Welding Handbook

Ninth Edition

Volume 3

WELDING PROCESSES, PART 2

American Welding Society

Welding Handbook, Ninth Edition

Volume 1 *Welding Science and Technology*

Volume 2 *Welding Processes, Part 1*

Volume 3 *Welding Processes, Part 2*

Volume 4 *Materials and Applications, Part 1*

Volume 5 *Materials and Applications, Part 2*

Welding Handbook

Ninth Edition

Volume 3

WELDING PROCESSES, PART 2

Prepared under the direction of the
Welding Handbook Committee

Annette O'Brien, Editor
Carlos Guzman, Associate Editor

American Welding Society
550 N.W. LeJeune Road
Miami, FL 33126

Library of Congress Control Number: 2001089999
ISBN: 978-0-87171-053-6

The *Welding Handbook* is the result of the collective effort of many volunteer technical specialists who provide information to assist with the design and application of welding and allied processes.

The information and data presented in the *Welding Handbook* are intended for informational purposes only. Reasonable care is exercised in the compilation and publication of the *Welding Handbook* to ensure the authenticity of the contents. However, no representation is made as to the accuracy, reliability, or completeness of this information, and an independent substantiating investigation of the information should be undertaken by the user.

The information contained in the *Welding Handbook* shall not be construed as a grant of any right of manufacture, sale, use, or reproduction in connection with any method, process, apparatus, product, composition, or system, which is covered by patent, copyright, or trademark. Also, it shall not be construed as a defense against any liability for such infringement. Whether the use of any information in the *Welding Handbook* would result in an infringement of any patent, copyright, or trademark is a determination to be made by the user.

Printed in the United States of America

DEDICATION

Dr. John M. Gerken

This book is dedicated to Dr. John M. Gerken in recognition of his distinguished work in the field of metallurgical engineering and the transfer of its technology, and for his steadfast commitment to the *Welding Handbook*. Dr. Gerken served on the Welding Handbook Committee beginning in 1992 during the preparation of the two Eighth Edition volumes of the *Welding Handbook, Materials and Applications, Part 1 and Materials and Applications, Part 2*. He served as the Chair of the Volume 1 Committee, which prepared Volume 1 of the Ninth Edition, *Welding Science and Technology* for publication in 2001. He is known nationally for his dedication to the welding industry and his service to the American Welding Society as a volunteer.

Dr. Gerken had a distinguished career with TRW, directing several welding research and engineering development programs involving the use of newly developed aerospace metals and requiring expertise in the major welding processes. After retiring from TRW, he extended his career at the Lincoln Electric Company as Manager of Technology Transfer, where he revised the *Welding Procedure Handbook*.

Dr. Gerken is a past president of the American Welding Society and served on the Board of Directors. In 1995 he was made a Fellow of the American Welding Society. He has been an active member of various national technical standards committees for AWS, currently including the Subcommittee on Nickel Alloys, the Committee on Joining Metals and Alloys, the U.S. Tag for ISO/TC 44/SC6, Resistance Welding; the Committee on Resistance Welding, and the Subcommittee on Stainless Steel Alloys.

He began his education at Newark College of Engineering, but interrupted his studies to serve in the United States Army. He resumed his education at Rensselaer Polytechnic Institute, where he earned a doctorate in metallurgical engineering.

In addition to his public service, Dr. Gerken has privately served as a mentor to many in the welding community over the years. The Welding Handbook Committee is grateful for his guidance and friendship.

CONTENTS

ACKNOWLEDGMENTS

The American Welding Society Welding Handbook Committee and the editors recognize the contributions of the volunteers listed below who have created, developed, and documented the technology of welding and shared it in past editions of the *Welding Handbook,* beginning with the first edition published in 1938. The enthusiasm and meticulous dedication of the authors and technologists reflected in the previous eight editions of the *Welding Handbook* are continued in this volume of the Ninth Edition.

This volume was compiled by the members the Welding Handbook Volume 3 Committee and the Chapter Committees, with oversight by the Welding Handbook Committee. Chapter committee chairs, chapter committee members, and oversight persons are recognized on the title pages of the chapters.

The Welding Handbook Committee and the editors recognize and appreciate the AWS technical committees who developed the consensus standards that pertain to this volume, and acknowledge the work of R. L. O'Brien, Editor of Volume 2, Eighth Edition. The Welding Handbook Committee is grateful to members of the AWS Technical Activities Committee and the AWS Safety and Health Committee for their reviews of the chapters. The Editors express their appreciation to the AWS Technical Division staff for their assistance during the preparation of this volume.

Welding Handbook Committee Chairs, 1938–2007

1938–1942	*D. S. Jacobus*
Circa 1950	*H. L. Boardman*
1956–1958	*F. L. Plummer*
1958–1960	*R. D. Stout*
1960–1962	*J. F. Randall*
1962–1965	*G. E. Claussen*
1965–1966	*H. Schwartzbart*
1966–1967	*A. Lesnewich*
1967–1968	*W. L. Burch*
1968–1969	*L. F. Lockwood*
1969–1970	*P. W. Ramsey*
1970–1971	*D. V. Wilcox*
1971–1972	*C. E. Jackson*
1972–1975	*S. Weiss*
1975–1978	*A. W. Pense*
1978–1981	*W. L. Wilcox*
1981–1984	*J. R. Condra*
1984–1987	*J. R. Hannahs*
1987–1990	*M. J. Tomsic*
1990–1992	*C. W. Case*
1992–1996	*B. R. Somers*
1996–1999	*P. I. Temple*
1999–2004	*H. R. Castner*
2004–2007	*P. I. Temple*

PREFACE

This is Volume 3 of the *Welding Handbook*, the third of the five-volume series in the Ninth Edition. Following the publication of *Welding Handbook* Volume 1, *Welding Science and Technology*, the Welding Handbook Committee determined that one volume could no longer contain the process technology described in the 29 chapters of Volume 2, Eighth Edition; thus Volume 2 and Volume 3 of the Ninth Edition were designed to accommodate the expanded information. *Welding Processes, Part 1* presented updated information on arc welding and cutting, the gas processes, brazing, and soldering. This volume, *Welding Processes, Part 2* is devoted to information on resistance welding, solid state processes and other joining and cutting methods. Volumes 4 and 5 of the Ninth Edition of the *Welding Handbook* will address welding materials and applications.

Volume 3 contains updated resistance welding chapters: spot and seam welding, projection welding, flash and upset welding, high-frequency welding, and resistance welding equipment. The chapter on friction welding has been updated, and new in this edition, a separate chapter on the developing process variation, friction stir welding, has been added. Other chapters are ultrasonic welding, explosion welding, adhesive bonding, thermal spraying and cold spraying, diffusion welding and diffusion brazing. The chapters on electron beam welding and laser beam welding and cutting contain significantly expanded technology. The last chapter, Other Welding and Cutting Processes, contains information on two new or revitalized processes, magnetic pulse welding, and electro-spark depositing. New information is presented on water jet cutting, which is reappearing in many current applications as a modern, efficient process.

A table of contents of each chapter is outlined on the cover page, and a subject index with cross-references appears at the end of the volume. A major subject index of this volume and previous editions of the *Welding Handbook* is included. Appendices A and B are lists of safety standards, and Appendix C presents a list of SI/inch-pound conversions of commonly used pressure units.

The chapters in this volume reflect the dramatic changes brought into welding processes over the past decade by the precise control of welding parameters made possible with digital controls and microprocessors as applied to new techniques and advanced materials. To meet the challenge of including this expanded technology, each chapter was prepared by a committee made up of highly qualified experts enthusiastic about the subject process, and headed by a chapter chair with an admirable dedication to the details of infusing state-of the-art information into the basics of the processes. All committee members are volunteers who generously devoted countless hours of personal time to the chapters.

One hundred and ten American Welding Society volunteers contributed to this book, representing university, government, and private welding research and development institutions, manufacturers of welding equipment and materials, and manufacturers, fabricators, and welders who use this technology.

The Welding Handbook Committee welcomes your comments and suggestions. Please address them to the Editor, *Welding Handbook*, American Welding Society, 550 N. W. LeJeune Road, Miami, FL 33126.

Phillip I. Temple, Chair
Welding Handbook Committee

Richard C. DuCharme, Chair
Welding Handbook Volume 3 Committee

Annette O'Brien, Senior Editor
Welding Handbook

REVIEWERS
AMERICAN WELDING SOCIETY
SAFETY AND HEALTH COMMITTEE AND
TECHNICAL ACTIVITIES COMMITTEE

D. E. Clark	*Idaho National Engineering Laboratory*
L. C. Connor	*Consultant*
D. A. Fink	*The Lincoln Electric Company*
S. R. Fiore	*Edison Welding Institute*
R. L. Holdren	*Applications Technologies, LLC*
P. Howe	*American Welding Society*
D. J. Landon	*Vermeer Manufacturing Company*
M. J. Lucas	*GE Aviation*
S. O. Luke	*Black & Veatch*
K. A. Lyttle	*Praxair Incorporated*
R. B. Maust	*Raytheon Integrated Defense Systems*
H. Mikeworth	*Dukane Corporation*
D. D. Kautz	*Sandia National Laboratories*
A. F. Manz	*Consultant*
A. Odermatt	*Hobart Brothers*
A. W. Sindel	*Alstom Power Incorporated*
D. H. Sliney	*U.S. Army Center for Health Promotion and Preventive Medicine*
P. T. Vianco	*Sandia National Laboratories*
J. L. Warren	*CNH America, LLC*

CONTRIBUTORS

WELDING HANDBOOK COMMITTEE

P. I. Temple, Chair	*Detroit Edison*
C. E. Pepper, First Vice Chair	*URS Corporation*
B. J. Bastian	*Benmar Associates*
H. R. Castner	*Edison Welding Institute*
R. C. DuCharme	*Consultant*
J. H. Myers	*Welding Inspection & Consulting Services*
W. Lin	*Pratt & Whitney*
A. O'Brien, Secretary	*American Welding Society*
W. L. Roth	*Procter & Gamble, Incorporated*
R. W. Warke	*LeTourneau University*

WELDING HANDBOOK VOLUME 3 COMMITTEE

Richard C. DuCharme, Chair	*Consultant*
D.W. Dickinson	*The Ohio State University*
S. P. Moran	*Miller Electric Company*
A. O'Brien, Secretary	*American Welding Society*
W. L. Roth	*Procter & Gamble*
P. F. Zammit	*Brooklyn Iron Works, Incorporated*

CHAPTER CHAIRS

Chapter 1	J. W. Dolfi (Ret.)	*Ford Motor Company*
Chapter 2	J. C. Bohr	*General Motors*
Chapter 3	L. E. Moss	*Automation International, Incorporated*
Chapter 4	M. Siehling	*RoMan Manufacturing, Incorporated*, and
	D. M. Beneteau	*Centerline (Windsor)*
Chapter 5	M. Kimchi	*Edison Welding Institute*
Chapter 6	T. V. Stotler	*Edison Welding Institute*
Chapter 7	T. J. Lienert	*Los Alamos National Laboratory*
Chapter 8	D. K. Graff	*Edison Welding Institute*
Chapter 9	G. A. Young	*Dynamic Materials Corporation*
Chapter 10	G. W. Ritter	*Edison Welding Institute*
Chapter 11	M. F. Smith	*Sandia National Laboratories*
Chapter 12	S. C. Maitland	*Goodrich Landing Gear*
Chapter 13	D. D. Kautz	*Los Alamos National Laboratory*
Chapter 14	D. D. Kautz	*Los Alamos National Laboratory*
Chapter 15	J. E. Gould	*Edison Welding Institute*

CHAPTER 1

RESISTANCE SPOT AND SEAM WELDING

Photograph courtesy of Soudronic Automotive AG

Prepared by the Welding Handbook Chapter Committee on Resistance Spot and Seam Welding:

J. W. Dolfi, Chair
Ford Motor Company (Ret.)

J. E. Gould
Edison Welding Institute

M. J. Karagoulis
General Motors Corporation

B. G. Kelly
Kelly Welding Solutions, P.C.

C. J. Orsette
RoMan Engineering Services

W. Urech
Soudronic Automotive

Welding Handbook Committee Member:

S. P. Moran
Miller Electric Manufacturing Company

Contents

CHAPTER 1

RESISTANCE SPOT AND SEAM WELDING

INTRODUCTION

Spot welding and seam welding are resistance welding processes. Resistance welding includes a group of processes that produce coalescence of the faying surfaces with the heat obtained from the resistance of the workpieces to the flow of the welding current in a circuit of which the workpieces are a part, and by the application of pressure.

A resistance spot weld is made between or on overlapping members (workpieces) in which coalescence may start and occur on the faying surfaces, or may proceed from the outer surface of one member. The weld cross section is approximately circular.

A seam weld is a continuous weld made between or on overlapping members, in which coalescence may start and occur on the faying surfaces, or may have proceeded from the outer surface of one member. The continuous weld may consist of a single weld bead or a series of overlapping spot welds.[1,2]

Spot welding and seam welding adapt well to automation. They are mainstay processes in the automotive industry and in other industries that manufacture products involving the welding of similar thicknesses of the same metal. Examples include automobile fuel tanks, catalytic converters, mufflers, and roof joints. Seam welding is an important process for manufacturers of cans and containers of all types. Other products of spot and seam welding are furnace heat exchangers and storage tanks.

This chapter covers the fundamental principles of these processes and their variations, equipment, and mechanical systems. The advantages and limitations of the processes are discussed, and welding conditions as applied to various metals are presented. Other topics include joint design, welding schedules, surface preparation, weld quality, and economics. The chapter concludes with a section on safe practices specific to the spot and seam welding processes.

FUNDAMENTALS

In resistance spot welding and seam welding, the coalescence of metals is produced by the heat generated in the workpieces due to resistance to the passage of electric current through each workpiece. Force, exerted on the joint through the electrodes, is always applied before, during, and after the application of current to confine the weld contact area at the faying surfaces and, in some applications, to forge the weld metal during postheating.

A resistance welding electrode is the part of a resistance welding machine through which the welding current and (in most cases) force are applied directly to the workpiece. The electrode may be in the form of a rotating wheel, rotating roll, bar, cylinder, plate, clamp, chuck, or a modification of one of these.

Figure 1.1 illustrates the mechanisms of the two processes.

1. Welding terms and definitions used throughout this chapter are from *Standard Welding Terms and Definitions*, AWS A3.0:2001, Miami: American Welding Society.

2. At the time of the preparation of this chapter, the referenced codes and other standards were valid. If a code or other standard is cited without a date of publication, it is understood that the latest edition of the document referred to applies. If a code or other standard is cited with the date of publication, the citation refers to that edition only, and it is understood that any future revisions or amendments to the code or standard are not included; however, as codes and standards undergo frequent revision, the reader is encouraged to consult the most recent edition.

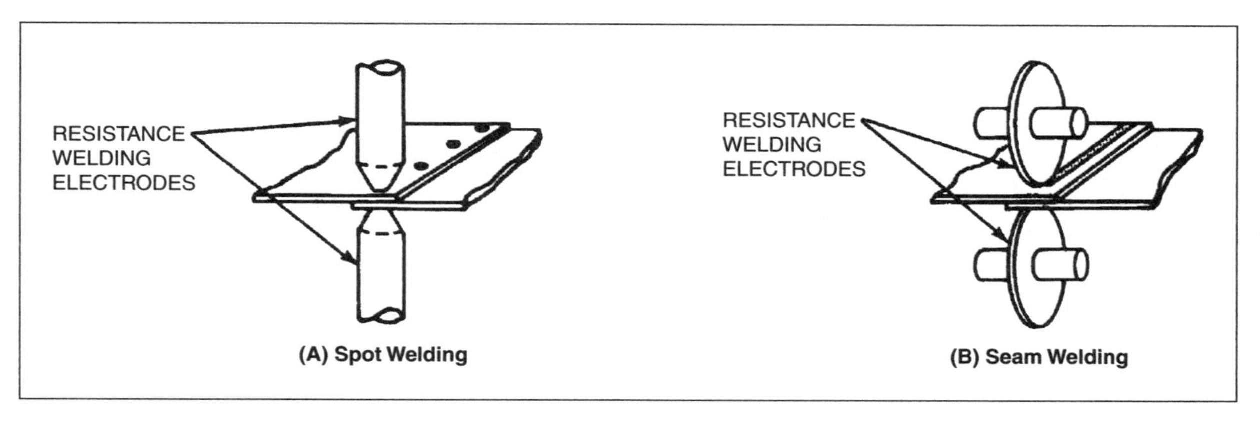

Figure 1.1—Simplified Diagrams of Basic Resistance Spot Welding (A), and Seam Welding (B)

In spot welding, a nugget (the weld metal that joins the workpieces) is produced at the electrode site, but two or more nuggets may be made simultaneously using multiple sets of electrodes. Seam welding is a variation of spot welding in which a series of overlapping nuggets are produced to obtain a continuous, leak-tight seam. One or both electrodes generally are in the form of wheels that rotate as the workpieces pass between them or between one wheel and a flat copper electrode. A seam weld can be produced with spot welding equipment, but the operation is much slower. A series of separate spot welds may be made with a seam welding machine and wheel electrodes by suitably adjusting the travel speed and the time between welds.

PRINCIPLES OF OPERATION

Spot and seam welding operations involve the coordinated application of electric current and mechanical pressure of sufficient magnitude and duration. The welding current must pass from the electrodes through the workpieces. The continuity of weld current is promoted by force applied to the electrodes. The first requirement in the sequence of operations is to develop sufficient heat to raise a confined volume of metal to the molten state. The fused metal is then allowed to cool while under pressure until it has adequate strength to hold the workpieces together. The current density and pressure must be high enough to form a nugget, but should not be so high that they cause molten metal to be expelled from the weld zone. The duration of weld current must be accurately timed to prevent excessive heating of the electrode faces that could fuse an electrode to a workpiece and greatly reduce the service life of the electrode.

As noted, heat in a resistance welding process is produced by electrical current flowing through the electrical resistance in the workpiece. Because metals have relatively low resistance, welding current must be relatively high to produce sufficient heat to develop welding temperatures at the desired location. Weld current also must exceed heat loss by thermal conduction in the workpiece and loss to the relatively cool electrodes in contact with the workpiece.

Heat Generation

The amount of heat generated in an electrical conductor depends on the following factors:

1. Amperage,
2. Resistance of the conductor (including interface resistance), and
3. Duration of current.

These three factors affect the heat generated, as expressed in the following equation:

$$Q = I^2Rt \tag{1.1}$$

where

Q = Heat generated, joules (J);
I = Current, amperes (A);
R = Resistance of the workpieces, ohms (Ω); and
t = Duration of current, seconds (s).

The heat generated is proportional to the square of the welding current and directly proportional to the resistance and the time. Some of the heat is used to make the weld and some is lost to the surrounding metal.

The welding current required to produce a given weld is an approximation inversely proportional to the square root of the time. Thus, if the time is extremely short, the current required will be high.

The secondary circuit of a resistance welding machine and the workpieces constitute a series electrical circuit. The total electrical impedance of the current path affects the current magnitude for a given applied voltage. The current in a series circuit will be the same in all parts of the circuit. The heat generated at any location in the circuit will be directly proportional to the resistance at that point.

A very important characteristic of resistance welding is the rapidity with which welding heat can be produced. The temperature distribution in the workpieces and electrodes in spot and seam welding is illustrated in Figure 1.2.

In effect, there are at least seven resistances connected in series that account for the temperature distribution in a weld. The resistances numbered from 1 through 7 shown in Figure 1.2 apply to a two-thickness joint. Numbers 1 and 7 denote the electrical resistance of the electrode material.

Numbers 2 and 6 represent the contact resistance between the electrode and the base metal. The magnitude of this resistance depends on the surface condition of the base metals and each electrode, the size and contour of each electrode face, and the electrode force. (Resistance is approximately inversely proportional to the contacting force.) This is a point of high heat generation, but the surface of the base metal does not reach fusion temperature during the passage of current due to the high thermal conductivity of the electrodes (1 and 7) and the fact that they are usually water-cooled. For free-zinc galvanized steels (when zinc is in elemental form as opposed to zinc alloys such as zinc-nickel or zinc iron [galvanneal]), the contact and faying surface resistance is not significantly higher than the bulk resistance of the material. Therefore, bulk heating of the material contributes substantially to the development of the weld.

Planes 3 and 5 are locations of the total resistance of the base metal, which is directly proportional to its resistivity and thickness, and inversely proportional to the cross-sectional area of the current path.

Plane 4 is the location of the base metal interface resistance where the weld is to be formed. This is the point of highest resistance and therefore the point of greatest heat generation. Since heat is also generated at Points 2 and 6, the heat generated at Plane 4 is not readily lost to the electrodes.

Heat is generated at all of these locations, not at the base metal interface alone. The flow of heat to or from the base metal interface is governed by the temperature gradient established by the resistance heating of the various components in the circuit. This in turn assists or retards the creation of the proper localized welding heat.

Heat will be generated in each of the seven locations shown in Figure 1.2 in proportion to the resistance of

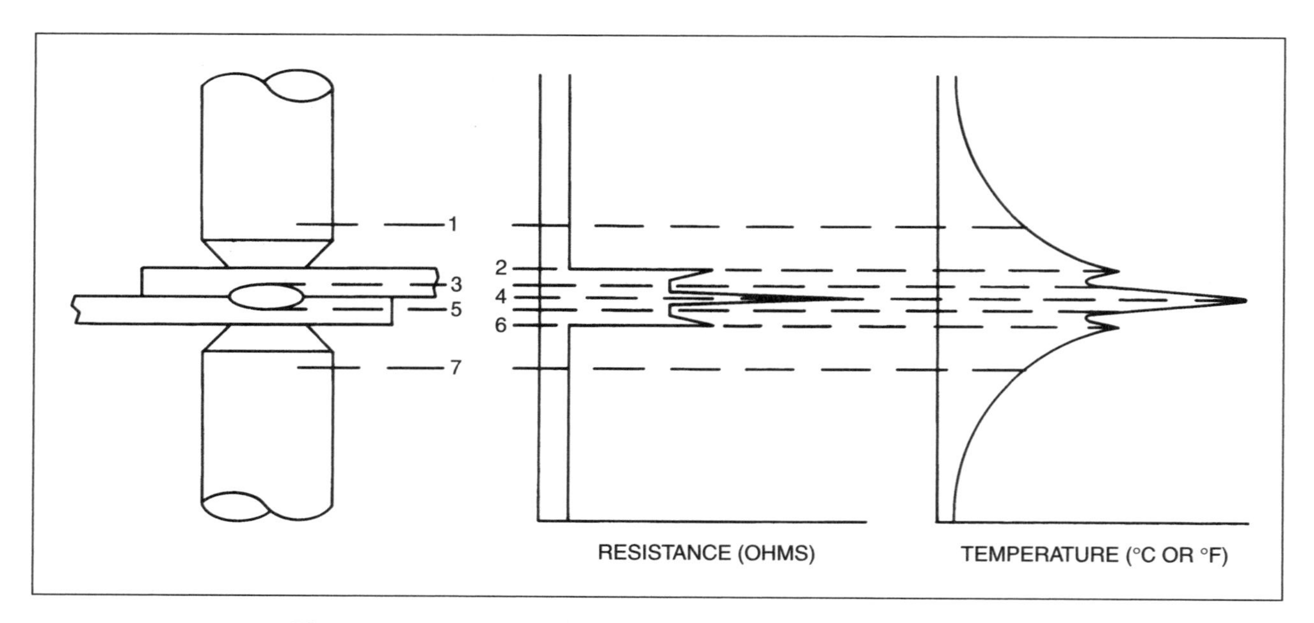

Figure 1.2—Relationship of Resistance and Temperature as a Function of Location in the Diagrammed Circuit

each. However, welding heat is required only at the base metal interface, and the heat generated at all other locations should be minimized. Since the greatest resistance is located at Plane 4, heat is most rapidly developed at that location. Points of next-lower resistance are Planes 2 and 6. The temperature rises rapidly at these points also, but not as fast as at Plane 4. After about 20% of the weld time has elapsed, the heat gradient may conform to the profile shown in Figure 1.2. Heat generated at Planes 2 and 6 is rapidly dissipated into the adjacent water-cooled electrodes at Planes 1 and 7. The heat at Plane 4 is dissipated much more slowly into the base metal. Therefore, while the welding current continues, the rate of temperature rise at Plane 4 will be much more rapid than at Planes 2 and 6. The welding temperature is indicated on the chart at the right in Figure 1.2 by the number of short horizontal lines within the drawing that lead to the matching curve. In a well-controlled weld, the welding temperature is first reached at numerous contact points at the weld interface; then melting occurs, and with time, the contact points quickly grow into a nugget.

The following factors affect the amount of heat generated in the weld joint by a given current for a unit of weld time:

1. The electrical resistances within the workpieces and the electrodes,
2. The contact resistances between the workpieces and between the electrodes and the workpieces, and
3. The heat lost to the workpieces and the electrodes.

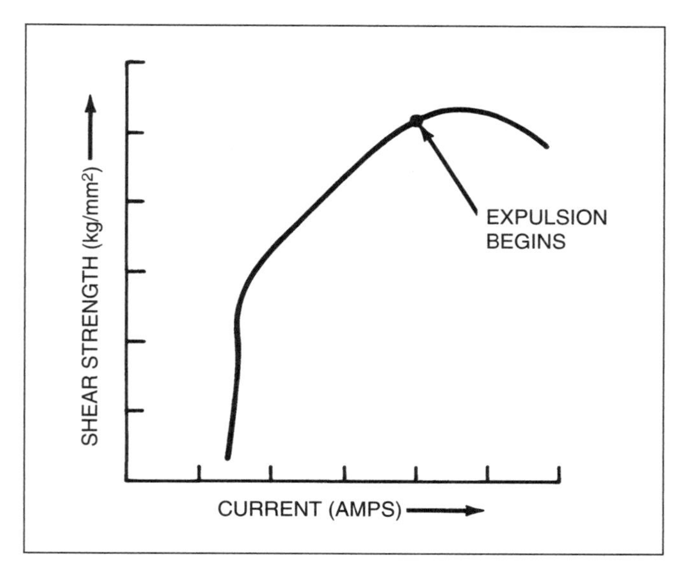

Figure 1.3—Effect of Welding Current on Shear Strength of Spot Welds

Effect of Welding Current

In the formula, $Q = I^2Rt$, current has a greater effect on the generation of heat than either resistance or time. Therefore, it is an important variable to be controlled. Two factors that cause variation in welding current with alternating current (ac) welding machines are fluctuations in power-line voltage and variations in the impedance of the secondary circuit. Impedance variations are caused by changes in circuit geometry or by the introduction of varying masses of magnetic metals into the secondary loop of the machine.

In addition to variations in the magnitude of welding current, current density may vary at the weld interface. This change in current density can result from the shunting of current through preceding welds and contact points other than those at the weld. An increase in electrode face area will decrease current density and welding heat. This may cause a measurable decrease in weld size. Minimum current density for a finite time is required to produce fusion at the interface. Sufficient heat must be generated to overcome the losses to the adjacent base metal and the electrodes.

Weld nugget size and strength increase rapidly with increasing current density. Excessive current density will cause molten metal expulsion (resulting in internal voids), weld cracking, and lower mechanical strength properties. Typical variations in the shear strength of spot welds as a function of current magnitude are shown in Figure 1.3. In spot and seam welding, excessive current overheats the base metal and results in deep indentations in the workpieces and will cause overheating and rapid deterioration of the electrodes.

Effect of Weld Time

The rate of heat generation must be adjusted so welds with adequate strength will be produced without excessive heating and rapid deterioration of the electrode. The total heat developed is proportional to weld time. Essentially, heat is lost by conduction into the surrounding base metal and the electrodes; a very small amount is lost by radiation. These losses increase in proportion to weld time and metal temperature.

Given suitable current density, some minimum time is required to reach melting temperature during a spot welding operation. If current is continued, the temperature at Plane 4 (refer to Figure 1.2) in the weld nugget will far exceed the melting temperature, and the internal pressure may expel molten metal from the joint. Generated gases or metal vapor and minute metal particles may be expelled. If the workpiece surfaces are scaly

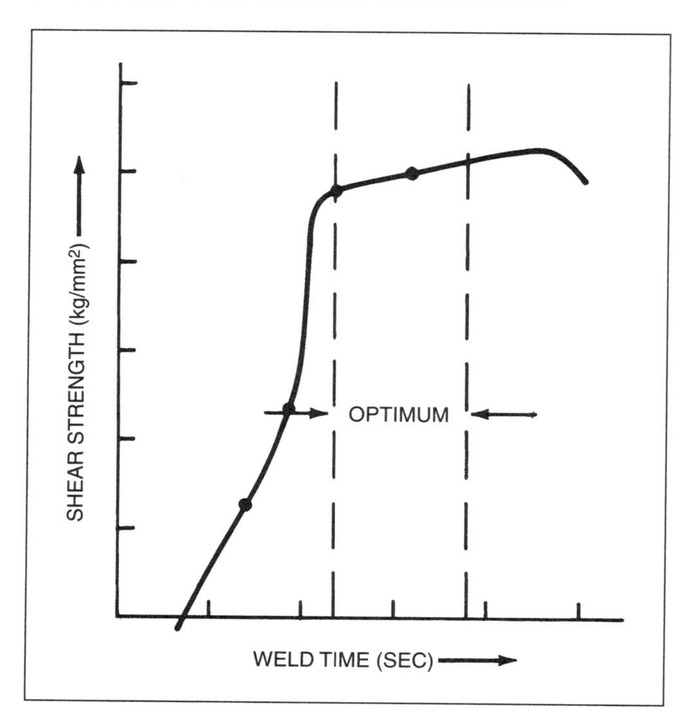

Figure 1.4—Tensile Shear Strength as a Function of Weld Time

or pitted, gases and particles also may be expelled at Planes 2 and 6.

In most cases, at some point during an extended welding interval the heat losses will equal the heat input and temperatures will stabilize. An example of the relationship between weld time and spot weld shear strength is shown in Figure 1.4, assuming all other conditions remain constant.

To a limited extent, weld time and amperage may be complementary. The total heat may be changed by adjusting either the amperage or the weld time. Attempts to speed up the process by increasing current and reducing weld time eventually will produce a runaway thermal condition at the faying surface accompanied by severe expulsion. The heat must be applied long enough for the faying surface heat to be conducted into the surrounding metal to grow the nugget and produce the desired weld size.

When spot-welding heavy plate, welding current commonly is applied in several relatively short impulses without removal of electrode force. The purpose of pulsing the current is to gradually build up the heat between the workpieces at the weld interface. The amperage needed to accomplish welding can rapidly melt the metal if the heat pulse time is too long, resulting in expulsion.

Effect of Welding Pressure

The resistance R in the heat formula (Equation 1.1) is influenced by welding pressure through its effect on contact resistance at the weld interface. Welding pressure is produced by the force exerted on the joint by the electrodes. Electrode force is considered to be the net dynamic force of the electrodes on the workpieces, and it is the pressure produced by this force that affects the contact resistance.

Workpieces to be spot or seam welded must be clamped tightly together at the weld location to facilitate passage of the current. For a given set of parameters, including welding circuit, welding equipment, electrode geometry, and workpiece surface condition, the amperage will rise to some limiting value as the electrode force or welding pressure is increased. The effect on the total heat generated in the workpiece, however, may be the reverse. As the pressure is increased, the contact resistance and the heat generated at the interface will decrease. To increase the heat to the previous level, amperage or weld time must be increased to compensate for the reduced resistance.

On a microscopic scale, the surfaces of metal components are a series of peaks and valleys. When they are subjected to light pressure, the actual metal-to-metal contact will be only at the contacting peaks, which is a small percentage of the area. Thus, with light pressure on the workpieces, contact resistance will be high. As the pressure is increased, the high spots are depressed and the actual metal-to-metal contact area will be increased, causing a decrease in the contact resistance. In most applications, the electrode material is softer than the workpieces; consequently, the application of a suitable electrode force will produce lower contact resistance at the electrode-to-workpiece interfaces than at the interface between the workpieces.

Of the variables in the resistance welding process, surface resistance may be the most difficult to control. Increasing the weld force frequently will reduce this resistance and cause the process to be more dependent on the bulk material resistance, which is generally consistent. This reduces process variation at the expense of higher current, increased wear and tear on the machine, and other negative factors associated with higher force.

Influence of Electrodes

Electrodes perform a vital function in the generation of heat because they conduct the welding current to the workpieces. In spot and seam welding, the electrode contact area largely controls the welding current density and the resulting weld size. Electrodes must have good electrical conductivity, but they also must have adequate strength and hardness to resist deformation caused by repeated applications of high electrode force.

Deformation or "mushrooming" of the electrode face increases the contact area and decreases both current density and welding pressure. Weld quality deteriorates as deformation of the electrode face proceeds; therefore, the electrodes must be reshaped or replaced at intervals to maintain adequate heat generation for acceptable weld properties. A similar increase in the electrode contact area is caused by erosion and pitting of the electrodes when welding coated materials.

When the electrodes are slow in following a sudden decrease in total thickness of the workpieces, a momentary reduction in pressure occurs. If this happens while the welding current is on, interface contact resistance at Locations 2, 4, and 6 (refer to Figure 1.2) and the rate of heat generation will increase. An excessive heating rate at the three contacting surfaces tends to cause overheating and violent expulsion of molten metal. Molten metal is retained at each interface by a ring of unfused metal surrounding the weld nugget. A momentary reduction in electrode force permits the internal metal pressure to rupture this surrounding ring of metal that is not fused. Internal voids or excessive electrode indentation may result. Weld properties may fall below acceptable levels, and electrode wear will be greater than normal.

Influence of Surface Condition

The surface condition of the workpieces influences heat generation because contact resistance is affected by oxides, dirt, oil, and other foreign matter on the surfaces. The most uniform weld properties are obtained when the surfaces are clean.

The welding of workpieces with an irregular coating of oxides, scale, or other contamination causes variations in contact resistance. This produces inconsistencies in heat generation. Heavy scale on the work surfaces also may become embedded in the electrode faces, causing rapid deterioration of the electrode. Oil and grease on the surface may pick up dirt that will contribute to electrode deterioration.

Influence of Metal Composition

The electrical resistivity of a metal directly influences resistance heating during welding. In high-conductivity metals such as silver and copper, little heat is developed even under high current densities. The small amount of heat generated is rapidly conducted into the surrounding workpiece and the electrodes.

The composition of a metal determines its specific heat, melting temperature, latent heat of fusion, and thermal conductivity. These properties govern the amount of heat required to melt the metal and produce a weld. However, the amount of heat necessary to raise a unit-mass of most commercial metals to the fusion temperature is very nearly the same. For example, stainless steel and aluminum require the same number of energy units to reach fusion temperature, even though they widely differ in spot welding characteristics. As a result, the electrical and thermal conductivities become dominant. The conductivities of aluminum are about ten times greater than those of stainless steel. Consequently, the heat lost in the electrodes and surrounding metal is greater with aluminum. Accordingly, the welding current used for aluminum must be considerably greater than that used for stainless steel.

Heat Balance

Heat balance occurs when the depth of fusion (penetration) in the two workpieces is approximately the same. The majority of spot and seam welding applications involves the welding of similar thicknesses of the same metal, with electrodes of the same alloy, shape, and size. In these cases, heat balance is automatic. The heat generated in the workpiece is unbalanced, however, for applications in which welding is performed on different gauges and grades of materials.

Heat balance may be affected by the following conditions:

1. Relative electrical and thermal conductivity of the workpieces,
2. Relative geometry of the workpieces at the joint,
3. Thermal and electrical conductivity of the electrodes, and
4. Geometry of the electrodes.

It should be noted that if dissimilar electrodes are used in high-volume production operations, the design must be error-proofed by using electrode (tip) designs that cannot be interchanged. The use of the incorrect electrode tends to cause an unbalanced heat condition leading to discrepant welds.

Heating will be unbalanced when workpieces have significantly different composition or different thickness, or both. The imbalance can be minimized in many cases by the design of the workpiece or the design and material of the electrode. Heat balance also can be improved by using the shortest weld time and lowest current that will produce acceptable welds.

Heat Dissipation. During welding, heat is lost by conduction into the adjacent base metal and the electrodes, as shown in Figure 1.5. This heat dissipation continues at varying rates during the application of current and afterward until the weld has cooled to room temperature. It may be divided into two time phases: during the time of current application and after the cessation of current. The extent of the first phase depends on the composition and mass of the workpieces, the

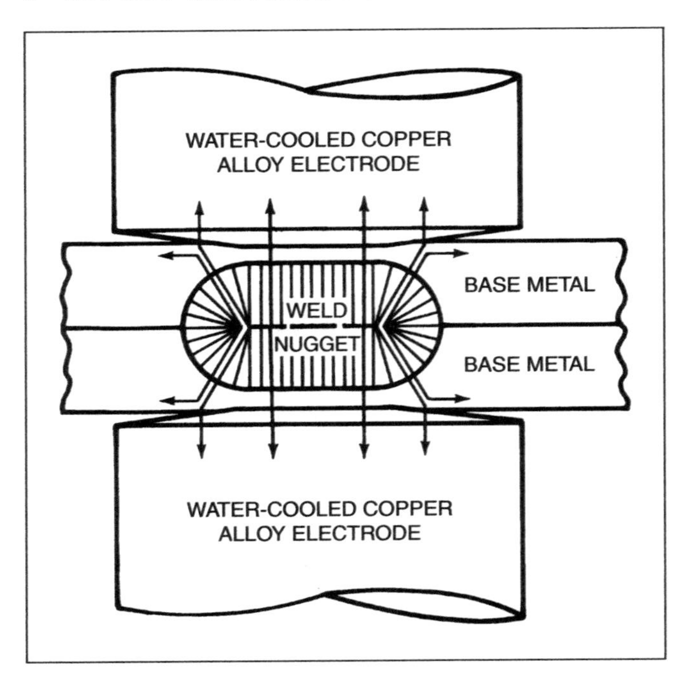

Figure 1.5—Heat Dissipation into Surrounding Base Metal and Electrodes During Resistance Welding

welding time, and the means of external cooling. The composition and mass of the workpieces are determined by the design. External cooling depends on the welding setup and the welding cycle.

The heat generated by the welding current is inversely proportional to the electrical conductivity of the base metal. The thermal conductivity and temperature of the base metal determine the rate at which heat is dissipated or conducted from the weld zone.[3] In most cases, the thermal and electrical conductivities of a metal are similar. In a high-conductivity metal, such as copper or silver, high amperage is needed to produce a weld and compensate for the heat that is dissipated rapidly into the adjacent base metal and the electrodes. Spot and seam welding of these metals is difficult.

If the electrodes remain in contact with the workpieces after weld current ceases, they rapidly cool the weld nugget. The rate of cooling decreases with longer welding times because a larger volume of base metal would have been heated. This reduces the temperature gradient between the surrounding base metal and the weld nugget. The cooling rates are slower for thick sheets of metal that generally require long welding times than for thin sheets and short weld times.

Problems may result if the electrodes are removed from the weld too quickly after the welding current is turned off. This action may cause excessive warping when welding thin sheets. With thick sheets, adequate time is needed to cool and solidify the large weld nugget while under pressure. On coated materials, such as galvanized steel containing zinc, the removal of the electrodes too soon may allow the surface temperature to increase to the level that zinc penetration into the grain boundaries of the steel may occur. Zinc penetration can reduce weld strength and performance. It usually is best, therefore, to maintain the electrodes in contact with the workpieces until the weld cools to a temperature that is strong enough to sustain any loading that will be imposed when the pressure is released.

The cooling time for a seam weld nugget is short when the electrodes are rotated continuously. Therefore, welding is commonly done with water flowing over the workpieces to remove the heat as rapidly as possible. The flowing water develops a steam bubble over the hot area, and more effective cooling is achieved by directing a stream of water from a nozzle at the point where the welded material exits from under the resistance welding electrode. It is not always good practice to cool the weld zone rapidly. When welding quench-hardenable steel alloys, it usually is best to retract the electrodes as quickly as possible to minimize heat dissipation to the electrodes, and thus reduce the cooling rate of the weld.

WELDING CYCLE

The welding cycle for spot and seam welding consists of the four basic phases: squeeze time, weld time, hold time, and off time. Off time generally is used only for manually initiated repetitive welding cycles. The phases of the welding cycle are described as follows:

1. *Squeeze time*—the time interval between initiating the timer and the first application of current; the time interval added to ensure that the electrodes contact the workpieces and establish the desired electrode force before welding current is applied;
2. *Weld time*—the time that welding current is applied to the workpieces in making a weld in single-impulse welding;
3. *Hold time*—the time during which force is maintained on the workpieces after the last impulse of current ends, allowing the weld nugget to solidify and cool until it has adequate strength; and

3. For additional information, see “Heat Flow in Welding,” Chapter 3 in American Welding Society (AWS) Welding Handbook Committee, Jenney, C. L, and A. O'Brien, Eds., 2001, *Welding Science and Technology*, Vol. 1, *Welding Handbook*, 9th ed., Miami: American Welding Society.

4. *Off time*—the time during which the electrodes are off the workpiece and the workpiece is moved to the next weld location. The term is generally applied when the welding cycle is repetitive.

Figure 1.6 shows a basic welding cycle. One or more of the following features may be added to this basic cycle to improve the physical and mechanical properties of the weld zone:

1. Precompression force to seat the electrodes and workpieces together,
2. Preheat to reduce the thermal gradient in the metal at the start of weld time,
3. Forging force to consolidate the weld nugget,
4. Quench and temper times to produce the desired weld strength properties in hardenable alloy steels,
5. Postheat to refine the weld grain size in steels, and
6. Current decay to retard cooling in aluminum.

In some applications, the welding current is supplied intermittently during a weld interval time; it is on during heating time and ceases during cooling time. Figure 1.7 shows the sequence of operations in a more complex welding cycle than the basic cycle.

WELDING CURRENT

Alternating current (ac) or direct current (dc) can be used to produce spot and seam welds. The transformer in the welding machine transforms high-voltage line power to low-voltage, high-amperage welding power. Many applications use single-phase ac of the same frequency as the power line, usually 60 hertz (Hz). Direct current is used for applications that require very high welding current. Direct current is preferred for these applications because the load power can be drawn from a three-phase power line. The use of dc welding current also reduces power losses in the secondary circuit. Direct current may be essentially constant for a timed period or may be in the form of a high-peaked pulse. The latter normally is produced from stored electrical energy.

PROGRAMMING THE CURRENT

Welding current used to produce the heat in the workpiece can be manipulated by electronic control systems. Wave forms can be alternating current, direct current, and combinations of these. In addition, current can be obtained from the power lines and stored for later use. When a stored-energy machine is used, the demand from the power lines is reduced. Power is

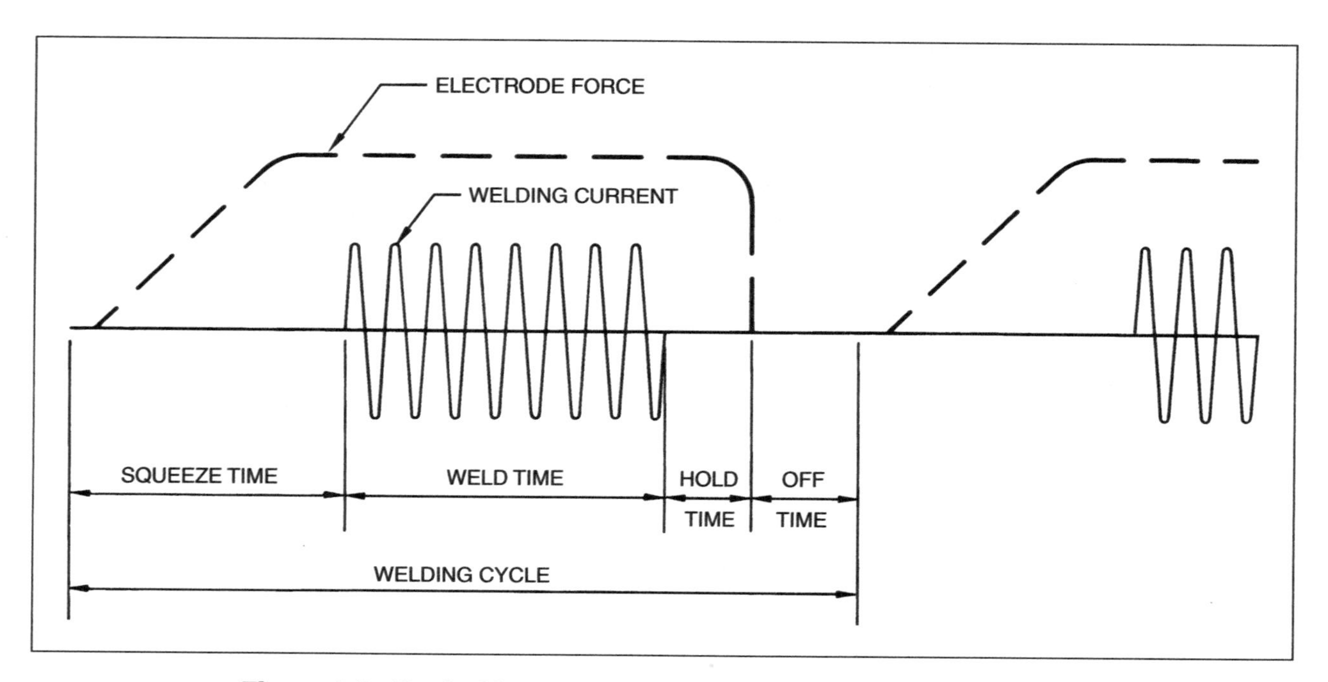

Figure 1.6—Basic Single-Impulse Welding Cycle for Spot Welding

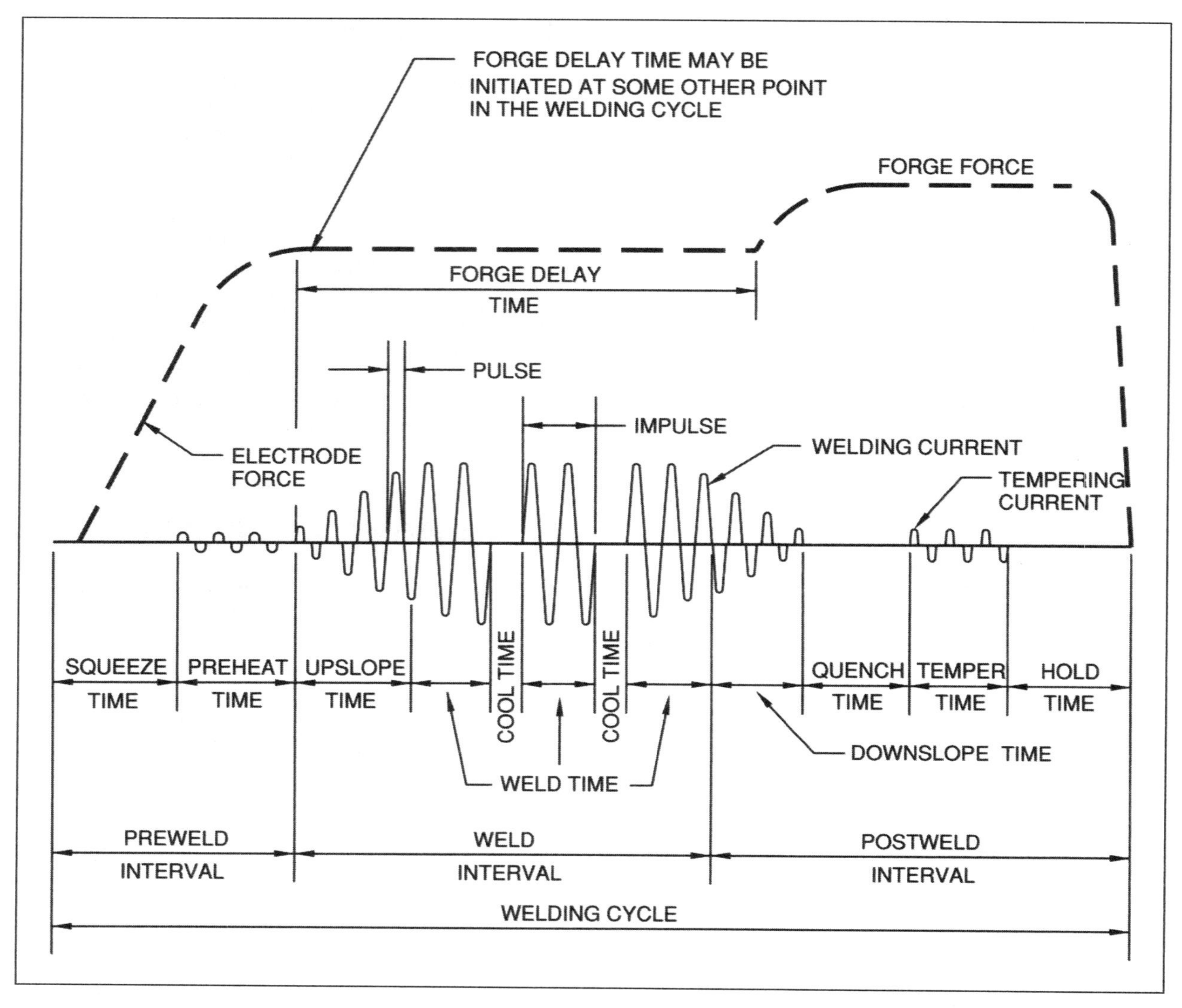

Figure 1.7—Enhanced Welding Cycle with Preheat, Upslope, Downslope, Quench and Temper Times, and Hold Time at Forging Force (Alternating Current)

acquired over longer periods of time than it takes to make a weld. When the primary power source is not capable of delivering sufficient power, stored-energy systems may offer the only way to make some resistance welds. The next section describes some of the systems in common use for controlling welding current.

Direct-Energy and Stored-Energy Welding Machines

The term *direct-energy welding machine,* as applied in this instance, refers to a machine that obtains energy directly from the utility line supply as the weld is being made. A stored-energy machine obtains and stores the energy from the utility before the weld is made and then discharges the stored energy into the weld system. Typically the weld time of a stored-energy machine is fixed by the physical characteristics of the system and only the magnitude of the current can be adjusted.

The rate of current rise and fall can be programmed when using direct-energy welding machines. The period during which the current continuously rises or increases is called *upslope time.* The period during which the current continuously falls or decreases is called *downslope time.* These cycles are illustrated in Figure 1.7. These features are available on machines equipped with electronic control systems.

Upslope generally is used to avoid overheating and expulsion of metal at the beginning of weld time, when the base metal interface resistance is high. Downslope is used to control weld nugget solidification, to avoid cracking in metals that are quench-hardenable or subject to hot tearing.

Prior to welding, the base metal can be preheated using a low current. Following the formation of the weld nugget, the current can be reduced to some lower value for postheating of the weld zone. This may be the downslope interval, as shown in Figure 1.7, or a separate application of current following a period of quench time.

WELD TIME

Weld time is the duration of the flow of welding current through the workpieces while making a weld. The time of applied current for machines other than stored-energy power sources is controlled by electronic, mechanical, manual, or pneumatic means. Weld times commonly range from one-half cycle (1/120 s) for very thin sheet metal to several seconds for thick plate metal. For the capacitor or magnetic type of stored-energy welding machines, the weld time is determined by the electrical constant of the system.

Single-Impulse Welding

The use of one continuous application of current to make an individual weld is called *single-impulse welding* (refer to Figure 1.6). Upslope or downslope current may be included in the time period.

Multiple-Impulse Welding

Multiple-impulse welding uses two or more pulses of current separated by a preset cooling time, as shown in Figure 1.7. This sequence is used to control the rate of heating at the interface while spot welding relatively thick steel sheet.

ELECTRODE FORCE

Completion of the electrical circuit through the electrodes and the workpieces is accomplished by applying electrode force. This force is produced by hydraulic, pneumatic, magnetic, or mechanical devices. The pressure developed at the interfaces depends on the area of the electrode faces in contact with the workpieces. This force, or pressure, performs the following functions:

1. Brings the various interfaces into intimate contact,
2. Reduces initial contact resistance at the interfaces,
3. Suppresses the expulsion of molten weld metal from the joint, and
4. Consolidates the weld nugget.

Forces may be applied during the welding cycle to accomplish the following:

1. Maintain a constant weld force;
2. Regulate precompression and weld forces to supply a high initial level to reduce initial contact resistance and bring the workpieces into intimate contact, followed by a lower level for welding;
3. Supply precompression at several levels of weld force and forging force at the first two levels as described in No. 2, followed by a forging force near the end of the weld time; and
4. Apply required weld and forging forces, the latter used to reduce porosity and hot cracking in the weld nugget.

EQUIPMENT

Spot and seam welding equipment consists of three basic elements: an electrical circuit, the control equipment, and a mechanical system.[4]

ELECTRICAL CIRCUIT

The electrical circuit is comprised of a welding transformer, a primary power control system, and a secondary circuit. The secondary circuit is made up of the electrodes that conduct the welding current to the workpiece, and includes the workpieces as part of the circuit. In some cases, a means of storing electrical energy is also included in the circuit. Either alternating current or direct current can be used. The welding machine converts 60-Hz line power to low-voltage, high-amperage power in the secondary circuit of the welding machine.

Alternating-Current Machines

Some resistance welding machines produce single-phase alternating current of the same frequency as the power line, usually 60 Hz. These machines contain a single-phase transformer that provides the high welding currents required at low voltage. Depending on the thickness and type of material to be welded, currents

4. Chapter 4 of this volume, "Resistance Welding Equipment," provides more detailed information on welding controls.

may range from 1000 amperes (A) to 100 000 A. A typical electrical circuit designed for this type of machine is shown in Figure 1.8.

Direct-Current Machines

Welding machines may produce direct current of continuous polarity, pulses of current of alternating polarity (frequency-converter machines), or high-peak pulses of current. The high-peak pulses are produced by stored electrical energy.

Rectifier Machines

Rectifier machines are direct-energy types, in which ac power from the electric utility plant distribution system (line power) passes through a welding transformer and is then rectified to dc power. Silicon-diode rectifiers are widely used in secondary circuits because of their inherent reliability and efficiency. The system can be single phase; however, one of the advantages of direct-current systems is the capability of using a three-phase transformer to feed the rectifier system in the secondary circuit. This makes it possible to use balanced three-phase line power.

Frequency-Converter Machines

Frequency-converter machines have a special welding transformer with a three-phase primary and a single-phase secondary. The primary current is controlled by ignitron tubes or silicon-controlled rectifiers. Half-cycles of three-phase power, either positive or negative, are conducted to the transformer for a timed period that depends on the transformer design. The transformer output is a pulse of direct current. By switching the polarity of the primary half-cycles, the polarity of the secondary current is reversed. A weld may be made with one or more dc pulses.

Stored-Energy Machines. Stored-energy machines are electrostatic designs. They draw power from a single-phase system, store it, and then discharge it in a very short pulse period to make the weld. These machines draw power from the supply line over a relatively long time between welds, accumulating power to deliver to the electrodes during a short weld time.

The equipment for electrostatic stored-energy welding consists primarily of a bank of capacitors, a circuit for charging the capacitors to a predetermined voltage, and a system for discharging the capacitors through a suitable welding transformer. High-voltage capacitors generally are used, the most common varying from 1500 volts (V) to 3000 V.

ELECTRODES

Resistance welding electrodes perform the following functions:

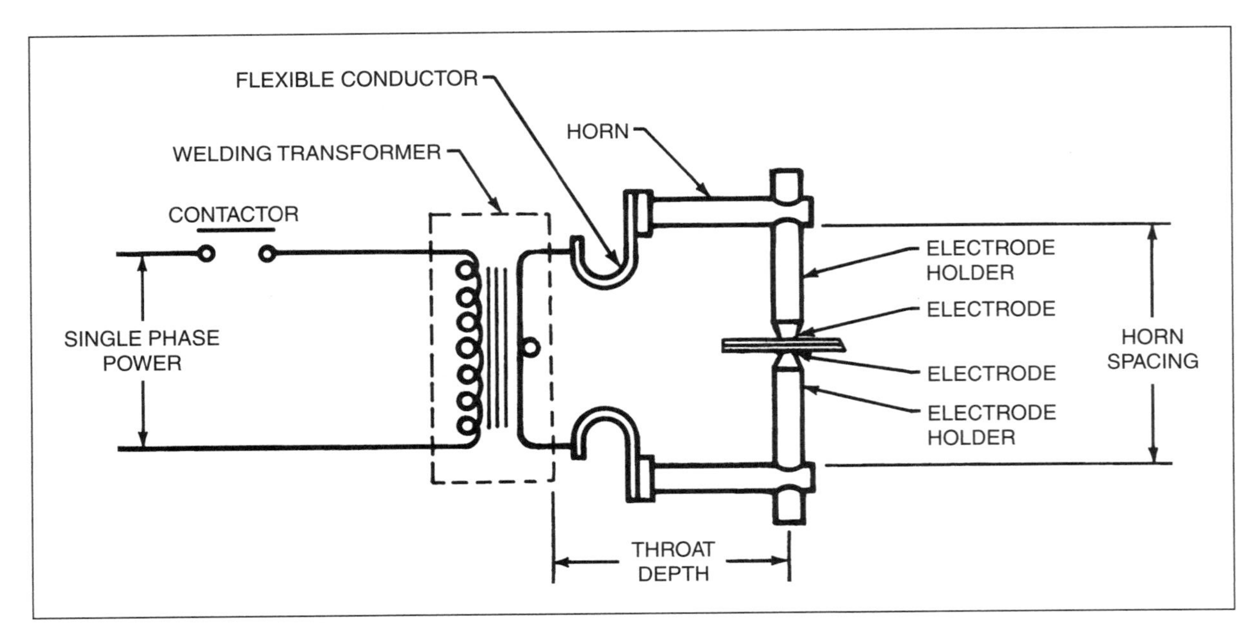

Figure 1.8—Typical Single-Phase Spot Welding Circuit

1. Conduct the welding current to the workpieces, and for spot or seam welding, provide stable electrical contact area at the weld zone;
2. Transmit a force to the workpieces;
3. Dissipate part of the heat from the weld zone; and
4. Maintain relative alignment and position of the workpieces.[5]

If the application of pressure were not involved, electrode material could be selected almost entirely on the basis of electrical and thermal conductivity. However, since the electrodes are often subjected to forces of considerable magnitude, they must be capable of withstanding the imposed stresses at elevated temperatures without excessive deformation. Proper electrode shape is important because the current must be confined to a fixed area to achieve the needed current density.

When only one spot or seam weld is to be made at a time, only one pair of electrodes is required. In this case, the force and current are applied to each weld by shaped electrodes.

Electrodes made from copper alloys with satisfactory physical and mechanical properties are available commercially. Generally, the harder the alloy, the lower are its electrical and thermal conductivities. The choice of a suitable alloy for any application is based on a compromise between its electrical and thermal properties and its mechanical qualities. Electrodes selected for aluminum welding, for instance, should have high conductivity at the expense of high compressive strength to minimize the sticking or fusing of electrodes to the workpieces. Conversely, electrodes for welding stainless steel should sacrifice high conductivity to obtain enough compressive strength to withstand the required electrode force.

Resistance of the electrode to deformation or mushrooming depends on the proportional limit and hardness of the electrode alloy. Hardness is established by heat treatment or cold working during the forming process. The temperature of the electrode face is a major factor in electrode service life because this is where softening takes place. The sizes and shapes of electrodes usually are determined by the type and thickness of the metal sheet to be welded.

CONTROL EQUIPMENT

Welding controls usually provide one or more of the following principal functions:

1. Initiate and terminate current to the welding transformer,
2. Control the magnitude of the current, and
3. Actuate and release the electrode force mechanism at the proper times.

Welding controls are divided into three functional groups: welding contactors, timing and sequencing controls, and other current controls and regulators.

A welding contactor, often simply referred to as a *weld timer*, connects and disconnects the primary power and the welding transformer. Electronic contactors using phase-shift heat control may use silicon-controlled rectifiers, ignitron tubes, thyratron tubes, power transistors or saturable reactors to regulate the primary current. The majority of weld controls use solid-state devices for regulating the weld current.

Adjustments to the timing and sequence control establish the welding sequence and the duration of each function of the sequence. This includes the application of electrode force and current, and the timing of intervals following each function.

The welding current output of a machine is controlled by transformer taps or an electronic heat control, or both. An electronic heat control is used to regulate the current conduction in power-switching devices. It controls current by delaying the conduction of the power switches (in the case of phase-shift heat control) during each half-cycle (1/120 s for 60 Hz power). Varying the firing delay time can be used to gradually increase or decrease the effective primary voltage. This provides control of upslope and downslope welding current.

Transformer taps are used to change the number of primary turns connected across the ac power line. This changes the turns ratio of the transformer, and produces an increase or decrease in open-circuit secondary voltage. Decreasing the turns ratio will increase the open-circuit secondary voltage, the primary current, and the welding current.

MECHANICAL SYSTEMS

Resistance spot welding and seam welding machines essentially have the same types of mechanical operation. The electrodes approach and retract from the workpieces at controlled times and rates. Electrode force is applied by hydraulic, pneumatic, magnetic, or mechanical means. The rate of electrode approach must be rapid, but controlled, so that the electrode faces are not deformed from repeated impact with the workpieces. The locally heated weld metal expands and contracts rapidly during the welding cycle and the electrodes must follow this movement to maintain welding pressure and electrical contact. The ability of the machine to follow motion is influenced by the moving mass of the machines force control system and by friction.

5. Resistance welding electrodes are discussed in more detail in Chapter 4, Resistance Welding Equipment.

If the pressure between the electrodes and workpieces drops rapidly during weld time, the contact surfaces of the electrodes and workpieces may overheat and result in the burning or pitting of the electrode faces. The electrodes may stick to the workpieces, and in some instances, the surfaces of the workpieces may vaporize from the very high energy concentration.

The electrode force used during the melting of the weld nugget may not be adequate to consolidate the weld metal and to prevent internal porosity or cracking. Multiple-level force machines may be employed to provide a high forging pressure during weld solidification. The magnitude of this pressure should be appropriate to the composition and thickness of the metal and the geometry of the workpieces. The forging pressure applied often is two to three times greater than the welding pressure. Since the weld cools from the periphery inward, the forging pressure must be applied at or close to the time the current terminates.

SURFACE PREPARATION

For all types of resistance welding, the condition of the surfaces of the workpieces largely controls how consistent weld quality will be. The contact resistance of the faying surfaces has a significant influence on the amount of welding heat generated; hence, the electrical resistance of these surfaces must be highly uniform for consistent results. They must be free of high-resistance materials such as paint, scale, thick oxides, and heavy oil and grease. If it is necessary to use a primer paint on the faying surfaces prior to welding, as is sometimes the case, the welding operation must be performed immediately after applying the primer, or special conducting primers must be used. For best results, the primer should be applied as thinly as possible so that the electrode force will displace it and allow metal-to-metal contact.

Paint should never be applied to outside base metal surfaces before welding because the paint will result in poor surface appearance and reduced electrode life. Heavy scale should be removed by mechanical or chemical methods. Light oil on steel is not harmful unless it has picked up dust or grit. Remnants of drawing compounds containing mineral fillers should be removed before welding.

Some seam welding and special processes that use continuously fed foil or wire between the electrode and workpiece can be used for welding coated materials. This is discussed in more detail in the seam welding section.

PREPARATION METHODS

The methods of preparing surfaces for resistance welding differ for various metals and alloys. The manufacturer of the alloy should be consulted for recommendations on preparing the surfaces for welding. The use of surface preparation processes or compounds may involve materials that present specific risks to the user. Procedures for the safe use of the cleaning materials recommended by the suppliers of the cleaners, solvents, and etchants should be carefully and consistently followed.

Some base metals may be prepared for welding by the metal supplier and shipped with protective coating or wrapping to preserve the clean surface required for good weldability. This service may be available to high-volume users.

Aluminum

The chemical affinity of aluminum for oxygen causes the aluminum to become coated with a thin film of oxide when it is exposed to air. The thin oxide film that forms on a freshly cleaned aluminum surface does not cause sufficient resistance to be troublesome for resistance welding. The permissible holding period, or elapsed time between cleaning and welding, may vary from 8 hours to 48 hours or more, depending on the cleaning process used, the cleanliness of the shop, the alloy, and the application. Heavy oxide, which is an excellent electrical insulator, and resulting hydrated surface compounds that form on aluminum alloys prevent the successful welding of aluminum that has been stored over long periods of time.

An aluminum surface may be mechanically cleaned for resistance welding using a fine grade of abrasive cloth, fine steel wool, or a fine wire brush. Clad aluminum also may be cleaned by mechanical means, but damage to the cladding must be carefully avoided. Numerous commercial chemical cleaners are available for aluminum. Chemical cleaning usually is preferred for large-volume production for reasons of economy as well as for uniformity and control.

Magnesium

The cleaning of magnesium alloys is particularly important because they readily alloy with copper at elevated temperatures. The contact resistance between the electrode and the workpiece must be kept as low as possible. Manufacturers of magnesium alloys supply the material with a coating of oil, or the alloys are chrome-pickled to protect them from oxidation during shipment and storage. The protective coating must be removed to facilitate the removal of residual magnesium oxide to obtain sound and consistent welds.

Copper

The proper cleaning of copper alloys is important. The beryllium-coppers and aluminum-bronzes are particularly difficult to clean by chemical means. Mechanical means are preferred. In some instances, a flash coating of tin is applied to produce a uniformly higher surface resistance than that of pure copper.

Nickel

Nickel and nickel alloys demand high standards of cleanliness for successful resistance welding. The presence of grease, dirt, oil, and paint increases the probability of sulfur embrittlement during welding and will result in defective welds. Oxide removal is necessary if heavy oxides are present from prior thermal treatments. Machining, grinding, blasting, or pickling may be employed. Wire brushing does not produce a satisfactory surface.

Titanium

The surfaces of titanium workpieces should be scrupulously clean before welding. Contaminants such as oil, grease, dirt, oxides and paint can adversely affect both weld consistency and chemical composition. Titanium and titanium alloys react with many elements and compounds at welding temperatures. Contamination by oxygen, hydrogen, nitrogen and carbon, which enter the microstructure interstitially, can significantly reduce weld ductility and toughness. Scale-free surfaces may be welded either after degreasing or after degreasing and acid pickling. The surfaces may be degreased with acetone, methylethylketone, or a dilute solution of sodium hydroxide. Chlorinated solvents should not be used. Titanium and titanium alloys are susceptible to stress corrosion. Pickling may be used to remove light oxide scale before welding.

Steel

Plain carbon steels and low-alloy steels have relatively low resistance to corrosion in ordinary atmosphere; hence, these metals usually are protected by a thin film of oil during shipment, storage, and processing. This oil film has no harmful effects on the weld, provided the oily surfaces are not contaminated with shop dirt or other poorly conductive or dielectric materials.

Steels are supplied with various surface finishes. Some of the more common finishes are the following:

1. Hot-rolled;
2. Hot-rolled, pickled, and oiled; and
3. Cold-rolled with or without annealing.

Hot-rolled steel must be pickled or mechanically cleaned prior to welding. Hot-rolled pickled steel is weldable in the as-received condition, possibly needing only wiping to remove loose dirt. Cold-rolled steel presents the best welding surface and, if properly protected by oil, requires no cleaning prior to welding other than wiping to remove loose dirt.

High-alloy steels and stainless steels are noncorrosive and usually require only minimum cleaning before resistance welding. When exposed to elevated temperatures, stainless steels acquire an oxide film; the thickness of the oxide depends on the temperature and the length of time of exposure. The film is an oxide of chromium, which is effectively removed by pickling. Oil and grease should be removed with solvents or by vapor degreasing prior to welding.

Coated Steels. The various coatings and types of plating applied to carbon steel sheet metal to provide corrosion resistance or to enhance appearance lend themselves satisfactorily to welding with resistance welding processes, with few exceptions. In general, good results may be obtained without special cleaning processes. However, the surfaces of aluminized steel should be wire brushed before welding to minimize expulsion and electrode pickup.

Phosphate coatings can increase the electrical resistance of the surfaces to a degree that welding current cannot pass through the sheets with low welding pressures. Higher pressures will produce welds, but slight variations in coating thickness may prevent welding.

SURFACE PREPARATION CONTROL

Surface preparation control can be maintained by periodically measuring the room temperature contact resistance of the workpieces immediately following cleaning. The measurement is most readily taken tip-to-tip between two resistance welding electrodes through two or more thicknesses of metal. The unit surface resistance varies inversely with pressure, temperature, and area of contact. To make the measurements significant in the control of surface cleanliness, the test conditions must be specified.

RESISTANCE SPOT WELDING

Resistance spot welding (RSW) produces a weld at the faying surfaces of a joint by the heat obtained from resistance to the flow of welding current through the workpieces from electrodes that serve to concentrate the welding current and pressure at the weld area. Figure 1.9

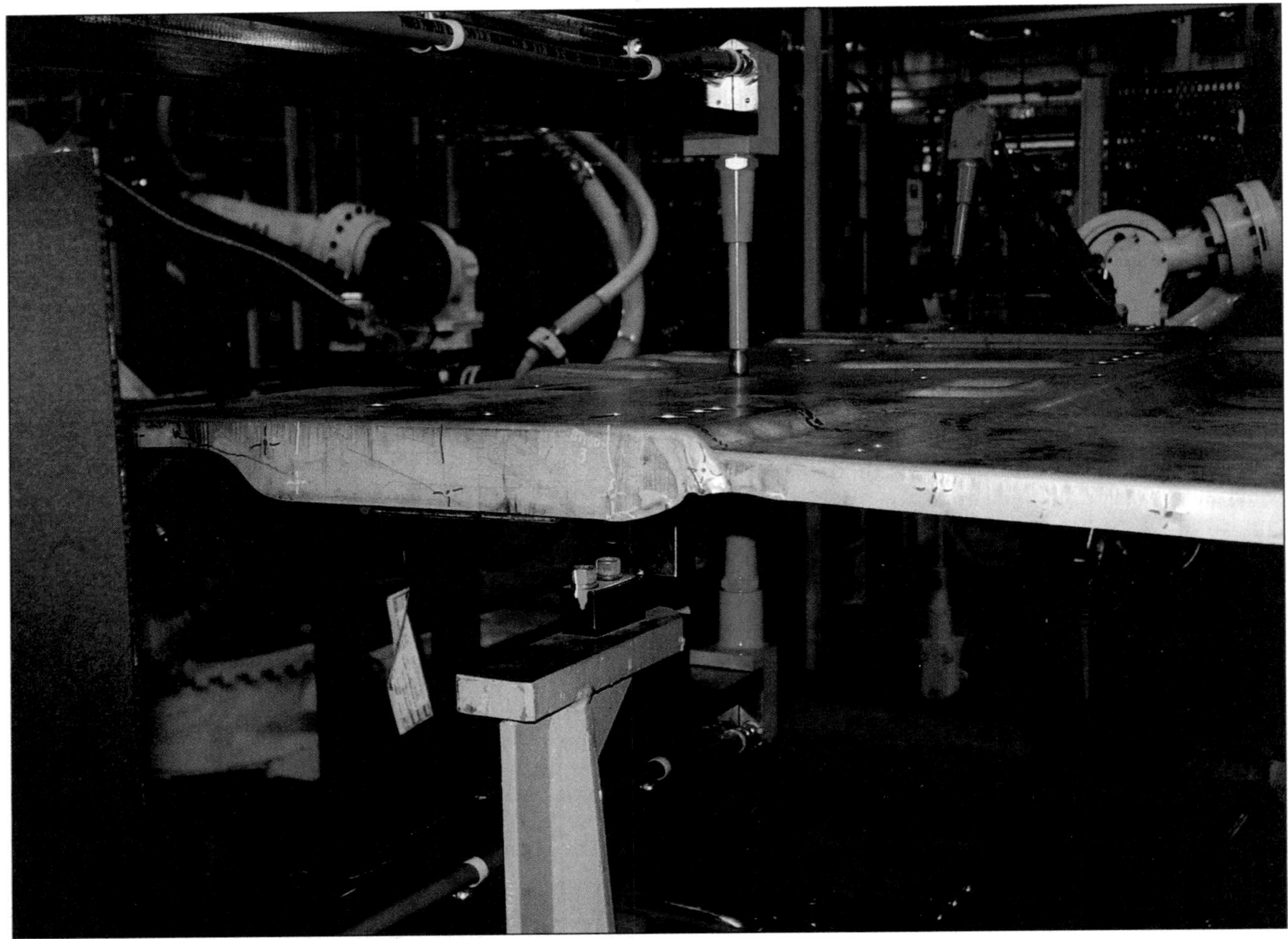

Photograph courtesy of KUKA Flexible Production Systems

Figure 1.9—Resistance Spot Welding Machine

shows a robotically controlled resistance spot welding system, used for manufacturing automotive underbodies.

APPLICATIONS

Resistance spot welding is used to fabricate sheet metal assemblies up to about 3.2 mm (0.125 in.) in thickness when the design permits the use of lap joints and leak-tight seams are not required. Occasionally the process is used to join steel plates 6.35 mm (1/4 in.) or more in thickness; however, the loading of such joints is limited and the joint overlap adds weight and cost to the assembly when compared to the cost of an arc-welded butt joint.

Stainless steel, aluminum, and copper alloys are commonly used metals in commercial spot welding. The process is used extensively for joining low-carbon steel sheet metal components for automobiles, cabinets, furniture, and similar products.

Resistance spot welding is used in preference to mechanical fastening, such as riveting or screwing, when disassembly for maintenance is not required. It is much faster and more economical because separate fasteners and additional filler materials are not needed for assembly.

ADVANTAGES AND LIMITATIONS

The major advantages of resistance spot welding are its high speed and adaptability for automation in the high-rate production of sheet metal assemblies. Spot welding is also economical in many job shop operations because it is faster than arc welding or brazing and requires less skill to perform.

Some of the limitations of the process are the following:

1. Disassembly for maintenance or repair is very difficult;
2. A lap joint adds weight and material cost to the product compared to a butt joint;
3. The equipment costs generally are higher than the costs of most arc welding equipment;
4. The short time, high current, and high power requirements produce unfavorable line power demands, particularly with single-phase machines; and
5. Heat produced by the spot weld tends to develop in the center of a stack of sheets; therefore, welding a thin outside sheet to two thick sheets becomes difficult and is not recommended where gauge ratios exceed 2:1.

SPOT WELDING VARIATIONS

Variations in the methods of performing resistance spot welds include direct and indirect welding, series welding, and parallel welding. The methods differ in the application of welding current and pressure, and in the arrangement of the secondary circuit.

Direct and Indirect Welding

Direct welding involves a resistance welding secondary circuit variation in which welding current and electrode force are applied to the workpieces by directly opposed electrodes, wheels, or conductor bars for spot or seam welding.

Arrangements for direct single spot welds are shown schematically in Figure 1.10(A), (B), and (C). In Figure 1.10(A), direct welding uses electrodes with similar geometry. In Figure 1.10(B), a larger electrode against one workpiece provides an increased contact surface area for applications requiring welding heat balance, or to reduce the indentation on the lower sheet by the electrodes. Figure 1.10(C) is similar to 1.10(B) except individual spot welds can be produced by sequenced weld guns.

Indirect welding involves a resistance spot or seam welding secondary circuit variation in which the welding current flows through the workpieces in locations away from, as well as at the point of welding.

Figures 1.10(D) through 1.10(G) represent indirect resistance welding arrangements. A backing plate arrangement in Figure 1.10(D) provides a current path and pressure when the backing plate is made of a conducting material. If the backing plate is non-conductive, it provides support for only the welding force, and the current path will be from the top electrode, through the faying surface location, along the lower workpiece to the return connection further down the joint. Figure 1.10(E) is similar to (D) except that the electrode that does not exert pressure is away from the lap joint. The current path may be through a conducting backing plate between the electrodes, or through the base material between the electrodes. Figures 1.10(F) and (G) are similar to the joints in (D) and (E), but are for the spot welding of high-resistance materials that require higher voltages. The two secondary circuits are in series, and are connected to two transformers. The primary circuits may be connected either in series or in parallel. The two secondary circuits provide the sum of their respective voltages at the spot welds. This transformer and electrode arrangement is commonly called "push-pull," or "over-under" and typically is used when welds are needed in the center of large workpieces. With this arrangement there is no need for a circuit to wrap around the workpiece.

Parallel and Series Welding

Parallel welding and series welding are secondary-circuit variations of resistance welding used for multiple spot welding applications. In parallel welding, the secondary current is divided and conducted through the workpieces and electrodes in parallel electrical paths, simultaneously forming multiple spot welds. Examples of parallel welding are shown schematically in Figure 1.11(A) and (B); series welding arrangements are shown in Figure 1.11(C) and (D).

In Figure 1.11(A), the welding current is from a single transformer with multiple electrodes in the secondary circuit in parallel. Figure 1.11(B) represents a parallel welding system operating with a three-phase primary. This particular system is limited to three workstations.

In series welding, Figures 1.11(C) and (D) show that the secondary circuit current is conducted through the workpieces and the electrodes in a series electrical path, simultaneously forming multiple spot welds at the electrode locations. A series-welding arrangement requires equal resistance values at the faying surfaces to obtain uniform heating at each spot weld. When spot welding with two electrodes in series, a portion of the current travels through the adjacent area of the workpiece from one electrode to the other, bypassing the faying surfaces. This shunted current does not contribute to the spot weld, and this must be taken into account when developing a procedure for series spot welding. It should be noted that on galvanized steel, this current raises the

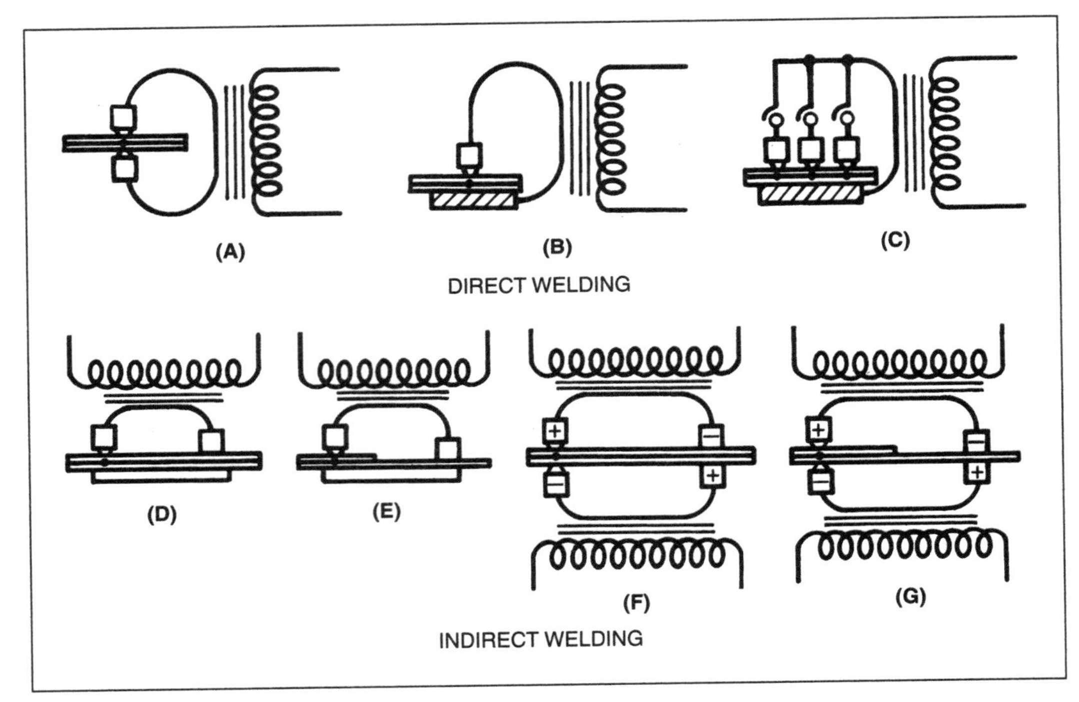

Figure 1.10—Typical Arrangements for Single Spot Welds

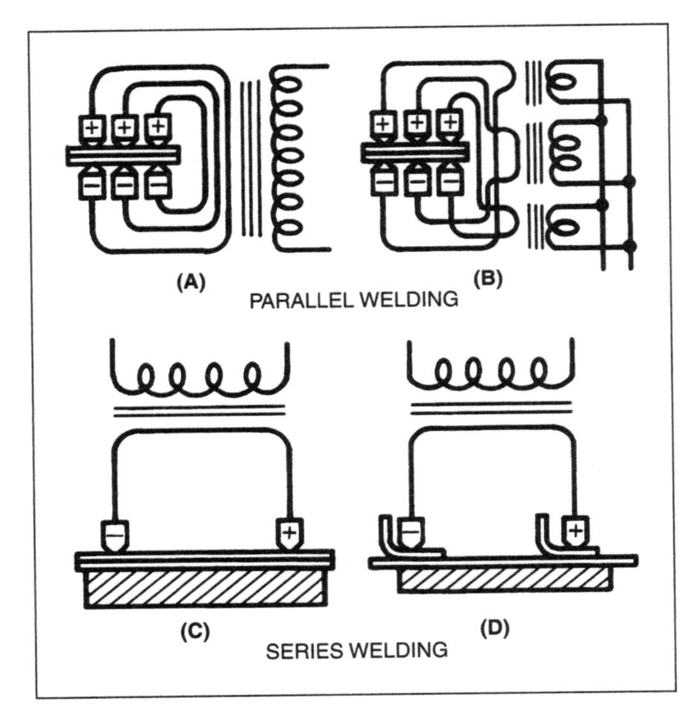

Figure 1.11—Typical Secondary Arrangements for Direct Multiple-Spot Welding

temperature of the steel surrounding the weld to the level at which zinc penetrates into the grain structure of the steel and may substantially reduce strength. Therefore, series welding should be not be used on galvanized steel.

HEAT BALANCE

Heat balance in a spot weld occurs when the depths of fusion in the two workpieces are approximately the same. Problems with heat balance arise when joints are made between metals of different thickness, different electrical conductivity, or a combination of both. Variations in electrode configurations and compositions can be used to overcome unbalanced heating to some extent, as shown in Figure 1.12. This sketch illustrates general methods for overcoming improper joint heating involving base metals with different electrical conductivity.

In Figure 1.12(A), an electrode with a smaller face area is used on the metal with the higher conductivity. The smaller contact area increases the current density in the higher conducting metal. Less heat is conducted away from the joint by the base metal and the electrode. More heat is generated in the workpiece, and the fusion area shifts from the less conductive metal toward the metal with higher conductivity. An alternative would be

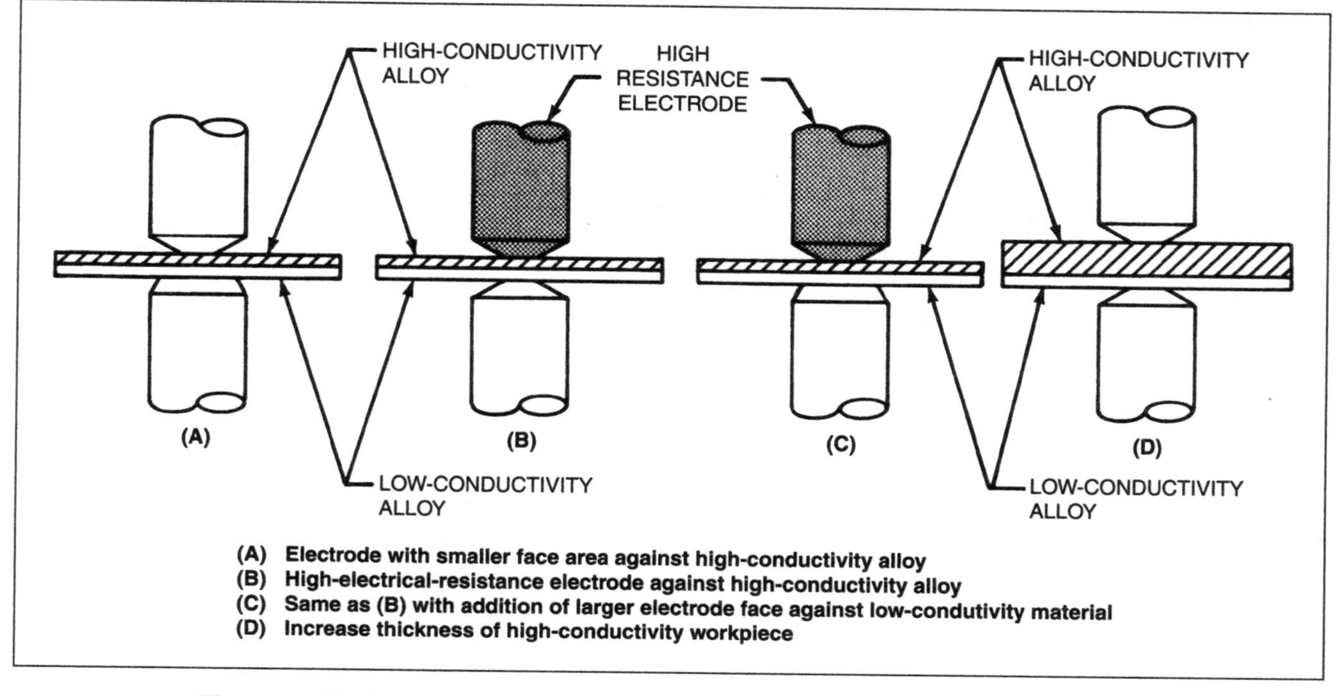

Figure 1.12—Typical Techniques for Improving Joint Heat Balance When Spot Welding Metals with Different Electrical Conductivities

to apply an electrode with higher resistance against the metal with higher conductivity to limit the heat loss through that electrode, as shown in Figure 1.12(B). Figure 1.12(C) shows the combination of a higher-resistance electrode and a smaller electrode face area applied to the more conductive metal. Better heat balance also can be obtained by increasing the thickness of the more conductive metal, as shown in Figure 1.12(D). This results in an increase in the effective resistance of the sheet.

The spot welding of metals with similar electrical characteristics but differing thickness also will result in uneven joint heating. The thicker workpiece will exhibit a higher resistance (lower conductivity) than the thinner sheet, resulting in deeper penetration into the thicker sheet. The heat balance can be improved by decreasing the current density in the thicker sheet, by decreasing the heat loss from the thinner sheet, or by a combination of both. Applying a large-diameter electrode on the thicker sheet will concentrate the current density into the thinner metal, shifting the nugget penetration deeper into the thinner sheet. It should be noted that when dissimilar electrodes are used to achieve heat balance, the design should include an error-proofing function that prevents the use of incorrect electrodes.

In order to effectively make spot welds in two or more dissimilar thicknesses of the same metal, a maximum section-thickness ratio of the outer sheets is suggested. For carbon steels, the suggested maximum section-thickness ratio is 4 to 1. (If the same electrodes are used on both sides, the ratio must not exceed 3 to 1.) To join three different thicknesses of carbon steel using Type A or E electrodes (classified by the Resistance Welding Manufacturing Alliance [RWMA]),[6] an outer sheet thickness ratio of 2.5 to 1 is suggested. Zinc-coated materials require a reduced ratio of about 2 to 1. To accommodate joints with higher thickness ratios, altering the electrode face diameter and the electrode composition are important methods of balancing the heat produced in each member of the joint.

In multiple layers of dissimilar thickness, a long weld time permits more uniform distribution of heat in the asymmetrical resistance path between the electrodes. Correct heat balance may be obtained by using multiple-impulse (pulsed) welding, or a single impulse of continuous current for an equivalent time.

6. Resistance Welding Manufacturing Alliance, a standing committee of the American Welding Society, 550 N.W. LeJeune Road, Miami, Florida 33126. www.rwma.org.

LAP JOINT DESIGN

The joint design in all applications of spot welding consists of a lap joint. One or more of the welded members may be flanges, or formed sections such as angles and channels. The use of standard resistance welding machines, portable welding guns, and special-purpose machines must be considered when designing the lap joint configuration. The joint design for direct welding must allow access to both sides of the joint by the electrodes.

Design Requirements

Factors that should be considered when designing for spot welding include:

1. Edge distance,
2. Joint overlap,
3. Fitup,
4. Weld spacing,
5. Joint accessibility,
6. Surface marking, and
7. Weld strength.

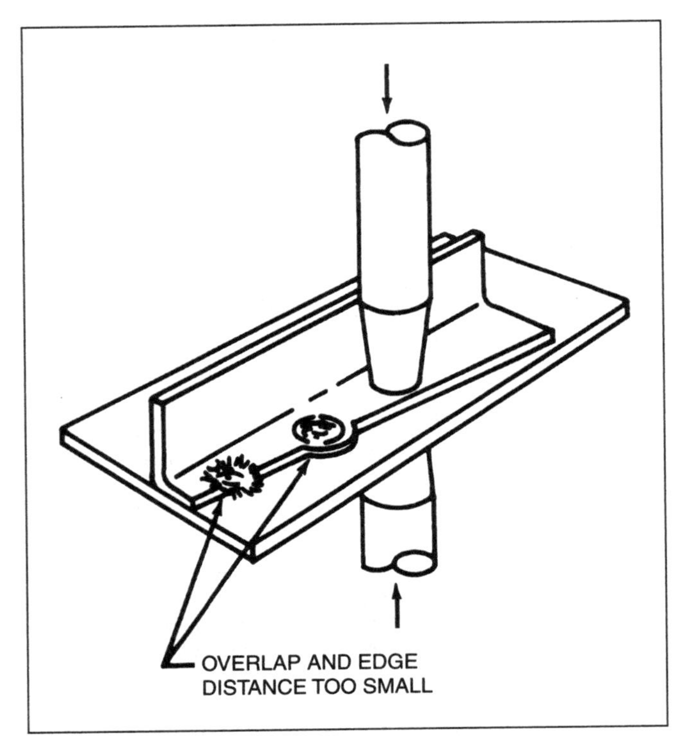

Figure 1.13—The Effect of Improper Overlap and Edge Distance

Edge Distance. The edge distance is measured from the center of the weld nugget to the edge of the sheet. The location of the spot weld must ensure that enough base metal is available to resist the expulsion of molten metal from the joint. Spot welds made too close to the edge of one or both workpieces will cause the base metal at the edge of the workpiece to overheat and upset outward, as shown in Figure 1.13. The restraint by the base metal at the edge of the molten nugget is reduced and expulsion of molten metal may occur due to the high internal pressure in the nugget. The weld nugget may be unsound, the electrode indentation excessive, and the weld strength low. The required minimum edge distance is a function of base metal composition and strength, section thicknesses, electrode face contour, and the welding cycle.

Joint Overlap. The minimum permissible joint overlap in sheet metal is calculated at two times the minimum edge distance. The overlap must include the base metal requirement for avoiding edge overheating and expulsion for both sheet metal workpieces. Other factors, such as electrode clearance, positioning tolerance of the weld tip and the workpieces, may require a larger overlap to provide for consistent weld quality. If the overlap is too small, the edge distance will automatically be insufficient, as sketched in Figure 1.13.

Fitup. The faying surfaces of the workpieces should fit together along the joint with little or no gap between them. Any force required to overcome gaps in the joint will reduce the effective welding force. The force required to close the joint may vary as welding progresses, and consequently may change the actual welding force. The ultimate result may be significant variations in the strength of the individual welds.

Weld Spacing. When numerous spot welds are made successively along a joint, a portion of the secondary current shunts through the adjacent welds. This shunting of the current must be considered when establishing the distance between adjacent spot welds and when establishing the welding machine settings. Typical weld current and minimum spacing shown in general welding charts do not provide compensation for this shunt effect. Some welding machines are capable of providing a different weld schedule for the second weld when spacing would be less than the recommended minimum. If this is the case, spacing can be somewhat below the minimum, but greater electrode wear and surface marking can be expected because of the higher current required to make the weld. The effect of shunting current also can be influenced by presence of sealers, adhesives and other coatings at the faying surfaces.

The division of current depends primarily on the ratio of the resistances of the two paths, one through the adjacent welds and the other across the interface between the sheet metal workpieces. If the path length through the adjacent weld is longer than the thickness of the joint, the resistance will be high compared to the resistance of the joint, and the shunting effect will be negligible.

The minimum spacing between spot welds is related to the conductivity and thickness of the base metal, the diameter of the weld nugget, and the cleanliness of the faying surfaces. For example, thick sections or metals with high conductivity will require greater spacing between spot welds. The suggested minimum spacing between adjacent spot welds is increased when joining three or more sheets. The spot spacing for a weld joining three thicknesses is generally 30% greater than the spacing required for welding two sections of the thicker outer sheet. Current levels may be increased in order to provide more current to the weld and thus offset the shunting effects; however, the higher heat inputs may cause expulsion if applied to the first spot weld, which is not shunted. An auxiliary weld timer or current control may be provided to produce the first spot weld using lower heat input. Modern electronic welding controls frequently provide several different welding schedules that can be selected in this situation. This is not a recommended practice when using a manual welding gun with a standard control, because the operator will be required to select alternate triggers to get the alternate weld schedule. Even the most conscientious worker sometimes may select the incorrect one. Most weld controllers cannot compensate automatically for shunting effects.

Joint Accessibility. The joint should be designed in consideration of the size and shape of commercially available electrodes and electrode holders, as well as the type of spot welding equipment that will be used. Each side of the weld joint should be accessible to the electrodes mounted on the welding machine or to backup electrodes in the case of indirect welding. Chapter 4, "Resistance Welding Equipment," contains information on electrodes and electrode holder designs. It should be noted that while offset points are shown on standard electrode designs, the electrodes must be installed with the correct orientation, which may be difficult to maintain in a production environment.

SPOT WELD QUALITY

Spot weld quality involves a combination of positive attributes related to the strength and fatigue life of the weld and negative attributes such as surface marking, workpiece distortion and adherent expulsion that results from a poorly controlled resistance welding process. Maintenance practices and proper setup can reduce marking and distortion and assure that the weld strength is consistent.

Indentation

Indentation, or surface marking, is a depression on the exterior surface of the workpieces. It is an undesirable effect resulting from workpiece shrinkage, which is caused by a combination of the heat of welding and electrode penetration into the surface of the workpiece.

When the welding current is on, the workpiece is resistance-heated locally and has a tendency to expand in all directions. Because of the pressure exerted by the electrodes, expansion transverse to the plane of the sheets is restricted. As the weld cools, contraction takes place almost entirely in the transverse direction and produces concave surfaces or marks at the electrode locations, as shown in Figure 1.14. This type of weld shrinkage is different from excessive electrode indentation into the workpieces caused by improper welding procedures. The contraction shrinkage is almost certain to occur to some degree and seldom exceeds a few tenths of a millimeter.

A circular ridge around the spot weld concavity occurs when the expanding workpiece upsets in the plane of the sheets around the electrode face, as can be observed in Figure 1.14. This ridge is caused by the relatively high electrode force and occurs to some extent with all shaped electrodes. The ridge is even more obvious with coatings such as galvanizing, where the coating melts during the weld process and is squeezed out from under the electrodes.

After some finishing operations, such as painting, the marks may be very conspicuous. It is difficult to eliminate the marks completely, but they can be reduced materially by modifying the welding procedure. For example, the indentation into the sheet can be minimized by welding in the shortest time practical.

Various techniques are used to minimize these markings. A common method is to use a large flat-faced electrode against the "show side" of the joint (the side that is visible when the assembly is in use). The electrode

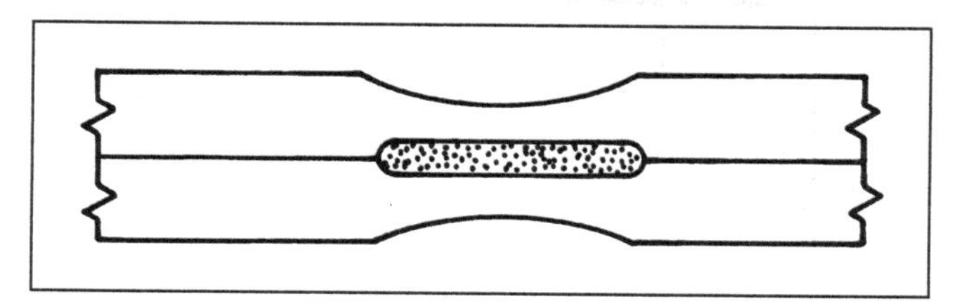

Figure 1.14—Indentation Produced by Shrinkage During Spot Welding

used for this technique should be made of a hard copper alloy to reduce wear. This method can be used in manual welding applications with galvanized steel; however, in automated welding, the weld is made at precisely the same location each time. Therefore, precision placement results in more rapid wear of the backup electrode and will require frequent maintenance to preserve surface shape. Another technique is to use indirect welding arrangements such as those shown in Figure 1.15 (also refer to Figure 1.10). Attempts to minimize weld markings are associated with higher maintenance costs and the process becomes less robust in achieving correct weld size. Achieving balance between acceptable weld size and acceptable appearance raises issues between the competing requirements of appearance and product performance. Some projection welding machines use stored energy and high-performance follow-up systems to make welds with very little marking because the short weld time minimizes the heat input. Additional information is presented in Chapter 2 of this volume, "Projection Welding."

Surface marking also may occur when an electrode or the electrode holder accidentally contacts the workpiece adjacent to where the spot weld is to be made. The resultant arcing may produce a small pit in the workpiece, which is undesirable in some applications. If localized melting occurs as a result of the contact, cracks may result in some materials.

Electrode misalignment, skidding, or deflection of the supporting machine component under load also may result in undesirable surface marking. Localized overheating and electrode deflection will not be a problem if the correct joint design, electrodes, and equipment are used.

Weld Strength

The strength of a single spot weld in shear is determined by the cross-sectional area of the nugget in the plane of the faying surfaces. Strength tests for spot welds are published in *Recommended Practices for*

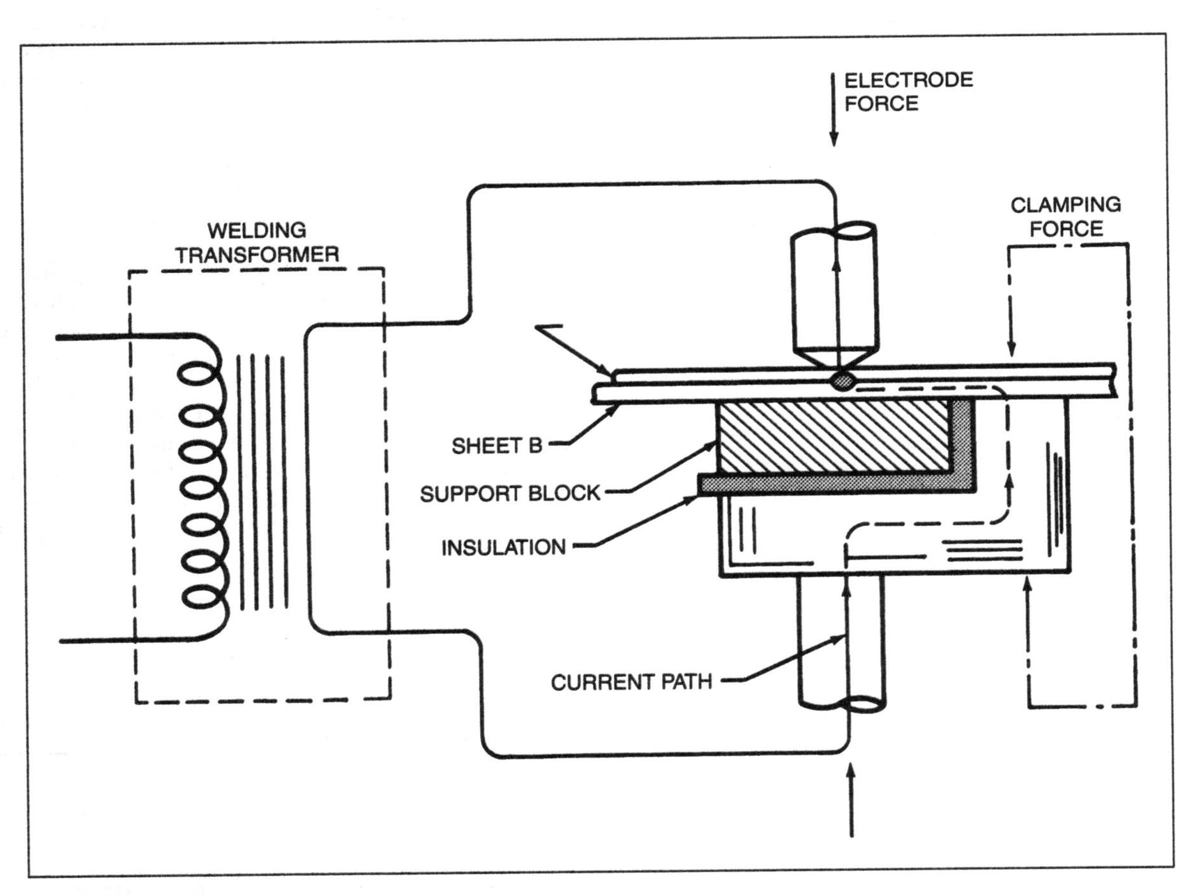

Figure 1.15—Application of Indirect Welding to Minimize Weld Marking on One Side

Resistance Welding, AWS C1.1.[7] Additional information on spot weld test procedures can be found in Chapter 12, Welding Inspection and Nondestructive Examination, in Volume 1 of the *Welding Handbook*.[8]

Lap joints tested with the weld in shear will undergo an eccentricity of loading resulting in rotation of the joint at the weld as the test load increases. Resistance to joint rotation increases in proportion to increasing sheet thickness. The joint may fail either by shear through the nugget, or by tearing of the base metal adjacent to the weld nugget, as shown in Figure 1.16. Normally, in mild steels with tensile strength below approximately 550 megapascals (MPa) (80 kips per square inch [ksi]), low weld strengths are associated with weld nugget shear failure, and high strengths are associated with base metal tearing. As the base metal tensile strength approaches the weld nugget strength, more nugget shear can be expected for a specific weld size. A specific minimum-size nugget diameter is required to obtain failure by base metal tearing. The specific nugget diameter necessary to produce base metal tearing is directly related to the type of base material, surface condition, and, if applicable, coating type.

7. American Welding Society (AWS) Committee on Resistance Welding, 2000, *Recommended Practices for Resistance Welding,* Miami: American Welding Society.

8. American Welding Society (AWS) Welding Handbook Committee, Jenny, C. L. and A. O'Brien, eds., 2001, *Welding Handbook,* 9th Ed., Vol. 1, *Welding Science and Technology,* Miami: American Welding Society.

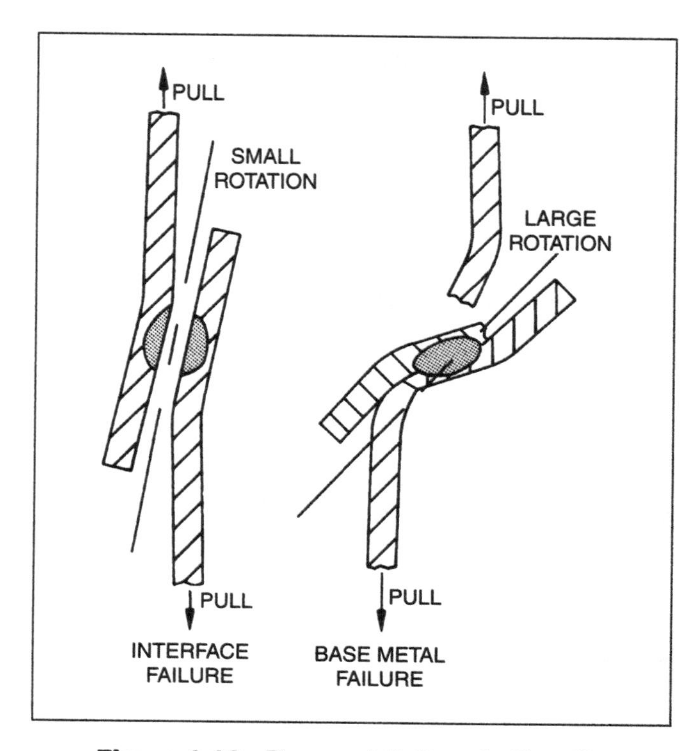

Figure 1.16—Types of Failure in Tensile Shear Test as a Function of Rotation

Increasing the nugget diameter above this minimum value may result in some increase in the weld strength. Figure 1.17 shows the slight increase in strength values for low-carbon steel as the nugget size increases.

Spot welds have relatively low strength when stressed in tension by loading transverse (normal) to the plane of the sheets. This is due to the sharp notch between the sheets at the periphery of the weld nugget; consequently, when the weld will be subjected to loading of this type, the weld must be designed to compensate for this lower mechanical performance.

The strength of multiple spot-welded joints is dependent on material thickness, weld spacing, and weld pattern. The spacing between adjacent spot welds may alter the weld strength of the joint due to current shunting through previous welds. As the spacing between adjacent spot welds decreases, the joint shear strength may decrease.

Figure 1.18 shows the effect of shunt distance (weld spacing) on the tensile shear strength of spot welds. Data were taken from welds made on 6.35 mm (1/4 in.) thick by 76 mm (3 in.) wide strips of mild steel. All welds were made with one shunt circuit. The average shear strength of twenty-four welds was 8000 kg (17 570 lb).

To obtain a desired joint strength, the number of required welds must satisfy minimum spacing requirements in order to reduce the effects of current shunting. A staggered weld pattern of multiple rows provides better strength than a rectangular pattern because the staggered pattern distributes the load more efficiently among the spot welds.

Table 1.1 summarizes the relationship between resistance spot welding variables and joint strength and suggests resistance spot welding schedules for welding two equal-thickness uncoated low-carbon steel sheets, showing the resulting minimum shear strengths and the diameters of the nugget and button (a button is the part of a weld, including all or part of the nugget, that tears out during destructive testing).

Electrode Maintenance

Proper maintenance of electrodes is necessary for the production of consistently good welds. An abnormal increase in the size of the electrode faces contacting the workpieces is detrimental to weld strength and quality. For example, if an electrode face 6.3 mm. (1/4 in.) in diameter is allowed to increase to 7.9 mm (5/16 in.) by mushrooming, the contact area will increase 50%, with a corresponding decrease in current density and

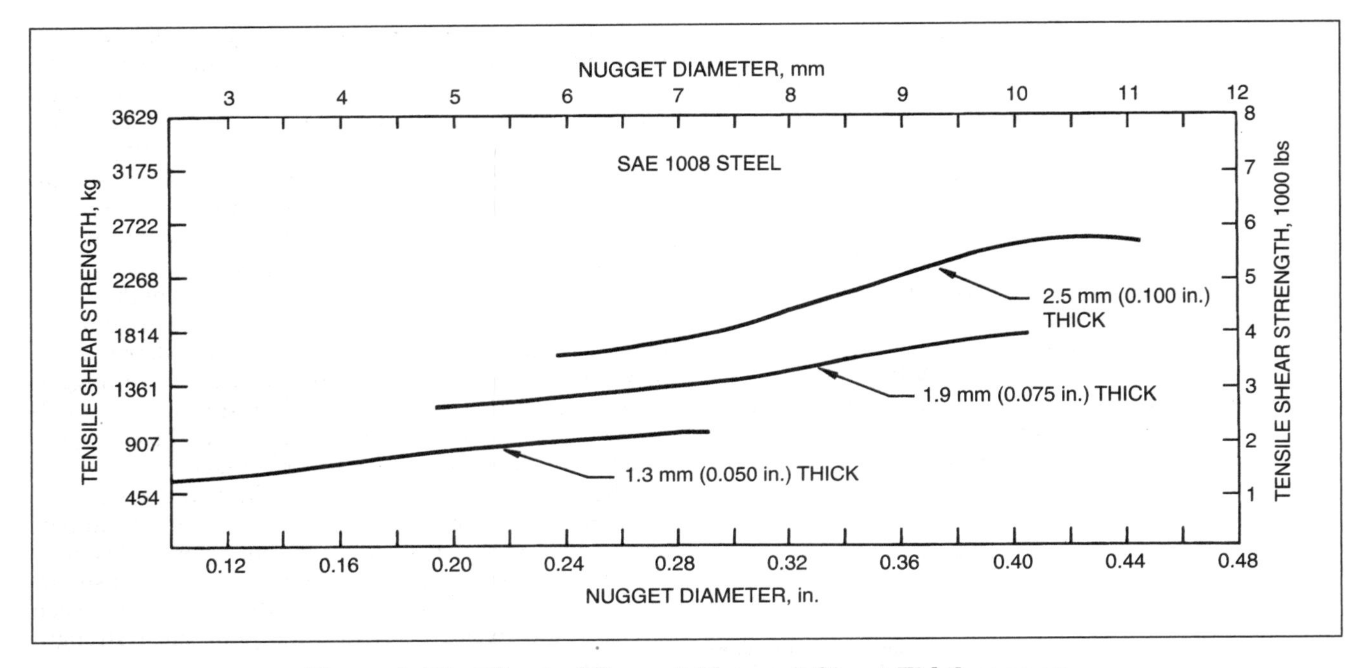

Figure 1.17—Effect of Nugget Size and Sheet Thickness on Tensile Shear Strength, Failure Occurring by Base Metal Tear-Out

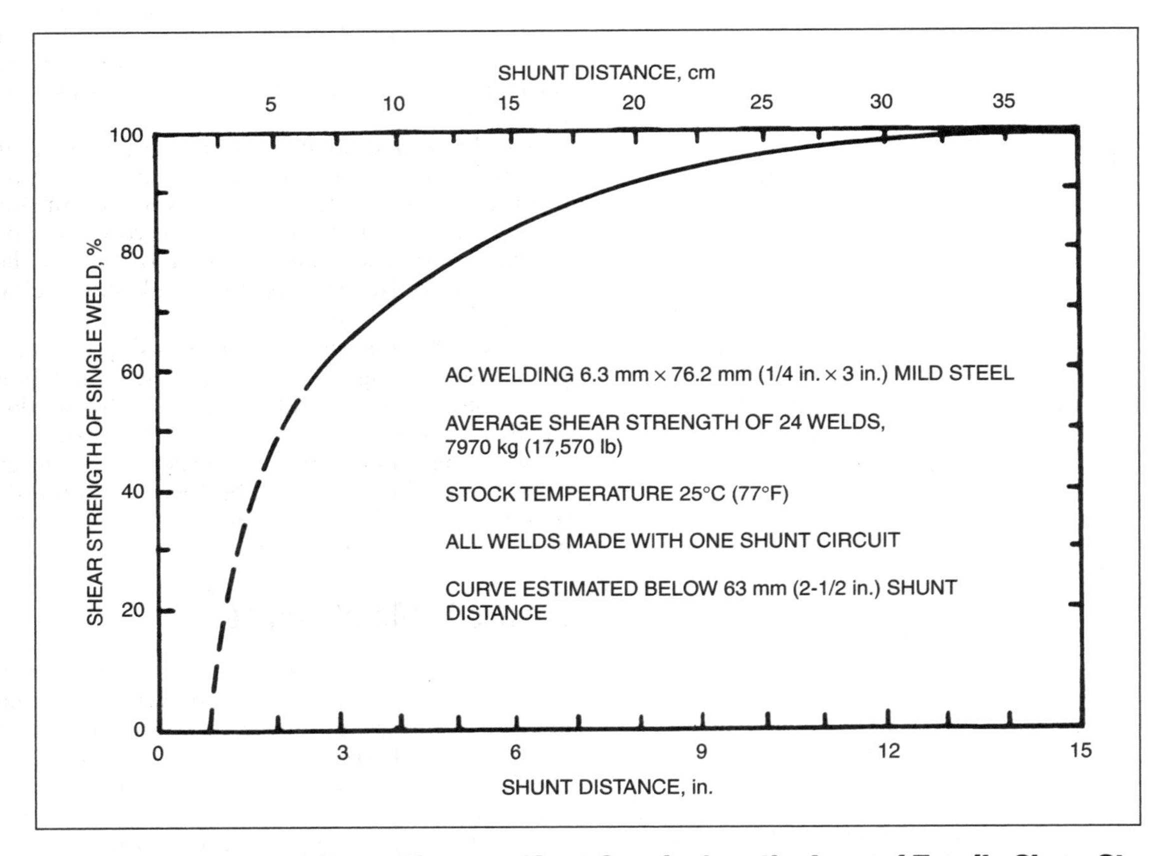

Figure 1.18—The Effect of Shunt Distance (Spot Spacing) on the Loss of Tensile Shear Strength

Table 1.1
Suggested Schedules for Spot-Welding Two Equal-Thickness Uncoated Low-Carbon Steel Sheets

Sheet Thickness		Electrode				Force		Weld Time (60-Hz)	Welding Current (Approx.)	Minimum Contact Overlap		Minimum Weld Spacing				Minimum Shear Strength		Button Diameter	
		Face Dia.			Bevel Angle**							2 Stack		3 Stack					
mm	in.	mm	in.	Shape*	Degrees	kN	lbf	Cycles	A	mm	in.	mm	in.	mm	in.	kN	lb	in.	mm
0.5	0.02	4.7	0.188	E, A, B	45	1.8	400	7	8500	11.0	0.44	9.5	0.38	15.5	0.62	1.4	320	0.10	2.5
0.6	0.03	4.7	0.188	E, A, B	45	2.0	450	8	9500	11.8	0.47	15.5	0.62	22.0	0.88	2.0	450	0.12	3.0
0.8	0.03	6.3	0.25	E, A, B	45	2.2	500	9	10 500	11.8	0.47	15.5	0.62	22.0	0.88	2.6	575	0.14	3.6
0.9	0.04	6.3	0.25	E, A, B	45	2.7	600	9	11 500	13.3	0.53	18.8	0.75	26.5	1.06	3.3	750	0.16	4.1
1.0	0.04	6.3	0.25	E, A, B	45	3.1	700	10	12 500	13.3	0.53	18.8	0.75	26.5	1.06	4.1	925	0.18	4.6
1.1	0.05	6.3	0.25	E, A, B	45	3.3	750	11	13 000	14.8	0.59	23.5	0.94	29.5	1.18	5.1	1150	0.19	4.8
1.3	0.05	7.8	0.312	E, A, B	30	3.6	800	12	13 500	14.8	0.59	23.5	0.94	29.5	1.18	6.0	1350	0.20	5.1
1.4	0.06	7.8	0.312	E, A, B	30	4.0	900	13	14 000	15.8	0.63	26.5	1.06	32.8	1.31	7.5	1680	0.21	5.3
1.5	0.06	7.8	0.312	E, A, B	30	4.4	1000	14	15 000	15.8	0.63	26.5	1.06	32.8	1.31	8.2	1850	0.23	5.8
1.8	0.07	7.8	0.312	E, A, B	30	5.3	1200	16	16 000	16.5	0.66	29.5	1.18	37.5	1.50	10.2	2300	0.25	6.3
2.0	0.08	7.8	0.312	E, A, B	30	6.2	1400	18	17 000	18.0	0.72	34.5	1.38	40.0	1.60	12.0	2700	0.26	6.6
2.3	0.09	9.4	0.375	E, A, B	30	7.1	1600	20	18 000	19.5	0.78	39.0	1.56	47.0	1.88	15.3	3450	0.27	6.9
2.6	0.11	9.4	0.375	E, A, B	30	8.0	1800	23	19 500	21.0	0.84	42.0	1.68	50.0	2.00	18.5	4150	0.28	7.1
3.0	0.12	9.4	0.375	E, A, B	30	9.3	2100	26	21 000	22.0	0.88	45.3	1.81	62.5	2.50	22.2	5000	0.30	7.6

*Shape Definitions:
E = Truncated cone
A = "A-Nose" pointed
B = 3-in. radius

**Applies to truncated-cone electrodes only and is measured from the plane of the electrode face.

Notes:
1. For metals of intermediate thickness, force and weld time may be interpolated.
2. Minimum weld spacing is measured from centerline to centerline.
3. The data in this table were supplied by the AWS D8 Subcommittee on Sheet Steel and represent an average of typical variables used by the automotive industry. Refer to AWS D8.6, *Standard for Automotive Resistance Spot Welding Electrodes*, for electrode specifications.

pressure. Depending somewhat on the weld schedule, the result may be weak or defective welds. A visual indicator is the production of poorly shaped spots, which may be caused by the following conditions:

1. Noncircular electrode faces,
2. Too large a flat face on the electrode,
3. Concavity or convexity of the electrode face, and
4. Misalignment of the electrodes with respect to the workpiece.

Correct electrode alignment is relatively easy to maintain with stationary welding machines and proper supporting fixtures; however, misalignment is common with portable gun-type machines. The seriousness of this condition depends on the ease with which the equipment can be manipulated and correctly positioned for welding. It is likely that the electrodes will have longer life between maintenance dressings when welding positioned workpieces with stationary machines than when welding non-positioned workpieces with portable welding guns.

WELD BONDING

Weld bonding is a resistance spot welding process variation in which the spot-weld strength is augmented by an adhesive at the faying surfaces. Weld bonding is a combination of resistance spot welding and adhesive bonding. An adhesive in the form of paste or film is placed between the workpieces; resistance welds are then made through the adhesive layer. The adhesive is allowed to cure either at ambient temperature or by heating in an oven, as required by the adhesive manufacturer. The principal purpose of the spot welds is to hold the joint together during curing, since the welds do not contribute greatly to the strength of the joint. When large bond areas are used, fewer spot welds may be used than otherwise would be required. Adhesive around the weld also greatly increases the fatigue life of the joint. Chapter 10 of this volume, "Adhesive Bonding," provides more detailed information.

Structures common to the aerospace and transportation industries are joined by weld bonding. For example, weld bonding is used to attach beaded panels to aircraft skins, and aircraft or truck skins to channels, angles, and other types of reinforcement.

The adhesive, whether paste or film, can be applied to one or both joint surfaces. The electrode force during welding squeezes out the adhesive at the spot weld locations, creating a current path through the sheets. The adhesive must have good wetting and flow characteristics in order to create a secure bond at the faying surfaces. Premature curing of the adhesive, prior to or during spot welding, may hamper proper adhesive movement and result in high resistance between the faying surfaces. High resistance may impede weld current or result in excessive heating and subsequent metal expulsion. The application of a precompression electrode force prior to the welding cycle may help displace the adhesive at each weld site.

Advantages and Limitations

Among the advantages of weld bonding are improved fatigue life and durability of the joint compared to joints made with spot welding alone. The process also may improve stress distribution and joint rigidity and result in higher resistance to buckling in thin sheets. The adhesive in the joint dampens vibration and noise and provides some corrosion resistance. In some aircraft components, greater cost effectiveness is obtainable with weld bonding than with mechanical fastening or adhesive bonding alone.

In most applications, the limitations of weld bonding include the additional costs of the adhesive, added time required for the curing operation, and the time and labor costs of cleaning the components. Additional cost factors include treating the workpiece surfaces, applying the adhesive, and maintaining the desired bond line thickness in the joint. Bond line thickness is sometimes maintained by the addition of glass beads or conductive fillers of a specified diameter. Used as fillers, these beads may interfere with weld nugget development. Another limitation is that operating temperatures in service for the component are limited to the effective service temperature of the adhesive.

The presence of adhesive in the joint makes welding more difficult, and may contribute to significant variations in weld quality. Regardless of welding conditions, not all of the adhesive will be displaced from between the sheets; therefore the contact resistance will be higher than with clean sheets.

RESISTANCE SEAM WELDING

A resistance seam weld is made on overlapping workpieces and is a continuous weld formed by overlapping weld nuggets, by a continuous weld nugget, or by forging the joint as it is heated to the welding temperature by resistance to the flow of welding current.

ADVANTAGES AND LIMITATIONS

Resistance seam welding has the same advantages and limitations as resistance spot welding. An additional advantage is the ability to produce a continuous leak-tight weld. Seam welds must be made in a straight or uniformly curved path. Corners can be welded by interrupting the weld and crossing the end of the first with the start of the next weld. Leak-tight joints must be made with a continuous weld seam. A continuous weld seam cannot be produced if abrupt changes occur in welding direction or in joint contour along the path. This limits the design of the assembly.

The strength properties of seam-welded lap joints are generally lower than those of fusion-welded butt joints due to the eccentricity of loading on lap joints and the built-in notch along the nugget at the sheet interface.

TYPES OF SEAM WELDS

Details of three types of seam welds in sheet metal: the lap seam weld, mash seam weld, and metal finish seam weld are shown in cross section in Figure 1.19. The joint in Figure 1.19(A) is classified as a lap seam weld, a joint in which the workpieces are overlapped sufficiently to prevent the sheet edges from becoming part of the weld. Lap seams are commonly used in automotive applications such as fuel tanks, catalytic converters, mufflers, and roof joints. Other applications include the manufacturing of furnace heat exchangers, water tanks, and cans of all types. The lap-seam welding of multiple stackups and materials of dissimilar thickness is possible.

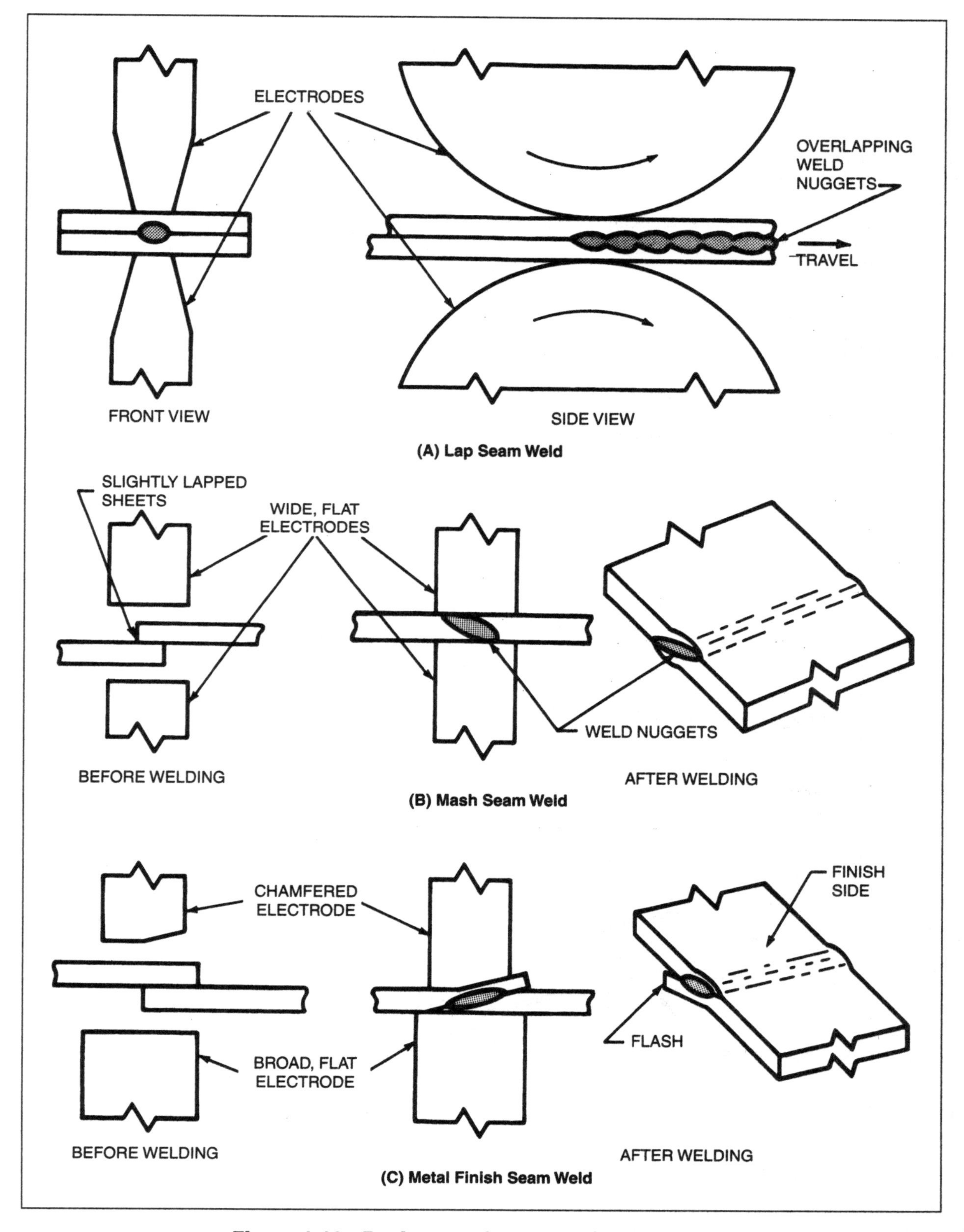

Figure 1.19—Resistance Seam Welding Variations

Lap joints can be seam-welded using two wheel electrodes, as shown in Figure 1.20, or with one wheel and a mandrel. The minimum joint overlap is the same as for spot welding: two times the minimum edge distance (the distance from the center of the weld nugget to the edge of the sheet).

When surface appearance is not a concern, radius-face wheels generally will provide a wider current range because as the weld becomes hotter the wheels sink in more, thus increasing the area and reducing the current density. The wheels generally will require continuous dressing or rolling to maintain shape.

APPLICATIONS

Seam welds typically are used to produce continuous gas-tight or liquid-tight joints in sheet assemblies, such as automotive gasoline tanks. The process also is used to weld longitudinal seams in structural tubular sections that do not require leak-tight seams. In most applications, two wheel electrodes or one translating wheel and a stationary mandrel are used to provide the current and pressure for resistance seam welding. The position of the electrode wheels on a resistance seam welding machine is shown in Figure 1.20. Leak-tight seam welds also can be produced using spot welding electrodes. This requires the purposeful overlapping of the spot welds to obtain the leak-tight weld. Overlapping spot welding requires an increase in power after the first spot weld to offset the shunting effect in order to obtain adequate nugget formation as welding progresses. Figure 1.21 shows a resistance seam welding machine making a leak-tight joint on a coated-steel fuel tank. This machine does not use a knurl wheel as an electrode but uses a wheel to deliver a wire electrode, which is wrapped around the wheel so that it always presents a new contact surface to the workpiece.

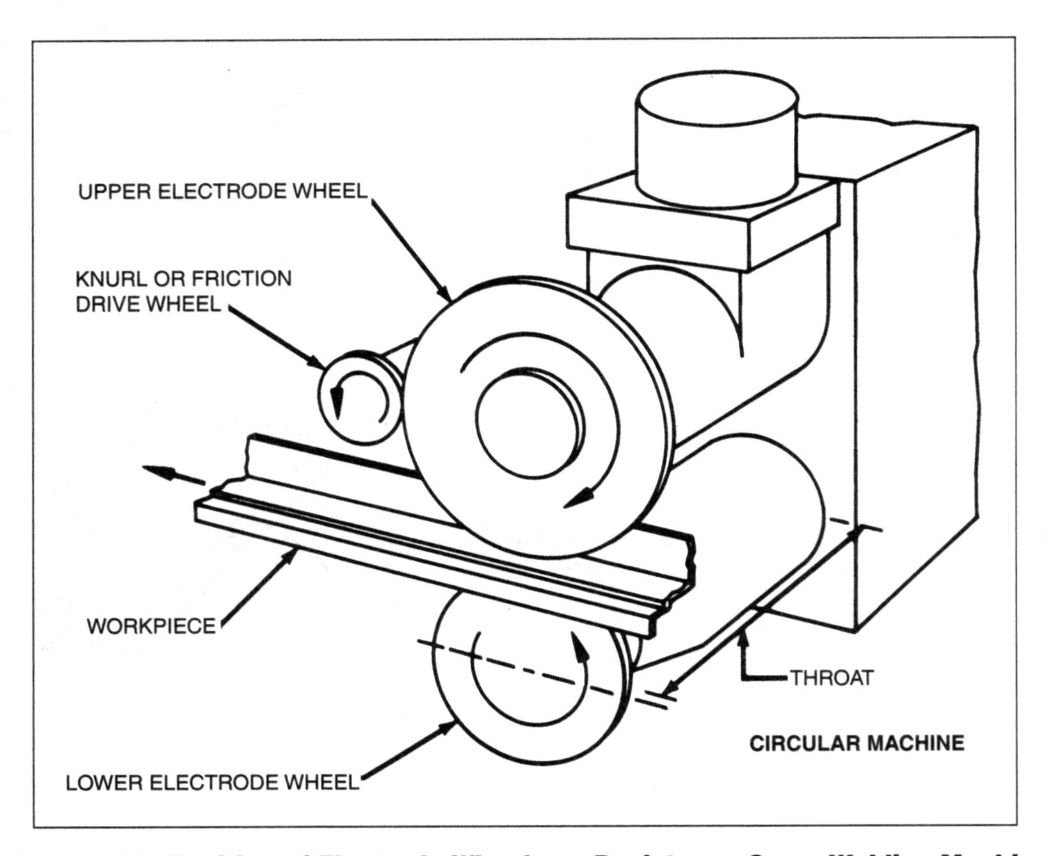

Figure 1.20—Position of Electrode Wheels on Resistance Seam Welding Machine

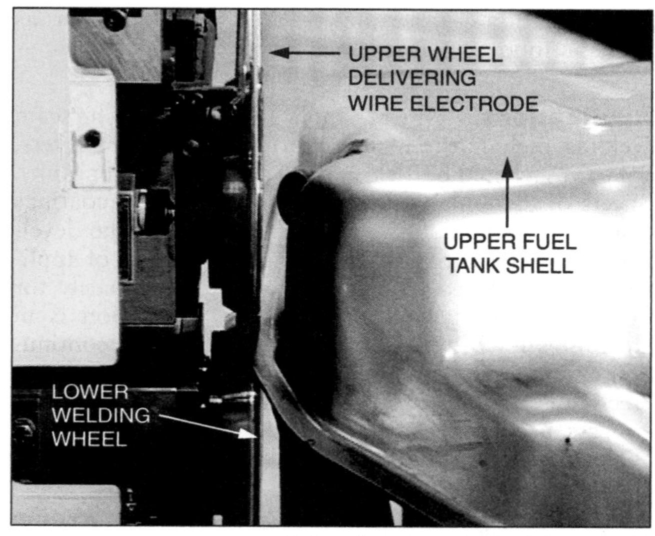

Photograph courtesy of Soudronic Automotive AG

Figure 1.21—Resistance Seam Welding of a Coated-Steel Fuel Tank

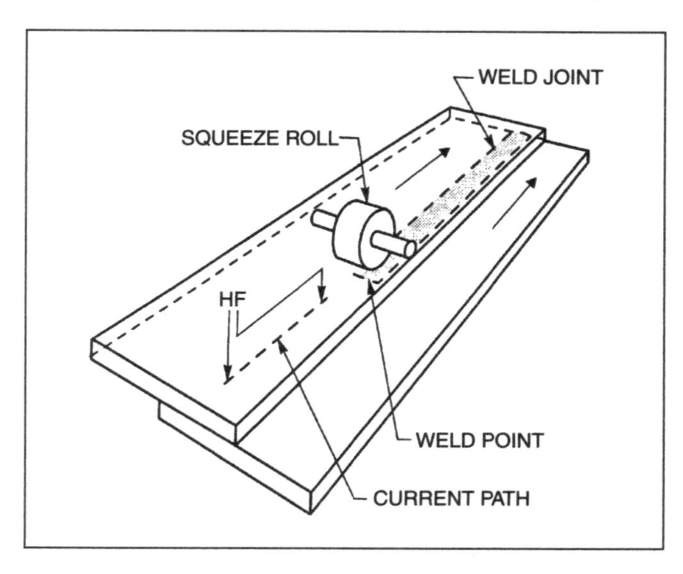

Figure 1.22—Configuration for a Lap Joint Made by High-Frequency Seam Welding

SEAM WELDING PROCESS VARIATIONS

High-frequency seam welding, induction seam welding, and mash seam welding are variations of the resistance seam welding process. Other modes of seam welding operations described in this section are metal finish seam welds, electrode wire seam welds, and roll spot welding.

High-Frequency Seam Welding

High-frequency seam welding (RSEW-HF) is a resistance seam welding process variation in which high-frequency welding current is supplied through electrodes into the workpieces. The configuration for a lap joint using high-frequency seam welding is shown in Figure 1.22. High-frequency seam welding is discussed in Chapter 5 of this volume.

Induction Seam Welding

Induction seam welding (RSEW-I) is a resistance seam welding process variation in which high-frequency welding current is induced in the workpieces. This process is used mainly in the production of tubes. Another application is to provide annealing of weld joints. Induction seam welding is discussed in detail in Chapter 5, "High Frequency Welding."

Mash Seam Welding

Mash seam welding (RSEW-MS) is a resistance seam welding process variation that makes a lap joint primarily by high-temperature plastic working and diffusion as opposed to melting and solidification. The joint thickness after welding is less than the original assembled thickness.

A mash seam weld is a joint in which the contact overlap of the two metals is only one to two times the thickness of the thinner sheet. Two exceptions are welds in tailored blank welds, in which the recommended overlap is the sum of the two sheet gauges but with a 3.5 mm (0.14 in.) maximum, and the welding of tinplate cans with a sheet overlap of two to four times the sheet thickness. The weld area is forged or mashed down during welding to a total thickness of 5% to 25% greater than the original starting thickness of a single sheet. Even in this instance, welds in tailored blanks and in can bodies are exceptions, with a thickening of up to 50% of a single sheet or the thinner sheet if the two sheets are different. It is necessary to have some overage of thickness so the current delivered from the electrode wheels can be confined to the weld area. This is needed because the resistance welding electrodes used in mash seam welding are flat-faced wheels and usually are wider than the required weld. The final thickness and surface appearance of the finished weld can be improved and controlled by planishing. Planishing wheels are hardened steel or carbide wheels that cold-work the joint after welding. Because of hardening induced by cold-working of the planishing process, postweld heat treatment may be necessary.

Mash seam welding, shown in Figure 1.19(B), requires considerably less overlap than the conventional lap joint. Wide, flat-faced wheel electrodes that completely cover the overlap are used. Mash seam welding requires high electrode force, continuous or pulsed welding current, and accurate control of force, current, welding speed, overlap, and joint thickness in order to obtain consistent welding characteristics. Overlap is maintained at close tolerances, usually by rigidly clamping or tack-welding the workpieces.

In certain applications, the exposed surface or show side of the welded component is placed against a mandrel, which acts as an electrode and supports the workpieces. A wheel electrode is applied to the side of the joint that does not show. The show surface of the joint must be mashed as nearly flat as possible so that it will present a good appearance. Proper positioning of the wheel with respect to the joint is required to obtain a smooth weld face. When the appearance of the finished product is important, some polishing of the weld area may be required before painting or coating.

Mash seam welding produces continuous seams that have good appearance and are free of crevices. Crevice-free joints are necessary in applications that have strict contamination or cleanliness requirements, such as joints in food containers or refrigerator liners.

Applications. Typical applications of mash seam welding include the manufacturing of drums, buckets, vacuum-jacketed bottles, aerosol cans, and water tanks. Sheets of dissimilar thickness with or without coatings also can be successfully mash seam welded. The development of this capability has created a new set of applications in tailored blank manufacturing, primarily for automotive use. Another important application is in steel mills for the joining of sheets to produce continuous coils of sheet metal. Figure 1.23 shows tin-coated steel cans manufactured with the mash seam welding process using intermediate copper wire electrodes. A typical mash seam welding machine can weld up to 1000 can bodies per minute.

Disadvantages of the mash seam welding variation include the following:

1. Offset occurs at the joint due to the inability of the process to completely flatten the seam,
2. Distortion caused by the inherent lateral flow of metal as it is welded requires restraining by fixturing or by tack welding, and
3. Very rigid fixturing is required to resist weld distortion.

To obtain acceptable welds, the materials to be joined by mash seam welding must have wide plastic temperature ranges. Low-carbon steels or stainless steels can be joined by mash seam welding in certain applications.

Photograph courtesy of Soudronic Automotive AG

Figure 1.23—Tin-Coated Steel Can Bodies Welded with the Mash Seam Welding Process

Metal Finish Seam Welding

Metal finish seam welds, lap welds, and mash seam welds differ with respect to the amount of forging, or "mash down" that takes place during welding. Practically no forging occurs in lap welding while the thickness of a mash seam weld approaches that of one sheet thickness. Metal finish seam welding is a compromise between lap and mash seam welding: forging occurs on only one side of the joint, as shown in Figure 1.19(C).

The amount of deformation, or mash, is affected by the geometry of one electrode wheel face and the position of the joint relative to the face. The wheel face is beveled on one side of the midpoint, as illustrated in Figure 1.24. This configuration varies the amount of deformation across the joint. Good surface finish can be produced on the side of the joint against the flat wheel when proper welding procedures are used.

The location of the edge of the sheet contacting the flat-faced electrode relative to the bevel on the other electrode must be held within close tolerance, as shown in Figure 1.24. For example, with 0.8 mm (0.031 in.) low-carbon steel sheet, the edge must be within 0.4 mm (0.016 in.) of center. The overlap distance is not critical.

Higher amperage and electrode force are required for metal finish seam welding than for mash seam welding because of the greater overlap distance. Materials that are easily joined with mash seam welding (materials with wide plastic temperature ranges) also are easily welded using the metal finish seam welding technique.

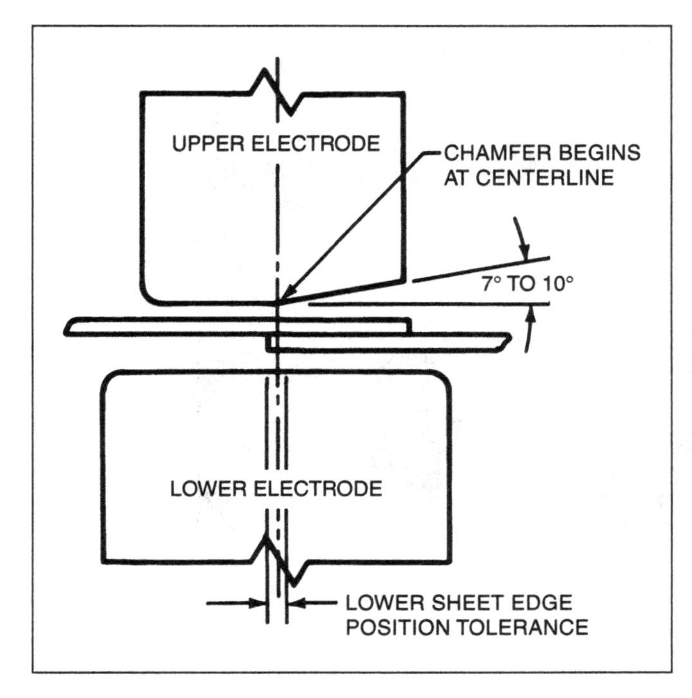

Figure 1.24—Electrode Face Contour and Joint Position for Metal Finish Seam Welding

Electrode Wire Seam Welding

Electrode wire seam welding is a technique that uses an intermediate copper wire electrode between each wheel electrode and the workpiece, as shown in Figure 1.25. The technique is used almost exclusively for mash seam or overlap welding of tin mill products to fabricate cans and as overlap welding for the welding of fuel tanks made from metallic-coated or organic-coated steel sheets.

Figure 1.26 shows a seven-axis, programmable wire electrode seam welding machine designed to weld fuel tanks using intermediate copper wire electrodes. Travel speed in this type of application is 5 to 8 meters per minute (m/min) (16.4 to 26.2 feet per minute [ft/min]), depending on the sheet quality, thickness, and coating. Figure 1.27 shows a close-up of the wheel electrode assembly. In this system, the copper wire electrode travels around the wheel electrodes at the selected welding speed. The copper wire electrode provides a continuously renewed surface, but it is not consumed in the welding operation. This avoids the buildup of coating material that would occur on a copper wheel electrode. The copper wire electrode may have a circular, flat, or other cross section profile.

Figure 1.28 shows profiles of copper wire electrodes used to overlap welds on sheet metal with coatings of varying thicknesses. A cross section of the weld and recommended wire profiles and diameters also are shown.

Electrode wire seam welding requires specially designed welding systems. The seam welds may be made with two wheel electrodes, or one wheel electrode and a mandrel electrode.

The temperature range of a mash seam weld made with electrode wire should provide good solid-phase bonding and should not exceed the melting point of the base metal. If the surface of the seam weld reaches temperatures greater than the melting point of the base metal, spikes or splashes of molten metal will be expelled from the seam weld. Splashes of this material can lead to corrosion of the weldment, and therefore are undesirable.

Electrode wire seam welding has little tolerance for temperature variation, but very high welding speeds are obtained by the very constant resistance maintained between the weld wheel, the intermediate copper wire electrode, and the sheet surface. As examples, in the welding of can bodies, the welding speed is up to 100 m/min (328 ft/min). The welding speed in fuel tank welding is in the range of 5 m/min to 8 m/min (16 ft/min to 26 ft/min). Variations in welding temperature resulting from

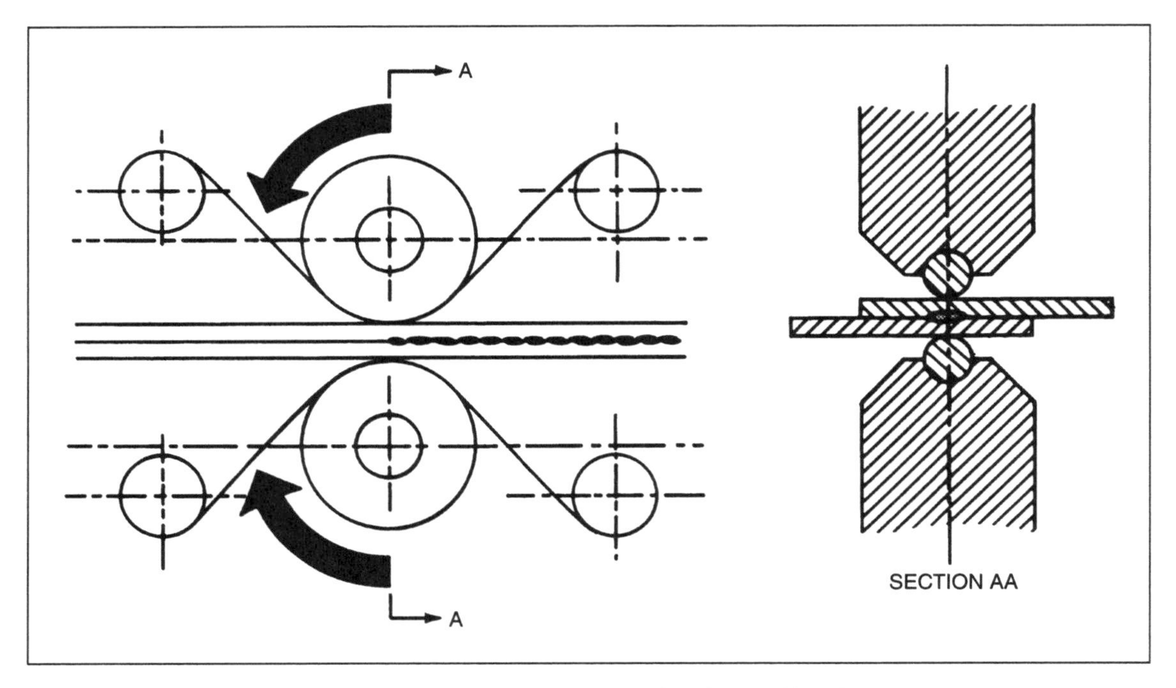

Figure 1.25—Electrode Wire Seam Weld

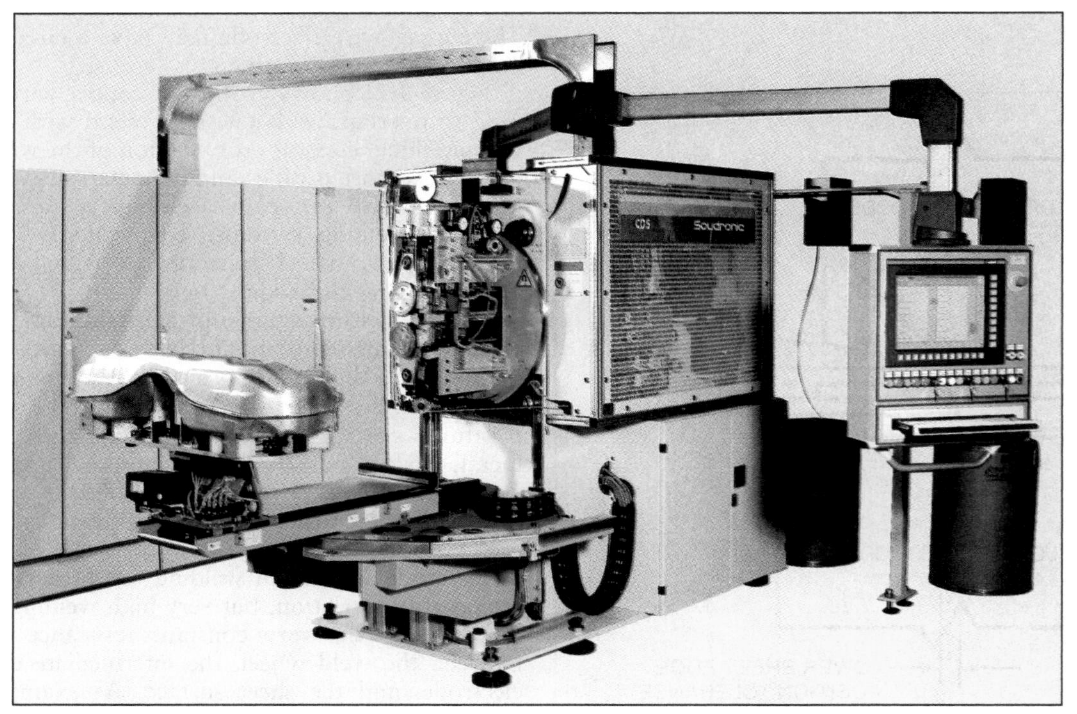

Photograph courtesy of Soudronic Automotive AG

Figure 1.26—Seven-Axis, Programmable Wire-Electrode Seam Welding Machine Customized for Welding Fuel Tanks

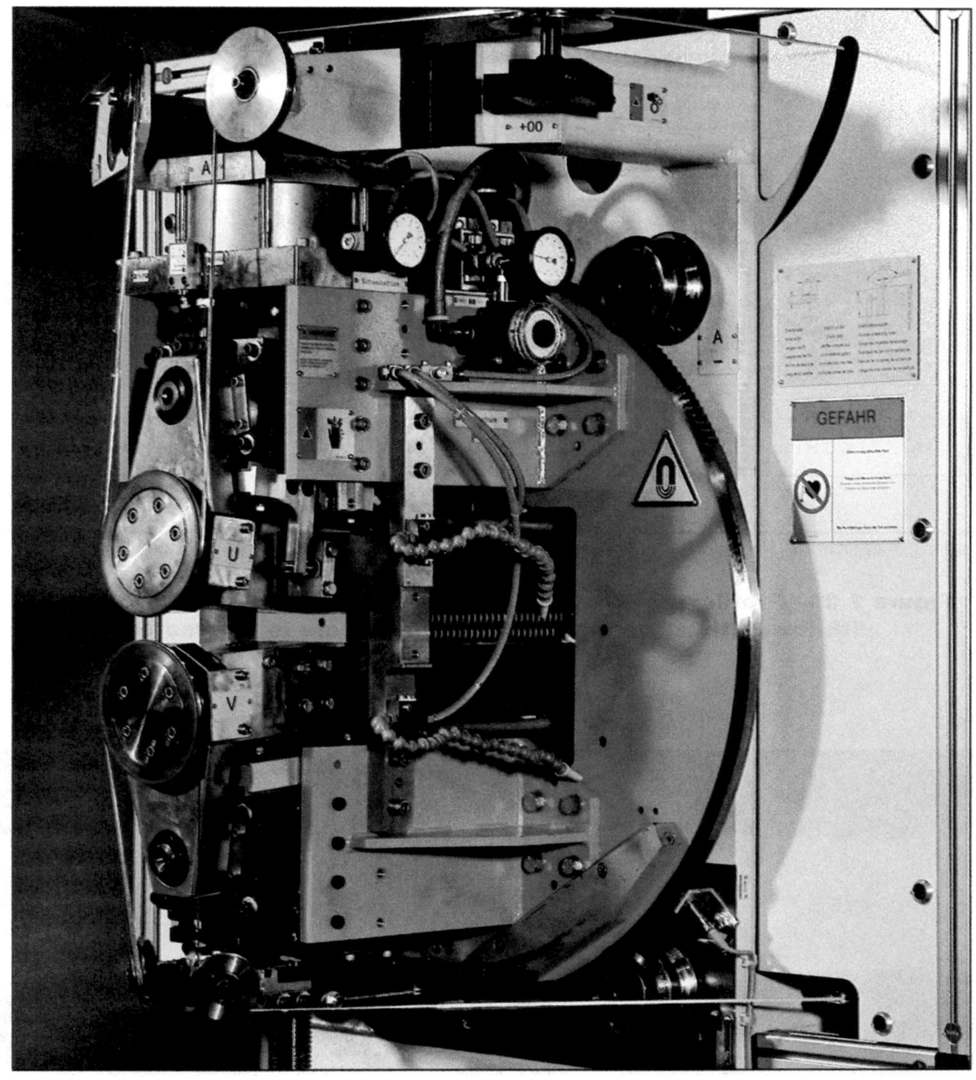

Photograph courtesy of Soudronic Automotive AG

Figure 1.27— Close-up of the Wheel Electrode and Wire-Electrode Assembly in a Seam Welding Machine

fluctuations in electrical power or electrode pressure and changes in overlap distances usually are acceptable.

Figure 1.29 shows a micrograph of transverse and longitudinal sections of the overlap of a seam weld in a fuel tank. The material is steel sheet with an electrolytic zinc-nickel coating and an additional coating of aluminum-rich paint. The seam is welded with copper wire electrodes Type E, shown in Figure 1.28.

Butt Seam Welds. Two types of butt joints, autogenous and exogenous, can be made with electrode wire seam welding. A joint in which two abutting edges are welded without a filler metal is classified as an autogenous butt seam joint. The thickness of the weld should be approximately the same or slightly less than the thickness of the sheet. Autogenous butt seam welding typically is reserved for applications in which other butt-welding processes cannot be used, for example, for tube welding and for welding sheet metal in railroad cars.

A joint in sheet metal in which two abutting edges are welded using a strip of filler metal (usually called a

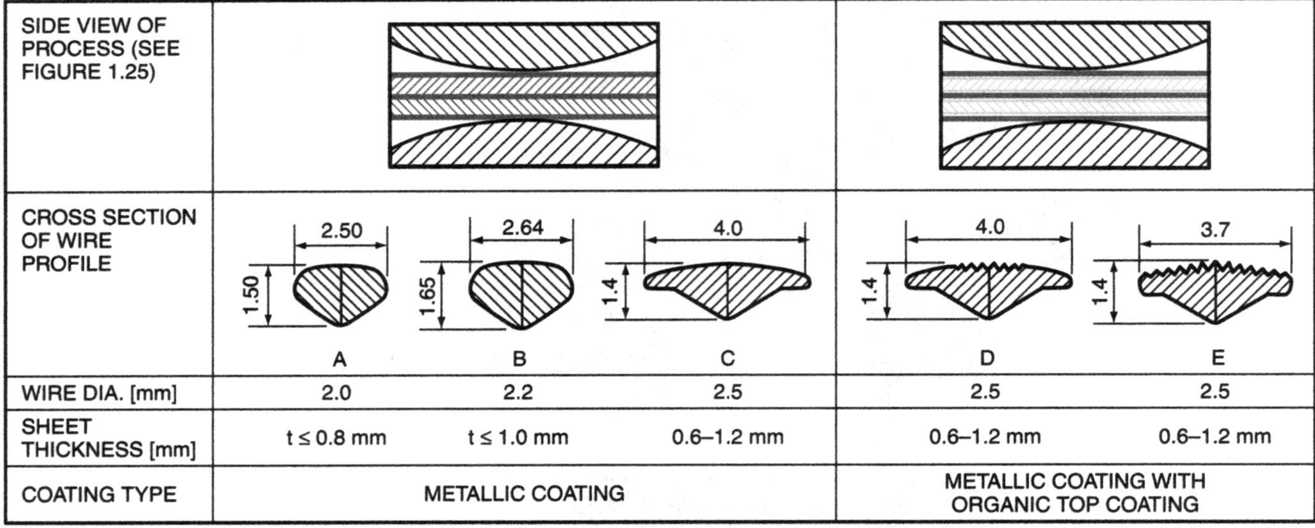

SIDE VIEW OF PROCESS (SEE FIGURE 1.25)					
CROSS SECTION OF WIRE PROFILE	2.50; 1.50 A	2.64; 1.65 B	4.0; 1.4 C	4.0; 1.4 D	3.7; 1.4 E
WIRE DIA. [mm]	2.0	2.2	2.5	2.5	2.5
SHEET THICKNESS [mm]	t ≤ 0.8 mm	t ≤ 1.0 mm	0.6–1.2 mm	0.6–1.2 mm	0.6–1.2 mm
COATING TYPE	METALLIC COATING			METALLIC COATING WITH ORGANIC TOP COATING	

Source: Adapted from Soudronic Automotive AG.

Figure 1.28—Profiles of Wire for Use with Various Types of Sheet Metal with Protective Coatings (Measurements are in Millimeters)

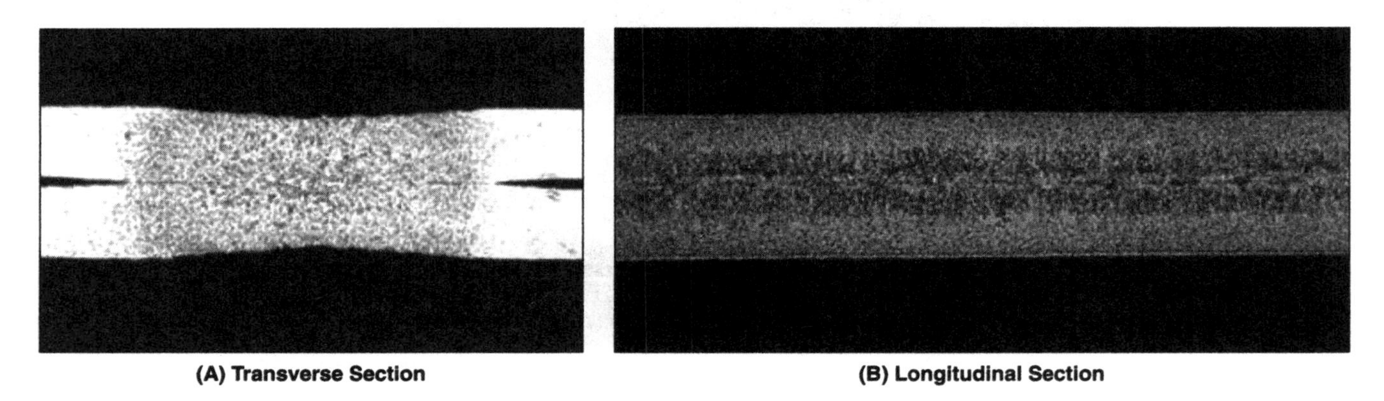

(A) Transverse Section

(B) Longitudinal Section

Figure 1.29—Seam Weld in Prepainted Zn/Ni-Coated Steel Using an Electrode Wire System

foil seam weld) that becomes part of the weld is classified as an exogenous butt seam weld. An exogenous weld is made by positioning the edges of the sheets to form a butt joint. The filler, usually a thin, narrow strip of metal that also serves as an electrode, is fed between the workpieces and the wheel electrode and welded to one or both sides of the joint. The metal strip bridges the root opening, distributes the welding current to both sheet edges, provides added electrical resistance, and contains the molten weld nugget as it forms. This produces a flush or slightly reinforced weld joint, as shown in Figure 1.30.

Strip electrode configurations can be circular, triangular, or flat. The metal strip must be guided accurately and centered on the joint to ensure even distribution of current to both sheet edges. The strip may be welded to the sheets by roll spot welding at low power before the joint is seam-welded. Very little forging of the joint during welding is required; therefore, less distortion of the joint results compared to lap-seam welding. The travel speed for butt-joint

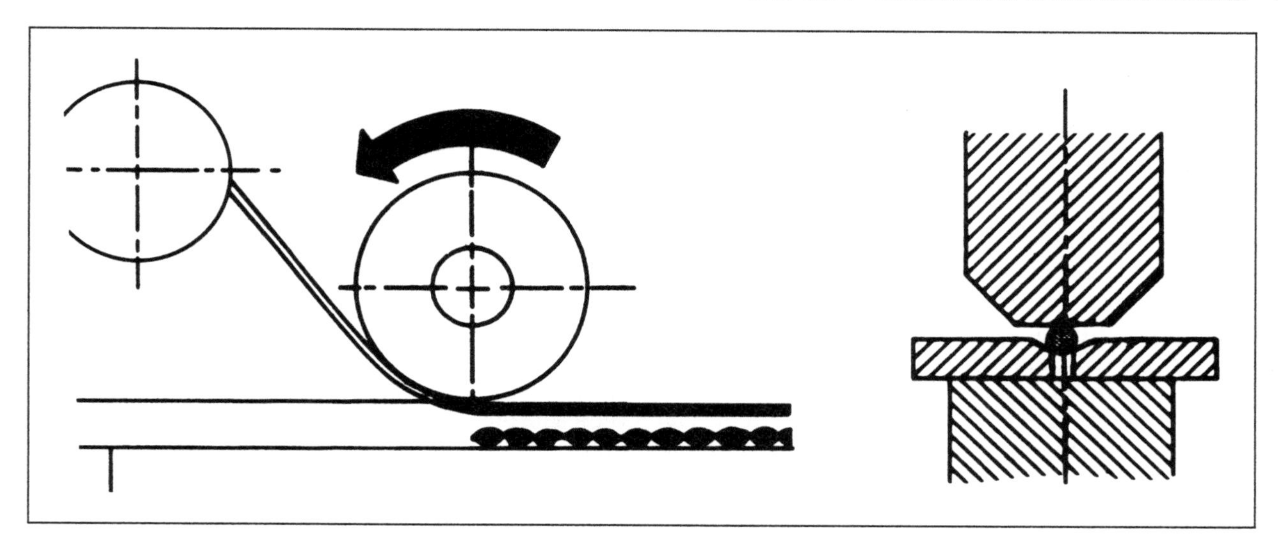

Figure 1.30—Exogenous Seam-Welded Butt Joint

seam welding of low-carbon steel is comparable to that for lap-seam welding. To achieve good corrosion resistance, welding procedures must ensure that welding of the strips over the entire width of the joint is maintained.

Roll Spot Welding

Roll spot welding is a resistance seam welding mode that makes intermittent spot welds using one or more rotating circular electrodes. The rotation of the electrode wheels may or may not be stopped during the weld cycle. In this method, a row of spaced spot welds can be made with a seam welding machine without retracting the electrode or removing the electrode force between welds. The radius of the wheel electrode, the contour of the face, and the weld time influence the shape of the nugget. The nugget usually is oval-shaped.

Weld spacing is obtained by the adjustment of cooling time with the wheel electrodes continuously rotating at a set speed. Holding time is effectively zero. Roll spot welding also may be done with interrupted electrode rotation when a hold-time period is needed to consolidate the weld nugget as it cools.

When continuously moving electrodes are employed, as commonly is the case, weld time usually is shorter and welding amperage higher than those used for conventional spot welding. The higher amperage employed sometimes may require the use of a higher electrode force. Otherwise, the recommended practices for spot welding apply.

Continuous Current. To weld low-carbon steel, welding current can be applied continuously along the length of the seam at high travel speeds if the waveform of the available current will produce the proper nugget size and spacing. In this case, the level of weld quality required may be secondary to high-production requirements. Continuous current can be used for sheet up to 1 mm (0.040 in.) thick. Above this thickness, the condition of the surface has a significant detrimental effect on welding and electrode life is short. Continuous-current welds in a particular thickness can be made over a wide range of speeds. For example, two sheets of 1-mm (0.040-in.) steel stock can be welded at speeds ranging from 44 mm/s to 131 mm/s (105 in./min to 310 in./min). Amperage requirements increase with travel speed.

A problem that may arise when operating with continuous alternating current is arcing between the wheel electrode and a localized region of the weld assembly on the exit side of the electrode. The arcing may produce superficial melting of the sheet surface and the electrode. In steels, the rapid cooling of molten metal that results from workpiece-to-electrode arcing can result in the formation of brittle martensite. The localized martensitic microstructure may provide initiation sites for the formation of cracks.

Welding Speed. The speed of welding depends on the metal being welded, stock thickness, and the weld strength and quality requirements. In general, permissible welding speeds are much lower when welding stainless steels and nonferrous metals because of restrictions on heating rate necessary to avoid weld metal expulsion.

In some applications, it is necessary to stop the movement of the electrodes and workpieces as each weld nugget is made. This usually is the case for sections over 4.78 mm (0.188 in.) thick, and for metals that require postheating or forging cycles to produce the desired weld properties. Interrupted motion significantly reduces welding speed because of the relatively long time required for each weld.

When using continuous motion, welding machine adjustments for current, time, and speed must take place to maintain weld quality and joint strength: the welding current must be increased and the heating time decreased as welding speed is increased. There is a speed beyond which the required welding current may cause undesirable surface burning and electrode pickup. This will accelerate electrode wear.

Tandem Seam Welding

Two seam welds can be made in series by using two weld heads mounted side by side or in tandem. The two seams can be welded with the same welding current, and power demand will be only slightly greater than for a single weld.

A tandem wheel arrangement can reduce welding time by 50%, since both halves of a joint can be welded simultaneously. Thus, for a joint 182 cm (72 in.) long, two welding heads can be placed 91 cm (36 in.) apart, with the welding current path directed through the workpiece from one wheel electrode to the other. A third continuous electrode is used on the other side of the joint. The full length of the joint can be welded with only 91 cm (36 in.) of travel.

Electrodes. Seam welding electrodes normally are wheels ranging in diameter from 50 mm to 600 mm (2 in. to 24 in.). Typical wheel sizes are 175 mm to 300 mm (7 in. to 12 in.) in diameter and widths are 10 mm to 19 mm (3/8 in. to 3/4 in.).

The width of the cross section at the weld interface should range from 1-1/2 to 3 times the thickness of the thinner member. The ratio of weld width to sheet thickness normally decreases as the thickness increases. The weld width is always slightly less than the electrode face width when commercial welding schedules are used.

Seam welding electrodes are discussed further in Chapter 4, "Resistance Welding Equipment."

SEAM WELDING VARIABLES

Several variables must be considered when planning a seam welding application, including heat balance, welding cycle, current requirements, welding speed, electrodes, external cooling, and joint design.

Heat Balance

The seam welding of dissimilar types of metals or metals of unequal thickness presents the same heat balance problems as those of spot welding. The techniques for improving joint heat balance are similar. On workpieces requiring a low current density or fast cooling, the contact area between the workpiece and the electrode can be enlarged by increasing the diameter or width of the wheel electrode. As an alternative, one of the electrode wheels or mandrels may be made of an alloy with higher thermal conductivity to facilitate heat conduction from the workpiece via the electrode.

Welding Cycle

Resistance seam welds typically require higher welding currents than resistance spot welds due to the shunting of the welding current through previously made welds.

The resistance welding current normally is supplied in timed pulses (heating times) separated by periods of cooling times (hold times). A weld nugget is produced during each pulse of current. For a given welding speed and heating-cooling cycle, the welding current determines the depth of weld penetration. The weld nugget overlap is controlled by the welding current schedule and weld speed. As the welding speed increases, the ratio of heating time to cooling time must be increased in order to maintain weld overlapping of the nugget. The heating time controls the size of the weld nugget. Figure 1.31 shows the effect that cooling time has on weld nugget penetration (macrosection perpendicular to weld centerline) and weld overlap (macrosection parallel to weld centerline).

The use of welding current higher than required to obtain proper joint properties may result in excessive indentation or burning of the weldment. Short heating times or fast welding speeds require higher current levels for proper weld nugget formation, but may produce greater electrode wear.

Either pulsed (intermittent) current or continuous alternating current is used for resistance seam welding. As an example of the lap seam welding process, Table 1.2 provides suggested resistance seam welding conditions for welding uncoated low-carbon steel sheet.

Pulsed Current

Pulsed current usually is selected for most seam welding operations for the following reasons:

1. Good control of the heat can be maintained,

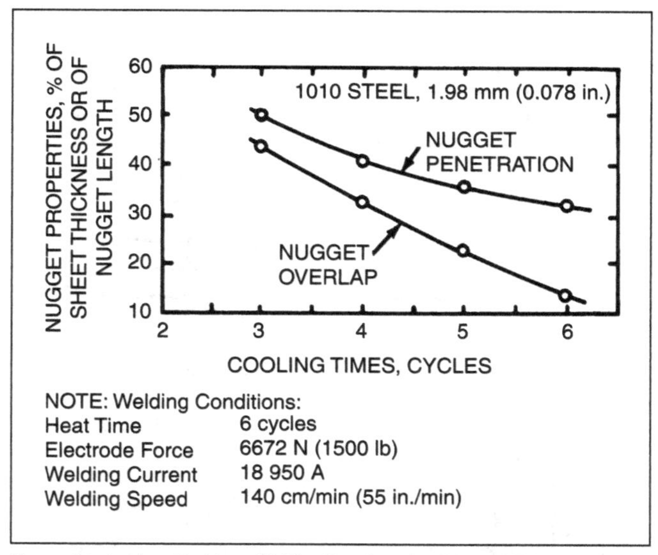

Source: Adapted from Resistance Welding Manufacturing Alliance (RWMA) Bulletin 23.

Figure 1.31—Effects of Cooling Time on Nugget Penetration and Nugget Overlap in Seam Welding

2. Each weld nugget in the seam is allowed to cool under pressure,
3. Distortion of the workpieces is minimized,
4. Expulsion or burning is easily controlled, and
5. Sound welds with good surface appearance are possible.

For a leak-tight seam, the nugget overlap should be 15% to 20% of the nugget diameter. For maximum strength, the overlap should be 40% to 50%. The size of the nugget depends on the heating time for a given welding speed and current. The amount of overlap depends on the cooling time.

The number of welds (nuggets) per unit of length that can be produced economically falls within a range specific to the type and thickness of the sheet metal. In general, as the sheet thickness decreases, the number of welds per unit of length must increase to obtain a strong, leak-tight seam weld. The ratio of welds per millimeter (in.) to welds per minute will establish the welding speed in millimeters (or inches) per minute. The number of welds per minute is the number of cycles of

Table 1.2
Suggested Schedules for Seam Welding Uncoated Low-Carbon Steel Sheet

Thickness		Electrode Width (W) and Shape (E)* W, Min.		E, Max.		Force		On Time	Off Time	Weld Speed		Welds	Welds	Current	Minimum Contact Overlap	
mm	in.	mm	in.	mm	in.	kN	lbf	(Cycles)**	(Cycles)**	mm/min	in./min	per cm	per in.	(Amp.)	mm	in.
0.3	0.01	9.5	0.38	4.5	0.18	1.8	400	2	1	2032	80	6	15	8000	9.5	0.38
0.5	0.021	9.5	0.38	4.8	0.19	2.4	550	2	2	1905	75	5	12	11 000	11.0	0.44
0.8	0.031	12.5	0.5	6.3	0.25	4.0	900	3	2	1829	72	4	10	13 000	12.5	0.5
1.0	0.04	12.5	0.5	6.3	0.25	4.4	980	3	3	1702	67	4	9	15 000	12.5	0.5
1.3	0.05	12.5	0.5	7.8	0.31	4.7	1050	4	3	1651	65	3	8	16 500	14.0	0.56
1.6	0.062	12.5	0.5	7.8	0.31	5.3	1200	4	4	1600	63	3	7	17 500	15.5	0.62
2.0	0.078	1.6	0.062	9.5	0.38	6.7	1500	6	5	1397	55	2	6	19 000	17.3	0.69
2.4	0.094	1.6	0.062	11.0	0.44	7.6	1700	7	6	1270	50	2	5.5	20 000	18.8	0.75
2.7	0.109	18.8	0.75	12.5	0.5	8.7	1950	9	6	1219	48	2	5	21 000	20.3	0.81
3.1	0.125	18.8	0.75	12.5	0.5	9.8	2200	11	7	1143	45	2	4.5	22 000	22.0	0.88

W
R = 76.2 mm (3 in.)
E
L

*Shape (E) is the width of a 75-mm (1/4-in.) radius on the electrode wheel face.
**On time and off time for 60-Hz power.

Notes:
1. Type of steel: SAE 1010.
2. Material should be free from scale, oxides, paint, grease and oil.
3. Welding conditions determined by thickness (T) of outside piece.
4. Data for total thickness of pile-up not exceeding 4 T. Maximum ratio between thicknesses: 3 to 1.
5. Electrode material: Class 2; minimum conductivity: 75% of copper; minimum hardness: 75 Rockwell B.
6. For large assemblies, minimum contacting overlap indicated should be increased 30%.

ac per minute, divided by the sum of the heating and cooling times (in cycles) for a single weld.

To obtain the minimum number of welds per unit of length that will produce the specified seam at a given welding speed, the heating time and welding current should be adjusted to provide the required weld nugget geometry. The cooling time should then be set to provide the necessary nugget overlap. Since decreasing cooling time may increase the heat buildup, nugget penetration may increase. With 50 Hz or 60 Hz weld controls, the number of spots per minute is limited by the value of line power, and thus the maximum travel speed. Inverter-type weld controls that allow heating and cooling times to be programmed in milliseconds (ms) may allow greater flexibility and greater speed. High-speed can-body welding machines operate with up to 1000 Hz.

External Cooling

Flood cooling, immersion cooling, or mist cooling commonly is used with seam welding. This generally is in addition to any internal cooling of the components in the secondary circuit of the welding machine. Electrode wear and distortion of the workpieces may be excessive when external cooling is not used and the electrode wheels are not internally water-cooled. When welding nonferrous metals and stainless steel, clean tap water for cooling is satisfactory. For ordinary steels, a 5% borax solution commonly is used to minimize corrosion.

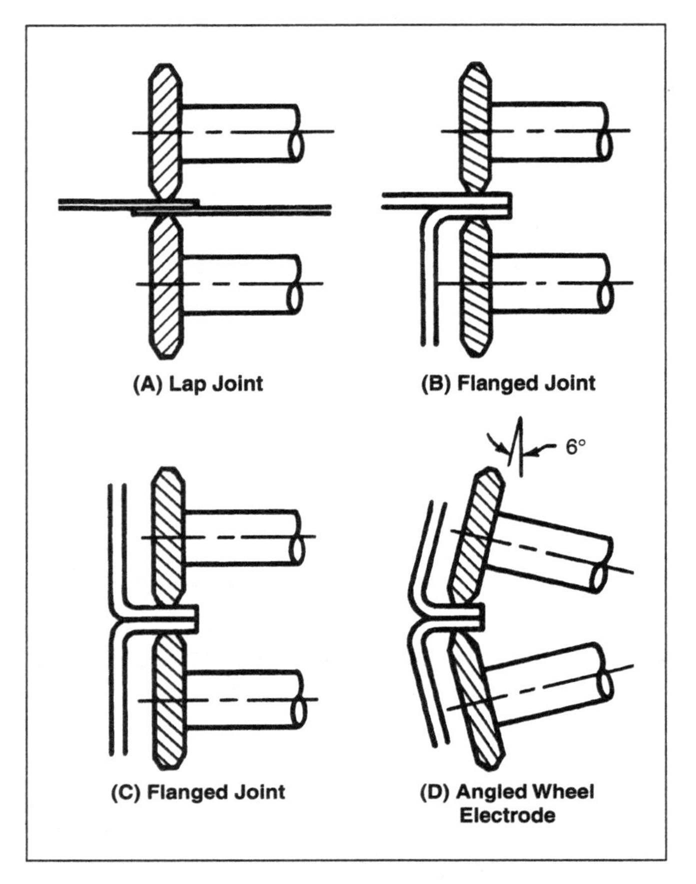

Figure 1.32—Examples of Resistance Seam-Welded Joints

Seam Welding Joint Designs

The various requirements that must be met in designing spot-welded joints also apply to seam-welded joints. With seam welding, the requirements of the electrode configuration, attachment method, and leak-tight requirements place some limitations on the workpiece design.

Wheel electrodes are relatively large and require unobstructed access to the joint. Since the electrodes rotate during welding, they cannot be inserted into small recesses or internal corners. External flanges must change direction over large radii in order to produce a strong, leak-tight seam weld. Joint designs that incorporate corners having small radii may result in welding problems when using the resistance seam welding processes. Decreased welding speeds are sometimes required in order to maintain weld quality.

The commonly used joint designs for seam welding shown in Figure 1.32 are similar to those used for spot welding applications. The lap joint in Figure 1.32(A) is the most common design. The workpiece edges must overlap sufficiently to prevent expulsion of the weld metal from the edges of the workpiece. However, excessive overlapping may entrap dirt or moisture within the joint and may cause subsequent manufacturing or service problems. Lap seam welds are used for the longitudinal seams in cans, buckets, water tanks, mufflers, and large-diameter, thin-walled pipes.

Flange joints are forms of lap joints. The design in Figure 1.32(B), in which one workpiece is straight, is commonly used to weld flanged ends to containers of various types. In Figure 1.32(C), both pieces are flanged. This design is used to join the two sections of automotive gasoline tanks. Often the flanged pieces are dished to obtain added strength or improve performance of the final product, in which case it is necessary to mount one or both wheels at an angle to clear the workpieces, as shown in Figure 1.32(D). A practical limit is 6° because greater angles cause excessive bearing thrust.

Special workpiece designs may require the adjustment of the shape and contour of the wheel electrode. Workpieces that contain regularly spaced contours may be welded with notched or segmented wheel electrodes, as shown in Figure 1.33.

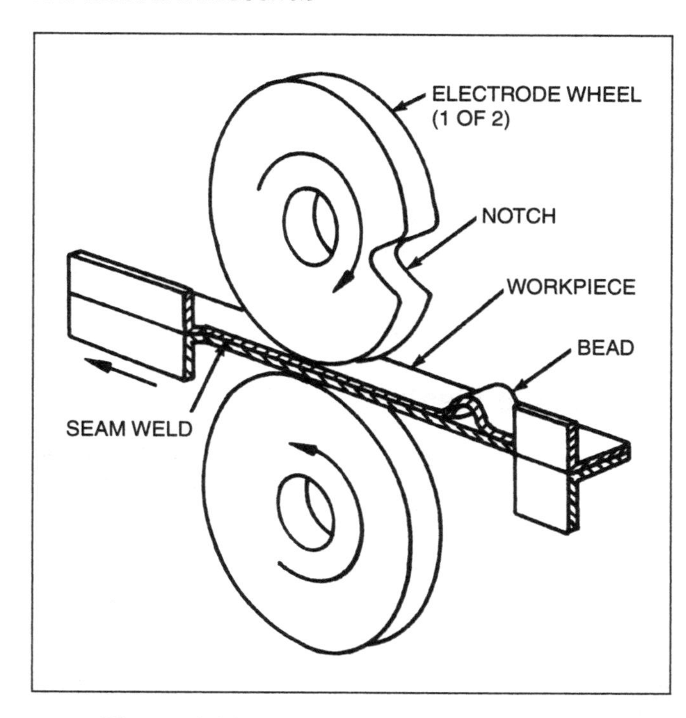

Figure 1.33—Notched Electrode Wheel for Seam Welding of a Workpiece with Obstructions in the Path of the Wheel

METAL PROPERTIES AND WELDABILITY

Various metals and their properties influence weldability by resistance spot and seam welding processes. The properties of these metals present variables that must be considered when planning a specific application.

PROPERTIES OF METALS

Following are properties of metals that determine weldability:

1. Electrical resistivity,
2. Thermal conductivity,
3. Thermal expansion,
4. Hardness and strength,
5. Oxidation resistance,
6. Plastic temperature range, and
7. Metallurgical properties.

Electrical Resistivity

From a resistance welding standpoint, workpiece resistivity probably is the most important welding property, since the heat generated by the welding current is directly proportional to resistance. More current is required to generate heat for a metal of low resistivity than one of high resistivity. A metal such as pure copper has low electrical resistivity and is difficult to join by resistance welding. In addition, current shunting through adjacent welds is more significant in low-resistivity metals than in those with high resistivity. Therefore, metals with high electrical resistivity are considered more weldable than those with low resistivity. It should be noted that low electrical resistivity also is a factor in power requirements. High currents require large transformers and high-capacity power lines, which increase equipment operating costs.

Thermal Conductivity

Thermal conductivity is important because part of the heat generated during resistance welding is lost by conduction into the base metal. This loss must be overcome by greater power input. Therefore, metals with high heat conductivity are less weldable than those with low conductivity. Thermal conductivity and electrical conductivity of the various metals closely parallel one another. Aluminum, for instance, is a good conductor of both heat and electricity, while stainless steel is a poor conductor of both.

Thermal Expansion

The coefficient of thermal expansion is a measure of the change in dimensions that takes place as a result of a temperature change. When the coefficient of thermal expansion is large, warping and buckling of welded assemblies can be expected.

Hardness and Strength

The hardness and strength of metals are important variables in resistance welding. Soft metals can be indented easily by the electrodes. Hard, strong metals require high electrode forces, which in turn require electrodes with high hardness and strength to prevent rapid deformation of the electrodes. Metals that retain their strength at elevated temperatures may require the use of welding machines capable of applying a forging force to the weld.

Oxidation Resistance

The surfaces of all commonly used metals oxidize in air, some more readily than others. Surface oxide films generally have high electrical resistance, which reduces the weldability of metals by resistance welding processes. In spot and seam welding, these oxides can cause surface flashing, pickup of metal on the electrode, and poor surface appearance. If the thickness of the oxide film varies from one workpiece to another, inconsistent weld strength may result.

Aluminum alloys form surface oxides rapidly. Therefore, welding must take place within a short time after deoxidation cleaning to avoid significant variations in surface contact resistance. In contrast, preweld deoxidation cleaning usually is not necessary for stainless steel if it is cleaned at the mill prior to packaging and shipping. Whether preweld deoxidation cleaning is necessary depends on the amount of oxide present on the surfaces and how it will affect weld properties. Surface resistance measurements may be used to confirm cleanliness. In general, all mill scale, heavy oxide from prior heat treatment, and extraneous material such as paint, drawing compounds, or grease should be removed before welding.

Plastic Temperature Range

If a metal melts and flows in a narrow temperature range, the welding variables must be more closely controlled than the variables for metals that melt and flow over a wide plastic temperature range. This property may have considerable bearing on the selection of welding procedures and equipment. Aluminum alloys have narrow plastic ranges and require precise control of welding current, electrode force, and electrode follow-up during welding. Low-carbon steels have a wide plastic range and thus are easily joined by resistance welding.

Metallurgical Properties

In resistance welding, a small volume of metal is heated to forging or melting temperature in a short time. The heated metal is then cooled rapidly by the electrodes and surrounding metal. Cold-worked metal will be annealed in the areas exposed to this thermal cycle. In contrast, the rapid cooling will cause hardening in some steels. High-carbon steel may harden so rapidly that the welds crack. A tempering cycle following the welding cycle is needed to avoid cracking. For optimum mechanical properties in the weld region, the heat-treatable alloys must be properly heat-treated after welding.

METALS WELDED

The metals commonly joined by the resistance spot welding and seam welding processes include low-carbon steels, hardenable steels, stainless steels and coated steels; also nickel-base alloys, copper alloys, aluminum and magnesium alloys, and titanium.

Low-Carbon Steels

Low-carbon steels generally contain less than 0.25% carbon. The overall weldability of these steels by the resistance welding process is good. Their electrical resistivity is average. Hardenability is low. Welds with good strength can be obtained over a wide range of current, electrode force, and weld time settings.

Hardenable Steels

High-carbon steels may contain 0.55% to 1.0% carbon; medium-carbon steels may contain from 0.25% to 0.55% carbon. Low-alloy steels contain up to 5.5% total alloying elements, including cobalt, nickel, molybdenum, chromium, vanadium, tungsten, aluminum and copper.

Alloying additions produce certain desirable properties in steels. The steels may respond to heat treatment and may be hard and brittle unless a postheat tempering cycle is employed. Process controls for welding operations that include post-weld heat treatment also should include recognition that a programming error might eliminate the post-weld heat treatment and go unnoticed because the workpiece will appear to be properly welded. Seam welding process controls require that travel must stop after each weld nugget is formed to apply the postheat tempering cycle. Standard machines with special controls are available to perform this function.

In general, hardenable steels are less weldable because of their hardenability properties than low-carbon steel.

Stainless Steels

Stainless steels contain relatively large amounts of chromium or chromium and nickel as alloying elements. Stainless steels are divided into three groups: martensitic, ferritic, and austenitic. Whether a stainless steel is hardenable depends on the amounts of carbon, chromium, and nickel it contains.

Ferritic and Martensitic. Ferritic steels contain at least 10% chrome and consist of ferrite and carbide microstructures. These materials sometimes are considered nonhardenable; however, the rapid cooling caused by the chilled resistance welding electrodes can produce martensite in these materials. Martensitic stainless steels contain sufficient amounts of carbon and alloying elements to promote hardening after rapid cooling in air. Weldability is poor for both types. When welding the martensitic types, the procedures given for high-carbon and low-alloy steels should be followed. The ferritic steels have low ductility and a characteristic coarse-grained structure in the weld region. These steels are generally not suitable for applications in which a ductile weld is required. With martensitic steels, a postweld heat treatment improves weld ductility. However, postweld heat treatment of the ferritic types is not beneficial.

Austenitic. There are a number of austenitic stainless steels, each having suitable properties for particular uses. The most commonly used austenitic steels contain 18% chromium, 8% nickel, and approximately 0.10% carbon. The unstabilized austenitic steels are susceptible to carbide precipitation if heated for an appreciable time between 427°C to 871°C (800°F to 1600°F), but when short weld times are used they can be resistance-welded without producing harmful carbide precipitation.

These alloys require less current than that required for low-carbon steels, since the electrical resistance is about seven times greater. Relatively high electrode force is needed because of the high strength of these alloys at elevated temperatures. Austenitic stainless steels have higher coefficients of thermal expansion than carbon steels. As a result, seam-welded assemblies may warp excessively. Distortion may be reduced by using welding schedules that lower the total heat input.

Coated Steels

Many plated and coated steels can be joined by spot welding or seam welding, but the weld quality usually is affected by the composition and thickness of the coating. Coatings on steel are usually applied for corrosion resistance, enhanced appearance or decoration, or a combination of these. Welding procedures should assure reasonable preservation of the coating function as well as produce welds of adequate strength. Strength requirements usually entail machine settings similar to those for bare carbon steel. Adjustments to compensate for the coating are determined by a number of factors, including the effect of the coating on contact resistance, acceptable electrode indentation, the tendency of the coating to alloy with the base metal, and the tendency of the electrode to fuse to the workpiece.

As previously noted, the heat generated by the welding current is directly proportional to resistance; thus, the thickness of the coating is the most important variable affecting the weldability of coated steels. When coating thickness presents problems in welding, better quality welds often can be obtained by decreasing the coating thickness. A weld nugget of the desired size may be obtained without too much disturbance of the outside surfaces by using higher welding current, greater electrode force, and shorter welding time than that used for the same thickness of bare steel. However, it is difficult to prevent alloying and metal pickup around the periphery of the electrode face, particularly when welding coatings with low melting points such as lead, tin, and zinc. Short welding times, good electrode tip maintenance, and attention to electrode cooling are the best preventive measures.

Nickel-Base Alloys

Generally, nickel-base alloys are readily joined by the resistance welding processes. However, the cast nickel-base alloys (for example, precipitation-hardenable, low-ductility Alloy 713C) normally are difficult to join by resistance welding because of cracking. High electrode forces are needed because of the high strength of nickel-base alloys at elevated temperatures. These alloys are subject to embrittlement by sulfur, lead, and other low-melting-point metals when exposed to high forces at high temperatures. Oil, grease, lubricant, marking material, and other foreign substances that might contain sulfur or lead must be removed from the workpieces prior to welding, or weld cracking may occur. Pickling prior to welding will be necessary only if a significant amount of oxide is present, indicated by surface discoloration.

Pure nickel can be welded rather easily. Some mechanical sticking of electrodes may be experienced because of the high electrical conductivity of nickel. A dome electrode with a cone angle of 170° is recommended for spot welding.

Monel 400™ is an alloy approximately two-thirds nickel and one-third copper. It has higher electrical resistivity and strength than low-carbon steel; therefore, somewhat lower welding current and higher electrode force are required for this alloy than for low-carbon steel.

Monel K-500,™ which can be age-hardened at 538°C (1000°F), has higher electrical resistivity and strength but lower thermal conductivity than Monel 400. Therefore, lower welding currents but higher electrode forces are required for Monel K-500 than for Monel 400. Monel K-500 will crack in the age-hardened condition if subjected to appreciable tensile stress at 595°C (1100°F). Spot and seam welding should be done on annealed materials.

Inconel 600™ contains approximately 78% nickel, 15% chromium, and 7% iron. Also, it has higher electrical resistivity and strength but lower thermal conductivity than Monel 400. Therefore, lower welding currents and higher electrode forces are required for this alloy than for Monel 400. Inconel 600 can be readily joined by resistance welding using procedures similar to those for stainless steels.

Inconel X-750,™ Inconel 718,™ and Inconel 722™ are age-hardenable alloys. They possess high strength at elevated temperatures, and have high electrical resistance. Relatively low welding current and high electrode force are needed for these alloys. These materials should be welded in the solution-annealed condition.

Copper Alloys

Copper alloys have a wide range of weldability that varies almost inversely with their electrical resistance. When the resistance is low, they are difficult to weld; when the resistance is high, they are rather easy to weld. Welding machines that provide adequate current capacity and moderate force are necessary. Because of the narrow plastic range of these alloys, machines with low-inertia heads should be used to provide a faster follow-up of the upper electrode to maintain pressure on the joint, which will prevent metal expulsion. The machines should be capable of accurate control of welding current, time, and electrode force because of the sensitivity of these alloys to variations in welding conditions. Shorter welding times are recommended to prevent metal expulsion and the incidental welding of the electrode to the workpiece. The fusing of the electrodes with the workpiece can be reduced by using electrodes faced with a refractory metal.

Copper-zinc alloys (brasses) become easier to weld with increasing zinc content because the electrical resistivity increases. The red brasses are difficult to weld, while the brasses with high zinc content can be welded throughout a range of welding conditions, even though the required energy input is high compared with that required for carbon steel.

Copper-tin alloys (phosphor-bronze), copper-silicon alloys (silicon-bronze), and copper-aluminum (aluminum-bronze) are relatively easy to weld because they have relatively high electrical resistance. These alloys, particularly phosphor-bronze, have a tendency to be hot short, which may result in cracking in the weld.

Aluminum and Magnesium Alloys

Most commercial aluminum and magnesium alloys produced in sheet or extruded form may be spot welded and seam welded, provided the thickness involved is not too great. Proper welding equipment, correct surface preparation, and suitable welding procedures are necessary to produce satisfactory welds.

Aluminum and magnesium alloys have high thermal and electrical conductivities. Therefore, high welding current and short welding time are needed. Machines with low-inertia heads should be used for spot and seam welding because aluminum and magnesium alloys soften rapidly at welding temperature. Rapid acceleration of the welding head is necessary to maintain contact between the electrodes and workpieces to prevent expulsion.

Titanium Alloys

Titanium and titanium alloys are readily welded by the resistance welding processes. Resistance welding is facilitated by the relatively low electrical and thermal conductivity of these materials. Although titanium and titanium alloys are very sensitive to embrittlement caused by reaction with air at fusion welding temperatures, these alloys can be joined by resistance welding without shielding with an inert gas. During resistance welding, the molten weld metal is completely surrounded by the base metal, thus protecting it from contamination. The welding time is short.

WELDING SCHEDULES

When setting up for welding a particular metal and joint design, a schedule must be established to produce welds that meet the design specifications. Schedules from previous experience can provide a starting point for the initial setup. If the application is new, reference to published information on the welding of the material by the designated process will serve as a guide for the initial setup.

Sample welds should be made, tested, and recorded while changing one process variable at a time within a range to establish an acceptable value for the specific variable. It may be necessary to establish the effect of one variable at several levels of another. For example, weld or heat time and electrode force may be evaluated at several levels of current. Visual examination and destructive test results can be used to select an appropriate welding schedule. Finally, samples of parts from first production or simulations of the product should be welded and destructively tested after welding. Final adjustments are then made to the welding schedule to meet design or specification requirements.

Starting schedules for many commercial alloys may be available from the equipment manufacturer. Some may be found in the following publications:

1. *Recommended Practices for Resistance Welding*, AWS C1.1M/C1.1:2000;[9]
2. *Specification for Resistance Welding of Carbon and Low-Alloy Steels*, C1.4M/C1.4:1999;[10]
3. *Recommended Practices for Automotive Portable Gun Resistance-Spot Welding on Electrode and Force Recommendation for Two and Three*

---

9. American Welding Society (AWS) Committee on Resistance Welding, 2000, *Recommended Practices for Resistance Welding*, AWS C1.1M/C1.1:2000, Miami: American Welding Society.

10. American Welding Society (AWS) Committee on Resistance Welding, 1999, *Specification for Resistance Welding of Carbon and Low-Alloy Steels*, C1.4M/C1.4:1999, Miami: American Welding Society.

Loose Metal Thickness, Supplement to AWS D8.5-66;[11]

4. *Recommended Practices for Automotive Weld Quality—Resistance Spot Welding*, D8.7M:2005;[12]
5. *Resistance Welding Manual*, Revised 4th Edition;[13] and
6. *Metals Handbook: Welding, Soldering and Brazing*, Volume 6.[14]

ECONOMICS

The cost of a process is determined by the choice of joining methods and the materials to be joined. Once the joining method has been determined, the volume of assemblies required and the production rate will determine the cost of joining. The ongoing cost of energy per weld usually is the least concern for resistance welding processes. However, the initial investment required for the power supply system used in resistance welding can be substantial due to the requirement for relatively high currents compared to many other welding processes.

The cost of consumables used in resistance welding generally is much lower compared to other welding processes. The cost of maintaining the consumables is an additional consideration. Generally, the weld contact tips and surfaces require the most maintenance and may require servicing at intervals as frequently as once every 200 welds for certain coated steels and aluminum resistance welding applications. For uncoated steel, the resistance welding electrodes may last through more than 14,000 welds.

The surface requirements for the weld also may require the use of specially contoured contact backups that require frequent dressing and maintenance. Aluminum- and zinc-coated steels require more frequent electrode tip maintenance than uncoated steel. Pilot trials in the laboratory frequently are used to forecast weldability and electrode tip life for specific combinations of materials.

Positioning and locating the workpieces has an impact on welding cost. Resistance spot welding process equipment can be stationary and the workpieces moved to position the weld at the required location, or the equipment may be moved to place the electrodes at the welding location. Both of these methods can be used for high- or low-volume production requirements. The decision on which method is most favorable must take the following costs into account:

1. Manual handling of workpieces;
2. Automation of workpiece handling;
3. Flexibility of handling a variety of workpiece configurations;
4. Tooling requirements, maintenance and service life;
5. Accessibility to equipment for maintenance; and
6. Anticipated maintenance intervals.

The economic decisions required for resistance welding processes are similar to those of any other industrial assembly process: they focus on personnel, tooling costs, production volume and production rate. However, the resistance welding cost estimate also must contain the cost of maintenance of electrical components in the secondary circuit. When the joint produced is suitable for the intended service and the production volume reaches thousands, resistance welding for the joining of steels almost always is less costly than other processes.

Chapter 4 of this volume contains more detailed information on resistance welding equipment. Valuable information also can be obtained from commercial suppliers of high-volume production equipment listed with the Resistance Welding Manufacturing Alliance.[15]

SAFE PRACTICES

Spot and seam welding may involve hazardous situations that can be avoided by taking the precautions outlined in this section. Safe practices are designed to address potential mechanical and electrical risks, and measures for personal protection.

Safety requirements for the welding industry are published by the American National Standards Association (ANSI) in *Safety in Welding, Cutting, and Allied Processes*, ANSI Z49.1.[16] Recommendations for the safe use of resistance welding equipment provided in

11. American Welding Society (AWS) Committee on Resistance Welding, 1966, *Recommended Practices for Automotive Portable Gun Resistance-Spot Welding on Electrode and Force Recommendation for Two and Three Loose Metal Thickness*, Supplement to AWS D8.5-66, Miami: American Welding Society.
12. American Welding Society (AWS) Committee on Resistance Welding, 2005, *Recommended Practices for Automotive Weld Quality—Resistance Spot Welding*, D8.7M:2005, Miami: American Welding Society.
13. Resistance Welding Manufacturing Alliance (RWMA), 2003, *Resistance Welding Manual*, Revised 4th Ed., Miami: Resistance Welding Manufacturing Alliance.
14. ASM International, 1993, *Metals Handbook, Welding, Soldering and Brazing*, Volume 6, Materials Park, Ohio: ASM International.

15. Resistance Welding Manufacturing Alliance (RWMA), 550 N.W. LeJeune Road, Miami, Florida. rwma@aws.org and www.aws.org/rwma.
16. American National Standards Institute (ANSI) Accredited Standards Committee Z49, 2005, *Safety in Welding, Cutting, and Allied Processes*, ANSI Z49.1:2005. Miami: American Welding Society.

the manufacturers' operating manuals should be carefully followed. The suppliers of materials provide Materials Safety Data Sheets (MSDSs) for each product. The MSDS should be read and understood by all involved in welding operations before the product is used. Additional resources, pertinent safety associations, and publishers of standards are listed in Appendix A and B of this book.

MECHANICAL GUARDING

Power-initiating controls on welding equipment, such as push buttons or switches, should be arranged or guarded to prevent the operator from inadvertently activating them.

In some multiple-gun welding machine installations, the operator's hands can be expected to pass under the point of operation. These machines should be effectively guarded by a suitable device, such as proximity-sensing gates, latches, blocks, barriers, or dual hand controls.

One or more emergency stop buttons should be provided on all welding machines, with a minimum of one at each operator position.

ELECTRICAL

Resistance welding equipment should be designed to avoid accidental contact with the components of the system that are electrically hazardous. High-voltage components must have adequate electrical insulation and be completely enclosed. All doors, access panels, and control panels of resistance welding machines must be kept locked or interlocked to prevent access by unauthorized persons. The interlocks must effectively interrupt power and discharge all high-voltage capacitors into a suitable resistive load when the door or panel is open. In addition, a manually operated switch and suitable device should be provided to give a visual indication that discharge of all high-voltage capacitors has taken place.

All electrical equipment must be suitably grounded and the transformer secondary must be grounded or provided with equivalent protection. The only exception to grounding the secondary is in the situation in which the secondary and the workpiece are not accessible to operators while the weld is in process. In this case a mechanical contactor may be used to disconnect the primary power from the transformer, except while welding. It should be noted that this does not provide protection for maintenance persons who may need to make welds on coupons to confirm the proper operation of the welding machine. The publication, Occupational Safety and Health Administration (OSHA) *Title 29—Labor*, In *Code of Federal Regulations (CFR)*, Chapter XVII, Parts 1910.255b(9) and c(6) should be consulted.[17] In addition to the items on electrical safety covered in this code, the Resistance Welding Manufacturing Alliance (RWMA) *Bulletin 16* provides requirements for welding machine grounding in *Methods for Grounding Resistance Welding Circuits*.[18]

External weld-initiating control circuits should operate at low voltage for portable equipment. Circuits that are initiated by grounding are not recommended because they can be initiated by a failed or damaged conductor or device. Safeguards against excessive noise must be provided.

PERSONAL PROTECTIVE EQUIPMENT

The personal protective equipment needed depends on the particular welding application. The following equipment is generally required for resistance welding:

1. Eye protection in the form of face shields (preferred) or hardened-lens goggles;
2. Ear protection;
3. Skin protection provided by nonflammable gloves, and clothing with a minimum number of pockets and cuffs in which hot or molten particles can lodge; and
4. Protective footwear.

OTHER HAZARDS

A stringent fire prevention protocol must be implemented in the workplace, as recommended in the National Fire Protection Association (NFPA) publication *Fire Prevention During Welding, Cutting, and Other Hot Work*, NFPA 51B.[19]

Adequate ventilation must be ensured, as recommended in the Threshold Limit Values (TLVs®) and Biological Exposure Indices (BEIs®) guidelines published periodically by the American Conference of Governmental Industrial Hygienists (ACGIH).[20]

17. Occupational Safety and Health Administration (OSHA), *Title 29—Labor*, In *Code of Federal Regulations (CFR)*, Chapter XVII, Parts 255b(9) and c (6), Washington D.C.: Superintendent of Documents, U.S. Government Printing Office.
18. Resistance Welding Manufacturing Alliance (RWMA) *Bulletin 16*, Section 1.7.20, *Methods for Grounding Resistance Welding Circuits*, Miami: American Welding Society.
19. National Fire Protection Association (NFPA), 2003, *Fire prevention during welding, cutting, and other hot work*, NFPA 51B, Quincy, Massachusetts: National Fire Protection Association.
20. American Conference of Governmental Industrial Hygienists (ACGIH), 1330 Kemper Meadow Drive, Cincinnati, Ohio 45240. http://www.acgih.org.

Measures must be in place for the safe use and correct disposal of hazardous materials, such as alkaline solutions, pickling acids and solvents used in resistance spot and seam welding operations.

Safeguards against excessive noise must be integrated in the workplace.

CONCLUSION

This chapter has provided the reader with basic information on resistance spot and seam welding, with insights drawn from the many years of experience of the contributors. The fundamentals presented in this chapter provide solid direction for users of the resistance spot and seam welding processes by guiding the choices of equipment, welding parameters, workpiece preparation, welding schedules, maintenance, and other operating strategies. Although the resistance welding industry relies mainly on computer-controlled equipment, which allows manipulation of virtually any welding parameter during the welding cycle, manually operated equipment is still in use.

Resistance spot and resistance seam welding are adaptable to mass production, small-lot, or one-of-a-kind production. Evidence of the effectiveness of resistance welding can be observed in the wide variety of applications in automotive, home appliance, hand tool, and furniture manufacturing. Many resistance-welded items that may go unnoticed include diverse products such as electronic assemblies, containers, food processing equipment, medical equipment and devices, and sports equipment.

The ability to optimize the performance of resistance spot and seam welding systems can be broadened not only by hands-on experience, but also by a comprehension of electrical circuits and welding metallurgy. The reader will benefit from continued study of the resistance welding processes by consulting pertinent publications of the American Welding Society[21] and the Resistance Welding Manufacturing Alliance,[22] as well as the resources listed at the end of this chapter.

BIBLIOGRAPHY

American Conference of Governmental Industrial Hygienists (ACGIH). TLVs® and BEIs®: Threshold limit values for chemical substances and physical agents in the workroom environment. Cincinnati: American Conference of Governmental Industrial Hygienists.

American National Standards Institute (ANSI) Accredited Standards Committee Z49. 2005. *Safety in welding, cutting, and allied processes*, ANSI Z49.1:2005. Miami: American Welding Society.

ASM International. 1993. *Metals handbook: welding, soldering and brazing. Volume 6*. Materials Park, Ohio: ASM International.

American Welding Society (AWS) Committee on Definitions and Symbols. 2001. *Standard welding terms and definitions*, AWS A3.0:2001. Miami: American Welding Society.

American Welding Society (AWS) Committee on Resistance Welding. 2000. *Recommended practices for resistance welding*, AWS C1.1M/C1.1:2000. Miami: American Welding Society.

American Welding Society (AWS) Committee on Resistance Welding. 1999. *Specification for resistance welding of carbon and low-alloy steels*, C1.4M/C1.4:1999. Miami: American Welding Society.

American Welding Society (AWS) Committee on Resistance Welding. 1966. *Recommended practices for automotive portable gun resistance-spot welding on electrode and force recommendation for two and three loose metal thickness*. Supplement to AWS D8.5-66. Miami: American Welding Society.

American Welding Society (AWS) Committee on Resistance Welding. 2005. *Recommended practices for automotive weld quality—resistance spot welding*. D8.7M:2005, Miami: American Welding Society.

American Welding Society (AWS) Welding Handbook Committee. Jenney, C. L. and A. O'Brien, Eds. 2001. *Welding science and technology*. Vol. 1 of *Welding handbook*, 9th ed. Miami: American Welding Society.

National Fire Protection Association (NFPA). 2003. *Fire prevention during welding, cutting, and other hot work*. NFPA 51B. Quincy, Massachusetts: National Fire Protection Association.

Occupational Safety and Health Administration (OSHA). 1999. *Title 29—Labor*. In *Code of Federal Regulations (CFR)*, Parts 1910.255b(9) and c(6). Washington D.C.: Superintendent of Documents. U.S. Government Printing Office.

Peterson, R. E. 1974. *Stress concentration factors*. New York: John Wiley and Sons.

Resistance Welding Manufacturing Association. 2003. *Resistance welding manual*, Revised 4th Ed. Miami: American Welding Society.

21. American Welding Society, 550 N.W. LeJeune Road, Miami, FL 33126.
22. Resistance Welding Manufacturing Alliance, 550 N.W. LeJeune Road, Miami, FL 33126.

SUPPLEMENTARY READING LIST

Adams, T. 1985. Nondestructive evaluation of resistance spot welding variables using ultrasound. *Welding Journal* 64(6): 27–28.

Agashe, S. and H. Zhang. 2003. Selection of schedules based on heat balance in resistance spot welding. *Welding Journal* 82(7): 179-s–183-s.

Anon. 1983. Flexible controller helps "turn the corner" in resistance welding. *Welding Journal* 62(11): 68–69.

Aidun, D. K., and R. W. Bennett. 1985. Effect of resistance welding variables on the strength of spot welded 6061-T6 aluminum alloy. *Welding Journal* 64(12): 15–25.

AWS D8 Committee on Automotive Welding. 2002. *Recommended practices for test methods for evaluating the resistance spot welding behavior of automotive sheet steel materials,* AWS/SAE D8.9M:2002. Miami: American Welding Society.

Brown, B. M. 1987. A comparison of ac and dc resistance welding of automotive steels. *Welding Journal* 66(1): 18–23.

Bowers, R. J., C. D. Sorensen, and T. W. Eagar. 1990. Electrode geometry in resistance spot welding. *Welding Journal* 69(2): 455.

Brown, B. M. 1987. A comparison of ac and dc resistance welding of automotive steels. *Welding Journal* 66(1): 18–23.

Chang, H. S., and Y. J. Cho. 1990. A study on the shunt effect in resistance spot welding. *Welding Journal* 69(8): 308-s–317-s.

Cho, H. S., and Y. J. Cho. 1989. A study of thermal behavior in resistance spot welds. *Welding Journal* 68(6): 236-s.

Dickinson, D. W., J. E. Franklin, and A. Stanya. 1980. Characterization of spot welding behavior by dynamic electrical parameter monitoring. *Welding Journal* 59(6): 170-s–176-s.

Fukumoto, S. I. Lum, E. Biro, D. R. Boomer, and Y. Zhou. 2003. Effects of electrode degradation on electrode life in resistance spot welding of aluminum alloy 5182. *Welding Journal* 82 (11): 307-s–312-s.

Gedeon, S. A. 1987. Measurement of dynamic electrical and mechanical properties of resistance spot welds. *Welding Journal* 66(12): 378-s–382-s.

Gedeon, S. A., D. Schrock, J. LaPointe, and T. W. Eagar. 1988. Metallurgical and process variables affecting the resistance spot weldability of galvanized sheet steels. SAE Technical Paper Series No. 840113. Warrendale, Pennsylvania: Society of Automotive Engineers.

Gould, J. E. 1987. An examination of nugget development during spot welding, using both experimental and analytical techniques. *Welding Journal* 66(1): 1-s–5-s.

Gould, J. E. 2003. Theoretical analysis of welding characteristics during resistance mash seam welding of sheet steels. *Welding Journal.* 82(10): 263-s–267-s.

Hain, R. 1988. Resistivity testing of spot welds challenges ultrasonics. *Welding Journal* 67(5): 46–50.

Hall, P. M., and W. R. Hain. 1987. Nondestructive monitoring of spot weld quality using a four-point probe. *Welding Journal* 66(5): 20–24.

Han, Z., J. Orozco, J. E. Indacochea, and C. H. Chen. 1989. Resistance spot welding: a heat transfer study. *Welding Journal* 68(9): 363-s–368-s.

Hawkins, R., and Rabinovich, S. 2005. *Inspection Trends.* Miami: American Welding Society. 8(2): 30–33.

Howe, P. and S. C. Kelley. 1988. Coating-weight effect on the resistance spot weldability of electrogalvanized sheet steels. *Welding Journal* 67(12): 271-s–275-s.

Howe, P. and S. C. Kelley. 1988. A comparison of the resistance spot weldability of bare, hot-dipped, galvannealed, and electrogalvanized DQSK sheet steels. SAE Technical Paper Series No. 880280. Warrendale, Pennsylvania: Society of Automotive Engineers.

Kanne, R. 1986. Solid-state resistance welding of cylinders and spheres. *Welding Journal* 65(5): 33–38.

Kim, E. W., and T. W. Eagar. 1989. Measurement of transient temperature response during resistance spot welding. *Welding Journal* 68(8): 303-s–307-s.

Kimichi, M. 1984. Spot weld properties when welding with expulsion—A comparative study. *Welding Journal* 63(2): 58-s–63-s.

Lane, C. T., C. D. Sorensen, G. B. Hunter, S. A. Gedeon, and T. W. Eagar. 1987. Cinematography of resistance spot welding of galvanized sheet steel. *Welding Journal* 66(9): 260-s–264-s.

Linnert, G. E. 1994. *Welding metallurgy: carbon and alloy steels.* Miami: American Welding Society.

Nied, H. A. 1984. The finite element modeling of the resistance spot welding process. *Welding Journal* 63(4): 123-s–132-s.

Rathburn, R. W., D. K. Matlock, and J. G. Speer. 2003. Fatigure behavior of spot welded high-strength sheet steels. *Welding Journal* 82(8): 207-s–218-s.

Savage, W. F., E. F. Nippes, and F. A. Wassell. 1977. Static contact resistance of series spot welds. *Welding Journal* 56(11): 365-s–370-s.

Savage, W. F., E. F. Nippes, and F. A. Wassell. 1978. Dynamic contact resistance of spot welds. *Welding Journal* 57(2): 43-s–50-s.

Sawhill, J. M., Jr., H. Watanabe, and J. W. Mitchell. 1977. Spot weldability of Mn-Mo-Cb, V-N, and SAE 1008 steels. *Welding Journal* 56(7): 217-s–224-s.

Song, Q., Zhang, W., and Bay, N. 2005. An experimental study determines the electrical contact resistance in resistance welding. *Welding Journal.* 84(5): 73-s–76-s.

Sun, X., E. V. Stephens, R. W. Davies, M. A. Khaleel, and D. J. Spinella. 2004. Effects of fusion zone size on failure modes and static strength of aluminum resistance spot welds. *Welding Journal.* 83(11): 308-s–318-s.

Zhou, M., and S. J. Hu. 2003. Relationships between quality and attributes of spot welds. *Welding Journal* 82(4): 72-s–77-s.

CHAPTER 2

PROJECTION WELDING

Photograph courtesy of CenterLine (Windsor) Limited

Prepared by the Welding Handbook Chapter Committee on Projection Welding:

J. C. Bohr, Chair
General Motors

W. H. Brafford
Tuffaloy Products

G. J. Daumeyer III
Resistance Welding Equipment & Supply Company

B. G. Kelly
Kelly Welding Solutions, P.C.

D. P. Kelly
Fusion Welding Solutions

M. Kimchi
Edison Welding Institute

H. Zhang
University of Toledo

Welding Handbook Volume 3 Committee Member:

S. P. Moran
Miller Electric Manufacturing Company

Contents

CHAPTER 2

PROJECTION WELDING

INTRODUCTION

Projection welding (PW) is an electric resistance welding process that produces welds by the heat obtained from the resistance to the flow of the welding current. The resulting welds are localized at predetermined points by projections, embossments, or intersections.[1,2]

Force is always applied before, during, and after the application of current to confine the weld contact area at the faying surfaces and, in some applications, to forge the weld metal during postheating. The localization of heating is obtained by a projection or embossment on one or both of the workpieces.

Projection welding is primarily used to join a stamped, forged, or machined component to another component. One or more projections are produced on the workpiece during the forming operations or other prior operations. A common application of projection welding is the use of specially designed nuts that have projections on the portion of the workpiece to be welded to the assembly. Fasteners or mounting devices, such as bolts, nuts, pins, brackets, and handles, can be welded to sheet metal, providing a variety of design options for countless applications.

Projection welding generally is used for section thicknesses ranging from 0.5 millimeters (mm) to 3.2 mm (0.02 inch [in.] to 0.125 in.). The process is used to join various carbon steels, alloy steels, and some nickel alloys in a large variety of applications, from automotive mounting brackets to the cross-wire welding of grocery carts. Projection welding is easily automated for the high-speed welding of multiple projections.

This chapter presents the fundamentals of the projection welding process, its advantages, limitations, and applications. Topics include the equipment and materials associated with the process, welding variables, welding schedules and weld quality. A discussion of a variation of projection welding, cross wire welding, is presented. The chapter continues with a brief discussion of the economics of projection welding, and concludes with a section on safe practices.

FUNDAMENTALS

Projection welding, like spot and seam welding, involves the coordinated application of electric current and mechanical force of the proper magnitude and duration.[3] In projection welding, embossed dimples, machined projections, or component features are used to produce the concentration of localized current and heating when two workpieces are placed in contact. Weld position is determined by the shape of the component and the projections rather than the shape of the electrode, as is the case in resistance spot welding.

The electrode is the part of the resistance welding machine through which the welding current passes, and in most cases, it is used to apply force directly to the workpiece. The electrode may be in the form of a rotating wheel, rotating roll, bar, cylinder, plate, clamp, chuck, or a modification of these. The welding current still must pass from the electrodes through the workpiece with continuity assured by forces applied to the electrodes through the projections. The projections are shaped to provide the necessary current density.

1. Welding terms and definitions used throughout this chapter are from *Standard Welding Terms and Definitions*, AWS A3.0:2001, Miami: American Welding Society.

2. At the time this chapter was prepared, the referenced codes and other standards were valid. If a code or other standard is cited without a publication date, it is understood that the latest edition of the document referred to applies. If a code or other standard is cited with the publication date, the citations refer to that edition only, and it is understood that any future revisions or amendments to the code or other standard are not included. As codes and standards undergo frequent revision, the reader is advised to consult the most recent edition.

3. The fundamentals of resistance welding processes are covered in Chapter 1 of this volume, "Spot and Seam Welding."

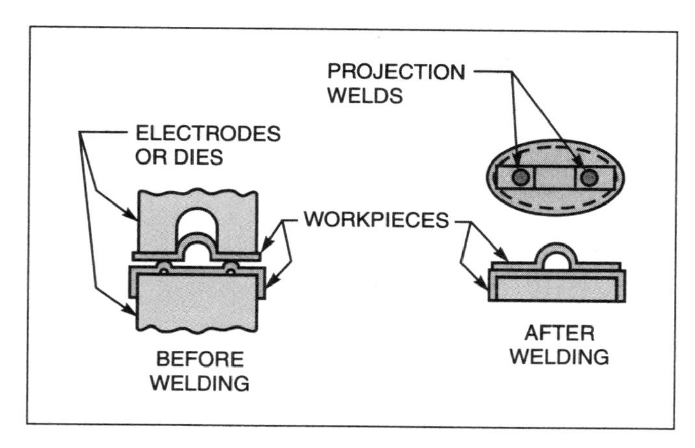

Figure 2.1—Simplified Diagram of the Projection Welding Process

The sequence of operations begins with the development of sufficient heat to raise a confined volume of metal to the plastic or molten state. This metal is then allowed to cool while under pressure until it develops adequate strength to hold the workpieces together. Figure 2.1 illustrates the projection welding process.

Several types of projections commonly are used: embossed button or dome type, usually round, as shown in Figure 2.2(A); the stud (shoulder) type, as shown in Figure 2.2(B); and the annular type, as shown in Figure 2.2(C). Other types include elongated projections, radius projections, and wire projections used in cross wire welding.

When joining sheet metal with projection welding, a melted weld zone similar to that of a resistance spot weld is produced. When a machined or solid-formed projection is used, a solid-phase forge weld is produced without melting at the weld interface. The bond across the weld interface is produced by the plastic deformation of the heated workpieces in contact with one another.

ADVANTAGES AND LIMITATIONS

In general, projection welding can be used instead of spot welding to join small components to one another and to larger components. The selection of one method over another depends on the advantages, limitations, and economics of the two processes. The major advantage of projection welding is that electrode service life is increased because larger contact surfaces are used. Other advantages of projection welding include the following:

1. A number of welds can be made simultaneously in one welding cycle of the machine, limited only by the ability of the welding machine to apply uniform electrode force and welding current to each projection;
2. Less overlap and closer weld spacing is possible because the current is concentrated by the projection and does not shunt through adjacent welds;
3. Thickness ratios of at least 6 to 1 are possible because of the flexibility in projection size and location;

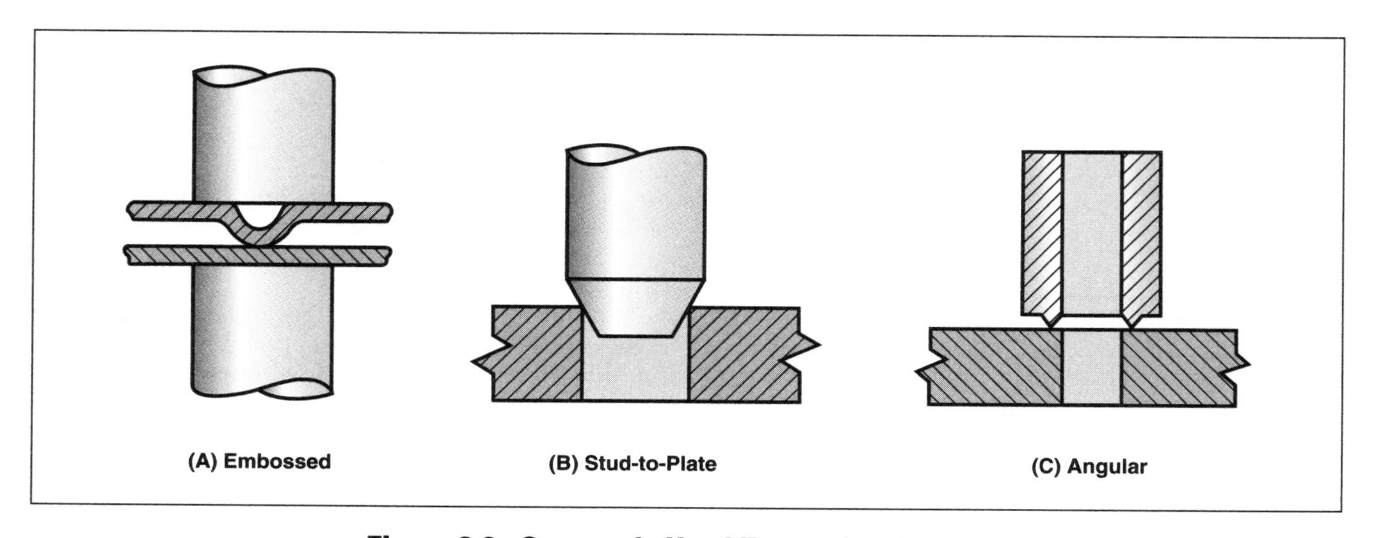

Figure 2.2—Commonly Used Types of Projections

4. Projection welds can be smaller than spot welds, because the projections are uniform and can be located with greater accuracy and consistency;
5. Projection welds generally are better in appearance on the side without the projection than spot welds, because most deformation and greatest temperature rise occur in the workpiece with the projection, leaving the exposed surface of the other workpiece relatively cool and free of distortion;
6. Electrode maintenance costs are lower because the large, flat-faced electrodes used in projection welding incur much less wear than spot welding electrodes;
7. When welding small components, fixturing costs may be reduced because fixtures or component locators sometimes can be combined with the welding electrodes; and
8. Although weld quality will be better if surfaces are clean, oil, rust, scale, and coatings are less of a problem than with spot welding because the tip of the projection tends to break through the foreign material early in the welding cycle.

The most important limitations of projection welding are the following:

1. The forming of projections may require an additional operation unless the workpieces are press-formed to the design shape;
2. When multiple welds are to be produced, accurate control of projection height and precise alignment of the welding electrodes are necessary to equalize the electrode force and welding current;
3. When welding sheet metal, the process is limited to thicknesses in which projections with acceptable characteristics can be formed, and for which suitable welding equipment is available; and
4. Multiple welds must be made simultaneously, which requires higher-capacity equipment than spot welding and limits the practical size of the component that contains the projections.

EQUIPMENT

Projection welding equipment consists of three basic groups: an electrical circuit, the control equipment, and a mechanical system.[4] These components are integrated into projection welding systems such as those shown in Figure 2.3, a pedestal welding machine designed to weld nuts to a surface and equipped with a nut feeder; and in Figure 2.4, a machine with a heavy-duty frame adaptable for either low-volume or high-volume production.

4. For more detailed information see Chapter 4 of this volume, "Resistance Welding Equipment."

ELECTRICAL CIRCUIT

The electrical circuit is comprised of a welding transformer, a primary contactor, and a secondary circuit. The secondary circuit includes the workpieces and the electrodes that conduct the welding current to the workpiece. In some cases, a means of storing electrical energy is also included in the circuit. Alternating current (ac) or direct current (dc) may be used for projection welding. The welding machine converts 60-Hz line power to low-voltage, high-amperage power in the secondary circuit of the machine.

CONTROL EQUIPMENT

Welding control equipment provides one or more of the following functions:

1. Initiates and terminates current to the welding transformer,
2. Controls the magnitude of the current, and
3. Actuates and releases the electrode force mechanism at the proper times.

Controls may be divided into three groups based on function: welding contactors, timing and sequencing controls, and other current controls and regulators.

A welding contactor connects and disconnects the primary power and the welding transformer. Electronic contactors use silicon-controlled rectifiers (SCRs), ignitron tubes, or thyratron tubes to interrupt the primary current. The timing and sequence control establishes the welding sequence and the duration of each function of the sequence. This includes the application of electrode force and current, and also the time intervals following each function.

Transformer taps or an electronic heat control, or both, control the welding current output of a machine. An electronic heat control is used in conjunction with ignitron tubes or SCRs. It controls current by delaying the firing of the ignitron tubes or SCRs during each half-cycle (1/120 second [s]). Varying the firing delay time can gradually increase or decrease the primary root means square (rms) amperage. This provides upslope and downslope control of welding current.

Photograph courtesy of Dengensha America

Figure 2.3—Projection Welding Machine for a Nut-Welding Application, with Nut Feeder

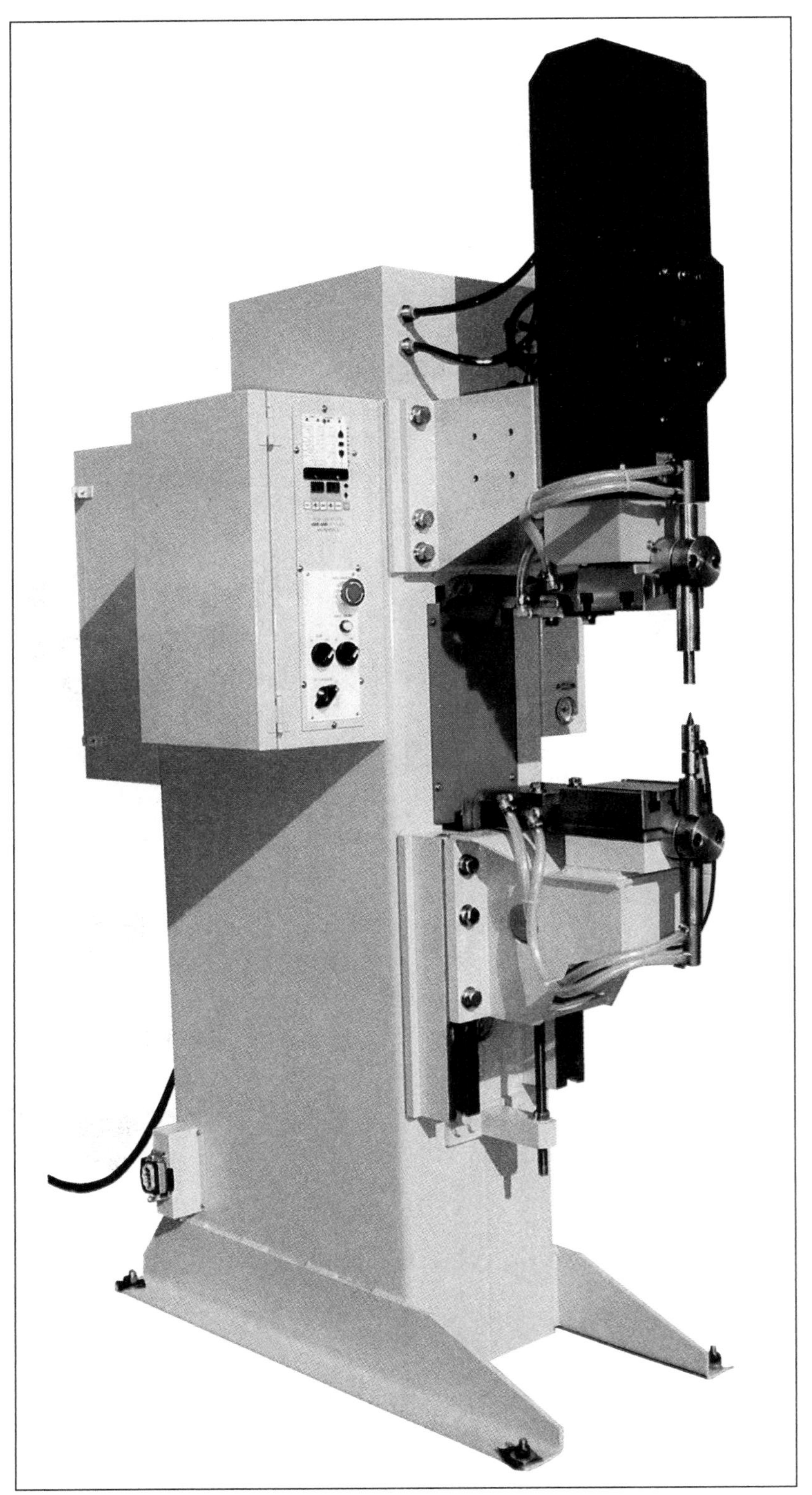

Photograph courtesy of Dengensha America

Figure 2.4—Projection Welding Machine with Heavy-Duty Frame

Transformer taps are used to change the number of primary turns connected across the power line. This changes the turns ratio of the transformer, with an increase or decrease in open-circuit secondary voltage. Decreasing the turns ratio increases the open-circuit secondary voltage, the primary current, and the welding current.

MECHANICAL SYSTEM

Projection welding machines essentially have the same types of mechanical operation as spot and seam welding equipment. The electrodes approach and retract from the workpieces at controlled times and rates. Electrode force is applied by hydraulic, pneumatic, magnetic, or mechanical means. The rate of electrode approach must be rapid but controlled, so that the electrode faces are not deformed from repeated blows. When welding projections are stamped in thin materials, it is important to manage the impact of the electrodes as they close on the workpiece. Too much impact can flatten the projection and make welding much more difficult or impossible. The locally heated weld metal expands and contracts rapidly during the welding cycle and the electrodes must follow this movement to maintain welding pressure and electrical contact. The ability of the machine to follow motion is influenced by the mass of the moving components, or their inertia, and by friction between the moving components and the machine frame. Machines designed for projection welding typically have low-friction moving systems and frequently are designed to separate the electrode from large moving masses. They provide springs to increase the speed of electrode movement as the projections collapse.

If the pressure between the electrodes and the workpiece drops rapidly during weld time, the contact surfaces of the electrodes and workpieces may overheat and result in the burning or pitting of the electrode faces. The electrodes may stick to the workpiece, and in some cases the surfaces of the workpieces may vaporize as a result of the very high energy. The electrode force used during the melting of the weld nugget (the weld metal joining the workpieces in projection welds) may not be adequate to consolidate the weld metal and to prevent internal porosity or cracking.

Multilevel-force machines may be employed to provide a high forging pressure during weld solidification. The magnitude of this pressure should be compatible with the composition and thickness of the metal and the geometry of the workpieces. The forging pressure often is two to three times the welding pressure. Since the weld cools from the periphery inward, the forging pressure must be applied at or close to the time current is terminated.

WELDING VARIABLES

Projection welding variables include surface preparation, the type of joint, joint design, projection design, type of component (machined or forged), heat balance, welding cycle, and the type of electrodes and welding dies required.

SURFACE PREPARATION

For projection welding and all types of resistance welding (for example, flash, high-frequency, seam, spot, and upset welding), the condition of the surfaces of the workpieces affects how consistent weld quality will be. The contact resistance of the faying surfaces has a significant influence on the amount of welding heat generated; hence, the electrical resistance of these surfaces should be uniform for consistent results. The surfaces must be free of highly resistant materials such as paint, scale, thick oxides, and heavy oil and grease. If it is necessary to use primer paint on the faying surfaces prior to welding, as sometimes is the case, the welding operation must be performed immediately after applying the primer, or special conducting primers must be used. For best results, the primer should be applied as thinly as possible so that the electrode force will displace it and allow metal-to-metal contact.

Paint should never be applied to outside base metal surfaces before welding because it will reduce electrode life and produce a poor surface appearance. Heavy scale should be removed by mechanical or chemical methods. Light oil on steel surfaces is not harmful unless it has picked up dust or grit. Remnants of drawing compounds containing mineral fillers should be removed before welding.

The methods used for preparing surfaces for projection welding differ for various metals and alloys. The surface conditions and methods of cleaning and preparation for welding described in Chapter 1, "Spot and Seam Welding" are applicable to projection welding.

TYPES OF JOINTS

In addition to resistance spot and seam welding, projection welding also can be used to produce lap joints. The number and shape of the projections depend on the requirements for joint strength in each application.

Annular or ring projections can be used in applications requiring either gas-tight or water-tight seals, or to obtain a larger-area weld than dome-type projections can provide.

JOINT DESIGN

Lap joint designs for projection welding are similar to those used in spot welding; however, joint overlap and edge distances for projection welding generally can be less than for spot welding. Most applications use multiple projections. The minimum distance between projections should be two times the diameter of the projection.

The design of the component at the joint location may be significantly limited because the welding electrodes normally contact several projections simultaneously. The electrodes must be mounted rigidly on the welding machine, and the supporting members must be strong enough to minimize deflection when electrode force is applied. Press-type welding machines commonly are used for projection welding applications.

Fitup is important with multiple-projection welding. Each projection must be in contact with the mating surface to accomplish a weld. Uniformity of projection height is a factor in good fitup. The resistance welding electrodes must be carefully designed and accurately manufactured to mate with the workpieces at the weld locations. A resistance welding electrode usually is shaped to the workpiece contour to clamp the workpieces and to conduct the welding current. The electrodes should not force the deformation of the workpieces to obtain good fitup.

When good surface appearance on one component is a requirement, the marking left by the projection must be minimized. The projections should be placed on the other component. Using a large, flat electrode on the "show" or presentation side of the joint should prevent electrode marking, although slight shrinkage may occur at each projection weld. This may become visible after some finishing operations.

When projection welds are used to attach other fasteners such as weld nuts or bolts, the fasteners must contain a sufficient number of projections to carry the design load. Applicable mechanical testing should be applied to prove the integrity of the design. Production quality control should be programmed to ensure that weld quality does not drop below the design standards.

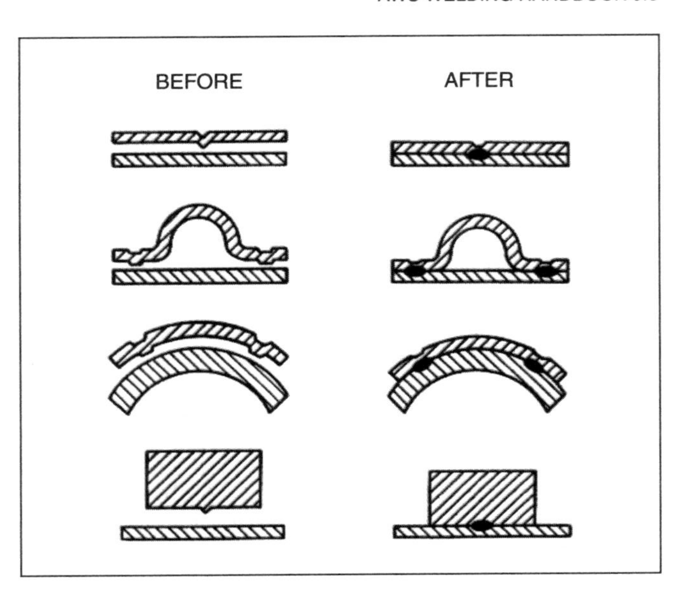

Figure 2.5—Examples of Various Projection Designs Before and After Welding

PROJECTION DESIGNS

The purpose of a projection is to concentrate the heat and pressure at a specific location on the joint. The projection design determines the current density requirement. Various types of projection designs are shown before and after welding in Figure 2.5.

The means of producing projections depends on the material in which they are to be produced. Projections in sheet metal components generally are made by embossing. Failure to maintain projection size and shape generally has been the biggest cause of weld quality problems for projections on thin sheet metal components. Projections in thick metal pieces are formed either by machining or forging. When welding stamped workpieces, projections generally are located on the edge of the stamping.

Sheet Metal

A projection design for sheet metal requires the following characteristics:

1. Sufficient rigidity to support the initial electrode force before welding current is applied;
2. Adequate mass to heat a spot to welding temperature on the other surface to prevent the collapse of the projection before the other surface is adequately heated;
3. Collapse of the projection without metal expulsion between the sheets or sheet separation after welding;
4. Easy formation that does not allow partial shearing from the sheet during the forming operation; and
5. Minimum distortion of the workpiece during forming or welding.

The general design of a projection suitable for steel sheet is shown in Figure 2.6. This design prevents the tendency to shear or to significantly thin the projection wall during the forming operation. If not designed to withstand the forming operation, the projections may be weak and the resulting welds may be easily torn from the sheet on loading. The designs of the punch and die that form this projection shape are illustrated in Figure 2.7. The projection sizes recommended for various sheet thicknesses and the punch and die dimensions to produce the projections are listed in Table 2.1.

Projections may be elongated to increase nugget size, and thus the strength of the weld. In this case, the contact between the projection and the faying surface is linear. Elongated projections generally are used for metal sheet gauges thicker than approximately 2.0 mm (0.08 in.).

On thin sheet metal, a small-diameter annular projection may be used instead of a round projection. The annular projection has greater stiffness and resists collapse when electrode force is applied. Thinner sections require special welding machines capable of following the rapid collapse of the projections.

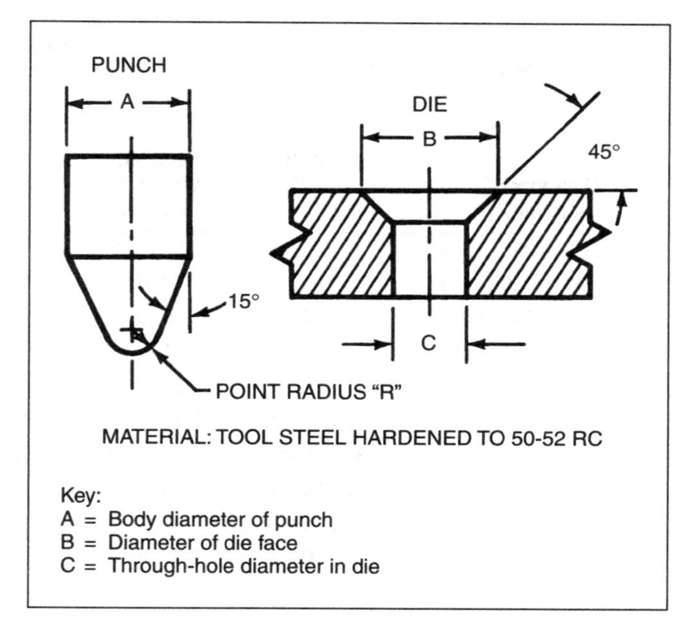

Figure 2.7—Basic Design of a Punch and Welding Die Used to Form Spherical Dome Projections in Sheet Steel (See Figure 2.6)

Machined or Forged Components

Annular projections frequently are used on forged components to carry heavy loads and for applications that require a pressure-tight joint around a hole between two workpieces. Annular projections also produce high-strength welds when large bosses or studs are welded to thin sheet metal. Figure 2.8 shows two applications of annular projections: a threaded boss and a shoulder stud. The summit of the circular ridge should be rounded, as shown in Figure 2.8(A). This shape improves heat balance, particularly with heavy sections. The completed weld is shown in (B). As shown in the shoulder stud in Figure 2.8(C), a relief path should be provided at the base of the projection for the upset metal to fill in as the projection collapses and completes the weld (D). This assures a tight joint without a gap.

Various designs of weld fasteners are available commercially for projection welding applications. Typical examples are shown in Figure 2.9. The design of the projections and the number of projections required depend on the specific application. Additional information on forming annular and recessed projections, including sample die and punch designs, are published by the Resistance Welding Manufacturing Alliance, a standing committee of the American Welding Society, in the *Resistance Welding Manual*.[5]

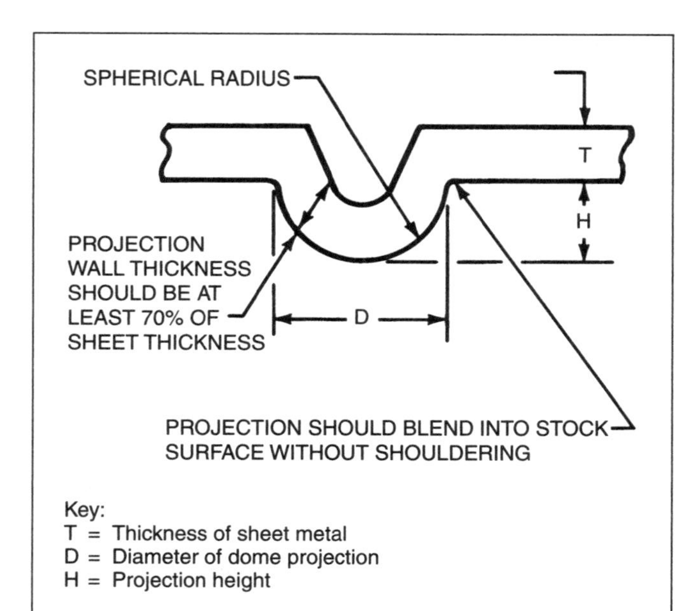

Figure 2.6—Basic Design of a Spherical Dome Projection Placed in Steel Sheet

5. Resistance Welding Manufacturing Alliance (RWMA), a standing committee of the American Welding Society, 2003, *Resistance Welding Manual*, Revised 4th Ed., Miami: Resistance Welding Manufacturing Alliance, 3-12–3-15.

Table 2.1
Punch and Die Dimensions for Spherical Dome Projections (Refer to Figure 2.7)

	Projection		Punch		Die	
Thickness, T mm (in.)	Height, H Within 2% mm (in.)	Diameter, D Within 5% mm (in.)	Diameter, A mm (in.)	Point Radius, R Within 0.002 mm (in.)	Hole Diameter, B Within 0.005 mm (in.)	Chamber Diameter mm (in.)
0.56–0.86 (0.022–0.034)	0.64 (0.025)	2.29 (0.090)	9.53 (0.375)	0.79 (0.031)	1.93 (0.076)	2.29 (0.090)
0.91–1.09 (0.036–0.043)	0.89 (0.035)	2.79 (0.110)	9.53 (0.375)	1.19 (0.047)	2.26 (0.089)	2.79 (0.110)
1.24–1.37 (0.049–0.054)	0.97 (0.038)	3.56 (0.140)	9.53 (0.375)	1.19 (0.047)	2.64 (0.104)	3.30 (0.130)
1.55–1.70 (0.061–0.067)	1.07 (0.042)	3.81 (0.150)	9.53 (0.375)	1.57 (0.062)	3.05 (0.120)	3.81 (0.150)
1.96 (0.077)	1.22 (0.048)	4.57 (0.180)	9.53 (0.375)	1.57 (0.062)	3.66 (0.144)	4.57 (0.180)
2.34 (0.092)	1.27 (0.050)	5.33 (0.210)	12.7 (0.500)	1.98 (0.078)	4.37 (0.172)	5.33 (0.210)
2.72 (0.107)	1.40 (0.055)	6.10 (0.240)	12.7 (0.500)	1.98 (0.078)	4.98 (0.196)	6.10 (0.240)
3.12 (0.123)	1.47 (0.058)	6.89 (0.270)	12.7 (0.500)	2.39 (0.094)	5.61 (0.221)	6.86 (0.270)
3.43 (0.135)	1.57 (0.062)	7.62 (0.300)	12.7 (0.500)	27.69 (01.09)	6.35 (0.250)	7.62 (0.300)
3.89 (0.153)	1.57 (0.062)	8.38 (0.330)	12.7 (0.500)	3.18 (0.125)	6.86 (0.270)	8.38 (0.330)
4.17 (0.164)	1.73 (0.068)	8.89 (0.350)	12.7 (0.500)	3.58 (0.141)	7.54 (0.297)	9.14 (0.360)
4.55 (0.179)	2.03 (0.080)	9.91 (0.390)	12.7 (0.500)	3.96 (0.156)	8.33 (0.328)	9.91 (0.390)
4.95 (0.195)	2.13 (0.084)	10.41 (0.410)	12.7 (0.500)	3.96 (0.156)	8.59 (0.338)	10.41 (0.410)
5.33 (0.210)	2.34 (0.092)	11.18 (0.440)	12.7 (0.500)	4.75 (0.187)	9.09 (0.358)	11.18 (0.440)
5.72 (0.225)	2.54 (0.100)	11.94 (0.470)	12.7 (0.500)	4.75 (0.187)	9.35 (0.368)	11.94 (0.470)
6.22 (0.245)	2.84 (0.112)	13.46 (0.530)	12.7 (0.500)	4.75 (0.187)	10.31 (0.406)	13.46 (0.530)

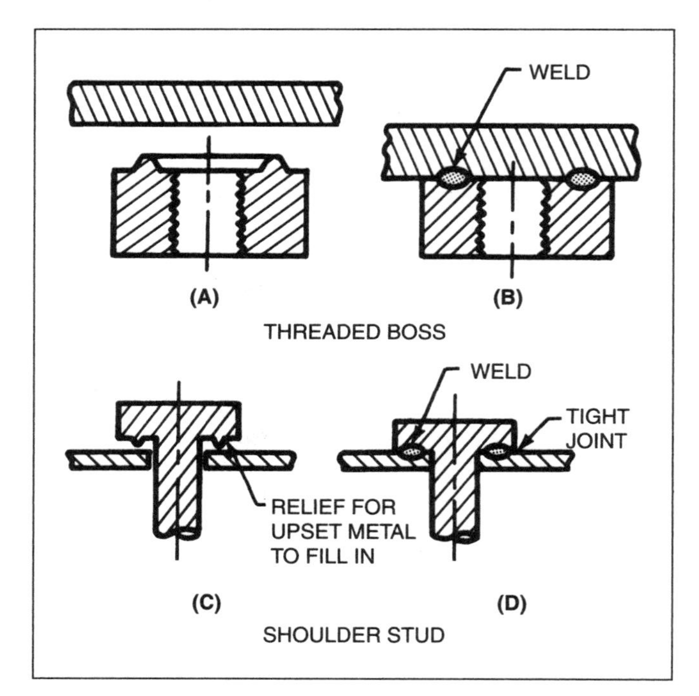

Figure 2.8—Projection Welding Applications Using Annular Projections

HEAT BALANCE

As in resistance spot welding, the distribution of heat in the two sections being welded with the projection welding process must be reasonably uniform to obtain strong welds. The following factors affect heat balance:

1. Design and location of the projection,
2. Thickness of the sections,
3. Thermal and electrical conductivity of the sections,
4. Heating rate, and
5. Electrode alloy.

The major portion of the heat develops in the projections during welding. Consequently, heat balance is generally easier to obtain in projection welding than in resistance spot welding; however, heat balance may be complicated when making multiple simultaneous projection welds. Uniform allocation of welding current and electrode force is necessary to obtain the even heating of all projections. Since the current paths through the projections are in parallel, any variation in resistance between the projections will cause the current to be distributed unequally.

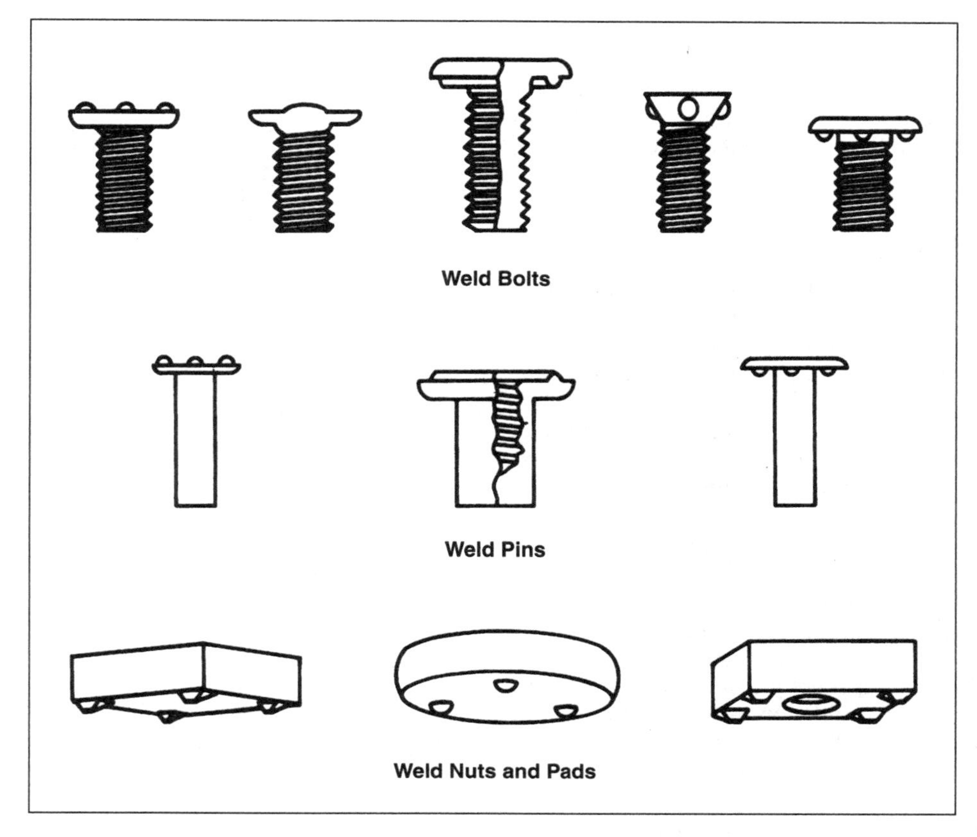

Figure 2.9—Typical Commercially Available Fasteners Used in Projection Welding

Projections must be designed to support the electrode force needed to obtain good electrical contact with the mating workpieces, and to collapse when heated. With multiple projections, slight variations in projection heights can affect heat balance. This may occur as a result of wear of the projection-forming punches. Heat balance in metals of dissimilar thickness is maintained by placing the projection in the thicker of the two workpieces. The size of the projection is based on the requirements for heating the thinner section. Likewise, to maintain heat balance in materials of dissimilar conductivity, the projection is located in the workpiece with the higher conductivity (lower resistivity). The electrode must be chosen on the basis of its composition because the electrode alloy determines the conductivity of the electrode, which also can affect heat balance.

WELDING CYCLE

Variables in the projection welding process include welding current, weld time, electrode force, and electrodes.

Welding Current

The current for each projection generally is less than that required to produce a spot weld in the same thickness of the same metal. If excessive current is used, the projection will heat rapidly, melt, and result in expulsion. However, the current must be at least high enough to create fusion before the projection has completely collapsed. For multiple projections, the total welding current should approximately equal the current for one projection multiplied by the number of projections. Some adjustment may be required to account for normal projection tolerances, component design, and the impedance of the secondary circuit.

Weld Time

Weld time is about the same for single or multiple projections of the same design. Although a short weld time may be desirable from a production standpoint, it will require correspondingly higher amperage. This may cause overheating and metal expulsion. In general, longer weld times and lower amperages are used for projection welding than those for spot welding.

In some cases, multiple welding pulses may be advantageous to control the heating rate. This is helpful with thick sections and metals with low thermal conductivity.

Electrode Force

The electrode force used for projection welding depends on the metal being welded, the projection design, and the number of projections in the joint. The force should be adequate to flatten the projections completely when they reach the welding temperature, bringing the workpieces in contact. Excessive force will prematurely collapse the projections and the weld nuggets will be ring-shaped, with incomplete fusion in the center.

The welding machine must be capable of mechanically following the workpiece with the electrodes as the projections collapse. Slow follow-up will cause metal expulsion before the workpieces are together.

The sequence of events during the formation of a projection weld is shown schematically in Figure 2.10. In Figure 2.10(A), the projection is shown in contact with the mating sheet. In (B), the current has started to heat the projection to welding temperature. In (C), the electrode force causes the heated projection to collapse rapidly and then fusion takes place. The completed weld is shown in (D).

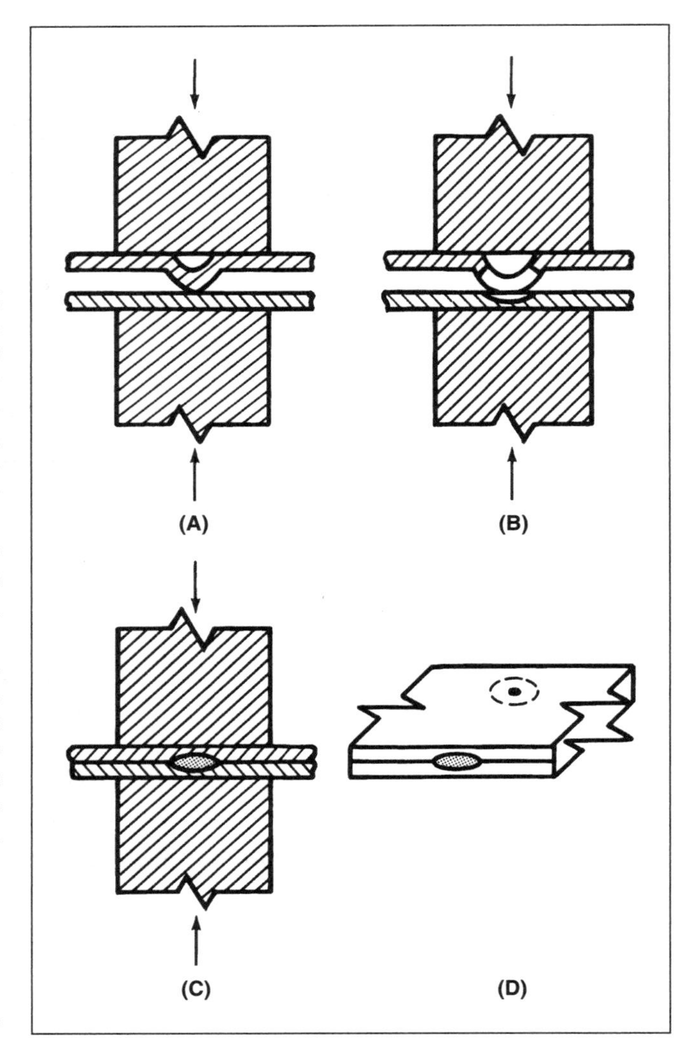

Figure 2.10—Sequence of Events During the Formation of a Projection Weld

ELECTRODES

The areas of the workpieces at the joint interface frequently are flat except for the projections. In such cases, large flat-faced electrodes are used. If the surfaces to be contacted are contoured, the electrodes are fitted to them. With contoured electrodes, the electrode force can be applied without distorting the workpieces and the welding current can be introduced without overheating the contact areas.

For a single projection, the face diameter of the electrode should be at least twice the diameter of the projection. With multiple projections, the electrode face should extend a minimum space equal to one projection diameter beyond the boundary of the projection pattern.

The best electrode material is one that is hard enough to minimize wear but not so hard that it causes cracking or surface burning of the workpiece. If burning or cracking is encountered, an electrode made of a softer alloy of higher conductivity should be used. With multiple projections, electrode wear can disturb the balance of welding current and electrode force on the projections. Then the strength and quality of the welds may become unacceptable.

Electrodes for large production requirements often have inserts of Resistance Welding Manufacturing Alliance (RWMA) Group B material at the points of greatest wear. In some cases, it is more economical and equally satisfactory to use one-piece electrodes of RWMA Group A, Class 3 alloy.[6]

Welding electrodes and locating fixtures for projection welding usually are combined. With the proper fixtures, it is possible to obtain accuracy with projection welding equal to that of any other assembly

6. American Welding Society (AWS), *Specification for Automotive Resistance Spot Welding Electrodes*, D8.6:2005 Miami: American Welding Society.

process. The welding fixtures should meet the following requirements:

1. Provide accurate positioning of the workpieces;
2. Permit rapid loading and unloading;
3. Allow no alternative path for the welding current;
4. For ac welding, be made of nonmagnetic materials; and
5. Be properly designed for operator safety.

The electrodes must be mounted solidly on the welding machine. The workpieces are correctly positioned in one electrode and all the welds are made at one time in one operation of the machine. The workpieces may be positioned and aligned with one another by punching holes in one and semi-punching the other to match. The projections usually can be embossed or forged in the same operation.

In some designs, insulated pins or sleeves may be used in the electrode to position and align the workpieces. Basic examples are shown in Figures 2.11 and 2.12.

When the small component of an assembly can be placed on the bottom and the large component on top, it is a simple matter to hold the small component in a recessed lower electrode, as shown in Figure 2.13. However, a problem is created when the design requires placing a small part on top of a larger part. Sometimes the small component can be located and held by a removable device and then welded with a flat upper electrode. Components that nest into the upper electrode may be held by spring clips attached to the electrode. Figure 2.14 shows a spring-loaded retainer entered through the side of the electrode holding a bolt for welding. A vacuum also may be used to hold small components in the upper electrode when component design permits.

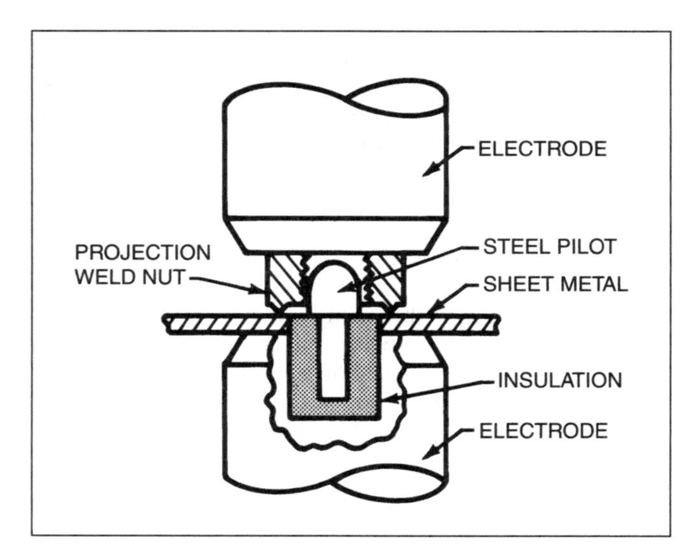

Figure 2.11—Insulated Pin Used to Position a Weld Nut

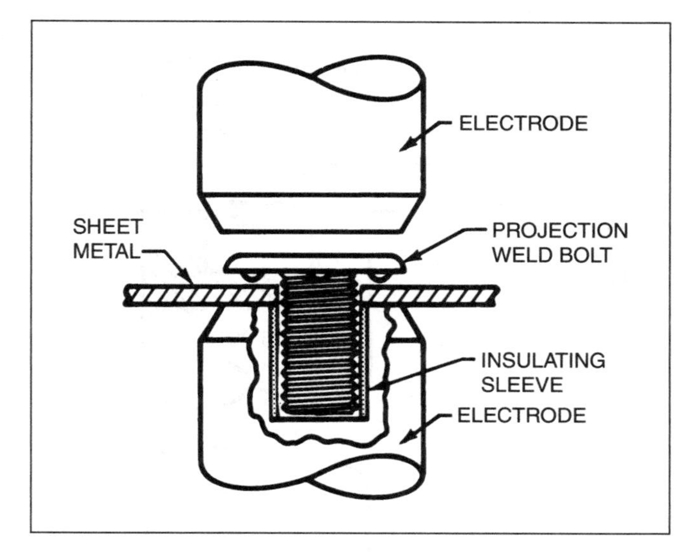

Figure 2.12—Positioning a Weld Bolt with an Insulating Sleeve

The success of projection welding operations in production largely depends on the proper selection, installation, and maintenance of the electrodes. If the electrodes are correctly designed and constructed, installation is next in importance. First, the platens of the welding machine must be parallel to one another and perpendicular to the motion of the ram. The platens should be smooth, clean, and free of nicks and pit marks. If they are not, the platens should be removed and machined smooth and flat before installing the electrodes. The test for parallelism of the platens should be made under the operating forces to be used during welding. This can best be done by placing a steel block with smooth parallel faces between the platens, applying the intended electrode force, and then checking the platens for gaps with "feeler" gauges.

The next step is to check the bases of the electrodes. They must be clean, smooth, flat, and free from burrs and nicks. If they are not suitable, the electrode bases should be machined. The electrodes are then installed on the machine. Most machines have T-slots at right angles to one another in the two platens to permit universal alignment of the electrodes. After the electrodes are properly aligned, they should be clamped securely to

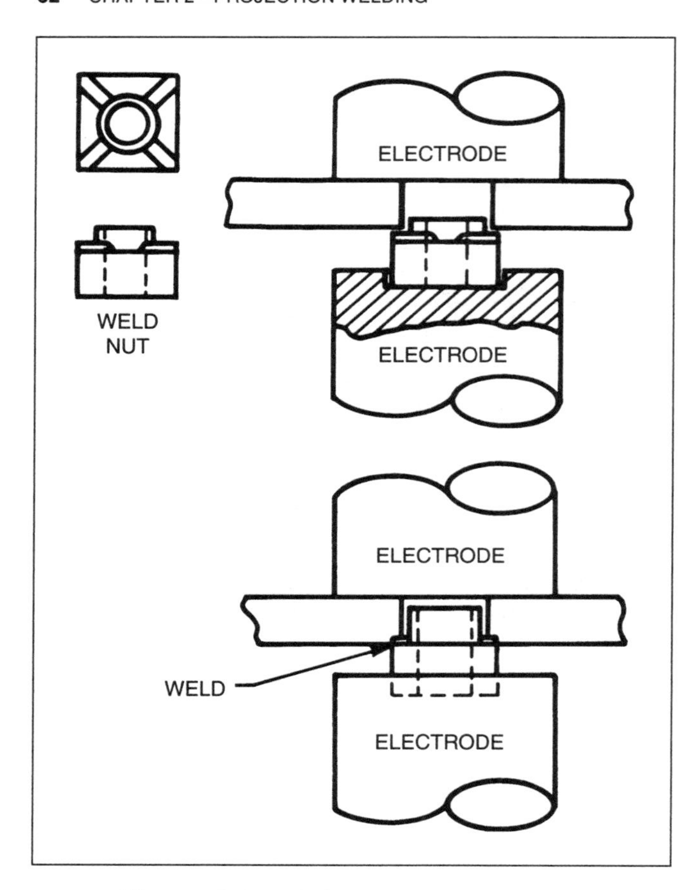

Figure 2.13—A Recessed Electrode Used to Position a Weld Nut

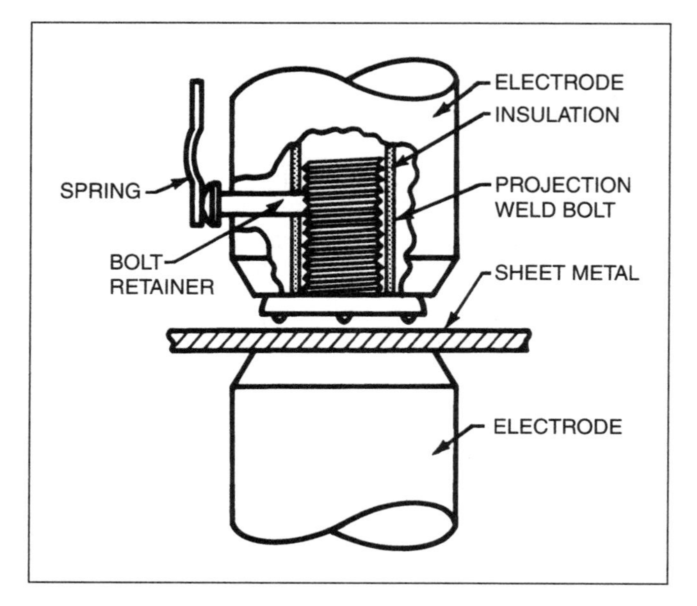

Figure 2.14—Holding a Bolt in the Upper Electrode with a Spring Retainer

the platens. With the workpieces in place in the electrodes, the position of the ram or "knee" of the machine should be adjusted for the proper stroke, including the necessary allowance for upset of the projections.

If the tips of the projections are in one plane and of uniform height, the setup is ready for trial welds. The following conditions may cause inconsistent current or force on the projections:

1. Shunting of the current through locators,
2. Unequal secondary circuit path lengths,
3. Excessive play in the welding head, and
4. Too much deflection in the knee of the machine.

The use of shims between electrode components or between electrodes and platens should be avoided. If shims must be used they should be made of clean, annealed, pure copper sheets of sufficient area to carry the secondary current.

If projections are located on curved or angled surfaces, accurate templates should be provided for checking the electrodes. It should be noted that when curved workpieces are welded, or two or more components are welded to others components, the mill tolerances for the metal thicknesses involved might cause problems. Mill tolerances must be provided for in the design of the components and the arrangement of the projections.

WELDING SCHEDULES

When setting up a welding process for a particular metal and joint design, a schedule must be established to produce welds that meet the design specifications. Previous experience can provide a starting point for the initial setup. If the application is a new one, reference to published information on the welding of the material by the designated process can serve as a guide for the initial setup.

Sample welds should be made and tested while changing one process variable at a time within a range to establish an acceptable value for that variable. It may be necessary to establish the effect of one variable at several levels of another. For example, weld or heat time and electrode force may be evaluated at several levels of current. Visual examination and destructive test results can be used to select an appropriate welding schedule. Finally, the first components off the production line, or simulations, should be welded and destructively tested. Final adjustments to the welding schedule then can be made to meet design or specification requirements.

Starting schedules for many commercial alloys may be available from the equipment manufacturer. Some also may be found in the following publications:

1. *Recommended Practices for Resistance Welding,* AWS C1.1M/C1.1;[7]
2. *Specification for Resistance Welding of Carbon- and Low-Alloy Steels,* C1.4M/C1.4;[8] and
3. *Resistance Welding Manual.*[9]

METALS WELDED

Projection welding typically is restricted to low-carbon steels or microalloyed steels. Some applications in aluminum are successfully welded with the process; however, the initial weld force may crush or collapse the aluminum projection prior to the application of current. Therefore, projection welds in aluminum are very difficult to make.

The following metals commonly are joined by projection welding:

1. Low-carbon steel;
2. Hardenable steels;
3. Stainless steels—ferritic, martensitic, and austenitic types;
4. Nickel-base alloys;
5. Copper alloys;
6. Aluminum and magnesium alloys;
7. Titanium alloys; and
8. Coated and plated steels.

Many plated and coated steels can be welded with the projection welding process, but weld quality usually is affected by the composition and thickness of the coating. Coatings on steel usually are applied for corrosion resistance, decoration, or a combination of these. Welding procedures should assure the production of welds with adequate strength and also should reasonably preserve the coating function. Strength specifications usually require welding machine settings similar to those used for bare carbon steel. Adjustments to compensate for the coating can be determined by a number of factors, including the effect of the coating on contact resistance, acceptable electrode indentation, the tendency of the coating to alloy with the base metal, and the tendency of the electrode to fuse to the workpiece.

METAL PROPERTIES INFLUENCING WELDABILITY

The following properties of metals have a bearing on weldability by the projection welding process:

1. Electrical resistivity,
2. Thermal conductivity,
3. Thermal expansion,
4. Hardness and strength,
5. Oxidation resistance,
6. Plastic temperature range, and
7. Metallurgical properties.

Chapter 1 of this volume, which covers resistance spot and seam welding, provides information on the properties of metals that also applies to projection welding.

WELD QUALITY

The weld quality required depends primarily on the application. Generally, the quality of projection welds is determined by the following basic factors:

1. Surface appearance,
2. Weld size,
3. Penetration,
4. Strength and ductility, and
5. Internal discontinuities.

SURFACE APPEARANCE

Normally, the surface appearance of a projection weld should be relatively smooth. It should be round or oval in the case of a contoured workpiece, and free from surface fusion, electrode deposit, pits, cracks, excessive electrode indentation, or any other condition that indicates improper electrode maintenance or operation. Table 2.2 lists some undesirable surface conditions, their causes, and the effects on weld quality.

WELD SIZE

The diameter of the nugget or the width of the fused zone must meet the weld size requirements of the appropriate specifications or design criteria. In the absence of such requirements, the following general rule should be used: projection welds should have a

7. American Welding Society (AWS) Committee on Resistance Welding, *Recommended Practices for Resistance Welding*, AWS C1.1M/C1.1:2000 (or latest edition), Miami: American Welding Society.
8. American Welding Society (AWS) Committee on Resistance Welding, (latest edition), *Specification for Resistance Welding of Carbon- and Low-Alloy Steels*, C1.4M/C1.4:1999 (or latest edition), Miami: American Welding Society.
9. Resistance Welding Manufacturing Alliance (RWMA), 2003, *Resistance Welding Manual*, Revised 4th Ed., Miami: American Welding Society.

Table 2.2
Undesirable Surface Conditions for Projection Welds

Type	Cause	Effect
1. Deep electrode indentation	Improperly dressed electrode face Improper control of electrode force Excessively high rate of heat generation due to high contact resistance (low electrode force)	Loss of weld strength due to reduction of metal thickness at the periphery of the weld area Bad appearance
2. Surface fusion (usually accompanied by deep electrode indentation)	Scaly or dirty metal Low electrode force Misalignment of workpieces High welding current Electrodes improperly dressed Improper sequencing of pressure and current	Undersize welds due to heavy expulsion of molten metal Large cavity in weld zone extending through to surface Increased cost of removing burrs from outer surface of workpiece Reduced electrode life and increased production time due to frequent electrode dressings
3. Irregular shaped weld	Misalignment of workpieces Excessive electrode wear or improper electrode dressing Badly fitting workpieces Electrode bearing on the radius of the flange Skidding Improper cleaning of electrode surfaces	Reduced weld strength due to change in interface contact area and expulsion of molten metal
4. Electrode deposit on workpieces (usually accompanied by surface fusion)	Scaly or dirty material Low electrode force or high welding current Improper maintenance of electrode contacting face Improper electrode material Improper sequencing of electrode force and weld current	Bad appearance Reduced corrosion resistance Reduced weld strength if molten metal is expelled Reduced electrode life
5. Cracks, deep cavities, or pin holes	Removing the electrode force before welds are cooled from liquidus Excessive heat generation resulting in heavy expulsion of molten metal Poorly fitting components requiring most of the electrode force to bring the faying surfaces into contact	Reduction of fatigue strength if weld is in tension or if crack or imperfection extends into the periphery of weld area Increase in corrosion due to accumulation of corrosive substances in cavity or crack

weld size equal to or larger than the diameter of the original projection.

There is a maximum limit to the size of a projection weld. Since this limit usually is controlled by the workpiece configuration, cost, or practicality of making the weld, the limit should be established based on the design requirements of the application and prevailing shop practices.

PENETRATION

Penetration is the depth to which the weld nugget extends into the workpiece. Generally, the acceptable minimum penetration is 20% of the thickness of the thinner outside workpiece. If penetration is less than 20%, the weld is considered to be cold, because not enough heat was generated in the weld zone, or the joint interface was too small.

STRENGTH AND DUCTILITY

Structures joined by projection welds usually are designed so that the welds are loaded in shear when the components are exposed to tension or compression loading. In some applications, the welds are loaded in tension, or in a combination of tension and shear, but only when the direction of loading is perpendicular to the plane of the joint.

The strength requirements for projection welds normally are specified in kilograms (pounds) per weld. The strength generally increases in proportion to increases in nugget diameter or weld size, even though the average unit stress decreases. The unit stress decreases because of the greater tendency for failure to occur at the edge of the nugget as its size increases. In low-carbon steel sheet, for example, the calculated average shear stress in good welds at fracture will vary from 69 MPa to 414 MPa (10 ksi to 60 ksi). Low values apply to relatively large welds, and high values to relatively small welds. In both instances, the actual tensile stress in the sheet at the weld periphery is at or near the ultimate tensile strength of the base metal. For this reason, the shear strength of circular welds tends to vary linearly with nugget diameter.

Single projection welds are not strong in torsion when the axis of rotation is perpendicular to the plane of the welded components. The torsional strength tends to vary with the cube of the nugget diameter (D^3). Little torsional deformation occurs in low-ductility welds prior to failure. Angular displacements may vary from 5° to 180°, depending on weld metal ductility.

Test Methods

The standard methods of measuring ductility, such as those that measure the percentage of elongation or reduction of area in a tensile test, are not adaptable to projection welds. Hardness testing is the closest alternative to ductility testing for these welds. It should be noted that although the ductility for a given alloy decreases with increasing hardness, different alloys of the same hardness do not necessarily possess the same ductility.

Another method of indicating the ductility of a projection weld is to determine the ratio of direct tension-shear strength.[10] A weld with good ductility has a high ratio; a weld with poor ductility has a low ratio. Where this ratio is specified, 0.25 is usually the minimum for hardenable steel welds after tempering.

Various methods are available to minimize the hardening effect of rapid cooling in the welds. Following are some of these methods:

1. Use long weld times to put heat into the workpiece,
2. Preheat the weld area with a preheat current,
3. Temper the weld and heat-affected zones with a temper current at some interval after the weld time, and
4. Anneal or temper the welded assembly in a furnace.

10. These tests are described in American Welding Society (AWS) Welding Handbook Committee, Jenney, C. L. and A. O'Brien, Eds., 2001, *Welding Handbook*, 9th ed., Vol. 1, *Welding Science and Technology*, Miami: American Welding Society, Chapter 6.

INTERNAL DISCONTINUITIES

Internal discontinuities in resistance welds include cracks, porosity or spongy metal, large cavities, and, in some coated metals, metallic inclusions. Generally, these discontinuities will not have a detrimental effect on the static or fatigue strength of a resistance weld if they are located entirely in the central portion of the weld nugget. This is true because the stresses essentially are zero in the central portion of the nugget. Conversely, no discontinuities should occur at the periphery of a weld, where the load stresses are highly concentrated. The high stresses at the weld periphery can be attributed to the high stress concentration factor associated with the overlapping joint geometry, as described in *Peterson's Stress Concentration Factors*.[11] Since high stress concentration can greatly reduce the fatigue strength or service life of a metal, projection welding and other resistance welding processes generally are not used for applications in which the joint is subject to high cyclic load stresses.

Projection welds in metals approximately 1 mm (0.040 in.) thick and greater may have a small amount of shrinkage porosity in the center of the weld nugget, as illustrated in Figure 2.15. these discontinuities are less pronounced in some welds than in others due to the

11. Stress concentration factors are discussed in detail in Pilkey, W. D., 1997, *Peterson's Stress Concentration Factors (2nd Edition)*, New York: John Wiley and Sons, Chapter 6. This book is an essential addition to the professional libraries of engineers and designers working in the automotive, aerospace, and nuclear industries, civil and mechanical engineers and students and researchers in these fields.

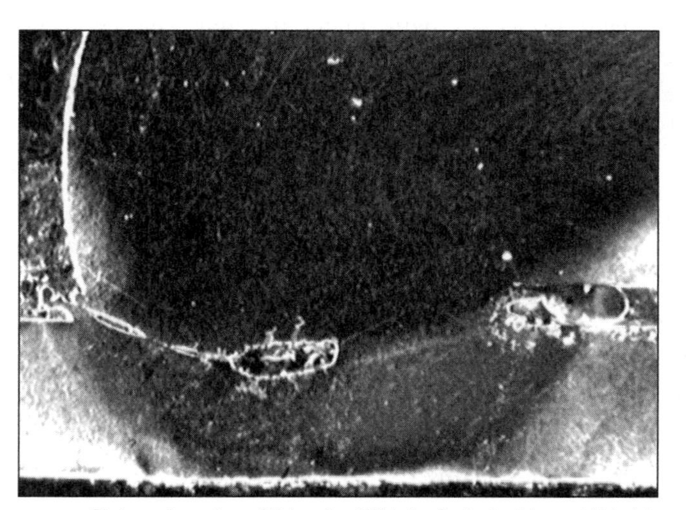

Photograph courtesy of University of Waterloo Centre for Advanced Materials

Figure 2.15—Discontinuities in Projection Welds

difference in forging action of the electrodes on the hot metal. Porosity or cavities that result from expulsion of molten metal are much larger than shrinkage cavities. A certain number of expulsion cavities usually are expected in production welding of various metals. These discontinuities are shown in Figure 2.15.

Low electrode force, high welding current, and poor fitup generally are the causes of internal discontinuities in projection welds. Removing the electrode force too soon after the welding current stops also causes internal discontinuities. When these conditions occur, the weld nugget will not be properly forged during cooling.

If crack-like indications are observed in the heat-affected zone at low magnification, these indications should be reexamined at higher magnification to determine whether they are actual cracks or "coring." A cored area is filled with material that has a dendritic structure. Coring may result from incipient melting or backfilling of cracks by the molten metal in the heat-affected zone, based on the dendritic structure. Based on service experience, coring did not appear to affect the serviceability of the welded joints of various resistance-welded, nickel-base alloy jet engine components, such as nozzles and combustor housings.

Photograph courtesy of InterMetro Industries

Figure 2.16—Steel Utility Shelves Fabricated with Cross Wire Welding

CROSS WIRE WELDING

Cross wire welding is a commonly used variation of projection welding in which the localization of the welding current is achieved by the intersection contact of wires, and is usually accompanied by considerable embedding of one wire into another. This section covers general principles, wire materials, and welding techniques for cross wire welding.

Cross wire welding applications usually consist of welding a number of parallel wires at right angles to one or more other wires or rods to manufacture the following types of products:

1. Stove and kitchen racks,
2. Grills of all kinds,
3. Lampshade frames,
4. Poultry processing equipment,
5. Wire baskets,
6. Fencing,
7. Grating,
8. Concrete reinforcing mesh, and
9. Wire shelving.

The steel wire utility shelves shown in Figure 2.16 represent a typical application of cross wire welding. The close-up view of the welds in Figure 2.17 illustrates the high-quality appearance of welds required by many cross-wire welding applications.

FUNDAMENTALS

There are several specific ways to perform the welding operation, depending on production requirements, but the quality of the finished product is essentially the same regardless of the method used. Figure 2.18 shows a section of a typical cross wire weld.

Wire racks can be welded in a press-type projection welding machine or in special automatic indexing machines with hopper feeding systems and a separate gun for each weld.

Concrete reinforcing mesh is made on continuous machines. The stay wires are fed either from wire reels

Photograph courtesy of InterMetro Industries

Figure 2.17—Close-Up View of Cross Wire Welds with Appearance as a Factor

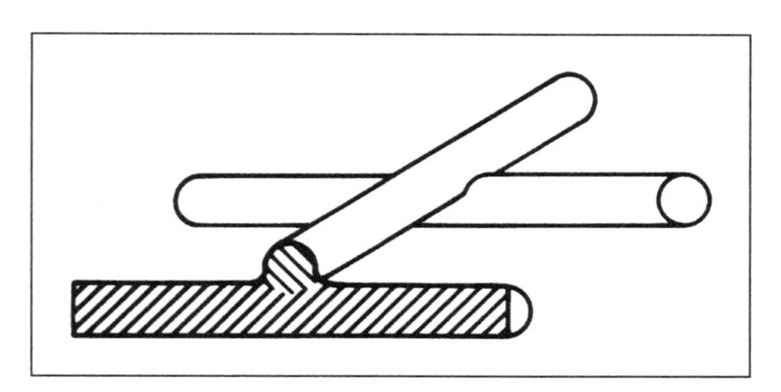

Figure 2.18—Section of a Typical Cross Wire Weld

on the side of the machine or from magazines of cut wire. The welded mesh is either rolled into coils, like fencing, or cut into mats and then stacked and bundled.

As in spot and projection welding, the wire or rod should be clean and free from scale, rust, dirt, paint, heavy grease, or other highly resistant coatings. Plated or galvanized wire or rods may be used, but the coating at the weld would be destroyed.

Wire Materials

Low-carbon steel wire is the material most commonly welded. Typical machine settings for the cross wire welding of low-carbon steel are listed in Table 2.3. Next in importance are stainless steel and copper-nickel alloy wires. Copper-nickel alloy wires require about the same weld time and amperage as carbon steel wires, and about two times the electrode force. Stainless steel wires also require about the same weld time, but use 60% of the amperage and two and one-half times the electrode force.

Welding Technique

Normally, cross wire welds are not dressed after welding. Therefore, the major consideration may be appearance, with strength secondary in importance for some applications.

In setting up the welding machine, consideration must be given to the following variables:

1. Design strength,
2. Appearance,
3. Welding electrodes,
4. Electrode force,
5. Weld time, and
6. Welding current (heat).

The specifications of the particular application determine which is most important, strength or appearance, when setting up welding conditions for cross wire welding. It is normally assumed that high-strength welds with an acceptable appearance are desired.

The required electrode force, welding current, and weld time depend greatly on the specified amount of compression that is applied to the wires or rods. The amount of compression is correctly called *upset* (but often called *setdown*). Upset is the ratio of the decrease in joint height to the diameter of the smaller wire. Weld strength generally increases as upset percentage increases. Upset is defined and illustrated in Figure 2.19, which shows how the percentage of upset is calculated.

The welding electrodes must be of the proper material and shape, with provision for water-cooling. RWMA Class II alloy electrodes usually have an acceptable service life, although electrode facings of harder alloys sometimes are used for special applications.[12] Although flat electrodes commonly are used for cross wire welding, certain advantages may be gained by shaping the electrodes to conform to the wires or rods

12. Resistance Welding Manufacturing Alliance (RWMA) *Bulletin 16:(1)*, 1996, Miami: American Welding Society.

Table 2.3
Conditions for Cross Wire Welding of Low-Carbon Steel Wire

	Cold-Drawn Wire				Hot-Drawn Wire			
Wire Diameter mm (in.)	Weld Time, Cycles	Electrode Force J (lb)	Weld Current A	Weld Strength J (lb)	Weld Time, Cycles	Electrode Force J (lb)	Weld Current A	Weld Strength J (lb)
	15% Upset				15% Upset			
1.59 (1/16)	5	135.58 (100)	600	610.12 (450)	5	135.58 (100)	600	474.54 (350)
3.18 (1/8)	10	169.48 (125)	1800	1321.92 (975)	10	169.48 (125)	1850	1016.86 (750)
4.76 (3/16)	17	488.09 (360)	3300	2711.64 (2000)	17	488.09 (360)	3500	2033.73 (1500)
6.35 (1/4)	23	786.37 (580)	4500	5016.53 (3700)	23	786.37 (580)	4900	3796.29 (2800)
7.94 (5/16)	30	1118.55 (825)	6200	6914.67 (5100)	30	1118.55 (825)	6600	6236.76 (4600)
9.53 (3/8)	40	1491.40 (1100)	7400	9083.98 (6700)	40	1491.40 (1100)	7700	8406.07 (6200)
11.11 (7/16)	50	1898.15 (1400)	9300	13 015.85 (9600)	50	1898.15 (1400)	10000	11 931.20 (8800)
12.7 (1/2)	60	2304.89 (1700)	10300	16 540.98 (12200)	60	2304.89 (1700)	11000	15 591.91 (11500)
	30% Upset				30% Upset			
1.59 (1/16)	5	203.37 (150)	800	677.91 (500)	5	203.37 (150)	800	542.33 (400)
3.18 (1/8)	10	352.51 (260)	2650	1525.30 (1125)	10	352.51 (260)	2770	1152.45 (850)
4.76 (3/16)	17	813.49 (600)	5000	3253.96 (2400)	17	813.49 (600)	5100	2304.89 (1700)
6.35 (1/4)	23	1152.45 (850)	6700	5694.44 (4200)	23	1152.45 (850)	7100	4067.45 (3000)
7.94 (5/16)	30	1965.94 (1450)	9300	8270.49 (6100)	30	1965.94 (1450)	9600	6779.09 (5000)
9.53 (3/8)	40	2792.98 (2060)	11300	11 321.08 (8350)	40	2792.98 (2060)	11800	9219.56 (6800)
11.11 (7/16)	50	3931.87 (2900)	13800	15 320.74 (11300)	50	3931.87 (2900)	14000	13 015.85 (9600)
12.7 (1/2)	60	4609.78 (3400)	15800	18 439.12 (13600)	60	4609.78 (3400)	16500	16 812.14 (12400)
	50% Upset				50% Upset			
1.59 (1/16)	5	271.16 (200)	1000	745.70 (550)	5	271.16 (200)	1000	610.12 (450)
3.18 (1/8)	10	474.54 (350)	3400	1694.77 (1250)	10	474.54 (350)	3500	1220.24 (900)
4.76 (3/16)	17	1016.86 (750)	6000	3389.54 (2500)	17	1016.86 (750)	6300	2440.47 (1800)
6.35 (1/4)	23	1681.21 (1240)	8600	5965.60 (4400)	23	1681.21 (1240)	9000	4203.04 (3100)
7.94 (5/16)	30	2711.64 (2000)	11400	8812.82 (6500)	30	2711.64 (2000)	12000	7185.83 (5300)
9.53 (3/8)	40	4067.45 (3000)	14400	11 931.20 (8800)	40	4067.45 (3000)	14900	9761.89 (7200)
11.11 (7/16)	50	6033.39 (4450)	17400	16 134.23 (11900)	50	6033.39 (4450)	18000	13 829.34 (10200)
12.7 (1/2)	60	7185.83 (5300)	21000	19 794.94 (14600)	60	7185.83 (5300)	22000	17 625.63 (13000)

Note: Upset % = Decrease in joint height × 100 ÷ diameter of smaller wire.

being welded. Shaped electrodes provide better contact between the electrode and the workpieces.

The electrode force depends on the wire diameter, the specified upset, the desired appearance, and the strength of the weld design. The electrode force affects the appearance of the weld. The values listed in Table 2.3 will produce welds with good appearance. Lower-strength welds than those shown in Table 2.3 will result if higher electrode forces are used without decreasing the weld time and increasing the welding current. The weld time needed depends on the diameter of the wire to be welded. The welding current depends on the diameter and the specified upset. Current should be slightly less than that which would result in "spitting" or expulsion of hot metal. For best results, the values shown in Table 2.3 should be used.

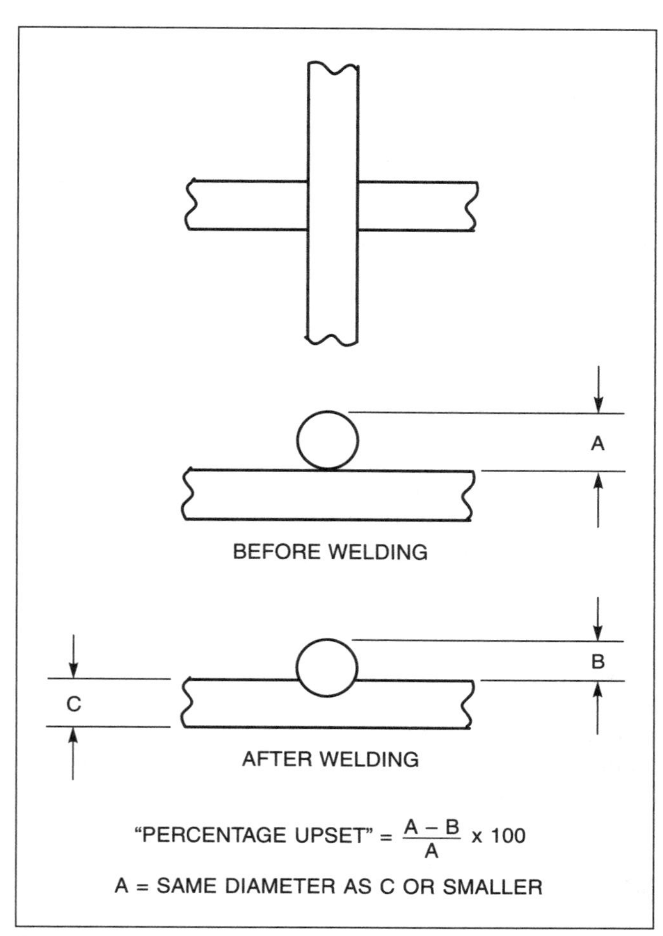

Source: Adapted from American Welding Society (AWS) Committee on Resistance Welding, 2000, *Recommended Practices for Resistance Welding*, AWS C1.1M/C1.1:2000, Miami: American Welding Society, p. 77.

Figure 2.19—Method of Calculating Upset for a Low-Carbon Steel Application

ECONOMICS

The advantages of projection welding over other welding processes are numerous, but the following benefits are important because they impact the economics of the application:[13]

1. A row or group of projection welds are readily made, especially in heavy materials;
2. Electrode maintenance is lower in projection welding than other resistance welding processes;
3. Because there is no mushrooming of the electrodes, current density remains unchanged regardless of electrode wear, although repeated welds eventually will wear a depression in the electrode surface that may have an effect on the weld process;
4. Projection welds can be readily made on angled or curved surfaces, particularly when two or more welds are made at an angle to one another;
5. Projections can be spaced closer together than conventional spot welds;
6. For multiple spots, projection welding equipment and tooling is simpler and less expensive in most cases; and
7. Projection welds require less current per weld than other processes because less metal is heated during welding.

SAFE PRACTICES

Projection welding and other resistance welding processes may involve hazardous situations that can be avoided by taking the proper precautions. Recommendations for safe practices in resistance welding are provided in *Recommended Practices for Resistance Welding*, AWS C1.1M/C1.1.[14] *Safety in Welding, Cutting, and Allied Processes*, ANSI Z49.1, accredited by the American Standards Institute, provides general information on safety in welding and has a section on resistance welding.[15] The Occupational Safety and Health Administration (OSHA) of the United States government publishes mandatory safety requirements in *Welding, Cutting and Brazing*, Subpart Q of *Occupational Safety & Health Standards for General Industry.*[16] Additional safety standards and publishers are listed in Appendices A and B of this volume.

13. Resistance Welding Manufacturing Alliance (RWMA), a standing committee of the American Welding Society AWS), 2003, *Resistance Welding Manual,* Revised 4th Edition, Miami: American Welding Society.

14. American Welding Society (AWS) Committee on Resistance Welding, *Recommended Practices for Resistance Welding*, AWS C1.1M/C1.1, Miami: American Welding Society.

15. American National Standards Institute (ANSI) Accredited Standards *in Welding, Cutting, and Allied Processes*, ANSI Z49.1, Miami: American Welding Society.

16. Occupational Safety and Health Administration (OSHA), latest edition, *Occupational Safety and Health Standards for General Industry*, in *Code of Federal Regulations (CFR)*, Title 29 CFR 1910, Subpart Q, Washington D.C.: Superintendent of Documents, U.S. Government Printing Office.

MECHANICAL HAZARDS

The safety recommendations in equipment manuals and material safety data sheets (MSDSs) provided by the manufacturers of equipment and materials must be carefully followed. Machines and mechanical equipment must be properly installed and routine inspection and maintenance must be performed. Operators must be well trained in the safe use of the equipment and must use personal protective equipment.

Guarding

Initiating controls on welding equipment, such as push buttons or switches, should be strategically arranged or guarded, as detailed in OSHA 29, CFR 1910, Subpart S, *Electrical*, to prevent the operator from inadvertently activating them.[17]

In some multiple-gun welding machine installations, the operator's hands can be expected to pass under the point of operation. Effective guarding devices such as proximity-sensing gates, latches, blocks, barriers, or dual hand controls should be provided in these machines.

Stop Buttons

One or more emergency stop buttons should be provided on all welding machines, with a minimum of one at each operator position.

PERSONAL PROTECTIVE EQUIPMENT

The protective equipment needed depends on the particular welding application. The following equipment generally is needed for resistance welding:

1. Eye protection in the form of face shields (the preferred form of protection) or hardened-lens goggles, with visitors to the welding area required to wear, as a minimum, approved safety glasses with side shields;
2. Skin protection provided by nonflammable gloves and clothing with a minimum number of pockets and cuffs in which hot or molten particles could lodge; and
3. Protective footwear.

ELECTRICAL

Resistance welding equipment should be designed to avoid accidental contact with components of the system that are electrically hazardous. High-voltage components must have adequate electrical insulation and must be completely enclosed. All doors, access panels, and control panels of resistance welding machines must be locked or interlocked to prevent access by unauthorized persons. The interlocks must effectively interrupt power and discharge all high-voltage capacitors into a suitable resistive load when the door or panel is opened. A manually operated switch or suitable positive device should be provided to assure complete discharge of all high-voltage capacitors. Three types of controls may be used to determine that voltage is completely discharged:

1. A device with two resistors and timed discharge, with failure indicators;
2. A single-discharge resistor with on-off lamp to verify presence or absence of voltage; and
3. A device for use in traction applications (electric rail) with parallel redundant systems and two lamps for indicating high voltage.

All electrical equipment must be suitably grounded and the transformer secondary must be grounded or provided with equivalent protection. For portable equipment, external weld-initiating control circuits should operate at low voltage.

NOISE

Projection welding equipment may cause excessive noise. If engineering methods such as reducing the intensity of the source, shielding the source, providing sound-deadening enclosures and sound deflectors do not reduce noises to acceptable levels, ear muffs or earplugs may be required.

FUMES

Fumes and airborne particulate may result from the metals being welded and from the composition of the electrodes. Ventilation should be provided so that hazardous concentrations of airborne contaminants are maintained below the allowable levels specified in OSHA *Code of Federal Regulations* and American Conference of Industrial Hygienists (ACGIH) TLVs® and BEIs:® *Threshold Limit Values for Chemical Substances and Physical Agents in the Workroom Environment.*[18]

17. Occupational Safety and Health Administration (OSHA), latest edition, *Occupational Safety and Health Standards for General Industry*, in *Code of Federal Regulations (CFR)*, Title 29 CFR 1910, Subpart S, Washington D.C.: Superintendent of Documents, U.S. Government Printing Office.

18. American Conference of Industrial Hygienists (ACGIH) TIVs® and BEIs:® *Threshold Limit Values for Chemical Substances and Physical Agents in the Workroom Environment*, Cincinnati: American Conference of Industrial Hygienists.

BIBLIOGRAPHY

American National Standards Institute (ANSI) Accredited Standards Committee. 2005. *Safety in welding, cutting, and allied processes*, ANSI Z49.1. Miami: American Welding Society. 38–41.

American Welding Society (AWS) Committee on Definitions and Symbols. 2001. *Standard welding terms and definitions*. AWS A3.0:2001. Miami: American Welding Society.

American Welding Society (AWS) Committee on Resistance Welding. 2000. *Recommended practices for resistance welding*. AWS C1.1M/C1.1:2000. Miami: American Welding Society.

American Welding Society (AWS) Welding Handbook Committee. Jenney, C. L. and A. O'Brien, eds. 2001. *Welding Handbook*, 9th ed., Vol. 1. *Welding science and technology*. Miami: American Welding Society. Chapter 6.

American Welding Society (AWS). *Specification for automotive resistance spot welding electrodes*, AWS D8.6:2005. Miami: American Welding Society.

American Welding Society (AWS) Committee on Resistance Welding. 2000. *Recommended practices for resistance welding*, AWS C1.1M/C1.1:2000. Miami: American Welding Society.

American Welding Society (AWS) Committee on Resistance Welding. 1999. *Specification for resistance welding of carbon- and low-alloy steels*, C1.4M/C1.4:1999. Miami: American Welding Society.

Occupational Safety and Health Administration (OSHA). 1999. *Occupational safety and health standards for general industry* in *Code of federal regulations* (CFR). Title 29 CFR 1910. Subpart Q. Washington D.C.: Superintendent of Documents. U.S. Government Printing Office.

Occupational Safety and Health Administration (OSHA). 1999. *Occupational safety and health standards for general industry* in *Code of federal regulations* (CFR). Title 29 CFR 1910. Subpart S. Washington D.C.: Superintendent of Documents. U.S. Government Printing Office.

Pilkey, W. D. 1997. *Peterson's stress concentration factors* (2nd Ed.). New York: John Wiley and Sons. Chapter 6.

Resistance Welding Manufacturing Alliance (RWMA). Resistance Welding Manufacturing Alliance (RWMA), a standing committee of the American Welding Society. 2003. *Resistance Welding Manual,* Revised 4th Edition, Miami: American Welding Society. 3-12–3-15.

SUPPLEMENTARY READING LIST

Kuntz, M. 2006. Modeling projection welding of fasteners to advanced high-strength steel (AHSS) sheet using finite element method. *Sheet Metal Welding Conference XII. May 2006. Conference Proceedings.* Detroit: AWS Detroit Section.

Sun, X. 2001. Effect of projection height on projection collapse and nugget formation—a finite element study. *Welding Journal* 80(9): 211–216.

Wu, P., W. Zhang, and N. Bay. 2005. Characterization of dynamic mechanical properties of resistance welding machines. *Welding Journal*. 84(1): 17-s–21-s.

Zhang, H., and J. Senkara. 2006. *Resistance Welding Fundamentals and Applications*. Boca Raton, Florida: CRC Press.

CHAPTER 3

FLASH AND UPSET WELDING

Photograph courtesy of Holland Company

Prepared by the Welding Handbook Chapter Committee on Flash and Upset Welding:

L. E. Moss, Chair
Automation International, Inc.

B. J. Bastian
Consultant

S. Maitland
Goodrich Landing Gear

Welding Handbook Volume 3 Committee Member:

W. L. Roth
Procter & Gamble Inc.

Contents

CHAPTER 3

FLASH AND UPSET WELDING

INTRODUCTION

Flash welding (FW) and upset welding (UW) are resistance welding processes generally used to make butt joints in components of similar cross section by making a weld simultaneously across the entire joint area. Filler metal is not used. To form the solid-state weld characteristic of both processes, upset force is applied during or shortly after the resistive heating cycle is complete. Butt joints between workpieces with similar cross section can be made by both flash welding and upset welding. Upset welding is similar to flash welding except that no flashing action occurs in upset welding.

The two variations of upset welding are high-frequency upset welding (UW-HF) and induction upset welding (UW-I). The method of heating, the closing rate, and the timing of the application of force are the distinguishing characteristics of these process variations.

The flash welding and upset welding processes are a mainstay of many industries requiring high-integrity welds, such as those used in the automotive, aerospace, railroad, petroleum, and construction industries.

Flash welding and upset welding are discussed in separate sections in this chapter. Each section includes the fundamentals of the process, applications, equipment, welding procedures, and process variables. However, both flash welding and upset welding are represented in the sections on weld quality, methods of inspection and examination, and economics. The chapter concludes with a section on safe practices applicable to both processes.

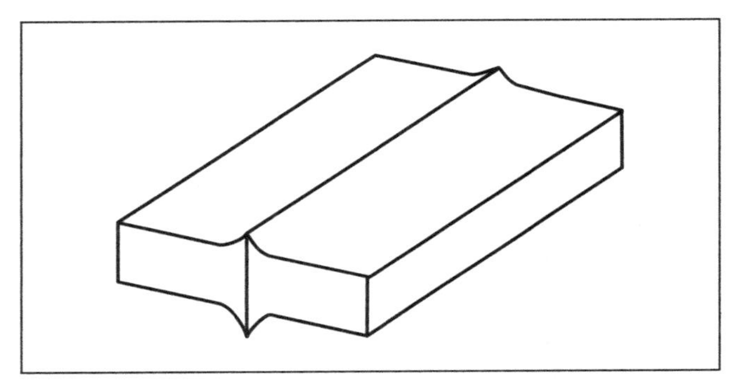

Figure 3.1—Cross Section of a Flash Weld

FLASH WELDING

Flash welding (FW) is a resistance welding process that produces a weld at the interface of a butt joint by heat produced during the flashing action and by the sudden application of high pressure (created by the upset force) after heating is substantially completed. The flashing action, caused by very high current densities at small contact points between the workpieces, forcibly expels material from the joint as the workpieces are slowly moved together. The weld is completed by a rapid upsetting of the workpieces.[1,2] A cross section of a flash weld is shown in Figure 3.1.

1. Welding terms and definitions used throughout this chapter are from *Standard Welding Terms and Definitions*, AWS A3.0:2001, Miami: American Welding Society.

2. At the time of the preparation of this chapter, the referenced codes and other standards were valid. If a code or other standard is cited without a date of publication, it is understood that the latest edition of the document referred to applies. If a code or other standard is cited with the date of publication, the citation refers to that edition only, and it is understood that any future revisions or amendments to the code or standard are not included; however, as codes and standards undergo frequent revision, the reader is encouraged to consult the most recent edition.

FUNDAMENTALS

The flash welding process starts as the two components to be joined (the workpieces) are clamped in resistance welding dies (the electrodes), which are connected to the secondary circuit of a resistance welding transformer. Voltage is applied as one workpiece is advanced slowly toward the other. When contact occurs between the workpieces at the high points of surface irregularities, resistance heating occurs. The high current used in the process causes rapid melting and vaporization of the metal at the points of contact, and subsequently causes minute arcs to form. This action is called *flashing*. As the workpieces are moved together at a suitable closing rate, flashing continues until the weld interface is covered with molten metal, and a short length of each workpiece component reaches forging temperature; this area becomes the heat-affected zone (HAZ). A solid-state weld is then created by the rapid application of a high upset force that brings the faying surfaces in full contact and forges the workpieces together. Flashing voltage and current are either reduced or terminated at the start of upset. The metal and any oxides formed during flashing are expelled from the weld interface. The expelled material is called *flash*.

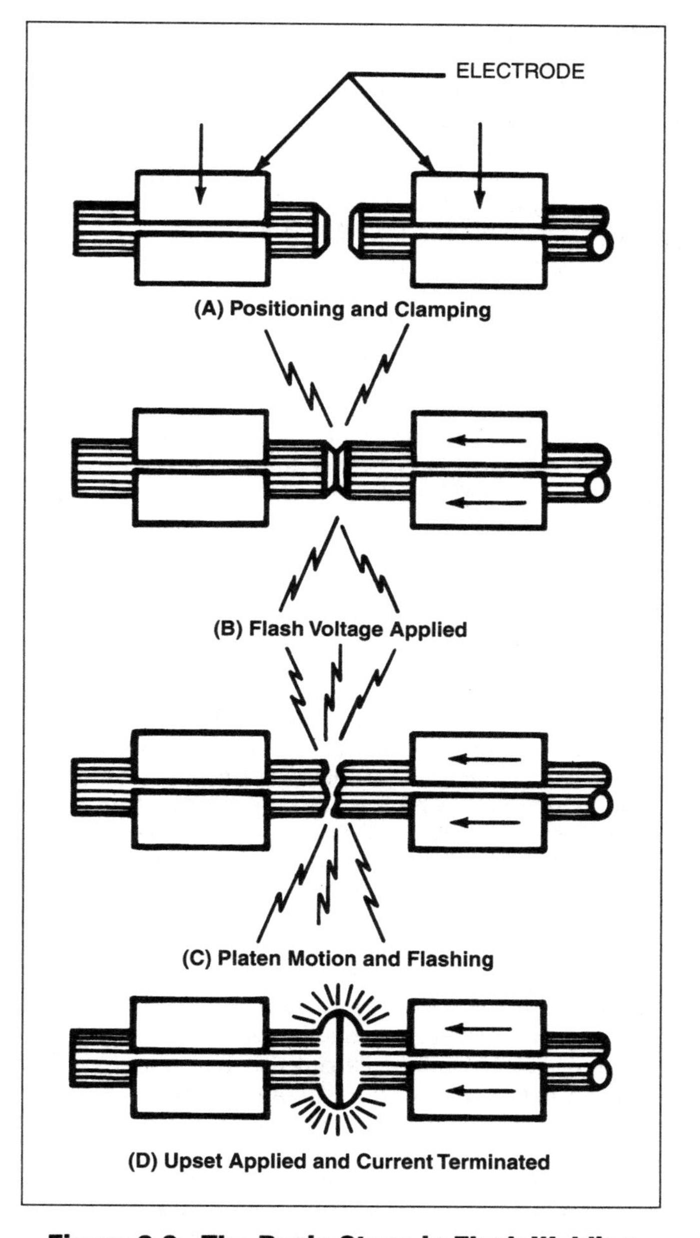

Figure 3.2—The Basic Steps in Flash Welding

Principles of Operation

Flash welding is an automated process; the welding machine performs the tasks essential to producing an upset-welded joint. The welding operator is required only to position the workpieces in the machine, clamp them in the resistance welding dies (electrodes), and initiate the welding sequence. The automated welding sequence then takes over to perform the following actions:

1. Applies the flashing voltage;
2. Slowly moves the platens holding the welding dies to bring the components into contact;
3. Initiates welding current;
4. Continues to move the platens together as flash material is expelled;
5. Applies the upset force after the joint has been sufficiently heated;
6. Terminates welding current at some point just before, during, or after upset; and
7. Trims off the flash with cutting dies so the weld area is flush with the surface of the weldment (optional for some applications).

The welding operator then unclamps the weldment and returns the platen and unloads (or unloads the weldment and returns the platen).

Figure 3.2 illustrates these basic steps. In Figure 3.2(A) the workpiece is positioned and clamped; in Figure 3.2(B) the flashing voltage is applied; in Figure 3.2(C) platen motion is started and flashing occurs; and in Figure 3.2(D) upset is applied and terminated. Additional steps such as preflash, preheat, variable voltage flashing, postheat, annealing, and trimming of the flash may be added as the application requires.

Flashing takes place between the faying surfaces as the movable workpiece is advanced toward the stationary one. Heat is generated at the joint and the temperature increases with time. The flashing action (which results in metal loss) increases with workpiece temperature.

A graph relating workpiece motion with time is known as the *flashing pattern*. In most cases, a flashing pattern should show an initial period of constant-velocity linear motion of one workpiece toward the other to facilitate the start of flashing. This motion should then merge into an accelerating motion, which should closely approximate a parabolic curve. This pattern of motion is known as *parabolic flashing*.

To produce a strong joint with uniform upset, the temperature distribution across the joint should be uniform and the average temperature of the faying surface should be the melting temperature of the metal. Once these conditions are reached, further flashing is not necessary.

The steepness of the temperature gradient corresponding to a stable temperature distribution is a function of the workpiece acceleration during parabolic flashing. In general, the higher the rate of acceleration, the steeper is the stable temperature gradient produced. Thus, the shape of the temperature distribution curve in a particular application can be controlled by the appropriate choice of a flashing pattern. Since the compressive yield strength of a metal is temperature-sensitive, the behavior of the metal during the upsetting portion of the welding cycle is markedly dependent on the flashing pattern. Therefore, the choice of flashing pattern is extremely important for the production of sound flash welds. The minimum flashing distance is the amount of flashing required to produce a stable temperature distribution. From a practical standpoint, the flashing distance should be slightly greater than the minimum acceptable amount to ensure that a stable temperature distribution is always achieved.

Upset occurs when a stable temperature distribution is achieved by flashing and the two workpieces are rapidly brought together. The movable workpieces should be accelerated rapidly so that the molten metal on the flashing surfaces will be extruded before it can solidify in the joint. Motion should continue with sufficient force to upset the metal and weld the two pieces together in a solid-state bond.

Upset current is often applied as the joint is being upset to maintain temperature by resistance heating. This permits upset of the joint with lower force than would be required without resistance. Upset current and duration normally are adjusted by electronic heat control on the basis of either previous experience or the results of welding tests.

ADVANTAGES AND LIMITATIONS

Some important advantages of flash welding are the following:

1. Cross-sectional shapes other than circular, such as angles, H-sections, and rectangles can be welded;
2. Workpieces of similar cross section can be welded with the axes aligned or at an angle to one another, within limits;
3. The expulsion of the molten metal film at the weld interface during the upset stage of welding removes impurities from the interface and allows for the formation of a sound solid-state bond;
4. Preparation of the faying surfaces is not critical (within limits) if the proper welding parameters are used;
5. Rings of various cross sections can be welded; and
6. The heat-affected zones of flash welds are much narrower than those of upset welds.

Some limitations of the process are the following:

1. The high single-phase power demand produces imbalance on three-phase primary power lines; however, three-phase direct-current (dc) rectification commonly is used, especially on larger welding machines that reduce the power demand;
2. The molten metal particles ejected during flashing may present a fire hazard, may injure the operator, and may adhere to various welding machine components, requiring periodic cleaning;
3. Removal of flash and upset metal from the weldment generally is necessary and may require special equipment for deburring, scarfing, or grinding;
4. Alignment of workpieces with very small cross sections is required; and
5. The workpieces require similar or identical cross sections, depending on the application.

The limitations in No. 2 of this list require the operator to wear face and eye protection and use a barrier or shield to block flying sparks. Welding machine components can be protected by designing the equipment to shield it from the expelled material.

FLASH WELDING APPLICATIONS

Many ferrous and nonferrous alloys and some dissimilar metals can be joined by flash welding. Base metals welded by the process and typical products manufactured are discussed in this section.

Base Metals

Typical metals welded by the process are carbon steels and low alloy steels, stainless steels, aluminum alloys, nickel alloys, and copper alloys. Titanium alloys can be flash-welded, but it is necessary to use an inert shielding gas to displace air from around the joint to minimize embrittlement. Dissimilar metals may be flash-welded if their upsetting characteristics are similar. Some

Figure 3.3—Copper-to-Steel Joined by Flash Welding

of the dissimilar characteristics can be overcome by changing the initial extensions between the clamping dies, adjusting the flashing distance, and the selection of welding variables. Typical examples are the welding of aluminum to copper or welding a nickel alloy to steel. A copper-to-steel weld is shown in Figure 3.3.

Typical Products

The automotive industry uses wheel rims produced from flash-welded cylindrical blanks that are formed from flat cold-rolled steel stock and aluminum alloys. The electrical industry uses motor and generator frames produced by flash welding plate and bar stock previously rolled into cylindrical form. Cylindrical transformer cases, circular flanges, and seals for power transformer cases are other examples. The aerospace industry uses flash welds in the manufacture of landing gear struts, control assemblies, hollow propeller blades, and rings for jet engines and rocket casings.

The petroleum industry uses oil-drilling pipe with fittings attached by flash welding. Railroads use flash welding to join sections of relatively high-carbon steel track. Tracks are welded using stationary machines or in the field using mobile machines and portable generating equipment mounted on railroad cars or specially designed trucks. A truck designed for the field welding of rail tracks is shown in Figure 3.4.

The construction industry uses angle iron truss sections joined by the flash welding process. Miter joints commonly are used in the production of rectangular frames for windows, doors, and other architectural trim. These products are made of plain carbon steels or stainless steels, aluminum alloys, brasses, and bronzes. Usually the service loads are limited, but appearance requirements of the finished joints are stringent.

EQUIPMENT

This section presents an overview of the machines, controls, auxiliary equipment and flash welding electrodes (dies) required for flash welding. The equipment and materials used in flash welding and other resistance welding processes are discussed in detail in Chapter 4 of this volume.

Typical Welding Machines

A typical flash welding machine consists of six major components, including the following:

1. The welding machine bed, which has platen ways (guiding surfaces) attached;
2. The platens, which are mounted on the ways;
3. Two clamping assemblies, one of which is rigidly attached to each platen to align and hold the workpieces;
4. A means for controlling the motion of the movable platen;
5. A welding transformer with adjustable tap settings; and
6. Control systems to initiate workpiece motion and flashing current.

Flash welding machines are available for manual, semi-automatic, or fully automatic operations; however, the majority of modern flash welding machines are either semi-automatic or fully automatic. In manual operations, the operator controls the speed of the platen from the time flashing is initiated until upset is completed. In semi-automatic operation, the operator usually initiates flashing manually, which in turn energizes an automatic cycle that completes the weld. In a fully automatic operation, the workpieces are loaded into the machine, either manually or by an automatic loading system, and the welding cycle is then completed automatically. The platen motion of small flash welding machines is provided mechanically by a cam driven by an electric motor through a speed reducer, or through pneumatic or servo-cylinder units. Large machines normally are hydraulically operated.

Photograph courtesy of Holland Company

Figure 3.4—Mobile Welding Machine Used in the Production of Flash-Welded Joints in Rail Track

Operating personnel should be instructed in operating the machinery in a safe manner, as specified in operation manuals supplied by the equipment manufacturer and recommended in the Safe Practices section of this chapter. In general, the hands of the operator must be kept clear of moving machinery, and contact with electrically charged surfaces must be avoided.

System Controls and Auxiliary Equipment

Electronic or computer controls on flash welding machines are integrally designed to sequence the machine, control the welding current, and precisely control the platen position during flashing and upsetting. Silicon-controlled rectifier (SCR) contactors are widely used on large machines drawing line power up to 5000 amperes (A).

Preheat and postheat cycles are controlled either by electronic timers or phase-shift heat controls. Although these functions sometimes may be initiated manually, a majority are automatically initiated and terminated in proper sequence during the welding period.

Resistance Welding Electrodes

Resistance welding electrodes (dies) used in flash welding are not in direct contact with the welding area, unlike spot and seam welding electrodes. Resistance welding dies may serve as workpiece-holding clamps or current-conducting clamps, or both. A common arrangement consists of upper clamping dies and lower current-conducting clamping dies. Since the current density in these dies normally is low, they may be made of relatively hard materials with low electrical conductivity. To avoid overheating, water-cooling of the dies is common in high-production applications.

There are no standardized designs for these electrodes, since they must fit the contour of the workpieces. The size of the dies largely depends on the geometry of the workpieces and the mechanical rigidity needed to maintain proper alignment of workpieces during upsetting. The dies usually are mechanically fastened to die holders (fixtures) that are then fastened to the welding machine platens.

The electrode contact area should be as large as practical to avoid hot spots or local die burns to the workpiece. The contact surfaces may be incorporated in small inserts attached to larger dies for low-cost replacement and convenient detachment for redressing.

If the workpieces are backed up or reinforced so that the clamping dies do not need to carry the upset force, clamping pressures need to be sufficient only to provide good electrical contact. If the workpiece cannot be reinforced, clamping force arrangements must be made in the machine design to supply sufficient pressure to eliminate slippage of the workpiece during upset. In extreme cases, it may be necessary to use serrated clamp inserts. In this case, the inserts usually are made of hardened tool steel.

Flash welding dies tend to wear, but do not plastically deform, or "mushroom." As wear takes place, the contact area may decrease and cause local hot spots or die burns. The dies should be kept clean. Dirt and flash particles tend to become embedded in the dies. All bolts, nuts, and other die-holding devices should be tight.

Fixtures and Backups

Fixtures perform the following functions for flash welding:

1. Rapidly and accurately position two or more workpieces relative to one another,
2. Hold workpieces in proper location while they are being welded, and
3. Permit easy release of the welded assembly.

A fixture is either fastened to the machine or built into it. The workpieces are loaded directly into the fixture and welded.

Resistance welding processes are very rapid compared to other methods of joining. If maximum production is to be attained, fixtures must be designed to be easily loaded and unloaded.

The following factors should be considered when designing a fixture:

1. Quick-acting clamps, toggles, and similar devices should be employed, including the ejector pins sometimes used to facilitate removal of the finished assembly;
2. The fixture must be designed so that welding current is not shunted through any fixturing or positioning devices, which may require insulation of pins and locating strips;
3. Nonmagnetic materials usually are preferred, because any magnetic material located in the throat of the machine will increase the electrical impedance and limit the maximum current the machine can deliver;
4. The operator should be able to load and unload the workpieces safely, which may require the use of swivel devices or slides to move the fixture out of the machine;
5. A combination guard and flash shield that swings in place should be provided to prevent the hand of the operator from reaching between the platens and protects the operator from flash;
6. A fixture must provide for movement of the workpieces as they are being clamped in the dies; and
7. All bearings, pins, slides, and other moving parts should be protected from the expelled material.

Backups. Backups are needed if the clamping dies cannot prevent slippage of the workpieces when the upsetting force is applied. Slippage usually occurs when the section of the workpiece in the die is too short for effective clamping, or the workpiece cannot withstand the required clamping force without damage. A backup often consists of a steel bracket that can be bolted to the platen in various positions. Brackets may have either fixed or adjustable stops against the workpieces.

WELDING PROCEDURES

Like other welding processes, flash welding involves numerous variables that affect the quality of the resulting weld. To ensure consistent weld quality, a welding procedure for each application should be developed that prescribes the welding machine settings for the variables. A sample data sheet for flash welding, upset welding, and other resistance welding processes is reproduced in Figure 3.5. The latest edition of the American Welding Society AWS C1.1M/C.1.1, *Recommended Practices for Resistance Welding*, should be consulted.[3]

Welding Parameters

Flash welding involves dimensional, electrical, force, and time parameters, or variables. The dimensional variables shown in Figure 3.6 are for the flash welding of tubing and flat sheets. Table 3.1 is coordinated with Figure 3.6 and provides the recommended data for welds in steel tubing and sheet material. (The first section of Table 3.1 shows measurements in Standard International units and the second section in U.S. Customary units.)

Figure 3.7 provides similar variables for solid round, hexagonal, square and rectangular bars. Recommended data to support these welds is presented in Table 3.2. The paths of the movable platen and the faying surfaces during flashing and upsetting are also shown in Figure 3.8. The current, force, and time variables are shown in Figure 3.9. Most operations do not involve all of the variables in Figure 3.9. A simple flash welding cycle involves flashing at one voltage setting followed by upset.

3. American Welding Society (AWS) Committee on Resistance Welding, latest edition, *Recommended Practices for Resistance Welding*, AWS C1.1M/C.1, Miami: American Welding Society.

RESISTANCE WELDING DATA SHEET

MACHINE DESCRIPTION ______________________

____

MFGR. ______________________

kVA ____________ SERIAL ____________

SIDE B (1) (6) SIDE A

(13) (3) (2) (12)

MATERIAL DATA	SIDE A	SIDE B
Approx. Analysis (type)		
Thickness		
Width		
Diameter		
Area		
Surface Cond.		
Ult. Strength		
Yield Strength		
Elongation %		
Red. in Area %		
Hardness		
Shape		

MACHINE DATA (in.)	
1. Initial Die Opening	
2. Material Extension A	
3. Material Extension B	
4. Total Flash Off	
5. Total Upset	
6. Final Die Opening	
7. Total Material Loss	
ELECTRODE DATA	
8. Electrode Material A	
9. Electrode Material B	
10. Clamping Force A kN (lb)	
11. Clamping Force B kN (lb)	
12. Contact Length A mm (in.)	
13. Contact Length B mm (in.)	

	PREHEAT	INITIAL FLASHING	FINAL FLASHING	UPSET	POSTHEAT
Time					
Voltage (Open Circuit)					
Distance at					
Force					
Current					

COMMENTS ______________________

____

PLATEN TRAVEL INFORMATION ______________________

TESTING METHODS	REMARKS:
Tension ____________	
"U" Bend Test ____________	
Hardness ____________	
Macro ____________	

Figure 3.5—Sample Resistance Welding Data Sheet for Flash Welding and Upset Welding

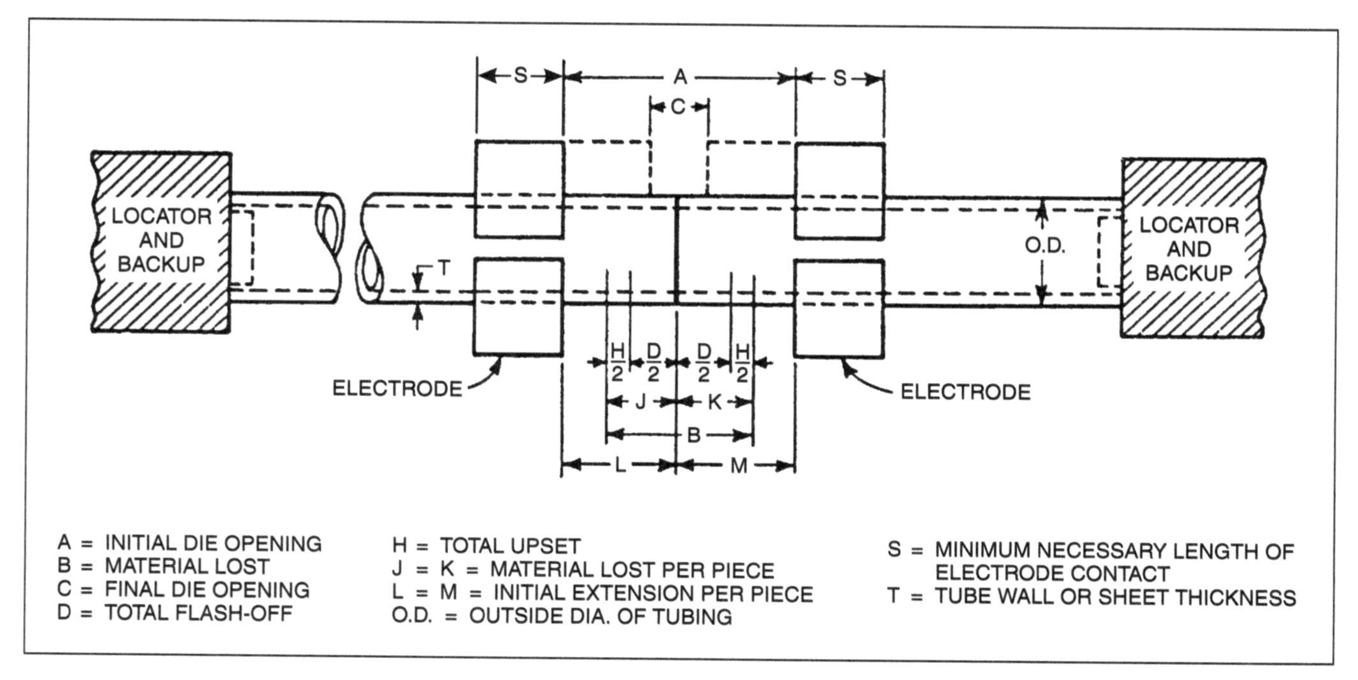

Figure 3.6—Flash Welding Variables for Tubing and Flat Sheets

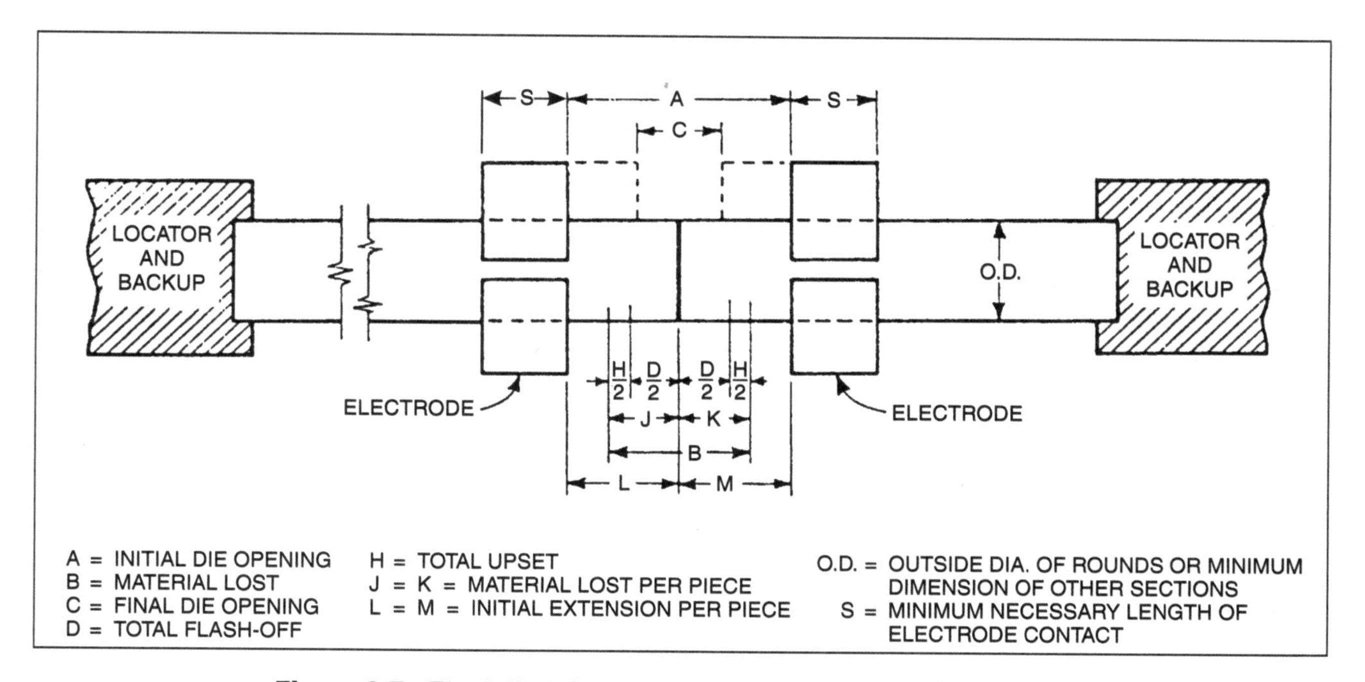

Figure 3.7—Flash Welding of Solid, Round, Hexagonal, Square, and Rectangular Bars (see Table 3.2 for Recommended Data)

Table 3.1
Data for Flash Welding of Tubing and Flat Sheets (Millimeters)

(See Figure 3.6 for Assembly of Workpieces)										Minimum Length of Electrode Contact (S)	
Tube, Wall, or Sheet Thickness (T)	Initial Die Opening (A)	Material Lost (B)	Final Die Opening (C)	Total Flash Loss (D)	Initial Extension per Piece (M)	Material Loss per Piece (J = K)	Initial Extension per Piece (L = M)	Flash Time Seconds	O.D.	With Locator	Without Locator
0.25	2.79	1.52	1.27	1.02	0.51	0.76	1.40	1.00	6.35	9.53	25.40
0.51	5.46	2.92	2.54	2.03	0.89	1.47	2.74	1.50	7.92	9.53	25.40
0.76	8.26	4.45	3.81	3.18	1.27	2.24	4.14	2.00	9.53	9.53	38.10
1.02	10.92	5.84	5.08	4.19	1.65	2.92	5.46	2.50	12.70	9.53	44.45
1.27	13.46	7.11	6.35	5.21	1.91	3.56	6.73	3.25	19.05	12.70	50.80
1.52	15.75	8.38	7.37	6.10	2.29	4.19	7.87	4.00	25.40	19.05	63.50
1.78	18.16	9.78	8.38	7.11	2.57	4.90	9.09	5.00	38.10	25.40	76.20
2.03	20.45	11.05	9.40	8.00	3.05	5.54	10.24	6.00	50.80	31.75	*
2.29	22.48	12.07	10.41	8.76	3.30	6.05	11.25	7.00	63.50	44.45	*
2.54	24.64	13.21	10.92	9.53	3.68	6.60	12.32	8.00	76.20	50.80	*
2.79	26.92	14.48	12.45	10.41	4.06	7.24	13.46	9.75	88.90	57.15	*
3.05	28.96	15.49	13.46	11.18	4.32	7.75	14.48	10.25	101.60	63.50	*
3.30	31.12	16.51	14.61	11.94	4.57	8.26	15.57	11.00	114.30	69.85	*
3.56	33.53	17.78	15.75	12.95	4.83	8.89	16.76	12.75	127.00	69.85	*
3.81	35.31	18.54	16.76	13.46	5.08	9.27	17.65	13.50	139.70	76.20	*
4.06	37.34	19.56	17.78	14.22	5.33	9.78	18.67	14.00	152.40	82.55	*
4.32	39.12	20.32	18.80	14.73	5.59	10.16	19.56	15.75	165.10	88.90	*
4.57	41.15	21.34	19.81	15.49	5.84	10.67	20.57	16.50	177.80	95.25	*
4.83	42.93	22.10	20.83	16.00	6.10	11.05	21.46	17.25	190.50	101.60	*
5.08	44.70	22.86	21.84	16.51	6.35	11.43	22.35	18.00	203.20	107.95	*
6.35	51.05	25.65	25.40	18.54	7.11	12.83	25.53	24.00	215.90	114.30	*
7.62	57.02	28.45	28.58	20.57	7.87	14.22	28.52	30.00	228.60	120.65	*
8.89	62.48	30.73	31.75	22.35	8.38	15.37	31.24	36.00	241.30	127.00	*
10.16	67.06	32.77	34.29	23.62	9.14	16.38	33.53	42.00	254.00		*
11.43	70.61	34.29	36.32	24.64	9.65	17.15	35.31	48.00			*
12.70	73.91	35.81	38.10	25.91	9.91	17.91	36.96	54.00			*
13.97	77.22	37.21	40.01	26.24	10.41	18.62	38.61	60.00			*
15.24	79.63	38.23	41.40	27.56	10.67	19.13	39.83	66.00			*
16.51	82.42	39.50	42.93	28.58	10.92	19.76	41.22	73.00			*
17.78	85.34	40.89	44.45	29.46	11.43	20.45	42.67	80.00			*
20.32	89.54	42.55	46.99	30.73	11.81	21.29	44.78	92.00			*
22.86	92.96	43.94	49.02	31.75	12.19	21.97	46.48	104.00			*
25.40	96.52	45.72	50.80	33.02	12.70	22.86	48.26	116.00			*

*Not recommended for welding without use of locators.

Note: Data are based on welding without preheat, and for two workpieces with same welding characteristics.

Table 3.1 (Continued)
Data for Flash Welding of Tubing and Flat Sheets (Inches)

(See Figure 3.6 for Assembly of Workpieces)										Minimum Length of Electrode Contact (S)	
Tube, Wall, or Sheet Thickness (T)	Initial Die Opening (A)	Material Lost (B)	Final Die Opening (C)	Total Flash Loss (D)	Initial Extension per Piece (M)	Material Loss per Piece (J = K)	Initial Extension per Piece (L = M)	Flash Time Seconds	O.D.	With Locator	Without Locator
0.010	0.110	0.060	0.050	0.040	0.020	0.030	0.055	1.00	0.250	0.375	1.000
0.020	0.215	0.115	0.100	0.080	0.035	0.058	0.108	1.50	0.312	0.375	1.000
0.030	0.325	0.175	0.150	0.125	0.050	0.088	0.163	2.00	0.375	0.375	1.500
0.040	0.430	0.230	0.200	0.165	0.065	0.115	0.215	2.50	0.500	0.375	1.750
0.050	0.530	0.280	0.250	0.205	0.075	0.140	0.265	3.25	0.750	0.500	2.000
0.060	0.620	0.330	0.290	0.240	0.090	0.165	0.310	4.00	1.000	0.750	2.500
0.070	0.715	0.385	0.330	0.280	0.101	0.193	0.358	5.00	1.500	1.000	3.000
0.080	0.805	0.435	0.370	0.315	0.120	0.218	0.403	6.00	2.000	1.250	*
0.090	0.885	0.475	0.410	0.345	0.130	0.238	0.443	7.00	2.500	1.750	*
0.100	0.970	0.520	0.430	0.375	0.145	0.260	0.485	8.00	3.000	2.000	*
0.110	1.060	0.570	0.490	0.410	0.160	0.285	0.530	9.75	3.500	2.250	*
0.120	1.140	0.610	0.530	0.440	0.170	0.305	0.570	10.25	4.000	2.500	*
0.130	1.225	0.650	0.575	0.470	0.180	0.325	0.613	11.00	4.500	2.750	*
0.140	1.320	0.700	0.620	0.510	0.190	0.350	0.660	12.75	5.000	2.750	*
0.150	1.390	0.730	0.660	0.530	0.200	0.365	0.695	13.50	5.500	3.000	*
0.160	1.470	0.770	0.700	0.560	0.210	0.385	0.735	14.00	6.000	3.250	*
0.170	1.540	0.800	0.740	0.580	0.220	0.400	0.770	15.75	6.500	3.500	*
0.180	1.620	0.840	0.780	0.610	0.230	0.420	0.810	16.50	7.000	3.750	*
0.190	1.690	0.870	0.820	0.630	0.240	0.435	0.845	17.25	7.500	4.000	*
0.200	1.760	0.900	0.860	0.650	0.250	0.450	0.880	18.00	8.000	4.250	*
0.250	2.010	1.010	1.000	0.730	0.280	0.505	1.005	24.00	8.500	4.500	*
0.300	2.245	1.120	1.125	0.810	0.310	0.560	1.123	30.00	9.000	4.750	*
0.350	2.460	1.210	1.250	0.880	0.330	0.605	1.230	36.00	9.500	5.000	*
0.400	2.640	1.290	1.350	0.930	0.360	0.645	1.320	42.00	10.000		*
0.450	2.780	1.350	1.430	0.970	0.380	0.675	1.390	48.00			*
0.500	2.910	1.410	1.500	1.020	0.390	0.705	1.455	54.00			*
0.550	3.040	1.465	1.575	1.033	0.410	0.733	1.520	60.00			*
0.600	3.135	1.505	1.630	1.085	0.420	0.753	1.568	66.00			*
0.650	3.245	1.555	1.690	1.125	0.430	0.778	1.623	73.00			*
0.700	3.360	1.610	1.750	1.160	0.450	0.805	1.680	80.00			*
0.800	3.525	1.675	1.850	1.210	0.465	0.838	1.763	92.00			*
0.900	3.660	1.730	1.930	1.250	0.480	0.865	1.830	104.00			*
1.000	3.800	1.800	2.000	1.300	0.500	0.900	1.900	116.00			*

*Not recommended for welding without use of locators.

Note: Data are based on welding without preheat, and for two workpieces with same welding characteristics.

Table 3.2
Data for Flash Welding of Solid Round, Hexagonal, Square, and Rectangular Bars (Millimeters)

(See Figure 3.7 for Assembly of Workpieces)

O.D.	Initial Die Opening (A)	Material Lost (B)	Final Die Opening (C)	Total Flash Loss (D)	Total Upset (H)	Material Loss per Piece (J = K)	Initial Extension per Piece (L = M)	Flash Time Seconds	O.D.	Minimum Length of Electrode Contact (S) With Locator	Minimum Length of Electrode Contact (S) Without Locator
1.27	2.54	1.27	1.27	1.02	0.25	0.64	1.27	1.00	6.35	9.53	25.40
2.54	4.62	2.08	2.54	1.57	0.51	1.04	2.31	1.50	7.92	9.53	25.40
3.81	6.86	3.05	3.81	2.29	0.76	1.52	3.43	2.00	9.53	9.53	38.10
5.08	8.89	3.81	5.08	2.79	1.02	1.91	4.45	2.50	12.70	9.53	44.45
6.35	10.92	4.57	6.35	3.30	1.27	2.29	5.46	3.25	19.05	12.70	50.80
7.62	12.95	5.33	7.62	3.81	1.52	2.67	6.48	4.00	25.40	19.05	63.50
8.89	15.24	6.35	8.89	4.57	1.78	3.18	7.62	5.00	38.10	25.40	76.20
10.16	17.40	7.24	10.16	5.21	2.03	3.63	8.71	6.00	50.80	31.75	*
11.43	19.56	8.13	11.43	5.84	2.29	4.06	9.78	7.00	63.50	44.45	*
12.70	21.59	8.89	12.70	6.35	2.54	4.45	10.80	8.00	76.20	50.80	*
13.97	23.88	9.91	13.97	7.11	2.79	4.95	11.94	9.00	88.90	57.15	*
15.24	26.04	10.80	15.24	7.75	3.05	5.41	13.03	10.00	101.60	63.50	*
16.51	27.94	11.43	16.51	8.26	3.18	5.72	13.97	11.00	114.30	69.85	*
17.78	29.97	12.19	17.78	8.89	3.30	6.10	14.99	12.00	127.00	69.85	*
19.05	32.00	12.95	19.05	9.53	3.43	6.48	16.00	13.00	139.70	76.20	*
20.32	34.04	13.72	20.32	10.16	3.56	6.86	17.02	14.00	152.40	82.55	*
21.59	36.07	14.48	21.59	10.80	3.68	7.24	18.03	15.00	165.10	88.90	*
22.86	38.10	15.24	22.86	11.43	3.81	7.62	19.05	16.00	177.80	95.25	*
24.13	40.13	16.00	24.13	12.07	3.94	8.00	20.07	17.00	190.50	101.60	*
25.40	42.16	16.76	25.40	12.70	4.06	8.38	21.08	18.00	203.20	107.95	*
26.67	44.20	17.53	26.67	13.34	4.19	8.76	22.10	20.00	215.90	114.30	*
27.94	46.23	18.29	27.94	13.97	4.32	9.14	23.11	22.00	228.60	120.65	*
29.21	48.26	19.05	29.21	14.61	4.45	9.53	24.13	24.00	241.30	127.00	*
30.48	50.29	19.81	30.48	15.24	4.57	9.91	25.15	27.00	254.00		*
31.75	52.32	20.57	31.75	15.88	4.70	10.29	26.16	30.00			*
33.02	54.36	21.34	33.02	16.51	4.83	10.67	27.18	33.00			*
35.56	58.42	22.86	35.56	17.78	5.08	11.43	29.21	36.00			*
38.10	62.48	24.38	38.10	19.05	5.33	12.19	31.24	42.00			*
40.64	66.55	25.91	40.64	20.32	5.59	13.72	33.27	48.00			*
43.18	70.61	27.43	43.18	21.59	5.84	13.72	35.31	56.00			*
45.72	74.68	28.96	45.72	22.86	6.10	14.48	37.34	66.00			*
48.26	78.74	30.48	48.26	24.13	6.35	15.24	39.37	77.00			*
50.80	82.80	32.00	50.80	25.40	6.60	16.00	41.40	92.00			*

*Not recommended for welding without use of locators.

Note: Data are based on welding without preheat, and for two workpieces with same welding characteristics.

Table 3.2 (Continued)
Data for Flash Welding of Solid Round, Hexagonal, Square, and Rectangular Bars (Inches)

(See Figure 3.7 for Assembly of Workpieces)										Minimum Length of Electrode Contact (S)	
O.D.	Initial Die Opening (A)	Material Lost (B)	Final Die Opening (C)	Total Flash Loss (D)	Total Upset (H)	Material Loss per Piece (J = K)	Initial Extension per Piece (L = M)	Flash Time Seconds	O.D.	With Locator	Without Locator
0.050	0.100	0.050	0.050	0.040	0.010	0.025	0.050	0.250	0.375	1.000	0.050
0.100	0.182	0.082	0.100	0.062	0.020	0.041	0.091	0.312	0.375	1.000	0.100
0.150	0.270	0.120	0.150	0.090	0.030	0.060	0.135	0.375	0.375	1.500	0.150
0.200	0.350	0.150	0.200	0.110	0.040	0.075	0.175	0.500	0.375	1.750	0.200
0.250	0.430	0.180	0.250	0.130	0.050	0.090	0.215	0.750	0.500	2.000	0.250
0.300	0.510	0.210	0.300	0.150	0.060	0.105	0.255	1.000	0.750	2.500	0.300
0.350	0.600	0.250	0.350	0.180	0.070	0.125	0.300	1.500	1.000	3.000	0.350
0.400	0.685	0.285	0.400	0.205	0.080	0.143	0.343	2.000	1.250	*	0.400
0.450	0.770	0.320	0.450	0.230	0.090	0.160	0.385	2.500	1.750	*	0.450
0.500	0.850	0.350	0.500	0.250	0.100	0.175	0.425	3.000	2.000	*	0.500
0.550	0.940	0.390	0.550	0.280	0.110	0.195	0.470	3.500	2.250	*	0.550
0.600	1.025	0.425	0.600	0.305	0.120	0.213	0.513	4.000	2.500	*	0.600
0.650	1.100	0.450	0.650	0.325	0.125	0.225	0.550	4.500	2.750	*	0.650
0.700	1.180	0.480	0.700	0.350	0.130	0.240	0.590	5.000	2.750	*	0.700
0.750	1.260	0.510	0.750	0.375	0.135	0.255	0.630	5.500	3.000	*	0.750
0.800	1.340	0.540	0.800	0.400	0.140	0.270	0.670	6.000	3.250	*	0.800
0.850	1.420	0.570	0.850	0.425	0.145	0.285	0.710	6.500	3.500	*	0.850
0.900	1.500	0.600	0.900	0.450	0.150	0.300	0.750	7.000	3.750	*	0.900
0.950	1.580	0.630	0.950	0.475	0.155	0.315	0.790	7.500	4.000	*	0.950
1.000	1.660	0.660	1.000	0.500	0.160	0.330	0.830	8.000	4.250	*	1.000
1.050	1.740	0.690	1.050	0.525	0.165	0.345	0.870	8.500	4.500	*	1.050
1.100	1.820	0.720	1.100	0.550	0.170	0.360	0.910	9.000	4.750	*	1.100
1.150	1.900	0.750	1.150	0.575	0.175	0.375	0.950	9.500	5.000	*	1.150
1.200	1.980	0.780	1.200	0.600	0.180	0.390	0.990	10.000		*	1.200
1.250	2.060	0.810	1.250	0.625	0.185	0.405	1.030			*	1.250
1.300	2.140	0.840	1.300	0.650	0.190	0.420	1.070			*	1.300
1.400	2.300	0.900	1.400	0.700	0.200	0.450	1.150			*	1.400
1.500	2.460	0.960	1.500	0.750	0.210	0.480	1.230			*	1.500
1.600	2.620	1.020	1.600	0.800	0.220	0.540	1.310			*	1.600
1.700	2.780	1.080	1.700	0.850	0.230	0.540	1.390			*	1.700
1.800	2.940	1.140	1.800	0.900	0.240	0.570	1.470			*	1.800
1.900	3.100	1.200	1.900	0.950	0.250	0.600	1.550			*	1.900
2.000	3.260	1.260	2.000	1.000	0.260	0.630	1.630			*	2.000

*Not recommended for welding without use of locators.

Note: Data are based on welding without preheat, and for two workpieces with same welding characteristics.

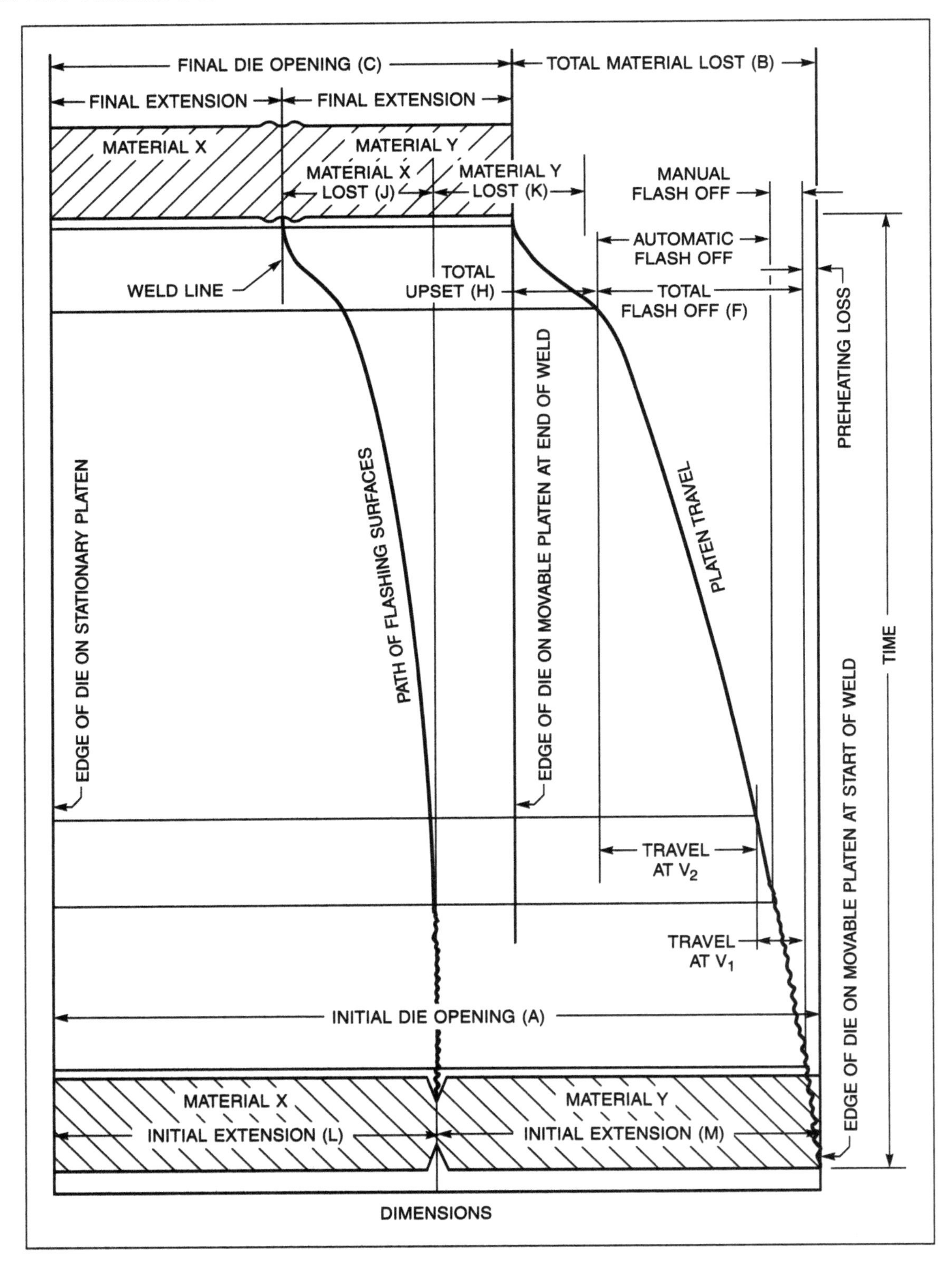

Figure 3.8—Flash Welding Dimensional Variables and Motions

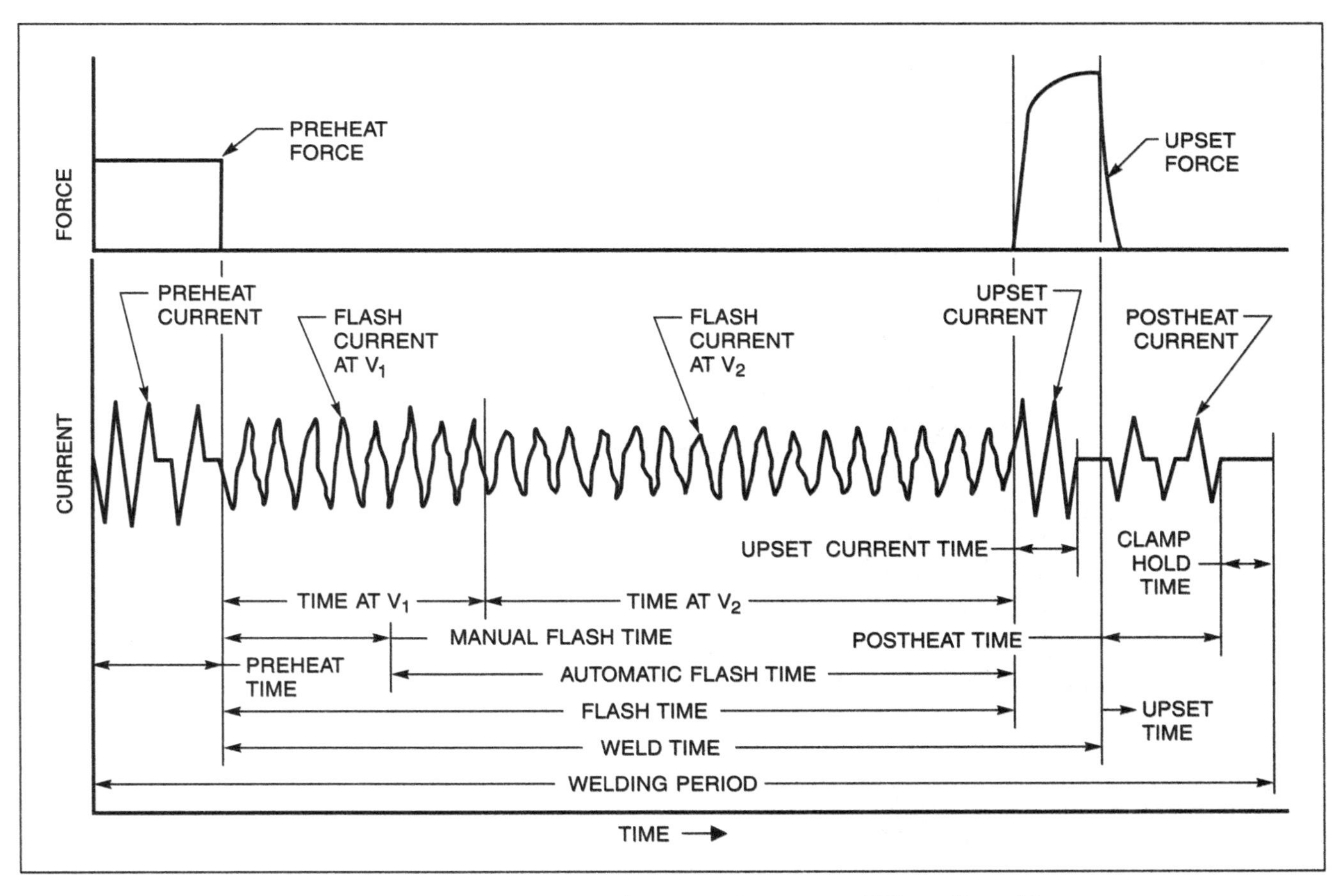

Figure 3.9—Flash Welding Current, Force, and Time Variables

Welding of Steel

Tables 3.1 and 3.2 are applicable to steels with low and medium forging strength. The tables provide the recommended dimensions for setting up a flash welding machine to weld sections of various thicknesses. Total flashing time is based on welding without preheating.

When setting up a welding schedule, the dimensional variables and flashing time may be selected from Tables 3.1 and 3.2. The welding machine is adjusted to the lowest secondary voltage at which steady and consistent flashing can be obtained. The secondary voltages available are dependent on the electrical design of the welding transformer.

The upsetting force used for a particular application depends on the alloy and the cross-sectional area of the joint. The selection of equipment for steels should be based on the values of recommended upset pressures given in Table 3.3. These values are based on welding without preheat.

Joint Design

Figure 3.10 illustrates three common joint designs for welds made by flash welding: the axially aligned weld (A), the miter weld (B), and the ring weld (C). Following are several basic design requirements:

1. The design should provide for an even heat balance in the workpieces so that the ends of the workpieces will have nearly equal compressive strength at the finish of the flashing time;
2. Compensation for the loss of metal lost during flashing (flash loss) and upset must be anticipated and included in the initial length when designing the workpiece;

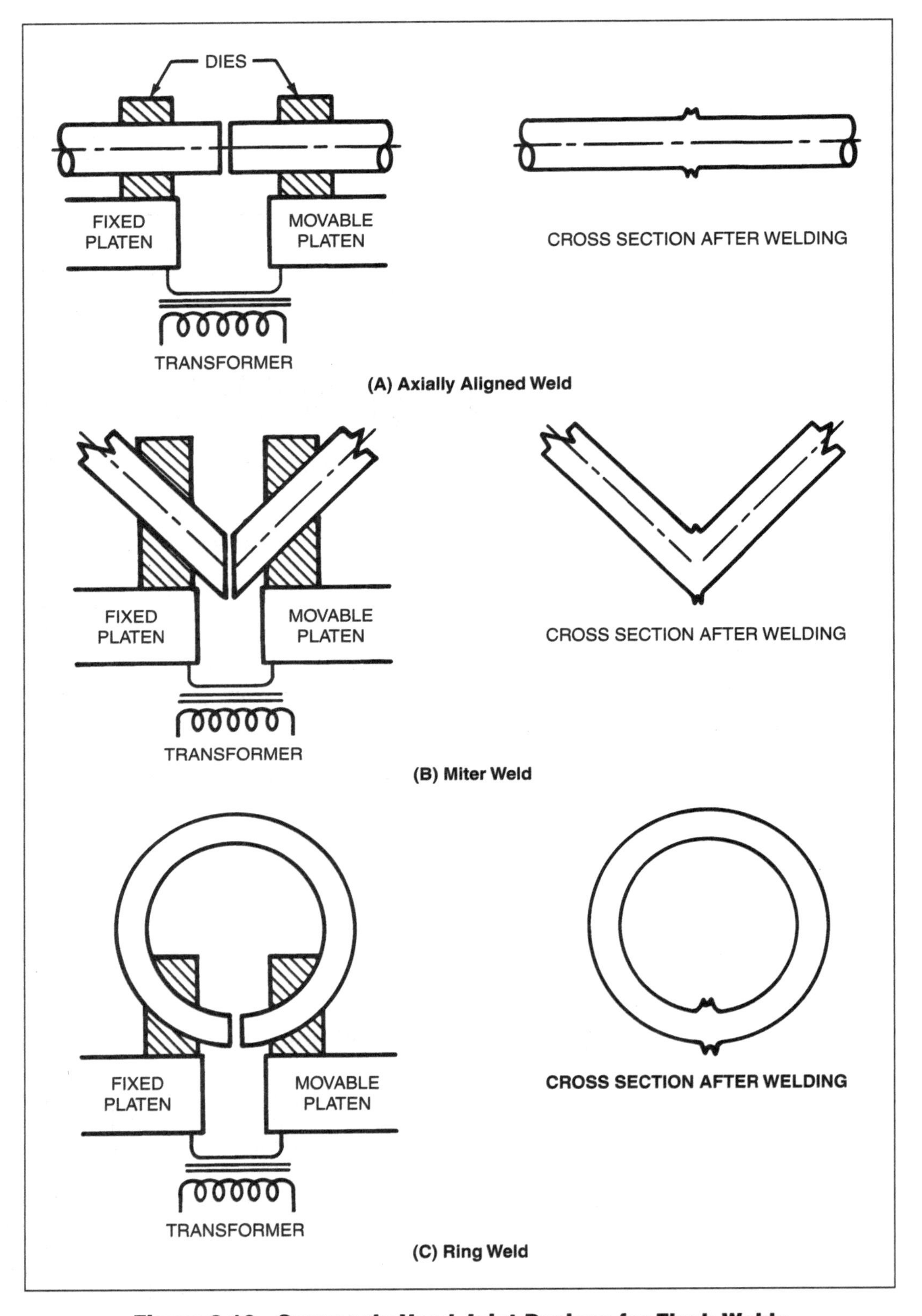

Figure 3.10—Commonly Used Joint Designs for Flash Welds

3. With miter joints, the angle between the two members must be taken into account in the design;
4. The workpieces must be designed so that they can be suitably clamped and held in accurate alignment during flashing and upsetting, with the joint perpendicular to the direction of upset force; and
5. The preparation of the workpiece end should be designed to allow flash material to escape from the joint, and so that flashing starts at the center or the central area of the workpieces.

In general, the two workpieces should have the same cross section at the joint. Bosses may have to be machined, forged, or extruded on the workpieces to meet this requirement. Based on the particular application, successful flash welds have been made in workpieces with dissimilar cross sections using computer servo systems that employ sophisticated parameter settings.

When welding extruded or rolled shapes of different thickness within the cross section, the temperature distribution during flashing will vary with section thickness. This tendency often can be counteracted by proper design of the clamping dies, provided the ratio of the thicknesses does not exceed about 4 to 1.

The recommended maximum joint lengths for several thicknesses of steel sheet are given in Table 3.3. The maximum diameters for steel tubing of various wall thicknesses are listed in Table 3.4. The limits can be exceeded in some cases if special equipment and procedures are used.

When welding rings with the flash welding process, the ratio of circumference to cross-sectional area should be determined; the shunting of current below this ratio becomes a problem because of power loss, which can be high. The minimum ratio will depend on the electrical resistivity of the metal to be welded. The ratio can be lower with metals of high resistivity, such as stainless steels, than with low-resistivity metals, such as aluminum.

When heavy sections are welded, it is often advisable to bevel the end of one workpiece to facilitate the start of flashing. Beveling the one end may eliminate the necessity for preheating or initially flashing at a voltage higher than normal. Suggested dimensions for beveling rods, bars, tubing and plate are shown in Figure 3.11.

Table 3.3
Recommended Maximum Joint Lengths of Flat Steel Sheet for Flash Welding

Sheet Thickness		Maximum Joint Length		Sheet Thickness		Maximum Joint Length	
mm	in.	mm	in.	mm	in.	cm	in.
0.25	0.010	25	1.00	1.5	0.060	6.35	25.00
0.50	0.020	125	5.00	2.0	0.080	8.90	35.00
0.75	0.030	250	10.00	2.5	0.100	11.45	45.00
1.0	0.040	375	15.00	3.2	0.125	14.50	57.00
1.3	0.050	500	20.00	4.8	0.187	22.35	88.00

Table 3.4
Recommended Maximum Diameters of Steel Tubing for Flash Welding

Wall Thicknesses		Maximum Tubing Diameter		Wall Thicknesses		Maximum Tubing Diameter	
mm	in.	mm	in.	mm	in.	mm	in.
0.5	0.020	13.0	0.50	3.2	0.125	102	4.00
0.8	0.030	19.0	0.75	4.7	0.187	152	6.00
1.3	0.050	32.0	1.25	6.4	0.250	230	9.00
1.6	0.062	38.0	1.50				
2.0	0.080	51.0	2.00				
2.5	0.100	76.0	3.00				

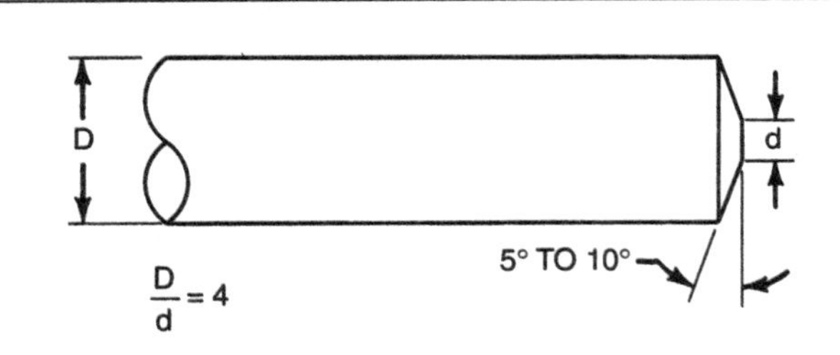

(A) Rods and Bars 6.3 mm (0.25 in.) and Larger

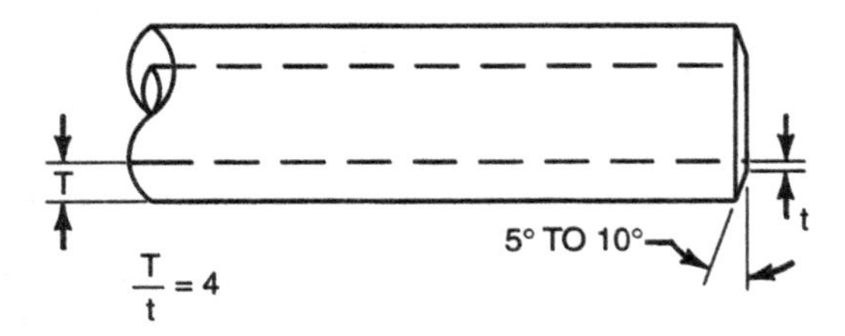

(B) Tubing 4.8 mm (0.188 in.) Wall and Larger

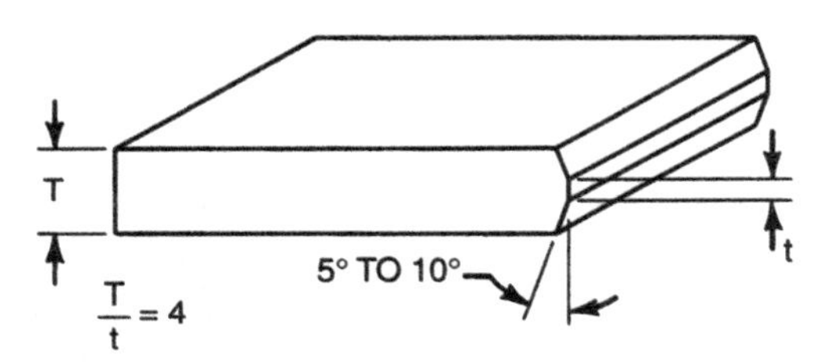

(C) Flat Plate 4.8 mm (0.188 in.) and Thicker

Note: Only one workpiece should be beveled when D is 6.3 mm (0.25 in.) or larger and T is 4.8 mm (0.188 in.) or greater.

Figure 3.11—End Preparation of One Workpiece to Facilitate the Flashing of Large Sections

Heat Balance

In axially aligned joints, when the two workpieces are of the same alloy and cross section, the heat generated in each of the workpieces during the welding cycle will be the same, provided the physical arrangement for welding is uniform. Flash loss and upset loss also will be equal in each workpiece. In general, the heat balance between workpieces of the same alloy will be adequate if the two cross-sectional areas do not differ by more than normal manufacturing tolerances.

Dissimilar Metals

When two dissimilar metals are flash-welded, the metal loss during flashing may differ for each metal. This behavior can be attributed to differences in electrical conductivity, thermal conductivity, or melting temperature, or all three. To compensate for this, the extension from the clamping die of the more rapidly consumed workpiece should be greater than that of the other workpiece. When welding aluminum to copper, the extension of the aluminum workpiece should be twice that of the copper workpiece.

Flash welding of nonaligned sections, such as miter joints, may produce varying properties across the joint because of heat imbalance across the joint. Since the faying surfaces are not perpendicular to the respective lengths of the workpieces, the volume of metal decreases across the joint to a minimum at the apex. Consequently, flashing and upset at the apex may vary significantly from that which occurs across the remainder of the joint.

Miter joints between round or square bars should have a minimum groove angle of 150°. At smaller angles, the weld area at the apex will be of poor quality because of the lack of adequate backup metal. Satisfactory miter joints can be made between thin rectangular sections in the same plane with a groove angle as small as 90°, provided the width of the stock is greater than 20 times its thickness. If service loading produces a tensile stress at the apex, the outside corner should be trimmed to remove the poor-quality weld metal from the joint area.

Surface Preparation

Surface preparation for flash welding is of minor importance and in most cases none is required. Clamping surfaces usually require no special preparation unless excessive scale, rust, grease, or paint is present. The abutting surfaces should be reasonably clean to accomplish electrical contact. Once flashing starts, dirt or other foreign matter will not seriously interfere with the completion of the weld.

Initial Die Opening

The initial die opening is the sum of the initial extensions of the two workpieces. (Refer to Figures 3.6, 3.7, and 3.8.) The initial extension for each workpiece must provide for metal loss during flashing and upset, and for some undisturbed metal between the upset metal and the clamping die. Initial extensions for both workpieces are determined from available welding data or from welding tests. The initial die opening should not be too large; otherwise, inconsistent upset and joint misalignment may occur.

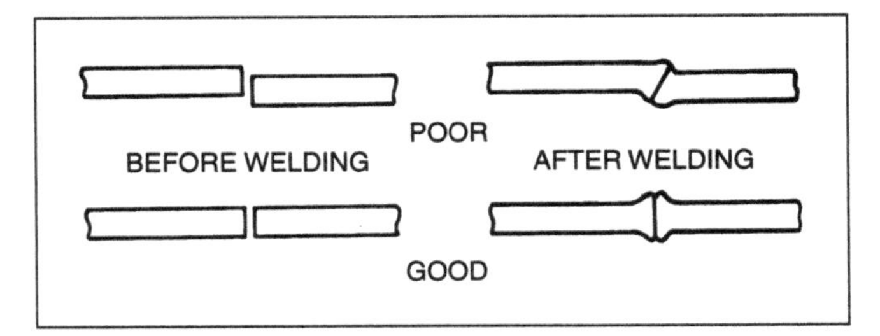

Figure 3.12—Effect of Poor Alignment on Joint Geometry

Alignment

It is important that the workpieces are properly aligned in the welding machine so that flashing on the faying surfaces is uniform. If the workpieces are misaligned, flashing will occur only across opposing areas and heating will not be uniform when upset, thus causing the workpieces to slip past each other. This is illustrated in Figure 3.12. Alignment of the workpieces should be given careful consideration when designing the welding machine, the workpieces, and the tooling. This is especially true for sections with a large width-to-thickness ratio.

Metal Loss

The final length of the weldment will be less than the sum of the lengths of the original workpieces because of the loss of metal during flash and upset. The extent of these losses must be established for each assembly and then added to the workpiece length so that the weldment will meet design requirements. Changes in welding procedures may require modification of workpiece lengths.

Shielding Gas

In some applications, displacement of air from the joint area by an inert or reducing gas shield may improve joint quality by minimizing contamination by oxygen or nitrogen, or both. However, gas shielding cannot compensate for improper welding procedures, and it should be used only when required by the application. Shielding gas is rarely used in actual production applications.

Argon or helium is particularly effective when flash welding reactive metals, such as titanium. At high temperatures, reactive metals are embrittled when they are exposed to air. Dry nitrogen may be effective with stainless steels and heat-resisting steels. The value of a protective atmosphere depends on the effectiveness of the device delivering the shielding gas. The material ejected during flashing may be deposited on the shielding gas apparatus and will interfere with its operation. Provisions for platen movement must be incorporated in the design.

If gas cylinders are used to provide the shielding gas, they must be protected from damage by plant traffic. Cylinder storage racks must have securing devices. If the shielding gas is provided by a piping system, the piping should be properly labeled.

Preheating

During preheating, the workpieces are brought into contact under light pressure and then the welding transformer is energized. In some applications, the platen is oscillated forward and backward at a variable rate during preheat. The resistance heating effect of high-density current flow heats the metal between the dies. The temperature distribution across the joint during preheating approximates a sinusoidal waveform with the peak temperature point at the weld interface.

A preheating operation serves the following useful functions:

1. It raises the temperature of the workpieces, which makes flashing easier to start and maintain;
2. It produces a temperature distribution with a flatter gradient that persists throughout the flashing operation and distributes the upset over a greater length than is the case when no preheat is employed; and
3. It may extend the capacity of a machine and permit the joining of larger cross sections than otherwise would be possible (modern control systems automate this process through platen control, variable voltage and timing).

Welding Cycle

Most commercial flash welding machines are operated automatically. A welding schedule is established for the particular application by a series of test welds that are evaluated for quality. The machine is then set up to reproduce the qualified welding schedule for the particular application.

The operator loads and unloads the machine and observes the welding cycle for consistency of operation. In some cases, flash weld monitors are integrated in the machine to monitor data from each weld for quality and mechanical operation. The monitors also can show the welding parameters graphically and archive the data for future needs. Other applications, such as automatic feed and ejection devices, may be incorporated into the machine.

Postheating

Steels with extremely high alloy or carbon content may crack if the weld is cooled too rapidly. In some cases, this condition may be avoided by preheating large workpieces, which will decrease the subsequent cooling rate. Postweld heating the joint in the welding machine by resistance heating or by immediately placing the weldment in a furnace operating at the desired temperature may prevent cracking when preheating is ineffective.

A postweld heat cycle also may be incorporated in a flash welding machine by using control systems and phase-shift heat control. Postweld heat timing can be initiated at the end of upset (distance) or after a time delay. The desired temperature can be obtained by adjusting the heat control. However, heat will be transferred from the weldment to the clamping dies during the postweld heat cycle. This must be considered in the design and selection of materials of the die. Water-cooling is usually necessary.

Flash Removal

Frequently, it is necessary to remove flash residue from the welded joint. In some cases, this is done only to enhance appearance. A joint is somewhat stronger in tension if the flash is not removed because of the larger cross section provided by upset material. However, the notch effect at the weld interface may then cause a reduction in fatigue strength. The notched portion of the upset material should be removed, but the balance may be left in place when the design of workpieces indicates that reinforcement is beneficial.

It generally is easier to remove the flash immediately after welding while the metal is still hot. This can be done by a number of methods, including machining, grinding, high-speed abrasion wheels, die trimming, oxyfuel gas cutting, high-speed sanding, and pinch-off clamping dies. Trimming machines specifically designed to remove flash also may be incorporated in the process. With some alloy steels, flash removal with cutting tools often is difficult because of the hardness of the steels. In these cases, either grinding or oxyfuel gas cutting usually is employed.

With soft metals such as aluminum and copper, the flash may be partially sheared off using pinch-off dies. These dies have sharp tapered faces, which cut almost through the metal as upsetting takes place. The final die opening is small. The partially sheared flash is then easily removed by other means, and the joint can then be smoothed by filing or grinding.

WELDING MACHINE SETTINGS

The principal variables of the flash welding process that are controlled by settings on the welding machine include flashing voltage, flashing time, and upset rate, distance, and current.

Flashing Voltage

Flashing voltage is determined by the welding transformer tap setting. The lowest setting possible consistent with good flashing action should be selected. The use of electronic phase-shift heat control always has been discouraged as a means of reducing the flash voltage; however, in most practical applications, it is still widely used. When using heat control, a transformer tap should be selected that maintains the phase angle as high as possible. Heat settings below 60% are not desirable when using alternating current (ac). On dc power systems, the minimum heat value is even lower, as the three phases will still overlap well below 50%. Normally, the lowest transformer tap and the highest heat setting possible produces the best results when using either ac or dc, because near-full waveform voltages are preferred to enhance the flashing action.

Flashing Time

Flashing is carried out over a time interval to obtain the required flash loss of metal. The time required is related to the secondary voltage and the rate of metal loss as flashing progresses. Since a flashing pattern generally is parabolic, the variables are interrelated. In any case, smooth flashing action for a specified minimum flashing distance during a specified time interval is necessary to produce a sound weld.

Upset

To produce a satisfactory flash weld, the flash and upset variables are interrelated and must be considered together. The upset variables include the following:

1. Flashing voltage cutoff,
2. Upset rate,
3. Upset distance and upset time, and
4. Current magnitude and duration.

Table 3.5
Upset Pressures for Various Classes of Alloys*

Strength Classification	Examples	Upset Pressure MPa	Upset Pressure ksi
Low forging	SAE 1020, 1112, 1315 and steels commonly designated as high-strength low-alloy.	69	10
Medium forging	SAE 1045, 1065, 1335, 3135, 4130, 4140, 8620, 8630.	103	15
High forging	SAE 4340, 4640, 300M, tool steel, 12% Cr and 18-8 stainless steel, titanium, and aluminum.	172	25
Extra-high forging	Materials exhibiting extra-high compressive strength at elevated temperature, such as A286, 19-9 DL, nickel-base, and cobalt-base alloys.	241	35

*Classified by Society of Automotive Engineers (SAE).

Flashing Voltage Cutoff. In the past, systems were designed to provide two voltage ranges using two primary contactors, each connected to a separate transformer tap. This allowed for a higher secondary voltage (V_1) for the initial stage to assist in starting flashing action and the second contactor was initiated to provide a normal secondary voltage (V_2) as the first contactor is de-energized. A newer approach is to vary the speed of the platen and monitor secondary voltage. Control circuitry signals the drive system to vary the forward speed in relation to voltage so that the best flashing action is obtained.

In some cases, flashing voltage is terminated at the moment that upset of the weld commences. If flash current is maintained long after upset has occurred, excessive heating, electrode (die) burning, and deformation will occur. However, with modern control systems, voltage and current can be sustained at an accurate rate during upset to produce general material heating, aid in upset, retard cooling rates in hardenable material, and reduce compressive yield strength.

Rate. Upset is initiated by rapidly increasing the acceleration of the workpieces to bring the faying surfaces together quickly. The molten metal and oxides present on the surfaces are expelled from the joint and the hot weld zone is upset. The upset rate must be sufficient to expel the molten metal before it solidifies and to produce the optimum upset while the metal has adequate plasticity.

The welding machine must apply a force to the movable platen to properly accelerate the workpiece and overcome the resistance of the workpieces to plastic deformation. The force required depends on the cross-sectional area of the joint, the yield strength of the hot metal, and the mass of the movable platen. Table 3.5 provides the approximate minimum upset pressures for the flash welding of various classes of alloys. These values may be used as a first approximation in determining the size of the welding machine required to flash-weld a particular joint area in one of these alloys.

Distance. The upset distance must be sufficient to accomplish two actions: the oxides and molten metal must be expelled from the faying surfaces, and the two surfaces must be brought into intimate metal-to-metal contact over the entire cross section.

The amount of upset required to obtain a sound flash weld depends on the metal and the section thickness. If the flashing conditions produce relatively smooth flashed surfaces, smaller upset distances than those needed for roughly flashed surfaces will be satisfactory for most metals. Some heat-resistant alloys may require upset distances as great as 1 to 1.25 times the section thickness. Satisfactory welds are made in aluminum with upset distances about 50% greater than those employed with steels of similar thickness. (Refer to Tables 3.1 and 3.2 for typical flash welding dimensions, including upset distances and material losses for low- and medium-strength forging steels.)

Current. As discussed in the section on postweld heating, in some cases the weld zone may tend to cool too rapidly after flashing is terminated. This may result in inadequate upset or cold cracking of the upset metal. The joint temperature can be maintained during upset by resistance heating with current supplied by the welding transformer. The magnitude of the current generally is controlled electronically.

Upset current normally would be terminated at the end of upset. If the flash is to be mechanically trimmed immediately after welding, upset current may be maintained for an additional period to achieve the desired temperature for trimming.

UPSET WELDING

Upset welding (UW) is a resistance welding process that produces solid-state coalescence over the entire area of the faying surfaces or progressively along a butt joint by the heat achieved by resistance to the flow of welding current through the area where those surfaces are in contact. Pressure is used to complete the weld. High frequency upset welding and induction upset welding are two variations of the process.

FUNDAMENTALS

Upset welding is done in the solid state. As a light force is applied to the joint to bring the faying surfaces into intimate contact, the metal at the joint is resistance-heated to a temperature at which recrystallization can take place rapidly across the weld interface. After sufficient heating, a high upset force is applied. The upset pressure hastens recrystallization at the interface and forms a solid-state bond. During upsetting, some metal is expelled outward from weld interface; this purges the joint of oxides and other contaminants that may weaken the bond line. Figure 3.13 shows a cross section of an upset weld in bar stock or plate.

PROCESS VARIATIONS

The two variations of upset welding are high-frequency upset welding (UW-HF) and induction upset welding (UW-I).

High-Frequency Upset Welding

High-frequency upset welding is a process variation in which high-frequency welding current is supplied through electrodes into the workpieces.

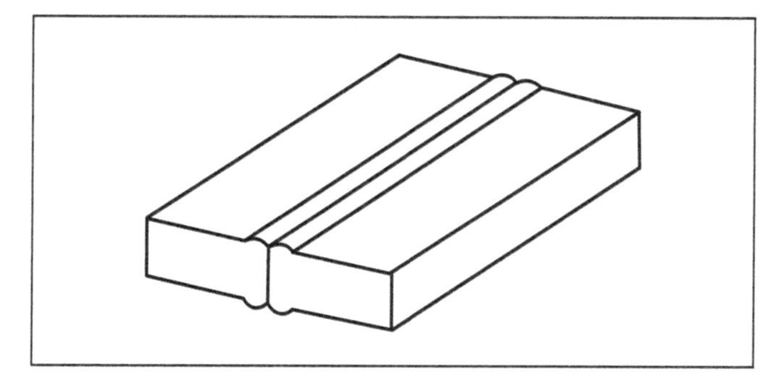

Figure 3.13—Cross Section of an Upset Weld in Bar Stock or Plate

Figure 3.14 shows typical joints made by high-frequency upset welding. Two techniques for making butt joints in bar stock or plate are illustrated: one uses a squeeze roll, shown in Figure 3.14(A), and the other uses a pressure pad (B) to apply upset force. Figure 3.14(C) shows a lap joint used in high-frequency seam welding. A seam weld in a tube welding is shown in Figure 3.14(D).

Induction Upset Welding

Induction upset welding is an upset welding process variation in which high-frequency welding current is induced in the workpieces. The induction upset welding of a tube seam is illustrated in Figure 3.15.

UPSET WELDING APPLICATIONS

Upset welding infers the use of butt joints and often the process is simply called *butt welding*. The process has two major applications: the end-to-end joining of two sections of the same cross section and the continuous welding of butt-joint seams in roll-formed products such as pipe and tubing.

Butt joints in two workpieces of the same cross section can be accomplished by upset welding, flash welding, and friction welding. Continuous butt joints in pipe and tubing can be welded with upset welding and high frequency welding. However, upset welding provides a number of advantages when used in the manufacturing of wire, pipe, and tubing.

Upset welding is used in wire mills for the production of wire and by other industries for manufacturing products made from wire. Wire and rod with diameters from 1.30 mm to 31.80 mm (0.05 in. to 1.25 in.) can be joined with the upset welding process. In wire mill applications, the process is used to join coils of wire to one another to facilitate continuous processing. Upset welding also is used to fabricate a wide variety of products from bar, strip, and tubing.

Typical examples of mill forms and products made with upset welding are shown in Figure 3.16. Wire and rod with diameters from 1.30 to 31.80 mm (0.05 to 1.25 in.) can be joined with the upset welding process.

Metals Welded

A wide variety of metals in the form of wire, bar, strip, and tubing can be joined end-to-end by upset welding, including the following:

1. Carbon steels,
2. Stainless steels,
3. Aluminum alloys,
4. Nickel alloys, and
5. Alloys with high electrical resistance.

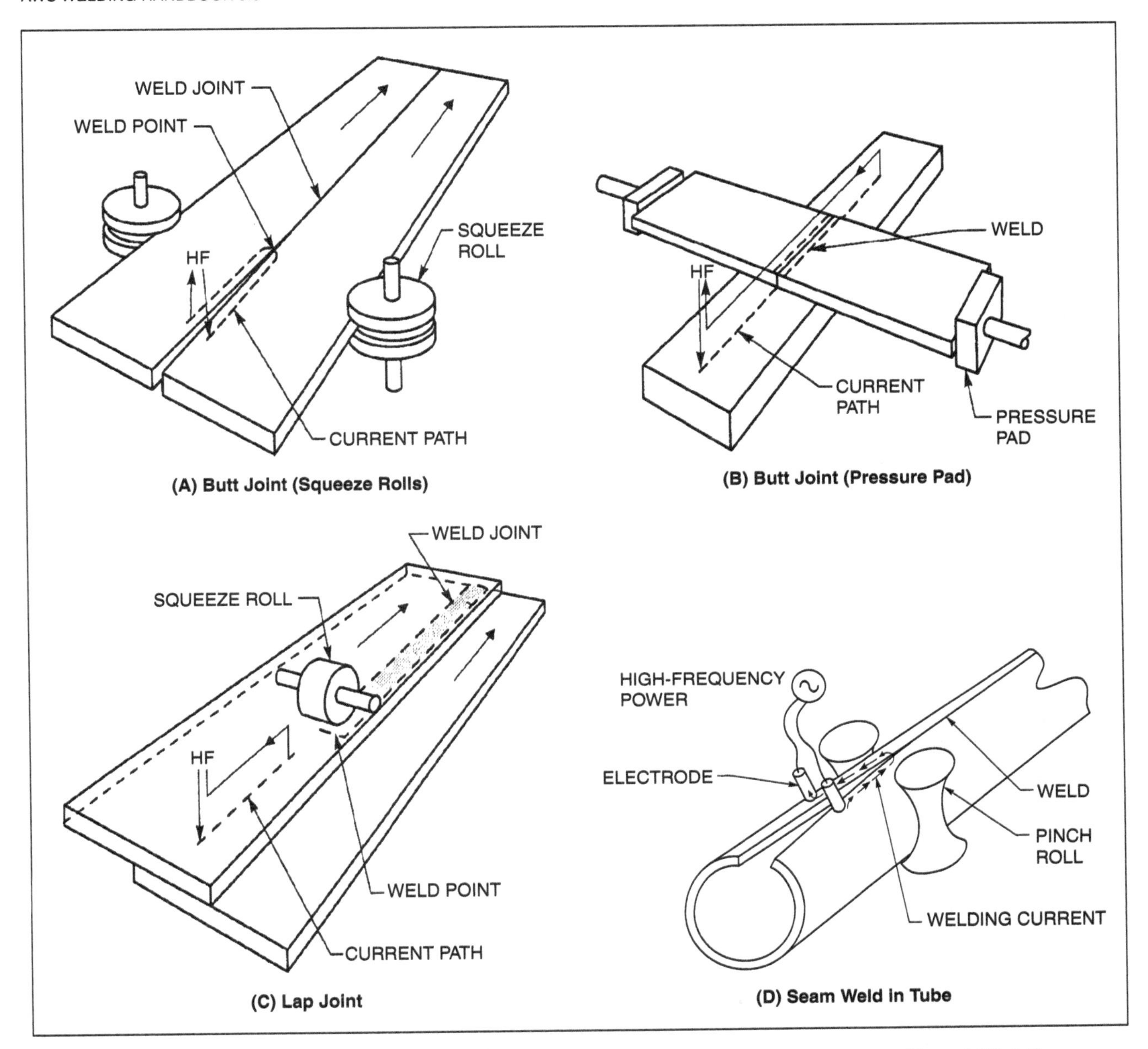

Figure 3.14—Typical Resistance-Welded Joints Made by High-Frequency Upset Welding

SHEET, STRIP, AND WIRE WELDING

The most common applications of upset welding include the fabrication of innumerable metal products requiring butt joints in sheet, strip, and wire.

Advantages and Limitations

The upset welding process can produce butt welds in sheet and strip materials at a higher rate of speed than a comparative flash welding process. Typical upset welding machines use dc power sources to maximize current density. The speed advantage comes at the cost of great sensitivity to joint condition, clamping forces, resistance welding die quality, and upset forces. To reduce the sensitivity to these variations, typical butt welding machines employ automated loaders to consistently load the workpieces and automated machines to trim the weldments as quickly as possible once they are complete. Since very little heat is generated outside the

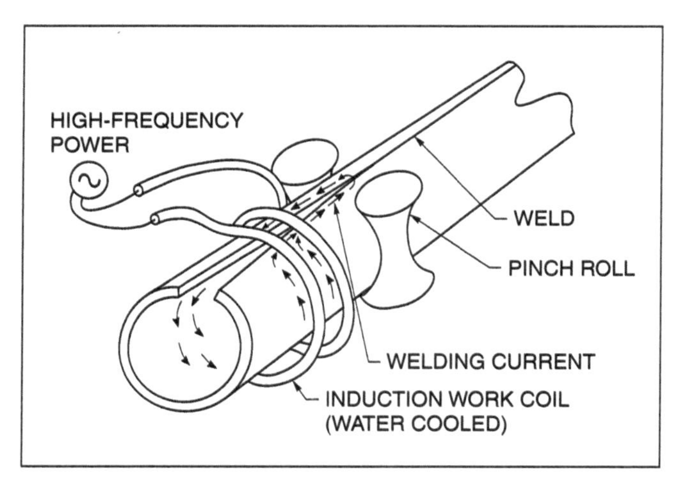

Figure 3.15—Induction Upset Welding of Tube

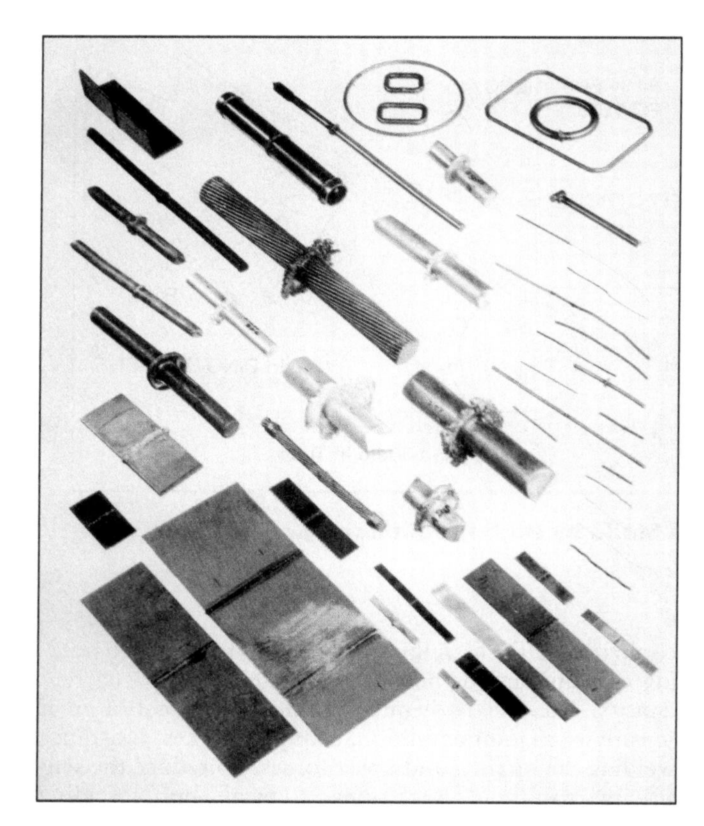

Figure 3.16—Typical Mill Forms and Products Joined by Upset Welding

welding zone, the heat will rapidly dissipate into the base metal and welding dies. Because no material is ejected from the machine, upset welding is inherently clean and does not produce metal fumes, eliminating the need for extensive exhaust systems. Flash splatter buildup is also nonexistent, which allows elaborate automation equipment to be located near the machine without mechanical shielding.

Sequence of Operations

In an automated upset welding sequence for an upset welded butt joint in sheet or strip, the welding operator loads the welding machine with the workpieces aligned end-to-end and clamps them securely, with even die contact along the length of the workpiece. The welding machine performs the following tasks:

1. Initiates the sequence;
2. Applies welding force;
3. Initiates current;
4. Applies upset force (if required);
5. Terminates current, and
6. Releases upset force.

The operator quickly unclamps the weldment and trims the upset material, then returns the movable platen and unloads the weldment (or unloads the weldment and returns the movable platen).

The general arrangement for upset welding is shown in Figure 3.17. One clamping die is stationary and the other is movable to accomplish upset. Upset force is applied through the moveable clamping die or a mechanical backup, or both.

Joint Preparation

For uniform heating in sheet and strip applications, the faying surfaces of the workpieces should be flat, comparatively smooth, and perpendicular to the direction of the upsetting force. Prior to welding, they should be cleaned to remove any dirt, oil, oxidation, or other contaminants that will impede welding and degrade joint integrity.

Heating Process and Heat Gradient

The contact resistance between the faying surfaces is a function of the smoothness and cleanliness of the surfaces and the contact pressure. This resistance varies inversely with the contact pressure, provided the other factors are constant. When current is applied, the base metal between the dies also will be heated by the welding current. Higher joint forces will generate a wider heating zone; lower forces will concentrate the heating to the joint itself. At a certain force, typically about 900 kilogram-

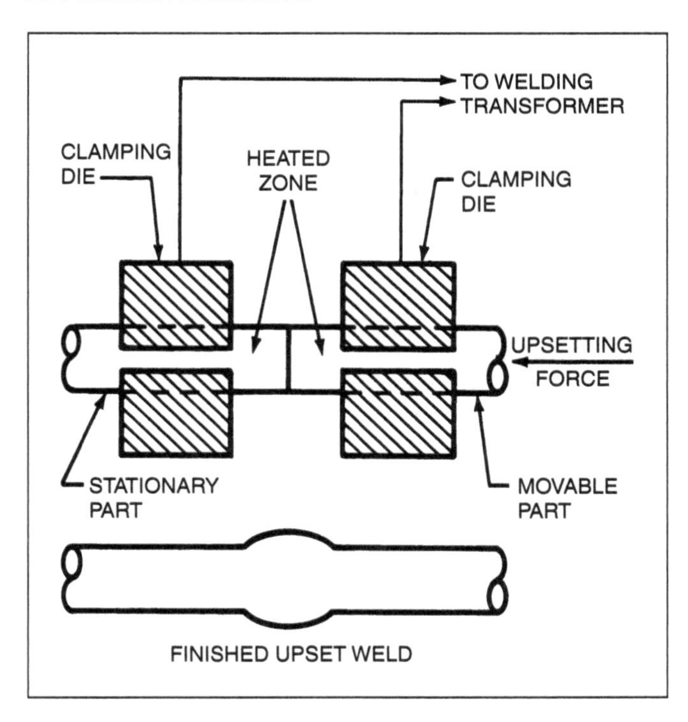

Figure 3.17—General Arrangement for Upset Welding of Sheet

force (kgf) (2000 pound-force [lbf]) or less, the faying surfaces will superheat and produce the weld. It should be noted that the upset speed is governed by the heat gradient produced within the workpieces. The heat gradient is shaped by the thermal, electrical, and mechanical characteristics of the workpieces, die contact and temperature, and the force, current, and weld time specified in the welding schedule.

Base Metal Properties

In metal sheet and strip applications, the heat gradient in the workpiece is created in response to the electrical resistance of the workpiece material and the faying surfaces coupled with the thermal resistance of the metal. This governs the rate at which the heat moves across the workpieces and into the resistance welding dies. The mechanisms of electrical and thermal resistance are highly time-dependent and therefore require that welding conditions be extremely well controlled and repeatable in order to generate a consistent heat gradient. The sensitivity of upset welding to deviation is a function of the characteristics of the base metal, with electrical and thermal resistances shaping the gradient and the melting point controlling the collapse rate of the metal. For example, steel that has a high melting point and relatively high electrical and thermal conductivity would have a high temperature gradient and relatively slow collapse rate compared to copper, which has even higher electrical and thermal values. Copper will heat quickly and then collapse very quickly, thus requiring a welding machine design that can accommodate the physical characteristics of various metals.

Electrode (Die) Opening

Since the welding equipment exerts a multi-ton force against the faying surfaces, the die opening for sheet and strip applications should be kept as small as possible to prevent deflection of the workpiece material. It should be noted that uneven loading and any pre-existing deflection of the workpieces would result in a skewed weld, or even a lapped joint.

Resistance Welding Dies

As previously noted, the upset welding process for sheet and strip applications requires the workpieces to be held together at high pressure to prevent slipping and must have uniform clamping force across the workpieces. On sheet welding machines, these two requirements can be problematic because of the high ratio of surface area to cross-sectional area of the workpieces. Additional aesthetic requirements, such as those of automotive applications, limit the die marks that may be left in the metal after welding. This requirement rules out serrated or other machined dies that would provide additional grip. Typical machines use lower dies made of copper for electrical contact and upper dies made of steel or other hardened materials. The lower dies normally need to be changed periodically due to wear. Generally, the clamping force for upset welds has to be much higher than that typically used for flash welds because of the higher pressures required for upset welding.

Die Cooling

As noted, the typical upset weld for sheet and strip metal applications has a small die opening, with a correspondingly short thermal path to the dies. Therefore, the temperature of the dies has an important function in the shaping of the heat gradient in the workpiece. For this reason, the dies must be maintained at a reasonably constant temperature to produce repeatable results. The dies usually are water-cooled and may be connected to chillers to maintain stable temperatures. An often-overlooked problem occurs when the dies are allowed to become too cold; typical plant cooling systems have no provision for maintaining cooling water at a set temperature, and possibly may even heat it. This can lead to unacceptable variations in the welds and a number of defective weldments after periods of inactivity, such as shift changes, breaks, and long shutdowns. Other reasons for variations in workpiece temperature are the result of environmental sources, such as winter-to-summer variations.

Power Systems

Power systems used for sheet and strip applications generally consist of a 220 volts (V) or 480 V three-phase dc power source that can produce a low-voltage, high-current dc power for the weld. The dc power system usually has an SCR-based heat control and possibly transformer taps for setting the desired range of welding voltage. Since the heat gradient of the process is based on the current (and hence the voltage of the power source), it is important that the power source is able to provide the desired welding current despite variations in line power voltage. The control of power fluctuations usually is accomplished by various feedback controls, such as constant voltage, which will vary the phase angle of the SCR to maintain a constant power output as the line voltage varies, and constant current, which monitors and maintains constant welding current. Power is conveyed to the welding machine via large water-cooled copper buss bars, which are sized for minimal heat rise at the welding currents required.

Upset Welding Machines

Upset welding machines for strip and sheet metal usually are designed to weld a particular group of alloys, such as steels, within a size range based on cross-sectional area. The mechanical capacity and electrical characteristics of the welding machine are matched to the specific type of application. Machines designed for aluminum and other metals with high conductivity and low melting points require much higher current input and faster force control systems than those required for steel applications. Generally, this allows a welding machine designed for aluminum to weld steel if the required parameters are achievable by that machine. However, a welding machine designed for steel cannot be used for aluminum.

An upset welding machine has two platens: one is stationary and the other is moveable. The clamping dies are mounted on these platens. The clamps operate either in a straight-line motion or through an arc about an axis, depending on the application. Since clamp contact is so crucial, it is desirable to have a method of adjusting the clamping force front-to-back to compensate for welding-die and workpiece irregularities. The forces required for upset welding are produced by a hydraulic system with controls capable of rapidly applying a desired force at the faying surfaces of the workpieces. The controls must be capable of maintaining the force in stationary and moving sequences of operation as the workpieces heat and collapse. Failure to maintain force during the collapse will result in weld irregularities and possibly spontaneous flashing during the weld.

Heat Balance

The upset welding process generally is used to join two workpieces of the same alloy and same cross-sectional geometry. In this case, heat balance should be uniform across the joint. If the workpieces are similar in composition and cross section but of unequal mass, the workpiece with the larger mass should project from the clamping die somewhat farther than the other workpiece. When welding dissimilar metals, the metal with higher electrical conductivity should extend farther from the clamp than the other. When the upset welding process is used for large workpieces that do not make good contact with one another, it is sometimes advantageous to interrupt the welding current periodically to allow the heat to distribute evenly into the workpieces.

CONTINUOUS UPSET WELDING

Continuous upset welding is used in the manufacture of pipe and tubing. In these applications, a coiled strip is fed into a set of forming rolls. The rolls progressively form the strip into a cylindrical shape. The edges to be joined approach one another at an angle and culminate in a longitudinal *V* groove at the point of welding. A wheel electrode contacts both edges of the tube a short distance from the apex of the *V*. Current from the power source travels from one electrode along the adjacent edge to the apex, where welding takes place, and then back along the other edge to the second electrode. The edges are heated to welding temperature by resistance to this current flow. The hot edges are then upset, brought together by a set of pinch rolls to consummate the weld.

Tube-Welding Equipment

Upset welding may use either ac or dc power. Alternating-current machines may be operated with either 60-Hertz (Hz) single-phase power or with higher frequency produced by a single-phase alternator. Direct-current machines are powered by a three-phase transformer-rectifier unit.

Welding Procedures

As the formed tube passes through the zone between the electrodes and the pinch rolls, a variation in pressure across the joint takes place. If no heat were generated along the edges, this pressure would be maximum at the center of the squeeze rolls. However, since heat is generated in the metal ahead of the centerline of the squeeze roll, the metal gradually becomes plastic and the initial point of contact with the edge is slightly ahead of the squeeze roll axes. The point of maximum upset pressure is somewhat ahead of the squeeze roll centerline.

The current across the seam is distributed in inverse proportion to the electrical resistance between the two electrodes. This resistance, for the most part, is the electrical contact resistance between the edges to be welded. Pressure levels directly impact contact resistance (higher pressures equal lower contact resistance and lower contact pressure equal higher contact resistance). As the temperature of the joint is increased, the pressure decreases due to material at the joint softening, and the electrical resistance increases.

When using ac welding power, a very sharp thermal gradient caused by resistance heating at the peaks of the ac cycle produces an intermittent or "stitch" effect. The stitch is normally circular in cross section, lying centrally in the weld area and parallel to the line of initial closure of the seam edges. It is the hottest portion of the weld. The stitch area is molten while the area between stitches is at a lower temperature. The patches of molten metal are relatively free to flow under the influence of the motor forces (current and magnetic flux) acting on them. Consequently, the molten metal may be ejected from the stitch area or forced outward at full upset. If the welding heat is excessive, too much metal is ejected and pinhole leaks may result. With too little heat, the individual stitches will not overlap sufficiently, resulting in interruptions or gaps in the seam weld.

The longitudinal spacing of the stitches must have some limit. The spacing is a function of the power frequency and the travel speed of the tubular workpieces. With 60-Hz power, the speed of welding should be limited to approximately 0.45 m/sec (90 ft/min). To weld tubing at higher speeds than this requires welding power of higher frequency. Table 3.6 shows typical welding speeds using various sizes of 180-Hz power sources for longitudinal welds in steel tubing of several wall thicknesses.

It is good practice to close the outside corners of the edges first as the formed tube moves through the machine so that the stitches will be inclined forward. This condition is known as an *inverted V*. The advantages of using an inverted *V* are twofold. First, the angle deviation from vertical reduces the forces tending to expel any molten metal in the joint. Second, the major portion of the solid upset metal is extruded to the outside where it is easily removed. The tubing normally is formed so that the angle of the *V*-groove is about 5° to 7°.

Surface Burns

As in spot and seam welding, the current that provides the welding heat in upset welding must enter the workpiece through electrode contacts. The electrical resistance of these contacts must be kept to a minimum to avoid resistance heating that is sufficient to produce surface burns on the tube. Burns are actually surface portions of the tube that are heated to the melting point. The molten metal from the burned area may stick to the face of the wheel electrode or become embedded in it. If these areas of metal pick-up are large enough and become embedded in the electrode face, the contact resistance at these locations will increase signifi-

Table 3.6
Typical Speeds for the Seam Welding of Steel Tubing Using 180 Hz Power Sources

Wall Thickness		Speed, m/min (ft/min)			
mm	in.	125 kVA m (ft)	200 kVA m (ft)	300 kVA m (ft)	500 kVA m (ft)
1.27	0.050	45.7 (150)	70.0 (200)	—	—
1.65	0.065	33.5 (110)	42.7 (140)	70.0 (200)	—
2.11	0.083	21.9 (72)	32.0 (105)	44.0 (145)	—
2.41	0.095	—	26.0 (85)	35.0 (115)	—
2.77	0.109	—	20.1 (66)	27.4 (90)	—
3.17	0.125	—	15.2 (50)	21.3 (70)	42.7 (140)
3.40	0.134	—	—	18.3 (60)	38.1 (125)
3.96	0.156	—	—	—	26.0 (85)

cantly and cause more severe surface burning of the tube. This action continues to worsen with each revolution of the electrode. To stop surface burning, the operation must be interrupted and the electrodes cleaned or replaced, or an automatic electrode dressing device must be installed.

To prevent burns, the area of contact and the pressure between the electrode and the tube must be optimum. As a rule of thumb, each electrode should have sufficient contact area so that the current density will be less than 78 amperes per square millimeter (A/mm^2) (50 000 amperes per square inch [A/in.^{2}]). The relative shapes of the formed tube and the electrode should ensure that the maximum contact pressure occurs next to the seam.

Without the aid of some reinforcing support, electrode contact pressure is limited by the ability of the tube to resist the forces being applied. The maximum permissible pressure in the weld throat is a function of the yield strength of the metal and the ratio of the tube diameter to wall thickness (D/T ratio). In extreme cases in which the D/T ratio is high, a backup mandrel must be used to prevent distortion of the tube wall and misalignment of the joint.

FLASH AND UPSET WELD QUALITY

Weld quality is significantly affected by the specific welding variables selected for the application. Table 3.7 indicates the effects of several variables on weld quality when they are not optimum. Each variable is considered individually, although more than one variable may produce the same result or there may be an interaction of the variables. Common discontinuities found in flash welds and upset welds are discussed in this section.

Base Metal Structure

Metallurgical discontinuities that originate from conditions present in the base metal usually can be minimized by specifying necessary qualities when selecting the base metal. The inherent fibrous structure (such as centerline segregation) of wrought mill products may cause an isotropic mechanical behavior. An out-turned fibrous structure at the weld interface often results in some decrease in mechanical properties compared to the base metal, particularly concerning ductility.

The decrease in ductility caused by flash welding or upset welding usually is insignificant except in the following two cases:

1. When the base metal is not homogenous, for example, severely banded steels, alloys with excessive stringer-type inclusions, and mill products with seams and cold laps produced during the fabrication process; and
2. When the upset distance is excessive.

If banding, centerline segregation, or other stringer-type inclusions are present on a welded tube, the inclusions may be forged to the surface of the weld joint and produce discontinuities that are very brittle. This could cause the weld to fail during bending or flairing, such as when a fitting is created at the tube end. If the upset distance is excessive, it may cause these undesirable stringer-type microstructures to be completely reoriented transverse to the original structure and would produce a weak, brittle band prone to opening under even light pressure.

Table 3.7
Effect of Variables on Flash Weld and Upset Weld Quality

	Voltage	Platen Rate	Time	Current	Distance or Force
Excessive	Deep craters are formed that cause voids and oxide inclusions in the weld; cast metal in weld.	Tendency to freeze.	Metal too plastic to upset properly.	Molten material entrapped in upset; excessive deformation.	Tendency to upset too much plastic metal; flow lines bent perpendicular to base metal.
Insufficient	Tendency to freeze; metal not plastic enough for proper upset.	Intermittent flashing, which makes it difficult to develop sufficient heat in the metal for proper upset.	Not plastic enough for proper upset; cracks in upset.	Longitudinal cracking through weld area; inclusions and voids not properly forced out of the weld.	Failure to force molten metal and oxides from the weld; voids.

Oxides

Another source of discontinuities is the entrapment of oxides at the weld interface. Such defects are rare because proper upset should expel any oxides formed during the flashing operation. However, if tap or heat control settings are too high and cause flashing action to be excessive, the tendency to create discontinuities in the joint increases significantly, even if optimum upset conditions are used. This emphasizes the fact that interactions between variables is an important aspect to consider when initial parameters are set and when troubleshooting problems.

Flat Spots

Flat spots are metallurgical discontinuities that usually are limited to ferrous alloys. Their exact cause is not clear. They appear on a fractured surface through the weld interface in the form of smooth, irregularly shaped areas.

Excellent correlation exists between the location of flat spots and localized regions of carbon and silicon segregation in steels. In many cases, the cooling rates associated with flash welds are rapid enough to produce brittle, high-carbon martensite in areas on the flashing interface where the carbon content happens to be greater than the nominal composition of the alloy. Microhardness tests and metallographic examination have confirmed the presence of high-carbon martensite in the region surrounding a flat spot in almost every case, even in plain carbon steels. High-silicon steels, such as electrical steels used in transformers, also may encounter flat spots where the presence of brittle silicon oxides have not been properly forged out during upsetting. In addition, steels with banded microstructures appeared significantly more susceptible to this type of defect than unbanded steels.

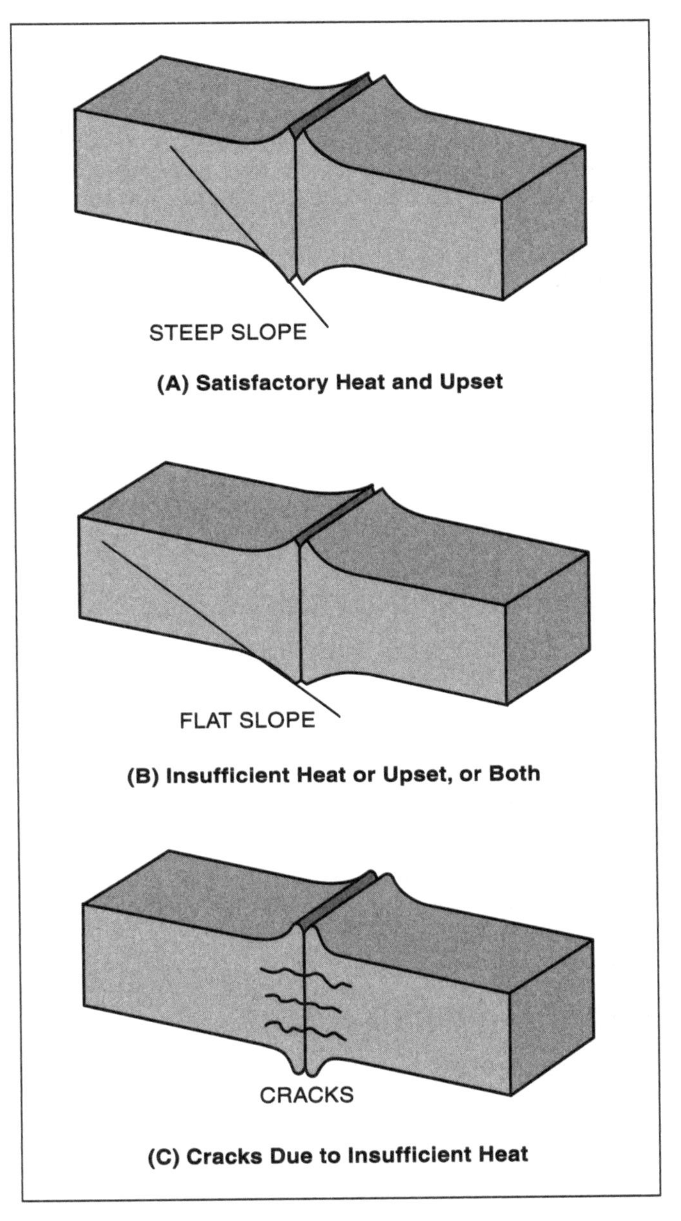

Figure 3.18—Visual Indications of Flash Weld Quality

Die Burns

Die burns are discontinuities produced by local overheating of the base metal at the weld interface between the clamping die and the workpiece surface. Die burns usually can be avoided by keeping the workpieces and die surfaces clean and having the workpieces fit properly with the dies.

Voids

Voids usually are the result of either insufficient upset or excessive flashing voltage. Deep craters produced on the faying surfaces by excessive flashing voltage may not be completely eliminated during upset. These discontinuities usually are discovered during qualification of the welding procedure. Voids and craters in flash welds can be readily avoided by decreasing the flashing voltage or increasing the upset distance. As shown in Figure 3.18(A) and (B), visual inspection can reveal welds with satisfactory and unsatisfactory upset.

Cracking

A crack is a type of discontinuity that may be internal or external. It may be related to the metallurgical and compositional characteristics of the metal. Alloys that exhibit low ductility over some elevated temperature range may be susceptible to internal hot cracking. Such alloys, known as hot-short alloys, are somewhat difficult to join with flash welding, but usually can be successfully welded when the proper welding conditions are used.

Cold cracking may occur in hardenable steels. It usually can be prevented by using welding conditions that moderate the cooling rate of the weld, coupled with postweld heat treatment as soon as possible after welding.

Insufficient heating prior to or during upset is the usual cause of cracking in the external upset metal, as shown in Figure 3.18(C). This can be eliminated by resistance heating during upset.

Mechanical Discontinuities

Mechanical discontinuities include misalignment of the faying surfaces of the workpieces prior to welding and inconsistent upset during welding. These discontinuities are easily detected by visual inspection. Misalignment of the workpieces is corrected by adjustment of the clamping dies and fixtures. Inconsistent upset may be caused by workpiece misalignment, insufficient clamping force, or excessive die opening at the start of upset. The latter can be corrected by decreasing the initial die opening and then adjusting the welding schedule, if necessary.

EXAMINATION AND TESTING

The nondestructive evaluation of flash welds and upset welds is conducted with methods similar to those used for resistance welding. Computer controls and monitoring devices are used to collect data for a multitude of parameters.

A major advantage of flash welding and upset welding is that the processes can be highly mechanized and automated. Therefore, consistent quality level is readily maintained after optimum welding conditions are established. The fact that no filler metal is employed means that the strength of the weld is primarily a function of the base metal composition and properties. Consequently, properly made flash welds should exhibit satisfactory mechanical properties.

Both flash and upset welds can be inspected and tested in the same manner. They may be evaluated by tension testing, in which the tensile properties are compared to those of the base metal, bend testing, and bulge testing, where a metal ball is forced up through the weld in sheet product while the weld area is clamped in a die. Bulge testing is one of the fastest means of testing upset and flash welds of sheet product. Metallographic and dye-penetrant inspection techniques also are used.

In commercial practice, both destructive and nondestructive tests are employed to ensure that the desired quality level is maintained in critical flash welded or upset welded products. The process control procedure usually includes the following functions:

1. Material certification,
2. Qualification of welding procedure and data monitoring,
3. Visual inspection of the product, and
4. Destructive testing of random samples.

When the product is used in a critical application, this process-control procedure is supplemented by other tests, such as magnetic particle and dye-penetrant examination. When the welded joint subsequently is machined, routine measurement of the hardness of the weld area also may be specified.

Base Metal Certification

Since defects in base metals may cause discontinuities in flash and upset welds, each lot of base metal should be carefully inspected on delivery to ensure that it meets specifications. Certified chemical analysis, mechanical property tests, macroscopic examinations, and magnetic particle inspection may be applicable.

Procedure Qualification

Each different combination of metals and size of sections to be welded by flash welding or upset welding normally require the qualification of an individual welding procedure. This usually involves welding a number of test specimens that duplicate the material, section size, welding procedure, and heat treatment to be used in producing the product. All of these specimens are visually inspected for cracks, die burns, misalignment and other discontinuities. When specified, weld hardness is measured. To verify weld strength, a tensile specimen should be machined from a test weld using the entire welded cross section when possible. The test results should be compared to base metal properties and design requirements. Bend testing and bulge testing also may be required to verify ductility and weld integrity. In sheet products, the bulge test is the quickest, cheapest, and easiest test to perform to determine weld integrity and ductility. Conducting this test first may preclude the need to machine expensive tensile specimens if the bulge test samples fail to produce the required results.

All pertinent welding conditions used in producing the qualification test should be recorded. The production run is then made using the qualified welding procedure.

Nondestructive Examination

Each completed weld in the production run should be visually examined for evidence of cracks, die burns, misalignment, and other external weld defects. When specified, magnetic particle or fluorescent penetrant inspection is performed on random samples to assist in detecting flaws not visible to the unaided eye. In critical applications, random radiographic examination also may be specified.

Destructive Testing

Depending on the size of the production run, a specified number of randomly chosen weldments may be selected for destructive testing of the welds. All of the results of these destructive tests must meet the same criteria specified in the welding procedure qualification test. Additional tests are required if any of the welds fail. A report of the results of all destructive tests is then prepared to certify the maintenance of the required average quality level for the lot.

Bend Tests for Sheet and Strip

Notched bend tests may be used in sheet and strip applications to force a fracture to occur along the weld interface for visual examination. A bend test may be useful as a qualitative means for establishing a welding schedule. However, such tests usually are not used for specification purposes.

Bulge Tests for Sheet and Strip

Bulge tests can be used in sheet and strip applications. A metal ball is forced up through the weld in sheet product while the weld area is clamped in a die. This action causes the weld to fracture along the weld interface if the weld has poor ductility or has discontinuities. If fracture occurs in the base metal and transverse to the weld, the joint usually is deemed acceptable. Typically, a series of tests spaced at regular intervals are conducted across the full width of the sheet or strip (if it is wide) to get an adequate sampling rate. This test is useful as a qualitative means for establishing welding schedules and random monitoring of weld quality during production.

Bend Test for Wire

The bend test is a common method for evaluating an upset weld in wire. A welded sample is clamped in a vise with the weld interface located a distance of one wire diameter from the vise of the jaws. The sample then is bent back and forth until it breaks in two. If the specimen fractures through the weld interface and shows complete fusion or if it occurs outside the weld, the weld quality is considered satisfactory.

Tension Tests

When strength testing is required, the tension test specimen should be machined to include the entire cross section of the weld, as described in the Procedure Qualification section, and properties of the weld metal should be compared to those of the base metal.

ECONOMICS

Flash and upset welding provide many advantages for welding applications involving butt welds in sheet, bar, tube, pipe, and wire. Economic effectiveness is important among these advantages.

Cost savings are inherent to flash welding and upset welding because filler metal is not required and shielding gas normally is not used. The processes are generally clean—they do not produce excessive metal fumes and spatter, eliminating the need for costly exhaust systems and mechanical shields.

Labor costs for joint preparation are minimal. Little if any joint preparation is necessary because any contaminants present at the weld face are expelled during the upsetting stage. The processes are automated, thus high production can be achieved while labor costs remain low. Automated systems require minimal skill of the welding operator. Excellent weld quality can be achieved economically.

SAFE PRACTICES

Many of the safe practices applicable to other welding and fabricating processes also apply to the safe use of flash and upset welding. General safe practices for the welding industry are published by the American Welding Society in *Safety in Welding, Cutting and Allied Processes*, ANSI Z49.1. The latest edition of this document is available electronically to users.[4] Mandatory safety standards are provided from the United States government Occupational Safety and Health Administration in *Occupational Safety and Health Standards for General Industry.*[5] Other safety standards and publishers are listed in Appendices A and B of this volume. Internet addresses of the publishers are included.

Mechanical, electrical, and operator safety is discussed briefly in this section.

MECHANICAL

The welding machine should be equipped with appropriate safety devices to prevent injury to the operator's hands or other parts of the body. Initiating devices, such as push buttons or foot switches, should be arranged and guarded to prevent them from being inadvertently actuated.

Machine guards, fixtures, or operating controls should prevent the hands of the operator from entering between the work-holding clamps or the workpieces. Dual hand controls, latches, presence-sensing devices, or similar devices must be employed to prevent operation in an unsafe manner. Safe operating procedures specified by the equipment manufacturer must be carefully followed.

ELECTRICAL

All doors and access panels on machines and controls must be kept locked or interlocked to prevent access by unauthorized personnel. When the equipment utilizes capacitors for energy storage, the interlocks should interrupt the power and discharge all the capacitors through a suitable resistive load when the panel door is open. A manually operated switch also should be provided in addition to the mechanical interlock or contacts. The use of this device assures complete discharge of the capacitors.

A lock-out procedure should be followed prior to working with electrical or hydraulic systems.

PERSONAL PROTECTION

Flash guards made of fire resistant material should be provided to protect the operator from sparks and to avoid fires. The operator must wear personal eye protection with suitable shaded lens, and must wear flame-resistant gloves capable of providing protection from hot metal. The gloves must be insulated to protect from electric shock.

When welding operations produce high levels of noise, operating personnel should be wear ear protection. Allowable noise exposure levels are published by the Occupational Safety and Health Administration (OSHA) in *General Industry Standards,* Title 29 CFR1910.95.[6]

Although flash welding and upset welding usually do not produce metal fumes, if fumes are created, for example, when welding stainless steel, fumes must be removed by an adequate exhaust system.

CONCLUSION

Flash welding and upset welding are viable joining processes for a variety of applications and materials. Technological advances in computer control systems, power sources, hydraulics and servo controls have greatly improved the process and broadened the scope of applications. In many cases, dissimilar materials and cross sections can now be welded based on individual requirements of the application. Technological advances in flash welding have increased production, reduced power consumption and made the process environmentally cleaner.

Both flash welding and upset welding produce high-quality solid-state welds using no filler materials or gases. Both welding processes are readily adapted to automation. Tooling designs and improvements in parameter control and options have enabled flash welding and upset welding machines to become more versatile and reliable, and have increased the range of the products weldable with these processes.

Applications include various automotive parts, such as wheel manufacturing in steel alloys and aluminum, coil joining and splicing in the basic metals industries, and engine components and landing gears in the aerospace industry. The railroad industry uses flash welding to welds rails and the construction industry builds trusses and related structural items using these welding processes. The petroleum industry use these processes to weld pipelines and various oil well applications. Smaller applications of these processes include the welding of band saw blades, wire joining, and related electronic micro-welding in the electronic industry. As distinctive resistance welding processes in the current

4. American National Standards Institute (ANSI) Accredited Standards Committee Z49, latest edition, *Safety in Welding, Cutting, and Allied Processes*, ANSI Z49.1, Miami: American Welding Society. This document can be accessed electronically at http://www.aws.org.
5. Occupational Safety and Health Administration (OSHA). *Title 29—Labor. In Code of Federal Regulations (CFR)*, Title 29, CFR 1910, Subpart Q, Washington D.C.: Superintendent of Documents, U.S. Government Printing Office.

6. See Reference 5.

marketplace, flash and upset welding have evolved into accurate and reliable process for the joining of metals. Future advances will continue to improve the processes and diversify their application.

BIBLIOGRAPHY

American National Standards Institute (ANSI) Accredited Standards Committee Z49. 2005. *Safety in welding, cutting, and allied processes,* ANSI Z49.1:2005. Miami: American Welding Society.

American Welding Society (AWS) Committee on Definitions and Symbols. 2001. *Standard welding terms and definitions.* AWS A3.0:2001. Miami: American Welding Society.

American Welding Society (AWS) Committee on Resistance Welding. 2000. *Recommended practices for resistance welding.* AWS C1.1M/C.1:2000. Miami: American Welding Society.

Occupational Safety and Health Administration (OSHA). 1999. Title 29—Labor. In *Code of Federal Regulations (CFR), Title 29, CFR 1910, Subpart Q.* Washington D.C.: Superintendent of Documents, U.S. Government Printing Office.

SUPPLEMENTARY READING LIST

Anon. 1976. Union Pacific used flash welding to take clickity-clack out of its tracks. *Welding Journal* 55(11): 961–962.

Holko, K. H. 1970. Magnetic force upset welding dissimilar thickness stainless steel tee joints. *Welding Journal* 49(9): 427s–439s.

Kotecki, D. J., D. L. Cheever, and D. G. Howden. 1974. Capacitor discharge percussion welding; microtubes to tube sheets. *Welding Journal* 53(9): 557–560.

MIL-W-6873, Military Specification, Welding; Flash, Carbon and Alloy Steel.

Petry, K. N., et al. 1970. Principles and practices in contact welding. *Welding Journal* 49(2): 117–126.

Savage, W. F. 1962. Flash welding: the process and application. *Welding Journal* 41(3): 227–237.

Savage, W. F. 1962. Flash welding: process variables and weld properties. *Welding Journal* 41(3): 109s–119s.

Sullivan, J. F. and W. F. Savage. 1971. Effect of phase control during flashing on flash weld defects. *Welding Journal* 50(5): 213s–221s.

Thompson, E. G. 1982. Attachment of thermocouple instrumentation to test components by all-position percussion welding. *Welding Journal* 61(6): 31–33.

Turner, D. L., et al. 1982. Flash butt welding of marine pipeline materials. *Welding Journal* 61(4): 17–22.

CHAPTER 4

RESISTANCE WELDING EQUIPMENT

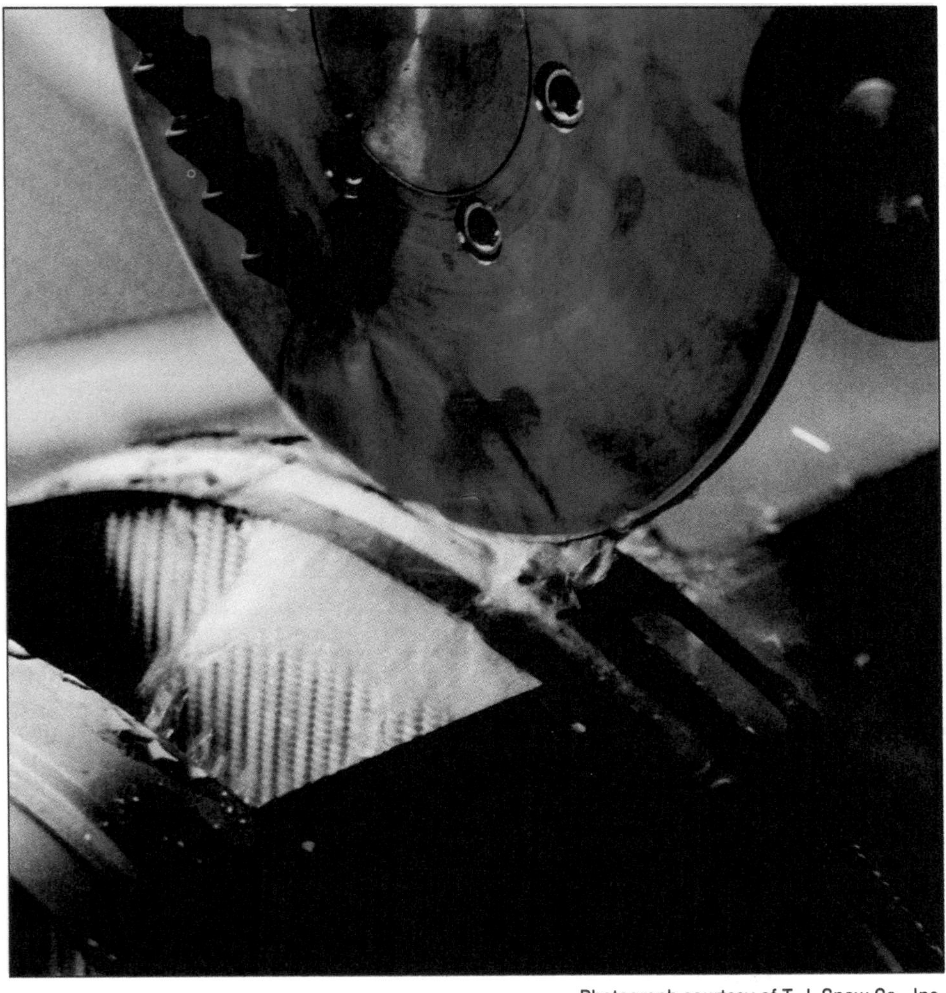

Photograph courtesy of T. J. Snow Co., Inc.

Prepared by the Welding Handbook Chapter Committee on Resistance Welding Equipment:

M. B. Siehling, Chair
RoMan Manufacturing, Inc.

D. M. Beneteau, Co-Chair
CenterLine (Windsor) Ltd.

P. L. Haynes
Miyachi Unitek Corporation

R. B. Hirsch
Unitrol Electronics

B. G. Kelly
Kelly Welding Solutions P.C.

R. P. Matteson
Taylor-Winfield Corp.

R. G. Van Otteren
Newcor Bay City Division

E. Waelchli
International Business Development & Technology Transfer

Welding Handbook Volume 3 Committee Member:

S. P. Moran
Miller Electric Manufacturing Co.

Contents

CHAPTER 4

RESISTANCE WELDING EQUIPMENT

INTRODUCTION

Resistance welding includes a group of welding processes that produce coalescence of the faying surfaces with the heat obtained from resistance of the workpieces to the flow of the welding current in a circuit of which the workpieces are a part, and by the application of pressure.[1,2] Commonly used resistance welding processes include flash welding, pressure-controlled resistance welding, projection welding, seam welding, spot welding, and upset welding. These processes and allied processes such as resistance soldering, resistance brazing, and hot upsetting are highly important to many industries.

Resistance welding provides a fast and economical way to weld all types of steels: coated or uncoated steels, low-carbon and high-carbon steels, dual-phase or heat-treatable steels, and stainless steels. Other materials, including aluminum, titanium, and copper alloys also are commonly joined by resistance welding processes.

Resistance spot welding has the advantages of high speed and adaptability to automation for the production of sheet metal components used in automobiles, cabinets, furniture, and other products. Resistance seam welding is used to produce continuous leak-tight joints in sheet assemblies such as hot water tanks, automotive gasoline tanks, and food containers, and also for longitudinal seams in tubular structures.

1. Welding terms and definitions used throughout this chapter are from *Standard Welding Terms and Definitions*, AWS A3.0:2001, Miami: American Welding Society.
2. At the time of the preparation of this chapter, the referenced codes and other standards were valid. If a code or other standard is cited without a date of publication, it is understood that the latest edition of the document referred to applies. If a code or other standard is cited with the date of publication, the citation refers to that edition only, and it is understood that any future revisions or amendments to the code or standard are not included; however, as codes and standards undergo frequent revision, the reader is encouraged to consult the most recent edition.

Projection welding contributes valuable design advantages to manufacturers. Fasteners or mounting devices, such as bolts, nuts, pins, brackets, threaded bosses, and handles can be welded to sheet metal with this process. Crossed wires are commonly joined with the projection welding process in a variety of common products, such as concrete reinforcing mats, shopping carts, and wire shelving.

Flash welding is used to join products of rod, bar, tube, pipe, and wire made of carbon steels, low-alloy steels, stainless steels, aluminum, nickel, and copper alloys. Upset welding makes end-to-end joints in two components of the same cross section, and makes continuous seams in pipe and tubing and in sheet and strip metals.

These resistance welding processes are described in Chapter 1, "Spot and Seam Welding," Chapter 2, "Projection Welding," and Chapter 3, "Flash and Upset Welding." This chapter covers the machines and equipment required to perform the varied joining applications of resistance welding.

EQUIPMENT SELECTION

The selection of resistance welding equipment usually is determined by the requirements of joint geometry, workpiece materials, quality, production schedules, and economic considerations. The equipment configurations run the gamut from standard machines to specially designed, complex resistance welding systems. It is common to incorporate other processes into the equipment to minimize workpiece handling, improve process reliability, or to conserve floor space.

The standard resistance welding machine typically has the following principal elements:

1. An electrical circuit consisting of a resistance welding transformer and a connected secondary circuit, which includes electrodes that conduct the current to the workpiece;
2. A mechanical system consisting of a machine frame and associated mechanisms to hold the workpiece and apply the welding force; and
3. Control equipment that initiates and times the duration of welding current, and may also control the magnitude of the welding current and the sequencing of other aspects of the machine cycle.

With respect to electrical operation, resistance welding machines are classified in two basic groups: direct energy and stored energy. Direct-energy machines convert energy from the electric supply lines for use in the process. Stored-energy machines draw power from the electric supply lines and accumulate it so that it can be applied after the charging interval. Machines in both groups may be designed to operate on either single-phase or three-phase power.

While the use of three-phase resistance welding machines continues to increase, the most commonly employed machines are the single-phase, direct-energy type. Single-phase machines are the simplest form and the least expensive in initial cost, installation, and maintenance. The mechanical systems and secondary circuit designs are essentially the same for all types of welding machines, but transformer designs and control systems differ considerably.

A single-phase welding machine has a larger volt-ampere (V-A) demand than a three-phase machine of equivalent rating. The demand of a single-phase machine causes an imbalance on a three-phase power line. Also, the power factor is relatively low because of the inherent inductive reactance in the welding circuit of a single-phase machine. Single-phase demand may not be a problem if the welding machine is a small part of the total electrical system load, or if a number of single-phase welding machines are operating across different supply phases to balance the load.

A three-phase, direct-energy machine draws power from all three phases of the power line. The inductive reactance of the welding circuit is low because direct current (dc) is used for welding; thus, the required secondary circuit voltage for a given welding current is reduced and the kilovolt ampere (kVA) demand of a three-phase machine is lower than that of an equivalent (equal current) single-phase machine. This is a definite advantage when a large-capacity machine is needed and power line capacity is limited.

The operating principle of a stored-energy machine is to accumulate electrical energy and then discharge it to make the weld. The energy normally is stored in a capacitor bank. Single-phase power generally is used for small bench-model stored-energy machines. The power demand is low because charging time is relatively long compared to the weld time.

The configurations of resistance welding machines are as varied as the applications of the processes. The need for distinctive machine configurations is driven by the unique requirements of each resistance welding process variation. In general, the machines are classified as either standard or custom. Many of the standard machine configurations can be traced to the origin of the process in the late 1800s, but continue to be used in present-day applications because the machines are cost effective and offer maximum versatility.

SPOT WELDING AND PROJECTION WELDING MACHINES

Machine configurations are available in the greatest variety for spot welding and projection welding because they are the most commonly applied resistance welding processes. In many instances, standard machines can be used interchangeably for either spot welding or projection welding. The exceptions are identified in the descriptions in the following sections.

Rocker-Arm Type

The simplest and most commonly used spot welding machine is the rocker-arm design, so called because of the pivoting movement of the upper horn. A horn is essentially an arm or an extension of an arm of a resistance welding machine that transmits the electrode force and, in most cases, the welding current. This type of machine is readily adaptable for spot welding of most weldable metals. Several choices of actuators are available for providing force: pneumatic (referred to as air-operated), foot-controlled, and motor-controlled.

Pneumatically operated machines, such as the one shown in Figure 4.1, use compressed air and are the most popular. Compressed air cylinders are the most common choice due to availability, low cost, and simplicity. Hydraulic cylinders sometimes are used when high forces are required. In both cases, the machine cycle generally is controlled automatically by the resistance welding control system. These machines can operate rapidly and are easily set up for welding.

Foot-operated machines incorporate a linkage between a foot pedal and the moveable arm, along with a mechanism, such as a spring, that regulates the electrode force and triggers the welding control sequence. Foot-operated machines are chosen for low cost and for low production runs, because they require an operator.

Photograph courtesy of Standard Resistance Welder Co.

Figure 4.1—Rocker-Arm Spot Welding Machine Operated with Compressed Air

They are not widely used, but may have applications for sheet metal fabrication, particularly for short production runs or when making prototype parts.

Motor-operated machines are another type that is not commonly used, but may be used when compressed air is not readily available. Electric motors are chosen because of energy efficiency, high speed, and operating capability; they produce repeatable welds and have programmable force and movement.

Standard rocker-arm machines generally are available with throat depths of 30.5 centimeters (cm) to 91.4 cm (12 inches [in.] to 36 in.) and transformer capacities of 5 kVA to 100 kVA. The general construction of these machines is the same for all three systems of operation.

Mechanical Design. The machine frame houses the transformer, the transformer tap switch, and also supports the mechanical and electrical components, as shown in Figure 4.1.

For rocker-arm machines, the stroke of the actuator must be proportional to the ratio of length from the electrode to the pivot, and the length from the pivot to the point where the actuator is attached. The force of the actuator must be similarly proportioned. If the length of the horns or electrodes is changed, the ratios affecting the electrode force and stroke ratios will be altered. Changes to the length of the horns or electrodes can be beneficial in cases where the actuator does not have sufficient adjustment to provide the desired electrode force.

When air-operated, the electrode force is in direct proportion to the air pressure as controlled by a pressure regulator. With foot-operated and motor-operated machines, the amount of force is determined by the stiffness of the spring and the compression distance.

Application Considerations. The upper electrode of the rocker arm machine operates on a fulcrum, and its path of travel follows an arc. Because of this, these machines are not recommended for projection welding.

To ensure electrode alignment and to minimize electrode skidding, the electrodes should be arranged so that a plane of the weld passes through the fulcrum as closely as possible. The two horns should also be parallel when the electrodes are in contact with the workpiece. Even with parallel horns, electrode skidding may occur because of the flexing of the mechanical components. In this case, electrode skidding can be reduced by changing to more rigid electrode holders, adjusting the position of the electrodes, or providing support to the lower horn.

On alternating current (ac) machines in which inductive reactance is a factor in electrical efficiency, it is prudent to pay attention to the size of the electrical secondary circuit area and the placement of magnetic materials, since these will affect the current available to make the weld.

Press-Type Machines

Press-type machines are used for a wide range of standard projection welding operations and for many spot welding applications. The movable welding head of a press-type machine travels in a straight line in guide bearings, or ways. The bearings must be of sufficient proportions to withstand any eccentric loading on the welding head.

Standard press-type welding machines, as defined by the Resistance Welder Manufacturers' Association (RWMA),[3] are available with capacities of 5 kVA to 500 kVA and throat depths up to 137 cm (54 in.). A standard press-type machine is shown in Figure 4.2. The

3. The Resistance Welder Manufacturers' Association (RMWA) ceased operations as an independent organization in 2005 and became the Resistance Welding Manufacturing Alliance, a standing committee of the American Welding Society. RWMA documents are available through the American Welding Society, 550 N.W. LeJeune Road, Miami, FL 33126, www.aws.org and www.rwma.org.

Photograph courtesy of Taylor-Winfield Corporation

Figure 4.2—Press-Type Welding Machine with T-Slotted Top and Bottom Platens

smaller versions, known as *bench welders*, are widely used for welding components in electronics, jewelry and medical devices. Because bench welders are used to weld very small workpieces made from a wide range of materials, these machines commonly incorporate inverter and capacitor-discharge controls that provide fine control of the weld current.

Mechanical Design. The machine frame of a press-type welding machine houses the power source and tap switch. The weld control, filter, lubricator, regulator, accumulator, surge tank, and coolant manifold usually are mounted on the outside of the machine frame. The machine frame must be strong enough to withstand the force delivered by the actuator to the workpiece.

Force may be applied by the complete range of welding actuators. A general rule is that machines rated above 500 kVA incorporate hydraulic actuators and for those below this value, pneumatic operation is preferred. The kVA rating is not the only factor to consider when determining if the machine should be equipped for air operation, since the high kVA rated system may be supplied to meet a high duty-cycle requirement and not a high kVA demand.

In large systems it is necessary to consider the effect of system dynamics on the weld. Large rams or a slow actuator may prevent the machine from keeping the electrodes in proper contact with the workpiece during the welding process. This is especially true in projection welding and upset welding, in which considerable workpiece displacement may occur. In such cases, it may be necessary to incorporate low-mass springs or a diaphragm air cylinder between the ram and the electrode to provide improved dynamic response. These devices also may improve the welding of thin sections or soft materials, such as aluminum, because they will reduce the likelihood that excessive mechanical force will damage the workpieces when the electrodes close.

If the slow response of machine is related to the capacity of the power source, an accumulator may be introduced into the system. Either hydraulic or pneumatic systems can be fitted with an accumulator that holds a volume of a pre-charged fluid that is quickly available to the actuator. This volume of fluid will support the system demand if the supply lines cannot maintain the instantaneous flow demands. The pneumatic version of the precharged air accumulator is commonly referred to as a *surge tank*.

Application Considerations. Standard press-type welding machines are designed and built on the module principle for economy in manufacturing. The same frame size is used with two or three transformers of different kVA ratings and with a range of throat depths.

Projection welding machines have platens on which electrode holders, fixtures, and other tooling devices are mounted. In most cases, the platens are a direct part of the secondary circuit. Platens have flat surfaces and usually have standard T-slots on which to bolt attachments.

Machines designed for spot welding are equipped with horns and electrode holders. A combination unit shown in Figure 4.3 has platens and horns, with one throat depth when used as a projection welding machine and a greater throat depth when used as a spot welding machine.

The platens, ram, and actuator are all on the same centerline. The distance from the centerline to the face of the secondary plate is the depth of the projection-welding throat. On standard machines with horns, the spot welding electrodes are located 15 cm (6 in.) or more from the face of the secondary plate. This is true whether or not platens are used.

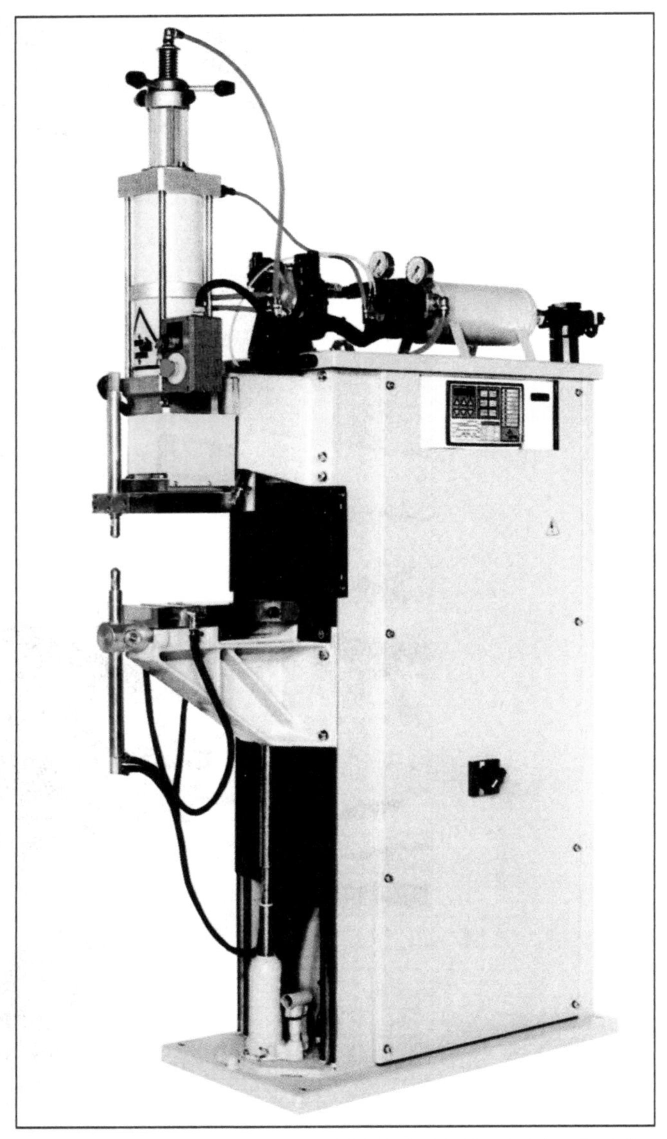

Photograph courtesy of Lors Machinery, Inc.

Figure 4.3—Combination Projection Welding and Spot Welding Machine with Platens, Horns, and Electrode Holders

On projection and combination machines, the lower platen is mounted to the knee (or it may be a part of the knee) and can be adjusted vertically. The knee may be made of copper, bronze, steel, or cast iron.

Multiple-Spot Welding Machine

A multiple-spot welding machine, generally referred to as a *multi-spot welding machine*, is designed to weld

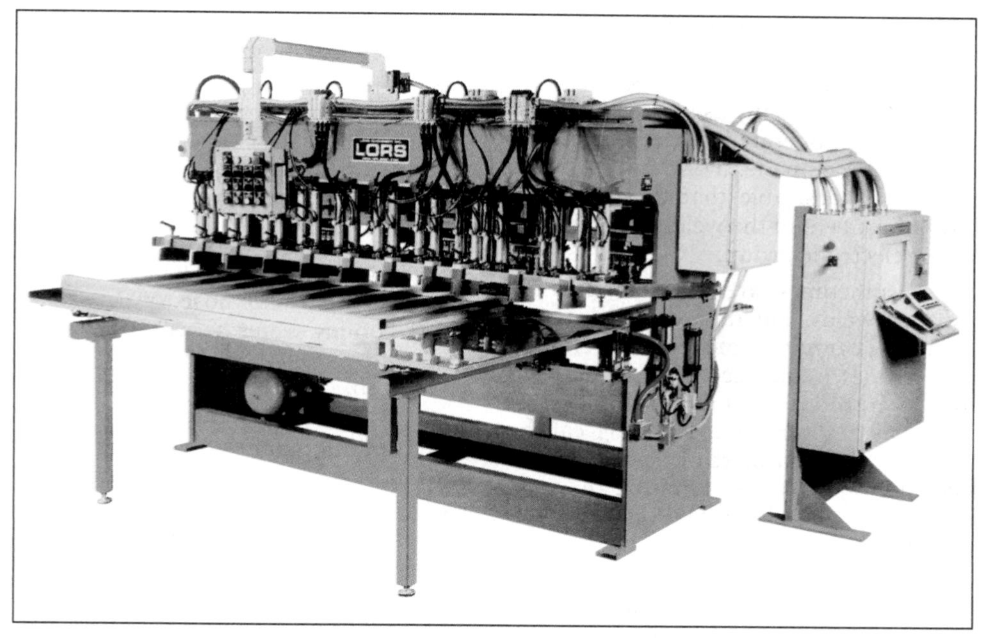

Photograph courtesy of Lors Machinery, Inc.

Figure 4.4—Multiple-Spot Welding Machine

a specific assembly. The machine shown in Figure 4.4 is configured to weld a series of stiffeners to elevator door panels. A multiple-spot welding machine should be considered when production requirements are high and when the spot or projection welds on an assembly are so numerous that welding with a single-point machine is not economical. The principal advantages of these machines are the following:

1. A number of welds can be made at the same time,
2. Workpiece dimensions and weld locations are consistent after welding, and
3. The equipment is reliable and easy to maintain.

Mechanical Design. Multiple-spot welding machines range from generic models to customized machines that are adapted to workpiece or application requirements. Generally they consist of an upper and lower platen, one of which may be moveable, to which in-line welding guns and electrodes are mounted. In-line (I-type) guns consist of an actuator with a rod-mounted electrode holder, an electrode, and a connector for a flexible conductor, such as an air-cooled cable or a laminated shunt. The connector typically acts as a clamp to secure the holder to the actuator. A number of welding transformers, usually of dual secondary circuit design, are connected by various methods, depending on factors such as workpiece geometry, number of welds, and cycle time requirement. Electrode force is applied directly to the electrodes through the electrode holder by welding actuators of various types. The upper electrodes typically are moveable and the workpieces rest on the lower stationary electrodes.

Application Considerations. Many designs of multiple-spot machines are available because of their broad usage and special requirements. The machines may be designed as welding stations in large automated high-production, assembly lines, or they may be used independently. Independently used machines may be loaded and unloaded either manually or automatically. They are commonly interfaced with robots for both welding and material handling.

If cycle time permits, gun sequencing and cascade sequencing are two control methodologies that can be can be employed to reduce the number of weld controls and the instantaneous demand on the electrical supply. In gun sequencing, one weld control is cycled repeatedly as each electrode or set of electrodes advances in turn to complete the secondary circuit. The workpiece and fixturing must be properly secured so that the shock of electrode contact and magnetic field in the secondary circuit of the welding machine do not dislodge components away from the weld location.

With the addition of welding contactors, a single control can be used to do the welding in a cascade sequence. In this configuration, all of the electrodes are closed at the same time and as many as six welding contactors per weld control are operated one at a time until all of the welds have been made. For both of these sequencing methods it is desirable that the control is configured to allow adjustment of the welding parameters for each electrode or electrode group.

Offset electrodes sometimes are considered when making closely spaced welds, but they should be used only when the actuators cannot be moved close enough together. Offset electrodes induce eccentric loading on the actuator and may introduce a skidding component if the electrode force is high enough. Skidding can be minimized by limiting the actuator stroke or otherwise reducing the electrode impact force. Limiting the electrode opening is also beneficial for visual confirmation of electrode alignment.

If multiple lower electrodes are fixed, it also will be necessary to consider how their height will be maintained. Common strategies are to use a large contact face on the electrode to minimize wear, use materials with superior performance, and electrode configurations that have a consistent height, or provide for independent adjustment of electrode height. For these reasons, fixed-height electrodes typically use threaded attachment instead of tapers.

A transformer with two insulated secondary circuits is commonly used to power two separate welding circuits. Because of different impedances in each secondary circuit, there can be significant variance in the output current of each secondary, even when great care is taken to provide duplicate secondary loops. If this is a problem, the sequencing previously described can be used. A current-balancing transformer also may help to achieve equivalent welding currents in each circuit.

Self-equalizing gun designs often are used when standard electrodes are needed on both sides of the weld to obtain good heat balance, or when variations in workpieces will not permit consistent contact with a stationary lower electrode. The same basic welding gun is used for these designs but is mounted on a special C-frame similar to that for a portable spot welding gun. The entire assembly can move as electrode force is applied at the weld locations.

Portable Machines

In some applications it is more convenient to manipulate the welding machine instead of the workpiece. Examples are very large workpieces such as rail cars, and complicated automotive parts that require extensive tooling and welding on many planes. There are few truly standard portable welding machines, but manual resistance welding guns have a basic architecture and quite a few common components.

Similar to other machines in this class, manual spot welding machines consist of four basic components:

1. A manual welding gun or tool;
2. A welding transformer and, in some cases, a rectifier;
3. An electrical contactor and sequence timer; and
4. A cable and hose system to carry power and cooling water between the stationary system elements and the manually manipulated welding gun.

In addition, some sort of superstructure is required to support the weight of the portable unit and give it an acceptable range of motion relative to the workpiece. Devices are added to aide in manipulation such as a balancer to reduce the effort required to move the unit, handgrips to safely position the welding gun, and some form of initiating switch. Safety devices are required to isolate the electrical and water supplies in the event of fault.

There are two basic configurations of manual welding guns. One is the scissor type, which is analogous to a rocker-arm spot welding machine. The other is the C type, which is analogous to the press-type welding machine.

Manual welding guns may incorporate all types of actuators to provide the necessary electrode force. Design consideration must be given to the motion of the actuator during the welding sequence, because a large motion of the actuator or a change in the center of gravity can make manual manipulation cumbersome.

The welding transformer may be a remotely mounted portable gun transformer or an integral (transgun) transformer. Portable gun transformers are connected to the manual welding gun with a dual conductor cable. The introduction of this cable into the secondary circuit has the following three fundamental effects:

1. It increases the total impedance on an ac system. On a dc system, the cable increases the resistance. Therefore, whether ac or dc systems are used, considerably higher secondary voltage is required in a portable-gun welding machine to produce a given secondary current than is required on a stationary welding machine. Portable gun transformers typically have open-circuit secondary voltages that are 2 to 4 times greater than transformers for stationary machines;
2. It increases the resistance component of the impedance so that the power factor is much higher than in a stationary welding machine; and

3. It minimizes the effect of workpiece impedance on the welding current and the power factor of the load. Electrical calculations for maximum welding amperes, kVA demand, secondary voltage requirements, and power factor are similar to other welding machines.

When the transformer is mounted integrally to the resistance welding gun, the assembly is called a *transgun*. An example of a manual transgun is shown in Figure 4.5. The advantage of a transgun over a manual welding gun with a remote portable gun transformer is the elimination of the voltage drop of the cable. Transguns also tend to be ergonomically better, since the gun is not tethered by the large and cumbersome dual-conductor cable, and the transformer adds no additional load if long lateral moves exceed the limits of the cable movement. Transguns have power factors that may exceed 85%. The workpiece, however, is the primary resistance component of the secondary circuit and must be taken into account when calculating impedances to size the transformer. When a transgun is applied as a manual welding gun, additional safety protection for the operator is required because of the presence of line voltage in the cable and transformer. RWMA *Bulletin 5*, Section 5-015.68.04 describes special considerations for portable (manual) transguns.[4]

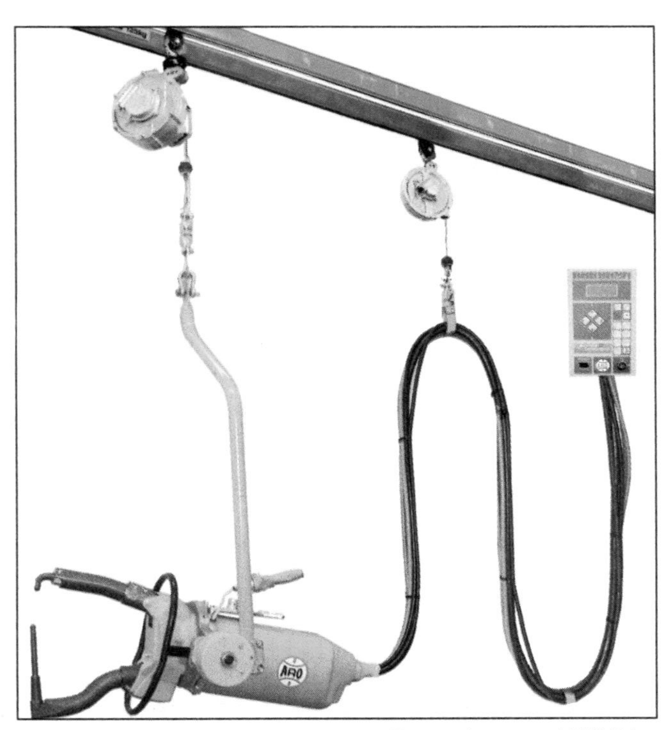

Photograph courtesy of SAVAIR Inc.

Figure 4.5—Manual Transgun

Custom Machines

When high production volume or large numbers of welds are required, or the assembly is too cumbersome to weld with a standard machine, a custom machine should be considered. The multiple-spot welding machines previously described may range from standard configurations to custom machines. The machines described in this section are uniquely designed to fabricate a specific assembly or range of products. The highest level of versatility can be achieved with robotic resistance welding equipment of the type shown in Figure 4.6. The sequence and location of welds, and sometimes the welding gun, can be selected to suit the requirements of a specific assembly. Figure 4.6 shows a robot cell (under construction) and the robot, which will manipulate a large servo welding gun that attaches components to an inverted floor pan assembly for an automobile.

Custom welding machines are configured to optimally suit the production task and may incorporate other processes to facilitate the production of complex assemblies with minimal or no operator intervention. Custom machines have the following advantages:

1. A number of welds can be made at the same time. These welds can be of different types, in different weld planes, and executed in a specific sequence;
2. Workpiece dimensions and weld locations are consistent after welding. Gauging or tooling can be incorporated to prepare the workpiece for welding or subsequent operations;
3. Automation can be incorporated to increase productivity; and
4. Machines can be stand-alone or incorporated into a large, high-production, automated assembly line. They can be interfaced with robots for both welding and material handling.

When designing the machine for a particular weldment, the following factors must be considered:

1. Shape, size, and complexity of the workpiece or weldment;
2. Dimensional consistency of the workpieces;
3. Composition and thickness of the workpieces;
4. Required appearance of the weld;
5. Production rate requirements;

4. Resistance Welding Manufacturing Alliance (RWMA), *Bulletin 5*. Miami: RWMA/American Welding Society.

Photograph courtesy of CenterLine (Windsor) Ltd.

Figure 4.6—Robotic Resistance Welding Machine

6. Availability of auxiliary equipment (presses, frames, and dial tables);
7. Changeover time for different assemblies;
8. Additional processes that must be incorporated into the machine, such as date and part number stamps, other welding processes, piercing or forming operations, and assembly and validation processes; and
9. Economic factors, including initial equipment cost, operating, labor, and maintenance costs.

SEAM WELDING MACHINES

A seam welding machine is similar in principle to a spot welding machine, except that circular electrodes in wheel form are used instead of the electrode tips used in spot welding. The circular electrodes are commonly called *welding wheels*. Welds can be performed in a continuous mode to create gas-tight welds, or intermittently to perform a line of spot welds, known as *roll spot welding*.

The essential elements of a standard seam welding machine are similar to the press-type machine shown in Figure 4.2, except that the horns are replaced by current-carrying bearings and circular electrodes. A drive assembly for one or both of the wheels is commonly provided. The other difference from a standard press-type welding machine is that the transformer normally is designed for heavier duty because the continuous nature of seam welding is more demanding than spot welding.

Most of the seam welding of thin-gauge metal is done with continuous-drive systems. Intermittent-drive systems must be used for thick-gauge metals to maintain electrode force on the weld nugget as it solidifies. The thickness range that can be welded with each drive system depends on the type of metal being joined.

The majority of continuous-drive mechanisms use a constant-speed, ac electric motor with a variable-speed drive. The speed range depends on the drive design and the electrode diameter. Good flexibility also may be obtained with a constant-torque, variable-speed dc drive or servo motor.

Seam Welding Machine Configurations

The three general configurations of seam welding machines are circumferential, longitudinal, and universal.

In the circumferential type shown in Figure 4.7, the axis of rotation of each electrode is perpendicular to the front of the machine. This type is used for long seams in flat work and for circumferential welds, such as welding the heads into containers.

In longitudinal seam welding machines, the axis of rotation of the electrodes is parallel to the front of the machine. This type is used for applications such as the welding of side seams in cylindrical containers and short seams in flat workpieces. Figure 4.8 shows a typical longitudinal seam welding machine.

In universal machines, electrodes may be set in either the circular or longitudinal position. This is accomplished with a swivel-type upper head in which the electrode and its bearing can be rotated 90° around a vertical axis. Two interchangeable lower arms are used, one for circular operation and the other for longitudinal operation.

Electrode Drive Mechanisms

The two types of electrode drive systems generally used in standard seam welding machines are the knurl or friction roller and the gear-drive system.

Knurl or Friction Roller Drive. In a knurl or friction-roller type welding machine, either the upper or the lower electrode, or both, is driven by a friction wheel on the periphery of the electrode. If the friction rolls have knurled teeth, they are known as *knurls* or *knurl*

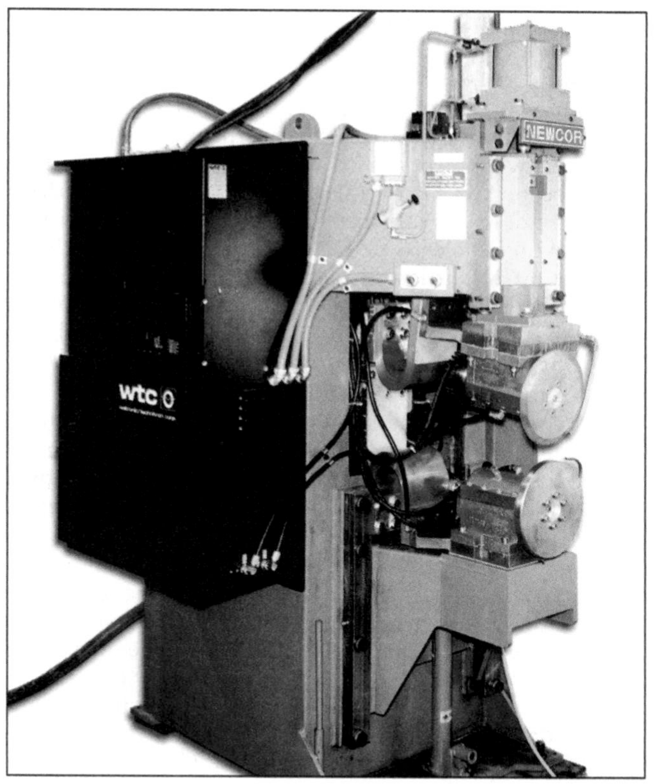

Photograph courtesy of Newcor Bay City Division

Figure 4.7—Circumferential Seam Welding Machine

Photograph courtesy of T. J. Snow Co., Inc

Figure 4.8—Longitudinal Seam Welding Machine

drives. A knurl or friction roller drive is designed to maintain a constant welding speed as the diameter of the electrode decreases from wear.

A knurl drive commonly is used on machines for the seam welding of galvanized steel, terne plate, scaly stock, or other workpieces with surface contaminants that might be picked up by the electrodes. The knurl drive wheel tends to break up the transferred contaminant material on the electrode face that can interfere with the welding process if left in place. When the nature of the work permits, both electrodes should be knurl-driven to provide a more positive drive and lessen the possibility of skidding.

A knurl drive also may function to control the shape of the contact face of the wheel electrode. This can be accomplished by using knurls designed with a radius in the wheel contact area, or by using a flat knurl designed with side cutters that constantly trim the wheel contact face to maintain a specific width.

Gear Drive. In a gear-driven machine, the electrode shaft is driven by a gear train powered by a variable-speed drive. Only one electrode should be driven to avoid skidding, otherwise, a differential gear box is necessary. This type of drive generally is less desirable than a knurl drive because the welding speed decreases as the electrode wears and decreases in size. This can be overcome by gradually increasing the drive speed.

The most important applications for a gear-driven machine are the welding of aluminum and magnesium and the fabrication of small-diameter containers. Standard seam welding machines are designed with a specific minimum distance between electrode centers for each machine size. If one of the electrodes must be small to fit inside a container, the other must be correspondingly larger to maintain the required center distance. If the ratio of the two electrode diameters exceeds about 2 to 1, the smaller electrode should be driven and the large one should idle to minimize electrode skidding.

Current-Carrying Bearings

Current-carrying bearings are required in seam welding machines to carry welding current from stationary secondary conductors to the circular electrodes. These bearings also must support the welding force in most cases. Current-carrying bearings generally are divided into two types, straddle bearings and silver contact heads.

The straddle bearing consists of a copper-alloy shaft with a hub in the center. The circular electrode normally is bolted to the hub. The shaft is supported on either side of the hub by plain bearings normally made from bronze and are lubricated with grease. The welding force creates the electrical connection through the lubricant between the rotating shaft and the fixed bearings. Straddle bearings typically are used on longitudinal seam welders such as container welders and coil joiners or strip welders.

Silver contact heads are a sealed assembly that supports the circular electrodes at the end of a copper alloy shaft. A pair of bearings mounted in a single housing supports the shaft. The housing also incorporates a series of silver contact shoes that are forced into contact with the shaft and the housing by spring force. These heads normally are filled with castor oil for lubrication. The castor oil can be pumped through the head, which allows the oil to be filtered and cooled. Silver contact heads typically are used on circular seam welding machines such as those designed to weld gas tanks and roll-formed bumpers. Silver contact heads have a much lower voltage drop and a longer service life than straddle bearings, but also are much more expensive.

Cooling Methods

Seam welding machines sometimes involve unique cooling schemes to achieve proper cooling of the machine, the electrodes, the current-carrying bearings, and other components of the secondary circuit. Rising temperatures in any of these components will cause an increase in electrical resistance in the secondary circuit, resulting in lower welding current and increased system degradation.

Cooling the workpieces also is important in most applications to minimize warping caused by localized heating. Indirect cooling of the electrode by way of a water-cooled hub or arbor is the preferred method. Cooling with water or cold gas jets spraying on both the workpiece and the welding electrodes also is quite satisfactory for most common workpiece materials. Welding under water may be done in special cases. Another method of cooling the weldment is to provide a water mist that removes heat by evaporation. A mist can be produced by mixing air and water in proper proportions and delivered through a nozzle.

Special-Purpose Seam Welding Machines

Special-purpose seam welding machines are available for specific applications and generally can be grouped in three types: traveling electrode, traveling fixture, and portable machines.

Traveling-Electrode Machine. When the traveling-electrode machine is used, the seam to be welded is clamped or otherwise positioned on a fixed mandrel or shoe of some type and the ram and wheel electrodes are moved along the seam. The mandrel or shoe is the lower electrode. The ram and electrode are moved by a pneumatic or hydraulic cylinder or by a motor-driven screw. Sometimes two upper electrodes operating in series are used side by side or in tandem. Figure 4.9

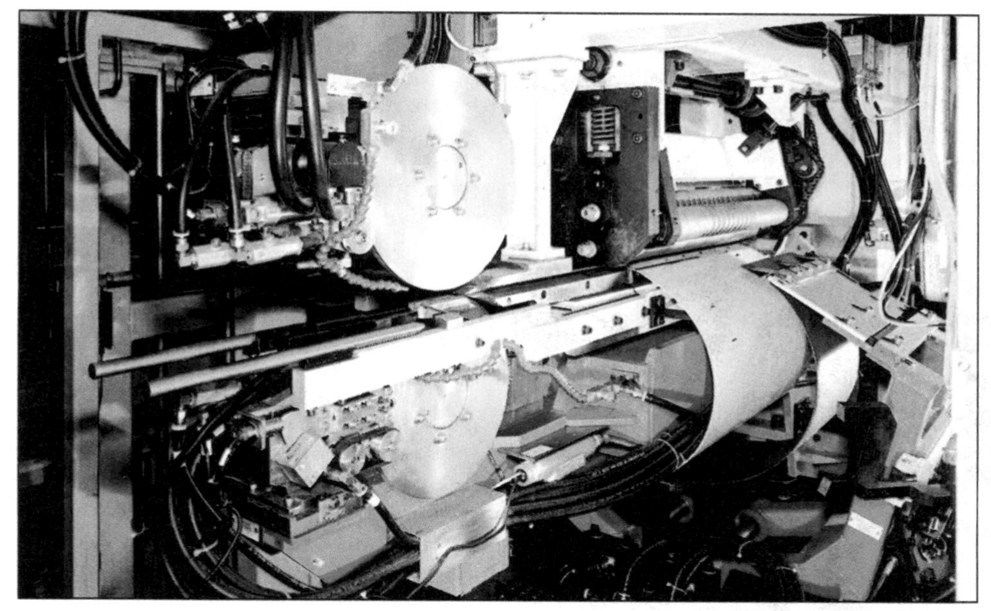

Photograph courtesy of RWC Inc.

Figure 4.9—Traveling-Electrode Seam Welding Machine

shows a traveling-electrode machine designed to weld clothes washer tub blanks.

Traveling Fixture. In the traveling-fixture machine, the one electrode remains in a fixed position. The fixture and workpieces are moved over the electrode by a suitable drive system. An example of a custom machine of this type is shown in Figure 4.10. This machine was constructed to weld steel barrel blanks.

Portable. Portable seam welding machines (guns) may be used for workpieces that are too large and bulky to be fed through a standard machine. The gun consists of a pair of motor-driven wheel electrodes and bearings, used with an air cylinder and associated mechanisms for applying the electrode force. Welding current is supplied in the same manner as for portable spot welding machines. A variable-speed dc drive may be used when a wide range of welding speeds is desirable. The motor and speed reducer are mounted directly on the welding gun frame.

FLASH WELDING AND UPSET WELDING MACHINES

Flash welding uses expulsion and arcing to clean and prepare the faying surfaces, followed with an upsetting action aided by pressure to form the weld. Upset welding does not employ expulsion and arcing, but achieves coalescence only by heat generated by the resistance of the workpiece to the welding current and the application of pressure. These processes are discussed in Chapter 3.

Flash and upset welding machines are similar in construction; the major differences are the motion of the movable platen during welding and the mechanisms used to impart the motion. The most visible difference is the expulsion and arcing of the flash weld, whereas the upset weld has no expulsion or arcing.

The process differences between flash welding and upset welding will make one process more appropriate than the other for a given application. Flash welding generally is preferred for the end-to-end joining of components of equal cross section and when the faying surfaces are very rough or contaminated. Upset welding normally is used to weld wire, rod, or bar stock of small cross section and to make leak-proof joints in pipe or tubing. Typically, the faying surfaces for upset welding must have a better finish than workpieces for flash welding. Flash welding machines usually have the capacity to weld workpieces with much larger cross sections than upset welding machines. The flash welding process has a notably longer weld time than upset welding.

Flash Welding Machines

A typical flash welding machine consists of a main frame, a stationary and movable platen (or two around-

Photograph courtesy of Newcor Bay City Division

Figure 4.10—Custom Traveling-Fixture Seam Welding Machine

a-centerline platens), synchronized movable platens, clamping mechanisms and fixtures, a transformer, a transformer tap switch, electrical controls, and mechanisms for flashing and upsetting. One platen is fixed or has a limited amount of adjustment for electrode and workpiece alignment. The other platen is mounted on slide ways for movement. This moveable platen is connected to an actuator for the flashing and upsetting operations. Both platens usually are made of cast or fabricated steel, although some small welding machines may have cast bronze, cast iron, or copper platens. The platens are connected to the secondary circuit of the transformer. Electrodes that hold the workpieces and conduct the welding current to them are mounted on the platens. The transformer and tap switch generally are located within or immediately behind the frame with short, heavy-duty copper shunts or a bus bar connecting to the platens.

The depth of the frame, and consequently the width of the platens, depend on the size of the workpieces and on the design of the clamping mechanism. Upsetting force should be aligned as nearly as possible with the geometric center of the workpieces to minimize deflection of the workpieces and the machine frame. Dual flashing and upsetting cylinders or cams sometimes are used with wide platens to provide uniform loading or clearance for long pieces to extend over the mechanism.

Transformers and Controls. Flash welding transformers essentially are the same as those used for other types of single-phase resistance welding machines. A transformer tap switch in the primary circuit normally is used to adjust flashing voltage. The primary power to the transformer is switched with an electronic contactor. Reduced power can be applied for preheating or post-weld heat treatment within the machine.

The transformer tap switch should never be used to control the secondary voltage during flashing; only the weld contactor should be used. If the tap switch is adjusted during flashing, the secondary voltage would become uncontrolled. This would result in inconsistent times of no secondary voltage followed by an instantaneous voltage, which could be quite high and could cause deep craters and entrapped oxides in the weld zone. With certain equipment, the flash voltage can be changed (reduced) during the final flashing phase to create a finer flash with a smaller arc-gap (ionized gas protection) and less surface cratering. This normally is accomplished by switching transformer taps by means of a paired and synchronized silicon-controlled rectifier (SCR) switch arrangement. With ignitron contactors, auxiliary load resistors must be connected in parallel with the transformer primary for proper operation of the ignitrons.

Flashing and Upsetting Mechanisms. A flash welding machine operates with one workpiece on the moveable platen electrode and the other workpiece (or the workpiece end when welding a ring) on the stationary platen electrode. The movement of the workpiece must be carefully controlled to produce consistently sound welds. The faying surfaces are brought into light contact and voltage is applied to initiate arcing. After the appropriate flashing time, the workpieces are rapidly brought into contact and upset. The upsetting action must be accurately synchronized with the termination of flashing.

The type of mechanism used for flashing and upsetting depends on the size of the welding machine and the requirements of the application. Some mechanisms permit the faying surfaces to be butted together under pressure and then preheated. After the appropriate temperature is reached, the pieces are separated and then the flashing and upsetting sequence is initiated. The movable platen may be actuated by one of three mechanisms: a motor-driven cam with integrated upset block, a motor-driven cam and upset cylinder (pneumatic or hydraulic), or with a servo-controlled hydraulic cylinder.

Motor-operated machines use an ac or dc motor with a variable-speed drive, which in turn drives a rotary or wedge-shaped cam. The cam is designed to produce a specific flashing pattern. It may contain an insert block to upset the joint at the end of flashing. The speed of the cam determines the flashing time. The platen may be moved directly by the cam or by a lever system. The motor may operate intermittently or continuously for each welding cycle. With continuous operation, the drive is engaged by a clutch on the output shaft of the speed reducer. The motor speed may be electronically controlled to produce a specific flashing pattern. A typical motor-operated flash welding machine is shown in Figure 4.11.

A motor-driven flashing cam may be used in combination with a pneumatic or hydraulic upsetting mechanism, particularly on larger machines. This combination of movements provides adjustment of upset speed, distance, and force independently of the flashing pattern. Current is synchronized with the motion of the platen by limit switches or electronic sequence controls.

Large and medium flash welding machines use hydraulically operated flashing and upsetting mechanisms. These machines are capable of applying high upsetting forces for large sections. They are accurate in operation and are readily set up for a wide range of weld requirements. A servo system is used to control the platen motion for flashing and upsetting. The servo system may be actuated by a pilot cam mechanism or by an electrical signal generated from the secondary voltage or the primary current. The choice of the operating mode depends on the application. The control may be programmed to include preheat and post heat treatments. An accumulator generally is required to provide an adequate volume of hydraulic fluid from the pumping unit during upsetting.

Two types of electro-hydraulic servo systems generally are used. In one design, the servo valve meters the fluid directly to the hydraulic cylinder for position control. In the other design, the servo valve meters the fluid to a small control cylinder that operates a follower valve on a separate hydraulic system. The first design is

Photograph courtesy of Hess Engineering, Inc.

Figure 4.11—Automatic Motor-Operated Flash Welding Machine

simple and straightforward; the second system is more complex, but has two distinct advantages. First, it has two separate hydraulic circuits for improved valve life. Second, the speed of response is fast and the control of the platen position is accurate.

Clamping Mechanisms and Fixtures. Several designs of clamping mechanisms are available to accommodate different types of workpieces. Clamping mechanisms generally are designed for operation in either the vertical or the horizontal position. In special cases, the mechanisms may be mounted in other positions.

In vertical clamping, the movement of the electrode may be in a plane perpendicular to the platen ways. The electrode may move either through a slight arc or in a straight line. If operating through an arc, a clamping arm pivots around a trunnion. This design generally is known as the *alligator* type. A machine with this clamping arrangement is shown in Figure 4.12. Clamping force may be applied by a pneumatic or hydraulic actuator operating directly or through a leverage or cam-operated mechanism. Vertical clamping is commonly used for bar stock and other compact sections.

In horizontal clamping, the motion of the electrodes is parallel to the platen and generally in a straight line, as shown in Figure 4.13. The major advantage of this type of clamping mechanism is that the welding transformer secondary can be connected to both halves of the electrodes for uniform transfer of welding current into the workpieces. This arrangement is highly desirable for welding components with large cross sections. Clamping force can be applied with one of the mechanisms described for vertical clamping.

Fixtures. Fixtures are used to support and align the workpieces for welding and to back up the workpieces to prevent slippage of the electrodes during upsetting. They can be adjusted to accommodate the cross section and length of the workpieces. The fixture must be sturdy enough to withstand the upsetting force without deflecting. When the workpieces can be supported, the clamping force on the electrodes can be limited to only the amount needed to ensure good electrical contact and maintain satisfactory joint alignment.

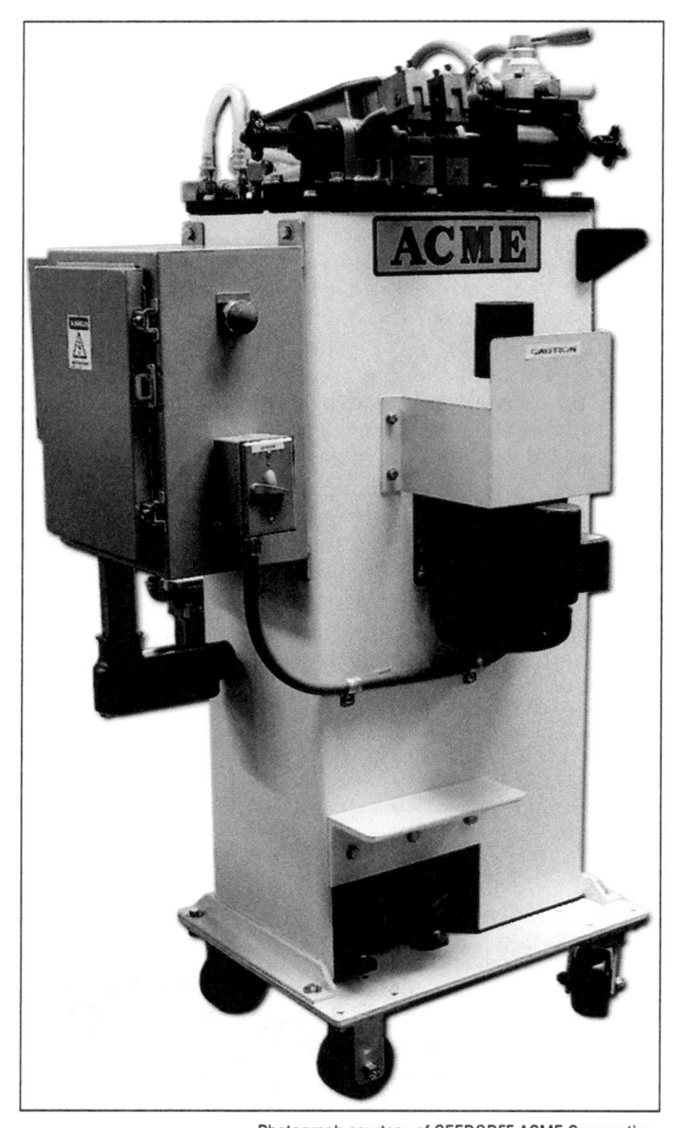

Photograph courtesy of SEEDORFF ACME Corporation.

Figure 4.12—Automatic Flash Welding Machine with Vertical "Alligator" Clamping

Upset Welding Machines

Upset welding machines are quite similar to flash welding machines in principle, except that no flashing mechanism is required.

The simplest type of upset welding machine is manually operated. In this machine, the workpieces are clamped in position in the electrodes. Force is exerted on the movable platen with a hand-operated leverage system. Welding current is applied, and when the abutting workpieces reach the proper welding temperature and force, the weld is accomplished. The current is manually shut off at the proper time during the welding cycle. The weldment is then removed from the electrodes. A limit switch or a timing device may be used to terminate the welding current automatically after the weld has achieved a predetermined length.

Photograph courtesy of Automation International Inc.

Figure 4.13—Automatic Hydraulically Operated Flash Welding Machine with Horizontal Clamping

Automatic machines may use springs, air cylinders, or servo-controlled hydraulic cylinders to provide upset force. Any of these devices can provide uniform force consistently and can be controlled in conjunction with the heat input. Springs or air cylinders are particularly adapted for welding nonferrous metals that have narrow plastic ranges.

Upset welding is used extensively for the welding of small wires, rods, and tubes in the manufacture of items such as chain links, refrigerator and stove racks, automotive seat frames, wheel rims and for joining coils of wire for further processing. The upset welding process is often selected for applications in which the upset (bulk deformation resulting from the application of pressure) is not objectionable in the context of product design. It is most easily adapted for joints with relatively small cross sections in which uniformity of welding current is not a problem or for uniform and even thickness over the cross sections.

For joints in large workpieces, attention must be paid to proper current and heat distribution. A die may be used to remove objectionable upset. Further finishing operations may be introduced to achieve desired surface finish. Automatic, servo-controlled upset welding machines have replaced most of the flash welding machines in the high-volume production of short tubular parts, such as wheel rims and containers. For this type of application, the welding machines typically are integrated into mechanical systems that subsequently trim the extruded material (flash) from the workpieces.

ELECTRODES

The consumable tools used in resistance welding are the electrodes, which may be in the form of a wheel, roll, bar, plate, clamp, chuck, or a variation of these. A welding electrode performs one or more of the following functions:

1. Conducts welding current and transmits force to the workpieces,
2. Fits the workpieces into proper alignment and in a fixed position, and
3. Removes heat from the weld metal or adjacent metal.

The electrode design always should provide sufficient mass to transmit the required welding force and current, and provide adequate cooling when needed. High-production applications sometimes involve thick sections that require special electrode designs. If it is necessary to compromise the design, it can be expected that electrode life, weld quality, the production rate, or all three may be affected. Consequently, the selection of the correct electrode material is very important for good performance.

ELECTRODE MATERIALS

Resistance welding electrode materials have been classified by the RWMA into three groups: A, copper-base alloys; B, refractory metal compositions, and C, specialty materials.[5] In addition to these materials, a number of proprietary alloys are available from electrode manufacturers. Table 4.1 lists the minimum mechanical and physical properties copper-base alloys must have to meet the various RWMA classification requirements. The specific alloy compositions are not classified, as they vary among manufacturers.

5. Standard electrode materials are described in the *Resistance Welding Manual*, published by the Resistance Welding Manufacturing Alliance, a standing committee of the American Welding Society, Miami: American Welding Society.

Table 4.1
Minimum Mechanical and Physical Properties of Copper-Base Alloys for RWMA Electrodes

Group A[a] Copper Base Alloys		Hardness Rockwell			Conductivity %IACS[b]			Yield Strength,[c] ksi (5% Ext. Under Load)			Ultimate Tensile Strength, ksi			Elongation % in 2-in. or 4-in. Diameters		
Size Range		Class			Class			Class			Class			Class		
in.	mm	1	2	3	1	2	3	1	2	3	1	2	3	1	2	3
Diameter—Round Rod Stock (Cold Worked)																
Up to 1	Up to 25	65 HRB	75 HRB	90 HRB	80%	75%	45%	45	55	90	60	65	95	13%	13%	9%
Over 1 to 2	Over 25 to 51	60 HRB	70 HRB	90 HRB	80%	75%	45%	45	55	90	55	59	92	14%	13%	9%
Over 2 to 3	Over 51 to 76	55 HRB	65 HRB	90 HRB	80%	75%	45%	45	55	90	50	55	88	15%	13%	9%
Thickness—Square, Rectangular, and Hexagonal Bar Stock (Cold Worked)																
Up to 1	Up to 25	55 HRB	70 HRB	90 HRB	80%	75%	45%	45	45	90	60	65	95	13%	13%	9%
Over 1	Over 25	50 HRB	65 HRB	90 HRB	80%	75%	45%	45	40	90	50	55	90	14%	13%	9%
Thickness—Forgings																
Up to 1	Up to 25	55 HRB	65 HRB	90 HRB[d]	80%	75%	45%	45	45	50	60	55	94	12%	13%	9%
Over 1 to 2	Over 25 to 51	50 HRB	65 HRB	90 HRB[d]	80%	75%	45%	45	45	50	50	55	90	13%	13%	9%
Over 2	Over 51	50 HRB	65 HRB	90 HRB[d]	80%	75%	45%	45	40	50	50	55	88	13%	13%	9%
Castings																
All	All	NA	55 HRB	90 HRB	NA	70%	45%	NA	20	45	NA	45	75	NA	12%	5%

[a] All materials are in fully heat treated condition unless otherwise specified. Round rod up to 1 in (25 mm) diameter is fully heat treated and cold worked.

[b] International Annealed Copper Standard (IACS). Conductivity = $\frac{1}{0.0058\ (\text{Resistivity})}$ where: Conductivity is expressed in %IACS and resistivity is in the units μΩ cm.

[c] Yield strength listed is typical—actual can be higher or lower.
[d] Hot worked and heat treated—but not cold worked.

Source: Adapted from American Welding Society (AWS), *Standard for Automotive Resistance Spot Welding Electrodes*, AWS D8.6:2005, Miami: American Welding Society, Table 2, Page 4.

Group A: Copper-Base Alloys

The copper-base alloys are divided into five classes. Class 1 alloys are general-purpose materials for resistance welding applications. These electrodes may be used for spot and seam welding for applications in which electrical and thermal conductivity is of greater importance than mechanical properties. Class 1 alloys are recommended for use in spot and seam welding electrodes for aluminum, brass, bronze, magnesium, and metal-coated steels, because this alloy class has high electrical and thermal conductivity.

Class 1 alloys are not heat treatable. Strength and hardness in these alloys is increased by cold working. Therefore, they have no advantage over unalloyed copper for castings, and are rarely used or fabricated in this form.

Class 2 alloys have higher mechanical properties but somewhat lower electrical and thermal conductivity than Class 1 alloys. Class 2 alloys have good resistance to deformation under moderately high pressures, and are the best general-purpose alloys. This alloy class is suitable for high-production spot welding and seam welding of mild steels and low-alloy steels, stainless steels, low-conductivity-copper-base alloys, and nickel alloys. These materials are used in the majority of resistance welding applications. Class 2 alloys also are suitable for shafts, clamps, fixtures, platens, gun arms, and various other current-carrying structural parts of resistance welding equipment. Class 2 alloys are heat treatable and may be used in both wrought and cast forms. Maximum mechanical properties are developed in wrought form by cold working after heat treatment.

Class 3 alloys also are heat treatable, but have higher mechanical properties and lower electrical conductivity than Class 2 alloys. The main application for spot or seam welding electrodes made of a Class 3 alloy is for welding heat-resistant alloys that retain high strength properties at elevated temperatures. The welding of these materials requires high electrode force, which in turn requires a strong Class 3 electrode alloy. Typical heat-resistant alloys are some low-alloy steels, stainless steels, and nickel-chromium-iron alloys. Class 3 alloys are especially suitable for use in many types of electrode clamps and current-carrying structural members of resistance welding machines. The properties of these alloys are similar in both the cast and wrought condition because they develop most of their mechanical attributes from heat treatment.

Class 4 alloys are age-hardenable types that develop the highest hardness and strength of the Group A copper alloys. The low conductivity of Class 4 alloys and their tendency to become brittle when heated make them unsuitable for spot or seam welding electrodes. They generally are recommended for components that have a relatively large contact area with the workpiece, which would include flash and projection welding electrodes and inserts. Other uses for these alloys are in workpiece backup devices, heavy-duty seam welding machine bearings, and other machine components for which resistance to wear and high pressure are important. Class 4 alloys are available in both cast and wrought forms. Because of their high hardness after heat treatment, they are frequently machined in the solution-annealed condition.

Class 5 alloys are available principally in the form of castings. They have high mechanical strength and moderate electrical conductivity, and are recommended for large flash welding electrodes, backing material for other electrode alloys, and many types of current-carrying structural members of resistance welding machines and fixtures.

Group B: Refractory Metal Compositions

Electrodes made from Group B materials contain a refractory metal in powder form, usually tungsten or molybdenum. These materials are made by powder metallurgy processes. Their chief attribute is resistance to deformation in service. They function well for achieving heat balance when two different electrode materials are needed to compensate for a difference in the thickness or composition of alloys being welded.

Class 10, 11, and 12 compositions are mixtures of copper and tungsten. The hardness, strength, and density increase and the electrical conductivity decreases with increasing tungsten content. These compositions are used as facings or inserts when exceptional wear resistance is required in various projection, flash, and upset welding electrodes. It is difficult to establish guidelines for the application of each grade, and commercial availability is a large factor in the material selection. The electrode design, welding equipment, opposing electrode material, and workpiece composition and condition are some of the variables that should be considered for each application.

Class 13 is commercially pure tungsten, and Class 14 is commercially pure molybdenum. They generally are considered to be the only electrode materials that will give good performance when welding nonferrous metals that have high electrical conductivity. The welding of braided copper wire to braided copper wire or copper and brass wire to copper and brass wire, or to various types of terminals are typical uses for Class 13 and 14 materials.

Group C: Other Materials

A number of unclassified copper alloys and other materials may be suitable for use in resistance welding electrodes. Aluminum-oxide dispersion-strengthened copper is grouped in RWMA Class 20 and is commonly used in the form of caps or circular electrodes. It is high-purity copper that contains small amounts of microscopic aluminum oxide uniformly distributed in the matrix. The aluminum oxide significantly strengthens the copper matrix and raises the recrystallization temperature of cold-worked material. The high recrystallization temperature of wrought material provides excellent resistance to softening and mushrooming of electrodes when the contacting surfaces are heated. This significantly contributes to long service life of the electrode.

The mechanical properties and electrical conductivity of dispersion-strengthened copper bars meet the requirements for RWMA Group A, Class 1 and Class 2 alloys, but they are not classified as such. Class 20 electrodes have shown improved resistance to sticking when welding galvanized steel with inadequate cooling or low electrode force, or both, compared to conventional Class 2 electrodes. The purchase cost penalty is offset by savings in production down-time and maintenance labor.

The suitability of a particular material for electrodes depends on the application. Although most requirements are met by materials meeting RWMA standards, there are cases in which other materials will function as well or better. For example, steel may be used for flash welding electrodes for certain aluminum applications.

SPOT WELDING ELECTRODES

A spot welding electrode has four features: the face, shank, a means of attachment, and a provision for cooling. The electrode face (the end of the electrode that contacts the workpiece) is the segment that is pointed, domed, flat, or shaped. The shank is the straight section between the face and the opposite end of the electrode; the tapered end provides the means of attaching the electrode. The provision for cooling is the hole in the electrode that allows water to be directed to the underside of the face.

Face

The face design is influenced by the composition, thickness, and geometry of the workpiece. In turn, the electrode face geometry determines the current and pressure densities in the weld zone. Figure 4.14 shows standard RWMA electrode face and taper designs.

Figure 4.14 shows two dimensions for the RWMA #4 taper because although the electrode nominally is 12.7 millimeters (mm) (1/2 in.) in diameter, in practice, the point at which the taper attains the proper length is a diameter of 12.2 mm (0.482 in.). Material with the nominal dimension of 0.482 in. commonly is used because it eliminates the requirement to do any special machining to blend the end of the taper into the outside diameter. There is no metric standard corollary to these standard RWMA electrodes. (Refer to Figure 4.20, which provides ISO nose style designators for ISO standard electrode caps.)

The most commonly used electrode contours are the radius, dome, and flat-faced styles. The flat-faced electrode is used to minimize surface marking or to maintain heat balance. The electrode face may be concentric to the axis of the electrode, as shown in Figure 4.14(A), (B), (C), (E), and (F); eccentric or offset, as shown in (D); or at some angle to the axis, as illustrated in Figure

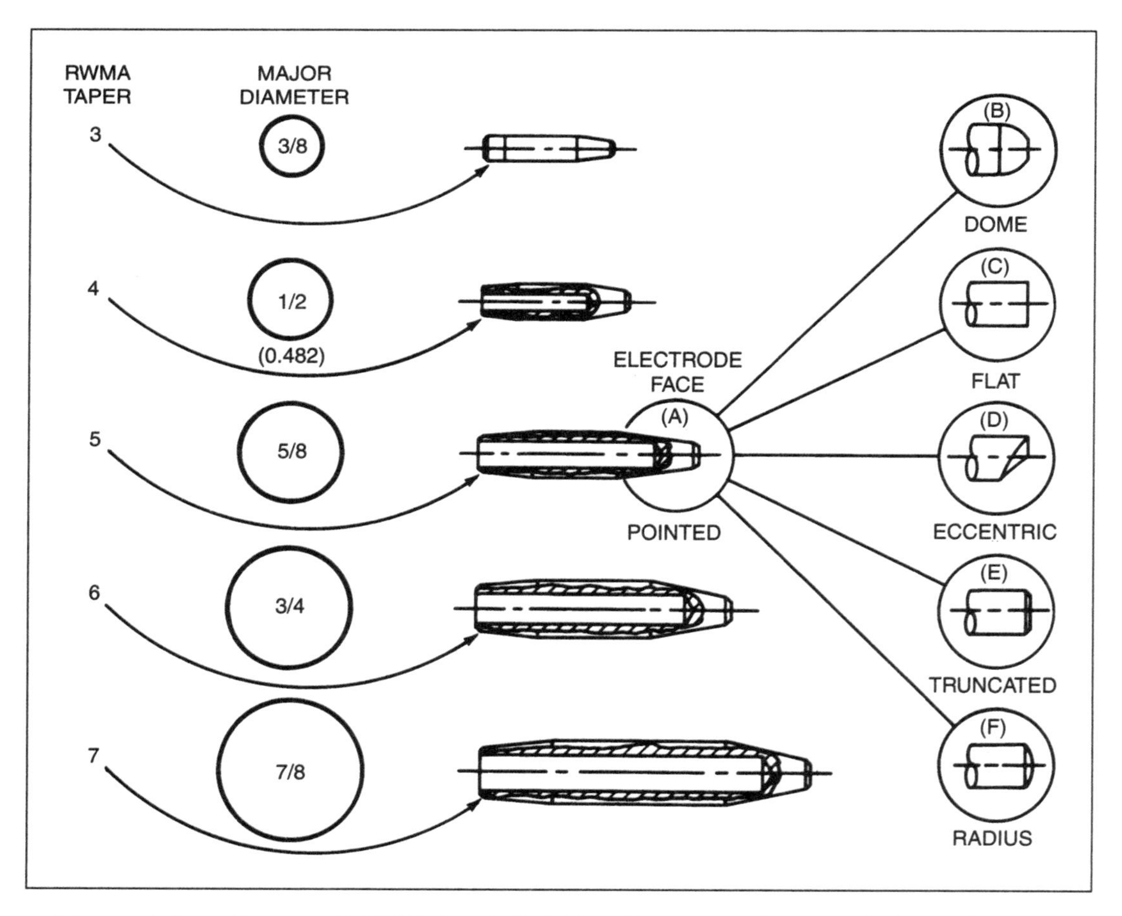

Figure 4.14—Standard RWMA Spot Welding Electrode Face and Taper Designs

4.15. So-called *offset electrodes* with eccentric faces are used to make a weld near a corner or in other less accessible areas. This is illustrated in Figure 4.16. A facing of Group B material may be brazed to a shank of an alloy from Group A to produce composite electrodes for special applications, as illustrated in Figure 4.17.

Shank

The shank of an electrode must have sufficient cross-sectional area to support the electrode force and carry the welding current. The shank may be straight or bent. Typical bent-shank electrodes are shown in Figure 4.18. (Refer to Figure 4.14 for standard shank diameters.)

To avoid incurring the expense of a new shank when the electrode face is worn out or damaged, an electrode cap and adapter combination can be used on the same shank to replace the damaged electrode face. Electrode caps with nose configurations corresponding to those illustrated in Figure 4.14 are provided with male or female tapers, as shown in Figure 4.19. The cap adapters are available in the same variety of straight and bent forms as the shanks. The taper used to secure the electrode cap to the cap adapter has mechanical load limits to avoid overloading the connection. The joint geometry imposed by the taper also will determine some

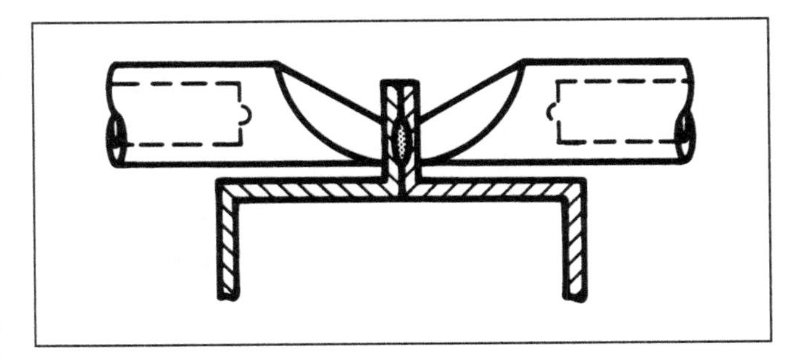

Figure 4.16—An Application of Type D Offset Spot Welding Electrodes

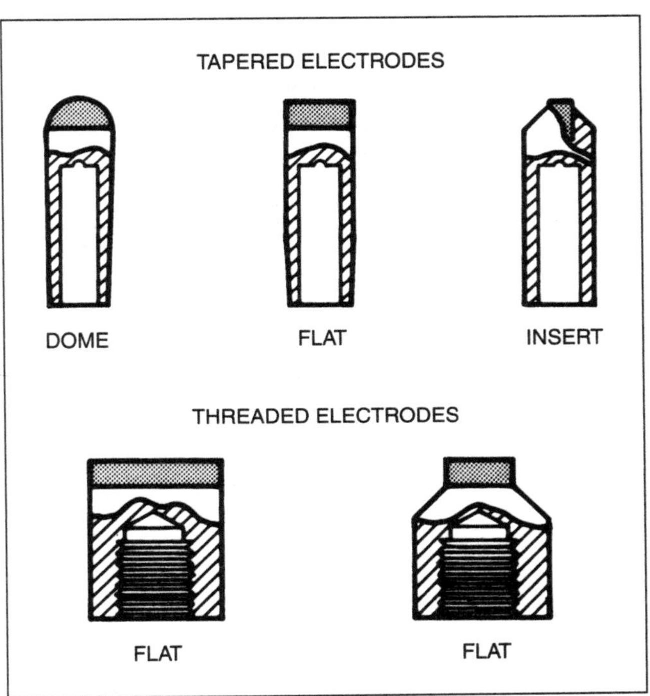

Figure 4.17—Typical Group B Electrode Faces Brazed to Group A Alloy Shanks and Threaded Electrodes

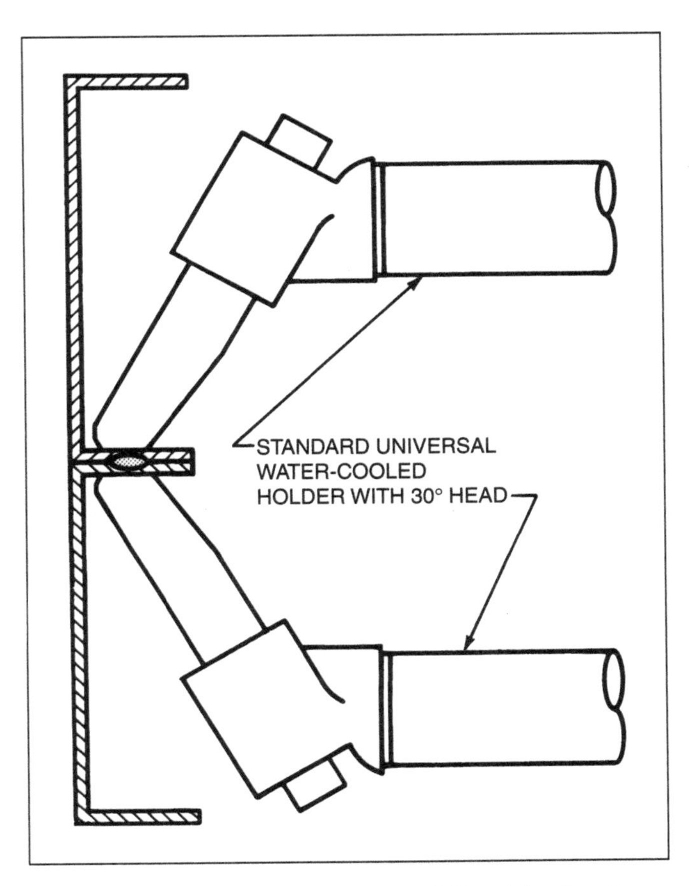

Figure 4.15—Special Spot Welding Electrodes with the Faces Angled at 30°

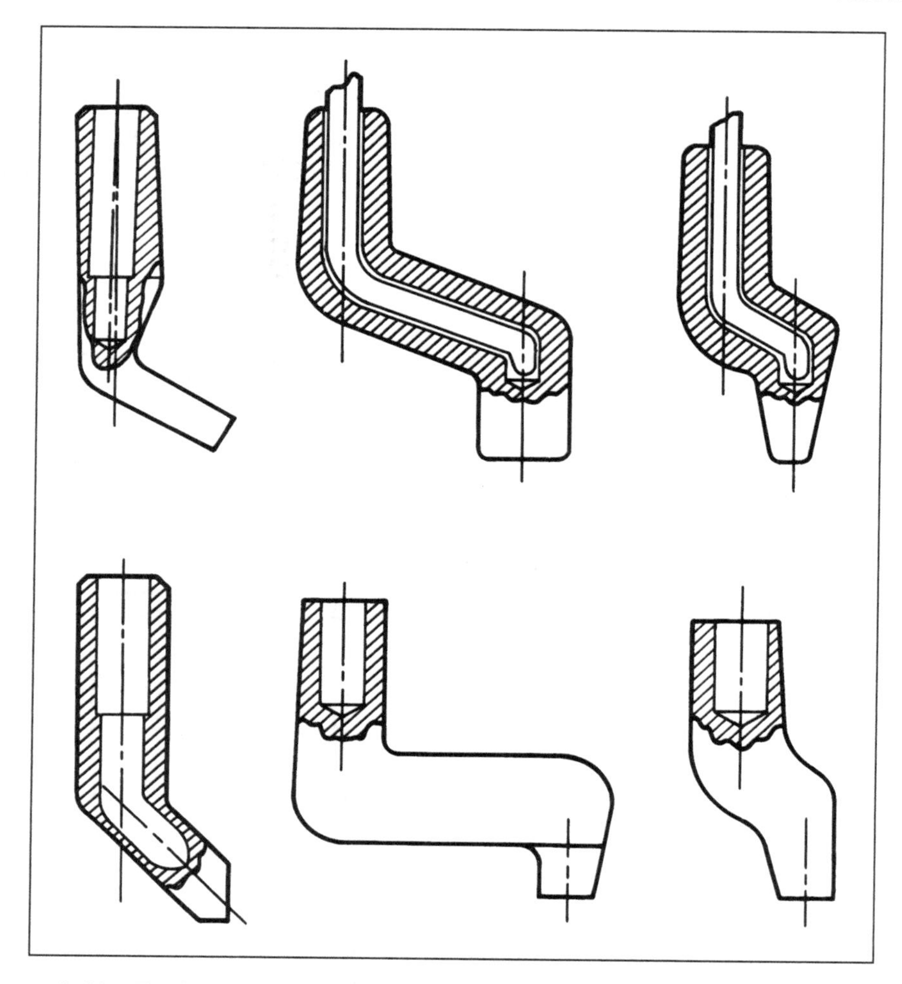

Figure 4.18—Typical Single-Bent and Double-Bent Spot Welding Electrodes

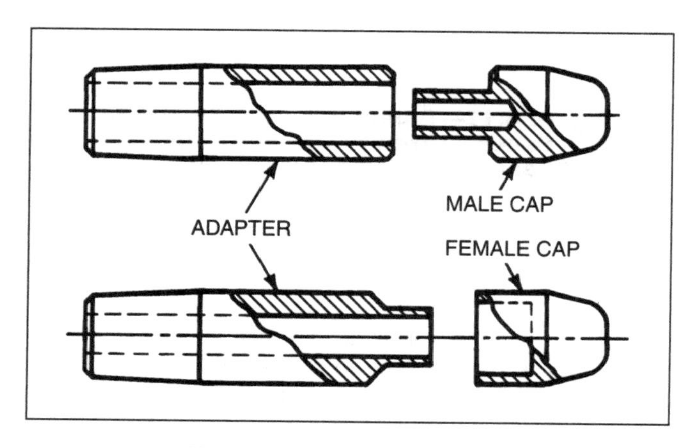

Figure 4.19—Adapter and Cap Designs of Two-Piece Spot Welding Electrodes

aspects of the configuration of the electrode cap and cap adapter combination.

The choice of using male or female electrode caps depends on each application. Male caps typically require more material to manufacture than an equivalent female cap and usually cost more. Male caps typically provide better heat transfer than equivalent size female caps and thus may be better suited for more difficult thermal conditions. Male cap tapers also resist side loading that may otherwise cause deformation of the relatively thin-walled female tapered caps, causing the latter to fall off. The choice of the taper also may relate to the exposure of the taper or to the space or length required to incorporate it in the electrode design. Sometimes the application of a variety of tapers is used as a means of error proofing, so that unique electrodes are used in the proper location.

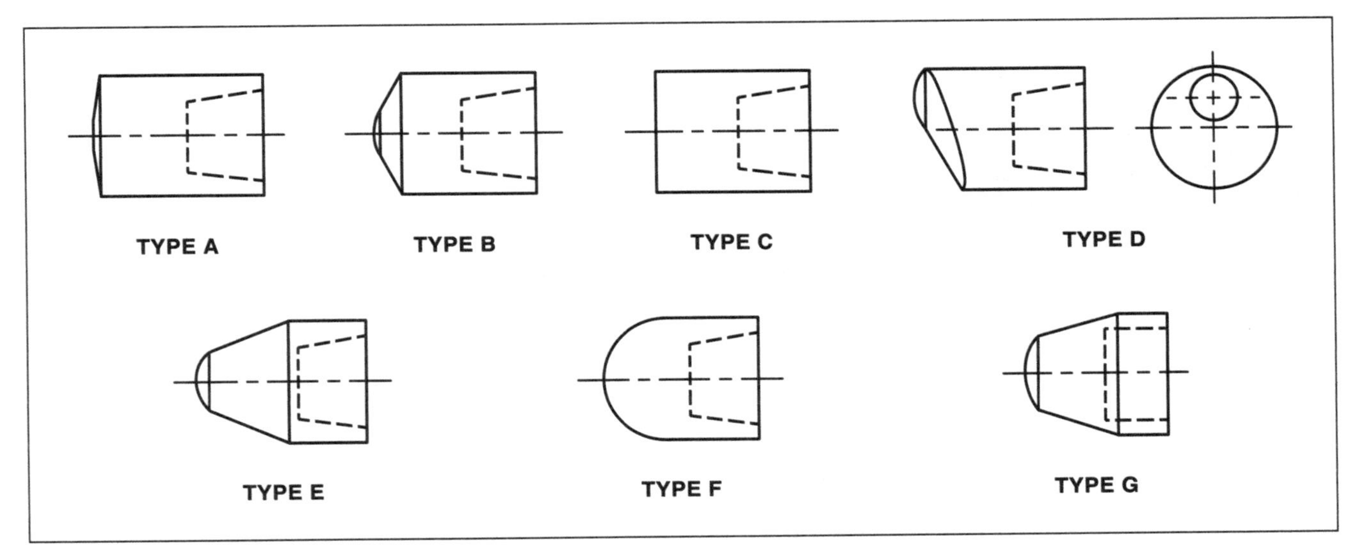

Figure 4.20—Standard ISO Female Cap Electrodes

The ISO standard female electrode caps shown in Figure 4.20 have significantly different designations. They are available with diameters of 13 mm (0.5 in.), 16 mm (0.6 in.), and 20 mm (0.8 in.). The metric electrode caps have a relatively short history and as a result, metric male caps were not fully developed, and rarely are used.

Attachment Mechanisms

A means of attachment mechanism must be provided to secure the electrode to the welding machine. Standard mounting arrangements employed to secure the shank end of the electrode in the electrode holder are tapers, threaded attachments, or an abutting surface.

RWMA and ISO standard tapers commonly are employed for electrode attachment or fixing. The tapers are available in a variety of sizes for electrodes of different diameters and lengths for different loading conditions. Because of the mechanical principles involved in engaging the male and female tapers, there is a practical limit to the load that can be applied. To illustrate these principles, recommended maximum electrode forces for the various sizes of RWMA standard male tapered electrodes are provided in Table 4.2. While not included in Table 4.2, similar specifications should be considered when applying RWMA and ISO standard female tapered electrodes caps.

Threaded attachments are used when high welding forces would make the removal of tapered electrodes difficult, or when electrode position is critical. Typical threaded electrodes are shown in Figure 4.21.

Cooling

When practical, spot welding electrodes should have an internal water-cooling passage extending close to the

Table 4.2
Recommended Maximum Electrode Force for RMWA Standard Male Tapered Spot Welding Electrodes

Taper No.	Shank Diameter	Face Diameter	Maximum Electrode Force
4	12.2 mm (0.482 in.)	4.8 mm (0.19 in.)	363 kgf (800 lbf)
5	15.9 mm (0.625 in.)	6.4 mm (0.25 in.)	680 kgf (1500 lbf)
6	19.1 mm (0.750 in.)	7.1 mm (0.28 in.)	907 kgf (2000 lbf)
7	22.2 mm (0.875 in.)	7.9 mm (0.31 in.)	1089 kgf (2400 lbf)

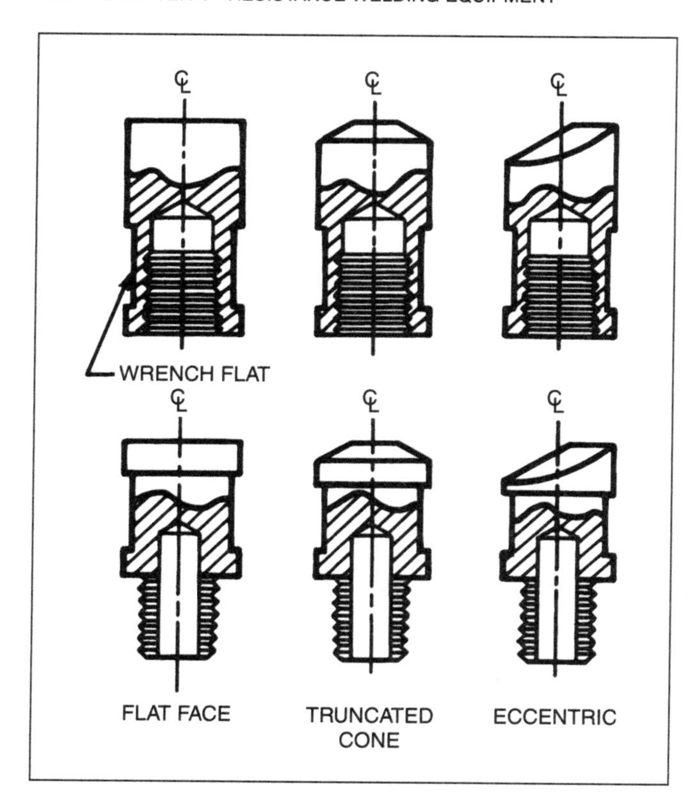

Figure 4.21—Typical Threaded Spot Welding Electrodes

welding electrode face. To prevent the generation of steam during the welding process, water should be delivered as closely as possible to the end of this passage. A water tube of copper, stainless steel, or polymer usually is provided for this purpose as either part of the electrode holder or fitted inside the holder if it would be difficult to install, for example, an installation in the field. Checking the direction of water flow is important to ensure that the cool water is directed at the inside of the tip. When internal cooling is not practical, external cooling of bent electrodes by immersion, flooding, or attached cooling coils should be considered.

Method of Electrode Manufacture

Electrodes can be made from cast, consolidated, forged, and wrought materials. As described previously, the properties of a number of the Group A copper-base alloys are improved by cold working. Some electrodes are produced from cold-worked rods, while others are bent, extruded, flattened, and shaped. Forging (in the cold or hot state) is used to accomplish the desired shape while at the same time adding grain refinement and thereby improving the electrode function. Complicated shapes can be created economically by the casting process. In some cases, the electrodes are subjected to heat treatment after cold working to achieve the desired operational performance.

Electrode Maintenance

The electrode face contour and finish, and surface cleanliness have an effect on welding performance. Electrode maintenance may vary from cleaning or machining to scheduled replacement.

The electrode face can be contaminated by interaction with the workpiece coating or base metal, and any applied lubricant, adhesive, sealer or paint. A small amount of contamination may not be harmful, but will increase surface resistance, which causes additional heating and results in electrode degradation. Intermittent cleaning may extend the service life of the electrodes.

In some cases, the concentrated heat and pressure during welding will cause "mushrooming" (plastic deformation) of the electrodes. Mushrooming is common in resistance spot welding. The result is increased contact area that causes a reduction in current and force density. Without compensation or other remedial action, the weld will get smaller or fail to form. The two responses to mushrooming are to compensate by increasing the current setting, or to reshape the electrode tip by machining. Machining to re-establish the tip profile is known as *electrode dressing;* this can be accomplished in-situ with hand tools or fixed apparatus, or the electrode can be removed from the welding machine to be processed.

In automated systems, it is usually possible to establish a reproducible pattern of electrode maintenance. The required frequency may vary widely with the welding environment, parameters, and workpiece conditions. The maintenance requirement usually is established by a periodic check of weld quality or through experience.

The electrodes in manual welding systems may be subjected to twisting, skidding and mechanical damage that will alter the electrode wear pattern or rate. It may be necessary to supplement scheduled maintenance with visual observation of the welding process or inspection of the weldment.

A minor amount of manual redressing of electrodes is permissible. In the laboratory this might entail using an abrasive cloth-wrapped paddle contoured on both sides to match the electrode face contour. The electrodes are brought against the abrasive cloth under a light load. The paddle is then rotated to redress the electrode faces. A file should never be used for redressing electrodes in the machine because the resulting electrode faces may become irregular in size and contour.

Heavier redressing of the electrode may be done in the machine with a manual or power-operated dressing tool. In robotic applications, dedicated tip dressers frequently are used to maintain the electrode during the material-handling or idle time. In this case, it is preferable to do frequent light cuttings to maintain the desired electrode face contour. The objective rarely is to re-establish the entire welding electrode tip profile, because this might entail too much unnecessary material removal.

To maximize the electrode service life and minimize the electrode cost per weld the following suggestions may be helpful:

1. Standard electrodes and holders should be used whenever possible,
2. The electrode should be made of the material recommended for the application,
3. Adequate water-cooling should be used, with attention paid to the proper direction of water flow to ensure that the water impinges on the electrode as closely as possible to the weld face,
4. The electrodes should be aligned so they will not skid against the workpieces and will maintain alignment when they are in contact with the workpieces,
5. Only rawhide or rubber mallets should be used for tapping electrodes into position and only ejector-type holders or the proper tools should be used to remove electrodes from the welding machine, and
6. The welding machine must be set up properly so the electrodes contact the workpieces with minimum impact and develop the proper weld force before current flows. Adequate force must be maintained until the current ceases.

PROJECTION WELDING ELECTRODES

Projection welding electrodes are similar to the previously described spot welding electrodes, but projection welding electrodes are predominantly flat faced. The flat face usually is much larger than the workpiece projection, which reduces the need for accurate alignment and reduces the current and pressure density that would otherwise accelerate electrode wear.

The other distinctive characteristic of projection welding electrodes is that many are made from hard Group A alloys or faced with Group B refractory metals. These materials can resist the mechanical loads better over very long service intervals. An example of this type of design is shown in Figure 4.22.

Projection welding electrodes eventually will become pitted or deformed at the weld locations. When this deterioration interferes with proper electrode contact or weld quality, the electrodes or inserts must be redressed or replaced. Regular cleaning of the electrodes to remove grease, dirt, flash, or other contamination will prolong electrode life.

SEAM WELDING ELECTRODES

Circular seam welding electrodes may be in the form of a wheel, ring, or disk. The basic considerations are face contour, width, diameter, cooling, and method of mounting. The diameter and width of the electrode usually are dictated by the thickness, size, and shape of the workpieces. The face contour depends on the requirements for current and pressure distribution in the weld nugget and the type of drive mechanism in use. The four basic face contours in common use are flat, single-bevel, double-bevel, and radius, as shown in Figure 4.23.

The electrodes usually are cooled by either flooding or directing jets of water on both of the electrodes and the workpieces from top and bottom. In applications in which those methods of cooling are unsatisfactory, the electrodes and shafts should be designed for internal cooling.

Cooling by simple flooding alone is not always adequate. A steam pocket may develop at the point where the electrode meets the workpiece, which will block cooling water from the immediate area. When flood cooling is unsuitable, water mist or vapor cooling may be effective.

A seam welding electrode must be attached to the shaft with a sufficient number of bolts or studs to withstand the driving torque. The contact area with the shaft must be large enough to transmit the welding current with minimum heat generation.

Peripheral-drive mechanisms, such as knurl or friction drives running against the electrode require adequate workpiece clearance. A knurl drive will leave marks on the electrode face, which in turn will mar the surface of the weld. However, a knurl drive wheel tends to remove surface pickup from the electrode face.

Although the workpiece and the drive method may require flat-faced electrodes with or without beveled edges, these electrode types are more difficult to set up, control, and maintain than radius-faced electrodes. Of the two types, radius-faced electrodes produce welds with the best appearance. Radius-faced electrodes have an additional benefit. As the weld temperature increases, they sink into the weld metal a little more deeply, and in doing so, increase the area of the weld. This provides a small amount of self-regulation of the process and effectively increases the process capability.

Like spot-welding electrodes, seam welding electrodes have a predetermined area of contact with the workpieces that must be maintained within limits if consistent weld quality is to be maintained. Electrode

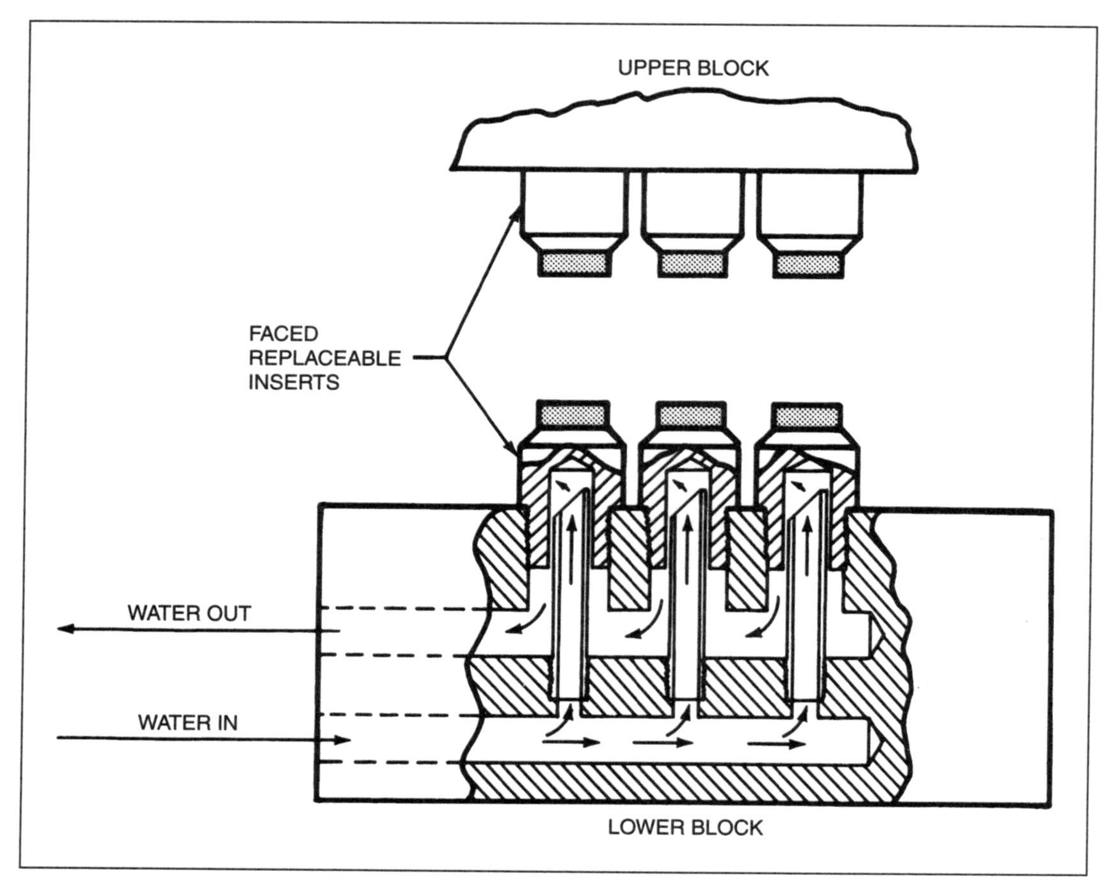

Figure 4.22—Typical Construction of a Multiple-Electrode Projection Welding Machine

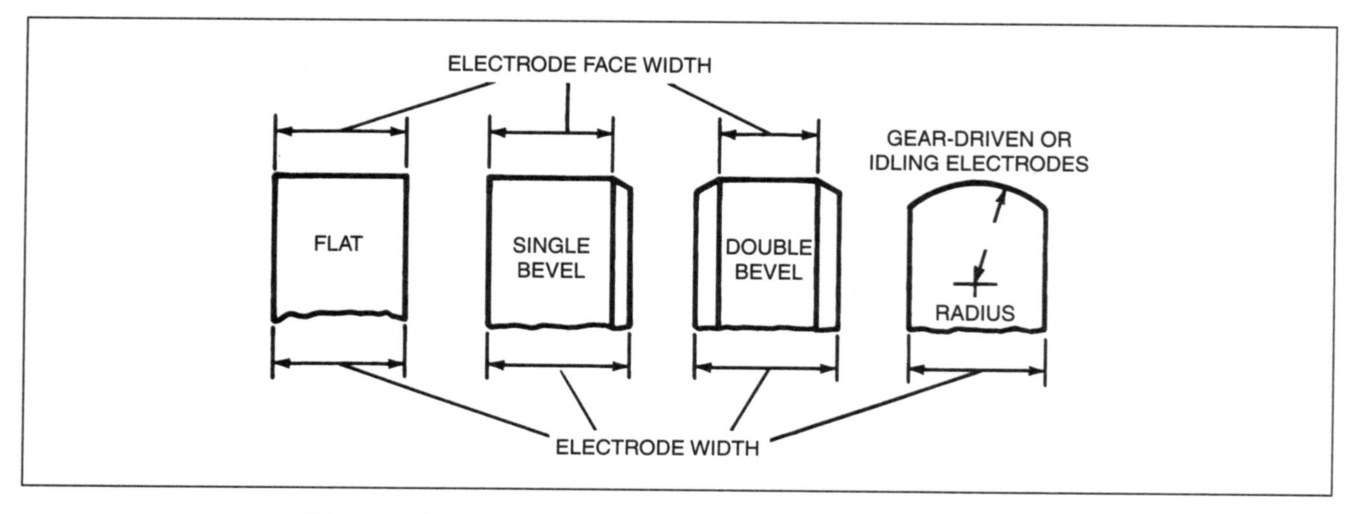

Figure 4.23—Face Contours of a Seam Welding Electrode

maintenance while the electrode is in the machine should be attempted only for minor dressing or touchup with light abrasives. Wheel dressers may be used for continuous electrode maintenance. Machining in a lathe is the preferred method of dressing an electrode to its original shape.

Precautions must be taken to prevent foreign materials from becoming embedded in the electrode wheel or the workpieces. Rough faces do not improve traction. Welding should be stopped while electrodes are still on the workpieces.

FLASH WELDING AND UPSET WELDING ELECTRODES

Unlike spot welding and seam welding electrodes, flash welding and upset welding electrodes usually are not in direct contact with the weld area. These electrodes function as workpiece-holding and current-carrying clamps, and often are referred to simply as clamps. The electrode also may function as a die to shape the workpiece, and in this case would appropriately be called a *die*. They normally are designed to contact a large area of the workpiece; thus the current density in the contact area is relatively low. Accordingly, relatively hard electrode materials with low conductivity give satisfactory performance.

Since the electrodes must conform to the workpieces, there are no standard designs. Two important requirements are that the materials have sufficient conductivity to carry the current without overheating, and that the electrodes have sufficient rigidity to maintain workpiece alignment and minimize deflection.

The electrodes are mechanically fastened to the welding machine platen. They can be solid, one-piece construction of one of the RWMA Group A electrode materials in Classes 1 through 5. Service life sometimes can be increased by using Class 2, 3, and 5 materials with replaceable inserts of Class 3 or 4, or one of the Group B materials at the wear points.

A varying amount of wear inevitably occurs, and this may result in decreased contact area and localized burning of the workpieces. For good service, the electrodes should be kept cool, clean, and free of dirt, grease, flash, and other foreign particles. An anti-spatter compound may help prevent flash adherence. All fasteners and holding devices should be tight and properly adjusted, and gripping surfaces should be properly maintained to avoid workpiece slippage during welding.

Filling sockets in the head of cap screws with modeling clay or putty will make it easier to remove them later for maintenance in high-moisture or weld-flash environments. It is much easier to scrape the modeling clay or putty out of the socket than to remove the accumulation of corrosion and weld flash. When assembling equipment, exposed threads beyond the nut should be covered with an extra nut; exposed threads tend to accumulate flash that would be difficult to remove.

ACTUATORS

Welding actuators include a broad range of devices from fluid power cylinders using pneumatic or hydraulic pressure to electric actuators using a variety of electric motors and magnetic drivers. Combinations of any of these operating principles may be used to take advantage of the properties of each.

PNEUMATIC

Pneumatic actuators have replaced most manually operated welding pressure systems because they are adjustable, reliable, and produce repeatable welds. Pneumatic actuators are available in diverse sizes, shapes and methods of construction.

Press-type welding machines traditionally incorporate a direct-acting air cylinder actuator coupled to the ram type to exert electrode force and retract the electrode. Four general types of double-acting air cylinders are employed, as illustrated in Figure 4.24.

In all cases, compressed air for the pressure stroke enters at Port A. During the pressure stroke, Port B must be allowed to dump remnant air in Chamber M to minimize loss of force. For the return stroke, the compressed air enters at Port B. During the return stroke, remaining air in Chamber L must be allowed to dump through Port A. Figure 4.24(B) shows an adjustable-stroke cylinder with stroke adjustment. The stroke-adjusting Screw K limits the travel of Piston P and therefore the electrode opening.

An adjustable-stroke cylinder incorporates an adjustable piston that provides a mechanical limit for the stroke, shown as "dummy" Piston R in Figure 4.24(B). The dummy piston is used to reduce the volume of compressed air and reduce the time necessary to achieve desired force. The dummy Piston R is attached to Adjusting Screw, K, which fixes this piston at the desired location. Chamber L is connected to Port A through the hollow adjusting screw. The stroke of the force on Piston P is determined by the position of the adjustable Piston R above it. This cylinder design responds faster than a fixed-stroke cylinder because the volume L above Piston P can be made smaller than that of an adjustable-stroke cylinder of the same size.

The adjustable-stroke cylinder can be modified to provide a retraction feature as shown in Figure 4.24(C). This feature provides two distinct stages of stroke for the piston P. These two stages can accommodate a weld

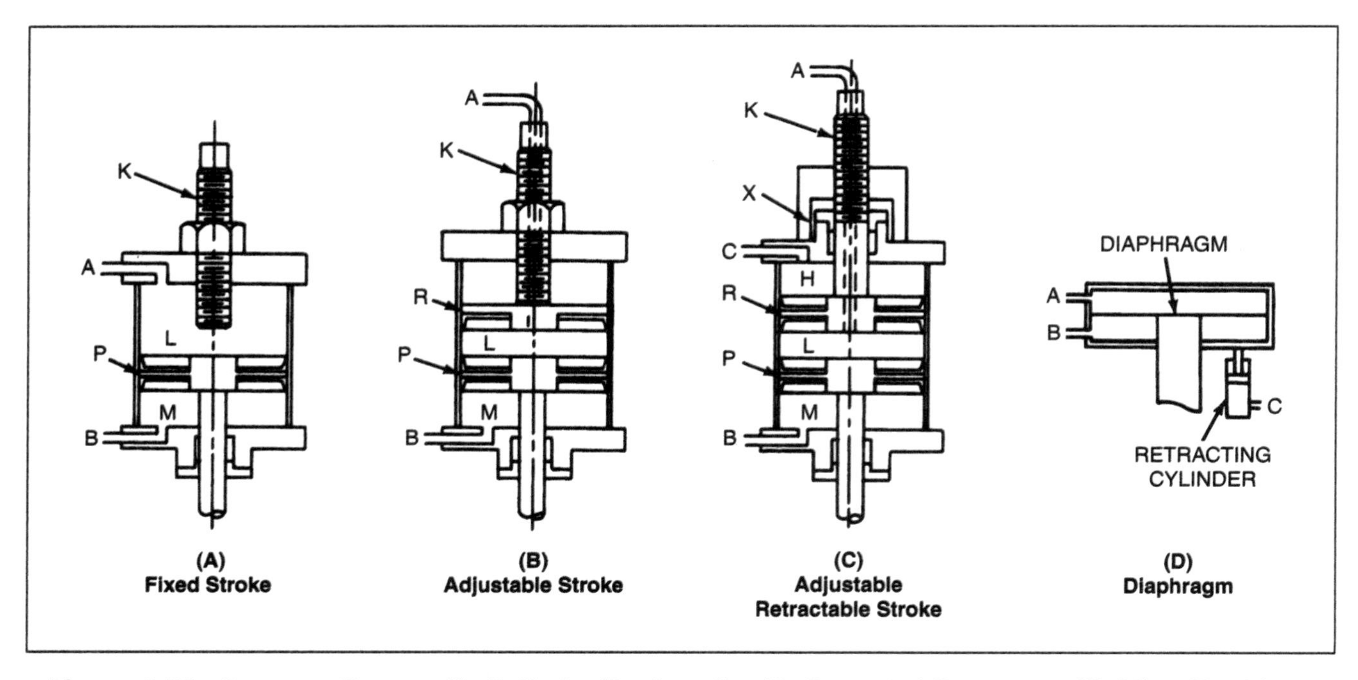

Figure 4.24—Common Pneumatic Cylinder Designs for Air-Operated Press-type Welding Machines

stroke and additional electrode opening called retract stroke, for loading and unloading the workpiece, or may be used for electrode maintenance. With the adjustable retractable-stroke cylinder, a third Port C is connected to Chamber H above the adjustable Piston R. If air is admitted to Chamber H at a pressure slightly higher than the operating pressure in Chamber L, Piston R will move down to a position determined by the adjustable Stop X. This piston movement and location is the retracting stroke, which determines the returned position for Piston P and the electrode opening for the weld stroke. When the air from Chamber H is exhausted to atmosphere, Piston P will lift Piston R with it until Stop X contacts the cylinder head. This will increase the electrode opening for loading and unloading workpieces. The readmission of air to Chamber H will return Pistons P and R to welding position when the pressure in Chamber H is slightly higher than that in Chamber M. Flow-control valves or cushions generally are used to control the operating speed of an air cylinder.

There may be safeguarding advantages to limiting stroke to no more than required. If the gap is too small to allow an operator to get a finger between the electrode and the workpiece, less guarding may be acceptable.

Figure 4.24(D) shows a diaphragm cylinder. In this design, separate cylinders are used to retract the entire cylinder and ram to allow workpiece loading. The deflection of the diaphragm by the pressure differential on either side of it provides the electrode movement. This system responds very rapidly due to its inherent low friction and inertia, providing fast follow-up of the electrodes as the weld nugget is formed. Dual electrode force is easily attained by alternately pressurizing and depressurizing Chamber B while Chamber A is held at a constant pressure.

To make welds on close centers, the weld cylinder must have a small diameter. This can be accomplished by connecting multiple pistons on the same shaft of a pneumatic actuator. While the area of the rod must be taken into consideration, the contribution of each successive piston area added will add proportionally to the output force. The penalty, of course, is cylinder length.

Air pressure below 140 kPa (20 psi) generally is not used because the static friction of the cylinder seals might cause erratic electrode movement and the pressure control valve may not operate consistently at low pressure values.

HYDRAULIC

Hydraulic actuators are employed when space is at a premium or when high forces are required. The design of these hydraulic cylinders is similar to that of air-operated cylinders. (Refer to Figure 4.24.) Hydraulic cylinders generally are smaller in diameter than air cylinders because higher pressures can be developed with a liquid system.

In the simplest type of hydraulic system, a constant-speed motor drives a constant-pressure, constant-delivery pump. The output pressure of the pump is controlled by an adjustable relief valve. Pressurized hydraulic oil delivery is controlled with a four-way valve similar to that employed in an air system. Auxiliary devices include a sump, filter, heat exchanger, gauge, and sometimes an accumulator.

Equipment cost and operating expense are primary considerations for selection of hydraulic cylinders. The application may be practically limited to instances when many small cylinders are required, or in high-force situations in which other actuators become impractical.

INTENSIFIER CYLINDERS

Air-over-oil intensifier cylinders exploit piston area differences to amplify the available pressure for high welding pressures. Generally, an incompressible hydraulic fluid trapped inside the cylinder is acted upon by an air-driven rod connected to a much larger piston that is being acted upon by regulated supply pressure. Since the viscous fluid does not have an escape route during the welding stroke, the intensifier cylinder is very effective in maintaining electrode follow-up. A cylinder diameter of 50 mm (2 in.) may be capable of delivering 1800 kilogram-force (kgf) (4000 pound-force [lbf]) over a short stroke distance with an input of 550 kPa (80 psi).

SERVO ACTUATORS

Pneumatic, hydraulic, and electric servo-controllers are all employed as actuators in resistance welding machines. The servo-controller uses feedback of available energy or output force, rod position, and sometimes velocity to predict adjustments required to maintain the desired position or force. The actuators are described as servo-pneumatic, servo-hydraulic and servo-electric; however, the generic term *servo actuator* usually refers to servo-electric actuators, since they are the most commonly used.

The servo-electric actuator employs a simple but powerful electric motor, generally used with an apparatus that converts rotary motion to linear motion. In many cases, this mechanical device is a screw of the ball-bearing or planetary-roller design. The servo controller excites the motor windings in the sequence and with the amplitude required to achieve the required position, velocity, and force in the prescribed time.

The advantage of all of these servo systems is consistency of programmed weld force and operating time. Setting the exact position of the electrodes greatly benefits some applications. The welding force and electrode position can be easily adjusted to suit the welding application.

For most applications it is not necessary to have direct feedback measurement of the welding force, but some applications require closed-loop force control.

POWER CONVERSION EQUIPMENT

The electrical characteristics of resistance welding equipment are heavily influenced by the characteristics of the welding circuit. In this section, the welding circuit is defined as all components through which electrical welding current flows. These components may include the workpieces, electrodes, fixturing, platens, bus bars, shunts, cables and the energy-conversion device. The energy-conversion device is the means through which electricity is changed into a form suitable for resistance welding. Energy-conversion devices may be classified as direct-energy and stored-energy devices.

Direct energy devices draw electricity for welding during weld time only. Direct energy-devices include transformers, transformer-rectifiers and frequency converters. The most common direct-energy device is the single-phase transformer that uses alternating current. This transformer has specially designed secondary terminals that also can be used with a rectifier for direct-current welding. Direct-energy systems may use three-phase input power to achieve smoother output dc current and reduce energy consumption. Three-phase input devices include medium-frequency, direct-current (MFDC), three-phase, full-wave dc, and frequency converters.

Stored-energy devices employ chemical, electrical, or electromechanical storage of accumulated electric energy that is discharged during the weld. These devices include chemical batteries, capacitors, and dynamos. Stored-energy welding machines are available with single-phase or three-phase input power. Welding current is unidirectional and weld times generally are shorter than those of direct-energy equipment. The choice of energy conversion is dictated by equipment costs, available utilities, and the characteristics of the welding circuit.

SINGLE-PHASE EQUIPMENT

The typical electrical system of a single-phase resistance welding machine consists of a transformer, a tap switch, and a secondary circuit, which includes the electrodes and workpieces.

The welding transformer, in principle, resembles any other iron-core transformer. The main differences in a welding transformer are the use of water cooling to minimize size and the use of a secondary circuit with only one or two turns. Transgun transformers for robotic applications are optimized for weight and output. Machine transformers are optimized for higher amperages and higher production rates. Fixture transformers are designed for connection flexibility and a wide range of secondary current. Portable gun transformers are designed with higher secondary voltage for use with a low-reactance dual-conductor water-cooled cable, commonly referred to as a *kickless cable.*

The dual-secondary transformer is noted for its flexibility. If desired, only one of the dual secondaries can be used at one time. Each of the secondaries can be used to power separate welding circuits, although there can be no individual current control in each circuit. For higher secondary voltage, the two secondaries can be connected in series to feed one secondary welding circuit. For high numbers of welds per minute, the secondaries can be connected in parallel to feed one secondary circuit.

Transformer Rating

Resistance welding transformers normally are rated on the basis of the temperature-rise limitations of its insulation materials. The standard rating in kVA is based on the capability of a transformer to produce the rated power at a 50% duty cycle without exceeding the temperature-rise limitations of the design. This means that, if properly cooled, a transformer can produce the rated power for a total time of 30 seconds during each minute of operation without exceeding temperature limitations. The kVA rated at 50% duty cycle is simply a base from which comparison can be made with other transformers.

The maximum permissible kVA demand for a standard resistance welding transformer at a particular duty cycle can be determined using the following formula:

$$\text{kVA demand} = 7.07 \text{ kVA-rated}/(\text{DC})^{1/2} \qquad (4.1)$$

where

kVA demand = Maximum permissible input power;

kVA rated = Standard transformer power rating at 50% duty cycle; and

DC = Operating duty cycle, %.

For example, a welding transformer rated at 100 kVA may be operated at 223 kVA demand at 10% duty cycle without overheating.

Duty cycle is the percentage of time that the transformer is actually on during a 1-minute integrating period. For 60 Hz power, the percent duty cycle can be expressed by the following formula:

Percent duty cycle =

$$\frac{\text{Welds/min} \times \text{weld time in cycles}}{(60 \text{ cycles/s})(60 \text{ s/min}} \times 100 \qquad (4.2)$$

For example, if a machine is producing 30 welds per minute with a weld time of 12 cycles (60 Hz), the operating duty cycle is the following:

$$\frac{30 \times 12}{3600} \times 100 = 10\%$$

Transformer Tap Switches

Transformer tap switches are devices used to connect various primary taps on the transformer to the output of the weld control. They usually are a rotary type and are designed for flush mounting in an opening in the machine frame or on the transformer. The switches are designed to accommodate the arrangement of the transformer taps. Straight rotary designs normally are used on transformers with 3 to 8 taps. To provide a range of secondary voltages, taps are placed at various turns on the primary winding. The taps are connected to the tap switch, and thus the turns of the transformer primary can be changed to produce different secondary voltages, as illustrated in Figure 4.25. In addition, there may be a series-parallel switch that connects two sections of the primary in series or in parallel. This provides a wider range of secondary voltages.

Most switch handles have lock buttons so that the contacts can be centered in each operating position. In addition, some switches have an off position that disconnects a line to the transformer. A tap switch should not be operated while the transformer is energized; doing so will cause arc-over between contacts and damage the switch.

Alternating Current Secondary Circuit

The geometry of the ac secondary circuit (loop), the size of the conducting components, the material from which they are made, and the presence of magnetic material in the secondary circuit will affect the electrical characteristics of the welding machine. Available welding current and kVA demand are influenced by the impedance of the secondary circuit. The electrical arrangement of a single-phase ac machine with an ac secondary circuit is shown in Figure 4.26.

The electrical impedance of an alternating current welding machine should be minimized to permit the delivery of the required welding current at minimum kVA demand. The electrical impedance is reduced when the following conditions are implemented:

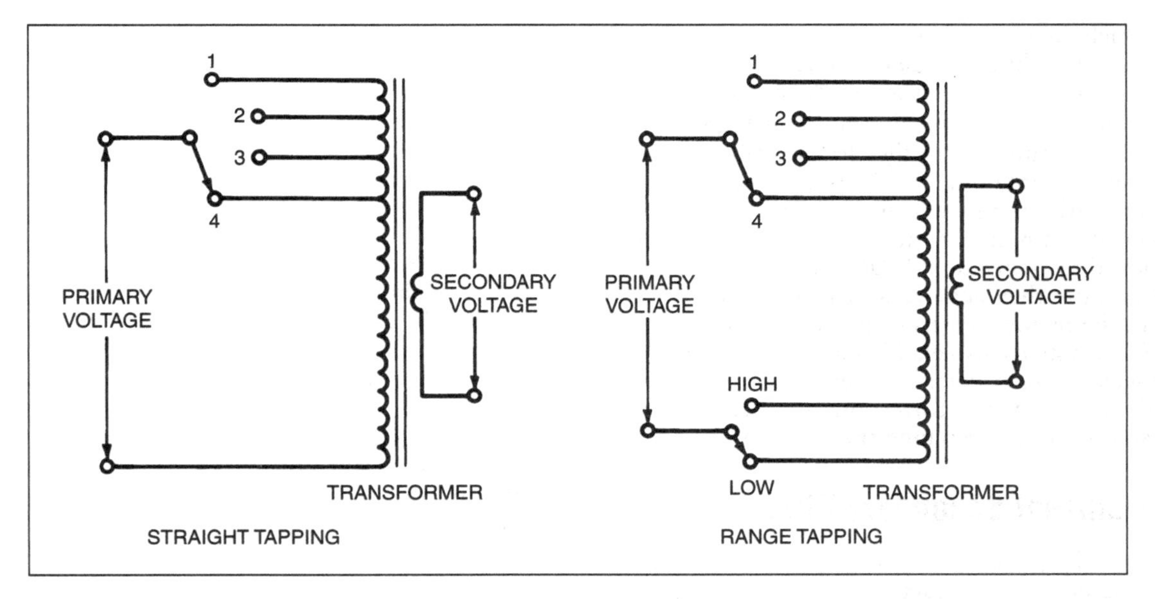

Figure 4.25—Rotary Tap Switches Used to Provide a Range of Secondary Voltage

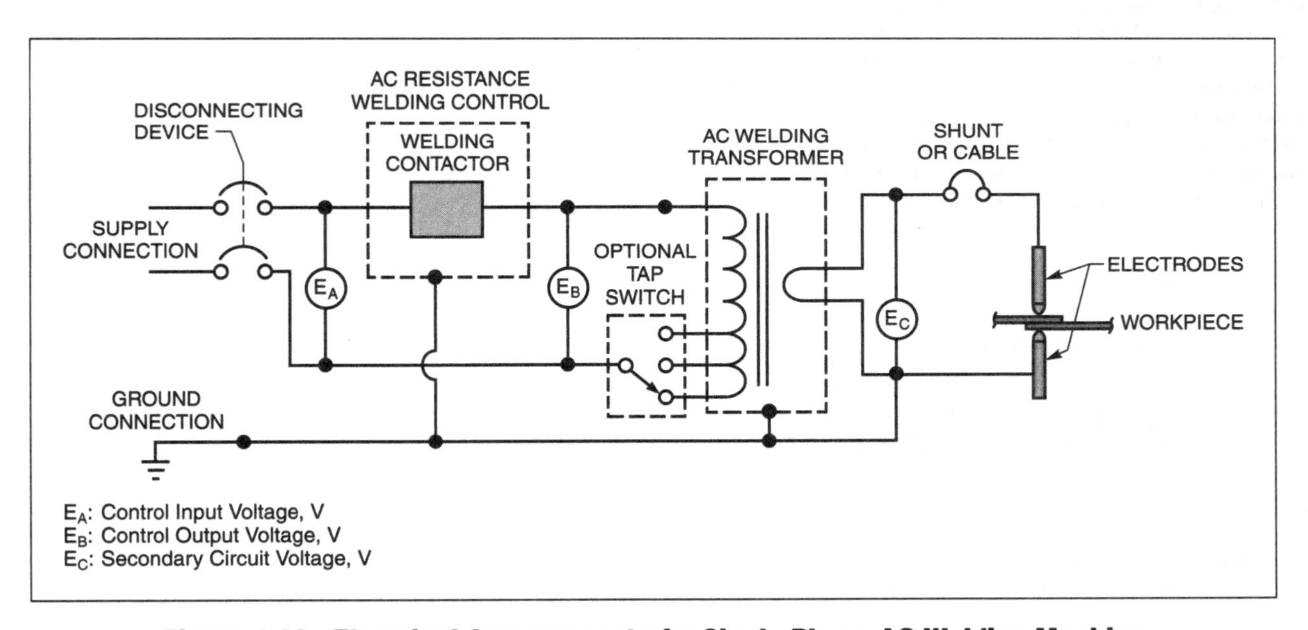

Figure 4.26—Electrical Arrangement of a Single-Phase AC Welding Machine

1. The throat area of the welding machine is decreased,
2. The amount of magnetic material in or near the throat of the machine is decreased,
3. The length or the secondary conductors are decreased,
4. The size and conductivity of the secondary conductors are increased, and
5. Electrical joints are tight and free of corrosion.

Well designed welding machines effectively minimize the impedance of the secondary circuit. However, the size of the workpieces and associated fixturing may require a large throat depth or throat height. This requirement may add considerable inductance to the secondary circuit. The increased inductance causes a reactive voltage drop, which in turn, decreases the power factor. To compensate for this, a higher secondary voltage is required to achieve the necessary electrical current.

A low power factor and intermittent high electrical demand are detrimental to the electric utility company, which must maintain a stable power supply for all customers. Alternating-current systems continue to be the most common for critical applications. Medium-frequency, direct-current (MFDC) systems are often deployed because they achieve a near-unity power factor without a large and expensive power factor correction. The MFDC systems also enable more sophisticated control of welding parameters. A benefit of the increased use of these systems is a corresponding decrease in cost. As a result, they are becoming the de facto standard.

Direct-Current Secondary Circuit

In ac welding machine secondary circuits, a component of the power (the reactive power) is consumed to maintain a magnetic field around the conductors and other nearby magnetic materials, such as the workpiece and tooling. The induced current losses add to the electrical resistance of the secondary circuit, resulting in a total apparent opposition to the welding current. This is referred to as the *impedance*. One method of decreasing impedance losses in the secondary circuit is to rectify the secondary power to dc. Single-phase dc resistance welding machines have a center-tapped secondary and a full-wave silicon diode rectifier. With this system, the kVA rating of a machine need not be increased much to provide for a larger throat area. For a given size and application, the kVA demand of a dc machine is significantly lower than that of an ac machine. For this reason, the power factor is about 0.90 for dc machines, compared to 0.25 to 0.30 for ac machines. This is advantageous for spot and seam welding operations during which the amount of magnetic material in the welding machine throat increases or decreases as welding proceeds. Figure 4.27 shows the electrical arrangement of a single-phase dc resistance welding machine.

When welding galvanized steel or aluminum, the positive electrode deteriorates substantially faster than the negative electrode. In some applications, it may be possible to select the polarity to place the greater deterioration on the electrode that is most easily changed or dressed.

THREE-PHASE EQUIPMENT

The power conversion devices described in this section are classified by the requirements of the following:

1. Direct-energy usage,
2. Three-phase power input, and
3. Unidirectional (dc) output.

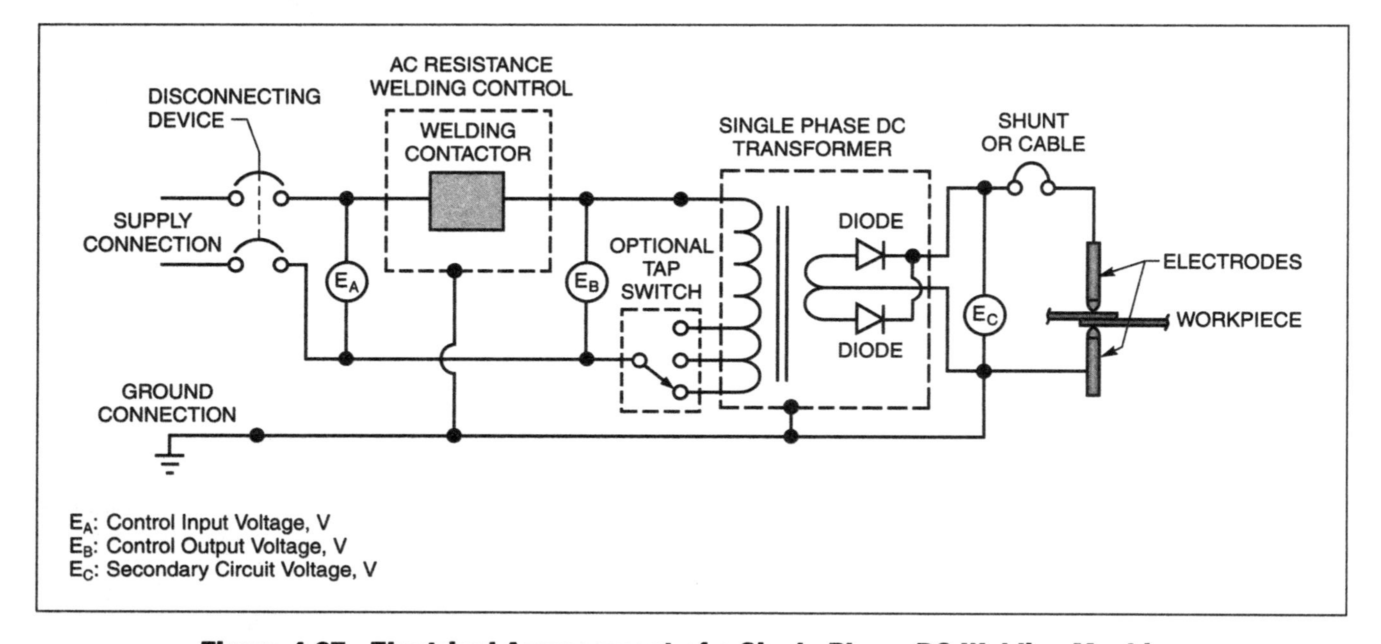

Figure 4.27—Electrical Arrangement of a Single-Phase DC Welding Machine

Direct-energy devices draw input power from the electrical power lines on demand. However, in this case, power is drawn from three phases instead of a single phase, as required by the single-phase ac equipment described in the previous section. Output current for this type of equipment is unidirectional, and therefore is defined as direct current.

The following three-phase power conversion devices are the most commonly used:

1. Medium frequency, direct current, also known as *inverter dc*;
2. Three-phase, full-wave rectified dc; and
3. Frequency converter.

While other power conversion devices, such as three-phase, half-wave, and Scott-T have been applied, the majority of devices in use are the MFDC inverter power source, the direct-current rectifier, and the frequency converter power source.

Medium-Frequency, Direct-Current Power Sources

Medium-frequency, direct-current power sources (inverters) have been in use since the mid 1980s, their popularity driven by the increased use of robots. Close-coupled transformer gun combinations on robots have increased payload limitations and allowed for larger and heavier welding guns. The MFDC system accommodates the heavier power conversion equipment, and has resulted in significant improvement over single-phase, close-coupled transformers.

The technology for MFDC became available through the inverter drives used to control electric motors, hence the term *inverter*. Inverters provide efficient output frequencies higher than the 50-Hz or 60-Hz power line frequency. The transformer designed for frequencies greater than 400 Hz achieves a substantial weight reduction over a comparable 50-Hz or 60-Hz transformer.

Inverter control components and functions are described in the section titled "Resistance Welding Controls" in this chapter. The focus of this section is on the transformer rectifier components and their functions in MFDC systems.

The favored frequency for MFDC systems has been established at 1000 Hz. A major benefit of 1000 Hz compared to 50 Hz or 60 Hz is the reduction of transformer core weight. However, a problem develops when operating at 1000 Hz: it results in increased inductive reactance losses. These losses are minimized by the rectification of the secondary voltage using a full-wave center tap rectifier and a transformer with a specially designed center tap secondary circuit.

The transformer for a center tap rectifier must be designed to allow connection to the electrical center of the secondary, hence the name *center tap*. A center transformer tap rectifier usually has a minimum of two diodes, i.e., one diode for each half-cycle pulse, as shown in Figure 4.28.

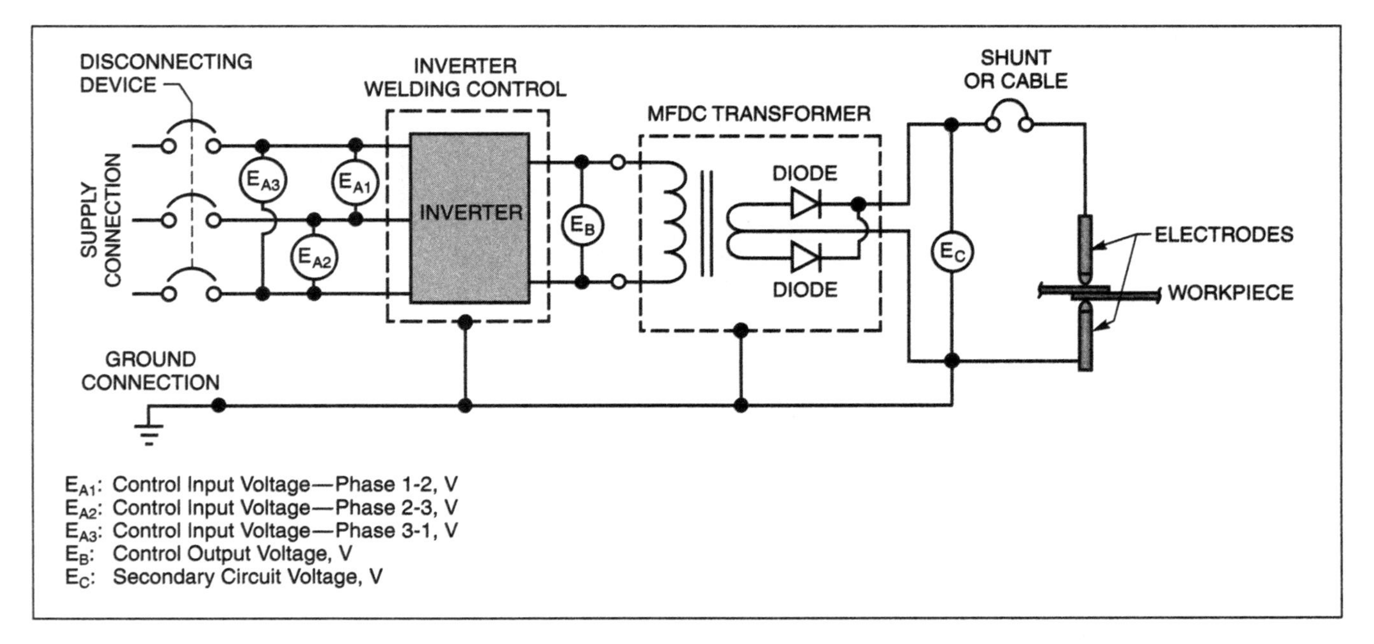

Figure 4.28—Electrical Arrangement of an MFDC Welding Machine

In a rectifier, the check-valve action of a diode allows the passage of unidirectional (dc) pulses only. The MFDC system substantially reduces current ripple, a common characteristic of ac or 60-Hz dc welding circuits, thereby minimizing losses due to inductive reactance. For low-voltage applications, the center tap rectifier is favored over the bridge rectifier because the voltage drop of one diode in a center tap rectifier is preferable to the voltage drop of two diodes in a bridge rectifier. Proper cooling of diodes is very important to diode longevity; inadequate cooling may cause premature diode failure.

Construction of the transformer secondary typically has two turns in series. The electrical center of the secondary carries both half-cycles of current. The two outer secondary terminals carry, in an alternating sequence, a positive half-cycle current pulse.

The electrical benefits of MFDC systems are summarized as follows:

1. Reduced weight and size;
2. Balanced three-phase load;
3. Power factor at or near unity;
4. Reduced primary demand;
5. Reduced wire size, breaker, bus bar, and substation for new facilities;
6. Increased current on large guns;
7. Reduced current loss due to magnetic materials; and
8. Functions such as the weld interval can be timed in 1-millisecond (ms) increments for an MFDC unit operating at 1000 Hz.

Direct-Current Rectifier

A three-phase dc rectifier-type welding machine is similar to the single-phase type in that each welding transformer powers a rectifier bank. The output of the rectifiers is fed into the welding circuit. Full-wave rectification, as shown in Figure 4.29 is the most commonly used. The transformer secondary is connected with a 6-phase star diode connection.

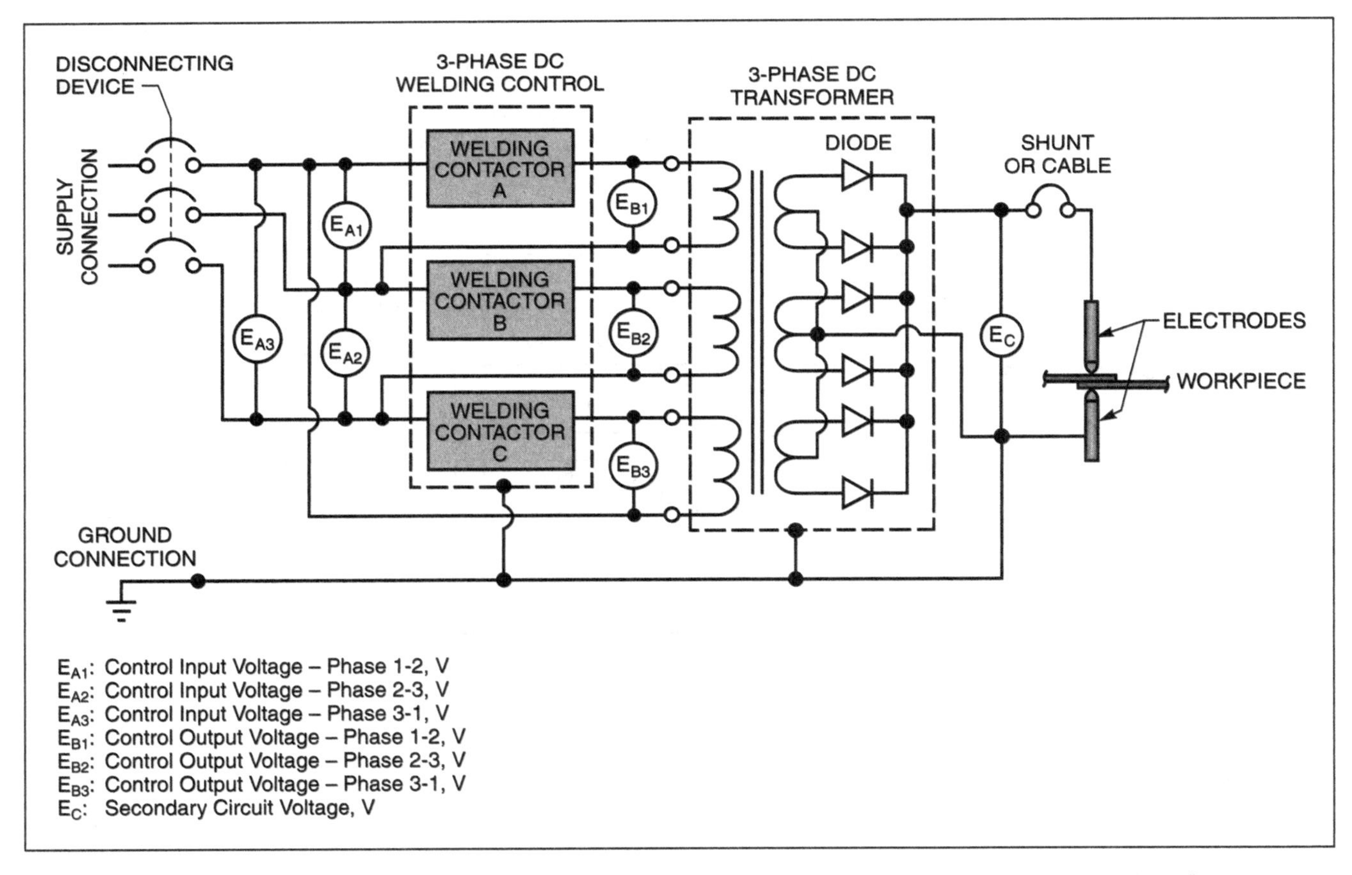

Figure 4.29—Electrical Arrangement of a Three-Phase DC Rectifier Welding Machine

Welding current is controlled by electronic heat control, sometimes in conjunction with a transformer tap switch. The design of the primary circuit and control device varies among equipment manufacturers. The secondary current output of a three-phase machine is much smoother than that of a single-phase machine. Power demand is balanced on the input line.

The three-phase rectifier consists of silicon diodes mounted on water-cooled conductors. The arrangement of conductors and diodes is electrically symmetrical. The impedance of each diode circuit and their other electrical characteristics must be similar so that the diodes will share the current load equally. Diodes can have a long service life if properly applied and used. Welding current may be supplied continuously, provided that the thermal rating of the welding machine is not exceeded.

Frequency Converters

Two types of frequency converter systems exist: the classic half-wave system and the full-wave type, which uses three-phase input to a rectifier to supply a low-frequency converter. Both of these systems perform in a similar fashion, but the half-wave type uses a large-core, three-phase transformer; the full-wave type uses a large-core, single-phase transformer.

The half-wave frequency converter has a specially designed transformer with three primary windings, each of which is connected across one of the three input phases. One secondary winding is interleaved among the primary windings and connected to the conductors of the welding circuit, as illustrated in Figure 4.30.

The transformer primary windings shown in Figure 4.30 are connected to the electric supply lines though three electronic contactors. Ignitron tubes or SCRs may be used as contactors. A welding control causes Contactors A, B, and C, to conduct in sequence. With the correct sequence and conduction time, current is passed through the three primary windings in the same direction. This causes unidirectional current in the secondary circuit. Contactors A, B, and C are then shut off at the end of a preselected time. Then Contactors A, B, and C

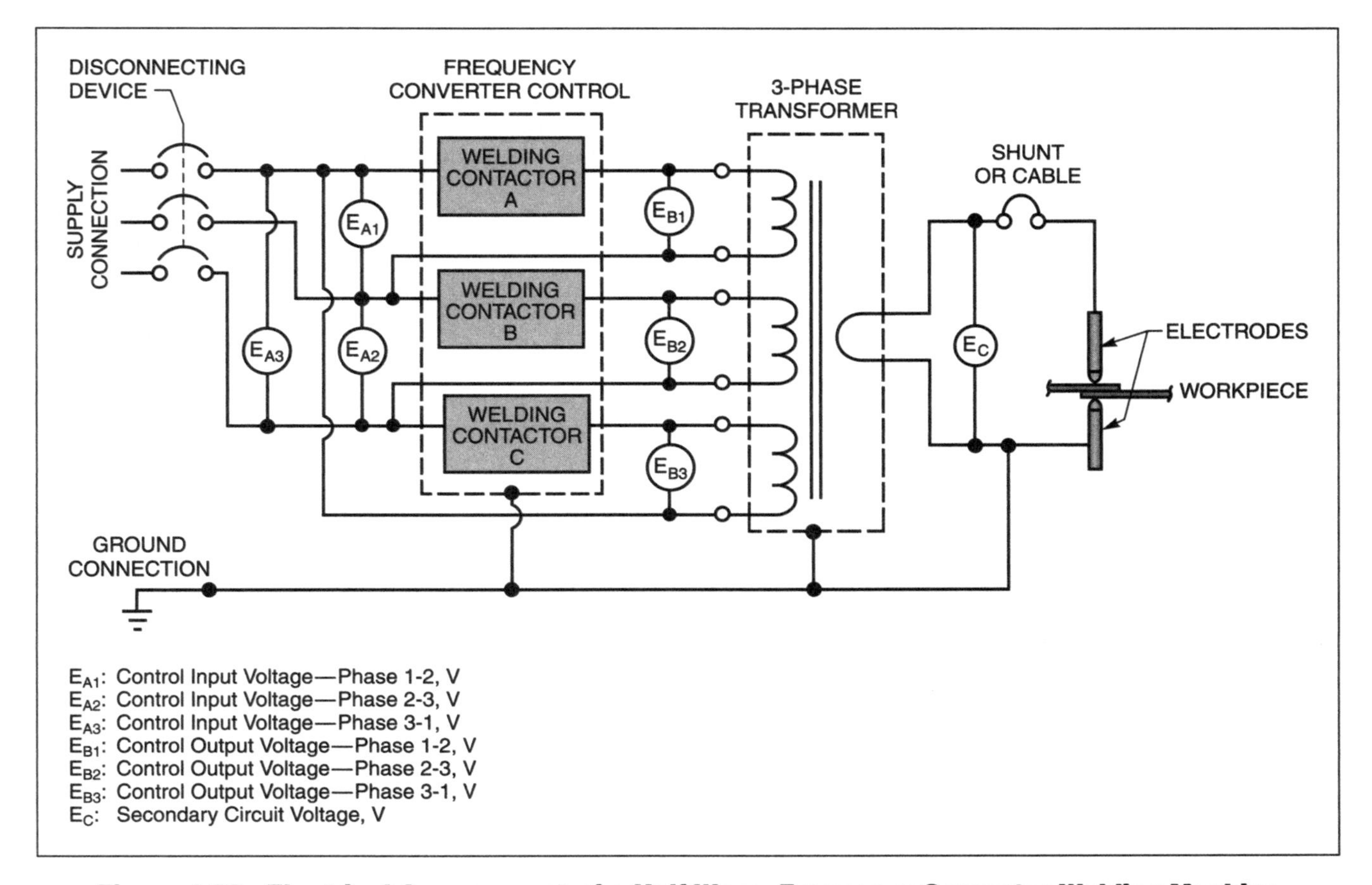

Figure 4.30—Electrical Arrangement of a Half-Wave, Frequency-Converter Welding Machine

are caused to conduct again, with the correct sequence and conduction time, and current goes in the opposite direction through the primary windings and the secondary circuit. This action effectively applies a reversing dc voltage to the primary windings.

The maximum duration of unidirectional primary current is governed mainly by the size of the transformer and its saturation characteristics. It is common practice to have two maximum dc pulse lengths. One is a short time of about 5 cycles (60 Hz) for high-current applications, and the other usually is 10 cycles with welding current limited to about 50% of maximum. Massive, specially designed transformers may permit the use of high current for the longer time period.

STORED-ENERGY EQUIPMENT

Stored-energy systems are well suited to the welding of materials that either have a high thermal conductivity or are heat sensitive. Stored-energy machines are no longer in common use because of their complexity and safety requirements, but some are used specifically because of their relatively low electrical power demand. The machines can be made in all sizes but are most commonly configured as small units suitable for bench mounting. Larger units usually are powered from a three-phase source, and smaller machines are effective when supplied by a single-phase source. Welding heads or portable tongs are used to connect to the power units with cables. This equipment serves a wide variety of applications, including stud welding, the assembly of small electrical components of nonferrous alloys, and the spot welding of foils.

Electrode force may range from a few grams (ounces) to several kilograms (pounds). Calibrated springs are used in a manual force system to apply the electrode force. Stored energy is used to produce a single pulse of welding current. The amplitude, duration, and wave shape of the welding current are determined by the electrical characteristics of the power source, including capacitance, reactance, resistance, and capacitor voltage. Welding times often are substantially shorter than one half-cycle of 60 Hz.

Figure 4.31 shows the electrical arrangement of a capacitive discharge spot welding machine, a common type of stored-energy welding machine. In this case, a bank of capacitors is used to supply the welding current when the welding contactor is switched on.

RESISTANCE WELDING CONTROLS

The principal functions of resistance welding controls are to provide signals to control machine actions, control electrode movement, start and stop the current to the welding transformer, and control the magnitude of the current. Welding controls also regulate the movement of machine components and provide dynamic welding data through various communication systems.

Historically, resistance welding controls were classified by welding function, such as slope and pulsation. In these older controls, each function required the use of a unique hardware circuit. The contemporary use of

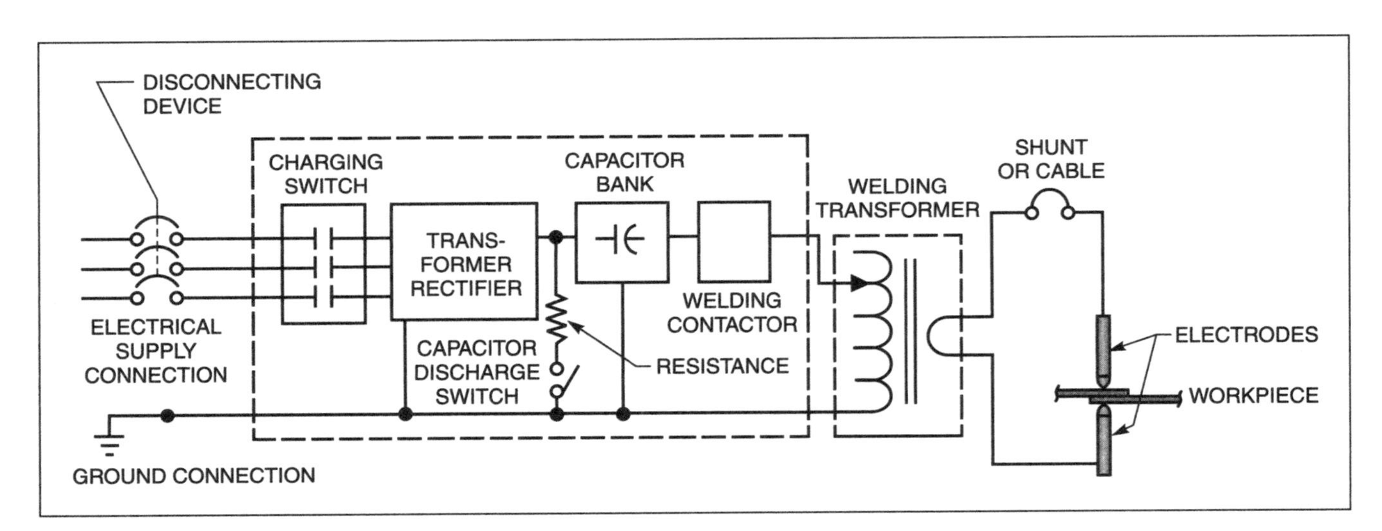

Figure 4.31—Electrical Arrangement of a Capacitive Discharge Spot Welding Machine

microprocessors in resistance welding controls has virtually eliminated these classifications.

HIERARCHY OF WELDING CONTROLS

The range of contemporary welding controls include simple heat controls, general heat and electrode movement controls, and sophisticated heat, electrode movement, and machine function controls.

Heat Controls

Heat controls are used to supply the desired welding current by sequencing the welding contactor. They are typically used when another device, such as a programmable logic controller (PLC) or a machine cam controls the electrode movement. Heat controls typically are digital, but analog heat controls are available to handle non-precision welding applications.

Heat and Electrode Movement Controls

These controls combine heat control and control of the electrode movement. They typically use an internal microprocessor or they can be integrated into a module that plugs into the backplane of a PLC.

Heat, Electrode Movement, and Machine Function Controls

These controls add machine functions to heat and electrode controls and are adaptable to spot, projection, or seam welding. The welding controller for the flash welding process may include functions to regulate the contact and flashing gap positions, squeeze time, preflash time, preheat level and time, final flash level and time, heat level, upset distance, postheat level and time, and hold time.

Machine functions may include the operation of multiple welding actuators, machine table movement, and the monitoring of machine sensors. They usually are able to control all functions of a complex welding machine, eliminating the need for a PLC or other machine control system. A benefit is the reduced number of interconnected systems necessary and the resulting simplification of machine maintenance and troubleshooting.

CONTROL CONFIGURATIONS

Resistance welding control configurations are designed to match the requirements of the variety of power conversion equipment previously discussed in this chapter.

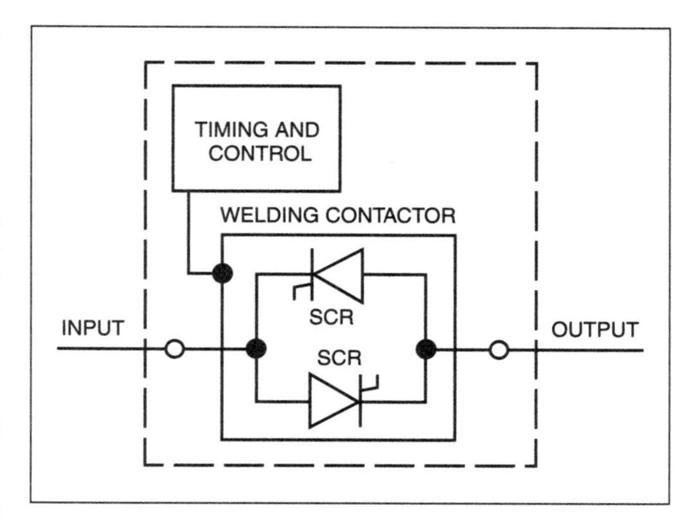

Figure 4.32— Single-Phase AC Resistance Welding Control

Single-Phase AC Welding Controls

Single-phase ac resistance welding controls use an SCR contactor, as shown in Figure 4.32, to provide an adjustable voltage input to the welding transformer, as shown in Figure 4.26. With ac technology, timing is required to be based on ac line frequency and is usually determined in cycles. At a 60-Hz line frequency, the interval of each time increment is 16.7 milliseconds (ms). Control can be as fine as one-half line cycle at heat settings from 1% to 99% of the available line voltage. These controls are used in conjunction with both ac and dc single-phase power sources.

Alternating-current controls use a pulse-width modulation scheme to control the amount of energy flowing to the transformer. The switching of the SCR contactor is delayed to provide slices of the input voltage waveform that correspond to the desired welding current. An illustration of this concept is shown in Figure 4.33, where the voltage traces correspond to the signal measurement locations shown in Figure 4.26.

Medium-Frequency, Direct-Current Welding Control

Medium-frequency, direct-current (MFDC) controls, also called inverter controls, incorporate the use of insulated-gate, bipolar-transistor (IGBT) devices to switch the current to the MFDC welding transformer at a nominal frequency of 1000 Hz. Figure 4.34 shows the basic design of the welding control corresponding to the system in Figure 4.28.

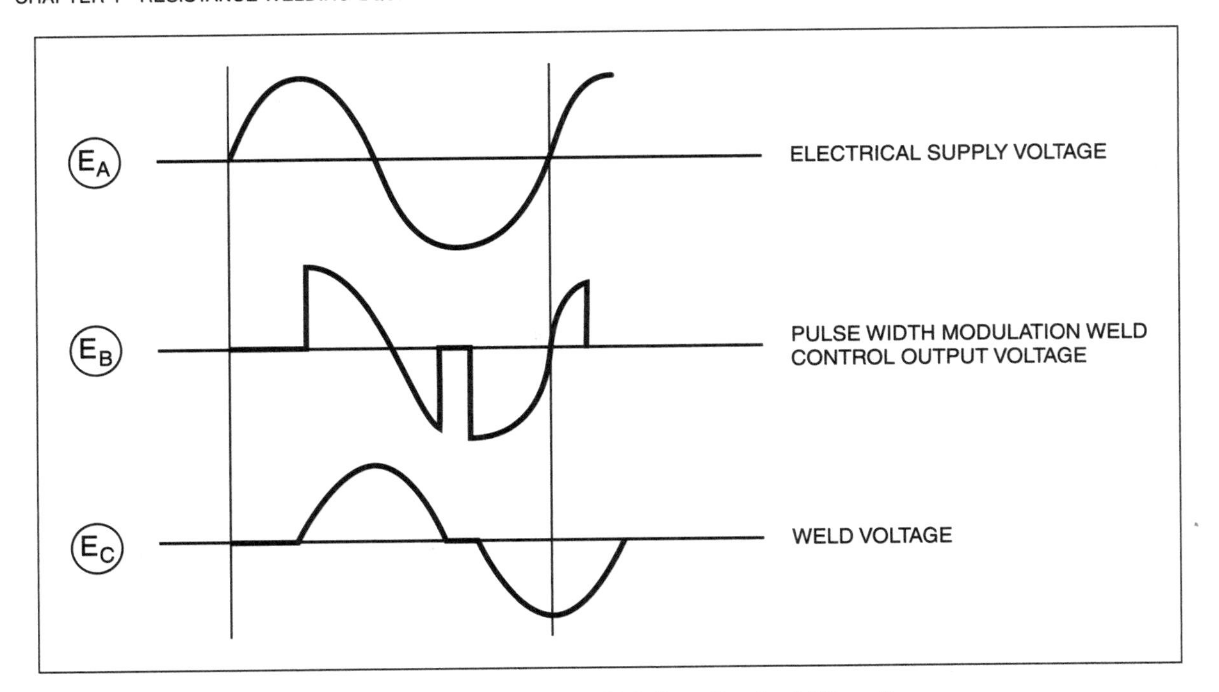

Figure 4.33—Typical AC Welding Control Waveform

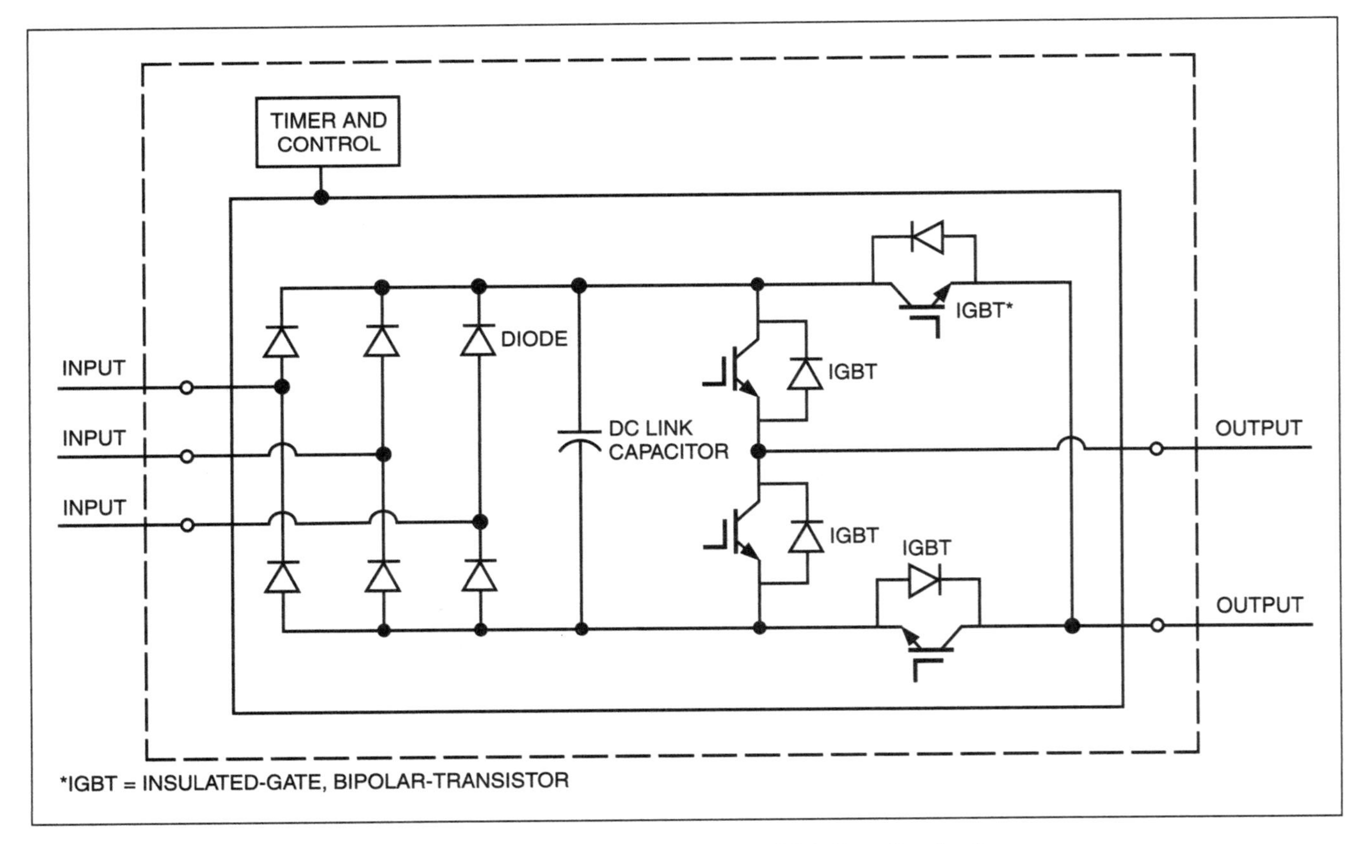

Figure 4.34—Typical Inverter Welding Control

In this system, the three-phase ac line power from the electric utility is first rectified by a full-wave bridge rectifier. Service voltage of 480-V, 60-Hz ac (common in the United States) will generate an internal dc bus voltage of approximately 670 V. A group of capacitors is connected across the dc bus to stabilize the bus voltage. The dc bus is connected across an H-bridge of IGBT devices. The IGBT devices are switched at a nominal frequency of 1000 Hz. This produces an ac voltage to the transformer primary at a nominal frequency of 1000 Hz. The secondary voltage of the transformer is full-wave rectified to dc voltage in the nominal range of 6 V to 13 V dc. Figure 4.35 shows the input voltage, the inverter output voltage, and the resulting welding current waveform corresponding to the measurement points referenced in Figure 4.28.

Controls of this type operating at 1000 Hz can provide control in increments of 1 ms. This eliminates the interruption in welding current inherent in ac welding systems during changeover from one half-cycle to the next half-cycle. This provides more consistent current to the weld and eliminates heating and cooling cycles during nugget formation.

Welding power is controlled by pulse width modulation (PWM), as illustrated in Figure 4.36, or by frequency modulation (FM) or a combination of both. In the PWM mode, the "on" duration is increased to raise the effective primary voltage into the welding transformer. This results in an increase in the secondary voltage out of the transformer. Welding current in the secondary circuit is the result of Ohm's Law, $I = E/R$, where an increase in secondary voltage results in an increase in weld current. In the FM mode, the frequency of the output used to change the power density of the transformer input is varied across a range of 400 Hz to 1800 Hz.

When MFDC technology was initially introduced, the cost of the controls and transformers was extremely high compared to conventional ac systems. Technology advances and increased production volumes have lowered the cost of these components. The ability of MFDC to provide a higher degree of control over the welding current has made them the control of choice.

Three-Phase DC Controls

As shown in Figure 4.37, three-phase dc controls operate as three synchronized ac welding systems of the type shown in Figure 4.29. The control regulates the supply of power to three single-phase transformers connected in DELTA or WYE. By operating on all three phases, this control minimizes the load on each line phase. Each transformer secondary is center-tapped and includes diodes to provide unfiltered dc to the welding electrodes.

Figure 4.38 shows a typical three-phase dc welding system waveform corresponding to the referenced locations in Figure 4.29.

Three-Phase Frequency Converter Controls

These controls turn on SCRs for all three phases in the same polarity for 1 to 12 cycles, depending on the size of the transformer core and windings. If more heat is required, the control stops conducting for a half-cycle and then fires all three phases for multiple cycles in the opposite direction. The result is a group of positive and negative voltage impulses that can be used as either unfiltered dc or may be used to imitate low-frequency ac. The three-phase frequency converter control of the

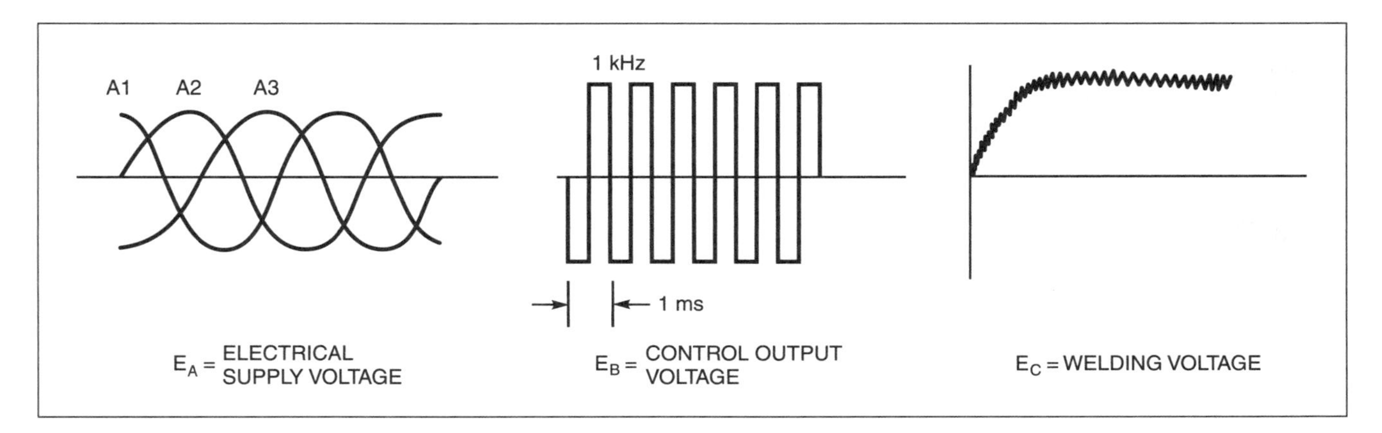

Figure 4.35—Medium-Frequency, Direct-Current Forms of Primary Voltage and Secondary Current

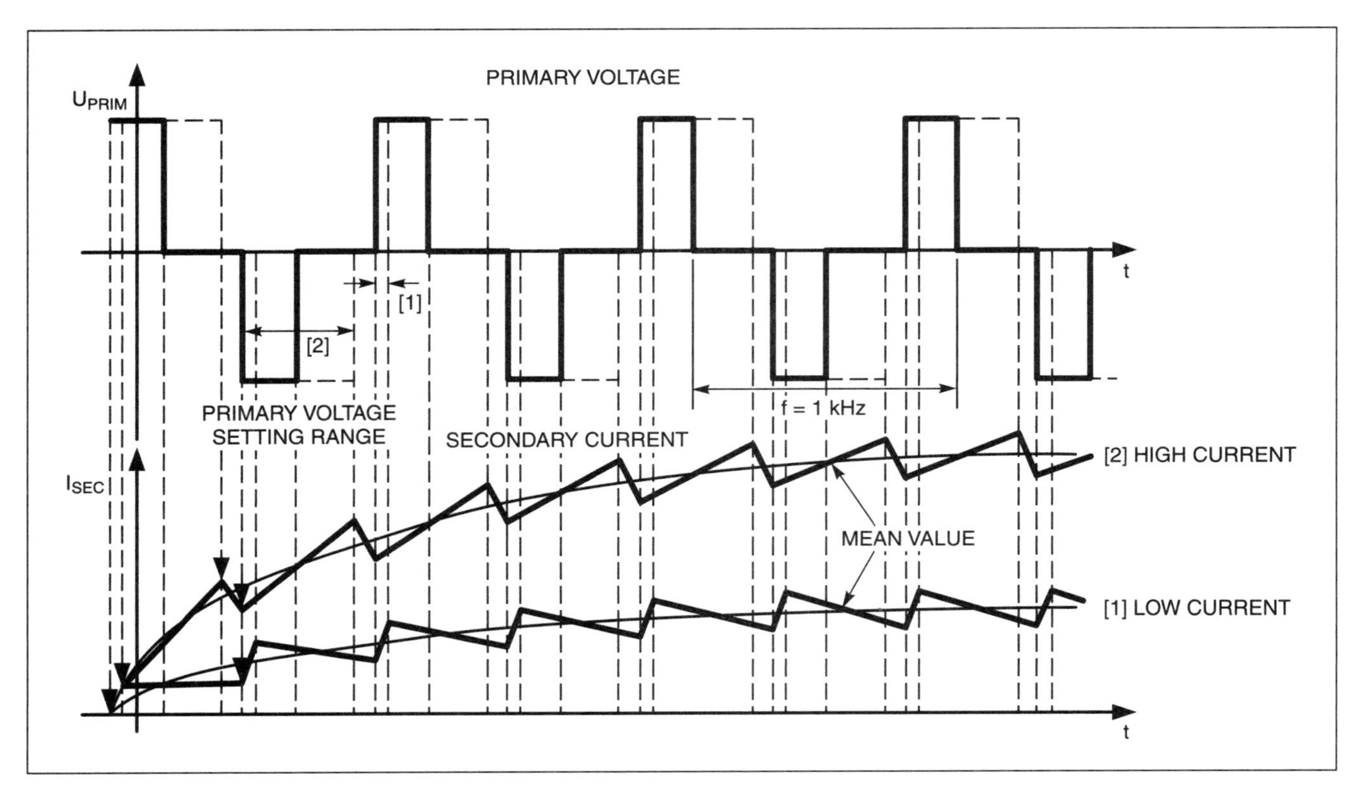

Figure 4.36—Primary Voltage and Secondary Current in Pulse Width Modulation (PWM) Control

system shown in Figure 4.30 and its waveform are illustrated in Figure 4.39 and Figure 4.40.

CONTROL COMPONENTS

Resistance welding controls include standard design elements and components to produce a range of inputs to the resistance welding transformer. Principally, these elements control timing of the welding current or equipment function and may regulate contactors of various forms that convert control signals to power levels suitable for the welding operation.

Time Mechanisms

Several types of timers have been developed to control the duration of various functions during the welding cycle. The availability and efficiency of inexpensive microprocessors and associated digital circuits have facilitated their use in most, if not all, of the welding controls being manufactured. Many older designs used resistor-capacitor timers for determining intervals.

Digital counters, with or without microprocessor control, provide accurate measurement and control of welding cycles or even segments of cycles (as needed for heat controls). These counters may be used to time conduction intervals or other actions associated with the welding process.

Some operations, such as the postheating of flash welds or upset welds, are not critical with respect to accuracy of timing. Pneumatic or motor-operated timers may be suitable for these applications. Timing ranges may vary from a few seconds to several minutes.

Welding Contactors

A contactor is used within the resistance welding control to connect and disconnect the electrical service to the welding transformer. The term *contactor* is a carryover from the mechanical (magnetic) contactors previously used to control welding transformer conduction in non-synchronous welding controls. Four basic types of resistance welding contactors are discussed in this section: ignitron contactors, SCR contactors, IGBTs, and isolation contactors.

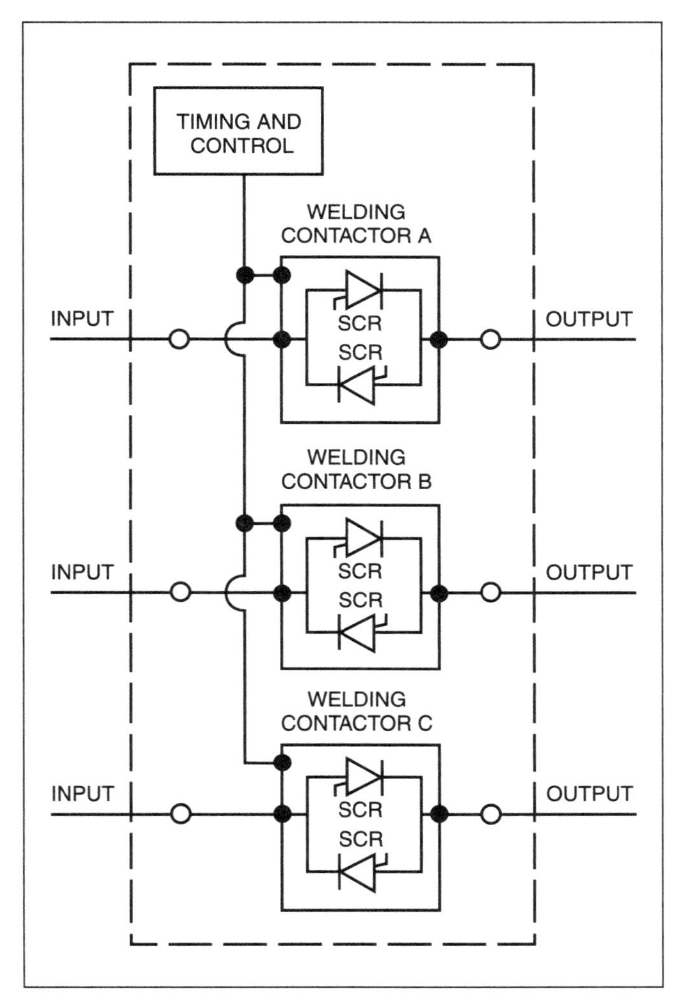

Figure 4.37—Basic Design of a Three-Phase DC Resistance Welding Control

Ignitron Contactors. Ignitron contactors utilize two inversely connected sealed glass tubes; each contains a pool of mercury at the bottom and utilizes a small igniter to initiate conduction. Ignitron tubes were the first non-mechanical contactors used in resistance welding, and the design allowed for precise use of selected portions of each half-cycle of line voltage to vary the heat presented to the process. These contactors were replaced by the SCR contactor in the 1960s and are not used in any controls produced today. SCR replacement kits are readily available to upgrade older controls. Care must be taken in the handling and disposal of the ignitron tubes since they contain a significant amount of mercury.

Silicon-Controlled Rectifier Contactors. Silicon-controlled rectifier contactors consist of two SCRs inversely connected and mounted to a heat sink. These solid-state switches use a very small current at the gate to conduct very high currents to the welding machine transformer primary. Like the ignitron, an SCR contactor can precisely switch voltage on at any point in a line voltage half-cycle to adjust the voltage output of the welder transformer secondary. Once a short gate pulse has been sent to an SCR, the SCR will remain conductive until the voltage of the current being conducted through the switch goes to zero.

Good water cooling is essential to ensure that the SCR contactor will have a long service life. Two different water cooling arrangements are available: direct and indirect water cooling. Direct cooling uses water-cooled blocks in contact with the SCR contact surfaces on one or both sides. The blocks are connected with hoses that provide water flow. Because there is a voltage difference between the cooling blocks, a voltage appears across the length of the hose. The type and length of hose are critical. If a replacement hose is needed on this type of device, the hose must be made of a nonconducting material, and the length must not be shorter than the original. Applications using this type of cooling should not have water flow turned off if power is applied. Water conductivity is a consideration with this design. The second type is indirect cooling, which are assembled with an insulator between the SCR and the cooling plate. This eliminates the voltage applied to the cooling plates, and eliminates concerns about water conductivity and the application of power without water flow. Modular packages typically use this design. A disadvantage is that the electrical insulation between the SCR and the cooling plate also reduces the heat flow. Therefore, the SCR current rating is lower than it would be for the same device with direct cooling. SCR contactors are available sizes as small as 75 A and as high as 5000 A or more.

Insulated-Gate Bipolar Transistors. Insulated-gate bipolar transistors (IGBTs) are switches used in MFDC (inverter) controls. They are used to create a square-wave voltage that is sent to the primary side of an MFDC welding transformer. Unlike an SCR contactor, an IGBT switch will stop conducting when the current being sent to the gate of a switch is turned off. A typical MFDC control uses four IGBT contactors to alternately switch the positive and the negative voltage from the previous rectifier stage of the control. The size of IGBT contactors ranges from 100 A to 2000 A or higher.

Isolation Contactor. An isolation contactor is an electromechanical device used to provide a means of opening and closing the physical connection between the output of the SCR or IGBT contactor and the

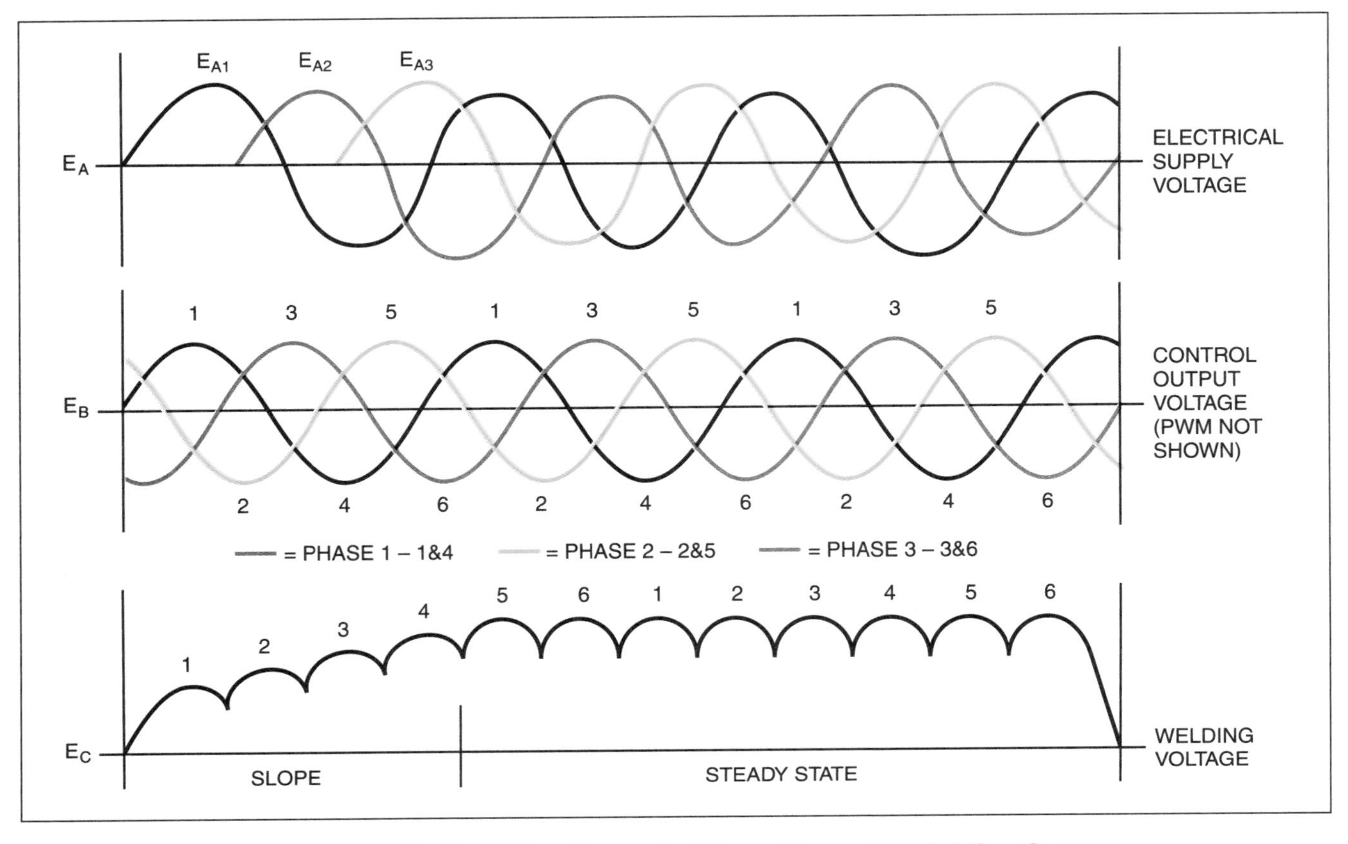

Figure 4.38—Typical Waveform of a Three-Phase DC Welding System

primary of the welding transformer. Isolation contactors are used to provide supplemental protection to personnel by isolating the welding transformer supply conductors and the transformer from any input voltage except during the actual welding sequence.

HEAT CONTROL METHODS

Coarse adjustment of the heat or current output of a welding machine can be accomplished with adjustable taps of the welding transformer. The tap switch changes the ratio of transformer turns for major adjustment of welding current to increase or decrease the output voltage of the transformer (refer to Figure 4.25). Precise control is accomplished with electronic heat control devices.

In electronic heat control circuits used for single-phase or three-phase control, the firing time of each SCR relative to the start of each half-cycle can be delayed to produce the desired heat setting. As the delay angle of the firing pulses is decreased, the SCRs begin to fire late in the half-cycle and the root mean square (rms) value of the welding transformer primary voltage will be low. As the delay angle is further decreased, the SCRs will fire earlier, and will conduct current for a greater part of the half-cycle. The rms current will increase. When the delay angle equals the power factor of the load, 100% rms primary current will be conducted to the welding transformer. Figure 4.41 illustrates this concept.

The reduction in heat or energy varies as the square of the current. Thus, if the rms current can be varied from 100% to 20%, the heat will vary from 100% to 4%. SCRs allow the control of heat over the entire range from 0% to 100%.

To minimize variations in welding current, the heat control should be operated in a range between 60% and 90%. At low settings, a small change of the heat percentage setting can significantly change the rms current. Line voltage disturbances, such as the operation of another welding machine, can sufficiently distort the line voltage waveform to produce such a change. At high settings, controls with automatic voltage compensation systems will not have the ability to react to line voltage drops. Also, controls that use constant-current control may not have the ability to increase heat when needed to compensate for conditions between the electrodes.

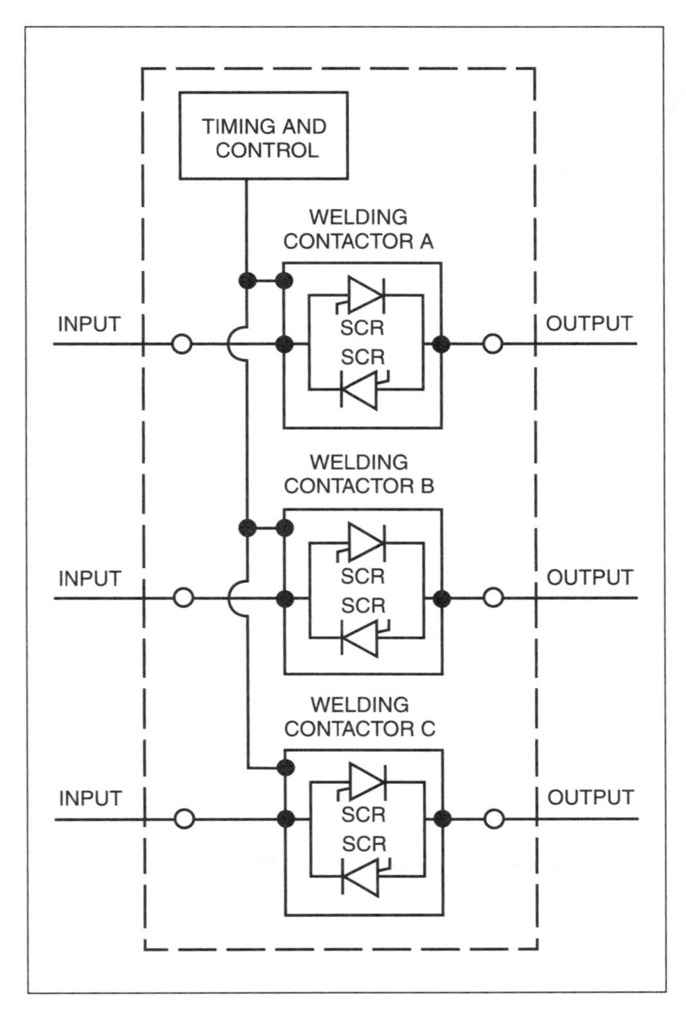

Figure 4.39— Three-Phase Frequency Converter Resistance Welding Control

Constant Current Control

A control with a constant current function is designed to maintain a constant welding current under changing conditions. This device will make corrections for either line voltage fluctuations or impedance changes caused by the insertion of magnetic material into the throat of the welding machine. The constant-current regulator compares primary current to a program-selected current setting so that it can vary the phase-shift heat control to maintain the desired value. The current measurement may be taken by a current transformer on the primary, or a Rogowski coil or Hall Effect transistor on the secondary of the welding transformer. The primary or secondary current is measured in each half-cycle for ac controls, or every square-wave cycle for MFDC controls. Regulation is active through all cycles of the weld time, with the result that the current programmed in the weld schedule is maintained.

The current regulation for ac controls is accomplished by measuring current on a half-cycle or full cycle, then adjusting the phase shift for the next cycle to maintain the correct value. The correction is at its best on the following cycle, thus, large changes may require more than one cycle to make the correction. For this reason, constant-current controls may not completely correct for variations on welds with very short weld times. If current regulation is needed, the MFDC control has a much faster reaction time.

Automatic Voltage Compensation

Controls with automatic voltage compensation (also known as line voltage compensation) are capable of dynamically adjusting the SCR firing angle to maintain the rms voltage at the primary of the welding transformer. This function compensates for decreases (sags) in ac line voltage and increases (surges) in the ac line voltage. Since heat also varies as the square of voltage, a 10% drop in line voltage will result in a 19% heat reduction. The welding transformer tap switch should then be set so that the desired weld is obtained at a controlled heat setting of 81% or less, if a10% line sag is expected. Reaction time for heat controls employing this compensation feature can be as fast as one half- cycle.

BASIC WELDING CONTROL FUNCTIONS

The basic resistance welding sequence consists of three functions: squeeze time, weld time, and hold time.

Squeeze Time

Squeeze time is the interval between the weld control initiation signal and the start of current flow through the electrodes. If squeeze time is set correctly, it will allow enough time for the electrodes to come to optimum force for the type of metal being welded. It is also possible to monitor actuator pressure or electrode force to determine when conditions are suitable to start the flow of current.

Weld Time

Weld time is the duration of current delivered to the welding transformer during the weld interval. In single-impulse welding, this is the total time welding current is on. In multiple-impulse welding, this is the time welding current is on in each impulse.

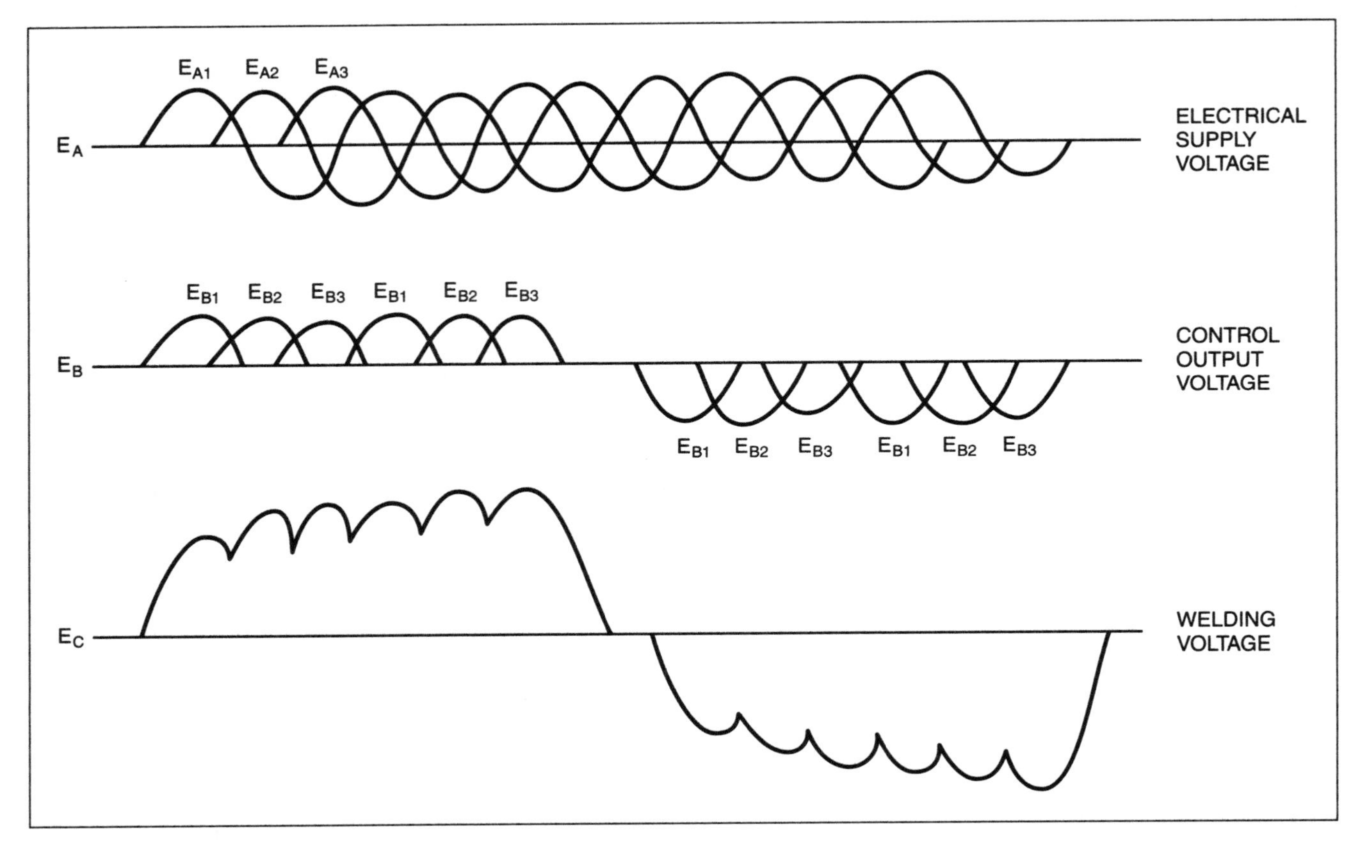

Figure 4.40—Typical Three-Phase Frequency Converter Welding Control Waveform

Hold Time

Hold time is the period during which the electrode force is maintained on the weld after current ceases. During this interval the weld is allowed to solidify and cool.

ADDITIONAL WELDING FUNCTIONS

Many controls designed for stand-alone operation regulate various welding functions in a preset format. Other controls use a free programmable format that allows the configuration of an almost unlimited combination of welding functions. This section covers the most common welding functions.

Repeat Off Time

In addition to squeeze time, weld time and hold time, most controls operating a single welding electrode provide a fourth function, "repeat off time." With the control in repeat mode, this function times the interval between the end of the welding sequence and the initiation of another sequence as long as the control initiation is maintained. This time is provided to allow the electrodes to be retracted to a position that permits moving to the next weld location. For seam welding machines, this time corresponds to the spacing between adjacent roll spot welds.

For manual operation of a welding gun, the off time can be adapted to the operator movement of the gun, so that while the gun is open during off time, the operator can move to the next weld position. Where large numbers of welds are required and when spacing between welds is not critical, manual welding in this manner can make many welds in a short time.

Preheat

This function occurs at the beginning of the weld sequence and typically provides a low level of heat to either preheat heavy metal or to fully seat the electrodes

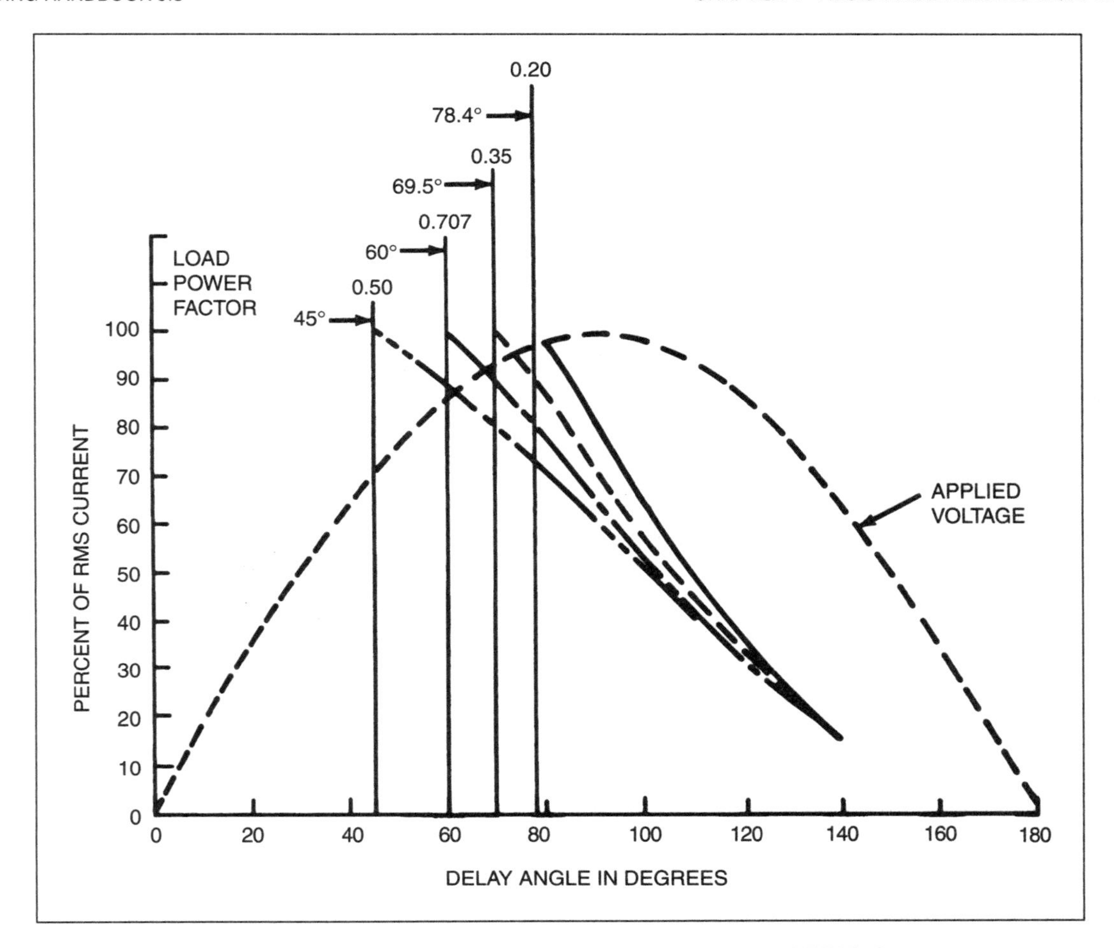

Figure 4.41—Relationship between Percentage of RMS Current and Firing-Delay Angle for Various Power Factors

on the workpiece prior to the application of full welding current. Preheat can be used also to liquefy a coating, such as zinc on galvanized steel, and allow the coating to be removed from the weld nugget area prior to the application of welding heat.

Upslope

This function applies heat that starts at a low level and increases in a linear fashion. Upslope serves the same function as preheat, but the ramping of heat values provides more ability for matching the heat input capability of the workpiece.

Downslope

During the downslope function, heat decreases in a linear fashion. The gradual decrease in current reduces the cooling rate of the weld. When welding hardenable steels, it may be useful to decrease the current to minimize the cooling rate and also the tendency for cracking.

Postheat

The postheat function occurs after downslope and typically provides a low level of heat to control the cooling of a weld nugget in thermally sensitive materials.

Quench and Temper

This dual function occurs at the end of a welding sequence. The quench time has no heat flow and allows the nugget to cool. The temper portion typically provides a low level of heat for a relatively long time. This sequence is used to remove brittleness from weld nuggets in high-carbon steels.

Forge Delay

Forge delay applies a high force between the electrodes, typically during the last cycle of a weld on thick material or heat-sensitive material. This high force is applied as the nugget cools and can be helpful in eliminating or reducing the occurrence of stress cracks and voids in the weld nugget.

Pulsation

Pulsation is the repetition of a number of weld intervals that are spaced by a cooling interval (cool time). This allows the application of additional current to increase the size of the weld nugget while the cooling interval permits cooling of the contact surface of the electrode. This function is used when welding very thick metals that require high thermal input, or when making welds near the edge of a flange.

AUXILIARY CONTROLS

Additional control devices work in conjunction with resistance welding controls to maximize reliability and productivity.

Cascade Controls

A single weld control can incorporate up to six separate welding contactors and operate them in sequence. This type of firing sequence is referred to as *cascade firing*. The advantage of cascade firing is in the reduction of weld controls required. However, the disadvantage to cascade firing is the loss of cycle time while waiting for the completion of each weld.

Since each contactor is operated independently in rapid succession, the demand on the electrical service is less than would be required for simultaneous operation. Independent heat and time control are maintained, as is the ability to monitor each weld made. The use of a cascade control for multiple welding sequences provides the fastest possible welding sequence with one control on a single-phase electrical service.

Valve Sequence Controls

A valve sequence control operates a number of actuators, such as pneumatic weld cylinders, that are connected to a single welding control. A different actuator is advanced to close the welding secondary each time the welding contactor is cycled. This control allows a reduced number of resistance welding transformers to be used and results in a lower line-voltage demand. This system is somewhat slower than cascade firing because additional time is required to allow the electrodes to separate from the workpiece.

Load Distribution Controls

A load distribution control manages the number of welding contactors or machines that can operate simultaneously. The electric service load distribution can be controlled through sequence interlocking or by the sensing of the applied load. Sequence interlocking ensures that only a prescribed amount of load is applied at any time to the electrical service.

If the load distribution control uses line voltage as the enabling parameter, load scheduling also may be required. For example, if several welding machines are waiting for the electrical service voltage to recover from a sag condition, it is not good practice to let all of the welding controls sequence simultaneously when the voltage returns to the desired range. A common solution is to provide unique recovery delay timer settings for different machines or groups of machines. One machine may be set to wait 10 cycles after 460 V and the next will be programmed to wait 20 cycles after 460 V. The second welding machine will either continue to wait if the first machine also pulls down the line voltage, or it will initiate normally after its programmed delay.

QUALITY CONTROL FUNCTIONS

A number of factors affect the consistency of resistance spot welds during a production run. These include line voltage variations, electrode deterioration, changes in workpiece surface resistance, shunting current paths, and variations in the electrode force system. Several systems are available to monitor specific welding variables or actions that occur during the welding cycle. If the monitor detects a fault, it can activate one or more of the following functions:

1. Turn on an alarm or signal light,
2. Document the information,
3. Reject or identify the faulty component,
4. Interrupt the process until the problem is corrected,
5. Alter time or current for the next weld, and
6. Change a variable during the weld cycle to ensure a good weld.

Variables that affect process stability and weld consistency include weld time, welding current, impedance, welding energy, and electrode force. Physical changes that take place in the weld zone are temperature, expansion and contraction, electrical resistance, and in some cases, metal expulsion. Monitoring devices can compute either weld energy or impedance by measuring welding voltage, current, resistance, or time. When the computed value falls below the acceptable limit, the monitoring devices can notify the operator or automatically adjust one or more of the variables prior to the next weld.

Several systems of adaptive feedback have been developed to improve the capability of making acceptable welds consistently. Adaptive feedback systems, whether used singly or in combination, have certain limitations. As examples of limitations, the systems may require frequent calibration; they may work on single-point welding machines only; or they may add significantly to machine maintenance. While the systems described below are available, they are not widely used in general industry.

The relationship between electrode indentation and weld strength is sometimes used as a control parameter. By controlling electrode indentation it is believed that welds of consistent strength can be obtained before welding current is terminated. A welding control based on this principle has been developed.

A relationship between nugget expansion and weld strength also has been established. Instruments are available to monitor nugget expansion and either increase or decrease the welding current in real time with reference to a baseline nugget/time expansion curve. The objective is to obtain acceptable welds consistently. This type of feedback control can compensate for any shunting effect, even in aluminum alloys.

Manual welding guns are widely used in the automotive industry. With a specially designed welding gun, adaptive feedback control can be achieved using the electrode indentation method. However, nugget expansion feedback control is difficult to achieve when a portable gun is used. For this reason, the resistance method and the acoustic emission analysis method are used to improve the performance of a manual gun.

In the resistance method, a resistance/time curve for a good weld is established. If the resistance/time curve of a subsequent weld deviates from this baseline curve, thereby indicating that expulsion is imminent, welding current is terminated.

The acoustic method detects metallurgical actions such as melting, expulsion, solidification, phase transformation, and cracking by the acoustic waves they emit. Each has a distinguishable waveform and amplitude. By detecting the acoustic waves at the threshold of expulsion, welding current can be terminated to obtain a strong weld.

As an alternative to either taking immediate action (terminating current) or passive monitoring (notifying the operator), some controls are capable of analyzing the data from many welds and detecting trends that can be used to maintain high quality welds. Trend analysis allows the control to compensate for the lowering of weld strength and slow deviations from the desirable weld results by varying process conditions.

ELECTRIC SERVICE EQUIPMENT

Electric service demand depends on the welding process and the design of the welding machine. An adequate power supply is one of the prerequisites for high-production resistance welding. The major part of a line power system for an industrial plant is within the plant itself. The plant electrical service generally consists of the incoming power, switching apparatus, transformers, and conductors.

ELECTRICAL SERVICE TRANSFORMERS

When planning the installation of a resistance welding machine, it is necessary to determine if the power supplied to the plant is adequate. This includes the kVA rating of the electrical service transformer and the size of the electric service conductors. The electric service transformer is connected to a primary feeder capacity of 2300 V, 4800 V, 7500 V, or 13 000 V and produces 230 V or 460 V power. The electric service conductors are the leads between the electric service transformer and the welding machine.

The adequacy of the electric service transformer and conductors is governed by two factors: the permissible voltage drop and permissible heating. The permissible voltage drop is the determining factor in the majority of installations, but consideration also must be given to heating.

The size of the electric service transformer for single-point welding machines should at least equal the value of the kVA demand during welding. Electric service transformers have impedance that is generally in the area of 5%. This means that at their kVA rating, the voltage drop on the secondary circuit will be 5% of the rated voltage. Also, the electric service conductors generally are sized to have no more than a 5% voltage drop. This will add up to a 10% voltage drop at the welding machine, which is the maximum that most machine manufacturers recommend for their products.

To determine the size of the electrical service transformer required to serve a welding machine on the basis of voltage drop, it is first necessary to determine the maximum permissible voltage drop specified by the welding machine manufacturer. If this information is not readily available, then a voltage drop of 5% or less should be used. When the same electric service transformer is used with two or more machines, the voltage drop caused by one machine will be reflected in the operation of the second machine. In this case, it is advisable to confine the total voltage drop to not more than 10% to ensure consistent weld quality. The voltage drop should be measured at the welding machine location. The percentage of voltage drop is calculated by the following formula:

Voltage drop, % =

$$\frac{\text{(No-load voltage)} - \text{(full-load voltage)}}{\text{(No-load voltage)}} \times 100 \qquad (4.3)$$

BUS OR FEEDER SYSTEM

In general, the bus or feeder conductors from the electric service transformer to the machines always should be as short as possible and should be designed for low reactance to minimize the voltage drop in the line. The simplest and most economical power line consists of insulated wires taped together in a conduit. When only two or three machines are to be served at a common location, this construction is effective and economical. Bus duct construction that permits easy welding machine supply connections at frequent intervals along the length of the duct is desirable in production plants where manufacturing layouts are continually changing.

As previously described, load distribution controls are available to interlock two or more machines to prevent simultaneous firing and the accompanying excessive voltage drop. Any scheme of interlocking will cause some curtailment in production.

INSTALLATION

Resistance welding machines should be connected to the electric service according to electrical codes and the recommendations of the welding machine manufacturer. The primary cable should be sized to allow for both thermal drop and voltage drop.

A means for disconnecting the resistance welding control from the service equipment also should be provided. It is desirable that this means of disconnecting incorporate suitable protection against overload or fault, such as a short-circuit. The protection device must have adequate interrupting capacity to safely disconnect the machine if the welding circuit is shorted and the machine is at full load. The required interrupting value may be from two to four times the nameplate rating of the machine.

When fuses are used, the fault values of the fuses should be in accordance with the machine manufacturer's instructions. The means of disconnecting must be suitably sized to prevent the welding of the contacts and to prevent arc flash. If a circuit breaker is used, it may be possible to incorporate a pushbutton to remotely trip the breaker in the event of a fault.

ECONOMICS

Resistance welding is considered to be one of the most efficient joining processes because the process occurs very rapidly, and the fabrication and maintenance of tooling and equipment is relatively straightforward. The heat and metal required by the process are provided by the workpieces.

The economic benefits of resistance welding are based on the availability of modern equipment capable of achieving very high production rates. The resistance welding process is easily automated through the use robots or dedicated tooling. The high levels of automation minimize direct labor cost. In highly automated resistance welding systems with electrode cap changers, the only labor cost is periodic maintenance.

Preventive or periodic maintenance costs may include checking guards or safety features for proper functioning; replacement of cables or shunts; checking tightness of bolted joints, or inspection for leaking fluids.

At lower levels of automation, an operator is required to load and unload the workpieces, but training costs are minimal. Various levels of automation may incorporate clamping, fixtures, component feeders, or additional operations on surface preparation of the workpiece. The operator may be trained for operation and periodic maintenance of the equipment.

Because typical weld times are a fraction of a second, the cost of electricity per weld is typically calculated as a fraction of a cent. Studies on a typical automobile having 2000 to 3000 resistance spot welds have concluded that the cost of electricity ranges from 5 dollars to 10 dollars. When determining the total electrical cost, consideration must be given to the operating cost of pumps, chillers, compressed air and robot motors.

Filler metals or shielding gases generally are not required in resistance welding, thus the associated cost is eliminated. Although no consumables are used in resistance welding, the cost of periodic replacement of welding electrode caps must be considered. At less frequent intervals, flexible secondary conductors such as shunts or cables may need to be replaced.

Resistance welding requires a high capital cost for equipment to achieve high production rates. This cost varies with the degree of automation and the cost is generally spread across the number of weldments planned for the equipment.

Cost justification for resistance welding equipment requires evaluation of high production rates balanced against equipment and utilities costs, the service life expectancy of the equipment, and the number of welded assemblies produced.

SAFE PRACTICES

Resistance welding processes are widely used in high-speed production operations, especially in the automobile and appliance manufacturing industries. The hazards encountered during resistance welding require protection against electrical shock, hot materials, sharp metal edges, expulsion of hot metal, fumes, pinch points, electromagnetic fields, and stored energy devices.

SAFETY STANDARDS

The American Welding Society publishes the standard, *Safety in Welding, Cutting, and Allied Processes,* ANSI Z49.1,[6] which covers the general safety and health hazards of welding. *Recommended Practices for Resistance Welding,* AWS C1.1M/C1.1 has a section on safe practices specific to resistance welding.[7] Additional information covering health and safety topics can be obtained from the publishers of safety standards listed in Appendices A and B of this volume.

Mandatory health and safety standards for industrial operations are contained in *Occupational Safety and Health Standards for General Industry,* published by the United States government.[8]

Resistance welding processes include spot, seam, flash, upset, and projection welding, in which a wide range of machine types are used. The main hazards that may arise with the processes and equipment are the following:

1. Electric shock due to contact with hazardous voltage;
2. Eye injury or fires caused by ejection of small particles of molten metal from the weld;
3. Crushing of some part of the body between the electrodes or other moving components of the machine; and
4. Welding fumes from the workpieces or from oil, lubricant, or other contaminants on the workpieces.

Each installation must be evaluated individually to accommodate the variations in resistance welding operations. General considerations include electrical shock and flash guards.

Personnel must be protected from electrical shock. The proper use of direct grounding conductors, isolation contactors, ground reactors or ground impedance must be ensured. Further information is published in the American National Standard Institute publication, ANSI Z49.1; the Occupational of Safety and Health Act (OSHA) publication, *General Industry Safety Standards,* Part 12,[9] and by the International Engineering Consortium (IEC),[10] National Electrical Manufacturers Association (NEMA),[11] and the Resistance Welding Manufacturing Alliance (RWMA) *Bulletin 16.*[12]

Flash guards of suitable fire-resistant material must be provided to control flying sparks and molten metal, especially on flash welding equipment.

Installation

All equipment should be installed in conformance with *National Electric Code,* NFPA No. 70.[13] and other applicable state and local codes. The equipment should be installed and maintained by qualified personnel. Prior to use in production, the equipment should be inspected by personnel responsible for safety to ensure that it is safe to operate.

PERSONAL PROTECTION

Personal protective equipment must be employed to protect the operator and others who may be in the area. Eye protection against expelled metal particles must be provided by a guard made of suitable fire-resistant material, or by using approved personal protective

6. American National Standards Institute (ANSI) Accredited Standards Committee Z49.1, 2005, *Safety in Welding, Cutting, and Allied Processes,* Miami: American Welding Society. May be downloaded from the Internet at: www.aws.org
7. American Welding Society (AWS) Committee on Resistance Welding, 2000, *Recommended Practices for Resistance Welding,* AWS C1M/C.1, Miami, American Welding Society.
8. Occupational Safety and Health Administration (OSHA), *Occupational Safety and Health Standards for General Industry, in Code of Federal Regulations (CFR),* Title 29 CFR 1910, Subpart Q, Washington, D.C.: Superintendent of Documents, U.S. Government Printing Office. Also available on the Internet at: www.osha.gov.
9. See Reference 11.
10. International Electrotechnical Commission (IEC), IEC Central Office, 3, rue de Varembé, CH-1211, Geneva 20, Switzerland.
11. National Electrical Manufacturers Association ((NEMA) 1300 N. 17 Street, Suite 1752, Rossyln, VA 22209.
12. Resistance Welding Manufacturing Alliance (RWMA) *Bulletin 16,* Miami: RWMA/American Welding Society.
13. National Fire Protection Association (NFPA), *National Electric Code,* NFPA 70, Quincy, Massachusetts: National Fire Protection Association.

eyewear. Safety glasses with side shields are recommended in all work areas. Shields or curtains should be installed to protect against metal expulsion. Gloves made of a heat-resistant and cut-resistant material should be worn to protect against the residual heat of the resistance welding process and workpieces with sharp edges. Safety footwear, aprons, and other equipment also may be required.

PINCH HAZARDS

Protection of personnel requires that operator remain safe from pinch points. In addition to the movement of welding electrodes, various automated devices for clamping, fixturing, or material handling may pose pinch-point conditions. The operator must be protected from pinch points by the appropriate use of guards, barriers, sensors, and other safety devices such as anti-tie-down palm buttons. Controls must be in place to prevent the unintended operation of machinery.

All chains, gears, operating linkages, and belts associated with the welding equipment should be protected in accordance with the American Society of Mechanical Engineers standard, *Safety Standard for Mechanical Power Transmission Apparatus*, ASME B15.1.[14]

ELECTRICAL HAZARDS

Precautions must be taken to address hazardous voltage that exists in the welding environment.

Grounding

The welding transformer secondary should be grounded by one of the following methods:

1. Permanent grounding of the welding secondary circuit by means of a suitably sized conductor or impedance, or
2. Connection of a grounding reactor across the secondary winding with a reactor tap to ground.

An OSHA-permitted alternative on stationary machines, an isolation contactor, may be used to open all of the primary lines. This applies only when operators cannot touch the workpiece or any of the secondary conductors while welds are in progress. It should be noted that it is not acceptable for an operator to load workpieces onto a turntable while a robot is welding on the other side. The turntable moves on lubricated bearings and is not considered to be grounded, so an electrical failure in the transformer on the robot could produce hazardous voltage on the turntable.

This secondary grounding should be done with an understanding of how undesirable transient circulating currents can flow between transformers in the system. When multiple-phase primary supplies or different secondary voltages, or both, are used for the several adjacent transformers, the use of ground reactors or isolation contactors may be necessary.

Capacitors

Resistance welding equipment and control panels containing capacitors involving high voltages (over 550 volts rms) must have adequate electrical insulation and be completely enclosed. All enclosure doors must be provided with suitable interlock switches, and the switch contacts must be wired into the control circuit.

The interlocks must effectively interrupt power and discharge all high voltage capacitors into a suitable resistive load when the door or panel is open. In addition, a manually operated switch or suitable positive device should be provided to assure complete discharge of all high-voltage capacitors. Supplemental visual indication also may be provided to verify that all hazardous voltage has been discharged.

Other Stored-Energy Sources

Lock-out procedures must include the elimination of the hazard that may arise from other stored energy sources. In addition to charged capacitors, these sources may include such things as compressed air, pressurized cooling water, gravity, and springs. There should be a way to relieve the stored energy or contain it by the application of blocks, locks or similar devices.

Locks and Interlocks

All doors, access panels, and control panels of resistance welding machines must be kept locked or interlocked. This is necessary to prevent access by unauthorized persons.

Stop Buttons

One or more emergency stop buttons should be provided on all welding machines, with a minimum of one at each operator position.

ELECTROMAGNETIC FIELDS

While there has been no link established between exposure to electromagnetic fields (EMF) and undesirable health effects, it is considered prudent to minimize

14. American Society of Mechanical Engineers (ASME), *Safety Standard for Mechanical Power Transmission Apparatus*. ASME B15.1, New York: American Society of Mechanical Engineers.

exposure whenever possible. The source of EMF is the electrical current flowing through conductors. The intensity relates directly to the value and frequency of the current. Electromagnetic field strength diminishes rapidly with increasing distance away from current-carrying conductors. The effect of electromagnetic fields can be observed in the movement of flexible conductors or the attraction to any magnetic objects such as tools, or nuts and bolts left loose in the vicinity of the electromagnetic field.

Personnel with pacemakers or other medical devices that may be susceptible to electromagnetic fields should consult their doctor to verify that it is safe to operate the resistance welding equipment.

FUMES AND METALLIC DUST

Heat generated during welding generally causes fumes. The fumes result from heat reacting with the base metal and any coatings on the base metal. The coatings can be inherent to the metal, such as galvanizing, primer, corrosion-resistant coatings, and plating, and also byproducts of handling and forming, such as forming lubricants and corrosion inhibitors, drawing compounds and oils. Adequate ventilation must be provided so that personnel will not be exposed to harmful levels of fumes. When necessary, monitoring worker fume exposure should be conducted by a trained industrial hygienist.

Fine particles of metal are created by expulsion during the resistance welding process and some may be suspended in the air. The particles also can coat surfaces and come in contact with the operator or contaminate the assembly. Adequate ventilation and dust collection should be provided to minimize personnel exposure. Ventilation systems should be rated for this purpose, since finely divided metal particles can be extremely combustible.

Another potential source of dust arises when maintenance personnel are doing work on electrodes and other metallic materials in the welding machine environment. Adequate personnel protective equipment must be used whenever this type of work is done. Special precautions may be necessary if the metal contains chromium, beryllium, or other elements that have restricted exposure limits.

SAFE OPERATION OF EQUIPMENT

The safety devices that may be required must be determined after careful consideration of the hazards involving equipment. Some considerations include the following:

1. Initiating devices on welding equipment, such as push buttons and foot switches, should be arranged or guarded to prevent the operator from inadvertently activating them;
2. In installations in which the operator's hands can be expected to pass under the point of operation. These machines should be effectively guarded by suitable devices such as proximity-sensing devices, latches, blocks, barriers, or dual hand controls;
3. On press-type machines and flash and upset welding machines, static safety devices such as pins, blocks, or latches should be provided to prevent movement of the platen or head during maintenance or setup for welding. More than one device may be required, but each device should be capable of sustaining the load; and
4. All suspended manual welding gun equipment, with the exception of the gun assembly, should have a support system that can withstand the total shock load in the event of failure of any component of the system. The system should be fail-safe. The use of adequate devices such as cables, chains, or clamps is considered satisfactory.

SAFETY TRAINING

This OSHA code requires that resistance welding machine operators be properly instructed and judged competent to operate the equipment. Training should include an understanding of the hazards and safe practices associated with the resistance welding process. In addition to an understanding of the requirements of safe operation of the specific machine, operators must know what actions should be taken if a problem or fault should occur.

CONCLUSION

In the 100 years that resistance welding has been used in commercial applications, the development and refinement of equipment has been a constant. Recently, innovation has been driven by high energy costs and the need to weld new and challenging materials. Sophisticated algorithms and electronics have enabled unprecedented process reliability. This chapter has provided a general overview of the design and application of resistance welding equipment.

BIBLIOGRAPHY

American National Standards Institute (ANSI) Accredited Standards Committee Z49.1. 2005. *Safety in welding, cutting, and allied processes*. ANSI Z49.1:2005. Miami: American Welding Society.

American Society of Mechanical Engineers (ASME). *Safety standard for mechanical power transmission apparatus*. ASME B15.1. New York: American Society of Mechanical Engineers.

American Welding Society (AWS) and Society of Automotive Engineers (SAE). *Standard for automotive resistance spot welding electrodes*. AWS D8.6/SAE HS-J1156. Miami: American Welding Society.

American Welding Society (AWS) Committee on Definitions and Symbols. 2001. *Standard welding terms and definitions*. AWS A3.0:2001. Miami: American Welding Society.

American Welding Society (AWS) Committee on Resistance Welding. 2000. *Recommended practices for resistance welding*. AWS C1M/C.1. Miami: American Welding Society.

National Fire Protection Association (NFPA). *National electric code*. NFPA 70. Quincy, Massachusetts: National Fire Protection Association.

Occupational Safety and Health Administration (OSHA). *Occupational safety and health standards for general industry*, in *Code of federal regulations (CFR)*. Title 29 CFR 1910. Subpart Q. Washington D.C.: Superintendent of Documents. U.S. Government Printing Office.

Resistance Welding Manufacturing Alliance (RWMA), *Bulletin No. 5*. Miami: American Welding Society.

Resistance Welding Manufacturing Alliance (RWMA), *Bulletin No. 16, Resistance welding standards*. Miami: American Welding Society.

Resistance Welding Manufacturing Alliance (RWMA). 2003. *Resistance welding manual, revised fourth edition*. Miami: American Welding Society.

SUPPLEMENTARY READING LIST

Brown, B. M. 1987. A comparison of ac and dc resistance welding of automotive steels. *Welding Journal* 66(1): 18–23.

Bowers, R. J., C. D. Sorensen, and T. W. Eagar. 1990. Electrode geometry in resistance spot welding. *Welding Journal* 69(2): 455.

J. Darrah. 1995. Controllers, SCR contactors vital to resistance welding. *Welding Journal* 74(8): 53.

K. Hofman, M. Soter, C. Orsette, S. Villaire, and M. Prokator. 2005. AC or DC for resistance welding dual-phase 600. *Welding Journal* 84(1): 46.

M. Malberg, N. Bay. 1998. Electrical systems of resistance welding machines, methods for characterizing. *Welding Journal* 77(4): 59.

M. J. Karagoulis. 1994. Nuts-and-bolts approach to the control of resistance spot welding. *Welding Journal* 76(7): 27.

P. Wu, W. Zhang, and N. Bay. 2005. Characterization of dynamic mechanical properties of resistance welding machines. *Welding Journal* 84(1): 17-s–21-s.

R. L. Stuefen. 1980. AC or DC for aluminum resistance welding in consideration of electrical demand. *Welding Journal* 59(1): 39.

Zang, Hongyang, and J. Senkara. 2006. *Resistance welding: fundamentals and applications*. Boca Raton, Florida: Taylor and Francis CRC Press.

CHAPTER 5

HIGH-FREQUENCY WELDING

Prepared by the Welding Handbook Chapter Committee on High-Frequency Welding:

Menachem Kimchi, Chair
Edison Welding Institute

W. A. Peterson
Edison Welding Institute

P. F. Scott
Thermatool Corporation

R. D. Czerski
Names Group

Welding Handbook Volume 3 Committee Member:

S. P. Moran
Miller Electric Manufacturing Company

Contents

Photograph courtesy of Thermatool Corporation

CHAPTER 5

HIGH-FREQUENCY WELDING

INTRODUCTION

High-frequency resistance welding (HFRW) includes a group of resistance welding process variations that use high-frequency welding current to concentrate the welding heat at the desired location.[1,2] The heat generated by the electrical resistance of the workpiece to high-frequency currents produces the coalescence of metals, and an upsetting force usually is applied to produce a forged weld. High-frequency resistance welding is an automated process and is not adaptable to manual welding.

The two high-frequency resistance welding process variations discussed in this chapter are contact resistance welding and induction welding. Both variations use high-frequency current to produce the heat for welding. The current for contact high-frequency resistance welding is conducted into the workpiece through electrical contacts that physically touch the workpiece. For induction welding, the current is induced in the workpiece by magnetic coupling with an external induction coil. There is no physical contact with the workpiece.

High-frequency resistance welding was developed during the late 1940s and early 1950s to fill the need for high-integrity butt joints and seam welds in pipe and tubing. The progressive development of this technology, detailed in Figure 5.1, served as the basis for the modern high-speed pipe and tube welding systems in current use.

The principal application of high-frequency welding continues to be in the manufacture of seam-welded pipe and tube. The processes also are used in the manufacture of products such as spiral-fin boiler tubes, closed roll form shapes, and welded structural beams. Other examples of manufacturing applications are the welding of the metal shielding layer that surrounds high-frequency coaxial cable, the manufacture of butt joints in strip material and for solar panels, and for the butt joints in pipe and tubing.

Topics covered in this chapter include the fundamentals of high-frequency welding, equipment and controls, process variations, economics of the process and safe practices. Additional sources of information are listed in the Supplementary Reading List at the end of the chapter.

1. Welding terms and definitions used throughout this chapter are from *Standard Welding Terms and Definitions*, AWS A3.0:2001, Miami: American Welding Society.

2. At the time of the preparation of this chapter, the referenced codes and other standards were valid. If a code or other standard is cited without a date of publication, it is understood that the latest edition of the document referred to applies. If a code or other standard is cited with the date of publication, the citation refers to that edition only, and it is understood that any future revisions or amendments to the code or standard are not included; however, as codes and standards undergo frequent revision, the reader is encouraged to consult the most recent edition.

FUNDAMENTALS

High-frequency welding processes rely on the properties of high-frequency electricity and thermal conduction, which determine the distribution of heat in the workpieces. High-frequency contact welding and high-frequency induction welding are used to weld products made from coil, flat, or tubular stock with a constant joint symmetry throughout the length of the weld. Figure 5.2 illustrates basic joint designs used in high-frequency welding. Figure 5.2(A) and (B) are butt seam welds; Figure 5.2(C) is a mash seam weld produced with a mandrel, or backside/inside bar. Figure 5.2(D) is a butt joint design in strip metal; and Figure 5.2(E) shows a T-joint. Figure 5.2(F) and (G) are examples of helical pipe and spiral-fin tube joint designs. Figure 5.2(H) is an example of a seam weld in a closed roll shape; Figure 5.2(J) illustrates a butt joint in pipe, showing the placement of the coil. Figure 5.2(K) shows a butt joint in bar stock.

High-frequency current in metal conductors tends to flow at the surface of the metal at a relatively shallow

History of High-Frequency Resistance Welding of Pipe and Tube

1940 to 1950—Butt Joints

Welds in butt joints were made with 10-kHz motor-generator power sources equipped with induction coils that could be opened for removal from the pipe or tube after the weld was formed. A municipal utility was the first to use mobile high-frequency welding equipment in the streets of New York City for the welding of butt joints in pipe. Pipes as large as 305 mm (12 in.) in diameter with wall thicknesses up to 8 mm (5/16 in.) were welded. The pipe ends were fitted and pressed together, then were induction-heated to the forge-welding temperature in about 60 seconds.

1949—Forge Welding of Continuous Seams

The first high-frequency induction welding system was introduced for the continuous forge welding of the longitudinal seam in small steel tubes. The system used a 10-kHz motor generator equipped with a split-return induction coil suspended over the seam, similar to those used for continuously normalizing the weld seam in pipe welding. These systems were operated for many years, but had the disadvantage of heating a large portion of the pipe.

1952—Induction Welding of Continuous Seams

Thomas Crawford tested the continuous induction welding of longitudinal seams in various metal tubes, including some with electrical cable inside, using the induction welding process. Tests were made at an electrical frequency of 400 kHz and used an induction coil surrounding the tube. This technique for the seam welding of pipe and tube was highly successful and continues to be employed for pipe with outside diameters up to 406 mm (16 in.).

1952—Butt, Lap, and T-Joints

Wallace Rudd and Robert Stanton invented a high-frequency contact welding process for welding a large variety of continuous joints. The Rudd and Stanton system operated at 400 kHz and introduced the high-frequency welding current directly into the workpiece by means of sliding contacts. The sliding contacts permitted the production of butt joints, lap joints, and T-joints in pipe, strip, and structural products.

1954—Continuous Seam Welding

Successful tests were made for welding the longitudinal seam in aluminum tubing, leading to the introduction of the first commercial welding of aluminum irrigation tubing. This high-frequency welding equipment used motor-generator sets with a maximum output frequency of about 10 kHz. By the early 1960s these systems had been superseded by high-frequency vacuum tube oscillators that typically operated around 400 kHz.

1986—Solid-State Induction Welding

The first solid-state induction welding machine was installed and operated at the Société Meusienne in France. This equipment operated at a welding frequency of about 100 kHz.

1990—Solid-State Inverter Power Sources

Solid-state inverters replaced vacuum-tube oscillators in many new equipment installations during the 1990s. These units were more easily configured to operate at a variety of welding frequencies between 80 kHz and 800 kHz. Experience, supported by mathematical theory, showed that selecting the correct welding frequency could greatly improve weld quality, improve mill yield for difficult-to-weld metals, and adapt the properties of the weld to the specific application. Seams in pipes up to 660 mm (26 in.) in diameter were welded using solid-state inverter power sources operating between 80 kHz and 150 kHz with power outputs of up to 1800 kW.

2000s—Selectable-Parameter Power Sources

High-frequency welding power sources with selectable parameters became available that allowed the continuous adjustment of the parameters of this process. This increased the range of pipe and tube products that could be produced by a single tube mill.

Figure 5.1—Historical Overview of High-Frequency Resistance Welding of Pipe and Tubing

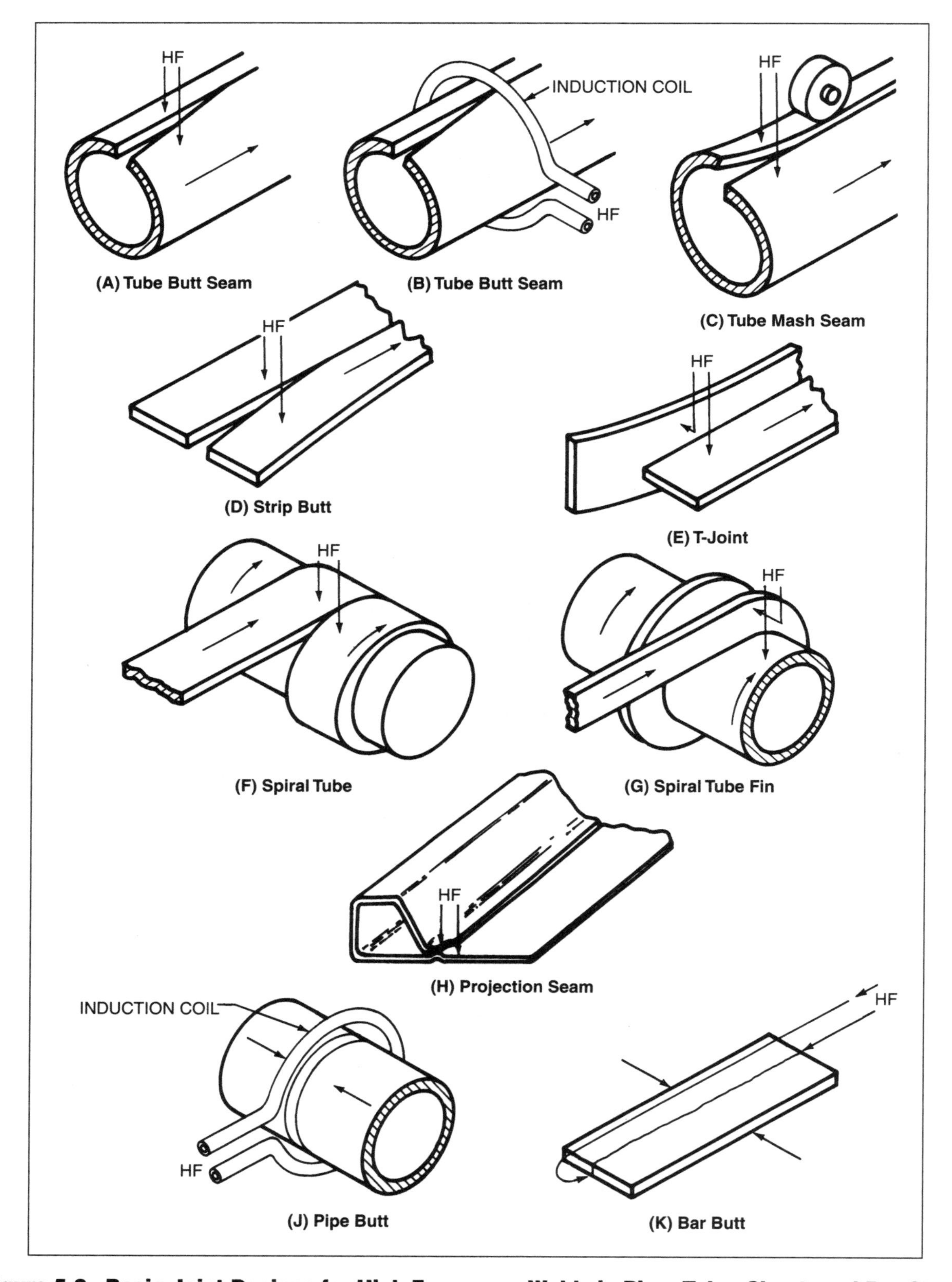

Figure 5.2—Basic Joint Designs for High-Frequency Welds in Pipe, Tube, Sheet, and Bar Stock

depth, which becomes shallower as the electrical frequency of the power source is increased. This commonly is called the *skin effect*. The depth of electrical current penetration into the surface of the conductor also is a function of electrical resistivity and magnetic permeability, the values of which depend on temperature. Thus the depth of penetration also is a function of the temperature of the material. In most metals, the electrical resistivity increases with temperature; as the temperature of the weld area increases, so does the depth of penetration. For example, the resistivity of low-carbon steel increases by a factor of five between room temperature to welding temperature. Metals that are magnetic at room temperature lose the magnetic properties above the Curie temperature.[3] When this happens, the depth of penetration increases drastically in the portion of metal that is above the Curie temperature while remaining much shallower in the metal that is below the Curie temperature. When these effects are combined in steel heated at a frequency of 400 kilohertz (kHz), the depth of current penetration is 0.05 millimeter (mm) (0.002 inch [in.]) at room temperature, while it is 0.8 mm (0.03 in.) at 800°C (1470°F). The depth of current penetration for several metals as a function of frequency is shown in Figure 5.3.

The second important physical effect governing the high-frequency welding process is thermal conduction of the heat generated by the electric currents in the workpiece. Control of the thermal conduction and the penetration depth provides control of the depth of heating in the metal. Because thermal conduction is a time-dependent process, the depth to which the heat will conduct depends on the welding speed and the length of the electrical current path in the workpiece. If the current

3. Curie temperature is the point at which a transition in substance occurs; for example, the temperature at which a transition occurs between the ferromagnetic and paramagnetic phases.

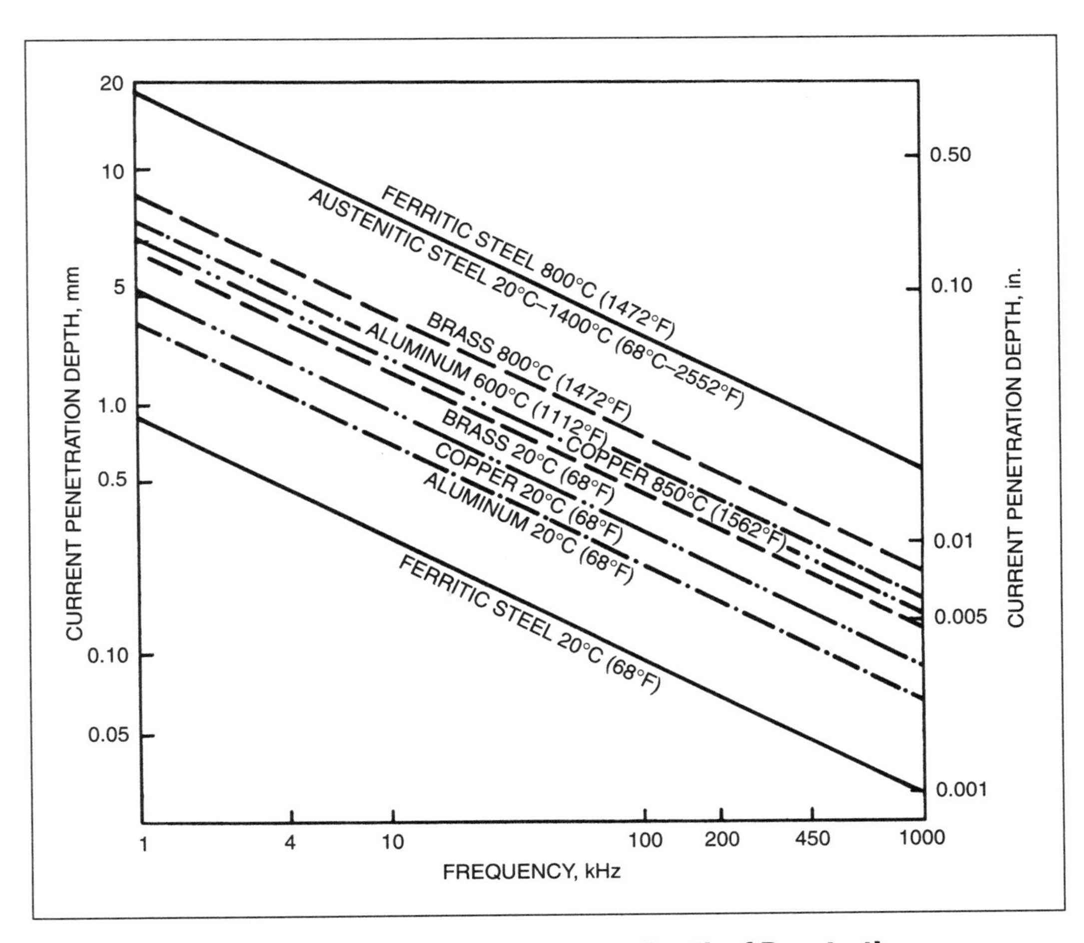

Figure 5.3—Effect of Frequency on Depth of Penetration into Various Metals at Selected Temperatures

path is shortened or the welding speed is increased, the heat generated by the electric current in the workpiece will be more concentrated and intense. However, if the current path is lengthened or the welding speed is reduced, the heat generated by the electric current will be dispersed and less intense. The effect of thermal conduction is especially important when welding metals with high thermal conductivity, such as copper or aluminum. It is not possible to weld these materials if the current path is too long or the welding speed is too slow. Changing the electrical frequency of the high-frequency current can compensate for changes in welding speed or the length of the weld path, and the choice of frequency, welding speed, and path length can adapt the shape of the heat-affected zone (HAZ) to optimize the properties of the weld metal for a particular application.

It should be noted that the high-frequency current path at the surface of the workpiece is controlled by how close it is, or its proximity to its own return path. This phenomenon, called the *proximity effect*, is illustrated in Figure 5.4. The proximity effect becomes more pronounced as the frequency is increased. Increased frequency reduces the penetration depth by confining the current to a shallower and narrower path. This property is illustrated in Figures 5.5(A) and (B), where current patterns are shown for workpieces with the same geometry when the frequency of the electric current is 60 hertz (Hz) and 10 kHz.

Referring to Figure 5.5, a 60-Hz current in the steel plate flows in the opposite direction of the current flowing in the adjacent proximity conductor. In this case, the size and shape of the proximity conductor have negligible effect on the distribution of current in the steel plate. This example also shows that penetration at 60 Hz is deep compared to the thickness of the plate. As a result, the current flows fairly uniformly throughout the cross section of the plate. When the 10-kHz current is applied to the same system, the current flowing in the steel plate is confined to a relatively narrow band immediately beneath the proximity conductor. This narrow band is the path of lowest inductive reactance for the current in the plate. The shape and magnetic surroundings of the proximity conductor have a considerable effect on the distribution of the current in the steel plate, but have no effect on the depth of penetration.

Controlling the concentration of current in a workpiece can achieve extremely high heating rates and high temperatures in a localized area, thereby heating only the region of the workpiece to be welded. By controlling the relative position of the surfaces to be welded, the heat can be positioned at the weld interface where it is needed without excessively heating the rest of the metal. The effect of changing the spacing and geometry of the conductors is shown in Figure 5.5(C). The closer proximity conductor develops a more confined current path. A rectangular proximity conductor with the narrow edge at the same distance from the steel plate as the close round conductor exhibits a broader current distribution in the plate. If a magnetic core were placed around the proximity conductor, the current would be further concentrated and heating would take place directly below the proximity conductor, as shown in the figure.

If the two conductors, with currents flowing in opposite directions, are metal sheets placed edge to edge in a plane with a small gap between them, the proximity effect will cause the current to concentrate at the two adjacent edges, thus causing them to heat. The skin effect will confine the currents to a shallow depth at those edges. This is the situation that characterizes all high-frequency welding applications.

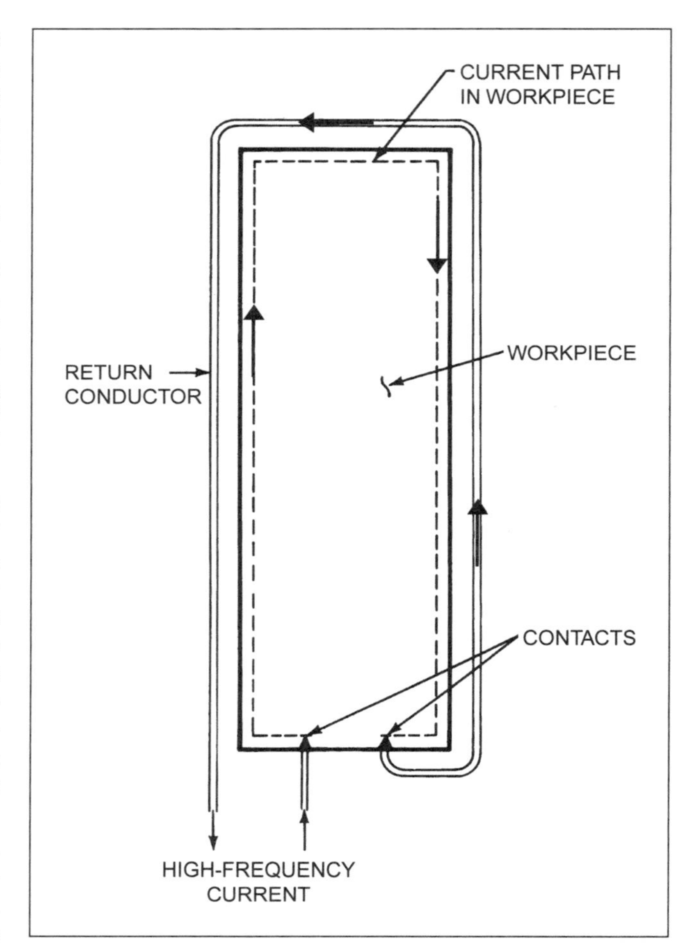

Figure 5.4—Restriction of the Flow Path of High-Frequency Current by the Proximity Effect of the Return Conductor

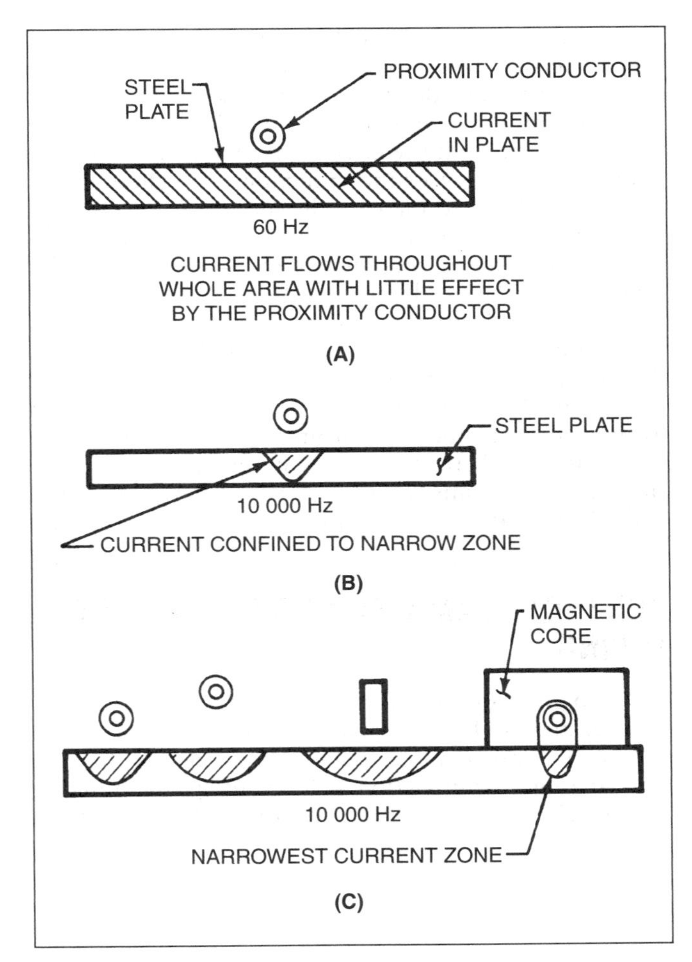

Figure 5.5—Depth and Distribution of Current Adjacent to Various Proximity Conductors

ADVANTAGES OF HIGH-FREQUENCY WELDING

A wide range of commonly used metals can be welded, including low-carbon and alloy steels, ferritic and austenitic stainless steels, and many aluminum, copper, titanium, and nickel alloys.

Because the concentrated high-frequency current heats only a small volume of metal at the weld interface, the process can produce welds at very high welding speeds and with high energy efficiency. High-frequency resistance welding can be accomplished with a much lower current and less power than is required for low-frequency or direct-current resistance welding. Welds are produced with a very narrow and controllable heat-affected zone and with no superfluous cast structures. This often eliminates the need for postweld heat treatment.

Oxidation and discoloration of the metal and distortion of the workpiece are minimal. Discoloration may be further reduced by the choice of welding frequency.

Maximum speeds normally are limited by mechanical considerations of material handling, forming and cutting. Minimum speeds are limited by material properties, excessive thermal conduction such that the heat dissipates from the weld area before bringing it to sufficient temperature, and weld quality requirements.

The fitup of the surfaces to be joined and the manner in which they are brought together are important if high-quality welds are to be produced. However, high-frequency welding is far more tolerant in this regard than some other processes.

Flux is almost never used, but can be introduced into the weld area in an inert gas stream. Inert gas shielding of the weld area generally is needed only for joining highly reactive metals such as titanium or for certain grades of stainless steel.

APPLICATIONS

The high-frequency induction and contact welding processes have been applied to a myriad of metal products made from bar, sheet, strip, pipe, and tubing.

INDUCTION SEAM WELDING OF PIPE AND TUBING

The welding of continuous-seam pipe and tubing is the predominant application of high-frequency induction welding. The pipe or tube is formed from metal strip in a continuous-roll forming mill and enters the welding area with the edges to be welded slightly separated. In the weld area, the open edges of the pipe or tube are brought together by a set of forge pressure rolls in a vee shape until the edges touch at the apex of the vee, where the weld is formed. The weld point occurs at the center of the mill forge rolls, which apply the pressure necessary to achieve a forged weld.

An induction coil, typically made of copper tubing or copper sheet with attached water-cooling tubes, encircles the tube (the workpiece) at a distance equal to one to two tube diameters ahead of the weld point. This distance, measured from the weld point to the edge of the nearest induction coil, is called the *vee length*. The induction coil induces a circumferential current in the tube strip that closes by traveling down the edge of the vee through the weld point and back to the portion of the tube under the induction coil. This is illustrated in Figure 5.6.

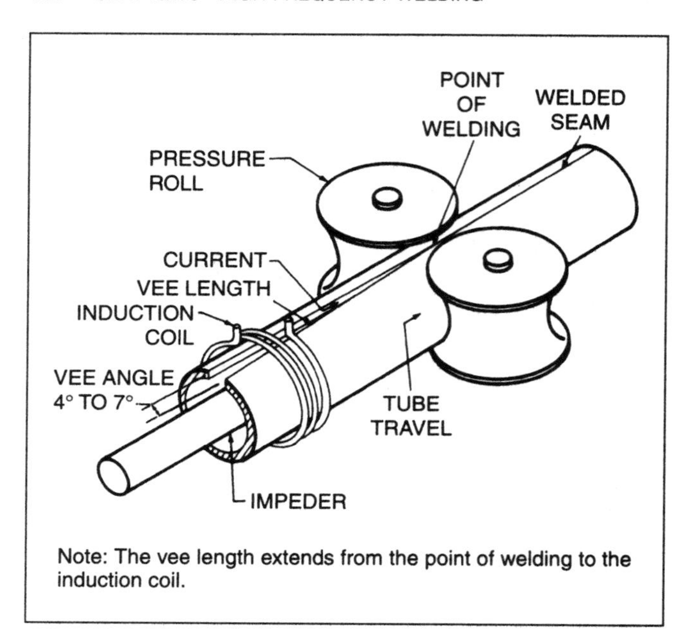

Note: The vee length extends from the point of welding to the induction coil.

Figure 5.6—High-Frequency Induction Seam Welding of a Tube

The high-frequency current flows along the edge of the weld vee due to the proximity effect (see Fundamentals), and the edges are resistance-heated to a shallow depth due to the skin effect.

The geometry of the weld vee is such that its length usually is between one and one half to two tube diameters long. The included angle of the vee generally is between 3° and 7°. If this angle is too small, arcing between the edges may occur, and it will be difficult to maintain the weld point at a fixed location. If the vee angle is too wide, the proximity effect will be weakened causing dispersed heating of the vee edges, and the edges may tend to buckle. The best vee angle depends on the characteristics of the tooling design and the metal to be welded. Variations in vee length and vee angle will cause variations in weld quality.

The welding speed and power source level are adjusted so that the two edges are at the welding or forge temperature when they reach the weld point. The forge rolls press the hot edges together, applying an upset force to complete the weld. Hot metal containing impurities from the faying surfaces of the joint is squeezed out of the weld in both directions, inside and outside the tube. The upset metal normally is trimmed off flush with the base metal on the outside of the tube, and sometimes is trimmed from the inside, depending on the application for the tube being produced.

An impeder, which is made from a magnetic material such as ferrite, generally is required to be placed inside the tube. The impeder is positioned so that it extends about 1.5 mm to 3 mm (1/16 in. to 1/8 in.) beyond the apex of the vee and the equivalent of 1 to 2 workpiece diameters upstream of the induction coil. The purpose of the impeder is to increase the inductive reactance of the current path around the inside wall of the workpiece. This reduces the current that would otherwise flow around the inside of the tube and cause an unacceptable loss of efficiency. The impeder also decreases the magnetic path length between the induction coil and the tube, further improving the efficiency of power transfer to the weld point. The impeder must be cooled to prevent its temperature from rising above its Curie temperature, where it becomes nonmagnetic. For ferrite, the Curie temperature typically is between 170°C and 340°C (340°F and 650°F).

CONTACT SEAM WELDING OF PIPE AND TUBING

The high-frequency contact welding process provides another means of welding continuous seams in pipe and tubing. The process essentially is the same as that described above for induction welding and is illustrated in Figure 5.7. The major difference is that sliding contacts

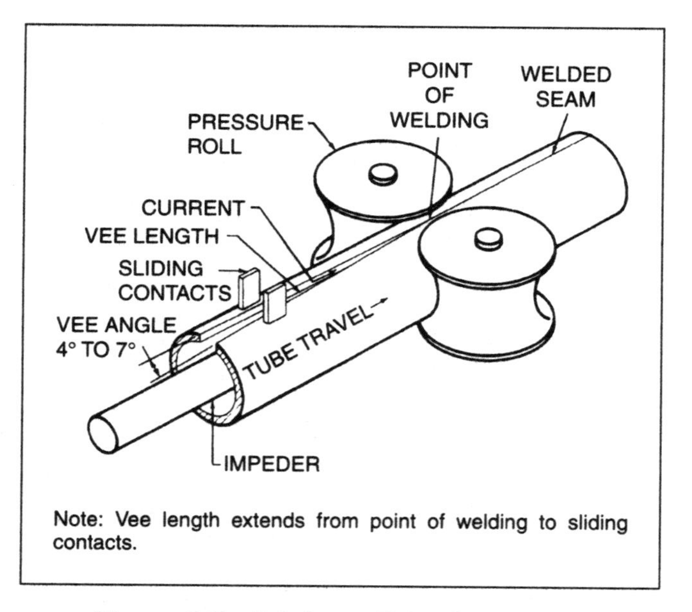

Note: Vee length extends from point of welding to sliding contacts.

Figure 5.7—Joining a Tube Seam with High-Frequency Resistance Welding Using Sliding Contacts

are placed on the tube adjacent to the unwelded edges at the vee length. With the contact process, the vee length generally is shorter than that used with the induction process. This is because the contact tips normally can be placed within the confines of the forge rolls where the induction coil must be placed sufficiently behind the forge rolls, so that the forge rolls are not inductively heated by the magnetic field of the induction coil. Because of the shorter vee lengths achievable with the contact process, an impeder often is not necessary, particularly for large-diameter tubes where the impedance of the current path inside the tube has significant inductive reactance.

Comparison of the Induction and Contact Processes for Pipe and Tube Seam Welding

The choice between the induction and contact processes for pipe and tube seam welding is the result of practical trade-offs between the following factors that differentiate the two processes.

High-frequency contact welding is a more efficient process than induction welding because of the shorter lengths and because there are no losses in the induction coil or in the path around the tube under it. For the seam welding of large-diameter pipe, the contact process can use as little as half the power required by the induction process.

Induction welding has the disadvantage that it is strictly a continuous process and the roll form mill generally has an accumulator to ensure a continuous supply of strip to the mill. Large-diameter tube is often welded coil to coil, requiring the roll form mill to be rethreaded between each coil run. In this situation, contact welding is preferred because the strip does not have to be threaded through the tight induction coil each time a tube is produced.

The major disadvantage of the contact process over the induction process is contact wear. This requires replacement of the contact tips at regular intervals, generally once every shift. The service life of the contact tips degenerates with increasing welding power level and for this reason, conventional single-shoe contact assemblies often are not operated above 600 kilowatts (kW) welding machine output power. Special dual-shoe contact assemblies that have two separate contacts at each side of the weld vee have proven to operate at 800 kW with acceptable contact tip wear.

When using the contact welding process under some conditions, arcing between the sliding contacts and the tube can occur. This may cause surface marking or "arc marks." In applications such as the manufacture of pipe to API standards for oil and gas transport, any arc marks must be removed by a subsequent grinding operation.

The choice between the induction and contact processes is often resolved by employing dual welding machines that allow operation with both the induction and contact processes. With these units, induction welding can be used to weld the smaller-diameter tube sizes in the tube mill product range, and contact welding can be used to weld the larger-diameters sizes.

Special Pipe and Tube Applications

Helically wound pipe and tube can be made with high-frequency welded lap or butt joints. Refer to Figure 5.2(F) and (G) for examples of helical pipe and tube joint designs. Helically wound corrugated culvert pipe with diameters between 600 mm (24 in.) and 2400 mm (72 in.) can be welded with the high-frequency process.

Zinc-coated or aluminum-coated tubing can be welded from precoated strip. After the weld has been made and the outside weld bead removed, a galvanized metal spray can be applied over the weld area to provide a complete coating. Alternatively, a steel strip can be welded and then cleaned, full-body heated by induction, and passed though a molten zinc bath in the production line to produce a galvanized tube directly from an uncoated strip. This process is used to manufacture most electrical conduit.

Tubes can also be hot-stretch reduced or cold stretch-reduced in line after welding. Induction heating generally is used to heat the tube prior to hot stretch-reducing. Alternatively, the tubes can be cold stretch-reduced inline and subsequently stress relieved by induction heating prior to coiling or cutting to length.

Coaxial Cable Shields. The concentric shield over the inside conductors of coaxial electrical cable is welded using the induction welding process. The shield is formed in a cable mill from very thin aluminum or copper strip as a covering over the center electrical conductors. The seam is then inductively welded using the same process used to manufacture seamed tubing. Although the center cable conductors preclude the use of an impeder, the shield is thin enough so that minimal weld power is required. Thus, the loss in efficiency is acceptable and the resulting heat levels are low enough to prevent damage to the center insulation of the conductor.

Seam Welding of Closed Roll Form Shapes. Applications requiring the welding of closed roll form shapes, such as automobile bumpers, store shelving supports and office partition columns, are a straightforward extension of pipe-and-tube seam welding. The difficulty is that the geometry of the workpiece often complicates the forge roll design. Thus, care must be exercised in assembling a system that presents the edges parallel to one another without waviness, and that applies sufficient forge pressure. (Refer to Figure 5.2[H], which shows an example of a joint design for closed roll form shapes.)

The cross-sectional geometry of the workpiece often determines the choice between the contact and induction processes. Lap welds often are made in roll form components and the theory of operation is similar to the joining process described above for butt joints. However, lap joints must be designed with consideration for the proximity effect, and the two faying surfaces must be brought together to form a vee.

Other processing may be integrated within the high-frequency welding mill either before or after welding. For example, patterns of holes may be punched in the strip prior to forming and welding, or a welded rectangular tube may be curved in a die sweep after welding to form an automobile bumper. A flying-shear cutoff system can be synchronized with these operations to provide a finished product requiring no subsequent machining operations.

Contact Welding of Structural Beams

Continuous high-frequency contact welding is not confined to welding closed shapes such as tube and pipe. It can be adapted to many other products such as the fabrication of structural I-beams, T-beams, and H-beams (refer to Figure 5.2[E]).

High-frequency-welded structural beams can be produced from low-carbon steels or high-strength, low-alloy metals. High-strength, low-alloy steel beams are used in commercial vehicle and trailer frames, in which the high strength-to-weight ratio is particularly valuable. Structural stock typically is made from low-carbon steels and high-strength, low-alloy steels, but nonferrous beams can be produced for special applications.

Welding Spiral Fins to Boiler Tube

Two different geometries of finned tube can be fabricated by high-frequency welding. In one, the fin is helically wound on edge around a tube and simultaneously welded to the surface of the tube. The fin and the tube may be made of the same material or different materials, as shown in Figure 5.2(G). This type of tube is used in power plant heat exchangers.

Fins can be welded longitudinally to a tube on one or both sides. Again, the materials may be the same, or dissimilar materials may be used for the fins and tube. This type of tube is used to manufacture water walls in boilers.

Tube also can be welded to strip or sheet metal for products such as solar absorber plates and freezer liners.

Induction Welding of Pipe Butt Joints

The high-frequency induction process has been used to make butt joints between sections of pipe or tubing. A narrow induction coil is placed around the joint. High-frequency current in the coil induces a circulating current concentrated in the area of the pipe butt joint, which is heated very rapidly. When the metal reaches welding temperature, upset force is applied to produce a forge weld. The placement of the coil in this application is shown in Figure 5.2(J).

Contact Welding of Finite-Length Plate Butt Joints

Techniques are available for welding two strips or plates together in a butt joint (refer to Figure 5.2[D] and [K]). The strips can be of the same or different metals, and can be of different thicknesses, as is the case with welded tailored blanks. These joint designs also can be used to weld the ends of two strips or to join the ends of pipes and tubes.

The weld is accomplished by passing a high-frequency current through the joint area. The current is introduced at each end of the joint by small contacts and is confined to the area of the joint by a proximity conductor. Generally, a magnetic core is used to assist in narrowing the current path (refer to Figure 5.4[C]). Low frequencies between 1 kHz and 10 kHz generally are used to perform these finite length welds. By selecting the proper frequency, the depth of current penetration can be adjusted to heat the joint throughout the thickness of the joint. When the joint reaches the welding temperature, a forging force is applied and the hot metal is upset. Welds of this type can be made at rates up to 1000 joints per hour. The use of a proximity conductor for butt joints in strips or plates is shown in Figure 5.8.

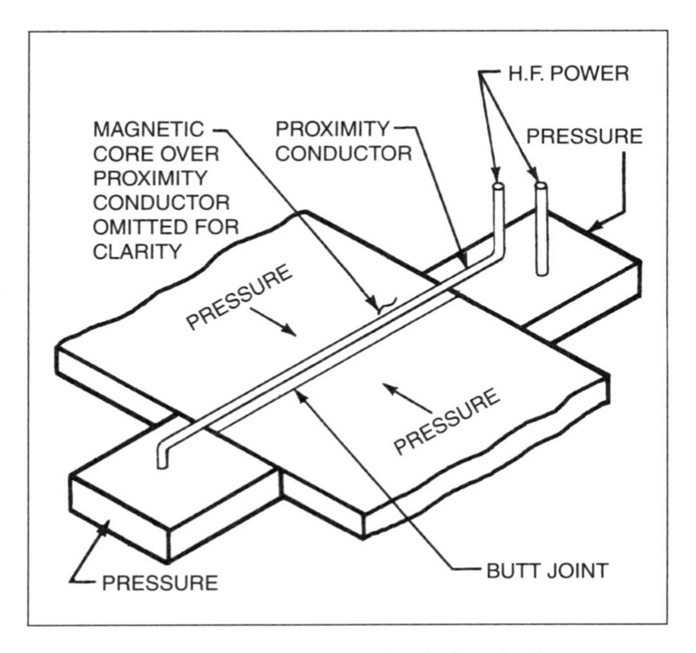

Figure 5.8—Method of Joining Strip or Plate with High-Frequency Welding

WELDING PARAMETERS

Table 5.1 is a master chart of typical welding parameters for high-frequency seam welds and butt joints for applications of induction welding and contact welding. Recommendations for process selection, welding speed, power range and welding frequency are shown for various metals and metal product forms.

EQUIPMENT

Equipment for high-frequency resistance welding includes a power source (usually a solid-state inverter type), induction coils, contacts, impeders, control devices, and mechanical equipment for preparing and aligning the workpieces. Equipment is sometimes required for postweld heat treatment.

Solid-State Power Sources

The predominant power source in modern installations of high-frequency welding equipment is the solid-state inverter power source. These units provide welding output power ranging from 50 kW to 1800 kW and operate at frequencies from 80 kHz to 800 kHz. A basic circuit for a typical solid-state inverter power source used for high-frequency induction welding is shown in Figure 5.9; the sequence of operation is shown in Figure 5.10.

Solid-state power sources are smaller in size than traditional vacuum-tube units and typically demonstrate efficiencies over 80%, while vacuum-tube units operate at efficiencies between 50% and 65%. Economically efficient operation results in a significant decrease in power consumption and cooling-water requirements. Units can be manufactured for special line voltages, but typical input line voltages are 480 V, 60 Hz, 3-phase; or 380 V, 50 Hz, 3-phase.

Referring to Figure 5.9, the operation of a solid-state power source can be described as follows. A power transformer (if required) steps the line voltage from the incoming power line circuit breaker transformer to the level that is compatible with the voltage breakdown rating of the solid-state power switching devices. The power transformer also provides galvanic isolation between the induction coil or contacts and the incoming power line. If a power transformer is not present, this is accomplished by a radio frequency (rf) transformer that is positioned in the circuit following the placement of the power switching devices. The line voltage is then rectified to produce a direct-current link voltage that is used to supply the high-frequency inverter section of the power source. Filtering also must be provided for the direct-current link voltage because any ripple on this voltage caused by the rectifier will modulate the welding power and this will manifest itself as unevenness or "stitching" in the weld. Typically, a ripple level of less than 1% is adequate to ensure a smooth weld. However, very thin-walled pipe or tube welding applications require ripple levels less than 0.3%.

The direct-current link voltage is used to power a high-frequency inverter. The inverter is constructed from a set of solid-state switching devices. These devices essentially behave like electrical switches, as illustrated in Figure 5.9. In practice, these generally are implemented using metal oxide semiconductor field effect transistors (MOSFET) or insulated gate bipolar transistor (IGBT) solid-state devices. The MOSFET types can switch at higher frequencies than IGBT devices, but IGBTs can handle more current per unit area of silicon than MOSFET devices.

Several circuit topologies for the inverter section may be used; each has advantages and limitations, depending on the requirements of the application. Circuit topologies include a current-fed inverter that switches current and a voltage-fed inverter that switches the voltage. For either of these topologies, the switching devices can be arranged in a full-bridge or half-bridge configuration. The diagrams in Figures 5.9 and 5.10 show a full-bridge, current-fed inverter. This example serves to illustrate basic inverter operation; other configurations operate in analogous ways.

In the current-fed inverter, a constant current choke is placed in series between the direct-current power source section and the solid-state high-frequency inverter section. The purpose of the choke is to supply a constant current from the dc power source. The high-frequency voltage variations created by the inverter appear across the constant-current choke. In a voltage-fed inverter, the function of the constant current choke is replaced by a parallel capacitor, the purpose of which is to hold the voltage constant and to absorb the high-frequency current variations created by the inverter.

The direct current provided to the inverter is converted to high-frequency alternating current by switching the solid-state power devices on and off in sequence. The resulting high-frequency current drives a resonant tank circuit composed of the tuning capacitor and the inductance of the induction coil and workpiece. The resonant frequency of this tank circuit is tuned to the welding frequency and converts the on and off current (created by the solid-state switching device) to a sine wave.

Figure 5.10 shows the switching sequence for the solid-state devices required by a current-fed inverter. It can be noted that Stage 1 of the switching sequence occurs when power devices *A* and *B* are on and power devices *C* and *D* are off. In Stage 1, the current flows through the induction coil in the clockwise direction.

Table 5.1
Typical High-Frequency Welding Parameters

Process and Application	Metal	Principal Workpiece Dimension	Thickness mm (in.)	Welding Speed	Welding Power Source Range	Welding Frequency
Induction seam welding: pipe and tube	Low-carbon steel (can be aluminum- or zinc-coated)	Outside diameter: 10 mm to 75 mm (0.4 in. to 3 in.)	Wall thickness: 0.6 mm to 2 mm (0.025 in. to 0.080 in.)	25 m/min to 300 m/min (80 ft/min to 1000 ft/min)	50 kW to 500 kW	300 kHz to 400 kHz
	Carbon steel less than 0.05% carbon (can be aluminum- or zinc-coated)	Outside diameter: 50 mm to 200 mm (2 in. to 8 in.)	Wall thickness: 1.5 mm to 12.7 mm (0.060 in. to 0.5 in.)	20 m/min (60 ft/min) to 150 m/min (500 ft/min)	200 kW to 1000 kW	200 kHz to 400 kHz
	Carbon steel less than 0.05% carbon	Outside diameter: 150 mm to 400 mm (6 in. to 16 in.)	Wall thickness: 2 mm to 12.7 mm (0.080 in. to 0.5 in)	15 m/min (50 ft/min) to 90 m/min (300 ft/min)	400 kW to 1200 kW	150 kHz to 300 kHz
	Carbon steel less than 0.05% carbon	Outside diameter: 300 mm to 600 mm (12 in. to 24 in.)	Wall thickness: 6 mm to 25 mm (0.25 in to 1 in.)	10 m/min (30 ft/min) to 50 m/min (165 ft/min)	800 kW to 1200 kW	100 kHz to 200 kHz
Contact seam welding: pipe and tube	Low-carbon steel (can be aluminum- or zinc-coated)	Outside diameter: 10 mm to 75 mm (0.4 in. to 3 in.)	Wall thickness: 0.6 mm to 2 mm (0.025 in. to (0.080 in.)	25 m/min (80 ft/min) to 300 m/min (1000 ft/min)	50 kW to 350 kW	300 kHz to 400 kHz
	Carbon steel less than 0.05% carbon (can be aluminum- or zinc-coated)	Outside diameter: 50 mm to 200 mm (2 in. to 8 in.)	Wall thickness: 1.5 mm to 12.7 mm (0.060 in. to 0.5 in.)	20 m/min (60 ft/min) to 150 m/min (500 ft/min)	150 kW to 450 kW	300 kHz to 400 kHz
	Carbon steel less than 0.05% carbon	Outside diameter: 150 mm to 400 mm (6 in. to 16 in.)	Wall thickness: 2 mm to 12.7 mm (0.080 in. to 0.5 in)	15 m/min (50 ft/min) to 90 m/min (300 ft/min)	250 kW to 600 kW	200 kHz to 400 kHz
	Carbon steel less than 0.05% carbon	Outside diameter: 300 mm to 600 mm (12 in. to 24 in.)	Wall thickness: 6 mm to 25 mm (0.25 in. to 1 in.)	10 m/min (30 ft/min) to 50 m/min (165 ft/min)	300 kW to 800 kW	150 kHz to 300 kHz
Induction seam welding: pipe and tube	Stainless steel	Outside diameter: 10 mm to 75 mm (0.4 in. to 3 in.)	Wall thickness: 0.6 mm to 2 mm (0.025 in. to 0.080 in.)	10 m/min (30 ft/min) to 90 m/min (300 ft/min)	50 kW to 300 kW	250 kHz to 300 kHz
Induction seam welding: pipe and tube	Aluminum	Outside diameter: 10 mm to 38 mm (0.4 in. to 1.5 in.)	Wall thickness: 0.3 mm to 1 mm (0.010 in. to 0.040 in.)	40 m/min (120 ft/min) to 150 m/min (500 ft/min)	50 kW to 150 kW	400 kHz to 800 kHz
	Aluminum	Outside diameter: 25 mm to 75 mm (1 in. to 3 in.)	Wall thickness: 1 mm to 2 mm (0.040 in. to 0.080 in.)	40 m/min (120 ft/min) to 90 m/min (300 ft/min)	50 kW to 300 kW	400 kHz to 600 kHz
Induction seam welding: pipe and tube	Copper	Outside diameter: 10 mm to 75 mm (0.4 in. to 3 in.)	Wall thickness: 0.3 mm to 2 mm (0.010 in. to 0.080 in.)	40 m/min (120 ft/min) to 150 m/min (500 ft/min)	50 kW to 300 kW	400 kHz to 600 kHz
Induction welding: butt joints in pipe and tube	Low-carbon steel	Outside diameter: Up to 300 mm (12 In.)	Wall thickness: 2 mm to 12 mm (0.08 in to 0.5 in.)	10 seconds to 60 seconds per joint	100 kW to 500 kW	1 kHz to 10 kHz
Induction welding: butt joints in beams	Low-carbon steel	Height: 75 mm to 500 mm (3 in. to 20 in.) Width: 50 mm to 300 mm (2 in. to 12 in.)	Web thickness: 2 mm to 10 mm (0.080 in. to (0.375 in.) Flange thickness: 3 mm to 12 mm (0.125 in. to 0.5 in.)	10 m/min (30 ft/min) to 40 m/min (120 ft/min)	100 kW to 350 kW	200 kHz to 400 kHz
Induction welding: spiral-fin boiler tubes	Low-carbon steel, stainless steel or in combination	Outside diameter: 25 mm to 75 mm (1 in. to 3 in.)	Wall thickness: To 10 mm (0.374 in.)	10 m/min (30 ft/min) to 40 m/min (120 ft/min)	100 kW to 350 kW	300 kHz to 400 kHz

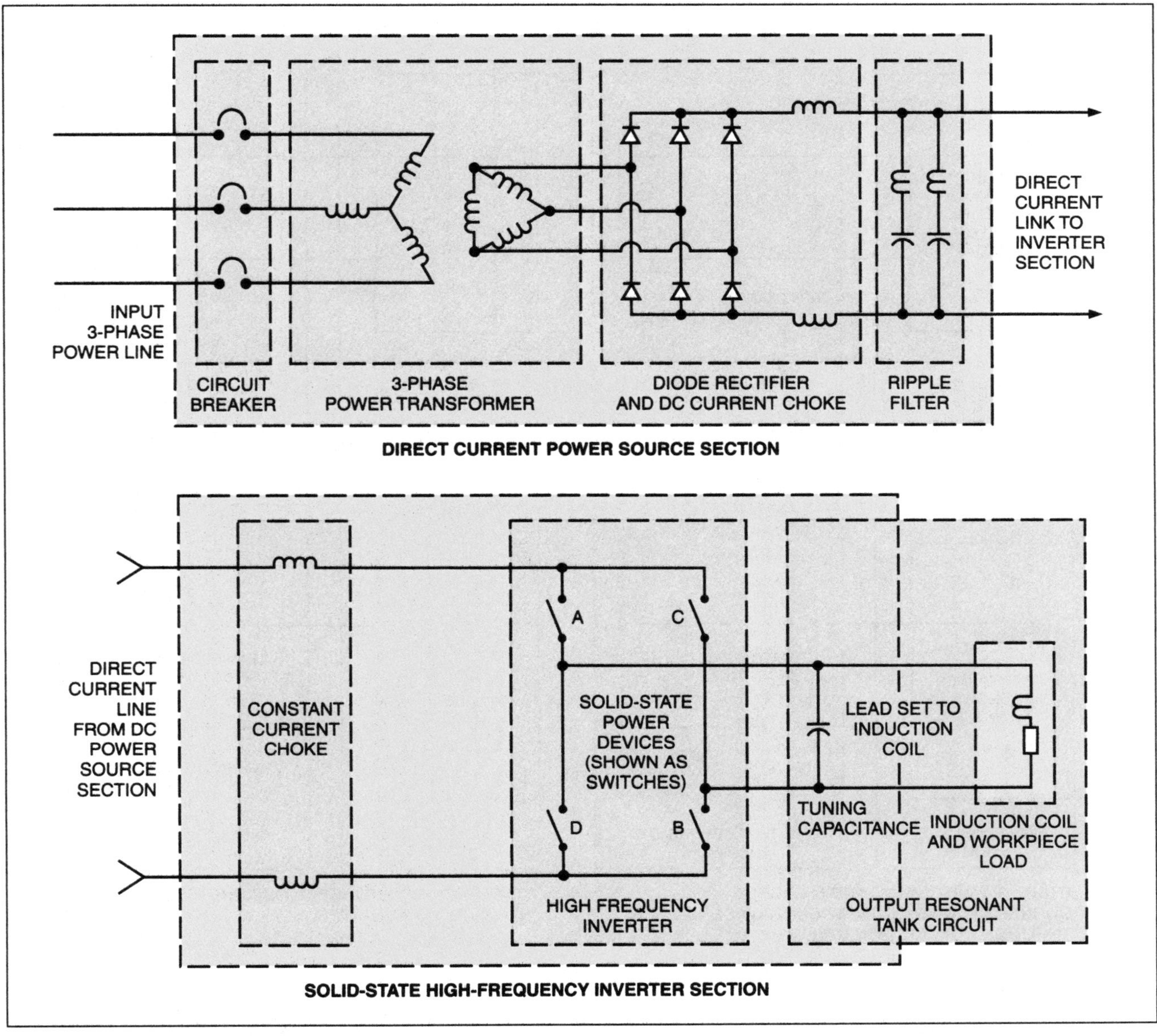

Source: Adapted from Thermatool Corporation.

Figure 5.9—Basic Circuit of a Solid-State Inverter Power Source

This part of the sequence creates the first half of the high-frequency sine-wave current in the induction coil. The devices then sequence to Stage 2, when all of the switching devices are turned on for a brief instant. (If using the voltage-fed topology, the devices would all briefly turn off.) The purpose of Stage 2 of the switching sequence is to ensure that the steady current provided by the constant-current choke is not interrupted. The solid-state power devices are then switched to Stage 3 of the switching sequence, where devices C and D are on and devices *A* and *B* are off. The current now flows counterclockwise through the induction coil; this action creates the second half of the output sine wave. The devices are then switched to the pattern shown as Stage 4. This has the same purpose as Stage 2, and prepares for the switching of the devices again to Stage 1. The complete switching sequence repeats once for every cycle of the high-frequency output sine wave.

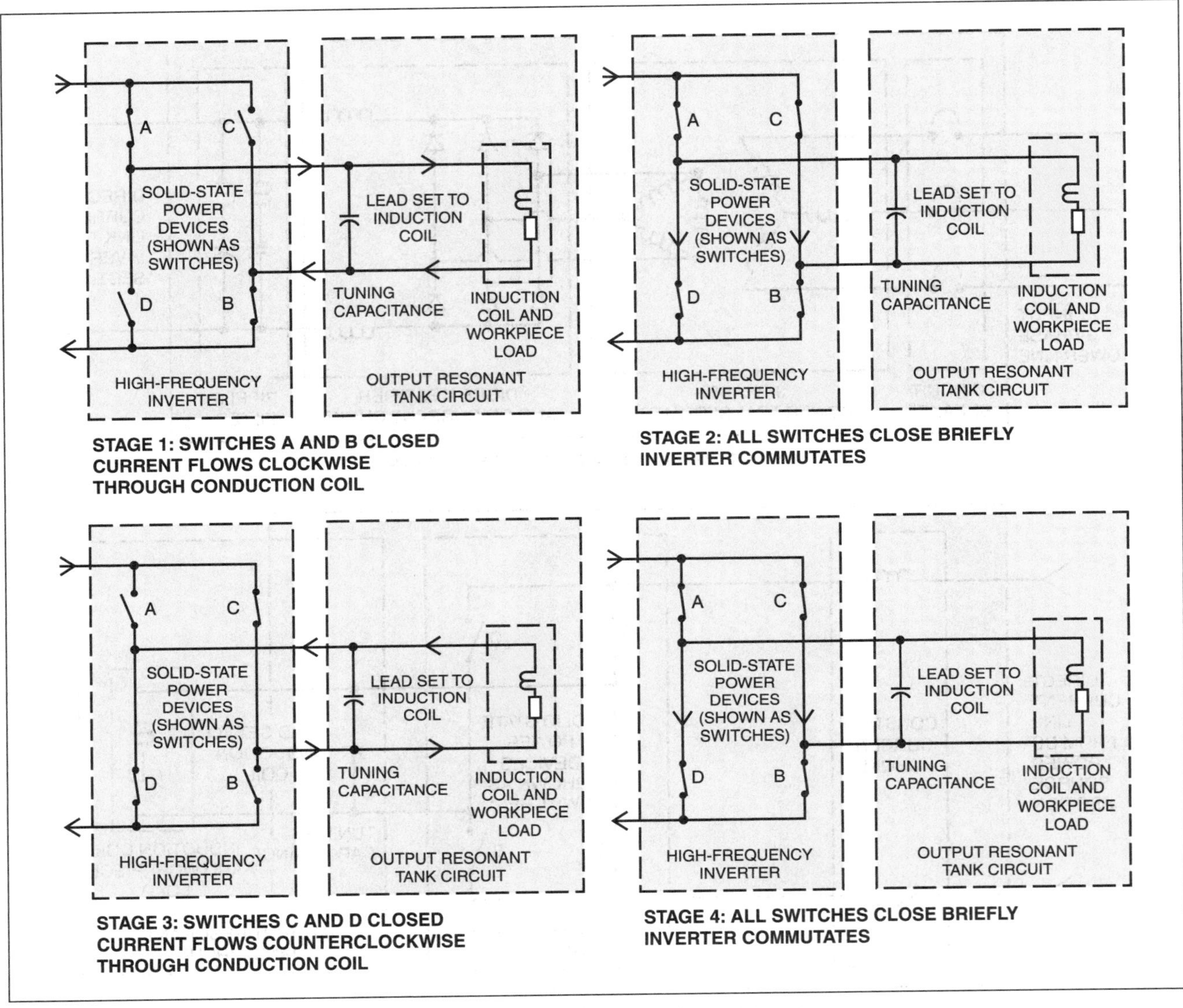

Source: Adapted from Thermatool Corporation.

Figure 5.10—Current-Fed Solid-State Inverter Switching Sequence

Power control of the inverter is accomplished either by manipulating the solid-state power device switching sequence or by controlling the output voltage of the direct-current power supply section.

Vacuum-Tube Oscillator Power Sources

Although most modern installations are solid-state power sources, many existing installations and some new installations use vacuum-tube oscillator power sources instead of solid-state inverters. The schematic in Figure 5.11 illustrates the circuit operation for a typical high-frequency vacuum-tube oscillator.

Incoming power flows through a circuit breaker and contactor (not shown in the drawing), then through a 3-phase SCR switch voltage regulator. The main purpose of the regulator is to provide a means for control of output power by varying the primary voltage supplied to the high-voltage plate transformer. The plate transformer converts its primary voltage to a much higher voltage (typically around 10 000 V), which is then rectified to provide the direct current required by

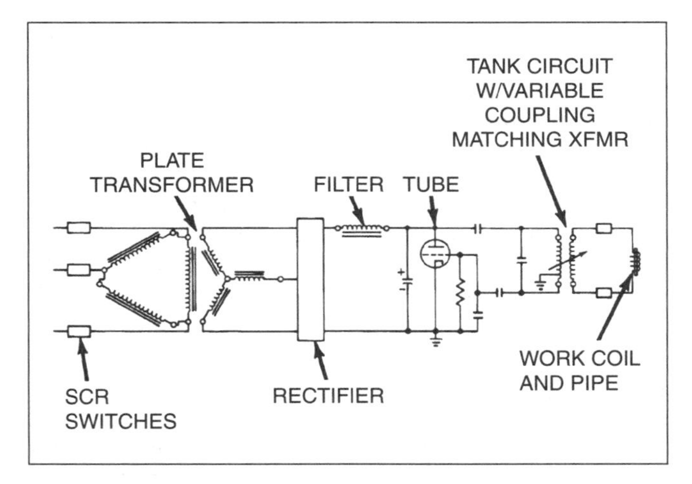

Figure 5.11—Circuit for a Typical High-Frequency Oscillator

the vacuum-tube oscillator circuit. The filter choke and capacitor reduce the ripple in the direct current to an acceptable level—typically less than 1%.

The oscillator circuit converts the direct current to high-frequency alternating current for the output transformer. The output transformer converts the high-voltage, low-current power provided by the vacuum tube of the oscillator to the low-voltage, high-current power required for welding. For induction welding, variable impedance-matching transformers often are used in which the primary winding can be moved relative to the secondary winding to match the relatively wide impedance range typical of induction applications.

The secondary winding of the impedance-matching transformer is in series with the induction coil or contacts and the workpiece. This forms the low-voltage, high-current system. The connecting leads should have the lowest possible impedance to obtain high efficiency and to minimize the voltage drop in the leads. This may be achieved by using short, wide leads made of flat copper plate separated by approximately 1.6 mm (1/16 in.) of insulation. Power losses in poorly designed leads or incorrectly matched transformers can seriously degrade the performance of a high-frequency welding system.

Induction Coils

The induction coils used in the high-frequency induction welding process generally are fabricated from copper tubing, hollow copper bar or copper sheet. They normally are water-cooled, where the cooling water is piped through the tubing or bar, or in the case of sheet-fabricated coils, is pumped through tubes brazed to the sheet or is sprayed on the coil. The highest efficiency is obtained when the induction coil completely surrounds the workpiece. The coil may have one or more turns; the number of turns is selected to provide the best match between the impedance of the workpiece and the output impedance of the power source. The strength of the magnetic field that induces the heating current in the workpiece diminishes rapidly as the distance between the coil and the workpiece is increased. Typical spacing between the coil and the workpiece ranges from 3 mm (1/8 in.) for small-diameter products to up to 25 mm (1 in.) for large-diameter products. Some typical induction coils are shown in Figure 5.12.

Photograph courtesy of Thermatool Corporation

Figure 5.12—Typical Induction Coils

Contacts

Sliding contacts transfer the current from the high-frequency power source to the workpiece in high-frequency contact welding. The contacts usually are made of copper alloy or may be composed of hard metallic or ceramic particles in a copper or silver matrix. The contacts are silver-brazed to heavy, water-cooled copper mounts, often called *contact shoes*. Replacement of the contacts is made by exchanging the mount and contact tip assembly. The area of contact tips range from 160 square millimeters (mm^2) to 650 mm^2 (0.25 square inch [$in.^2$] to 1 $in.^2$), depending on the magnitude of current that must be carried. Welding current usually ranges from 500 amperes (A) to 5000 A. Consequently, both internal and external cooling is required for the contact tip and mounts.

The level of force of the contact tip against the workpiece usually is in the range of 20 newtons (N) to 220 N (5 pounds [lb] to 50 lb) for a continuous welding system. The force required depends on the contact size and placement, the surface condition of the workpiece, the composition of the contact material, and the welding current required. Welding current is determined by the width of the surfaces being joined, the welding frequency, and the welding speed.

The service life of contacts depends on a number of factors, including the composition of the contact material, contact pressure, the metals being joined, and the welding power required. Contact service life can be as low as 300 meters (m) (1000 feet [ft]) under very severe conditions to over 90 000 m (300 000 ft) if circumstances are optimal. A typical contact system with contact regulation provided by pneumatic "springs" is shown in Figure 5.13.

Impeders

When welding closed workpieces such as pipe and tubing, current can flow on the inside surface as well as on the outside surface. The inside surface current decreases welding current that flows in the weld vee resulting in a substantial power loss at the weld point. Because the lost power does not heat the joint edges, the welding temperature cannot be reached unless the welding speed is reduced or power increased. If this condition is severe enough, it will be impossible to obtain a weld at any power or speed level.

Photograph courtesy of Thermatool Corporation

Figure 5.13—Typical Contact Assembly for High-Frequency Resistance Welding

To eliminate this loss, an impeder is placed inside the workpiece in the weld area and generally positioned so that in extends about 1.5 mm to 3 mm (1/16 in. to 1/8 in.) beyond the apex of the vee and the equivalent of 1 to 2 workpiece diameters upstream of the induction coil or contacts. The impeder, which is made from a magnetic material such as ferrite, increases the inductive reactance of the current path around the inside wall of the workpiece. The higher inductive reactance reduces the undesirable inside current and more current then flows through the weld vee.

Impeders typically are made of one or more ferrite rods or bars, and are cooled with water or mill coolant to keep the operating temperature below the Curie point, above which the ferrite loses magnetic properties.

Because the amount of magnetic flux the impeder can carry is determined by its cross-sectional area, the total impeder cross section should be made as large as practical. If the amount of impeder material is insufficient, the impeder will reach magnetic flux levels greater than the saturation flux and it will become inoperative at high power levels.

Impeders are particularly important when a mandrel must be run through the workpiece in the weld area to perform an inside weld bead treatment, such as inside bead scarfing or bead rolling. Although made of a nonmagnetic material such as austenitic stainless steel, a mandrel without impeders will reduce the inside reactance and will experience induction heating due to the magnetic field inside the workpiece. This will greatly reduce the efficiency of the process and can cause the mandrel to melt due to overheating. Therefore, impeders should be placed on top of a mandrel, or preferably around it. Typical impeders are shown in Figure 5.14.

Impeders generally are not required when welding large-diameter workpieces with the high-frequency contact welding process because the inductive reactance of the inside current path is generally large enough to limit losses at the shorter contact vee lengths.

Control Devices

In order to maintain proper welding conditions at different mill speeds and especially to minimize scrap material resulting when the mill is started and stopped, the weld power can be automatically adjusted as a function of mill speed. This system is most effective when welding low-carbon steel tubing, and can virtually eliminate any unwelded seam when the mill is stopped and restarted. The system also will reduce scrap when welding stainless and alloy steels and nonferrous metals, but typically a small length of product remains unwelded when starting and stopping the mill.

Photograph courtesy of Electronic Heating Equipment Company

Figure 5.14—Typical Impeders Used for Tube and Pipe Welding

Variations in weld temperature can be caused by variations in strip thickness and welding speed, and also by deterioration of the impeder. These variations can be controlled by the addition of a weld temperature control system. This reads the output of a pyrometer or analyzes the image obtained by an infrared camera aimed at the weld vee, and automatically adjusts the welding power to maintain a constant, preset temperature. When using a temperature control system, it is important to minimize the water or mill coolant used in the weld area, as this will interfere with obtaining an accurate temperature measurement.

Mechanical Equipment

Mechanical equipment is required in both the continuous and finite-length high-frequency welding processes. The edges to be welded must be properly aligned and brought together in parallel alignment by the mechanical equipment. This alignment must be maintained as the upset pressure is applied to forge the weld. Non-parallel edges will cause uneven heating of the inside and outside vee edges because the proximity effect will be unbalanced. This effect can sometimes be used to advantage by providing control over the size of the inside and outside weld beads.

The condition of the edges to be welded also is important. Mill-slit edges normally are satisfactory if the edges are not damaged during transport or forming. When manufacturing precision thin-wall tubing and high-quality, heavy-wall pipe, the strip edges should be trimmed in the production line. The edges can be trimmed with a stationary cutting tool or with a milling cutter. When welding large-diameter pipe made of single-strand rolled material, the edges often are slit in the production line immediately ahead of the forming section of the pipe mill.

Upset material on both the outside and inside weld bead can be removed using single-point cutting tools, or scarfing tools, arranged closely behind the weld point. The weld bead must be removed while it is red hot and the metal is still soft. Too much distance between the weld point and the scarfing tools, or too slow a mill speed can result in ragged scarfing and tool chatter. The inside bead may be left in the as-welded condition or rolled smooth, as required by product specifications. An induction tube welding machine in operation is shown in Figure 5.15.

Photograph courtesy of EFD Induction

Figure 5.15—High-Frequency Induction Tube Welding Machine

Postweld Heat Treating Equipment

Pipe or tube products made of certain materials, such as medium–carbon steel, high-carbon steel or steel alloys, may need post-weld heat-treatment. This is a requirement for welded pipe products such as oil pipelines or oil well casings made according to American Petroleum Institute (API) standards.[4] Low-carbon steel may be annealed to remove untempered martensite or may be stress-relieved to restore ductility that was lost during forming, welding and sizing. In most cases, only the weld zone is heat-treated. This is called *seam annealing*, but generally the seam is only normalized and not fully annealed. Seam annealing is performed inline by induction heating immediately following the upset-removal operation. A special linear inductor is used to anneal the seam at a frequency of 1 kHz to 3 kHz, depending on the thickness of the pipe wall. Low-frequency (1 kHz) is more effective on wall thicknesses over 10 mm (3/8 in.) while pipes less than 13 mm (0.5 in.) in wall thickness are often seam-annealed at 3 kHz.

4. American Petroleum Institute (API), 1220 L Street S.W., Washington, D.C. 20005-4070.

An annealing application of large pipe is shown in Figure 5.16, a 2400-kW induction seam annealer running on a 61 cm (24 in.) pipe. Each individual inductor is capable of 400 kW. The white stripe on the side is being painted on at the weld station. The vision system shown in the photo is used to track this white stripe, which allows the seam annealing equipment to orbitally track the weld seam.

Some applications require the complete tube to be heat treated. Heat treatment may be performed in the production line using induction heating after the welding and sizing procedures or may be performed off-line as a secondary operation. The frequency for induction heating depends on the tube material, wall thickness, and the required temperature. Medium frequencies of 1 kHz to 10 kHz usually are applied to full-body annealing low-carbon steel. Frequencies as high as 300 kHz are used for stress-relieving small-diameter, thin-wall tubes before coiling, and 30 kHz is often used for annealing austenitic stainless steel. An inert atmosphere may be provided during heating and cooling to prevent surface oxidation. This process is often called *bright annealing*.

Photograph courtesy of EFD Induction

Figure 5.16—Induction Seam Annealing of 61-cm (24-in.) Pipe with In-Line Tracking System

INSPECTION AND QUALITY CONTROL

Many standard inspection techniques are available to insure the production of quality high-frequency welded products. The selection and combination of techniques employed depend on the nature of the product and requisite product standards.

PRODUCT STANDARDS

Requirements for the manufacture of tubular products are described in a standard published by the American Society for Testing and Materials (ASTM), *Standard Specification for General Requirements for Carbon, Ferritic Alloy and Austenitic Alloy Steel Tubes*, A450/A450M.[5] This specification covers mandatory and non-mandatory requirements for a large variety of tubular products. Requirements that may not be mandatory for some products become mandatory if they are specified in the product specification or purchase documents. ASTM A450/A450M covers requirements for chemical and mechanical testing, dimensional tolerances, hydrostatic testing and nondestructive testing of the product.

The ASTM publication, *Standard Specification for Electric Resistance Welded Steel Shapes*, ASTM A769/A769M, covers the requirements for structural shapes such as I-beams and T-sections produced by high-frequency welding.[6] This specification describes the intended classes of application for structural products, the manufacturing, chemical, and mechanical property requirements, dimensional tolerances, test methods and frequency of testing, and requirements for inspection and testing.

The API documents, *Specification for Line Pipe*, API 5CT, and *Specification for Casing and Tubing*, API 5L provide the necessary manufacturing, testing and quality control requirements for producing oil line pipe and oil well casing.[7]

5. American Society for Testing and Materials (ASTM), *Standard Specification for General Requirements for Carbon, Ferritic Alloy and Austenitic Alloy Steel Tubes*, A450/A450M, West Conshohocken, Pennsylvania: American Society for Testing and Materials.
6. American Society for Testing and Materials (ASTM), *Standard Specification for Electric Resistance Welded Steel Shapes*, ASTM A769/A769M, West Conshohocken, Pennsylvania: American Society for Testing and Materials.
7. American Petroleum Institute (API), *Specification for Line Pipe*, API 5L, and *Specification for Casing and Tubing*. API 5CT, Washington, D.C.: American Petroleum Institute.

Visual Examination and Dimensional Inspection

A specification for a typical product should include acceptance criteria for outside diameter, wall thickness, accuracy of shape, straightness, and general appearance. Other high-frequency welded products are required to meet specified dimensional tolerances. In most cases, the dimensional checks are done manually on a small sample of the total number of items produced. For products with critical specifications, non-contact gauging systems based on ultrasonic, laser beam or similar monitoring techniques can be used to provide continuous measurement of wall thickness and outside diameter.

Metallurgical Examination

Metallographic examination commonly is used to evaluate weld quality. It is used both for inspecting the base material and for examination of the high-frequency weld. Transverse weld cross sections generally are used. A typical weld cross section from a 219-mm (8.625-in.) diameter pipe with 10-mm (0.395-in.) wall thickness is shown in Figure 5.17. In this sample, neither the outside nor inside upset has been removed, so that the amount of material squeezed from the weld area during forging is clearly visible. Other physical property tests normally performed include hardness tests, tensile tests, Charpy toughness tests, bend tests, and hydrostatic burst tests.

Nondestructive Examination

Requirements for nondestructive inspection of high-frequency welded tubes are provided in *Standard Specification for General Requirements for Carbon, Ferritic Alloy and Austenitic Alloy Steel Tubes*, ASTM A450/450M, and applicable documents referenced in that specification.

Ultrasonic testing is described in *Standard Recommended Practice for Ultrasonic Inspection of Longitudinal and Spiral Welds of Welded Pipe and Tubing*, ASTM E213.[8]

Eddy current testing procedures are described in *Standard Recommended Practice for Eddy-Current Examination of Steel Tubular Products Using Magnetic Saturation*, ASTM E309[9] and *Standard Recommended Practice for Electromagnetic (Eddy Current) Testing of Seamless and Welded Tubular, Austenitic Stainless Steel*

8. American Society for Testing and Materials (ASTM), *Standard Recommended Practice for Ultrasonic Inspection of Longitudinal and Spiral Welds of Welded Pipe and Tubing*, E213, West Conshohocken, Pennsylvania: American Society for Testing and Materials.

Photograph courtesy of Thermatool Corporation

Figure 5.17—Cross Section of a Weld in a Small-Diameter Tube

and Similar Alloys, ASTM E426.[10] Coils encircling the tube usually are used for small-diameter tubes, and sector coils located over the weld seam are used for large diameters. A typical system consists of an exciting coil that induces eddy currents into the tube and a sensor coil that reads the resulting magnetic flux created by the induced currents. The exciting and sensor coils typically are packaged together as a single unit. A discontinuity in the welded seam will disturb the normal current flow pattern. The disturbed current will create a magnetic field that differs from that produced without a discontinuity. The sensor coil detects the difference in magnetic field.

Flux-leakage testing is described in *Standard Recommended Practice for Flux Leakage Examination of Ferromagnetic Steel Tubular Products*, ASTM E570.[11] The tube is first magnetized to a level approaching its magnetic saturation. Discontinuities cause a leakage of the magnetic flux that is found by a magnetic detector.

In all of these methods, calibration standards are prepared by testing tubes of the same size and material as the one to be examined. These specimen tubes contain known discontinuities such as drilled holes or transverse, tangential or longitudinal notches; they are used to simulate the type of discontinuity that may occur in the welding process. When a signal occurs that exceeds the magnitude of that required by the calibration procedures, a marking system (typically a paint spray) identifies the location of the potential defect. In many cases, because of the high speed of the high-frequency welding process, it is difficult to discriminate between real weld defects and various discontinuities that may not be cause for rejection. Therefore, the marked areas may subsequently be retested off-line to verify the on-line test.

Weld testing procedures such as X-ray, magnetic particle, and liquid penetration generally are not applicable to high-frequency welding.

Ultrasonic testing, eddy-current testing, flux-leakage testing, or all three also may be performed after subsequent processing such as stretch-reduction, drawing, or cold expansion has been performed. These subsequent procedures may enlarge a discontinuity and make it easier to detect.

ECONOMICS

When high-frequency welding processes are applicable to the product, and when a sufficiently large number of welded products are required, high-frequency welding is far more economical than other forms of welding. The speeds at which products can be welded typically are 5 to 10 times faster than other welding methods, such as low-frequency resistance welding, gas tungsten arc welding or gas metal arc welding, or laser beam welding.

9. American Society for Testing and Materials (ASTM), *Standard Recommended Practice for Eddy Current Examination of Steel Tubular Products Using Magnetic Saturation*, E309, West Conshohocken, Pennsylvania: American Society for Testing and Materials.
10. American Society for Testing and Materials (ASTM), *Standard Recommended Practice for Electromagnetic (Eddy Current) Testing of Seamless and Welded Tubular, Austenitic Stainless Steel and Similar Alloys*, E426, West Conshohocken, Pennsylvania: American Society for Testing and Materials.
11. American Society for Testing and Materials (ASTM), *Standard Recommended Practice for Flux Leakage Examination of Ferromagnetic Steel Tubular Products*, ASTM E570, West Conshohocken, Pennsylvania: American Society for Testing and Materials.

The cost of capital equipment for high-frequency resistance welding is more than the cost of equipment for gas tungsten arc welding or gas metal arc welding, but typically is less than equipment for laser beam welding The increased production rates afforded by high-frequency welding generally translate directly into increased productivity, often making it the most cost-effective welding process for the particular application.

SAFE PRACTICES

The health and safety of the welding operators, maintenance personnel, and other persons in the area of the welding operations must be considered when establishing operating practices. The design, construction, installation, operation, and maintenance of the equipment, controls, power sources, and tooling should conform to the requirements of the United States Department of Labor in *Occupational Safety and Health Standards for General Industry,* (29) CFR Part 1910, Subpart Q.[12]

The high-frequency power source must also conform to the requirements of the Federal Communication Commission (FCC) as stated in Title 47, Part 15 concerning the radio frequency emissions from industrial, scientific and medical sources.[13] Responsibility for complying with FCC standards is undertaken by the power source manufacturer and does not pose a problem for the end user of the equipment, if the power source is installed following the manufacturer's recommendations. Information manuals provided by the manufacturers of equipment must be consulted, and recommendations for safe practices must be strictly followed. State, local, and company safety regulations also must be followed.

The American Welding Society document *Safety in Welding, Cutting, and Allied Processes*, ANSI Z49.1: 2005 covers safe practices specifically for the welding industry.[14]

Voltages produced by high-frequency power sources with solid-state inverter power sources (as high as 3000 V) and voltages produced vacuum-tube oscillators (as high as 30 000 V) can be lethal. Proper care and precautions must be taken to prevent injury while working on high-frequency generators and related control systems.

Modern power sources are equipped with safety interlocks on access doors and automatic safety grounding devices that prevent operation of the equipment when access doors are open. The equipment should never be operated with panels or high-voltage covers removed or with interlocks and grounding devices bypassed.

High-frequency currents are more difficult to ground than low frequency currents, and ground lines should be as short as possible to minimize inductive reactance. All leads between the power source and the contacts or induction coil should be totally enclosed in an insulated or grounded structure and constructed in a way that minimizes electromagnetic interference (EMI). Also, care should be taken to prevent the high-frequency magnetic field around the coil and leads from induction heating of the adjacent metal mill components.

The weld area should be protected so that operating personnel cannot come in contact with any exposed contacts or induction coils while these devices are energized.

Injuries to personnel from direct contact with high-frequency voltages, especially at the upper range of welding frequencies, may produce severe local tissue damage. Additional safety standards and publishers are listed in Appendices A and B of this volume.

CONCLUSION

High-frequency welding is a versatile process that can be used to join most weldable metals at high production rates. It is commonly used to produce seam-welded pipe and tubing from 10 mm (0.4 in.) in diameter through 610-mm diameter (24 in.), with wall thicknesses up to 25.4 mm (1 in.). It is also used to produce spiral fin boiler tubes, welded structural beams, and closed roll formed shapes. The process is particularly amenable to automated production. When high-frequency welding can be applied, it usually is the most productive welding process available, typically achieving welding rates 5 to 10 times faster than alternative welding methods.

BIBLIOGRAPHY

American National Standards Institute (ANSI) Accredited Standards Committee Z49. 2005. *Safety in welding, cutting, and allied processes*, ANSI Z49.1:2005 Miami: American Welding Society.

American Petroleum Institute (API). (Latest edition). *Specification for line pipe*, API 5L. Washington, D.C.: American Petroleum Institute.

12. Occupational Safety and Health Administration (OSHA), latest edition, Occupational safety and health standards for general industry in Code of Federal Regulations (CFR), Title 29 CFR 1910, Subpart Q. Washington, D.C.: Superintendent of Documents, U.S. Government Printing Office.
13. Federal Communications Commission, Title 47 of the *Code of Federal Regulations* (CFR), Part 15, *Radio Frequency Devices*, Washington, D.C.: United States Government Printing Office.
14. American National Standards Institute (ANSI) Accredited Standards Committee Z49, 2005, *Safety in Welding, Cutting, and Allied Processes*, ANSI Z49.1:2005, Miami: American Welding Society.

American Petroleum Institute (API). (Latest edition). *Specification for casing and tubing*, API 5CT. Appendix C. Washington, D.C.: American Petroleum Institute.

American Society for Testing and Materials (ASTM). *Standard recommended practice for ultrasonic inspection of longitudinal and spiral welds of welded pipe and tubing*. E213. West Conshohocken, Pennsylvania: American Society for Testing and Materials.

American Society for Testing and Materials (ASTM). *Standard recommended practice for eddy-current examination of steel tubular products using magnetic saturation*. E309. West Conshohocken, Pennsylvania: American Society for Testing and Materials.

American Society for Testing and Materials (ASTM). *Standard recommended practice for electromagnetic (eddy current) testing of seamless and welded tubular, austenitic stainless steel and similar alloys*, E426. West Conshohocken, Pennsylvania: American Society for Testing and Materials.

American Society for Testing and Materials (ASTM). *Standard specification for general requirements for carbon, ferritic alloy, and austenitic alloy steel tubes*. ASTM A450/A450M. West Conshohocken, Pennsylvania: American Society for Testing and Materials.

American Society for Testing and Materials (ASTM). *Standard recommended practice for flux leakage examination of ferromagnetic steel tubular products*, ASTM E570. West Conshohocken, Pennsylvania: American Society for Testing and Materials.

American Society for Testing and Materials (ASTM). *Standard specification for electric resistance welded steel shapes*, ASTM A769/A769M. West Conshohocken, Pennsylvania: American Society for Testing and Materials.

American Welding Society (AWS) Committee on Definitions and Symbols. 2001. *Standard welding terms and definitions*, AWS A3.0:2001. Miami: American Welding Society.

American Welding Society (AWS) Committee on Resistance Welding. 2000. Reaffirmed 2006. *Recommended practices for resistance welding*. AWS C1.1M/C1.1:2000 (R2006). Miami: American Welding Society.

Occupational Safety and Health Administration (OSHA). *Occupational safety and health standards for general industry*, in *Code of Federal Regulations* (CFR), Title 29, CFR 1910, Subpart Q. Washington, D.C.: Superintendent of Documents, U.S. Government Printing Office.

SUPPLEMENTARY READING LIST

Aceralia, F., and J. Ripidas. 2003. Manufacturing tube for hydroforming applications. *The Tube and Pipe Journal*: 5–6.

Asperheim, J. L., B. Grande, L. Markegard, J. E. Buser, and P. Lombard. Temperature distribution in the cross-section of the weld V. *Tube International*: (11).

Brown, G. H., C. N. Hoyler, and R. A. Bierwith. 1957. *Theory and applications of radio frequency heating*. New York: D. Van Nostrand Co., Inc.

Dailey, R. F. 1965. Induction welding of pipe using 10,000 cycles. *Welding Journal* 44(6): 475–479.

Haga, H., K. Aoki, and T. Sato. 1980. Welding phenomena and welding mechanisms in high-frequency electric resistance welding. *Welding Journal*. 59(7): 208s–212s.

Haga, H. et al. 1981. *Intensive study for high quality ERW pipe*. Document Number 3101, ERW-01-81-0. Nippon Steel Corporation.

Harris, S., and G. Butt. 1961. Welding of steel pipe using induction heating. *Welding Journal* 40(2): 57s–65s.

Johnstone, A. A., F. J. Trotter, and H. F. Brassard. 1960. Performance of the thermatool high-frequency resistance welding process. *British Welding Journal* 7(4): 238–249.

Koppenhofer, R. L., et al. Induction-pressure welding of girth joints in steel pipe. *Welding Journal* 39(7): 685–691.

Martin, D. C. 1971. High-frequency resistance welding. *Bulletin No. 160*. New York: Welding Research Council.

Oppenheimer, E. D. 1967. Joining helical and longitudinally finned tubing by high-frequency resistance welding. *ASTME Technical Paper AD67–197*. Dearborn, MI: Society of Manufacturing Engineers.

Oppenheimer, E. D., R. G. Kumble, and J. T. Berry. 1975. The double ligament tensile test: its development and application. *Journal of Engineering Materials and Technology* (4)107–112.

Osborn, H. B., Jr. 1956. *High-frequency continuous seam welding of ferrous and non-ferrous tubing. Welding Journal*. 35(12): 199-s–206-s.

Rudd, W. C. 1957. High-frequency resistance welding. *Welding Journal*. 36(7): 703-s–707-s.

Rudd, W. C. 1965. High-frequency resistance welding. *Metal Progress*. (10): 239–244.

Rudd, W. C. 1967. Current-penetration seam welding—a new high speed process. *Welding Journal*. 46(9): 762–766.

Scott, P. F. 1995. Introduction to solid state pipe and tube welding systems. *Proceedings of FMA tube 95 conference*. Rockford, Illinois: Fabricators and Manufacturers Association, Intl.

Scott, P. F. 1996. The effects of frequency in high-frequency welding. *Proceedings of ITA Tube Toronto 2000 Conference*. Leamington Spa, U.K.: International Tube Association.

Scott, P. F. 1999. Weld area design for high-frequency welding pipe and tube—a systems approach. *Proceed-*

ings of ITA Tubemaking for Asia's recovery conference. Singapore. Leamington Spa, U.K.: International Tube Association.

Scott, P. F. and W. Smith. 1996. The key parameters of high-frequency welding. *Tube International* (3)1996.

Scott, P. F. 2003. Selecting a welding frequency. *The Tube and Pipe Journal* (10):11.

Udall, H. N. 1986. Metallographic techniques—their contribution to quality high-frequency welded products. *Proceedings of 1986 International Conference—Tomorrow's Tube*, 10–12 June 1986. Leamington Spa, U.K.: International Tube Association.

Udall, H. N. and J. T. Berry. 1977. High-frequency welding of HSLA steel structurals. *Metal Progress.* 112(3).

Udall, H. N., J. T. Berry, and E. D. Oppenheimer. 1978. A high-speed welding system for the production of custom-designed HSLA structural sections. *Proceedings of International Conference on Welding of HSLA (Microalloyed) Structural Steels.* Rome. *9–12 Nov. 1976.* Materials Park, Ohio: ASM International.

Wolcott, C. G. 1965. *High-frequency welded structural shapes. Welding Journal.* 44(11): 921-s–926-s.

CHAPTER 6

FRICTION WELDING

Prepared by the Welding Handbook Chapter Committee on Friction Welding:

T. V. Stotler, Chair
Edison Welding Institute

B. J. Girvin
Edison Welding Institute

D. L. Kuruzar
Manufacturing Technology, Inc.

M. D. Pezzutti
GE Transportation

D. J. Walsh
Walsh Friction Welding, Inc.

Welding Handbook Volume 3 Committee Member:

W. L. Roth
Procter and Gamble, Inc.

Contents

Photograph courtesy of Thompson Friction Welding

CHAPTER 6

FRICTION WELDING

INTRODUCTION

Friction welding (FRW) is a solid-state process that produces a weld when two or more workpieces, rotating or moving relative to one another, are brought into contact under pressure to produce heat and plastically displace material from the faying surface (weld interface).[1,2]

Although friction welding is a solid-state welding process, under certain circumstances a molten film may form at the weld interface during the heating stage. Localized melting is seldom a problem, however, because this layer is extruded during the extensive hot working that takes place during the final stage of the process. Filler metal, flux, and shielding gas are not required with this process.

The two main variations of friction welding are direct drive friction welding (FRW-DD) and inertia friction welding (FRW-I). In direct drive friction welding, the energy required to make the weld is supplied by the welding machine through a direct motor connection for a preset period of the welding cycle. The energy required to make an inertia friction weld is supplied primarily by the stored rotational kinetic energy of the welding machine.

Friction stir welding (FSW) is classified by the American Welding Society as a variation of friction welding, but is not included in this chapter because of the substantial differences in the mechanics of the processes. Friction stir welding is discussed in detail in Chapter 7.

1. Welding terms and definitions used throughout this chapter are from *Standard Welding Terms and Definitions*, AWS A3.0:2001, Miami: American Welding Society.
2. At the time this chapter was prepared, the referenced codes and other standards were current. If a code or other standard is cited without a date of publication, it is understood that the latest edition of the document referred to applies. If a code or other standard is cited with the date of publication, the citation refers to that edition only, and it is understood that any future revisions or amendments to the code or standard are not included; however, as codes and standards undergo frequent revision, the reader is advised to consult the most recent edition.

Friction welding and its variations are used in nearly all industries for many different applications. The processes are used for high-volume production applications, when solid-state joining is a preference or a necessity, or when the geometries of the workpieces (the components) are such that other welding processes are not applicable. Friction welding can be used to weld nearly all combinations of metals and their alloys, including many dissimilar material combinations. The processes are important to fabricators who serve the aerospace, automotive, defense, transportation, agricultural, chemical, construction, and other industries. The photograph on the title page of this chapter shows an example of a common application: a threaded steel joint being welded to the end of a steel drill pipe used in the oil industry. The equipment is a direct drive friction welding machine with a 300-ton forge force.

The fundamentals of friction welding, variations of the process, and modes of operation are covered in this chapter. Other topics include the advantages and limitations of friction welding, applications, equipment, weldability of various materials, weld quality, inspection, and economics. Important information on safe practices is presented in the last section of the chapter. The supplementary reading list provides sources of additional technical information.

FUNDAMENTALS

Direct drive and inertia friction welds are produced when a compressive force is applied to workpieces rotating or moving relative to one another. The heat resulting from friction is used to assist plastic displacement of material from the weld interface. Friction welds are characterized by a narrow heat-affected zone (HAZ), the presence of plastically deformed material (flash) around the weld, and the absence of both fusion and partially melted zones.

In general, the process can be divided into two distinct stages: the friction stage and the forge stage. The friction stage can be further divided into two stages, first friction and second friction, which are described in more detail under the Direct Drive and Inertia Welding section.

The basic steps in the friction welding process are illustrated in Figure 6.1. As shown in Figure 6.1(A), one workpiece is rotated and the other is held stationary. When the appropriate rotational speed is reached, the two workpieces are brought together (B) under axial force. Abrasion at the weld interface heats the workpiece locally and upsetting (axial shortening) starts, as shown in (C). These two steps occur during the friction stage. Finally, rotation of the workpiece ceases and upset (forge) force (D) is applied to consolidate the joint. This occurs during the forge stage.[3]

The following terms and definitions relate to friction welding:

Friction speed—the relative velocity of the workpieces at the time of initial contact.

Friction force—the compressive force applied to faying surfaces during the time there is relative movement between the workpieces from the start of welding until the application of the forge force.

Friction time—the duration of time from the application of friction force until the application of forge force.

Friction upset distance—the decrease in length of workpieces during the time of friction welding force application.

Forge (upset) force—the compressive force applied to the weld after the heating portion (friction stage) of the welding cycle is essentially complete.

Forge (upset) distance—the total reduction in the axial length of the workpieces from the initial contact to the completion of the weld.

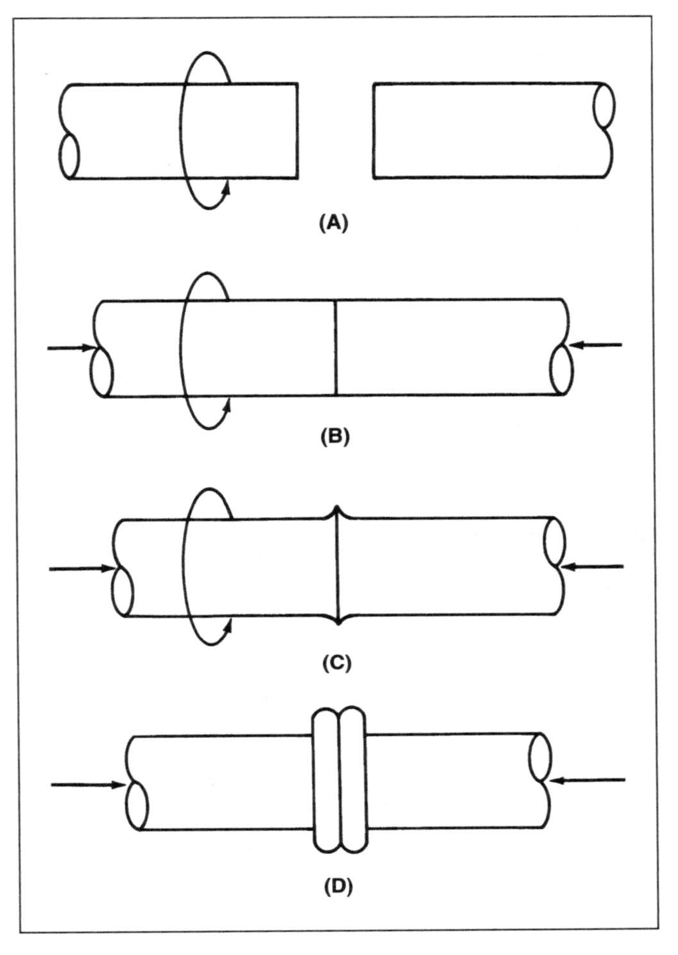

Figure 6.1—Basic Sequence of Friction Welding

ADVANTAGES AND LIMITATIONS

Following are several operational and economic advantages of friction welding:

1. No filler metal is required for all similar and most dissimilar material joints;
2. Flux and shielding gas normally are not required;
3. Solidification defects and porosity normally are not a concern;
4. The process is environmentally clean due to the minimization of sparks, smoke, or fumes;
5. Surface cleanliness is not as critical compared to other welding processes;
6. Heat-affected zones are narrow;
7. Most engineering materials and dissimilar metal combinations are well suited for joining;
8. In most cases, the weld is at least as strong as the weaker of the two materials being joined (high joint efficiency);
9. Operators are not required to have manual welding skills;
10. The process is easily automated for mass production;
11. Cycle times are short; and

3. American Welding Society (AWS) Committee on Friction Welding, Reaffirmed 1998, *Recommended Practices for Friction Welding*, ANSI/AWS C6.1-89, Miami: American Welding Society.

12. Plant requirements, such as space, electric power, and special foundations are minimal.

Following are some limitations of friction welding:

1. In general, one workpiece must have an axis of symmetry and be capable of rotation about that axis;
2. Alignment of the workpieces may be critical to developing uniform frictional heat;
3. Preparation of the interface geometry may be critical to achieving proper heat balance; and
4. Capital equipment and tooling costs are high, but payback periods typically are short for high-volume production.

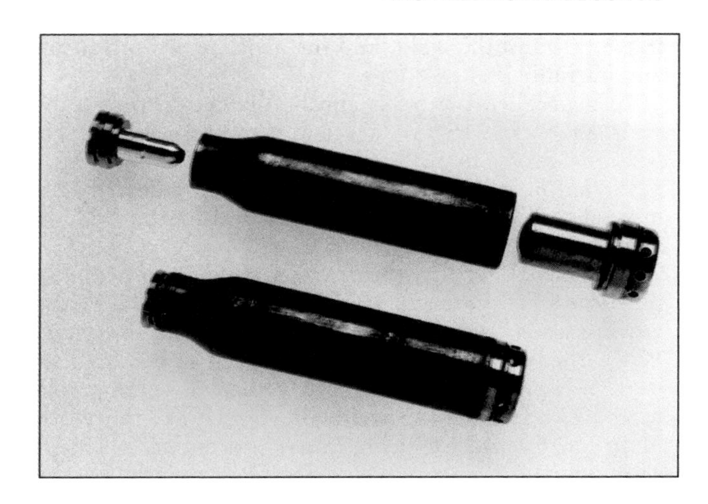

Figure 6.2—Automotive Airbag Inflator Welded with Direct Drive Friction Welding

DIRECT DRIVE AND INERTIA FRICTION WELDING

The friction welding cycle typically takes place in three stages: the first two comprise the friction stage, and the third involves the forging or upsetting stage. The heat for welding is developed during the friction stage, and the weld normally is consolidated with the forging or upsetting stage.

Typical examples of high-integrity welds made with direct drive and inertia friction welding are shown in Figures 6.2 and 6.3. Figure 6.2 depicts an airbag inflator made with direct drive friction welding; Figure 6.3 depicts a transmission gear made with inertia friction welding.

Figure 6.3—Transmission Gear Welded with Inertia Friction Welding

Friction Stage

The friction stage is the portion of the weld cycle when the heat is generated for forging. As the workpieces move in relation to one another and make contact, as shown in Figure 6.1(B), frictional heating takes place at the faying surfaces. When the workpieces first come into contact, the torque exerted on them and on the tooling is high; the first of two torque peaks occurs at this time during the welding cycle. To minimize the friction torque peak, the friction stage is often subdivided into "first" and "second" friction levels. The first friction level, often referred to as *scrub force*, uses a slightly lower axial pressure than the second to minimize the torque on the workpieces. As heat generation progresses, the axial pressure is subsequently increased to a higher level (the second friction) to more efficiently displace the plasticized material. This helps maintain a narrow heat-affected zone.

As the metals heat, bonding takes place at discrete points across the interface. These bonds may be stronger than the surrounding metal. Shearing takes place as the bonds break apart, transferring fragments of metal from one surface to the other. The transferred fragments accumulate until they form a continuous layer of plasticized metal. It is unlikely that a liquid film would form, but if so, it would occur at this point. As the friction stage continues, shown in Figure 6.1(C), the interfacial temperature increases, thus resulting in a reduction in the torque. As the torque becomes reasonably constant, the metal entering the interface is heated and forced from the interface as axial shortening continues.

Forging Stage

Toward the end of the heating process (friction stage), forging pressure is applied to the workpiece to rapidly displace the remaining plasticized material. This produces a flash "curl," shown at the weld joint in Figure 6.1(D). As the speed decreases, a second torque peak occurs, when the interface establishes a bond and cools from maximum temperature. This second torque peak typically is more pronounced with inertia friction welding. The torque then decreases as the spindle speed (rpm) drops to zero.

Energy Input Methods

Direct drive and inertia friction welding can be identified by the methods by which energy is supplied. Direct drive friction welding uses continuous energy input from a motor, whereas inertia friction welding uses stored energy from a flywheel.

Direct Drive Friction Welding. In direct drive friction welding, one of the workpieces is attached to a motor-driven chuck while the other is restrained from rotation. The motor-driven workpiece is rotated at a predetermined constant speed. The workpieces are advanced into contact with one another, and a friction force is applied. Heat is generated from friction as the faying surfaces abrade. This continues for a predetermined time, or until a preset amount of displacement occurs. In some cases, a preset rate of displacement is desirable and also can be programmed. Once an appropriate amount of heat is generated, the rotational driving force is disengaged, and the rotating workpiece is stopped by the application of a braking force (mechanical or electrical, or both). The friction welding force (forge force) is maintained or increased for a predetermined time or distance after rotation ceases. The relationship of the direct drive friction welding parameters is shown in Figure 6.4.

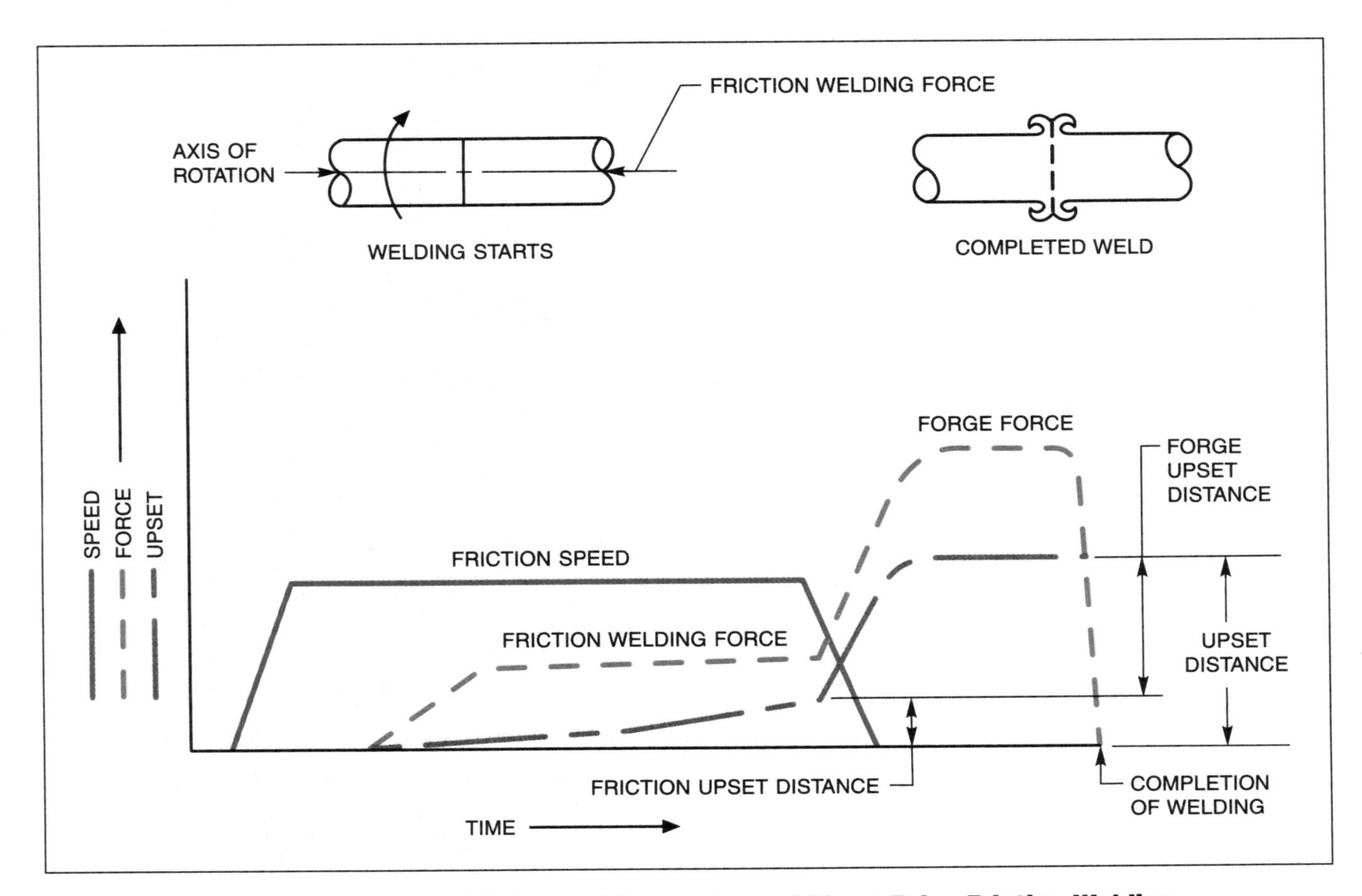

Figure 6.4—Characteristics and Parameters of Direct Drive Friction Welding

Inertia Friction Welding. With inertia friction welding, one of the workpieces is connected to a spindle which has one or more flywheels mechanically fixed to it. The other workpiece typically is restrained from rotating. The spindle is accelerated to a predetermined rotational speed, storing kinetic energy. When the desired speed is reached, the drive motor is disengaged and the workpieces are brought together under a preset amount of axial force. This causes the faying surfaces to rotationally abrade under pressure. The kinetic energy stored in the rotating spindle is dissipated as heat as the cycle progresses. A portion of the heat generated at the interface conducts into the base material of the workpieces and, depending on the clamp distance, into the tooling itself. The remaining thermal energy is extruded from the joint in the weld flash. Axial force is sometimes increased during the cycle before the rotation terminates. This is regarded as a forge force. The forge force is maintained for a predetermined time after rotation ceases. The characteristics of inertia friction welding parameters and their relationships are shown in Figure 6.5.

Types of Relative Motion

In most direct drive and inertia friction welding applications, one of the two workpieces is rotated about an axis of symmetry with the faying surfaces perpendicular to that axis. In normal cases, one of the two workpieces must therefore be circular or tubular in cross section at the joint location. Typical arrangements for single- and multiple-welding operations are shown in Figures 6.6(A) through 6.6(E).

Figure 6.6(A) depicts the conventional and most commonly used mode, in which one workpiece rotates while the other remains stationary. Figure 6.6(B) shows the counter-rotation mode, in which the workpieces are

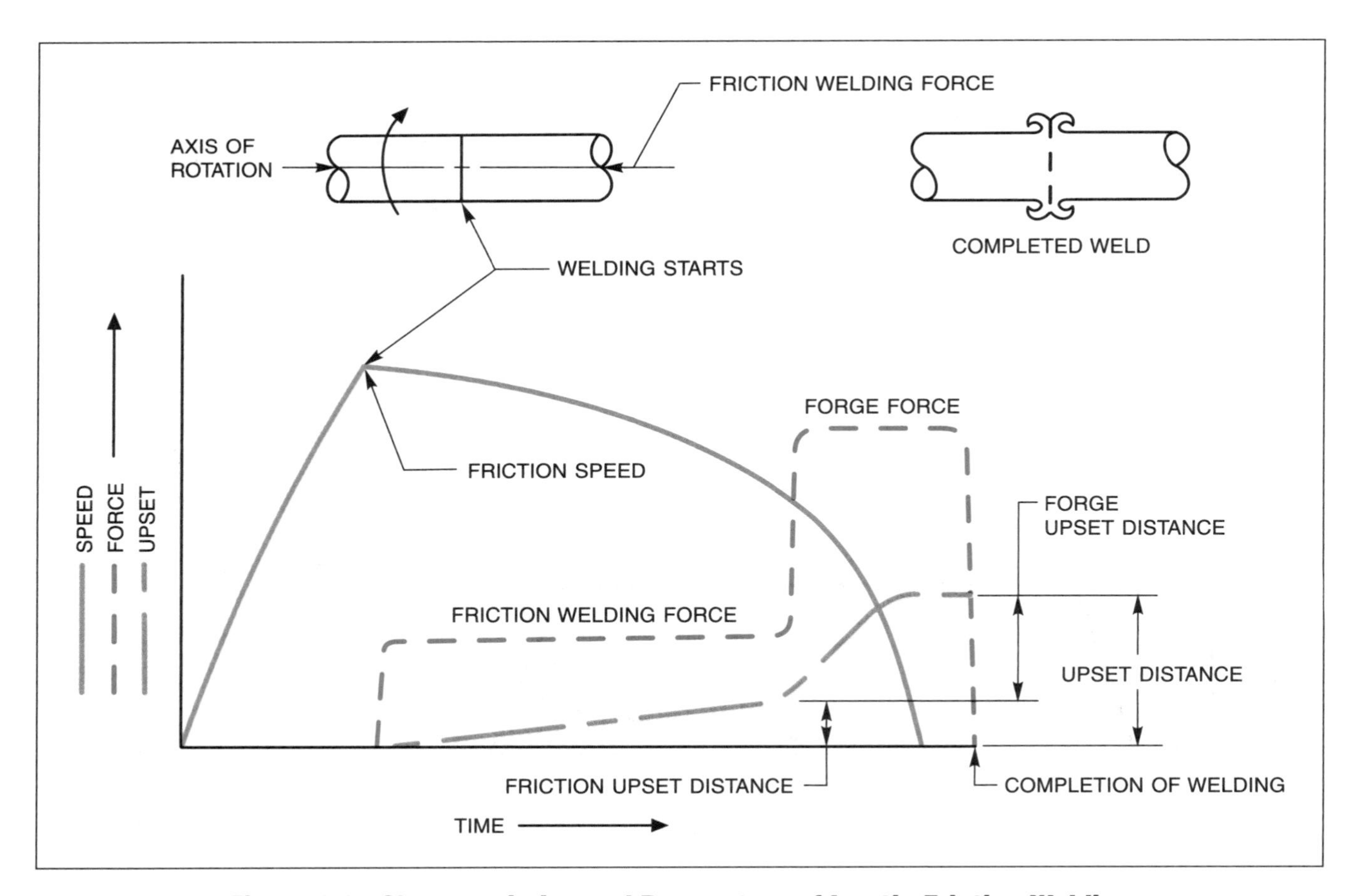

Figure 6.5—Characteristics and Parameters of Inertia Friction Welding

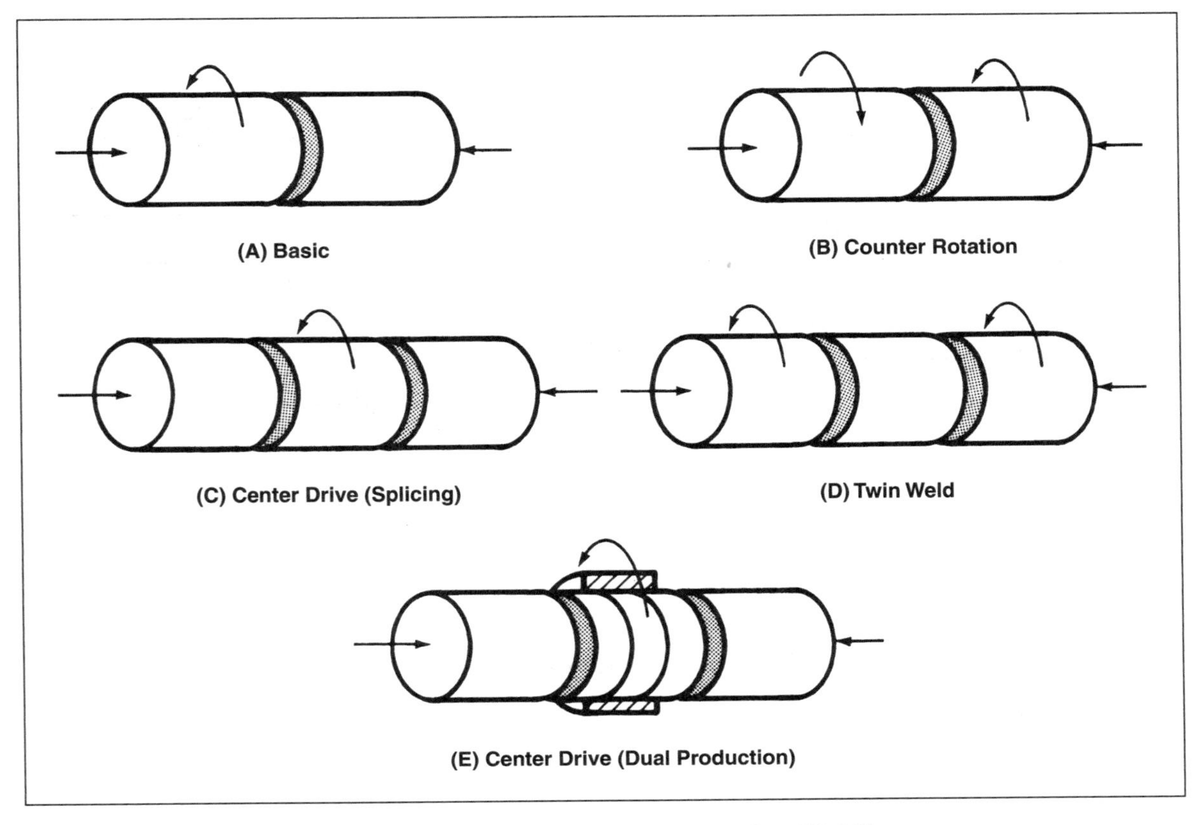

Figure 6.6—Typical Modes of Friction Welding

rotated in opposite directions. This procedure would be suitable for producing welds that require very high relative speeds. Figure 6.6(C) shows the center drive (splicing) mode where two stationary workpieces push against a rotating piece positioned between them. This setup might be desirable if the two end workpieces are very long or have such an awkward shape that rotation would be difficult or impossible by the other modes. This is one method of achieving rotational orientation to the center between the two workpieces. The twin-weld mode, shown in Figure 6.6(D), involves two rotating workpieces that contact a stationary workpiece at the middle. The center drive (dual production) mode, shown in Figure 6.6 (E), applies the same principle as the twin-weld mode to make two back-to-back welds using one rotating spindle at the center, thus improving productivity.

Other forms of friction welding (radial, orbital, angular reciprocating, and linear) and friction surfacing are unique in that each uses a different system to provide relative motion. These methods of friction welding are described in the following sections.

Radial. Radial motion can be used to weld circular sections for applications in which it is undesirable to rotate one or both of the workpieces. This method is often used to weld collars to shafts and tubes. As illustrated in Figure 6.7, the applied force on the rotating band is perpendicular to the axis of rotation. The collar is rotated and compressed as heat is generated. An internal expanding mandrel supports the pipe walls and prevents penetration of upset metal into the bore of the pipe.

Orbital. Friction welding with orbital motion is a technique in which one workpiece rotates (orbits) around the other. As illustrated in Figure 6.8, neither workpiece actually rotates around its axis. Consequently, the workpieces do not need to be circular or tubular in cross section. Orbital motion is a process option when member-to-member angular orientation is necessary.

Orbital Friction Surfacing. Orbital friction surfacing is achieved by the rotational motion of one consumable workpiece as it traverses across the surface of another work piece. The surfaced workpiece is perpendicular to the axis of rotation of the consumable work-

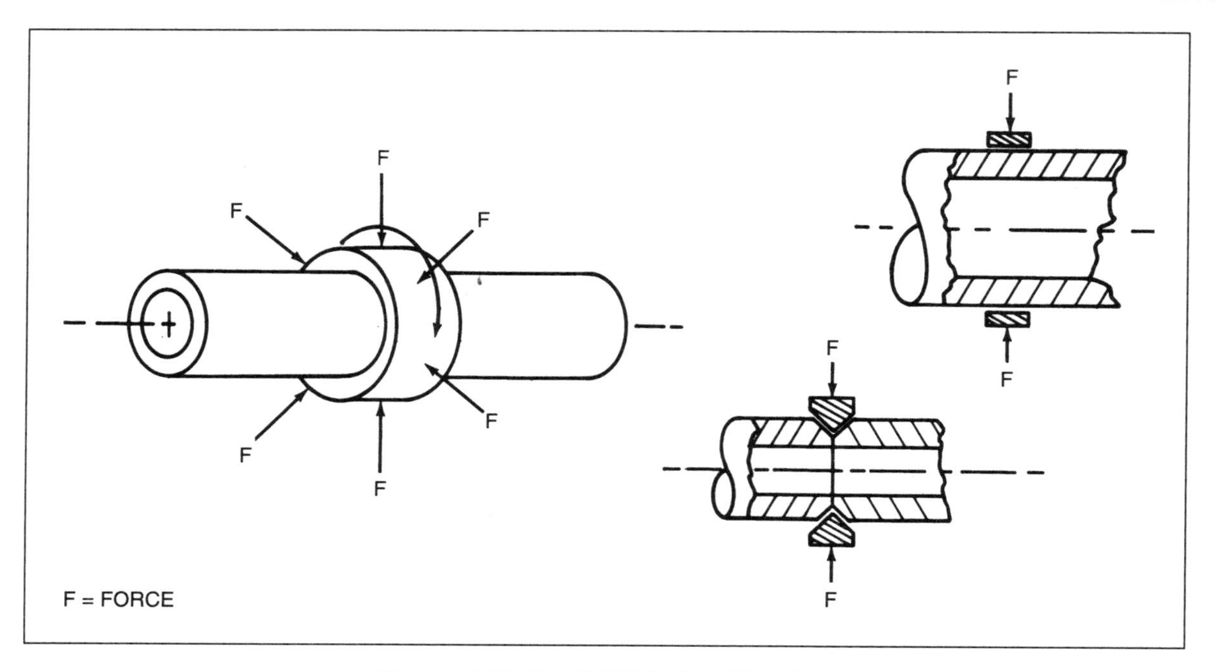

Figure 6.7—Radial Friction Welding

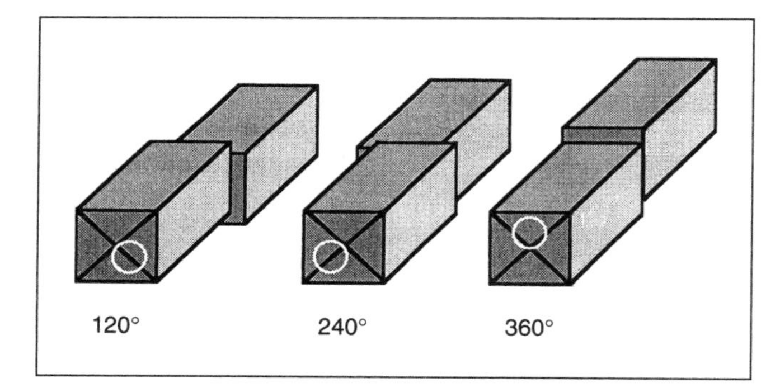

Figure 6.8—Three Consecutive Views of Orbital Friction Welding at 120° Intervals

piece. This technique is used to deposit material in a solid-state mode to a variety of configurations from flat plates to circular or cylindrical shapes. Orbital friction surfacing is illustrated in Figure 6.9.

Angular Reciprocating. Angular reciprocating motion primarily is used for joining plastics. This mode of friction welding employs a cyclic reversing rotational motion in which one or both of the moving workpieces rotate through a given angle that is less than one full rotation. This mode is shown in Figure 6.10.

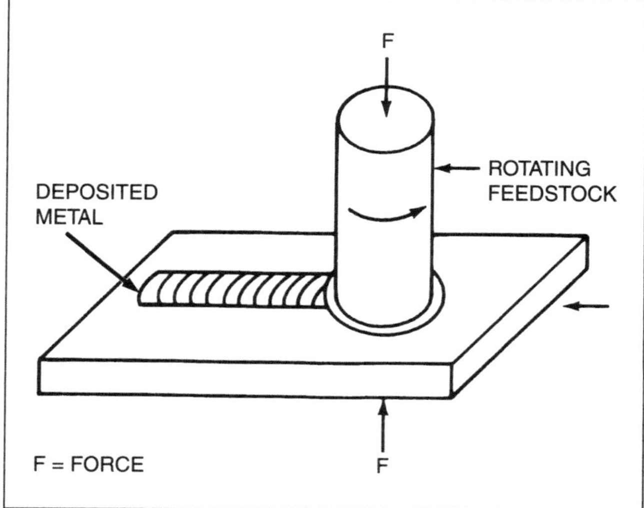

Figure 6.9—Orbital Friction Surfacing

Linear Friction Welding. Linear friction welding is accomplished with a straight-line, back-and-forth motion between the two workpieces, as illustrated in Figure 6.11. An advantage of this technique is that rotational symmetry of the workpieces is not required.

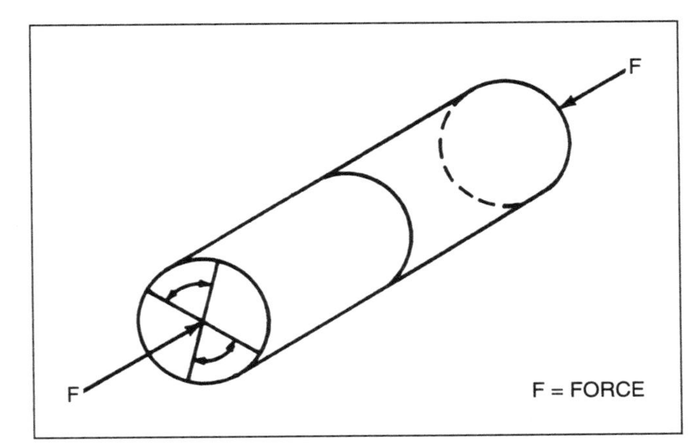

Figure 6.10—Angular Reciprocating Friction Welding

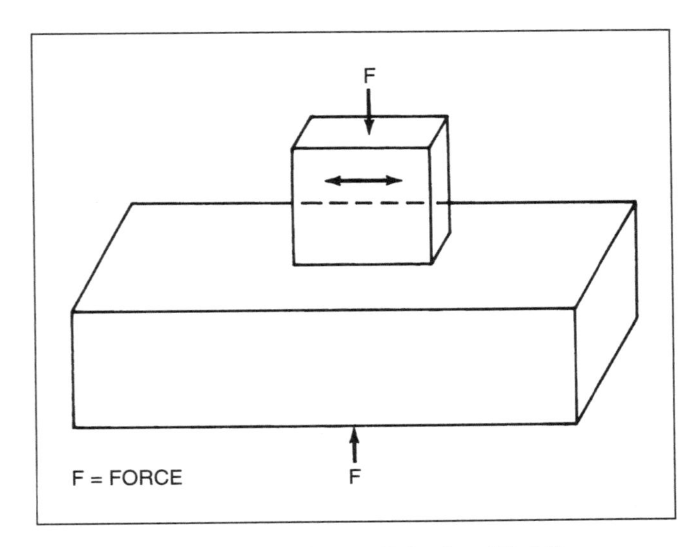

Figure 6.11—Linear Friction Welding

WELDING VARIABLES

In addition to the variables common to most welding processes, such as composition of alloys or materials, geometry of the components to be welded, and service requirements, several other variables are specific to the various modes of friction welding.

DIRECT DRIVE FRICTION WELDING VARIABLES

Rotational or tangential speed, pressure at the weld interface, and heating time are the variables that must be considered in direct drive friction welding.

Speed

The function of rotation is to produce a relative velocity at the faying surfaces. From the standpoint of weld quality, speed generally is not a critical variable; that is, it can vary within a fairly broad tolerance band and still provide sound welds. For steels, the tangential velocity should be in the range of 1.3 m/s (250 ft/min). This is true for both solid and tubular workpieces. Tangential speeds below 1.3 m/s (250 ft/min) produce very high torques that can cause problems with workpiece clamping, inconsistent upset, and metal tearing. Production machines normally are designed to operate with speeds of 300 rpm to 650 rpm. For example, a friction welding machine with a spindle speed of 600 rpm can be used to weld steel from 50 mm to 100 mm (2 in. to 4 in.) in diameter at velocities from 1.6 m/s to 3.2 m/s (310 ft/min to 620 ft/min).

High rotational speeds and the lower heat inputs associated with them, as shown in Figure 6.12, can be used to weld hardenable steels. A longer heating time provides additional heat input to the workpiece to help slow the cooling rate and avoid quench cracking. Conversely, for certain dissimilar metal combinations, low velocities (and the associated shorter heating times) can minimize the formation of brittle intermetallic compounds. For production welding, heating time for a given amount of upset usually is controlled by varying the friction welding pressure rather than altering the spindle speed.

Pressure

A broad range of weld pressures may be applied for any specific operation. Pressure, however, interacts with other process variables, so it should be ensured that the selected pressures are reproducible. Pressure determines the rate of upset, and consequently, the temperature gradient in the weld zone, the required power necessary to drive the spindle, and the amount of axial shortening during the forge stage. The pressure employed for a specific application depends on the materials and the joint geometry. Pressure also can be used to compensate for heat imbalance due to dissimilar mass, as is the case in tube-to-plate welds.

Heating (friction) pressure must be high enough to hold the faying surfaces in intimate contact to avoid oxidation. For a set spindle speed, low pressure limits the temperature at which material can be extruded while causing little or no axial shortening. High pres-

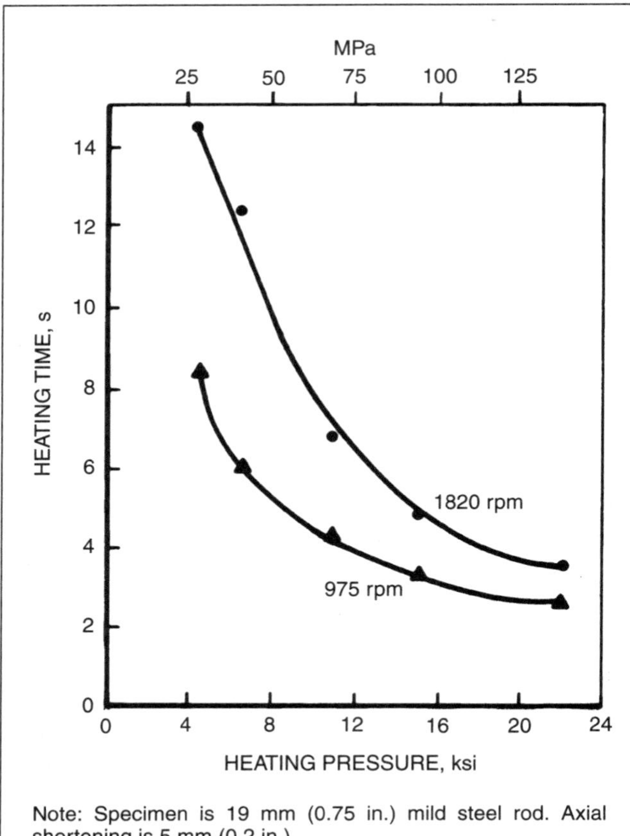

Note: Specimen is 19 mm (0.75 in.) mild steel rod. Axial shortening is 5 mm (0.2 in.).

Figure 6.12—Relationship Between Heating Time and Heating Pressure for Direct Drive Friction Welding of Mild Steel

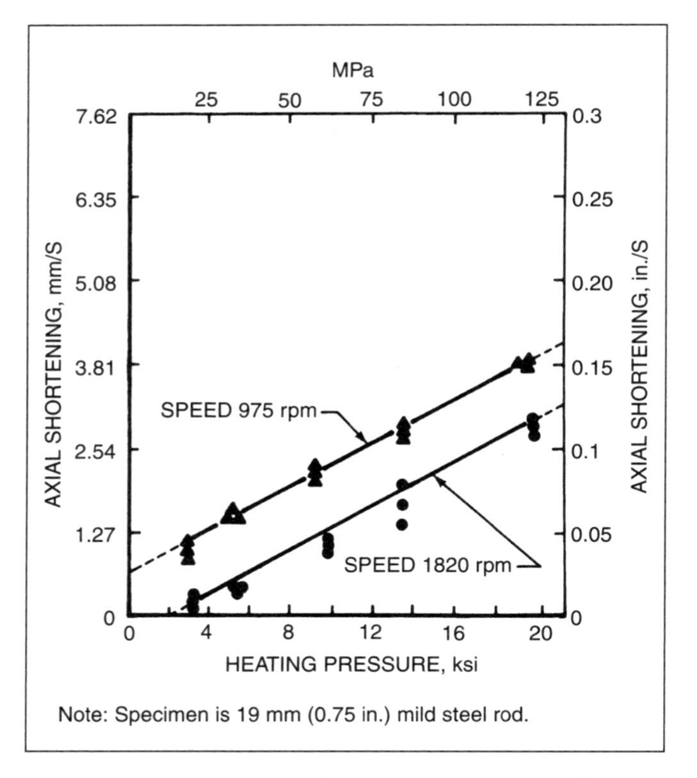

Note: Specimen is 19 mm (0.75 in.) mild steel rod.

Figure 6.13—Relationship Between Axial Shortening and Heating Pressure for Direct Drive Friction Welding of Mild Steel

sure enables the displacement of material at much lower temperatures and increases axial shortening. With mild steel, the rate of axial shortening is approximately proportional to heating pressure, as illustrated in Figure 6.13. It also shows that for a given pressure during the heating phase, axial shortening is greater at low speed than at high speed. Joint quality is improved in many metals, including steels, by applying an increased forging force at the end of the heating period.

Heating Time

For a particular application, heating time is determined during setup or from previous experience. Excessive heating time limits productivity, wastes material, and may have a negative effect on joint properties. Insufficient time may result in uneven heating, entrapped oxides, and unbonded areas at the weld interface. Uneven heating is typical of friction welds in bar stock when insufficient friction time is used. Welds produced on bar stock using lower surface velocities also may be affected by short friction times, particularly at the center of the bar. This is due to the reduced surface velocities in this location. Hence, thermal diffusion from the outer portion of the faying surface must take place in order to ensure a sound weld throughout.

Heating time can be controlled in three ways. The first method is to use a suitable timing device that stops rotation at the end of a preset time. Preheat and forging functions can be incorporated with heating time using a sequence timer. The second method is to stop rotation after a predetermined axial shortening. This control for axial shortening is set to consume a sufficient length to ensure adequate heating prior to upsetting. Variations in surface condition can be accommodated, to some degree, without sacrificing weld quality. The third method is to control the rate of axial shortening. Direct drive friction welding control devices often have a burn-off rate parameter that controls the pressure applied to the workpieces, within set limits, to achieve the desired rate.

In summary, for a given amount of axial shortening, the heating time is significantly influenced by heating pressure and rotation speed. Heating time is reduced as heating pressure is increased. Heating time also decreases proportionally with rotation speed at the same heating pressure.

Direct Drive Friction Welding of Mild Steel

A wide range of pressures may be used to make sound welds in steel. For mild steel, heating (friction) pressures of 30 MPa to 60 MPa (4350 psi to 8700 psi) and forging pressures of 75 MPa to 150 MPa (10 800 psi to 21 750 psi) are acceptable. Commonly used values are 55 MPa (8000 psi) for heating and 140 MPa (20 000 psi) for forging. High-strength alloys for high-temperature use, such as stainless steels and nickel-base alloys, require higher forging pressures.

If a preheat effect is desired in order to achieve a slower cooling rate or reduce the load on the spindle motor, a pressure of about 20 MPa (3000 psi) can be applied for a brief period at the initiation of the weld cycle. This is referred to as *scrub*. After a set time or distance, the pressure can be increased to that required for welding. (See Appendix C for pressure unit conversions.)

INERTIA FRICTION WELDING VARIABLES

In inertia friction welding the speed continuously decreases with time during both the friction and the forging stages. This contrasts to direct drive friction welding, in which the friction stage proceeds at constant speed. Throughout the welding cycle, the thickness of the plasticized layer is related to the surface velocity. As the speed decreases the rate of heat generation decreases, thereby causing a reduction in the thickness of the hot plasticized layer. The torque on the workpieces reaches a second peak when the weld enters the forging stage. The increased axial pressure that occurs during this stage further assists the extrusion of plasticized metal from the joint. Axial shortening continues until there is insufficient energy to displace material. Rotation eventually ceases. Forge pressure is maintained for some duration as the joint cools.

Relationship of Variables

The three welding variables specific to inertia friction welding are the following:

1. Moment of inertia of the flywheel,
2. Initial flywheel speed, and
3. Axial pressure.

The first two variables determine the total kinetic energy available to accomplish welding. The amount of pressure generally is based on the cross-sectional area of the material to be welded.

The energy in the flywheel at any instant during the welding cycle is defined by the following equation:

$$E = \frac{I\,S^2}{C} \tag{6.1}$$

where

E = Energy, joules (J) (ft-lb);
I = Moment of inertia, kg-m^2 (lb-ft^2);
S = Speed, rpm;
C = 182.4, when the moment of inertia is expressed in kilograms-meters^2 (kg-m^2); and
C = 5873, when the moment of inertia is expressed in pound-foot^2 (lb-ft^2).

For mathematical modeling and parameter calculations, the derived value of unit energy is defined by the following equation:

$$E_u = \frac{E}{A} \tag{6.2}$$

where

E_u = Unit energy, J/mm^2 (ft-lb/in.^{2});
E = Energy, J (ft-lb); and
A = Faying surface area, mm^2 (in.^{2}).

Unit energy can be used to scale or extrapolate data from one type of material, size, or geometry to another. This extrapolation often can serve as a first approximation.

With a particular flywheel system, the energy in the flywheel is determined by its rotational speed. If the mass of the flywheel is changed, the available energy at any particular speed will change. Therefore, the capacity of an inertia friction welding machine can be modified by changing the flywheel, within the limits of the machine capability.

During welding, energy is extracted from the flywheel, causing the speed to decrease. The total time used by the flywheel to stop depends on the rate at which the kinetic energy in the flywheel is converted to heat.

The shape and width of the heat-affected zone can be adjusted by varying the flywheel moment of inertia, heating pressure, and speed. These variables also are used to control the cooling rate of the weldment. The effect of flywheel energy, heating pressure, and tangential velocity on the heat pattern and flash formation of welds in steel are shown in Figure 6.14.

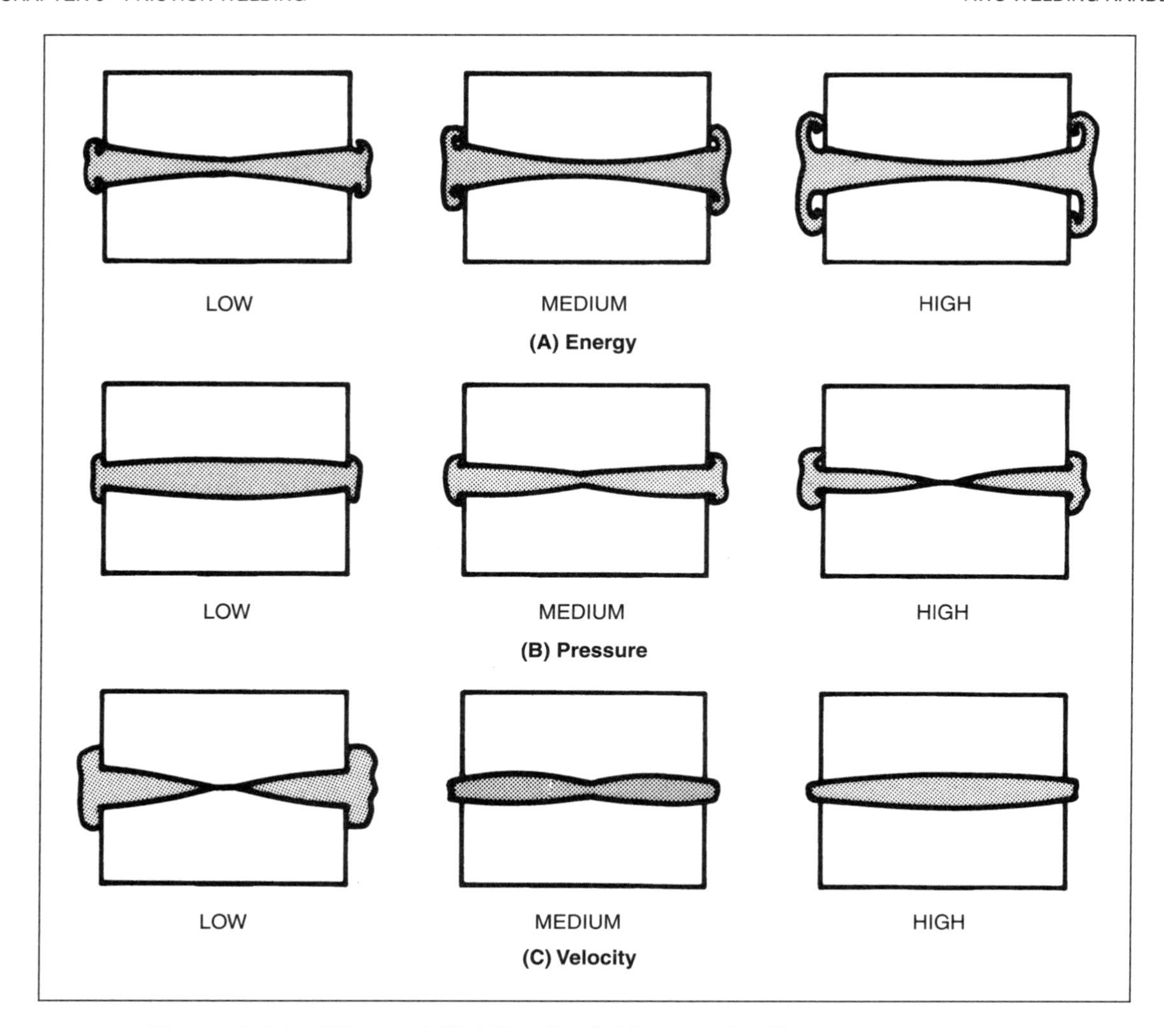

Figure 6.14—Effect of Welding Variables on the Heat Pattern at the Weld Interface and Flash Formation of Inertia Welds in Steel

Flywheel Effect. The moment of inertia of the flywheel depends on its section shape, diameter, and mass. For a specific application and initial speed, the energy available for welding can be increased by changing to a flywheel with a larger moment of inertia. The product of flywheel moment of inertia and the square of its initial velocity determine the total amount of energy produced.

The amount of upset occurring near the end of the welding cycle depends on the remaining energy in the flywheel and the heating or forging pressure. For low-carbon steel, forging usually starts at a peripheral velocity of about 1.0 m/s (200 ft/min). Large flywheels can prolong the forging or upsetting phase. If the flywheel is too small, the upset may be insufficient to eject impurities from the interface and consolidate the weld.

For a given initial velocity and heating pressure, a larger flywheel increases the available energy. The effect of this is shown in Figure 6.14(A). As the available energy is increased, the amount of material displacement (upset) becomes greater. The heating pattern remains reasonably uniform, but the excessive energy wastes metal in the form of flash.

Heating Pressure. The effect of varying heating pressure generally is opposite to that of velocity. As shown in Figures 6.14(B) and (C), welds made at low heating pressure are similar to welds made at high velocity with regard to the formation and appearance of weld upset and heat-affected zones. Excessive pressure produces a weld that lacks good bonding at the center and exhibits a large amount of weld upset. The effective heating pressure range for a solid bar of medium-carbon steel is 152 MPa to 207 MPa (22 000 psi to 30 000 psi).

Velocity. The instantaneous tangential velocity varies directly with the radius and rotational speed according to the following equation:

$$V_t = K\,r\,s \qquad (6.3)$$

where

V_t = Tangential velocity, ft/min;
r = Radius, m (in.);
s = Instantaneous speed, rpm;
K = 0.1, when r is expressed in meters; and
K = 0.52, when r is expressed in inches.

When the workpiece is a rotating solid rod, the velocity varies linearly from zero at the center to a maximum at the periphery. This contrasts with the behavior of a thin-wall tube where the change in velocity across the faying surface is minor. Hence, the energy required for welding a rod and a tube of the same alloy and equal faying surface area will be different.

For each type of metal there is a range of peripheral velocities that produces the best weld properties. For welding solid bars of steel, the recommended initial peripheral velocity of the workpiece ranges from 2.5 m/s to 7.5 m/s (500 ft/min to 1500 ft/min); however, welds can be made at velocities as low as 1.5 m/s (300 ft/min). If the surface velocity is too low, whether at the required energy level or not, the heating at the center will be too low to produce a bond across the entire interface and the flash will be rough and uneven. This is illustrated in Figure 6.14 (C). At medium velocities of 1.5 m/s to 4.1 m/s (300 ft/min to 800 ft/min), the heating pattern in steel has an hourglass shape at the lower end of the range and gradually flattens at the upper end of the range. For steel, the weld becomes rounded and is thicker at the center than at the periphery at initial velocities above 6.1 m/s (1200 ft/min).

WELDABILITY OF MATERIALS

Friction welding can be used to join a wide range of similar and dissimilar materials, including metals, some metal matrix composites, ceramics, and plastics. Combinations of materials that can be joined by friction welding are shown in Figure 6.15. The information in this chart should be used only as a guide. Weldability may depend on a number of factors, including specific alloy compositions, applicable process variation, component design and service requirements.

Any metal that can be hot forged and is not suitable for dry bearing applications can be joined by friction welding. Some metals may require postweld heat treatment. Free-machining types of alloys should be welded with caution because redistribution of inclusions may create planes of weakness in the weld zone. These welds exhibit low strength, decreased ductility, and reduced notch toughness.

In general, a consequence of reorienting inclusions into the weld plane is that the ductility and toughness across the joint tend to approach the wrought short-transverse properties of the materials. If these properties are critical, it is essential to use microstructurally clean materials.

The bonding mechanism of dissimilar metals is more complex. A number of factors, including physical and mechanical properties, surface energy, crystal structure, mutual solubility and intermetallic compounds may influence the bonding mechanism. It is likely that some alloying will occur in a very narrow region at the weld interface as a result of mechanical mixing and diffusion. The properties of this layer may have a significant effect on overall joint properties. Mechanical mixing and interlocking also may contribute to bonding. The complexity of the bonding mechanism makes it very difficult to predict the weldability of dissimilar metals. The suitability of a particular combination should be established prior to each application with a series of tests designed for that purpose.

A number of dissimilar metal combinations are marginally weldable. They may involve combinations that have high and low thermal conductivities, a large difference in forging temperatures, or the tendency to form brittle intermetallic compounds. Examples are aluminum alloys welded to copper or steel, and titanium alloys welded to steel or stainless steel.

The metallurgical structures produced by friction welding generally result from elevated-temperature deformation. Time at temperature is short, and the temperatures achieved generally are below the melting point. With nonhardenable metals such as mild steel, changes in properties are negligible in the weld zone. Conversely, structural changes may occur in the heat-affected zone of hardenable steels. Hardenable steels should be welded with a relatively long heating time to achieve a slower cooling rate and preserve toughness.

The interface structures of dissimilar metal combinations are significantly affected by the particular welding conditions employed. The longer the welding time, the greater is the consideration that must be given to diffusion across the weld interface. Proper welding conditions usually will minimize undesired diffusion or intermetallic compound formation. The interface between an aluminum and carbon steel friction weld is shown in Figure 6.16. A very narrow diffusion zone is apparent.

Examination of joints between dissimilar metals may show mechanical mixing at the interface in some cases. This action in a joint between Type 302 stainless steel and tantalum is shown in Figure 6.17.

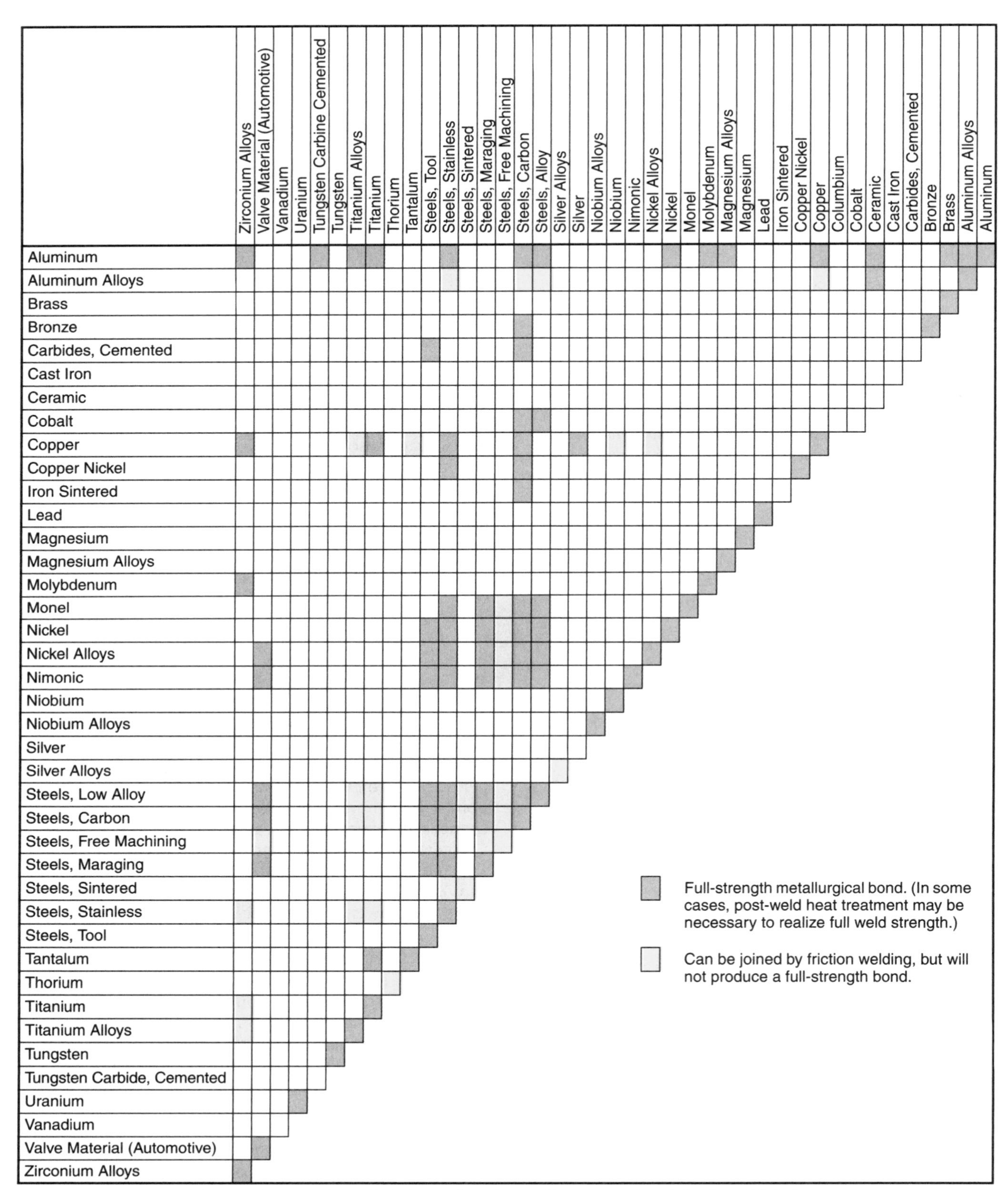

Figure 6.15—Material Combinations Weldable by Friction Welding

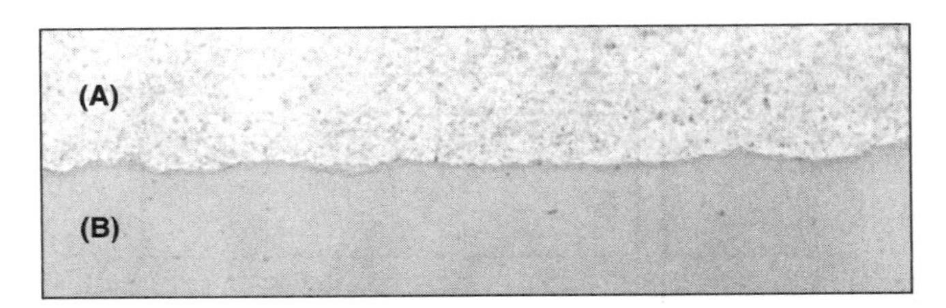

Figure 6.16— Micrograph of the Interface of a Friction Weld Between (A) Aluminum and (B) Carbon Steel (x1000)

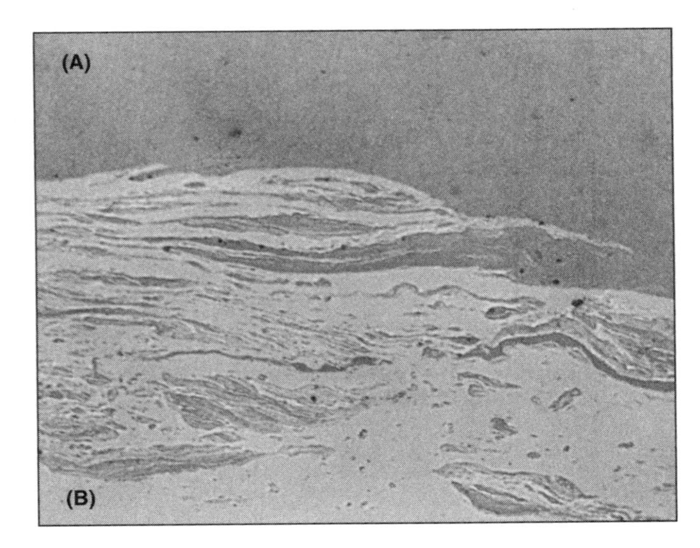

Figure 6.17— Micrograph of the Interfacial Mixing of a Friction Weld Between (A) Tantalum and (B) Type 302 Stainless Steel (X200 and Reduced 66%)

JOINT DESIGN

The nature of direct drive and inertia friction welding requires that the joint face of at least one of the two workpiece be essentially round, with the exception of orbital and linear friction welding. The rotated workpiece should be somewhat balanced in shape because it must revolve at relatively high speed. Preparation of the joint surfaces normally is not critical except when welding alloys with distinct differences in mechanical or thermal properties, or both.

The basic joint designs for combinations of bar, tube, and plate materials are illustrated in Figure 6.18. When bars or tubes are welded to plate, most of the flash comes from the bar or tube because there is less mass in that section and the heat penetrates deeper into it. This effect can be usefully employed in joints between dissimilar metals with widely different mechanical or thermal properties. For these joints, the material with lower forging strength or lower thermal conductivity should have the larger cross-sectional area.

Conical joints usually are designed with the weld faces at 45° to 60° to the axis of rotation, as shown in Figure 6.19. For low-strength metals, large angles are preferred because they will support the axial thrust required to produce adequate heating pressure. Certain applications may require the use of conical joints; however, experience has shown that a butt weld (perpendicular geometry) is superior. Butt-weld geometry results in less residual stress and distortion.

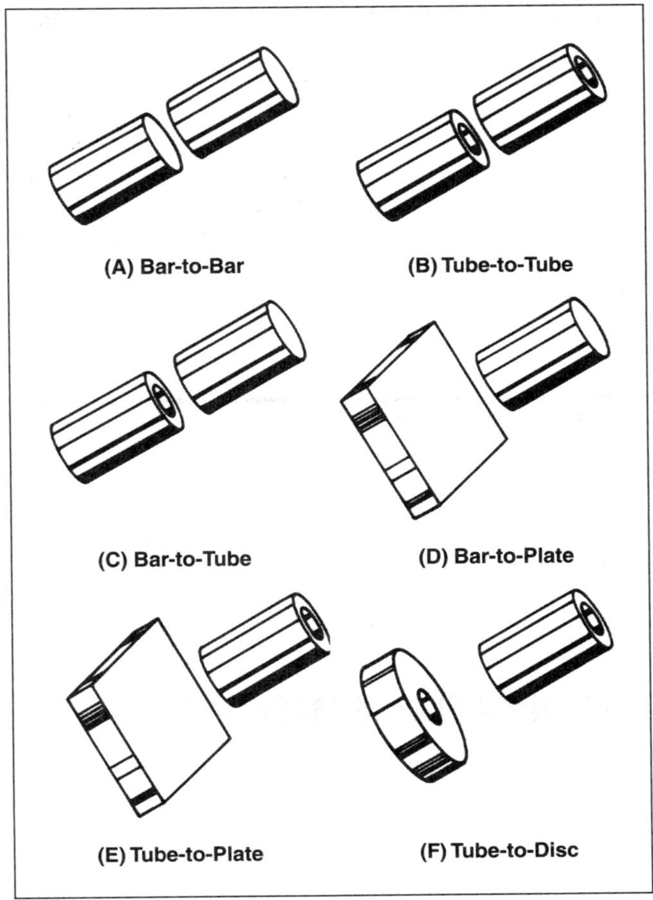

Figure 6.18—Typical Friction Weld Joint Applications

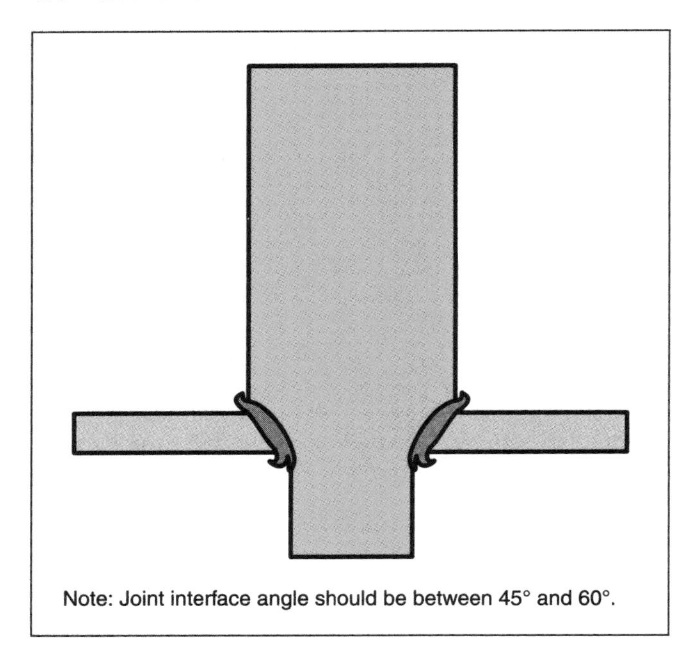

Figure 6.19—Typical Conical Weld Joint Design

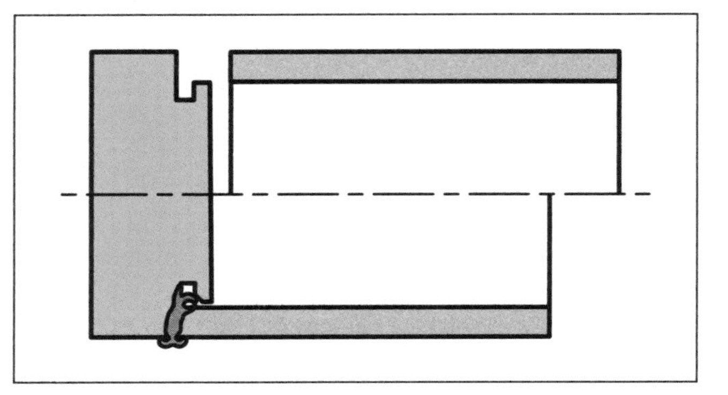

Figure 6.20—Typical Flashtrap Joint Design

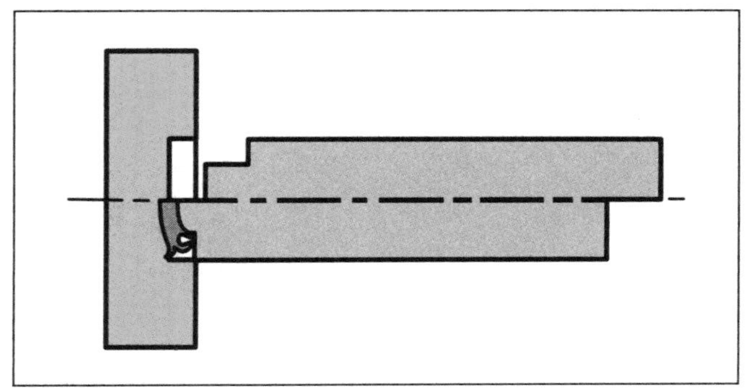

Figure 6.21—Typical Flashtrap Joint Design for Bar-to-Plate Weld

For applications in which the flash cannot be removed but must be isolated for cosmetic or functional reasons, clearance for flash can be provided in one or both workpieces with a flashtrap configuration. Two flashtrap designs are illustrated in Figures 6.20 and 6.21.

EQUIPMENT

Friction welding machines generally resemble machining lathes with modifications made for the high loads on the spindle and frame. An exception to this resemblance is the portable friction welding machine designed to be used in the field.

FRICTION WELDING MACHINES

A typical friction welding machine consists of the following components:

1. Head,
2. Base,
3. Clamping arrangements,
4. Rotating and upsetting mechanisms,
5. Power source,
6. Controls, and
7. Optional monitoring devices.

These components are used in both direct drive and inertia friction welding; however, the machines for each variation differ somewhat in design and method of operation.

Typical equipment designs allow for forces ranging from 890 newtons (N) (200 lbs) to 250 metric tons (275 tons) in direct drive friction welding machines, and up to 2040 metric tons (2250 tons) maximum forge force in inertia friction welding machines. This means that workpieces with diameters ranging from 1.5-mm (0.06-in.) bar stock to 600-cm (24-in.) tubes can be welded. The cross-sectional area ranges from 2 mm^2 to 160 000 mm^2 (0.003 $in.^2$ to 250 $in.^2$). Based on manufacturers' recommendations for mild steel, a machine generally can make welds that have a range of cross-sectional area of 8 to 1. For example, a 27-metric ton (30-ton) maximum forge force machine can make welds in bar stock ranging from 13 mm (1/2 in.) diameter to approximately 38 mm (1-1/2 in.) diameter.

Direct Drive Welding Machines

When using direct drive friction welding machines, one of the workpieces is clamped in a vise and the other is held in a centering chuck mounted on a spindle. The spindle is driven by a motor with a single- or variable-speed drive.

To make a weld, the rotating workpiece is thrust against the stationary workpiece to produce frictional heat at the contact surfaces, as illustrated in the schematic of a direct drive welding machine in Figure 6.22. Friction generated by the combination of speed and pressure raises the contact surfaces to a suitable temperature, and deformation (upset) occurs. Rotation is then stopped and the pressure is maintained or increased to further upset the interface and complete the weld. (Refer to Figure 6.4 for an illustration of a typical direct drive weld cycle.)

The machine spindle can be driven directly by a motor and allowed to stop by natural deceleration characteristics and the retarding torque exerted by the weld. A common practice is to include a fast-acting brake (mechanical or electrical, or both) on the spindle. The function of the brake is to rapidly terminate rotation at the end of a specified heating time or after a preset axial shortening of the weldment. This feature provides good control of overall weldment length and broadens the acceptable range of welding variables for critical applications. Dynamic braking also is used, performed electrically within the motor drive system to allow for tighter control of rotational orientation or final weldment length, or both.

Two variables are used to control the friction heating phase: axial shortening during friction (displacement), and heating (friction) time. Using displacement feedback control, the friction heating phase continues until a given workpiece length is achieved. Distance control is used to compensate for variations in preweld workpiece length, or to consistently achieve an appropriate range of weldment lengths. Displacement control should not be used if material cleanliness cannot be guaranteed. If workpieces become contaminated with lubricants, for example, friction times may become excessive. In these cases, the joint may overheat, causing degradation of weldment properties. This is particularly true for dissimilar metal welding in which intermetallic formation is a concern.

The friction-time mode of feedback control is intended to provide repeatable energy input. It also is possible to combine both options: a preweld distance can be set, after which control changes to a time-based mode, or the time can be set before preweld distance. For critical applications in which the workpieces normally would be of uniform length before welding, the time mode is preferred. A variation is the use of burnoff rate control. In this instance, the friction speed is held constant and the force is varied within certain boundaries to yield a certain rate (mm/min or in/sec). This method is used to further refine the heat generated during friction. In all cases, a minimum loss in length must occur between the components to ensure the removal of contaminants at the interface and the production of a sound weld.

The variables associated with this method are the following:

1. Rotation speed (rpm);
2. Preheat (scrub) pressure;
3. Preheat (scrub) distance or time;
4. Friction pressure;
5. Friction distance, time, or rate;

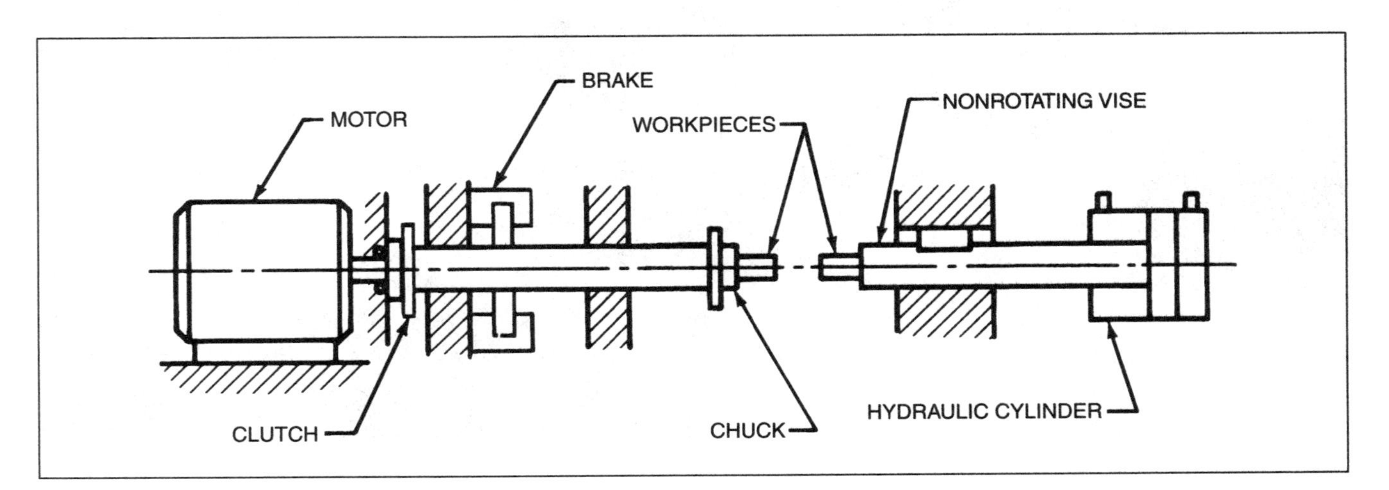

Figure 6.22—Basic Arrangement of a Direct Drive Welding Machine

6. Braking time (including delay and rate);
7. Forge pressure; and
8. Forge time (including delay and rate).

This list is comprehensive and it should be noted that not all machines or weld schedules will require settings for all variables. A direct drive friction welding machine is shown in Figure 6.23.

Inertia Welding Machines

An inertia welding machine (see Figure 6.24) has a flywheel mounted on the spindle between the drive and the rotating chuck, as illustrated in Figure 6.25. The flywheel, spindle, chuck, and rotating workpiece are accelerated to a select speed corresponding to a specific energy level. When the appropriate speed is attained, the drive motor is disengaged from the flywheel, allowing the workpiece to spin freely. The two workpieces are then brought together and a specific axial thrust is applied. The kinetic energy of the flywheel is transferred to the weld interface and converted to heat. As a result, speed decreases and the flywheel finally comes to rest. Simultaneously, the tangential velocity is decreasing to zero, with time in an essentially parabolic mode.

In the majority of applications, inertia friction welding machines use a single axial thrust to produce heating and forging pressure. However, the machines are normally capable of applying two levels of thrust. When forging pressure is used, it is triggered near the end of the cycle by a previously selected speed setting. (Refer to Figure 6.5 for an illustration of a typical iner-

Photograph courtesy of Thompson Friction Welding

Figure 6.23—Direct Drive Friction Welding Machine

Photograph courtesy of Manufacturing Technology, Inc. (MTI)

Figure 6.24—Typical Inertia Friction Welding Machine

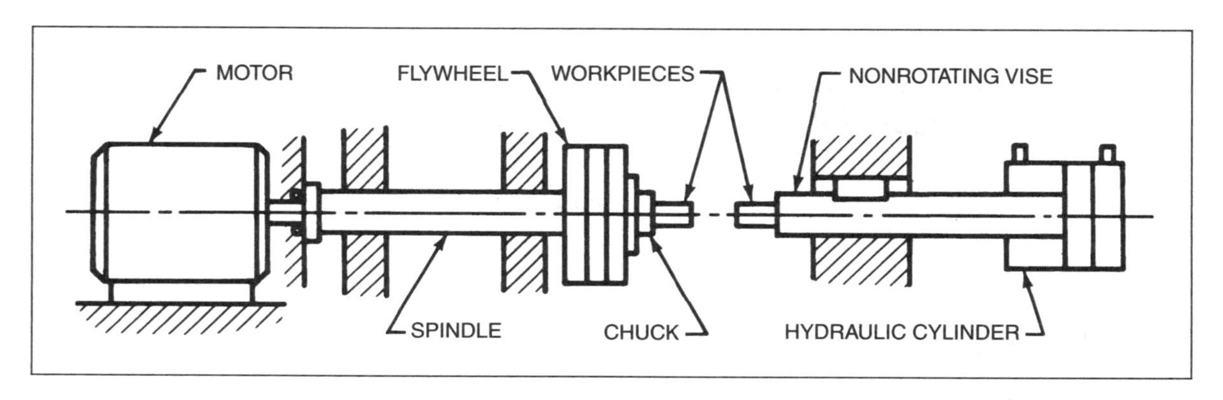

Figure 6.25—Basic Arrangement of an Inertia Friction Welding Machine

tia weld cycle.) This multiple-force technique also can be used to provide preheating before welding takes place, similar to the method used in direct drive systems.

A distance-setting control mode also can be achieved in inertia friction welding. Energy input is varied by adjusting rpm, depending on initial workpiece lengths. Most welds are performed by varying the speed and pressure only. A correlation between energy and upset must be established experimentally before employing this mode.

Typical inertia friction welding machines can be adjusted for the following variables that control weld quality:

1. Total moment of inertia,
2. Weld speed (initial rpm),
3. Weld pressure,
4. Upset (forge) speed (rpm at which upset pressure is applied), and
5. Upset (forge) pressure.

TOOLING AND FIXTURES

The two basic types of tooling are rotating and stationary devices. The friction welding machines in Figures 6.24 and 6.25 are equipped with both types. Each type can be designed for either manual or automatic operation. As a rule, manually actuated tooling is used only for small-quantity production. Rotating tooling must be well balanced, have high strength, and provide good gripping power.

The most commonly used stationary gripping device is a vice-like fixture with provision for absorbing thrust. This device permits reasonable tolerance in the stationary workpiece diameter and yet maintains concentricity with the other piece in the chuck. More accurate devices may be necessary when concentricity is critical. The effective mating of the faying surfaces and the concentricity of the workpieces depend on the accuracy of manufacture, projecting length from the clamping fixture, and the rigidity of the tooling.

WELDING PROCEDURES

Friction welding procedures include correct surface preparation, accurate fitup, and the requirements of the application, such as heat treatment.

SURFACE PREPARATION

Weld quality and consistency are best when the faying surfaces are free of dirt, oxide or scale, grease, oil, or other foreign materials. In addition, the faying surfaces should fit together with very little angularity. In some applications, a certain amount of contamination and angular contact of the faying surface may be tolerated. This holds true if sufficient axial shortening is used to account for the angular contact and to extrude sufficient plasticized metal from the interface to carry away any contaminants. Sheared, flame-cut, or sawed surfaces may be used with adequate axial shortening, provided the surfaces are essentially perpendicular to the axis of rotation. If the surfaces are not perpendicular, joint mismatch could result. For best practice, the squareness (angularity) should be within 0.01 mm/mm (0.010 in./in.) of joint diameter.

Thick layers of mill scale should be removed from the workpieces prior to welding to avoid unstable heating. A thin layer of scale may not be detrimental with adequate axial shortening. Center projections left by cutoff tools are not harmful. However, pilot holes or concave surfaces should be avoided because they may entrap air or impurities at the interface.

The surface cleanliness of both workpieces is critical for dissimilar-metal welds between materials with large differences in hot-forging behavior. The squareness of the harder material also is critical. Examples include steel-to-aluminum, steel-to-copper, and copper-to-aluminum welds.

FITUP

Friction welding requires the workpieces to be in a rigidly fixed position. All chucks or other gripping devices used for holding the workpieces must be reliable. Slippage of a workpiece in relation to the chuck results either in a defective weld or damage to the gripping device or the workpiece.

The gripping mechanism of the chucking devices must be rigid enough to resist the applied thrust. The extension of the workpiece from the device should be as short as practical to minimize deflection, eccentricity, and misalignment. The diameter of the grip must be as large as the diameter of the weld interface or larger, otherwise the workpiece may shear at the gripping point. Serrated gripping jaws are recommended for maximum clamping reliability.

HEAT TREATMENT

Prior heat treatment of the workpieces generally has little effect on the weldability of specific alloys by friction welding. However, heat treatment may affect the mechanical properties of the heat-affected zone and the gripping of the workpieces.

Postweld heat treatment can be employed to produce the desired properties in the base metal, the welded joint, or both. Postweld annealing may be used to soften or stress-relieve the joint.

For the welding of dissimilar metals, it should be ensured that postweld heat treatment does not contribute to the formation or expansion of an intermetallic layer at the interface, which may lower joint ductility or strength. The postweld heat treatment should be evaluated for the application by destructive testing.

APPLICATIONS

Friction-welded applications span the aerospace, agricultural, automotive, defense, marine, and oil industries. Friction welding is used for everything from tong-holds on forging billets to critical components of aircraft engines. Some of the automotive parts manufactured by friction welding include gears, engine valves, axle tubes, driveline components, strut rods and shock absorbers. Manufacturers of agricultural equipment commonly use friction welding to fabricate hydraulic piston rods, track rollers, gears, bushings, axles and similar components. Friction-welded aluminum-to-copper joints are in wide usage in the electrical industry. Stainless steels are welded to carbon steels in various sizes to make components for marine drive systems and to manufacture water pumps for residential and industrial use. Friction-welded assemblies often are used to replace expensive castings and forgings.

Some typical automotive and aircraft applications are shown in Figures 6.26 and 6.27. Figure 6.28 shows three weldments of dissimilar metals.

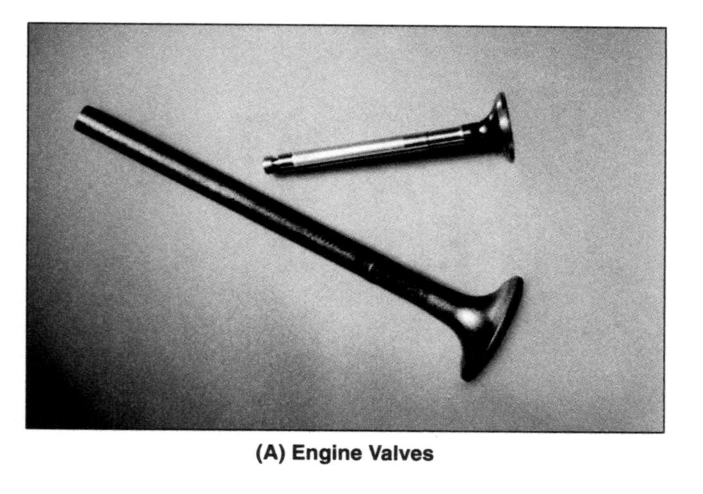

(A) Engine Valves

(B) Piston Assembly

(C) CV Joints

Figure 6.26—Typical Automotive Applications

WELD QUALITY

Weld quality is dependent on the proper selection of the type and quality of material and other welding variables. Good welds can be made between similar metals with a wide range of speeds, pressures and times. The welding parameters of dissimilar metal combinations are more critical than those of similar metals.

DISCONTINUITIES

Discontinuities characteristic of fusion welds, such as gas porosity and slag inclusions, are not encountered in friction welding; however, other types of discontinuities may occur. Discontinuities in friction welds are associated with improper surface preparation, incorrect welding conditions, defective material, or combinations of these. Discontinuities at the center of a weld may occur

(A) Titanium Compressor Rotor

(B) Titanium Compressor Rotor Section

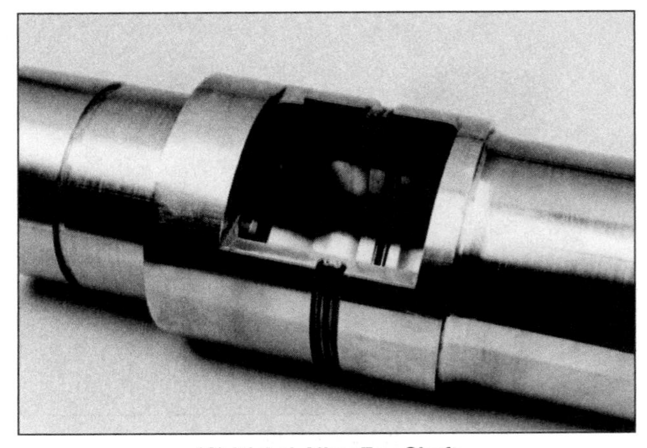

(C) Nickel Alloy Fan Shaft

Figure 6.27—Typical Aircraft Applications

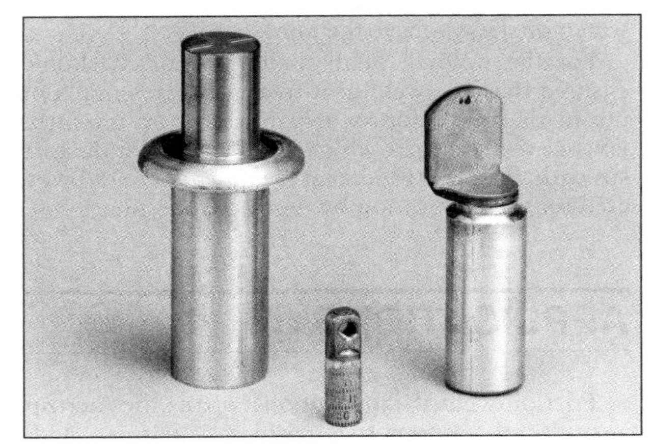

(A) Aluminum-to-Copper Electrical Connectors

(B) Aluminum-to-Stainless Steel Fuser Roller

(C) Aluminum-to-Inconel™ Transition Joint

Figure 6.28—Bi-Metal Combinations

for various reasons, such as welding conditions that do not create sufficient heating for coalescence.

Discontinuities in inertia friction welds made with the same speed and inertial mass but with a decreasing heating pressure (axial shortening) are shown in Figure 6.29(A) through (D). The two cross sections shown in Figures 6.29(E) and (F) exhibit center discontinuities due to insufficient pressure.

Incomplete bonding also may occur in direct drive friction welds when inadequate speed, heating time or heating pressure is used.

PROCESS MONITORING

Computerized data acquisition and analysis systems have revolutionized process monitoring in friction welding. Microprocessor-controlled welding machines are capable of maximizing both output and quality. Particularly useful is their agility in documenting each weld and manipulating data for statistical process control (SPC) purposes.

Factors that are monitored and documented include friction and forge pressures, friction speed, friction upset distance and time. Other parameters such as torque and energy also may be monitored in specific applications.

INSPECTION AND TESTING

Examination and testing are applied to incoming materials and the resulting weldments. Production sampling for testing to guarantee quality, in-process monitoring, and nondestructive examination usually are employed. Depending on the quality level needed, this may range from simple visual inspection and mechanical tests to advanced nondestructive examination techniques. Testing standards published by the American Welding Society include *Guide for the Visual Examination of Welds*, B1.11;[4] *Guide for the Nondestructive Examination of Welds*. B1.10;[5] and *Standard Methods for Mechanical Testing of Welds*, B4.0M.[6]

A photograph of a friction-welded automotive half-shaft is shown in Figure 6.30(A). Peak temperature is used as a process control, and the infrared image of the weld is shown in Figure 6.30(B). Figure 6.31 represents a series of visual and destructive examination techniques for an aluminum-to-copper inertia friction weld.

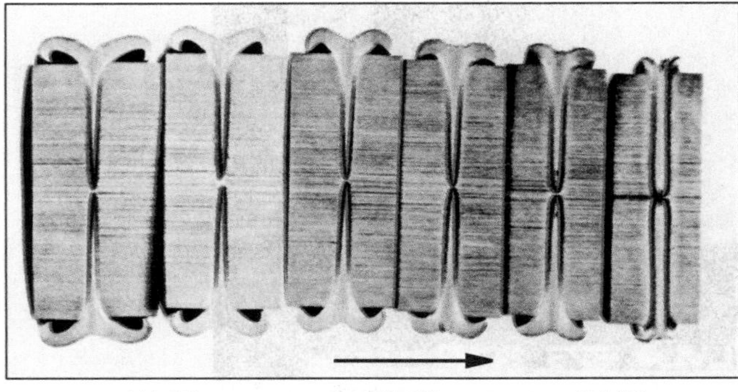

Figure 6.29—Effect of Axial Shortening on Friction Weld Joints

ECONOMICS

Friction welding has several economic advantages over other processes. Fluxes, shielding gases, and filler metals are not required, systemically reducing the cost of associated supplies and consumables. Less skill is needed to operate a friction welding machine than is required for some of the other processes. Most friction welding machines operate with little or no operator input, other than feeding the workpieces into the machine.

Although the initial capital cost of friction welding machines often is higher than machines used in other welding processes, return on investment typically can be realized in one to two years, depending on production volume. For low-volume production, welding can be outsourced to job shops or vendors who specialize in friction welding, thus lowering the cost of production.

SAFE PRACTICES

Friction welding machines are similar to machine tool lathes in that one workpiece is rotated by a drive system. They are also similar to hydraulic presses in that one workpiece is forced against the other with high loads. Safe practices for lathes and power presses

4. American Welding Society (AWS) Committee on Methods of Inspection, *Guide for the Visual Examination of Welds*, B1.11, Miami: American Welding Society.
5. American Welding Society (AWS) Committee on Methods of Inspection, *Guide for the Nondestructive Examination of Welds*, B1.10, Miami: American Welding Society.
6. American Welding Society (AWS) Committee on Mechanical Testing of Welds, *Standard Methods for Mechanical Testing of Welds*, B4.0M, Miami: American Welding Society.

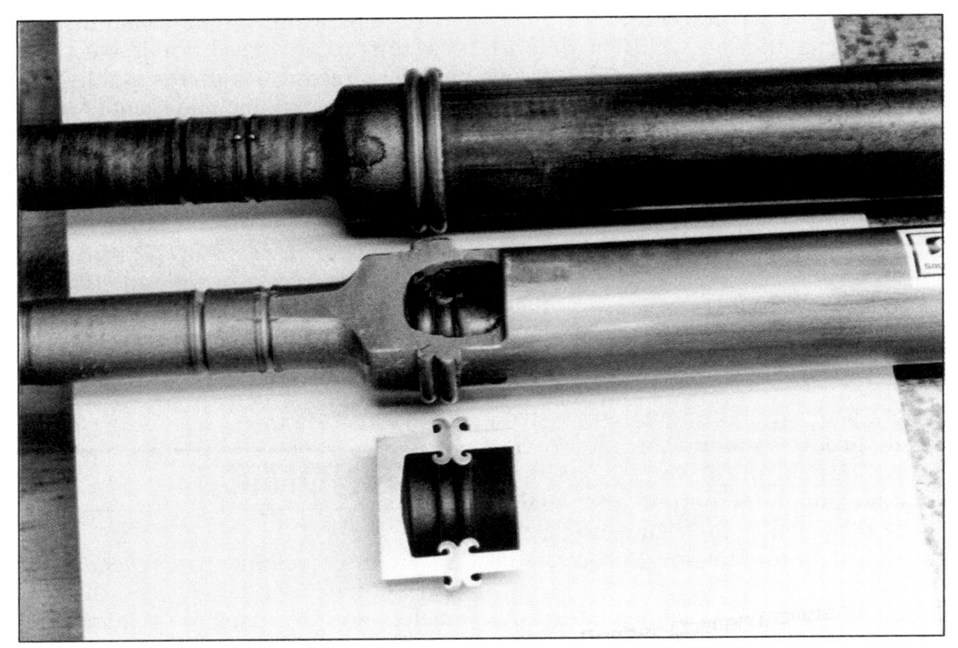

(A) Friction-Welded Joint in an Automotive Halfshaft

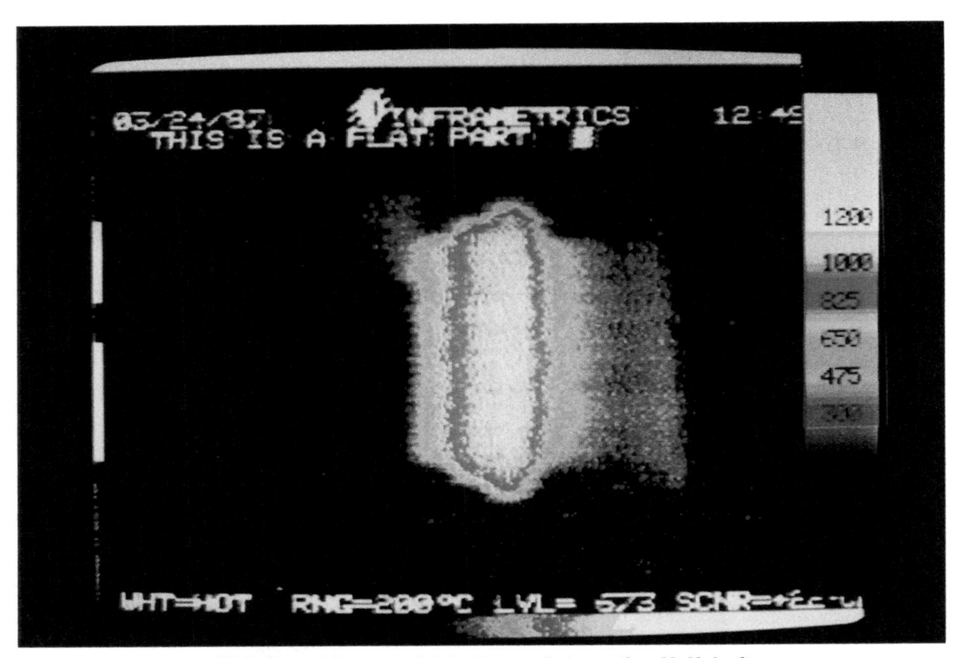

(B) Infrared Image of Joint in an Automotive Halfshaft

Figure 6.30—Friction-Welded Joint (A) and Infrared Imagery (B) Used to Measure Peak Weld Temperature

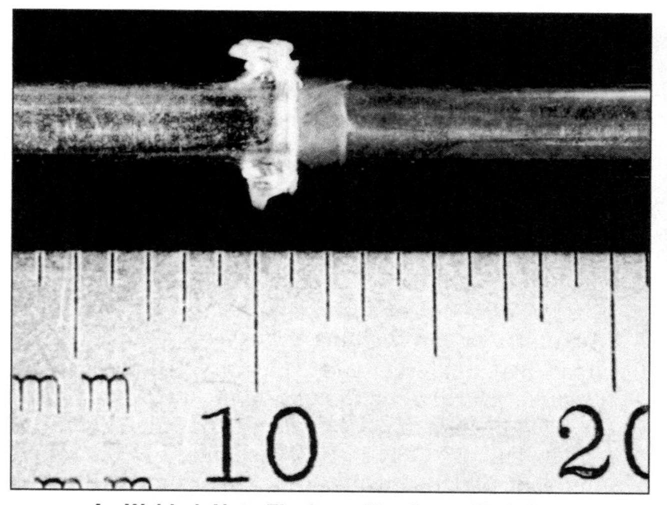

As-Welded; Note Flash on Aluminum 4043 Only

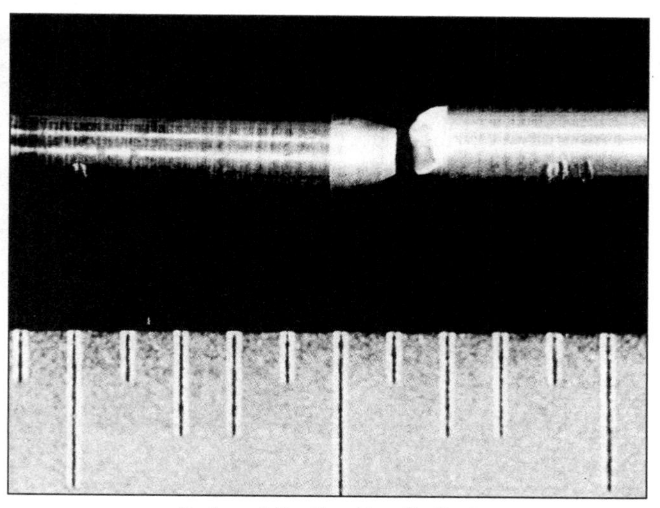
Reduced-Section Tensile Test

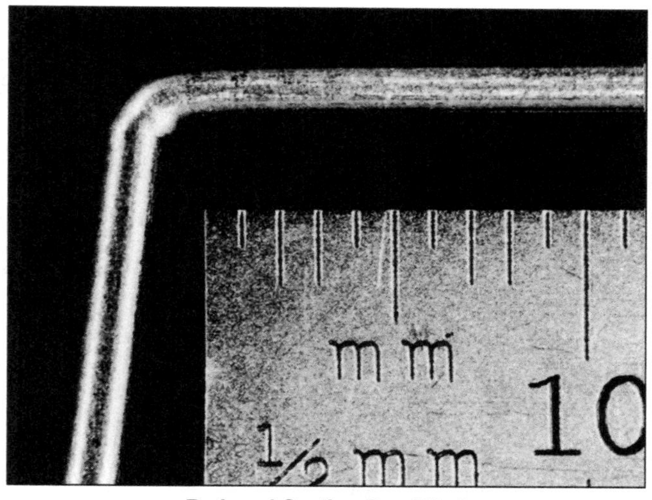

Reduced-Section Bend Test

Figure 6.31—An Inertia Friction-Welded Transition Joint Between OFHC Copper and 4043 Aluminum to Facilitate Simultaneous Solderability and Weldability of a Ground Pin

should be used as guides for the operation of friction welding machines. This information is published by the Occupational Safety and Health Administration (OSHA), United States Department of Labor, in the *Code of Federal Regulations (CFR)*, Title 29, Part 1910. 210, Subpart O.[7] Typical hazards are those associated with high rotational speeds of components, flying particles, and high noise levels.

Machines should be equipped with appropriate mechanical guards and shields, two-hand operating switches, and electrical interlocks. These devices should be designed to prevent operation of the machine when the work area, rotating drive, or force system is accessible to the operator or other personnel. Operation manuals and recommendations for the safe installation, operation, and maintenance of machinery are supplied by the equipment manufacturers and should be strictly followed.

7. Occupational Safety and Health Administration (OSHA), *Occupational Safety and Health Standards for General Industry, in Code of Federal Regulations (CFR)*, Title 29 CFR 1910.217, Subpart O, Washington D.C.: Superintendent of Documents, U.S. Government Printing Office.

Operating personnel should wear appropriate eye and ear protection and safety apparel commonly used with machine tool operations. Ear protection should be provided to guard against the high noise levels produced during friction welding.[8] Applicable codes published by OSHA in the *Code of Federal Regulations (CFR)*, Title 29, Part 1910 should be strictly observed.[9]

Appendices A and B of this volume provide lists of health and safety codes, specifications, books, and pamphlets. The publishers and the facts of publication are also listed.

CONCLUSION

Friction welding is a safe and economical solid-state welding process used to join a wide variety of material and joint combinations. Millions of weldments requiring high standards of quality and reliability have been welded using the friction welding process and its variations. Highly sophisticated control devices and techniques used to verify weld quality are ever evolving, assuring continued and enthusiastic use of the process.

Friction welding machines are easy to maintain and can be easily automated, reducing labor cost, facilitating weld repeatability, and ultimately increasing productivity.

BIBLIOGRAPHY

American National Standards Institute (ANSI) Accredited Standards Committee Z49. 2005. *Safety in welding, cutting, and allied processes*, ANSI Z49.1. Miami: American Welding Society.

American Welding Society (AWS) Committee on Definitions and Symbols. 2001. *Standard welding terms and definitions*, AWS A3.0:2001. Miami: American Welding Society.

American Welding Society (AWS) Committee on Friction Welding. 1989 (Reaffirmed 1998). *Recommended practices for friction welding*, AWS C6.1-89. Miami: American Welding Society.

8. American National Standards Institute (ANSI) Accredited Standards Committee Z49, *Safety in Welding, Cutting, and Allied Processes*, ANSI Z49.1, Miami; American Welding Society.

9. Occupational Safety and Health Administration (OSHA), *Occupational Safety and Health Standards for General Industry*, in *Code of Federal Regulations (CFR)*, Title 29 CFR 1910, Subpart Q, Washington D.C.: Superintendent of Documents, U.S. Government Printing Office.

American Welding Society (AWS) Committee on Methods of Inspection. 2000. *Guide for the visual examination of welds*, B1.11:2000. Miami: American Welding Society.

American Welding Society (AWS) Committee on Methods of Inspection. 1999. *Guide for the nondestructive examination of welds*, B1.10:1999. Miami: American Welding Society.

American Welding Society (AWS) Committee on Mechanical Testing of Welds. 2000. *Standard methods for mechanical testing of welds*, B4.0M:2000. Miami: American Welding Society.

Occupational Safety and Health Administration (OSHA). *Occupational safety and health standards for general industry*, in *Code of federal regulations (CFR)*, Title 29 CFR 1910. Washington D.C.: Superintendent of Documents, U.S. Government Printing Office.

SUPPLEMENTARY READING LIST

American Society for Materials. 1993. ASM Handbook, Volume 6, *Welding, brazing, and soldering*. Materials Park, Ohio: American Society for Materials.

American Welding Society (AWS) Committee on Friction Welding. 2006. *Specification for friction welding of metals*, AWS C6.2/C6.2M:2006. Miami: American Welding Society.

Baeslack, W. A. III, and K. S. Hagey. 1988. Inertia friction welding of rapidly solidified powder metallurgy aluminum. *Welding Journal* 67(7): 139-s.

Bell, R. A., J. C Lippold, and D. R. Adolphson. 1984. An evaluation of copper-stainless steel inertia friction welds. *Welding Journal*. 63(11): 325-s.

Dinsdale, W.O. and S. B. Dunkerton. 1981. The impact properties of forge butt welds in carbon-manganese steels, Part I: Continuous drive friction welds. Welding Institute Research Report, (9)159/1981. Cambridge, UK: The Welding Institute.

Dinsdale, W. O. and S. B. Dunkerton. 1981. The impact properties of forge butt welds in carbon-manganese steels, Part II: Orbital Friction and Inertia Welds. Welding Institute Research Report, (9)160/1981. Cambridge, UK: The Welding Institute.

Dunkerton, S. B. 1985. Properties of 25 mm diameter orbital friction welds in three engineering steels. *Welding Institute Research* Report, (4) 272. Cambridge, UK: The Welding Institute.

Dunkerton, S. B. 1985.Toughness properties of friction welds in steels. *Welding Journal*. 65(8): 193-s.

Eberhard, B. J., B. W. Schaaf, Jr., and A. D. Wilson. 1983. Friction weld ductility and toughness as influenced by inclusion morphology. *Welding Journal* 62(7): 171-s.

Ellis, C. R. G. 1972. Continuous drive friction welding of mild steel. *Welding Journal*. 51(4): 183-s–197-s.

Ellis, C. R. G. and J. C. Needham. 1972. Quality control in friction welding, IIW Document III. Miami, Florida: American Welding Society. 460-s–472-s.

Fu, L., Y. Duan, and S. G. Du. 2003. Numerical simulation of inertia friction welding process by finite element method. *Welding Journal* 82(3): 65-s–70-s.

Jessop, T. J. 1975. Friction welding of dissimilar metal combinations: aluminum and stainless steel. The Welding Institute Research Report, Cambridge, UK: The Welding Institute. 73–75.

Jessop, T. J., et al. 1978. Friction welding dissimilar metals. *Advances in welding processes*, 4th International Conference, Harrogate, England, 23-36. Cambridge, UK: The Welding Institute.

Kuruzar, D. L. 1979. Joint design for the friction welding process. *Welding Journal*. 58(6): 31-5.

Kyusojin, A., et al. 1980. Study on mechanism of friction welding in carbon steels. *Bulletin of the ASME*. 23(182). *New York:* American Society of Mechanical Engineers.

Lebedev, V. K., et al. 1980. The inertia welding of low carbon steel, *Avt Svarka* (7): 18-2.

Lippold, J. C. and B. C. Odegard. 1985. Microstructural evolution during inertia friction welding of austenitic stainless steels. *Welding Journal* 64(12): 327-s.

Mortensen, C., G. Jensen, C. Conrad and F. Losee. 2001. Mechanical properties and microstructure of inertia-friction-welded 416 stainless steel. *Welding Journal* 80(10): 268-s–273-s.

Murti, K. G. K., and S. Sundaresan. 1985. Thermal behavior of austenitic-ferritic joints made by friction welding. *Welding Journal* 64(12): 327-s.

Needham, J. C., and C. R. G. Ellis. 1972. Automation and quality control in friction welding. The Welding Institute Research Bulletin, 12(12), 333–339 (Part 1), 1971 December; 13(2), 47-51 (Part 2), 1972 February. Cambridge, UK: The Welding Institute.

Nessler, C. G., et al. 1971. Friction welding of titanium alloys. *Welding Journal* 50(9): 379-s–85s.

Nessler, C. G., et al. 1983. Radial friction welding. *Welding Journal* 62(7): 17–29.

Nicholas, E. D. and W. M. Thomas. 1986. Metal deposition by friction welding. *Welding Journal* 65(8).

Nicholas, E. D. 1987. Friction welding noncircular sections with linear motion: a preliminary study. Welding Institute Research Report, 337. Cambridge, UK: The Welding Institute.

Ochi, H., K. Ogawawa, Y. Yamamoto, and Y. Suga. 2004. Friction welding using insert metal. *Welding Journal* 83(3): 36–40.

Ruge, J., K. Thomas, and S. Sundaresan. Joining copper to titanium by friction welding. *Welding Journal* 65(8): 28.

Sassani, F., and J. R. Neelam. 1988. Friction welding of incompatible materials. *Welding Journal* 67(11): 264-s.

Searl, J. 1971. Friction welding noncircular components using orbital motion. *Welding and Metals Fabrication*. 39(8): 294–297.

Stotler, T., and J. E. Gould. 1992. Inertia welding of steel to a structural ceramic, 73rd AWS Meeting, (Chicago). Miami: American Welding Society.

Tumuluru, M. D. 1984. A parametic study of inertia friction welding for low alloy steel pipes. *Welding Journal*. 63(9): 289-s.

Vill, V. I. 1962. *Friction welding of metals*. Translated from Russian. Miami: American Welding Society.

Wang, K. K. 1975. Friction welding. Bulletin 204 (4). New York: Welding Research Council.

Wang, K. K. and G. Rasmussen. 1972. Optimization of inertia welding process by response surface methodology. *Trans—ASME Journal Engrg. Ind.* 94, Series B (4): 999–1006.

Wang, K. K., G. R. Reif, and S. K. Oh. 1982. In-process quality detection of friction welds using acoustic emission techniques. *Welding Journal*. 61(9): 312-s.

Yashan, D., S. Tsang, W. L. Johns, and M. W. Doughty. 1987. Inertia friction welding of 1100 aluminum to Type 316 stainless steel. *Welding Journal*. 66(8): 27.

CHAPTER 7

FRICTION STIR WELDING

Photograph courtesy of NASA, Marshall Space Flight Center

Prepared by the Welding Handbook Chapter Committee on Friction Stir Welding

T. J. Lienert, Chair
Los Alamos National Laboratory

M. W. Mahoney
Consultant

R. Nandan
The Pennsylvania State University

M. P. Posada
Naval Surface Warfare Center

T. V. Stotler
Edison Welding Institute

J. M. Thompson
General Tool Company

Welding Handbook Volume 3 Committee Member:

W. L. Roth
Procter & Gamble, Inc.

Contents

CHAPTER 7

FRICTION STIR WELDING

INTRODUCTION

Friction stir welding (FSW) is a variant of friction welding that produces a weld between two (or more) workpieces by the heating and plastic material displacement caused by a rapidly rotating tool that traverses the weld joint.[1,2] Heating is believed to be caused by both frictional rubbing between the tool and workpiece and by visco-plastic dissipation of the deforming material at high strain rates. Like conventional friction welding processes, the FSW process is solid state in nature.

Both friction welding and friction stir welding produce a volume of hot-worked metal along the bond line. However, friction stir welding differs from friction welding in one important aspect: in friction welding, the relative motion is between the workpieces that are held in compression, whereas the relative motion in FSW is between the workpieces and a rotating tool.

Friction stir welding has been in existence since the early 1990s and was developed mainly for the welding of aluminum alloys. The significant benefits of this process were quickly recognized, motivating considerable research and development during the next decade that expanded the technology to other materials.

Information in this chapter provides a basic understanding of the friction stir welding process and a variation of the process, friction stir spot welding. Topics include equipment and controls, process variables, materials and applications, mechanical properties, weld quality, economics, and safe practices. Additional topics discussed in this chapter are heat transfer, material flow, residual stress, nondestructive examination, and corrosion.

1. Welding terms and definitions are from American Welding Society (AWS) Committee on Definitions and Symbols, 2001, *Standard Welding Terms and Definitions,* AWS A3.0:2001, Miami: American Welding Society.
2. At the time this chapter was prepared, the referenced codes and other standards were current. If a code or other standard is cited without a date of publication, it is understood that the latest edition of the document referred to applies. If a code or other standard is cited with the date of publication, the citation refers to that edition only, and it is understood that any future revisions or amendments to the code or standard are not included; however, as codes and standards undergo frequent revision, the reader is advised to consult the most recent edition.

Lists of technical papers that have been used in compiling this chapter are provided in the Bibliography. An important feature of the Bibliography is the grouping of these papers according to the specific alloys welded with the friction stir process.

FUNDAMENTALS

Friction stir welding uses a nonconsumable, rotating welding tool to create heat locally. The rotating tool hot-works the material surrounding the weld interface to produce a continuous solid-state weld. A common tool design has the shape of a rod with a concave area (the shoulder) with a pin (or probe) that is coaxial with the axis of rotation. The workpieces are rigidly clamped and are supported by a backing plate, or anvil, that bears the load from the tool and constrains deforming material at the backside of the joint. In most cases, the pin is designed to be slightly shorter than the thickness of the weld joint to prevent contact with the backing plate and to promote complete penetration without defects. The process is illustrated in Figure 7.1.

To make a linear weld in a butt joint configuration, the workpieces are positioned on the backing plate with the edges in contact. To start the process, the rotating friction stir welding tool is plunged into the weld joint until the shoulder of the tool makes contact with the top surfaces of the workpieces. Frictional rubbing and visco-plastic dissipation cause the heated material to soften and plastically flow. The motion of the tool promotes displacement of the softened material to produce the weld. The hot-worked material is swept around the tool to produce recoalescence behind the tool. The tool shoulder provides constraint against the escape of hot-worked material, while applying a forging force to the top surface of the weld.

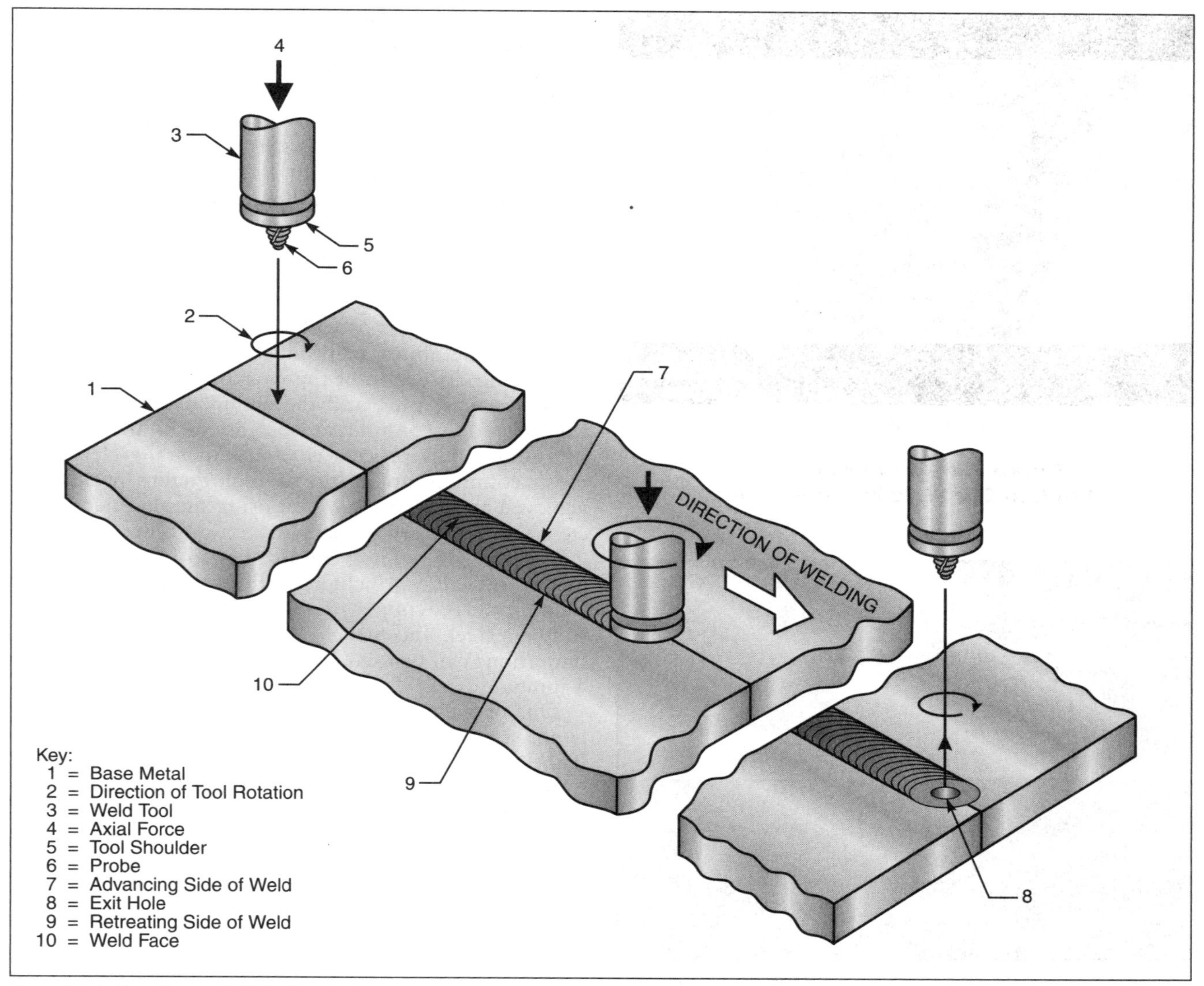

Source: Adapted from Thomas, W. T., Tool Drawing No. WMT104/04Lr1, Cambridge: TWI.

Figure 7.1—Schematic of the Friction Stir Welding Process

A particular nomenclature has been adopted to account for the asymmetry of the FSW process relative to the weld centerline. As indicated in Figure 7.1, the side of the weld where the tool traverse vector is parallel to the vector of tool rotation is called the *advancing side*. The opposite side of the weld, where the tool traverse vector and the tool rotation vector are anti-parallel, is referred to as the *retreating side*.

The tool continues to rotate while traveling along the joint, completing the weld as travel progresses. When the desired length has been achieved, the tool is removed. A macrograph of a transverse section of a high-integrity weld in a butt joint configuration is shown in Figure 7.2. The smooth and regular spacing of the toolmarks along the crown of a friction stir weld is shown in Figure 7.3.

Friction stir spot welding (FSSW) is a variation of friction stir welding commonly used to join sheet metal or other thin materials. A modified friction stir welding tool is used to penetrate the top sheet of the joint to a set depth. Friction stir spot welding typically is used for lap joints, so the tool normally penetrates only a small depth into the lower workpiece. The tool is retracted once it reaches its final position.

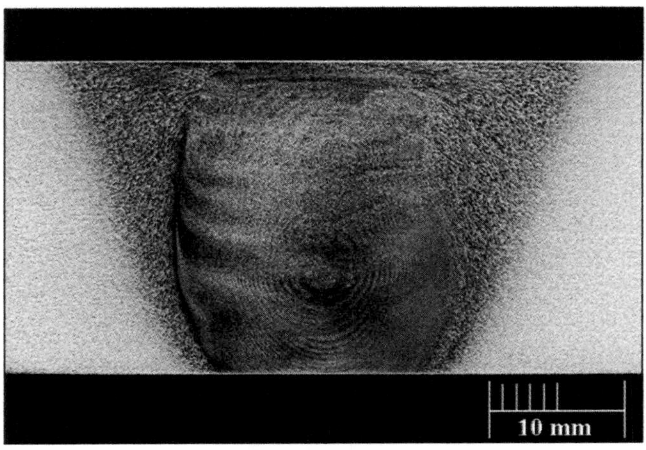

© Photomicrograph courtesy of Edison Welding Institute

Figure 7.2—Micrograph of a Friction Stir Weld in a Butt Joint

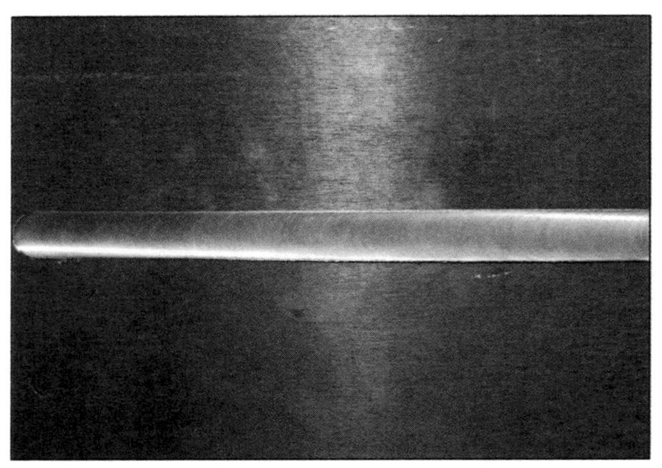

Photograph courtesy of NASA

Figure 7.3—Sample of a Friction Stir Weld in a Butt Joint

ADVANTAGES AND LIMITATIONS

Since friction stir welding is a solid-state forge welding process, it shares many of the principal advantages and disadvantages of other solid-state welding processes, particularly forge welding processes, addressed in this volume.

Friction stir welding normally is done in a single pass with full penetration and with little or no joint preparation. Depending on the workpiece material and thickness, minimal distortion occurs during welding, provided proper clamping is used. The welds typically exhibit as-welded mechanical properties superior to the properties of fusion welds. With the possible exception of materials with high flow stress at hot-working temperatures, such as steel and titanium alloys, friction stir welding can be done at relatively high processing speeds. Higher travel speeds can be achieved with friction stir welding than those attained with arc welding, but FSW travel speeds may not be competitive with laser beam welding. Friction stir welding is a machine-tool process; this aspect facilitates repeatable welds in production applications with little operator input.

The FSW process provides several other advantages, including the ability to produce solid-state welds with little or no distortion of the workpieces, the avoidance of fumes or spatter, and the elimination of solidification-related discontinuities, such as cracks and porosity. In addition, the process is environmentally clean. For most common applications, high-quality welds are achieved at relatively low cost using simple, energy-efficient mechanical equipment.

Friction stir welding can be used to join a variety of metals and alloys, including alloys of aluminum (Al), titanium (Ti), copper (Cu), magnesium (Mg), steel, stainless steel, and nickel (Ni). A number of joint configurations can be used, including butt, lap, corner, and T-joints.

One of the main limitations of FSW is that the joint is not self-supporting and must be properly restrained. If the workpiece is designed in a manner that requires support of the joint, tooling costs might be significant. However, the costs of a friction stir welding machine and associated tooling are much the same as equipment for other solid-state forge welding processes. The initial high equipment cost typically yields a quick return on investment due to the high-volume welding capability of the process.

EQUIPMENT AND CONTROLS

The equipment used in friction stir welding includes the welding machine, the friction stir welding tool, and welding fixtures to hold components of the weldment in the correct orientation for proper fitup during welding. These topics and methods for controlling the process are discussed in this section.

FRICTION STIR WELDING MACHINES

Friction stir welding machine designs typically are supported by one of three different machine platforms: C-frames, gantries, or vertical structures similar to those used in boring mills. Two of the various styles of machines are shown in Figure 7.4(A) and (B). Each style usually has an X, Y, and Z axis with a single-spindle motor. The friction stir welding tool may be tilted with either a manual or powered axis. More complex machines allow for gimballing and tilting of the spindle by the addition of A and B axes. These axes provide the means for welding corner joints and out-of-plane joints. Two-piece friction stir welding tools and bobbin tools (illustrated in the Friction Welding Tool Design section) require another motion axis coaxial with the spindle to move the pin relative to the shoulder. Some designs incorporate a separate spindle motor for the pin, allowing the shoulder and pin to rotate at different rotational speeds, which in turn permits an increase in the fraction of heat generated by the pin.

The movement of each travel axis normally is powered by electric motors and gearboxes to provide adequate torque, although some machines are driven hydraulically. Devices for force measurements typically are provided on each axis to allow for process control and process performance documentation. The magnitude of forces on the traveling axes X and Y ranges up to 22 000 newtons [N] (5000 lbf) for thin materials and up to 44 500 N (10 000 lbf) for thicker materials. The Z axis, used with single-piece and two-piece tools, typically has a load capacity ranging from 44 500 N (10 000 lbf) for thin materials and from 156 000 N (35 000 lbf) to 222 000 N (50 000 lbf) for thicker materials.

Friction stir welding typically is performed on machine-tool-like equipment. The welding of thin sections with small-to-moderate forces often is done on machine tools. In heavier work, the force levels and thermal phenomena associated with the process normally require equipment specifically designed for the FSW process.

Friction stir welding machines typically must retain positioning accuracy while operating under high process loads. Positioning accuracy is achieved by either compensating for deflection with internal deflection control algorithms or by making compensating measurements of position that are unaffected by the loads, such as a head-to-workpiece displacement measurement. Alternatively, the machine may be designed so that the process forces are loaded onto other machine features in a manner that prevents distortion of the coordinate system of the machine by process loads.

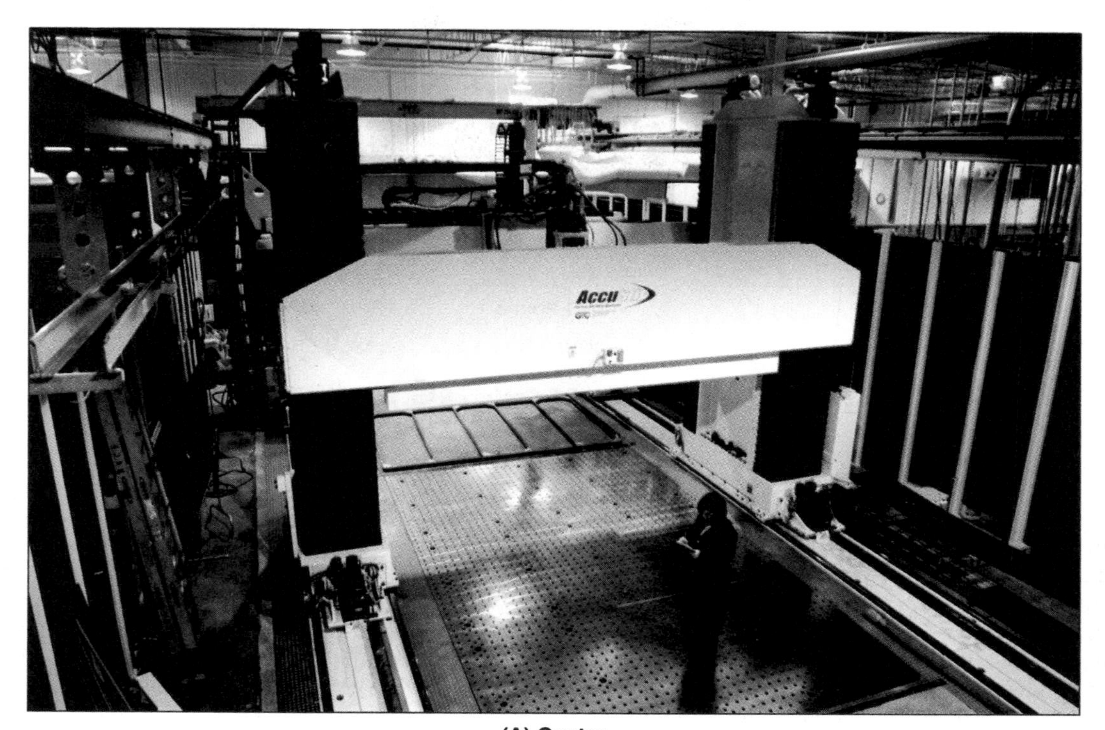

(A) Gantry

© Photograph courtesy of Edison Welding Institute. Used with permission.

Figure 7.4—Examples of Friction Stir Welding Machine Designs

(B) C-Frame

Figure 7.4 (Continued)—Examples of Friction Stir Welding Machine Designs

Friction stir welding machines are equipped with spindle bearing systems designed to achieve the required design function at the desired process load. Additional spindle features may be designed into the machine for the use of adjustable-length pin tools and bobbin tools. The spindle may be hollow to allow a second coaxial spindle to work inside the shoulder spindle. Force transducers typically are integrated into the structural load path of the machine to monitor plunge and traverse loads for process control and for process monitoring for quality assurance.

Friction stir welding equipment often incorporates features for the cooling of the backing plate and/or the FSW tool. A spindle cooling system can be used to protect the spindle bearings from heat, keep process heat from significantly changing important length dimensions of the FSW tool, or to actively cool the tool. The backing plate and the clamping system may be cooled when necessary to maintain thermal boundary conditions and keep the process in a steady-state condition.

Friction stir welding equipment, like all other industrial equipment, is available with a range of kinematic capabilities. Equipment designs may range from simple straight line welding of a single weld configuration with manually adjustable tilt angles to machines with 5 degrees or more of freedom to weld a complex path. Equipment capable of complex paths typically is fitted with three-dimensional fixturing to keep the workpieces properly constrained. The complexity of the workpiece is reflected in the tooling. If a bobbin FSW tool is used, the need for a backing plate is eliminated, but the tooling still has to constrain the workpieces from separating and distorting under load.

Force feedback control methods are used as a way of ensuring the position of the FSW tool in the workpiece material and/or controlling the travel speed. Force control, or load control, is necessary since FSW machines cannot be made infinitely stiff and workpieces do not always have the exact same thickness and strength level. Hence, given the variations in incoming material and machine design limitations, force control is a powerful aid to minimizing the problems associated with machine compliance to compensate for thermal boundary conditions. For example, the thermal boundary conditions at the start and end of a weld are different from the conditions in the center of a welded plate. If a machine flexes away from a fixed position of the FSW tool in the material, force control seeks to move the tool toward the weld to re-establish the force, yielding a machine or control system that compensates for machine flex.

Controls

Friction stir welding machines are controlled either with software based on machine tool codes, or with integrated functions designed specifically for FSW. Data acquisition functions typically are built into the control system. Some FSW control systems are based on computer numerical controls (CNC). Other machines use teaching, seam-tracking, or vision systems to define or control a weld path. Workpiece tolerances, path complexity, tooling, machine deflections under load, and thermal distortions of the workpiece need to be considered in assessing the type of control to be employed.

Any control method employed for friction stir welding must have sufficient accuracy to meet the intended use of the welding machine. Most friction stir welding machines are capable of controlling either force or position as the primary depth control variable, while the non-primary variable is monitored relative to a range of acceptable limits. For example, in load control, a load set point is specified, and a range of allowable position travel is specified, precluding tool advancement into the backing plate and accommodating the expected maximum thickness of the material. The converse is sometimes used, with a target position specified, but with limits placed on minimum and maximum force to protect the machine, FSW tools, and tooling. When the secondary variable reaches or exceeds the limits, the process can be terminated, the operator can be warned, or the pursuit of the primary set point can be stopped. Most software for machine control allows for the setting of these limits and the actions to be taken as the limit values are reached or exceeded.

A similar type of load control method can be used for the traveling axis. A minimum and maximum load value is set. While traveling to make the weld, the speed of advancement along the weld is allowed to increase or decrease as necessary to maintain the desired load value.

TOOL DESIGN

The friction stir welding tool is the most significant component of the system and usually is designed for a specific type of weld joint. Tools are manufactured from wear-resistant materials with good static and dynamic properties at welding temperature. The strength and wear resistance of the tool must be superior to the base metal used in the weldment. For example, a friction stir welding tool made of tool steel such as H13 is commonly used to weld aluminum alloys.

The three most common variations of tool design are a single-piece tool, a two-piece retractable-pin tool, and a bobbin tool. A single-piece tool is a monolithic design consisting of a pin and a shoulder, as shown in Figure 7.5. A retractable-pin tool, shown in Figure 7.6, enables independent motion of the pin and shoulder in the Z

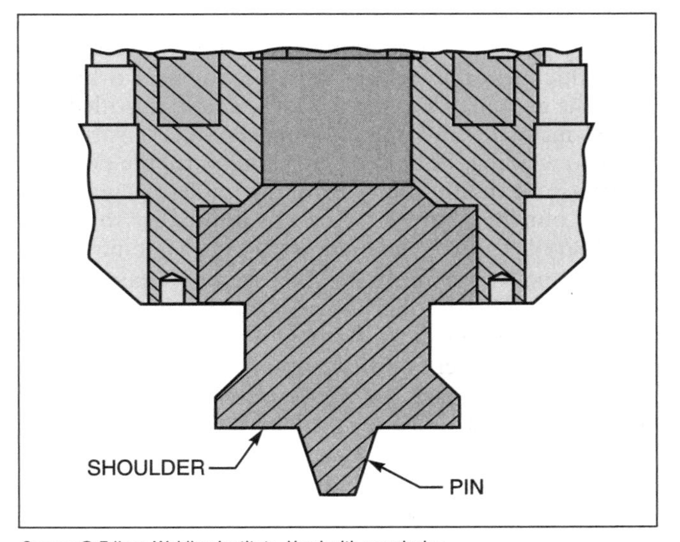

Source: © Edison Welding Institute. Used with permission.

Figure 7.5—Single-Piece Tool Used in Friction Stir Welding

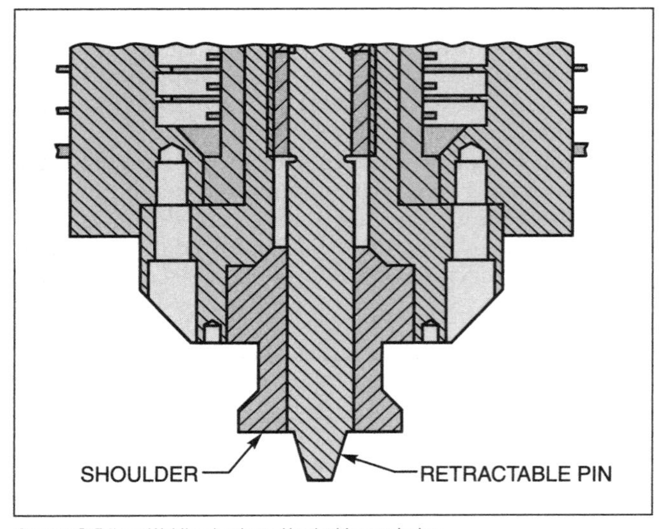

Source: © Edison Welding Institute. Used with permission.

Figure 7.6—Retractable Pin Tool Used in Friction Stir Welding

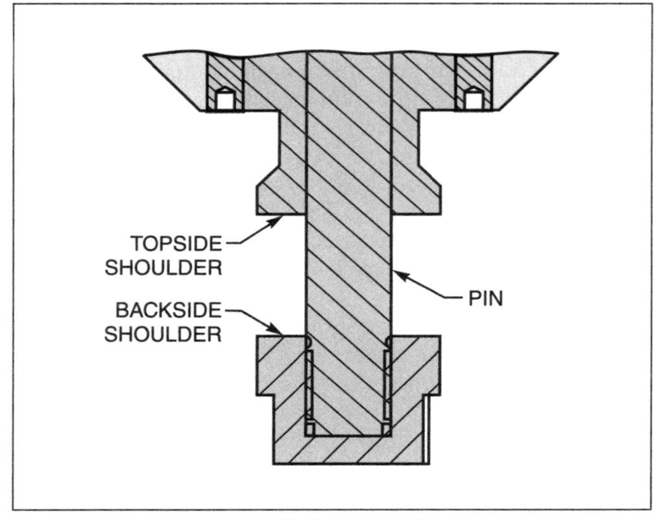

Source: © Edison Welding Institute. Used with permission.

Figure 7.7—Bobbin-Type Friction Stir Welding Tool

direction. This arrangement allows the pin to retract at the end of the weld, and eliminates the keyhole from the plate. A retractable-pin tool allows the pin to be positioned relative to a backing plate to ensure full penetration, and the shoulder can be operated in load control to ensure proper process hydrostatic pressure, all relatively independent of workpiece thickness.

A bobbin tool, as shown in Figure 7.7, has a top shoulder and a backside shoulder. This tool allows for faster welding speeds and eliminates the need for a backing plate. In some applications for a given workpiece thickness the bobbin is run onto the workpieces from the end; in other applications, a hole is drilled, the pin is inserted, and the bottom shoulder is attached prior to welding. Bobbin tooling eliminates the possibility of a root discontinuity from incomplete penetration by the pin, but bobbin tools are limited in thickness capability because of bending stresses in the pin.

An issue that complicates the use of retractable pin tools and bobbin tools is the problem of workpiece material extruding into the narrow clearance between pin and shoulder and eventually binding relative motion between the tool components. These tool configurations need to be cleaned periodically to remove any extruded material.

Tool shoulders normally have either a cupped (concave) design as shown in Figure 7.8(A), or a scrolled design as shown (B). The concave tool design normally is used with a one- degree to three-degree tilt angle (pushing angle), while the scrolled designs normally are employed at zero tilt. With the tilted arrangement, the rear edge of the tool provides increased compressive force to aid in recoalescence of the material. The function of the scroll is to force the material in the stir zone[3] to flow inward toward the pin in order to provide the compressive force.

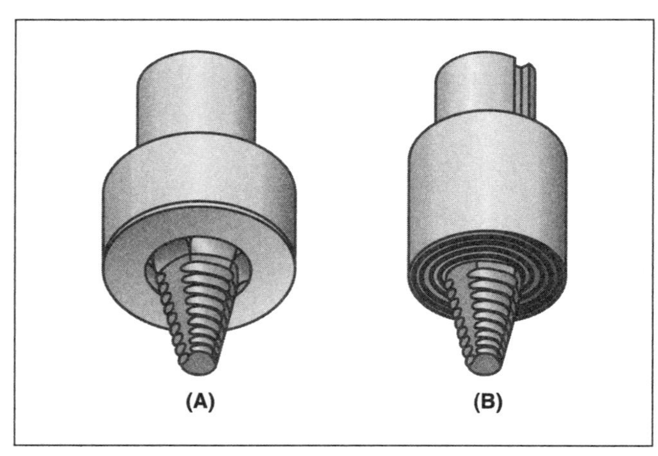

Source: © Edison Welding Institute. Used with permission.

Figure 7.8—(A) Concave Tool Design and (B) Scrolled Design

3. The term *stir zone* is used in this chapter to describe the weld metal zone in a friction stir weld. The term *nugget* is preferred by The Welding Institute (TWI), patent holder for the invention of the process.

The pin of the tool typically has some type of profiling, such as threaded features, fluted features or flats. Pin designs vary, depending on the material and joint configuration. Properly designed friction stir welding tools have the capability of making more than 1000 meters (m) (3280 feet [ft]) of welds, for example, in 5-mm (0.20-in.) thick series 5xxx and 6xxx aluminum alloys, before a tool change is required.

FIXTURING

Fixturing for friction stir welding usually is the most complicated and critical aspect of the process. The workpieces must be clamped to a rigid backing plate (anvil) and secured to resist the perpendicular and side forces that develop during welding. These forces tend to lift and push the workpieces apart. Fixtures are designed to restrain the workpieces and keep them from moving. A root opening (gap) of less than 10% of the material thickness can be tolerated for thicknesses up to about 13 mm (1/2 in.). The fixtures that hold the materials to the backing plate should be placed as close to the joint as possible to reduce the clamping load.

Various clamping configurations are possible, but always depend on the joint configuration. Some joints are designed to allow the workpieces to provide the necessary backing, for example, a T-joint. If the vertical leg of a T-joint is of sufficient thickness, tooling would be needed only to hold it in a perpendicular position.

WELDING PARAMETERS

Friction stir welding has four primary parameters: tool position (or load, depending on the type of control), spindle rpm, travel speed, and tilt angle of the tool. Other variables include joint design, surface preparation, and the characteristics of the materials to be joined. The important material characteristics are discussed in the section Factors Affecting Weldability.

TOOL POSITION

The position of the tool (typically called the Z-axis position) is an important parameter. The thickness of the base metal and the alloy being welded dictate the type and level of adjustments to be made to the parameters and to the tool design. The tool must be in the correct location to ensure sufficient contact with the top surfaces of the workpieces. Insufficient shoulder pressure permits loss of constraint, which promotes improper recoalescence of the hot-worked metal under the tool. The formation of a running pore on the top side of the weld is evidence of insufficient pressure. If the tool is positioned too deeply, three results typically occur: the weld area may be thinner than the base metal, the tip of the pin may drag on the backing plate and areas in the stir zone might reach the solidus temperature. The solidus temperature is defined as the temperature at which melting begins during heating. Further discussion of the solidus temperature is provided in a section on phase diagrams in Chapter 4, Volume 1 of the *Welding Handbook*, ninth edition.[4] Although the tendency is to err by positioning the tool too deeply, caution must be taken to avoid overheating the hot-worked material in the stir zone. If the material in this zone reaches the solidus temperature, it extrudes from under the stirring tool and may cause a running pore (surface or subsurface).

SPINDLE RPM, TRAVEL SPEED, AND TOOL ANGLE

Spindle rpm, travel speed, and tool angle interact with the tool position. The proper spindle rpm and travel speed typically decrease with increases in base metal thickness and strength. For example, high-strength aluminum alloys typically use lower spindle rpm and travel speeds.

The tool angle normally is set either to zero (perpendicular to the weld path) or at a slight pushing angle (1° to 3°). This tilt promotes recoalescence of the material in the stir zone at the rear of the tool. In contrast to the cupped shoulder used on a tilted tool, FSW tools that operate with zero tilt typically have a scroll feature on the shoulder where it contacts the base metal. The scroll restrains the deforming material and directs it inward toward the pin.

JOINT DESIGN

Friction stir welding can be used on a range of basic joint designs, as shown in Figure 7.9. In any joint design, fixturing must be designed to react to the forces from the FSW tool. For some applications, the joint design may allow for reaction to these forces, thus reducing fixturing requirements.

4. American Welding Society (AWS) Welding Handbook Committee, 2001, C. Jenney and A. O'Brien, ed. *Welding Handbook*, Vol. 1, ninth edition, *Welding Science and Technology*, Chapter 4, Miami: American Welding Society.

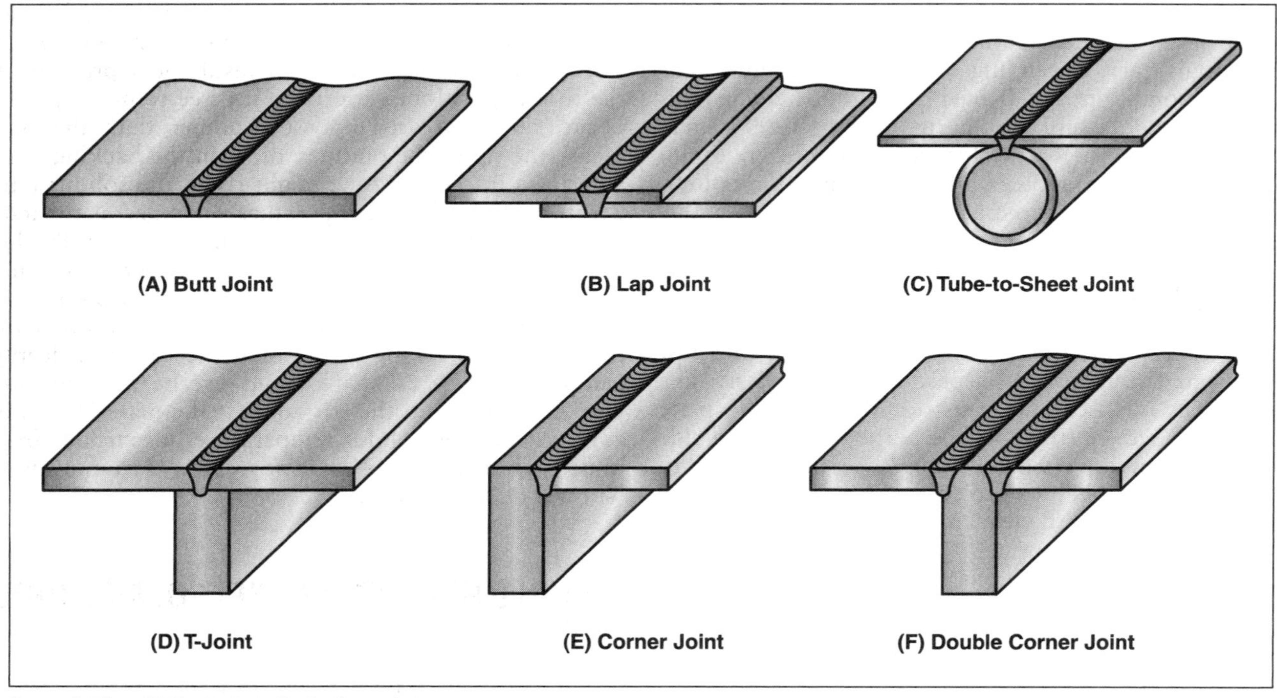

Source: © Edison Welding Institute. Used with permission.

Figure 7.9—Basic Joint Designs Used in Friction Stir Welding

Joint Preparation

Friction stir welding generally requires minimal joint preparation prior to welding. This requirement varies, however, depending on the base metal and the mechanical requirements of the joint. Cleaning the joint area by wiping usually is sufficient. For some applications of welds in butt joints, the heavy oxides on the joint surfaces of the workpieces must be removed, since these oxides could be swept into the joint area. For lap joints, the oxides on the overlapping surfaces must be removed.

REGIONS OF A FRICTION STIR WELD

Since FSW has been used primarily on aluminum alloys, the various regions of the weld have been named according to those found in optical macrographs of welds in aluminum alloys. However, it should be understood that similar terms are often are used with friction stir welds on other materials, and that some of the regions may be absent in welds on other materials. Different mechanisms may give rise to variations in the microstructure of similarly named regions due to differences in the metallurgical properties of the workpieces.

The various regions of a friction stir weld are illustrated in Figure 7.10. Figure 7.10(A) shows an optical macrograph of a transverse section of a friction stir weld on 6.35-mm (~0.25-in.) thick 6061-T651 aluminum alloy. The various regions of the weld discussed below are indicated by numbers. The origin of the microstructures in the various weld regions is discussed in the subsection Factors Affecting Weldability.

Region 1, the stir zone shown in Figure 7.10(A), contains relatively equiaxed grains that are refined relative to the grains of the base metal. The stir zone contains material that has interacted most closely with the tool during welding and is generally believed to experience the highest peak temperatures and the greatest strains and strain rates. As a result of etching, the stir zone often will display a series of concentric rings of constant spacing sometimes called the *onion ring pattern.* A

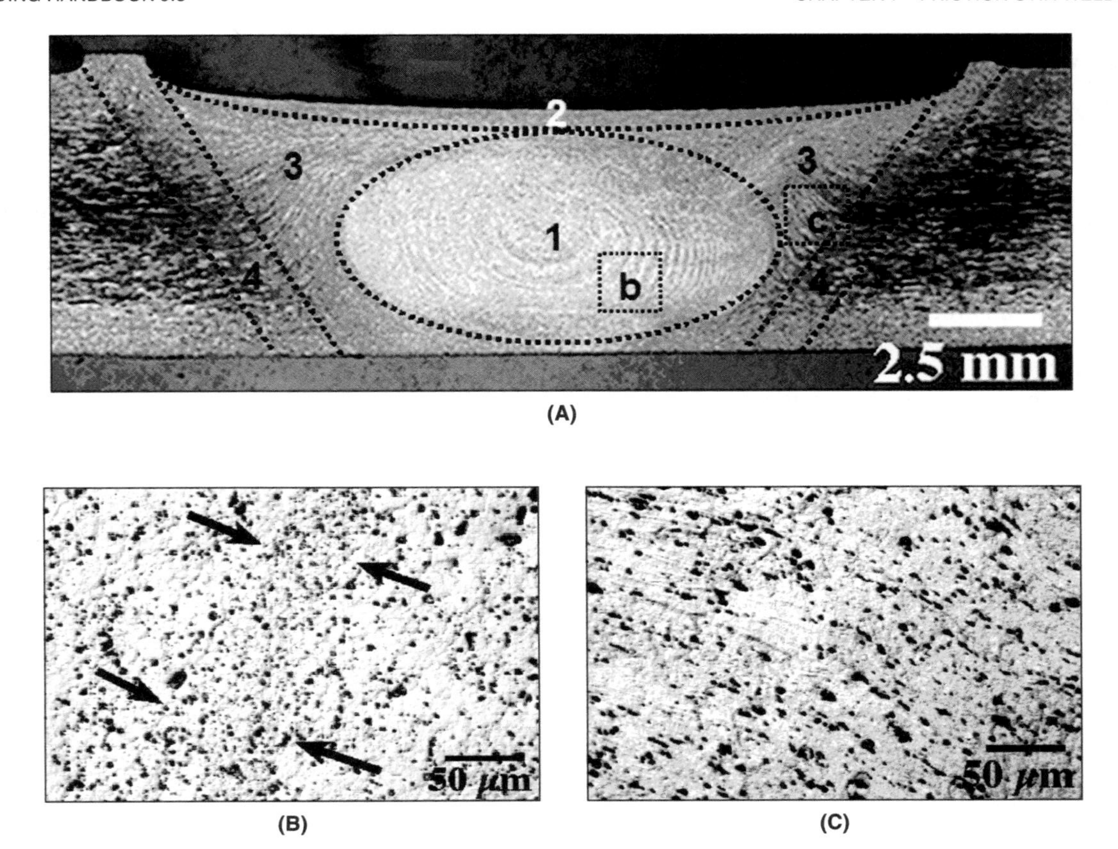

Figure 7.10—(A) Optical Macrograph of a Transverse Section of a Friction Stir Weld on 6.35-mm (0.25-in.) Thick 6061-T651 Aluminum, (B) Banding in the Stir Zone, and (C) Evidence of Material Flow in the TMAZ

higher magnification micrograph of the ring pattern in the stir zone is shown in Figure 7.10(B), also indicated as *b* in Figure 7.10(A).

Region 2, called the *flow arm*, is considered to be part of the stir zone. It contains material that has been strongly influenced by direct interaction with the tool shoulder.

Region 3, the thermomechanically affected zone (TMAZ) or heat-and-deformation-affected zone (HDAZ), surrounds the stir zone and flow arm. As shown in Figure 7.10(A), the grains of the TMAZ, which are initially oriented in the rolling plane of the plate, are swept upward by the deformation of material in the stir zone. However, the grains are not refined with respect to those of the base metal. A higher-magnification micrograph illustrating the deformation and bending of the grains in the TMAZ is shown in Figure 7.10(C) (refer to *c* in Figure 7.10(A). The TMAZ is believed to have undergone peak temperatures and strains below those found in the stir zone.

Region 4, the furthest from the stir zone, is the true heat-affected zone (HAZ). This region has experienced temperatures high enough to cause some metallurgical changes but has undergone little or no deformation. The presence of the HAZ in precipitation-strengthened aluminum alloys generally is not obvious in optical macrographs except for small differences in etching contrast. In these alloys, the HAZ derives from coarsening and/or overaging of the precipitates. Evidence of these changes can be detected only by using transmission electron microscopy (TEM) or detecting microhardness changes. In aluminum alloys that have been strengthened by cold working, the grains of the HAZ may undergo recrystallization and grain growth.

The final region is the unaffected base metal region. This region has been subjected to no appreciable temperature rise and has experienced no deformation. Consequently, the properties of the base metal have not been altered by welding.

FACTORS AFFECTING WELDABILITY

This section provides a review of the friction stir welding of specific alloys, including alloys of Al, Mg, Cu, Ti, steel, and stainless steel. Topics such as weldability, types of tools required, and microstructural evolution are discussed for each alloy group. Process parameters for the successful friction stir welding of some alloy groups are provided.

Although the concept of weldability as it relates to friction stir welding has not been well defined, it is clear that some alloys are easier to weld than others. Several factors may be included in the definition of weldability for friction stir welding: the size of the process window (the range of acceptable welding variables that produce quality welds), the maximum welding rate possible, the magnitude of resultant forces, and the tool requirements for producing successful friction stir welds.

Friction stir welding is best understood when viewed as a hot-working process used for joining. Hot-working requires reaching temperatures in excess of 50% to 60% of the absolute melting temperature of the base metal in order to lower flow stresses.

The friction-stir weldability of a given material appears to scale closely with the ease of extrusion of the same material. Hence, aluminum and magnesium alloys have relatively good weldability, since rapid welding speeds can be used, forces on the tool and FSW machine are relatively low, and steel tools can be utilized due to the low working temperatures. Copper alloys are slightly more difficult to weld with the friction stir process because they have a higher hot-working temperature range and require the use of tools made from tungsten or nickel alloys. Titanium, steel, stainless steel, and nickel alloys clearly are more challenging to weld. They require tools made of materials with much higher temperature capabilities, such as molybdenum (Mo), tungsten-rhenium (W-Re) or cubic boron nitride (CBN), and typically require slower traverse rates. Refractory alloys like molybdenum and tungsten alloys have not been successfully welded with the friction stir process because of the lack of suitable tool materials.

Many technical papers have been published on the microstructural evolution in a variety of aluminum alloys. A review of the specifics of each alloy is prohibited by space considerations. However, the reader is referred to the Bibliography, which categorizes the papers specific to each alloy. A general review of the microstructural evolution of friction stir welds in aluminum alloys follows.[5]

5. The titles, authors, and facts of publication of technical papers and other resources represented in this section are listed in the Bibliography in the Aluminum Alloys section. This group includes Alloys 1100 through 7075.

ALUMINUM ALLOYS

Aluminum alloys possess a high strength-to-density ratio and provide good corrosion resistance in many environments. Consequently, they are often used in weight-critical structural applications in the aerospace and transportation industries. The face-centered-cubic (fcc) crystal structure of these alloys provides good ductility and toughness. Further details of the physical metallurgy of these alloys and a discussion of the alloy classes and temper designations can be found in the *Welding Handbook,* Volume 3 of the 8th edition, *Materials and Applications*, Chapter 1.[6]

Many aluminum alloys have been successfully welded using friction stir welding, including the 1xxx, 2xxx, 5xxx, 6xxx, and 7xxx series of alloys, several casting alloys, rapidly solidified aluminum-iron-silicon-vanadium (Al-Fe-Si-V) alloys, aluminum-beryllium (Al-Be) alloys, and Al-metal matrix composites. Friction stir welding has the potential to join aluminum alloys that cannot be successfully welded by fusion processes because of problems with solidification cracking and loss of properties in the HAZ.

Aluminum alloys can be welded using inexpensive and common tool materials such as tool steels. As mentioned previously, the weldability with FSW scales closely with the ease of extrusion for a given alloy. Consequently, the 1xxx alloys are more tolerant of changes in welding variables: they have a large processing window and can be welded rapidly at low forces. The 5xxx and 6xxx alloys are slightly more difficult to weld with the friction stir process, while the 2xxx and 7xxx alloys are the most difficult of the aluminum alloys. This ranking follows closely with the extrudability ranking.

Microstructures in the various regions of friction stir welds on aluminum alloys develop and evolve in accordance with the local thermomechanical cycle. In general, aluminum alloys can be strengthened by precipitation hardening during aging and/or by cold working. The final microstructures depend on the effects of the thermomechanical cycle on the original microstructures and may develop through a variety of processes, including dissolution, coarsening and reprecipitation of precipitates as well as recrystallization, recovery, and grain coarsening. Changes in microstructure are reflected in changes in microhardness and other mechanical and corrosion properties.

Figure 7.11 is a schematic drawing that summarizes some of the key microstructural features of a friction stir weld on the aluminum alloy 7075-T651. Local temperatures in the various weld regions also are indicated. The stir zone undergoes the highest peak temperatures

6. American Welding Society (AWS) Welding Handbook Committee, W. R. Oates, Editor, 1996, *Welding Handbook*, 8th edition, Volume 3, *Materials and Applications, Part 1,* Chapter 1, Miami: American Welding Society.

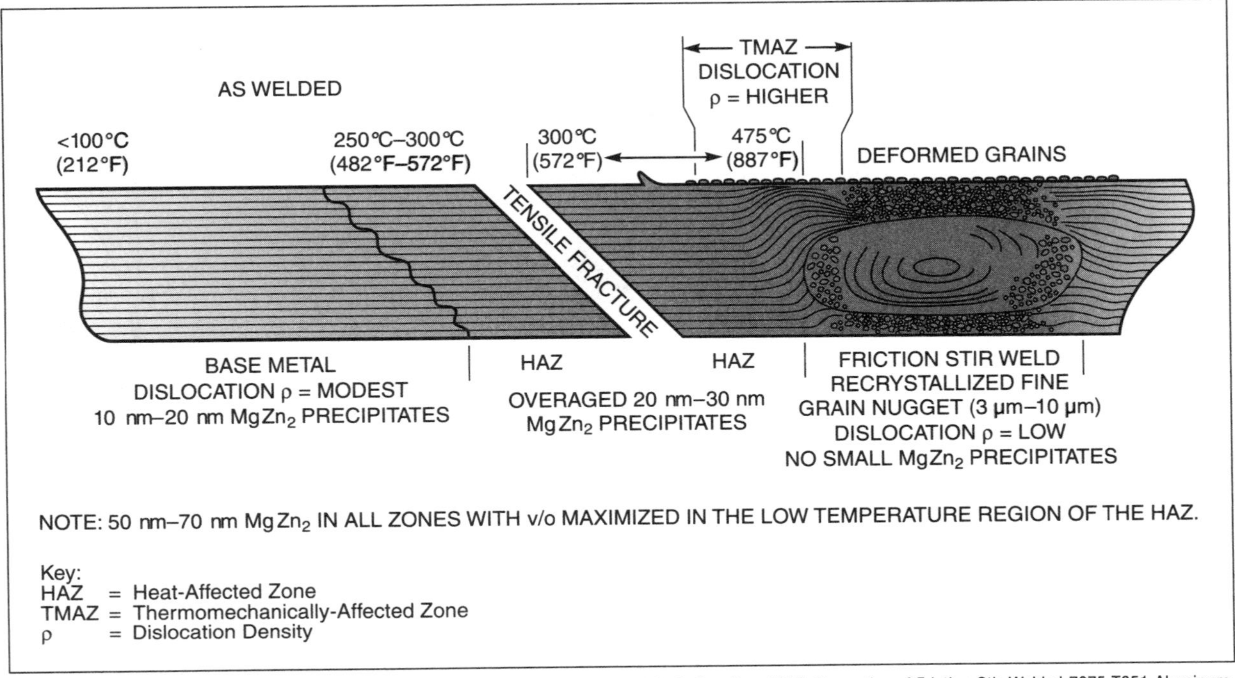

Source: Adapted from Mahoney, M. W., C. G. Rhodes, J. G. Flintoff, W. H. Bingel, and R. A. Spurling, 1998, Properties of Friction Stir Welded 7075 T651 Aluminum, *Metallurgical and Materials Transactions*, 29A: 1955–1964.

Figure 7.11—Microstructures, Precipitate Distributions, and Temperature Ranges for a Friction Stir Weld on 7075-T651 Alloy

and the greatest strains and strain rates. Peak temperatures in the stir zone usually are in excess of the solvus temperature for the strengthening precipitates, and consequently, the precipitates in the stir zone normally are dissolved during the friction stir process. The TMAZ experiences lower temperatures <475°C (<887°F) relative to the stir zone and much lower strain. The dislocation density in the TMAZ is higher than that in the stir zone, as indicated in Figure 7.11. The HAZ experiences even lower temperatures, 250°C to 350°C (482°F to 662°F). The HAZ thermal cycle causes local overaging of the strengthening precipitates, as noted in the figure. The fracture path for tensile tests on these welds is also indicated in Figure 7.11. Fracture always occurred through the HAZ region due to overaging of the precipitates.

As shown in Figure 7.12, peak temperatures measured in the stir zone approached 475°C (887°F) and decreased with increasing distance from the edge of the stir zone to approximately 250°C (482°F) in the HAZ. Heat flow and peak temperatures are discussed in further detail in the sections, Heat Transfer and Material Flow.

Cooling rates normally are rapid enough to retain most of the solute in solution to room temperature. As a result, the stir zone is initially supersaturated with solute after cooling to room temperature. In 6xxx alloys, and especially in 2xxx and 7xxx alloys, the stir zone subsequently can undergo natural aging at room temperature by the formation of new precipitates. The strength of the metal in the stir zone also can be increased via artificial aging at temperatures above ~100°C (~212°F).

During friction stir welding, the grains in the stir zone also are refined by some restorative process involving dynamic recovery (DRV) and/or dynamic recrystallization (DRX), as a result of the deformation imposed by the welding tool. However, the initial restored grain size may increase as a result of grain growth during cooling after the tool has passed. The spacing between the rings in the ring pattern, which may be evident in the stir zone, is closely related to the distance the tool advances per revolution. The rings appear to involve alternate regions of fine grains interspersed between regions of even finer grains, as shown in Figure 7.10(b). The precipitate density in the fine grain bands also may differ from adjacent regions. The variations in grain size may stem from differences in strain and strain rates that occur in the different regions of the material. As shown in Figure 7.10, the intermetallic dispersoids in the finest-grained regions are broken up and formed into alternating bands.

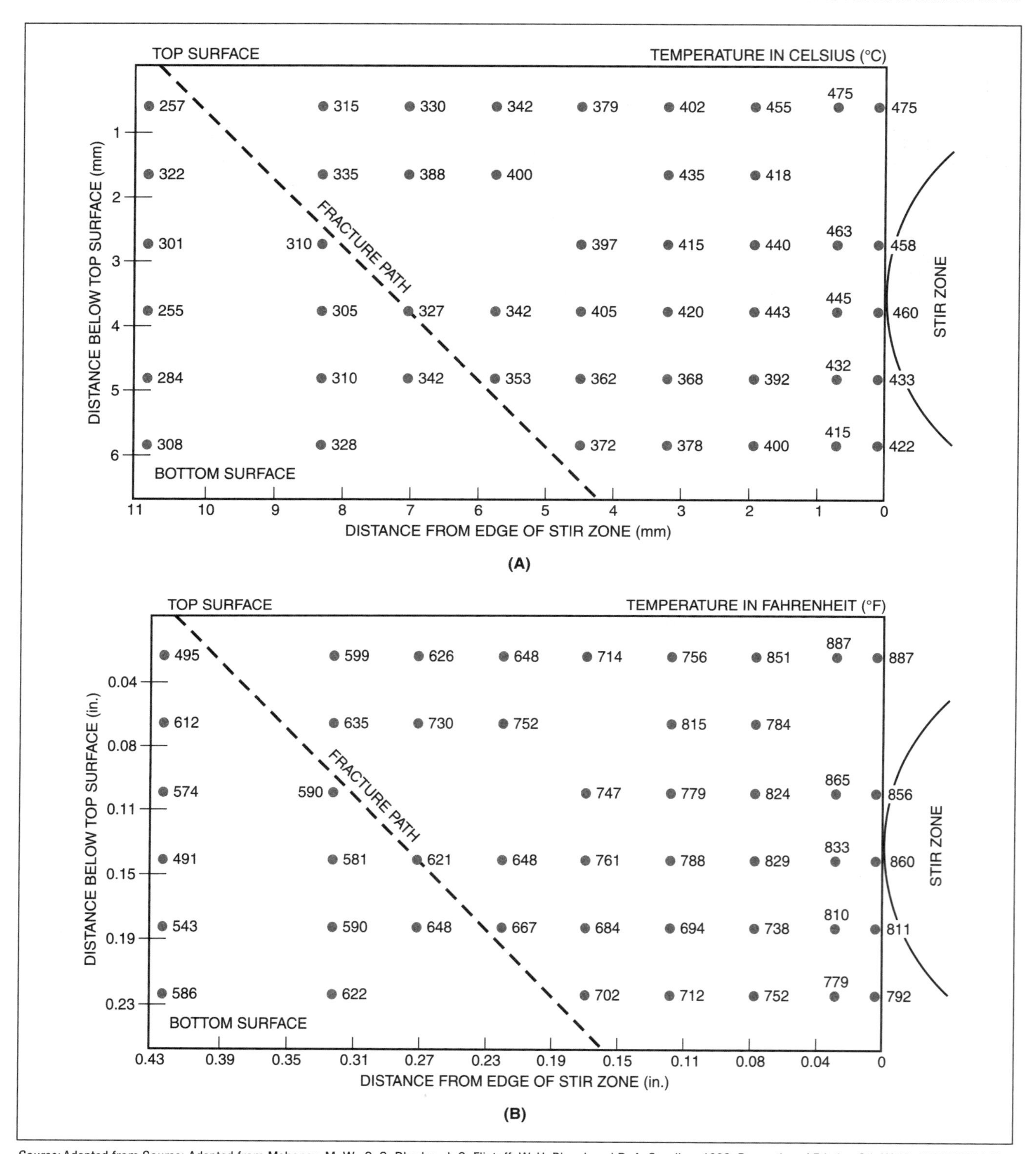

Source: Adapted from Source: Adapted from Mahoney, M. W., C. G. Rhodes, J. G. Flintoff, W. H. Bingel, and R. A. Spurling, 1998, Properties of Friction Stir Welded 7075 T651 Aluminum, *Metallurgical and Materials Transactions*, 29A: 1955–1964.

Figure 7.12—Summary of Peak Temperature Distribution in a Friction Stir Weld in Aluminum Alloy 7075-T651

The TMAZ undergoes peak temperatures too low to completely dissolve the precipitates, but high enough to cause their coarsening and partial dissolution. The grains of the TMAZ do not appear recrystallized but may undergo some softening due to recovery. These grains are swept upward under the influence of deformation in the stir zone. Precipitates in the heat-affected zone undergo coarsening and overaging, while cold-worked grains in the heat-affected zone are recrystallized. Both processes result in a commensurate loss of hardness and strength. Applying artificial aging to restore the hardness of the stir zone may result in further loss of hardness in the HAZ as a result of further overaging.

The pattern of microhardness in a transverse section of a friction stir weld in 6061-T651 aluminum can be observed in Figure 7.13. This pattern has a "W" shape and is typical of all precipitation-strengthened aluminum alloys. The base metal has a hardness value between 110 VHN (Vickers hardness number) and 115 VHN. The hardness drops to a value between 65 VHN and 70 VHN in the heat-affected zone due to coarsening and overaging of the strengthening precipitates. The hardness rises to a value between 80 VHN and 90 VHN in the weld zone because of a short artificial aging treatment subsequent to welding. Tensile properties of friction stir welds on aluminum alloys are discussed further in the section Mechanical Properties of Friction Stir Welds.

MAGNESIUM ALLOYS

The demand for magnesium alloys is increasing due to the useful combination of low density, tensile strength, and elastic modulus of these alloys. These alloys provide a good strength-to-weight ratio and are popular choices in the aerospace and transportation industries for weight-critical applications. They are mainly alloyed with either aluminum or zinc (Zn) in addition to rare earth elements, and can be strengthened by precipitation hardening and/or cold working. More information on the metallurgy of magnesium alloys is provided in Volume 3 of the *Welding Handbook*, eighth edition, *Materials and Applications, Part 1*.[7]

Magnesium alloys normally are cast or produced by some type of semi-solid casting process. The cast structure usually contains brittle eutectic and intermetallic

7. American Welding Society (AWS) Welding Handbook Committee, W. R. Oates, ed., 1996, Welding Handbook, Vol. 3, eighth edition, Materials and Applications, Part 1, Chapter 2. Miami: American Welding Society.

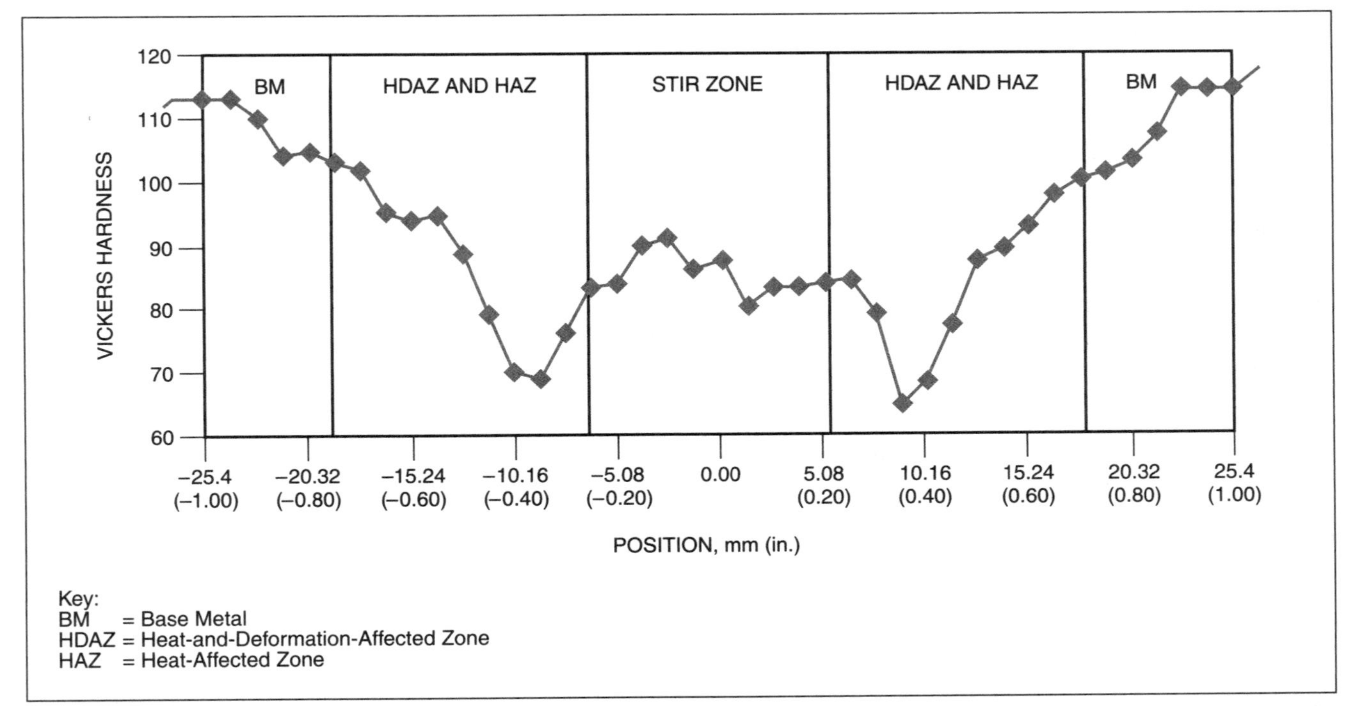

Source: Adapted from Lienertt, T. J., J. E. Grylls, and H. L. Fraser, 1998, Deformation Microstructures in Friction Stir Welds on 6061-T651. Second Symposium on Hot Deformation of Aluminum Alloys, Oct. 11–15, Rosemont, Illinois: 159–167.

Figure 7.13—Traverse of Microhardness in a Friction Stir Weld on 6061-T651 Alloy

phases. The hexagonal close-packed (hcp) crystal structure combined with brittle phases make the forming and shaping of Mg alloys difficult. These alloys can be fusion welded but porosity and coarse intermetallics may be present in the fusion zone. Problems resulting from the hcp structure and brittle phases also complicate the friction stir welding of magnesium alloys.

Many types of magnesium alloys are welded using the friction stir welding process, including AM50, AM60B, AZ31B, AZ61, AZ91D and experimental alloys.[8] Peak temperatures of these alloys during welding are reported to range from 325°C to 450°C (617°F to 842°F), depending on rpm and tool traverse rates. Magnesium alloys can be welded with tools made of tool steel.

In one study, the stir zone regions of magnesium alloys showed a refined grain size without coarse intermetallics. The refined grain size was believed to result from dynamic recrystallization, but no proof has been provided for this claim. The intermetallics were believed to be dissolved or broken up by interaction with the tool. Porosity in the zone of cast material was closed as a result of friction stir welding, and the primary solids in semi-solid cast alloys also were blended into the stir zone. The grains in the stir zone have been observed to show a high dislocation density, and the stir zone shows a preferred texture with certain crystallographic planes called *basal planes* aligned with the direction of tool rotation.[9] The transition zone (TMAZ) may be recrystallized and may show some dissolution of intermetallics. Hardness plots were either flat or showed an increase in the stir zone, due to the refined grain size and higher dislocation density of the alpha phase. Typical yield strengths were reported to be approximately 60% to 90% of the base metal, with tensile strengths of 75% to 90% of the base metal and lower ductilities. The texture has been reported to have an impact on tensile properties.

COPPER ALLOYS

Copper alloys have high thermal and electrical conductivity and are corrosion resistant in many environments. They have good ductility due to their face-centered cubic crystal structure and typically are strengthened by cold working, although some alloys can be precipitation strengthened.[10] The metallurgy of copper alloys is discussed in greater detail in the Volume 3 of the *Welding Handbook, Materials and Applications,* Chapter 3.[11]

Copper alloys can be fusion welded but the welds have problems with hot cracking and porosity. These alloys have been successfully joined using friction stir welding. A notable application involves the fabrication of nuclear waste storage containers. However, much less research has been reported on the friction stir welding of copper alloys than has been reported on aluminum alloys.

Peak temperatures in the stir zones have been reported to range up to 800°C (1472° F). Hence, tool materials must be chosen that will survive these temperatures without wear or fracture. Tool steels, cemented carbides, ceramics, and tungsten, molybdenum and nickel super-alloys have been tested as tool materials for welding copper with the following results:

1. Friction stir welding tools made of tool steel are not recommended because of rapid and excessive wear;
2. Carbide and ceramic stirring tools tend to fracture easily;
3. The refractory metals showed wear, but did not fracture; and
4. Nickel alloys performed best, particularly Nimonic 105.™

Microstructures in the stir zone were characterized by a refined grain size, probably due to classical discontinuous dynamic recrystallization. Transmission electron microscope observations of the stirred metal region revealed a lower density of dislocations relative to the base metal. Hardness in the stir zone has been shown to either increase or decrease, depending on welding conditions. Tensile testing showed that friction stir welds on copper alloys have very high ultimate tensile strengths relative to the base metal.

TITANIUM ALLOYS

Titanium alloys have densities intermediate to aluminum and steel alloys. They provide a good strength-to-weight ratio and can withstand higher temperatures in service than aluminum alloys. Titanium alloys also possess excellent corrosion resistance in many environments, including marine service.[12] These alloys are classified according to their room-temperature microstructure as alpha (α), alpha-beta (α-β), metastable (β) alloys, or (β) alloys. The α alloys have an hcp crystal structure, while

8. The titles, authors, and facts of publication of technical papers and other resources represented in this section are listed in the Bibliography in the Magnesium Alloys section.
9. The basal planes are those planes normal to the C-axis of the hpc crystal structure on which all of the dislocations glide and consequently on which all deformation takes place.
10. The titles, authors, and facts of publication of technical papers and other resources represented in this section are listed in the Bibliography in the Copper Alloys section.
11. American Welding Society (AWS) Welding Handbook Committee, W. R. Oates, ed., 1996, Welding Handbook, Vol. 3, Materials and Applications, Part 1, Chapter 2. Miami: American Welding Society.
12. The titles, authors, and facts of publication of technical papers and other resources represented in this section are listed in the Bibliography in the Titanium Alloys section.

the β alloys have a body-centered cubic (bcc) structure. The metallurgy of titanium alloys is discussed in greater detail in Chapter 9 of the *Welding Handbook*, eighth edition, Volume 4, *Materials and Applications*.[13]

Regardless of the welding process used, the joining of titanium alloys is complicated by problems associated with their high reactivity. These alloys rapidly dissolve oxygen, nitrogen, and hydrogen at temperatures above 500°C (932°F), resulting in subsequent embrittlement. Dissolution of these gases is especially rapid in the liquid phase. As a result, welding processes must be carried out in inert or vacuum environments to avoid embrittlement. Moreover, the workpieces and filler metals must be cleaned with solvents to remove hydrocarbon-based oils and moisture to prevent embrittlement. Heating of material in the heat-affected zone above the beta transus temperature can cause grain growth and produce coarse columnar grains in the fusion zone, resulting in a loss of ductility.

Several titanium alloys have been successfully joined using friction stir welding, including Ti-6Al-4V, Ti-15-3, and Beta-21S. The reactivity of titanium, the high working temperatures, and large forces require the use of specialized tool materials, and wear and deformation of the tool are prevalent. Commonly used tool materials for friction stir welding of titanium alloys include molybdenum alloy (TZM), tungsten, tungsten-rhenium alloys, and W-Re containing hafnium-carbide (HfC).

13. American Welding Society (AWS) Welding Handbook Committee, W. R. Oates, Ed., 1998. eighth edition, Welding Handbook, Volume 4, Materials and Applications, Part 2, Chapter 9. Miami: American Welding Society.

The beta alloys are easiest to join by friction stir welding, followed by the alpha-beta alloys. Commercial-purity (CP) titanium (alpha) appears to be the most difficult to weld because it possesses a hexagonal close-packed (hcp) structure and higher thermal conductivity relative to the other alloy classes. The process window for defect-free welds on CP titanium sheet material appears to be very small.

Ti-6Al-4V Alloys

Most studies on the friction stir welding of titanium alloys were conducted on Ti-6Al-4V alloys. This alloy is classified as an alpha-beta alloy and is the "workhorse" of the Ti alloys. Defect-free welds have been produced over a range of parameters using tools made of commercially pure tungsten; however, tungsten-rhenium tools also have been used. The tool and workpieces were protected from surface oxidation by making the weld in an inert gas chamber with a sliding top section.

In one study, peak temperatures in the stir zone were reported to exceed 1045°C (1913°F) and may have ranged up to 1150°C (2102°F). These temperatures are clearly in excess of the beta transus for this alloy, ~1000°C (~1832°F). Cooling rates were estimated to be about 40°C/s (72°F/s), which was too slow for the formation of alpha prime martensite. Peak temperatures in the TMAZ and the HAZ are believed to be below the beta transus temperature.

Figure 7.14 shows an optical macrograph of a transverse section of a friction stir weld on Ti-6Al-4V. Several microstructurally distinct weld regions can be observed, including the stir zone and the heat-affected zone, each

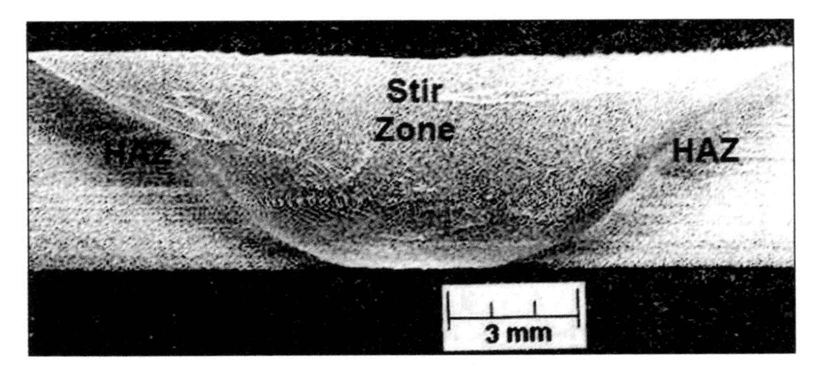

Source: Adapted from Lienert, T. J., K. V. Jata, R. Wheeler, and V. Seetharaman, 2001, Friction Stir Welding of Ti-6Al-4V Alloys, presented at Materials Solutions '00: Joining of Advanced and Specialty Materials III, Proceedings of the Conference, St. Louis, Missouri. Cambridge: TWI. 9–11 Oct. 2000, pp. 160–166.

Figure 7.14—Optical Macrograph of a Transverse Section of a Friction Stir Weld on Ti-6Al-4V

showing a different etching response. The stir zone in these welds contained equiaxed grains and had a refined grain size relative to the base metal. Consistent with the reported peak temperatures, the equiaxed grains in the stir zone exhibited grain-boundary alpha phase outlining prior beta grains, with fine colonies of alpha and beta formed by a nucleation and growth mechanism throughout the grain interiors. In a second study, the presence of secondary alpha within the beta was reported and is believed to form by a diffusion process on cooling. In accordance with the estimated cooling rates established in previous research, no evidence of alpha prime in the weld zones was found.

Results of microhardness tests across the various weld regions in a transverse section of a friction stir weld in Ti-6Al-4V are shown in Figure 7.15. The pattern of microhardness does not show the "W" shape typical of precipitation-strengthened aluminum alloys with a minimum of hardness in the heat-affected zone. Rather, the hardness profile is quite flat, with the exception of an increase in hardness in regions corresponding to the HAZ. The base metal had a Vickers hardness number between 320 and 340. Vickers hardness increased to about 370 in the heat-affected zone, and fell again to 350 in the stir zone. Tensile properties of friction stir welds on titanium alloys are summarized in the section Mechanical Properties of Friction Stir Welds.

Ti-15V-3Cr-3Al-3Sn Alloy

Defect-free friction stir welds also were produced in the metastable beta alloy Ti-15V-3Cr-3Al-3Sn (hereafter referred to as *Ti-15-3*) using tungsten-rhenium tools. Again, the tool and workpieces were protected from surface oxidation by welding in an inert gas chamber. Chemical analysis of the weld metal showed no measurable pickup of oxygen during welding. Tool wear and deformation were monitored before and after each weld by inspecting the tool with an optical comparator. Results indicated no measurable tool wear, deformation, or pickup of tungsten or rhenium in the workpiece after several welds totaling about 1 m (3 ft.) in length.

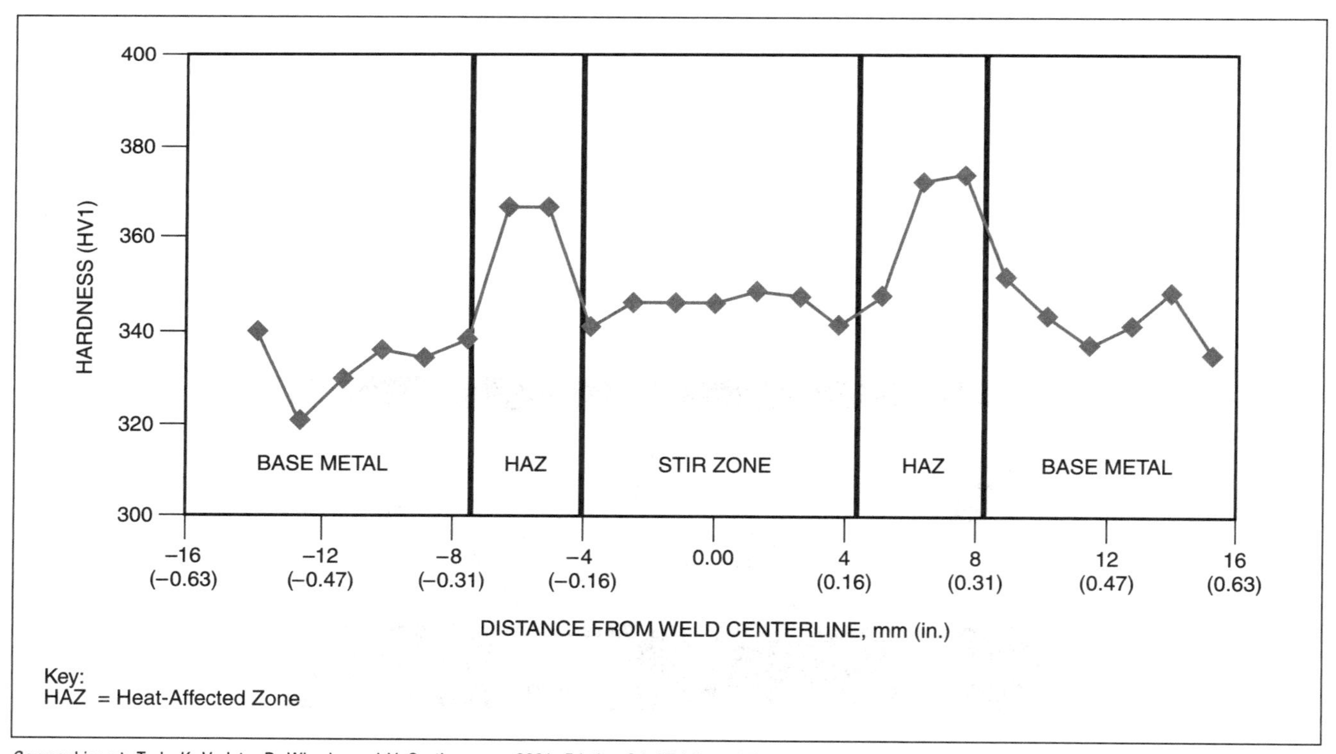

Source: Lienert, T. J., K. V. Jata, R. Wheeler and V. Seetharaman, 2001, Friction Stir Welding of Ti-6Al-4V Alloys, Materials Solutions '00: Joining of Advanced and Specialty Materials III; Proceedings of Joining of Advanced and Specialty Materials, St. Louis, Missouri, 9–11 Oct. 2000, Materials Park, Ohio: ASM International. pp. 160–166.

7.15—Microhardness Across the Various Weld Regions in a Transverse Section of a Friction Stir Weld in Ti-6Al-4V

Microstructural characterization of the welds on annealed Ti-15-3 revealed considerable grain refinement in the stir zone. No alpha phase was observed optically in the stir zone, the TMAZ or the HAZ of the welds on the annealed material. Welds on the annealed Ti-15-3 exhibited high yield joint efficiencies with acceptable ductility. Properly made welds failed through the base metal away from the weld region.

Little difference was observed in mechanical properties between welds produced on the annealed base metal and those given a postweld aging treatment at 635°C (1175°F) for 8 hours. The postweld aged samples exhibited grain boundary alpha phase along the beta grain boundaries and alpha phase dispersed throughout the grain interiors. Similar FSW success was achieved in Beta 21S™ alloys. A summary of the tensile properties of these welds is provided in the section Mechanical Properties of Friction Stir Welds.

CP Ti Alloy

Sound friction stir welds also were produced on a commercially pure titanium alloy using a titanium-carbide tool. The stir zone contained a considerable amount of twinning and a high dislocation density. The distribution of twins was found to vary throughout the stir zone, and the HAZ region was characterized by slight grain growth. Hardness results showed scatter with higher hardness in the regions with twins and higher dislocation densities.

STEEL ALLOYS

Steel alloys represent one of the most common groups of engineering materials. Steels are inexpensive to produce and are used extensively by the auto and shipbuilding industries.[14] Detailed information on the metallurgy of various steels is presented in the *Welding Handbook,* 8th edition, Volume 4, *Materials and Applications,* Chapters 1 through 5.[15]

Friction stir welding studies have been conducted on a variety of steels, including 1018, S355, DH-36, HSLA-65, and quenched and tempered steels. Tools must be made from materials that are able to withstand the high hot-working temperatures and that exhibit high strength and toughness at operating temperatures. Examples of tool materials that have been used successfully include tungsten-rhenium (W-Re) and polycrystalline cubic boron nitride (PCBN). Inert gas shielding is needed with W-based tools to limit embrittlement due to pickup of oxygen and nitrogen. Inert gas shielding also limits pickup of gases by the steel workpiece and improves the aesthetics of the weld. Tool wear and deformation may be common with metallic tools. The higher flow stresses and lower thermal conductivity associated with hot deformation of steels normally produce higher forces on the tool and the friction stir welding machine, thus mandating slower tool travel rates than those used for aluminum alloys. Studies have indicated that ferritic steels appear easier to join by friction stir welding than steels that are austenitic at welding temperature, because travel speeds may be faster.

1018 Steels

Friction stir welds of 1018 steel made with a tungsten tool were studied, and comparisons were made of tool dimensions before and after welding. These measurements indicated that most of the deformation and wear occurred during the tool plunging stage. Partial-penetration, partial-diameter pilot holes were used to minimize tool wear during the plunge period.

Thermocouple measurements and microstructural characterization indicated that peak temperatures in the stir zone reached the austenite phase field. Peak temperatures were estimated to exceed 1100°C (2012°F) and may approach 1200°C (2192°F). Cooling rates were approximately 5°C/second (s) to 10°C/s (ΔT_{8-5} = ~50 s to 60 s).

Figure 7.16 shows an optical macrograph of a friction stir weld on 1018 steel. Friction stir welds in carbon-manganese (C-Mn) steels, such as 1018, may exhibit several microstructurally distinct regions, including the stir zone (along the weld centerline), a grain-coarsened region surrounding the stir zone, a grain-refined region (encompassing the grain-coarsened region), as well as an intercritical region and a subcritical region containing

Source: Adapted from Lienert, T. J., W. L. Stellwag Jr., B. B. Grimmett, and R. W. Warke, 2003, Friction Stir Welding Studies on Mild Steel, *Welding Journal* 82(1): 1-s–9-s.

Figure 7.16—Optical Macrograph of a Friction Stir Weld on 1018 Steel

14. The titles, authors, and facts of publication of technical papers and other resources represented in this section are listed in the Bibliography in the Steel Alloys section.

15. American Welding Society (AWS) Welding Handbook Committee, W. R. Oates and A. M. Saitta, Eds. 1998, *Welding Handbook,* 8th Ed., Vol. 4, *Materials and Applications,* Chapters 1–5, Miami: American Welding Society.

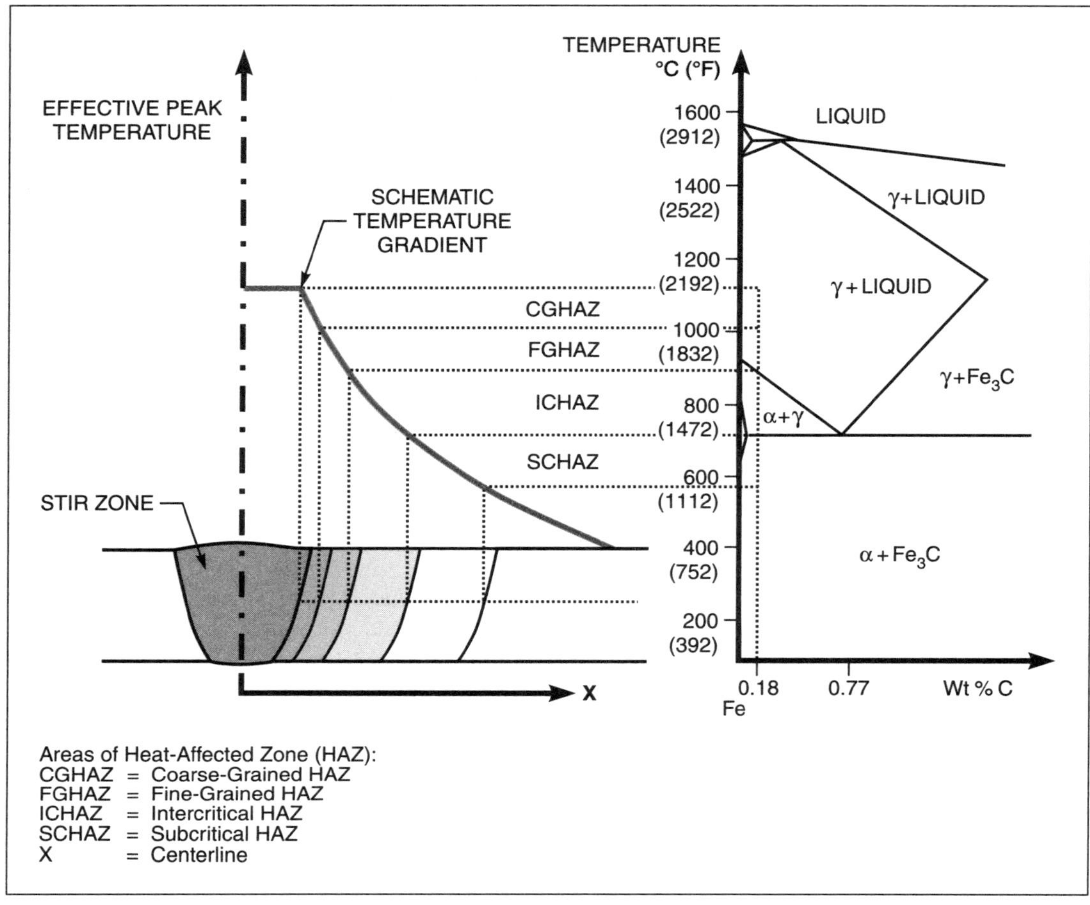

Source: Adapted from Lienert, T. J., W. L. Stellwag, Jr., B. B. Grimmett, and R. W. Warke, 2003, Friction Stir Welding Studies on Mild Steel, *Welding Journal* 82(1): 1-s–9-s.

Figure 7.17—Various Regions of the Heat-Affected Zone and Their Relation to Peak Temperatures in a Friction Stir Weld on 1018 Steel

partially spheroidized carbides. The various regions develop in accordance with the local thermomechanical cycle. A schematic illustration of the various regions of the heat-affected zone of a friction stir weld on 1018 steel and their relation to peak temperatures in the Fe-Fe_3C binary phase diagram is shown in Figure 7.17.

The distribution of microhardness across the various weld regions in a transverse section of a friction stir weld on 1018 steel is illustrated in Figure 7.18. Consistent with the slow cooling rate and limited hardenability of these steels, no evidence of martensite was found in the stir zone or the heat-affected zone. However, the stir zone and HAZ normally were harder than the base metal due to a finer grain size and to the formation of fewer transformation products during cooling. Welded samples failed in regions corresponding to the base metal and demonstrated yield and ultimate tensile strengths comparable to or in excess of those of the base metal. Further information on tensile properties of friction steel welds on steel is presented in the section Mechanical Properties of Friction Stir Welds.

DH-36

Friction stir welds have been produced on DH-36 steel using W-Re tools by several researchers. The stir zone microstructure contained grains of polygonal ferrite about 5 micrometers (μm) in size. No regions of martensite were observed in the stir zone. The heat-affected zone contained regions of partially spheroidized (fuzzy)

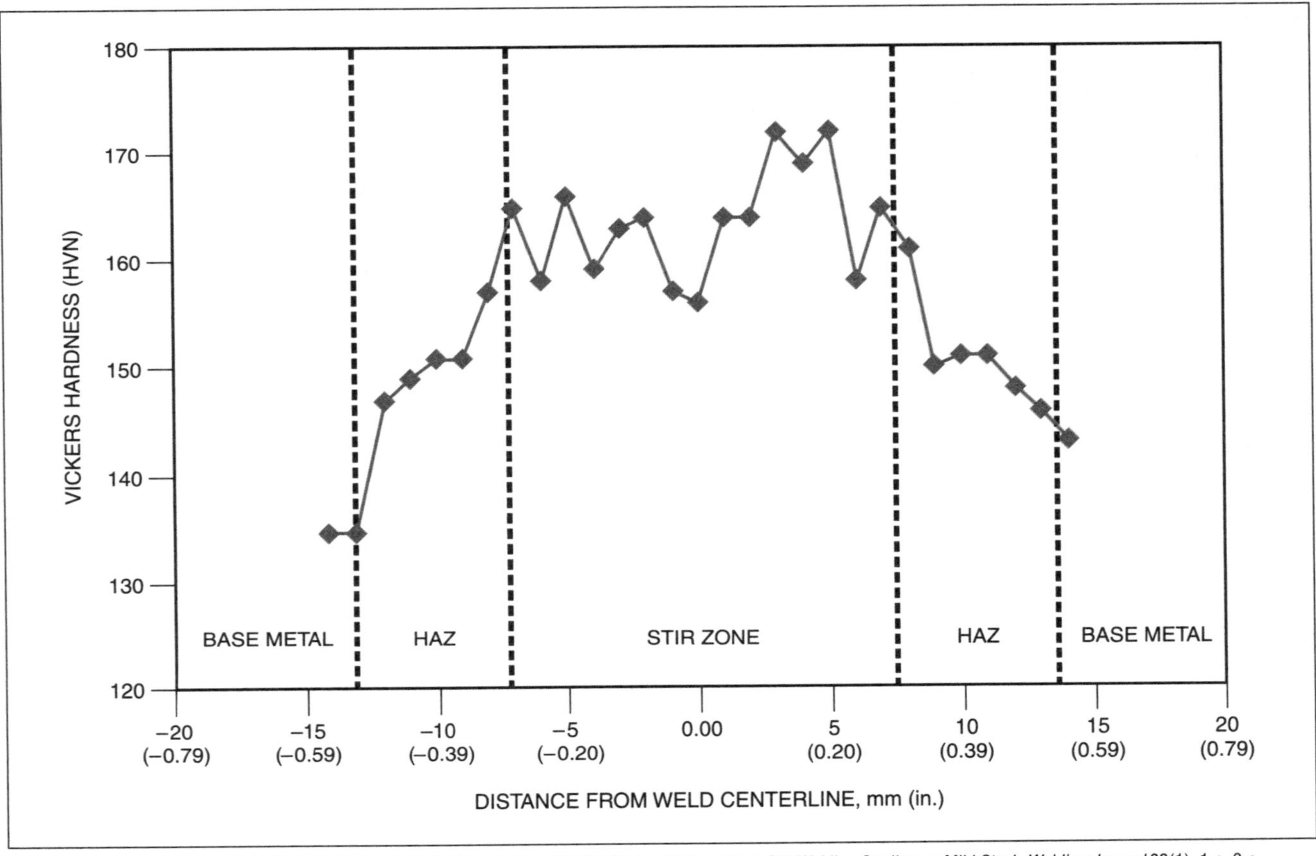

Source: Adapted from Lienert, T. J., W. L. Stellwag, Jr., B. B. Grimmett, and R. W. Warke, 2003, Friction Stir Welding Studies on Mild Steel, *Welding Journal* 82(1): 1-s–9-s.

Figure 7.18—Schematic Illustration of the Various Regions of the Heat-Affected Zone in a Friction Stir Weld on 1018 Steel

pearlite but no grain-coarsened or grain-refined regions. The microhardness of the stir zone averaged a Vickers hardness number of approximately 215, while that of the base metal was between 190 VHN and 195 VHN. Consistent with the absence of martensite, no evidence of extensive hardening was found in the microhardness results. The weld samples showed acceptable strain-to-failure with yield and tensile strengths exceeding those of the base metal.

HSLA-65

Studies of friction stir welding on HSLA-65 steels provided similar results to those of the DH-36 tests. Acceptable values of Charpy V-notch (CVN) toughness were shown for these welds, but for reasons unknown, large amounts of scatter were observed in the data. Corrosion testing with salt spray tests showed no evidence of preferential corrosion on these welds.

The heat-affected regions in two-pass friction stir welds on a quenched and tempered steel showed lowered hardness due to over-tempering. Hardness in the first pass was lower due to tempering during the second pass. Tensile tests showed ~70% joint efficiency for tensile strength. The CVN test showed that toughness was acceptable in these welds.

Stainless Steels

Stainless steel alloys provide a useful combination of strength and corrosion resistance. There are five major categories of stainless steels, classified according to their microstructures at room temperature. Most work on the friction stir welding of stainless steel was conducted on alloys from the austenitic class, including 304, 316, and AL6-XN. This class is not strengthened by heat treatment. The metallurgy of stainless steel alloys is discussed in "Stainless and Heat-Resisting

Steels," in Volume 4, eighth edition, of the *Welding Handbook, Materials and Applications, Part 2.*[16]

Austenitic stainless steels can be readily welded with fusion processes, but may have problems with hot cracking and distortion in large sections. The friction stir welding of these materials requires the use of tools made of W, W-Re, or PCBN using higher forces and slower travel speeds relative to Al alloys.[17] Peak stir-zone temperatures have not been reported, but likely exceed 1000°C (1832°F). As with most high-temperature materials, tool wear and deformation is a problem with metallic tools, while the service life of CBN tools may be limited by their low toughness.

Results of one study revealed that the stir zone of these welds was characterized by a refined grain size with equiaxed grains due to dynamic recrystallization. The stir zone also exhibited a banded region on the advancing side of the weld, often called the *swirl zone*. The formation of sigma phase has been confirmed in these bands by TEM examination with a local loss in corrosion properties. The accelerated formation of sigma phase was attributed to the intense deformation that occurred locally during friction stir welding. Microhardness results normally showed relatively flat profiles across all of the weld zones. Tensile results indicated increased yield strength and tensile strength with respect to the base metal, with a slight loss in elongation to failure.

A few studies were completed on the friction stir welding of duplex and ferritic stainless steels using cubic boron nitride tools, with the result that the ferritic alloy showed increased stir-zone hardness, suggesting the formation of martensite.

HEAT TRANSFER IN FRICTION STIR WELDS

The thermal cycles experienced by the material in and around the weld area have a profound influence on the final microstructures, mechanical properties, residual stresses and distortion of the weldment. Key features of the local thermal cycle that have the greatest effect include the peak temperature and the cooling rate. In this section, various models that have been used to simulate heat flow and thermal cycles during FSW are reviewed. Published research on experiments involving temperature measurements during FSW also are examined.[18]

MODELING OF HEAT TRANSFER

As a first approximation, temperature distribution around the FSW tool when welding thin plate can be calculated using Rosenthal's equation for moving-point heat source. As with laser beam welding or arc welding, the temperature contours are parabolic, compacted ahead of the tool and expanded behind it. However, the heat generation is not localized to a small area in a friction stir weld. It occurs over the entire interface between the tool and the workpiece. The thermal field is inherently asymmetric along the centerline of the weld due to differences in heat generation between the advancing and retreating sides. Hence, heat flow inherently is three dimensional. In addition, heat is carried by the considerable plastic flow that takes place around the tool, and consequently, heat transfer is both conductive and convective. Figure 7.19 shows computed temperature profiles in the X-Z, Y-Z and X-Y planes for a friction stir weld on AA6061 using an H-13 tool, with a pin radius of 6 mm (0.24 in.) and a shoulder radius of 25 mm (1 in.). The axial pressure was 12.7 MPa (1.842 ksi). The welding velocity was 1.59 mm/s (0.063 in./s) and the tool rotational speed was 344 rpm. The asymmetry in temperature distribution was not pronounced because of the high thermal conductivity of aluminum.

To understand heat transfer during friction stir welding, the heat generation must be quantified. Heat generation depends on the dynamics of the interface between the tool and the workpiece. A better understanding of heat generation and heat transfer in FSW has evolved over the years and is best represented by the increasing sophistication of numerical models for FSW.

Initial models for friction stir welding were analytical and were based on models for rotary friction welding. Local heat generation rate was a product of tool rpm, axial pressure, distance from the axis, and an effective coefficient of friction, μ. In one model, a value of two was chosen for μ, based on results from rotary friction welding experiments, and the thermal properties of the workpiece were assumed to be independent of temperature. The heat source was constructed by integrating the Rosenthal equation, using a series of point sources symmetrically disposed around the periphery of the shoulder. The model inherited the temperature singularity at the shoulder periphery from the Rosenthal formulation.

16. American Welding Society (AWS) Welding Handbook Committee, W. R. Oates and A. M. Saitta, ed., 1988, *Welding Handbook, Materials and Applications, Part 2,* eighth edition, Volume 3, Chapter 5, Miami: American Welding Society.

17. The titles, authors, and facts of publication of technical papers and other resources represented in this section are listed in the Bibliography in the Stainless Steel Alloys section.

18. The titles, authors, and facts of publication of technical papers and other resources represented in this section are listed in the Bibliography in the Heat Transfer section.

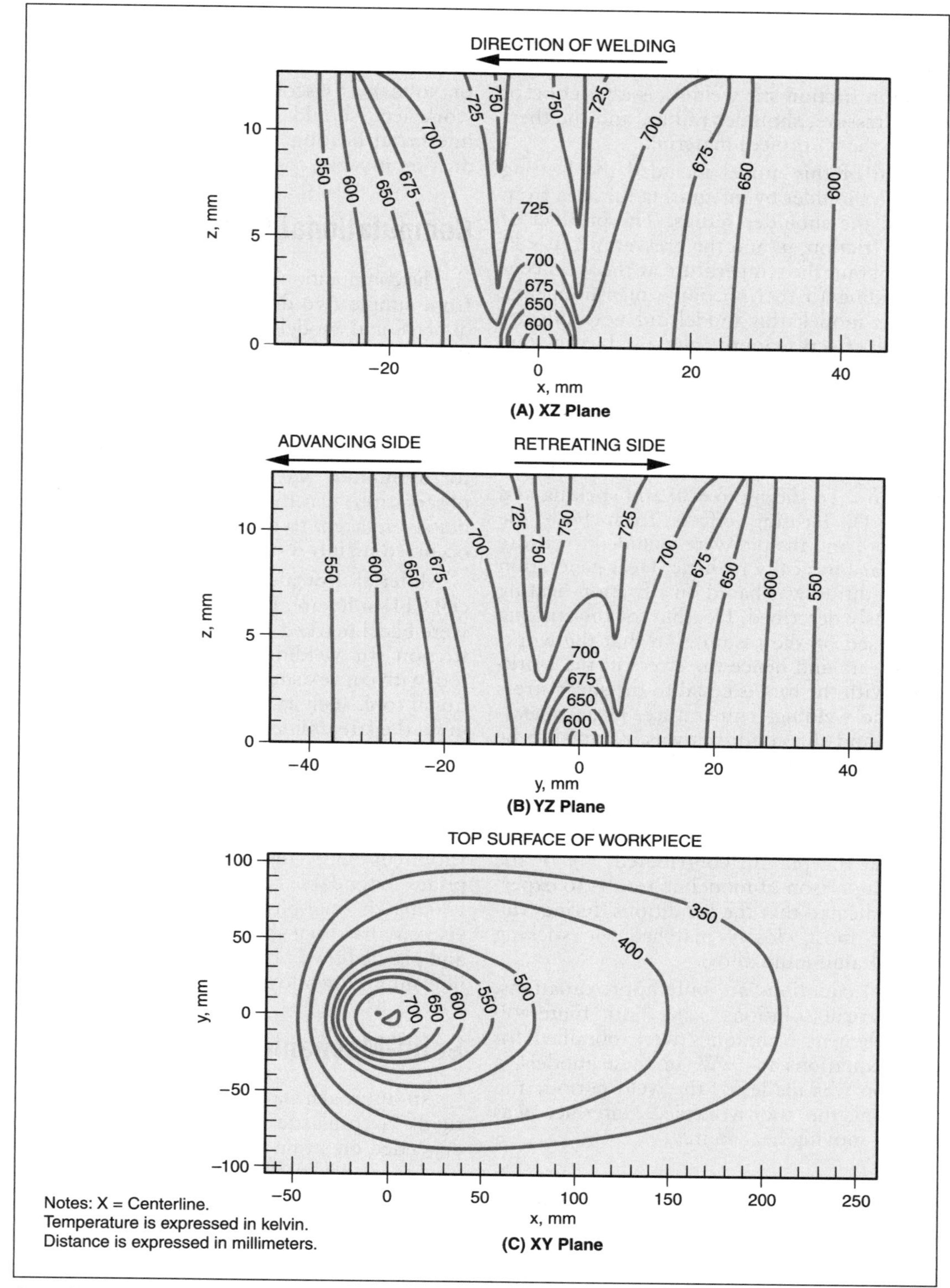

Source: Adapted from Nandan, R., G. G. Roy and T. DebRoy, 2006, Numerical Simulation of Three-Dimensional Heat Transfer and Plastic Flow During Friction Stir Welding, *Metallurgical and Materials Transactions A*, 37(4): 1247–1259.

Figure 7.19—Temperature (K) Distribution Profiles for a Friction Stir Weld on AA6061 at (A) XZ Plane, (B) YZ Plane, and (C) XY Plane

Mass flow, heating effects from the pin, and heating due to visco-plastic dissipation were not considered in this model. Nonetheless, the model confirmed that temperature distribution in friction stir welding is a function of tool rpm, axial pressure, shoulder radius, and the thermal properties of the workpiece material.

A modification of this model included the heating effect of the entire shoulder by integrating the area from the pin radius to the shoulder radius. The product of the coefficient of friction, μ, and the pressure, P, ($\mu \times P$) was adjusted to obtain the temperature at the weld centerline corresponding to thermocouple measurements. Like the previous model, this model did not consider mass flow, heating effects from the pin, and heating due to visco-plastic dissipation.

However, this approach using a coefficient of friction is too simplistic and is not adequate. Hence, in a subsequent model, a state variable, δ, was introduced to describe three different tool and workpiece contact conditions: sticking ($\delta = 1$), sliding ($\delta = 0$) and sticking and slipping ($0 < \delta < 1$). Heating effects from both the shoulder of the tool and the pin were included to ensure that the model was physically realistic. Heat generation for the sliding condition was based on a friction-heating approach previously described. Heating for the sticking condition was based on the assumption that the workpiece material shears and hence the stress in the workpiece in contact with the tool is equal to the shear stress for yielding at the welding temperature. Heat generation for the stick-and-slip condition was assumed to be a linear combination of the two, with δ describing the fraction of sticking. Results of the model suggested that the tool shoulder contributed 86% of the total heating, the vertical surface of the pin contributed 11% of the total heating, and the pin tip contributed 3% of the total heating. Comparison of modeling results to experimental results indicated that the conditions during friction stir welding most closely matched the sticking condition in 2024 aluminum alloy.

Since analytical equations are only approximations, several full numerical solutions using both finite-volume and finite-element techniques were obtained for the conduction equations in FSW. In these models, a steady assumption was made for the weld period, and heat generation at the tool-workpiece interface was simulated using a moving heat source.

Conductive and Convective Heat Transfer

Since large-scale plastic deformation occurs near the tool, heat transfer near the tool is predominantly convective. Hence, the effects of deformation must be taken into account for more accurate modeling. Primarily, there are two approaches to model deformation. One is to use solid mechanics, which involves finite-element analysis in most cases. Another approach is to treat the deformation as visco-plastic flow of high viscosity material; this falls under the purview of computational fluid dynamics (CFD) problems. Deformational heating (in solids) or viscous dissipation (in fluids) within the workpiece should be taken into account apart from interfacial heating, which was considered in heat-conduction models.

Computational Fluid Dynamics Models

The computational fluid dynamics models have evolved from simple two-dimensional to comprehensive three-dimensional models able to predict asymmetry of the thermal fields with respect to the joint line. One early CFD effort used a two-dimensional visco-plastic model for aluminum alloy based on laminar, viscous, and non-Newtonian flow around a circular cylinder. The non-Newtonian viscosity based on the constitutive equations for aluminum was calculated. One of the important observations was that beyond the rotational zone immediately adjacent to the FSW tool, the material transport occurred mainly along the retreating side.

Material flow also has been modeled using commercial CFD software, FLUENT.[19] Complex tool geometries were benchmarked for mechanical efficiency during the friction stir welding of a 7075 Al alloy. A triangular tool with convex surfaces, such as Trivex™ and a conventional tool, such as Triflute™ were compared by examining the streamlines around these tools.[20]

Two-dimensional models do not consider the vertical mixing during friction stir welding, which is observed experimentally. A comprehensive three-dimensional heat and material flow model solving the equations of conservation of mass, momentum, and energy, with appropriate boundary conditions and employing spatially variable thermal-physical properties and non-Newtonian viscosity has been shown to reliably predict heat transfer and plastic flow in friction stir welds of different systems, like aluminum alloys, mild steels, and stainless steels.

Solid-Mechanics Models

An approach based on solid mechanics, assuming a rigid visco-plastic material where the flow stress depended on strain rate and temperature, also has been used to model friction stir welding. The heat generation rate, expressed as the product of the effective stress and the effective strain rate, provided the boundary condition for calculation of temperature distribution in the workpiece and the tool, using a three-dimensional finite analysis code.

19. Computational fluid dynamics simulation software, Canonsburg, Pennsylvania: FLUENT.
20. Trivex and Triflute are trade marks of TWI.

One practical application involved the use of solid-mechanics-based models with adaptive boundary conditions for void-free welds. A fully coupled thermomechanical three-dimensional finite-analysis model was developed using the FE package ABAQUS[21] with material flow modeled using Johnson-Cook law. The contact forces were modeled by Coulomb's Law of friction. Results of the simulations indicated that the development of the sticking contact condition at the pin-workpiece interface is important for the success of processes involving recoalescence at the rear of the tool. The key to producing defect-free welds is the development of a high-temperature region close to the tool-workpiece interface surrounded by a zone of stiffer material at lower temperature.

Shock-Wave Model

An Eulerian shock wave physics code also has been used to model friction stir welding in aluminum. The three-dimensional code solves time-dependent equations for continuum mechanics and is well suited for modeling very large deformations at high strain rate, such as in a ballistic impact. The model predicted peak temperatures near 500°C (932°F) under the shoulder adjacent to the pin, where the largest deformation gradients were found. It should be noted that this result contrasts with those of other researchers who found peak temperatures near the shoulder periphery due to the greater relative local tool velocity.

Other Models

A force-balance approach has been employed to determine temperature profiles during friction stir welding; however, temperature fields found in this research were erroneous. Uncertainties in thermal conductivity were cited as possible causes for the discrepancy.

A model based on the torque input was also developed for friction stir welding. This model used the tool rotational speed and a shear stress of 14 MPa (2.03 ksi) in conjunction with tool dimensions to estimate the heat flux with and without backing plates by employing suitable convective heat transfer coefficients. Model results showed close qualitative agreement with measurements.

EXPERIMENTAL OBSERVATIONS OF TEMPERATURES

Many experiments aimed at determining temperatures at various locations of friction stir welds have been reported. Most of the experiments have involved thermocouple measurements. In this section, selected thermocouple measurements are discussed along with a method to infer peak temperatures in Al alloys from microstructural observations.

Thermocouple Measurements

In one study, peak welding temperatures reported during friction stir welding experiments on Al alloys were less than the solidus temperature of the alloy. Readers may wish to refer to the section on phase diagrams in Chapter 4 of Volume 1 of the *Welding Handbook*, ninth edition, for further discussion of the solidus temperature.[22] Peak temperatures of ~450°C (~842°F) were measured for the stir zone of friction stir welds made at 400 rpm on 6.4-mm (0.25-in.) 6061 aluminum. Figure 7.20 shows representative thermal cycles. Peak temperatures in the stir zone near the pin were nearly isothermal, and the peak temperature gradient through the plate thickness was small. No temperature difference was found between the advancing sides and the retreating sides. Peak temperatures were found to increase with increasing tool rpm and increasing tool pressure.

In another study, peak temperatures of 475°C (887°F) were reported near the boundaries of the stir zone and TMAZ in 7075 aluminum. Temperatures in the heat-affected zone ranged from 250°C (482°F) to 475°C (887°F), causing overaging and coarsening of the strengthening precipitates. (Refer to Figure 7.11 for an illustration of the weld zone microstructures, precipitate distributions, and temperature ranges for a friction stir weld on 7075 aluminum.)

In-Situ Inferences of Temperature

The θ (Al_2Cu) particles present in 2219 aluminum were used as in-situ sensors for eutectic melting to bound temperatures during the friction stir welding of this aluminum-copper alloy. This alloy contains at least 6.3% copper, which is above the maximum solid solubility for this system. The eutectic temperature for this system is 548°C (1018°F). No evidence of eutectic melting was observed in the stir zone or the TMAZ of these welds, confirming that peak temperatures were below the eutectic temperature and that no liquid phases existed during friction stir welding.

21. ABACUS, a finite element analysis program, Providence, Rhode Island: ABAQUS.

22. American Welding Society (AWS) Welding Handbook Committee, C. L. Jenney and A. O'Brien, ed., 2001, *Welding Handbook*, 9th ed., *Welding Science and Technology*, Vol. 1, Chapter 4, Miami: American Welding Society.

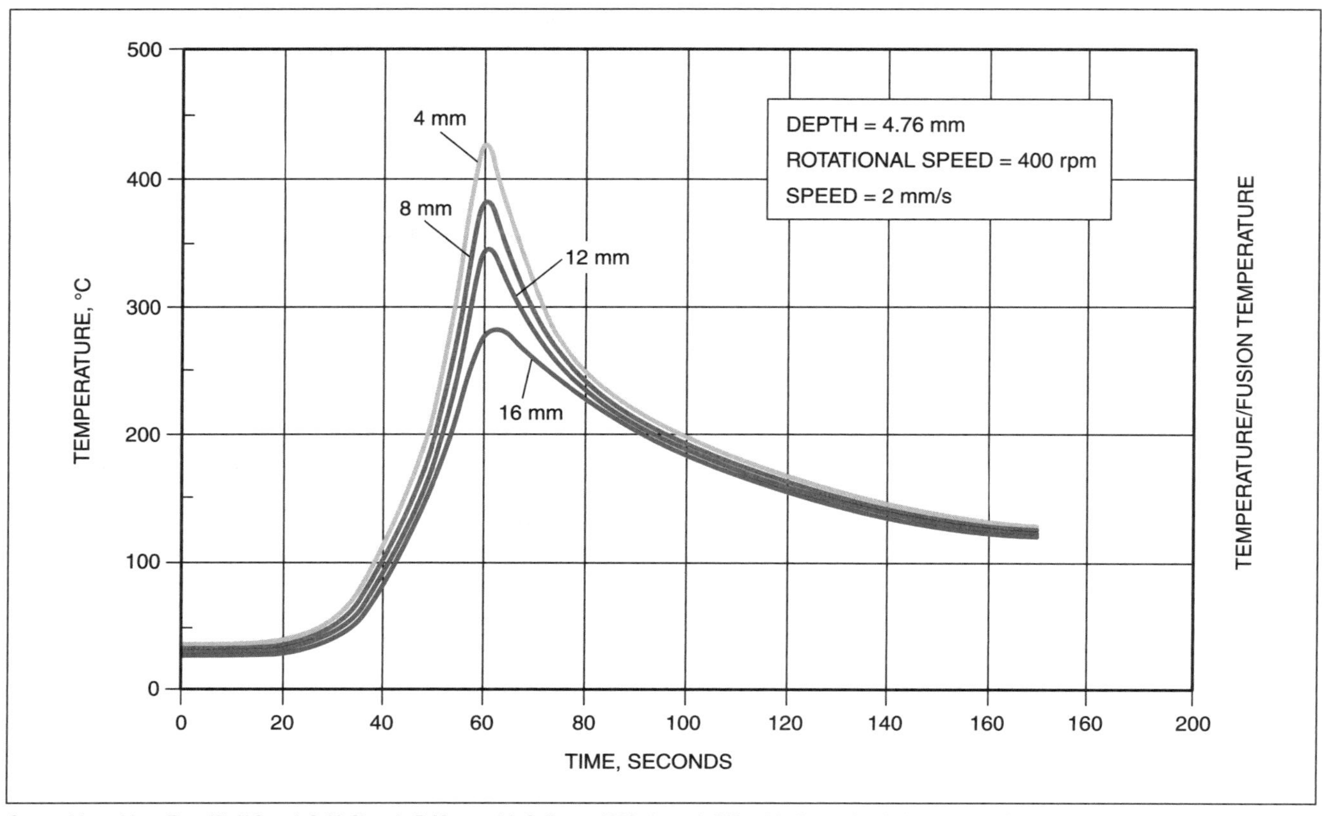

Source: Adapted from Tang, W., X Guo, J. C. McClure, L. E. Murr, and A. C. Nunes, 1988, *Journal of Materials Processing & Manufacturing Science* (USA), Vol. 7, No. 2 (10): 163–172.

Figure 7.20—Heat Input and Temperature Distribution in Friction Stir Welding on 6061 Aluminum

MATERIAL FLOW IN FRICTION STIR WELDING

Friction stir welding involves the movement of considerable amounts of viscous material around the tool followed by recoalescence of material near the back surface of the tool. An understanding of material flow during FSW is important for at least two reasons. First, welding conditions that produce certain undesirable flow patterns may result in the formation of discontinuities such as "wormhole" discontinuity.[23] Second, moving material also carries heat and may impact the thermal cycle. Models aimed at an improved understanding of material transport are discussed in the first part of this section. Subsequently, experimental observations of material flow associated with friction stir welds are reviewed.[24]

MODELING OF MATERIAL FLOW

In one set of simulations, the flow patterns around a friction stir welding tool were modeled using numerical solutions of coupled Navier-Stokes and heat transfer equations. The heat source was modeled by considering viscous dissipation of mechanical energy. Results of the simulations suggest that the following three distinct flow regimes exist as a function of distance from the shoulder:

1. A region of rotation just under the shoulder, where flow occurs in the direction of tool rotation;

23. The term *wormhole* as used in friction stir welding is a mechanical lack-of-bonding cavity discontinuity and is not to be confused with the nonstandard term *wormhole porosity* (elongated piping porosity) used with gas-shielded arc welding processes.

24. The titles, authors, and facts of publication of technical papers and other resources represented in this section are listed in the Bibliography in the Material Flow section.

2. A region near the base of the pin, where material is extruded past the pin; and
3. A transition area between the two regions, where flow is chaotic.

Of particular interest in the transition area was the report of an unstable region, in which flow reversal occurred at a location where the rotational and translational velocities of the tool were equal in magnitude and opposite in direction.

Solid-mechanics models also have been used to analyze the deformation of plasticized metal during friction stir welding. One solid-mechanics model was developed to analyze the flow of material during friction stir welding. This model closely reproduced the experimental marker studies that used dissimilar Al alloy markers, as discussed below. Results of the simulations also suggested that material tended to pass mainly around the retreating side of the pin.

Streamline plots of velocity obtained using a CFD technique gave insight into the nature of material flow around the tool pin. The pattern of streamlines plotted in Figure 7.21 illustrates a tool moving to the right and rotating in a counterclockwise direction when viewed from above. The streamlines show the paths taken by the material as it flows around the pin and is re-coalesced behind the pin. The streamlines split in front of the pin on the advancing side. Material within the shoulder diameter was transported around the retreating side of the pin in the direction of rotation, following arched paths shaped like concentric horseshoes (or the Greek letter "Ω"). The streamlines subsequently rejoined behind the pin on the advancing side.

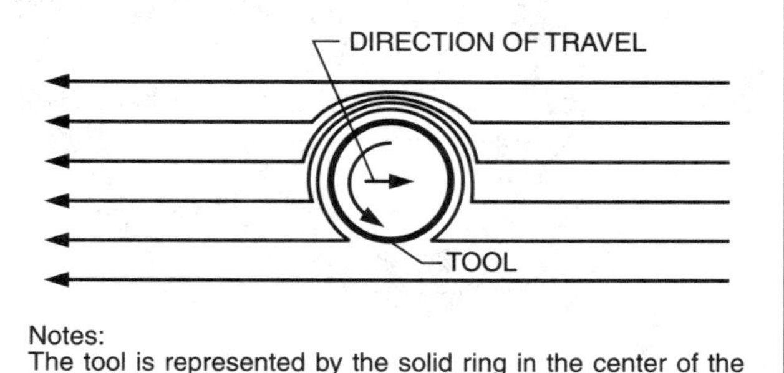

Source: Adapted from Seidel, T. U., and A. P. Reynolds, 2003, *Science and Technology of Welding and Joining*, Warrendale, Pennsylvania: The Minerals, Metals and Materials Society (TMS), 8(3): 175.

Figure 7.21—Plot of Streamlines Showing Metal Flow around the Friction Stir Welding Tool

The material that passed around the retreating side of the pin was deposited behind the pin in the same lateral position relative to the joint line as it originated. Consequently, the material originally located far on the advancing side (but within the pin diameter), was transported a greater distance than the material initially located closer to the retreating side, and thus experienced greater strains. Since the same material traveled a greater distance in the same time period (~1 revolution of the pin), it also was subjected to greater strain rates. In addition, results indicated that excessive rotational speeds (rpm) may allow some material to flow around the advancing side and that this condition may lead to the formation of incomplete bonding discontinuities.

EXPERIMENTAL OBSERVATIONS OF MATERIAL FLOW

Experiments with friction stir welds on both similar and dissimilar alloys between aluminum and copper and between different aluminum alloys have shown that the extensive plastic deformation during FSW creates complex vortex and swirl-like structures, especially near the interface of the weld zone and base metal. In welds of dissimilar aluminum alloys, striations and patterns of compositional difference were observed that corresponded to the spacing of the threads on the pin. The formation of these patterns was influenced by the geometry of the pin and the rotational speed.

Several researchers have used experimental tracer techniques to investigate the nature of plastic deformation and metal flow during friction stir welding. One investigation used steel shot markers placed at various plate depths and lateral distances from the weld centerline in an aluminum alloy. The position of the steel shot after welding was recorded using radiography. The markers initially placed near the top of the plate were lifted as they approached the tool and were deposited in a chaotic and random fashion behind the tool. Some of the material originally near the top surface subsequently was carried downward by the threads on the pin. Markers originally located near the mid-thickness of the plate also were lifted in front of the tool, and those positioned in the path of the pin were transported around the retreating side of the tool and deposited in the same lateral position relative to the centerline (refer to Figure 7.21). Markers positioned near the bottom of the plate passed under the pin without lifting.

Other efforts using markers in the form of plugs of dissimilar aluminum alloys with different responses to etching revealed that material was carried forward on the advancing side of the pin, and backward around the retreating side. In addition, the stir zone took on a more rounded shape when the diameter of the pin was increased, with the maximum stir-zone width occurring near the mid-thickness of the plate. The pin with the largest diameter produced a large bulge on the advancing side of the stir zone.

Composite markers of aluminum-silicon carbide (Al-SiC) and tungsten placed near the plate mid-plane on both the advancing and retreating sides also have been utilized. Results of this study showed that markers are lifted in front of the tool and subsequently carried downward by the threads. In addition, the markers initially on the advancing side were distributed over a wider region.

APPLICATIONS

Friction-stir welding has focused mainly on aerospace and marine applications, with a growing number of applications in the automotive, marine, and rail transportation industries. Some of the aerospace and aircraft projects include construction of the main fuel tank of the space shuttles, the booster core tanks of Delta rockets, floor decking for the C-17 Globemaster cargo aircraft, and fuselage and wing sections of the Eclipse 500 Business Class jet. Marine and rail transportation applications include the welding of fast-ferry decking and sections of subway and rail cars.

The first commercial application of friction stir welding was the joining of 20-m (65.6-ft) sheets of aluminum for the construction of fast ferries in Norway. In the United States, the Boeing Company was the first to use the friction stir welding process, developing its own technology and standards and cooperating with government, university, and industry resources to implement the construction of the Delta series of aerospace launch vehicles. The booster core of the Delta IV, or first-stage launch vehicle, includes a liquid oxygen tank 12-m (44-ft) high, an 8.4-m (28-ft) high fuel tank and a 4.8-m (216-ft) high inter-stage cylinder welded with friction stir welding. Figure 7.22(A) shows a weld qualification tank during production. The tanks are formed using three sheets of 22.22-mm (7/8-in.) thick 2014-T6 aluminum alloy. Tank panels are positioned for welding from the inside, as shown in Figure 7.22(B). Figure 7.22(C) shows the friction stir weld in progress.

Several aircraft builders use friction stir welding, including Eclipse Aviation and Airbus. Airbus used the friction stir process for their models A350, A340-500 and A340-600. Figure 7.23 shows the circumferential aluminum fuselage stiffeners and door doublers of

(A) Weld Qualification Tank in Production

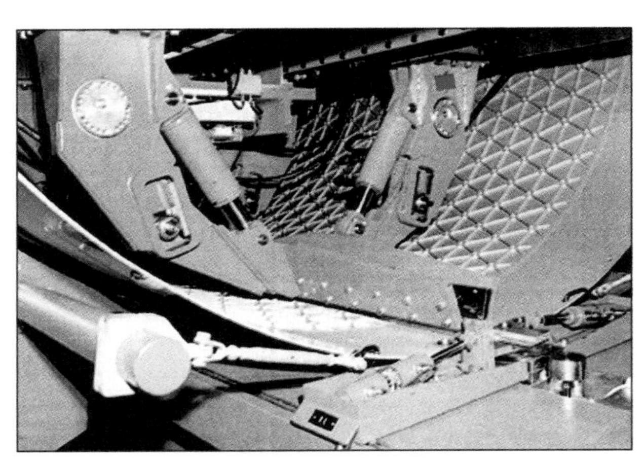

(B) Tank Panels Positioned for Welding from the Inside

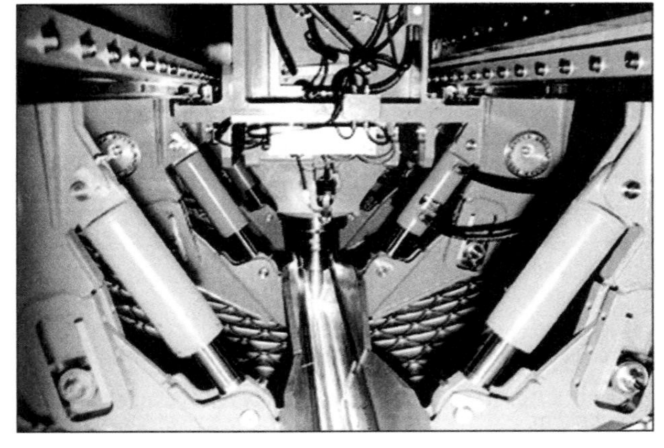

(C) The Friction Stir Weld in Progress

Photographs courtesy of the Boeing Company

Figure 7.22—Friction Stir Welding of 22.22-mm-Thick (7/8 in.) 2014-T6 Aluminum Alloy Panels for the Booster Core Tank of a Space Launch Vehicle

Photographs courtesy of Eclipse Aviation

Figure 7.23—(A) The Eclipse Jet with (B) Circumferential Aluminum Fuselage Stiffeners and (C) Door Doublers Attached with Friction Stir Welding

the Eclipse Business Class jet attached by friction stir welding.

Automotive applications of friction stir welding include suspension components and auto body weldments. Friction stir welding is used for these weldments because it does not react adversely to the coating used for this type of assembly. The Ford Motor Company has used the process to weld the central tunnel assembly of several thousand Ford GT automobiles. The Mazda Motor Corporation used friction stir spot welding to manufacture the automobile doors shown in Figure 7.24.

MECHANICAL PROPERTIES OF FRICTION STIR WELDS

Table 7.1 presents tensile data for base metal and friction stir weld samples in the transverse orientation for a number of common aluminum alloys.[25] The alloys

25. From a database compiled by T. J. Lienert. Base metal properties from ASM Metals Handbook. Materials Park, Ohio: ASM International.

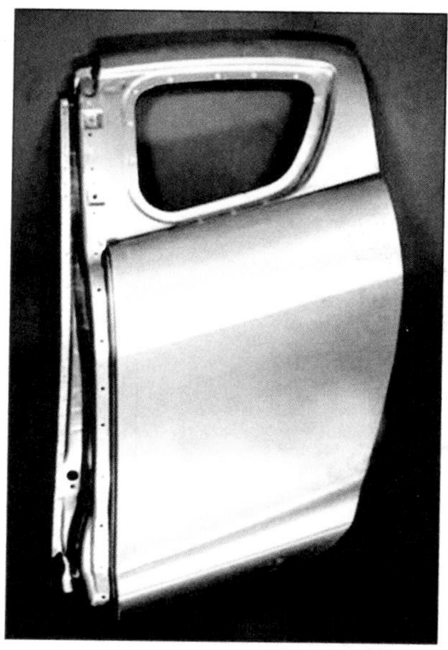

(A) Door Structure Spot-Welded for Impact Stability

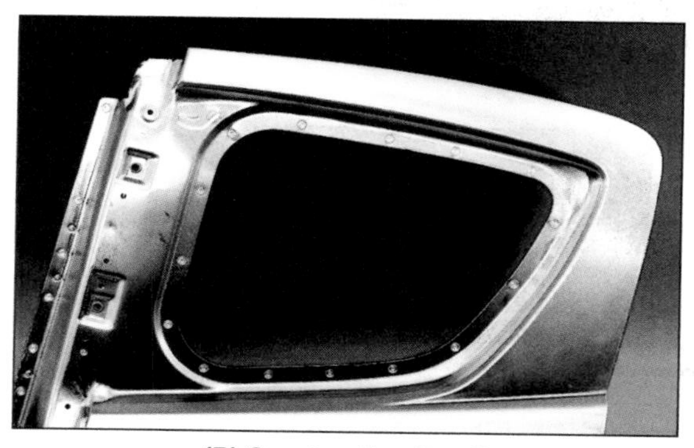

(B) Construction Detail

Photographs courtesy of Mazda Motor Corporation

Figure 7.24—Friction Stir Spot Welds in an Automobile Door

are listed along with their original temper conditions before welding. Yield strength, ultimate tensile strength and the percentage of elongation are provided for the base metal and friction stir weld metal for each alloy. Data for the friction stir welds were taken from reports in open literature and compiled into a database. Friction stir weld properties are shown for the average, standard deviation, and the number of data points (number of welds or tests) used to calculate the average and standard deviation for each alloy.

Comparisons of base metal and weld metal data for a given alloy indicate that yield and tensile strengths for friction stir welds generally exceed 60% to 70% of the yield and tensile strengths of the base metal. It should be noted that the strength data covers a considerable time span and includes data generated early in the development of friction stir welding. Markedly better results can be achieved with current tool designs and modern FSW machines that are stiffer and allow faster tool travel rates and lower heat input.

Referring to Table 7.1, it can be noted that the percent elongation for friction stir welds is always less than the values reported for the base metal. The lower values for friction stir welds (and nearly all welds in general) result from non-uniform elongation in the gauge length during testing. In turn, non-uniform elongation results from differences in microstructure across the weld region, and differences in microstructure result from variations in local thermomechanical histories developed during welding. In aluminum alloys, most of the strain during tensile testing is carried by the heat-affected zone. Partitioning of the strain across smaller lengths shows that the heat-affected zone actually undergoes considerable elongation. Hence, the low elongation values are artificial and should be viewed with these caveats in mind. As an example, Figure 7.25 shows the distribution of strains across the weld region for a tensile sample of a friction stir weld made on 7075-T6. The stir zone and base metal (not shown) undergo little straining, while the heat-affected zone carries considerable strain.

Fatigue data collected for friction stir welds is rather limited. Most fatigue studies have focused on aluminum alloys. In general, friction stir welds perform better in fatigue than fusion welds on the same material, but do not perform as well as that of the base metal. Figure 7.26 presents fatigue data for a friction stir weld on a 2024-T351 aluminum alloy. The testing was performed at a constant stress ratio of 0.1 and stress ranges up to 350 MPa (50.76 ksi).

Data for base metal and tensile properties of friction stir welds for three titanium alloys are presented in Table 7.2. Values are given for commercially pure Ti, the Ti-6Al-4V (Ti-6-4) alloy, and the Ti-15-3 alloy. In Table 7.2, the average standard deviation and number of data points used were included in the friction stir weld data where available. The loss of tensile properties in titanium alloys due to welding is much less than the loss incurred when welding aluminum alloys. These differences stem from the differences in the strengthening mechanisms between titanium and aluminum alloys. In other words, high-yield-strength and high-tensile-strength joint efficiencies can be achieved in Ti alloys with friction stir welds. In certain cases, the joint efficiency may appear to be greater than 100%. It should be noted that the caveats discussed previously for percentage of elongation also apply to titanium alloys.

Table 7.1
Test Data for Tensile Properties of Base Metals and Friction Stir Welds for Common Aluminum Alloys

Base Metal—Typical Properties*					Friction Stir Welds—Sample Properties								
					Yield Strength			Ultimate Tensile Strength			%Elongation		
Alloy	Original Temper	Yield Strength MPa (ksi)	Ultimate Tensile Strength MPa (ksi)	% Elongation	Average MPa (ksi)	Standard Deviation MPa (ksi)	Number of Data Points	Average MPa (ksi)	Standard Deviation MPa (ksi)	Number of Data Points	Average	Standard Deviation	Number of Data Points
2014	T6	414 (60)	483 (70)	13	243 (35.2)	25 (3.6)	2	355 (51.4)	36 (5.2)	9	N/A	N/A	N/A
2024	T3	345 (50)	483 (70)	18	308 (44.7)	34 (5)	15	416 (60.4)	27 (3.9)	23	7.8	3.8	14
2195	T8	N/A	N/A	N/A	253 (36.7)	10 (1.5)	29	396 (57.5)	11 (1.6)	35	8.5	1.6	33
5083	0	145 (21)	290 (42)	22	132 (19.1)	7 (1)	4	306 (44.4)	19 (2.8)	15	22.5	0.7	2
5454	0	117 (17)	248 (36)	22	101 (14.6)	8 (1.1)	3	245 (35.5)	17 (2.4)	5	18.9	4.4	5
5454	H32/H34	207 (30)/ 241 (35)	276 (40)/ 303 (44)	10	119 (17.3)	8 (1.1)	9	251 (36.4)	9 (1.3)	9	19.6	3.8	9
6013	T4	145 (21)	276 (40)	20	191 (27.7)	40 (5.8)	3	306 (44.4)	12 (1.8)	3	4.8	3.1	3
6013	T6	303 (44)	365 (53)	5	265 (38.5)	46 (6.6)	5	316 (45.9)	21 (3.1)	5	6.4	1.4	3
6061	T6	276 (40)	310 (45)	17	161 (23.4)	48 (6.9)	7	221 (32.1)	23 (3.3)	13	10	3.4	7
6082	T4	N/A	N/A	N/A	119 (17.2)	23 (3.3)	5	211 (30.6)	32 (4.6)	7	14.5	6.2	5
6082	T6	262 (38)	310 (45)	6	150 (21.7)	10 (1.4)	8	238 (34.6)	12 (1.7)	9	10.2	6.1	8
7050	T73/T7451	434 (63)	496 (72)	12	444 (64.4)	63 (9.1)	6	473 (68.6)	45 (6.5)	11	6.5	0.5	6
7075	T651	503 (73)	572 (83)	11	334 (48.4)	16 (2.3)	3	439 (63.6)	30 (4.3)	7	5	2.1	3
7075	T73	393 (57)	474 (69)	7	N/A	N/A	N/A	444 (64.4)	54 (7.8)	4	N/A	N/A	N/A

Notes:
— The test configuration was in the transverse orientation.
— N/A = Not available for this table.
— Standard deviation normally applies only to Gaussian distributions and is applicable only for distributions with sample sizes greater than 8; however, standard deviations are included in this table for small sample sizes to provide some indication of the distribution about the mean.

*Typical properties of base metals are from *ASM Metals Handbook*, Materials Park, Ohio: ASM International. Properties of 6013 and 6075 are from Aluminum Standards and Data, 2006, Aluminum Association. Properties of 6082 T6 are from ASTM B-209, ASTM International.

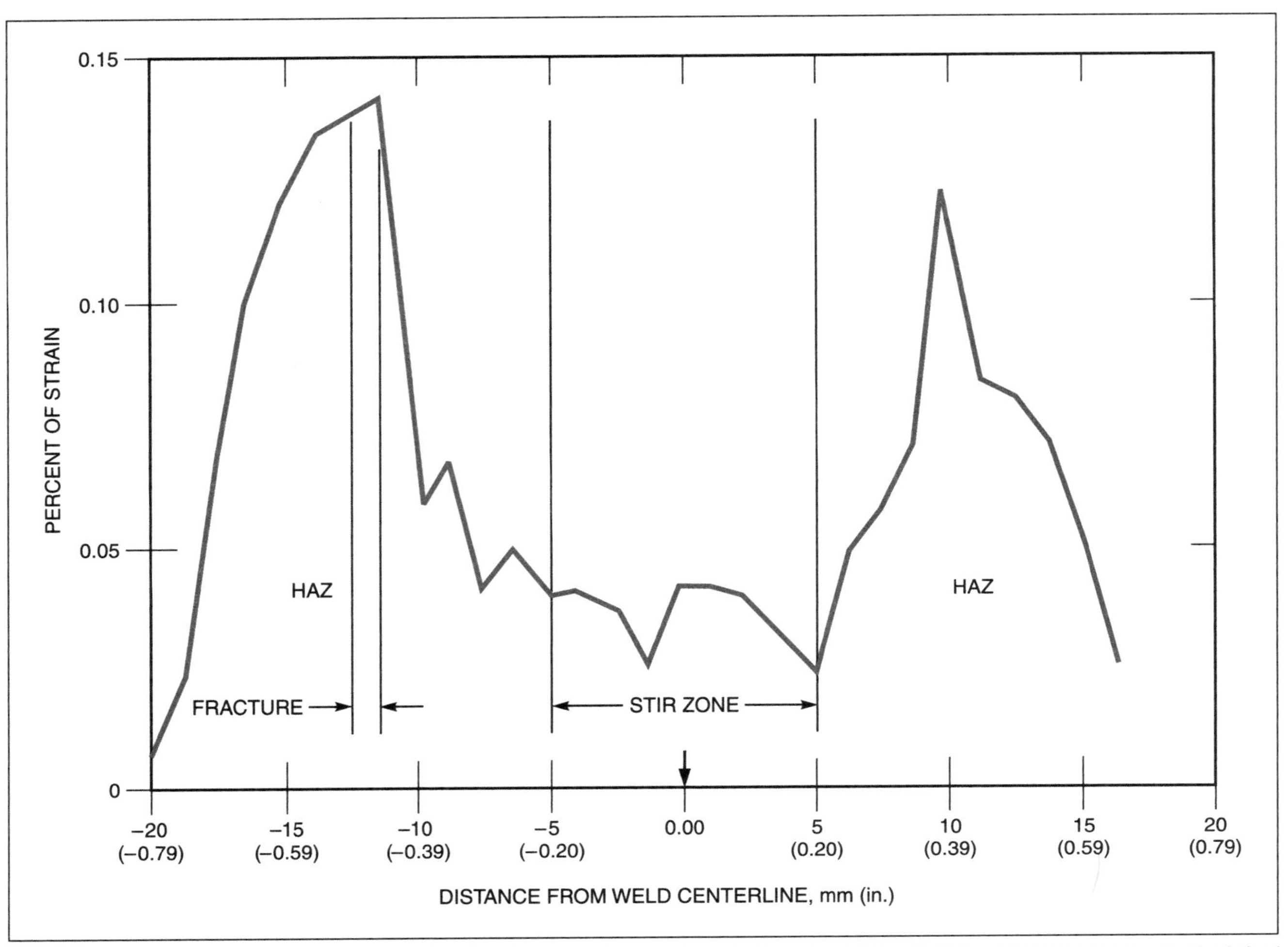

Source: Adapted from Mahoney, M. W., C. G. Rhodes, J. G., Flintoff, W. H, Bingel, and R. A. Spurling, 1998. Properties of Friction Stir Welded 7075-T651 Aluminum. *Metallurgical and Materials Transactions* 29A: 1955–1964.

Figure 7.25—Distribution of Tensile Strains Across the Weld Region in a Sample Friction Stir Weld in 7075-T651 Aluminum Alloy

Table 7.3 lists tensile data for the base metal and friction stir weld metal for 304 stainless steel and 2507 duplex stainless steel. The variations in base metal properties for the 304 alloys are believed to stem from differences in the amount of cold work. (These alloys are not heat-treatable.) It should be noted that the base metal for the 304 alloy has been tested in both transverse and longitudinal orientations. The average standard deviation and number of data points used are included in the friction stir weld data where available. Very good yield and tensile efficiencies can be achieved when welding stainless steel. Concerns discussed previously regarding the interpretation of percent elongation for friction stir welds also apply to stainless steels.

Acceptable yield and tensile joint efficiencies in steels can be achieved with friction stir welding. Tensile data for base metal and friction stir welds of several steels are reported in Table 7.4. The base metal for the DH-36 steel was tested in two orientations; the other steels were tested only in the transverse orientation. If base metal values were not available, minimum values were listed in this table. Charpy V-notch test data for friction stir welds on two steels are reported in Table 7.5.

Comparative data for this test were limited. In general, the average CVN data showed that friction stir welds in these steels provided acceptable toughness. However, in some cases, the scatter in the test data was large, and further study is required to understand the scatter.

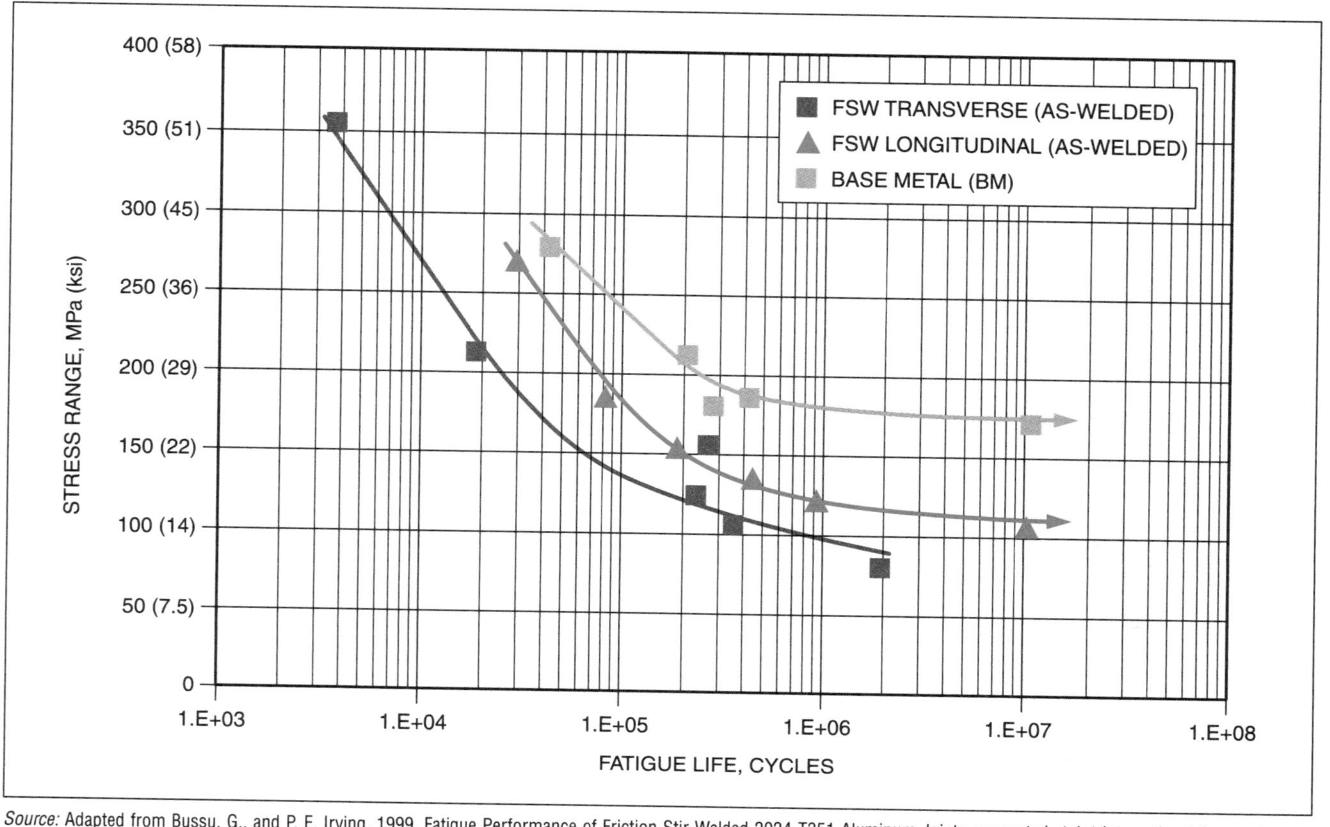

Source: Adapted from Bussu, G., and P. E. Irving, 1999, Fatigue Performance of Friction Stir Welded 2024-T351 Aluminum Joints, presented at 1st International Symposium on Friction Stir Welding, Thousand Oaks, California, Cambridge: TWI.

Figure 7.26—Fatigue Data for a Friction Stir Weld in 2024-T-351 Aluminum Alloy

RESIDUAL STRESSES IN FRICTION STIR WELDS

Friction stir welding normally involves lower heat input relative to fusion welding but requires higher clamping forces. The process also involves more deformation than fusion welding. Residual stresses develop in welded structures due to thermal gradients generated by the transient thermal cycle that occurs as the heat source moves. The workpieces may distort if they are not constrained by fixturing; however, if they are rigidly constrained, they cannot distort and residual stresses develop. The presence of residual stresses may impact subsequent mechanical and corrosion properties of the weldment.[26]

26. The titles, authors, and facts of publication of technical papers and other resources represented in this section are listed in the Bibliography in the Residual Stress section.

Residual stress measurements have been made on friction stir welds, mainly on aluminum alloys. The measurements normally are performed with neutron diffraction, but x-ray diffraction also has been used. In one study, residual stress distributions exhibited a symmetric double-peak profile, or "M" shape, with the greatest tensile stress peaks located in the HAZ regions with lowest hardness. An example of the residual stress distribution in a friction stir weld on an aluminum alloy is shown in Figure 7.27.

To balance these tensile stresses, compressive stresses must exist away from the weld area in the base metal. A lower magnitude of tensile residual stress, or sometimes compressive stresses of a low magnitude, occurred in the stir zone. The greatest residual stresses developed in the longitudinal direction (parallel to the direction of welding) and usually were equal to 50% to 75% of the room-temperature yield stress of the base metal. This condition contrasts with residual stresses found in fusion

Table 7.2
Test Data for Tensile Properties of Base Metals and Friction Stir Welds for Three Titanium Alloys

Base Metal—Typical Properties*				Friction Stir Welds—Sample Properties								
				Yield Strength			Ultimate Tensile Strength			% Elongation		
Alloy	Yield Strength MPa (ksi)	Ultimate Tensile Strength MPa (ksi)	% Elongation	Average MPa (ksi)	Standard Deviation MPa (ksi)	Number of Data Points	Average MPa (ksi)	Standard Deviation MPa (ksi)	Number of Data Points	Average	Standard Deviation	Number of Data Points
CP Ti (Grade 2)	275 (40)	441 (63.9)	25	N/A	N/A	N/A	430 (62.4)	14 (2)	N/A	20	2	N/A
Ti-6-4	913 (132.4)	1014 (147)	12.7	897 (130.1)	0.69 (0.1)	3	958 (138.9)	3 (0.5)	3	12.7	30.5	3
Ti-15-3	765 (111)	769 (111.5)	28	817 (118.5)	28.0 (4)	N/A	822 (119.2)	28 (4)	N/A	6.4	4	N/A

Notes:
— The test configuration was in the transverse orientation.
— N/A = not available for this table.
— Standard deviation normally applies only to Gaussian distributions and is applicable only for distributions with sample sizes greater than 8; however, standard deviations are included in this table for small sample sizes to provide some indication of the distribution about the mean.

*Typical properties of base metals are from *ASM Metals Handbook*, Materials Park, Ohio: ASM International.

Table 7.3
Test Data for Tensile Properties of Base Metals and Friction Stir Welds for 304 Stainless Steel and 2507 Duplex Stainless Steel

Base Metal—Typical Properties*					Friction Stir Welds—Sample Properties								
					Yield Strength			Ultimate Tensile Strength			%Elongation		
Alloy	Sample Orientation	Yield Strength MPa (ksi)	Ultimate Tensile Strength MPa (ksi)	% Elongation	Average MPa (ksi)	Standard Deviation MPa (ksi)	Number of Data Points	Average MPa (ksi)	Standard Deviation MPa (ksi)	Number of Data Points	Average	Standard Deviation	Number of Data Points
304	Transverse	689 (100)	724 (105)	28	689 (100)	N/A	N/A	738 (107)	N/A	N/A	30	N/A	N/A
304	Transverse	296 (43)	669 (97)	NA	395 (57.3)	50 (7.2)	2	707 (102.5)	40 (5.8)	2	N/A	N/A	N/A
304	Transverse	172 (25)	483 (70)	NA	340 (49.3)	32 (4.6)	3	621 (90)	21 (3)	3	N/A	N/A	N/A
304	Longitudinal and all weld metal	186 (27)	483 (70)	40	352 (51)	0	2	652 (94.5)	5 (0.7)	2	65	1.4	2
2507	Transverse	717 (104)	951 (138)	34	740 (107.4)	N/A	N/A	930 (134.9)	N/A	N/A	18	N/A	N/A

Notes:
— N/A = not available for this table.
— Standard deviation normally applies only to Gaussian distributions and is applicable only for distributions with sample sizes greater than 8; however, standard deviations are included in this table for small sample sizes to provide some indication of the distribution about the mean.

*Typical properties of base metals are from *ASM Metals Handbook*, Materials Park, Ohio: ASM International.

Table 7.4
Test Data for Tensile Properties of Base Metals and Friction Stir Welds for Several Steels

Base Metal—Typical Properties*					Friction Stir Welds—Sample Properties								
					Yield Strength			Ultimate Tensile Strength			% Elongation		
Alloy	Sample Orientation	Yield Strength MPa (ksi)	Ultimate Tensile Strength MPa (ksi)	% Elongation	Average MPa (ksi)	Standard Deviation MPa (ksi)	Number of Data Points	Average MPa (ksi)	Standard Deviation MPa (ksi)	Number of Data Points	Average	Standard Deviation	Number of Data Points
1018	Transverse	310 (45)	463 (67.2)	40	332 (48.1)	21 (3)	3	476 (69.1)	21 (3)	3	22	3	3
DH-36[a]	Transverse	345 (50*)	483 (70*)	20*	439 (63.6)	24 (3.5)	7	506 (73.4)	41 (5.9)	7	5.8	2.3	7
DH-36[a]	Transverse	427 (62)	579 (84)	N/A	565 (82)	29 (4.2)	2	624 (90.5)	24 (3.5)	2	N/A	N/A	N/A
DH-36[a]	Longitudinal and all weld metal	345 (50*)	483 (70*)	20*	610 (88.5)	5 (0.7)	2	800 (116.1)	39 (5.7)	2	9.3	6	2
HSLA-65	Transverse	448 (65*)	538 (78*)	20*	470 (68.1)	45 (6.5)	4	571 (82.8)	2 (0.3)	4	24.1	6.8	4
HSLA-65	Transverse	448 (65*)	538 (78*)	20*	527 (76.5)	14 (2.1)	2	645 (93.5)	14 (2.1)	2	25.9	1.4	2
S-355	Transverse	343 (49.7)	545 (79.1)	31	345 (50.1)	10 (1.4)	5	533 (77.3)	4 (0.6)	5	18.8	0.8	5
Q&T2[b]	Transverse	1427 (207)	1731 (251)	11.3	1041 (151)	21 (3)	N/A	1234 (179)	21 (3)	N/A	2.6	3	N/A
Q&T3[c]	Transverse	1489 (216)	1765 (256)	13	1124 (163)	N/A	N/A	1213 (176)	N/A	N/A	2.8	N/A	N/A

Notes:
a. Samples contained small discontinuities.
b. 500 BHN quenched and tempered steel (1/4 in. thick plate).
c. 500 BHN quenched and tempered steel (1/2 in. thick plate).
— N/A = not available for this table.
— Standard deviation normally applies only to Gaussian distributions and is applicable only for distributions with sample sizes greater than 8. However, standard deviations are included in this table for small sample sizes to provide some indication of the distribution about the mean.

*Typical properties of base metals are from *ASM Metals Handbook*, Materials Park, Ohio: ASM International.

Table 7.5
Charpy V-Notch Toughness Data for Friction Stir Welds on Selected Steel Alloys

	Energy Absorbed in Joules (J [ft-lb]) at –40°C (–40°F)		
Alloy	Base Metal	Weld Metal	Heat-Affected Zone
HSLA-65[a]	9 (7)	11 (8)	15 (11)
HSLA-65[b]	28 (21)	26 (19)	41 (30)
HSLA-65[b]	207 (153)	47 (35)	N/A
S355	17 (13)	18 (13)	19 (14)

Notes:
a. 6.4 mm (1/4 in.) plate.
b. 12.7 mm (1/2 in.) plate.
c. N/A = not available for this table.

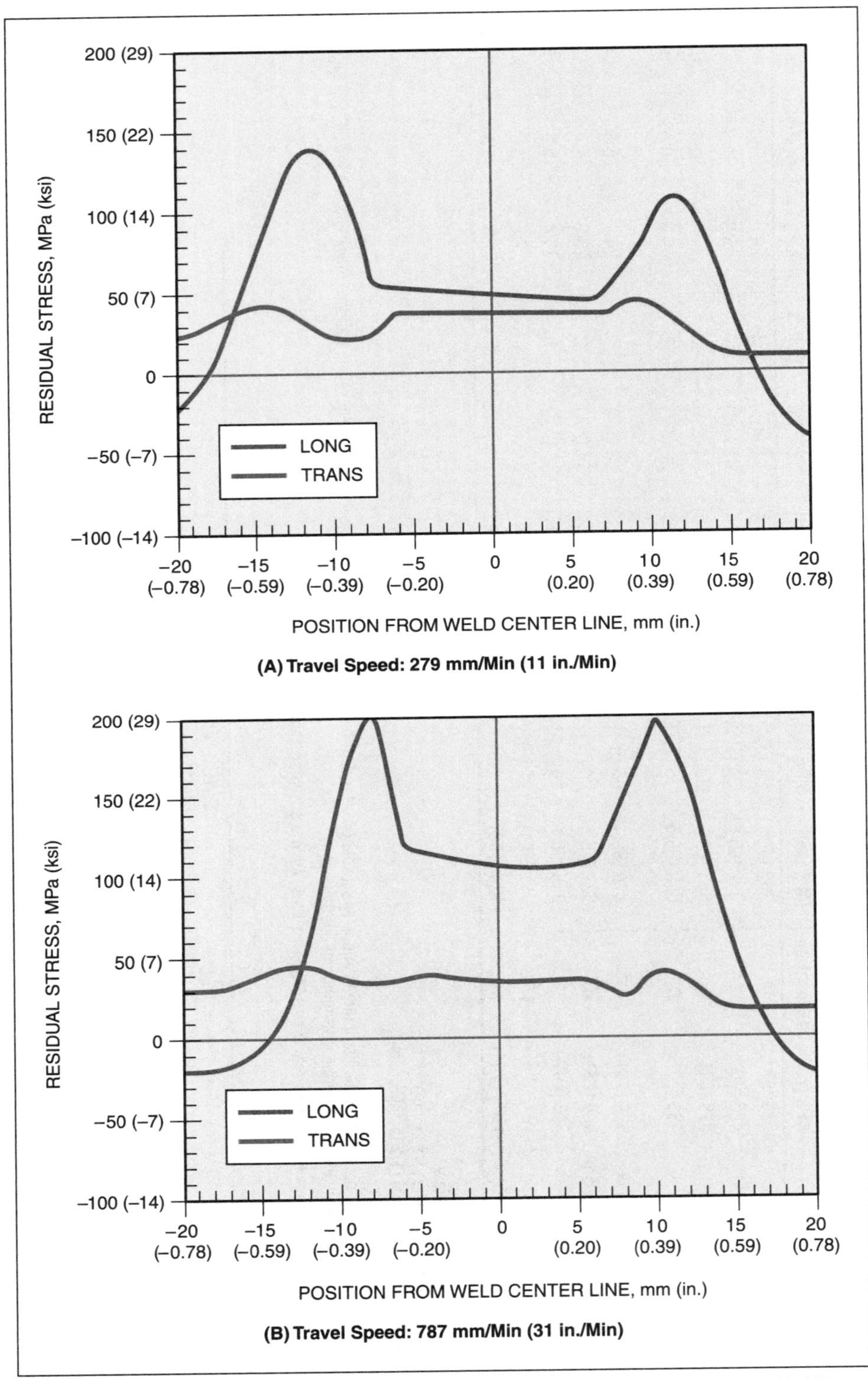

Source: X. L. Wang, Z. Feng, S. A. David, S. Spooner, and C. R. Hubbard, 2000, Nettron Diffraction Study of Residual Stresses in Friction Stir Welds, *Proceedings of Sixth Annual Conference on Residual Stresses*, Oxford, UK, London: IOM Communications, Ltd. pp. 1408–1414.

Figure 7.27—Residual Stress Distribution in Friction Stir Welds on AA6061-T6 at Varied Travel Speeds

welds, where longitudinal stresses equal to the room-temperature yield stress of the base metal are the norm. Transverse residual stresses in friction stir welds (perpendicular to the welding direction) were typically much lower than the longitudinal stresses, with the largest transverse stresses occurring at the mid-thickness of the plate.

The magnitude of the residual stresses is affected by welding parameters, with higher stresses associated with lower heat input. Welds with higher heat input (for example, low travel speeds and high rpm) widened the extent of the heat-affected zone, causing redistribution of stresses with lower magnitudes for a given location. Although residual stresses were lower than those found in fusion welds, they were high enough to affect fatigue crack growth rates.

A few residual stress measurements have been made on friction stir welds in metals other than aluminum alloys. Results of measurements of residual stress in friction stir welds on Type 304 stainless steel were similar to those usually found in fusion welds, with large longitudinal tensile stresses in the stir zone close to yield stress, and compressive stresses below the yield point in the heat-affected zone and base metal. The double-peak stress profile common to friction stir welds on aluminum alloys was not found in the 304 steel. Measurements of friction stir welds of Ti-6Al-4V showed that the greatest longitudinal stresses were in the stir zone near the top and bottom surfaces. These stresses were compressive in nature but well below the yield stress. A lower magnitude of tensile stresses was found within the plate thickness. The greatest magnitude of residual stress was found in the stir zone.

CORROSION ISSUES

Results of corrosion tests have been reported for friction stir welds in 2xxx, 5xxx, 6xxx, and 7xxx aluminum alloys. The bulk of the testing was on the 7xxx-series alloys due to the importance of these alloys in aerospace applications. Techniques used for corrosion evaluations included testing for susceptibility to stress corrosion cracking, pitting corrosion, exfoliation corrosion, polarization, salt spray, and corrosion fatigue. The corrosion resistance of friction stir welds normally was superior to that of fusion welds due to the elimination of solidification segregation.[27]

The friction stir welding of precipitation-hardened aluminum alloys involves heating and cooling under non-equilibrium conditions, which produce microstructures different from those of the base metal. Modifications to the microstructure involve coarsening and/or dissolution of the precipitates, usually in the HAZ or TMAZ. These changes can lead to increased corrosion susceptibility in different regions of the weld, depending on the alloy, starting temper, testing environment, and heat input. Generalization of the results of corrosion testing on all alloys is difficult because of metallurgical differences between the alloys as well as differences in heat input.

As an example, corrosion in friction stir welds on 7xxx alloys has been reported to occur due to the formation of precipitate-free zones along grain boundaries and copper depletion near the boundaries. This condition led to local compositional differences and galvanic potential between the grain boundaries and the matrix. Postweld heat treatment was effective in restoring corrosion resistance by allowing the diffusion of copper back to the precipitate-free regions, but also led to reduced overall strength, possibly with large reductions that were unacceptable. Tool designs influence the heat input and thermal cycles and have an effect on subsequent corrosion properties. References in the Bibliography of this chapter are arranged in topical order and will provide resources for detailed information for specific alloys and testing methods.

WELD QUALITY

The types of discontinuities found in FSW are quite different from those encountered in fusion welds. Because FSW is a solid-state joining process, solidification-related discontinuities, such as hot cracking, porosity, and slag inclusions are eliminated. However, other types of discontinuities may be produced. The discontinuities found in friction stir welds are defined in this section, and nondestructive examination methods for detecting them are discussed.

DISCONTINUITIES

The American Welding Society defines a discontinuity as an interruption of the typical structure of a material, such as a lack of homogeneity in its mechanical, metallurgical or physical characteristics. A defect is defined as a discontinuity or discontinuities that by nature or accumulated effect render a part or product unable to meet minimum applicable acceptance standards or specifications; thus a discontinuity is not necessarily considered a defect. The use of the term *defect* indicates that the part or product should be rejected. The acceptance criteria for welded parts or products typically are outlined in pertinent standards and specifi-

27. The titles, authors, and facts of publication of technical papers and other resources represented in this section are listed in the Bibliography in the Corrosion section.

cations documents. At the time of publication of this chapter, no nationally or internationally recognized standards or specifications related to friction stir welding have been established, although many corporations and government agencies have developed their own internal specifications. However, a document entitled *Specification for Friction Stir Welding of Aluminum Alloys for Aerospace Applications* is currently being prepared for publication by the American Welding Society.[28]

Several types of FSW discontinuities have been identified; some of these are illustrated in Table 7.6. Section 1 (7.6-1) of this two-part table shows discontinuities that are related to the welding process or procedure; Section 2 (7.6-2) shows discontinuities that are metallurgical in nature.

Process and Procedure Discontinuities

Incomplete root penetration in a friction stir weld, as shown in Table 7.6-1(A) and 7.6-1(B), results from insufficient pin length of the tool or insufficient plunge depth. As a consequence, the deformation in the stir zone does not extend deeply enough, and an unbonded region remains at the weld root. Small unbonded regions, such as those shown in 7.6-1(B), may be difficult to detect with NDE methods.

Joint mismatch, 7.6-1(C), can occur if fixturing is not properly made, if clamping pressures are too low or if the workpieces are warped. Excess weld flash, shown in 7.6-1(D), occurs if too great a tool shoulder plunge depth is used (with position control) or if too large a plunge force is used (with load control). An example of excess flash is shown in Figure 7.28. The underfill discontinuity, 7.6-1(E) often occurs simultaneously with excess weld flash and leaves the stir zone thinner than the base metal.

Metallurgical Discontinuities

Most metallurgical discontinuities involve incomplete bonding of the joint surfaces. Incomplete bonding is difficult to detect with most NDE methods and may be the initiation site for fatigue cracks and corrosion. Referring to Table 7.6-2, the so-called *kissing bond* discontinuity illustrated in 7.6-2(A) also involves incomplete bonding along a thin ligament at the weld root. An example of a kissing bond discontinuity is shown in Figure 7.29. It should be noted, however, that the kissing bond differs from incomplete penetration in that the tool may be sufficient in length and the deformation in the stir zone may extend completely through the plate thickness. It occurs when oxide films from the joint surfaces are not completely dispersed during welding, leaving a surface on which bonding is interrupted intermittently by a trail of oxide particles. It can also occur as a result of insufficient forge pressure or insufficient machine stiffness, which allows the tool to be deflected from the joint line leaving undispersed oxide films.

A joint-line remnant discontinuity, shown in Table 7.6-2(B) is sometimes referred to as a *lazy-S* discontinuity because of its shape. It is essentially an extended kissing bond and forms for the same reasons as the kissing bond. This discontinuity can be avoided by removal of surface oxides prior to welding or by a change in tool design.

Voids or internal cavities, shown in Table 7.6-2(C), (sometimes called *wormholes*),[29] are discontinuities caused by incomplete bonding that stems from insufficient forging pressure or incorrect tool design. A cross section of an internal cavity discontinuity is shown in Figure 7.30(A). Under extreme conditions, such as very high tool rotation speeds, an internal wormhole discontinuity may extend to the surface of the weld leaving a trench. An example of this discontinuity is illustrated in Figure 7.30(B).

The hooking discontinuity shown in Table 7.6-2(D) occurs in lap welds when the joint interface adjacent to the stir zone is pulled up or down, thereby decreasing the effective sheet thickness. This effect may impact the mechanical response of the weld.

One type of surface discontinuity not included in Table 7.6 is surface galling. Surface galling, as shown in Figure 7.31, may occur during the friction stir welding of some 2xxx and 7xxx aluminum alloys as a result of partial liquation. This discontinuity results from excessive heat input and can be eliminated by decreasing the tool rpm.

NONDESTRUCTIVE EXAMINATION OF FRICTION STIR WELDS

The various discontinuities found in friction stir welds require modifications of the examination techniques used for fusion welds. Large incomplete-bonding discontinuities such as voids can be readily found using standard radiography or ultrasonic examination methods. Unfortunately, kissing bonds and tight bonds with incomplete joint penetration are closed and are difficult to detect using standard nondestructive examination (NDE) methods such as radiography, die-penetrant and ultrasonic techniques. Some progress in identifying these discontinuities has been made using a phased-array ultrasonic technique. A relatively new technique using

28. American Welding Society, D17 Committee on Welding in the Aircraft and Aerospace, Forthcoming, *Specification for Friction Stir Welding of Aluminum Alloys for Aerospace Applications,* D17.3/D17.3M: 200x, Miami: American Welding Society.

29. See Reference 23.

Table 7.6
Types of Discontinuities in Friction Stir Welds

I. Welding Process or Procedure Related Discontinuities

(A) Incomplete Penetration

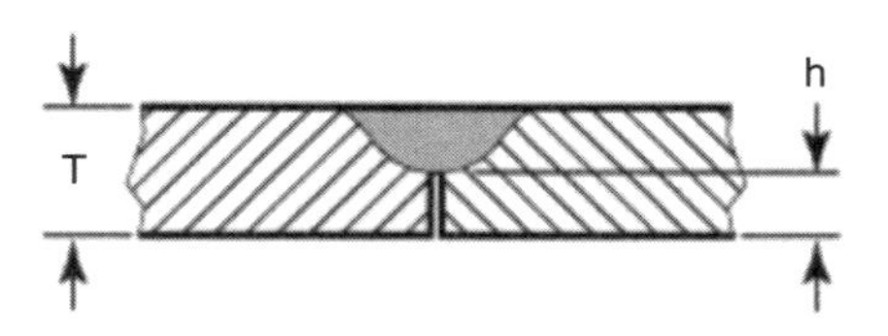

(B) Incomplete Penetration

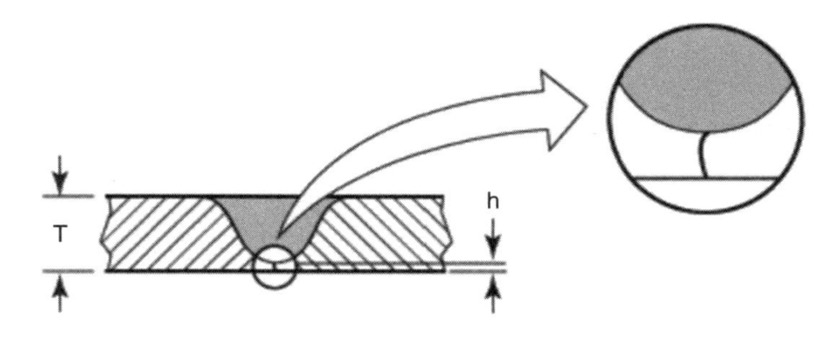

(C) Joint Mismatch

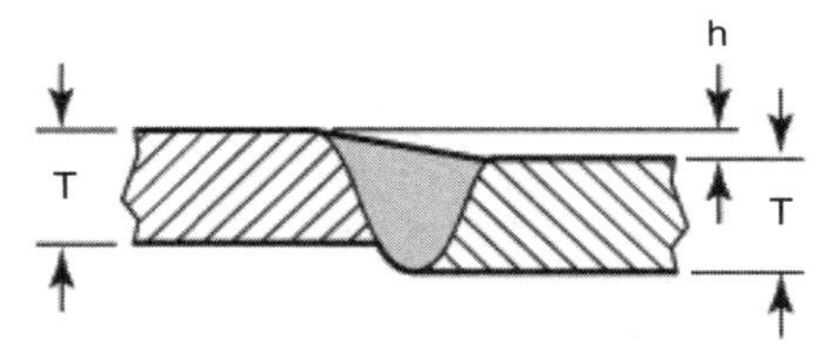

(D) Excess Weld Flash

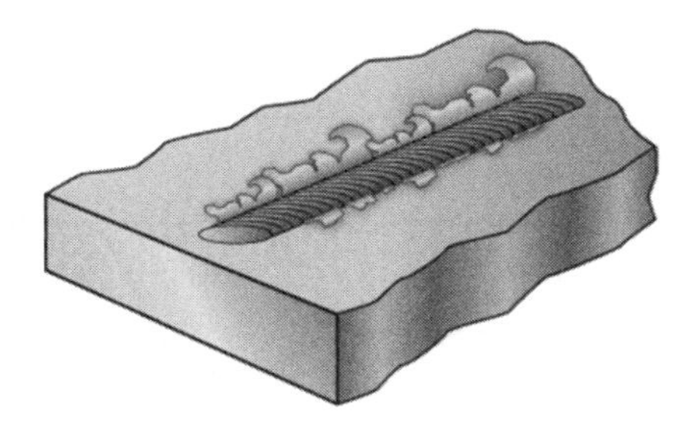

(E) Underfill

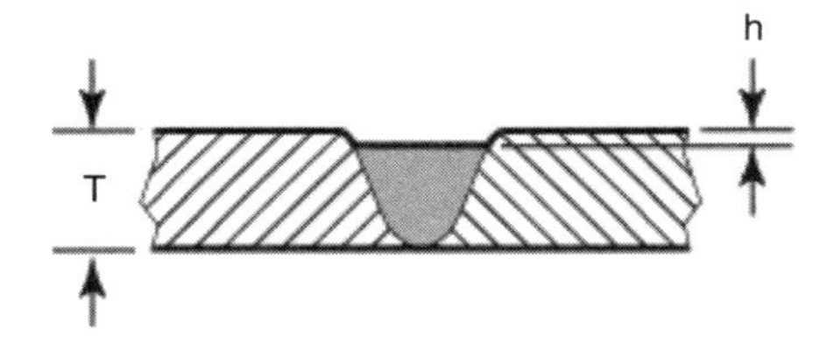

(Continued)

Table 7.6 (Continued)
Types of Discontinuities in Friction Stir Welds

II. Metallurgical Discontinuities

(A) Kissing Bond

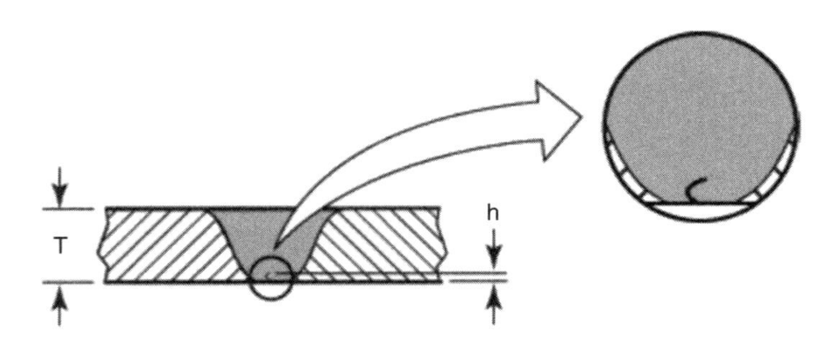

(B) Joint Line Remnant

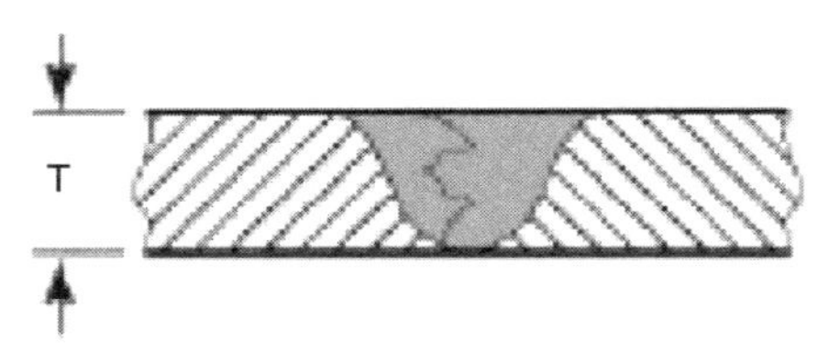

(C) Cavity

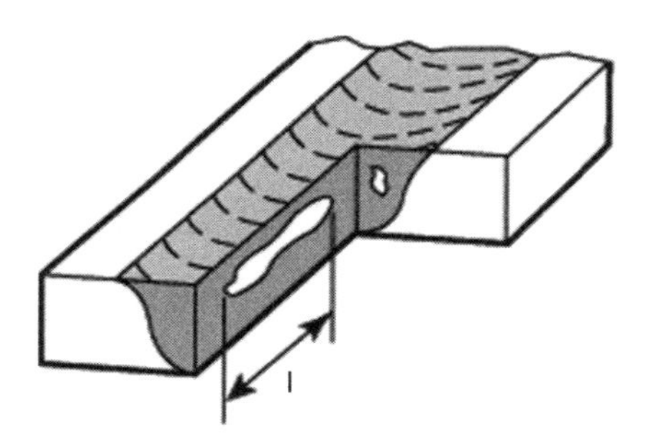

(D) Hooking

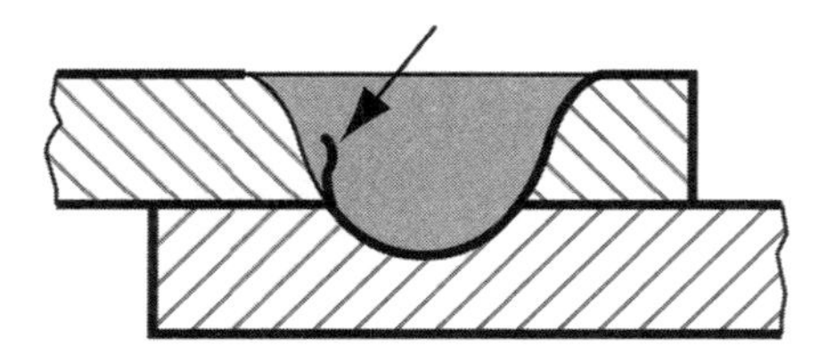

Key:
T = nominal thickness of the base metal
h = height of a discontinuity
l = length of a cavity in the longitudinal direction of the weld

Note: A cavity can also break through the workpiece surface.

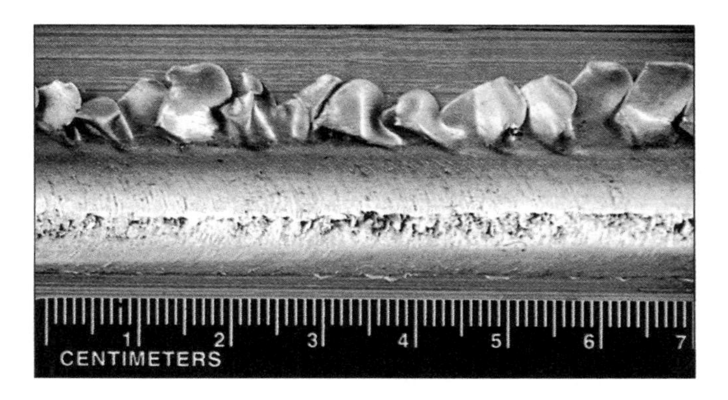

Figure 7.28—Excess Flash on a Friction Stir Weld on 6061-T6 Aluminum Alloy

Source: Lamarre, A., M. Moles, and O. Pupuis, 2004. Materials Solutions '03: Joining of Advanced and Specialty Materials VI; Pittsburgh, 13-15 Oct. 2003. Materials Park, Ohio: ASM International pp. 96–103.

Figure 7.29—Incomplete Bonding ("Kissing Bond") Discontinuity on a Friction Stir Weld

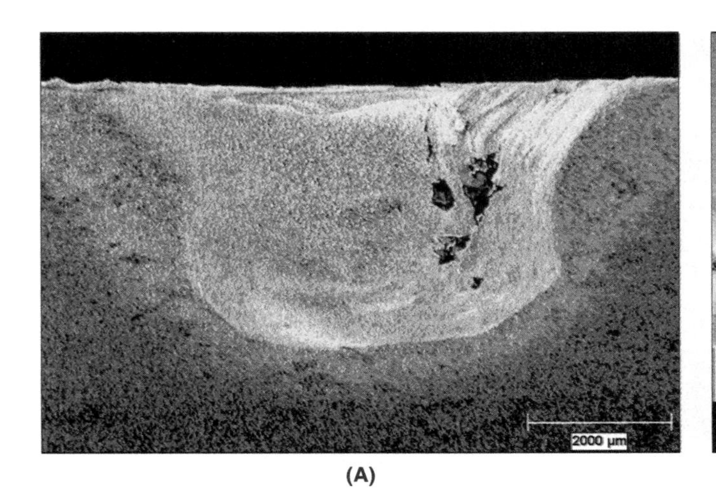

(A)

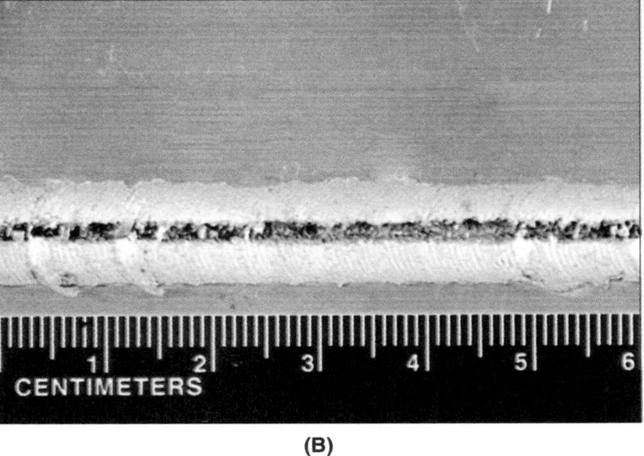

(B)

Figure 7.30—(A) Internal Cavity ("Wormhole") Discontinuity in a Friction Stir Weld on 6061-T6 and (B) Surface-Breaking Discontinuity Produced Using Very High Tool Revolution Speeds

Photograph courtesy of NASA Marshall Space Center

Figure 7.31—Galling Discontinuity in a Friction Stir Weld

ultrasonic attenuation measurements with the phased-array ultrasonic examination has shown greater success in finding kissing bonds. In addition, an eddy current method using a magneto winding magnetometer has been employed to map conductivities and find incomplete-penetration discontinuities in friction stir welds. Considerably more work is needed to develop viable inspection methods for friction stir welds.[30]

ECONOMICS

While cost-benefit analyses of friction stir welding probably have been compiled by major users of the process during the 1990s, no hard data on the subject is known to have been published. Hence, it is difficult to state actual savings associated with friction stir welding. However, friction stir welding provides several advantages over other fabrication methods. For example, the process allows the welding of high-strength aluminum alloys that are not readily weldable with fusion processes. This capability allows the use of greater amounts of high-strength aluminum alloys and permits down-gauging of materials, with commensurate weight and cost savings. Friction stir welding also can be used to fabricate wide panels of aluminum that are not available from suppliers of extruded aluminum. An important consideration is that friction stir welding normally provides better mechanical and corrosion properties for a given alloy than fusion welding.

30. The titles, authors, and facts of publication of technical papers and other resources represented in this section are listed in the Bibliography in the Nondestructive Inspection section.

Welds produced with friction stir welding have lower values of residual stresses and less distortion, thus reducing costs associated with stress relief and straightening. Friction stir welding uses no fluxes or filler metals, systemically reducing the cost of associated supplies and consumables required by other processes. Shielding gases typically are not used to weld aluminum alloys. The cost of FSW tools used for aluminum alloys is minimal, and a single tool can be used to make more than 1000 m (3280 ft) of welds.

Labor costs are reduced because less skill is needed to operate a friction stir welding machine than is required for other welding processes. Most friction stir welding machines can be run with little or no operator input, other than loading the workpieces into the machine. Fitup requirements are less stringent, and joint preparation is economical because friction stir welding requires less machining, cleaning, and etching than other processes.

When compared to riveting, the cost of rivets and the labor-intensive riveting process are eliminated. Startup costs and inspection costs for friction stir welding are lower than for riveting. Inspection procedures also are less involved with FSW with respect to riveting.

From an energy standpoint, the friction stir welding process is more efficient for aluminum alloys than resistance spot welding due to the low resistance of these alloys, and is more energy-efficient than laser beam welding due to the high reflectivity of aluminum. Higher travel speeds can be achieved with friction stir welding than those attained with arc welding.

Friction stir welding is often selected as the joining process for a given application, with the economics of its selection driven by savings in other manufacturing processes. Cost items such as preweld and postweld machining are lower due to reduced distortion. Material costs are often lower when using FSW compared to fusion welding. This is in part due to the higher mechanical properties in the as-welded condition, which allows the use of thinner material for a given design load.

Although the initial capital cost of friction stir welding machines often is higher than machines used in other welding processes, such as arc welding, the return on investment typically can be realized in one or two years, depending on production volume.

However, several drawbacks currently are associated with friction stir welding. The process is still proprietary and licensing costs will continue until the patent expires. The initial machine cost is high and there are a limited number of machine manufacturers. Moreover, the knowledge base is limited among manufacturing engineers. Finally, specification and standards for friction stir welding need to be established before the process is more broadly accepted. These problems will resolve as the process matures and gains greater popularity.

SAFE PRACTICES

Friction stir welding machines are similar to milling, drilling, and machining equipment. In some cases, they are similar to hydraulic presses, with the common feature that one workpiece is forced against the other with high loads. Thus, the safe practices for machining equipment and power presses should be used as guides for the operation of friction stir welding machines. This information is published by the Occupational Safety and Health Administration (OSHA), United States Department of Labor, in the Code of Federal Regulations (CFR), Title 29, Part 1910.210, Subpart O.[31] Typical hazards of friction stir welding are associated with high rotational speeds of components, flying particles, and high levels of noise.

Appendices A and B of this volume contain lists of health and safety standards, codes, specifications, books and pamphlets. The publishers and the facts of publication also are listed. The American National Standards Institute safe practices guide, *Safety in Welding, Cutting, and Allied Processes* ANSI Z49.1, published by the American Welding Society is available on the Internet.[32]

Machines should be equipped with appropriate mechanical guards, shields, and electrical interlocks. These devices should be designed to prevent operation of the machine when the work area, rotating drive, or force system is accessible to the operator or other personnel. Operation manuals and recommendations for the safe installation, operation, and maintenance of machinery are supplied by the equipment manufacturers and should be strictly followed.

Operating personnel should wear appropriate eye protection. Ear protection should be worn to guard against the potentially high noise levels produced during friction stir welding. Welding operators must wear safety apparel commonly used with machine tool operations. Applicable codes published by OSHA in the Code of Federal Regulations (CFR), Title 29, Part 1910 should be strictly observed.[33]

31. Occupational Safety and Health Administration (OSHA), 2005, *Occupational Safety and Health Standards for General Industry, in Code of Federal Regulations (CFR)*, Title 29 CFR 1910.217, Subpart O, Washington D.C.: Superintendent of Documents, U.S. Government Printing Office.
32. American National Standards Institute (ANSI) Z.49 Committee, 2005, *Safety in Welding, Cutting, and Allied Processes,* Miami: American Welding Society, available on line at http://www.aws.org
33. See Reference 34.

CONCLUSION

Friction stir welding is a newly developed solid-state process that is safe and economical for making welds that can be used to join a wide variety of joints and material combinations. Most of the advantages and disadvantages that apply to other solid state forge welding process also apply to friction stir welding. As FSW is a new process, new industrial applications and capabilities are being discovered even as this book goes to press. In the future, FSW may be the process of choice for many applications.

BIBLIOGRAPHY

American National Standards Institute (ANSI) Accredited Z49 Committee. 2005. *Safety in welding, cutting, and allied processes.* ANSIZ41.1:2005. Miami: American Welding Society. Available on line at www.aws.org.

American Welding Society (AWS) Committee on Definitions and Symbols. AWS A3.0. 2001. *Standard welding terms and definitions.* Miami: American Welding Society.

American Welding Society (AWS) Welding Handbook Committee, W. R. Oates, ed. 1996. *Welding handbook*, 8th edition, Volume 3, *Materials and applications, Part 1.* Miami: American Welding Society. Chapter 1.

American Welding Society (AWS) Welding Handbook Committee. 1996. *Welding handbook*, Vol. 3, *Materials and applications, Part 1.* Miami: American Welding Society. Chapter 3.

American Welding Society (AWS) Welding Handbook Committee. W. R. Oates, ed. 1998. *Welding handbook,* Volume 4, *Materials and applications, Part 2.* Miami: American Welding Society. Chapter 9.

American Welding Society (AWS) Welding Handbook Committee, W. R. Oates and A. M. Saitta, ed. 1998. *Welding handbook,* 8th ed., Vol. 4. *Materials and applications,* Miami: American Welding Society. Chapters 1–5.

American Welding Society, D17 Committee on Welding in the Aircraft and Aerospace. Forthcoming. *Specification for friction stir welding of aluminum alloys for aerospace applications.* Miami: American Welding Society.

Linnert, G. E. 1994. A. Saitta, ed. *Welding metallurgy,* Vol. 1, 4th ed. Miami: American Welding Society. 81.

Occupational Safety and Health Administration (OSHA). 2005. *Occupational safety and health standards for general industry,* in *Code of Federal Regulations (CFR),* Title 29 CFR 1910.217, Subpart O. Washington D.C.:

Superintendent of Documents, U.S. Government Printing Office.

Suits, M. W., J. Leak, and D. Cox. 2003. *Materials solutions '03: Joining of advanced and specialty materials VI*. Pittsburgh, Pennsylvania: ASM International. 104–111.

Seidel, T. U., and A. P. Reynolds. 2003. *Science and technology of welding and joining* 8(3): 175.

Vill, V. I. 1962. *Friction Welding of Metals. Translated from Russian.* Miami: American Welding Society. 9–38.

ALLOYS

Aluminum Alloys

Feng, Z., and J. E. Gould. 1998. Heat flow model for friction stir welding of aluminum alloys. *Journal of materials processing & manufacturing science.* 7(10)2: 185–194.

Feng, Z., J. E. Gould, and T. J. Lienert. 1998. *Second symposium on hot deformation of aluminum alloys II,* Rosemont, Illinois. Warrendale, Pennsylvania: The Minerals, Metals and Materials Society (TMS). 159–167.

Frigaard, O., O. Grong, and O. T. Middling. 2000. A process model for friction stir welding of age hardening aluminium alloys. *Metallurgical and materials transactions.* Vol. 32A, No. 5. 1189–1200.

Strangwood, M., J. E. Berry, D. P. Cleugh, A. J. Leonard, A. J., and P. L. Threadgil. 1999. Characterization of the thermomechanical effects on microstructural development in friction stir welded age hardening aluminium based alloys. *Proceedings of the 1st international conference on friction stir welding*. Session 11, Paper 3. Cambridge, U.K.: TWI.

1100 Alloys

Murr, L. E., G. Liu, and J. C. McClure. 1997. Dynamic recrystallization in friction stir welding of aluminum alloy 1100. *Journal of materials science* 16: 1801–1803.

Sato, Y. S., Y. Kurihara, S. H. C. Park, H. Kokawa, and N. Tsuji. 2004. Friction stir welding of ultrafine grained Al alloy 1100 produced by accumulative roll-bonding. *Scripta materialia* 50(1): 57–60.

2014 and 2024 Alloys

Charit, I., and R. S. Mishra. 2003. High strain rate superplasticity in a commercial 2024 Al alloy via friction stir processing. *Materials science and engineering: a.* 359(1–2): 290–296.

Heurtier, P., M. J. Jones, C. Desrayaud, J. H. Driver, and F. Montheillet. 2003. Thermomechanical conditions and resultant microstructures in friction stir welded 2024 aluminum. *Materials science forum.* Vol. 426–432, Part 4. 2927–2932.

Sutton, M. A., B. Yang, A. P. Reynolds, and J. Yan. 2004. Banded microstructure in 2024-T351 and 2524-T351 aluminum friction stir welds, part 1, metallurgical studies. *Materials science and engineering: a.* 364(1–2): 55–65.

Sutton, M. A., B. Yang, A. P. Reynolds, and J. Yan. 2004. Banded microstructure in 2024-T351 and 2524-T351 aluminum friction stir welds, part 2, metallurgical characterization. *Materials science and engineering: a.* 364(1–2): 66–74.

Sutton, M. A., B. Yang, A. P. Reynolds, and R. Taylor. 2002 Microstructural studies of friction stir welds in 2024-T3 aluminum. *Materials science and engineering: a.* (Switzerland). 323(1–2): 160–166.

2219, 2519, and 2195 Alloys

Cao, G., and S. Kou. 2005. *Welding journal.* Friction stir welding of 2219 aluminum: behavior of θ (Al2Cu) particles. 84(1). 1-s–8-s.

Colligan, K., P. J. Konkol, J. R. Pickens, I. Ucok, and K. McTernan. 2002. Friction stir welding of thick sections 5083-H131 and 2195-T8P4 aluminum plates. *Proceedings of the 3rd international symposium on friction stir welding.* 27–28 Sept. 2001. Kobe, Japan. Cambridge: TWI.

Fonda, R. W. and S. G. Lambrakos. 2003. Microstructural analysis and modeling of the heat-affected zone in Al 2519 friction stir welds. *Proceedings, 6th international conference: trends in welding research.* Pine Mountain, Georgia. Cambridge: TWI. 241–246.

Kinchen, D. G., Z. Li, and G. P. Adams. 1999. Mechanical properties of friction stir welds in Al-Li 2195-T8. *Proceedings of the first international conference on friction stir welding.* Session 9, Paper 2. Cambridge: TWI.

Li, Z., W. J. Arbegast, P. J. Hartley, and E. I. Meletis. 1999. Microstructure characterisation and stress corrosion evaluation of friction stir welded Al 2195 and Al 2219 alloys. *Trends in welding research: proceedings of the 5th international conference on trends in welding research.* Pine Mountain, Georgia. (June 1998): 568–573. Materials Park, Ohio: ASM International.

5083 Alloys

Charit, I and R. S. Mishra. 2004. Evaluation of microstructure and superplasticity in friction stir processed 5083 Al alloy. *Journal of materials research* 19(11): 3329–3342.

Colligan, K., P. J. Konkol, J. R. Pickens, I. Ucok, and K. McTernan. 2002. Friction stir welding of thick sections 5083-H131 and 2195-T8P4 aluminum plates. *Proceedings of the 3rd international symposium on friction stir welding*. 27–28 Sept. 2001. Kobe, Japan. Cambridge: TWI.

Shigematsu, I., Y.J. Kwon, K. Suzuki, T. Imai, and N. Saito. 2003. Joining of 5083 and 6061 aluminum alloys by friction stir welding. *Journal of materials science letters* 22(5): 353–356.

Svensson, L. E., L. Karlsson, H. Larsson, B. Karlsson, M. Fazzini, and J. Karlsson. Microstructure and mechanical properties of friction stir welded aluminium alloys with special reference to AA 5083 and AA 6082. *Science and technology of welding and joining* 5(5): 285–296.

Vaze, S. P., J. Xu, R. J. Ritter, K. J. Colligan, J. J. Fisher, and J. R. Pickens. 2003. Friction stir processing of aluminum alloy 5083 plate for cold bending. *Materials science forum* 426–432, Part 4: 2979–2986.

6061, 6063 and 6082 Alloys

Chao, Y. J., and X. Qi. 1998. Thermal and thermomechanical modeling of friction stir welding of aluminum alloy 6061-T6. *Journal of materials processing & manufacturing science*. 7(10). 215–233.

Karlsson, L., L. E. Svensson, and H. Larsson. 1998. Characteristics of friction stir welded aluminum alloys. *Proceedings of the fifth international conference on trends in welding research*. Materials Park, Ohio: ASM International. 574–579.

Lee, W-B., Y-M. Yeon, and S-B. Jung. 2004. Mechanical properties related to microstructural variation of 6061 Al alloy joints by friction stir welding. *Materials transactions* 45(5): 1700–1705.

Lee, W-B., H-S. Jang, Y-M. Yeon, and S-B. Jung. Effect of PWHT on behaviors of precipitates and hardness distribution of 6061 Al alloy joints by friction stir welding method. *Materials science forum* 449–452, Part 1: 601–604.

Lienert, T. J., R. J. Grylls, J. E. Gould, and H. L. Fraser. 1998. Deformation microstructures in friction stir welds on 6061-T651. *Proceedings, second symposium on hot deformation of aluminum alloys II*. Rosemont, Illinois. Warrendale, Pennsylvania: The Minerals, Metals and Materials Society (TMS). 159–167.

Lim, S., S. Kim, C-G. Lee, and S. Kim. 2004. Tensile behavior of friction-stir-welded Al 6061-T651. *Metallurgical and materials transactions a*. 35A(9): 2829–2835A.

Liu, L., H. Nakayama, S. Fukumoto, A. Yamamoto, and H. Tsubakino. 2004. Microscopic observations of friction stir welded 6061 aluminum alloy. *Materials transactions* 45(2): 288–291.

Liu, L., H. Nakayama, S. Fukumoto, A. Yamamoto, and H. Tsubakino. 2004. Microstructure of friction stir welded 6061 aluminum alloy. *Materials forum* 28: 878–882.

Liu, H., H. Fujii, M. Maeda, and K. Nogi. 2003. Tensile properties and fracture locations of friction-stir welded joints of 6061-T6 aluminum alloy. *Journal of materials science letters* 22(15): 1061–1063.

Liu, G., L. E. Murr, C-S. Niou, J. C. McClure, and F. R. Vega. 1997. Microstructural aspects of the friction-stir welding of 6061-T6 aluminum. *Scripta Materialia* (USA) 37(3): 355–361.

Murr, L. E., G. Liu, and J. C. McClure. 1998. A TEM study of precipitation and related microstructures in friction stir welded 6061 aluminum. *Journal of materials science* 33: 1243–1251.

Reynolds, A. P. 1998. Mechanical and corrosion performance of GTA and friction stir welded aluminum for tailor welded blanks: alloys 5454 and 6061. *Proceedings of the fifth international conference on trends in welding research*. Materials Park, Ohio: ASM International. 563–567.

Sato, Y. S., H. Kokawa, M. Enomoto, and S. Jogan. 1999. Microstructural evolution of 6063 during friction stir welding. *Metallurgical and materials transactions a*. 30A: 2429–2437.

Soundararajan, V., S. Zekovic, and R. Kovacevic. 2005. Thermomechanical model with adaptive boundary conditions for friction stir welding of Al 6061. *International journal of machine tools & manufacture*. 45 14(11): 1577–1587.

Svensson, L. E. and L. Karlsson. 1999. Microstructure, hardness and fracture in friction stir welded AA 6082. *1st international symposium on friction stir welding*. Thousand Oaks, California. Cambridge: TWI.

7050 Alloys

Jata, K. V., K. K. Sankaran, and J. J. Ruschau. 2000. Friction-stir welding effects on microstructure and fatigue of aluminum alloy 7050-T7451. *Metallurgical and materials transactions* 31A: 2181–2192.

Mahoney, M. W., R. Mishra, and T. Nelson. 2002. High strain rate superplasticity in thick section 7050 aluminum created by friction stir processing. *Proceedings, third international symposium on friction stir welding*. Kobe, Japan. Cambridge: TWI. 10.

Sankaran, K. K., R. J. Lederich, K. Jata, K. Rajan, and D. Schwartz. 2003. Metallurgical characterization of friction stir welded 7050-T74 and C458-T3 aluminum alloys. *Proceedings, 6th international conference: trends in welding research*. Pine Mountain, Georgia. Cambridge: TWI. 293–296.

Sharma, S. R., R. S. Mishra, J. A. Baumann, R. J. Lederich, and R. Talwar. 2003. Microstructural characterization of a FSW 7050 Al alloy. *Friction stir welding and processing II* held at the 2003 TMS annual meeting. San Diego, California. Warrendale, Pennsylvania: TMS. 209–217.

7075 Alloys

Baumann, J. A., R. J. Lederich, and R. Mishra. 2003. Evolution of thick-section friction stir welding of 7075 Al. *Proceedings, 4th international symposium on friction stir welding,* Park City, Utah. (May) Session 4A, Paper S04A-P1. Cambridge: TWI.

Goloborodko, A., T. Ito, X. Yun, Y. Motohashi, and G. Itoh. 2004. Friction stir welding of a commercial 7075-T6 aluminum alloy: grain refinement, thermal stability and tensile properties. *Materials transactions* 45(8): 2503–2508.

Mahoney, M. W., C. G. Rhodes, J. G. Flintoff, W. H. Bingel, and R. A. Spurling. 1998. Properties of friction stir welded 7075 T651 aluminum. *Metallurgical and materials transactions* 29A: 1955–1964.

Rhodes, C. G., M. W. Mahoney, W. H. Bingel, R. A. Spurling, and C. C. Bampton. 1997. Effects of friction stir welding on microstructure of 7075 aluminum. *Scripta materialia* 36(1): 69–75.

Shibayanagi, T., and M. Maeda. 2004. Characteristics of microstructure and hardness in friction stir welded 7075 aluminum alloy joints. *Transactions of the JWRI* 33(1): 17–23.

Copper Alloys

Cederqvist, L. and R. E. Andrews. 2003. A weld that lasts for 100,000 years: FSW (friction stir welding) of copper canisters. *Proceedings, 4th international symposium on friction stir welding,* Park City, Utah. Session 3B, Paper S03B-P1. Cambridge: TWI.

Hautala, T., and T. Tiainen. 2003. Friction stir welding of copper. *6th international conference: trends in welding research*. April 15–19, 2002. Pine Mountain, Georgia. Materials: ASM International. 324–328.

Kallgren, T. and R. Sandstrom. 2003. Microstructure and temperature development in copper welded by the FSW process. *Proceedings, 4th international symposium on friction stir welding,* Park City, Utah. Session 3B, Paper S03B-P2. Cambridge: TWI.

Lee, W-B, and S-B. Jung. The joint properties of copper by friction stir welding. 2004. *Journal of materials science letters*. 58(6): 1041–1046.

Park, S. H. C., T. Kimura, T. Murakami, Y. Nagano, K. Nakata, and M. Ushio. 2004. Microstructures and mechanical properties of friction stir welds of 60% Cu-40% Zn copper alloy. *Materials science and engineering A*. 371(1–2): 160–169.

Polar, A., T. Shah, and J. E. Indacochea. 2004. Friction stir welding of copper plates. *ASM materials solutions '03: joining of advanced and specialty materials,* Pittsburgh, Pennsylvania. October 2003. Materials Park, Ohio: ASM International. 20–27.

Magnesium Alloys

Chang, C. I., C. J. Lee, and J. C. Huang. 2004. Relationship between grain size and Zener-Holloman parameter during friction stir processing in AZ31 Mg alloys. *Scripta materialia* 51(6): 509–514.

Esparza, J. A., W. C. Davis, and L. E. Murr. Microstructure-property studies in friction-stir welded, thixomolded magnesium alloy AM60. *Journal of materials science,* 38(5): 941–952.

Esparza, J. A., W. C. Davis, E. A. Trillo, and L. E. Murr. 2002. Friction-stir welding of magnesium alloy AZ31B. *Journal of materials science letters*. 21(12): 917–920.

Johnson, R. 2003. Friction stir welding of magnesium alloys. *Proceedings, 4th international symposium on friction stir welding, Park City, Utah*. Session 3B, Paper S03B-P32003. Cambridge: TWI.

Nakata, K., I. Seigo, N. Yoshitaka, H. Takenori, J. Shigetoshi, and U. Masao. 2001. Friction stir welding of AZ91D thixomolded sheet. *Proceedings, 3rd international symposium, Kobe, Japan*. Cambridge: TWI. Zhang, D., M. Suzuki, and K. Maruyama. 2005. Microstructural evolution of a heat-resistant magnesium alloy due to friction stir welding. *Scripta materialia* 52(9): 899–903.

Park, S. H. C., Y. S. Sato, and H. Kokawa. 2003. Microstructural evolution and its effect on Hall-Petch relationship in friction stir welding of thixomolded Mg alloy AZ91D. *Journal of materials science* 38(21): 4379–4383.

Park, S. H. C., Y. S. Sato and H. Kokawa. 2003. Effect of micro-texture on fracture location in friction stir weld of Mg alloy AZ61 during tensile test. *Scripta materialia* 49(2): 161–166.

Park, S. H. C., Y. S. Sato and H. Kokawa. 2003. Basal plane texture and flow pattern in friction stir weld of a magnesium alloy. *Metallurgical and materials transactions*. 34A(4): 987–994.

Sato, Y. S., S. H. C. Park, A. Matsunaga, A. Honda, and H. Kokawa. 2005. Novel production for highly formable Mg alloy plate. *Journal of materials science* 40(3): 637–642.

Somasekharan, A. C., and L. E. Murr. Microstructures in friction-stir welded dissimilar magnesium alloys and magnesium alloys to 6061-T6 aluminum alloy. *Materials characterization* 52(1): 49–64.

Stainless Steel Alloys

Klingensmith, S., J. N. DuPont, and A. R. Marder. 2005. Microstructural Characterization of a double-sided friction stir weld on a superaustenitic stainless steel. *Welding journal* 84(5): 77s–85s.

Okamoto, K., S. Hirano, M. Inagaki, S. H. C. Park, Y. S. Sato, H. Kokawa, T. W. Nelson, and C. D. Sorensen. 2003. Metallurgical and mechanical properties of friction stir welded stainless steels. *Proceedings, 4th International Symposium on friction stir welding*. Park City, Utah. Session 10A, Paper S10A-P1. Cambridge: TWI.

Park, S. H. C., Y. S. Sato, H. Kokawa, K. Okamoto, S. Hirano, and M. Inagaki. 2004. Corrosion resistance of friction stir welded 304 stainless steel. *Scripta materialia* 51(2): 101–105.

Park, S. H. C., Y. S. Sato, H. Kokawa, K. Okamoto, S. Hirano, and M. Inagaki. 2003. Rapid formation of the sigma phase in 304 stainless steel during friction stir welding. *Scripta materialia* 47(12): 1175–1180.

Park, S. H. C., Y. S. Sato, H. Kokawa, K. Okamoto, S. Hirano, and M. Inagaki. 2005. Microstructural characterisation of stir zone containing residual ferrite in friction stir welded 304 austenitic stainless steel. *Science and technology of welding and joining* 10(5): 550–556.

Posada, M., J. Deloach, A. P. Reynolds, M. Skinner, and J. P. Halpin. 2001. Friction stir weld evaluation of DH-36 and stainless steel weldments. *Proceedings, friction stir welding and processing symposium*. Indianapolis, Indiana. Warrendale, Pennsylvania: TMS. 159–171.

Reynolds, A.P., M. Posada, J. DeLoach, M.J. Skinner, J. Halpin and T. J. Lienert. 2001. Friction Stir Welding of Austenitic Stainless Steels. *Proceedings of the 3rd international conference on friction stir welding*. Session 2, Paper 1. Cambridge: TWI.

Reynolds, A. P., W. Tang, T. Gnaupel-Herold, and H. Prask. 2003. Structure, properties, and residual stress of 304L stainless steel friction stir welds. *Scripta materialia* 48(9): 1289–1294.

Sato, Y. S., T. W. Nelson, C. J. Sterling, R. J. Steel, and C. O. Pettersson. 2005. Microstructure and mechanical properties of friction stir welded SAF 2507 super duplex stainless steel. *Materials science and engineering a*. 397(1–2): 376–384.

Steel Alloys

Johnson, R. J. Dos Santos, and M. Magnasco. 2003. Mechanical properties of friction stir welded S355 C-Mn steel plates. *Proceedings, 4th International Symposium on friction steel welding*. Park City, Utah. Session 9B, Paper S09B-P1. Cambridge: TWI.

Konkol, P. J., J. A. Mathers, R. Johnson, and J. R. Pickens. 2003. Friction stir welding of HSLA-65 steel for shipbuilding. *Proceedings, 4th international symposium on friction stir* welding. *Park City, Utah*. Session 10A, Paper S10A-P2. Cambridge: TWI.

Konkol, P. J. 2003. Characterisation of friction stir weldments in 500 Brinell hardness quenched and tempered steel. Session 10A. Paper S10A-P2. *Proceedings of the 4th international symposium on friction stir welding*. Park City, Utah. Cambridge: TWI.

Lienert, T. J. 2004. Friction stir welding of DH-36 steel. *Proceedings, materials solutions '03 conference: joining of advanced and specialty materials VI*. Pittsburgh, Pennsylvania. Materials Park, Ohio: ASM International. 28–34.

Lienert, T. J., W. L. Jr. Stellwag, B. B. Grimmett, R. W. Warke. 2003. Friction stir welding studies on mild steel. *Welding journal* 82(1): 1s–9-s.

Lienert, T. J., and J. E. Gould. Friction stir welding of mild steel. 1999. *1st international symposium on friction stir welding*. June 14–16. Thousand Oaks, California. Cambridge: TWI.

Lienert, T. J., W. Tang, J. A. Hogeboom, and L. G. Kvidahl. 2003. Friction stir welding of DH-36 steel. *Proceedings, 4th international symposium in friction stir welding*. Park City, Utah. Session 9B, Paper S09B-P2. Cambridge: TWI.

Posada, M., J. Deloach, A. P. Reynolds, R. Fonda, and J. P. Halpin. 2003. Evaluation of Friction Stir Welded HSLA-65 *Proceedings, 4th international symposium on friction stir welding*. Park City, Utah. Session 10A, Paper S10A-P3. Cambridge: TWI.

Reynolds, A. P., W. Tang, M. Posada, and J. DeLoach. 2003. Friction stir welding of DH36 steel. *Science and technology of welding and joining* 8(6): 455–460.

Thomas, W. M., P. L. Threadgill, and E. D. Nicholas. 1999. Feasibility of friction stir welding steel. *Science and technology of welding and joining* (UK) 4(6): 365–372.

Titanium Alloys

Juhas, M. C., G. B. Viswanathan, and H. L. Fraser. 2000. Microstructural evolution in Ti alloy friction stir welds. *Proceedings of the 2nd international symposium on friction stir welding*. Session 5, Paper 4. Gothenburg, Sweden. Cambridge: TWI.

Lee, W-Bae., C-H. Lee, W-S. Chang, Y-M. Yeon, and S-B. Jung. 2005. Microstructural investigation of friction stir welded pure titanium. *Journal of materials science letters*. 59(26): 3315–3318.

Lienert, T. J., K. V. Jata, R. Wheeler, and V. Seetharaman. 2000. Friction stir welding of Ti-6Al-4V alloys. *Proceedings, materials solutions '00 conference: joining of advanced and specialty materials III*. St. Louis, Missouri. Materials Park, Ohio: ASM International. 160–166.

Ramirez, A. J. and M. C. Juhas. 2003. Microstructural evolution in Ti-6Al-4V friction stir welds. *Materials science forum* 426–432(Part 4): 2999–3004.

Reynolds, A. P., E. Hood, and W. Tang. 2005. Texture in friction stir welds of Timetal 21S. *Scripta materialia* 52(6): 491–494.

CORROSION

Biallas, G., R. Braun, C. Dalle Donne, G. Staniek, and W. A. Kayser. 1999. Mechanical properties and corrosion behavior of friction stir welded 2024-T3. *Paper presented at 1st international symposium on friction stir welding*. Thousand Oaks, California. Cambridge: TWI.

Corral, J., E. A. Trillo, Y. Li, and L. E. Murr. 2000. Corrosion of friction-stir welded aluminum alloys 2024 and 2195. *Journal of materials science letters*. 19(23): 2117–2122.

Dunlavy, L. M. and K. V. Jata. 2003. High-cycle corrosion fatigue of friction stir welded 7050-T7451. *Friction stir welding and processing II, 2003 TMS annual meeting*. San Diego, California. Warrendale, Pennsylvania: TMS. 91–98.

Frankel, G. S., and Z. Xia. 1999. Localised corrosion and stress corrosion cracking resistance of friction stir welded aluminium alloy 5454. *Corrosion* 55(2): 139–150.

Hannour, F., A. J. Davenport, and M. Strangwood. 2000. Corrosion of friction stir welds in high strength aluminium alloys. *Proceedings, 2nd international symposium on friction stir welding*. Gothenburg, Sweden. Session 10, Paper 2. Cambridge: TWI. 26–28.

Li, Z. X., W. J. Arbegast, P. J. Hartley, and E. I. Meletis. 1999. Microstructure characterization and stress corrosion evaluation of friction stir welded Al 2195 and Al 2219 alloys. *Proceedings, 5th international conference: trends in welding research*. June 1998. Materials Park, Ohio: ASM International. 568–573.

Lumsden, J., G. Pollock, AND M. Mahoney. 2003. Effect of post weld heat treatments on the corrosion properties of FSW AA7050. *Friction stir welding and processing II, 2003 TMS annual meeting*. San Diego, California. Warrendale, Pennsylvania: TMS. 99–106.

Lumsden, J. M. Mahoney, G. Pollock, D. Waldron, and A. Guinasso. 1999. Stress corrosion susceptibility in 7075 T7541 aluminium following friction stir welding. *Paper presented at 1st international symposium on friction stir welding*. Thousand Oaks, California. Cambridge: TWI.

Lumsden, J., G. Pollock, and M. Mahoney. 2005. Effect of tool design on stress corrosion resistance of FSW AA7050-T7451. *Proceedings, friction stir welding and processing III, 2005 TMS annual* meeting. San Francisco, California. Warrendale, Pennsylvania: TMS. 19–25.

Meletis, E. I., P. Gupta, and F. Nave. 2003. Stress corrosion cracking behavior of friction stir welded high-strength aluminum alloys. *Friction stir welding and processing II as held at the 2003 TMS annual* meeting. San Diego, California. Warrendale, Pennsylvania: TMS. 107–112.

Merati, A., K. Sarda, D. Raizenne, and C. Dalle Donne. 2003. Improving corrosion properties of friction stir welded aluminum alloys by localized heat treatment. *Friction stir welding and processing II, 2003 TMS annual meeting*. San Diego, California. Warrendale, Pennsylvania: TMS. 77–90.

Padgett, B. N., C. Paglia, and R. G. Buchheit. 2003. Characterization of corrosion behavior in friction stir weld Al-Li-Cu AF/C458 alloy. *Friction stir welding and processing II as held at the 2003 TMS annual meeting*. San Diego, California. Warrendale, Pennsylvania: TMS. 55–64.

Paglia, C. S, L. M. Ungaro, B. C. Pitts, M. C. Carroll, A. P. Reynolds, and R. G. Buchheit. 2003. The corrosion and environmentally assisted cracking behavior of high strength aluminum alloys friction stir welds: 7075-T651 vs. 7050-T7451. *Friction stir welding and processing II, 2003 TMS annual* meeting. San Diego, California. Warrendale, Pennsylvania: TMS. 65–75.

Pao, P. S., R. W. Fonda, H. N. Jones, C. R. Feng, B. J. Connolly, and A. J. Davenport. 2005. Microstructure, fatigue crack growth, and corrosion in friction stir welded Al 5456. *Proceedings, friction stir welding and processing III as held at the 2005 TMS* annual meeting. San Francisco, California. Warrendale, Pennsylvania: TMS. 27–34.

Pao, P. S., E. Lee, C. R. Feng, H. N. Jones, and D. W. Moon. 2003. Corrosion fatigue in FSW welded Al 2519. *Friction stir welding and processing II, 2003 TMS annual meeting*. San Diego, California. Warrendale, Pennsylvania: TMS. 113–122.

Pao, P. S., S. J. Gill, C. R. Feng, and K. K. Sankaran. 2001. Corrosion-fatigue crack growth in friction stir welded Al 7050. *Scripta materialia* 45(5): 605–612.

Park, S. H. C., Y. S. Sato, H. Kokawa, K. Okamoto, S. Hirano, M. Inagaki. 2004. Corrosion resistance of friction stir welded 304 stainless steel. *Scripta materialia* 51(2): 101–105.

Sankaran, K. K., H. L. Smith, and K. Jata. 2003. Pitting corrosion behavior of friction stir welded 7050-T74 aluminum alloy. *Proceedings, 6th international conference: trends in welding research*. Pine Mountain, Georgia. April 2002. Cambridge: TWI. 284–286.

Srinivasan Bala, P., W. Dietzel, R. Zettler, J. F. dos Santos, and V. Sivan. 2005. Stress corrosion cracking susceptibility of friction stir welded AA7075-AA6056 dissimilar joint. *Materials science and engineering a*. 392(1–2): 292–300.

Subramanian, P. R., N. V. Nirmalan, L. M. Young, P. Sudkamp, M. Larsen, P. L. Dupree, and A. K. Shukla. 2003. Effect of microstructural evolution on mechanical and corrosion behavior of friction stir-processed aluminum alloys. *Friction stir welding and processing II as held at the 2003 TMS annual meeting*. San Diego, California. Warrendale, Pennsylvania: TMS. 235–242.

FRICTION STIR WELDING MACHINES

General Tool Company: http://frictionstirwelding.com/frictionstirwelding.html.

MTS Systems: http://www.mts.com/aesd/friction_stir1.htm.

Nova-Tech Engineering: http://www.ntew.com/products/fsw.html.

HEAT TRANSFER

Chen, C. M. and R. Kovacevic. 2003. Finite element modeling of friction stir welding—thermal and thermomechanical analysis. *International journal of machine tools and manufacture*. 43(13): 1319–1326.

Khandkar, M. Z. H., J. A. Khan, and A. P. Reynolds 2003. Input torque based thermal model of friction stir welding of Al-6061. *Proceedings, 6th international conference: trends in welding research*. Pine Mountain, Georgia. April 2002. Cambridge: TWI. 218–223.

Khandkar, M. Z. H., J. A. Khan, and A. P. Reynolds. 2003. Prediction of temperature distribution and thermal history during friction stir welding: input torque based model. *Science and technology of welding and joining* 8(3): 165–174.

Nandan, R., G. G. Roy, and T. DebRoy. 2006. Numerical simulation of three dimensional heat transfer and plastic flow during friction stir welding. *Metallurgical and materials transactions* (2) 2006.

Schmidt, H., J. Hattel, and J. Wert. 2004. An analytical model for the heat generation in friction stir welding. *Modeling and simulation in materials science and engineering* 12(1): 143–157.

Smith, C. B., G. B. Bendzsak, T. H. North, J. F. Hinrichs, J. S. Noruk, and R. J. Heideman. 2000. Heat and material flow modeling of the friction stir welding process. *Proceedings, ninth international conference on computer technology in welding, Detroit, Michigan*. September 1999. Miami: American Welding Society. 475–486.

Song, M., and R. Kovacevic. 2004. Heat transfer modeling for both workpiece and tool in the friction stir welding process: a coupled model. *Proceedings of the institution of mechanical engineers B. Journal of engineering manufacture*. 218(1): 17–33.

Song, M., and R. Kovacevic. 2003. Thermal modeling of friction stir welding in a moving coordinate system and its validation. *International journal of machine tools and manufacture* 43(6): 605–615.

Song, M., R. Kovacevic, J. Ouyang, and M. Valant. 2003. A detailed three-dimensional transient heat transfer model for friction stir welding. *Proceedings, 6th international conference: trends in welding research*. Pine Mountain, Georgia. April 2002. Cambridge: TWI. 212–217.

Tang, W., X. Guo, J. C. McClure, L. E. Murr, and A. C. Nunes Jr. 1998. Heat input and temperature distribution in friction stir welding. *Journal of materials processing & manufacturing science*, Vol. 7. 2(10): 163–172.

Wang, K. K., and P. Nagappan. 1970. Transient temperature distribution in inertia welding of steels. *Welding journal*. 49(9): 419-s–426-s.

MASS FLOW

Chao, Y. J., and X. Qi, and W. Tang. Heat transfer in friction stir welding—experimental and numerical studies. *Journal of manufacturing science and engineering* 125(1): 138–145.

Chao, Y. J., and X. Qi. 1998. Thermal and thermomechanical modeling of friction stir welding process. *Journal of materials processing and manufacturing science* 7(2): 215–233.

Chen, C. M., and R. Kovacevic. Finite element modeling of friction stir welding—thermal and thermomechanical analysis. *International journal of machine tools and manufacture* 43(13): 1319–1326.

Colegrove, P. A., and H. R. Shercliff. 2004. Development of trivex friction stir welding tool, part 1—two-dimensional flow modeling and experimental validation. *Science and technology of welding and joining* 9 (4): 345–351.

Colegrove, P. A., and H. R. Shercliff. 2004. Development of trivex friction stir welding tool, part 2—three-dimensional flow modeling. *Science and technology of welding and joining* 9 (4): 352–361.

Colegrove, P. A., and H. R. Shercliff. 2004. Two-dimensional CFD modeling of flow around profiled FSW tooling. *Science and technology of welding and joining* 9(6): 483–492.

Colligan, K. 1999. Material flow behavior during friction stir welding of aluminum. *Welding journal* 78(7): 229-s–327-s.

Deng, X., and S. Xu. 2001. Solid mechanics simulation of friction stir welding process. *Transactions of NAMRI/SME*. SME Vol. XXIX: 631–638. Dearborn, Michigan: Society of Manufacturing Engineers

Feng, Z., J. E. Gould, and T. J. Lienert. 1998. A heat flow model for friction stir welding of aluminum alloys. *Proceedings, second symposium on hot deformation of aluminum alloys II*. Rosemont, Illinois. Warrendale, Pennsylvania: TMS. 149–158.

Guerra M., J. C. McClure, L. E. Murr, and A. C. Nunes Jr. 2002. *Proceedings, symposium on friction stir welding and processing, 2001 TMS fall meeting*. Edited by K. V. Jata, M. W. Mahoney, R. S. Mishra, S. L. Semiatin, and D. P. Field. Page 25. Warrendale, Pennsylvania: TMS.

Khandkar, M. Z. H., J. A. Khan, and A. P. Reynolds. 2003. Prediction of temperature distribution and thermal history during friction stir welding: input torque based model. *Science and technology of welding and joining* 8(3): 165–174.

Kazi, S. H., and L. E. Murr. 2002. *Proceedings, symposium on friction stir welding and processing, 2001 TMS fall meeting*. Edited by K. V. Jata, M. W. Mahoney, R. S. Mishra, S. L. Semiatin, and D. P. Field. Warrendale, Pennsylvania: TMS.

Li, Y., L. E. Murr, and J. C. McClure. 1999. Structural materials: properties, microstructure and processing. *Materials science and engineering a*. A271(1–2): 213–223.

Li, Y., L. E. Murr, and J. C. McClure. 1999. Solid-state flow visualization in the friction-stir welding of 2024 Al to 6061 Al. *Scripta materialia* 40(9): 1041–1046.

London B., M.W. Mahoney, W. Bingel, M. Calabrese, R.H. Bossi, and D. Waldron. 2003. *Proceedings: symposium on friction stir welding and processing II*. Edited by K. V. Jata, M. W. Mahoney, R. S. Mishra, S. L. Semiatin, and T. J. Lienert. Page 3. Warrendale, Pennsylvania: TMS.

McClure, J. C., W. Tang, L. E. Murr, X. Guo, Z. Feng, and J. E. Gould. 1999. A thermal model of friction stir welding. *Trends in welding research: proceedings, 5th international conference*. Pine Mountain, Georgia. June 1998. Materials Park, Ohio: ASM International. 590–595.

Murr, L. E., R. D. Flores, O. V. Flores, J. C. McClure, G. Liu, and D. Brown. 1998. Friction-stir welding: microstructural characterization. *Materials research innovations* 1(4): 211–223.

Murr, L. E., Y. Li, R. D. Flores, E. A. Trillo, and J. C. McClure. 1998. Intercalation vortices and related microstructural features in the friction-stir welding of dissimilar metals. *Materials research innovations* 2(3): 150–163.

Reynolds, A.P., T.U. Seidel, and M. Simonsen. 1999. *Proceedings, 1st international symposium on friction stir welding*. Edited by P. Threadgill, Session 4, Paper 1. Warrendale, Pennsylvania: TMS.

Reynolds, A.P. 2000. Visualisation of material flow in autogenous friction stir welds. *Science and technology of welding and joining* 5(2): 120–124.

Schmidt, H., and J. Hattel. 2005. A local model for the thermomechanical conditions in friction stir welding. *Modeling and simulation materials science and engineering* 13(1): 77–93.

Schmidt, H., J. Hattel, and J. Wert. 2004. An analitical model for the heat generation in friction stir welding. *Modeling and simulation in materials science and engineering* 12: 143–157.

Schmidt, H. N. B., and J. Hattel. 2004. Heat source models in simulation of heat flow in friction welding. *International journal of offshore and polar engineering* 14(4): 296.

Seidel, T. U., and A. P. Reynolds. 2001. Visualization of the material flow in AA2195 friction-stir welds using a marker insert technique. *Metallurgical and materials transactions a*. 32A(November): 2879–2884.

Seidel, T. U., and A. P. Reynolds 2003. Two–dimensional friction stir welding process model based on fluid mechanics. *Science and technology of welding and joining* 8(3): 175–183.

Seidel, T. U., and A. P. Reynolds. 2003. Two–dimensional friction stir welding process model based on fluid mechanics. *Science and technology of welding and joining* 8(3): 175–183.

Song, M., and R. Kovacevic. 2003. Thermal modeling of friction stir welding in a moving coordinate system and its validation. *International journal of machine tools and manufacture* 43(6): 605–615.

Song, M., and R. Kovacevic. 2004. Heat transfer modeling for both workpiece and tool in the friction stir welding process: a coupled model. *Proceedings of the institution of mechanical engineers, part b: journal of engineering manufacture* 218(1): 17–33.

Xu, S., X. Deng, A.P. Reynolds, and T.U. Seidel. 2001. Finite element simulation of material flow in friction stir welding. *Science and technology of welding and joining* 6(3): 191–193.

Zhu, X. K., and Y. J. Chao. Numerical simulation of transient temperature and residual stresses in friction stir welding of 304L stainless steel. *Journal of materials processing technology* 146(2): 263–272.

MODELING

Askari, A., S. Silling, B. London, and M. Mahoney. 2001. Modeling and analysis of friction stir welding processes. *Friction stir welding and processing*. 4-8 Nov. 2001. Indianapolis, Indiana. Warrendale, Pennsylvania: TMS. 43–54.

Colegrove, P., M. Painter, D. Graham, and T. Miller. 2000. 3-dimensional flow and thermal modeling of the friction stir welding process. *Proceedings of the 2nd international symposium* on *friction stir welding*. Session 4, Paper 2. Gothenburg, Sweden. Cambridge: TWI.

Colegrove, P. A., and H. R. Shercliff. 2005. 3-Dimensional CFD modeling of flow round a threaded friction stir welding tool profile. *Journal of materials processing technology*. 169(2): 320–327.

G. J. Bendzsak, T. H. North, and C. B. Smith. 2000. An experimentally validated 3D model for friction stir welding. *Proceedings of the 2nd international sym-*

posium on friction stir welding, Gothenburg, Sweden. P. Threadgill, ed. Session 4, Paper 1. Cambridge: TWI. 26–28.

Stewart, M. B., A. C. Nunes Jr., G. P. Adams, and P. Romine. 1998. A combined experimental and analytical modeling approach to understanding friction stir welding. SECTAM XIX. 3–5 May. Deerfield Beach, Florida. *Proceedings of the 19th southeastern conference on theoretical and applied mechanics: developments in theoretical and applied mechanics.* Vol. 19. 472–484.

NONDESTRUCTIVE EXAMINATION

Bird, C. R. 2003. *Proceedings, 4th international symposium on friction stir welding.* Park City, Utah. Session 9A, Paper S09A-P1.2003. Cambridge: TWI.

Bird, C., O. Dupuis, and A. Lamarre. 2003. A new developments of the of ultrasound phased array for the evaluation of friction stir welds. *Friction stir welding and processing II.* TMS Annual Meeting. San Diego, California. Warrendale, Pennsylvania: The Minerals Metals and Materials Society (TMS). 135–141.

Goldfine, N., D. Grundy, V. Zilberstein, and D. G. Kinchen. 2003. Friction stir weld inspection through conductivity imaging using shaped-field MWM arrays. *6th international conference: trends in welding research,* Pine Mountain, Georgia. Materials Park, Ohio: ASM International. 318–323.

Goldfine, N., D. Grundy, V. Zilberstein, and D. G. Kinchen. 2004. Nondestructive inspection techniques for friction stir weld verification on the space shuttle external tank. *6th international conference: Trends in welding research.* 15–19 Apr. 2002, Pine Mountain, Georgia. Materials Park, Ohio: ASM International.

Lamarre, A., M. Moles, and O. Pupuis. 2004. Ultrasonic phased array inspection technology for the evaluation of friction stir welds. *Materials solutions '03: joining of advanced and specialty materials VI.* Pittsburgh, Pennsylvania. October 2003. Materials Park, Ohio: ASM International. 96–103.

RESIDUAL STRESSES

Bussu, G., and P. E. Irving. 2003. The role of residual stress and heat–affected zone properties on fatigue crack propagation in friction stir welded 2024-T351 aluminium joints. *International journal of fatigue* (UK) 25(1): 77–88.

Dalle Donne, C. D., E. Lima, J. Wegener, A. Pyzalla, and T. Buslaps. 2002. Investigation of residual stresses in friction stir welds. *Proceedings, third international symposium on friction stir welding.* Kobe, Japan. September 2001. Cambridge: TWI. 10.

Dalle Donne, C., G. Biallas, T. Ghindini, and G. Raimbeaux. 2000. Effect of weld imperfections and residual stresses on the fatigue crack propagation in friction stir welded joints. *Proceedings, 2nd international symposium on friction stir welding.* Gothenburg, Sweden. Session 8, Paper 2. Cambridge: TWI.

Dalle Donne, C., E. Lima, J. Wegener, A. Pyzalla and T. Buslaps. 2003. *Proceedings from the 3rd international friction stir welding* symposium. Kobe, Japan. Session 7, Paper D. Cambridge: TWI.

James, M, and M. W. Mahoney. 1999. Residual stress measurements in friction stir welded aluminium alloys. *Proceedings, 1st international symposium on friction stir welding.* Thousand Oaks, California. Cambridge: TWI.

John, R., K. V. Jata, and K. Sadananda, K. 2003. Residual stress effects on near-threshold fatigue crack growth in friction stir welds in aerospace alloys. *International journal of fatigue* 25(9–11): 939–948.

Peel, M., A. Steuwer, M. Preuss, and P. J. Withers. 2003. Microstructure, mechanical properties and residual stresses as a function of welding speed in aluminium AA5083 friction stir welds. *Acta materialia* 51(16): 4791–4801.

Reynolds, A. P., W. Tang, T. Gnaupel-Herold, and H. Prask. 2003. Structure, properties, and residual stress of 304L stainless steel friction stir welds. *Scripta materialia* 48(9): 1289–1294.

Staron, P., M. Kocak, and S. Williams. 2003. Residual stress distributions in friction stir welded Al sheets determined by neutron strain scanning. *Proceedings, 6th international conference: trends in welding* research. Pine Mountain, Georgia, April 2002. Cambridge: TWI. 253–256.

Sutton, M. A., A. P. Reynolds, D-Q. C. Wang, and R. A. Hubbard. 2002. Study of residual stresses and microstructure in 2024-T3 aluminum friction stir butt welds. *Journal of engineering materials and* technology. Transactions of the ASME. 124(2): 215–221. New York: American Society of Mechanical Engineers.

Wang X. L., Z. Feng, S. A. David, S. Spooner, and C. R. Hubbard. 2000. *Hubbard: proceedings, sixth annual conference on residual* stresses. Oxford, London: IOM Communications. 1408–1414.

CHAPTER 8

ULTRASONIC WELDING OF METALS

Prepared by the Welding Handbook Chapter Committee on Ultrasonic Welding:

K. F. Graff, Chair
Edison Welding Institute

J. F. Devine
Sonobond Ultrasonics, Incorporated

J. Keltos
Branson Ultrasonics

N. Y. Zhou
University of Waterloo

Welding Handbook Volume 3 Committee Member:

W. L. Roth
Procter and Gamble, Incorporated

Contents

Photograph courtesy of © Edison Welding Institute. Used with permission.

CHAPTER 8

ULTRASONIC WELDING OF METALS

INTRODUCTION

Ultrasonic welding (USW) is a solid-state welding process that produces a weld by the local application of high-frequency vibratory energy as the workpieces are held together under pressure.[1,2] The vibratory energy creates a relative transverse motion between the two surfaces, disperses interface oxides and contaminants from the interface to achieve metal-to-metal contact, and produces the weld.

The advantages of ultrasonic welding, a process originally developed for the welding of metals needed in critical applications, are complimentary to many of the requirements of joining plastic materials; thus the process is widely used in the welding of plastics. Because of the fundamental differences in the welding of metals and plastics, the scope of this chapter is limited to the ultrasonic welding of metals.

Ultrasonic welding is used for applications involving similar and dissimilar metallic joints. It is used to produce lap joints between metal sheets or foils, between wires or ribbons and flat surfaces, between parallel wires, and for joining other types of assemblies that can be supported on an anvil. The process is used as a production tool in the automotive, residential and commercial appliance, refrigeration, air conditioning, and other manufacturing industries. It is used for producing batteries and electrical contacts, fabricating small motor armatures, manufacturing aluminum foil, and joining aluminum components. Ultrasonic welding is used in the automotive and aerospace industries as a structural joining method. The process is uniquely useful for encapsulating materials such as explosives, pyrotechnics and reactive chemicals that require hermetic sealing but cannot withstand high-temperature joining processes. It is widely used in the semiconductor and microelectronics industries.

This chapter covers the fundamentals of ultrasonic welding and describes the equipment and operating techniques used in production. A separate section on ultrasonic microwelding is presented. Other topics are welding variables and procedures, weld quality, and applications. A section on the economics of ultrasonic welding is included. Safe practices specific to ultrasonic welding are discussed, and references are supplied for general information on safety in welding.

1. Welding terms and definitions used throughout this chapter are from American Welding Society (AWS) Committee on Definitions and Symbols, 2001, *Standard Welding Terms and Definitions*, Miami: American Welding Society.

2. At the time this chapter was prepared, the referenced codes and other standards were valid. If a code or other standard is cited without a publication date, it is understood that the latest edition of the document referred to applies. If a code or other standard is cited with the publication date, the citations refer to that edition only, and it is understood that any future revisions or amendments to the code or other standard are not included. As codes and standards undergo frequent revision, the reader is advised to consult the most recent edition.

FUNDAMENTALS

The components of an ultrasonic welding system are an electronic power source, a transducer, an acoustic coupling system, a sonotrode, an anvil and a clamping system. The power source provides high-frequency electrical power to the ultrasonic transducer. High-frequency mechanical vibrations are generated in the transducer and transmitted through an acoustic coupling system to the workpieces via a weld tool, called the *sonotrode*. The workpieces, supported by the anvil, are firmly held together during welding by a clamping force applied through the coupling system and anvil. Ultrasonic welding generally is accomplished with either a lateral drive or a wedge-reed system, as discussed in Principles of Operation in this chapter.

ULTRASONIC WELDING SYSTEMS

Based on the type of weld required for the specific application, several ultrasonic welding systems can be implemented to produce spot, seam, and torsion welds in thin-gauge materials (for example, foil and wire) and to produce spot welds in microelectronics.

Spot Welding

Ultrasonic spot welding produces individual welds by introducing a pulse of vibratory energy into the workpieces as they are held together under clamping force between the sonotrode and the anvil. The shape and size of the spot weld is determined by the shape and size of the sonotrode face when in contact with the workpiece and, to some extent, by the shape of the anvil.

Typical spot welds are circular in pattern but may be adjusted to have rectangular, elliptical or other irregular shapes, depending on the sonotrode face. Thus, elongating a rectangular pattern of the weld face permits the creation of a linear weld pattern. A typical spot weld is approximately 5 mm to 6 mm (0.20 inch [in.] to 0.23 in.) in diameter. Weld dimensions are dependent on the requirements of the application, the power capabilities of the welding machine, and the type and thickness of the workpieces. Depending on the application, the spacing between spot welds may range from the relatively wide spacing needed to meet strength requirements to overlapping welds, which can produce a leak-tight seal.

Seam Welding

In seam welding, a continuous joint is produced between workpieces that are passed between a rotating, disk-shaped sonotrode and either a roller or a flat anvil. The rotating transducer-anvil system may be either a fixed unit or it may be mechanized to travel along the joint of the workpieces.

Torsion Welding

Torsion welding, which produces a circular weld pattern, is created by a configuration of ultrasonic transducers and coupling system components that result in a twisting, or torsional, vibration of the sonotrode in the plane of the workpieces. In contrast to the linear motion produced by spot welding systems, torsional motion produces the natural ring, or circular pattern of the weld.

Microwelding of Electronic Components

The use of ultrasonic welding in the microelectronics industry is generally referred to as *microbonding*, although in the context of this chapter, the terms *welding* and *bonding* are not synonymous.[3] Modifications of spot welding systems, typically combining features of both lateral-drive and wedge-reed systems, are used in microwelding. While in principle, microwelding has similarities to ultrasonic spot welding, microwelding is significantly different in terms of power levels, vibrational parameters (such as frequency and amplitude), and in dimensions and types of workpiece materials. Further discussion of the features of this unique application of ultrasonic welding is presented in a separate section of this chapter under the title "Microwelding."

Ultrasonic Welding of Plastics

A major use of ultrasonic welding is the joining of plastics. It should be noted that the welding of plastics is fundamentally distinct from the ultrasonic welding of metals, with the differences predominant in two key areas. First, in plastics welding, the ultrasonic vibrations create a relative normal motion between the surfaces, in contrast to transverse motion for metal welding. Second, and most critical, is that plastic welding is a fusion-based process that requires heating, softening, and melting of the plastics at the adjoining surfaces to create a bond. Thus, in a strict sense, the ultrasonic plastic welding process bears its closest analogy to fusion-based metal welding processes, such as arc welding or laser beam welding. Further information on ultrasonic plastic welding can be found in the AWS publication, *Guide to Ultrasonic Assembly of Thermoplastics*, AWSG1.1M/G1.1,[4] and in the *Welding Handbook*, Volume 3, *Materials and Applications*.[5]

PROCESS ADVANTAGES AND LIMITATIONS

The primary advantages of ultrasonic welding are the inherent features of being a rapid, solid-state welding process. No melting of the metal occurs, in contrast to fusion welding processes. Thus, expelled molten metal,

3. The AWS Committee on Definitions and Symbols reserves the term *bonding* for the joining and allied processes where either an adhesive bond or a mechanical bond is predominant at the interface created by the process actions, for example, adhesive bonding and thermal spraying. When an atomic bond between the atoms at that interface is predominant, the resulting joint is called a *weld* and the process that produced that joint is called *welding*, without regard to whether the weld interface is created as a result of fusion or in the solid state. The interatomic bond existing between metal atoms at the weld interface of a fusion weld is no different from that at the weld interface of a solid-state weld.

4. American Welding Society (AWS) Committee on Joining of Plastics and Composites, 2006, *Guide to Ultrasonic Assembly of Thermoplastics*, AWS G1.1M/G1.1: 2006, Miami: American Welding Society.

5. American Welding Society (AWS) Welding Handbook Committee, W. R. Oates, ed., 1996, *Welding Handbook*, Volume 3, *Materials and Applications*, Chapter 6. Miami: American Welding Society.

heat-affected zones (HAZs), cast nuggets or brittle intermetallics do not occur in ultrasonic welding. Of special value is the capability of welding aluminum, copper, and other high-thermal-conductivity metals with little difficulty, typically using substantially less energy than alternative processes. Dissimilar metals with wide differences in melting points can be welded if appropriate welding machine settings are used.

Because it is a solid-state process, ultrasonic welding facilitates the welding of thin sections to thick sections, and readily welds a wide variety of dissimilar metals. Welds can be made through some types of surface coatings, such as thermal spray deposits or plating, although careful evaluation of the coating is important to ensure that adhesion of the workpiece to the substrate takes place rather than adhesion to the surface coat or plate.

Ultrasonic welding often provides cost and quality advantages over alternative processes. For example, wire harness manufacturers reported a significant improvement in weld quality at a cost reduction of 50% or more when switching from the crimp-and-solder method to the ultrasonic welding of stranded copper wires. The ultrasonic-welded joints were stronger, had better electrical conductivity, and also reduced the net weight by eliminating the need for the mechanical crimp. When compared to resistance welding, the energy requirement for an ultrasonic welding system is only about 5% of that needed for a resistance welding machine. The need for water-cooling systems required in some resistance welding operations is eliminated with ultrasonic welding.

A limitation of the ultrasonic welding process is that the application must allow the workpieces to be placed into a lap-joint configuration, somewhat similar to that required for most resistance spot welds. Thus, butt joints and T-joints common to arc welding and laser beam welding processes are not easily achieved with ultrasonic welding systems.

Joint thickness (the thickness of the component adjacent to the sonotrode) presents limitations, depending on the material being welded, the size of weld area and the power of the welding system available to make the weld. Joint thicknesses may range from foil-like dimensions of 0.1 mm (0.004 in.) to several millimeters. For example, spot welds on 5XXX- and 6XXX-series aluminums are restricted to the one-to-three millimeter thickness ranges for welding systems with 3 kW to 4 kW power ratings. Maximum joint thickness is closely related to material strength and hardness. While the weldability of nearly all metals has been demonstrated, some steels, aerospace alloys, and superalloys introduce special challenges to ultrasonic welding unless the application involves thin-gauge metal or the use of very powerful welding machines capable of exerting high clamping forces.

PRINCIPLES OF OPERATION

As noted, ultrasonic welding uses high-frequency mechanical vibrations to produce the welding action between two surfaces held together under pressure during the welding cycle. Details of the two most common types of ultrasonic welding systems, the lateral-drive and the wedge-reed systems, are illustrated in Figure 8.1.

The lateral-drive and wedge-reed ultrasonic welding systems have three components in common: a power source, an ultrasonic transducer, and a horn. The primary component is the electronic power source that converts line frequency to the ultrasonic frequency required by the transducer. The transducer normally uses power in the frequency range of 15 kilohertz (kHz) to 40 kHz (but may be as high as 300 kHz for microwelding). Depending on the power rating of the transducer and the size of the workpieces, the power source may need to provide output levels from 10 watts (W) to several kilowatts.

The second common requirement of both systems is the ultrasonic transducer. The transducer converts high-frequency electrical power to an ultrasonic vibration in the unit at the frequency of the electrical signal. The conversion of electrical energy to mechanical vibrations (and mechanical vibrations to electrical energy) is a property of piezoelectric materials. Transducers are fabricated with disks of piezoelectric materials in the form of ceramics, such as lead zirconate titanate. The conversion of electrical to mechanical energy occurs as a result of the expansion and contraction that piezoelectric materials undergo in the presence of an alternating electric field. (The piezoelectric materials used in ultrasonic welding are similar, although larger in size, to the materials used to create and detect ultrasonic waves in nondestructive examination transducers.) The design and dimensions of the transducer are engineered to produce a tuned, resonant vibration, creating a high-frequency, piston-like vibration at its end. The vibrations occur along the long axis of the transducer (longitudinal vibrations), and although high in frequency (e.g., 20 kHz), they are low in amplitude. Thus, vibration amplitude of 20×10^{-6} m, or 20 microns (μm), is typical; it is far below that detectable by the human eye and requires special instruments to measure.

A piezoelectric-based ultrasonic transducer is composed of the piezoelectric material, metal back and front masses, a bolt used to assemble and pre-compress the workpieces, and an enclosure case. Figure 8.2 shows a typical piezoelectric-based ultrasonic transducer with a representation of the nature of the tuned longitudinal vibration. In operation, the transducer resonates in a half-wavelength acoustical vibration pattern, with maximum amplitude at the ends of the transducer.

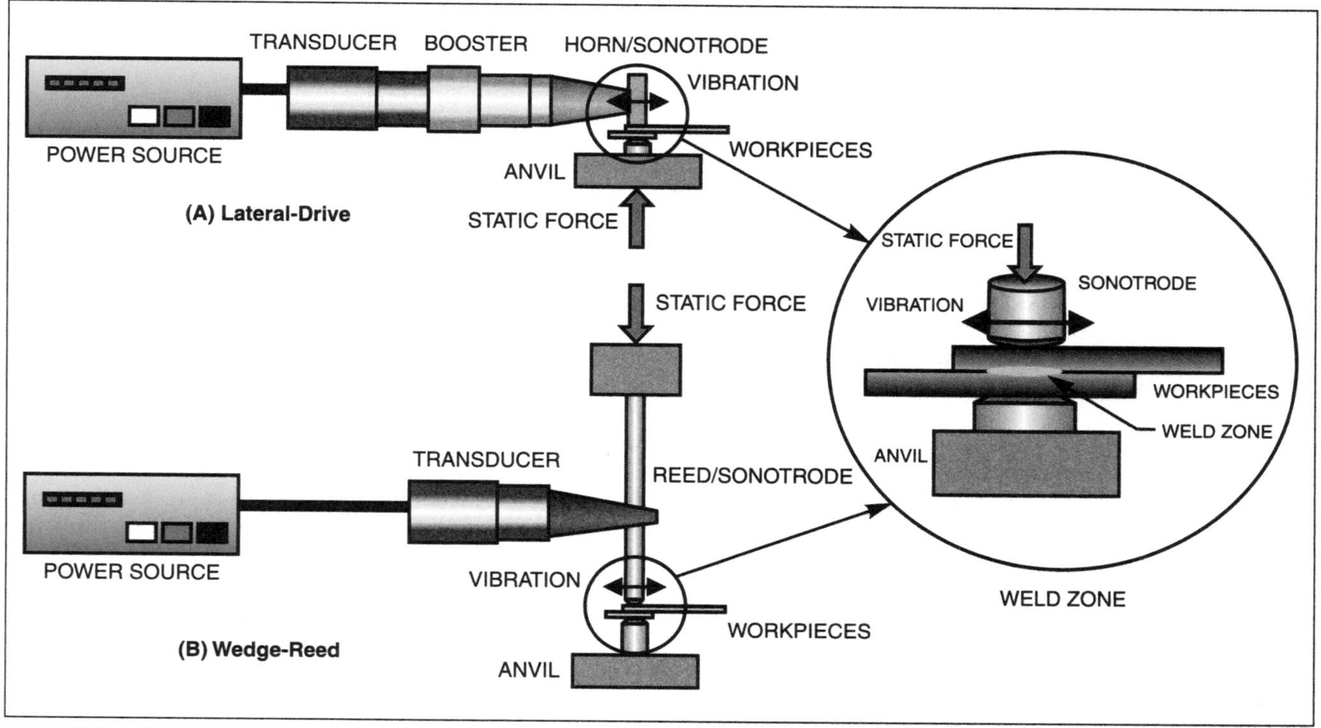

Source: © Edison Welding Institute. Used with permission.

Figure 8.1—(A) Lateral-Drive and (B) Wedge-Reed Ultrasonic Welding Systems, with Weld Zone Common to Both Systems

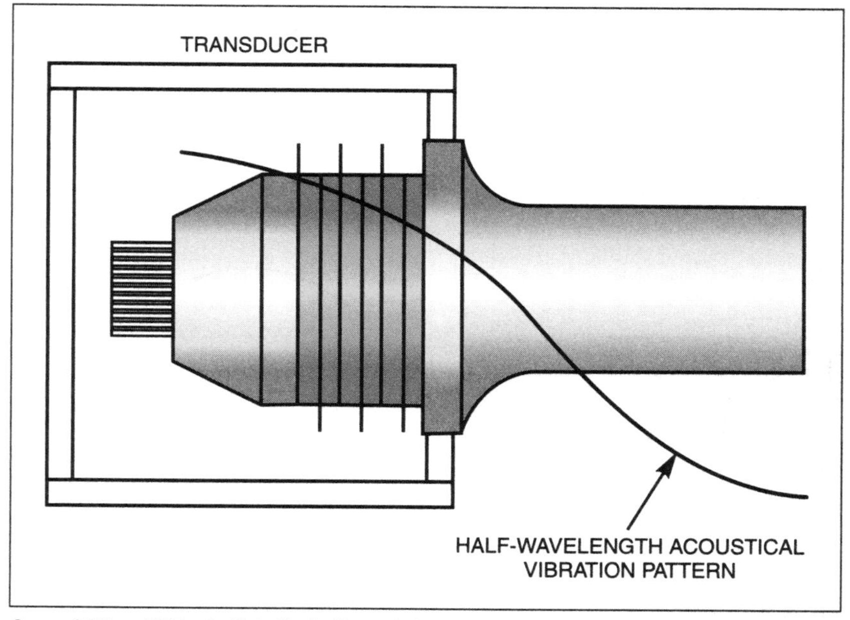

Source: © Edison Welding Institute. Used with permission.

Figure 8.2—Piezoelectric-Based Ultrasonic Transducer

While piezoelectric-based transducers are used almost exclusively in ultrasonic welding systems, it also is possible to use transducers based on the magnetostrictive phenomenon. Magnetostrictive materials, such as nickel, also have the property of expanding and contracting under an alternating magnetic field. These materials were used in the early years of ultrasonic welding, but because of their relatively low efficiency they were supplanted by the more efficient piezoelectric materials.

The third component common to both the lateral-drive and wedge-reed systems is the *horn*, which provides the acoustic coupling of the ultrasonic vibrations to the weld. The horn amplifies the vibrations and transmits them to the remaining components of the coupling system. In the lateral-drive system, the horn is usually called a *booster* because its usual function is to boost, or amplify, the vibrations. In the wedge-reed system, the horn is called the *wedge* because of its similar shape. The horn also is tuned to vibrate at the same frequency as the transducer and in an acoustic half-wavelength.

Beyond these similarities, the lateral-drive and the wedge-reed systems diverge slightly in the means of transmitting the ultrasonic vibrations into the balance of the acoustic coupling system and to the workpieces. The design of the anvils is different in the two systems. The anvil of a lateral-drive system is completely rigid, while in the wedge-reed system, the anvil may be rigid or it can be tuned to vibrate as part of the acoustic coupling system. Overall, both systems achieve a similar result: the creation of vibrational action of the workpieces in a direction parallel to surface of the workpieces and the weld interface. This feature is evident in Figure 8.1 in the encircled diagram of the weld zone, produced by either system.

Lateral-Drive System

In the lateral-drive system, the booster drives another horn also tuned to an acoustic half-wavelength to which the sonotrode is solidly attached. The sonotrode may be attached either by a threaded connection, or may be integrally machined into the horn. The sonotrode directly contacts and transmits vibratory energy into the workpieces, which are supported by the anvil. The anvil in a lateral-drive system is rigidly constructed to resist deflection and to support the static clamping force applied to the workpieces during welding.

Static force is applied to the workpieces through the coupling action of the leveraged forces. It should be noted that the direction of the vibration of the sonotrode, and hence the vibration transmitted into the workpieces, is parallel to the workpiece surfaces and the weld interface. (This is in contrast to the ultrasonic welding of plastics, in which the direction of transmitted vibrational energy is perpendicular to the workpiece surfaces and the weld interface.) Figure 8.3 shows the assembly of the transducer, booster, horn, and sonotrode, and shows the increasingly amplified longitudinal vibration pattern of the three units.

Wedge-Reed System

In the wedge-reed system, the transducer-wedge is solidly coupled to the reed. In most cases, the reed and transducer-wedge are welded together to ensure a solid interface. The reed, which may be considered as a beam, is driven by the wedge to vibrate in a bending, or flexural, mode. This is in contrast to the transducer-

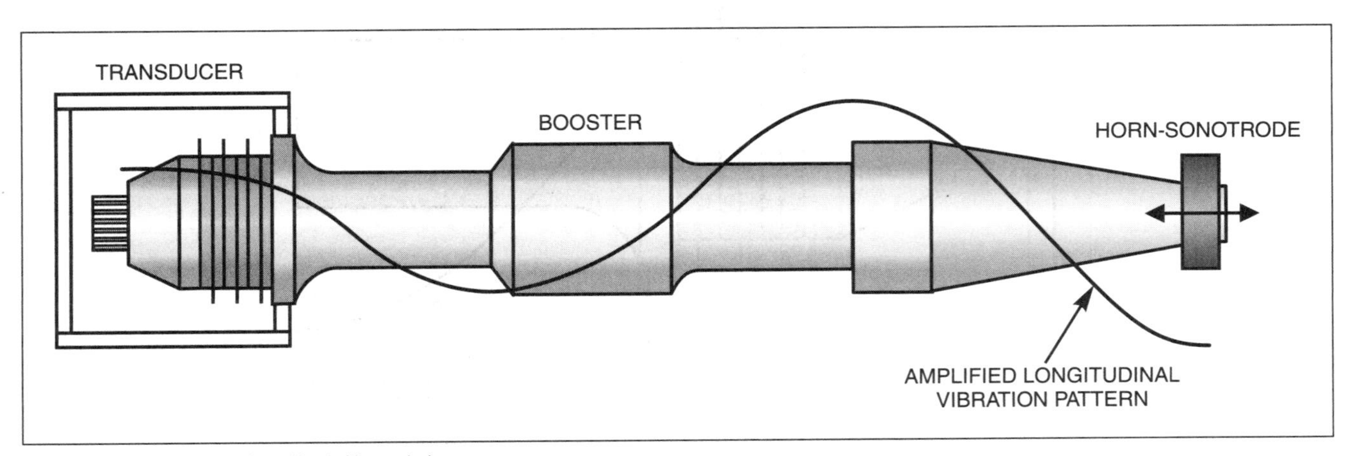

Source: © Edison Welding Institute. Used with permission.

Figure 8.3—Lateral-Drive Transducer, Booster and Horn-Sonotrode Assembly in Longitudinal Vibration

wedge and the components of the lateral-drive system, all of which vibrate longitudinally. The reed (beam) vibration is tuned to be in resonance with the transducer-wedge. The nature of this vibration is transverse to the axis of the reed, similar to the vibrations of a tightened string. The result of this arrangement is that the sonotrode, attached at the end of the reed, is vibrated parallel to the workpiece surface and the weld interface, exactly in the manner achieved by the lateral-drive system. Thus, while the two systems transmit energy to the workpieces by different means, both achieve similar vibrational action at the workpieces. These features are illustrated in Figure 8.4.

The workpieces are supported by an anvil in the wedge-reed system. The anvil may be rigid or it may be adjusted to vibrate as part of the acoustic coupling system. The clamping force is applied between a mass at the upper end of the reed and the anvil. The force is transmitted directly along the axis of the reed to maintain the workpieces in intimate contact for welding.

SYSTEM PARAMETERS

The key parameters of the ultrasonic welding process are vibration frequency, vibration amplitude, clamping force, weld power, weld energy, weld time, type of material, and tooling.

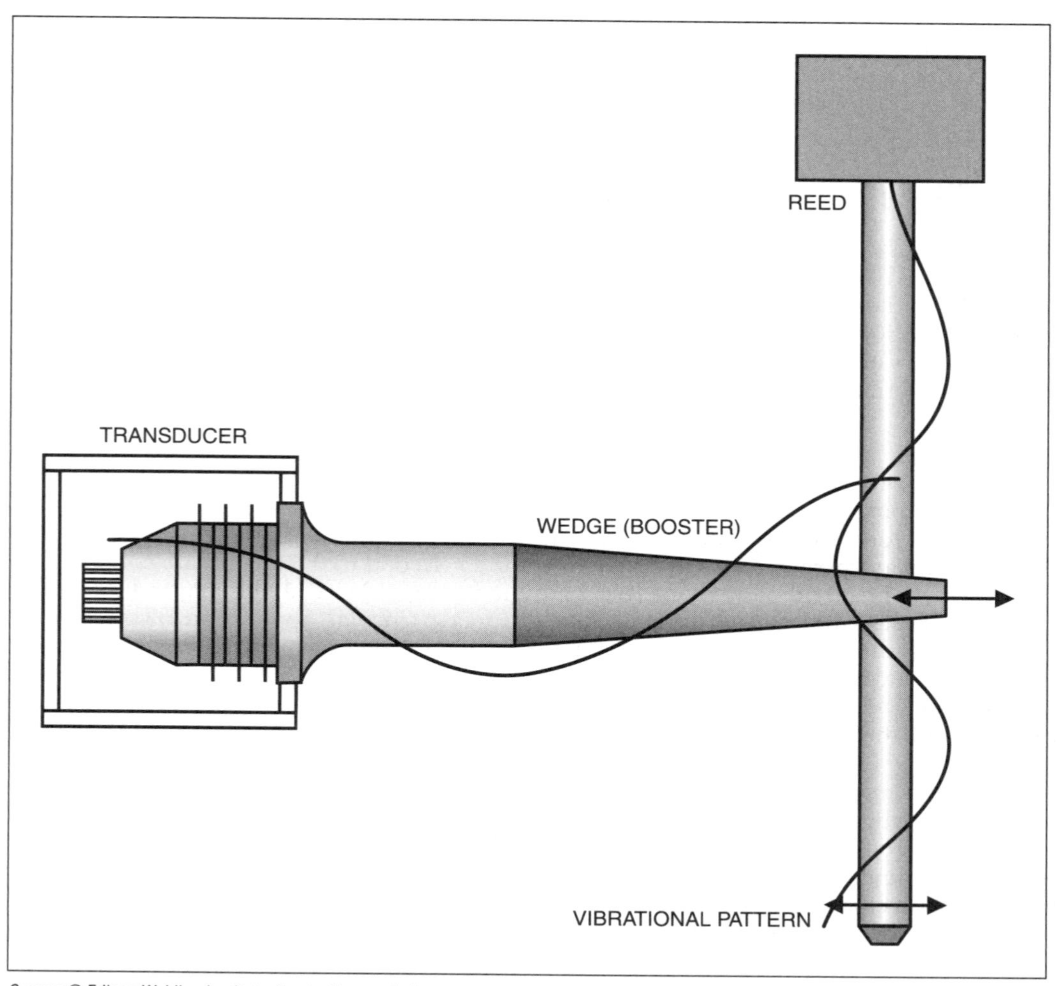

Source: © Edison Welding Institute. Used with permission.

Figure 8.4—Wedge-Reed Transducer, Wedge and Reed in Combined Longitudinal and Bending Vibration Pattern

Vibration Frequency

Ultrasonic transducers are designed and tuned to operate at a specific frequency. Operating frequencies of welding transducers may range from 15 kHz to as high as 300 kHz, with a typical frequency of 20 kHz used for welding. The nature of the application determines the best operating frequency to use. Frequency selection tends to be governed by practical matters of weld power requirements, which in turn are typically dictated by workpiece size, type of metal or material, and basic acoustic principles in the design of transducers and coupling components that generate and transmit ultrasonic vibrations.

While a transducer may be designed for a single frequency, in operation the nominal frequency actually may vary slightly above or below the rated frequency (normally ±2% to 3%) due to manufacturing and material variations, slight dimensional changes due to temperature effects, the level of clamping force and the effects of a changing weld load. Ultrasonic power sources are designed to automatically compensate for such shifts in resonant frequency.

Vibration Amplitude

The amplitude of vibration of the sonotrode at the weld strongly influences the vibratory energy delivered to the weld zone. Amplitude requirements (typically on the order of 10 µm to 80 µm) depend on the system design, the characteristics of the materials being welded, and the amount of power delivered to the transducer.

Clamping Force

The function of clamping (static) force is to hold the faying surfaces of the workpieces in intimate contact. The applied force is always perpendicular to the plane of the weld interface. The method of applying this load depends on the overall design of the welding machine. Hydraulic systems are satisfactory for large units. Intermediate-size units may incorporate pneumatically actuated or spring-loaded systems, and forces can range from 200 newtons (N) to 5000 N (50 pounds [lb] to 1000 lb).

Excessive force may produce surface deformation and increase the required welding power. Insufficient force may cause sonotrode slippage, causing surface damage, excessive heating, or unsatisfactory welds. Clamping force for a specific application is established in conjunction with ultrasonic power requirements.

Electrical Power

The electrical power delivered from the power source to the transducer transforms to ultrasonic power and energy. The efficiency of the power actually delivered to the weld zone depends on several factors, including the efficiency of the transducer and coupling system, the coupling of the sonotrode to the workpieces, and losses radiated from the weld into the anvil and surrounding workpieces or structures. The welding machine power setting may be indicated in terms of the high-frequency power input to the transducer, or the load power (the power dissipated by the transducer-sonotrode-workpiece assembly). As previously noted, the power requirement varies with the type and thickness of the materials being welded. The minimum effective power for a given application can be established by running a series of tests from which a threshold curve for welding is established. While the concept may seem simple, the accurate determination of weld power is complex, and typically can be estimated at best. Therefore, while electrical power input may be accurately measured, accurately determining the various losses between the electrical terminals and the actual weld zone is difficult. However, by using estimates based on overall system efficiencies it is possible to arrive at estimates of weld power in relation to input electrical power.

Weld Energy

The electrical energy delivered to the transducer is the integral of the electrical power to the transducer. Similarly, the total ultrasonic energy delivered to the weld is the integral of the weld power. Energy into the actual weld is a function of system efficiencies, subject to the uncertainties previously noted.

Weld Time

The weld time used for an ultrasonic weld is very short, typically less than one second; weld times of 0.25 second to 0.5 second are common. If longer welding times are needed, this may indicate insufficient power for a particular application. High power and a short welding time usually produce welds that are superior to those achieved with low power and long welding time. Excessive welding time can cause poor surface appearance, internal heating, and internal cracks. The same factors of power and unit time also apply to seam welding.

Materials Welded

The types of metals or materials being welded are of prime importance in the overall range of welding parameters. The physical properties of the materials, such as the nature of modulus, yield strength and hardness are first considerations. Next, surface characteristics such as finish, oxides, coatings and contaminants must be considered, since these would interfere with the vibratory interaction of the workpiece surfaces. The surfaces must be in intimate contact, as required by the ultrasonic welding process. Finally, the geometric shapes of the workpiece and the properties of the materials, particularly the thickness, are important.

Tooling

The geometry of the ultrasonic tooling components (the sonotrode and the anvil) that contact the workpieces act to transmit the vibratory energy and support the workpieces. These are important welding parameters that can have a significant effect on weld quality and repeatability.

The contact surfaces of the sonotrode and the anvil may have various patterns of grooves and lands (for example, knurl patterns) to better grip the workpieces. Tooling is designed and fabricated with materials of varying hardness, in some cases with special coatings or surfaces, depending on the application.

EQUIPMENT

The power sources and welding systems required for ultrasonic lateral-drive, wedge-reed, and torsion welding, including spot and seam welding, are described in this section. It should be noted that descriptions of power sources and controls in this section also apply to ultrasonic seam welding, torsion welding systems and automated ultrasonic welding equipment.

POWER SOURCES

The function of the power source is to change electrical line power of 50 Hz or 60 Hz to the frequency requirements of the welding system design. This is done in two stages through an oscillator and an amplifier. The output is the high-frequency, high-voltage electrical power needed to drive the ultrasonic transducer.

The power source must incorporate electrical impedance-matching networks between the amplifier output and the ultrasonic transducer. Typically, the output impedance of the power source is low (about 50 ohms), whereas the input impedance of the transducer tends to be high and reactive. Input impedances of hundreds of ohms for piezoelectric materials, capacitive in nature, are common. Electrical matching to this load requires transformers and induction coils following the amplifier output stage. Automatic frequency control is a standard feature of power sources. Circuits can be controlled either by free-running oscillators or operated through positive feedback derived from the load.

Some ultrasonic welding machines also incorporate constant amplitude control. This is accomplished by representing the electromechanical transducer as an equivalent electrical circuit and identifying and sensing a critical voltage or current that directly correlates to the vibration amplitude of the transducer. Using this concept, it becomes possible to control functions of the vibration amplitude during the weld cycle, such as maintaining constant amplitude or programming amplitude variations during the weld cycle. All of these functions, frequency conversion, amplification, impedance-matching and frequency control, typically are combined into a separate module detached from the welding hardware. In some low-power systems, these functions may be integrated into the hardware of the welding head.

Power and Energy in the Weld Zone

The flow and absorption of ultrasonic power and energy into the weld zone is affected by numerous factors. These include efficiencies of transducer and acoustic coupling components, radiation into surrounding workpiece structure, and transmission through the weld into the anvil structure. Since determining the exact amount of power and energy delivered into the weld is difficult, only approximations can be made. However, with a properly designed system, it can be estimated that as much as 80% to 90% of the input electrical power to the transducer will be delivered into the weld zone.

Weld Power. For practical usage, the power required for welding usually is measured in units of high-frequency electrical power delivered to the transducer. This power can be monitored continuously, providing a reliable average value that can be associated with equipment performance and with weld quality. To calculate the watt-seconds or joules (J) of energy used in welding, the power in watts is multiplied by the time in seconds.

ULTRASONIC SYSTEM COMPONENTS

The transducer contained in the various welding systems operate at a given rate, or duty cycle. As noted previously, piezoelectric transducers are highly efficient devices. Nevertheless, they are subject to some losses and heating during welding operations. When operated at high duty cycle or high power settings, the transducer should provide for air-cooling to avoid thermal damage.

The acoustic coupling systems (horns and reeds) are assembled by bolting the components together at the various interfaces. The transmission components may be made of steel, titanium or aluminum. In the wedge-reed system, the wedge and reed typically are made of steel. In the lateral-drive system, the booster may be titanium and the horns mounting the sonotrode may be titanium or steel. These various materials are selected to provide low energy losses and high fatigue strength under the applied static and vibratory stresses.

For highest reliability, the various joints in the transducer must have high integrity and excellent fatigue life. Because of the high-frequency vibration, brazed, welded, and mechanical junctions were used in the past, but most modern welding machines use threaded mechanical joints for easy interchangeability. Transducer-sonotrode systems usually have acoustically designed mounting arrangements to ensure maximum efficiency of energy transmission when static force is applied through the system. These force-insensitive mounts prevent any shift of the resonant frequency of the system, and minimize loss of vibratory energy into the supporting structure.

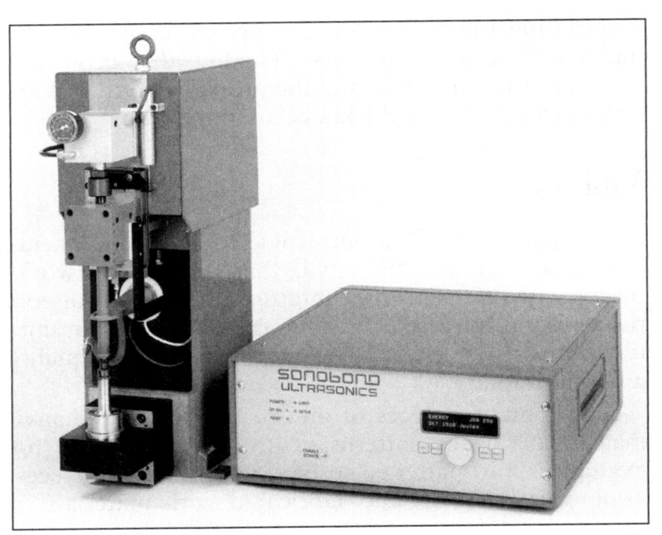

Photograph courtesy of Sonobond Ultrasonics, Incorporated

Figure 8.6—20-kHz, 3-kW Wedge-Reed Welding Machine

SPOT WELDING MACHINES

The schematic diagrams of lateral-drive and wedge-reed spot welding systems viewed in Figure 8.1 are actualized in Figures 8.5, 8.6, and 8.7. Figure 8.5 shows a 20-kHz, lateral-drive welding machine; Figure 8.6 and Figure 8.7 show two wedge-reed systems, one supplying 20 kHz, 3 kW and the other supplying 15 kHz, 4.5 kW.

SEAM WELDING MACHINES

The basic design of an ultrasonic seam welding machine (not including details of bearings and driving mechanisms) is shown in Figure 8.8. A continuously

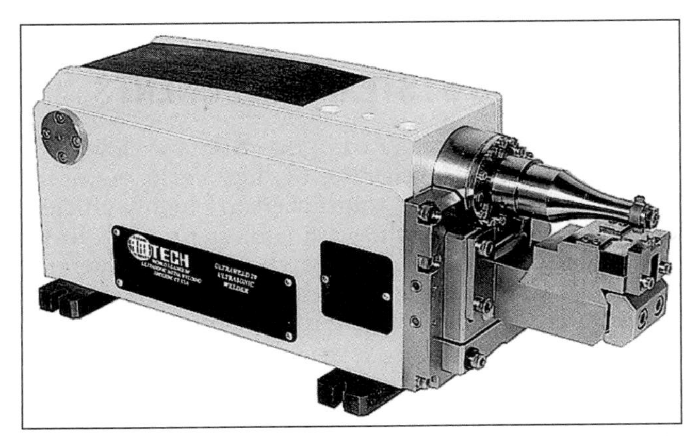

Photograph courtesy of Branson Ultrasonics

Figure 8.5—20-kHz Lateral-Drive Welding Machine

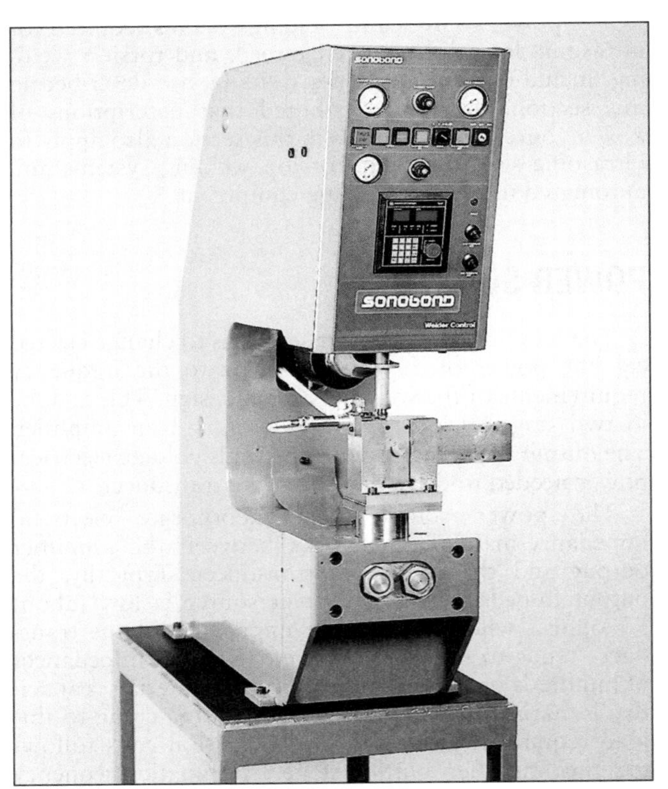

Photograph courtesy of Sonobond Ultrasonics, Incorporated

Figure 8.7—15-kHz, 4.5-kW Wedge-Reed Welding Machine

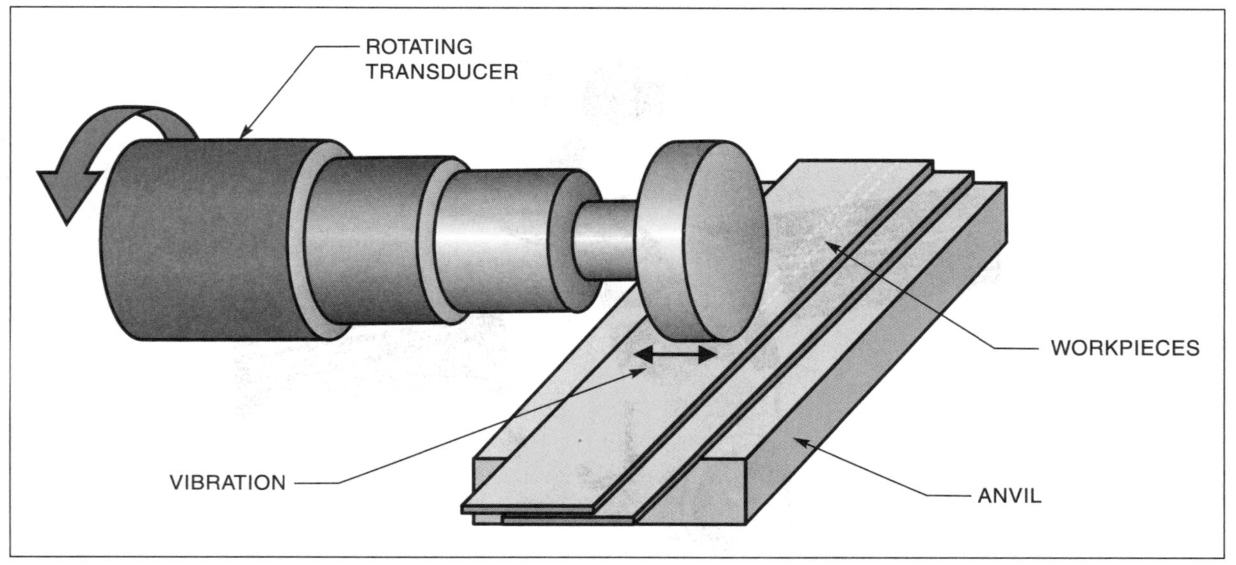

Source: © Edison Welding Institute. Used with permission.

Figure 8.8—Schematic of an Ultrasonic Seam Welding Machine (Not Including Bearings and Drive Mechanisms)

turning, lateral-drive ultrasonic transducer vibrates a circular disk sonotrode that travels along the top surface of the workpieces and produces a continuous weld. A fixed anvil also is shown in Figure 8.8, but in practice, the anvil may move in pace with the turning sonotrode disk. This synchronous motion may be produced by a turning-disk anvil, or by a laterally moving flat anvil base. In some equipment, the rotating ultrasonic transducer may be in a fixed position. In other models, it may be mechanized to traverse the joint of the workpieces. As with spot welding machines, the transducer must be leveraged to apply a static force to the workpieces as it turns.

Seam welding machines are used to join aluminum and copper foils, to form thin-wall tubing, and to weld aluminum contacts on photovoltaic cells. Figure 8.9 shows an example of a traversing-head seam welding machine. Figure 8.10 shows a fixed-head seam welding machine.

TORSION WELDING MACHINES

A schematic of an ultrasonic torsion welding machine is shown in Figure 8.11, in which two ultrasonic transducers operating in longitudinal vibration are attached (typically by welding) to ultrasonic boosters and tooling. These configurations provide a push-pull, twisting, or torsional vibration to the booster system. The key feature of this arrangement is that the booster and tooling is acoustically designed to be resonant in the torsional mode, in contrast to the longitudinal mode as used in spot welding and seam welding systems. The torsional resonance system naturally produces a circular vibration pattern at the sonotrode, which in turn creates a circular weld pattern when pressed against the workpieces. Although the vibration pattern is circular, it is important to note that the motion of the tool surface is still parallel to the workpiece surface, exactly as in the spot welding action.

The torsion welding machine illustrated in Figure 8.11 has two transducers. Depending on power requirements, from one to four transducers may be used. Figure 8.12 shows a high-power ultrasonic torsion welding machine, incorporating four transducers (two of which may be seen emerging from the front of the system), operating at 20 kHz and providing 10 kW of power.

AUTOMATED ULTRASONIC WELDING SYSTEMS

The basic controls for an ultrasonic welding system typically consist of the on/off switch for line power, adjustments for vibration amplitude, clamping force, power level, energy and welding time. The number of controls and the level of sophistication vary with the

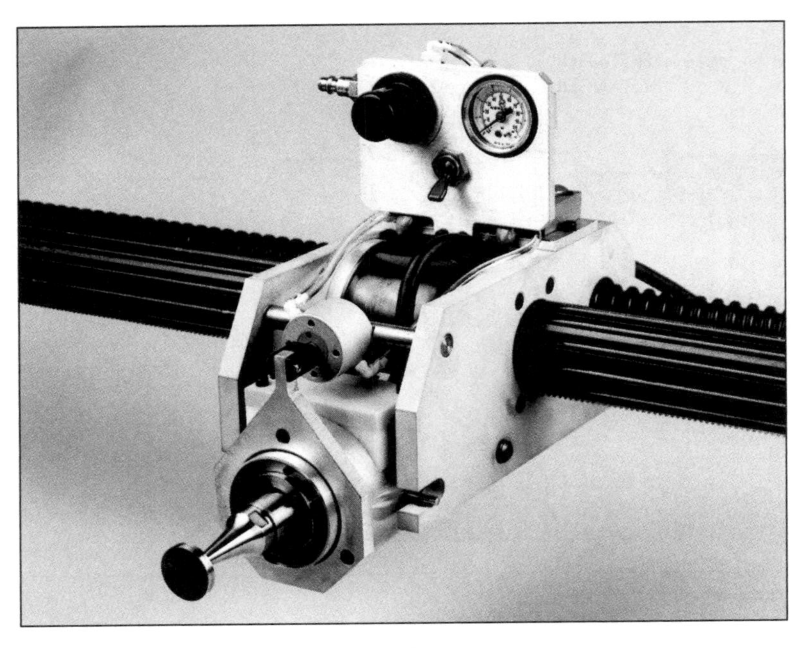

(A)

(B)

Photographs courtesy of Sonobond Ultrasonics, Incorporated

Figure 8.9—(A) Traversing-Head Ultrasonic Seam Welding Machine (B) Foil-Splicing Application

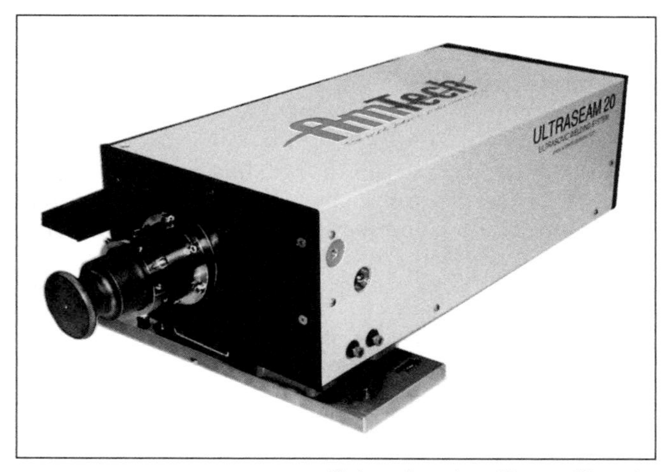

Photograph courtesy of Branson Ultrasonics

Figure 8.10—Fixed-Head Ultrasonic Seam Welding Machine

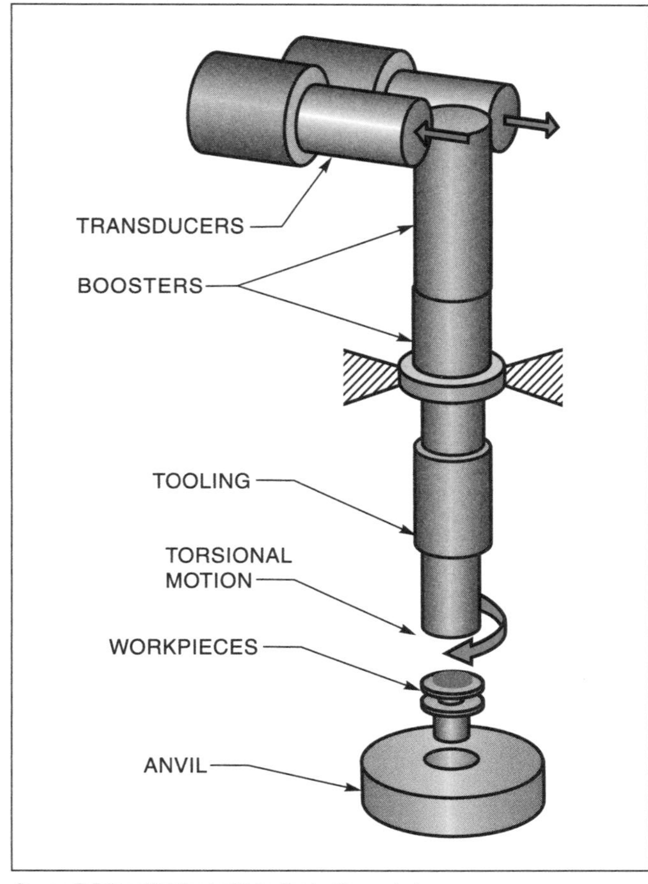

Source: © Edison Welding Institute. Used with permission.

Figure 8.11—Ultrasonic Torsion Welding Machine

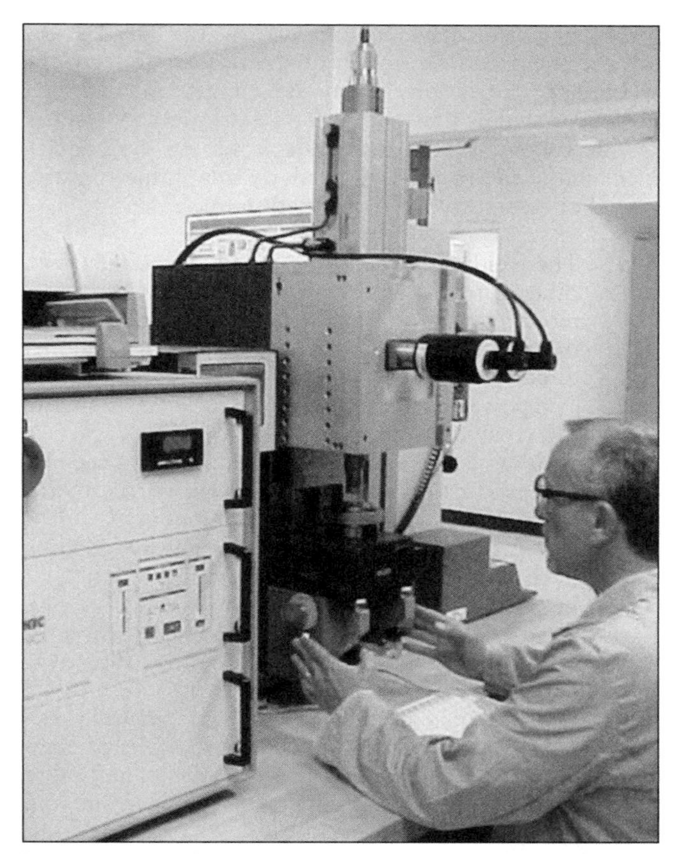

Photograph courtesy of © Edison Welding Institute and Telsonic, Inc. Used with permission.

Figure 8.12—10-kW Ultrasonic Torsion Welding System

type of welding system and by manufacturer. Of the several control variables, not all can be independently set (for example, weld energy or weld time can be set, but not energy and time simultaneously). The type and sophistication of readouts, data processing, data recording, and computer interfaces also vary by system and manufacturer. Most controls and output parameters ultimately can be related to weld quality.

The welding cycle in production applications generally is controlled automatically and usually is actuated by dual palm buttons or a foot switch on the welding machine. The automatic cycle consists of lowering the sonotrode tip or raising the anvil, applying clamping force, introducing the ultrasonic pulse, and retracting the sonotrode or anvil.

Other controls and indicators are included on some welding machines to monitor operation of the equipment or to provide flexibility in use. These machines may provide for the selection and control of energy in joules, height or distance, and also for the speed of the advancement and retraction of the sonotrode from the weldment. Deviations from set ranges may trigger alarms or signals that will reject a workpiece or weld.

The following features of ultrasonic welding equipment make the process particularly adaptable to automated or semiautomated production lines:

1. The welding head can be readily interfaced with other automatic processing equipment. It can be mounted on any rigid structure and in any position, with the tip contacting the work from any direction;
2. The power source may be located as far as 15.2 m (50 ft) away from the welding head;
3. Welding times are usually a fraction of a second, and production rates are limited primarily by the speed of the work-handling equipment;
4. The process does not involve extensive heating of the equipment or the workpiece;
5. In automatic filling and closing production lines, accidental spillage of the contents on the weld interface usually will not significantly affect weld quality; and
6. Automated equipment also may include frequency counters, weld counters, material handling actuators, indexing mechanisms, and other devices to minimize operator involvement and maximize production rates.

WELDING VARIABLES AND PROCEDURES

Setting up procedures for ultrasonic welding production includes determining welding machine settings for the basic variables of ultrasonic welding: vibration amplitude, power, energy, clamping force, and welding time. Other considerations are choosing a joint design appropriate to the application, choosing the welding machine with the power capacity suitable to the application and the thickness of the metal, selecting means of controlling workpiece resonance, determining the type of surface preparation, and providing for a welding atmosphere if it is needed.

Only a select few variables inherent to ultrasonic welding may be independently set on a given type of welding machine for a given application. One system may accommodate the selection of power, clamping force, and time, while another system may permit the selection of amplitude, force, and energy. Variables for a specific application are typically established experimentally, making use of experience gained from similar applications when possible. Once the parameters are determined and the welding machine controls are set, adjustments usually are not required unless alterations to the equipment have been made, such as sonotrode tip changes or changes in the workpieces.

An interaction between the welding variables is possible. For example, for a given application, there may be an optimum clamping force at which minimum power is required to produce acceptable welds. This condition can be established by plotting a threshold curve of the formation shown in Figure 8.13, which defines the conditions of the best dynamic coupling between the sonotrode tip and the workpiece and thus, the minimum power to produce strong welds.

Plotting a threshold curve consists of making welds at the selected power and clamping force settings and evaluating weld quality by appropriate criteria, for example, pulling a nugget in spot welds, or weld shear strength, or in some cases, visual examination of weld deformation.

The steps may consist of the following:

1. The welding time is set at a reasonable value. One-half second is a good starting point for most metals. For very thin metals, a shorter weld time is usually chosen.

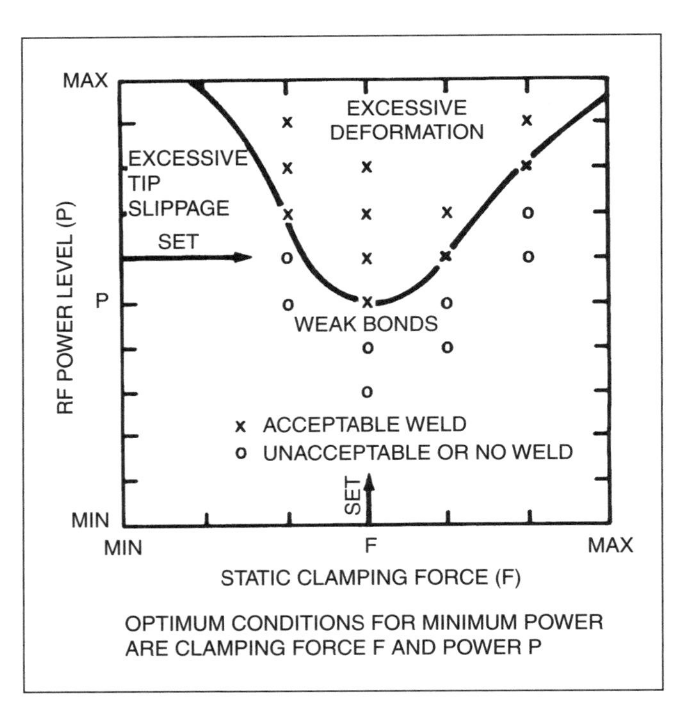

Figure 8.13—Typical Threshold Curve Relating Power and Clamping Force

2. Welding is started at low values of clamping force and power, and a series of test welds is made with incrementally increasing values of clamping force at a fixed power level. The welds are evaluated and the results plotted, as shown in Figure 8.13, indicating acceptable and unacceptable welds.
3. This procedure is repeated at other values of ultrasonic power until an inverted bell-shaped curve is generated.

The use of these data will generate a curve that separates the acceptable welds from the unacceptable welds. Welding is ordinarily done using the clamping force value for minimum acceptable power and a power level somewhat above the minimum. The total energy (E) required is the product of the selected power (P) and weld time (T), or $E = P \times T$. If welding time is decreased, power must be increased accordingly. The threshold curve is a practical and efficient method for determining proper welding machine settings for all types of ultrasonic welds.

Welding machines with different power levels typically have differing ranges of clamping force; consequently machines that deliver increased power are capable of applying greater clamping force required to achieve a given weld. Table 8.1 shows typical ranges for welding machines of various power capacities.

WELDING IN ENERGY MODE

Variations in surface conditions (oxides and contaminants) can result in varying the efficiency with which energy is coupled to the workpieces during a weld cycle. For example, an oil film might delay effective coupling in an interval during which the film is essentially burned off, after which coupling and welding proceed as they would for initially clean workpieces. If a weld cycle is set for time alone under such circumstances, some welds might be incomplete due to the time required for burnoff, while others, where less oil film is present, might be overwelded. Figure 8.14 shows the power curves for two welds with different surface conditions, and a third weld in which a part of the component (workpiece) is missing. While differing in peak powers, the two curves have the same net energies (the areas under the curves), with both resulting in sound welds.

Table 8.1
Typical Clamping Force Ranges for Ultrasonic Welding Machines of Various Power Capacities

Machine Power Capacity, W	Approximate Range of Clamping Force	
	Newtons	lbf
20	0.04 to 1.7	0.009 to 0.39
50 to 100	2.2 to 67	0.5 to 15
300	22 to 800	5 to 180
600	310 to 1780	70 to 400
1200	265 to 2650	60 to 600
4000	1100 to 14 250	250 to 3200
8000	3500 to 17 800	800 to 4000

POWER-FORCE PROGRAMMING

Power-force programming involves applying incremental variations in power and clamping force during the welding cycle. The cycle is initiated at low power and high clamping force. After a brief interval, power is increased and force reduced. The cycle is accomplished automatically with special logic circuitry. Certain materials, such as refractory metals and alloys, are more effectively welded when power-force programming is used. Weld strength is higher and cracking of the weld metal is minimized when these programming techniques are used.

GEOMETRY AND COMPOSITION OF SONOTRODES AND ANVILS

A wide variety of weld patterns can be achieved by varying the details of the weld surface of the sonotrode. For example, small-diameter 0.10 mm (0.004 in.) ring patterns can be made, and even larger diameters are achievable with special horn configurations. Linear pattern lengths of 25 mm to 38 mm (1 in. to 1.5 in.) may be made, and patterns as long as 200 mm (8 in.) are achievable for some foils.

The sonotrode usually is made of high-quality, heat-treated tool steel. The fit between the tip and the horn for the lateral-drive welding machine or the reed for the wedge-reed system must be of the highest integrity for efficient transmission of vibratory energy. A poor fit will create heat at the interface and will eventually damage the components of the welding machine and reduce the service life of the sonotrode. The welding sonotrode is a consumable item that must be repaired or replaced as required.

The wedge-reed system uses a locking taper on the sonotrode, making it easily removable. A replaceable sonotrode also may be used with a lateral-drive welding machine. Alternatively, lateral-drive systems may have a tip integrated with the welding horn and may have several surfaces for welding. As the horn rotates, it provides a new welding surface.

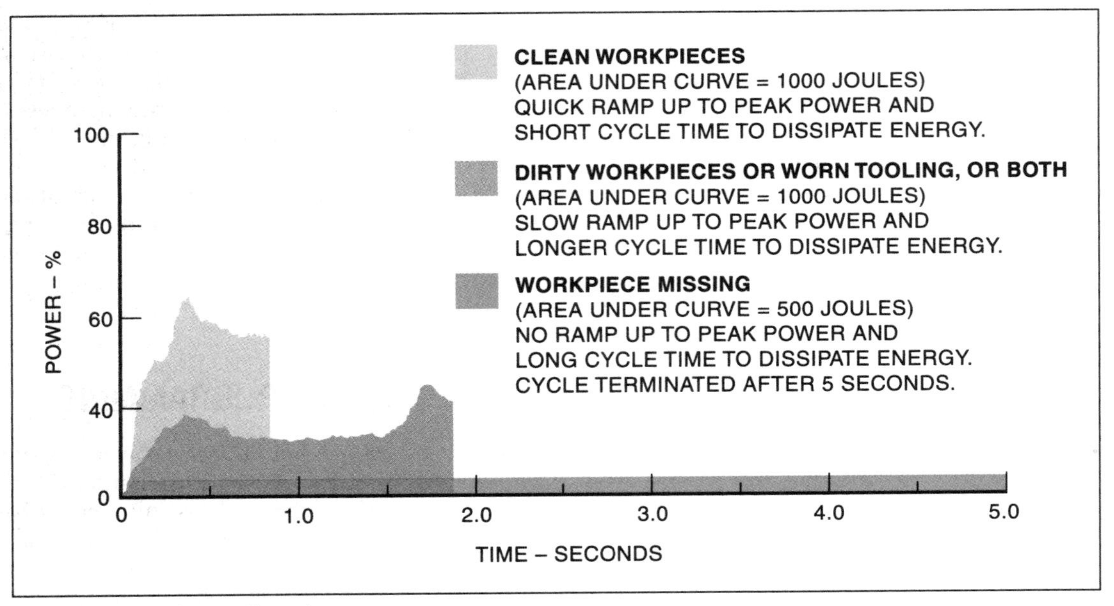

Source: Adapted from Branson Ultrasonics.

Figure 8.14—Weld Power Curves for Varying Surface Conditions

Sonotrode Sticking

When certain alloys are welded they may stick to the sonotrode. If the alloy sticks to the sonotrode, a mechanical stripper can be used, or sometimes a subsequent low-power ultrasonic pulse will be sufficient to free the workpiece from the sonotrode. If a nugget remains on the sonotrode it can be removed by operating the weld cycle into a thick piece of brass. This usually removes the stuck nugget better than mechanical abrasion. Some exotic alloys have been used in manufacturing sonotrodes to prevent particularly tenacious sticking, but with limited success. A repeated low-volume water spray on the sonotrodes can be used to reduce or eliminate the sticking. A steel shim with an oxidized surface is particularly effective in preventing both tip sticking and workpiece deformation when welding aluminum and titanium panels, for example, in an aircraft application.

Sonotrodes and anvils with various serrated or crosshatch patterns are common designs that are useful in preventing slippage between the sonotrode or anvil surfaces and the weldment. A typical crosshatch pattern would be 0.5 mm (0.020 in.) peak-to-peak, and about 0.2 mm (0.008 in.) deep. Alternatively, a roughened surface achieved by electrical discharge machining or grit blasting (for example, to a 200-micron finish) may be used to prevent slippage of the sonotrode or anvil. These various patterns and treatments of the weld face lead to longer service life of the sonotrode.

Torsion-welding sonotrodes are solid components made with the desired weld pattern integrally machined to the end of the sonotrode, typically in a circular raised, knurled pattern. Anvils may be flat or appropriately contoured to achieve intimate contact with the workpiece. For example, when welding a lid to a cylindrical container, the anvil usually is recessed to accommodate the container, with the flange in contact with the anvil surface.

Sonotrodes used in seam welding are resonant or mass-loaded disks. For welding flat surfaces, the disks are machined with a convex edge. The edge of the disk also may be contoured to achieve intimate contact with the workpieces; for example, the entire periphery of a disk can be grooved to the permit continuous seam welding of a rib to a cylinder.

Sonotrode Maintenance

When the sonotrode begins to show wear, erosion, or material pickup, it may be reconditioned by machining it to a flat surface and regrinding the serration pattern. The manufacturer often performs this service at less cost than replacing the tip. If excessively worn, the sonotrode should be replaced.

JOINT DESIGN

Joint designs for ultrasonic welding of sheet metal are less restrictive than designs used in some other types of welding. Edge distance is not critical. The main requirement is that the sonotrode tip does not crush or gouge the sheet edge. Welds in structural aluminum alloys of varied thickness have shown the same strength when welded at either 3 mm (0.125 in.) or 19 mm (0.750 in.) from the edge. Weight and material savings are achieved by using the minimum acceptable overlap.

A key factor in joint design is joint thickness. While comprehensive guidelines for combinations of all material thicknesses are not available, experience has shown that designs can be developed for various metals, thicknesses, and combinations of weld power. Figure 8.15 illustrates several joint designs for torsion welding and typical applications. Examples of hermetically sealed products and other joint designs created with ultrasonic torsion welding are shown in Figure 8.16.

Figure 8.17 lists the capacities of several ultrasonic welding machines for welding various thicknesses and types of metal.

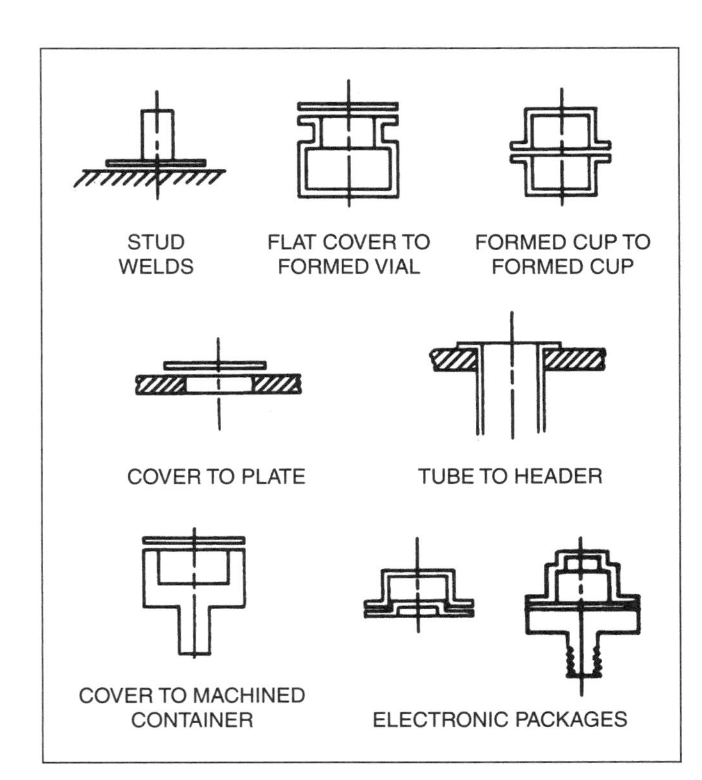

Figure 8.15—Typical Ultrasonic Torsion Welding Joint Designs and Applications

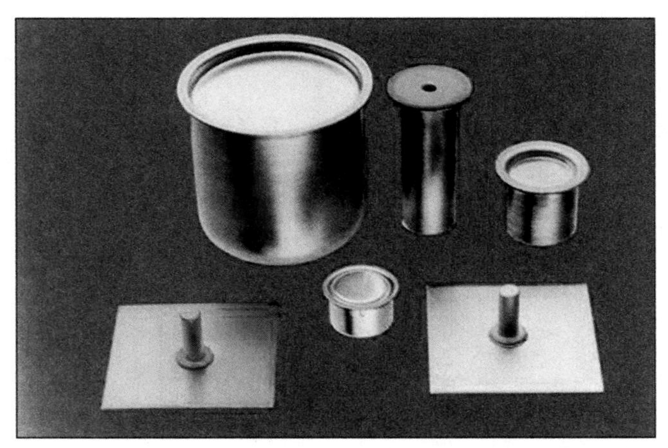

Photograph courtesy of Telsonic, Inc.

Figure 8.16—Examples of Hermetically Sealed Products and Other Applications Welded with Ultrasonic Torsion Welding System

Ultrasonic spot welding places no restrictions on weld spacing. Consecutive or overlapped welds have no effect on the quality of previously made welds, except under the resonance conditions described in the next section.

Ultrasonic torsion welding has unique capabilities for hermetic sealing, as indicated by the joint designs in Figure 8.15. Torsion welds may be preferred to spot welds for structural applications. The overlapping rings provide relatively uniform stress distribution with less stress concentration, less tendency toward cracking, and generally do not produce workpiece resonance.

CONTROL OF WORKPIECE RESONANCE

Ultrasonic welding is inherently a vibration-driven process and consequently may transmit vibrations into the workpieces. This can cause workpiece resonance, which, if not controlled, can cause fatigue damage to the workpieces or to previously made welds, or both. Several corrections can be applied separately or in combination. Resonance vibration can be eliminated by altering the workpiece dimensions or the orientation of the workpieces in the welding machine. Damping the vibration in thin sections frequently can be accomplished by applying pressure-sensitive tape to the workpieces. Clamping masses to the workpieces or clamping the workpieces into a comparatively massive fixture also will limit the resonance, even for the most difficult cases.

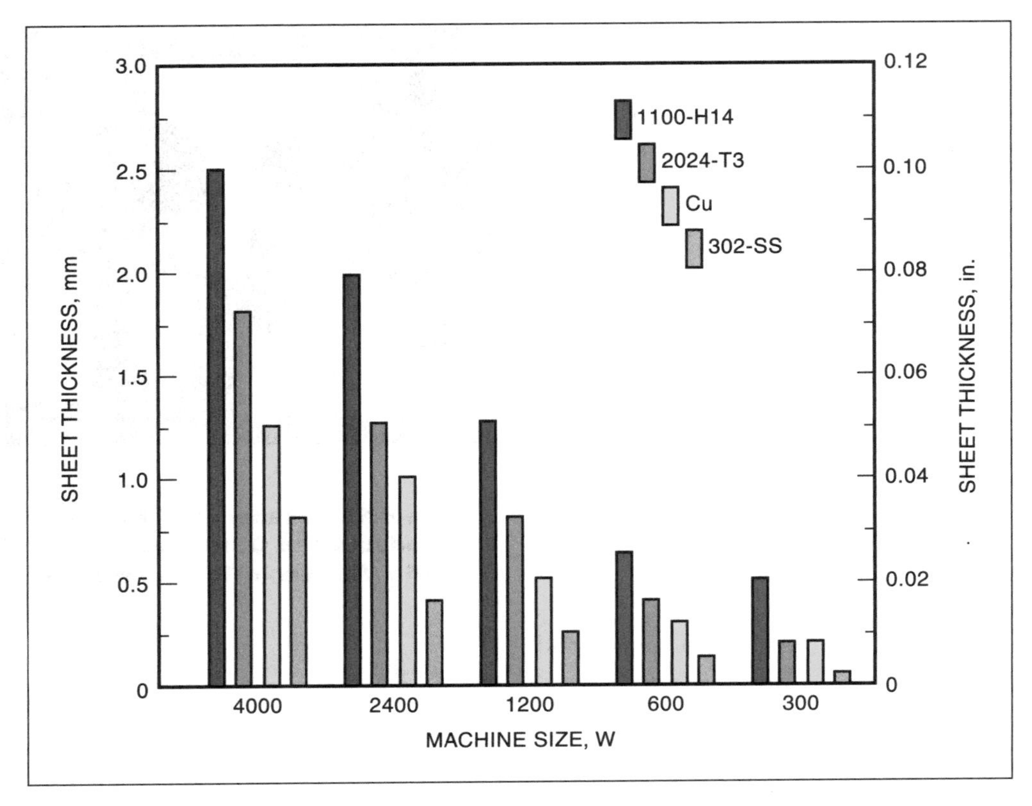

Figure 8.17—Capacities of Several Ultrasonic Spot Welding Machines Used to Join Selected Metals

SURFACE PREPARATION

A good surface finish contributes to the ease with which ultrasonic welds are made. Some of the more readily weldable metals, such as aluminum, copper, or brass can be welded in the mill-finish condition if they are not heavily oxidized. Normally, these materials require the removal of surface lubricants with only a detergent. Thin oxide films usually do not inhibit welding since they are disrupted and dispersed during the process. Metals that are heavily oxidized or contain surface scale require more careful surface preparation. Mechanical abrasion or descaling in a chemical etching solution sometimes is necessary to provide a clean surface for welding.

Once the surface scale is removed, the elapsed time before welding is not critical as long as the materials are stored in a noncorrosive environment. It is possible to weld some metals through certain surface films, coatings, or insulations, but somewhat higher ultrasonic energy levels are required. Some types of film cannot be penetrated and always must be removed prior to welding.

SPECIAL WELDING ATMOSPHERES

Ultrasonic welding usually does not require special atmospheres. When welding some metals, the process may produce discoloration of the surface in the vicinity of the weld. When such a surface is undesirable, it can be minimized by protecting the weld with inert gas, for example, with small jets of argon impinging around the contact area of the tip. For packaging applications in which sensitive materials must be protected from contamination, welding can be accomplished in a chamber filled with inert gas.

METALS WELDED

As with most other welding processes, there are no definite rules to determine which metals can or cannot be welded using the ultrasonic welding process, or the relative ease with which they may be welded. Nevertheless, some guidelines for weldability exist. The weldability of a metal is determined mainly on its tendency to deform, a property strongly tied to the crystal microstructure and the overall hardness of the metal. Weldability is known to decrease when going from a face-centered cubic to a body-centered cubic microstructure. This decrease is even more evident when welding metals with hexagonal microstructures. Weldability decreases also with increasing hardness of the base metal. When welding dissimilar metals, the properties of the softer and more easily weldable material are of great importance.

Another important factor is the hardness difference between the oxide and the base material. As an example, aluminum oxide is several times harder than the aluminum base metal, so it fractures easily when the base metal is plastically deformed, thus allowing cleaning for a metal-to-metal contact. Conversely, copper has a relatively soft oxide. It deforms with the copper base metal and therefore it is much more difficult for metal-to-metal contact to take place. Copper usually requires extensive cleaning prior to welding.

In general, most metals and alloys can be ultrasonically welded. Figure 8.18 identifies some of the monometallic and bimetallic combinations that can be welded on a commercial basis. Blank spaces in the chart indicate combinations of metals for which welds have not been attempted or are known to have been unsuccessful. Figure 8.17 also illustrates that various metals differ in weldability according to the composition and properties of the specific metal. Metals considered difficult to weld typically require either high power or long weld times, or both. These applications tend to incur operational problems such as sonotrode sticking or short tool life.

COMMONLY WELDED METALS

Specific information on several groups of metals and alloys commonly welded by ultrasonic welding is presented in this section.

Aluminum Alloys

All combinations of aluminum alloys are weldable. Photomicrographs show significant diffusion across the interface in ultrasonic welds of aluminum alloys. Aluminum in any available form can be welded: cast,

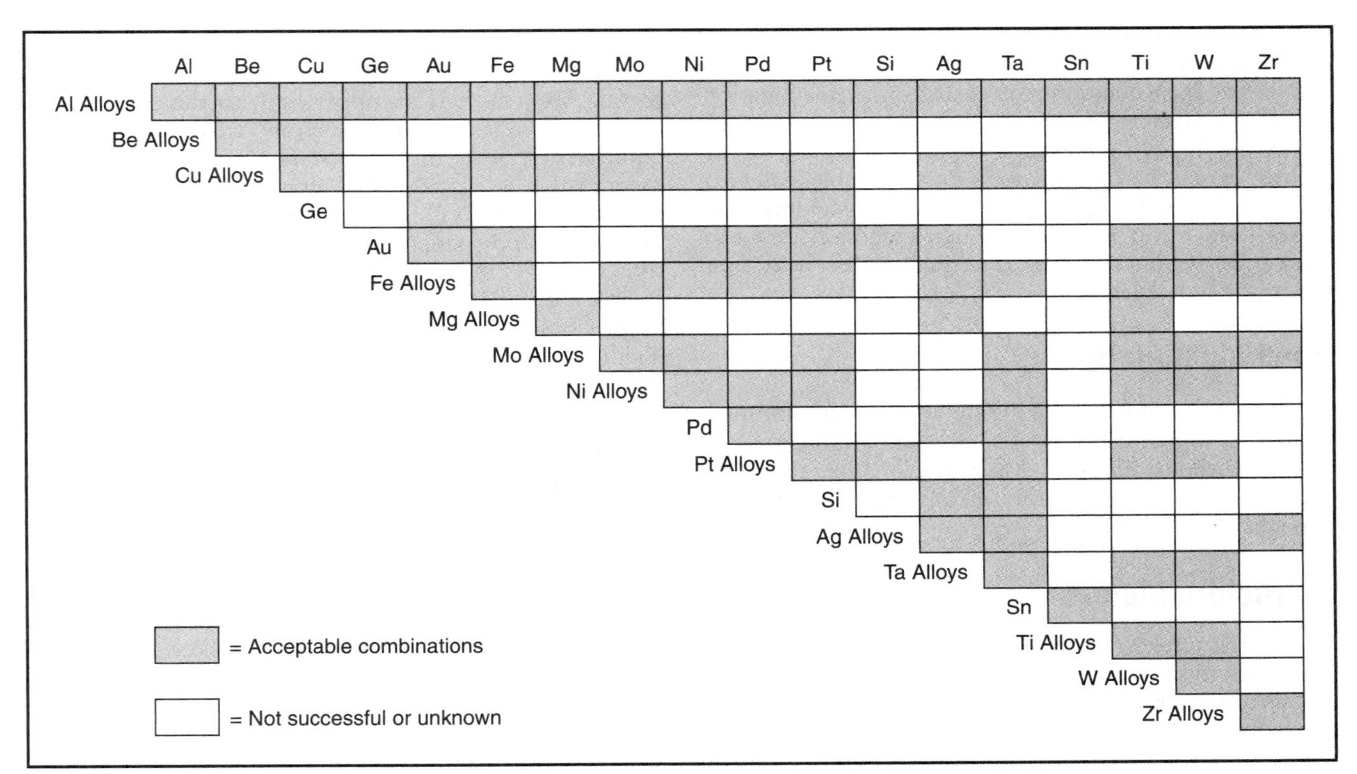

Figure 8.18—Metal Combinations Welded with the Ultrasonic Process

extruded, rolled, forged, or heat-treated. Soft aluminum cladding on the surface of these alloys facilitates welding. Aluminum can be welded to most other metals, including germanium and silicon, the primary semiconductor materials. Annealed materials exhibit high deformation. In general, the hardness factor of work-hardenable aluminum alloys should be 25% to 50% of the maximum hardness of the alloy.

In ultrasonic spot welding, some aluminum alloys, particularly the 1100 alloys and the 2036 alloy, tend to stick to the sonotrode. Sonotrode sticking sometimes can be alleviated by using high power and a short weld time. Sonotrode tips made of hardened tool steel produce the best results. Other remedies include the use of a mechanical stripper for light sticking, or placing a hard steel shim between the vibrating tip and the weldment.

Copper Alloys

Copper and copper alloys, such as brass and gilding metal, are relatively easy to weld. High thermal conductivity is not as deterring a factor in ultrasonic welding as it is in fusion welding. Surface condition is an especially important variable in the ultrasonic welding of copper alloys.

Iron and Steel

Satisfactory welds can be produced in various types of iron and steel, such as ingot iron, low-carbon steels, tool and die steels, austenitic stainless steels, and precipitation-hardening steels. The power requirements for these metals are higher than for aluminum and copper. Nevertheless, it may be difficult to justify the replacement of resistance welding and other common welding processes used for steels unless a particular advantage of ultrasonic welding is applicable, such as less heat damage to the workpieces.

Precious Metals

The precious metals, including gold, silver, platinum, and palladium and alloys of these can be ultrasonically welded without difficulty. Most precious metals can be satisfactorily welded to other metals and to germanium and silicon.

Refractory Metals

The refractory metals, including molybdenum, niobium, tantalum, and tungsten and some alloys of these metals are among the most difficult metals to join with ultrasonic welding. Thin foils of these metals can be joined if they are relatively free from contamination and surface or internal defects.

Other Metals

Thin-gauge nickel, titanium, zirconium, beryllium, magnesium, and many alloys of these metals can be ultrasonically welded to similar and dissimilar metals and to other materials. Metal foils and wires can be readily joined to thermally sprayed metals on glass, ceramics, or silicon. These welds are particularly useful in the semiconductor industry, as discussed in the section "Microwelding."

MULTIPLE-LAYER WELDING

Multiple-layer welding is feasible with ultrasonic welding. As many as 80 to 100 layers of 25 µm (0.001 in.) aluminum foil can be joined simultaneously with either spot welds or continuous seam welds. Several layers of dissimilar metals also can be welded together. A typical application is welding multiple layers of aluminum ribbon to a terminal to produce foil-wound capacitors.

THICKNESS LIMITATIONS

Different materials present different levels of difficulty in producing a weld. Varying thickness limitations for ultrasonic welds depend on the weldability of the metal. Also, there is an upper limit to the thickness of any metal that can be effectively welded because of equipment power limitations. For a readily weldable metal, such as Type 1100 aluminum, the maximum thickness in which reproducible high-strength welds can be made is approximately 2.5 mm (0.10 in.). The upper thickness limit of harder metals is in the range of 0.4 mm to 1.0 mm (0.015 in. to 0.040 in.). This limitation applies only to the workpiece that is in contact with the welding tip; the other workpiece may have greater thickness. Thus, thin-to-thick sections can be welded. Extremely thin sections can be welded successfully. For example, fine wire with diameters less than 0.01 mm (0.0005 in.) and foil as thin as 0.004 mm (0.00017 in.) can be welded.

When a weld is difficult to achieve with available power levels, good quality joints can be made by inserting a foil of another metal between the two workpieces. Examples of this technique are the use of 0.01 mm (0.0005 in.) nickel or platinum foil between molybdenum workpieces, and beryllium foil welded to AISI Type 310 stainless steel using an interleaf of thin Type 1100-H14 aluminum foil. The weldability range of Type 2014-T6 aluminum alloy can be extended by using a foil interleaf of Type 1100-O aluminum.

APPLICATIONS

The application of ultrasonic welding spans a wide range of industries. Major applications include electrical connections, foil and sheet products, encapsulation and packaging, the sealing of leak-tight containers, and structural welding.

ELECTRICAL CONNECTIONS

Electrical connections of various types are effectively made by ultrasonic welding. Both single and stranded wires can be joined in various combinations to other wires and to terminals. These joints are frequently uninhibited by anodized coatings on aluminum and certain types of electrical insulation, and the weld is made through these coatings to the base metal. Other current-carrying devices, such as electric motors, field coils, transformers, capacitors and auto, truck, and appliance harnesses, may be assembled with ultrasonically welded connections. Other examples include welding copper strip to copper strip for grounding networks, copper contacts to aluminum plates in ignition modules, and flat wires to starters, coils and transformers. Some typical electrical connections are shown in Figure 8.19. Figure 8.19(A) is a field coil assembly for automotive starter motors. Figure 8.19(B) shows a typical wire harness connection; Figure 8.19(C) is a truck-trailer wire harness interconnection, and Figure 8.19(D) shows thin-gauge aluminum (left) and nickel (right) tabs welded to 20 layers of expanded aluminum and copper mesh used in a battery intercell.

Photograph courtesy of Sonobond Ultrasonics, Incorporated

(A) Field Coil Assembly

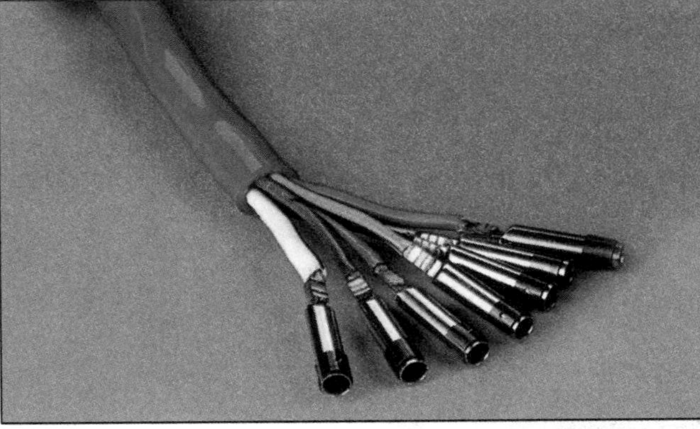

Photograph courtesy of Branson Ultrasonics

(C) Typical Wire Harness Connection

Photograph courtesy of Sonobond Ultrasonics, Incorporated

(B) Multiple-Wire Splice

Photograph courtesy of Branson Ultrasonics

(D) Aluminum and Nickel Tabs Welded to Multiple Layers of Expanded Metal Mesh for a Battery Intercell

Figure 8.19—Electrical Interconnections

One of the most extensive fields of application of ultrasonic welding is in microelectronics, by which fine aluminum and gold lead wires are attached to transistors, diodes, and other semiconductor devices. This application is discussed in further detail in the section "Microwelding."

Foil and Sheet

An important application of ultrasonic welding is in foil rolling mills, where broken and random lengths of aluminum foil are welded in continuous seams. Highly reliable splices capable of withstanding annealing operations are rapidly welded in foils up to 0.127 mm (0.005 in.) in thickness and 180-cm (72-in.) wide. The splices are almost undetectable after subsequent working operations. Aluminum and copper sheet up to about 0.5mm (0.020 in.) thick can be spliced by ultrasonic welding using special processing and tooling. Multiple layers of foil are readily welded to capacitor terminals with the ultrasonic welding process.

Encapsulating and Packaging

Ultrasonic welding is used for a wide variety of packaging applications that range from soft foil packets to pressurized cans. Leak-tight seals can be produced by torsion, seam, and line (straight-line closure) welding. The process is useful for encapsulating materials that are sensitive to heat or electrical current, such as primary explosives, slow-burning propellants and pyrotechnics, high-energy fuels and oxidizers, and living tissue cultures.

WELDING IN PROTECTIVE ATMOSPHERES

Ultrasonic welding can be accomplished in a protective atmosphere or vacuum. Thus, the process is frequently used to join instrument components and other items that must be protected from dust or contamination. This capability also permits the encapsulation of chemicals that react with air. Torsion welds can be made for closures of copper and aluminum canisters with diameters up to about 6.5 cm (2.5 in.). Ultrasonic welding is extensively used for tube closures in refrigeration systems.

Line welding is used for packaging applications with one or more straight-line seams, such as sealing the ends of squeeze tubes. Square or rectangular packets are produced by intersecting line welds on each of the four edges. Continuous seam welding is used to seal packages that cannot be accommodated with ring or line welding. Figure 8.20 shows a semi-portable ultrasonic welding system for creating a line weld to close copper tubing.

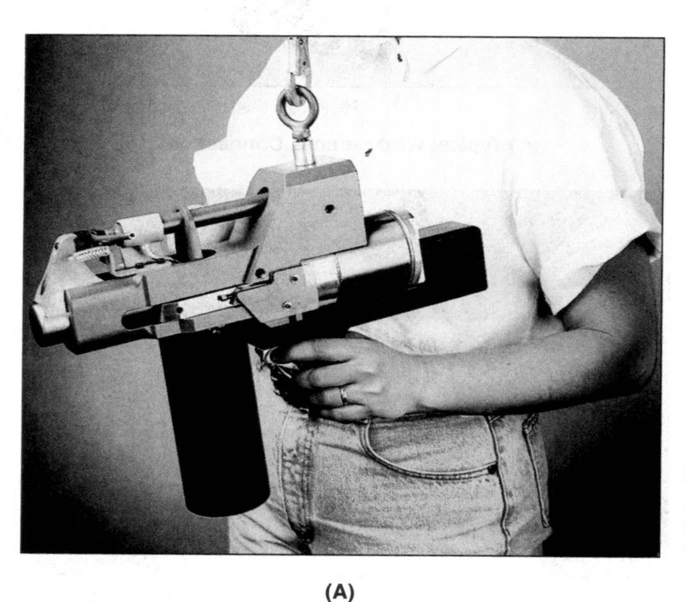

(A)

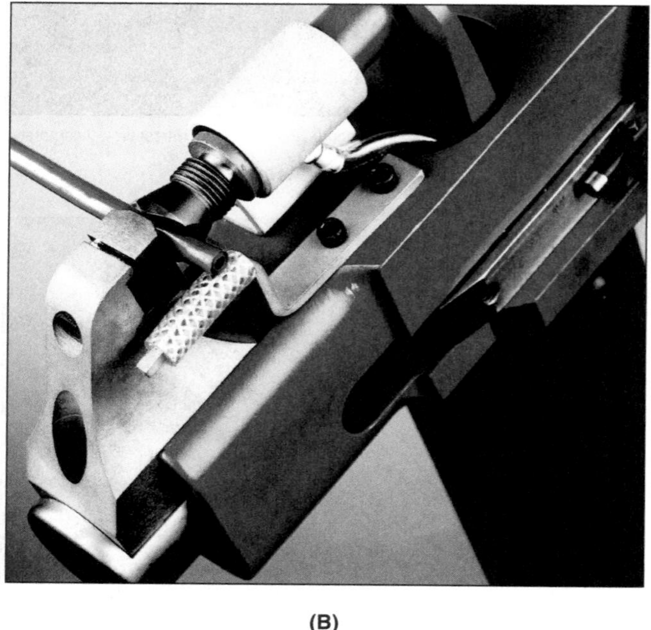

(B)

Photographs courtesy of Sonobond Ultrasonics, Incorporated

Figure 8.20—(A) Semi-Portable Ultrasonic Welding System for Tube Closure and (B) Tube Closure in Progress

STRUCTURAL WELDING

Ultrasonic welding provides high-integrity joints for structural applications within the limitations of weldable sheet thickness. The process has been used to assemble secondary aircraft structures, such as the helicopter access door shown in Figure 8.21. This assembly consists of inner and outer skins of aluminum alloy joined by multiple ultrasonic spot welds. Individual ultrasonic welds in this application had 2.5 times more average strength than the minimum required for resistance spot welds in the same metals and same thickness. Assembled doors sustained loads 5 to 10 times greater than the design load without weld failure in air load tests. Significant savings in fabrication and energy costs also were evident when compared to costs of adhesive bonding.

Solar Energy Systems

The use of ultrasonic welding has reduced fabrication costs for some solar energy conversion and collection systems. Systems for converting solar heat to electricity frequently involve photovoltaic modules of silicon cells joined by aluminum connectors. An ultrasonic seam welding machine, operating at travel speeds of up to 9.14 m/min (30 ft/min), is capable of joining all connectors in a single row in a fraction of the time required for hand-soldering or individual spot welding. After all connections are made on one side of the assembly, the process is repeated on the opposite side.

Solar collectors for hot-water heating systems may consist of copper or aluminum tubing attached to a collector plate. An ultrasonic seam welding machine or step-mode spot welding machine can be used to weld through the plate to the tube. Fabrication costs are lower than those of soldering, resistance spot welding, or roll welding.

Photographs courtesy of Sonobond Ultrasonics, Incorporated

Figure 8.21—Ultrasonic Spot Welds in a Helicopter Access Door

OTHER APPLICATIONS

Continuous ultrasonic seam welding is used to assemble components of corrugated heat exchangers. Strainer screens are welded without clogging the holes. Space radiation counters are fabricated using ultrasonic torsion welding to join beryllium foil windows to stainless steel frames to provide a leak-tight weld to contain helium. The few examples noted here demonstrate the versatility of the ultrasonic welding process.

MECHANISMS OF ULTRASONIC WELDING

Complex mechanical and metallurgical interactions between the surfaces of the two workpieces are involved in ultrasonic welding. These interactions have an effect on the degree of weldability of various metals.

DEFORMATION IN THE WELD ZONE

Prior to welding, contact between two workpieces occurs only between local surface asperities, as shown in Figure 8.22(A). Thus, it can be appreciated that no matter how smooth a surface may appear visually or by measurement, under sufficient magnification the surface is revealed as irregular, having asperities (peaks and valleys). When two surfaces are brought together, they will touch at local asperities. Oxides and contaminants on the surfaces will prevent intimate metal-to-metal contact and welding will not take place.

In the initial stages of ultrasonic welding, the introduction of ultrasonic vibration creates a relative motion between the faying surfaces, causing shearing and elastic-plastic deformation between the asperities. This effect is enhanced by the static force pressing the surfaces together. Highly localized interfacial slipping tends to break up the oxides and surface film, permitting metal-to-metal contact at many asperities, as shown in Figure 8.22(B). The original microstructure of the base metal may be substantially altered or almost completely destroyed and replaced by a fine-grained structure. Residual stresses due to elastic deformation may be released through atomic reorientation or diffusion.

The region of elastic-plastic deformation that is initially confined near the asperities may extend into the interior of the workpiece, depending on the levels of

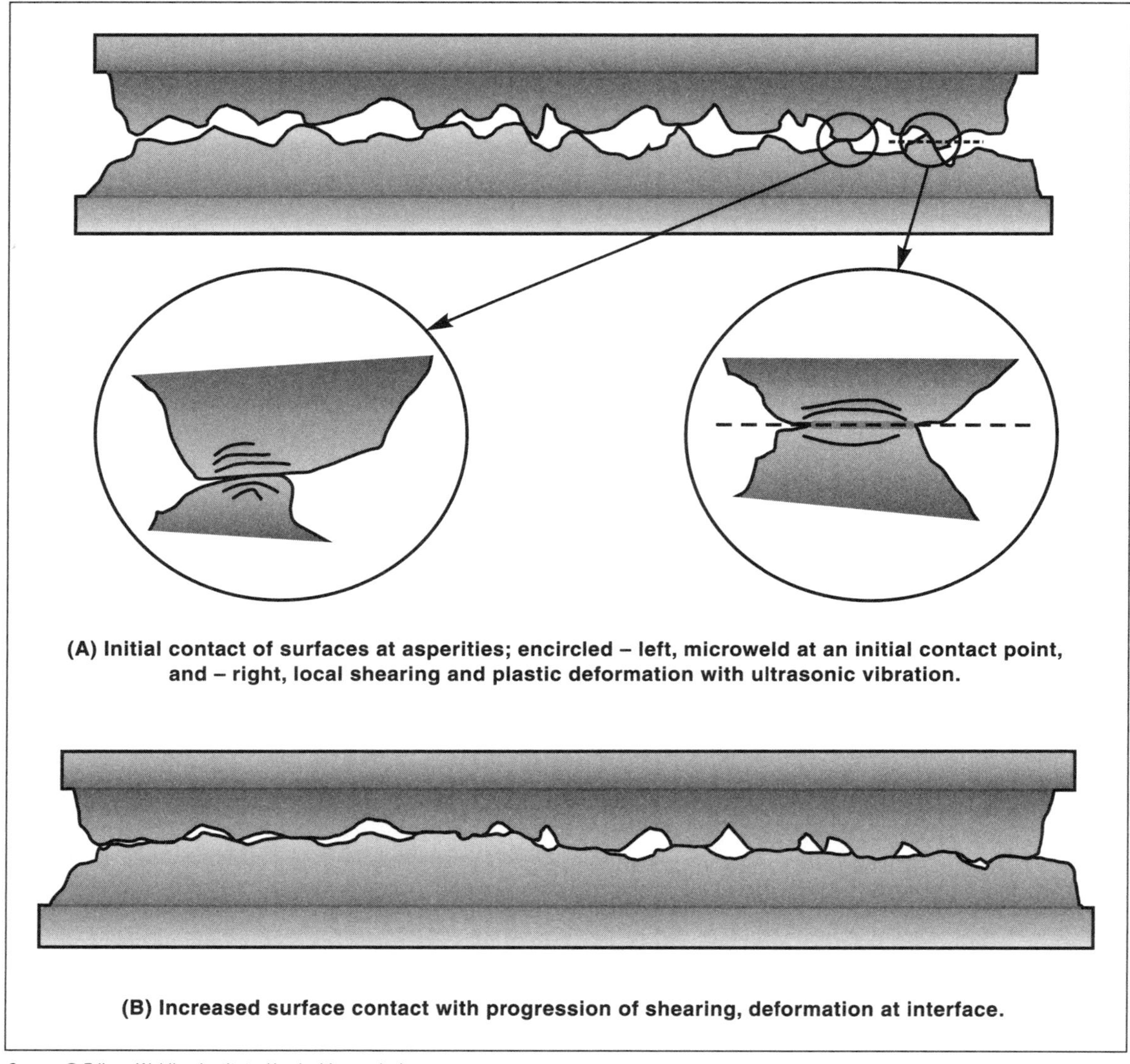

Source: © Edison Welding Institute. Used with permission.

Figure 8.22—Development of Contact Surfaces in Ultrasonic Welding

ultrasonic power and force involved in the weld. Microscopic examination of this region will show evidence of complex patterns of local plastic flow. The welding action will break up and disperse surface oxides, and in some cases, may create some extrusion of metal adjacent to the weld, as illustrated in Figure 8.23.

Stress Patterns

Microscopic examination of photoelastic stress models reveals the significant aspects of elastic-plastic stress patterns during welding. With applied static force only, the stress pattern is symmetrical around the axis of force application. When a lateral force is superimposed, such as that occurring during one half-cycle of vibration, the force shifts in the direction of this lateral force, and shear stress is produced on that side of the axis. When the direction of the lateral force is reversed, as in the second half of the vibratory cycle, the shear stress shifts to the opposite side of the axis. The shear stress changes direction thousands of times per second during ultrasonic welding.

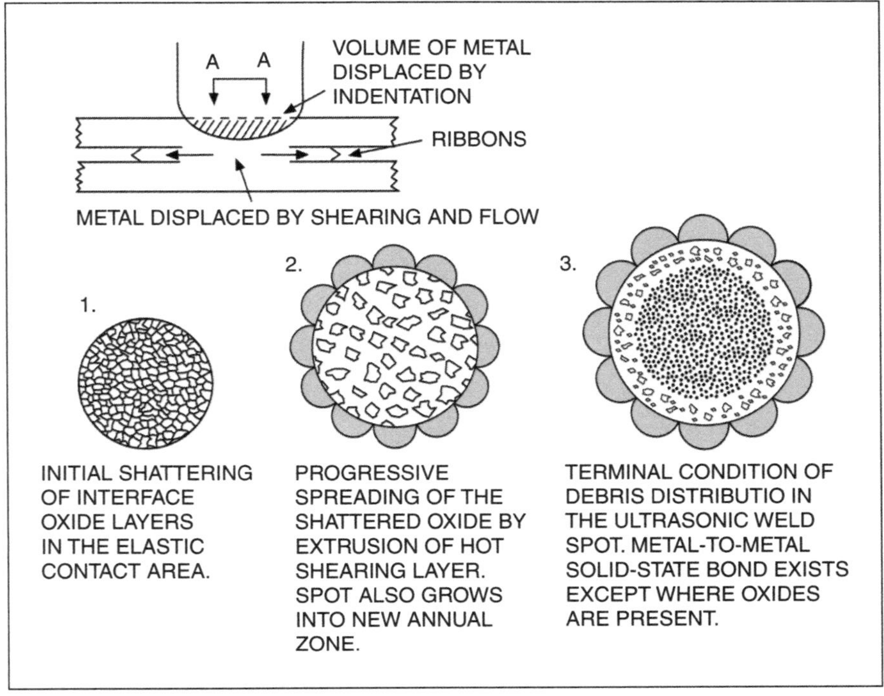

Source: Sonobond Ultrasonics, Incorporated.

Figure 8.23—Dispersion of Oxides and Extrusion of Metal Near the Weld

It should be noted that the stress pattern at the weld interface also changes with sonotrode shape. Flat sonotrode surfaces result in more even stress distribution at the weld interface than spherical shapes. The sonotrode shape thus influences the plastic deformation pattern in the weld zone.

Weld Zone Temperature

The ultrasonic welding of metals at room temperature produces a rise in localized temperature from the combined effects of elastic hysteresis, localized interfacial slip, and plastic deformation. However, base metals do not melt at the interface when the correct welding machine settings for clamping force, power, and weld time are used. Ultrasonic weld specimens examined with both optical and electron microscopy have shown phase transformation, recrystallization, diffusion, and other metallurgical phenomena, but no evidence of melting.

Interfacial temperature studies made with very fine thermocouples and rapid-response recorders show a high initial rise in temperature at the interface, followed by a leveling off. Increasing the power raises the maximum temperature achieved. Increasing the clamping force increases the initial rate of temperature rise but lowers the maximum temperature achieved. Thus, it is possible to control the temperature profile, within limits, by appropriate adjustment of welding machine settings.

The interface temperature rise also is related to the thermal properties of the metal being welded. The absolute temperature within the weld is mainly a function of the temperature-dependent mechanical properties of the harder and stronger material to be welded. For example, the temperature rise at the interface is significantly higher when copper is welded to Monel™ than when copper is welded to copper. The extent of the temperature increase also is dependent on the thermal conductivity of the materials.

When welding metals that have a wide range of melting temperatures, the maximum temperature in the weld is approximately 35% to 50% of the absolute melting temperature of the metal. When suitable welding machine settings are used, metals with wide differences in melting temperature can be welded. This capability is one of the major benefits of ultrasonic welding.

Power Requirements and Weldability

The energy required to make an ultrasonic weld can be related to the hardness of the workpieces and the thickness of the workpiece in contact with the sonotrode. Analysis of data covering a wide range of materials and thicknesses is expressed in the following empirical relationship:

$$E = K(HT)^{3/2} \quad (8.1)$$

where

E = Electrical energy, W × S(J);
K = Constant for a given welding system;
H = Vickers hardness number; and
T = Thickness of the sheet in contact with the sonotrode tip, mm (in.).

The constant K is a complex function that primarily involves the electromechanical conversion efficiency of the transducer, the impedance match into the weld, and other characteristics of the welding system. Different types of transducer systems will have substantially different K values.

Figure 8.24 shows the relationship between the energy required for sound spot welds and the hardness of various thicknesses of any weldable sheet metal, based on the above equation. It provides a convenient first approximation of the minimum electrical input energy required for a ceramic transducer-type spot welding machine based on the sheet thickness and the Vickers hardness evaluation of the metal. Similar data can be derived for torsion, line, and seam welds. For seam welds, the energy is expressed in terms of the unit length of the seam.

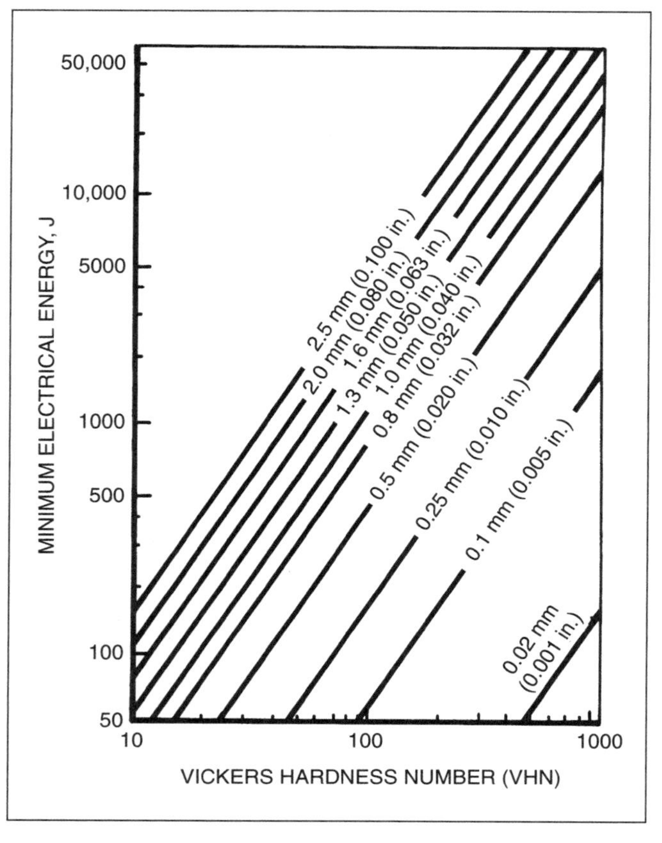

Figure 8.24—Relationship Between the Minimum Electrical Energy Required for Ultrasonic Spot Welding and Metal Hardness for Several Thicknesses of Sheet

MICROSTRUCTURAL PROPERTIES

Metallographic examination of ultrasonic welds in a wide variety of metals shows that a number of metallurgical phenomena result when vibratory energy is introduced into the weld zone. The following are three important types of metallurgical events:

1. Interfacial phenomena, such as interpenetration and surface film disruption;
2. Working effects, such as plastic flow, grain distortion, and edge extrusion; and
3. Heat effects, such as recrystallization, precipitation, phase transformation, and diffusion.

Ultrasonic welding is accompanied by local plastic deformation along the faying surfaces by interdiffusion or recrystallization at the weld interface, and by interruption and displacement of oxide or other barrier films. Surface films, which are broken up by the stress reversals and plastic deformations that occur along the interface, may be displaced in the vicinity of the interface or may simply be interrupted in continuity in random areas within the weld zone. The actual behavior of the surface film depends on several factors, including the welding machine settings, the properties of the film and the base metal, and the temperature achieved at the weld interface. Recrystallization of the metal frequently occurs in the weld nugget. Sufficient heat may be generated in certain alloys that exhibit precipitation behavior or phase transformation to induce these effects. Although diffusion may occur across the interface, the extent of the diffusion is limited by the short weld time. More than one of these metallurgical effects may be apparent in the same weld, and different effects may occur in welds in the same metal but produced at different welding machine settings.

Several typical examples are illustrated in Figure 8.25. An extreme example of interpenetration across a

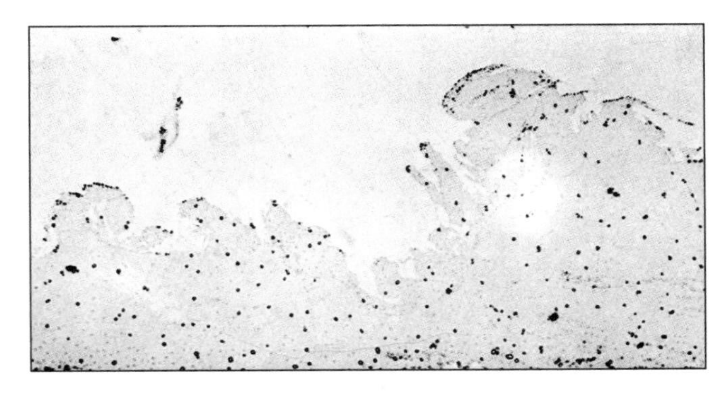

(A) 0.076 mm (0.003 in.) Nickel Foil (Top) to 0.076 mm (0.003 in.) Gold-Plated Kovar™ Foil (×150)

(B) 0.027 mm (0.005 in.) Nickel Foil (Top) to 0.501 mm (0.020 in.) Molybdenum Sheet (×100)

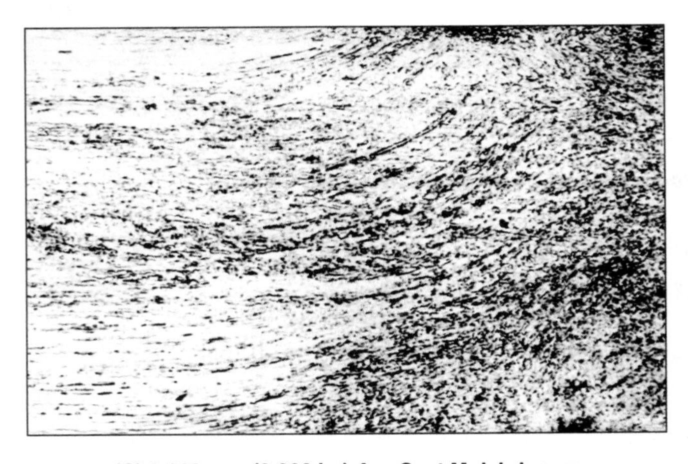

(C) 0.203 mm (0.008 in.) Arc-Cast Molybdenum Sheet to Arc-Cast Molybdenum Sheet (×70)

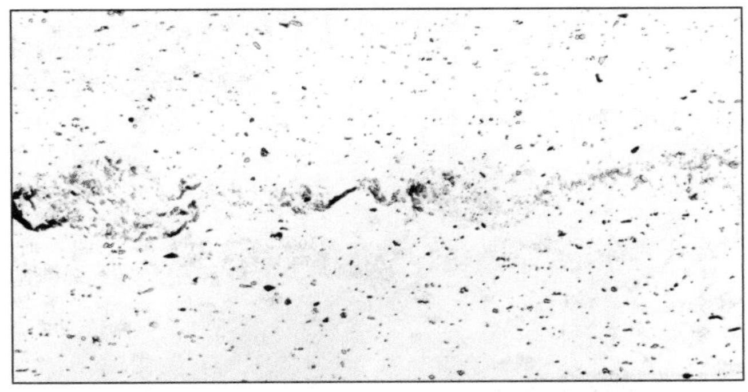

(D) 0.305 mm (0.012 in.) Type 1100-H14 Aluminum Sheet to Type 1100-H14 Aluminum Sheet (×250)

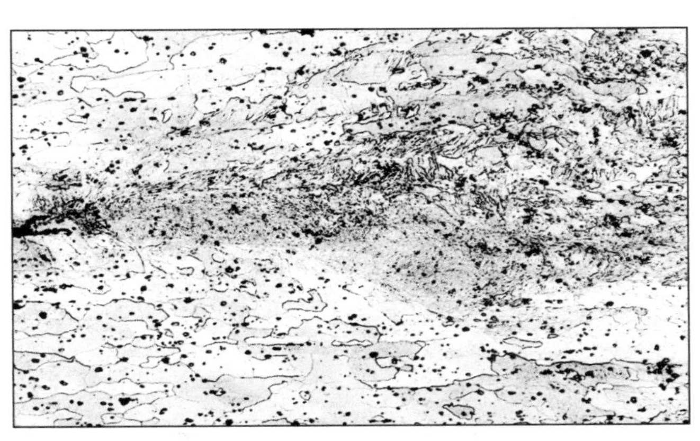

(E) 0.813 mm (0.032 in.) Type 2024-T3 Aluminum Alloy Sheet to Type 2024-T3 Aluminum Alloy Sheet (×75)

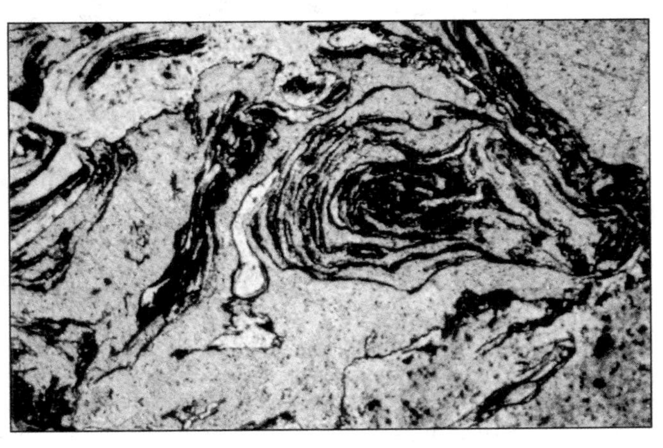

(F) Copper-to-Silver Weld Interface (×200)

Figure 8.25—Photomicrographs of Typical Ultrasonic Welds

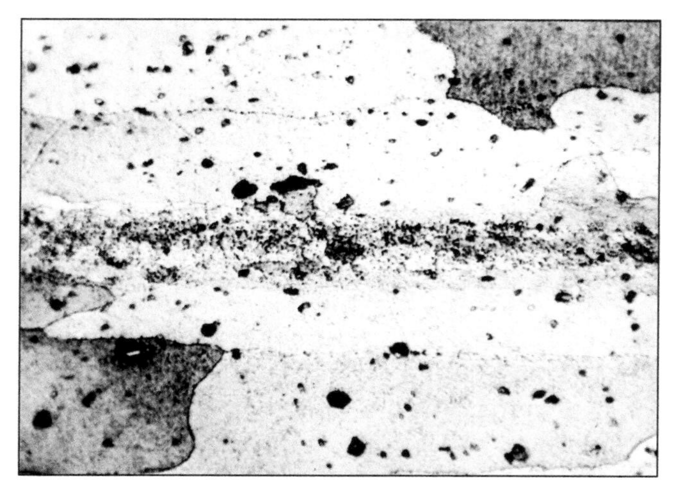

(G) 1.016 mm (0.040 in.) Type 2020 Aluminum Sheet to Type 2020 Aluminum Sheet (×375)

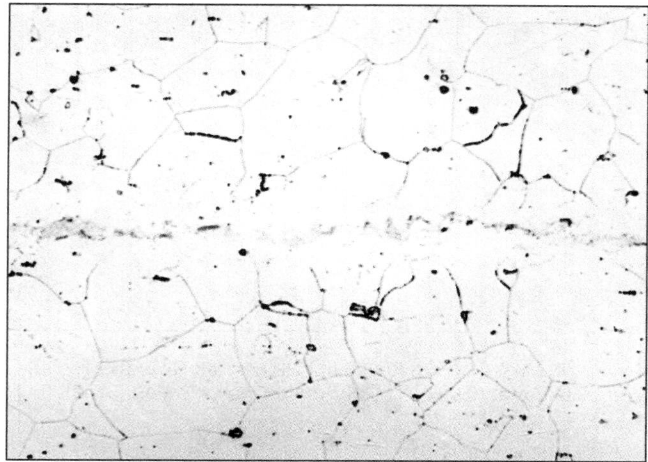

(J) 0.305 mm (0.012 in.) Solution Heat-Treated and Aged Inconel™ Sheet to Solution Heat-Treated and Aged Inconel (×750)

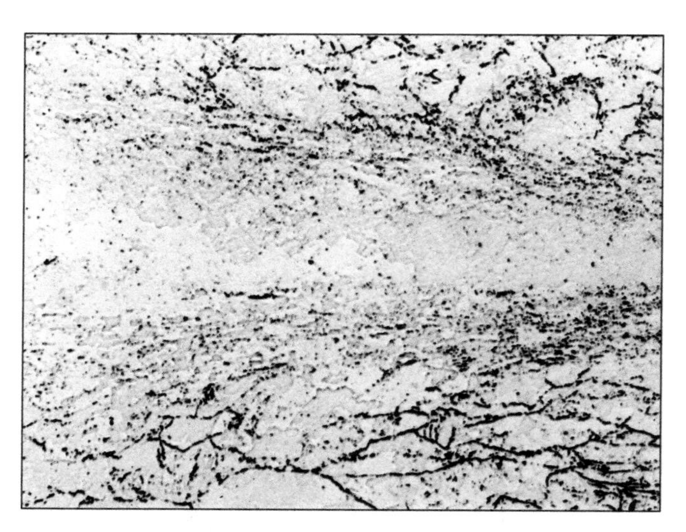

(H) 0.356 mm (0.014 in.) Half-Hard Nickel Sheet to Half-Hard Nickel Sheet (×250)

(K) 0.813 mm (0.032 in.) Die Steel (0.9% C) to 0.813 mm (0.032 in.) Ingot Iron (×500)

Figure 8.25 (Continued)—Photomicrographs of Typical Ultrasonic Welds

weld interface is shown in Figure 8.25(A), in which a Kovar™ foil (a low-expansion iron-base alloy of 29% nickel, 17% cobalt) has intruded into as much as 75% of the thickness of a nickel foil. A gold plate on the surface of the Kovar was dispersed throughout the highly worked region. In Figure 8.25(B), interfacial ripples in a nickel-to-molybdenum weld illustrate the plastic flow that occurs locally. Entrapped oxide is indicated by the dark patches on the extreme right of the photomicrograph.

The weld between two sheets of arc-cast molybdenum, Figure 8.25(C), shows very little interpenetration, and the bond line is thin. Figure 8.25(D) illustrates the surface oxide film dispersion that may occur during the welding of aluminum sheet. General plastic flow along the weld interface is observed in the Type 2024-T3 aluminum alloy weld in Figure 8.25(E), where the metal has recrystallized to a fine grain size. The copper–silver weld in Figure 8.25(F) illustrates the intense plastic flow and turbulence that often occurs at the weld boundary of ultrasonic welds.

Evidence of recrystallization may be observed in ultrasonic welds in several structural aluminum alloys, beryllium, low-carbon steel and other metals, even though they were not in the cold-worked condition prior to welding. For instance, the weld in Type 2020 aluminum alloy in Figure 8.25(G) shows evidence that mutual deformation of the surfaces and subsequent recrystallization has occurred. In Figure 8.25(H), the elevated temperature during welding resulted in recrystallization of prior cold-worked nickel.

An additional effect of interfacial heating is illustrated in Figure 8.25(J), which shows a weld in a solution-treated and aged nickel-base alloy. In the aged condition, a precipitate normally appears throughout the grains and in the grain boundaries. In the vicinity of this interface, the oxide scale is dispersed and the grain boundaries appear to stop short of the interface, indicating that the precipitate was dissolved during welding. An example of alloying that may occur in the weld between ferrous metals of different carbon content is shown in Figure 8.25(K).

ULTRASONIC MICROWELDING

The terms *ultrasonic microwelding* and *ultrasonic microbonding* are often used interchangeably, and although *microbonding* is the term most commonly used in the microelectronics industry. The most extensive application of ultrasonic welding is in microelectronics, where lead wires made of fine aluminum or gold are attached to substrates such as transistors, diodes, and other semiconductor devices to make electrical interconnections. The trillions of microwelds performed annually far exceed any other use of ultrasonic metal welding. The extensive range of materials and thicknesses that can be joined by microwelding processes are summarized in Table 8.2.

The two main systems used in the microelectronics industry are ultrasonic wedge microwelding and thermosonic microwelding. Ultrasonic wedge welding (different from the wedge-reed system) produces two wedge welds per connection. The thermosonic system produces a ball weld, or first weld, and a crescent weld, or second weld.

MICROWELDING SYSTEMS

Microwelding (microbonding) systems, as shown in Figure 8.26, are similar to ultrasonic spot welding systems. An ultrasonic transducer-horn vibrating in longitudinal motion is used to drive a flexible beam, called the *sonotrode* or *welding tool* in bending vibration. This results in a transverse vibration at the tip of the beam, and creates a weld between the thin wire and the substrate. The microwelding system shown in Figure 8.26 consequently has similarities to the wedge-reed system in that it transmits ultrasonic energy into the weld via bending vibrations. The difference is that the low clamping force required for microwelding permits the force to be applied to the tip of the welding tool simply by leveraging forces on the transducer system, as is used in the lateral-drive system. (Refer to Figure 8.1.) The welding tool is typically called the *capillary* in thermosonic welding, and the *wedge* in ultrasonic wedge welding.

The key difference between microwelding and other methods of ultrasound welding is not in the principles of operation, which generally are the same as those of lateral-drive and wedge-reed welding machines, but in the frequency ranges and power levels used. Thus, microwelding machine frequencies tend to start at 60 kHz and range up to 150 kHz, with frequencies in experimental systems reported at 300 kHz. Power levels of microwelding equipment usually are in the range of 0.1 W to 50 W, whereas the higher power systems discussed in previous sections operate in the range of 10 W to 1000 W.

MECHANISMS OF WEDGE MICROWELDING

The ultrasonic welds used in microelectronic applications are made with highly deformable materials such as aluminum, copper, and gold wire. The wire diameters used in these ultrasonic welds are usually 25 µm to 50 µm. Thus, aluminum-to-aluminum welds are made between fine aluminum: 1% silicon wire, 25 µm to 50 µm in diameter, and various aluminum alloys (for example, alloys with 1% silicon and 1% to 2% copper) bonding pads on semiconductor chips. Large-diameter wire, up to about 0.75 mm (0.03 in.) supplied in the fully annealed condition is used to connect power devices that require higher currents. Figure 8.27 shows the details of wedge welding at the wedge–wire interface in which the wire is fed through the tip of the wedge to the substrate.

The complex welding process for this application can be summarized as follows: The application of ultrasonic power results in increased wire flow and a gradual bonding action between the outer surface of the deformed wire and the substrate material. Thus, wire-to-substrate interfacial motion may occur during the first few milliseconds, resulting in only a minimal temperature rise of 50°C to 100°C (120°F to 210°F). After this time, small microwelds form along or just inside

Table 8.2
Metal Wire and Ribbon Leads That May Be Ultrasonically Welded to Thin Metal Surfaces on Stable Nonmetallic Substrates

	Metal Form		Lead Diameter or Thickness Range	
Substrate	Film	Wire/Ribbon	mm	in.
Glass	Aluminum	Aluminum wire	0.051 to 0.254	0.002 to 0.010
	Aluminum	Gold wire	0.076	0.003
	Nickel	Aluminum wire	0.051 to 0.508	0.002 to 0.020
	Nickel	Gold wire	0.051 to 0.254	0.002 to 0.010
	Copper	Aluminum wire	0.051 to 0.254	0.002 to 0.010
	Gold	Aluminum wire	0.051 to 0.254	0.002 to 0.010
	Gold	Gold wire	0.076	0.003
	Tantalum	Aluminum wire	0.051 to 0.508	0.002 to 0.020
	Chromel	Aluminum wire	0.051 to 0.254	0.002 to 0.010
	Chromel	Gold wire	0.076	0.003
	Nichrome	Aluminum wire	0.0635 to 0.508	0.0025 to 0.020
	Platinum	Aluminum wire	0.254	0.010
	Gold-platinum	Aluminum wire	0.254	0.010
	Palladium	Aluminum wire	0.254	0.010
	Silver	Aluminum wire	0.254	0.010
	Copper on silver	Copper ribbon	0.711	0.028
Alumina	Molybdenum	Aluminum ribbon	0.076 to 0.127	0.003 to 0.005
	Gold-platinum	Aluminum wire	0.254	0.010
	Gold on molybdenum-lithium	Nickel ribbon	0.051	0.002
	Copper	Nickel ribbon	0.051	0.002
	Silver on molybdenum-manganese	Nickel ribbon	0.051	0.002
Silicon	Aluminum	Aluminum wire	0.254 to 0.508	0.010 to 0.020
	Aluminum	Gold wire	0.051	0.002
Quartz	Silver	Aluminum wire	0.254	0.010
Ceramic	Silver	Aluminum wire	0.254	0.010

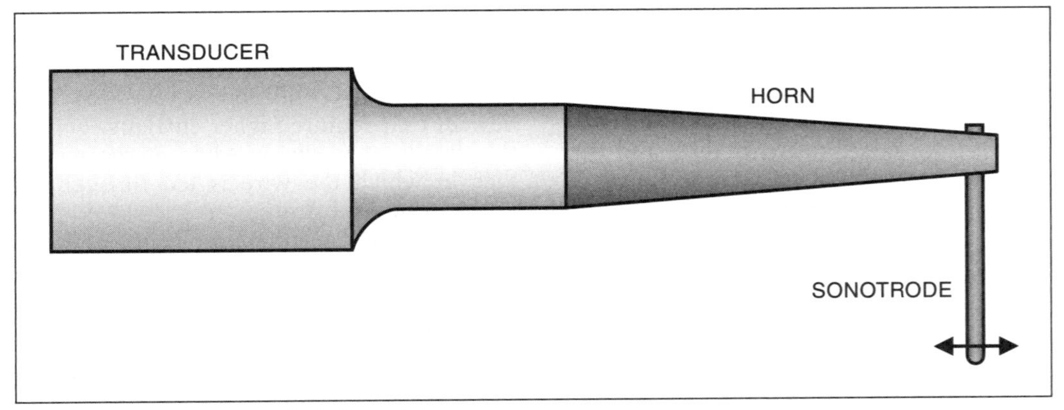

Source: © Edison Welding Institute. Used with permission.

Figure 8.26—Basic Features of an Ultrasonic Microwelding System

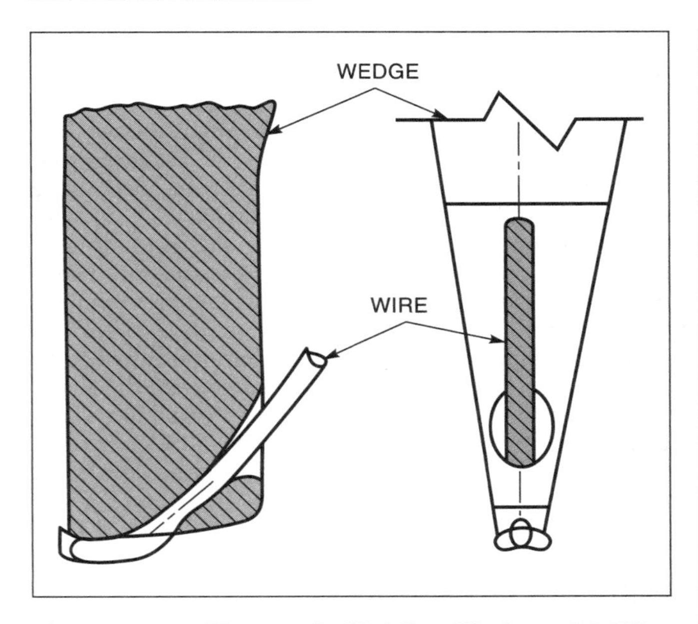

Figure 8.27—Ultrasonic Welding Wedge with Wire

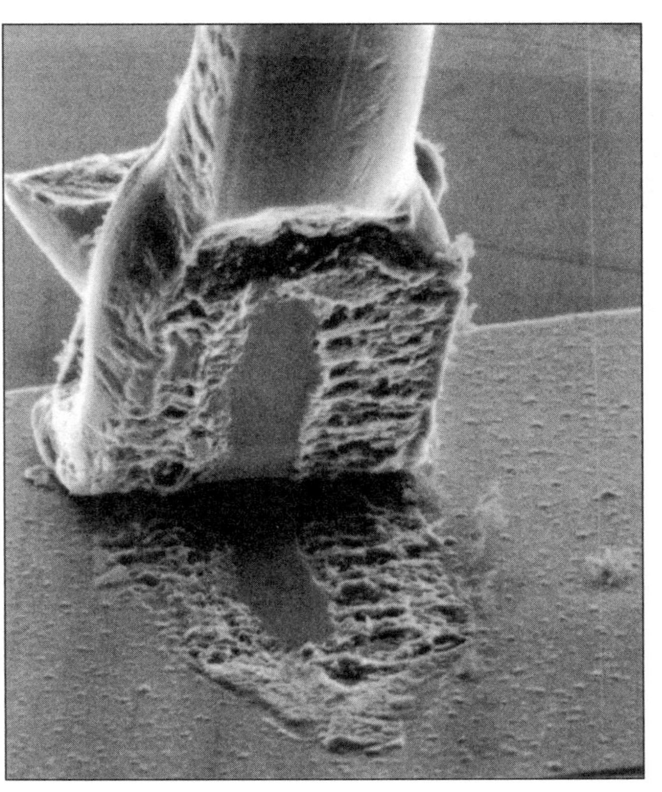

Figure 8.28—Typical Ultrasonic Microweld Between a 25-µm Diameter Aluminum, 1% Silicon Wire and an Aluminum Substrate

the perimeter of the faying surfaces. The vibratory action effectively removes surface contaminants and exposes fresh material for welding. Using a scanning electron microscope as a diagnostic tool, the disruption of surface contaminants can be readily observed below the heel of the weld tool on the weld substrate. The interfacial motion progresses into the interface of the welding tool and the wire, and more ultrasonic energy is absorbed into the weld area. It is assumed that this interfacial motion slows and ceases as these microwelds grow. The microwelds join together and grow toward the center, generally leaving the center unwelded, as shown in Figure 8.28 (the wire in Figure 8.28 has been lifted up so the weld pattern can be seen.) In this application, the weld was made with 50 ms weld time and 25 kg (55 lb) load.

An example of the initial stage of a weld of an aluminum-silicon ribbon to a substrate is shown in Figure 8.29. The ribbon has been removed, revealing microwelds beginning to form at the perimeter of the joint where deformation is greatest.

Transmission electron micrographs and scanning electron micrographs taken along the interface of monometallic welds have variously shown grain boundaries, no grain boundaries, debris zones of oxides and contaminants, and numerous crystallographic defects. Gold-to-aluminum and other bimetallic welds made at room temperature do not show the formation of intermetallic compounds, but a clear boundary similar to a grain boundary is always observed. It is of interest to note that the significance of the material flow induced by ultrasonic energy can easily be observed by placing the welding wedge on the wire to be joined to a substrate using the full welding force, but without applying ultrasonic energy. With this condition, the flow of the wire is almost unnoticeable.

Reflected sound and light are used to reveal the movements of the welding tool. The magnitude of the wedge movement increases with increasing ultrasonic energy. By gradually increasing the power and the time for a given load on the wedge, characteristics of the ultrasonic wire weld interface can be observed. For a given combination of wire and substrate materials, a range of power (welding tool movement), time, and welding machine load variables can be determined that will provide acceptable weld strength values. There is a trade-off between reducing the wire strength because of deformation and the strength of the weld interface. For reliable welding, the wire always should be weaker than the weld interface.

Figure 8.29—Initial Stage of a Weld of an Aluminum-1% Silicon Ribbon (12 µm × 37 µm Cross Section) to an Aluminum Substrate

THERMOSONIC MICROWELDING

In the microelectronics industry during the early 1970s, ultrasonic wedge welding was predominately applied to the joining of aluminum wires to aluminum metallized bond pads on semiconductor devices, and joining wires to either aluminum-clad or gold-plated leads on the package. The thermal compression welding mode of ultrasonic welding also was in use at that time for joining wire to a substrate. In this technique, fresh metal surfaces were exposed for welding by mechanical compressive action that disrupted the surface films. This effect was further promoted by heating the substrate, typically from 10°C to 350°C (50°F to 662°F). Modern thermosonic microwelding emerged from this early version of thermal compression welding.

Modern thermosonic microwelding (microbonding) involves the ultrasonic welding of metal wire to a heated substrate. Interface temperatures of about 100°C to 200°C (212°F to 400°F) normally are used; this is lower than temperatures needed for thermal compression welding and thus less damaging to the plastic materials used to attach chips to packages. This technique is the most popular method of wire-to-substrate welding, with trillions of joints produced annually in microelectronic applications.

Figure 8.30 shows a ball or capillary welding tool with an unwelded ball in position to begin the welding process. The ball is formed by a spark discharge that melts the tip of the wire. The heat-affected zone above the ball is fully annealed by the discharge. The wire feeds through the capillary, allowing the welding head to travel on the surface toward the crescent weld. High-purity gold wire (99.99% gold) is the predominant welding wire. Micro-alloying of the residual impurities of 100 parts per million must be carefully controlled to ensure acceptable ball formation and control the shape of the wire loop. The principal welding variables—weld force, weld time and ultrasonic energy—are controlled by computer software. Welding temperature is controlled independently.

EQUIPMENT

High-speed, automated microwelding machines are used for wire welding applications. Typically, these machines are capable of joining 6 to 8 wires (12 to 16 welds) per second. The two basic types of welding systems used are thermosonic ball welding machines and ultrasonic wedge welding machines. Both systems use ultrasonic energy, but they require different tooling and

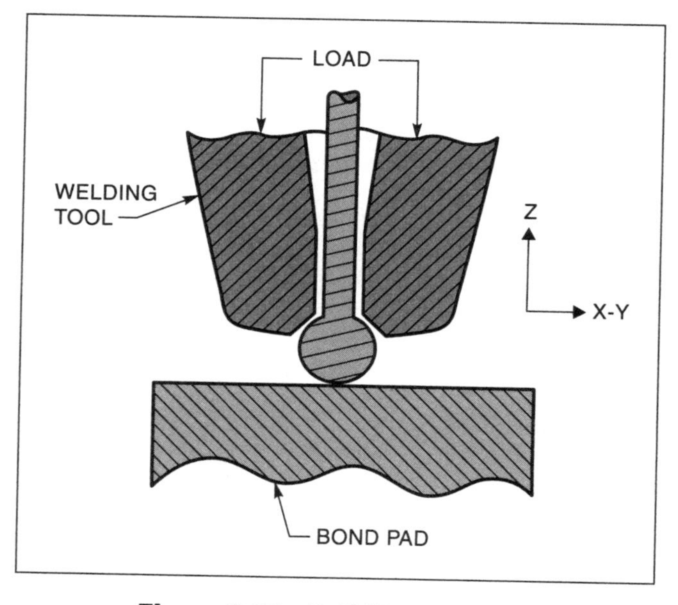

Figure 8.30—Ball-Welding Tool with Unwelded Ball

operating procedures and are used to weld different materials. Ultrasonic energy for both systems is generated by a piezoelectric transducer attached to a horn. The transducer design is tapered to provide mechanical gain. The system resonates at 60 kHz or higher. Phase-locked loop circuits are designed to match the electronic and mechanical systems for optimized output. Sensors that monitor changes in system impedance during welding are used to achieve real-time control of output during weld formation. Motion control is accomplished through software-controlled servo systems. The thermosonic ball welding machine requires three axes of motion for the welding head manipulator and one for the positioner. The wedge welding machine requires a rotational axis. The servos are required to position the tool with an accuracy of ±2.5 microns (63.5 µm) in all three axes.

WELD QUALITY

Variations in weld quality may result from several factors generally associated with the workpieces and the welding machine or its settings. Weld quality ordinarily is not affected by normal manufacturing variations in metal workpieces. Metals that meet the specification requirements usually can be consistently welded without varying the welding machine settings. Problems sometimes are encountered, however, if close tolerances are not held. For example, nickel, copper, and gold plating on metal surfaces frequently have thickness variations that affect weld quality. Surfaces for torsion welding must be flat and parallel to ensure uniform welding around the periphery.

If there is significant change in the workpieces during a production operation, the welding schedule usually must be adjusted to accommodate the change. Variations in weld quality during production runs are often traced to unauthorized changes in metal alloy, workpiece geometry, or surface finish. For example, magnetic wires that are lubricated to facilitate coil winding may be ultrasonically welded without cleaning. However, a change in the type of lubricant may cause unacceptable welds unless welding machine settings are appropriately adjusted.

A substantial change in load power may indicate a faulty weld, which could be due to changes in workpiece dimensions or surface finish, improper assembly of workpieces, or welding machine malfunction. On some machines, the acceptable high and low limits of power can be set, and deviations from this range can be programmed to trigger a rejection signal. Similar arrangements can be made for welding machines based on the constant-energy principle. Uniform-quality welding also depends on the mechanical precision of the welding machine. Lateral deflection of the sonotrode or looseness of the anvil can produce unacceptable aberrations in the welds. Sonotrode tips must be acoustically designed and precisely ground to the desired contour. The surfaces must be properly maintained to ensure reproducible welds.

EXAMINATION OF PHYSICAL AND METALLURGICAL PROPERTIES

The examination of physical and metallurgical properties of the workpieces may help determine ultrasonic welding procedures and ensure weld quality. Among the physical properties are the surface appearance and thickness deformation. Metallurgical properties include microstructure and weld stress.

Surface Appearance

The surface of the workpiece at a weld location is usually roughened or deformed slightly by the combined compressive and shear forces. This effect can be minimized with adjustments in the welding machine settings and with careful sonotrode tip maintenance. The surface contour depends primarily on sonotrode geometry. Spot-weld sonotrodes usually leave an elliptical impression, embossing the sonotrode pattern on the workpiece. This is a function of linear displacement of the tip. A weld impression is larger in soft, ductile metals, such as aluminum, than in hard metals of the same thickness. The weld impression is controlled by appropriate adjustment of welding machine settings. Spot size can be increased by using a tip with a larger radius. The actual weld area does not necessarily duplicate the surface impression, except in thin sheet. Sometimes spot welds have areas of incomplete fusion in the center, a condition that usually can be remedied by decreasing the tip radius or reducing the clamping force.

Thickness Deformation

A weld may show some thickness deformation because of the applied clamping force. The deformation in sheet metals usually is less than 20% of the total joint thickness, even with soft metals. When contoured workpieces such as wires are welded, deformation is somewhat greater unless the contour of the sonotrode tip is exactly matched with the workpiece. Deformation may exceed 50% in fine wires.

Microstructural Properties

The microstructural properties of ultrasonic welds, as discussed in the section "Mechanisms of Ultrasonic Welding" influence weld quality. Metallographic examination is an important means of determining the quality of ultrasonic welds, often showing the degree of bonding along the weld zone and the extent of local plastic deformation.

Mechanical Properties

A variety of mechanical tests may be used to evaluate the properties of ultrasonic welds. The property most frequently tested is shear strength. In addition, data is reported on tensile strength, peel strength, microhardness, corrosion resistance, and hermetic sealing properties. Properly performed ultrasonic welding produces welds of acceptable strength and integrity.

Shear Strength

Shear strength tests usually are conducted on simple lap joints containing single or multiple spot welds or predetermined lengths of seam or line welds. The preparation of test specimens and testing procedures essentially duplicate those used for resistance spot and seam welds. Figure 8.31 shows the increase in shear strength in proportion to sheet thickness for single spot welded specimens in two aluminum alloys. Usually in the thin gauges of aluminum sheet and often in the intermediate gauges, failure occurs by fracture of the base metal or by tear-out of the weld button rather than by shear of the weld itself. Similar data for several stainless steel and nickel alloys are shown in Figure 8.32 and for several refractory metals and alloys in Figure 8.33.

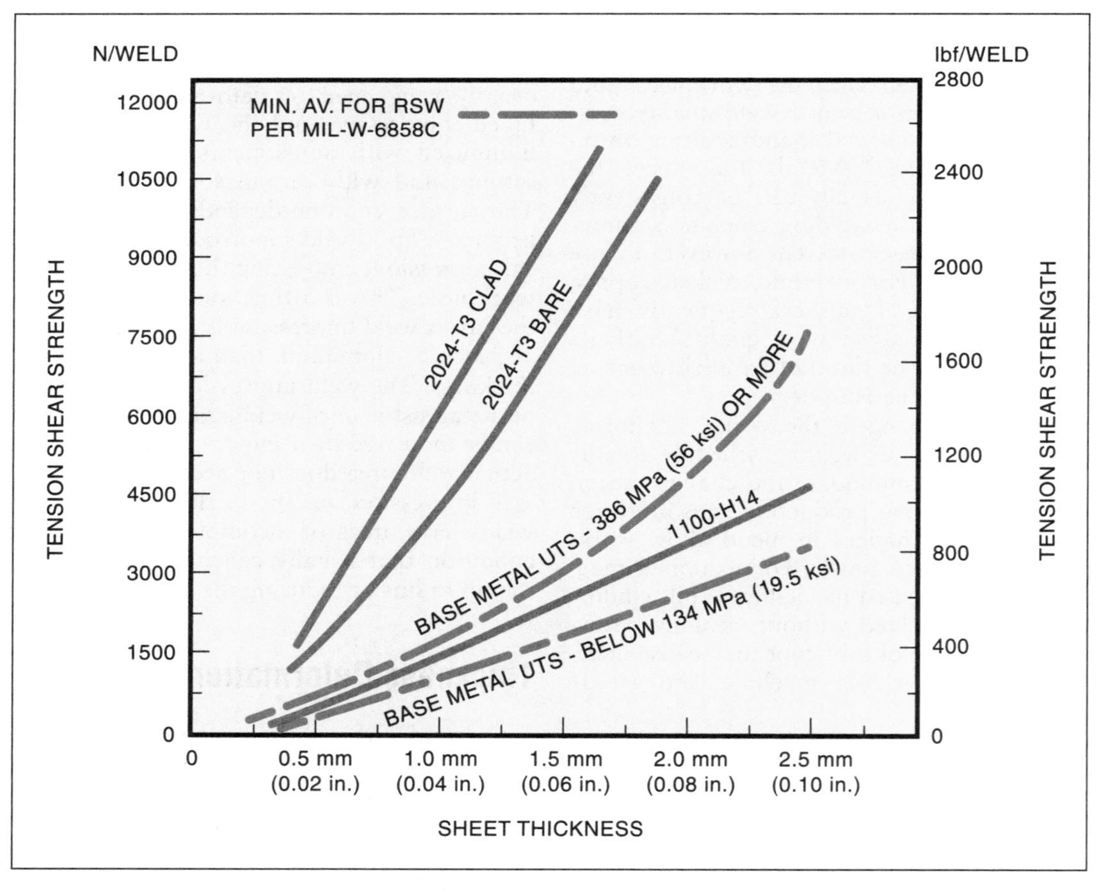

Figure 8.31—Typical Shear Strengths of Ultrasonic Spot Welds in Aluminum-Alloy Sheet

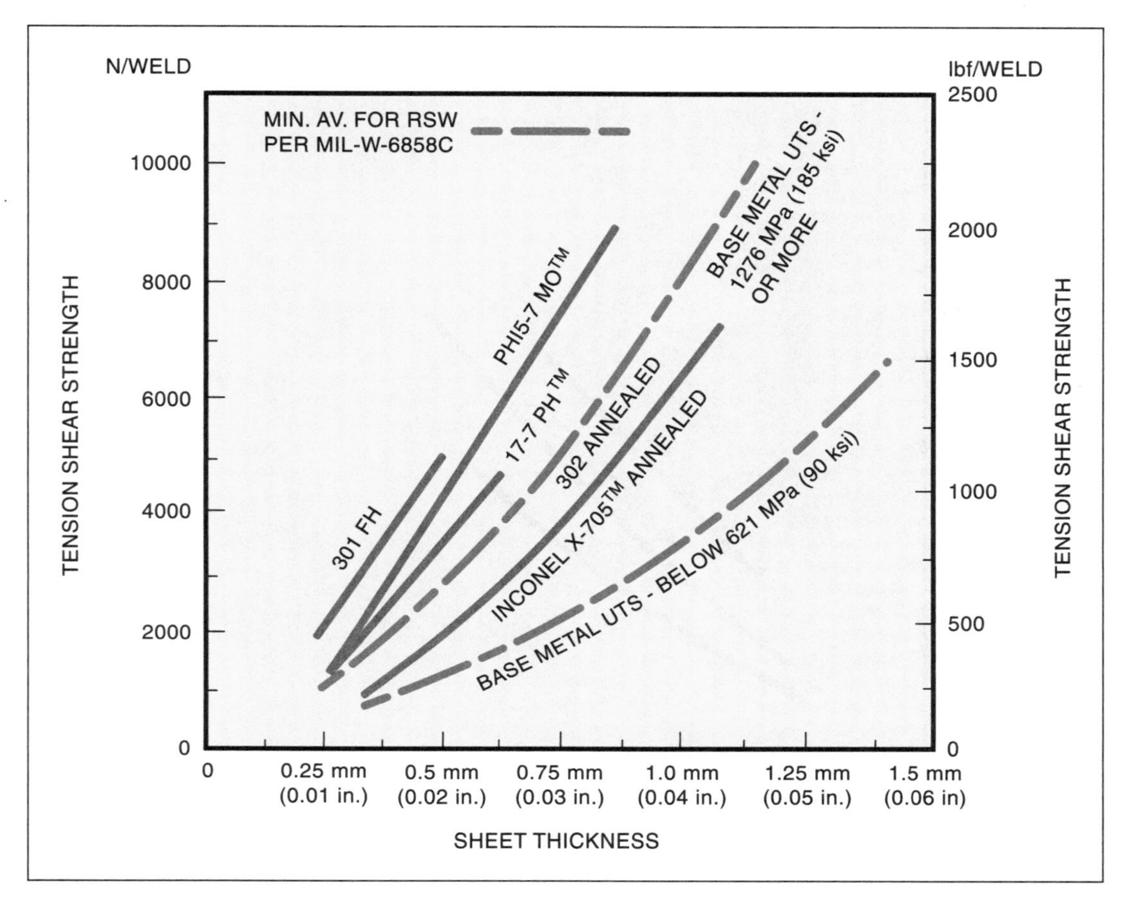

Figure 8.32—Typical Shear Strengths of Ultrasonic Spot Welds in Stainless Steel and Nickel-Base Alloys

Typical spot-weld strengths in a variety of metals are summarized in Table 8.3. Of particular interest is the low variability associated with the strength data. In most instances this is less than 10%.

Strength tests of line welds and seam welds show approximately the same strength as the base metal, particularly in thin-gauge metals. As examples, spot and seam welds in structural aluminum alloys have shown strengths equivalent to 85% to 95% of the ultimate tensile strength of the material under both shear and hydrostatic tests. Line welds in 0.025 mm (0.001 in.) Type 5052-H16 aluminum alloy average 85% to 92% of the base metal strength. Continuous seam welds in thin-gauge 1100 aluminum show 88% to 100% joint efficiency.

Elevated-temperature tests on weld specimens of several metals and alloys indicate that weld strength is no lower than that of the base material at the same temperature.

Corrosion Resistance

The cast nugget of a resistance spot weld is frequently the site of a localized corrosion attack when the weldment is exposed to an unfavorable environment. This is not true of ultrasonic welds, mainly because they are solid-state welds with no cast structure and thus no welding-induced metallurgical segregation. Weld specimens in aluminum alloys and stainless steels that were exposed to boiling water, sodium chloride solutions, and other corrosive materials have shown no preferential attack in the weld. However, when dissimilar metals are welded, the possibility of galvanic corrosion at the weld nugget must be recognized.

NONDESTRUCTIVE EVALUATION

Welding machine control functions, such as settings for energy or distance, provide an effective means of

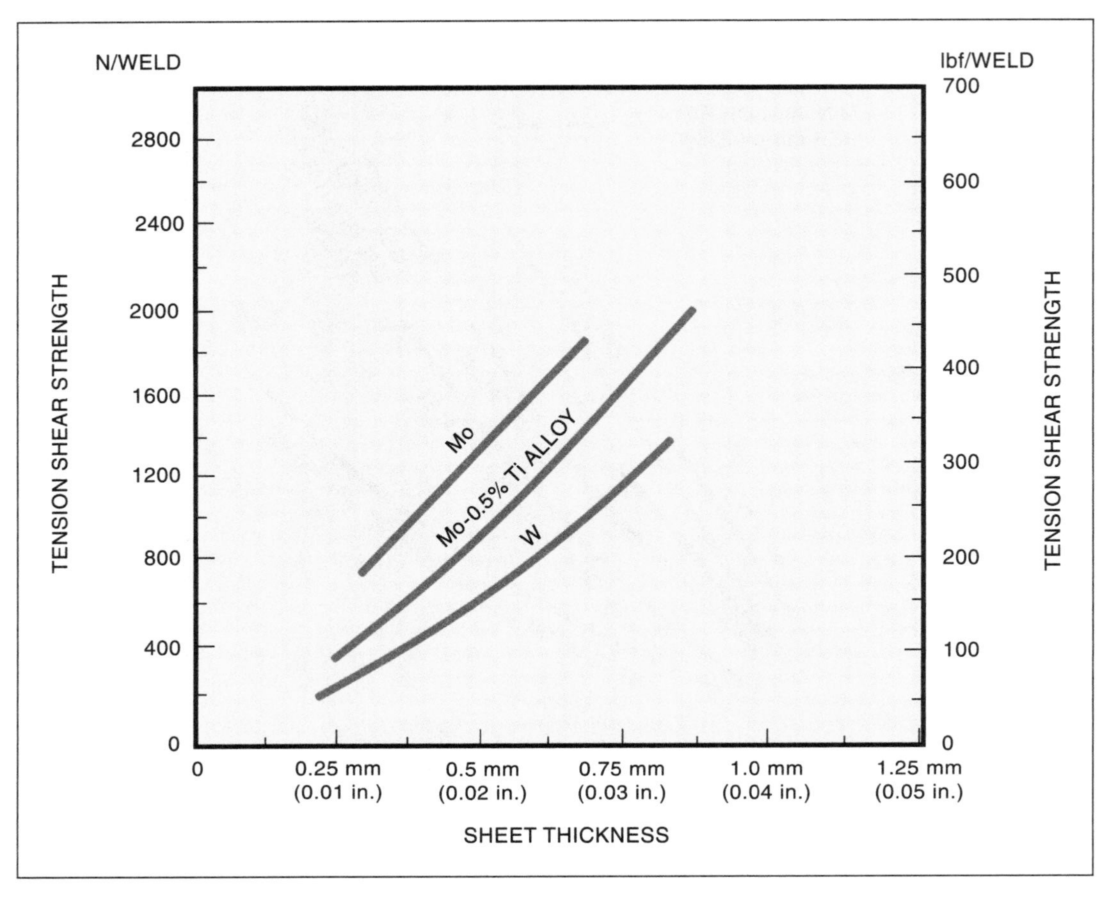

Figure 8.33—Typical Shear Strengths of Ultrasonic Spot Welds in Several Refractory Metals and Alloys

monitoring weld quality while the weld is in progress. The operator can immediately detect an improper cycle and reject the workpiece, or in automated systems, logic may be provided for automatic rejection of a weldment made at an unsatisfactory level of quality. Nearly all nondestructive examination techniques available for fusion welds are applicable to ultrasonic welds. Therefore, ultrasonic, radiographic, and infrared radiation testing techniques may be used in specific applications. If hermetic sealing is the primary requirement of the weld, helium leak tests are effective.

MECHANICAL TESTING

An approach to quality assurance in some applications involves mechanical (destructive) testing of randomly selected specimens during a production run. For welds in relatively thin ductile sheet, a peel test will indicate adequate weld strength if failure occurs by nugget tear-out or fracture of the base metal. Metallographic sectioning for examination provides a reliable indication of weld quality, but it is slow and expensive.

For most applications, shear testing is the most practical destructive test. Figure 8.34 shows typical variations in shear strength of random spot weld samples in 1.0 mm (0.040 in.) Type 2024-T3 aluminum alloy, produced with a specific welding machine setting at different times of the day for a number of days. The maximum, average, and minimum strength values for each set of weld samples are shown. The horizontal lines indicate the mean value and standard deviation range for the entire group. As illustrated in Figure 8.34, the process began to show poor control on the seventh and eighth days. Control was restored on the ninth day by making appropriate adjustments in amplitude.

Table 8.3
Typical Shear Strengths of Ultrasonic Spot Welds in Several Alloys

Metal	Alloy or Type	Sheet Thickness, mm (in.)	Mean Shear Strength with 90% Confidence Interval, Newton (lbf)
Aluminum	2020-T6	1.016 (0.040)	5515.8 ± 222.4 (1240 ± 50)
	3003-H14	1.016 (0.040)	3247.2 ± 177.9 (730 ± 40)
	5052-H34	1.016 (0.040)	3336.2 ± 133.4 (750 ± 30)
	6061-T6	1.016 (0.040)	3558.6 ± 177.9 (800 ± 40)
	7075-T6	1.27 (0.050)	6850.3 ± 400.3 (1540 ± 90)
Copper	Electrolytic	1.143 (0.045)	3780.1 ± 88.1 (850 ± 20)
Nickel	Inconel X-750™	0.813 (0.032)	6761.3 ± 444.8 (1520 ± 100)
	Monel K-500™	0.813 (0.032)	4003.4 ± 266.9 (900 ± 60)
	Rene 41™	0.508 (0.020)	1690.3 (380)
	Thoria dispersed	0.63 (0.025)	910
Steel	AISI 1020	0.635 (0.025)	2224.1 ± 88.1 (500 ± 20)
	A-286	0.381 (0.015)	3024.8 ± 311.4 (680 ± 70)
	AM-350	0.203 (0.008)	1378.9 ± 88.1 (310 ± 20)
	AM-355	0.203 (0.008)	1690.3 ± 11.4 (380 ± 70)
Titanium	8% Mn	0.813 (0.032)	7695.4 ± 889.6 (1730 ± 200)
	5% Al–2.5% Sn	0.711 (0.028)	8674.3 ± 533.8 (1950 ± 120)
	6% Al–4% V	1.016 (0.040)	10 052.1 ± 800.7 (2260 ± 180)

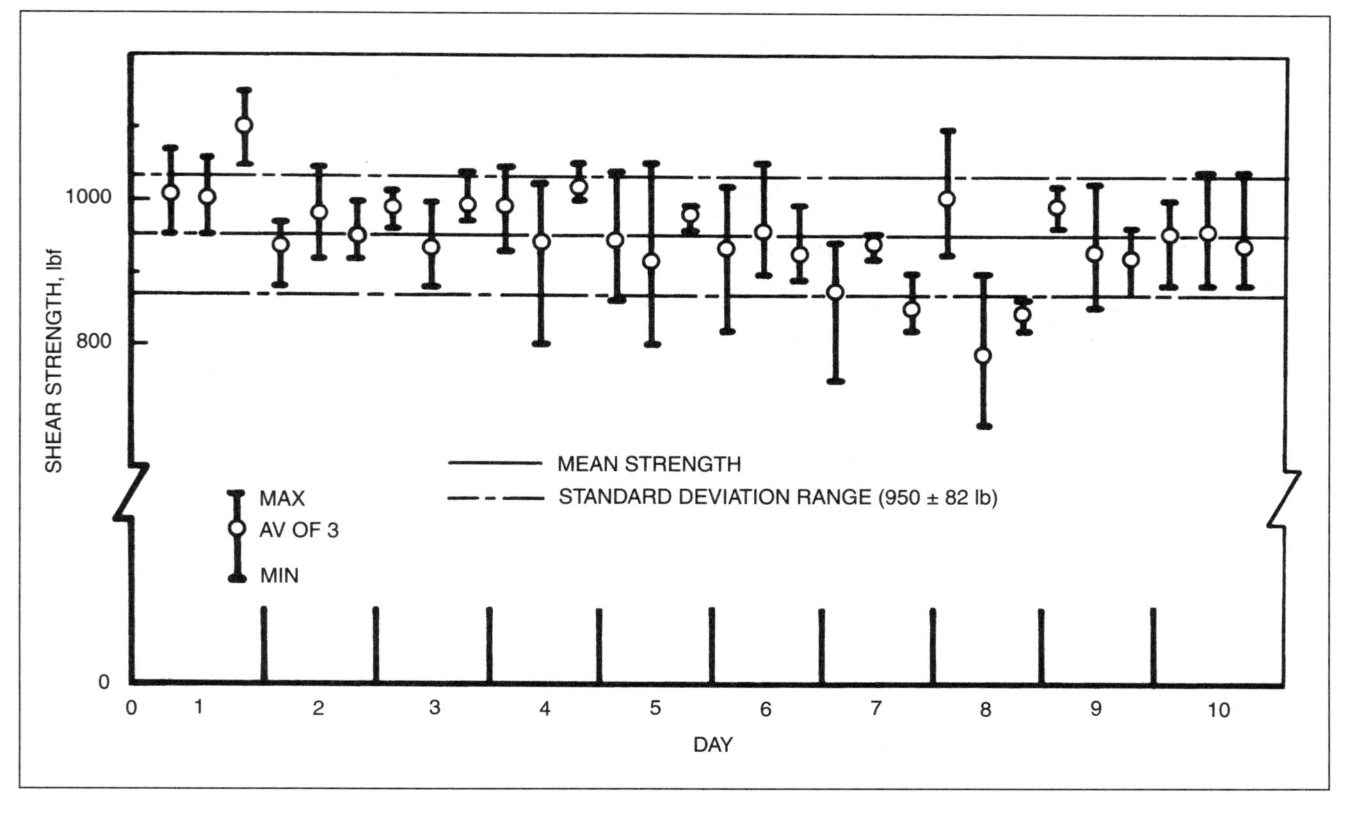

Figure 8.34—Typical Variance in Ultrasonic Weld Tensile Shear Strength in 1.0 mm (0.040 in.) Type 2024-T3 Aluminum Alloy

ECONOMICS

The economic motivation for the use of ultrasonic welding ultimately is to be found in the inherent advantages of the process. Because it is a solid-state welding process, heat generation is low and has minimal or no impact on material properties. The process is excellent for joining aluminum, copper, and other materials with high thermal conductivity. It has the capability of joining a number of dissimilar metal combinations, and can weld thin-thick material combinations. The initial stages of the process are frictional in nature, which permits welding through oxides and contaminants. Ultrasonic welding is fast, easily automated, and does not require filler metals or welding gases. Finally, the energy requirements of ultrasonic welding are low when compared to processes that are considered competitive, such as resistance spot welding or laser beam welding.[6]

Nearly all of these advantages, and hence the economic benefits, ultimately can be traced to the solid-state nature of the process. Thus, issues of economics relative to ultrasonic welding may resolve to that of being able to make the weld ultrasonically and being unable to make it by other methods without extensive materials and workpiece redesign. In making an economic assessment on the use of ultrasonic welding, the factors of equipment, energy, tooling and other costs must be considered.

Equipment costs depend on the type of ultrasonic system required (refer to Figures 8.1 and 8.2). Lateral-drive, wedge-reed or piezoelectric-based ultrasonic transducer systems are available at varying costs, depending on the selection of optional functions.

Ultrasonic welding machines cost anywhere from $18,000 to $45,000, depending on power capacity and various system options, such as statistical process control packages. A basic 20-kilohertz, 2.5- to 3.5-kilowatt ultrasonic metal welding system could be acquired for around $30,000. Resistance welding machines typically cost between $8,000 and $30,000. Laser welding machines were $50,000 and up.

Energy requirements for ultrasonic welding vary according to the application, but USW systems generally use less energy than alternative processes and often provide quality advantages as well as cost advantages. In addition to resistance spot welding and laser beam welding, alternative joining processes may include roll welding, soldering, adhesive bonding, and riveting. Cost savings are attributed to the use of small-capacity power sources and the small amount of energy required for each weld. A typical ultrasonic welding power source has a 3-kW capacity; the energy required for a weld ranges from 100 joules to several hundred joules.

Most welding processes incur expenses for tooling or fixtures that must be factored into the cost of the system. As noted in the preceding sections of this chapter, the sonotrode performs the function of tooling for ultrasonic welding. Sonotrodes are available in two forms. In one, the actual pattern of the weld tip is machined into the tip of a solid horn; in another, the weld tip is a small component that can be detached, either from the reed of a wedge-reed system, or from the sonotrode of a lateral-drive system. Thus, tooling costs remain low.

6. Weber, A., 2003, The Economics of Ultrasonics, ASSEMBLY Magazine, 46(8).

SAFE PRACTICES

The hazards specifically associated with operating ultrasonic welding equipment are identified and discussed in this section. Thus, the hazards associated with fusion welding or cutting processes in general are not addressed in this section. General precautions for welding are covered in detail in the latest edition of *Safety in Welding, Cutting, and Allied Processes,* ANSI Z49.1, which should be consulted.[7] Legal regulations established to protect personnel working on or around various forms of industrial machinery and welding equipment are intentionally omitted from this section, but must be followed. The general requirements outlined in applicable documents such as the United States Department of Labor document *Occupational Safety and Health Standards for General Industry, Title 29, in Code of Federal Regulations (CFR). 1910* must be adhered to at all times.[8] Additional sources of safety information are listed in Appendix B of this volume.

The ultrasonic welding operator may require eye and ear protection, depending on the specific application and equipment. Safety glasses are recommended for all workplace environments. Sound levels in the work area should be monitored to determine if the need for ear protection exists.[9]

Most ultrasonic welding equipment is designed with interlocks and other safety devices to prevent operation

7. American National Standards Institute (ANSI) Accredited Standards Committee Z49, *Safety in Welding, Cutting, and Allied Processes*, ANSI Z49.1, Miami: American Welding Society.
8. Occupational Safety and Health Administration (OSHA), *Occupational Safety and Health Standards for General Industry*, in *Code of Federal Regulations (CFR)*, Title 29 CFR 1910, Subpart Q, Washington D.C.: Superintendent of Documents, U.S. Government Printing Office.
9. American National Standards Institute (ANSI) Accredited Standards Committee, *Practices for Occupational and Educational Eye and Face Protection*, ANSI Z87.1, New York: American National Standards Institute.

under unsafe conditions. Nevertheless, additional consideration must be given to the health and safety of the operators, maintenance personnel, and other personnel in the area of the welding operations. Good engineering practice must be followed in the design, construction, installation, operation, and maintenance of equipment, controls, power sources, and tooling. This will ensure conformance to applicable United States Department of Labor standards, and American Welding Society standards, the safety recommendations of the equipment manufacturers, and the safety standards of the user.

With high-power electrical equipment, high voltages are present in the frequency converter, the welding head, and the coaxial cable connecting these components. Consequently, the equipment never should be operated with the electrical panel doors open or the housing covers removed. Door interlocks usually are installed to prevent the introduction of power to the equipment when the high-voltage circuitry is exposed. The electrical cables are fully shielded and present no hazard when properly connected and maintained.

Because of hazards associated with application of clamping force, the operator should never place hands or arms in the vicinity of the welding sonotrode tip when the equipment is energized.

Manually operated ultrasonic welding equipment usually is activated by dual palm buttons that meet the requirements of local laws. Both buttons must be pressed simultaneously to actuate a weld cycle, and both must be released before the next cycle is initiated. For automated systems in which the weld cycle is sequenced with other operations, guards should be installed for operator protection. Hazards can be further minimized by setting the sonotrode travel to the minimum compatible with workpiece clearance.

Torsion welding machines may be used for closing or hermetically sealing containers filled with detonable materials. While there are no known instances of premature ignition of explosive materials during ultrasonic welding, adequate provisions always should be made for remote operation by placing the welding machine either in a room separate from the control station or behind an explosion-proof barrier.

CONCLUSION

The ultrasonic welding process and the allied process of ultrasonic microwelding are implemented in an increasing range of uses, including those in the electrical, electronic, automotive, aerospace, medical products and other industries. The most significant advantage ultrasonic welding brings to its users is that it is a solid state joining process that imposes minimum impact on the properties of the materials being welded. It is a rapid process, achieving welds in fractions of a second, and is easily automated for advanced manufacturing. It is used in special applications with metals of high thermal conductivity, such as the light metals and copper alloys, when other welding processes sometimes encounter difficulty.

These positive characteristics of ultrasonic welding have resulted in a trend to use more powerful welding systems that permit the welding of thicker materials and tougher alloys. This trend, in addition to continued research on the underlying mechanisms and metallurgy of ultrasonic welding, shows promise for more widespread use and greater diversity of applications of ultrasonic welding in the future.

BIBLIOGRAPHY

American National Standards Institute (ANSI) Accredited Standards Committee, *Practices for occupational and educational eye and face protection*, ANSI Z87.1, New York: American National Standards Institute.

American Welding Society (AWS) Committee on Definitions and Symbols. 2001. *Standard welding terms and definitions*. Miami: American Welding Society.

American Welding Society (AWS) Committee on Joining of Plastics and Composites. 2006. *Guide to ultrasonic assembly of thermoplastics,* AWSG1.1M/G1.1: 2006. Miami: American Welding Society.

American Welding Society (AWS) Welding Handbook Committee. 1996. *Welding Handbook,* Volume 3, W. R. Oates, ed. *Materials and Applications*. Miami: American Welding Society. Chapter 6.

Occupational Safety and Health Administration (OSHA), *Occupational Safety and Health Standards for General Industry*, in *Code of Federal Regulations (CFR)*, Title 29 CFR 1910, Subpart Q, Washington D.C.: Superintendent of Documents, U.S. Government Printing Office.

SUPPLEMENTARY READING LIST

Anon. 1978. Ultrasonic welding sees growing use in small motor assembly. *Welding Journal* 57(9): 41–43.

Anon. 1980.Ultrasonic welding of silver electrical contacts. *Welding Journal* 59(5): 41–42.

Avila, A. J. 1964. Metal bonding in semiconductor manufacturing—a survey. *Semiconductor Products and Solid-State Technology* 7(11): 22–26.

Bilgutay, N., X. Li, and M. McBrearty. 1986. Development of non-destructive bond monitoring techniques for ultrasonic bonders. *Ultrasonics* 24(6): 307–317.

Chang, U., and J. Frisch. 1974. An optimization of some parameters in ultrasonic metal welding. *Welding Journal* 53(1): 24-s–35-s.

Devine, J. 1980. Joining electric contacts: ultrasonics works fast. *Welding Design and Fabrication*, (3).

Devine, J. 1984. Joining metals with ultrasonic welding. *Machine Design*, (9).

de Vries, E., 2000. Development of ultrasonic welding process for stamped 6000 series Aluminum. Diploma thesis, University of Applied Science, Emden.

de Vries, E. 2004. Mechanics and mechanisms of ultrasonic metal welding. Ph.D. Dissertation, The Ohio State University.

Estes, C. L., and P. W. Turner. 1973. Ultrasonic closure welding of small aluminum tubes. *Welding Journal* 52(8): 359-s–369-s.

Grigorashvili, Y., W. Sheel, and M. Thiede. 1984. Characterizing ultrasonic weld joints by measuring the electrical transition resistance. *Schweisstechnik* 34(10): 468–470.

Harman. G., and K. O. Keedy. 1972. An experimental model of the microelectronic ultrasonic wire bonding mechanism. *10th annual proceedings, reliability physics*. Las Vegas, Nevada. 49–56.

Harthoorn, J. 1973. Joint formation in ultrasonic welding compared with fretting phenomena for aluminum. *Ultrasonics International 1973 Conference Proceedings*. 43–51.

Harthoorn, J., 1978. Ultrasonic metal welding. Dissertation, Technical University, Eindhoven.

Harman, G. G. 1997. Wire bonding in microelectronics—materials, processes, reliability and yield. 2nd ed. New York: McGraw-Hill.

Hazlett, T. H., and S. M. Ambekar. 1970. Additional studies of interface temperature and bonding mechanisms of ultrasonic welds. *Welding Journal* 49(5): 196-s–200-s.

Hulst, A. P., and P. Lasance. 1978. Ultrasonic bonding of insulated wire. *Welding Journal* 57(2): 19–25.

Jones, J. B. 1967. Ultrasonic welding. *Proceedings of the CIRP International Conference on Manufacturing Technology,* 1387–1410. Ann Arbor, Michigan, September 1967.

Jones, J. B. et al. 1961. Phenomenological considerations in ultrasonic welding. *Welding Journal* 40(4): 289-s–305-s.

Joshi, K. C. 1971. The formation of ultrasonic bonds between metals. *Welding Journal* 11(50): 840–848.

Kelly, T. J. 1981. Ultrasonic welding of Cu-Ni to steel. *Welding Journal* 60(4): 29–31.

Kirzanowski, J. E. 1989. A transmission electron microscopy study of ultrasonic wire bonding. *Proceedings, 39th IEEE Electronic Components Conference,* 450–455. Houston, Texas, May 22–24, 1989.

Koziarski, J. 1969. Ultrasonic welding: engineering, manufacturing and quality control problems. *Welding Journal* 40(4): 349-s–358-s.

Langenecker, B. 1969. Effects of ultrasound on deformation characteristics of metals. *IEEE TransSonics and Ultrasonics,* SU-13. 1–8.

Littleford, F. E. 1976. Welding electronic devices by ultrasonics. *Industrial Electronics* 6(3): 123–126.

Lum, I., J. P. Jung, and Y. Zhou. 2005. Bonding mechanism in ultrasonic gold ball bonds on copper substrate. *Metallurgical and materials transactions A: Physical metallurgy and materials science.* 36A: 1279–1286.

Meyer, F. R. 1976. Assembling electronic devices by ultrasonic ring welding. *Electronic packaging and production* 16(7): 27–29.

Meyer, F. R. 1976. Ultrasonic welding process for detonable materials. *National Defense* 60(334): 291–293.

Meyer, F. R. 1977. Ultrasonics produces strong oxide-free welds. *Assembly Engineering* 20(5): 26–29.

Mitskevich, A. M. 1973. Ultrasonic welding of metals, from *Physical principles of ultrasonic technology,* Vol. 1, L. D. Rozenberg. Ed. New York: Plenum Press.

Pfluger, A., and X. Sideris. 1975. New developments in ultrasonic welding. *Sampe Quarterly* 7(1): 9–19.

Reuter, M., and Roeder, E. 1993. Ultrasonic welding of glass and glass-ceramics to metal. *Schweissen Schneiden* 45(4): E62–E65.

Shin, S. and H. T. Gencsoy. 1968. Ultrasonic welding of metals to nonmetallic materials. *Welding Journal* 47(9): 398s–403s.

Weare, N., J. Antonevich, and R. Monroe. 1960. Fundamental studies of ultrasonic welding. *Welding Journal* 49(8): 331–341.

Weber, Austin. 2003. The Economics of Ultrasonics. *Assembly Magazine* 29(8).

Welding Handbook Committee. 2001. C. Jenney and A. O'Brien, eds. *Welding science and technology.* 9th ed. Vol. 1 of *Welding Handbook*. Miami: American Welding Society.

CHAPTER 9

EXPLOSION WELDING

Prepared by the Welding Handbook Chapter Committee on Explosion Welding:

G. A. Young, Chair
Dynamic Materials Corporation

J. G. Banker
Dynamic Materials Corporation

A. Nobili
Nobelclad Europe

B. K. Sareen
Industrial Equipment & Engineering

Welding Handbook Volume 3 Committee Member:

W. L. Roth
Procter & Gamble Inc.

Contents

Photograph courtesy of Coek Engineering

CHAPTER 9

EXPLOSION WELDING

INTRODUCTION

Explosion welding (EXW) is defined as a solid-state process that produces a weld by high-velocity impact of the workpieces as the result of a controlled explosive detonation.[1,2] Although many in the explosion welding industry use the word *bonding* to describe a welded joint, *welding* is the preferred term for joining metals using explosives because an atomic bond occurs and is predominant at the interface.[3]

Explosion welding was developed by the DuPont Chemical Company as an outgrowth of their explosives business. Their research and development group and others worked on explosion metalworking following World War II; this led to the granting of a United States patent for explosion welding to the company in 1964.

Although there are many applications for explosion welding, this chapter emphasizes the manufacturing of clad plate, which is the most significant use of the process. An example of an explosion-welded product is shown on the first page of this chapter, a titanium-steel-clad pressure vessel 6 meters (20 feet) in diameter and 73 meters (240 feet) long.

1. Welding terms and definitions used throughout this chapter are excerpts from *Standard Welding Terms and Definitions*, AWS A3.0:2001, Miami: American Welding Society.
2. At the time this chapter was prepared, the referenced codes and other standards were valid. If a code or other standard is cited without a publication date, it is understood that the latest edition of the document referred to applies. If a code or other standard is cited with the publication date, the citation refers to that edition only, and it is understood that any future revisions or amendments to the code or other standard are not included. As codes and standards undergo frequent revision, the reader is advised to consult the most recent edition.
3. The AWS Subcommittee on Definitions and Symbols reserves the term *bonding* for the joining and allied processes, where either an adhesive bond or a mechanical bond is predominant at the interface created by the process actions, i.e., adhesive bonding and thermal spraying. When an atomic bond between the atoms at that interface is predominant, the resulting joint is called a *weld* and the process that produced that joint is called *welding*, without regard to whether the weld interface is created as a result of fusion or in the solid state. The interatomic bond existing between metal atoms at the weld interface of a fusion weld is not different from that of the weld interface of a solid-state weld.

Explosion-welded joints of dissimilar metals can be applied by a designer anywhere there is a need for a high-quality transition between metals. Typical uses include ultrahigh-vacuum joints between aluminum, copper, and stainless steel; corrosion-resistant claddings on mild steel substrates; and aluminum alloys joined to metals with low-expansion rates, for example, those used in electronic packages.

This chapter discusses the fundamentals of explosion welding and describes the equipment and facilities required. Other topics are applications, process variables, material properties, weld quality and testing, economics, and safe practices.

FUNDAMENTALS OF EXPLOSION WELDING

The three fundamental components of an explosion welding system include the prime metal, the base metal, and the explosive agent. The base metal, or backer plate, remains stationary as the prime metal is welded to it. A typical arrangement of the components for explosion welding is shown in Figure 9.1.

To make a weld, the workpiece (the prime metal) is accelerated across a short distance with the force of the explosion and collides with a stationary metal surface, the base metal. As the explosion progresses at high velocity across the surface of the prime metal, the force of the explosion causes this metal to conform to an advancing bend angle as it accelerates and collides with the backer plate. A jet is formed in the collapsing space preceding the collision point. The energy of the jet removes metal oxides from the faying surfaces. In this way, metallically pure surfaces are produced and brought together under enormous force, and microfusion occurs at the faying surfaces. Investigation has found micromelting in some metal combinations and it is believed to be present in all explosion welds, but

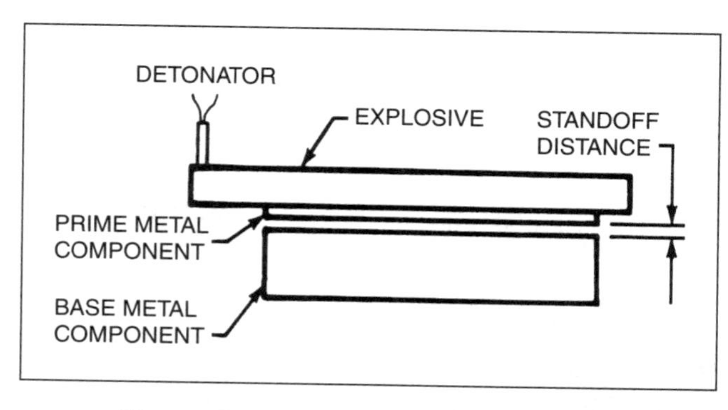

Figure 9.1—Typical Arrangement of Components for Explosion Welding

because of the speed of the explosion, little temperature rise takes place in the welded metals. Explosion welding is classified as a cold welding process. The joint strength of explosion welds generally is greater than the weaker of the prime metal and base metal.

Applications include welding and the bimetallic cladding of metals. The most important use of explosion welding is the manufacturing of clad plate. The cladding of dissimilar metals by explosion welding takes advantage of the best properties of both metals; one metal may provide protection from a harsh environment while the other may provide economic benefits. Explosion welds often are used for metal transition joints, either for electrical or mechanical purposes. Explosion welding can also be used as a joining process to increase mechanical strength between dissimilar metals that are not weldable by fusion processes. Metals that typically are combined include various grades of steel, stainless steel, nickel alloys, aluminum, copper alloys, titanium, zirconium and tantalum.

The prime metal component usually is positioned parallel to the base metal component; however, for special applications, the prime component may be placed at some small angle relative to the base component. In the parallel arrangement, the base metal and prime metal are separated by a specified spacing, called the standoff distance. In the angled position, the explosion locally bends and accelerates the prime component across the standoff distance at a high velocity so that it collides at a predetermined angle with the base component. The angular collision and collision point of the metal components progress across the joint as the explosion takes place and makes the weld.

The explosive agent typically is in granular form, or *pril,* and is distributed uniformly over the top surface of the prime component. The amount of force exerted on the prime component by the explosive depends on the detonation characteristics and the quantity of the explosive agent. The quantity and velocity of the explosive are two of the three essential process variables. A buffer membrane or coating may be used between the explosive and the prime component to protect the surface of the prime component from erosion by the detonating explosive.

EXPLOSIVE DETONATION

The action that occurs during explosion welding is illustrated in Figure 9.2. The manner in which the explosive is detonated is extremely important. Detonation must take place progressively across the surface of the prime component. The speed of the detonation front moving across the prime component establishes the velocity at which the collision progresses over the joint area. This is known as the *detonation velocity.* High-quality, reproducible explosion welds are dependent on the selection of the detonation velocity of the explosive and the correct preparation of the explosive. A uniform detonation front is required.

The three essential process variables are the following:

1. Standoff distance;
2. Explosive detonation velocity; and
3. Explosive load, that is, the quantity and thickness of the explosive layer or "bed" spread evenly on the prime metal, and the energy of the explosive agent.

These EXW variables are different for various combinations, types, and thicknesses of metals.

As the detonation front moves across the surface of the prime metal, both the intense pressure in the front and the pressure generated by the expanding gases immediately behind the front accelerate the prime metal toward the base metal. The velocity of the prime metal when it impacts the base metal is called the *collision velocity.* As the detonation front progresses across the prime metal, the force causes it to form an angular bend in the prime metal that travels ahead of the collision point. The collision angle is an important secondary parameter for explosion welding. It is secondary because the angle is a result of the combined parameters of the thickness and the mechanical properties of the prime metal, the velocity of the explosive, the load, and the standoff distance. As noted, the quantity and velocity of the explosive are essential variables of the process. The sequence of operations in explosion-welded clad plate manufacturing is illustrated in Figure 9.3.

The following are important interrelated variables of the explosion welding process:

1. Detonation velocity,
2. Collision angle, and
3. Collision velocity.

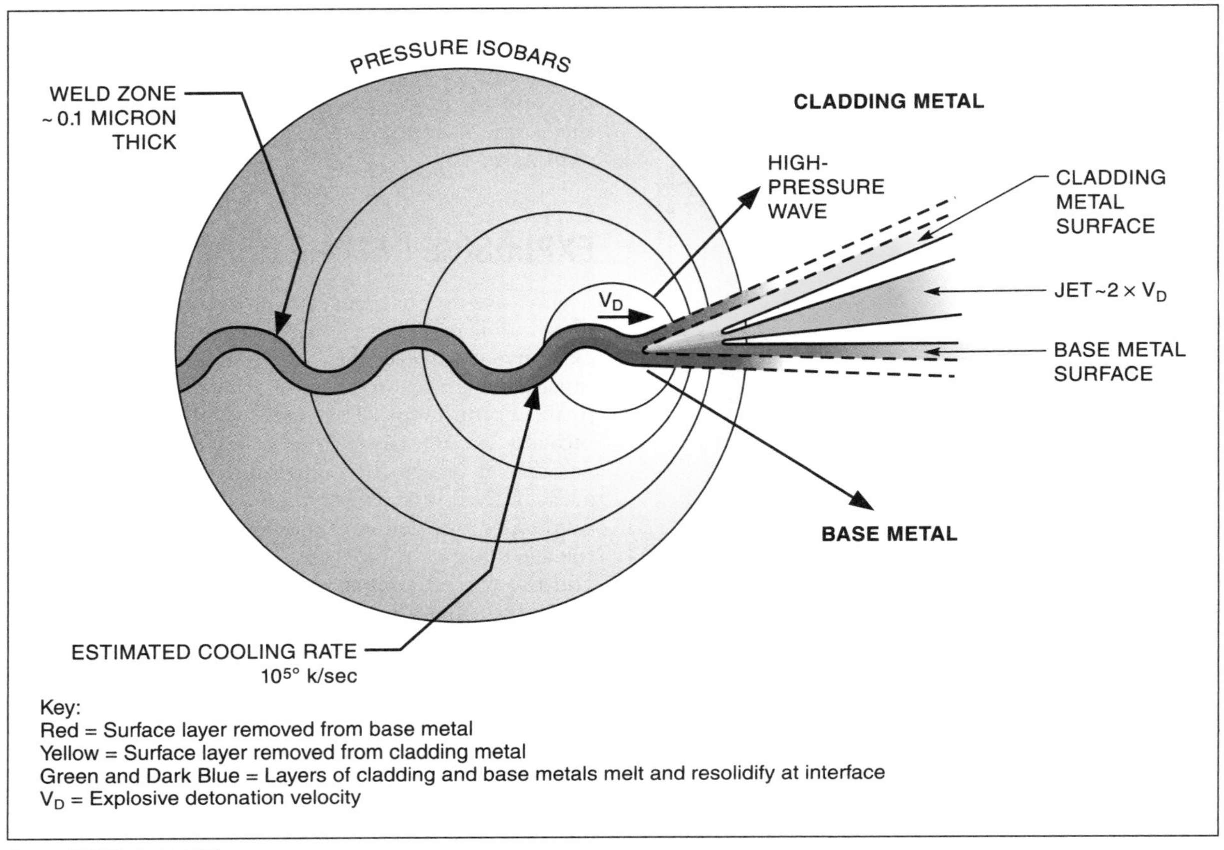

Source: DMC Clad Metal USA.

Figure 9.2—Action Between Components During Cladding with Explosion Welding

The intense pressure necessary to make a weld is generated at the collision point when any two of these variables are within well-defined limits. The limits are determined by the properties of the particular metals to be joined. Pressure forces the surfaces of the two components into intimate contact and causes localized plastic flow in the immediate area of the collision point. At the same time, a jet is formed preceding the collision point, as shown in Figure 9.3. The jet is formed as a result of momentum exchange when the two surfaces collide. A thin layer of the faying surfaces is spalled and expelled by the jet. This action also includes the removal of surface oxides, debris and standoff devices used in the manufacturing assembly, and exposes clean, oxide-free metal, which is required to make a metallurgically strong weld. The collision-velocity force acting on the metals behind the collision point applies residual pressure for a period of time long enough to avoid the release of the metal components and to complete the weld.

CHARACTERISTICS OF THE WELD

The interface between the two components of an explosion weld is almost always a sine-curve waveform. The wave size is dependent on the collision parameters and the properties of the metals. A typical wavy explosion weld interface is shown in Figure 9.4. Some metal combinations form welds with a wavy interface containing small pockets of re-solidified melt that normally are located on the front and back slopes of the waves. This region comprises a combination of the prime metal and base metal. The pockets will be ductile if the metal combinations can form solid solutions, but they may be brittle or may show discontinuities in metal combinations that form intermetallic compounds. Reactive metals such as titanium and zirconium may form these isolated pockets of intermetallic compounds when explosively welded to steel. The pockets typically are not detrimental to the intended service of the weld. Good welding practices will produce a weld with no intermetallics, or only very small pockets.

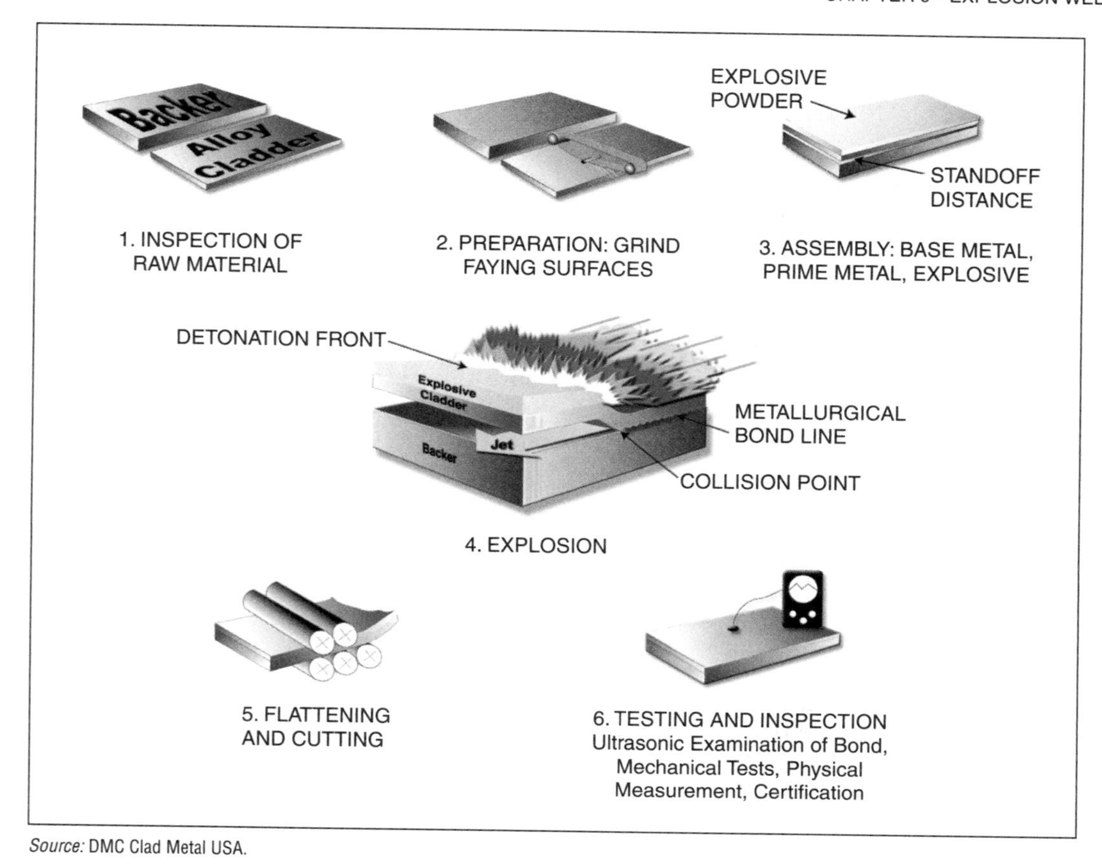

Source: DMC Clad Metal USA.

Figure 9.3—Sequence of Operations in Clad Plate Manufacturing with Explosion Welding Collision, Jetting, and Welding

Large areas of melt may occur if excessive collision energy (generated by detonation velocity, collision velocity, and collision angle) is applied. The large melt pockets or the continuous melted layers, or both, often contain a substantial number of shrinkage voids and other discontinuities that reduce strength and ductility. They usually are detrimental to the soundness and serviceability of the weld. For this reason, welding practices that produce excessively large pockets of melt or continuous melted layers normally are not used. Large waves are not necessarily bad unless they are associated with large melt pockets.

A flat weld interface can be formed when the collision velocity is below the critical value for the particular combination of metals being welded. Welds of this type may possess satisfactory mechanical properties, but as a rule are not practical. A well-shaped sine waveform usually indicates that the welding parameters are appropriately centered in the optimal range. A flat weld interface usually indicates that the weld parameters are

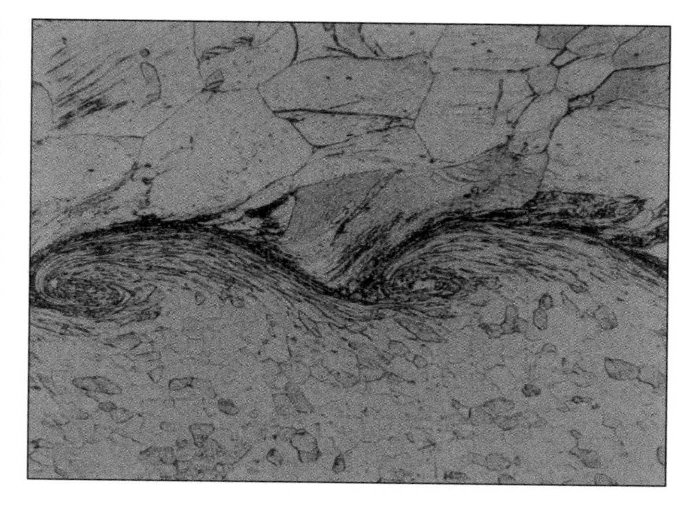

Figure 9.4—Typical Wavy Interface Formed Between Two Explosion-Welded Components, Tantalum to Copper

at the outer range of the requirements for a sound weld. A flat interface is indicative of low energy input, possibly bordering on a metallurgical discontinuity in the joined surface.

The range of explosion welding parameters that results in industrially acceptable weld characteristics is quite broad. Consequently, well-developed industrial explosion welding technology consistently yields high-quality welds for joining or cladding.

CHARACTERISTICS OF THE WELD INTERFACE

During the 1990s, three groups of researchers, two in Europe and one in Japan, attempted to characterize the microstructure of the weld interface zone. Even with modern, high-powered microscopy this proved to be elusive. The researchers chose to focus on a titanium-steel interface, since this is one of the most commonly welded combinations. A general consensus was reached through direct-observation electron microscopy and energy-balance theories and calculations.[4] Contrary to a previous and generally accepted theory, a very fine line of microfusion existed. For the titanium-steel combination it was indicated to be 0.05-microns (µm) to 2-µm wide in the studied sample, as illustrated in Figure 9.5. This zone appeared to be a metal atom mixture that had not recrystallized. In other words, the micro-fusion line is an amorphous structure, according to these researchers. Fine grains of the adjacent prime and base metal, apparently recrystallized, were found on both sides of the melt zone. Energy-balance equations validated the theory of the phase formation mechanism. Research in this area continues with work on weld zone characterization, not only with titanium-steel, but also with other metal combinations. The exact characterization previously has not been a priority within the industry because macrodestructive and nondestructive tests have proven satisfactory over the many years of successful industrial use of explosion welding.

4. Kreye, et al; Chiba, Onzawa, T. Ilyama, S. Koabayashi, A. Takasaki, and Y. Ujimoto, 1999, Microstructure of Explosively Bonded Interface between Titanium and Very Low Carbon Steel as Observed by TEM, *Proceedings of Reactive Metals in Corrosive Applications Conference*, Wah Chang, Albany, OR. 89–98.

EQUIPMENT AND FACILITIES

The required equipment for explosion welding varies according to the application and product. Since the primary commercial application of explosion welding is the manufacturing of large, flat, bimetallic plates used as a clad or a transition joint, that type of manufacturing facility is discussed in this section. Figure 9.6 shows a mechanized system for manufacturing large sections of clad plate with the explosion welding process. As evident in Figure 9.6, clad plates usually are distorted somewhat during explosion welding and must be straightened to meet standard flatness specifications. Straightening usually is done with a press or a roller leveler.

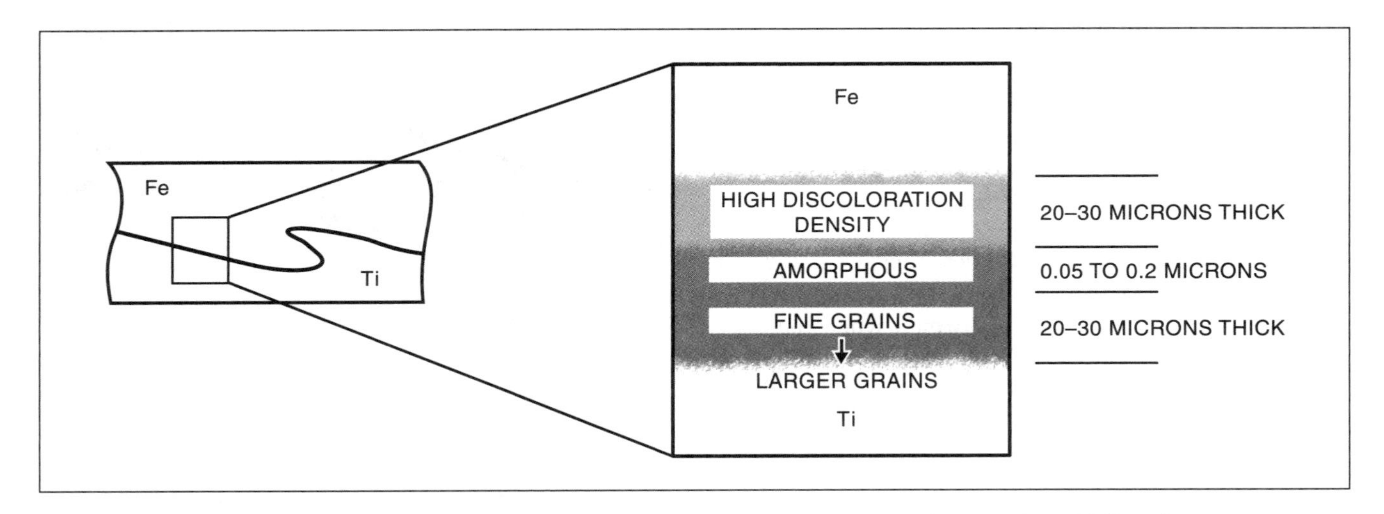

Figure 9.5—Microfusion Line in a Sample Explosion Weld in Steel and Titanium

Photograph courtesy of DMC Clad Metal USA

Figure 9.6—A Mechanized Factory for Manufacturing Large Explosion-Welded Plate

A typical manufacturing operation requires indoor and outdoor facilities and equipment. Metal preparation and processing is best performed indoors. Equipment and facilities are required for the following work functions:

1. Preparation of workpieces;
2. Fitup and assembly;
3. Handling, safety, and transportation;
4. Explosion process; and
5. Postweld treatment and testing.

Preparation of incoming base metal and prime metal may require a method of cutting components to size from larger pieces. Flame cutting and sawing often is used both before and after the explosion welding process. Mechanical descaling and abrasive grinding of the plate is required. The equipment for these functions varies from a small hand grinder to a large, mechanized gantry grinder, as shown in Figure 9.7. Wet grinding is required for reactive metals.

A gas tungsten arc seam-welding (GTAW) machine capable of making radiographic-quality seam welds of the prime metal plates often is required for large assemblies. The alloy prime metal frequently is seam-welded to match the size of the one-piece base plate. This welding booth for this operation is shown in Figure 9.8 and the GTAW seam welding operation is shown in Figure 9.9. The assembly operation requires general shop tools for woodworking and light-gauge metal fabrication, including a light-duty arc-welding machine.

An explosion shooting site such as an open field underground chamber, or vacuum chamber is required. The explosion shooting field, or chamber, usually is at a location separate from the metal processing equipment; therefore, sometimes the transportation of plates weighing 50 tons and more is necessary. Material handling equipment is necessary, as is equipment for the safe handling of high explosives and blasting agents.

Reactive clad metals and some other backer metals require postweld stress relief to achieve optimal properties. A furnace is required for these treatments. Since the explosion usually deforms the clad metal beyond a tolerance that is acceptable for subsequent processing, various flattening methods are employed. Two techniques, roller-leveler flattening and hydraulic pressing are shown in Figures 9.10 and 9.11.

Photograph courtesy of DMC Clad Metal USA

Figure 9.7—Mechanized Gantry with Wet Grinder Used in Plate Preparation

Photograph courtesy of DMC Clad Metal USA

Figure 9.8—Welding Booth for Automatic Gas Tungsten Arc Welding of Explosion-Clad Plate

Photograph courtesy of DMC Clad Metal USA

Figure 9.9—Gas Tungsten Arc Seam Welding of Prime Metal to Fit Base Metal Backer

Photograph courtesy of DMC Clad Metal USA

Figure 9.10—Roller Leveler Used to Flatten Explosion-Welded Plates

Photograph courtesy of DMC Clad Metal USA

Figure 9.11—Hydraulic Press Used to Flatten a Tube Sheet Blank for a Heat Exchanger

TESTING EQUIPMENT

Both destructive and non-destructive testing equipment is used to test explosion welds. Automatic and manual ultrasonic straight-beam inspection equipment is required for nondestructive testing. Destructive test equipment for tensile and impact testing, and equipment for flatness tests and surface inspection are required.

Explosion-welded clad plate usually is inspected with an ultrasonic beam to verify the continuity of the weld. A fully automatic, ultrasonic scanning head driven by a mechanized gantry is shown in Figure 9.12. It produces a permanent record for use in certification documents.

Mechanical tensile testing and other laboratory equipment, as shown in Figure 9.13, are required to certify that explosion-welded cladding has met industry standards. The testing laboratory of a well-equipped explosion welding company would also include a metallograph, as shown in Figure 9.14.

Further value-added manufacturing may require milling, machining, drilling, welding, or sawing or cutting. Cutting processes often used include oxygen cutting, plasma arc cutting or water jet cutting. Equipment to complete these manufacturing operations ranges from the most rudimentary to the mechanized and automated. Some explosion welding companies have started with minimal equipment by outsourcing many of the ancillary operations. High-quality reproducible welds require good equipment, adequate facilities and careful attention to welding procedures and manufacturing process controls.

WELDING VARIABLES

As discussed in the Fundamentals of Explosion Welding section, the three primary variables that must be considered are the standoff distance, the velocity of the detonation, and the quantity of the explosive. These variables are determined by the area of the weld, the thickness and properties of the prime and base metals, the desired weld quality features, and economic considerations.

Secondary variables include the following:

1. Surface preparation of the faying surfaces;
2. Type and spacing of standoff devices;
3. Prime metal overhang and metal sizing to accommodate usual incomplete edge bonding;
4. Preheating the metal prior to detonation;
5. Location of the detonation-initiating device and the booster explosive; and
6. The use, type, and placement of explosive containment.

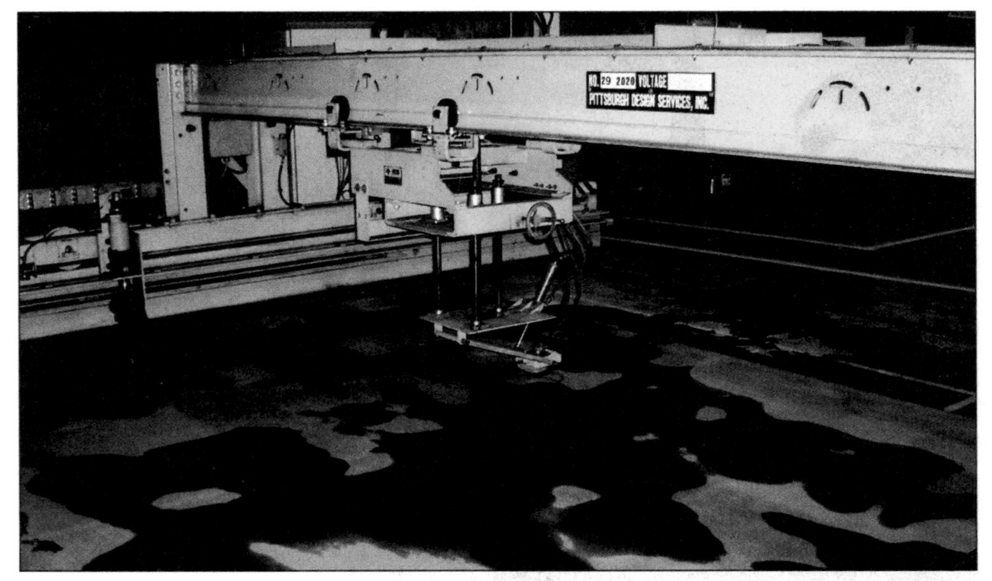

Photograph courtesy of DMC Clad Metal USA

Figure 9.12—Gantry-Mounted Ultrasonic Scanning System

Photograph courtesy of DMC Clad Metal USA

Figure 9.13—Mechanical Testing Equipment

Photograph courtesy of Nobelclad Europe

Figure 9.14—A Metallograph Used to Examine Weld Microstructure

Welding variables are significantly different when explosion welding is performed in a vacuum chamber. The use of a vacuum chamber for explosion welding is even more sophisticated in practice than open-air explosion welding. A vacuum chamber allows for the use of less explosive material, or load. The assembled metal and the explosive are placed in a specially constructed chamber that can be evacuated to a partial vacuum. As long as the chamber is properly sealed and evacuated, there is little chance for air over-pressure to damage the chamber. This technology has been proven for decades but is not widely practiced.

MATERIALS

The materials used for cladding by explosion welding are grouped into the following three categories:

1. Raw metal, also known as *plain material*;
2. Explosives (energetic materials); and
3. Expendable construction materials required to build fixtures needed to secure and position the workpieces prior to the explosion.

The raw metal consists of the base metal, the cladding metal, and an interlayer metal, if one is specified. The clad manufacturer receives metal in the form of plate or sheet from a producing mill, stocking warehouse or service center, or it may be provided by the buyer of the cladding service. The metal can be as small as a dinner plate or as large as the bed of a semi-truck trailer. Prime metals may come from coils and leveled for use. Cladding typically is applied to steel as the base metal. Typical prime metals used for cladding include stainless steels, nickel, copper, aluminum, titanium, zirconium, tantalum, and silver, and many alloys of these metals. Various combinations of the typical prime metals also are joined by explosion welding. Prime metal that is directly welded (without an interlayer) should be limited to a maximum yield strength. Depending on the welding parameters and practices of the welding company, an approximate maximum value is 276 000 kilopascals (kPa) (40 000 pounds per square inch [psi]). Flatness is a crucial quality of metals to be explosion welded. Flatness requirements over the entire area of the prime metal must be at least as stringent as those of commercial tolerances. Plate or sheet metal that is out-of-flat, wavy, or "oil-canned" affects the standoff distance, an essential process variable. Material tolerances must meet the requirements of national standards or those of the metal manufacturer.

The base metal, which is the thicker and heavier of the two components, can be obtained from the same sources as the prime metal. Plates weighing over 45 400 kg (50 tons) or more can be commercially welded by explosion welding. Commercial applications can use base metal as thick as 60 centimeters (cm) (24 inches [in.]) and even 90 cm (36 in.) thick for certain products. Generally, the base metal is carbon steel, alloy steel, or stainless steel. It usually is in plate form but also may be a forging. Flatness is a critical issue, but thick plates deform less than thin plates during shipping and handling. Steels are required to be virtually free of slag inclusions and laminations so they will sustain the force of the explosion.

Explosive Materials

The four basic components of the explosion welding system are the following:

1. The bulk explosive that provides the energy for welding,
2. An electric blasting cap or fuse used as an initiator,
3. A booster explosive that is easily ignited by the initiator and may be required to ignite the bulk material, and
4. A detonation cord that links the initiator to the booster and may link more than one assembled metal pack for simultaneous detonation.

Energetic materials used for the explosion welding process typically are granular and their composition usually is based on ammonium nitrate (AN) as the primary ingredient. The energy of the ammonium nitrate can be increased by mixing or coating the AN with other materials. One common additive is fuel oil or a similar petroleum-based fluid. The blend is then referred to as *ANFO*. This produces a detonation velocity of 2000 meters per second (m/s) to 3000 m/s (6500 feet per second [ft/s] to 9800 ft/s), which is the velocity range normally required to produce the collision-point conditions necessary for optimal welding. In general, the detonation velocity of an explosive depends on the chemical composition, an inert diluent addition, the depth of the explosive bed (thickness of the explosive covering), and packing or loading density.

Other important considerations that affect the characteristics of the ANFO explosive mixture include the following:

1. The size and dispersion of the granular material,
2. Moisture content (affected by duration and atmosphere of storage), and
3. Handling and containment methods (bags or other containers) that do not segregate size and density of the granular material.

Explosives other than AN and ANFO may be used, but because of the effectiveness, safety, and low cost, ANFO is commonly used.

Metals Welded

As a general rule, any metal can be explosion welded if it possesses sufficient strength and ductility to withstand deformation at the high velocity associated with the process.

Data published by various researchers and manufacturers over the years verified the successful joining of a large number of specific metal combinations, ranging from common industrial combinations like those presented in Table 9.1 to more exotic combinations, such as welding tungsten to molybdenum, tantalum to rhenium, and gold to platinum. Welding conditions for each metal combination are dependent on variables ranging from simple issues like availability, size, and thickness to complex metallurgical considerations. In many cases, interlayers of a third metal are beneficial in optimizing explosion welding viability. The precious metals perform well in explosion welding but are not significant in commercial products, primarily due to cost.

Metals with elongations of at least 5% to 6% in a 51-mm (2-in.) gauge length and Charpy V-notch impact strengths of 13.6 joules (J) (10 foot-pounds [ft-lb]) or more can be explosion welded. In special cases, metals with low ductility can be welded by preheating to a slightly elevated temperature to the point at which they will have adequate impact resistance. It should be noted that the use of explosives in conjunction with heated

Table 9.1
Common Industrial Explosion Welding Metal Types

Cladding Metals	Base Metals
Stainless steels (austenitic, martensitic, ferritic, and duplex)	Carbon and many low-alloy steels, plate, and forgings
Nickel and nickel alloys	Austenitic stainless steels (300 series)
Copper and copper alloys	Aluminum
Aluminum and aluminum alloys	
Titanium	
Zirconium	
Tantalum	

components requires special safety considerations. Metals such as beryllium, Al-Be alloys and rhenium that are difficult to weld can be joined with explosion welding. Powder metal products such as Glidcop™ and Al-SiC can be joined to wrought metal without thermal excursions.

Metals with high hardness, such as wear-resistant plate, do not possess adequate ductility to withstand the violent force of the explosion welding process.

Explosion welding does not produce major changes in the properties of the cladding metals or the base metals. However, the process can produce some changes in the mechanical properties and hardness of the metals, particularly in the immediate area of the weld interface, as indicated in Figure 9.15. Changes in the base metal properties most frequently are within the allowable tolerances according to materials specifications such as those published by the American Society for Testing and Materials (ASTM International). ASTM documents A516, A387, and A240 cover typical base metals.[5] In general, the severe localized plastic flow along the interface during welding increases the hardness and strength of the metal in this region. Accordingly, the ductility decreases.

5. ASTM International, 100 Barr Harbor Drive, West Conshohocken, PA 19428-2959.

Explosion-welded plate usually is supplied in the as-welded condition because the hardening that occurs immediately adjacent to the interface usually does not significantly affect the mechanical properties.

Hardening effects may be greatly reduced by a postweld heat treatment (refer to Figure 9.15), however, the particular heat treatment applied should be a type that will not reduce the ductility of the weld by undesirable atomic diffusion or the formation of brittle intermetallic compounds at the interface.

APPLICATIONS

The cladding of flat plate constitutes the major commercial application of explosion welding. The sections of heavy 304L stainless steel-clad plate shown in Figure 9.16 are examples of this application.

Clad metal produced by explosion welding is used to construct pressure vessels used in oil refineries and chemical processing plants. Components of heat exchangers also are a major use of corrosion-resistant, explosion-welded clad metal. Figure 9.17 shows an

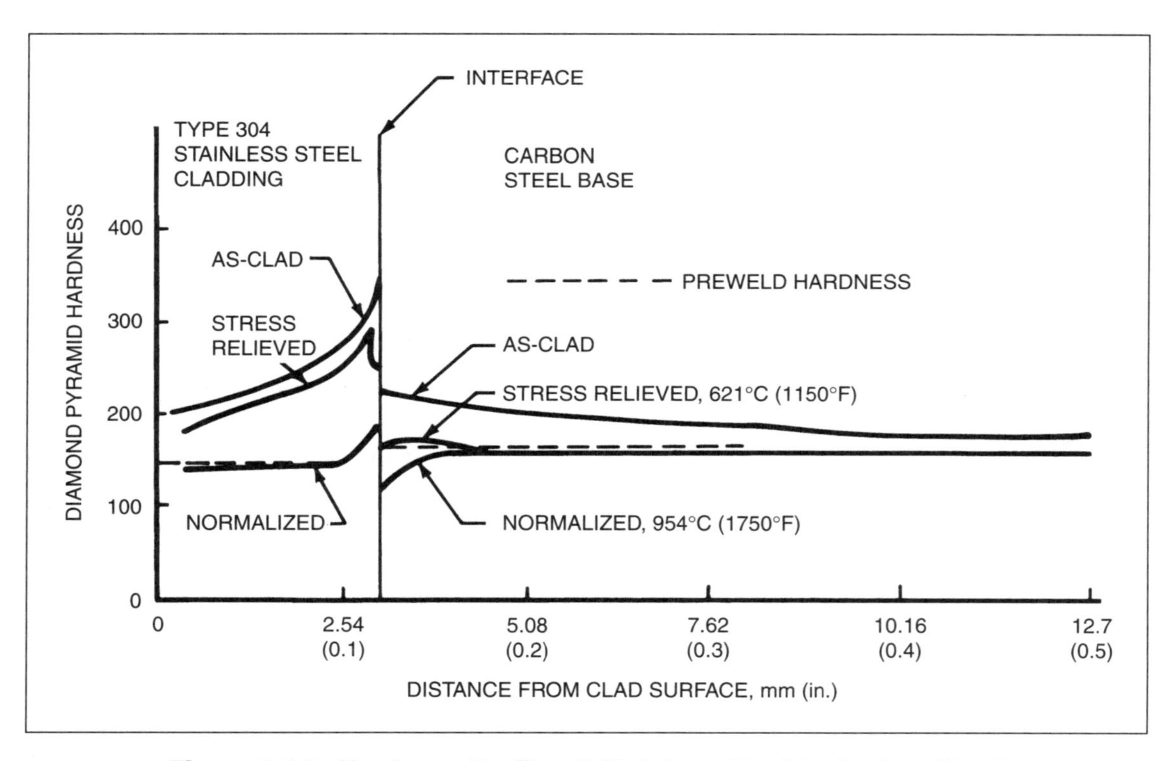

Figure 9.15—Hardness Profile of Stainless Steel-to-Carbon-Steel Clad Plate, As-Welded and After Heat Treatment

Photograph courtesy of Nobelclad Europe

Figure 9.16—Heavy 304L Stainless-Steel-Clad Plate

Photograph courtesy of Asahi Kasei

Figure 9.17—Explosion-Clad Titanium Tube-Sheet Blank (After Drilling) Used in a Condenser

explosion-clad titanium condenser tube-sheet blank after drilling. Pressure-vessel heads and other components frequently are made from explosion-clad plates by conventional hot-forming or cold-forming techniques. Explosion clad metal products can be formed by hot or cold spinning. The orange hemispherical pressure vessel head shown in Figure 9.18 was formed to finished shape by the hot spinning process. In this application, the material is 1/4-inch stainless steel clad to a low alloy, chromium-molybdenum alloyed steel base metal. The total thickness is over 3 inches. The finished product, a hydrotreater reactor column loaded for transporting to the refinery site is shown in Figure 9.19. Planning for hot forming must take into account the metallurgical properties of the base and prime metals and the possibility that undesirable diffusion may occur

Photograph courtesy Trinity Industries Head Division

Figure 9.18—Hemispherical Pressure Vessel Head Made of Explosion-Clad Plate Formed by Hot Spinning

Photograph courtesy of Beaird Industries

Figure 9.19—A Hydrotreater Reactor Column for an Oil Refinery Fabricated from Explosion-Clad Plate

at the interface. Compatible alloy combinations, such as stainless steel and carbon steel, may be formed by methods traditionally used for solid alloy heads. It must be carefully ensured that hot working or heat treatments, or both, are properly selected to be certain that both the prime and base metals retain the proper mechanical and corrosion-resistant properties in the completed head.

Alloys subject to carbon sensitization, and metals such as titanium or zirconium combined with steel require special consideration during cladding and various fabrication steps. Titanium-clad steel, for instance, should be hot formed at temperatures no higher than 760°C (1400°F) to prevent the formation of undesirable intermetallics, which could lead to brittle failure of the weld. Tantalum-clad steel plates, often specified when resistance to extreme corrosion is required, are shown in Figure 9.20. In another application, a titanium-clad tube sheet blank for a heat exchanger is shown in the as-machined condition in Figure 9.21.

Reducing the thickness of clad plate by hot rolling (called "*bang-and-roll*" or *conversion rolling*) provides a convenient and economical means of producing bimetal plate or sheet in thinner gauges than easily or economically produced by direct cladding by explosion welding. Thermal considerations for this process are similar to those used for forming pressure vessels or cylinder heads.

CYLINDERS

Explosion welding can be used to clad the outside surfaces of cylinders if they are limited to a few inches in diameter. Outside surface cladding is readily accomplished and used commercially. Cladding the inside of tubular sections has been successfully accomplished but has not been commercially viable. The primary application for which EXW was intended was welding nozzles for pressure vessels. Among other problems with this application, the high hoop stresses developed by the explosion are difficult to control without distorting the diameter of the tubular section.

Photograph courtesy of DMC Clad Metal USA

Figure 9.20—Tantalum-Clad Steel Plates (Before Edge Trimming) Specified for Resistance to Extreme Corrosion

Photograph courtesy of DMC Clad Metal USA

Figure 9.21—Titanium-Clad Tube Sheet Blank for a Heat Exchanger, in the As-Machined Condition

TRANSITION JOINTS

Bimetallic transition joints serve several needs in two primary categories: tubular transition joints for piping connections between dissimilar metals, and bimetallic flat transition joints specified for strength or electrical conductivity, or both, between dissimilar-metal structural members.

Structural transition joints generally are made from strips or small plates cut from a single larger piece of explosion-welded plate, as illustrated in Figure 9.22. A conventional fusion welding process, such as gas metal arc welding, gas tungsten arc welding, or plasma arc welding, then is used to attach the members of the transition joint to the similar-metal components. However, the arc-welding interpass temperature and subsequently the service temperature at the weld interface must be carefully limited to a level suitable for the combination of materials in the explosion-welded joint. Aluminum-steel is a typical combination for transition joints because explosion welding produces a strong joint that can be arc welded to both metals. A cross section of an explosion-welded transition joint in a typical shipboard application is shown in Figure 9.23.

Photograph courtesy of Nobelclad Europe

Figure 9.22—Transition Joint Made from Aluminum-Steel Clad Plate Welding Strips

Electrical Transition Joints

Aluminum, copper, and steel are the metals most commonly used in electrical systems because each provides special properties. Joints between these metals frequently are required to be continuous and of high quality if they are to conduct high current efficiently with minimum power loss. Poor conductivity causes a rise in temperature that can degrade poor joints. Explosion welding produces strong, heat-resistant bimetal transition joints that provide efficient conductors of electricity. The joints are cut from thick explosion-welded plates of aluminum-copper or aluminum-steel. This concept routinely is used in the fabrication of anodes for aluminum smelting.

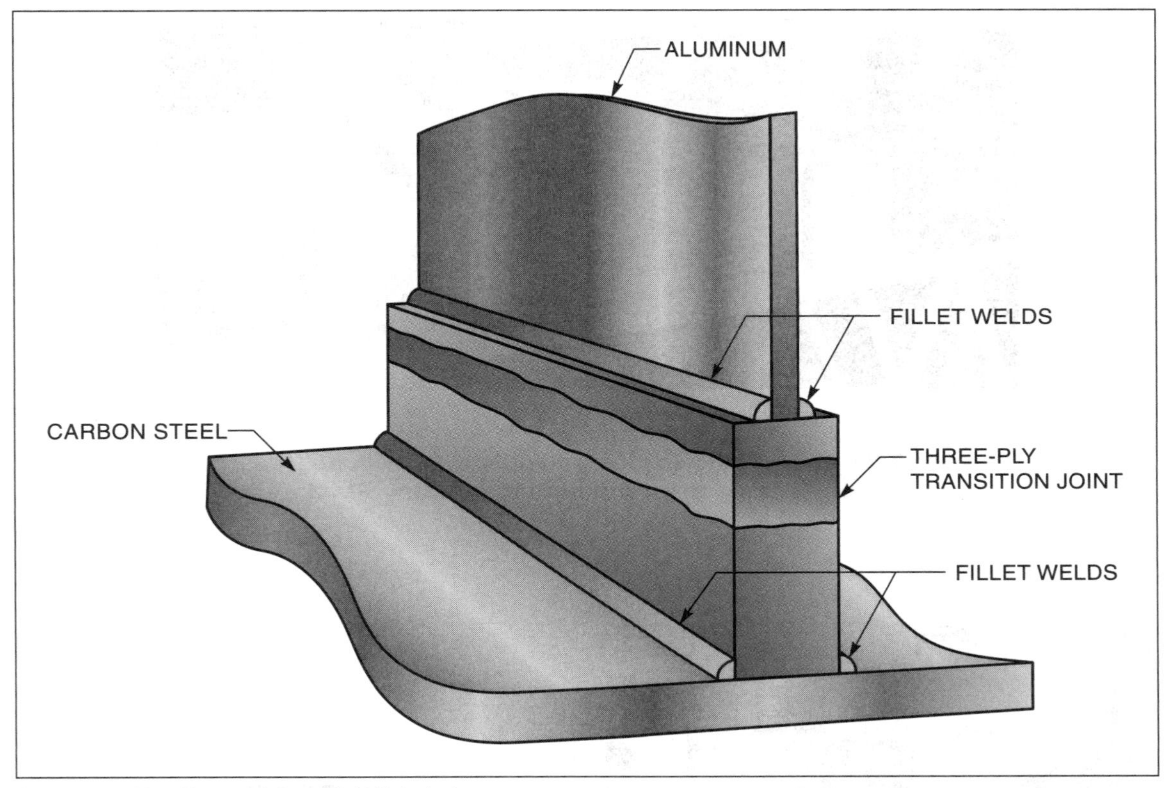

Source: Adapted from Merrem & la Porte BV, the Netherlands.

Figure 9.23—Explosion-Welded Aluminum-to-Steel Clad Metal Transition Joint Using Arc-Welded Fillet Welds to Join Steel to Aluminum for a Shipboard Application

The temperature limits for transition joints between aluminum and steel are 260°C (500°F) or less for long-term service. Copper-aluminum joints should be limited to 150°C (300°F). High-quality transition welds are unaffected by thermal cycling below these temperatures. Short-term exposure (10 minutes to 15 minutes) during attachment welding, for example, the joint may reach temperatures of 290°C to 315°C (550°F to 600°F) in aluminum-to-steel welds, and 200°C to 260°C (400°F to 500°F) for aluminum-to-copper welds without harm. These temperature limits can be increased significantly by the addition of an interlayer metal. The use of a titanium interlayer between aluminum and steel or aluminum and copper raises the functional operating temperature to over 500°C (1000°F).

Figure 9.24 shows bimetallic, explosion-welded electrical transition joints sawed from a large plate and ready for use as aluminum smelting yoke connections. The application is shown in Figure 9.25, in which the transition joint is used between the vertical rod and the yoke in aluminum smelting.

Marine Transition Joints

Ships and other marine structures often require the use of specific metals in different areas of the structure. A classic example is ship construction using a steel hull and an aluminum superstructure. Until the late 1960s, the connection between the aluminum and steel components was made with mechanical fasteners. The corrosion problems resulting from the use of galvanically dissimilar metals was further complicated by the crevice between the components. Maintenance was a major problem. Explosion-welded aluminum-to-steel transition joints were introduced to the shipbuilding industry and quickly became the industry-wide solution for making welds between aluminum and steel components. Although the explosion-welded transition joint does not eliminate galvanic-driven corrosion, it totally eliminates the dissimilar-metal crevice and creates a product that is easily protected by conventional corrosion management techniques. Similar aluminum-steel transition joints are used in rail and highway transportation equipment.

Photograph courtesy of Nobelclad Europe

Figure 9.24—Bimetallic Explosion-Welded Electrical Transition Joints Used in an Aluminum Smelting Plant

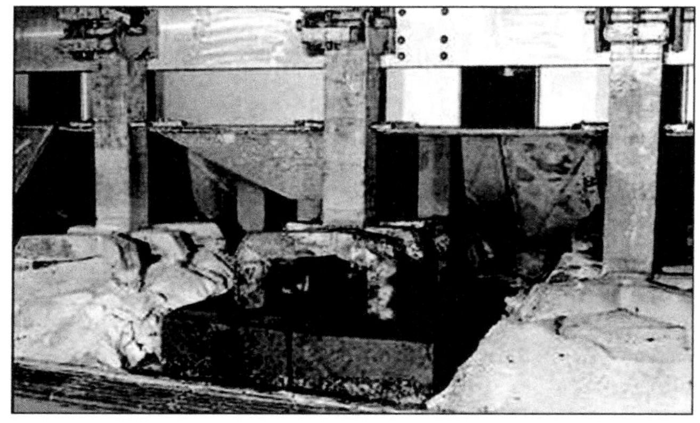

Photograph courtesy of Pechiney

Figure 9.25—Explosion-Welded Transition Joint in an Aluminum Smelter Anode

Photograph courtesy of Alstrom Naval

Figure 9.26—High-Speed Ferry Constructed with Explosion-Welded Transition Joints Between Dissimilar Metals

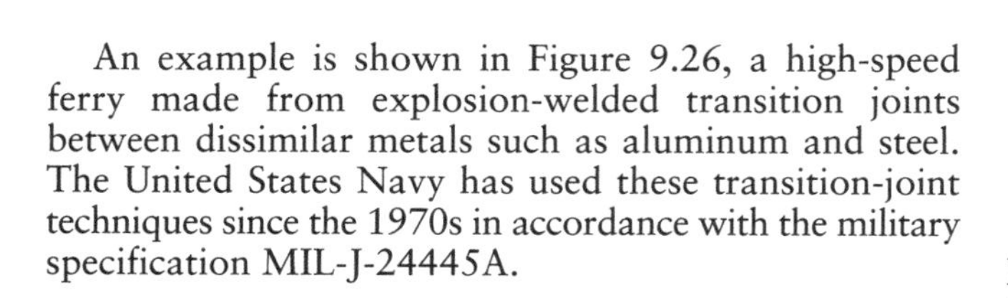

An example is shown in Figure 9.26, a high-speed ferry made from explosion-welded transition joints between dissimilar metals such as aluminum and steel. The United States Navy has used these transition-joint techniques since the 1970s in accordance with the military specification MIL-J-24445A.

Tubular Transition Joints

Tubular transition joints in various configurations can be machined from thick clad plate. The interface of the explosion weld is perpendicular to the axis of the tube. Examples of a variety of transition joints machined from explosion-clad plate are shown in Figure 9.27 and Figure 9.28. While the majority of explosion-welded tubular transition joints are aluminum to steel, other metal combinations for this type of joint include titanium to stainless steel, zirconium to stainless steel, zirconium to nickel-base alloys, and copper to aluminum.

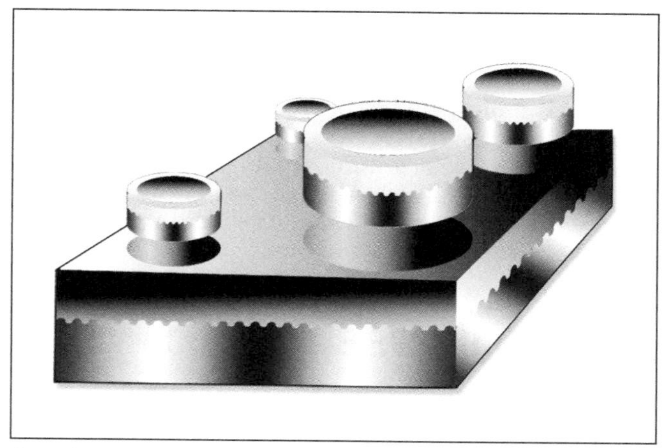

Source: DMC Clad Metal USA.

Figure 9.27—Tubular Transition Joints Sectioned from Explosion-Welded Stainless Steel-Aluminum Clad Plates for Cryogenic Use

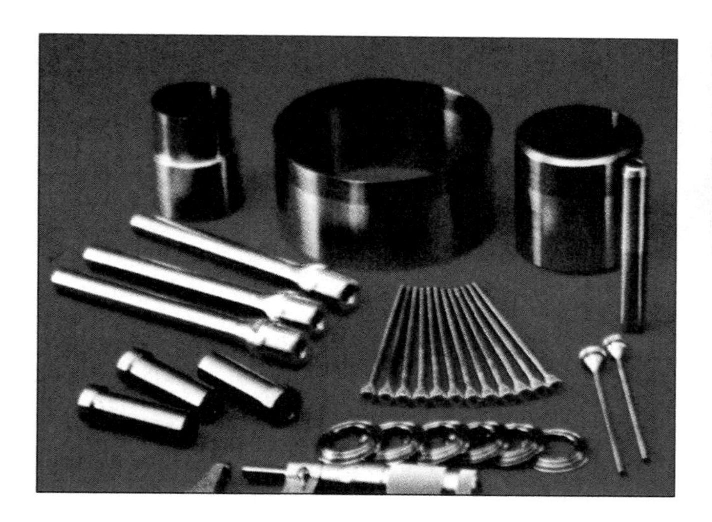

Figure 9.28—Aluminum-to-Steel, Titanium-to-Aluminum, and Titanium-to-Stainless Steel Tubular Transition Joints Machined from Explosion-Clad Plate

Joints can be fabricated directly by explosion welding in an overlapping or telescoping style similar to a cylindrical cladding operation. These joints have the advantage of a long overlap and frequently require little or no machining after welding. Figure 9.29A shows a direct explosion-welded tubular transition joint 30.5 cm (12 in.) in diameter made from 3003 aluminum explosion-welded to A106 Grade B steel.

Photograph courtesy of Batelle

Figure 9.29—A Tubular Transition Joint of 3003 Aluminum Explosion-Welded to A106 Grade B Steel

HEAT EXCHANGERS

Explosion welding can be used make tube-to-tube sheet welds in the fabrication of shell and tube heat exchangers for use in the manufacturing of chemicals, for oil refining, hydrocarbon processing and production, and other industrial processes.

As previously mentioned, commercial attempts to clad the inside of tubular sections and pressure vessel nozzles have not been successful. One of the reasons is that even a heavy-walled pipe section usually yields to the high hoop stress applied by the magnitude of explosive force necessary to produce a continuous surface weld. The distortion is not readily controlled.

A heat exchanger tube sheet may be 50 mm to 6 cm (2 in. to 24 in.) thick or more. Tube holes are drilled through the usually circular tube sheet in a hexagonal close-spaced pattern. The tubes usually are less than 25 mm (1 in.) in diameter. Tube wall thickness typically ranges from 2 mm to 3 mm (0.062 in. to 0.125 in.). Tube diameters range from 12 mm to 20 mm (0.5 in. to 1.5 in.). The tube wall is relatively thin and the tube sheet is massive, even with closely spaced holes. To seal the connection, the linear distance of the tube-to-tube-sheet weld needs to be only the equivalent of the wall thickness of the tube.

When explosion welding is used to join two flat pieces of metal, the standoff distance is maintained at a precisely set dimension. In tube-to-tube sheet explosion welding, the standoff distance is formed by an angle chamfered into a portion of the cylindrical hole drilled in the tube sheet, as shown in Figure 9.30(A), labeled "Machined Angle of the Tube Sheet." The explosive energy is provided by a detonator designed for a specific tube diameter and tube wall thickness. It is detonated from the inside and is sheared off as the explosion gases exit the hole. The process usually is performed inside a shop building with qualified technicians, shielding, and controlled space.

The standoff gap is an essential welding variable and the options of that parameter are limited. However, since the weld needs to be completed only on a short length of the tube wall, the optimal standoff parameter can be satisfied along the length of the sloping angle of the machined tube sheet hole. The availability of tube-to-tube sheet detonator kits and machine angles is well established.

Tube materials such as stainless steel, nickel alloys, copper alloys, aluminum, titanium and zirconium tubes have been welded, usually to a carbon-steel tube sheet. Explosion welding also can be used for plugging leaking tubes in heat exchangers. Electric power stations and petrochemical companies use the process because it is quick, easy and reliable. Although the process appears simple, only qualified, trained technicians should use the technique. An explosive handling permit is required in most states.

WELD QUALITY AND TESTING

The quality of an explosion weld can be verified by an examination of the joint properties, which depend on the initial properties of the prime metal, base metal, and the interface metal, if one is used. The properties of the metals include strength, toughness, and ductility. The effect that welding has on these properties can be determined by comparing the results of tension, impact, bend, and fatigue tests of welded and unwelded materials. The titles of ASTM testing procedures for clad plate are listed in Table 9.2.

The quality of the weld typically is evaluated by destructive and nondestructive tests. Special destructive tests sometimes are required because the size of test samples is limited by the thickness of the components. The test samples may be very small since the subject is an autogenous weld in thin metal. Subsized test specimens often are required. Also, the weld is planar and in essence has no thickness. The tests should reflect conditions the weld will withstand in service.

NONDESTRUCTIVE EXAMINATION

Due to the nature of explosion welds, nondestructive examination (NDE) is restricted almost totally to the ultrasonic method. Radiography generally is not an effective technique for determining of weld quality, strength, or continuity.

Ultrasonic Examination

Ultrasonic examination is the most widely used nondestructive method for the examination of explosion welds. Standard, straight-beam ultrasonic inspection verifies the continuity of the weld across the plate surface, but provides minimal information on weld properties.

Pulse-echo and straight-beam ultrasonic techniques, such as are those specified in the latest edition of *Standard Specification for Straight Beam Ultrasonic Inspection of Plain and Clad Steel Plates for Special Applications,* ASTM A578 (noted in Table 9.2), normally are applied for the examination of clad steels used in the construction of pressure vessels. An ultrasonic frequency in the range of 2.5 megahertz (MHz) to 10 MHz usually is recommended. Allowances must be made for the differences in the acoustical impedance of various metals.

The ultrasonic testing instrument can be calibrated on standard samples containing both welded areas and known unwelded areas. Unwelded areas reflect the signal prematurely, appearing less than the full height of the signal at the appropriate location on the oscilloscope display, or as a digital value less than the thickness of the complete welded section. An automatic, computer-controlled, gantry-driven transducer is the preferred apparatus used for testing explosion-clad plate and recording the results for use in certification documents (refer to Figure 9.12). For large clad plates where scanning 100% of the surface area typically is not necessary, the examination can be carried out on a rectangular grid pattern. Various industrial standards of ASTM and codes of the American Society of Mechanical Engineers (ASME)[6] address testing and acceptance standards.

6. American Society of Mechanical Engineers, 3 Park Avenue, New York, NY 10016-5990.

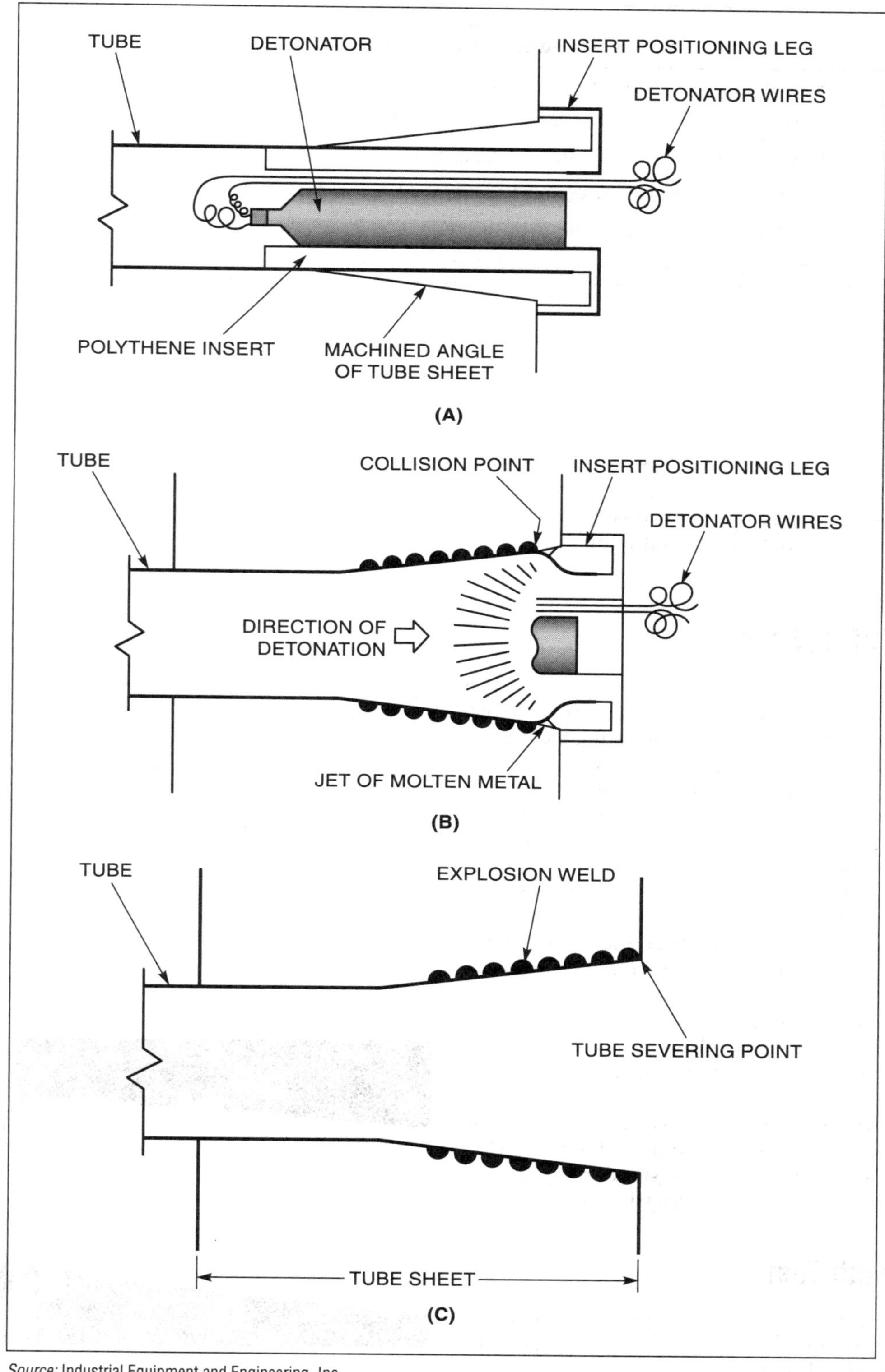

Source: Industrial Equipment and Engineering, Inc.

Figure 9.30—(A) Arrangement for Tube-to-Tube Sheet Explosion Weld, (B) Weld in Progress, and (C) Completed Weld

Table 9.2
ASTM Standards for Testing Clad Plate

ASTM A262	*Standard Practices for Detecting Susceptibility to Intergranular Attack in Austenitic Stainless Steels*
ASTM A263	*Standard Specification for Stainless Chromium Steel-Clad Plate*
ASTM A265	*Standard Specification for Nickel and Nickel-Base Alloy-Clad Steel Plate*
ASTM A578	*Standard Specification for Straight-Beam Ultrasonic Examination of Plain and Clad Steel Plates for Special Applications*
ASTM B432	*Standard Specification for Copper and Copper Alloy Clad Steel Plate*
ASTM B898	*Standard Specification for Reactive and Refractory Metal Clad Plate*

Source: ASTM International, West Conshohocken, Pennsylvania.

It should be noted that indications of rejectable weld discontinuities typically occur in less than 0.05% of the total surface area produced by professionally recognized industrial explosion welding companies.

DESTRUCTIVE TESTING

Destructive testing is used to determine the strength of the weld and the effect of the process on the properties of the component metals. Standard testing techniques can be used, but specially designed tests sometimes are required to determine weld strength for some configurations. For the base metal, the usual tests for tensile strength and impact strength are used to verify compliance with the original specifications. A thermal regimen that simulates conditions the clad metal may undergo during the complete manufacturing cycle frequently is required. The user may require that the prime metal alloy be tested for corrosion resistance after explosion welding before it is used in subsequent fabrication ASTM A262, *Standard Practice for Detecting Susceptibility to Intergranular Attack in Austenitic Stainless Steels* is an example of a corrosion test sometimes specified. (See Table 9.2.)

Testing requirements for carbon-steel clad plates and other commonly used metals are published in the ASTM documents listed in Table 9.2. These standards primarily use shear tests to determine the strength of the weld.

Shear Strength Test

To prepare a specimen, or coupon, for a shear-strength test, a small block of clad metal is cut and machined to specified dimensions and according to the appropriate ASTM clad product specification. All of the clad metal is removed except for a small lug. The coupon is fixtured in a die. A tensile testing machine is used to apply measured force until the clad metal lug shears from the base metal. A force-per-unit area is calculated and a value in MPa (psi) is recorded as the shear strength of the cladding weld to the base metal.

The jointly published materials code specifications of ASTM and ASME require minimum shear test values for several metal combinations used to construct pressure vessels. Explosion-welded clad material often exceeds the minimum shear values by two and even three times more than the values required. Figure 9.31 shows a standard ASME shear test for weld strength, in which failure occurred in the base metal rather than the weld, as expected. The bar of metal to the right was sheared from this area. The force was measured using a tensile machine. This shear value is a measure of the strength of the weld.

Photograph courtesy of DMC Clad Metal USA

Figure 9.31—Configuration of a Tensile Shear Test Specimen Showing Failure in the Base Metal

Tensile Strength Test

When testing thick explosion-welded products, typically thicker than 50 mm (2 in.) and when the thinner component is at least 13-mm (1/2-in.) thick, the tensile strength of the weld is measured with reduced-length, conventional tensile coupons. If the cladding metal is thin, it is very difficult to produce and test coupons of this type. A special ring-rupture tension test, or ram test, often is used for thin metals to evaluate the tensile strength of explosion welds. As shown in Figure 9.32, the specimen is designed to subject the weld interface to a tensile load. The cross-sectional area of the specimen is the annulus between the outside and inside diameters. The typical specimen is designed with a short gauge length that is intended to cause failure at the weld interface or immediately adjacent to it. If failure occurs in one of the base metals, the test shows that the weld is stronger than the prime metal. The ram tensile test is most often used with aluminum-to-steel weld cladding and conducted in accordance with specification MIL-J-24445A, *Military Specification Joint, Bimetallic Bonded, Aluminum to Steel.*[7] The explosion weld has a tensile strength greater than the value of the weaker of the two composite materials. Therefore, a tensile test generally is used only with aluminum combined with other metals because the aluminum is a relatively soft, weak metal and weld strength therefore is limited by the aluminum.

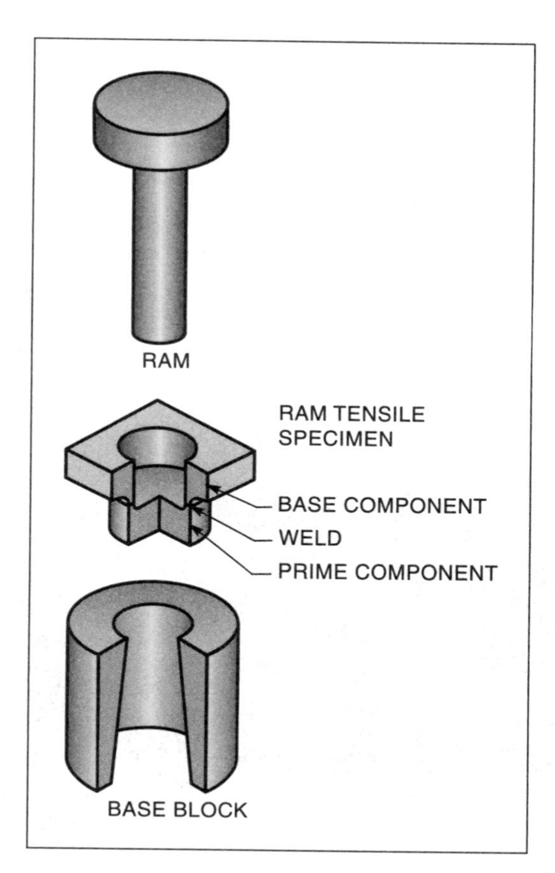

Figure 9.32—Configuration of a Typical Ring or Ram Tensile Test

The ram test is conducted by placing the specimen on the base block with the ram in the hole. A compressive load is then applied to the prime component through the ram supported by the base. The load at failure is recorded.

Metallographic Examination

Metallography can provide qualitative information about explosion welds. The visual weld characteristics, which are important when considering fitness for service, will vary greatly, depending on the metal types and the intended application of the product. Interpreting this information requires significant process insight and interpretive experience; consequently, metallography is not recommended as an evaluation tool unless the operator has training and experience in the characteristics of the microstructures of the individual metal combinations.

ECONOMICS

Although explosion welding is used for both joining and cladding, bimetal cladding makes up the vast majority of applications. Explosion welding is not cost effective when used for butt joints and generally will be considered only for metals that cannot be fusion welded. Procedures for lap joints in like metals have been developed but are not commonly used and do not provide an economic advantage. Conversely, explosion-welded cladding, generally chosen for applications in which corrosion resistance is required, competes with other solutions either by direct cost comparison or by the benefits of the unique features of the process.

The equipment designer has several cost-related options when planning explosion-welded cladding for corrosion-resistant applications. These include the following:

1. Providing an allowance for predicted corrosion by adding thickness to the base metal to enhance the metal of construction, such as carbon steel;

7. United States Department of Defense Index, latest edition, Washington, D.C.: United States Department of Defense.

2. Adding chemical corrosion inhibitors to the process to slow corrosion of the original metal of construction;
3. Specifying solid, through-wall thickness, corrosion-resistant alloy construction;
4. Specifying bimetallic clad construction made by explosion welding or roll bonding; and
5. Using arc-welded overlay (only with suitable metal combinations).

When the bimetallic clad option is selected, the designer must consider the engineering features and economic factors. The later usually consists of price and delivery time, and sometimes product form.

Weld overlay can be more cost competitive for very heavy thick plate. Weld overlay usually is applied to a pressure vessel component such as a shell plate or head after it is formed to shape. One or more fusion weld passes are applied to a base metal to obtain the minimum chemistry of the applied alloy. The weld quality is measured by shear tests or autoclave hydrogen-disbond testing, or both. It should be noted that explosion weld cladding of plate is only done in the flat position.

Hot-roll bond cladding has a price advantage in thin metal cross sections. When used in ASME-governed applications, bond strength must meet ASTM/ASME requirements detailed in previous references that also cover explosion weld cladding. (Refer to Table 9.2.)

When price is the only consideration, explosion-welded cladding has an advantage in plate thickness between the weld overlay and hot-roll bond cladding options. However, explosion-welded cladding has high weld shear strength similar to weld overlaid cladding. When fast delivery is needed, the explosion clad is the option of choice, regardless of thickness. For metal combinations of steel and aluminum, titanium, zirconium or tantalum, explosion-welded cladding is the first or only choice.

Other joining technologies, such as hot-roll bonding, friction welding, upset welding or diffusion welding may be used for bimetallic transition joints in electrical or mechanical connections. However, explosion welding provides a superior transition joint in almost all applications. Costs may vary between the products.

An example of an explosion welding application chosen for economic advantage is a titanium-clad steel autoclave used in nickel ore leaching, shown in Figure 9.33. The autoclave is fabricated from explosion weld-clad pressure-vessel plate. The alternative to the titanium lining was to install several courses of brick to

Photograph courtesy of Coek Engineering NV

Figure 9.33—Titanium-Clad Steel Autoclave to Be Used On-Site for Nickel Ore Leaching

serve as the thermal lining. Instead, a corrosion-resistant membrane was applied to the carbon steel before the brick was laid. The autoclave alone weighs approximately 450 metric tons.

SAFE PRACTICES

The explosives and explosive devices that are an integral part of explosion are inherently hazardous. Safe methods for handling them do exist. However, if the materials are misused, they can injure or kill anyone in the area and damage or destroy property.

Explosive materials must be handled and used by properly trained and experienced people. Handling and safety procedures must comply with all applicable federal, state, and local regulations. Federal jurisdiction over the sale, transport, storage, and use of explosives is administered through the U.S. Bureau of Alcohol, Tobacco, and Firearms;[8] the Hazardous Materials Regulation Board of the U.S. Department of Transportation;[9] the Occupational Safety and Health Administration,[10] and the United States Environmental Protection Agency.[11] Many states and local governments require a blasting license or permit, and some cities have special requirements for the handling and detonating of explosives. The Institute of Makers of Explosives (IME) provides educational publications to promote the safe handling, storage, and use of explosives.[12] Manufacturers performing explosion welding are obliged to develop their own safe practices and control methods that ensure the safety of employees and the public.

CONCLUSION

Explosion welding is one of the more unusual welding processes when compared to conventional fusion and solid state processes. Once initiated, explosion welding allows little room for error from the perspective of safety or process control. It happens fast and violently. Despite these realities, the process has the ability to weld many metallically incompatible metals that otherwise could not welded. It is most economically used to make relatively large flat plates. When used in cladding applications, explosion welding provides some features that are technically superior. Additionally, manufacturing time is short.

Although construction codes allow for certain repairs, explosion welding is an unforgiving process. It is well mastered and highly reproducible, but there are major impediments to the practice of the technology. One impediment is the requirement for a large area with controlled access, and another is the governmental permits and licenses required to handle and use explosives. Availability and storing of explosives also is a major consideration.

Since its inception, explosion welding has been further developed and used for many applications. Not all uses and applications have become or remained commercially viable. For example, in the early 1980s, the procedure for joining lengths of large-diameter gas and oil transmission pipelines by explosion welding was demonstrated as a field procedure. In 1984, the first application of this procedure was employed to join a section of pipeline 1067 mm (42 in.) in diameter and 6 kilometers (3.7 miles) long. The weld could be made quickly and economically, and although technically viable, the process has not gained significant commercial application.

Explosion welding is used by a relatively small number of businesses worldwide and is served by several competent companies. Development work continues on several fronts. Since it is a small, flexible industry, producers work with end users and fabricators to tailor the product for the intended use.

8. Bureau of Alcohol, Tobacco, Firearms and Explosives, Office of Public and Governmental Affairs, Title 27, in Code of Federal Regulations (CFR), Part 55, *Commerce in Explosives*. Washington, D.C. United States Government Printing Office.

9. United States Department of Transportation, 400 7th Street, S.W., Washington, D.C. 20590.

10. Occupational Safety and Health Administration (OSHA), Code of Federal Regulations (FR) Title 27, Alcohol, Tobacco Products and Fire Arms (ATF) Part 55—*Commerce in Explo*sives, Washington, D.C.: Superintendent of Documents, Government Printing Office.

11. United States Environmental Protection Agency, Ariel Rios Building, 1200 Pennsylvania Avenue, N.W., Washington, D.C. 20460.

12. Institute of Makers of Explosives, 1120 19th Street N.W., Suite 310, Washington, D.C. 20035-3605.

BIBLIOGRAPHY

American Society for Testing and Metals (ASTM International). *Standard specification for straight beam ultrasonic inspection of plain and clad steel plates for special applications*, ASTM A578. West Conshohocken, Pennsylvania: ASTM International.

American Society of Mechanical Engineers. 1984. High energy rate fabrication. Proceedings: 8th International Conference. San Antonio, Texas. 17–21 June 1984. J. Bermon and J. W. Schroeder, ed. New York: American Society of Mechanical Engineers.

American Welding Society (AWS) Committee on Definitions and Symbols, Standard Welding Terms and

Definitions, A3.0: 2001, Miami: American Welding Society.

Chiba, A., M. Nishida, and Y. Morizono. 2004. Microstructure of bonding interface in explosively-welded clads and bonding mechanism. Materials Science Forum. Switzerland: Trans Tech Publications. 465–474.

International Society of Explosives Engineers. 1998. Blasters' Handbook. Cleveland, Ohio: International Society of Explosives Engineers.

Kreye, H., and M. Hammerschmidt. 1982. Transmission electron microscope investigation of the microstructure affected by the bonding process during oblique collision of metallic surfaces. Proceedings of 5th international symposium on explosive working of metals, Oct. 12–14 1982. Gottwaldov, Czech Republic: Czechoslovak Scientific and Technical Society.

Occupational Safety and Health Administration (OSHA), Code of Federal Regulations (CFR) Title 27, Alcohol, Tobacco Products and Fire Arms (ATF) Part 55—*Commerce in explosives*, Washington, D.C.: Superintendent of Documents, United States Government Printing Office.

United States Department of Defense Index, latest edition, Washington, D.C.: United States Department of Defense.

Young, G. A., and J. G. Banker. 2004. Explosion welded, bi-metallic solutions to dissimilar metal joining. Proceedings of the 13th offshore symposium, Feb. 24, 2004. Houston: Society of Naval Architects and Marine Engineers, Texas Section.

SUPPLEMENTARY READING LIST

Banker, J. G. 2003. Recent developments in reactive and refractory metal explosion clad technology, NACE Paper 03459. Houston, Texas: NACE International.

Banker J. G. 2003. United States Patent #6,772,934. Kinetic energy welding process.

Banker, J. G., and A. Nobili. 2002. Aluminum-steel electric transition joints: effects of temperature and time upon mechanical properties. W. Schneider, ed. Light metals. Warrendale, Pennsylvania: The Minerals, Metals, and Materials Society. 439–445.

Banker J. G., and J. P. Winsky. 1999. Titanium/steel explosion-bonded clad for autoclaves and vessels. 1999. Proceedings of ALTA 1999 autoclave design and operation symposium. Melbourne, Australia ALTA Metallurgical Services.

Banker J. G. 1996. Try explosion clad steel for corrosion protection. Chemical engineering progress. American Institute of Chemical Engineers (AIChE) No. 7: 40–44.

Banker J. G. 1996. Commercial applications of zirconium explosion clad. Journal of testing and evaluation, West Conshohocken, Pennsylvania: American Society for Testing and Materials. 24(2) 91–95.

Banker J. G., and M. S. Cayard. 1994. Evaluation of stainless steel explosion clad for high temperature, high pressure hydrogen service. Proceedings of hydrogen in metals conference, Vienna. Austria. Houston, Texas: NACE International.

Bilmes, P. A., C. Gonzalez, and J. C. Cuyas. 1988. Barrier interlayers in explosive cladding of aluminum to steel. Metal construction. 20(3): 113–114.

Blazynski, T. Z. 1983. Explosive welding, forming and compaction. Essex, U. K.: Applied Science Publishers Ltd.

Chadwick, M. D., and P. W. Jackson. 1983. Explosion welding in planar geometries. Explosive welding, forming and compaction. T. Z. Blazynski, ed. Essex, U.K.: Applied Science Publishers Ltd. 219–287.

Cleland, D. B. 1983. Basic consideration for commercial explosive cladding processes T. Z. Blazynski, ed. Explosive welding forming and compaction. England: Applied Science Publishers Ltd. 159–188.

Cowan, G. R., J. J. Douglass, and A. H. Holtzman. 1964. United States Patent 3137937, Explosive bonding.

Crossland, B. 1982. Explosive welding of metals and its applications. Oxford University Press. 7–8.

Crossland, B. 1976. Review of the present state-of-the-art in explosive welding. Metals technology.

Fujita, M. 1982. An investigation of the combined underwater (explosive) bonding and forming process. High energy rate fabrication. Vol. 70. Proceedings: ASME Winter Meeting, Phoenix, Arizona, 14–19 November. New York: American Society of Mechanical Engineers. 29–37.

Hammerschmidt, M., and H. Kreye. 1980. Microstructure and bonding mechanism in explosive welding. Proceedings of international conference on metallurgical effects of high strain rate deformation and fabrication. Albuquerque, New Mexico. 961–973.

Holtzman, A. H. and G. R. Cowan. 1961. The strengthening of austenitic manganese steel by plane shock waves—response of metals to high velocity deformation. New York: Interscience Publishers. 447.

Holtzman, A. H. and G. R. Cowan. 1965. Bonding of metals with explosives. Bulletin 104 (4). New York: Welding Research Council.

Jamieson, R. M., A. Loyer, and W. D. Hauser. 1981. High-impact girth welds in large-diameter pipes. Steels for line pipe and pipeline fittings. Cambridge, England: The Metals Society and The Welding Institute. 342–453.

Johnson, T. E. and A. Pocalyko. 1982. Explosive welding for the 80s. High energy rate fabrication. Vol. 70. Proceedings: ASME Winter Meeting, Phoenix, Arizona, 14–19, November 1982, 63–82. New York: American Society of Mechanical Engineers.

Justice, J. T. 1986. Explosion welding proven for large-diameter gas lines. Oil and gas journal. 84(34): 44–50.

Justice, J. T., and J. J. OBeirne. 1986. Explosion welding of a large-diameter gas transmission pipeline. E. J. Seiders, ed. Proceedings: pipeline engineering symposium, 23–27 February, New Orleans. New York: American Society of Mechanical Engineers.

Kreye, H. 1977. Melting phenomena in solid state welding processes. *Welding Journal* 56(5): 156-s–158-s.

Linse, V. D. 1993. Procedure development and process considerations for explosion welding, ASM Handbook, Vol. 6, Welding, brazing and Soldering. Materials Park, Ohio: ASM International. 896–900.

Linse, V. D. 1985. The application of explosive welding to turbine components, 74-GT-85. New York: American Society of Mechanical Engineers.

Linse, V. D., and N. S. Lalwaney. 1984. Explosive welding. Journal of metals. 36(5).

Longstaff, G., and E. A. Fox. 14–19 November 1982. Fabrication and plugging of tube-to-tube-sheet joints using impact explosive welding technique. Proceedings: ASME Winter Meeting, Phoenix, Arizona. High energy rate fabrication, Vol. 70. New York: American Society of Mechanical Engineers: 39–53.

McKinney, C. R., and J. G. Banker. 1971. Explosion bonded metals for marine structural applications. Marine technology, Society of Naval Architects and Marine Engineers. 8(3), 285–292.

Nobili, A., T. Masri, and M. C. Latent. 1999. Recent developments in characterization of a titanium-steel explosion bond interface. Proceedings of conference on reactive metals in corrosive applications. J. Haygosth and J. Tosdale, ed. Wah Chang, Albany OR: 89–98.

Onzawa, T. IIyama, S. Koabayashi, A. Takasaki, and Y. Ujimoto. 1999. Microstructure of explosively bonded interface between titanium and very low carbon steel as observed by TEM, Proceedings of conference on reactive metals in corrosive applications. Wah Chang, Albany OR: 89–98.

Patterson, R. A. 1982. Explosion bonding: aluminum-magnesium alloys bonded to austenitic stainless steel. High energy rate fabrication, Vol. 70. Proceedings: ASME winter meeting, Phoenix, Arizona, 14–19 Nov. 1982. New York: American Society of Mechanical Engineers. 15–27.

Patterson R. A. 1993. Fundamentals of explosion welding. ASM Handbook, Vol. 6. Welding, brazing, and soldering. Materials Park, Ohio: ASM International. 1604.

Pocalyko, A. 1981. Explosively clad metals. Encyclopedia of chemical technology. Vol. 15, 3rd ed. New York: John Wiley & Sons. 275–296.

Richardson, Kirk. 2004. Explosion welding: from principle to practice. Outlook. First Quarter, 2004: 3–4.

Smith, L. M. and M. Celant. 1998. Practical handbook of cladding technology, Edmonton, Alberta: CASTI Publishing.

Tatsukawa, I. 1986. Interfacial phenomena in explosive welding of Al-Mg alloy/steel and Al-Mg alloy/titanium/steel. Tokyo: Japan Welding Society. 17(2): 110–116.

Yamashita, T., T. Onzawa, and Y. Ishii. 1975. Microstructure of explosively bonded metals as observed by transmission electron microscopy. Transactions of Japan Welding Society, Tokyo: Japan Welding Society. 9(2): 51–56.

CHAPTER 10

ADHESIVE BONDING OF METALS

Photograph courtesy of Dymax Corporation

Prepared by the Welding Handbook Chapter Committee on Adhesive Bonding:

G. W. Ritter, Chair
Edison Welding Institute

K. Y. Blohowiak
The Boeing Company

A. C. Pocius
3M Company

G. T. Schueneman
Henkel Loctite Corporation

Welding Handbook Volume 3 Committee Member:

D. W. Dickinson
The Ohio State University

Contents

CHAPTER 10

ADHESIVE BONDING OF METALS

INTRODUCTION

Adhesive bonding is a materials joining process in which a nonmetallic bonding material (an adhesive) is placed between the faying surfaces of the materials to be joined, (the adherends). The interposed adhesive hardens or solidifies by physical or chemical property changes to produce a bonded joint with useful strength.[1,2] During some stage of processing, the adhesive must become sufficiently fluid to wet the faying surfaces of the adherends.

Although adhesive bonding is used to join many nonmetallic materials, only the bonding of metals to metals or to nonmetallic structural materials is covered in this chapter.

The types of adhesives and the methods of applying them are discussed. A section on the advantages, limitations, and performance of adhesive-bonded joints will help the design engineer, fabricator or manufacturer determine whether the benefits of the process can be applied to a prospective joining project.

This chapter includes details of surface preparation for adhesive bonding, assembling and curing procedures, equipment, types of adhesives, joint design, process control, quality assurance, testing and inspection, economics, and safe practices.

Continuing research and the ongoing development of improved adhesives and materials have contributed to the dramatic advances in the properties of adhesives, the types of materials bonded, and the bonding techniques and applications discussed in this chapter.

1. Adapted from American Welding Society (AWS) Committee on Definitions and Symbols, 2001, *Standard Welding Terms and Definitions*, A3.0:2001, Miami: American Welding Society.
2. At the time of the preparation of this chapter, the referenced codes and other standards were valid. If a code or other standard is cited without a date of publication, it is understood that the latest edition of the document referred to applies. If a code or other standard is cited with the date of publication, the citation refers to that edition only, and it is understood that any future revisions or amendments to the code or standard are not included; however, as codes and standards undergo frequent revision, the reader is encouraged to consult the most recent edition.

FUNDAMENTALS

The general term *adhesive* includes such materials as cement, glue, mucilage, and paste. Although natural organic and inorganic adhesives are available, synthetic organic polymers generally are used to join metal assemblies. Various definitions specific to adhesive bonding are applied to the term *adhesive* to indicate certain characteristics, as follows:[3]

1. Physical form: solid, liquid or paste adhesive, film adhesive, and tape adhesive;
2. Chemical type: silicone, epoxy, phenolic, polyimide, acrylic, cyanoacrylate, urethane, and hybrid adhesive;
3. Application method: pump, in-line mixer, sprayed, film, and syringe;
4. Curing method: ambient, anaerobic (air expelling), primer-accelerated, heat, moisture ingress, ultraviolet, microwave, and induction; and
5. Materials bonded to metal: paper, metal, plastics, glass, ceramics, and wood.

The adhesive bonding of metals is similar to soldering and brazing in some respects, but a metallic bond does not take place. The surfaces being joined are not melted, although they may be heated. An adhesive in the form of a liquid, paste, or tacky solid is placed

3. Definitions and terms relating to adhesives can be found in ASTM International Committee D14.14, latest edition, *Standard Terminology of Adhesives*, D 907, West Conshohocken, Pennsylvania: ASTM International.

between the faying surfaces of the joint. After the faying surfaces are placed in intimate contact or mated with the adhesive in between, heat or pressure, or both, are applied to accomplish the bond.

An adhesive system must have the following characteristics:

1. At the time the bond is formed, the adhesive must become fluid so that it wets and comes into close contact with the surface of the adherends;
2. In general, the adhesive cures, cools, dries, or otherwise hardens during the time the bond is formed or soon thereafter;
3. The adhesive must have good mutual attraction with the metal surfaces, and have adequate strength and toughness to resist failure along the adhesive-to-metal interface under service conditions;
4. As the adhesive cures, cools, or dries, it must not be allowed to shrink excessively, which might cause undesirable internal stresses to develop in the joint;
5. To develop a strong bond, metal surfaces must be clean and free of dust, loose oxides, oil, grease, or other foreign materials;
6. Air, moisture, solvents, and other gases which may tend to be trapped at the interface between the adhesive and the metal must have a way of escaping from the joint; and
7. The joint design and cured adhesive must be capable of withstanding the intended service conditions.

PRINCIPLES OF BONDING

For wetting to occur in adhesive bonding, the free energy of the adherend surface must be greater than that of the adhesive. This usually is the case for metallic adherends and polymeric adhesives; however, contaminants adsorbed on the metal can lower the free energy of the surface and prevent the formation of a good adhesive bond. Contaminants can be removed from the surface by washing with solvent or by abrasion. The latter treatment, using grit blasting, abrasive paper, or abrasive pads, frequently is used to prepare metal surfaces before bonding. If the adhesive bond is to be exposed to heat and humidity in service, it is imperative that the best possible surface preparation is implemented. Good surface preparation requires a clean surface with sufficient profile to anchor the adhesive or any primer, if used.

In addition to the surface energetics, the viscosity of an adhesive must be low enough during the bond-forming process to readily spread a bondline over the surface of the adherend. The required viscosity depends on the surface roughness and the bondline. A smooth surface with a narrow bondline of <0.203 millimeters (mm) (<0.008 inches [in.]) may require an adhesive with <1000 centipoise (cps) viscosity. A grit-blasted surface with a bondline of >0.381 mm (>0.015 in.) may require several thousand centipoise viscosity or even a paste to ensure that the bond will not gap, but will still wet. The higher the viscosity, the greater the probability that the adhesive will not completely wet the surface and will entrap gases, liquids, or vapors in the bondline. This tendency can be reduced by applying pressure during the curing process. When bonding metal to plastic, two special considerations must be addressed. First, the relatively low surface energy (wetting) of the plastic usually requires an oxidative treatment, such as flame treatment, corona treatment, or plasma treatment. As an option, a suitable primer may be employed. Second, because of the poor thermal conductivity of plastics, a plastic surface heats much more slowly during a heated cure, resulting in poor bond strength at the plastic surface. Recently developed adhesives have been designed to bond plastics directly without surface treatments or primers.

ADVANTAGES AND APPLICATIONS

Adhesive bonding provides several advantages for joining metals when compared to resistance spot welding, brazing, soldering, or mechanical fasteners, such as rivets or screws. Adhesive bonding is selected for many applications that require one or more of the following capabilities:

1. Bonding dissimilar metals;
2. Bonding thin-gage materials;
3. Creating bonds at low temperatures;
4. Combining bonding and sealing in one operation;
5. Providing thermal and electrical insulation;
6. Distributing stress uniformly;
7. Producing a smooth surface appearance;
8. Providing fatigue, vibration, and sound damping;
9. Saving weight (in many applications); and
10. Simplifying the design of the adherend.

Bonding Dissimilar Metals

It is possible to bond dissimilar metals with minimal galvanic corrosion in service, provided the adhesive layer maintains electrical isolation between the metals. Many types of adhesive formulations are flexible enough to permit the bonding of dissimilar metals with widely different coefficients of thermal expansion. Such possibilities depend on the size of the pieces and the degree of joint strength required. A single adhesive may be used for joining a number of dissimilar metal combinations in a single assembly. Adhesive bonding also makes it possible to join metals to nonmetallic materials, such as various types of plastics.

Figure 10.1 shows a typical application in the aerospace industry, in which bonded composites are used with welded metal structures to produce an integrated

Photograph courtesy of Rolls-Royce plc

Figure 10.1—An Airplane Turbine Engine with Composites Bonded to Welded Metal Structures

assembly. Figure 10.2 shows the interior of an automobile, in which nonmetallic materials and welded metals are joined by adhesive bonding for integration.

Bonding Thin-Gauge Metals

Very thin metal components can be joined by adhesive bonding. Following are some examples:

1. Multiple layers of thin metal sheets can be bonded together to form electric motor laminates;
2. Various metal foils can be joined to one another or to other materials;
3. Thin-gauge metal sheets may be used as "sandwich" panel skins (two sheets with adhesive between them); and
4. Adhesive bonding may be required in cases in which normally weldable metals are too thin to weld, or when joining large areas of materials.

Low Processing Temperatures

The temperatures used for the heat curing of most adhesives are between 65° Celsius (C) and 176°C (150° Fahrenheit [F] and 350°F), which are temperatures often below the normal soldering range of 185°C (365°F). Room-temperature curing formulations provide sturdy structural bonds for service temperatures up to 82°C (180°F). High-performance epoxy adhesives cure at room temperature and maintain good strength up to 150°C (300°F). They can be used to join heat-sensitive components without damage, but it should be noted that a full cure may require 7 days to 30 days. These adhesives should be cured at room temperature for the time specified by the manufacturer, followed by a short post-bonding cure at the highest temperature to which the bond may be expected to be exposed when in service. Adhesive bonding should be considered when high-temperature joining operations would

Photograph courtesy of American Honda Motors

Figure 10.2—Interior of an Automobile Combining Metallic and Nonmetallic Materials Joined by Adhesive Bonding

cause microstructural changes or structural damage to the workpieces.

Unlike metal solders, many adhesives provide suitable performance at temperatures above their curing temperature. To perform at elevated temperature, the adhesive must experience that temperature sometime during the curing cycle, either by undergoing an elevated-temperature cure or by using the heat generated by the adhesive as it cures. This completes the necessary reaction chemistry to increase the glassy transition temperature (T_g) of the adhesive. (T_g is defined as the temperature at which a material changes from glassy or visco-elastic to plastic). Once all the reactive chemical species in the adhesive have reacted, the increased T_g provides additional temperature resistance.

Combined Bonding and Sealing

The adhesive that joins components also may serve as a sealant or coating to provide protection from oils, chemicals, moisture, or a combination of these. In Figure 10.3, an adhesive that cures at room temperature is applied to seal the ends of an antenna circuit. This adhesive "potting" compound is formulated to allow the transmission of radio frequency signals while the adhesive is in the uncured state, enabling quality control testing of the circuit to be performed immediately after application of the adhesive.

Figure 10.4 shows a robotic application of a hem-sealing adhesive. After the door panels are joined, the components may be held in correct alignment with welds.

Thermal and Electrical Insulation

Adhesives can provide thermal or electrical insulating layers between the two surfaces being joined. For example, almost all mass-produced printed circuits use adhesive bonding. In this application, the adhesive used to bond the copper conductor to the base material has electrical characteristics similar to those of the base material. An adhesive also may serve as an insulator between adjacent conductors. Most adhesives have the insulating capability of 300 volts (V) to 800 V per mil.

The addition of certain metallic or carbon fillers to adhesive formulations can make them electrically conductive. Before the decision is made to use a particular adhesive formulation, testing should be performed under simulated service conditions because corrosion

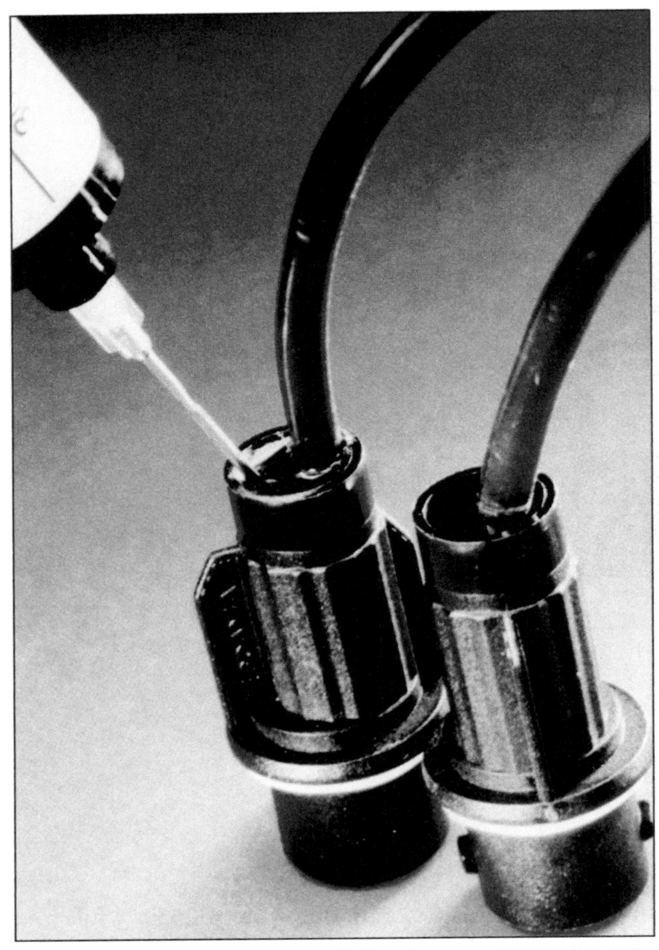

Photograph courtesy of Dymax Corporation

Figure 10.3—Application of an Adhesive Potting Compound to Seal an Antenna Circuit

Photograph courtesy of Sealant Equipment & Engineering

Figure 10.4—Robotic Application of a Hem-Sealing Adhesive on an Automobile Door Panel

may occur in some metal structures bonded with electrically conductive adhesives that are exposed to moisture. Modern formulations employ substantially chlorine-free resins that reduce the potential for corrosion. Metal powder additives also improve the thermal conductivity of adhesives. A typical conductive adhesive has a resistivity of 0.0005 to 0.002 ohm per centimeter (cm). Thermal conductivity approaches 2 watts per meter (W/m) per kelvin (°K). An unfilled adhesive has a resistivity of 10^{12} to 10^{16} ohms per centimeter and thermal conductivity of only 0.2 W/m per °K.

Uniform Stress Distribution

One of the great advantages of adhesive bonding is the ability to spread stress over large areas by using a weld-bonding technique. In weld bonding, adhesive bonding and resistance welding processes are combined in one operation. Joints can be designed to distribute the load over a relatively large bonded area to minimize stress concentrations. For example, in the construction of laminated wall panels, metal skin sheets are bonded to metal or paper honeycombs, or to many of the foamed polymers, such as polystyrene, polyurethane, polyacrylics, polyimides, and polyvinylchloride.

Smooth Surface Appearance

Adhesive bonding can ensure smooth, unbroken surfaces without protrusions, gaps, or holes. A typical example is the vinyl-to-metal laminate widely used in the production of television cabinets, housings for electronic equipment, and automotive trim. Figure 10.5 shows a trailer constructed with lightweight adhesive-bonded panels, where broad, smooth areas are required. Hood and roof stiffeners on automobiles are joined to the panels by adhesive bonding rather than by resistance spot welding to avoid marks on the panels that would be susceptible to rusting and might require filling, grinding, and polishing prior to painting.

Photograph courtesy of Featherlite Trailers

Figure 10.5—Trailer Constructed with Lightweight Adhesive-Bonded Panels

The bonding of thin metal to thicker metal can result in the adhesive "pulling" the thin metal, possibly causing *read-through* or *telescoping*, a condition in which the adhesive bead line becomes visible at shallow viewing angles because the adhesive has deformed the metal. Susceptibility to read-through is especially troublesome when the structure is heated for curing. Read-through can be remedied by using a more rubbery adhesive that has the ability to expand and contract during heating and cooling.

Vibration and Sound Damping

The capability of adhesives to absorb shock and vibration gives adhesive-bonded joints good fatigue life and sound-damping properties. The use of adhesives rather than rivets has increased joint fatigue life by a factor of ten or more in some applications. Increased fatigue performance is another implicit advantage to combining weld bonding with adhesive bonding. Figure 10.6 shows an instant-curing resin formulated to prevent the migration of sound vibrations between components.

A combination of adhesives and rivets for joints in very large aircraft structures has increased the fatigue life from 2×10^5 cycles for riveted joints to more than 1.5×10^6 cycles for bonded and riveted joints. The large bonded area also dampens vibration and sound. With rivet-bonding or bolt-bonding, the fastener must pass through a hole. The full advantage of the additional mechanical augmentation will not come into action until the adhesive fails or creeps. In weld bonding, the metal adherends are intimately joined by the weld. This allows the full advantage of the mechanical augmentation to function effectively.

Photograph courtesy of Dymax Corporation

Figure 10.6—Instant-Curing Resin Applied to a Component to Dampen Sound Vibrations

Weight Savings

The adhesive bonding process may permit significant weight savings in a finished product by using lightweight fabricated components. Honeycomb panel assemblies, used extensively in the aircraft industry and the construction field, are excellent examples of lightweight fabrications. Figure 10.7 shows a sheet-aluminum panel assembly with an aluminum honeycomb core. Not only is the honeycomb core joined to the metal face sheets by adhesive bonding in this application, but also the honeycomb configuration is fabricated by adhesive bonding. Weight reduction often is important in the functioning of the product and also may provide considerable cost savings in packing, shipping, and installation labor.

The uniform stress distribution typical of adhesive bonding eliminates many of the stress concentrations associated with spot welding or riveting. This allows more load to be carried by more area, potentially allowing thinner stock to be used.

Simplification of Design

Adhesive bonding often permits design simplification. For example, cast-metal components may be prone to porosity that may lead to rejection. In some instances, castings can be replaced by an assembly joined with adhesive bonding. In one application, aluminum die-cast pump sections were joined to a steel core by adhesive bonding. Previously in this application, the component was cast as one piece of steel, but porosity in the casting resulted in an excessive number of rejects. Redesigning the casting as an adhesive-bonded assembly reduced the number of rejects to nearly zero.

Design simplification may take the form of providing varied choices of materials for a given application. Rapid-cure adhesives are used to bond specific combinations of materials that cannot be welded. These may involve ceramics, magnets, and metals, or may combine plastics with other materials. Figure 10.8 shows adhesive-bonded magnet and metal components of a dc motor.

LIMITATIONS

Adhesive bonding has certain limitations that should be considered when selecting a joining process. The most important of these are limited peel strength, operational temperature limit, delayed completion because

Photograph courtesy of Canadian Commercial Vehicles

Figure 10.7—Aluminum-Sheet Panel Assembly with an Aluminum Honeycomb Core

Photograph courtesy of Dymax Corporation

Figure 10.8—Adhesive-Bonded Magnet and Metal Components of a DC Motor

of curing time, limited testing procedures, and restricted service conditions.

Low Peel Strength

Adhesives will not support high-peel loads at temperatures above 120°C (250°F), not even when the components are joined with adhesives that have high tensile and shear strengths at temperatures as high as 150°C (300°F). For applications in which high peel strength is essential, some mechanical reinforcement may be necessary. "Peel stoppers" in the form of strategically placed welds in combination with adhesive bonding are another effective use of weld bonding.

Operational Temperature Ceiling

Adhesives, including epoxy-phenolic, which are designed for minimal creep at elevated temperatures, have an operational temperature ceiling of about 260°C (500°F). Some new high-temperature adhesives derived from heat-stable polyamides, polybenzimidazoles, and related compounds can be used at temperatures up to 371°C (700°F), but they are costly and difficult to process.

In general, cyanoacrylates can be used up to about 70°C (180°F), urethanes up to about 120°C (250°F), acrylics up to about 180°C (350°F), and epoxies in the range of 230°C to 260°C (450°F to 500°F). Polyamides, bismaleimides, and polyetherether ketones (PEEK) can operate as high as 340°C (650°F), but their use is highly specific and requires advanced processing techniques. At these elevated temperatures, brazing or soldering might be the appropriate joining method in the absence of welding.

Curing Time

To develop full strength, joints must be fixtured and cured at temperature for some time. Conversely, mechanical fasteners provide design strength immediately and usually do not require extensive fixturing. Curing time must be factored into throughput requirements and fixture travel with the bonded component. Weld bonding addresses these requirements by providing the advantage of *in-situ* fixturing. Some adhesives are applied hot and have a green strength that is suitable for non-fixtured transport to the curing oven or other heating process that will fully cure the adhesive.

Testing Procedures

Nondestructive inspection methods normally used for other joining methods generally are not applicable to the evaluation of adhesive bonds. Both destructive and nondestructive testing must be used with process controls to establish the quality and reliability of bonded joints. Figure 10.9 shows an impact test for a bonded composite-steel assembly. The adhesive bond has survived an impact that deforms the steel.

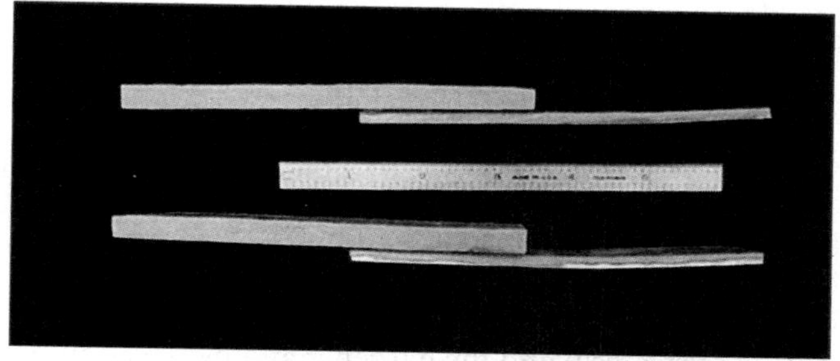

Photograph courtesy of Edison Welding Institute

Figure 10.9—Impact Test for a Bonded Composite-Steel Assembly

Limited Service Conditions

Service conditions may be restrictive. Many adhesive systems degrade rapidly when the joint is both highly stressed and exposed to a hot, humid environment. The effects of heat and humidity often can be ameliorated by the proper choice and application of surface preparation of the adherends. Specific service conditions must be considered with regard to the adhesive system. The adhesive system is composed of the substrate, interface, and adhesive. All three components must be stable in the service environment and loading to obtain an adhesive bond that will be successful for the long term.

MATERIALS BONDED

The metals most commonly joined by adhesive bonding include all steels, aluminum, titanium, magnesium, and to some extent, copper alloys. More difficulty is encountered when bonding nickel-base alloys, platinum, and gold.

ADHESIVES

Adhesives may be either thermoplastic or thermosetting. The principal ingredients of most adhesive formulations are a synthetic resin system, an elastomer or flexibilizer, inorganic materials, and colorants. Other possible ingredients include viscosity modifiers, adhesion promoters, and stabilizers.

THERMOSETTING ADHESIVES

Thermosetting resins, frequently referred to as structural adhesives, are the most important structural bonding materials used for metal adhesive formulations. Thermosetting adhesives harden, or cure, by chemical reactions that occur with the addition of a hardener or catalyst. The properties of these adhesives can be modified for specific applications by changing the chemical nature of the resin and the hardener and by the addition of modifying agents and fillers. Heat, pressure, radiation, or other energy can accelerate the curing rate. Once they cure, these adhesives cannot be remelted. A broken joint cannot be rebonded by heating. Depending on the composition, thermosetting adhesives may soften or weaken at high temperature and ultimately may decompose.

THERMOPLASTIC ADHESIVES

Thermoplastic resins are long-chain molecular compounds that soften on heating and harden on cooling. They undergo no chemical change on heating, so the cycle can be repeated. However, they can oxidize and decompose at excessively high temperatures, at about 180°C (350°F). Many thermoplastic resins also can be softened at room temperature with organic solvents. They harden again as the solvent evaporates.

Limited resistance to heat, solvents, and load-induced stresses generally make thermoplastic resins unsuitable as structural adhesives. However, some thermoplastic resins or elastomers are combined with thermosetting resins, such as epoxies and phenolics for improved flexibility, peel strength, and impact resistance. Elastomers or flexibilizers are added to adhesive formulations to add resiliency, improve peel strength, and increase resistance to shock and vibration.

Inorganic materials are added as fillers to improve the mechanical and physical properties of the adhesives. Fillers can add greatly to the stability of bonded joints by reducing shrinkage and thermal expansion and by increasing the modulus of elasticity of the adhesive. Unfortunately, the addition of fillers may embrittle an adhesive, so the correct balance between toughness and stiffness should be a property of the adhesive formulation.

TYPES OF ADHESIVES

The types of adhesives used to bond metals include solvent, hot-melt, pressure sensitive, and chemically reactive, as listed in Table 10.1.

Table 10.1
Types of Polymeric Adhesives Used to Bond Metals

Type	Adhesive
Solvent	Neoprene
	Nitrile
	Urethane (thermoplastic)
	Block copolymer
	Styrene-butadiene
Hot Melt (Thermoplastic)	Ethylene vinyl acetate
	Block copolymer
	Polyester
	Polyamide
Pressure Sensitive	Block copolymer
	Acrylic
	Silicone
	Urethane
Chemically Reactive (Thermoset)	Epoxy
	Phenolic
	Structural acrylic
	Anaerobic
	Cyanoacrylate
	Urethane

Solvent-Type Contact Adhesives

The solvent-type contact adhesives are predominantly elastomeric thermoplastics produced as solutions or suspensions. They build bond strength on removal of the solvent. The liquid adhesive is applied to the adherend surfaces and, for metal adherends, time is allowed for the solvent to evaporate. The adhesive-coated surfaces are then joined under contact pressure. Sometimes heat is applied to fuse the coated surfaces after drying. The use of these adhesives has been challenged because of concerns about the release of solvent constituents into the environment. However, newly developed solvent mixtures are less detrimental to the environment and nontoxic water-borne products are available.

Hot-Melt Adhesives

Hot-melt adhesives are thermoplastics. After the adherends have been coated with adhesive and mated, heat and pressure are applied to the assembly. The joint is then cooled to solidify the adhesive and achieve a bond. These adhesives normally are not used for structural applications, but they are popular for their low toxicity and relative ease in handling. Some hot-melt adhesives can withstand service temperatures close to 150°C (300°F). One problem associated with applying hot-melts to metals is the high thermal conductivity of the metal, which causes premature solidification of the adhesive-adherend interface and results in incomplete wetting. Preheating the metal adherend can eliminate or ameliorate this problem.

Pressure-Sensitive Adhesives

Pressure-sensitive adhesives (PSAs) are the most widely used adhesives in the world. Their formulations instantly provide a relatively low-strength bond with the brief application of pressure. Bond strength typically increases with time, which may range from a few minutes to a number of hours. They may be applied to any clean, dry surface. Conversely, dirty, oily, or moist surfaces do not accept most PSAs. They are capable of sustaining only very light loads because of retention of their flow characteristics, and most are not considered structural adhesives. However, a number of pressure-sensitive transfer tapes and double-sided tapes that exhibit high shear strengths are available. These types are used in the construction of trucks and trailers. Modern tape adhesives can be considered for certain low-load structural applications and offer several advantages, including very little waste, constant bondline thickness, and controlled application with no clean-up.

Chemically Reactive Adhesives

The chemically reactive adhesives consist primarily of thermosetting resins in liquid and solid forms, including films and tapes. They are activated either by the addition of a catalyst or hardener, or by the application of heat. Bond strength is achieved from the chemical reaction that takes place during the cure. Catalysts or hardeners may be incorporated by the adhesive manufacturer or may be added by the user immediately prior to application. Chemically reactive formulations generally must be used within a prescribed period of time after mixing to avoid premature setting. This period of time, called *open time*, may range from a few minutes to several days. One-part adhesives requiring elevated-temperature cure may have open times of months.

STRUCTURAL ADHESIVES

The ultimate objective of a structural adhesive is either to create a bond that is as strong as the material it joins or that forces the failure zone away from the joint. Since this goal is not always attainable, a structural adhesive can be defined as one that is used to transfer required loads between the adherends in a structure for the life expectancy of the structure when exposed to its service environment. Alternatively, a structural adhesive can be defined as an adhesive that will not creep significantly under the service conditions of the joint.

Although structural bonding has been successfully used for many aerospace applications since the 1950s, concerns about the long-term durability of structural adhesive bonds have limited the widespread use of this joining method. The combination of stress, even as low as 20% of the initial adhesive strength, and exposure to hot humid environments can cause significant degradation of bond performance, sometimes leading to failure. Many of the factors affecting the durability of adhesive joints are known and include the type of adhesive, the nature of the adherends, the surface preparation before bonding, and the service conditions.

The performance of a structural adhesive can be affected by the presence of moisture. The mechanical properties can change as water is absorbed due to the plasticizing action of water. Swelling stresses can lead to the formation of crazes and microcracks. In unfavorable situations, hydrolysis of the adhesive can occur. Water also can displace the adhesive from the metal surface and thus induce interfacial disbonding (failure at the bondline). Finally, water may hydrate and weaken the metal oxide surface layer of the adherend. In general, the relative importance of each of these factors is unknown, as are the details of the various mechanisms. For these reasons it is usually not possible to predict the durability of a given adhesive joint from first principles. However, there are several models available that can be used in combination with a test protocol that will provide some prediction of the durability of a joint under heat and humid conditions. Knowledge gained from experiments with the adhesive system exposed to the service environment and loading allows the discernment of specific surface treatments and adhesives that can be yield predictable long-term performance.

Phenolic Resins

Phenolic resins, being naturally brittle, are often modified with thermoplastics or elastomers for structural adhesive applications. These modified phenolics are available as solutions in organic solvents and also as films, both supported and unsupported. Such adhesives feature high peel strengths, and tensile and shear strengths in the range of 20 megapascals (MPa) to 34 MPa (3000 pounds per square inch [psi] to 5000 psi) on steel. Phenolic resins may emit water on cure. (See Appendix C for pressure unit conversions.)

Epoxy Resins

Epoxy resins are the most popular adhesive bases for structural applications. Epoxy resins combine the properties of excellent wetting action, low shrinkage, high tensile strength, toughness, and chemical inertness to produce adhesives noted for strength and versatility. The principal drawbacks of epoxy resins are slow curing and susceptibility to moisture. The so-called "five-minute epoxies" typically display low fracture toughness but are useable for quick repairs. Unlike phenolic adhesives, epoxies do not give off water while curing. They can be applied in liquid form without a solvent carrier. Because of this, entrapment of volatile products is minimized. Only low pressure is necessary to maintain intimate contact between the adherends during bonding, resulting in greatly simplified equipment requirements.

Epoxy adhesives are available as free-flowing liquids, films, powders, sticks, pellets, and mastics. This variety permits considerable latitude in the selection of application techniques and equipment. Fillers or plasticizers may be added to minimize stresses that can develop when the adhesive and adherends have different coefficients of thermal expansion.

The wide choice of hardeners available for epoxy formulations provides curing cycles ranging from a few seconds at elevated temperatures to several minutes or hours at room temperature. However, the heat-resistant formulations require high-temperature cures.

Unmodified epoxy-based adhesives show high shear and tensile strengths, but tend to be brittle and thus perform poorly in cleavage and peel. High peel strength can be obtained by adding an elastomer to the epoxy resin, but this reduces the modulus of the adhesive. Specially formulated "toughened" epoxy adhesives, in which a modifying rubber is present as a distinct phase evenly distributed throughout the cured adhesive, are widely available. These adhesives exhibit high shear strengths in addition to high peel strength and impact resistance. The phase-separated structure imparts high fracture toughness, or resistance to crack propagation, to the adhesive.

Epoxy Phenolics

An epoxy-phenolic resin cured with dicyandiamide is an adhesive that performs well in the 205°C to 260°C (400°F to 500°F) temperature range. It will produce a bond with good shear and creep properties at 260°C (500°F), but relatively poor peel resistance when cured in a representative cycle of 2 hours at 177°C (350°F) and 1 MPa (150 psi) pressure. Better overall strength properties often are obtained by using this adhesive in conjunction with a specially formulated primer that has high peel strength. The elastomer-modified phenolics that require curing at temperatures from 149°C to 177°C (300°F to 350°F) and pressures up to 690 kilopascals (kPa) (100 psi) provide good resistance to heat and water in service.

Other classes of high-temperature-resistant polymers have been evaluated as high-temperature adhesives. Of these, the polyamides are available as structural adhesives. Applications for these adhesives are limited, however, because of the difficult processing conditions they require.

Acrylic Adhesives

Structural acrylic adhesives cure by a chain-growth free-radical mechanism, which results in a range of cure times from about one minute to a few hours. The adhesives generally are supplied as two components that are mixed immediately prior to the assembly of the joint. Some formulations are available that require no mixing; cure is initiated on contact of the adhesive with a surface that has been coated with a special primer-activator.

Acrylics adhere well to a variety of metals and engineering plastics and are tolerant of surface contamination. Formulations are available that form strong bonds in metals even though the surfaces of the adherends have not been cleaned of mill oils and drawing compounds. However, the most durable bonds are produced when clean surfaces are used.

Anaerobic Adhesives

Anaerobic adhesives are shelf-stable, ready-to-use formulations that cure at room temperature. Their cure is inhibited by the presence of air (oxygen) in the package and during application. Once the joint is assembled and air is excluded from the liquid adhesive, curing begins. The major use of these adhesives is for sealing and securing threaded fasteners, although formulations are available for applications requiring high-strength bonds.

Cyanoacrylate Adhesives

Cyanoacrylate adhesives, called *super-glues,* are also shelf-stable and cure rapidly at room temperature when placed in contact with most surfaces. The cure is catalyzed by traces of basic compounds present on the surface of the adherends. Water, which is adsorbed on most materials, often acts as the effective catalyst. Like most rapid-cure adhesives, they have low inherent toughness. Cyanoacrylates work best with thin bondlines, so the surfaces should be well matched. The use of these adhesives in industrial applications was formerly limited because of their relatively poor moisture and heat resistance; however, some cyanoacrylate adhesives have been formulated to perform better in these respects.

Urethane Adhesives

Urethane adhesives are particularly useful for bonding to plastics. They may be used with metals but should not be used with unprimed aluminum, where moisture resistance is needed. Cure times typically are a few minutes to a few hours and temperature resistance of about 120°C (250°F) is common. This encompasses the operating ranges of most plastics. Urethane adhesives can also operate into a cryogenic range well below –18°C (0°F).

Silicone Adhesives and Sealants

Silicone adhesives usually are categorized as structural sealants. They have relatively low tensile strengths of about 6 MPa to 7 MPa (0.9 ksi to 1.0 ksi). They display excellent temperature operating ranges, typically from –50°C to 260°C (–60°F to 500°F). Silicone adhesives rarely are used in the absence of another joining method.

METHODS OF APPLICATION

Adhesives can be applied with rollers, brushes, caulking guns, trowels, spray guns, or by dipping. The form of the adhesive and method of application must be compatible.

Industrial adhesives are available in the following forms:

1. Liquids, ranging in viscosity from free-flowing to the consistency of thick syrup;
2. Pastes;
3. Mastics;
4. Solids;
5. Powders; and
6. Supported and unsupported films.

The method used for the application of a particular adhesive should be selected in consideration of the following factors:

1. Available forms of the selected adhesive,
2. Methods available for applying the various forms,
3. Joint designs and order of assembly,
4. Production rate requirements, and
5. Equipment costs.

Liquid Adhesives

Liquid adhesives commonly are applied by brush, short-napped paint roller, or dipping to join the components of small assemblies. The more viscous liquids are often applied with a trowel, mixing/extrusion guns, or plastic squeeze bottles or tubes. (Refer to Figure 10.6.) The polyethylene nozzles of bottles or tube containers should not be rubbed across prepared surfaces, such as etched aluminum. This action may deposit a wax-like coating to which the adhesive will not adhere. Small applicators, resembling a ballpoint pen or a hypodermic needle, may be used to deposit very narrow glue lines for spot application. The silk-screen process is also useful for applying an adhesive to selected areas. Automatic dispensing machines, which simplify the proportioning and the mixing of many two-component formulations, also are available.

For large areas, such as curtain wall panels, liquid adhesives may be applied by spraying, flow-coating, roller-coating, or troweling. Depending on the application method, consideration must be given to the viscosity and working life (pot life) of the adhesive formulation.

Paste and Mastic Adhesives

Paste and mastic adhesives may be applied with a smooth or serrated trowel, knife roller, or an extrusion device. Some paste formulations contain a thixotropic additive that inhibits sag or flow during application and cure. This feature permits their use on vertical or overhead surfaces and may eliminate or significantly reduce the need for special clean-up operations.

Solid Adhesives

One method of applying a solid adhesive is to heat the substrate to a temperature slightly above the melting point of the adhesive and then apply the adhesive to the hot surface as it melts. Some rod and powder forms of epoxy adhesives are applied in this manner. Specially developed flame spray guns also can be used, but this method may require powders with a particle size within narrow tolerances. Also, overheating the adhesive during application must be carefully avoided.

Film Adhesives

Mostly used in aerospace applications, adhesives in film or tape form are extremely simple to use and produce a bondline of relatively constant thickness and coating weight per unit area. These are important factors in most bonding applications. Films are made from adhesives that are thermoplastic, thermosetting, or pressure sensitive and are supplied in rolls or sheets that can be blanked or cut with scissors to the required shape. Film adhesives are particularly useful for bonding large areas, such as honeycomb sandwich panels.

(Refer to Figure 10.7.) Special films are available for use as adhesive backing on items such as nameplates and decals.

Generally, the film adhesive is placed in position between the adherends and activated with heat or solvent. In the case of pressure-sensitive film, pressure is applied to accomplish bonding. For applications demanding high strength at elevated temperatures, the films usually require both heat and pressure to create the bond.

Solvent-reactive and pressure-sensitive films are intended primarily for bonding large sheets where only nominal pressure is required. These types of films do not need special heating equipment. They are particularly useful for short-run production and for bonding components at room temperature.

Duplex bonding films, which combine the properties of elastomeric adhesives and epoxies, are sometimes used for honeycomb sandwich constructions. The elastomeric side, usually a nitrile phenolic, is bonded to the facing to provide peel resistance. The epoxy side forms fillets around the cell walls, and this increases the effective bonding area and the resulting joint strength.

ADHESIVE SELECTION

The selection of the proper adhesive for production normally depends on the following key variables:

1. Materials to be joined (possibly including a plastic),
2. Service requirements,
3. Method of adhesive application, and
4. Costs of adhesive bonding compared with other joining methods.

The service requirements of the completed assembly must be studied thoroughly. Several factors to be considered in determining the bonding application are the following:

1. Type and mode of loading (peel, tensile, shear, or combined loading; static or cyclic fatigue);
2. Operating temperature range;
3. Chemical resistance;
4. Weather and environmental resistance;
5. Flexibility;
6. Differences in thermal expansion rates;
7. Odor or toxicity problems; and
8. Color match.

Once the service requirements are known, adhesive systems with good durability potential can be selected. The desired form of the adhesive and method of application can then be chosen based on availability of production equipment and scheduling requirements.

In adhesive selection, the tendency to over-design should be avoided for economic reasons. However, some excess performance is necessary to allow for possible environmental attack and decrease in performance over time. Requirements for higher strength or greater heat resistance than is actually required for the specific application may exclude the consideration of many formulations that are adequate for the job, and perhaps are less costly or easier to handle in production.

In selecting an adhesive for a specific application, certain physical properties of the adhesive and the mechanical properties of bonded joints should be considered. These properties pertain to the behavior of an adhesive from the time it is made until the bond is accomplished, as well as its performance in service. Table 10.2 lists some of the properties and the applicable ASTM standards for the testing of adhesive bonds.[4] Individual applications may require in-house

4. American Society for Testing and Materials (ASTM International), 100 Barr Harbor Drive, P.O. Box C700, West Conshohocken, PA 19428.

Table 10.2
ASTM Standards for Determining Adhesive Properties

Property	ASTM Designation(s)
Physical	
Aging	D1151, D1183, D2918, D2919, D3236, D3762
Chemical	D896
Corrosivity	D3310, D3482
Curing rate	D1144
Flow properties	D2183
Storage life	D1337
Viscosity	D1084
Volume resistivity	D2739
Thermal T_g	E1356, E1545, E1824
Conductivity	D2739
Mechanical	
Cleavage strength	D1062, D3433
Creep	D1780, D2293, D2294
Fatigue	D3166
Flexural strength	D1184
Impact strength	D950
Peel	D903, D1781, D1876, D2918, D3167
Shear	D1002, D2182, D2295, D2919, D3528, D3983
Tensile	D897, D1344, D2095

modifications to these testing standards, but they are generally accepted throughout the industry.

For most applications, several candidate adhesives usually will be selected based on these considerations. The final choice generally is made after running a test program to determine the suitability of the adhesives for the particular application. Testing may involve various laboratory specimens or a prototype of the complete assembly. In either case, testing should include some estimate of the long-term durability of the adhesive bond in the service environment. The adhesive supplier often provides help technical support in the selection of adhesives, especially if a large quantity of adhesive is to be used. However, determining suitability for use and testing in support of suitability usually is the responsibility of the user.

JOINT DESIGN

To incorporate as many of the advantages of adhesive bonding as possible, joint design should be considered during the early stages of production planning. Joints that are common in welding operations are often undesirable for adhesive bonding. Weld joint designs typically feature high peel loads. If adhesive bonding is being considered as part of a redesign program, structural adhesives should not be substituted directly for other joining methods. The joint should be redesigned to take advantage of the properties of adhesive bonding and to allow for its limitations.

Although the primary objective is a strong assembly capable of meeting service requirements, proper joint design can often lead to other cost-saving benefits. Through good design, it may be possible to achieve satisfactory results with an economical adhesive formulation that will utilize a simple bonding process, and to minimize the quality control steps needed to ensure reliability.

Joint design often influences the form and characteristics of the bondline. The design must provide space for sufficient adhesive and a means of getting the adhesive into the joint area. There is an optimum bondline thickness for most adhesive-adherend combinations. Thinner bondlines generally result in higher strength, but there is a limit because the fracture resistance of an adhesive is somewhat dependent on how much adhesive is present. Filling the joint clearance (gap) is an advantage of adhesive bonding; therefore it is necessary to strike a balance between bondline thickness and manufacturing tolerance. Joint filling beyond 0.75 mm (0.030 in.) should be avoided. Acrylics and cyanoacrylates rarely tolerate joint filling beyond 0.25 mm (0.010 in.).

When considering a new design or a redesign for adhesive bonding, the following three rules should be observed:

1. The design loading should produce shear or tensile loading on the joint, and cleavage or peel loading should be minimized;
2. The joint design should ensure that the static loads do not exceed the adhesive plastic strain capacity; and
3. If it is anticipated that the joint will be exposed to low cyclic loads, the joint overlap should be increased sufficiently to minimize the possibility of creep in the adhesive. Creep is the primary fatigue mode for adhesives, comparable to the initiation and propagation of cracks in welds.

These rules may be difficult to achieve in practice. Some stress concentrations are unavoidable, and it is difficult to design a joint that will be stressed in one mode only.

The three main types of loading are illustrated in Figure 10.10. A fourth type, cleavage peel, is symmetrical peel or unsymmetrical tensile loading. An adhesive-bonded joint performs best when loaded in shear, that is, when the direction of loading is parallel to the plane of the faying surfaces. With thin-gauge metal bonds, joint designs can provide large bond areas in relation to the metal cross-sectional area. This makes it possible to produce joints that are as strong as the metal adherends.

The relationship between joint strength and overlap length for a double-lap shear joint is shown in Figure 10.11. The joint strength and overlap distance are

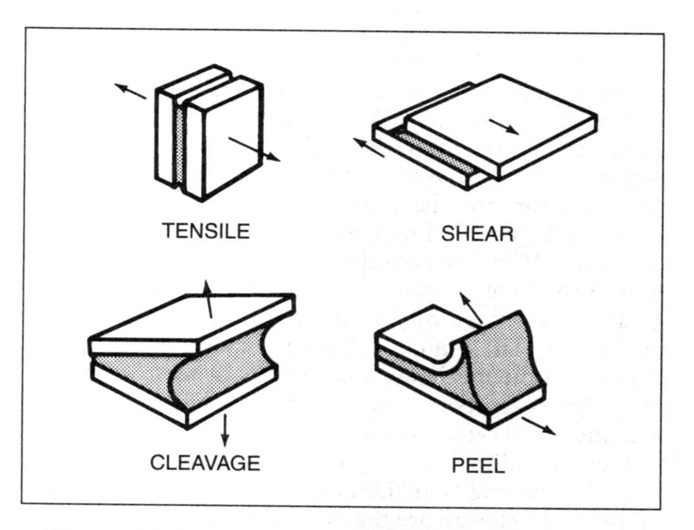

Figure 10.10—Four Principle Types of Loading

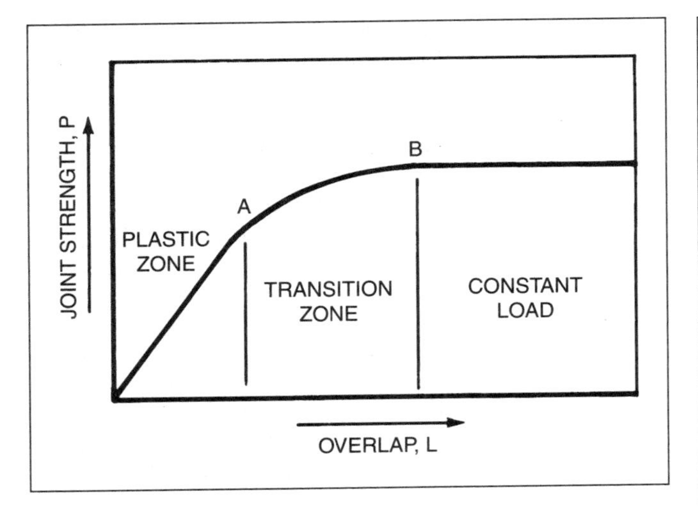

Figure 10.11—Relationship Between Joint Strength and Overlap in Shear

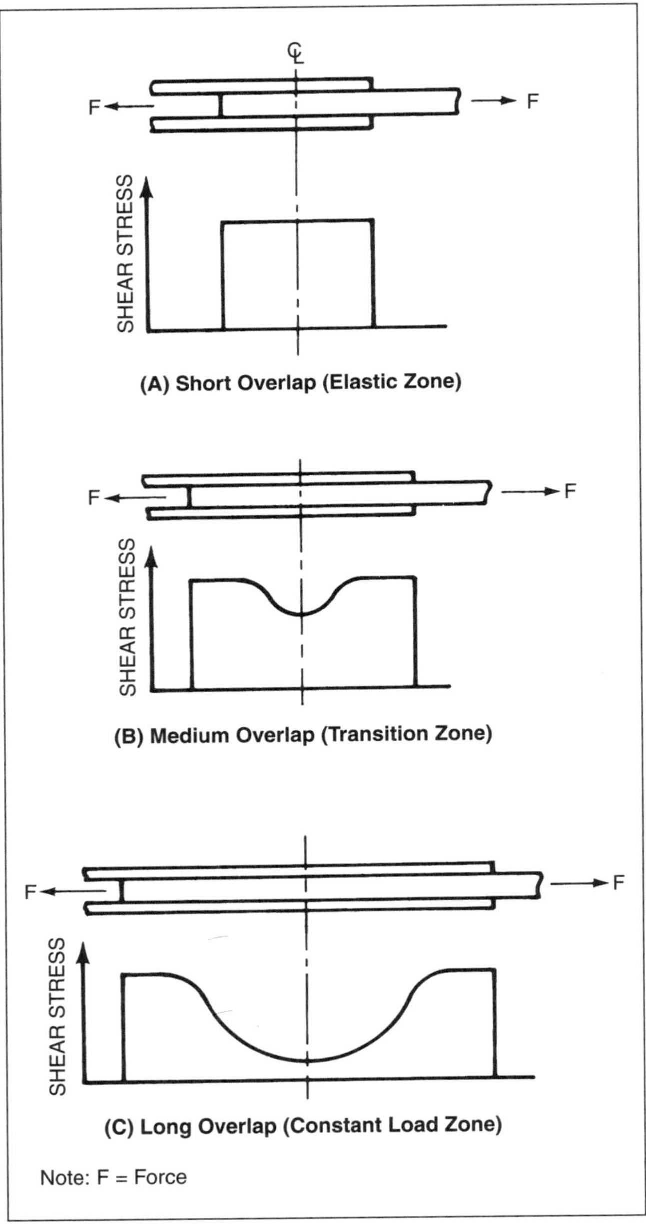

Figure 10.12—Change in Shear Stress Distribution with Overlap for Constant Load, F

proportional up to some limit (Point *A* in Figure 10.11). Then the unit increase in strength lowers as the overlap distance increases. Beyond some overlap (Point *B*), the failure load does not change significantly with overlap distance. If an adhesive is strained into the plastic region, it will not recover elastically (by definition), so the bond dimension will permanently change.

SHEAR LOADING

Figure 10.12 indicates the shear stress distribution across lap joints caused by load *F* with short, medium, and long overlaps. With short overlap, Figure 10.12(A), the shear stress is uniform along the joint. In this case, the joint can creep under load with time and failure may occur prematurely. When the overlap exceeds some value, the adhesive at the ends of the joint carries a larger portion of the load than the adhesive at the center. Therefore, the shear stress at the center is lower, as shown in Figure 10.12(B), and the likelihood of creep is decreased. With long overlap, Figure 10.12(C), the portion of the joint overlap that bears low shear stress is a greater percentage of the total, and creep potential is minimized. The joint overlap for minimum creep will depend on the mechanical properties of the base metal, the adhesive properties and thickness, the type of loading, and the service environment.

Weld bonding can improve overall stress distribution. Placing welds at the natural stress concentrations in the joint helps assure that stress loading is kept below the maximum allowable for the adhesive.

PEEL LOADING

Peel loading is the most challenging mode for an adhesive. Difficulties multiply when cleavage or peel-type loading is present. Cleavage loading produces non-uniform stress across the joint, and this causes failure to initiate at the edge of the adhesive. Thus, such a joint is considerably weaker than the same bonded area under uniform shear or tensile stress.

The situation is even more critical when the adhesive is subject to peel-type loading, in which a very narrow line of adhesive at one edge of the joint must withstand the load. Peel loading produces failure at only a fraction of the tensile load that would rupture a bond of the same area.

As previously noted, unidirectional loading is rarely accomplished. Most joints are subjected to multi-axial loads that combine cleavage or peel loading with tension or shear stress in the bond. One example is a straight butt joint that is designed to be stressed strictly in tension but is subjected to a bending moment that creates a cleavage load. Another example is a single lap joint that is designed to withstand expected shear stress but must bear cleavage or peel loads when the joint rotates slightly as the load forces tend to align, as shown in Figure 10.13. These problems usually can be minimized by selecting an adhesive designed to carry the type of loading expected and by employing the proper joint design.

Several of the more common types of joints for sheet metal are shown in Figure 10.14. A butt joint design, shown in Figure 10.14(A), is not recommended. They are rarely successful. Cleavage loading may develop if the applied loading is eccentric. A bevel or scarf joint, Figure 10.14(B), is a better design because the bonded area can be greater than that of a butt joint. Cleavage stress concentrations at the edges are minimized by the tapered edges of the adherends. Although widely used in wood bonding, this configuration is difficult to make in metals with regard to alignment and the application pressure during curing.

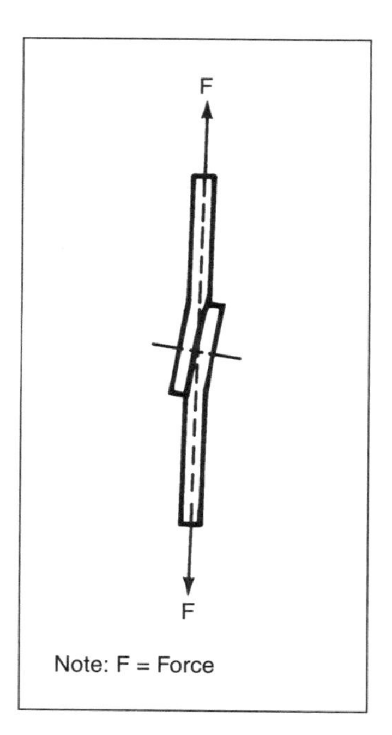

Figure 10.13—Lap Joint Rotation as a Result of Loading

The single lap joint, Figure 10.14(G), is probably the most commonly used type and is adequate for many applications. The beveled lap joint, Figure 10.14(H), has less stress concentration at the edges of the bond because of the beveled edges. The thin edges of the adherends deform as the joint rotates under load, and this minimizes peel action.

If joint strength is critical and the components are thin enough to bend under load, a "joggle" lap joint, shown in Figure 10.14(I), would be better. The load is aligned across the joint and parallel with the bond plane, thus minimizing the possibility of cleavage loading.

If the sections to be bonded are too thin to permit edge tapering, a double-strap joint, Figure 10.14(E), will provide good results. The best design is the beveled double-strap joint with tapered straps, as shown in Figure 10.14(F). For single lap joints, increasing the frontal joint area distributes tensile and shear stress, while increasing the joint depth increases peel resistance.

The butt joints shown in Figure 10.15 can be incorporated easily into machined or extruded shapes that are to be assembled and joined by adhesive bonding. The tongue-and-groove joint, shown in Figure 10.15(A), not only aligns the load-bearing interfaces with the plane of shear stress, but also provides good resistance to bending. The landed-scarf tongue-and-groove joint, shown in Figure 10.15(B), often has production advantages. The configuration of this joint automatically aligns the components to be mated, controls the length of joint, and establishes the thickness of the glue line. It is a good design for an assembly that is intended to withstand high compressive forces in service, and it adds the advantage of a clean appearance.

Corner and T-joint designs and evaluations are shown in Figure 10.16. The use of beveled or tapered reinforcing members requires a cost analysis to determine if the improved joint properties are justifiable. Joints requiring machined slots or complex corner fittings are seldom of interest in sheet metal designs.

TUBING

A tube joint can be described as a single-lap shear joint spun on an axis. Adhesive bonding is useful for tube joints, as shown in the examples in Figure 10.17. Large bonded areas have very strong joints with a clean appearance. Similar in concept to the single lap joint, increasing the joint depth defeats peel resistance from

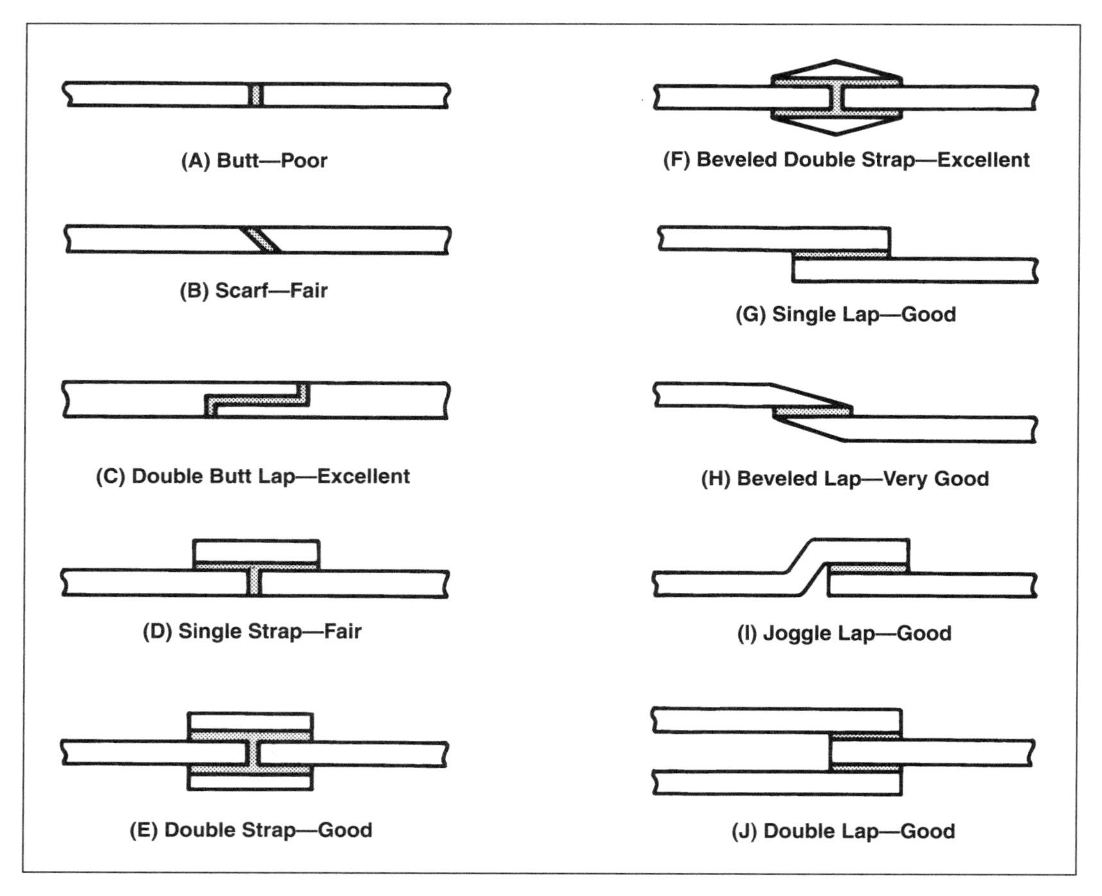

Figure 10.14—Joint Designs for the Adhesive Bonding of Sheet Metal

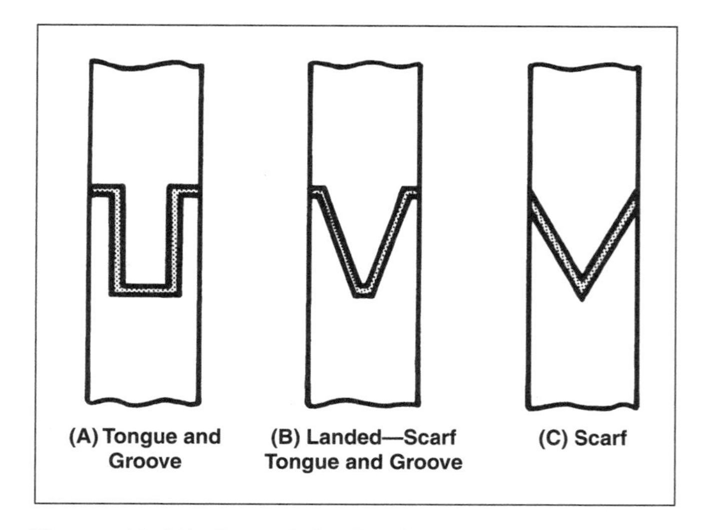

Figure 10.15—Butt Joint Designs for the Adhesive Bonding Machined or Extruded Shapes

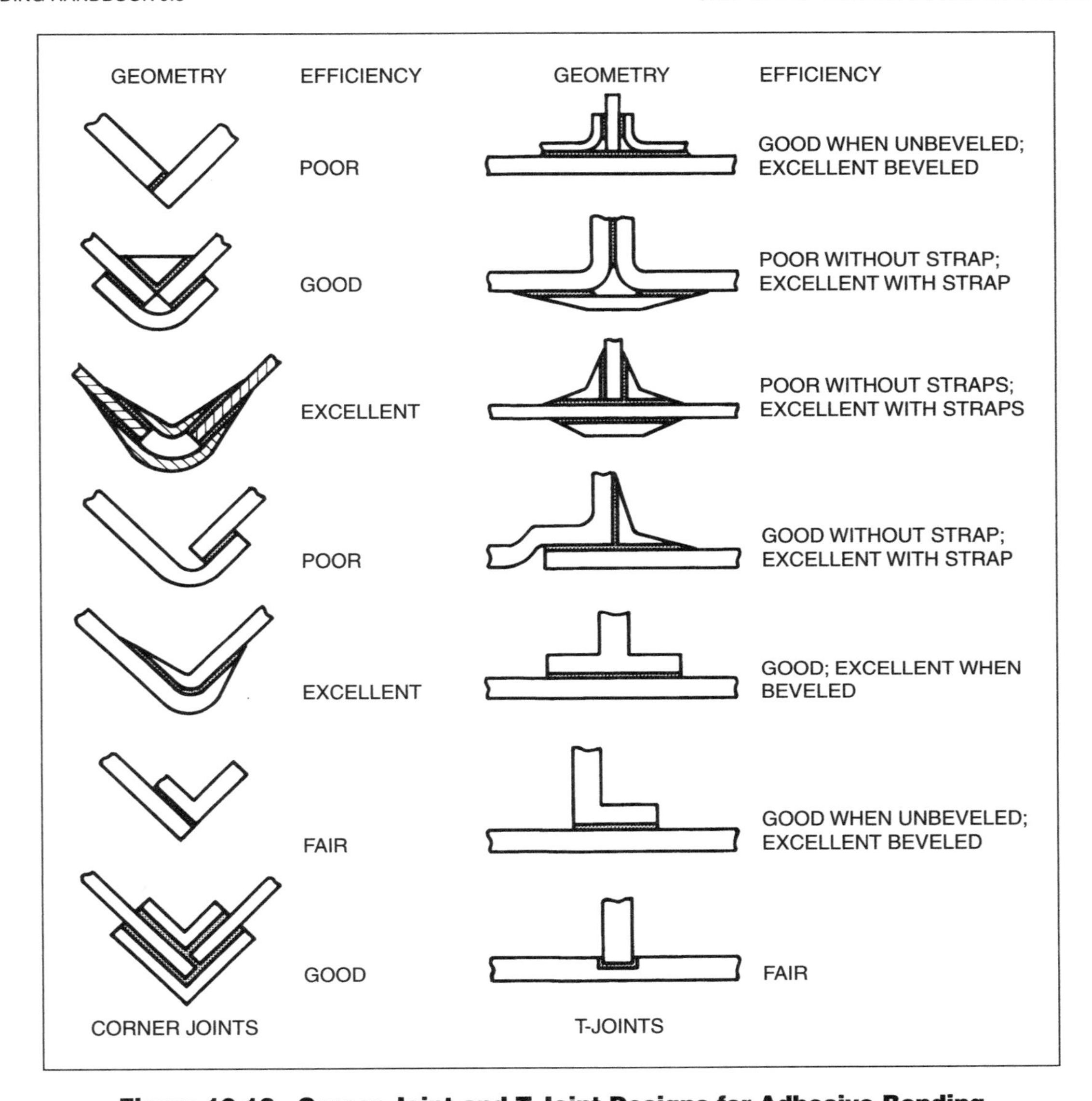

Figure 10.16—Corner-Joint and T-Joint Designs for Adhesive Bonding

off-axis loading and increases resistance to torsion shear.

Processing some designs may be complicated. During assembly of the designs shown in Figures 10.17(A) and (B), the adhesive may be pushed out of the joint. The design shown in Figure 10.17(C) partially overcomes this problem. Adhesive in the corners is forced into the joint by a positive-pressure filling action during assembly. Tapered or beveled tubular joint designs, as illustrated in Figures 10.17(D), (E), and (F), will produce a positive pressure on the adhesive during assembly that will completely fill the joint clearance, but they are costly to produce. The design pictured in Figure 10.17(G) shows a tubular sleeve joint that can be filled by injecting the adhesive under positive pressure through a hole in the sleeve. This technique results in completely filled and bonded joints at reasonable fabrication costs.

Using computers to develop analytical mathematical models makes it possible to optimize joint design by taking into account the geometry of the adherends and the properties of the adhesive. The thick adherend shear test (ASTM D3983) provides the shear modulus, the elastic shear stress limit, and the asymptotic shear stress

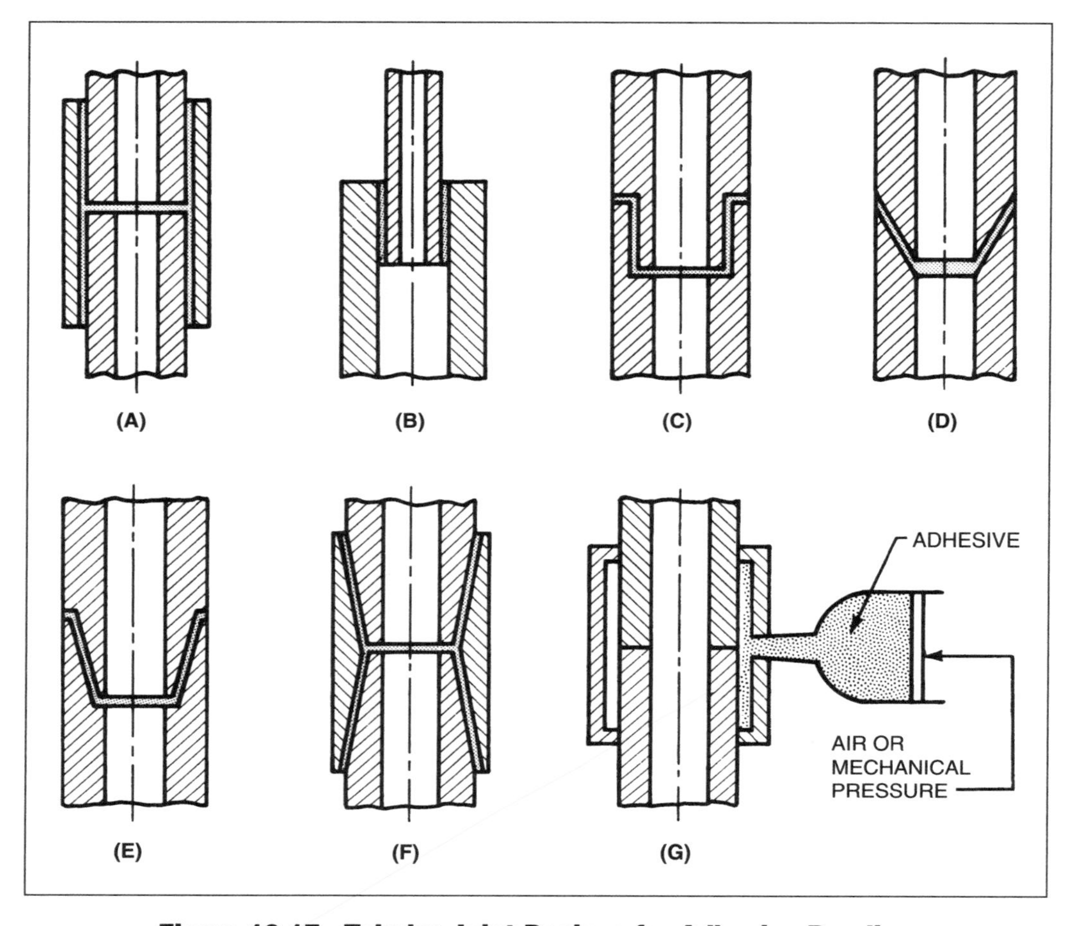

Figure 10.17—Tubular Joint Designs for Adhesive Bonding

of the adhesive that are useful in this respect.[5] Thick adherend specimens can test for "pure" adhesive properties but the bondlines must be kept relatively thin to reduce peel moment.

It eventually may be possible to include the effects of environmental exposure into the mathematical modeling of joint performance. This will allow the durability of adhesive-bonded structures to be predicted. Several models for predicting durability of adhesive joints exist. Their utility in the prediction of durability depends on a number of measurements that are used in combination with the model.

5. American Society for Testing and Materials (ASTM International). Subcommittee D14.70. 2004. *Standard Test Method for Measuring Strength and Shear Modulus of Nonrigid Adhesives by the Thick-Adherend Tensile-Lap Specimen*, ASTM D3983-98, West Conshohocken, Pennsylvania: ASTM International.

SANDWICH CONSTRUCTION

A major use of structural adhesive bonding is in "sandwich" construction, where two sheets of material are joined by adhesive bonding to a core material. First used in aircraft fabrication to meet the demand for a high stiffness-to-weight ratio, sandwich panels are widely used throughout general industry. The characteristics of these panels make them equally valuable in the fabrication of walls, truck bodies, refrigerators, cargo pallets, and many other commercial applications.

The sandwich skins, or facings, behave approximately as membranes stabilized by a lightweight core material that transmits shear between the skins. Basically, a high stiffness-to-weight ratio is achieved by placing the two load-carrying skins as far from the neutral axis as possible with the lightweight core bonded to them. In a sense, the face sheets perform the same functions as the flanges of an I-beam, and the core performs as the beam web.

The selection of skin and core materials for sandwich construction is dictated by service requirements and economic considerations, including the cost of fabrication and materials. Many different skin and core materials may be used as sandwich components. Sheets of metal, plastic, wood, and fire-resistant inorganic compositions are commonly used as skin materials.

The following three core materials are basic types:

1. Solids, such as hardwood, end-block balsa, or metal;
2. Honeycomb or corrugated cores made of various materials, typically metal foils, resin-impregnated paper, or reinforced plastics; and
3. Open-cell or closed-cell foamed materials, such as polystyrene, polyurethane, polyisocyanurate, and glass.

Metal-foil honeycomb cores are available with and without perforations in the cell walls. The perforations permit equalization of pressure in the completed panel and also vent gases produced by some types of adhesives during bonding. Cores are also available in truss and waffle configurations that are produced by corrugating and folding or by pressing the material to shape between matched dies.

Expanded plastic cores may be supplied in a preformed state that requires a separate bonding step, or the plastic may be foamed in place and thus simultaneously formed and bonded to the skins of the sandwich.

Special consideration should be given to selecting the material combination that will produce the optimum composite structure for a specific application. For example, a honeycomb core between thin metal skins can provide high strength and low weight; or a foamed plastic core can be used to assure high thermal insulation with almost any skin material.

Several basic types of adhesives with very similar properties are used in sandwich manufacturing. These basic formulations are modified to meet specific environmental conditions. A careful study of the design, environmental factors, and material properties should be made before selecting the adhesive.

SURFACE PREPARATION

Adhesives bond only to surfaces, therefore, the strength and durability of the bondline can be only as good as the quality of the surface preparation. On metal, that surface is an oxide layer. Frequently, the weakest link in an adhesive-bonded joint is the interfacial bond between the adhesive and the adherends. If the adherends are not properly cleaned and prepared to receive the adhesive, the bond will show less than optimum performance and the environmental resistance usually will be significantly reduced. Surfaces should be prepared by using procedures that ensure that the bond between the adhesive and metal surfaces is as strong as the adhesive itself. When the joint is tested under simulated service conditions, failure should occur within the adhesive rather than at the bondline. This is called *cohesive failure* and indicates that the surface preparation method is the best it can be.

The required degree of surface preparation depends mainly on the nature of the material and, to some extent, on the service requirements, bonding cycle, and the probable nature of contaminants. For some less critical applications, a solvent wipe or washing in a detergent solution may be adequate.

All cleaning agents must be carefully removed from the surface by rinsing and drying thoroughly prior to the application of the adhesive. For the best joint performance, the surface preparation procedure must be selected that will provide the best bond between the adhesive and adherend.

It is equally important to avoid recontamination of the clean surfaces during processing. Components should be handled with clean gloves, tongs, or hooks, and all contact with the bonding area should be avoided. Priming, bonding, or a combination of these should be accomplished as soon after surface cleaning as possible. In the interim, components should be stored in a clean, dry place.

PREPARATION OF METAL SURFACES

Metal surfaces always have an oxide layer and also may have significant levels of other contamination. It is necessary to remove or control these surface characteristics to ensure good bonding. The surface preparation of metals should accomplish the following:

1. Remove all contaminants from the surface;
2. Make the surface chemically receptive to the adhesive or primer and give satisfactory wetting characteristics;
3. Prevent poorly adhering or low-strength compounds from forming on the adherends;
4. Remove minimum amounts of metal;
5. Avoid embrittlement or corrosion, or forming a surface prone to environmental attack; and
6. Provide a surface resistant to moisture degradation.

Surface preparation methods for aluminum alloys, stainless steels, carbon steels, magnesium alloys, titanium alloys, and copper alloys are published in ASTM D2651, *Standard Recommended Practice for Preparation of Metal Surfaces for Adhesive Bonding.*[6] The best corrosion-resistant surface preparation method for aluminum alloys is stated in SAE Aerospace Recommended Practice (ARP) No. 1524, *Surface Preparation and Priming of Aluminum Alloy Parts for High Durability Structural Adhesive Bonding.*[7]

Different levels of surface preparation may be employed, depending on the end use of the bonded structure. An appropriate surface method may include some or all of the following methods of obtaining optimum strength characteristics:

1. Cleaning and degreasing;
2. Deoxidation (scouring, grit blasting, chemical etching); and
3. Anodizing, surface modification, or application of coatings with functional properties.

Cleaning and Degreasing

Cleaning usually begins with some form of degreasing with solvent-borne or waterborne degreasing systems. If immersion degreasing is used, the degreasing solution should be checked periodically for oily contaminants and decomposition products, and the solvent should be changed when necessary. When modern water-borne degreasing agents are used, solvent degreasing is rarely necessary, but if it is used, the adherends should be wiped with clean lint-free cloths or disposable tissues. Specific degreasing methods may be used, depending on the metal being degreased. Suppliers specializing in degreasing materials often can help select the proper degreasing system for a given metal. Degreasing may be followed by a water rinse with deionized water. Forced drying in an oven after water rinsing is preferred since this reduces the possibility of recontamination from dust and impurities while air drying at room temperature.

Deoxidizing

Following degreasing, mechanical abrading sometimes is additionally employed not only to clean, but more importantly, to deoxidize metal surfaces and increase the effective bonding area by roughening. Grinding, filing, wire brushing, sanding, and abrasive blasting are common methods. It is essential to degrease the surface prior to abrading so that contaminants are not ground into the surface. Abraded contaminants should be removed mechanically or with solvent before proceeding to additional treatment. Solvent wiping after mechanical cleaning has the potential to recontaminate the surface, particularly if a considerable surface profile has been added. Some operators simply brush the abraded dust off the surface using a clean brush. Air blasting the debris carries the danger of adding oil to the surface from compressed air lines.

If subsequent solvent cleaning is necessary, it must be carefully assured that the solvent does not contain contaminants that can remain on the surface after the bulk of the solvent has evaporated. Mechanical deoxidation alone is seldom sufficient preparation for the bonding of components to be used under harsh environmental conditions, such as extreme o heat and humidity.

Chemical deoxidation involves spraying or immersion with an acidic or basic reactant to remove oxide layers. If an immersion deoxidation treatment is employed, a thorough check of the composition of the treating solution should be made periodically. Special consideration should be given to the rate of metal processing over a given period of time. Failure to control the concentration of strong acid or alkali solutions may result in excessive metal loss. The frequency of monitoring depends on the rate of exhaustion of the treating solution and usually is determined from historical use patterns.

Anodizing and Etching

Depending on the end use of the bonded article, chemical deoxidation may be sufficient and the bonding operation may proceed to priming or bonding. However, in addition to mechanical or chemical deoxidation, surface modification might be required following the deoxidation processes. Aluminum or titanium often is anodized in the presence of sulfuric, phosphoric, chromic, or hydrofluoric acid to produce a durable, water-resistant oxide coating. Anodizing provides additional surface morphology that promotes primer or adhesive wetting as well as a degree of mechanical interlocking.

Surface conversion or etching treatments are similar to anodizing, but performed in the absence of applied electrical potential. Etching treatments include liquid or gelled topical etching solutions available from commercial suppliers, or phosphate-fluoride mixtures for use on aluminum or titanium. Stainless steels may be etched with ferric chloride solutions. These methods provide

6. American Society for Testing and Materials (ASTM International). Subcommittee D14.80. *Standard Recommended Practice for Preparation of Metal Surfaces for Adhesive Bonding,* ASTM D2651, West Conshohocken, Pennsylvania: ASTM International.

7. Society of Automotive Engineers (SAE). Ams-P Polymeric and Composite Materials Committee, 1990, *Surface Preparation and Priming of Aluminum Alloy Parts for High Durability Structural Adhesive Bonding*, ARP No. 1524, Warrendale: Pennsylvania: Society of Automotive Engineers.

some chemical etching and produce a residual surface chemistry that is amenable to bonding while providing some level of moisture resistance.

Anodizing treatments and etching treatments sometimes may carry heavy toxicity and environmental loads into the process. It is particularly important to reduce or eliminate the use of chromium-VI (Cr^{+6}) components, which are known carcinogens. However, remedial techniques have emerged that use a combination of grit blasting and silane treatment or solgel inorganic polymer treatments. These water-based systems feature low toxicity and have been successfully applied on aluminum, titanium, stainless steels, nickel-base alloys, and even carbon steels. Silanes and solgels provide chemical surface modifications that include reactive species specific to the anticipated primer or adhesive. This produces a chemically receptive surface for bonding.

PRIMERS

A primer is a surface modifier that changes the surface into one that accepts the adhesive or protects the interface in service. Certain service conditions may require the use of a primer for improved corrosion resistance, flexibility, shock resistance, or peel resistance of the adhesive bond. Primers also may be used to wet or penetrate the substrate or to protect a treated surface of a substrate prior to the application of adhesive.

Most primers are low-viscosity solutions commonly applied by spraying. Brushing may be satisfactory when relatively small areas are to be coated. In some cases, roll coating or dipping may be employed. Several prime coats may be required to build up the desired thickness, particularly if the adherends are porous. Air drying and full or partial curing generally is necessary prior to further processing.

PREPARATION OF OTHER MATERIALS

It can be anticipated that metals will be bonded to plastics, composites, glass, or ceramics in many structural designs. Rigid plastics can be lightly sanded to reduce gloss and remove mold-release compounds. Sanding or grit blasting increases surface area but does not increase surface energy. Cleaning, grit blasting, and solvent wiping are frequently employed in sequence to prepare composites for bonding. Once abraded, surfaces should be wiped and flushed with an oil-free solvent. Certain types of plastics, such as fluorocarbon, polypropylene and polyethylene, are difficult to bond and require extensive chemical treatment or surface oxidation, or both, by corona, flame, plasma, or ultraviolet-oxygen treatments. Overtreating a plastic material by any method can produce subsurface strain induced by the expanded surface. The amount of oxidative surface treatment on a plastic must be determined empirically.

Glass is easily cleaned by wiping with a suitable solvent. Joint durability can be greatly enhanced, particularly in moist environments, by first cleaning with a laboratory glassware cleaning solution or with a 30% hydrogen peroxide solution and then priming the surfaces with a silane finish. Light grit blasting also can improve bonding to glass and ceramics, both of which are inherently highly polar and wettable surfaces.

INSPECTION OF PREPARED SURFACES

The affinity of a clean surface for water is the basis of the most commonly used test for a chemically clean surface, the water-break test. This test involves cleaning the surface, rinsing well and running water over the surface. After the running water is stopped, the surface should be observed to assure that uniform wetting has occurred and that the uniform wetting film is maintained for 30 seconds. Uniform wetting means that the water film remains continuous, with no water breaks, no pulling away from areas, or beading. If water collects in droplets or forms a break in the continuous water film within the 30-second test, there is probably a residual contamination on the surface. The affected area should be cleaned again and the water-break test should be repeated.

If a contact angle method is used to test for surface contamination, the contact angle water drop should be greater than 12 degrees, indicating that the surface has been cleaned adequately to remove greases, oils, and other nonpolar contaminants. When contact angle measurement is used as a quality control method for cleanliness, inspection should be made immediately after the metal has been dried. If water remains in droplet form, the surface has not been suitably prepared.

PROCEDURES

The procedures for assembling, fixturing, applying pressure, heating and curing of adhesive-bonded components are discussed in this section.

ASSEMBLY AND CURE

The procedures for assembling and curing adhesive bonds depend on the following:

1. Type of adhesive;
2. Open time or work time allowed by the adhesive;

3. Type, size, and configuration of the assembly;
4. Service requirements of the completed assembly;
5. Affinity of the application for automation; and
6. Complexity, quality, and cost.

Drying time is an important factor when solvent-dispersed adhesives are used. Since this time varies with different formulations, it is essential that the recommendations from the adhesive manufacturer are followed. The solvent evaporation rate may be increased by moderate heating with infrared lamps, a hot-air oven, or other methods.

If there is sufficient porosity in one adherend to allow the solvent to escape, the components may be fitted together during the drying time. In any case, the assembly must not be heated for curing until the solvent has evaporated. It is also essential that coated components be fitted together before the tack range of the adhesive has expired. The components may be fitted together immediately after they are coated with chemically reactive adhesives. The mating of the components may be delayed, but it must be done before the adhesive starts to "body," or thicken excessively. Provision should be made for positioning the components and holding them in place while the adhesive cures or sets.

Fixtures

Assembly fixtures frequently are used for positioning. They may be simple jigs or self-contained equipment with provision for applying pressure or heat, or both. The fixture design depends on the size and configuration of the assembly and the amount of heat and pressure needed to cure the adhesive. When using a pressure-sensitive adhesive or tape, the components must be precisely positioned before mating because they may not come apart for repositioning.

Fixturing is particularly important when a contact adhesive is used. Extreme care should be taken to align the adherends accurately before they are mated, since a strong bond is created instantly on contact of the two coated surfaces. The use of release sheets, untreated kraft paper for example, is often helpful in avoiding premature contact. Positioning may not be as critical with some formulations with less aggressive tack periods if the assembly can be slightly adjusted after mating without damage to the bond.

The fixture should properly position the components to meet assembly tolerances and glue line thickness requirements. Fixtures should be lightweight for ease of handling and efficient heat transfer. A heavy fixture presents a large heat sink that may retard heating and cooling rates, which may be detrimental for some adhesive systems. In addition, the fixture must be strong enough to maintain dimensions under the curing conditions for the assembly. The expansion rate of the fixture material should nearly match that of the assembly to minimize distortion and subsequent stressing of the adhesive.

Pressure-sensitive tape may be used to hold the adherends in position if the tape can withstand the curing temperature. Tapes are particularly useful with epoxy formulations that cure at room temperature or slightly warmer and require only moderate pressure.

In weld bonding operations, adhesive bonding may be combined with resistance welding, laser welding, ultrasonic welding, or mechanical fasteners to improve the load-carrying capacity of the joint. The adhesive is applied to the adherends first. Then the components are fixtured and welds or mechanical fasteners are introduced to hold the joints rigid while the adhesive cures. Figure 10.18 illustrates typical design combinations. These techniques significantly reduce or eliminate fixturing requirements and decrease assembly time when compared to conventional adhesive bonding methods.

Application of Pressure

When pressure-sensitive adhesives, film, and some paste adhesives are used, it is necessary to apply and

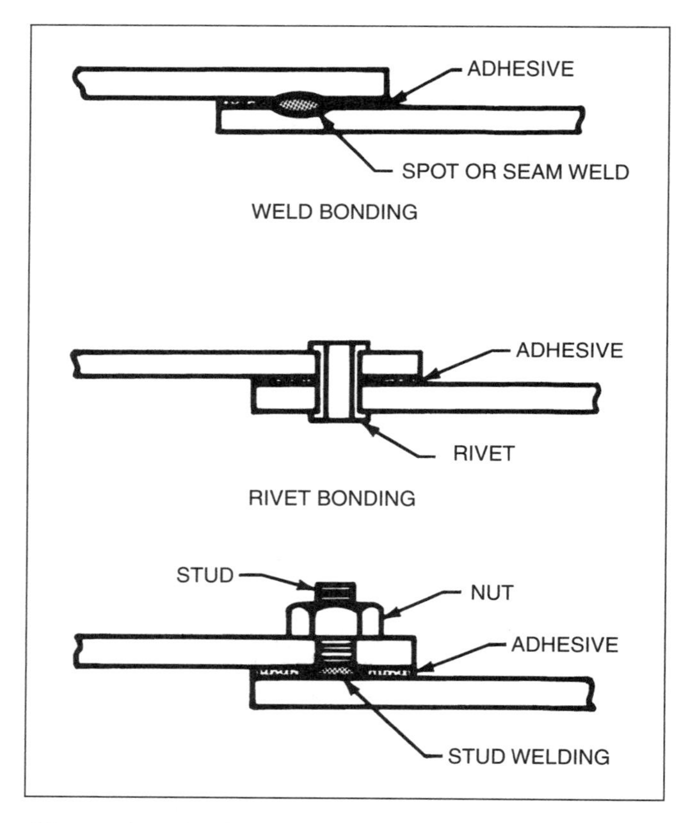

Figure 10.18—Adhesive Bonding in Combination with Resistance Welds and Mechanical Fasteners

maintain adequate pressure during cure to achieve the following:

1. Produce a uniformly thin glue line over the entire bonded area for optimum strength characteristics,
2. Facilitate flow or spreading of viscous adhesives,
3. Counteract any internal pressure caused by the release of volatile products,
4. Overcome minor imperfections in the faying surfaces, and
5. Compensate for solvent loss and dimensional changes.

Uniform pressure may be applied to the joint by several methods, including the following:

1. Dead weights, such as bags of sand or shot;
2. Mechanical devices, such as clamps, wedges, bolts, springs, and rollers;
3. Inflated tubes;
4. Air pressure bearing on an assembly that has been placed in a flexible, evacuated bag;
5. Mechanical or hydraulic presses; and
6. Autoclaves.

Inflated tubes are used in conjunction with a rigid backing fixture. When inflated, the tube presses uniformly along the bondline. Ambient air pressure is adequate for some applications. Air pressure is applied by enclosing the assembly in a thin, airtight bag and then evacuating the bag. Autoclaves are used in a similar fashion in that the assembly is placed in a thin, gas-tight bag vented to ambient pressure. The bagged assembly is placed in an airtight chamber, which is then pressurized to several levels of atmosphere. The pressure forces the bag to conform to the adherends and transmits the pressure to the assembly.

Phenol-based adhesives and film adhesives generally require curing pressures in the range of 400 kPa to 2070 kPa (50 psi to 300 psi). Flat panels coated with a neoprene contact-bond adhesive generally are mated by passing them through rollers under as much pressure as the components will withstand without crushing. A weighted hand roller or other pressure device also can be used.

For fabricating sandwich-panels, the upper pressure limit is governed by the compressive strength of the core material. The lower limit depends on the minimum requirements of the adhesive formulation. For sandwich panels containing solid inserts or edgings, special fixtures may be needed to apply higher pressure at the specific locations.

Throughout the curing cycle, pressure should be as uniform and constant as possible over the entire bond area. If necessary, irregular surfaces can be built up with pads of compressible material. In some cases, soft rubber pads are used to compensate for variations in the dimensions of sheet material and the fixtures. The mass of these materials should be minimized to avoid heat sink and insulating effects. Matched tooling is not often used for curved panels because of the high cost. A better method is to use a one-sided formed tool in conjunction with the vacuum bag or autoclave technique.

Curing Temperature

Since variations in the thermal conductivity of the components influence the amount of heat transmitted to the adhesive layer, curing temperature should be measured at the glue line. Otherwise, the adhesive may not develop the desired properties for the intended application because of improper curing temperature.

Epoxy and phenolic-based structural adhesives require curing at elevated temperatures, generally from about 150°C to 205°C (250°F to 400°F), for periods ranging from 0.5 hours to 2 hours. Many one-part epoxy adhesives can be cured at temperatures as low as 120°C (250°F). Many two-part epoxy systems cure at room temperature; however, they generally produce better properties when they are cured at elevated temperatures. A slightly elevated cure temperature of 50°C to 65°C (120°F to 150°F) provides a simple process control if the temperature in the building varies seasonally. The properties of two-part adhesives often can be improved by using a short higher-temperature post-cure.

When neoprene contact-bond adhesives are used, the adhesive-coated surfaces frequently are heated during the drying cycle and mated under pressure while still warm. When design requirements are not stringent, the adhesive may be dried and the components mated at room temperatures. However, joint properties tend to be more variable than those obtained when the hot contact bonding procedure is used.

As a general rule, curing time decreases as the curing temperature is increased, within limits. Even the epoxies designed to cure at room temperature will cure faster when heated to moderately elevated temperatures. Heating may reduce curing time from a number of hours to several minutes. In general, it is the time at temperature rather than the absolute temperature that governs cure and performance. Another consideration is that the adherends and all of the fixturing must be heated before the entire assembly can reach the desired temperature. Some room-temperature-curing adhesives will not cure properly below 16°C (60°F). This may be an important factor in field applications.

In many instances, longer curing times at elevated temperatures will improve the bond strength of the joint for service above room temperature. Post-curing of the bonded joint without pressure also can improve the heat resistance of the bond. It should be noted that the further reaction of chemical species on heating to temperatures higher than the initial cure will increase the cure of the adhesive until all possible reactions are complete and a maximum T_g (the temperature at which a material changes from glassy or visco-elastic to plastic) is achieved. Heating beyond this point will not be beneficial and may lead to degradation of the adhesive.

EQUIPMENT

In addition to the fixturing equipment described previously, equipment for adhesive bonding may require ovens, presses, or autoclaves to accommodate the curing of adhesive bonds.

Ovens

Ovens are the most widely used and least expensive method for heat-curing when only moderate pressure or simple positioning of the adherends is required. Ovens can be heated by gas, electricity, or steam. Adhesives that give off flammable vapor or solvent during cure should not be exposed to open flames or electrical elements. Ovens should be well vented, temperature-controlled, and fitted with an air-circulating fan for evacuation of flammable vapors prior to heating, and to ensure uniformity of heat throughout. Infrared lamps and ovens commonly are used for contact bond rapid-drying neoprene formulations.

Induction Curing. A common method for accelerating the cure of adhesives is induction curing, in which the entire joint area is placed within an induction field. When energized, the field may heat the metal, some plastics, or the adhesive itself. Carbon-filled adhesives and plastics heat faster than unfilled materials. With steel adherends, the temperature rise is typically quite fast, and curing may be complete in approximately 15 seconds. Induction-cured adhesives may outgas during cure, resulting in porosity, or they may become embrittled because of the tremendous energy release in a short period of time. The curing time at temperature also is important in assuring a good bond.

For field applications, an induction-curing generator that provides 1 kW to 5 kW can be used. Many coil-style units are designed to fit the specific field requirements for adhesive curing. Dimensional control by fixturing is necessary to maintain uniform field strength. If a thermoplastic adhesive has been used, induction heating equipment would be required to reheat the joint for disassembly.

Heated Presses

Hydraulic platen presses frequently are used for applying heat and pressure to flat assemblies. The platens usually are heated by electric heating elements, high-pressure steam, hot water, or some other heat-exchanging fluid.

When the adherends are placed in the press at temperatures below about 65°C (150°F), the procedure is called *cold entry.* Entry at the adhesive-curing temperature is known as *hot entry.* In general, adhesives that release volatile substances perform better when cold entry is employed. Certain adhesives are also affected by the rate of temperature rise or heat input. These factors influence the chemical reactions, the flow, and the density of cured adhesives of the volatile-releasing types. For example, cold entry and a rate of temperature rise of less than 16°C (0°F) per minute result in better shear strength at elevated temperatures for certain nitrile-phenolic film adhesives. Other adhesives, such as epoxy-phenolics, require either stepped heat input or the release of pressure (breathing) at specific temperatures to allow volatiles to escape. Nonvolatile adhesives, such as epoxies, are not affected to any great extent by entry temperature or by the rate of heat input.

Autoclaves

Large autoclaves are used for bonding aircraft assemblies and other extremely large parts. The typical operating range of an autoclave is 1380 kPa (200 psi) maximum pressure and 180°C (350°F) maximum temperature. Pressure generally is provided by compressed air or nitrogen, and curing temperature is achieved with steam-heated tubes or electrical elements.

QUALITY CONTROL

The guarantee of quality for an adhesive joint is the responsibility of the manufacturer and the user of the adhesive bonding process. No nondestructive testing method exists that measures the strength of an adhesive-bonded joint. The strength of the joint depends on the direction of the applied stress. A joint may show very good tensile performance, but poor peel performance. Adhesive-bonded joints are inspected and tested to determine quality and performance under the specific loading and environmental conditions that will be encountered in service. Based on the test results, quality requirements can be established, and inspection methods and procedures can be specified to assure that level of quality. The advantages and limitations of inspection and testing procedures must be understood in order to apply them successfully. A number of industry standard specifications for testing adhesive-bonded joints may be used. (Refer to Table 10.2.) Testing also may consist of accelerated, simulated, or actual use tests of the end product devised by the individual manufacturer or an industry group. For this reason, industry associations are good sources of information on testing procedures.

If an adhesive is to be used with a metal for which no performance data exists, or if it is to be used in an unusual environment, it should be subjected to testing that replicates the conditions of exposure. No new application should be released for manufacturing without adequate testing under all possibly perceived conditions. Single overlap shear specimens (ASTM D1002) can be used to evaluate the compatibility of a metal surface condition with an adhesive system and to evaluate the effect of any unusual environmental exposure.[8] If an adhesive is to be used in a structural joint under stress in a certain environment, test joints should be simultaneously subjected to both the stress and the environmental conditions expected in service.

PROCESS CONTROL AND QUALITY ASSURANCE

Process control remains the primary method for assuring a good adhesive bond. The assumption is that good bonding practices followed in order of requirements will assure good bonding. Good process control includes inspection and monitoring of all cleaning and processing procedures and equipment, evaluation of all materials, and control of storage time and conditions.

Adhesives and primers should be evaluated to assure conformance to the requirements of the design and the specifications of the manufacturer. Certified test reports from the manufacturer may be acceptable in lieu of actual performance tests.

Periodic tests should be performed to determine that cleaning, mixing, and bonding procedures are adequately controlled. The frequency of testing will depend on the volume of assemblies produced and the requirements for the application. However, many manufacturers who employ adhesive bonding in critical applications perform suitable quality control tests at least daily to verify that the process is within specifications. Any production assemblies rejected for dimensional reasons or structural damage should be destructively inspected for joint quality.

8. American Society for Testing and Materials (ASTM International). Subcommittee D1002, 2005, *Test Method for Apparent Shear Strength of Single-Lap-Joint Adhesively Bonded Metal Specimens by Tension Loading (Metal-to-Metal)*, ASTM D1002-05, West Conshohocken, Pennsylvania: ASTM International.

Lap shear tests published in ASTM D1002 generally are satisfactory for control of mixing, priming, and bonding. Peel tests should be performed to ascertain the adequacy of cleaning procedures. The climbing drum method, described in ASTM D1781,[9] and the crack extension (wedge) test in ASTM D3762,[10] also may be used for this purpose.

Crack Extension (Wedge) Test

The crack extension (wedge) test is designed for rapid screening of adhesive joint durability in a controlled humidity and temperature environment. A test specimen design for aluminum alloys is shown in Figure 10.19. One or more specimens are cut from an adhesive-bonded panel. The wedge is forced between the adherents and bends them apart. This separates the adhesive and produces cleavage loading at the apex of the separation. The location of the apex of the sheet separation is recorded.

The wedged specimens are then exposed at 49°C (120°F) to an air environment of 95% to 100% relative humidity for 60 minutes to 75 minutes. The water gener-

9. American Society for Testing and Materials (ASTM International). Subcommittee D14.80, 2004, *Standard Test Method for Climbing Drum Peel for Adhesives*, ASTM D1781-98, West Conshohocken, Pennsylvania: ASTM International.

10. American Society for Testing and Materials (ASTM International). Subcommittee D14.80, 2003, *Standard Test Method for Adhesive-Bonded Surface Durability of Aluminum (Wedge Test)*, ASTM D3762-03, West Conshohocken, Pennsylvania: ASTM International.

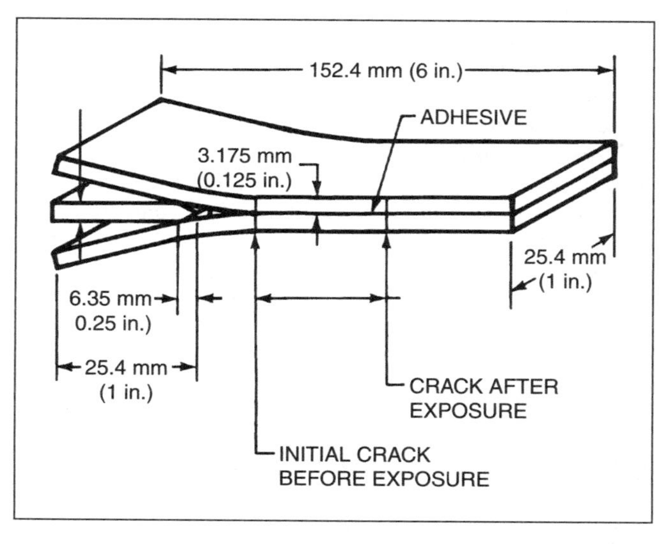

Figure 10.19—Crack Extension (Wedge) Specimen Designed for Aluminum Alloys

ated by the humidity should contain less than 200 parts per million total solids. The distance that the apex moved during exposure is measured two hours after exposure.

This test is used for surface preparation process control by comparing test results with a maximum acceptable increase in adhesive crack length. It also is used to determine adhesive durability characteristics and for surface preparation procedures. The test was originally designed for adhesive-bonded aluminum. However, it may be suitable for other metals with design modifications to account for differences in stiffness and yield strength. The crack-extension test is a good qualitative test to judge the relative merits of surface preparation methods, but is provides no quantification of bond strength. The duration of the test is 1 day to 7 days.

EVALUATION OF FABRICATED PRODUCTS

After the mechanical and processing properties of an adhesive system have been determined through destructive laboratory testing, the ability of manufacturing departments to duplicate these properties should be established. Therefore, complete testing of the first item produced or the first few items of the production run is recommended.

Test loads should be applied in the same manner in which the product item will be loaded in use. The dimensions of the test specimen should be scaled to those of the actual product. The test specimen should reflect that of the actual production item within the confines of economic considerations. During the design phase, at least one full-scale article should be tested under the most rigorous conditions envisioned. Actual loading conditions often are difficult to simulate. In cases involving multidirectional loads, particularly on full-scale test articles, design loads may be applied in each plane individually. The specimen then can be loaded to failure in the most critical load path to determine if it meets the minimum design strength.

When it is impractical to load a completed item for test because of geometry or difficulty in mounting, many companies fabricate test specimens that are either an integral part of the assembly or separate panels processed in the same manner as the production item. Mechanical properties of such specimens closely represent the actual strength of the product. This procedure can provide close control over materials and processing equipment.

NONDESTRUCTIVE INSPECTION

Several nondestructive inspection methods other than visual inspection may apply to adhesive bonding. One or more of the following methods may be appropriate:

1. Ultrasonic (most common),
2. Acoustic impact (tapping),
3. Liquid crystals,
4. Birefringent coatings,
5. Radiography,
6. Holography,
7. Infrared,
8. Proof test, and
9. Leak test.

Nondestructive testing methods for welded joints are discussed in Volume 1 of Welding Handbook, Chapter 14, *Welding Inspection and Nondestructive Examination.*[11] Many of these techniques are applicable to adhesive bonding.

The selection of a method that can be used for a specific application will depend on one or more of the following factors:

1. Design and configuration of the structure,
2. Materials of construction,
3. Types of joints,
4. Material thickness,
5. Type of adhesive, and
6. Accessibility of the joints.

In some cases, it may be necessary to incorporate features in the component design or the adhesive to utilize an inspection process. For example, a filler may be required in the adhesive to increase the thermal or electrical conductivity or the density of the adhesive. To determine the applicability of a particular inspection method, the manufacturers of the particular type of equipment should be consulted.

ECONOMICS OF ADHESIVE BONDING

Adhesive bonding methods combine many procedures, including surface preparations, priming and curing, adhesives application, fixturing, adhesive curing, and unloading the finished assembly from the fixtures. Additional procedures are associated with storage and disposal of adhesives. Each of these steps carries a cost. Thus, the economics of a bonding process depends on the specific assembly application and the contributing steps required to produce the boded article.

11. American Welding Society (AWS) Welding Handbook Committee, 2001, Jenney, C. L. and A. O'Brien, eds., *Welding Handbook*, Vol. 1, 9th ed., *Welding Science and Technology*, Miami: American Welding Society, pp. 580–636.

Bond durability is often related directly to the amount and type of surface preparation. Careful surface preparation results in higher installed cost. For a critical application with an expected long service life, that cost is implicit in the value of the component and will be reflected in the price. In the mid 2000s, high-performance adhesives cost $100 to $500 per gallon. However, a gallon of adhesive used in a joint that is 9.5-mm (3/8 in.) wide with a 0.39-mm (1/64-in.) thick bondline will produce almost 12.20 kilometers (40,000 feet) of bondline, bringing the material cost to 1.1/2 cents per 30.5 cm (1 foot) for even the most expensive adhesives.

Special packaging, such as cartridge kits, incurs additional cost but provides ease of mixing, thus less labor cost, and reduced waste. Most structural adhesives are competitively priced based on volume and packaging to provide a balance between performance, value, and ease of use. A more expensive adhesive may require less capital investment for dispensing and curing, or it may provide a faster cure time, which reduces labor time or work in processing.

Cyanoacrylate adhesives may be regarded as expensive, but they cure very quickly and require little capital investment for dispensing and no investment for curing ovens. Epoxy adhesives are less expensive but may require a curing oven for optimum performance or manufacturing speed. Urethanes are more expensive on average because of one of the chemical component types common to all urethanes. A urethane adhesive may be the only choice for bonding two plastics or for bonding a plastic to a metal. Structural tapes are expensive but require no mixing, metering, or curing. They have little storage or disposal cost and can be applied by anyone, without special training.

Equipment costs vary according to dispensing and curing requirements. A syringe dispenser may cost less than $1000 while a large drum pump may cost more than $50,000. A lamp for ultraviolet curing may cost from $2500 to $5000, while a large throughput oven may cost several tens of thousands of dollars. Induction and microwave equipment is expensive but will cure an adhesive in seconds at elevated temperatures. The trade-off in speed outweighs the capital investment to get that speed. Virtually all dispensing applications can be computer-controlled, which reduces labor cost.

Depending on the application, it has been found that adhesive bonding is cost-effective compared with mechanical fasteners requiring holes, such as bolts or screws. Hole drilling and matching require intensive care in assembly. The larger stress distribution area of an adhesive bond frequently means the metal or plastic thickness can be reduced, when compared with a mechanical fastener, because holes are stress concentrators.

Adhesive bonding rarely is cost competitive with welding. Weld bonding (discussed in Chapter 1) provides enhanced performance to the extent that the additional cost of adhesive provides considerable structural value to the end use. The use of welding combined with adhesive bonding can allow more freedom in surface preparation, which helps to offset the cost of the adhesive. The installed cost of an adhesive must be viewed in light of the end use, the manufacturing or design advantages, the implied value of weight savings, and end-use value to the customer.

SAFE PRACTICES

Adequate safety precautions must be observed when working with adhesives. Corrosive materials, flammable liquids, and toxic substances may be present during the adhesive bonding process. Therefore, manufacturing operations should be carefully supervised to ensure that proper safety procedures, protective devices, and protective clothing are being used. All federal, state, and local regulations should be complied with, including OSHA Regulation 29CFR 1900.1000, *Air Contaminants*.[12] The material safety data sheet of the adhesive should be carefully examined before the adhesive is handled to ensure that the appropriate safety precautions will be followed.

FLAMMABLE MATERIALS

All flammable materials, including solvents, should be stored in tightly sealed drums and issued in suitably labeled safety cans to prevent fires during storage and use. Solvents and flammable liquids should not be used in poorly ventilated, confined places, but only in adequately ventilated areas. Usually, ventilation is considered to be adequate if the offending substance does not emit a recognizable odor. Safety lids should be provided when containers of solvents are transported in trays. Flames, sparks, or spark-producing equipment must not be permitted in the area where flammable materials are being handled. Fire extinguishers should be readily available.

The National Fire Protection Agency (NFPA) publishes standards applicable to adhesive bonding, such as

12. Occupational Safety and Health Administration (OSHA) *Air Contaminants, Occupational Safety and Health Standards for General Industry, in Code of Federal Regulations (CFR)* Title 29 CFR 1900.1000, Washington, D.C.: Superintendent of Documents, U.S. Government Printing Office.

Standard for Fire Prevention During Welding, Cutting, and Other Hot Work, NFPA 51:2003.26.[13]

TOXIC MATERIALS

Severe allergic reactions can result from direct contact, inhalation, or ingestion of phenolics, epoxies, and some urethanes, and also some catalysts and accelerators. The eyes or skin may become sensitized over a long period of time even though no signs of irritation are visible. Once a worker is sensitized to a particular type of adhesive, he or she may no longer be able to work near it because of allergic reactions. Careless handling of adhesives by production workers may expose others to toxic materials if proper safety rules are not observed. For example, co-workers may touch tools, doorknobs, light switches, or other objects contaminated by careless workers.

For the normal individual, proper handling methods that eliminate skin contact with the adhesive should be sufficient. It is mandatory that protective equipment or protective creams, or both, be used to avoid skin contact with certain types of formulations. Proper disposable gloves should be worn by anyone who is in contact with the adhesive in a production or testing environment.

A number of factors must be considered in determining the extent of precautionary measures to be taken. All of the following elements should be evaluated in terms of the individual operation:

1. Frequency and duration of exposure,
2. Degree of hazard associated with the specific adhesive,
3. Solvent or curing agent used,
4. Temperature at which the operations are performed, and
5. Potential evaporation surface area exposed at the workstation.

Documents such as *Threshold Limit Values (TLVs®) for Chemical Substances and Physical Agents in the Workroom Environment*,[14] and *Methods for Sampling Airborne Particulates Generated by Welding and Allied Processes*, AWS F1.1,[15] should be consulted. General information on industrial hygiene practices in the presence of hazardous materials or circumstances can be found in *Safety in Welding, Cutting, and Allied Processes*, ANSI Z49.1.[16]

PRECAUTIONARY PROCEDURES

Precautionary procedures are recommended in this section for the handling and use of adhesives and auxiliary materials. The work area should be equipped and supplied to accommodate safe practices.

Personal Protective Hygiene

Personnel should read and understand material safety data sheets supplied by the vendors of potentially hazardous materials. Workers should be instructed in the proper procedures to prevent skin contact with solvents, curing agents, and uncured base adhesives. Showers, washbowls, mild soaps, clean towels, refatting creams, and personal protective equipment should be provided.

Curing agents should be removed from the hands with soap and water. Resins should be removed with soap and water, alcohol, or a suitable solvent. Any solvent used to remove curing agents from the hands or skin should be used sparingly and should be followed by washing with soap and water. In case of allergic reaction or burning, prompt medical aid should be obtained.

Plastic or rubber gloves should be worn at all times when working with potentially toxic adhesives. When contaminated, the gloves must not contact objects that others may touch with their bare hands. Contaminated gloves should be discarded or cleaned using procedures that will remove the particular adhesive. Cleaning may require solvents, soap and water, or both. Hands, arms, face, and neck should be coated with a commercial barrier ointment or cream. This type of material may provide short-term protection and facilitate removal of adhesive components by washing.

Full-face shields should be used for eye protection whenever the possibility of splashing exists, otherwise glasses or goggles should be worn. If splashed, the eyes should be immediately flushed with water for 10 minutes to 15 minutes and then promptly treated by a physician.

Protective clothing should be worn at all times by those who work with the adhesives. Shop coats, aprons, or coveralls may be suitable and they should be cleaned before reuse.

13. National Fire Protection Association (NFPA), latest edition, *Standard for Fire Prevention During Welding, Cutting, and Other Hot Work*, NFPA 51:2003, Quincy, Massachusetts: National Fire Protection Association.
14. American Conference of Industrial Hygienists, (latest edition) *Threshold Limit Values (TLVs®) for Chemical Substances and Physical Agents in the Workroom Environment*, American Conference of Governmental Industrial Hygienists, Cincinnati, Ohio: American Conference of Industrial Hygienists.
15. American Welding Society (AWS) Committee on Fumes and Gases, *Methods for Sampling Airborne Particulates Generated by Welding and Allied Processes*, AWS F1.1, Miami: American Welding Society.
16. American National Standards Institute (ANSI) Accredited Standards Committee Z49, *Safety in Welding, Cutting, and Allied Processes*, ANSI Z49.1 (latest edition), Miami: American Welding Society.

WORK AREAS

Areas in which adhesives are handled should be separated from other operations. In addition to proper fire extinguishing equipment, these areas should contain the following facilities:

1. A sink with running water,
2. An eye shower or rinse fountain,
3. First aid kit, and
4. Ventilating facilities.

Fumes and Radiation

Ovens, presses, and other curing equipment should be individually vented to remove gases and vapors. Proper guarding devices should be used to prevent thermal burns, radio frequency radiation, or microwave field exposure. Vent hoods should be provided at mixing and application stations. Fume control is always a good practice when working with materials that give off odors or gases. Adhesives containers will clearly warn of the need for ventilation control if contents or byproducts. Documents such as *Threshold Limit Values (TLVs®) for Chemical Substances and Physical Agents in the Workroom Environment,* and *Methods for Sampling Airborne Particulates Generated by Welding and Allied Processes,* AWS F1.1, should be consulted.

Anodizing treatments and etching treatments are another source of concern. They may carry heavy toxicity and environmental loads into the work area. It is particularly important to reduce or eliminate the use of chromium-VI (Cr^{+6}) components, which are known carcinogens.[17]

Appendix B of this volume, "Safety and Health Codes and Other Standards" lists health and safety standards, codes, specifications, books and pamphlets. The publishers and the facts of publication also are listed.

CONCLUSION

Because of the continuing development of sophisticated adhesives and the widening variety of fabricating and engineering materials currently available, a basic knowledge and understanding of the capabilities of adhesive bonding have become essential tools for the welding engineer.

Adhesive bonding processes employ a heterogeneous material as a bonding agent between two faying surfaces. Unlike the microstructural interactions frequently provided by brazes and solders, adhesives interact mechanically with surface roughness and polarities. Adhesives typically are organic materials with lower modulus than the materials being joined.

Adhesives can be used to bond any combination of metals, plastics, ceramics, and glasses. Few adhesives are useable above 345°C (650°F) and maximum bonded shear strength is 48.3 MPa (7000 psi), depending on the materials involved. Most structural adhesives are used with service temperatures below 90°C (200°F) and a shear requirement of 17.2 MPa (2500 psi). Adhesives offer significant advantages in stress distribution compared to discrete welds or fasteners.

Adhesives require good surface preparation to ensure maximum bonding and durability. Metals preparations focus on providing a hydrolytically stable, adherent oxide layer. Plastics bonding preparations emphasize increasing surface polarity, usually through an oxidative treatment, to provide interaction with the adhesive polymer chains. Primers may be employed on any surface to improve bonding and durability.

The function of the adhesive is to join the two surfaces and to transfer stress through the joint area. Good joint design allows the adhesive to function below its maximum limits under all operating and environmental conditions. In practice, a significant design factor of safety (FOS) is employed, ranging from 1.1 to 4 or more, depending on the criticality of the structure and the likelihood for abuse in service. Adhesives often show poor peel performance, thus any good joint design will be chosen or modeled to minimize peel. Adhesive joints have been designed to withstand nuclear blast. Good modeling of good joint design with allowance for the factor of safety can produce joints that will provide decades of survivability.

Adhesives may be combined with other joining processes including welding, riveting, bolting, and screw connectors. Weld bonding processes are often used with resistance welding, laser welding, ultrasonic welding, and friction-stir welding. Weld-bonded structures provide synergistic joining because the weld and adhesive are intimately coupled in stress transfer. Increased shear (over welds), increased peel (over adhesives and welds), and increased fatigue life (over adhesives or welds) are known. Bolts and screws require clearance holes so the full effect of the mechanical fastener may not be realized until significant shear strain has been imparted on the joint. Hence, bolting or "chicken bolts" will play a role only when the adhesive has failed or is near failure. Rivets and self-piercing rivets provide more fastener

17. Occupational Safety and Health Administration (OSHA), latest edition. *Occupational Safety and Health Standards for General Industry* in *Title 29, Labor,* in *Code of Federal Regulations,*(CFR)1910.1026. Washington D.C.: Superintendent of Documents. U.S. Government Printing Office.

participation because of reduced mechanical play in fastener holes.

Advanced designs will employ innovative combinations of materials. It is to be expected that the welding engineer will work with or around adhesives in processing of structural components, thus adhesive bonding will become an important technology for the welding engineer. A good knowledge base and understanding of adhesive bonding is becoming an essential tool for the welding engineer.

BIBLIOGRAPHY

American Conference of Industrial Hygienists. (Latest edition). *Threshold Limit Values (TLVs®) for Chemical Substances and Physical Agents in the Workroom Environment.* Cincinnati, Ohio: American Conference of Industrial Hygienists.

American Society for Testing and Materials (ASTM International) Subcommittee D14.70. 2004. *Standard test method for measuring strength and shear modulus of nonrigid adhesives by the thick-adherend tensile-lap specimen.* ASTM D3983-98. West Conshohocken, Pennsylvania: ASTM International.

American Society for Testing and Materials (ASTM International). 2002. *Test method for apparent shear strength of single-lap-joint adhesively bonded metal specimens by tension loading (metal-to-metal) adhesive bonds.* ASTM D1002-01. West Conshohocken, Pennsylvania: ASTM International.

American Society for Testing and Materials (ASTM International). 2004. *Standard test method for climbing drum peel for adhesives.* ASTM D1781-98. West Conshohocken, Pennsylvania: ASTM International.

American Society for Testing and Materials (ASTM International). 1998. *Standard test method for adhesive-bonded surface durability of aluminum (wedge test).* ASTM D3762-98. West Conshohocken, Pennsylvania: ASTM International.

American Society for Testing and Materials (ASTM International) Subcommittee D14.80. 2004. *Standard recommended practice for preparation of metal surfaces for adhesive bonding*, ASTM D2651. West Conshohocken, Pennsylvania: ASTM International.

American Welding Society (AWS) Committee on Definitions and Symbols. 2001. *Standard welding terms and definitions.* Miami: American Welding Society.

American Welding Society (AWS) Welding Handbook Committee. 2001. Jenney, C. L. and A. O'Brien, eds. *Welding Handbook*, Vol. 1, 9th ed., *Welding Science and Technology* Miami: American Welding Society.

American Welding Society (AWS) Committee on Fumes and Gases. 2006. *Methods for Sampling Airborne Particulates Generated by Welding and Allied Processes*, AWS F1.1M:2006, Miami: American Welding Society.

National Fire Protection Agency (NFPA). *Standard for fire prevention during welding, cutting, and other hot work*, NFPA 51:2003.26.

Occupational Safety and Health Administration (OSHA). (Annual) Air contaminants: permissible exposure limits (PELs): *Occupational safety and health standards for general industry,* in *Code of Federal Regulations (CFR)* Title 29 CFR 1910.1000. Washington, D.C.: Superintendent of Documents, U.S. Government Printing Office.

Society of Automotive Engineers (SAE). Ams-P Polymeric and Composite Materials Committee. 1990. ARP No. 1524, *Surface preparation and priming of aluminum alloy parts for high durability structural adhesive bonding*, Warrendale, Pennsylvania: Society of Automotive Engineers.

SUPPLEMENTARY READING LIST

American Society for Materials. 1990. *ASM engineered materials handbook*, Vol. 3. *Adhesives and sealants.* Materials Park, Ohio: American Society for Materials.

Hartshorn, S. R., ed. 1986. *Structural adhesives: chemistry and technology.* New York: Plenum Press.

Katz, I. 1971. *Adhesive materials, their properties and usage.* Long Beach, CA: Foster Publishing Co.

Kinloch, A. J., ed. 1983. *Durability of structural adhesives.* New York: Applied Science Publishers.

Kinloch, A. J. 1986. *Structural adhesives: developments in resins and primers.* London: Applied Science Publishers.

Kinloch, A. J. 1987. *Adhesion and adhesives: science and technology.* London: Chapman and Hall.

Landrock, A. H. 1985. *Adhesives technology handbook.* Park Ridge, NJ: Noyes Publications.

Landrock, A. H. 1987. *Aluminum adhesive bond permanence: treatise on adhesion and adhesives*, Vol. 5. New York: Marcel Dekker. 45–137.

Petrie, E. M. 2000. *Handbook of adhesives and sealants.* New York: McGraw-Hill.

Pocius, A. V. 2002. *Adhesion and adhesives technology: an introduction*, 2nd Ed., Munich: Hanser Publishers.

Pocius. A. V., D. A. Dillard, and M. K. Chaudhury. 2002. *Adhesion science and engineering* (in two volumes). Amsterdam: Elsevier.

Ritter, G. W. 1999. Adhesives are penetrating deeper into industry worldwide. *Welding Journal* 78(5): 46–48.

Rogers, N. L. 1966. Surface preparation of metals for adhesive bonding. Applied Polymer Symposium No. 3. New York: Interscience Publishers. 327–340.

Shields, J. 1984. *Adhesives handbook*, 3rd ed. London: Butterworths.

Skeist, I., ed. 1976. *Handbook of adhesives*, 2nd ed. New York: Van Nostrand Reinhold.

Snogren, R. C. 1974. *Handbook of surface preparation*. New York: Palmerton Publishing Co.

Note: Additional resources are available from the Adhesive and Sealant Council, 7979 Old Georgetown Road, Bethesda, MD 20814, http://www.ascouncil.org.

CHAPTER 11

THERMAL SPRAYING AND COLD SPRAYING

Prepared by the Welding Handbook Chapter Committee on Thermal Spraying and Cold Spraying:

M. F. Smith, Chair
*Sandia National Laboratories**

C. C. Berndt
Stony Brook University

D. E. Crawmer
Thermal Spray Technologies, Inc.

A. M. Kay
ASB Industries

W. J. Lenling
Thermal Spray Technologies, Inc.

R. A. Miller
R. A. Miller Materials Engineering

C. Moreau
National Research Council of Canada

R. A. Neiser
Sandia National Laboratories

L. D. Russo
Stony Brook University

Welding Handbook Volume 3 Committee Member:

P. F. Zammit
Brooklyn Iron Works Inc.

Photograph courtesy of Sulzer Metco Corporation

**Sandia is a multiprogram laboratory operated by Sandia Corporation, a Lockheed Martin Company, for the United States Department of Energy's National Nuclear Security Administration under Contract DE-AC04-94AL85000.*

Contents

CHAPTER 11

THERMAL SPRAYING AND COLD SPRAYING

INTRODUCTION

Thermal spraying and cold spraying are two closely related but fundamentally different spray-deposition processes. Thermal spraying encompasses a group of processes in which finely divided metallic or nonmetallic materials are spray-deposited in a molten or semi-molten condition onto a substrate to form a coating or to build up a free-standing shape on a mandrel.[1,2] Cold spraying is a comparatively recent and emerging process that is often included in the family of thermal spray processes because it also involves spraying finely divided materials onto a substrate to form a coating or free-standing shape. However, unlike traditional thermal spray processes, cold spray particles are not heated to a molten or semi-molten state, but arrive at the target surface in solid state, often at or near room temperature. The fundamental differences in bonding mechanisms and process characteristics related to the use of molten or semi-molten spray particles in thermal spraying versus solid, relatively low-temperature spray particles in cold spraying are discussed in this chapter.

Major commercial thermal spray processes include electric arc spraying, atmospheric plasma spraying (APS), vacuum plasma spraying (VPS), flame spraying (flame spray), detonation flame spraying, and high-velocity oxyfuel spraying (HVOF). The photograph on the title page of this chapter shows a rotary plasma spray gun as it is translated up and down through the cylinder bores of an aluminum automobile engine to quickly apply a wear-resistant coating to the cylinder walls.

Because traditional thermal spray processes are more widely used and have greater commercial importance, the majority of this chapter is devoted to the traditional thermal spray technologies. However, cold spraying offers important advantages for some applications, and the commercial use of this technology is increasing.

This highly versatile group of spray processes can rapidly deposit an exceptionally wide range of metals, ceramics, glasses, polymers, and composites on many different substrate materials. Feedstock for thermal spray processes may be in the form of powder, rod, wire, or cord. A few examples of the diverse variety of commercial applications of thermal spraying include anti-corrosion and anti-skid coatings on ships and bridges; laser-engraved ceramic coatings for analox rolls used in the printing industry; and wear-resistant coatings and thermal barriers used in industrial and aerospace turbine engine applications. Thermal spraying also is used in biomedical applications, such as titanium and hydroxy-apatite (artificial bone) coatings on human joint and dental implants.

Although there are many sophisticated applications, sprayed coatings applied to enhance or extend performance in severe thermal, wear-intensive, or corrosive environments remain the mainstay of the thermal spray industry. The thickness of the deposited layer for applications typically is in the range of 0.125 millimeter (mm) to 1.0 mm (0.005 inch [in.] to 0.040 in.). Much thicker spray deposits more than 25 mm (1 in.) can be produced with some spray materials, and thermal spraying sometimes is used to make freestanding shapes. For example, injection molding dies can be produced by spraying onto removable patterns or mandrels.

1. Welding terms and definitions used throughout this chapter are from *Standard Welding Terms and Definitions*, AWS A3.0:2001, Miami: American Welding Society.

2. At the time of the preparation of this chapter, referenced codes and other standards were valid. If a code or other standard is cited without a date of publication, it is understood that the latest edition of the document referred to applies. If a code or other standard is cited with the date of publication, the citation refers to that edition only, and it is understood that any future revisions or amendments to the code or standard are not included; however, as codes and standards undergo frequent revision, the reader is encouraged to consult the most recent edition.

Cold spraying currently (~2006) is limited to depositing ductile metals and a few metal-ceramic composites. Developed in Russia in the late 1980s, the earliest commercial applications exploited some of the special advantages of cold spraying for uses such as high-thermal-conductivity copper coatings for computer chip heat sink assemblies. The cold spray process often can produce much thicker coatings than thermal spray processes due to a more favorable compressive residual stress state inherent in cold-sprayed deposits. For example, copper has been successfully cold-sprayed to thicknesses of more than 10 centimeters (cm) (4 in.), with possible commercial use in the cold-spray forming of large rocket nozzles.

Highly sophisticated thermal spray applications are modern, but the process is not new. The first documented thermal spray patent in the English language was recorded by Max Ulrich Schoop, a Swiss scientist, in 1911. Patent literature shows that the fundamental principles of modern thermal spray processes were well established by the early 1920s. During this period and through the 1950s, most applications were concerned with reclamation of industrial parts, such as worn shafts. Exponential advances in technology since then have dramatically expanded the number and diversity of thermal spray processes, materials, and applications.

The cold spray process (originally called the *cold-gas dynamic spray method*) was developed in the mid-1980s by a group of Russian scientists researching gas dynamics in Siberia. Although this process was not widely known outside Russia until the early-to-mid 1990s, cold spraying continues to be studied in research laboratories around the world and evaluated for new commercial applications.

This chapter provides an overview of the thermal spray and cold spray processes and describes the equipment, materials, and techniques used for both. Sections on applications, post-treatments, quality control, economics, and safe practices are included. Sources of additional information are listed in the Supplementary Reading List at the end of the chapter.

FUNDAMENTALS

In a generic thermal spray process, electrical or chemical energy is used to create small molten or semi-molten droplets (typically 10 microns [μm] to 100 μm in diameter) from powder, wire, rod, or cord feedstock. As illustrated in Figure 11.1, the droplets are propelled

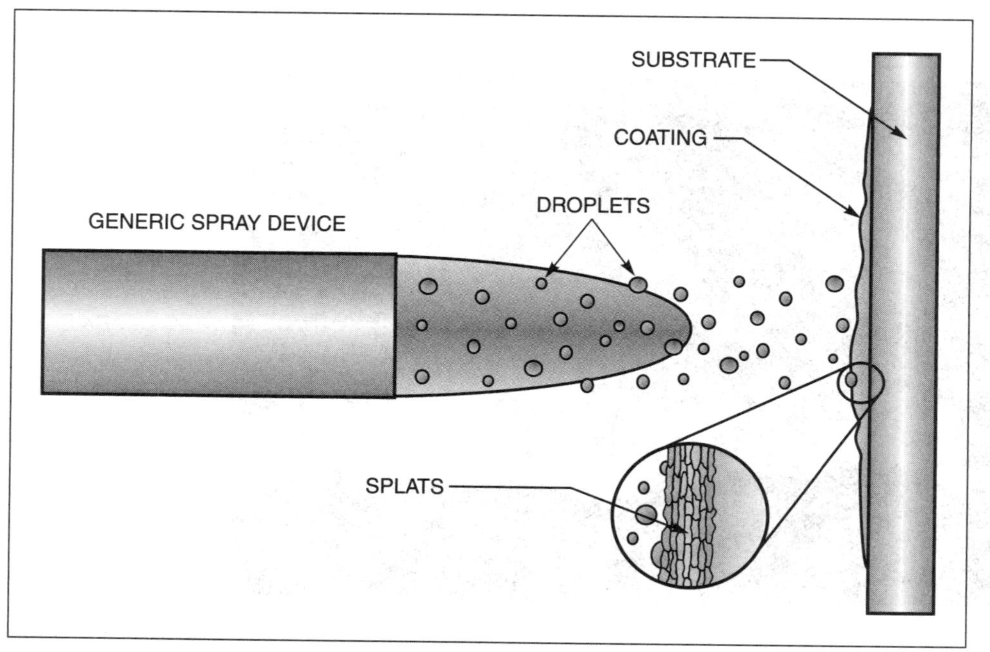

Source: Adapted from Sandia National Laboratories.

Figure 11.1—Generic Thermal Spray Device with Close-Up View of Spray Deposit

onto a workpiece surface (the substrate) by a subsonic or supersonic stream of gas. On impact, the droplets spread out and quickly solidify, cooling at rates ranging from 10^4 degrees per second to 10^8 degrees per second. The solidified droplets, called *splats*, randomly stack up on one another (much like randomly stacking pancakes or playing cards) to form the layered or lamellar microstructure that is characteristic of most thermal-spray-deposited materials. It should be noted that technically, only the first layer lands on the substrate; subsequently deposited material actually lands on previously deposited material, not on the original substrate surface.

The quality of the interfaces between the solidified splats, called *splat boundaries,* strongly influences the physical and mechanical properties of the spray-deposited material, much like grain boundaries influence the properties of wrought or cast materials. Deposition rates vary widely with different processes, materials, and applications, and typically are in the range of 2 kilograms per hour (kg/hr) to 20 kg/hr (4.4 lb/hr to 44 lb/hr). Figure 11.2 shows a composite wear coating of cemented carbide-nickel chromium plasma-sprayed onto a steel substrate.

As previously noted, cold spraying is a somewhat unusual process compared to other spray technologies: the spray particles are not melted, but rather are applied in the solid state at relatively low temperatures, normally well below the softening point of the feedstock material. Cold spraying can deposit ductile metal particles (typically 10 μm to 30 μm in diameter), which are accelerated to velocities approximately 500 meters per second (m/s) to 1200 m/s (1600 feet per second [ft/s] to 4000 ft/s) in a supersonic gas jet. The gas jet is created by passing compressed gas through a converging-diverging (DeLaval) nozzle. On impact with the underlying material, which may be the original surface or previously deposited coating, the solid particles plastically deform at very high rates, creating a hydrodynamic instability at the expanding interface between the spreading particle and the underlying material. Within the narrow region of this shear instability, solid material flows almost like a fluid, disrupting and sweeping away surface impurities. The nascent metal surfaces are forced together at high local pressures, thus creating a bond along the interface. This proposed model for the cold-spray bonding process is similar to explosion welding, but occurs on a micro scale.

Not all spray-deposited coatings of a particular material will result in identical performance. Differences in properties, performance, and adhesion to the substrate are influenced by many processing variables, such as the spray process, gas composition and flow rates, rate of deposition, coating thickness, substrate preparation, substrate geometry, substrate temperature, feedstock material, and gun-to-substrate (standoff) distance. The process and feedstock must be carefully optimized for each application.

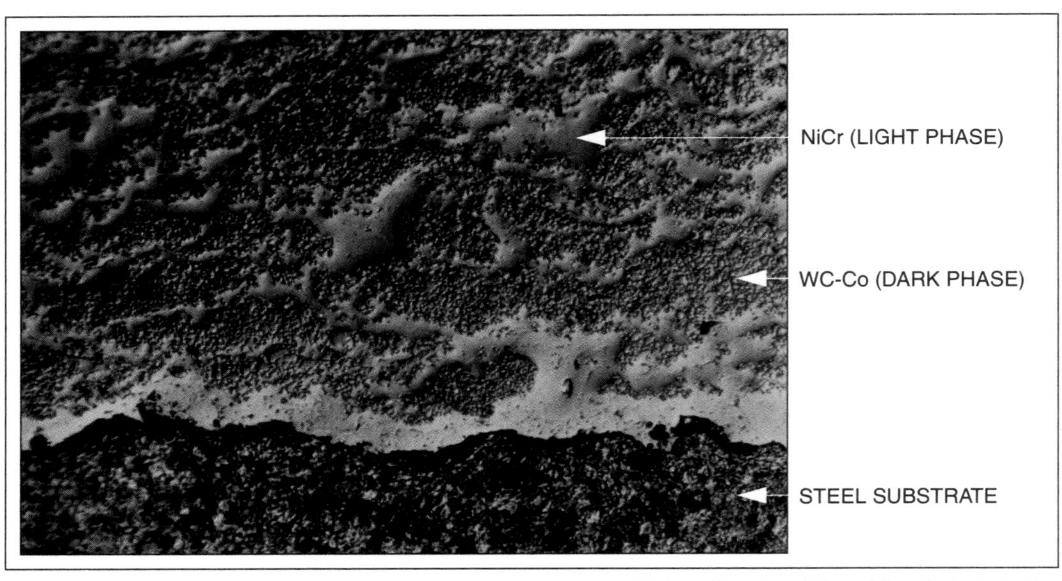

Photograph courtesy of Thermal Spray Technologies, Inc.

Figure 11.2—Metallographic Cross Section of the Lamellar Microstructure of a Typical Thermal-Sprayed Coating

Thermal spraying commonly is carried out in an ambient air environment, and a major advantage is the ability to spray-coat very large items *in situ*. However, inert atmosphere and vacuum-chamber processes can be used for special applications, such as applying premium-quality coatings on human dental or joint implants, and coatings on turbine blades for jet engines. These controlled-atmosphere processes, although considerably more expensive than spray-depositing in ambient air, can produce coatings of exceptionally high quality.

PROCESS VARIABLES

The primary factors that distinguish the various spray processes are the energy source (chemical or electrical), the temperature range of the process, and the velocity range of the spray droplets or particles. Figure 11.3 summarizes the approximate operating ranges of temperature and velocity for commonly used commercial thermal spray and cold spray processes. It should be noted that charts similar to Figure 11.3 have appeared in various forms in past years, and a comparison of different versions of the chart would reveal some significant differences. One reason for this inconsistency is that the temperature and velocity ranges can be defined in many different ways. For example, the gas jet temperature for a given process varies greatly at different locations within the jet. In arc spraying, the temperatures in the core of the electric arc may be several thousand degrees, while much of the rest of the jet is mainly composed of compressed air that may be near room temperature or perhaps even below, due to the expansion of the compressed gas. Similarly, gases in the high-pressure combustion chamber of a high-velocity oxyfuel system may approach the adiabatic (ideal maximum) flame temperature, but the gases cool significantly as they expand and accelerate to supersonic velocities while converting thermal energy into kinetic energy. Consequently, the chart showing the gas temperatures in Figure 11.3 should be considered only as

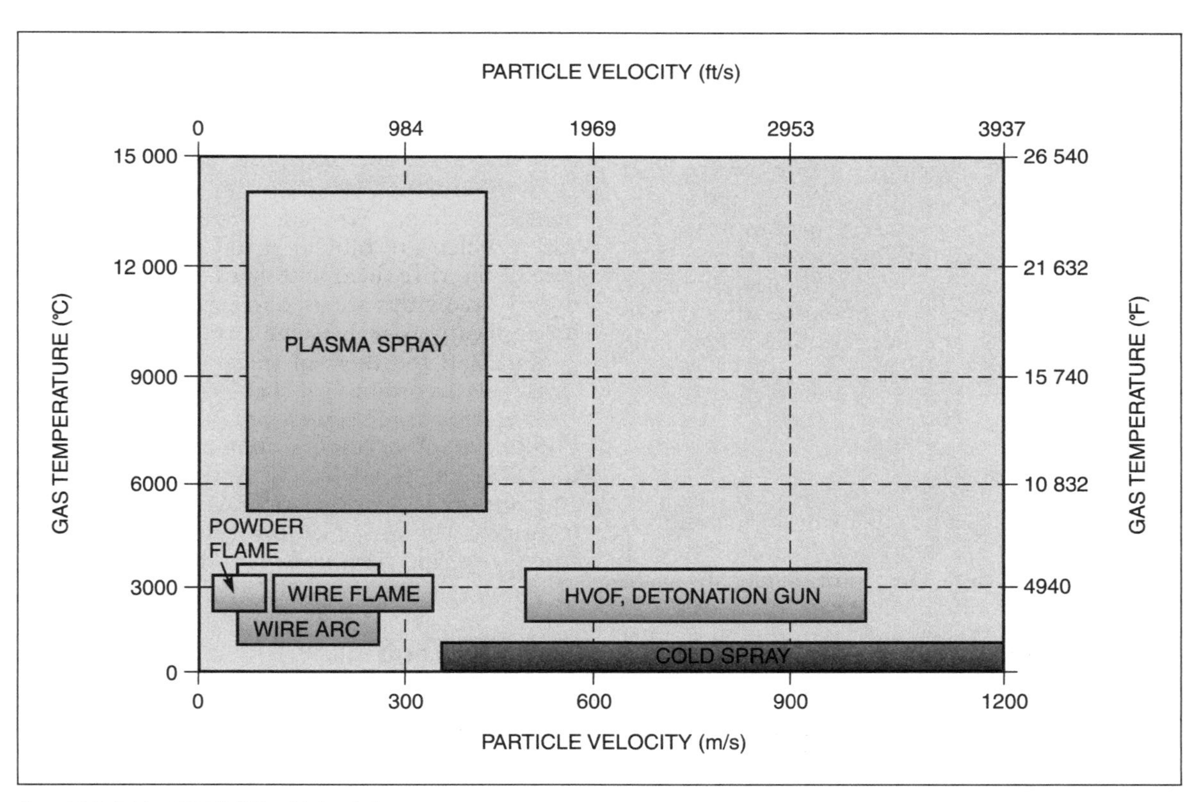

Source: Adapted from Sandia National Laboratories.

Figure 11.3—Approximate Temperature and Velocity Fields for Several Common Thermal Spray and Cold Spray Processes

an approximate guide to the range of gas temperatures that may be associated with a given process. The range of droplet or particle velocity shown in Figure 11.3 is somewhat easier to define, but even in this instance there are many variables, and the chart provides only a generalized comparison.

Despite these caveats, Figure 11.3 can be helpful as a preliminary screening tool to select candidate spray processes for new applications. As an example, if the application involves spraying a ceramic powder that has a melting point above approximately 2000° Celsius (°C) (3632° Fahrenheit [°F]), oxyfuel-based processes (for example, flame spraying, high-velocity oxyfuel spraying, and detonation gun spraying) may not provide sufficient heating to ensure complete melting. Although temperatures in the hottest regions of these devices theoretically may be sufficient to melt the powder, the time available to heat and melt the particle is brief, and depending on the specific flight path, not all particles may be exposed to the maximum temperatures in the spray gun. However, Figure 11.3 shows that plasma spraying produces much higher average jet temperatures, and therefore, is more likely to provide good melting.

In some applications, the higher velocities of HVOF, detonation gun, or cold spraying may be preferred because higher particle velocities generally tend to produce coatings with less porosity and less oxide. More favorable residual stress states also may be imparted to the coating due to the peening effect of high-velocity particles impacting the previously deposited material. There are some exceptions to these general tendencies; for example, even though particle velocities in vacuum plasma spray are significantly lower than those of the high-velocity processes, VPS can produce extremely dense coatings with little or no oxide because deposition is carried out at low ambient pressures inside a vacuum chamber.

FEEDSTOCK

Feedstock material (the metals or other materials to be spray-deposited) is another critical process variable in all spray deposition processes. In principle, any material that has a stable molten phase can be deposited by thermal spraying, and with special techniques, even materials that do not melt can be co-deposited simultaneously with sprayable materials to form a composite. Cold spraying requires that the impacting particles deform in a ductile manner on impact, so this process generally is restricted to ductile metals. However, like thermal spraying, some metal-ceramic composite coatings that incorporate non-ductile ceramic particles have been successfully produced with cold spraying by co-depositing a mixture of metal and ceramic particles or by using special composite powders in which each powder particle is a mechanical alloy of metal and ceramic.

Powder and wire are by far the most common forms of feedstock materials for thermal spraying, but ceramic rods are used with some devices; plastic cord filled with powder also is used for a few applications in Europe. Powder feedstock is used in all plasma spraying, high-velocity oxyfuel spraying, detonation gun spraying, powder flame spraying, and cold spraying. The chemistry of the feedstock material, the method of powder manufacture (for example, fused and crushed, spray-dried, or gas atomized), the average powder particle size, and size distribution of the powder particles all strongly influence the efficiency of the spray process, and also influence the quality and properties of the resulting deposited coating.

More than one composition of powder can be sprayed simultaneously to produce a composite coating or a graded coating. A graded coating is defined as a coating that has a systematically varied composition during the deposition procedure. For example, when there is a severe thermal expansion mismatch between a metal substrate and a ceramic coating, stress sometimes can be reduced by initially spraying a pure metal and then grading through a series of ceramic/metal composite mixtures, culminating in pure ceramic. This is accomplished by feeding only metal powder initially, then as the coating builds, progressively increasing the amount of the ceramic powder that is fed into the spray gun; the metal-powder feed rate gradually is reduced to zero, and at the end only pure ceramic is fed.

Wire feedstock is used in arc spray guns and wire flame spray guns. Wire spray devices also can use cored wires, which are tubular metal wires with the hollow core of the wire filled with hard particles or other additives. Cored wires are used to spray composite materials for applications such as enhanced wear resistance.

Wire and rod thermal spray feedstock materials are classified according to chemical composition in the *Specification for Thermal Spray Feedstock—Solid and Composite Wire and Ceramic Rods,* AWS C2.25/C2.25M, published by the American Welding Society and approved by the American National Standards Institute.[3] The classification system for feedstock wire, cored wire, and ceramic rods is shown in Figure 11.4.

SUBSTRATES

The substrates on which spray-deposited coatings can be applied include nearly all metals, many ceramics, most plastics, glass, wood, and in some instances, paper.

3. American Welding Society (AWS) Committee on Thermal Spray, 2002, *Specification for Thermal Spray Feedstock—Solid and Composite Wire and Ceramic Rods,* AWS C2.25/C2.25M:2002, Miami, American Welding Society.

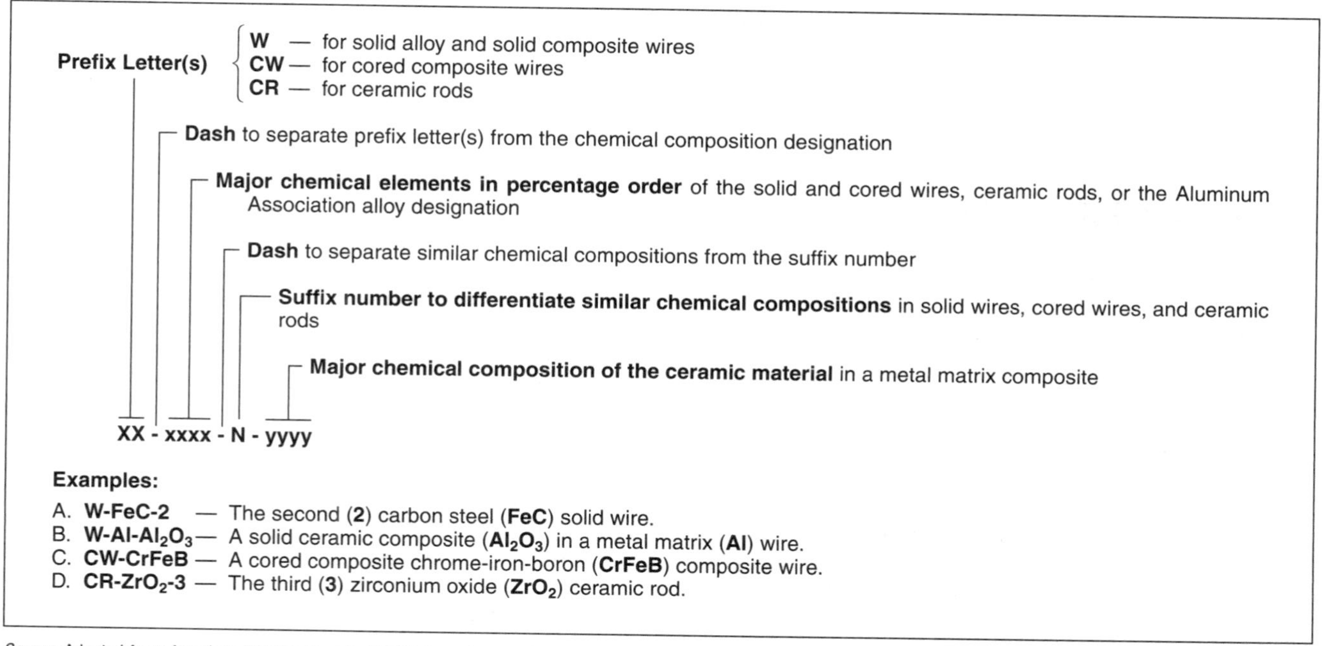

Source: Adapted from American Welding Society (AWS) Committee on Thermal Spray, 2002, *Specification for Thermal Spray Feedstock—Solid and Composite Wire and Ceramic Rods*, AWS C2.25/C2.25M:2002, Miami: American Welding Society, Figure 1.

Figure 11.4—Classification System for Wire, Cored Wire, and Ceramic Rod Feedstock

Not all spray materials can be applied to all substrates, however, the range of possible combinations probably is greater for thermal spraying than for any other surfacing technology. The selection of the appropriate spray process, feedstock material, substrate, surface preparation, and spraying parameters increases the probability of good adhesion of the coating to the substrate. The thermal properties of the substrate, such as conductivity and temperature before and during deposition, often are important because they affect the wettability of the splat and the tendency to build up residual stress.

It also is important to note that thermal spraying is a "line-of-sight" process. This means that if one were to view the substrate from the vantage point of the spray gun, only the surfaces that can be seen from that vantage point can be coated.

SPRAY TECHNIQUES

To achieve an optimized coating structure, the angle of the spray gun relative to the substrate generally should be kept as close to 90° as possible. Off-normal spraying generally will decrease deposition efficiency (that is, the percentage of the material sprayed that actually adheres to the substrate) and also may produce other undesirable effects in the microstructure, such as increased porosity. To achieve uniform deposition on large components, the spray gun or the substrate, or both, must be moved relative to one another in a suitable pattern. Areas that are not to be coated can be masked with appropriate single-use or reusable masking materials, similar to the masking tape used in spray painting. Commonly used thermal spray masking materials include special high-temperature masking tapes, rubber or plastic masks, and reusable metal masks.

If possible, sharp inside or outside corners on the substrate should be rounded before spray-coating, because sharp corners may increase local stresses in the coating, and thus may cause the coating to crack or spall. Sharp corners also may create local turbulence during deposition that disrupts the microstructure and properties of the deposited material in that area.

Since most thermal spray coatings tend to end up in a state of residual tensile stress, the overall shape of the surface that is to be spray coated can also affect coating adhesion. As a general rule, it is more difficult to achieve good adhesion to a concave surface than to a convex surface because the sprayed coating tends to shrink as is solidifies and cools. Good adhesion to the interior surface of a pipe or tube usually is more challenging than good adhesion to the exterior of the same pipe or tube.

SUBSTRATE PREPARATION

Proper preparation of the substrate is critical to good adhesion of the coating. Substrate preparation normally consists of an initial chemical cleaning operation followed by a mechanical surface roughening operation, such as grit blasting, and then a final cleaning. Cleaning methods such as a solvent rinse, vapor degreasing, or chemical etching are used to eliminate contamination (for example, dirt, oil, or grease) that may inhibit wetting and bonding of the deposits on the substrate.

Surface roughening is most commonly achieved by grit blasting. Grit blasting promotes better adhesion of the coating in at least two important ways. First, it removes surface impurities, such as oxide scale, that could interfere with the bonding of the coating to the substrate. Second, it creates a rough surface on the substrate, which significantly alters the stress distribution along the coating-substrate interface. Residual stresses are no longer highly concentrated along a large planar interface, but instead, are broadly distributed across short segments at different angles along a tortuous interface path.

Aluminum oxide is the most common grit media, and the grit particle size, geometry, angle, and impact velocity affect the resulting substrate surface profile. The grit should be kept clean and free of contaminants. The edges of the grit particles should remain sharp, so it is advisable to use only fresh blasting media (no recycled media) or, for less critical applications, at least to monitor blasting media quality and replace it as necessary.

After grit blasting, the substrate sometimes is wire-brushed or treated with a chemical etchant to remove residual grit that may become embedded in the workpiece surface during the blasting operation. Embedded grit is problematic because it can locally reduce adhesion of the coating to the substrate.

Mechanical roughening by special machining operations is an alternative to grit blasting to roughen a workpiece surface. Machining techniques are most commonly used with cylindrically shaped workpieces that can be mounted in a lathe for the roughening operation. With some softer metals such as aluminum, high-pressure water blasting, that is, water jet pressures exceeding 276 megapascals (MPa) (40,000 pounds per square inch [psi]), also has proven to be a very effective means to prepare a substrate surface prior to thermal spray depositing. In the right situations, these alternative surface preparation methods can provide excellent adhesion while avoiding concerns about embedded grit.

The precise nature of the bond formed between a sprayed coating and the substrate has been a subject of considerable debate, and it probably varies somewhat from one specific situation to another. Although it is commonly stated that mechanical interlocking is an important element of the bonding process, research on bonding mechanisms indicates that other mechanisms, such as metallic and chemical bonding, van der Waals forces,[4] and diffusion are primarily responsible for adhesion of the coating to the substrate. For example, it has been observed that aluminum metal can be cold-sprayed directly onto a perfectly smooth plate glass surface. The aluminum will adhere so tightly to the smooth glass that if sufficient force is applied to pull the aluminum coating off of the glass, the glass surface will fracture and fragments of glass will be found adhering to the aluminum coating. This example clearly illustrates that the adhesion of the aluminum to the very smooth glass surface involves something more than simple mechanical interlocking.

For some applications, the adhesion, or bond strength, of a sprayed material to a substrate can be significantly improved by first depositing an appropriately selected intermediary or prime coating, called a *bond coat*. Common bond coating materials are molybdenum, nickel-chrome alloys, or nickel-aluminum alloys, all of which provide excellent adhesion to a wide range of substrate materials. Bond coatings also can help mitigate stresses resulting from thermal expansion mismatch between the substrate and the coating, especially when a ceramic coating is applied to a metal substrate. When necessary, bond coat materials often can be built up to thicknesses of 1 mm (0.040 in.) or more. For example, a thick coating might be required to meet dimensional requirements when restoring a worn shaft, and it might be desirable to use a cemented-carbide coating to provide a more wear resistant surface. However, if a cemented-carbide coating is too thick, high residual stress may cause the coating to crack or spall. In this case, a bond coating material could be used to build up most of the required coating thickness before a cemented-carbide top coating is applied.

MICROSTRUCTURE OF SPRAY DEPOSITS

As previously noted, thermal spray coatings have a lamellar microstructure formed as incoming splats flatten out and pile up on one another (refer to Figure 11.2). Although the diameter and thickness of splats varies greatly according to the feedstock material and spray process, typical splats will have diameters on the order of 50 µm to 250 µm (0.002 to 0.01 in.) with typical thicknesses of roughly 2 µm to 20 µm (0.0001 to 0.001 in.).

4. Johannes Diderik van der Waals (1837–1923), Dutch physicist.

In most cases, when a molten spray droplet impacts the workpiece, it will flow out and completely solidify before the next molten droplet strikes that specific location. For this reason, although thermal spraying sometimes is described as "spray-painting" with molten metals or ceramics, this analogy is flawed. In spray painting, impacting paint droplets typically flow together to form a continuous liquid surface film. However, a continuous molten surface layer normally does not form as the thermal spray deposit is built up.

The physical properties of a sprayed material primarily are determined by the degree of bonding between individual splats, much as the properties of many wrought or cast materials are strongly influenced by grain boundaries. Splat boundaries can be strongly affected by impurities that may form between splats, such as oxides and pores (voids). Since many coatings are sprayed in an ambient air environment, metals or other reactive materials commonly will form a small percentage of oxide due to interaction with atmospheric oxygen. Cold spraying is an exception to this generalization, since the temperature of the incoming spray particles is so low that minimal or no reaction occurs with the ambient atmosphere with most materials. Oxidation can also be reduced or eliminated by spraying in a controlled atmosphere, as in vacuum plasma spraying.

It must be emphasized that most of the properties of a given material deposited by thermal spraying differ from the properties of the same material in a wrought or cast form. This is because the splat boundaries, not the grain boundaries, are the primary influence on mechanical, thermal, electrical, and other bulk properties of the coating material. As might be expected with the highly oriented structure of the splat boundaries, the properties of most sprayed coatings also tend to be highly anisotropic in the as-sprayed condition (that is, tensile strength, conductivity, and other variables measured parallel to the coating surface may differ greatly from the same properties measured perpendicular to the coating surface). Because the splat boundaries strongly influence coating properties, imperfections, such as pores and oxide, along the splat boundaries have a significant effect on properties.

The porosity for most thermal spray deposits typically falls in the range of 0.5% to 10%, depending on the process, spray parameters, and feedstock. As a general rule, low-velocity processes, such as arc spraying or flame spraying, tend to produce coatings with more porosity than high-velocity processes, such as HVOF and cold spraying. In some cases, porosity is desirable for coating function, for example, to reduce the thermal conductivity of ceramic thermal barrier coatings, and it can be intentionally pushed to even higher levels ranging up to approximately 20% in some cases.

Pores in a sprayed material often result from incomplete flow or splashing of the incoming material, and also from incomplete wetting due to local oxides or other surface imperfections. Hence, most of the porosity in a coating typically is concentrated between splats and along the splat boundaries. The size of pores within a coating can vary widely, but the pores usually are small in comparison to the size of the splats within the coating.

It is often difficult to accurately measure the porosity of a sprayed coating. For example, normal metallographic cutting and polishing processes may artificially increase or decrease the apparent porosity in a polished coating cross section specimen due to mechanisms such as pull-out of poorly adhered splat fragments or smearing of softer materials, such as aluminum. Similarly, pyncnometry may not yield accurate results, because some of the pores in the coating may be totally isolated from the surface. There is no universally best way to measure coating porosity in all situations, and testing experts should be consulted if an accurate measurement is needed for a specific situation.

In some applications, a sealant can be used to mitigate the effects of porosity; for example, a porous sprayed coating sometimes can be successfully used in a corrosive environment simply by applying an epoxy or other sealant to inhibit migration through the pores to the substrate-coating interface. A wide range of commercial sealants is available.

With some processes, such as high-velocity oxyfuel spraying, detonation gun spraying, vacuum plasma spraying, inert-atmosphere plasma spraying, and cold spraying, splat boundaries tend to be inherently cleaner with less oxide and porosity. In this case, the resulting properties are more uniform (isotropic) and the properties of the sprayed material are more similar to cast or wrought materials. Post-deposition thermal treatments also can be used to enhance properties. For example, a study showed that a post-spray anneal of cold-sprayed pure aluminum produced mechanical properties comparable to wrought 1100 aluminum in an H14 condition. A variety of post-spray treatments may be used, including heat treatment, flame fusing, laser glazing, and hot isostatic pressing.

ADVANTAGES

Compared to alternative deposition processes, such as electroplating, chemical vapor deposition (CVD), or physical vapor deposition (PVD), thermal spraying has the potential to provide several major advantages for appropriate applications, including the following:

1. The ability to deposit an extremely wide range of materials (in principle, anything that melts) onto an even wider range of substrate;

2. High deposition rates, ranging from roughly 2 kg/hr to 20 kg/hr (4.4 lb/hr to 44 lb/hr), and the ability to create thick deposits with maximum thicknesses of approximately 0.3 mm to 30 mm (0.01 in. to 1.2 in.), depending on the material;
3. The ability to produce customized microstructures and properties, such as strain-tolerant ceramics, and mixed, layered, or functionally graded composites; and
4. In some applications, thermal spraying is less expensive or less harmful to the environment, or both, than alternative surfacing technologies.

Essentially, all metals ranging from beryllium to uranium are process-compatible materials that can be successfully deposited by thermal spraying. Because cold spraying is a solid-state process, it is largely restricted to more ductile pure metals and alloys, but the variety of feedstock materials is still extensive. It is noteworthy that many simple and complex metal alloys are easily spray-deposited, unlike several competing technologies. For example, electroplating and chemical vapor deposition cannot deposit common engineering alloys, such as stainless steel or Inconel®.

Many ceramic materials with stable molten phases are routinely deposited with thermal spraying. In general, oxides are the most commonly sprayed pure ceramic coatings, because many carbides, nitrides, and borides tend to dissociate rather than melt when they are heated. However, a large volume of carbide is sprayed in the form of metal-carbide composite coatings, such as tungsten-carbide-cobalt (WC-Co) or chromium-carbide-nickel-chromium (Cr_3C_2-NiCr). Carbide coatings are commonly used for wear-resistance applications. The feedstock is a special composite powder that consists of hard carbide particles suspended in a metal matrix. During deposition, the particles are heated only enough to soften or partially melt the matrix metal, with the entrained carbide particles remaining solid. High-velocity processes, such as HVOF and detonation flame spraying, are preferred for high-quality cemented carbide coatings.

Some unusual mixtures of materials are commercially sprayed for special applications, such as nickel-graphite or aluminum-polyester composite coatings used for abradable gas path seals that help protect blade tips in the compressor sections of turbine engines. Many other materials that melt, such as glass and nylon, also have been successfully sprayed in commercial and research applications.

The variety of substrate materials suitable for thermal spraying is almost unlimited. While metal substrates are the most common, coatings have been successfully deposited on a wide array of solid materials, including ceramics, glass, plastic, reinforced composites, concrete, and even wood. Not all coating material and substrate combinations are compatible, but most are possible. Some spray processes are preferred over others for specific combinations, but numerous coating-substrate combinations are possible.

It is important to note that materials with relatively high melting points often can be deposited onto substrates that normally would not tolerate the temperatures necessary to melt the coating material. Highly dissimilar metals can be combined. For example, Figure 11.5 shows tungsten, with a melting point of 3387°C (6129°F), being plasma-sprayed onto aluminum, which has a melting point of only 660°C (1221°F), without melting the aluminum substrate. This is possible because the thermal energy contained in an individual molten tungsten spray droplet is limited and can be rapidly dissipated in the highly conductive aluminum substrate.

Material deposition rates for thermal spraying typically are several orders of magnitude higher than most alternative coating processes, such as electroplating or vapor deposition. This advantage may be partially offset in applications where competing processes can be adapted to simultaneous batch processing of large numbers of workpieces. However, thermal spraying usually affords significant processing speed advantages, and this advantage increases as coating thickness increases.

Photograph courtesy of Sandia National Laboratories

Figure 11.5—Tungsten Being Plasma-Sprayed Onto Aluminum

Another potential advantage of the high deposition rates inherent in most thermal spray processes is the capability to rapidly and economically deposit thick coatings or bulk deposits. Maximum thickness varies greatly, depending on factors such as the coating and substrate materials, the spray process, as well as the workpiece geometry and surface preparation. Most materials can be spray-deposited to a thickness of at least 0.3 mm (0.01 in.), and some materials can be sprayed to almost unlimited thickness exceeding 30 mm (1.2 in.).

The primary factor that limits ultimate thickness in thermal spraying is residual stress. After a molten droplet solidifies, it contracts as it cools. As more layers of material are added, the resulting thermal stress progressively increases, and it may eventually become high enough to cause the coating to crack or separate from the substrate. As a general guide, highly ductile metals with comparatively low melting points tend to have less residual stress, permitting greater coating thickness. In addition, greater coating thickness often can be built up using a high-velocity, relatively low-temperature deposition process, such as high-velocity oxyfuel spraying, detonation flame spraying, or cold spraying. The high-impact velocities inherent in these processes tend to produce an effect much like shot peening, which introduces compressive stress that can reduce residual tensile stress, and in some cases it may even produce residual compressive stress, thus permitting thicker coatings.

High deposition rates provide the capability of applying coatings rapidly and economically to extremely large surface areas. For example, aluminum- or zinc-based coatings are widely used to provide galvanic corrosion protection on large structures, such as bridges, ship decks, and high-capacity storage or processing tanks. In such systems, the aluminum- or zinc-based coating preferentially oxidizes (corrodes), sacrificially protecting the underlying steel. Because the protection is conferred by sacrificial oxidation of the coating material, the service life of galvanic protection increases in proportion to the thickness of the sacrificial coating. Thus, steel that is thermal-spray coated with a thick galvanic coating will far outlast steel coated by a process such as hot-dip galvanizing, because galvanizing provides a substantially thinner coating of sacrificial material.

The microstructure and properties of sprayed materials can be purposely varied over an extremely wide range, affording the materials engineer valuable opportunities to create engineered materials and surfaces for specific applications. Two or more similar or highly dissimilar materials can be co-deposited, layered, or graded to meet specific performance objectives. The high cooling rates of thermal spraying can be used to create amorphous materials or coatings that contain metastable phases. Also, the unique microstructures of sprayed materials often can be used to advantage. For example, fine-scale porosity in a ceramic coating can be used to limit crack propagation and thus create strain-tolerant ceramic coatings that will withstand significant amounts of strain without forming large cracks or spalling off the substrate.

The cost effectiveness of thermal spraying is another potential advantage. Costs vary considerably depending on the process and the materials to be deposited, but thermal spraying can afford significant savings over alternative technologies in appropriate applications.

With regard to the environment, in most instances the dusty overspray from thermal spraying is comparatively easy to collect and control using conventional methods, such as respirators equipped with appropriate filters and local exhaust vents connected to an approved dust filtration system. Collected overspray from some commercial operations can be reprocessed and recycled. Even if it is not recycled, the environmentally responsible disposal of solid dry powder from a thermal spray operation often is easier and less costly than large volumes of toxic liquid or gaseous effluents produced by some of the alternative coating processes.

LIMITATIONS

Thermal spraying or cold spraying may not be the optimum solution for all coating applications. Some potential limitations that should be considered when evaluating spray deposition technology for a given application include the following:

1. The application must allow line-of-sight deposition;
2. Only materials with a stable molten phase can be deposited by thermal spraying, and cold spraying can deposit only ductile metals or ceramic-ductile metal composites;
3. Oxides, porosity, or both may occur in sprayed deposits;
4. Physical and mechanical properties differ from wrought or cast materials; and
5. Residual stress may build up in the coating during deposition.

Because spray coating is a line-of-sight process, there are geometric limitations to the locations where a spray coating can be applied. For example, it is not possible to spray a coating on the interior surface of small-diameter tubes or holes due to physical access limitations. A tube or circular hole with an inside dimension (ID) of approximately 8 cm (3 in.) is the smallest inside dimension that can be spray-coated with commercially available spray guns specially designed for ID coating. In a few cases, somewhat smaller IDs have been successfully coated using custom-built miniature spray guns.

Thermal spraying generally is limited to applying materials that have a stable molten phase; that is, they melt rather than dissociate as they are heated. For example, many carbide, boride, and nitride ceramics dissociate rather than melt when heated. The exception to this general rule is that a material that does not melt, for example, graphite or carbide sometimes can be deposited by combining it with a sprayable matrix metal to create a composite coating, such as nickel-graphite (Ni-graphite) or tungsten-carbide-cobalt (WC-Co).

As previously mentioned, many thermal spray coatings deposited in an ambient air environment contain significant amounts of oxides and porosity, typically concentrated along splat boundaries. For this reason, the physical and mechanical properties of these sprayed materials often are very different from those of the same material in a conventional wrought or cast form. This can be either an advantage or a disadvantage, depending on the application and the properties desired. However, application engineers must be aware that the properties of thermally sprayed materials are quite different from materials created or deposited with other technologies and evaluate them accordingly.

Metallurgical and physical changes that occur during spray deposition can combine to create a complex residual stress state in the resulting coatings. In thermal spraying, as successive layers of molten droplets solidify and cool, thermal contraction can create substantial residual tensile stress in the spray deposited material. Differences in the thermal expansion coefficients of the coating and substrate materials also can result in high localized stresses. In cold spraying and other high-velocity spray processes, compressive stresses may result from high-velocity particles impacting (peening) previously deposited material. In coatings less than approximately 0.5 mm (0.020 in.) thick, residual stress generally is not a serious issue. However, as coating thickness increases or the service environment becomes more aggressive, residual stress can become a very important factor that must be considered and managed by the careful choice of coating material, spray process, and deposition parameters. In general, lower process temperatures, increased material ductility, and proper substrate temperature all tend to reduce residual stress in a sprayed coating. In thermal spraying, the compressive stresses induced by high-velocity particle impact sometimes can be used to reduce, or even eliminate, the tensile stresses resulting from thermal contraction as the coating cools to room temperature.

PROCESS VARIATIONS

All spray deposition processes can be grouped into three general categories, as described in Table 11.1. The first category includes thermal spray processes that burn combustible gases as the primary energy source to heat and accelerate spray particles. The second category includes thermal spray processes that use electrical power as the primary energy source. The third category utilizes compressed gas as the primary energy source to accelerate spray particles and involves much less thermal energy. Cold spraying is the only process currently in this third category.

COMBUSTION PROCESSES

In combustion spray processes, a fuel gas or combustible liquid fuel is mixed with an oxidizer, usually pure oxygen, and burned to provide thermal energy. The fuel gases most commonly used for combustion spraying are acetylene, propane, propylene, hydrogen, and MPS (C_3H_4—a mixture of liquefied petroleum gas and methylacetylene-propadiene). The most common liquid fuel is kerosene. Other fuels, such as methane and natural gas, also are used in some applications.

The flame temperature is a function of the type of fuel and the ratio of fuel to oxygen. Flame temperatures versus fuel-to-oxygen ratios for several fuel gases are summarized in Figure 11.6. Acetylene combined with oxygen produces the highest adiabatic flame temperatures. The distinct characteristics of an oxyacetylene flame make it easy to adjust the stoichiometry to pro-

Table 11.1
Basic Groups of Thermal Spraying Processes

Combustion Processes	Electrical Processes	Low Heat Input Processes
Flame spraying	Arc spraying	Cold spraying
High-velocity oxyfuel spraying	Atmospheric plasma spraying	
Detonation flame spraying	Vacuum plasma spraying	
	Induction-coupled plasma spraying	

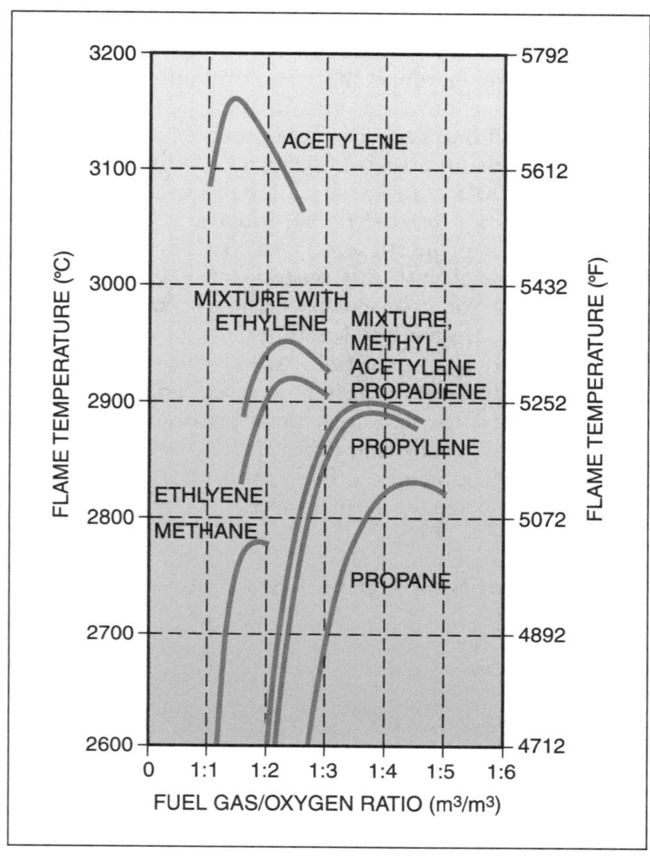

Source: Adapted from Linde AG Linde Gas Division.

Figure 11.6—Flame Temperature as a Function of Stoichiometry for Several Fuel Gases

duce oxidizing, neutral, or reducing conditions. However, safety-related restrictions on the maximum supply manifold pressure for acetylene, especially in the United States, can make it difficult to achieve and maintain optimum spray conditions for some materials. Proper supply manifold design is critical for safe operation and good performance with acetylene.

Flame Spraying

Flame spraying is a thermal spray process in which an oxyfuel gas flame is the source of heat for melting the surfacing material. In all flame spraying processes, the feedstock material is continuously fed into the oxyfuel flame where it melts. The flame typically is surrounded by a coaxial stream of compressed gas (most commonly air, but sometimes an inert gas) that helps cool the spray gun. In wire and rod guns, the high-flowing coaxial jet also may be used to help strip molten material from the wire or rod tip and atomize it into small droplets. Finally, the coaxial gas jet helps propel the molten spray droplets toward the substrate.

Flame spraying devices can be divided into two basic categories: spray guns that are designed for wire, rod, or cord feedstock and guns designed for powder feedstock. A cross section of a typical flame spray gun for wire, rod, or cord feedstock is shown in Figure 11.7.

Wire Flame Spraying. Wire flame spraying is a flame spray process variation in which the feedstock material is usually in the form of wire, but materials such as ceramics or low-ductility metals that cannot readily be made into wire commonly also can be fed

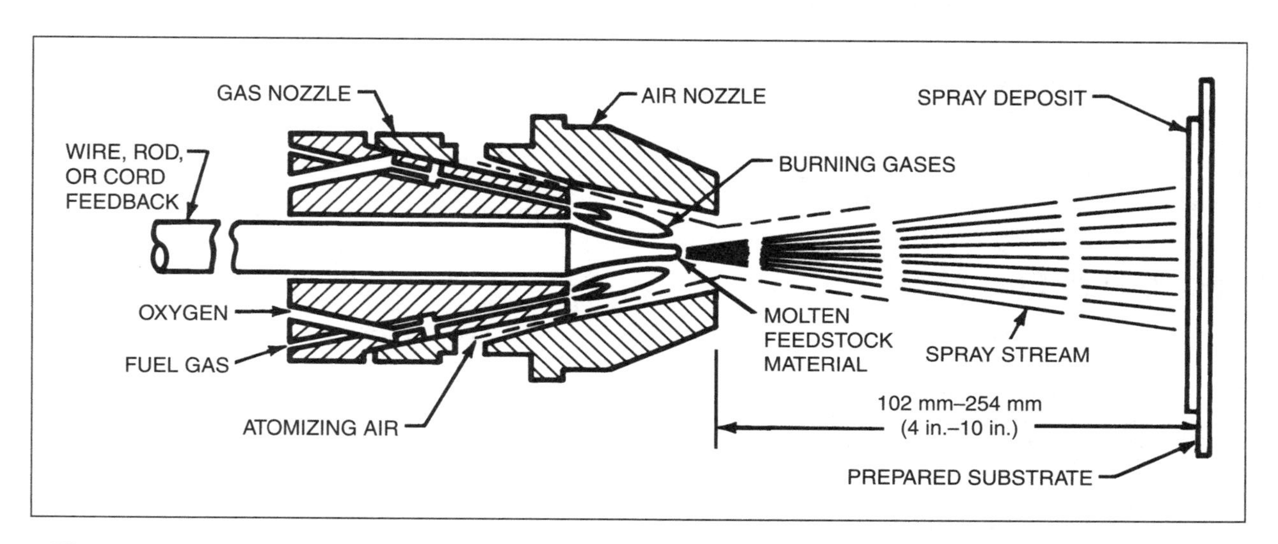

Figure 11.7—Cross Section of a Typical Flame Spray Gun for Wire, Rod, or Cord Feedstock

into the gun as straight rods. Cord feedstock (not widely used in the United States) typically consists of a combustible hollow tube made of a material that is consumed in the flame, such as plastic. The tube is filled with powder made from the material that is to be deposited. It is fed like wire into the flame, where the outer covering is burned away, thus releasing the powder into the flame.

The droplet velocity for wire and rod flame spraying can be somewhat higher, ranging up to 300 m/s or 400 m/s (980 ft/s or 1300 ft/s). The higher velocity primarily is due to smaller droplet size (lower droplet mass) under some wire or rod flame spray conditions. Conditions that produce high-velocity droplets tend to produce somewhat denser coatings with improved adhesion.

Powder Flame Spraying. Powder flame spraying is similar to wire and rod flame spraying. In powder flame spraying, however, a powder feeder is used to continuously meter powder feedstock material into the flame, often with compressed air or inert gas as the transporting medium. In most designs, the powder feeder is a separate piece of equipment external to the spray gun, although some spray guns have a powder-feeding mechanism as an integral part of the gun. Guns with integral powder feeders tend to be special-purpose units with limited powder hopper capacity.

The velocity of the spray droplets for powder flame spraying is among the lowest of all thermal spray processes—typically in the range of 30 meters per second (m/s) to 100 m/s (100 ft/s to 330 ft/s). This low-impact velocity tends to produce coatings with more porosity and less adhesion than coatings deposited with more energetic spray processes.

With a few exceptions, flame-sprayed coatings generally tend to have more porosity, higher oxide content, lower adhesion strength, and less corrosion resistance than coatings produced by more energetic processes, such as high-velocity oxyfuel spraying, detonation flame spraying, plasma spraying, or cold spraying. However, for suitable applications, flame spraying offers significant advantages as a relatively simple, highly portable, and comparatively inexpensive process.

Spray-and-Fuse Coatings

A technique known as *spray-and-fuse coating* can dramatically increase the density, adhesion, and performance of flame-sprayed coatings in selected applications. Spray-and-fuse coatings use self-fluxing metal alloys. These materials typically are nickel-base or cobalt-base metals with additions of fluxing agents, such as boron and silicon, which depress the melting point and enhance wetting. These materials are first deposited using conventional flame spray methods. As a post-treatment, the coating material is then reheated to form a continuous molten layer on the surface. The objective is to enhance the structural integrity of the coating, reduce porosity, and in many cases, create a true metallic bond with the substrate. The temperature used in this fusing post-treatment usually is in excess of 1040°C (1900°F). A variety of heating methods are used, such as an oxyfuel flame, induction heating, or a furnace. While properties and performance often can be greatly enhanced with this post-treatment, it should be noted that in some cases this technique may result in high residual stress in the fused material.

A wide variety of materials can be deposited with the flame spray process; however, many materials with high melting points (for example, most ceramics, refractory metals, and materials that burn, decompose, or become severely oxidized in an oxyfuel flame) should be applied with other thermal spray processes.

High-Velocity Oxyfuel Spraying

High-velocity oxyfuel spraying (HVOF) is a thermal spray process using a high-pressure oxyfuel mixture to heat and propel a powdered surfacing material to a substrate. Figure 11.8 shows a cross section of a generic HVOF spray gun.

Like flame spraying, the high-velocity oxyfuel spraying process uses a combustible gas or liquid that is mixed with pure oxygen and then burned in the combustion chamber of the spray gun. However, HVOF combustion occurs at a higher pressure in the combustion chamber and the resulting heated gas exits the gun through a converging-diverging nozzle, or other constricted flow path that produces a supersonic gas jet. Powder particles entrained in this high-velocity gas jet are accelerated up to velocities on the order of 500 m/s to 1 kilometer per second (km/s) (1600 ft/s to 3300 ft/s), depending on the spraying device, spray material, and operating conditions. These particle velocities are greater than those achieved in most other spray methods, except for detonation flame spraying and cold spraying. Hydrogen, propylene, natural gas, and propane are the HVOF fuel gases most commonly used. Kerosene is the most common liquid fuel. High-velocity air fuel spraying (HVAF) is a variant of the HVOF process that uses compressed air instead of oxygen.

Powder feedstock is brought to the thermal spray gun by an inert carrier gas, such as nitrogen or argon. The high kinetic energy of the spray particles contributes additional heat as the particles impact the substrate. This helps promote good bonding of the particles, and the peening effect imparts compressive residual stress and tends to produce relatively dense deposits with very low porosity. Since much of the thermal energy in the jet is converted to kinetic energy, the gas temperatures within the jet as it exits the spray gun

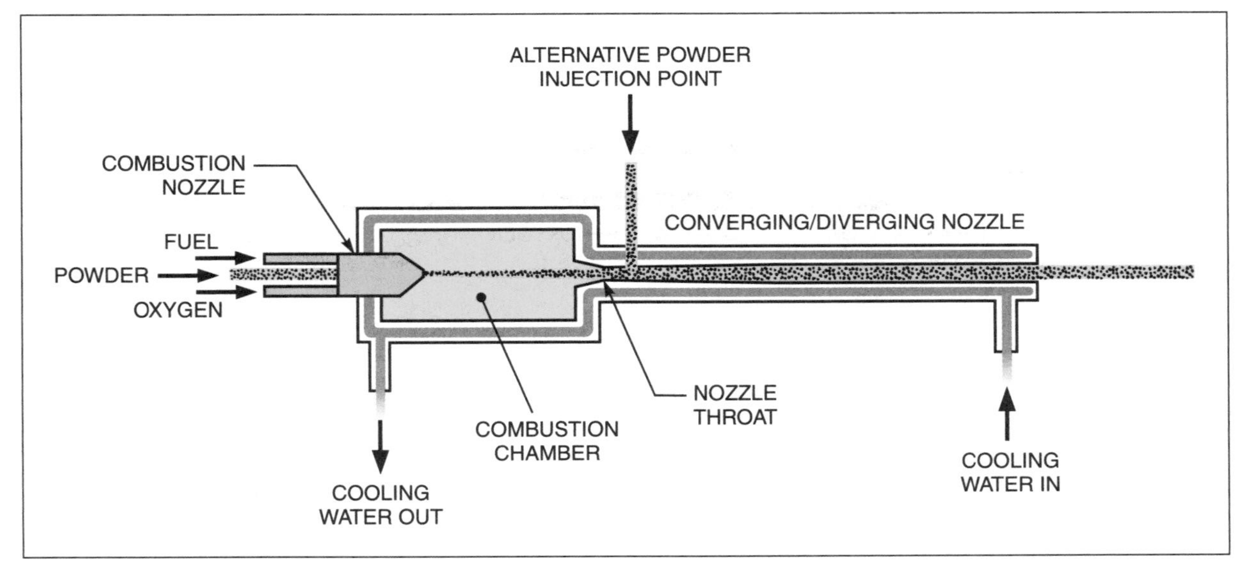

Source: Sandia National Laboratories.

Figure 11.8—Cross Section of a Generic HVOF Spray Gun

can be relatively low compared to other combustion processes. Low gas jet temperatures help prevent overheating of the sprayed material and decarburization of temperature-sensitive carbides, such as tungsten carbide (WC). For this reason, the HVOF process generally is preferred for high-quality cemented carbide wear-resistant coatings, such as WC-Co, and some metallic coatings. Conversely, the temperatures in most HVOF devices are not high enough to melt ceramic particles, so this process is seldom used to spray pure ceramics.

It should be noted that despite relatively moderate temperatures in the jet itself, heat input to the substrate with HVOF tends to be substantial. This occurs because deceleration of the high-velocity gas jet at the substrate surface converts the kinetic energy of the jet back into thermal energy, releasing a large amount of heat at the substrate surface. This heat load usually is manageable with auxiliary cooling jets and rapid movement of the spray jet over the workpiece surface; but the substrate heat load must be considered before selecting HVOF spraying for a specific application.

Detonation Flame Spraying

Detonation flame spraying is a thermal spraying process variation in which the controlled explosion of a mixture of fuel gas, oxygen, and powdered surfacing material is utilized to melt and propel the surfacing material to the substrate. This process differs from other thermal spraying methods in that it is a pulsed rather than continuous process. Individual charges of feedstock powder are heated and sequentially propelled onto the substrate by a series of detonations of an explosive mixture of oxygen and acetylene in the gun chamber.

A schematic of a detonation flame spray gun, (called a D-gun™), is shown in Figure 11.9. The gun consists of a long barrel into which is introduced a mixture of oxygen, acetylene, and powdered coating material suspended in nitrogen (the carrier gas). The oxygen-acetylene mixture is ignited by an electric spark, creating a controlled detonation wave with high localized temperatures and pressures. As the detonation wave moves down the barrel, the powder particles are accelerated and heated to the melting point. The molten material impinges the substrate and deposits the coating. After each shot, nitrogen is used to purge the gun prior to the next detonation event. Detonation frequencies are typically in the range of 3 Hertz (Hz) to 8 Hz.

Maximum particle velocities on the order of 700 m/s to 1000 m/s (2300 ft/s to 3300 ft/s) are possible. The high-velocity particle impingement results in a strong bond with the substrate, and excellent finishes can be achieved. Detonation flame spray coatings typically have very little porosity and low oxide content. Temperatures above 3315°C (6000°F) are achieved within the detonation wave, while the substrate temperature can be maintained below 150°C (300°F) with special techniques, such as using a carbon-dioxide (CO_2)

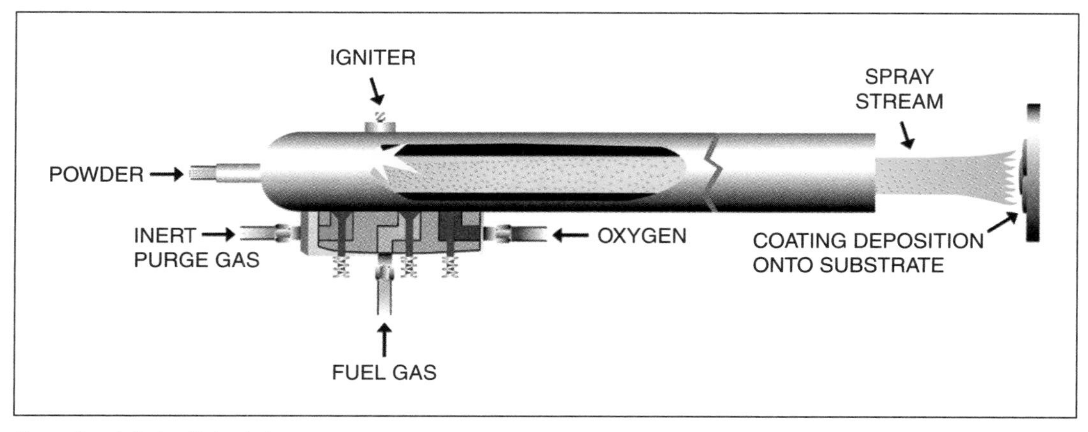

Source: Praxair Surface Technologies.

Figure 11.9—Schematic of a Generic Detonation Spray Gun

cooling system. Like HVOF, the detonation gun is a preferred method for depositing many metal and cemented carbide coatings. The resulting deposit is extremely hard, dense, and tightly bonded. Coating thickness typically ranges between 50 µm and 500 µm (0.002 in. and 0.020 in.).

ELECTRICAL PROCESSES

The electrical spray processes include arc spraying, plasma spraying, vacuum plasma spraying, and radio-frequency (RF) plasma spraying (also called induction-coupled plasma spraying). The heat source for these thermal spray processes is an electric arc.

Arc Spraying

The arc spray process (sometimes called *electric arc spraying*, *wire arc spraying*, or *twin-wire arc spraying*) uses a direct current (dc) electric arc drawn between two feedstock wires to create molten material. A jet of compressed gas is used to help atomize the molten material and accelerate the spray droplets toward the substrate.

The two feedstock wires are electrically isolated from one another, with one wire connected to the negative output terminal (cathode) of a welding-type power source and the other wire connected to the positive output terminal (anode). The wires are continuously fed by means of motorized drives through a guide mechanism that brings the wire tips together at a slight angle. An electric arc between wire tips continuously melts the tips, so the two wires are essentially consumable electrodes. An arc spray gun is shown schematically in Figure 11.10.

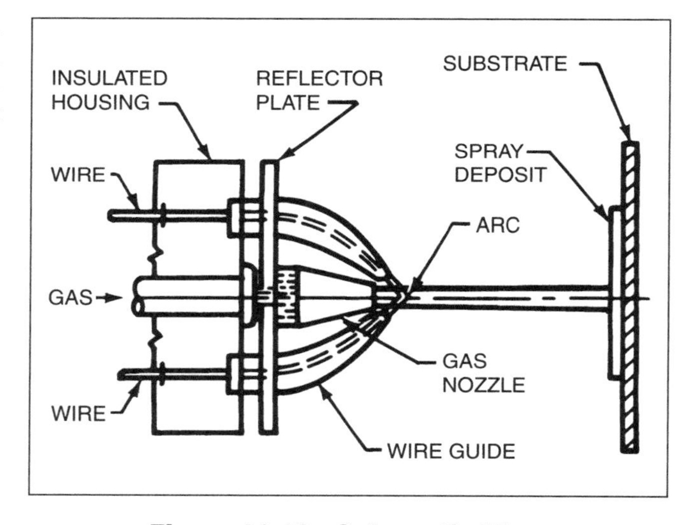

Figure 11.10—Schematic View of an Arc Spray Gun

An atomizing gas (usually compressed air) is directed across the arc, shearing off molten droplets to form an atomized spray stream. Inert gas can be substituted for the compressed air in order to reduce oxidation of the molten particles. In some instances, reactive gasses are used to create novel structures in the coating, for example, nitrogen used with titanium to produce titanium nitrides.

Particle velocities and temperatures of the arc spray process overlap those of powder and wire flame spraying, as indicated in Figure 11.3. The quantity of metal oxides is better controlled and spray rates are higher with arc spraying than with wire flame spraying. The feedstock materials used with arc spraying are metals and alloys in wire form. Powders contained within a metal sheath (cored wire) also are available. A limitation of arc spray is that it is restricted to feedstock that can be fabricated into electrically conductive wire.

Plasma Spraying

Plasma spraying is a highly versatile process able to melt and deposit an exceptionally wide range of materials. One mode of plasma spraying is called *atmospheric plasma* (APS) to differentiate this technique from vacuum plasma spraying (VPS). In atmospheric plasma spraying, a direct-current arc is used to create a high-temperature plasma jet that melts and propels the surfacing material.

As shown in Figure 11.11, a plasma spray gun has two water-cooled electrodes: a tungsten cathode surrounded by a cylindrical copper anode. To create the plasma jet, argon or nitrogen with optional additions of helium (and in some cases hydrogen or other gases) is introduced into the annular space between the anode and cathode. A high-amperage dc arc is then struck between the two electrodes, creating the plasma jet. The term *plasma* as it is used in this instance refers to a very hot gas in a partially ionized, electrically conductive state. Total power levels in currently available plasma spray guns typically fall in the range of 20 kilowatts [kW] to 200 kW.

The arc-driven plasma can reach temperatures on the order of 5000°C to 20 000°C (9000°F to 36 000°F), creating thermal energy sufficient to melt any feedstock material. The anode also serves as a nozzle to direct the flow of the hot plasma jet. The rapidly expanding, extremely hot plasma passes through the nozzle as a high-velocity jet with exit velocities that may be either subsonic or supersonic, depending on the nozzle design and the operating conditions. The surfacing material, in powder form, is injected into the plasma stream either inside the anode (internal injection) or just beyond the anode nozzle exit (external injection), where it melts and is propelled onto the substrate. A few plasma gun designs use axial injection, in which the powder is fed down the central axis of the spray gun.

The plasma spray process was developed in the mid-1950s and was used extensively in the early United States space programs. Because of the high temperature

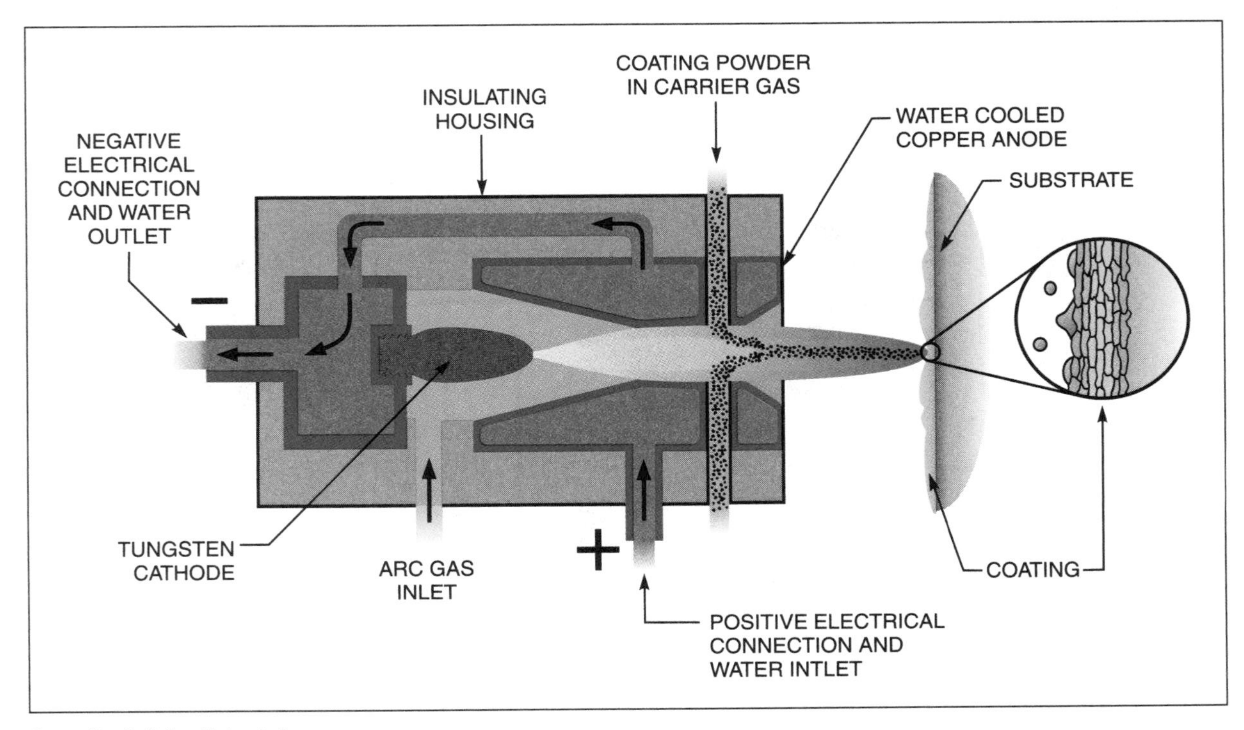

Source: Sandia National Laboratories.

Figure 11.11—Cross Section of a Generic Plasma Spray Gun with Internal Powder Injection

of the plasma jet, this process is particularly well suited for spray materials with high melting points, such as ceramics and refractory metals. It is still widely used for aerospace applications, especially coatings for aircraft and land-based turbine engines. Turbine and rocket engine components are exposed to extreme service conditions. Plasma-sprayed ceramic thermal barriers, for example, yttria-stabilized zirconia (YSZ), have been used to extend the service life of the components and increase performance in many critical applications.

Quality surfacing deposits require the introduction of powder at the proper point in the plasma jet and at the correct feed rate. Since the particles are in the plasma for a very short time, slight variations in the location of the feed point may significantly change the amount of heat transmitted to the powder. Particle velocity is an important variable with respect to bond strength, deposit density, and integrity. In general, higher particle velocities result in higher coating density and better adhesion.

Plasma Transferred Arc. The term *plasma spraying* normally refers to the use of a non-transferred arc plasma spray gun, such as the one illustrated in Figure 11.11. *Non-transferred arc* means that the plasma arc is contained entirely within the gun, and the substrate is not part of the electric circuit. There is another plasma surfacing technique in which an arc is established between an electrode of the plasma gun and the workpiece. This technique, known as *plasma transferred arc* (PTA) *surfacing*, differs from traditional thermal spray processes in that a pool of the molten coating material is created on the substrate surface. In the PTA process, the substrate is biased relative to the spray gun and electrical current flows via the conductive ionized plasma jet between the spray gun and the workpiece. Substantial additional heating of the substrate can be achieved with the transferred arc, melting the feedstock powder in a broad, shallow pool similar to a weld bead as it solidifies on the surface of the workpiece. Because of the molten pool on the substrate surface, the PTA process really has more in common with plasma arc welding than with thermal spray processes. Plasma transferred arc spray does produce a true metallic bond with the substrate, but at the expense of mixing some molten substrate material in with the coating material, possibly diluting the surfacing material or creating undesirable metallurgical phases.

Vacuum Plasma Spraying

Vacuum plasma spraying (VPS) is a variation of plasma spraying that uses a plasma spray gun operating inside a water-cooled vacuum chamber. The major components of a typical VPS system are illustrated in Figure 11.12. The term *vacuum* in this instance refers to a relatively soft dynamic vacuum; that is, the chamber is continuously pumped to balance the inflow from the arc gas and powder carrier gas, with operating pressures

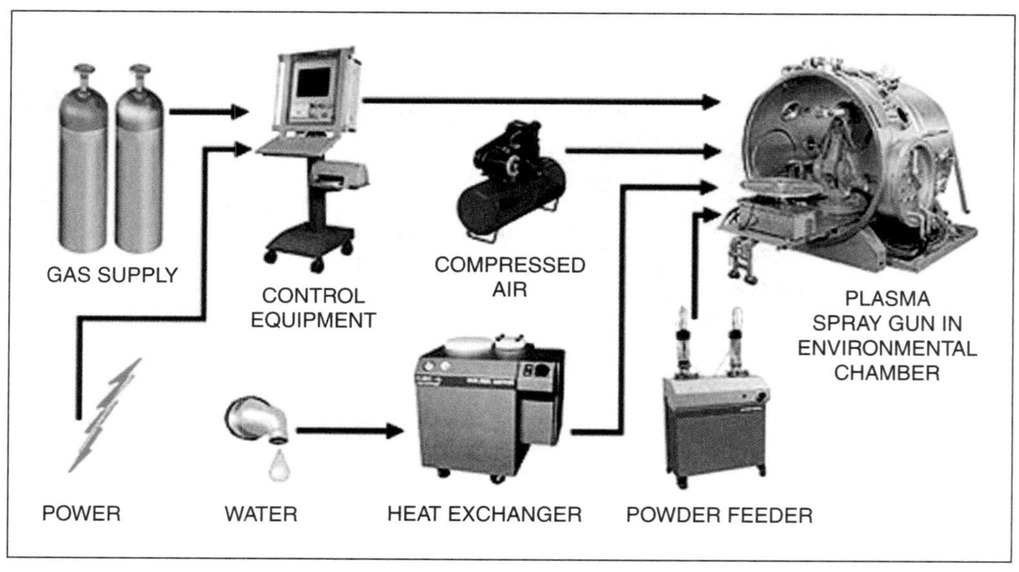

Source: Sulzer Metco Corporation.

Figure 11.12—Vacuum Plasma Spray System

typically above 6.7 kilopascals (kPa) (50 Torr). For this reason, the process sometimes is referred to as *low-pressure plasma spraying* (LPPS®) or *controlled-atmosphere plasma spraying* (CAPS). The process chamber is initially sealed and pumped down as far as possible with the spray gun off; subsequently, it is back-filled to the desired pressure, usually with an inert gas (such as argon) before coating deposition begins.

A secondary power source (not shown in Figure 11.12) sometimes is added to a VPS system to apply bias voltage to the substrate relative to the anode nozzle of the spray gun. With the use of this secondary power source, it is possible to pass substantial direct current through the conductive plasma plume, which is greatly extended and broadened at the low pressures inside the vacuum chamber. The secondary current passing through the plasma plume technically is a partial transferred arc, because the primary arc that generates the plasma is still contained within the plasma spray gun. However, in VPS, such secondary arcs are commonly referred as *transferred arcs*. A transferred arc is used for two reasons: to clean impurities (such as oxides) from the surface of the workpiece, and to provide additional heating of the coating and workpiece.

In the cleaning mode (also called *reverse transferred arc mode*) the workpiece is negatively biased and becomes a cathode. The transferred arc is turned on after the plasma spray gun is started, but before the powder feeder is turned on. Cathode spots move across the surface of the workpiece where the conductive plasma plume is in contact with it, resistively superheating and vaporizing surface impurities that have poor electrical conductivity, such as surface oxides. Figure 11.13 shows the reverse-transferred-arc cleaning of a turbine blade prior to applying the coating. The cathode spots can be seen as small bright regions on the surface of the workpiece in the lower part of the photograph.

If the polarity of the transferred arc is changed to positively bias the substrate, making it an anode, the current flowing through the transferred arc can be used to increase heating of the substrate and the coating material on the substrate. In many instances, this promotes better adhesion of the coating to the substrate with better inter-splat cohesion and lower porosity.

Because oxygen levels inside a vacuum plasma spray chamber are much lower than in a normal ambient air atmosphere, it is possible to make the following improvements in quality:

1. Coatings can be deposited with little or no oxidation of the sprayed material, resulting in much cleaner splat boundaries;
2. Coating porosity can be dramatically reduced; and
3. Coatings can be applied at much higher substrate temperatures without oxidizing the substrate material.

Photograph courtesy of Electro-Plasma, Inc.

Figure 11.13—Vacuum Plasma Spray Plume Used to Clean the Surface of a Turbine Blade

High-quality thermal spray coatings with properties nearly like wrought material and exceptional adhesion to the substrate can be achieved with this process. However, the capital cost of a VSP system may be approximately ten times that of standard atmospheric plasma spray equipment, and operating costs also are higher. For this reason, its use has been largely restricted to high-value-added applications, such as coating critical components for turbine engines and applying biomedical coatings for dental and orthopedic implants.

It is noteworthy that plasma spraying also has been carried out at atmospheric pressure in simple gas-tight enclosures that are purged and back-filled with argon or another inert gas to reduce the oxygen level in the spray environment. This approach provides some, though not all, of the advantages of vacuum plasma spraying at a lower cost. Such systems have been successfully used to spray highly reactive or hazardous metals, such as beryllium; however, commercial use is very limited.

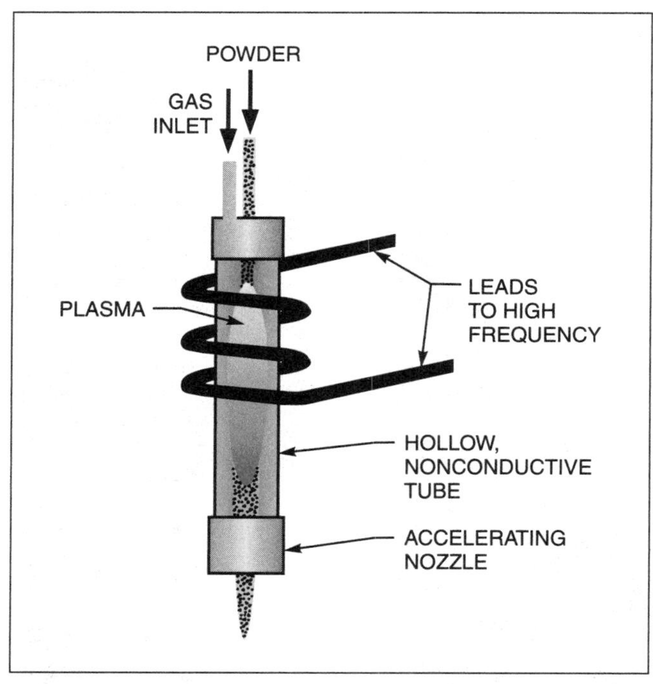

Source: Technar Automation LTEE.

Figure 11.14—Arrangement of a Radio-Frequency (RF) Plasma Spray Gun

Radio-Frequency Plasma Spraying

A radio-frequency (rf) plasma spray gun, shown in Figure 11.14, differs from a direct-current, arc-driven plasma spray gun in several important ways. The rf gun has no electrodes and consists of a hollow glass or ceramic tube about 50 mm (2 in.) in diameter and 150-mm (6-in.) long. Gas is injected into one end of the tube, and plasma is created inside the tube by an induction coil wrapped around the outside of the tube. For this reason, this process is also commonly referred to as *induction coupled plasma spraying*. The induction coil is powered by a radio-frequency (rf) generator. Powder injected into one end of the rf plasma spray gun moves at much lower velocity and is heated along a substantially longer path compared to a dc arc plasma spray gun. The longer heating period of the powder in the gun allows the complete melting of larger powder particles and can result in more consistent sprayed coatings.

However, rf spray guns generally are rigidly and closely affixed to a rather large rf power source, so the spray gun cannot be moved during the coating operation. Instead, the workpiece must be manipulated in front of the gun nozzle exit to spray the areas to be coated.

LOW HEAT INPUT PROCESSES

Cold spraying is a low heat input process in which coatings are applied at much lower temperatures compared to most traditional thermal spray processes, often at or near room temperature. The feedstock powder is not melted or even significantly heated in most cases. This provides the advantage that there is little or no reaction with atmospheric oxygen to form oxides in the coating, and no undesirable changes in the chemistry or phase composition of the coating material. For example, it is possible to spray pure copper in an ambient air environment with no measurable increase in oxide content of the coating compared to the feedstock powder. However, such processes generally are limited to highly ductile materials that will plastically deform and flow without shattering when striking a substrate at very high impact velocities.

Cold Spray Systems

In a typical cold spraying system, shown in Figure 11.15, a compressed gas is resistively pre-heated to temperatures normally in the range of 250°C to 800°C (480°F to 1470°F), then fed into a small high-pressure chamber in the spray gun, and finally allowed to exit through a converging-diverging nozzle, creating a supersonic low-temperature jet (gas jet cools as it expands). The primary reason for pre-heating the process gas is not to heat the spray particles, but rather to achieve higher gas jet flow velocities at a given inlet pressure. This is because the sonic velocity in the throat (the point of minimum diameter) of the converging-diverging cold spray nozzle increases as the temperature of the gas increases.

The process gas can be compressed air, nitrogen, or helium. Lighter, monatomic gasses provide higher jet velocities. For example, helium provides higher jet and particle velocities than nitrogen; however, the cost of helium is substantially higher. Compressed air produces the lowest particle velocities and it can be used only for a relatively limited range of spray materials.

Feedstock powder, typically 10 µm to 30 µm in mean diameter, is fed into the supersonic gas jet, and the particles are accelerated to velocities on the order of 500 m/s to 1000 m/s (1500 ft/s to 3000 ft/s) before impacting the substrate. A coating is formed only if particles impact at a velocity higher than a minimum velocity, known as the *critical velocity* or V_{crit}, which is material-dependent. Above the critical velocity, the kinetic energy is sufficient to plastically deform the particles and bond them to underlying material by a process believed to be similar to explosion welding, but on a micro scale. Increasing the particle velocity makes it possible to spray a larger variety of materials and to form denser coatings of higher quality.

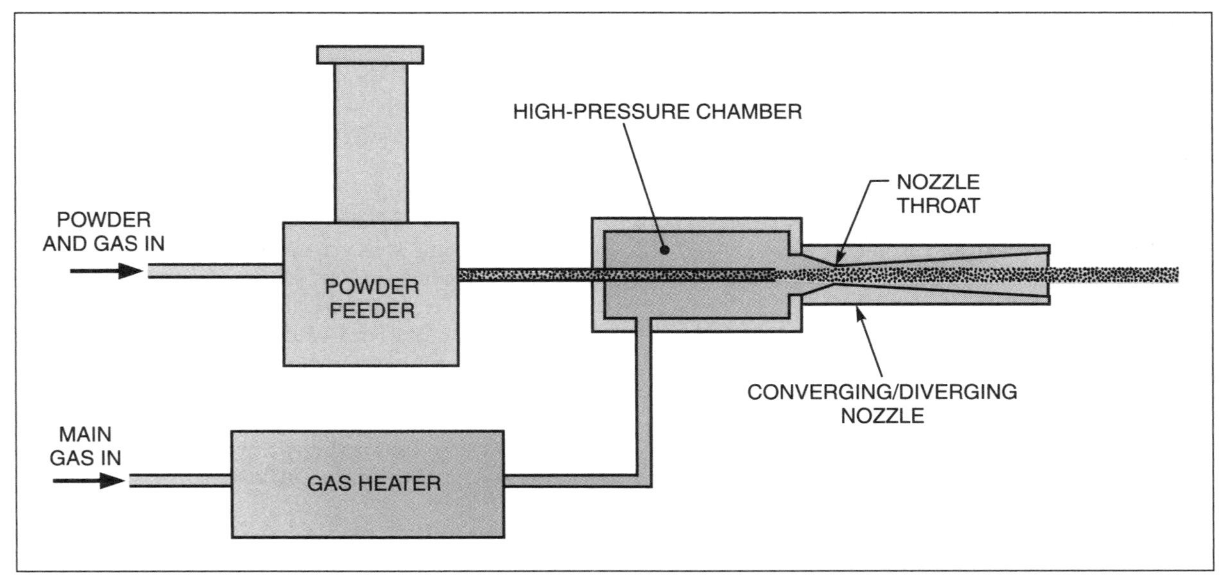

Source: Sandia National Laboratories.

Figure 11.15—Arrangement of a Cold Spray System

Only materials that are sufficiently ductile at high strain rates and low temperatures can be cold-sprayed. This means that the process generally is suitable only for ductile metals. A wide range of pure metals and metal alloys, such as aluminum (Al), copper (Cu), iron (Fe), nickel (Ni), stainless steel, and tantalum (Ta) have been successfully cold-sprayed. In some cases, it also is possible to cold-spray a non-ductile material by mixing it with a ductile matrix metal to form a composite coating material. In this case, the non-ductile phase becomes embedded and trapped in the more ductile matrix metal. Examples of coatings that have been successfully cold-sprayed include ceramic metal composites (cermets), such as aluminum plus silicon-carbide (Al-SiC), and aluminum plus boron-carbide ($Al\text{-}B_4C$).

In addition to selecting a suitable feedstock material, the substrate material for cold spraying must be sufficiently hard relative to the coating material to ensure that the impacting particles will plastically deform. If the substrate material is too soft, the impacting particles will simply embed themselves in the substrate material. In an extreme case, if a very soft substrate such as plastic is used, the high-velocity spray stream may rapidly erode or even cut through the substrate material.

It also is important that the coating material bonds well to the substrate material on impact. For example, atomic bonding mechanisms in metals are fundamentally different from those in ceramics or glasses. For that reason, metals typically do not adhere well when cold-sprayed directly onto ceramic or glass substrates. A notable exception is aluminum, which forms a very strong and stable oxide. Aluminum typically bonds well when cold-sprayed onto many oxide ceramics and glasses. For this reason, a thin layer of aluminum sometimes is cold-sprayed onto a ceramic or glass substrate before spraying some other metal on top of the aluminum. The aluminum serves as a bond coating layer and improves adhesion of the other metal to the ceramic or glass substrate, much like a primer coating applied to bare wood before applying house paint.

EQUIPMENT AND MATERIALS

As in many other industries, thermal spraying has been part of a strong trend toward automation. Increased sophistication of spray equipment has promoted the widespread implementation of robotic spray operations. Nevertheless, many simple hand-operated systems are still in use and are quite adequate for some applications. Because of the wide variety of options, the selection of a specific process and spray equipment must take many variables into consideration, including the size and complexity of the workpiece, the degree of uniformity required in the coating, and tradeoffs in cost versus quality of the coating for the intended application.

FLAME SPRAY EQUIPMENT

A typical flame spray system consists of the following:

1. Spray gun;
2. Feeding devices for wire, rod, or powder spray materials;
3. Pressure regulators, and flow meters for oxygen and fuel gas supplies;
4. Control unit for the compressed air source (when required);
5. Workpiece holding device;
6. Gun or workpiece manipulation device for semiautomatic or automatic processing (if required); and
7. Air cooling jets, or air siphon (when required).

Flame Spray Guns

Flame spray guns are available in various models and sizes designed for wire, rod, or powder feedstock. All commercial flame spray guns are similar in basic concept. Wire, rod, or powder is melted by an oxyfuel flame and propelled onto the substrate by a compressed gas jet or the combustion gases or both. Figure 11.16 shows a typical wire-fed flame spray system.

Small flame spray guns can be handheld and manipulated in much the same manner as paint spray guns. They often are used to apply protective coatings of aluminum or zinc to large objects, such as liquid storage tanks, ship hulls, and bridges. Larger, heavier units usually are designed to be mechanically manipulated. Various nozzle and air cap combinations are used to accommodate different feedstock materials and to alter the spray pattern and spraying conditions.

The arrangement of the oxyfuel gas jets and the atomizing gas orifices differs among various manufacturers, as do the mechanisms for feeding materials into the flame. Flame spray accessories in the form of air jets and air shrouds are available to change the flame characteristics. These accessories are used to adjust the shape of the flame and to control the velocity and temperature distribution of the spray droplets.

Materials are deposited in multiple layers, each of which can be as thin as 15 µm (0.0005 in.) per pass. The total thickness of the spray deposit depends on several factors, including the following:

1. Type and properties of the surfacing material;
2. Condition of the workpiece material, including geometry;
3. Service requirements of the coated product; and
4. Post-spray treatment of the coated product.

A wire flame spray gun consists of two subassemblies: a drive unit that feeds the wire; and a gas head that controls and mixes the flow of fuel gas, oxygen, and atomizing gas. The wire drive unit consists of a motor and drive rolls. The drive motor may be electric or pneumatic, with adjustable manual or automated

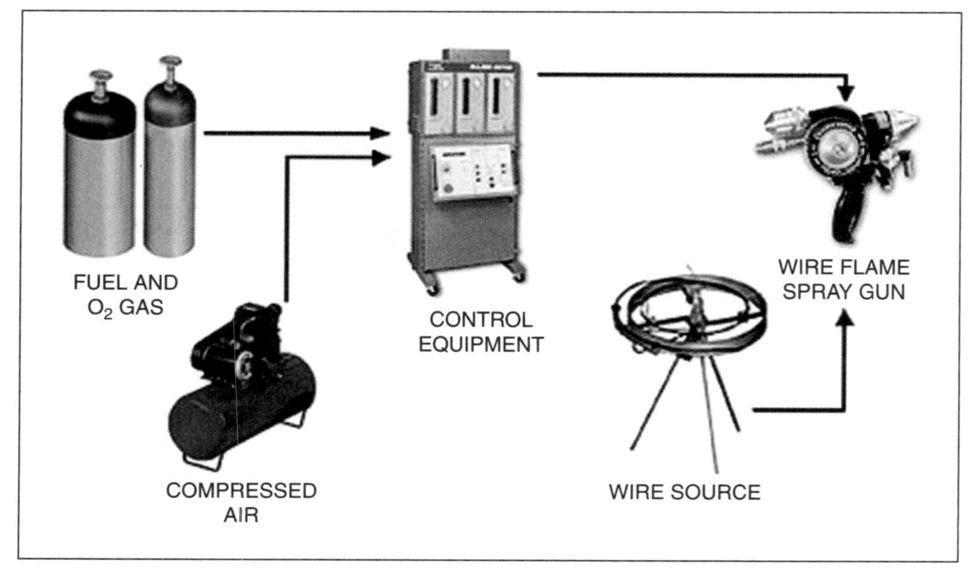

Source: Sulzer Metco Corporation.

Figure 11.16—Components of a Typical Wire Flame Spray System

speed controls. It should be noted that if the wire feed rate is excessive, the wire tip will extend beyond the hot zone of the flame and not melt or atomize properly. This produces very coarse deposits. If the wire feed is too slow, the metal will severely oxidize, and the wire may fuse to the nozzle. Such deposits have high oxide content because of the reduction in particle size that results.

Rod guns are very similar to wire guns, and some models can feed either wire or rod. Rod feedstock is used when the feedstock material cannot readily be made into wire. Common examples would be ceramics and metals with limited ductility. Straight lengths of rod are successively fed into the flame by drive rollers on the gun. Ceramic coatings applied by rod flame spray are characterized as being hard and dense, and they exhibit comparatively lower residual stress. However, spraying with ceramic rod requires greater care in adjusting the spray variables than wire spray because of the higher melting points and lower thermal conductivities of ceramics compared to metals. Following are some types of ceramics applied by this technique:

1. Alumina-titania,
2. Alumina,
3. Zirconia,
4. Rare earth oxides,
5. Zirconium silicate,
6. Magnesium zirconate,
7. Barium titanate,
8. Chromium oxide,
9. Magnesia-alumina,
10. Mullite, and
11. Calcium titanate.

The gas head of a powder flame spray gun is similar in basic design to the wire or rod guns, but it does not have a wire/rod drive mechanism. The powder to be sprayed is supplied to the gun from a hopper that may be remote from the gun or mounted directly on it. The powder may either be aspirated or injected into the flame by a carrier gas, such as air, nitrogen, or argon. Alternatively, it may be carried by one of the process gases (usually the oxygen), or it simply may be gravity-fed.

Metals, ceramics, and ceramic-metal blends can be flame sprayed in powder form. Some of the deposited metals are hard alloys designed for specific wear resistance or corrosion-resistance applications. Very hard materials, such as carbides and borides, often are blended with metal powders to form cermet composites that are tough and wear resistant. The degree to which spray powder particles melt depends on the melting point of the material and the time that the particles are exposed to the heat of the flame (called the *dwell time*). Powder materials with low melting points usually will become completely molten, and those with high melting points and low thermal conductivity, such as ceramics, may melt only on the outer surface of the particles.

Flame Spray Gas Controls

Oxygen and fuel gas flow meters are used to control the oxygen-to-fuel gas ratio and flame intensity. Since the molten particles are exposed to oxygen, some oxide typically will be formed on them, even when a reducing gas mixture is used. The amount of oxide does not vary greatly with changes in the oxygen-to-fuel gas ratio.

Compressed Air Source

Compressed air is used to propel, and in the case of wire and rod materials, to help atomize the molten particles. The cleanliness and dryness of the compressed air is important in producing a quality deposit. Oil or water in the compressed air will cause fluctuations in the flame, produce poor or irregular atomization of the spray material, reduce bond strength, and affect the quality of the deposit. To minimize problems associated with oil and air contamination, an after-cooler, a water extraction unit or a chemical filter, or both, should be installed between the air source and the spray unit. Accurate regulation of the air pressure also is important for uniform atomization.

HIGH-VELOCITY OXYFUEL SPRAY EQUIPMENT

High-velocity oxyfuel (HVOF) spraying technology was introduced commercially in the early 1980s and continues to be widely used for cemented carbide wear-coatings and other applications. Figure 11.17 shows an HVOF spray gun. The supersonic gas velocity of the flame jet causes the formation of bright spots, or "shock diamonds." Equipment for the HVOF process is similar to that used for the subsonic powder flame spray process. A fuel gas (for example, propylene, propane, hydrogen, or acetylene) or a liquid fuel (for example, kerosene) is burned with oxygen to provide both thermal and kinetic energy. The powder to be sprayed is entrained in a nitrogen or argon carrier gas and injected axially or radially into the spray gun. The converging-diverging nozzle on an HVOF gun produces supersonic exit gas velocities ranging up to approximately 1500 m/s (4900 ft/sec).

Liquid-fueled guns soon followed the original gas-fueled HVOF guns. These kerosene guns use either air or oxygen to combust the fuel. While fuel-gas driven HVOF devices can yield up to approximately 530 megajoules (MJ) (500,000 British thermal units [Btu]) of energy, kerosene guns can provide up to twice that amount of energy. This higher energy level creates

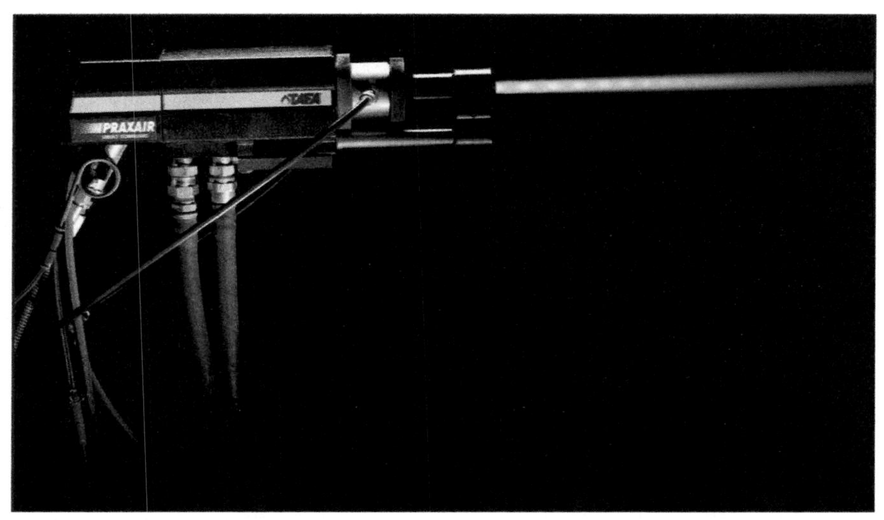

Photograph courtesy of Praxair Surface Technologies

Figure 11.17—High-Velocity Oxyfuel (HVOF) Spray Gun

higher particle velocities, the most significant benefit of which is compressive residual stress in the coating. Higher spray rates are achievable, but higher heat input to the workpiece can also be expected.

DETONATION GUN EQUIPMENT

Detonation gun spraying is accomplished with a specially designed gun, shown in Figure 11.18. Detonation spray guns produce extremely high levels of sound, on the order of 145 decibels (dB). For this reason, it requires sound mitigation measures beyond those used for most other thermal spray processes, and typically it is housed in a double-walled, sound-isolating room. The actual spray operation is completely automatic and remotely controlled.

Both coating services and detonation spray equipment are commercially available; however, due to the unique nature of this technology, most detonation gun coatings are applied by a specialized service provider.

ARC SPRAY EQUIPMENT

Arc spraying systems typically are small, self-contained units that include the electric power source, gas and electrical controls, wire feeder, and wire spool supports. Typical commercial systems, as shown in Figure 11.19, are capable of deposition rates ranging from 1 kilogram per hour (kg/hr) to 20 kg/hr (2.2 pounds per hour [lbs/hr] to 44 lbs/hr) for iron-based materials. Factors controlling the deposition rate include the electrical current rating of the power source and the permissible wire feed rate relative to the available power. Current, wire speed, and spray rate are all proportional and directly related to one another.

Direct-current, constant-voltage power sources normally are used for arc spraying. The power source typically provides adjustable voltage in the range of 18 volts (V) to 40 V, permitting the spraying of a wide range of metals and alloys. The arc length and vaporization increase when the voltage rises. Arc voltage should generally be kept at the lowest stable operating level to reduce vaporization of the feedstock.

Higher bond strength can be achieved with some materials by spraying the first pass using low arc current, low gas flow rate, and a short gun-to-workpiece distance. This method of producing a bond coat capitalizes on the tendency of arc spray to produce low-stress coatings by generating large, slow-cooling particles. After the first pass has been applied over the entire surface, subsequent spraying is done using standard gas flows, the lowest possible arc voltage consistent with good arc stability, and the normal spray gun-to-workpiece distance. These conditions ensure the following results:

1. Fine spray particle size,
2. Minimum loss of alloy constituents, and
3. A concentrated spray pattern.

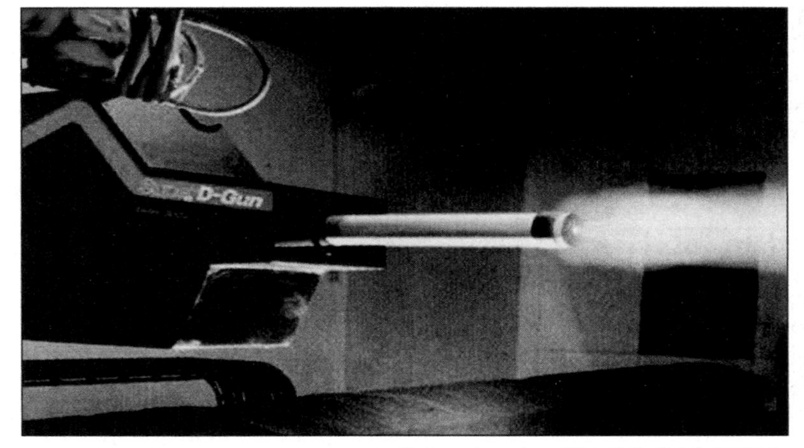

Photograph courtesy of Praxair Surface Technologies

Figure 11.18—Detonation Spray Gun

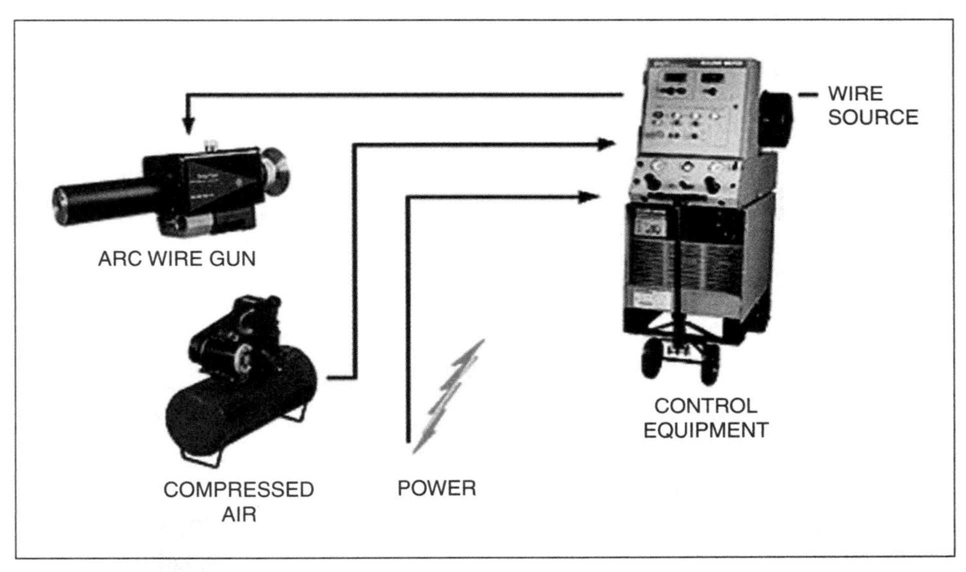

Source: Sulzer Metco Corporation.

Figure 11.19—Arc Spray Equipment

Controls are provided for the atomizing gas pressure, gas flow rate, wire feed speed, and arc power. On most equipment, switches or triggers are provided on the spray gun to energize the wire feed and the atomizing gas flow. Gas velocity through the atomizing nozzle can be regulated over a range of 4.0 m/s to 5.5 m/s (800 feet per minute [ft/min] to 1100 ft/min) to control deposit characteristics.

The wire control unit consists of two reel or coil holders, which are electrically isolated from one another and connected to the spray gun with flexible insulated wire guide tubes. Some designs supply the electrical current and wire through a single flexible conduit. Common wire sizes range from 1.6 mm to 3.2 mm (1/16 in. to 1/8 in.), although larger diameters are used for some high-volume applications.

Energy and labor costs tend to be lower for arc spraying because of higher thermal efficiency, higher deposition rates, lower equipment and maintenance costs, and lower gas costs. Due to the highly localized arc heating and the significant flow of cool atomizing gas, the heat input to the substrate can be lower than for many alternative processes; therefore, arc spray is one of the processes commonly used to apply metal coatings to heat-sensitive substrates such as plastic.

PLASMA SPRAY EQUIPMENT

A plasma spray system typically consists of a plasma gun, a direct current power source, process controls for gas, electricity, and cooling water, and a powder feeding system. The associated fixturing and traversing devices are specific to the application. A complete plasma spray system is shown in Figure 11.20.

Plasma Spray Gun

Several types of plasma spray guns are available; the traditional "stick cathode" gun illustrated in Figure 11.21 is the most common. Other plasma spray gun designs feature sophisticated multiple-electrode arrangements and some designs provide for axial injection of the powder feedstock down the central axis of the gun. Such guns typically are more complex and expensive, but may afford advantages for specific spray applications. Plasma spray gun power capacities vary widely from 20 kW to 200 kW. Direct current from 100 amperes (A) to 1200 A is used at voltages ranging from 25 V to 160 V.

Most plasma guns are light enough to be hand held; however, the trend in production applications of plasma spray is toward robotic gun manipulation. Robotic spraying provides greater uniformity and repeatability of the coating properties. It also minimizes exposure of the plasma spray operator to high levels of sound and other hazards of the immediate spray environment.

Power Sources

In general, plasma spray power sources closely resemble those used for arc welding. Power sources for plasma spraying should have the following characteristics:

1. Constant current dc output,
2. Variable open-circuit and load voltages,
3. Variable current control,
4. Low ripple, and
5. Good regulation.

Silicon-controlled rectifier (SCR) power sources generally meet these requirements. Some units are easily operated in parallel for operations requiring high power.

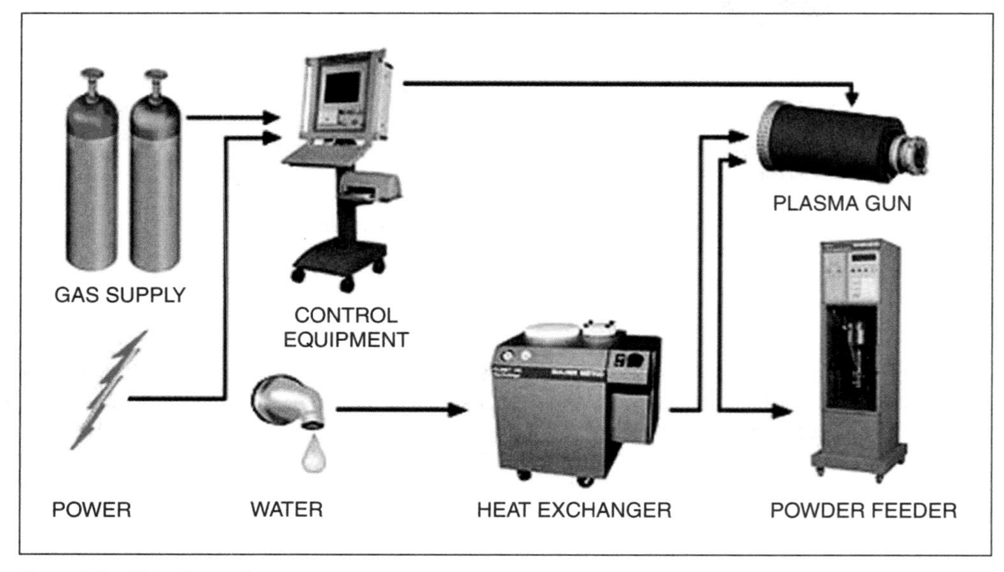

Source: Sulzer Metco Corporation.

Figure 11.20—Typical Plasma Spray System

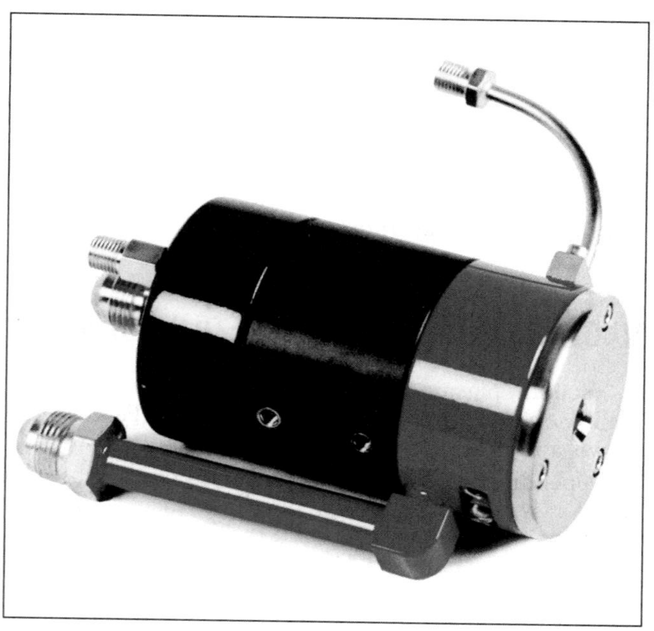

Photograph courtesy of Praxair Surface Technologies

Figure 11.21—A Typical 80-kW Direct-Current Plasma Spray Gun

Powder Feeding Devices

The three basic types of powder feeders used for plasma spraying are aspirator, volumetric (mechanical), and fluidized bed. These are the same powder feeders used for many other powder-fed thermal spray processes, with the exception that feeders for most high-velocity processes (for example, high-velocity oxyfuel spraying and cold spraying) tend to operate at significantly higher powder-canister pressures than typical plasma spray systems.

Gas Circuits

The following gas circuits commonly are required in a plasma spray system:

1. Primary arc gas,
2. Secondary arc gas, and
3. Powder-carrier gas.

In most systems, the primary and carrier gases are fed from the same source. Both monatomic gases (argon and helium) and diatomic gases (nitrogen and hydrogen) are used for plasma spray processes. Argon and helium produce the highest temperatures; nitrogen and hydrogen produce the highest enthalpy. The diatomic gases not only release ionization energy, but also the energy of molecular dissociation and recombination. The choice of gas greatly affects the plasma characteristics. Plasma arc gases should contain less than 50 parts per million (ppm) oxygen and methane (in the case of argon).

Argon and nitrogen are the most commonly used primary gases, in that order. Helium, hydrogen, and nitrogen are the common secondary gases, again in that order. These four gases have characteristics that are important to successful plasma spraying.

Argon. Argon provides high plasma flow velocity. It is used to spray materials that would be adversely affected if hydrogen or nitrogen were used. Carbides and high-temperature alloys are most commonly sprayed with argon, especially in aircraft applications.

Nitrogen. Nitrogen is widely used because it is inexpensive, diatomic, and permits high spray rates and deposit efficiencies. The service life of the nozzle usually will be considerably shorter with nitrogen than with monatomic gases.

Helium. Helium is commonly used as a secondary gas mixed with argon. Helium tends to raise the arc voltage and can be used up to a 2:1 argon-helium ratio. Helium significantly increases gas and particle velocities, and it has less effect on anode service life than either of the diatomic secondary gases.

Hydrogen. Hydrogen may be used as a secondary gas in mixes from 5% to 25% with nitrogen or argon. The addition of hydrogen raises the arc voltage, and thus the power and thermal energy of the arc. Hydrogen may have a detrimental affect on certain metals that tend to absorb hydrogen when in a molten condition or on substrate materials that readily absorb hydrogen, such as titanium. Like nitrogen, hydrogen can significantly reduce anode service life.

Plasma Spray System Controls

A complete plasma spray system, including the spray unit, can be remotely operated from a control console. The console provides adjustment of the primary and secondary plasma gas flow rates, powder carrier gas flow rate, plasma current, starting and stopping functions, and in some cases, operation of the powder feeding unit and any associated robotic manipulator systems. Plasma spray control systems range from basic and relatively inexpensive fully manual systems with analog gauges and controls to highly automated, computer-controlled systems that can provide a higher degree of automated process monitoring and control.

Plasma Spray Feedstock Materials

Plasma spraying is one of the most versatile of the thermal spray processes in terms of the variety of different materials that it can successfully deposit. Although the theoretical adiabatic flame temperatures of oxyfuel processes range up to approximately 3160°C (5720°F), it becomes increasingly difficult, from a practical standpoint, to achieve reasonable deposition rates and high-quality coatings as the melting point of the feedstock material moves toward this theoretical limit. Plasma spraying can deposit powdered materials that normally are applied by flame spraying, but it does so at a higher rate and often with lower porosity. The process also can deposit a wide range of refractory ceramics and metals that are difficult or impossible to deposit with the flame spray processes.

A partial list of surfacing materials applied by plasma spraying is shown in Table 11.2. Many additional commercial compositions are custom-formulated for specific applications and are proprietary.

Plasma-sprayed ceramic coatings generally exhibit higher densities and hardness than flame-sprayed deposits. Due to the higher bulk density, plasma-sprayed deposits may be thinner in some cases, but since there is less porosity to provide stress relief, residual stresses in the plasma-sprayed deposits also may be higher. A highly stressed thermal spray deposit may be more susceptible to cracking or spalling. Methods of mitigating residual stress include proper substrate temperature control and proper deposition procedures.

Bond Coating for Ceramics. When plasma-spraying ceramic coatings onto metal substrates, it is common to spray a bond coating first, such as molybdenum, NiCr, NiAl, or a proprietary alloy developed for this purpose. In some cases, graded mixtures of the ceramic and a suitable metal also have been used to provide a graded transition in thermal expansion between a ceramic coating and a metal substrate, thus reducing stress at the interface of the coating and substrate. This can be achieved by spraying various ceramic/metal mixtures, progressively increasing the amount of ceramic to produce the graded transition.

VACUUM PLASMA SPRAY EQUIPMENT

Vacuum plasma spray systems typically are large, highly sophisticated, and relatively expensive. Figure 11.22 shows a representative vacuum plasma spray system.

Depending on the capabilities and complexity of a specific system, the cost to purchase such equipment can range from several hundred thousand dollars up to several million dollars. The majority of these systems are used in the aerospace industry to apply coatings to aircraft turbine engine and rocket engine parts, and in the biomedical industry to apply coatings to dental or orthopedic implants.

It is important to note that great care must be exercised after spray operations with a reactive feedstock material. When venting any type of inert chambered thermal spray system back to ambient atmosphere, freshly generated, very fine reactive material with a large surface area that has not reacted during the process may react exothermically with atmospheric oxygen when it is exposed to air. It may spontaneously ignite, possibly in a violent manner. Experts knowledgeable in the safe operation of such equipment should be consulted to determine proper procedures and safety measures for a specific material and system.

Table 11.2
Examples of Materials Commonly Applied by Plasma Spraying

Metals	Carbides*	Oxides	Cermets
Aluminum	Chromium-carbide	Alumina	Alumina-nickel
Chromium	Titanium-carbide	Chromia	Alumina-nickel-aluminide
Copper	Tungsten-carbide	Magnesia	Magnesia-nickel
Molybdenum		Titania	Zirconia-nickel
Nickel		Zirconia	Zirconia-nickel-aluminide
Nickel-aluminum alloys			
Nickel-chromium alloys			
Tantalum			
Tungsten			

*Normally combined with a metal powder that serves as a binder.

Photograph courtesy of Sandia National Laboratories

Figure 11.22—Vacuum Plasma Spray System

RADIO-FREQUENCY PLASMA SPRAY EQUIPMENT

The use of radio-frequency (rf) inductively coupled plasma spray systems is limited, but growing, because they afford unique advantages for some applications. For example, significantly longer dwell times in the hot zone of the plasma provide opportunities for reactive spraying in which the spray powder (for example, titanium) could be chemically reacted with the plasma gas (for example, nitrogen) during the deposition process. Also, since there are no metal electrodes, continuous operation with reactive as well as inert plasma gases is possible without gun deterioration or deposit contamination due to erosion of metal electrodes. This permits the use of a wide range of plasma gases include air, argon, nitrogen, methane, and even oxygen. Plasma stability, power conversion efficiency, and maximum heat content are all related to the gas flow pattern, and this pattern varies with different gases. An inductively coupled plasma spray application is shown in Figure 11.23.

Control of the plasma effluent is obtained by varying the type and flow rate the plasma gas, the power input to the induction coil, and the design of the exit nozzle. Gas velocity, although generally lower than that used for traditional dc arc plasma spray, can be increased by changing the exit nozzle size and through the use of a low-pressure chamber.

The inductively-coupled plasma spray process has been used to spray inter-metallic powders, such as titanium aluminide for carbon-fiber-reinforced and metal-matrix composites (MMCs) with excellent results. Because an electrode (that might deteriorate during

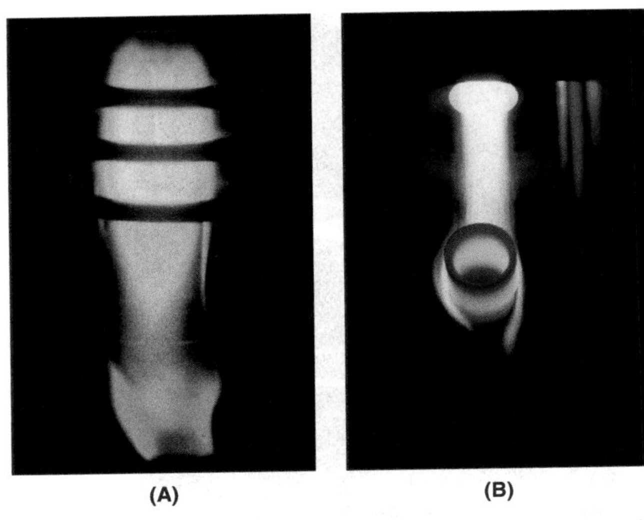

Photographs courtesy of Plasma Technology Research Center (Centre de Recherche en Technologie des Plasmas [CRTP]), University of Sherbrooke

Figure 11.23—(A) Inductively Coupled Plasma Spray Gun Showing Hot Plasma Region at the Core of the Induction Coil and (B) Workpiece Being Spray Coated

operation) is not used, a potential source of contamination is eliminated and deposits with a higher purity are possible. The predominant usage of induction plasma spray is in materials processing, that is, spheroidization, reactive spray, and chemical refinement.

COLD SPRAY EQUIPMENT

Commercially available cold spray equipment ranges from highly sophisticated, fully automated robotic spray systems to small, simple, and comparatively inexpensive hand-held systems. The smaller hand-held systems can spray only a limited range of materials that are relatively easy to deposit; however, they are useful for applications such as spraying aluminum or zinc for corrosion control.

A typical cold spray system, as shown in Figure 11.24, consists of a control console that regulates gas flow and overall system operation, an electrical resistance heater that heats the incoming process gas, a spray gun that may be hand held or mounted on a robot arm, and a powder feeder. The powder feeder usually is a high-pressure, volumetric feeder which also could be used for HVOF. The main process gas is typically heated to temperatures in the range of 250°C to 450°C (480°F to 840°F) to increase the velocity of both the gas jet and spray particles. Some cold spray systems are capable of heating the main process gas up to 800°C (1472°F), which allows them to achieve higher velocities and spray a much broader range of materials with nitrogen, thus reducing the need to use more expensive helium. Heating the process gas also significantly reduces gas consumption. Although the gas is hot as it enters the converging section of the spray gun nozzle, it rapidly cools as it expands in the diverging section of the nozzle, and the gas temperatures at the nozzle exit actually may be at or below room temperature in some spray operations.

The most common process gases are helium and nitrogen. Helium produces the highest particle velocity, which means that it can deposit the widest range of materials and generally yields the highest quality of spray-deposited material. However, helium is substantially more expensive than nitrogen, and the high cost of helium may constitute a barrier for some commercial applications. Gas consumption for the cold spray process is considerably higher than that of other thermal spray processes. However, if the production volume is high enough, a helium recycling system can be used to reclaim and recycle most of the expensive process gas, thus significantly reducing total operating costs. For a limited range of materials and applications, very inexpensive compressed air can be used as the cold spray process gas, but the coating quality with compressed air is usually inferior to that possible with nitrogen or helium.

Feedstock powders for cold spraying typically have mean diameters in the range of 10 µm to 30 µm (0.0004 in. to 0.0012 in.), somewhat finer than the powders normally used for other thermal spray processes. However, experimental gun designs have successfully cold-sprayed powders with much larger mean diameters (ranging above 100 µm [0.004 in.]), and it has been observed that the critical velocity decreases slightly with increasing particle size. Since the feedstock powder is not melted and re-solidified, cold spraying is much more sensitive to the precise physical characteristics of the feedstock powder particles, for example, phase composition and hardness, as compared to feedstock powders for traditional thermal spray technologies.

CHARACTERISTICS OF SPRAY-DEPOSITED MATERIALS

The basic premise for thermal spray processes is that all, or at least most, of the spray particles arrive at the substrate surface at a temperature and velocity that will produce the desired coating microstructure and proper-

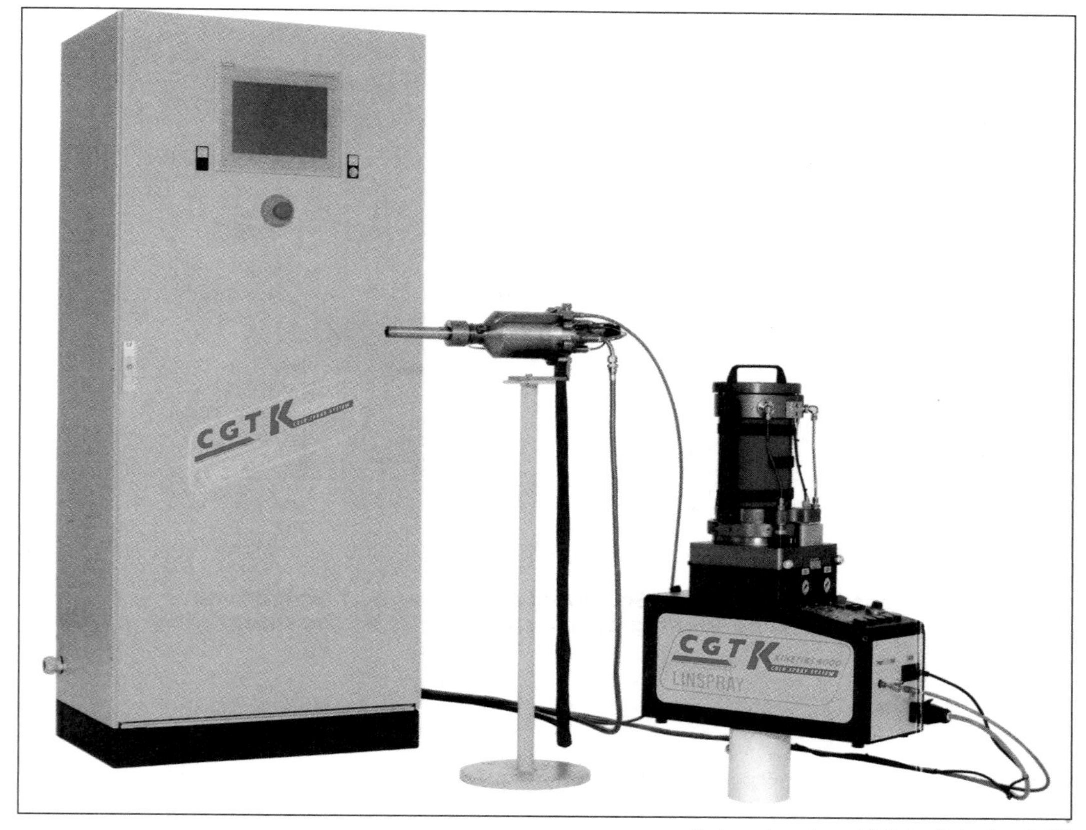

Photograph courtesy of Cold Gas Technology GmbH

Figure 11.24—Commercial Cold Spray System, Including Control Console, Gas Heater, Spray Gun, and Powder Feeder

ties. On impact with the substrate, molten droplets (or solid particles for processes such as cold spraying) flow out and bond to the underlying material to build up the coating. Individual flattened droplets or particles, called *splats,* are the basic building blocks of a sprayed coating microstructure, and the interfaces between the splats, called *splat boundaries,* strongly influence both the physical and mechanical properties of spray-deposited materials.

The characteristics of splat impact and flow directly affect the microstructure and properties of the coating. Relevant factors include the degree of spreading, or flattening ratio of the splat; the relative rates of solidification versus spreading (that is, whether solidification begins before or after droplet spreading is complete); splat morphology (fragmented or disc shaped); and the behavior of subsequent splats.

Achieving consistency and control of the coating structure requires an understanding of the relationships between spray process variables, coating microstructure development, coating properties, and ultimately, the performance of the coating. This is not a simple task, as these relationships often are complex and dependent on the materials and process.

MICROSTRUCTURE OF THE COATING

Visual observation of the microstructure of a coating is an effective method of gathering information about the overall quality. Typically, a metallographic specimen of a transverse (through-thickness) section of a coating is prepared to reveal its structure under optical or electron microscopy. Figure 11.25 shows a micrograph of a plasma-sprayed molybdenum coating. The layers apparent in this micrograph are individual splats that have solidified on top of one another. Because the heat flow is predominantly through the base of the splat as it flows out and solidifies, the small columnar grains within each splat are oriented parallel to the direction

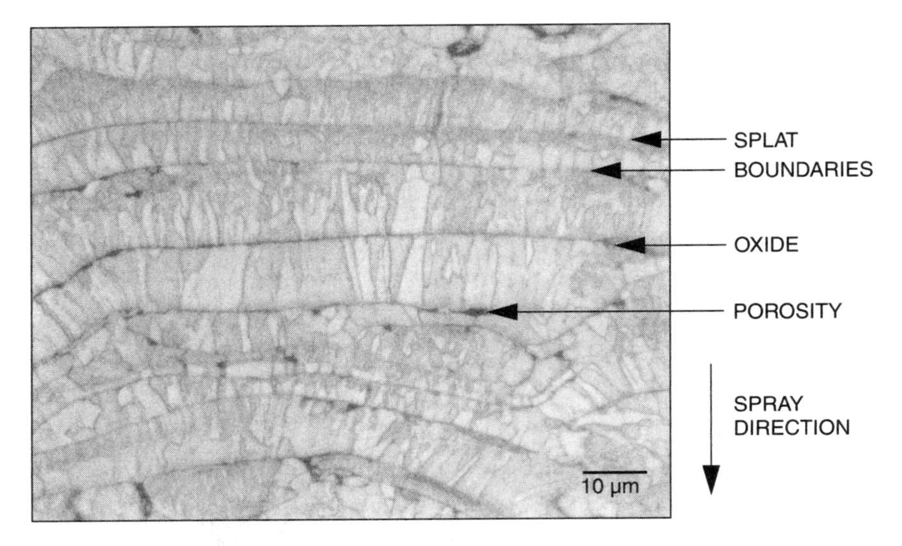

Micrograph courtesy of *Journal of Thermal Spray Technology*, 1996, Vol. 5(4), p. 452, ASM International

Figure 11.25—Cross-Sectional Micrograph of Molybdenum Applied by Plasma Spraying in an Air Environment

of heat flow during solidification (perpendicular to the substrate surface). The splat boundaries show evidence of pores (black areas) and oxides (gray areas). Unlike wrought or cast materials, the properties of individual grains and grain boundaries do not determine the physical and mechanical properties of the coating. Rather, the properties of the splat boundaries dominate the overall response of the coating and determine many of its physical and mechanical properties.

Much can be gathered from visual examination of metallographic cross sections, including information about the coating-substrate interface, unmelted or partially melted particles, oxide inclusions and the presence of porosity. The microstructure and properties for a given coating can vary dramatically according to the chosen process, feedstock material, and deposition parameters. For example, Figure 11.26 shows a cross-sectional micrograph of two copper coatings sprayed in an ambient air environment with the same feedstock powder, but with two different deposition processes, plasma spraying, shown in Figure 11.26(A), and cold spraying Figure 11.26(B). The plasma-sprayed sample (A) contains approximately 5% porosity (black areas) and 1.7 wt.% oxide (gray areas), with a conductivity only 15% of oxygen-free high-conductivity (OFHC) copper. The cold-sprayed sample (B) shows less than 1% porosity, 0.3 wt.% oxide, and has a conductivity that is 85% of OFHC copper. Thus, although the feedstock powder and ambient atmosphere were the same for both spray processes, the plasma-sprayed coating shows substantially more oxidation and porosity due to the much higher temperatures inherent in that process. In fact, due to the very low process temperatures of cold spraying, there was no measurable increase in the oxygen content of the sprayed copper coating material as compared to the oxygen content starting feedstock powder (also 0.3 wt.% oxide).

Proper techniques for the sectioning and polishing of a sample must be utilized in order to obtain an accurate representation of a sprayed coating structure. Thermal spray coatings are not homogenous structures, and therefore specimens cannot be prepared and polished in the same manner as wrought or cast structures. Subjecting a thermal spray coating to sectioning or polishing procedures that are too aggressive may cause hard, brittle oxides within the structure to be preferentially removed from the structure. This phenomenon is termed *pull-out*, which leaves behind empty pockets that appear to be pores. Due to the preferential removal of oxide, subsequent examination of this type of improperly prepared metallographic specimen will erroneously indicate a porosity content that is too high and an oxide content that is too low. The same applies to other hard or brittle features, such as carbides. Conversely, some coating materials, such as aluminum and the superalloys, are prone to smearing during polishing. If improperly prepared, the oxides and pores within the structure will be covered over by the smearing of softer phases and will not be visible. Recommended preparation procedures for specimens of specific thermal spray

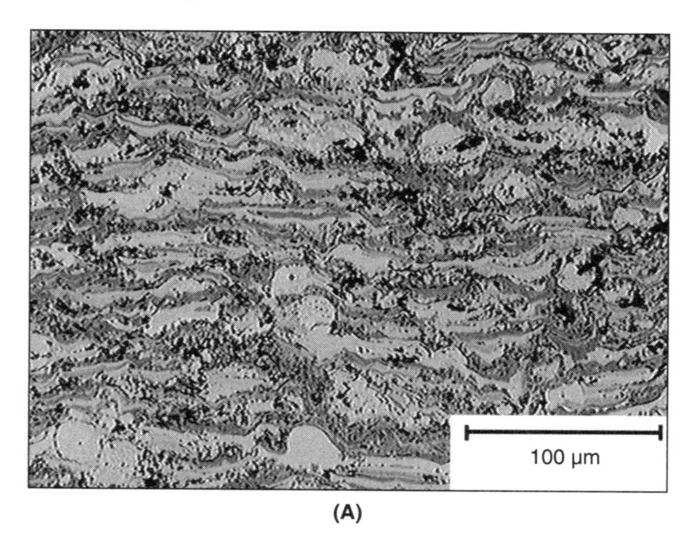

(A)

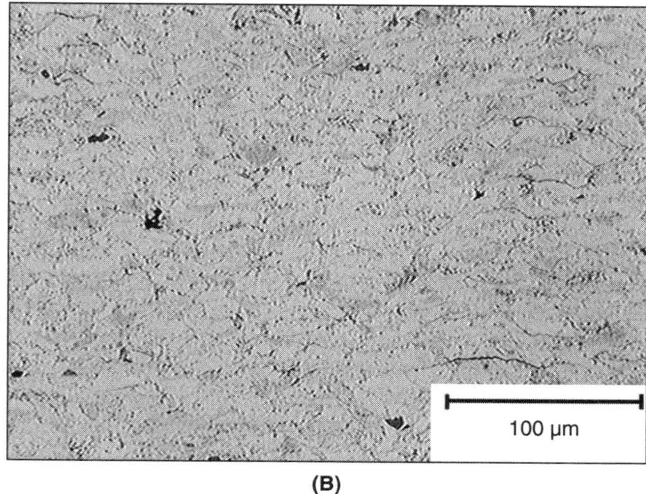

(B)

Micrographs courtesy of Sandia National Laboratories

Figure 11.26—(A) Copper Plasma-Sprayed in Ambient Air and (B) Copper Cold-Sprayed in Ambient Air

coating materials can be obtained from major metallographic equipment suppliers and from some professional organizations that specialize in thermal spraying and metallography.

CHEMISTRY OF THE COATING

Metals and alloys deposited by thermal spraying may not retain their exact original chemical composition. For example, when plasma and electric arc methods are used, appreciable amounts of high-vapor pressure constituents may be lost through vaporization due to the high temperatures produced by the arc. In some cases, the starting chemistry of the feedstock material is adjusted to compensate for these changes. The spray process or spray parameters also can be modified to minimize changes in chemistry during the deposition process. In addition to chemistry changes due to vaporization, the presence of oxidation may be significant for coatings when thermal-sprayed in an ambient air environment. This is especially true when air is used as the propellant or when pure oxygen is present, as in a combustion-based spray process.

Ceramics

Unlike metals, spraying an oxide ceramic such as alumina or zirconia in an ambient air environment has little effect on the chemical composition of the sprayed material. However, the high cooling rates associated with splat cooling often produce metastable phases in the coating. For example, most plasma-sprayed aluminum oxide (Al_2O_3) coatings contain a substantial amount of metastable gamma-phase alumina. Other commonly observed features in thermal spray coatings, especially ceramic coatings, are micro- and macro-cracking. Cracking within the structure results from quenching stresses during deposition. When a molten droplet impacts the relatively cold substrate, the splat solidifies and then shrinks as it cools, creating a large tensile stress, especially in relatively brittle, high-modulus materials like ceramics. Vertical micro-cracking between layers helps the structure to relieve these stresses. Large vertical cracks, sometimes extending throughout the entire thickness of the coating, are called macro-cracks. Macro cracking can be beneficial as a mechanism to relieve stress due to the thermal expansion mismatch between the substrate and the coating. However, it also can provide conduits for corrosive media to pass through the coating and reach the underlying substrate material.

Non-oxide ceramics (that is, borides, nitrides, and carbides) often are difficult or impossible to spray as pure materials, because many of these materials tend to dissociate or oxidize rather than melt when subjected to high temperatures. While it is true that cemented carbide coatings, such as WC-Co and Cr_3C_2-NiCr, commonly are sprayed to provide wear-resistant coatings, the deposition conditions are adjusted so that only the matrix metal is melted or softened, and the hard carbide phase remains in the solid state during deposition.

PHYSICAL AND MECHANICAL PROPERTIES

The physical and mechanical properties of a spray-deposited material usually differ substantially from the properties of that same material in a wrought or cast form. Defects (such as oxides and pores) in sprayed coating microstructures degrade the properties of the sprayed deposits, affecting properties such as tensile strength, ductility, and thermal and electrical conductivity. This contrasts with wrought or cast materials, which typically have fewer defects. Bulk densities for a given coating material are almost always lower than those of their wrought or cast counterparts due to the presence of porosity. The percentage of porosity is largely dictated by the process and parameters employed. As a general rule, processes that produce larger spray droplets with lower impact velocities, such as arc spraying, tend to produce more porosity than those that deposit smaller droplets with much higher impact velocities, such as high-velocity oxyfuel. Coating processes that minimize splat boundary defects, such as porosity and oxidation, tend to produce the best physical and mechanical properties in the coating. Post-deposition heat treatments also can be used to reduce residual stress and promote recrystallization or diffusion across splat boundaries in order to enhance mechanical and physical properties. For example, American Society of Testing and Materials (ASTM) tensile specimen bars were machined from a 1-cm (0.039-in.) thick cold-sprayed pure aluminum deposit; some of these bars were then given a post-deposition stress-relief heat treatment. Bars tested in the as-deposited condition were found to have no measurable ductility. However, the heat-treated bars had ductilities ranging up to 10%, comparable to wrought 1100 aluminum in an H14 condition.

Residual Stress

Stress begins to accumulate in molten thermal spray droplets as soon as the first splat impacts the substrate. The molten splat undergoes rapid cooling and begins to contract. The solid material under the splat (the substrate) restricts the ability of the splat to shrink completely. The stress resulting from droplet quenching is therefore tensile in the coating material. Other stresses due to a mismatch of thermal expansion between the substrate and the coating also may have an important influence. However, the stress due to a mismatch in thermal expansion coefficients can be either tensile or compressive, depending on the thermal expansion coefficients of the materials and their temperature history. Another type of stress can result from the peening effect of splats impacting at high velocity. This can be very significant for processes such as high-velocity oxyfuel spraying, detonation gun spraying, and cold spraying. Peening stress is compressive and, if properly controlled, can be used to offset or even overwhelm tensile stresses in the coating material.

Post-treatments of the coating, such as heat treatments, machining, and peening also affect the stress state of the coating. Peening will increase compressive stress, while machining may increase tensile stress. Heat treatments of metallic coatings usually help relieve stresses, whereas heat treatments of ceramic coatings may cause stress-relieving micro-cracks to sinter, possibly creating increased tensile stress within the coating.

The net residual stress of the coating is the sum of all stresses within the coating. Up to a point, net compressive residual stress usually is favorable because there are no driving forces to start or extend cracks within the coating. However, excessive compressive stress may cause the coating to spall (to separate from the substrate). Residual stress in the coating often is the primary factor that limits total coating thickness, because the stress progressively increases as the thickness is increased, ultimately rising to a level that will cause cracking or spalling of the coating. For this reason, coatings with lower tensile stresses (for example, coatings of ductile metals with low melting points) and coatings with residual compressive stresses (for example, coatings deposited with high-velocity oxyfuel spraying, detonation flame spraying or cold spraying) typically can be deposited to a much greater thickness.

HARDNESS

The hardness of a sprayed material varies greatly, depending on the nature of the material and the specific deposition process and conditions. Accurately measuring the hardness of a sprayed material also is more difficult than similar measurements of wrought or cast materials. This is due in part to the heterogeneous nature of sprayed materials that contain pores and oxides. In addition, the coating may be so thin that the substrate may influence hardness measurements. The selection of the appropriate hardness testing equipment, including the loading force as well as the type and shape of the indenter, is very important. Table 11.3 provides some approximate guidelines for minimum spray-deposit thickness for the various Rockwell scales.

Rockwell hardness measurements usually are taken on slightly polished coating surfaces with the indenter positioned perpendicularly to the top surface. Micro-hardness measurements, Vickers or Knoop, are usually performed on polished coating cross sections. In order to obtain an accurate micro-hardness measurement,

Table 11.3
Minimum Deposit Thickness for Rockwell Hardness Tests

Rockwell Scale	Minimum Thickness	
	mm	in.
15N	0.38	0.015
30N	0.64	0.025
45N	0.89	0.035
A	1.0	0.040
B	1.5	0.060
C	1.8	0.070
D	1.3	0.050

impressions should be taken in a zigzag pattern throughout the entire coating thickness. Specific recommendations for performing hardness tests on thermal sprayed coatings can be obtained from coating suppliers, metallographic equipment companies, and from the appropriate ASTM standards.

ADHESIVE AND COHESIVE STRENGTH

The strength of the bond between a sprayed coating and the substrate is called *adhesive strength* or *adhesion*. The strength of the bond between successive layers of deposited material within the coating itself is termed *cohesive strength* or *cohesion*. A standard test to determine the adhesive or cohesive strength of thermal-sprayed deposits is described in *Standard Test Method for Adhesion or Cohesive Strength of Flame Sprayed Coatings* ASTM C633-01.[5] In this test method, one end of a solid right cylinder is spray-coated, and the uncoated end of a second right cylinder is adhesively bonded to the coated surface of the first cylinder using standardized procedures and adhesives. The bonded assembly is then pulled apart in a tensile testing apparatus, and the failure stress as well as the mode of failure is determined. If the failure occurs between the coating and the substrate, adhesive strength is measured. If the failure is within the coating, cohesive strength is measured. In some instances, failure occurs in the adhesive bonding agent that was used to "glue" the coated and uncoated cylinders together. In this case, the test merely indicates that the ultimate adhesive and cohesive strengths of the sprayed coating must each be greater than the failure strength measured for the adhesive bonding agent.

While the strength values provided by the ASTM test are useful for relative comparisons, they typically do not reflect true bond strength. The true strength of the bond between a spray deposit and the substrate depends on many factors, including the following:

1. Substrate material and geometry;
2. Preparation of the substrate surface;
3. Spray angle relative to the substrate;
4. Substrate temperature during deposition;
5. Bond coating, if applicable;
6. Type of coating material and the nature of the microstructure;
7. Thickness of the coating; and
8. Post-spray thermal treatment, if applicable.

The coating should be applied to the end of a standard ASTM test specimen cylinder in a manner that simulates the real process as closely as possible, using the same substrate material, surface preparation, workpiece temperature, and spray conditions. The coating should be of uniform thickness, and an alignment fixture should be used during the bonding process to assure proper alignment of the test cylinders. The type of epoxy used is critical, because its viscosity may affect its ability to penetrate the coating. With a porous coating, penetration of the adhesive into the coating may produce invalid test results. Typically, the results of repetitive ASTM C633-01 testing will exhibit some scatter due to variables, such as edge effects or slight misalignment. Therefore, it is advisable to run multiple samples for a specific test condition in order to average the statistics over the scatter in the test results. Typical bond strengths for some common spray materials and processes are presented in Table 11.4.

DEPOSIT DENSITY

Thermal spray deposits have densities less than 100% compared to the same material in wrought form because the structures are porous and contain oxides. Porosity in thermal spray coatings either can be interconnected or isolated, and it is therefore difficult to accurately quantify. A simple, inexpensive method is to superimpose a grid over a polished cross section of the coating while examining it optically and to count the number of grid squares occupied by pores. A computerized version of this traditional method is available in standard image analysis software that commonly is used to quantify porosity levels in digitized micrographs. However, care must be taken to properly determine threshold levels for image analysis, as the gray levels for pores and oxides may overlap. Special care

5. American Society of Testing and Materials, *Standard Test Method for Adhesion or Cohesive Strength of Flame Sprayed Coatings*, ASTM C633-01, West Conshohocken, PA: American Society for Testing and Materials.

Table 11.4
Bond Strengths of Some Typical Thermal Spray Coatings

	Bond Strength, MPa (psi)			
Material	**Plasma-Sprayed**	**Arc-Sprayed**	**Low-Velocity Combustion-Sprayed**	**High-Velocity Combustion-Sprayed**
Molybdenum	24.0 (3500)	54.8 (8000)	Wire: 22.6 (3300) Powder: 24.7 (3600)	N/A
Nickel-Alumina	20.6 (3000)	54.8 (8000)	Wire: 21.6 (3150) Powder: 18.8 (2750)	54.8 (8000)
316 Stainless steel	13.7 (2000)	41.1 (6000)	Wire: 13.7 (2000) Powder: 10.3 (1500)	54.8 (8000)
Corundum	20.6 (3000)	N/A	Powder: 17.1 (2500)	13.7 (2000)
Tungsten carbide-12 cobalt	48.0 (7000)	N/A	N/A	54.8 (8000)

also must be taken in the preparation of metallographic samples for these measurements to avoid increasing apparent porosity due to pull-out, or reducing it due to smearing.

Other methods for determining density include water or toluene immersion, mercury porosimetery, and helium pyncnometry (pyncnometry determines pore volume based on precise measurements of gas pressure change in a known sample volume). However, all displacement and adsorption methods have limitations because some pores may be completely closed and isolated within the coating structure, while other pores may be directly or indirectly connected to the external surface of the sample. The various measurement techniques assume either that all pores are completely closed or that all pores are surface connected, so some degree of error results when a mixture of these two conditions exists.

The porous nature of sprayed coatings can be advantageous, especially for applications such as bearing surfaces and thermal barriers. The porosity within the structure permits oil absorption and retention as well as an escape for foreign material from actively loaded bearing surfaces. Porosity also is used to decrease the thermal conductivity of ceramic coatings, such as partially stabilized zirconia (PSZ), which is widely used for thermal barrier coatings to protect critical components in both aircraft and land-based turbine engines. For other applications, such as corrosion protection, porosity within a sprayed coating can be detrimental. In order to enhance performance in such cases, it is common to apply sealants that have been specially developed for this purpose after the sprayed coating is deposited.

As a general rule, dense coatings can be achieved through proper particle melting or high-impact velocity, or both. The vacuum plasma spraying and cold spraying processes can produce coatings close to wrought density with less than 0.5 vol.% porosity. The high-velocity oxyfuel spray and detonation spray processes also can produce very dense coatings, due to the high velocity of particle impact. Typically 1 vol.% to 3 vol.% porosity can be expected for HVOF coatings.

Plasma and arc spraying have similar porosity contents of 3 vol.% to 15 vol.%, varying with the type of material being sprayed and the parameters being used. Combustion processes using powder and wire are both low-temperature, low-velocity processes. They typically produce coatings with 8% to 20% porosity. Table 11.5 shows comparisons of the density of some wire flame-sprayed deposits for several metals to the density of the initial wire feedstock.

Table 11.5
Deposit Densities of Some Wire Flame-Sprayed Metals

	Density, g/cm^3 (lb/ft^3)	
Metal	**Wire Feedstock**	**Wire Flame-Sprayed Deposit**
1100 Aluminum	2.71 (169)	2.41 (150)
Copper	8.97 (560)	7.50 (468)
Molybdenum	10.2 (637)	9.02 (563)
AISI 1025 steel	7.86 (491)	6.75 (422)
304 Stainless steel	8.03 (501)	6.89 (430)
Zinc	7.14 (446)	6.84 (427)

POST-SPRAY TREATMENTS

Many products coated by means of thermal spraying can be placed directly into service, but some must undergo post treatments to meet product or performance specifications. Post treatments may include sealing and surface finishing, such as grinding or machining, and other treatments, such as buffing, polishing, abrasive tumbling, honing, or lapping.

SEALING

Sprayed deposits commonly are sealed to lengthen the service life of the coating or to prevent corrosion of the substrate, or both. Sprayed deposits of aluminum, zinc, or Al-Zn alloys may be sealed with vinyl sealant solutions, either clear or aluminum-pigmented. The sealer may be applied to fill only subsurface pores in the deposit, or both subsurface pores and surface irregularities. The latter technique provides a smooth coating that resists industrial atmospheres. Vinyl coatings may be applied with a brush or spray gun.

Sealers also are used on coated machine components. If the spray deposit will be exposed to acids in service, it is recommended that the surface be sealed with either a wax that has a high melting point or a phenolic-paint sealer. Spray deposits on high-pressure hydraulic rams, pump shafts, and similar parts should be sealed with air-drying phenolics to prevent the seepage of liquid through the coating around the packing. Thermal spray can be used to reclaim pressure cylinders of all types. Prior to finish grinding, the cylinder bore is sealed with a phenolic. This prevents grinding-wheel particles from becoming embedded in the pores of the sprayed metal and causing premature wear.

Epoxies, silicones, and similar materials are used as sealants for certain corrosive conditions. Vacuum impregnation with a plastic solution also is possible.

DIFFUSION SEALING

In some applications, a thin layer of aluminum may be diffused into a steel or silicon-bronze substrate at 760°C (1400°F). The diffused layer can provide corrosion protection against hot gases up to 870°C (1600°F). After depositing the aluminum, the workpiece can be coated with an aluminum-pigmented bitumastic sealer or another suitable material to prevent oxidation of the aluminum during the diffusion heat treatment. Similarly, in some aircraft applications the diffusion temperatures are dependent on the base material to which the aluminum is applied.

SURFACE FINISHING

Techniques for surface finishing thermal spray deposits differ somewhat from those commonly used for wrought or cast metals. Excessive pressure or heat generated in the coating during a finishing operation can cause damage, such as cracking, crazing, or separation from the substrate. Since the composition of an as-sprayed deposit is an aggregation of individual particles, improper finishing techniques can dislodge particles singly or in clusters. This may cause a severely pitted surface. The deposited particles should be cleanly cut and not pulled from the surface. Even so, if the coating is porous, the finished surface probably will not be shiny but may have a matte finish due to the porosity of the deposit. However, dense hard deposits can be polished or superfinished to mirror-like surfaces.

The selection of a finishing method depends on the type of deposit material, the hardness, and the coating thickness. Consideration should be given to the properties of the substrate material as well as dimensional and surface roughness requirements. Spray deposits of soft metals usually are finished by machining, especially those applied to machine components. The best finishes usually are obtained using carbide tools and high cutting speeds. More often, however, sprayed deposits are finished by grinding, particularly for hardfacing (for example, cemented carbide) and ceramic coatings. Various other finishing methods occasionally are used. These methods include buffing, tumbling, burnishing, belt polishing, lapping, and honing.

Machining

Tungsten carbide tools are commonly used for machining sprayed metal deposits and fused coatings. Proper tool angles are critical to successful machining of sprayed coatings. The surface speeds and the depth of cut are of equal importance. Improper tool angle and tool pressure can result in excessive surface roughness and the destruction of the bond between the coating and the substrate.

A cutting tool with a slightly rounded tip and a rake angle of 3° is recommended for many applications. On outside circumferences, the tip of the tool should be set 3° below center; in bores, this should be 3° above center. This will help limit the stress on the deposit. Peripheral speed should not exceed 400 mm/s (75 ft/min). The feed rate should be slow with light cuts for best surface finish.

Special cutting tools, such as oxide-coated carbide, cubic boron-nitride, ceramics, cermets, and diamonds may be used to machine very hard metal, ceramic, and cermet deposits. In many cases, machining with such tools may replace the grinding of intricate shapes and large pieces. The machining of flat deposits requires extreme care at the corners and edges to avoid damage. Shallow cuts and low feed rates should be used.

Grinding

It is common in many applications, especially cylindrical shafts, to deposit an extra thickness of material and then grind the surface to achieve a tight tolerance dimension or a specific surface finish. While it is possible to achieve excellent results with a wide range of materials and geometries, improper grinding techniques may severely damage a sprayed coating, so the proper selection of grinding wheels and speeds is very important.

Wet Grinding Metal Deposits. Wet grinding with large, wide wheels is the preferred method of grinding metal deposits. Wet grinding avoids heat buildup in the workpiece and permits closer tolerances than dry grinding. Grinding wheel manufacturers can recommend wheel types and grinding procedures for various metal deposits based on a particular type of grinding machine.

If it is necessary to dry-grind metal deposits, as is done with portable grinders mounted on a lathe, most of the material to be removed should be removed first by machining. The deposit is then ground to the required dimensions and final finish. Wheels used for dry grinding operations may be either aluminum oxide or silicon carbide, depending on the metal to be ground. The factors to be considered in selecting a wheel for grinding a spray deposit are similar to those for grinding the same metal in wrought or cast form. The grinding technique should be designed to minimize heat buildup in the deposit. The structure of the wheel should be as open as possible, and the grain size as coarse as possible, consistent with finish requirements. The wheel should be narrow, the infeed light, and the traverse as fast as possible without spiraling.

When grinding equipment is not available, metal deposits can be machined to within 50 μm to 150 μm (0.002 in. to 0.006 in.) of final size. Then they can be finished to size with a belt polishing unit. Close tolerances and fine finishes are possible using belt polishing with the proper selection of abrasive type and grit size.

Fused Deposits. Because most fused deposits (sprayed deposits that are re-melted, or fused, after they are deposited) are designed for hardfacing purposes, grinding is usually the most economical method of finishing them. Although most fused deposits can be machined with the proper type of cutting tool, close tolerance work is difficult because of rapid tool wear and the excessive heat generated. Dry grinding may be suitable for some operations, but the heat and fast wheel wear make it difficult to achieve close tolerances. Wet grinding can produce close tolerance components, fine finishes, and economical stock removal rates. Nickel-base alloys are best ground with silicon-carbide grinding wheels, and cobalt-base alloys with aluminum-oxide grinding wheels.

Grinding wheel manufacturers should be consulted for recommendations of the appropriate type for the job. Good practice usually suggests the coarsest wheel consistent with finish requirements, an open structure or soft bond, as large a wheel as possible, and good wheel-dressing techniques. The surface finish of fused coatings often can be improved after grinding by polishing with fine-grit belts.

Ceramic Deposits. Although individual particles of a thermally sprayed ceramic coating may have extreme hardness, sprayed ceramic coatings usually can be finished by conventional grinding techniques with standard equipment. Successful finishing requires selecting the proper grinding wheel, in some cases a diamond wheel, and following correct procedures. Flood-cooling should be employed during grinding and water cooling generally is preferred; however, some water-soluble coolant additives may stain light-colored ceramic deposits. General recommendations for grinding ceramic deposits are available from grinding wheel manufacturers.

Other Finishing

Other methods of surface finishing sometimes used for as-sprayed and fused deposits include the following:

1. Manual buffing or polishing,
2. Abrasive tumbling,
3. Honing, and
4. Lapping.

As-sprayed or as-machined deposits may be buffed or polished manually with abrasive stones, cloth, or paper. Abrasive tumbling can be used to polish the surface of small workpieces by removing high spots. An abrasive medium and cleaners, usually with a liquid, are vibrated or rotated in a drum in which the workpieces are tumbled until the correct finish is achieved.

Honing is done with abrasive stones mounted in a loading device. The workpiece normally moves in one direction or rotates, while the stones are oscillated under pressure transverse to the workpiece motion. Lapping is done with a fine, loose abrasive mixed with a carrier such as water or oil. The mixture is spread on lapping shoes or plates that are then rubbed against the spray deposit. The lap rides against the deposit, and the relative movements are continually changed.

APPLICATIONS

Much like welding, thermal spray processes are used in a broad range of applications in many major industries, including aerospace, power generation, petrochemical, automotive, marine, biomedical, primary metal producers, papermaking, printing, textiles, and many others. Thermal spray coatings are used to combat corrosive environments, to prevent high-temperature oxidation, to build-up or reclaim worn components, to minimize wear, to give specific surface properties to a bulk material, and for many other purposes.

SERVICE PROPERTIES

A few examples of industrial thermal spray applications are shown in Table 11.6, which illustrates the wide range of materials, processes, and applications in commercial use.

Table 11.6
Some Representative Examples of Industrial Thermal Spraying Applications

Industry	Application	Coating Purpose	Spray Process	Coating
Automotive	Seam welds	Filler	Arc	Silicon-bronze
	Piston cylinders	Wear resistance	Plasma	Steel
	Piston rings	Wear resistance	Plasma	Molybdenum
	Shift forks	Wear resistance	Flame	Molybdenum
	Oxygen sensors	Thermal barrier	Flame	Spinel
	Heat exchangers	Corrosion resistance	Arc	Zinc
Pulp and paper	Yankee dryer rolls	Wear resistance	Arc	FeCrBSi
	Center press rolls	Wear resistance	Plasma	Chrome-oxide
	Calendar rolls	Wear resistance	High-velocity oxyfuel (HVOF)	Chrome-carbide
	Boiler tubes	Wear, corrosion resistance	Arc	Nickel-chrome
	Corrugated rolls	Wear resistance	HVOF	Tungsten- carbide
	Digesters	Wear, corrosion resistance	Arc	Alloy 625
	ID fans	Wear resistance	HVOF	Tungsten-carbide
Aerospace	Aircraft engines	Thermal barrier	Plasma	Yttria-zirconia
	Aircraft engines	Abradable clearance control	Flame	Aluminum-polyester
	Aircraft engines	Wear resistance	HVOF	Tungsten-carbide
	Aircraft engines	Dimension restoration	Arc	Nickel-aluminum
	Landing gear	Wear, corrosion resistance	HVOF	Tungsten-carbide
	Airframe	Conductivity	Flame	Aluminum
	Airframe	Flap tracks	HVOF	Tungsten-carbide
Petrochemical	Ball valves	Wear, corrosion resistance	HVOF	Tungsten-carbide
	Gate valves	Wear, corrosion resistance	HVOF	Tungsten-carbide
	Choke stems	Wear, corrosion resistance	HVOF	Tungsten-carbide
	Piston rods	Wear resistance	Flame	Chrome-oxide
	Offshore oil rigs	Corrosion resistance	Flame	Aluminum
	Pump housings	Dimension restoration	Arc	Aluminum-bronze
	Compressor cylinders	Dimension restoration	Arc	420 stainless steel
	Processing tanks	Corrosion resistance	Flame	Aluminum

Corrosion and Oxidation Protection

Thermal spray coatings provide effective protection against many types of corrosive attack and are widely used to protect iron and steel. Both the coating material and application procedures are determined by service conditions. Undercoatings for organic materials, such as paints and plastic finishes, also are commonly applied by thermal spraying. For example, a thick layer of aluminum, zinc, or Al-Zn alloy can galvanically protect steel against oxidation, with the coating material itself sacrificially oxidizing, thus inhibiting oxidation of the underlying steel.

Specifications for the protection of steel are published in the American Welding Society (AWS) document, *Specification for the Application of Thermal Spray Coatings (Metallizing) of Aluminum, Zinc and Their Alloys and Composites for the Corrosion Protection of Steel,* AWS C2.23M/C2.23, NACE No. 12, SSPC-CS23.00.[6] It covers safety, job reference standards, equipment setup and preparation, surface preparation, aluminum and zinc application, and sealer and topcoat application. The document also specifies in-process quality control checkpoints.

Another AWS publication, Protection *of Steel with Thermal Sprayed Coatings of Aluminum and Zinc and Their Alloys and Composites,* AWS C2.18, covers safety, job and contract descriptions, background and requirements, selection of thermal spray coatings, operator qualification, materials and equipment; also application process methods with quality control checkpoints, job control records, maintenance and repair of thermal spray coatings, and debris containment and control.[7]

In many applications, these galvanic coatings can provide good protection alone, but even better protection often can be obtained if an organic seal coating, or paint, or both, is applied as a post-deposition treatment on top of the galvanic sprayed coating. Conversely, some other metals, such as nickel, nickel-copper alloys, stainless steels, and bronzes, are galvanically noble relative to steel. This means that spraying these metals onto steel in a corrosive environment could cause the steel to become the sacrificial material, accelerating oxidation (rusting) of the steel. In potentially corrosive environments, these materials should be used as a deposit on steel only if they are made impermeable to corrosive agents. This can be done by sealing the deposit with organic or inorganic liquids that penetrate the pores and form solid barrier layers, completely isolating the steel from the corrosive medium. It should be noted also that sealers are likely to entrap air bubbles in pores of the spray deposits, so the sealed component should not be heated excessively, because expansion of the air bubbles may rupture the sealer and thus provide a leakage path for corrosive agents.

Hard alloy deposits often are used on machine components, such as pump plungers, pump rods, hydraulic rams, packing sections of steam turbine shafts and valves. When sealed, hard-alloy deposits provide both corrosion protection and wear resistance.

Several different materials may be used to provide oxidation protection; the choice depends on the operating temperature anticipated in service. For service temperatures up to 870°C (1600°F), the workpiece can be aluminized by depositing a thin layer of aluminum. The aluminum is then diffused into the surface by a suitable heat treatment. For temperatures above 870°C (1600°F), a nickel-chromium alloy deposit may be used. This is followed by a coating of aluminum. Often, this deposit combination is then covered with an aluminum-pigmented bitumastic sealer. Finally, the workpiece is heat-treated by diffusion in a furnace, or it is placed directly in service if the operating temperature is above 870°C (1600°F). These deposits sometimes are used for cyanide pots, furnace kiln parts, annealing boxes, and furnace conveyors.

Zirconia and alumina ceramics sometimes are used for thermal barrier coatings, referred to as *TBCs*. When the workpiece will be exposed to thermal cycling, an alloy bond layer may help to minimize thermal stresses between the ceramic TBC and the substrate. The bond layer may contain nickel or cobalt, or both, and aluminum, often with additional elements such as chromium and yttrium.

The coating of steel-reinforced concrete represents another application of thermal spraying: the cathodic protection of chloride-contaminated concrete structures, such as substructures in water and tidal areas, from corrosion. Zinc thermal spray coatings are used as sacrificial anodes for passive cathodic protection and for distributed anodes used with active impressed currents. Detailed information is published in *Specification for Thermal Spray Zinc Anodes on Steel Reinforced Concrete,* AWS C2.20/C2.20M, which is formatted as an industrial process instruction and covers the application of zinc thermal spray coatings to concrete using arc spraying and flame spraying equipment.[8]

6. American Welding Society (AWS) Committee on Thermal Spray, *Specification for the Application of Thermal Spray Coatings (Metallizing) of Aluminum, Zinc and Their Alloys and Composites for the Corrosion Protection of Steel, AWS C2.23M/C2.23, NACE No. 12, SSPC-CS 23.00,* Miami: American Welding Society; an American National Standard published in cooperation with National Association of Corrosion Engineers (NACE International), Houston, Texas; and the Society for Protective Coatings (SSPC), Pittsburgh, Pennsylvania.

7. American Welding Society (AWS) Committee on Thermal Spray, *Guide for the Protection of Steel with Thermal Sprayed Coatings of Aluminum and Zinc and Their Alloys and Composites,* AWS C2.18, Miami: American Welding Society.

8. American Welding Society (AWS) Committee on Thermal Spray, 2002, *Specification for Thermal Spray Zinc Anodes on Steel Reinforced Concrete,* AWS C2.20/C2.20M, Miami: American Welding Society.

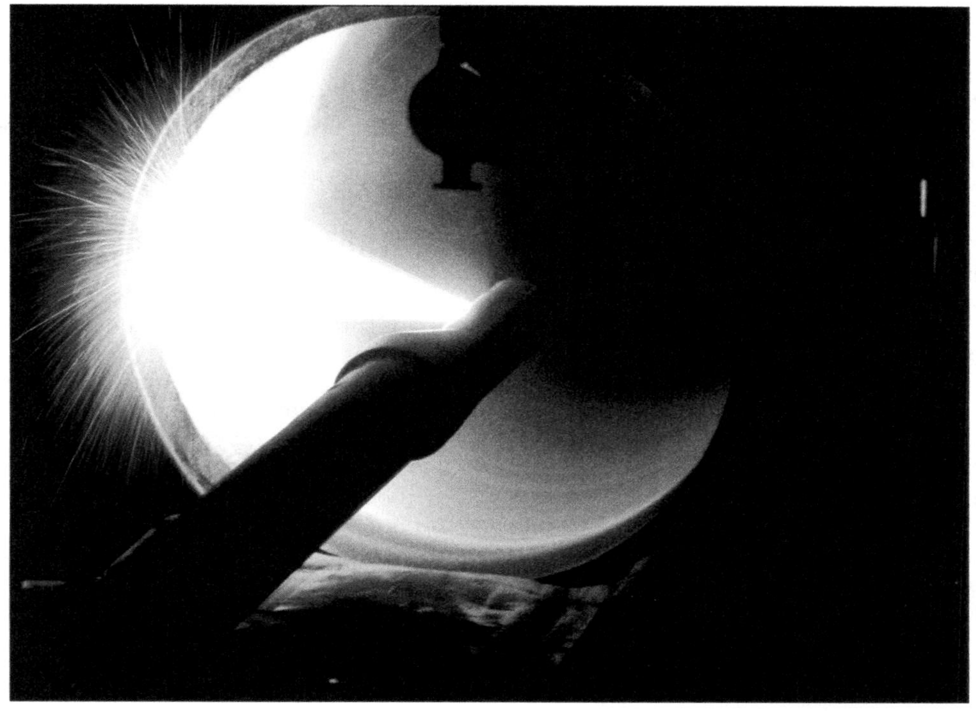

Photograph courtesy of Flame-Spray Industries, Inc.

Figure 11.27—Spraying a Wear-Resistant Coating onto the Inner Surface of a Cylinder with an Inner-Diameter Spray Gun

Wear Resistance

Thermal spraying materials for hardfacing can be used to protect mechanical components from many types of wear. The ability of somewhat porous metal spray deposits to absorb and maintain a film of lubricant is a distinct advantage in many applications. Spray-deposited surfaces often have a longer service life than the original surfaces, except when severe conditions of shock loading or abrasion are encountered. With proper equipment and sufficient working room, wear-resistant coatings can be applied to the inner surfaces of cylinder bores and other cylindrical workpieces. Figure 11.27 shows the spraying of a wear-resistant coating onto the inner surface of a cylinder with an inner-diameter (ID) spray gun. Post-deposition grinding or honing can be used to achieve a precise final internal diameter and surface finish.

With suitable masking, a low-cost base metal can be protectively coated with a high-quality, wear-resistant deposit only on the specific areas subject to wear.

Some metal deposits, such as nickel-copper alloys, nickel, and stainless steel, are virtually impervious to penetration by corrosives when they are applied in sufficient thickness and exposed only to moderate pressure. Vacuum impregnation of these surfaces with various phenol or vinyl sealant solutions or with fluorocarbon resins can further enhance performance in high-pressure service conditions. For applications that require extreme wear resistance or extreme corrosion resistance, or both, fused spray coatings can be used. Typically, these coating materials contain nickel, chromium, silicon, and boron, and are similar to nickel-base braze alloys.

ELECTRICAL APPLICATIONS

Because of factors such as oxide inclusions and microporosity, the electrical resistance of a metal spray deposit may be 50% to 100% higher than that of the same metal in cast or wrought form. These factors should be taken into consideration in design specifications of spray deposits for electrical conductors. Such applications include spraying copper onto electrical contacts, carbon brushes, glass automotive fuses, and silver or copper contacts. This is an area in which the cold spray process provides significant potential advantages, as the electrical and thermal conductivity of many cold-sprayed metals is much closer to that of wrought or cast metals.

Examples of other electrical applications of thermal spraying include electromagnetic shielding of computer cases and other electronic components by arc spraying zinc or tin-zinc onto a plastic electronics case or chassis, and condenser plates made by spraying aluminum onto both sides of a cloth tape. Various sprayed ceramic deposits also are used as electrical insulators.

FOUNDRY APPLICATIONS

Changes in the contours of expensive patterns and match plates used in foundries can be readily accomplished by the application of thermal spray deposits followed by appropriate finishing. Patterns and molds can even be repaired with wear-resistant deposits to minimize further wear in critical areas. Castings with porosity that becomes evident during machining also can be filled to reclaim the cast part that otherwise might be scrapped.

BRAZING AND SOLDERING

Thermal spraying can be used to pre-place many soldering or brazing filler metals. By spraying through a patterned mask, the filler material can be applied only in the regions where it is needed.

AIRCRAFT AND MISSILES

Thermal spray is used for gas path seals and wear-resistant surfaces for aircraft and missiles to remedy fretting, galling, and erosion. The process also is used to combat corrosion and oxidation at elevated temperatures. Deposits of alumina and zirconia are used for thermal barrier coatings to extend the safe operating range of critical metal components to higher ambient temperatures.

A robot set up for the plasma arc spraying of a bond coat onto a workpiece from the hot section of a gas turbine engine is shown in Figure 11.28. A thermal barrier coating is applied over the bond coat and forms an essential component of an engineered thermal barrier system.

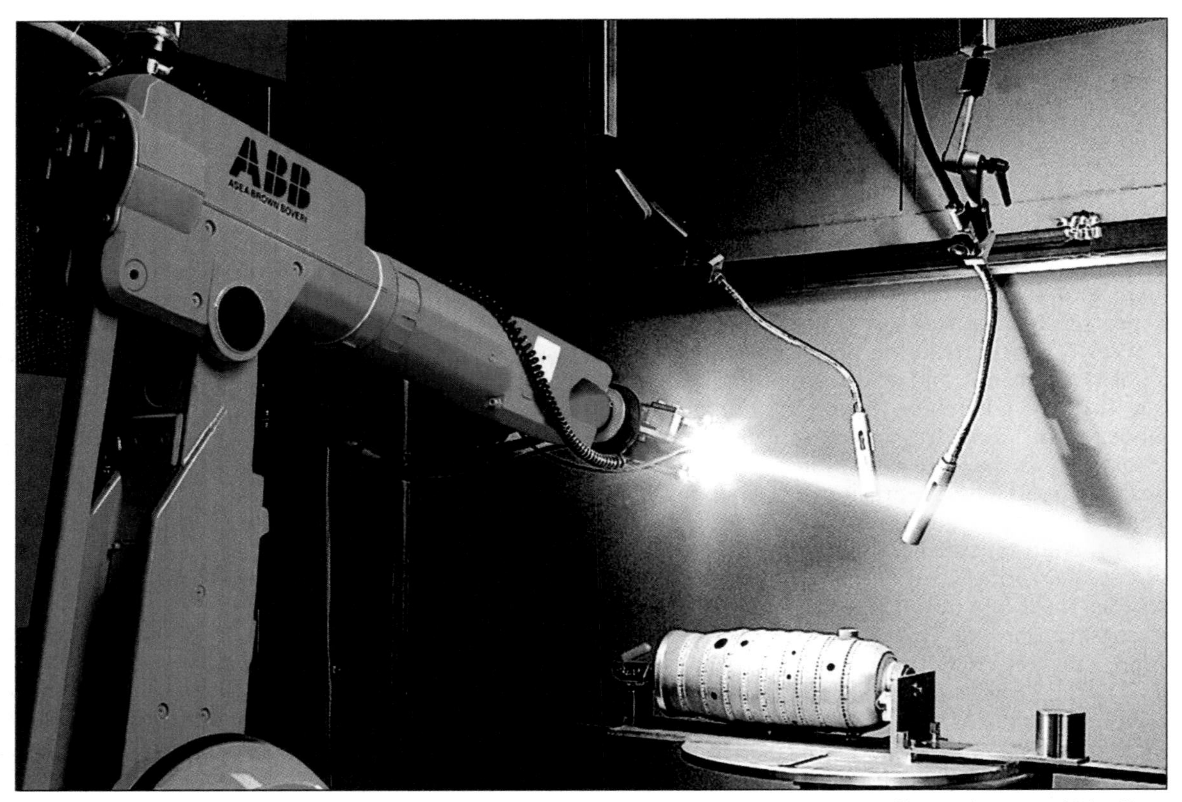

Photograph courtesy of Sulzer Metco

Figure 11.28—Robotic Spraying of a Bond Coating onto a Component of the Hot Section of a Gas Turbine Engine

QUALITY ASSURANCE

Worldwide emphasis on quality assurance and compliance with international certifications, such as those published by the International Organization for Standardization (ISO), influences broader implementation of the graded application of appropriate quality assurance practices and principles throughout the thermal spray industry. A properly designed quality assurance program covers every aspect of thermal spraying, including feedstock materials and process gases, surface preparation and masking, thermal spray equipment and accessories, operator training, process monitoring, documentation, measurement gauges, coating property tests, statistical analysis techniques, and customer feedback. However, it is important to strike a balance between quality control costs and benefits. Some of the considerations and tools that can be useful for a quality assurance program are described in this section.

MATERIALS

Quality control of materials includes obtaining material certification sheets for thermal spray feedstock materials and receiving gas certification sheets for the processing fuels and gases. For feedstock powder, the size distribution of the powder, the shape of the powder particles (particle morphology), and the method used to manufacture the powder (for example, spray-dried powder versus fused-and-crushed powder) all are important parameters that will affect the spray process and the quality of the coating.

Process gases often can be obtained in several grades of purity. It is important to ensure that the purity of the gas is appropriate for a specific spray process. A process gas that sometimes does not receive the attention that is deserves is compressed air. If compressed air is used for grit blasting to atomize droplets in a spray gun or to cool a component while it is being coated, it is vital that the air be free of contaminants. Proper conditioning of compressed air includes air driers to remove water and filters to remove oil and particulate.

The selection of the correct grit blasting media also is important, as proper surface preparation is crucial for good adhesion of the coating. Blasting media and blasting gun nozzles also should be periodically inspected and replaced, because the media loses sharp edges and breaks down into smaller particles with repeated use, and the diameter of the blasting gun nozzle increases with wear. This changes the impact velocity of the media.

EQUIPMENT

Thermal spray equipment, motion control equipment, and ventilation exhaust systems should be routinely inspected and maintained. Worn or damaged components not only may cause process parameters to stray, and thus potentially change the coating characteristics, but also may present serious safety hazards. The American Welding Society publication, *Specification for Thermal Spray Equipment Acceptance Inspection,* AWS C2.21M/C2.21, provides thermal spray equipment acceptance requirements for plasma, arc wire, flame powder, wire, rod and cord, and high-velocity oxygen fuel spray equipment. The publication includes inspection report forms which also serve as checklists.[9]

OPERATOR TRAINING

The value of effective training cannot be overstated as part of an overall quality assurance program. Good work practices embrace many small procedures that are seemingly inconsequential but are required for good quality. These include the handling of components, proper preheating and grit blasting, monitoring the spray consoles, monitoring the spray process, inspecting the coating, and maintaining accurate documentation. Recommendations for thermal spray operator qualification based on knowledge and skill testing are published by the American Welding Society in *Guide for Thermal-Spray Operator Qualification,* AWS C2.16/C2.16M. Individual thermal spraying operator qualification tests for various engineering and corrosion-control applications are included in the guide.[10] In some countries, notably in several European countries, formal thermal spraying operator certification programs have been implemented.

PROCESS MONITORING

Some thermal spray process control systems include sensors and software that monitors the operation of the spraying equipment and issues an alert if the process deviates beyond specified limits. High-tech accessory equipment is available to monitor the in-flight temperature and velocity of spray particles between the spray gun and the workpiece. However, not all applications require such sophisticated process monitoring equipment. A simple but effective means of monitoring the

9. American Welding Society (AWS) Committee on Thermal Spray, *Specification for Thermal Spray Equipment Acceptance Inspection*, AWS C2.21M/C2.21, Miami: American Welding Society.
10. American Welding Society (AWS) Committee on Thermal Spray, *Guide for Thermal-Spray Operator Qualification,* AWS C2.16/C2.16M, Miami: American Welding Society.

consistency of the process is to use witness coupons. Witness coupons are small samples of test substrates that are spray-coated simultaneously with the actual workpiece, or they may be coated immediately before or after the coating operation. For maximum effectiveness, witness coupons sometimes are coated three times: before, during, and after the primary coating operation. Simple non-destructive evaluations of witness coupons (such as weight gain and thickness) combined with destructive tests (such as metallographic sectioning) can monitor and track the reproducibility of the coating process and help determine whether the process has remained within acceptable limits. However, it should be noted that the geometry, material composition, and surface temperature of the witness coupon during deposition frequently differ from the actual workpiece, so the specific characteristics of the coating on the witness coupon probably will differ somewhat from the coating on the actual workpiece. Nevertheless, if the coating applied to the witness coupon remains highly consistent, the coating on the workpiece also should be consistent.

EXAMINATION TECHNIQUES

The advantages of nondestructive examination are that the workpiece is not damaged during examination, and it often can be performed *in situ*. However, nondestructive evaluation typically does not provide as much information as destructive testing. Three examination techniques commonly used to analyze thermal spray coatings are metallography, microhardness testing, and bond strength testing.

Metallography is an essential tool for in-depth analysis of coatings. The coating properties derived from metallographic examination may include porosity, oxide content, the distribution of carbides, the soundness of the coating-substrate interface, and the overall integrity of the coating. Combined with X-ray analysis, information about the coating chemistry and phase distribution can also be obtained.

Microhardness measurement is a standard tool used to reveal coating characteristics. The most popular test uses the Vickers square diamond pyramid. The Knoop test, which uses an elongated diamond pyramid, also is effective. Both of these tests should be performed according to *Standard Test Method for Micro-Indentation Hardness of Materials*, ASTM E384.[11] When coatings are sufficiently thick, superficial Rockwell tests may be used. There is no specific standard for Rockwell testing of sprayed coatings, but standardized Rockwell test methods are defined in the following standards: *Standard Test Methods for Rockwell Hardness and Rockwell Superficial Hardness of Metallic Materials*, ASTM E18;[12] *Metallic Materials—Rockwell Hardness Test* (Parts 1, 2 and 3), ISO 6508;[13] and *Standard Test Method for Rockwell Hardness of Plastics and Electrical Insulating Materials*, ASTM D785.[14]

Bond strength measurements, like witness coupons, can provide a measure of the consistency of the surface preparation and coating processes. Bond strength testing is most commonly performed according to *Standard Test Method for Adhesion or Cohesion Strength of Thermal Spray Coatings,* ASTM C633.[15] In this test, the face of a cylindrical plug is coated, glued to another blank cylindrical plug, and then pulled apart in a tensile testing machine.

Bond strength test results can provide a useful measure of the adhesion between a given coating and substrate; however, the results of this test cannot be directly correlated with the true bond strength on an actual workpiece due to differences such as workpiece geometry and temperature during deposition. Nevertheless, changes in bond strength can be used to evaluate and monitor intentional or unintentional process variations.

ECONOMICS

Thermal spraying is a versatile process that solves many surfacing challenges for engineers and designers. The technology allows a number of choices when selecting thermal spray processes, process parameters, and coating materials. Because of this diversity, the economics of thermal spraying can span a wide range. This can be a big advantage to the engineer because it provides many alternatives for the design of a coating appropriate to the product. Thermal spraying gives the engineer the ability to select a more costly coating that will provide the best properties, or alternatively, a more economical coating that will still provide sufficient properties required by the application.

11. American Society of Testing and Materials (ASTM) Subcommittee E04, *Standard Test Method for Micro-Indentation Hardness of Materials*, ASTM E384, West Conshohocken, Pennsylvania: American Society for Testing and Materials.

12. American Society of Testing and Materials (ASTM) Subcommittee E28, *Standard Test Methods for Rockwell Hardness and Rockwell Superficial Hardness of Metallic Materials*, ASTM E18, West Conshohocken, Pennsylvania: American Society for Testing and Materials.

13. International Organization for Standardization (ISO), *Metallic Materials — Rockwell Hardness Test* (Parts 1, 2, and 3), ISO 6508, Geneva: International Organization for Standardization.

14. American Society of Testing and Materials (ASTM) Subcommittee D20, *Standard Test Method for Rockwell Hardness of Plastics and Electrical Insulating Materials*, ASTM D785, West Conshohocken, Pennsylvania: American Society for Testing and Materials.

15. American Society of Testing and Materials (ASTM) Subcommittee B08, *Standard Test Method for Adhesion or Cohesion Strength of Thermal Spray Coatings*, ASTM C633, West Conshohocken, Pennsylvania: American Society for Testing and Materials.

Many components that are designed without a coating may require the selection of complex materials or the use of heat treatments, or both, to provide desired surface properties (such as high hardness for wear resistance) in specific areas.

Thermal spray coatings also can be used in the design of a component to reduce the overall production cost by using less expensive materials to make up the majority of the component. There are cases in which the surface requirements of a component are too great for a less expensive material to provide the necessary performance. Thermal spray coatings can provide the desired surface properties that a less expensive material alone cannot.

Determining the cost of a thermal spray coating can encompass a variety of factors. This section is not intended to cover all the contributing elements, but to provide some information on important topics to consider. Basic expenses include coating materials and gases. When estimating material costs, it is important to determine how much material will be processed through the thermal spray device to achieve the desired coating. It must be understood how the particular thermal spray system and process parameters affect the deposit efficiency of the material and how much overall time the system will need to spray the workpiece. Some of the other costs include consumables for the thermal spray device (such as nozzles, anodes, and cathodes) and the cost of surface preparation, which is grit blasting in most cases.

Masking the workpiece and post cleaning of overspray must be considered. It is common for many coatings to require post-coating machining or grinding process to meet surface finish or dimensional requirements, or both. Other factors are the quantity of workpieces to be processed, time for set-up, manipulation of the workpieces during the coating process, amount of labor required, and inspections.

The cost of thermal spray coating tasks can be estimated by using the following formulas:

1. Set up: number of operators × hourly wage × numbers of hours to set up;
2. Masking: Cost of masking materials + labor to apply masking;
3. Surface preparation (for example, grit blasting): time to blast workpiece × number of operators × hourly wage + cost of grit;
4. Coating material: process time per workpiece × feed rate of material × cost of material per unit weight;
5. Coating process labor: number of operators × hourly wage × workpiece processing time;
6. Post-coating operations: cost of materials + labor associated with the operations, (for example, removal of masking, grinding, polishing, overspray removal, and heat treating);
7. Allowances for materials, labor, scrap, and rework;
8. Manufacturing overhead: company hourly manufacturing overhead rate × total time to process workpiece;
9. Administrative overhead: company administration hourly rate × total time to process workpiece; and
10. Total cost: sum of Number 1 through Number 9.

There are also less obvious costs that also must be considered. Environmental controls are a good example. Costs are associated with protecting the operators from exposure to hazardous materials and process dangers such as ultraviolet light, noise, high temperature flames and plasmas. Overspray must be contained to protect the air from particulate and hazardous emissions.

The proper disposal of thermal spray waste needs to be considered. Other examples of environmental concerns that must be considered are the thermal spraying of materials that contain chromium. Many thermal spraying materials contain chromium in different forms, such as free chromium, chromium as part of an alloy, or chromium oxide. When any of these forms of chromium are processed through a thermal spraying device, there is a significant possibility that a small amount of hexavalent chromium will be produced. This form of chromium oxide is hazardous and is highly regulated by the EPA[16] and OSHA.[17] Fortunately, almost no hexavalent chromium is found in thermal spray coatings, but commonly can be present in the overspray materials generated during processing. Costs are associated with good air filtration equipment to protect the air, protecting the operators from exposure, and proper disposal of the overspray waste. Even though these environmental concerns do add some costs, there are very good methods to address all of these concerns. When comparing the methods of environmentally controlling thermal spraying with competing technologies such as electroplating, the challenges and costs of thermal spraying are typically much less. The ability to effectively control environmental concerns will continue to make the technology a favorable choice compared to many alternative coating technologies.

16. U.S. Environmental Protection Agency, Ariel Rios Building, 1200 Pennsylvania Avenue, N.W. Washington, DC 20460.

17. Occupational Safety and Health Administration (OSHA), *Occupational Safety and Health Standards for General Industry, in Code of Federal Regulations (CFR)*, Title 29 CFR 1910, Subpart Q, Washington D.C.: Superintendent of Documents, U.S. Government Printing Office.

SAFE PRACTICES

Thermal spraying and welding operations share many of the same health and safety risks, but thermal spraying also includes some issues that are specific to the process. This section presents an overview of some of the potential health and safety risks that should be considered when planning and performing thermal spray operations. This section is not intended to be a comprehensive treatment of the subject, but rather its purpose is to raise awareness and provide general safety concepts and sources of additional information. A list of safety standards, regulations, and publications is presented in Appendix B of this volume. Many of the publications on this list apply to the safe practices in welding that are shared with thermal spraying; others are related to general industrial safety.

It should be noted that the recommendations in the examples in this section might not satisfy local government regulations. Users of thermal spray equipment and processes must become familiar with safe practices and obtain the necessary information to safely operate the equipment at their location.

Personal safety ultimately is the responsibility of the user of the thermal spray process. The user must always be aware of the hazards that may be present. Proper training of the operator is essential. In the United States, all personal protective equipment (PPE) must be in accordance with OSHA Title 29 CFR 1910.132.[18] The manufacturer's safety standards and instructions for the handling of materials and equipment must be followed. Materials Safety Data Sheets (MSDS) are available from the manufacturers of materials used in thermal spraying, and should be read and understood prior to using the material. The American Welding Society publication *Safety in Welding, Cutting, and Allied Processes,* ANSI Z49.1 should be consulted.[19] The Thermal Spray Society has published a risk assessment document that is useful in planning thermal spray operations.[20]

Safe practices should be considered prior to beginning any thermal spray operation. The following five-step methodology can be used to assess and mitigate health and safety risks in thermal spraying:

1. Plan the work;
2. Identify and evaluate the associated risks;
3. Implement engineering controls, administrative controls, personal protective equipment, and training to mitigate risks;
4. Perform the work; and
5. Evaluate the process and make improvements.

In addition to proper operator training, the guiding safety philosophy is a three-tiered approach. Whenever possible, the best approach is to separate the operator from the hazards of the immediate spray environment and run the process remotely, for example by stationing the operator outside of a spray booth and using a robotic spray system. When this is not possible, then engineering controls, such as local exhaust, proper ventilation, and safety interlocks should be used to mitigate hazards to the extent possible. In addition, personal protective equipment, such as respirators and hearing protection, and administrative controls, such as checklists, represent important safeguards to help protect the operator.

A properly designed thermal spray facility can minimize many of the safety risks. In general, the hazards of thermal spray operations are associated with the following:

1. Gases;
2. Liquids;
3. Electrical equipment;
4. Feedstock material;
5. Fumes and dust;
6. Heat, sound, and light; and
7. Robotics and other mechanical equipment.

While the following discussion provides general information concerning some of the common hazards and mitigation measures, the reader is strongly encouraged to consult the references listed in this chapter and also with health and safety professionals.

GASES

Gases typically used in thermal spray operations include argon, hydrogen, helium, nitrogen, oxygen, acetylene, natural gas (methane), propane, and propylene; and synthetic fuel gases, such as methylacetylene-propadiene (stabilized) (MPS); also compressed air, and carbon dioxide (used as a coolant). Following are some hazards associated with gases in thermal spraying:

1. Damage or injury due to improper handling of high-pressure gases, cylinders, and apparatus;
2. Fire or explosion of combustible gases;
3. Displacement of air by asphyxiates;

18. Occupational Safety and Health Administration (OSHA), in *Code of Federal Regulations (CFR)* Title 29, Part 1910.132, (latest edition) *Occupational Safety and Health Standards*, Washington, D.C.: Superintendent of Documents, U.S. Government Printing Office.

19. American National Standards Institute (ANSI) Accredited Standards Committee Z49, *Safety in Welding, Cutting, and Allied Processes,* ANSI Z49.1:2005, Miami: American Welding Society.

20. ASM—Thermal Spray Society Safety Committee, *Safety Guidelines for Performing Risk Assessments,* Publication SG002-02, Cleveland: ASM, International, http://www. asminternational.org.

4. Spontaneous combustion or fire acceleration due to exposure to oxygen; and
5. Freezing and pressure hazards associated with cryogenic gases, for example, liquid argon or nitrogen.

Properly designed gas storage and handling systems can mitigate many of these hazards. The appropriate selection of piping material, manifold design, and the selection, placement, and maintenance of gas safety devices, such as pressure relief valves and flow limiters, are essential. The National Fire Protection Agency (NFPA) publishes several standards applicable to thermal spraying, such as *Standard for Fire Prevention During Welding, Cutting, and Other Hot Work*, NFPA 51:2003.26.[21] Standards published by the Compressed Gas Association should be consulted for codes relating to gas fittings and systems.[22] The training of personnel in safe handling and storage of pressurized gas containers and the proper operation and maintenance of gas systems also is important. The gas supplier usually is an excellent resource for assistance with gas system design, equipment selection, and training information and should be familiar with local fire and safety codes that may apply.

Gas Leaks

In addition to the fire hazard associated with open flames and hot spray particles generated during coating operations, some less obvious hazards must be considered. For example, a slow leak in a valve or fitting inside a closed spray booth could create an explosive atmosphere that could be easily ignited by something as simple as pressing a switch or turning on a light. The first line of defense against such hazards is a properly designed system that affords adequate separation of combustible gas lines from ignition sources, such as outlets and switches. Also, many modern spray booths incorporate interlock systems; for example, a booth may have an exhaust flow switch that will not permit any gas flow into a spray booth unless the booth ventilation system is operating. Many facilities also include monitors and alarm systems to detect the presence of hazardous concentrations of combustible gases in the booth atmosphere.

Like the spray booth, spray control consoles also represent an enclosed volume that can present a potential for explosion. Most modern spray control consoles are designed to physically isolate electrical components from gas lines, regulators, filters, and similar equipment. The consoles provide explosion-proof ventilation fans to prevent gas buildup within the cabinet in the event of a leak. However, some older spray equipment may not follow advanced design practices, and explosions, injuries, and even deaths have occurred due to leaks within a spray console. The equipment should be checked to ensure that it has an inherently safe design and that it meets local safety code requirements. Users should never attempt to modify commercially supplied consoles or other equipment, as inadvertent compromise of safety features may occur.

Another safety issue related to combustible gases is flashback in oxyfuel combustion spray guns. A flashback is a recession of the flame into or back of the mixing chamber of a combustion spray gun. It occurs when the velocity of the flame front exceeds the local flow velocity of the incoming gas, allowing the flame to migrate upstream into portions of the spray device or gas supply system that were not intended to contain it. This most commonly occurs because of improper gas settings or during shutdown, when the gas flow is ramping down. Mitigation measures include making certain that proper operating procedures are used, especially with regard to the shutdown sequence, and the use of backflow preventers, flashback arresters, and other safety measures to ensure that a flashback cannot propagate too far up the gas supply stream.

A gas leak, large or small, potentially can displace air in a closed spray booth to the point that there is insufficient oxygen to support life. Again, proper booth design, interlocks, flow limiters, and other control devices, such as an oxygen monitor, are important safeguards. Asphyxiation is an especially important concern when dealing with large-chambered spray systems. Some of these are vacuum tanks so large that a worker must physically enter them to service parts of the system. It is essential to sample the oxygen levels in all areas of such chambers in accordance with a defined confined-space safe entry procedure before allowing personnel to enter the chamber, even if the entry of a worker involves only placing his or her head inside the chamber. Procedures for working in confined spaces are published in *Safety Requirements for Confined Spaces*, ANSI Z117.1 by the American National Standards Institute (ANSI).[23]

As in welding, most combustion thermal spray devices use compressed pure oxygen as the oxidizer. Pure oxygen spontaneously ignites most organic liquids and can even cause metal piping to ignite during pressure surges if the local pressure (adiabatic compression) becomes too high. An oxygen system must be very carefully designed, and apparatus such as regulators, fit-

21. National Fire Protection Association (NFPA), latest edition, *Standard for Fire Prevention During Welding, Cutting, and Other Hot Work*, NFPA 51:2003, Quincy, Massachusetts: National Fire Protection Association.
22. Compressed Gas Association, 4221 Walney Road, Chantilly, VA 20151-2923.
23. American National Standards Institute (ANSI) Accredited Standards Committee Z117, (latest edition), *Safety Requirements for Confined Spaces*, ANSI Z117.1, New York: American National Standards Institute.

tings, and piping must be rated for service with pure oxygen and maintained in a clean condition. Workers must be careful not to expose clothing or other materials contaminated with grease or other organics to pure oxygen, as the contaminated materials may spontaneously ignite.

Due to the large volumes of gas used at some thermal spray facilities, some gases (argon or nitrogen) are commonly received and stored in cryogenic liquid form and converted into the gaseous state as they are withdrawn for use. In some cases, a liquefied gas such as carbon dioxide (CO_2) is used to cool the workpieces during coating operations. Liquefied petroleum gases also have potential cryogenic hazards. Serious injury and even death can result from improper handling and exposure to cryogenic liquids. Proper design and maintenance of the cryogenic system and thorough operator training are essential to ensuring safe operations.

LIQUIDS

In addition to liquefied gases, other liquids associated with thermal spraying include solvents used to clean and degrease surfaces prior to grit blasting; water used to actively cool many types of spray guns and occasionally to actively cool a workpiece; liquefied petroleum gases (notably propane and propylene); and kerosene, which is used to fuel high-enthalpy HVOF spraying systems.

Combustion hazards are associated with the solvents and fuels. Proper design of liquid storage and supply systems, for example, piping and valves, is as important for liquids as it is for gases. Provision also must be made for secondary containment of fluid that could leak from a primary containment vessel, such as a kerosene storage tank.

Cleaning solvents should be stored in cabinets approved for storage of flammable liquids. Solvent-soaked rags should similarly be kept in containers approved for this purpose, and care must be given to the proper disposal of excess or waste solvent. Cleaning solvents may also present health hazards due to inhalation or skin absorption, or both. Appropriate safety measures may include respiratory protection, safety goggles, face shields, gloves, aprons, and proper ventilation in the work area.

Primary hazards associated with water in a thermal spray environment are electrical shock and slips or falls that may result from spilled water. Safety measures may include ground fault interrupt (GFI) devices, insulating gloves, lock-out/tag-out (LOTO) procedures for equipment maintenance, spill kits, and rubber-soled footwear.

ELECTRICAL EQUIPMENT

Electrical hazards associated with thermal spray equipment may include exposure to high currents, high voltages, electric arcs, and flash or blast hazards. Sparks or arcs also could ignite flammable gases, liquids, or powders. In addition, tripping hazards may be associated with electrical cables and cords that are commonly found in the spray environment.

Electrical safety measures include emergency stop circuits, lock-out/tag-out procedures, the separation of electric and fuel sources, grounding, adequate access to electrical cabinets, appropriate lighting, overload protection (circuit breakers or fuses), ground fault interrupters, dust protection of electrical components, and adequate documentation of systems and procedures to ensure safe operation and maintenance of thermal spray equipment.

Power Sources

High-current dc power sources used in arc spraying and plasma spraying systems are an obvious source of electrical hazards. It must be carefully verified that cables, junction blocks, and similar apparatus are properly assembled and frequently inspected. The National Electrical Manufacturers Association (NEMA) publishes standards for safe use of electrical equipment.[24]

Corrosion on high-current junction blocks or damaged insulation could present serious hazards. The high-frequency unit used to initiate a plasma arc operates at very high voltages and may arc across unexpected pathways. Only properly trained personnel using appropriate safety measures, such as lockout/tagout, should service such equipment.

Some electrical equipment is located inside spray booths. This equipment potentially could serve as an ignition source for combustible gases or metal dust. Careful consideration should be given to providing adequate physical separation between electrical equipment and sources of combustible materials, liquids, gases, or vapors. Using well-designed electrical enclosures, properly locating them within the booth, and maintaining good housekeeping are important methods of preventing the gradual accumulation of metal powders in and around electrical equipment. Electrical cabinets, booth walls, and other metallic structures must be properly grounded according to local code requirements.

FEEDSTOCK

Of the available feedstock materials, that is, powders, wires, and rods, powders usually present the most sig-

24. National Electrical Manufacturers Association, 1300 North 17th Street, Suite 1847, Rosslyn, VA 22209.

nificant hazards. The hazards associated with thermal spray powders include inhalation, ingestion, skin or eye exposure, fire, and explosion.

A number of safe practices must be followed when handling thermal spray powders. Thermal spraying operators and other workers involved in the process must be familiar with the Material Safety Data Sheets provided by the manufacturers of the powders. Localized exhaust can reduce accumulation of fine powder particles in the air. To mitigate health hazards, the employer must have an appropriate hazard communication program such as that published in OSHA *Fact Sheet 93-26, Hazard Communication Standard* to communicate the risks associated with the materials used.[25] The operators must be provided with the necessary equipment, controls, personal protective equipment, (including respirator, gloves, and eye protection), and proper training.

For flammable or potentially explosive metal powders, such as aluminum or titanium, proper grounding of components such as the powder canister, hopper, and funnels is important to prevent the build-up of static electrical charges and associated sparks. Also, Class D metal fire extinguishers should be available in the spray area, and personnel should be trained in the challenges unique to fighting metal fires.[26]

Grit blast media, though not directly used in the spray operation, also presents safety issues. Silica-based media present an inhalation hazard, which may result in silicosis. In addition, the finely divided material removed from the workpiece during grit blasting operations may be hazardous and measures should be taken to prevent inhalation of these fine dusts and to protect skin and eyes.

FUMES AND DUST

Fumes and dust are hazardous byproducts of thermal spray operations. These hazards are similar to those encountered in many welding operations. Long-term exposure of the operator to fumes and dust can cause chronic and potentially serious health problems. In addition, acute illnesses, such as metal fume fever, can result from even a brief exposure to elevated concentrations of dust or fume, or both. The buildup of finely divided materials inside ductwork and exhaust filter systems can pose an additional safety threat.

In recent years, the air handling, filtration, and emissions requirements for thermal spray operations have become increasingly stringent, and related safety equipment has become more complex. The booth exhaust system must be designed to provide adequate air flow to accomplish multiple functions, including the removal of harmful particles from the air inside the spray booth; venting of thermal spray gases and byproducts from the booth; providing sufficient air motion to keep the booth clean and prevent dust contamination of the workpiece; preventing the buildup of combustible gases in the booth; and removing the heat generated by thermal spray guns.

Prior to 1980, water wash units were widely used in the thermal spray industry to purify exhaust air during spray operations. Contaminated air was passed through a water curtain to remove particulate, with a collection efficiency of approximately 75% to 90%. However, in most locations around the world, water wash units no longer meet government mandated environmental clean-air requirements. Modern thermal spray dust collection units use replaceable dry dust filters of various designs that can achieve collection efficiencies of 99% or higher.

The design of the overall exhaust ventilation system must accommodate the size and physical layout of the booth. Larger booths require higher air flows in order to maintain a clean environment. In general, spray booths should be just large enough to allow for safe processing with no additional space. The shape of the booth and the layout of equipment within the booth should be considered a part of the exhaust system design, because they greatly affect airflow and exhaust system performance. In concept, the spray booth should be treated as though it were a part of the overall exhaust ductwork system. An excellent publication entitled *Thermal Spray Booth Design Guidelines* has been published by the safety committee of the Thermal Spray Society.[27]

A suitable respiratory protection program must be in place to control exposure of the operator to thermal spray dust and fumes. It is recommended that a certified industrial hygienist be consulted when setting up a respiratory protection program and selecting appropriate personal protective equipment (PPE). Documents such as *Threshold Limit Values (TLVs®) for Chemical Substances and Physical Agents in the Workroom Environment*[28] and *Methods for Sampling Airborne Particulates Generated by Welding and Allied Processes*, AWS F1.1,[29] should be consulted.

25. Occupational Health and Safety Administration (OSHA), *OSHA Fact Sheet 93-26. Hazard communication standard. 29 CFR 1910.1200*, http://www.osha-slc.gov.
26. See Reference 21.
27. ASM—Thermal Spray Society Safety Committee, Thermal Spray Booth Design Guidelines, Publication SG003-03, Metals Park, Ohio: ASM, International, http://www.asminternational.org.
28. American Conference of Industrial Hygienists, *Threshold Limit Values (TLVs®) for Chemical Substances and Physical Agents in the Workroom Environment* American, Conference of Governmental Industrial Hygienists, Cincinnati, Ohio: American Conference of Industrial Hygienists.
29. American Welding Society (AWS) Committee on Fumes and Gases, *Methods for Sampling Airborne Particulates Generated by Welding and Allied Processes*, AWS F1.1/F1.1M, Miami: American Welding Society.

HEAT, NOISE, AND LIGHT

Thermal spray operations can emit dangerous levels of heat, sound, and light (radiation). The primary techniques of mitigating these hazards are the proper design of the work environment and use of personal protective equipment. The preferred approach is to place the operator at a workstation outside the spray booth. In situations where this is not possible, every measure must be taken to minimize worker exposure and to use the necessary PPE.

Heat

Thermal spray guns, especially HVOF devices, generate substantial quantities of heat. Without adequate ventilation, the booth can quickly become overheated. Excessive heat loads can damage the ductwork and the dust filter system. It is typically recommended that the total exhaust airflow be sufficient to keep the exhaust air temperature below 40°C (105°F). However, exceptionally high flow velocities exceeding 91 m/min (300 ft/ min) at the front face of the spray hood may interfere with spray operations. In addition to the obvious thermal hazards directly associated with spray devices, hot workpieces and tooling may require the use of thermally insulated gloves and handling tools, such as tongs or pliers. The use of auxiliary cooling jets may be required. Cotton clothing is preferred over synthetic fabrics, because it does not melt and does not support combustion as readily.

Noise

Sound levels associated with thermal spray operations vary greatly, but commonly exceed the limit of 85 dB for eight-hour occupational exposure without protective measures as prescribed by the Occupational Safety and Health Administration (OSHA) of the United States government.[30] OSHA rules concerning specific noise exposure limits and associated requirements should be consulted. In almost all cases, some form of hearing protection will be required. Some spray devices, for example, detonation spray guns, can produce extremely high sound levels ranging up to 145 dB. At these extremes, the operator must be removed from the environment because injury can occur by means of additional mechanisms such as bone conduction of the sound pressure waves. Table 11.7 lists representative sound levels for a variety of common thermal spray operations. The reader should be aware that actual sound pressure levels depend strongly on the specific spray environment, such as booth size, wall material, and construction details.

Table 11.7
Typical Sound Pressure Levels

Thermal Spray System	Decibel Range (dB)
Detonation flame spray gun	145
HVOF (liquid fuel)	133
HVOF (gaseous fuel)	125–135
HVAF (air-fuel system)	133
Combustion wire	118–122
Combustion powder	90–125
Ceramic rod gun	125
Arc wire	105–119
Air plasma (DC)	110–125
Vacuum plasma, low-pressure, or controlled atmosphere	Ambient
Cold spraying	110
Water-stabilized plasma	125
Inductively Coupled RF Plasma	95

Eye Protection from Light

Light (radiation) hazards associated with thermal spray operations include both visible and ultraviolet emissions. For example, plasma spray guns are well known for causing "sunburn" on unprotected skin. Excessive exposure to the bright emissions from thermal spray guns is similar to that associated with welding operations and can cause both acute injuries and chronic conditions, such as cataracts. Helmets, hand-held shields, face shields, and safety glasses or goggles should be used to protect the eyes, face, and neck during thermal spray operations. Protective measures are described in the American National Standards Institute documents *Practice for Occupational and Educational Eye and Face Protection*, ANSI Z87.1,[31] and *American National Standard for Industrial Head Protection*, ANSI Z89.1.[32]

Helmets, hand-held shields, and goggles must be equipped with suitable filter plates to protect the eyes from excessive ultraviolet and intense visible radiation. A guide for the selection of the proper filter shade number is shown in Table 11.8.

30. Occupational Safety and Health Administration (OSHA), latest edition, *Occupational Safety and Health Standards for General Industry* in *Title 29, Labor*, in *Code of Federal Regulations*, (CFR), Washington D.C.: Superintendent of Documents, U.S. Government Printing Office.

31. American National Standards Institute (ANSI), *Practice for Occupational and Educational Eye and Face Protection*, ANSI Z87.1, Des Plaines, Illinois: American Society of Safety Engineers (ASSE).

32. American National Standards Institute (ANSI) *National Standard for Industrial Head Protection*, ANSI Z89 1-97, Arlington, Virginia: International Safety Equipment Association.

Table 11.8
Recommended Lens Shade Plates for Thermal Spraying Operations

Thermal Spraying Operation	Shade Numbers
Wire flame spraying (except molybdenum)	5
Wire flame spraying of molybdenum	5 to 6
Flame spraying of metal powder	5 to 6
Flame spraying of exothermics or ceramics	5 to 8
Plasma and arc spraying	9 to 12
Fusing operations	5 to 6

Auto-darkening eye protection with variable filter lens shades is available. These devices have lenses that are light-activated, that is, they are clear when the spray device is off and immediately automatically darken when exposed to bright light from a spray gun. Other commonly used methods of reducing exposure to light hazards include remote process monitoring with cameras and the installation of curtains or dark shades on booth windows.

Robotics and Other Mechanical Hazards

Robotic spray gun and workpiece manipulation machines are commonly used in thermal spraying. At many facilities, the spray booth serves as the safeguarded space that keeps personnel from inadvertently entering the robot's envelope of motion. In addition to robots, many spray operations employ lathes, turntables, and other mechanical equipment to provide appropriate spray gun and workpiece manipulation. Hazards associated with such equipment include exposure of the thermal spray operator to entanglement in rotating machinery or being struck, pinched, or crushed by a powerful motorized assembly.

The Robotic Industries Association (RIA) publication *American National Standard for Safety Requirements for Industrial Robots and Robot Systems*, ANSI/RIA R15.06-1999 should be consulted regarding safeguards that should be designed into a robotic system to protect operators and maintenance personnel.[33] These safeguards may include presence-sensing devices, barriers, awareness signals, procedures, and training. Methods employed to warn and protect personnel, in order of priority, include the following:

1. Mechanical movement-restricting stops,
2. Warning and interlock devices,
3. Awareness barriers,
4. Warning signs, and
5. Written procedures.

Eliminating pinch points by mechanical means is given the highest priority. In general, personnel should not enter the safeguarded space when the robot is powered. If it is necessary for an operator to be present to program the system, the use of a speed-limited teach mode and a three-position motion enabling safety interlock is recommended. Speed-limited teach modes that allow only careful deliberate motion during an operation in which a mishap or involuntary action can occur are recommended. Three-position enabling devices permit operators to stop machine motion by simply releasing or tightening a grip on a safety switch. These have the advantage of stopping motion when the device is either released or tightly squeezed in a panic situation, or even if the operator is trapped or unconscious.

CONTINUING SAFETY PROGRAM

A broad range of skills and expertise is required to mitigate the hazards associated with thermal spraying and to provide for a safe work environment. Users of thermal spray processes are strongly urged to consult safety experts to address each area of thermal spraying that poses a safety concern. Safety must be a primary concern when a new facility is being set up, when an existing facility is being modified, when a new material is to be sprayed, or when new personnel are introduced into the work environment. A safe work environment includes careful analysis of the hazards, proper design of equipment and facilities, the use of personal protective equipment, and adequate training of all personnel. Additional safety-related information can be obtained from organizations at the Internet web sites listed in Table 11.9 and also from other resources listed in the Bibliography and Supplementary Reading List at the end of this chapter.

CONCLUSION

While simple in concept and broad in its applications, high-quality thermal spraying operations require knowledge, experience, and attention to detail. More detailed information on specific topics is available from the Internet resources and publications listed at the end of this chapter.

33. American National Standards Institute (ANSI) *American National Standard for Safety Requirements for Industrial Robots and Robot Systems*, ANSI/RIA R15.06-1999, Ann Arbor, Michigan: Robotic Industries Association.

Table 11.9
Electronic Access to Thermal Spraying Standards

Organization	Electronic Address
American National Standards Institute (ANSI)	www.ansi.org
American Society for Testing and Materials (ASTM)	www.astm.org
ASM International	www.asm-intl.org
American Welding Society (AWS)	www.aws.org
Compressed Gas Association (CGA)	www.cganet.com
International Organization for Standardization (ISO)	www.iso.ch
International Thermal Spray Association (ITSA)	www.thermalspray.org
National Fire Protection Association (NFPA)	www.nfpa.org
Society for Protective Coatings (SSPC)	www.sspc.org
Thermal Spray Society (TSS), an affiliate society of ASM International	www.asminternational.org
Occupational Health and Safety Administration (OSHA)	www.osha.gov
National Institute for Occupational Safety and Health	www.cdc.gov/niosh

The technology of the thermal spraying industry is supported by the referenced standards-writing groups, many of which are represented on Internet Web sites. Information on standards, recommended practices, safety, and other information can be obtained electronically from the major associations listed in Table 11.9.

BIBLIOGRAPHY

American Conference of Industrial Hygienists. *Threshold limit values (TLVs®) for chemical substances and physical agents in the workroom environment.* American Conference of Governmental Industrial Hygienists, Cincinnati, Ohio: American Conference of Industrial Hygienists.

American National Standards Institute (ANSI) Accredited Standards Committee Z117. *Safety requirements for confined spaces*, ANSI Z117.1. New York: American National Standards Institute.

American National Standards Institute (ANSI). *Practice for occupational and educational eye and face protection*, ANSI Z87.1. Des Plaines, Illinois: American Society of Safety Engineers (ASSE).

American National Standards Institute (ANSI). *National standard for industrial head protection,* ANSI Z89 1-97. Arlington, Virginia: International Safety Equipment Association.

American National Standards Institute (ANSI). *American national standard for safety requirements for industrial robots and robot systems*, ANSI/RIA R15.06-1999. Ann Arbor, Michigan: Robotic Industries Association.

American Society of Testing and Materials (ASTM). *Standard test method for micro-indentation hardness of materials*, ASTM E384. West Conshohocken, Pennsylvania: American Society for Testing and Materials.

American Society of Testing and Materials (ASTM). *Standard Test Method for Adhesion or Cohesive Strength of Flame Sprayed Coatings*, ASTM C633-01, West Conshohocken, PA: American Society for Testing and Materials.

American Welding Society (AWS) Committee on Definitions and Symbols. 2001. *Standard welding terms and definitions*, AWS A3.0:2001. Miami: American Welding Society.

American Welding Society (AWS) Committee on Thermal Spray. *Specification for the application of thermal spray coatings (metallizing) of aluminum, zinc and their alloys and composites for the corrosion protection of steel,* AWS C2.23M/C:2003, NACE No. 12, SSPC-CS 23.00. Miami: American Welding Society.

American Welding Society (AWS) Committee on Thermal Spraying. 2002. *Specification for thermal spray zinc anodes on steel reinforced concrete,* AWS C2.20/C2.20M:2002. Miami: American Welding Society.

American Welding Society (AWS) Committee on Thermal Spraying. 1985. *Thermal spray, practice, theory and application.* Miami: American Welding Society.

American Welding Society (AWS) Committee on Thermal Spraying. 1993. Reaffirmed 2001. *Guide for the protection of steel with thermal sprayed coatings of aluminum and zinc and their alloys and composites*, AWS C2.18. Miami: American Welding Society.

American Welding Society (AWS) Committee on Thermal Spray. 2003. *Specification for thermal spray equipment acceptance inspection*. AWS C2.21M/C2.21:2003. Miami: American Welding Society.

American Welding Society (AWS) Committee on Thermal Spray. 2002. *Guide for thermal spray operator qualification*, AWS C2.16/C2,16M:2002. Miami: American Welding Society.

American Welding Society (AWS) Committee on Fumes and Gases. *Methods for sampling airborne particulates generated by welding and allied processes*, AWS F1.1/F1.1M:2006. Miami: American Welding Society.

International Organization for Standardization (ISO). *Metallic materials — Rockwell hardness test* (Parts 1, 2, and 3), ISO 6508. Geneva: International Organization for Standardization.

National Fire Protection Association (NFPA). Latest edition. *Standard for fire prevention during welding. cutting, and other hot work*, NFPA 51B:1962. Quincy, Massachusetts: National Fire Protection Association.

National Fire Protection Association (NFPA. Latest edition. *Explosive materials code*, NFPA 495. Quincy, Massachusetts: National Fire Protection Association.

American National Standards Institute (ANSI) Accredited Standards Committee Z49.1, *Safety in welding, cutting and allied processes*, ANSI Z49.1. Miami: American Welding Society.

Occupational Health and Safety Administration (OSHA). OSHA Fact Sheet 93-26. *Hazard communication standard*, 29 CFR, standard number 1910.1200. Washington D.C.: Superintendent of Documents. U.S. Government Printing Office.

Occupational Safety and Health Administration (OSHA). Latest edition. *Occupational Safety and Health Standards for General Industry* in *Title 29, Labor*, in *Code of Federal Regulations*, (CFR). Washington D.C.: Superintendent of Documents. U.S. Government Printing Office.

SUPPLEMENTARY READING LIST

American Welding Society (AWS) Committee on Thermal Spraying. 1985. *Thermal Spraying: Practice, Theory, and Application, American Welding Society,* ISBN 0-87171-246-6. Miami: American Welding Society.

Berndt, C. C. 1992. *Preparation of thermal spray powders. Education module on thermal spray.* ASM International, Materials Park, Ohio: ASM International.

Berndt, C. C. Tensile adhesion testing methodology for thermally sprayed coatings, *Journal of materials engineering* 12(2):152.

Berndt, M. L., and C. C. Berndt. 2003. *Thermal spray coatings for corrosion protection*, ASM Corrosion Handbook, Volume 13A, Chapter 4d4, ISBN 978-0-87170-705-5. Materials Park, Ohio: ASM International.

Bernhardt, A. 1994. *Particle size analysis*. London: Chapman & Hall.

Davis, J. R., Ed. 2004. *Handbook of thermal spray technology*. Materials Park Ohio: ASM International.

Hall, A. C., D. J. Cook, R. A. Neiser, T. J. Roemer, and D. A. Hirschfeld, 2006, The effect of a simple annealing heat treatment on the mechanical properties of cold-sprayed aluminum, *Journal of thermal spray technology*, 15(2): 233–238.

Lowell, S., and J.E. Shields. 1991. *Powder surface area and porosity*. London: Chapman and Hall.

Matejka, D., and B. Benko. 1990. *Plasma spraying of metallic and ceramic materials*, 1st Ed. ISBN 0-47191-876-8. New York: John Wiley & Sons.

Pawlowski, L. 1995. *The science and engineering of thermal spray coatings*, ISBN 0-47195-253-2. New York: John Wiley and Sons.

Stafford, K. N. 1994. *Surface engineering: processes and applications*, ISBN 1-56676-154-9. CRC Press.

Suryanarayanan, R., Ed. 1993. *Plasma spraying: theory and applications*, ISBN 9-81021-363-8. World Singapore: Scientific Publishing Company, Inc.

Troczynski, T., and M. Plamondon. 1992. Response surface methodology for optimization of plasma spray. *Journal of thermal spray technology* 1(4): 293–300.

ADDITIONAL RESOURCES

Proceedings of the International Thermal Spray Conference is published annually by the ASM Thermal Spray Society, containing published papers on all aspects of thermal spray. Proceedings for specific years of interest and other thermal spray publications are available at the following ASM Thermal Spray Society website: http://www.asminternational.org.

Journal of Thermal Spray Technology is a peer-reviewed quarterly technical journal devoted to thermal spray published by the ASM Thermal Spray Society. Individual published papers are available at the following ASM-TSS website: http://www.asminternational.org.

CHAPTER 12

DIFFUSION WELDING AND DIFFUSION BRAZING

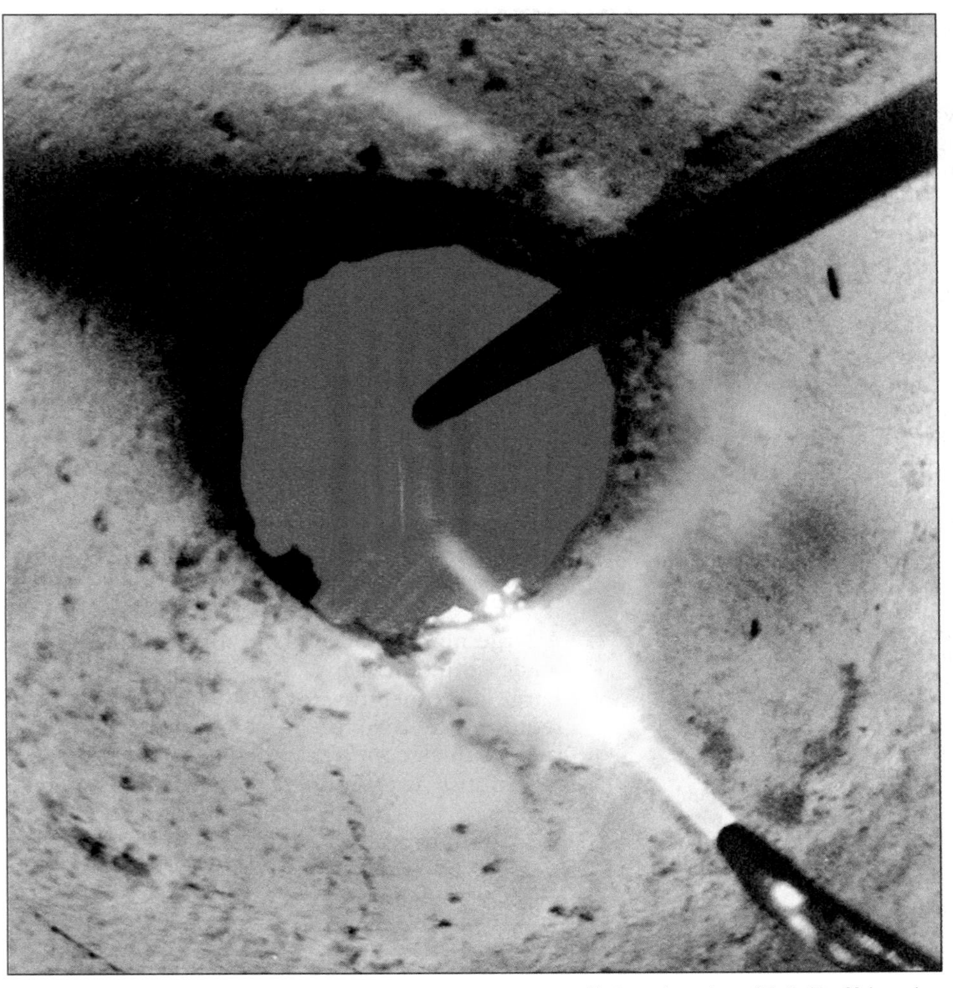

Photograph courtesy of Earthshine Mokume Inc.

Prepared by the Welding Handbook Chapter Committee on Diffusion Welding and Diffusion Brazing:

S. C. Maitland, Chair
Goodrich Landing Gear

P. Hall
Edison Welding Institute

Y. N. Zhou
University of Waterloo

Welding Handbook Volume 3 Committee Member:

D. W. Dickinson
The Ohio State University

Contents

CHAPTER 12

DIFFUSION WELDING AND DIFFUSION BRAZING

INTRODUCTION

Both diffusion welding (DFW) and diffusion brazing (DFB) are diffusion-controlled joining processes, but the former is a solid-state process while the latter involves a liquid phase in the process.

Diffusion welding produces a weld by the application of pressure at elevated temperature with no macroscopic deformation or relative motion of the workpieces. A filler metal may be inserted between the faying surfaces. A variation of diffusion welding, hot isostatic pressure welding (HIPW) produces coalescence by heating the metals and applying hot inert gas under pressure.[1, 2]

Diffusion brazing is distinguished from diffusion welding by the fact that liquid braze filler metal is introduced at the joint interface. When liquid filler metal is used, only moderate-to-low pressures are necessary versus the high pressures used for diffusion welding. The similarities and common features of diffusion welding and diffusion brazing are discussed in general in the first sections of this chapter. Information specific to each process is presented in separate welding and brazing sections. Both processes are discussed in the sections on Quality and Inspection, Economics, and Safe Practices. A brief section on a related process, diffusion soldering, also is included.

1. Terms that are sometimes used to define diffusion welding include *diffusion bonding*, *solid-state bonding*, *pressure bonding*, *isostatic bonding*, *hot press bonding*, *forge welding*, and *hot pressure welding*. The correct term, *diffusion welding*, should be used in all instances, except for hot pressure welding (HPW) and forge welding (FW), which have been classified by the American Welding Society (AWS) Committee on Definitions and Symbols as separate processes. Welding terms used throughout this chapter are from *Standard Welding Terms and Definitions*, AWS A3.0, Miami: American Welding Society.

2. At the time this chapter was prepared, the referenced codes and other standards were valid. If a code or other standard is cited without a date of publication, it is understood that the latest edition of the document referred to applies. If a code or other standard is cited with the date of publication, the citation refers to that edition only, and it is understood that any future revisions or amendments to the code or standard are not included. However, as codes and standards undergo frequent revision, the reader is advised to consult the most recent edition.

ADVANTAGES AND LIMITATIONS OF DIFFUSION WELDING AND DIFFUSION BRAZING

Diffusion welding and diffusion brazing have several significant advantages over the more commonly used welding and brazing processes, and also have some distinct limitations on applications of these processes.

Following are some of the advantages of the two processes:

1. Joints can be produced with properties and microstructures very similar to those of the base metal, which is particularly important for lightweight fabrications;
2. Components can be joined with minimum distortion, eliminating the need for subsequent machining or forming;
3. Dissimilar alloys that are not weldable by fusion processes or by processes requiring axial symmetry, such as friction welding, can be joined;
4. A large number of joints in an assembly can be made simultaneously;
5. Large joint members of base metals that would require extensive preheat for fusion welding, such as thick copper, can be more readily joined;
6. Assemblies with limited access can be joined; and
7. Discontinuities normally associated with fusion welding are not encountered.

Following are some limitations of the processes that should be considered:

1. The thermal cycle normally is longer than that of conventional welding and brazing processes;
2. Equipment costs usually are high, and this can limit the maximum size of weldments that can be economically produced;
3. The processes are not adaptable to high production rates, although a number of assemblies can be joined simultaneously;
4. Adequate nondestructive inspection techniques for quality assurance are limited, particularly those that assure design properties in the joint;
5. Suitable filler metals and procedures have not yet been developed for all structural alloys;
6. The faying surfaces and the fit of joint members generally require greater care in preparation than for conventional hot pressure welding or brazing processes;
7. In diffusion brazing, surface smoothness may be an important factor in quality control; and
8. The need to apply heat and a high compressive force simultaneously in the restrictive environment of a vacuum or protective atmosphere is a major limitation of diffusion welding.

APPLICATIONS OF DIFFUSION WELDING AND DIFFUSION BRAZING

A wide variety of similar and dissimilar metal combinations can be successfully joined by diffusion welding and diffusion brazing. Most applications involve titanium (Ti), nickel (Ni), and aluminum (Al) alloys, as well as several dissimilar metal combinations. The mechanical properties of the joint depend on the characteristics of the base metals. For example, the relatively low creep strength and the solubility of oxygen at elevated temperature contribute to the excellent properties of titanium-alloy diffusion weldments and brazements.

Nickel-base heat-resistant alloys are difficult to join because of the high creep strengths of these alloys, which require high pressures for diffusion welding. In addition, a thin, stable oxide film interferes with metal-to-metal contact because the oxygen is not soluble in nickel, unlike in titanium. These factors contribute to poor solid-state weldability of these nickel-base heat-resistant alloys. This problem can be overcome by the use of a relatively soft filler metal that provides more intimate contact.

Base metals strengthened by cold working will be irreversibly softened by the joining and heat-treating processes. However, heat-treatable alloys may be rehardened during joining and heat-treating or may be hardened with a postweld heat treatment.

Many diffusion welding and diffusion brazing applications involve titanium alloy components, the majority of which are made from Ti-6%Al-4%vanadium (V) alloy.[3] The popularity of the processes with alloys stems from the following factors:

1. Titanium is readily joined by both processes without special surface preparation or unusual process controls;
2. Diffusion-welded or diffusion-brazed joints may have better properties for some applications than fusion-welded joints;
3. Most titanium structures or components are used principally in aerospace applications in which weight savings or advanced designs, or both, are more important, within limits, than manufacturing costs.

Several well-established diffusion welding and brazing methods for joining titanium alloys are described separately in sections titled "Diffusion Welding" and "Diffusion Brazing."

SURFACE PREPARATION FOR DIFFUSION WELDING AND DIFFUSION BRAZING

The faying surfaces of joint members to be diffusion welded or diffusion brazed must be carefully cleaned and prepared before assembly. The primary surface finish is ordinarily obtained by machining, abrading, grinding, or polishing. In addition to cleanliness, surface preparation also requires the following:

1. Creation of an acceptable finish or smoothness;
2. Removal of surface oxides; and
3. Cleansing of gaseous, aqueous, or organic surface films.

Two properties of a correctly prepared surface are flatness and smoothness. A certain minimum degree of flatness and smoothness of the workpieces is required to ensure uniform contact. Conventional metal cutting, grinding, and abrasive polishing methods usually are adequate to produce the needed surface flatness and smoothness. A secondary effect of machining or abrading is the cold work introduced into the surface. Recrystallization of the cold-worked surfaces increases the diffusion rate across the weld or braze interface.

Chemical etching (pickling) is a commonly used form of surface preparation for welding or brazing that has two results. The first is the favorable removal of nonmetallic surface films, usually oxides. The second is less favorable since it removes all or part of the cold-worked layer that forms during machining. The need to

3. The weldability of titanium alloys is discussed in American Welding Society (AWS) Welding Handbook Committee, Oates, W. R. and Saitta, eds., 1998, Vol. 4 of *Welding Handbook*, 8th ed., 487–538.

remove oxide is apparent because it prevents metal-to-metal contact.

Degreasing is a universal part of any procedure for surface cleaning. Alcohol, acetone, detergents, and many other cleaning agents can be used. Frequently, the recommended degreasing technique is intricate and may include multiple rinse-wash-etch cycles using several solutions. Because some of these cleaning solvents are toxic or flammable, the safety precautions recommended by the manufacturer must be followed.

Many factors enter into selecting the treatment of the faying surfaces. In addition to the methods previously mentioned, the specific welding or brazing conditions may affect the selection. When high temperature or pressure is used, it becomes less important to obtain extremely clean surfaces. Increased atomic mobility, surface asperity deformation, and solubility of impurity elements all contribute to the dispersion of surface contaminants. When lower temperature or pressure is used, better-prepared and preserved surfaces are more important.

Preserving the clean faying surface is necessary following the surface preparation. One requirement is the effective use of a protective environment during diffusion welding or brazing. A vacuum environment provides continued protection from contamination. A pure hydrogen (H) atmosphere minimizes the amount of oxide formation, and at elevated temperature, it reduces the existing surface oxides of many metals. However, with titanium, zirconium (Zr), niobium (Nb), and tantalum (Ta), a pure hydrogen atmosphere forms hydrides that may be detrimental. High-purity argon (Ar), helium (He), and sometimes nitrogen (N) can be used to protect clean surfaces at elevated temperature. Many of the precautions and principles applicable to brazing atmospheres can be applied directly to diffusion brazing or welding.[4]

DIFFUSION WELDING

Two necessary conditions must be met before a satisfactory diffusion weld can be made. There must be mechanical intimacy of the faying surfaces, and interfering surface contaminants must be disrupted and dispersed to permit atomic bonding at the weld interface.

4. For additional information on brazing atmospheres, refer to Chapter 12 of American Welding Society (AWS) Welding Handbook Committee, A. O'Brien, Ed., 2004, *Welding Processes, Part 1*, Vol. 2 of *Welding Handbook*, 9th ed., Miami: American Welding Society. Also refer to American Welding Society (AWS) Committee on Brazing and Soldering, C. L. Jenney, ed., (forthcoming) *Brazing Handbook*, 5th ed., Miami: American Welding Society.

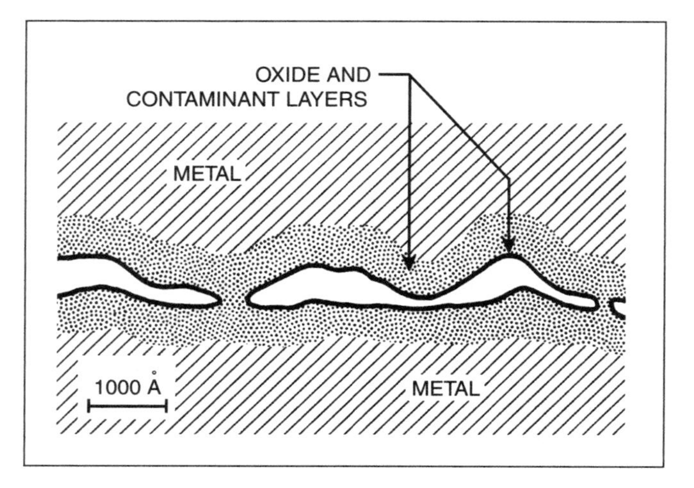

Figure 12.1—Characteristics of a Metal Surface Showing Roughness and the Presence of Contaminants

FUNDAMENTALS

As illustrated in Figure 12.1, metal surfaces have several general characteristics:

1. Roughness;
2. An oxidized or otherwise chemically reacted and adherent layer;
3. Other randomly distributed solid or liquid products such as oil, grease, and dirt; and
4. Adsorbed gas or moisture, or both.

For conventional diffusion welding without a diffusion aid (a solid filler metal applied to the faying surfaces), a three-stage mechanistic model, shown in Figure 12.2, describes the metallurgical stages that take place during weld formation.[5] The initial asperity contact is shown at left in Figure 12.2. In the first stage, (A) deformation of the contacting asperities occurs primarily by yielding and by creep deformation mechanisms to produce intimate contact over a large fraction of the interfacial area. At the end of this stage, the joint is essentially a grain boundary at the areas of contact with voids between these areas. During the second stage (B), diffusion becomes more prevalent than deformation, and many of the voids disappear as the grain-boundary diffusion of atoms continues. Simultaneously, the interfacial grain boundary migrates to an equilibrium configuration away from the original weld interface, leaving many of

5. This model is not applicable to diffusion brazing or hot pressure welding processes in which intimate contact is achieved through the use of molten filler metal in diffusion welding and bulk deformation in hot pressure welding.

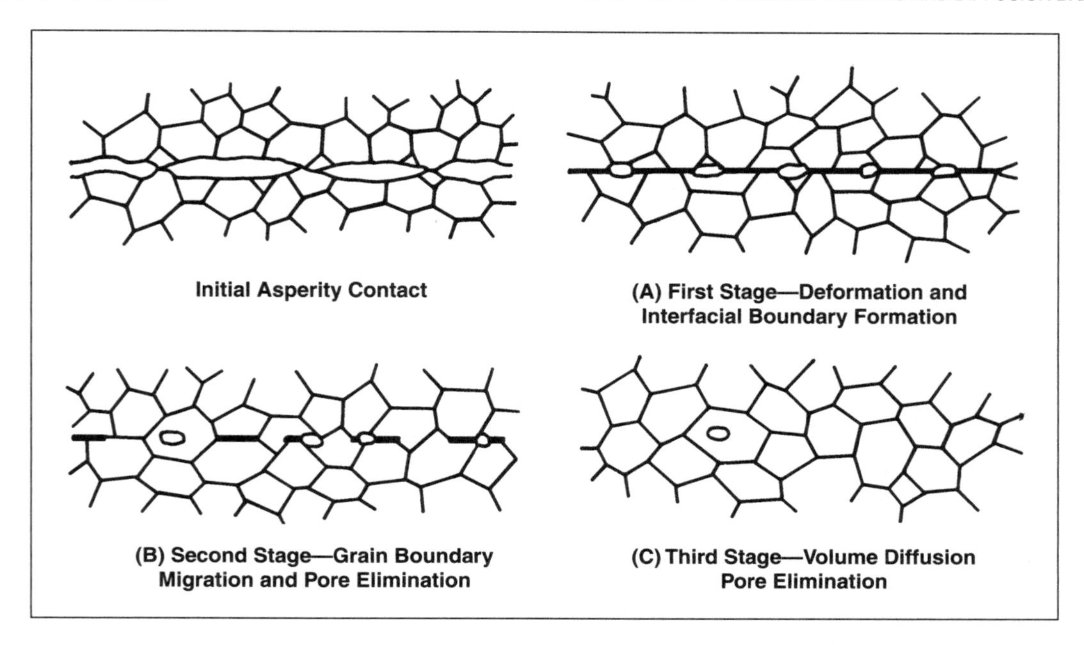

Figure 12.2—Three-Stage Mechanistic Model of Diffusion Welding

the remaining voids within the grains. In the third stage (C), the remaining voids are eliminated by volume diffusion of atoms to the void surface (equivalent to diffusion of vacancies away from the void). The stages overlap, and mechanisms that may dominate one stage also operate to some extent during the other stages.

This model is consistent with the following experimentally observed trends:

1. Temperature is the most influential variable, since the combination of temperature and pressure determines the extent of the contact area during the first stage and it alone determines the rate of diffusion that governs the elimination of voids during the second and third stages of welding;
2. Pressure is necessary only during the first stage of welding to produce a large area of contact at the welding temperature. Removal of pressure after this stage does not significantly affect joint formation. However, premature removal of pressure before completion of the first stage is detrimental to the process;
3. In general, rough initial surface finishes adversely affect welding by impeding the first stage and leaving large voids that must be eliminated during the later stages of welding; and
4. The time required to form a joint depends on the temperature and pressure used, as time is not an independent variable.

At the same time that intimate contact is being achieved, as previously described, various intervening surface films must be disrupted and dispersed so that metallic bonds can form. During the initial contact of the faying surfaces (Stage A in Figure 12.2), the films are locally disrupted and metal-to-metal contact begins at places where the surfaces move together under shear. The subsequent steps in the process (Stages B and C) involve thermally activated diffusion mechanisms that complete film dispersion and eliminate voids to achieve intimate metal contact.

The barrier film is largely an oxide. Proper cleaning methods reduce the other components of film to negligible levels. Two actions tend to disrupt and disperse the oxide film. The first is the solution of the oxide in the metal; the second is the spheroidization or agglomeration of the film. The surface oxides of titanium, tantalum, niobium, zirconium, and other metals in which interstitial elements are highly soluble will decompose at high temperatures because of the increased solubility of oxygen in these metals. If the oxide is relatively insoluble in the metal, as is the case for aluminum, the disruption action for the trapped film is spheroidization. This leaves a few oxide particles along the weld interface. However, if the weld is properly made, these particles are no more detrimental than the inclusions normally present in most metals and alloys.

Both the decomposition and spheroidization of the oxides require diffusion. Decomposition results in the diffusion of interstitial atoms of oxygen into the metal and spheroidization by diffusion as a result of the excessive surface energy of the thin films. The time of decomposition of a film of thickness X is proportional to X^2 over D, where D is the diffusion coefficient. The film must be kept very thin if diffusion-welding times are to be within acceptable limits. Spheroidization occurs more rapidly if the oxide films are thin; hence, control of the film thickness after cleaning and during heating to welding temperature is a critical factor in diffusion welding.

Once actual metal-to-metal contact is established, the atoms of both metals are within the attractive force fields of one another and a high-strength joint is generated. At this time, the joint resembles a grain boundary because the metal lattices on each side of the line have different orientations. However, the joint may differ slightly from an actual grain boundary because it may contain more impurities, inclusions, and voids that will remain if full asperity deformation has not occurred. For example, Stage B in the model for achieving intimate contact is not yet complete. As the process is carried to completion, this boundary migrates to a more stable non-planar configuration, and any remaining interfacial voids are eliminated through vacancy diffusion.

An intermediate filler metal is of significant practical importance in many systems, although the mechanisms described so far do not consider its use. When a filler metal is used or dissimilar base metals are welded, the mechanisms of diffusion of each metal into the other must be considered to develop a complete understanding of the DFW process.

DIFFUSION WELDING APPLICATIONS

Following are several categories of similar and dissimilar metal combinations that can be joined by diffusion welding:

1. Similar metals can be joined directly to form a solid-state weld. In this situation, the required pressures, temperatures, and times are dependent only on the characteristics of the base metals and surface preparation.
2. Similar metals can be joined with filler metal in the form of a thin layer of a different metal between them. In this case, the filler metal may promote more rapid diffusion or permit increased microdeformation at the joint to provide more complete contact between the surfaces. The filler metal may be diffused into the base metal by suitable heat treatment until it no longer remains a separate layer.
3. Two dissimilar metals can be joined directly where diffusion-controlled phenomena occur to form a metallic bond. The mechanisms are similar to those in Category 1, with the added effects that dissimilar metals create.
4. Dissimilar metals can be joined by using a third metal (for example, a filler metal) between the faying surfaces, to enhance weld formation either by accelerating diffusion or permitting more complete initial contact in a manner similar to Category 2 above.

PROCESS VARIABLES

Temperature, time, pressure, and metallurgical characteristics are welding variables that must be controlled for each diffusion welding application.

Temperature

Temperature is readily measured and controlled and is an important diffusion welding process condition for the following reasons:

1. In any thermally activated process, an incremental change in temperature will cause the greatest change in process kinetics when compared to most other process conditions;
2. Virtually all the mechanisms are temperature-dependent;
3. Physical and mechanical properties at elevated temperature, critical temperatures, and phase transformations are important reference points; and
4. Temperature must be controlled to promote or avoid certain metallurgical factors, such as allotropic transformation, recrystallization, and solution of precipitates.

Kinetic theory provides a means of understanding the quantitative effects of temperature in diffusion welding. Diffusivity can be expressed as a function of temperature as follows:

$$D = D_o e^{-Q/kT} \tag{12.1}$$

where

D = *Diffusivity*, the diffusion coefficient at temperature T;
D_o = A constant of proportionality;
e = An exponential value defined mathematically;
Q = Activation energy for diffusion;
T = Absolute temperature; and
k = Boltzmann's constant.

It is apparent from this equation that the diffusion-controlled processes vary exponentially with temperature. Thus, relatively small changes in temperature produce significantly large changes in process kinetics.

In general, the temperature at which diffusion welding takes place is above 0.5 *Tm*, where *Tm* is the melting temperature of the metal in degrees Kelvin or degrees Rankine. Temperatures between 0.6 *Tm* and 0.8 *Tm* are best for the diffusion welding of many metals and alloys. For a specific application, the temperature, pressure, time, and faying-surface preparation are interrelated and values must be determined.

Time

Time is closely related to temperature in that most diffusion-controlled reaction rates vary with time. The diffusion length, x, is the average distance traveled by migrating atoms during diffusion. It can be approximated as follows:

$$x = C(Dt)^{1/2} \tag{12.2}$$

where

x = Diffusion length,
C = A constant of proportionality (if undefined, use "1"),
D = Diffusion coefficient at T (absolute temperature), and
t = Time.

Thus, diffusion reactions progress with the square root of time (longer times become less and less effective), whereas they progress exponentially with temperature, as was previously shown.

Experience indicates that increasing both the time and the pressure at welding temperature increases joint strength up to a certain limit. Beyond this point no further gains are achieved. This limit is unknown. This illustrates that time is not a quantitatively simple condition. The simple relationship that describes the average distance traveled by an atom does not reflect the more complex changes in microstructure that result in the formation of a diffusion weld. Although atom motion continues indefinitely, microstructural changes tend to approach equilibrium. An example of similar behavior is the recrystallization of metals.

In a practical sense, time may vary over a very broad range, from seconds to hours. Production factors influence the practical time for diffusion welding. An example is the time necessary to attain the correct heat and pressure.

When the welding equipment has thermal and mechanical (or hydrostatic) inertia, welding times are long because of the impracticality of suddenly changing the conditions. When there are no inertial problems, welding time may be as short as 0.3 minute, as is the case when joining two thoria-dispersed nickel workpieces. Conversely, it may be as long as 4 hours, as when joining niobium to niobium with zirconium as a filler metal.

Pressure

Pressure is an important factor that affects several aspects of the diffusion welding process. Pressure is more difficult to deal with as a quantitative value than either temperature or time. The initial phase of metallic bond formation is certainly affected by the amount of deformation induced by the pressure applied. This is the most obvious single effect and probably the most frequently and thoroughly considered. Higher pressure invariably produces better joints when the other variables are fixed, within the limits of the welding range. Faying-surface deformation and asperity collapse are the most apparent reasons for this effect. The greater deformation also may lower the recrystallization temperature and accelerate the process of recrystallization at the welding temperature.

The welding equipment and the joint geometry place practical limitations on the magnitude of welding pressure. The pressure needed to achieve a good weld is closely related to the temperature and time. Pressure has additional significance when dissimilar metal combinations are considered. From an economic and manufacturing standpoint, low welding pressure is a desirable aspect. High pressure requires more costly equipment, better controls, and generally involves more complex production procedures.

The pressures and temperatures employed are largely interdependent, but the pressure need not exceed the yield stress of the base metal or filler metal at welding temperature. Thus, unless retaining dies are used, the pressure usually is kept slightly below the yield stress at welding temperature. The temperature and pressure normally are selected to produce a weld in an acceptable time.

Metallurgical Factors

In addition to the process conditions, a number of metallurgically important factors must be considered. Of particular importance for welds in similar metals are phase transformation and microstructural factors that tend to modify diffusion rates. Phase transformation (allotropic transformation) occurs in some metals and alloys. Steels are the most familiar of these, but titanium, zirconium, and cobalt (Co) also undergo phase transformation. During phase transformation, the metal is very plastic, and this promotes rapid deformation of the faying surfaces at lower pressures. Diffusion rates generally are higher during transformation, and also during recrystallization.[6]

Another means of increasing diffusion is alloying, or more specifically, introducing elements with high diffusivity at the faying surfaces. The function of a high-diffusivity element is to accelerate the elimination of voids. In addition to simple diffusion acceleration, these alloying elements may have secondary effects. The elements should have reasonable solubility in the base metal, but should not form stable compounds. Alloying should not promote melting at the weld interface.

When using an alloying element as a diffusion activator, it is desirable to hold the weldment at the diffusion temperature either during or after the welding process to reduce the high concentration of the alloying element at the weld interface. If this is not done, the high concentration may produce metallurgically unstable microstructures. This is particularly important for joints that will be exposed to elevated temperatures in service.

Filler Metals

It is sometimes advantageous to use some form of solid-state filler metal between the faying surfaces. The solid-state filler metal used in diffusion welding is distinguished from a brazing filler metal by the fact that it is not melted during the joining process. One purpose of a filler metal is to provide a layer of soft metal between the faying surfaces. A soft metal layer permits plastic flow to take place at lower pressures than would be required without the layer during the first stage of welding. After the joint is formed, the diffusion of alloying elements from the base metal into the filler metal reduces the compositional gradient across the joint.

Filler metals for diffusion welding may be necessary or beneficial in certain applications to achieve the following advantages:

1. Reduced welding temperature,
2. Reduced welding pressure,
3. Reduced process time,
4. Increased diffusivity, and
5. Scavenging of undesirable elements.

Filler metals can be applied in many forms. They can be electroplated, condensed, or sputtered onto the faying surface, or they can be a preplaced foil interlayer in the form of foil inserts or powder. The thickness of the filler metal should not exceed 0.25 mm (0.010 in.).

6. For additional information on welding metallurgy, refer to Chapter 4 of American Welding Society (AWS) Welding Handbook Committee, Jenney, C. L., and A. O'Brien, eds., 2001, *Welding Science and Technology,* Vol. 1 of *Welding Handbook*, 9th ed., Miami: American Welding Society.

The filler metal generally is a more pure version of the base metal. For example, unalloyed titanium is often used as a filler metal with titanium alloys, and nickel sometimes is used with nickel alloys. An exception to this general rule is the use of silver (Ag) as a filler metal in the diffusion welding of aluminum.

Aluminum Alloys

Aluminum alloys are among the most difficult metals to weld because of the rapid formation of a stable oxide film on bare aluminum surfaces. Most diffusion welding of aluminum is done at high temperatures. Lower temperature is required if foil interlayers or electroplated coatings are used as filler metals, and still lower pressures and deformations are needed in the presence of a transient liquid phase, as in diffusion brazing. However, these variables must be controlled, otherwise they can produce brittle intermetallic phases and result in low weld strength. Significant quantities of silver can dissolve in aluminum at 480°C to 530°C (896°F to 986°F), and silver oxides are unstable above 200°C (392°F). Thus, if the diffusion temperature is 480°C to 530°C (896°F to 968°F), silver oxides will not form, and the silver will dissolve in the aluminum base metal.

Any aluminum alloy can be welded by combining a silver coating on a surface clad with aluminum, and in practice, this could have the advantage of using a single welding procedure for all aluminum alloys. The diffusion welding of an aluminum assembly is illustrated in Figure 12.3.

Filler metals containing rapidly diffusible elements also can be used. For example, beryllium (Be) can be used with nickel alloys to decrease diffusion time. A properly selected diffusion aid (filler metal) will not melt at welding temperature or form a low-melting eutectic with the base metal. An improperly chosen diffusion aid can result in the following adverse conditions:

1. A decrease in the in-service temperature tolerance of the joint,
2. A decrease in the strength of the joint,
3. Degradation of the microstructure, and
4. Corrosion problems at the joint.

PROCESS VARIATIONS

Conventional diffusion welding involves the application of pressure and heat to accomplish a weld simultaneously along the entire length of one or more joints. Filler metal may or may not be used. Pressure may be applied using gas pressure or a mechanical or hydraulic press. Heat can be applied by any convenient means; however, electrical resistance heaters are the most commonly used. Forming the components of the weldment

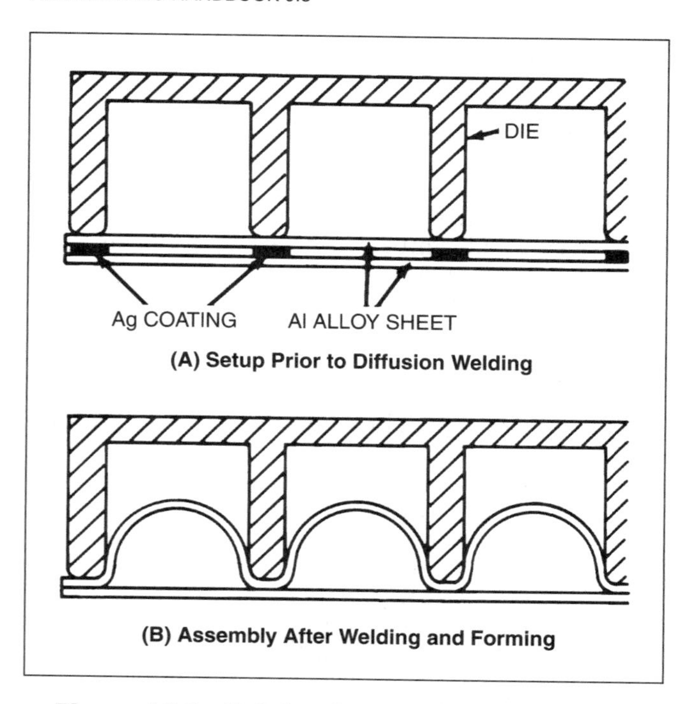

Figure 12.3—Fabrication of Diffusion-Welded Aluminum Assembly Using Silver as a Filler Metal

to shape is done prior to or after welding, using equipment designed for that purpose.

Although complex techniques may be required to create diffusion welds in some applications, others can be accomplished with the simplest of methods and tools. An example is a Japanese technique for the diffusion welding of several alloys to form a billet used to manufacture jewelry. The equipment consists of an oxyfuel gas welding torch, a C-clamp and a chamber created from refractory brick.[7,8] Material to be diffusion welded is cleaned and clamped tightly in the C-clamp and then transferred to the chamber. The chamber has two ports, one to allow the torch flame in and the other for observing the progress of the diffusion weld. During the firing, the billet must be constantly monitored for signs that the weld is taking place. Depending on the size of the billet and the firing, this technique can take as little time as one hour to complete a diffusion weld.

Hot Isostatic Pressure Welding

Hot isostatic pressure welding (HIPW) is a variation of diffusion welding in which the coalescence of metals is achieved by heating the workpieces and applying hot inert gas under pressure.

DIFFUSION WELDING TECHNIQUES

Diffusion welding techniques have been adapted to several critical applications, including the continuous seam welding of titanium and nickel-base superalloys, creep isostatic pressing of titanium sheet structures, and the superplastic forming and welding of titanium and aluminum in a combined procedure. These techniques are discussed in this section.

Continuous Seam Diffusion Welding

The components of continuous seams can be joined by yield-controlled diffusion welding. With this technique, the workpieces are positioned with tooling devices and then fed through a machine with four rollers. The top and bottom rollers are made of molybdenum and function much like resistance seam welding electrodes (wheel electrodes). The two side rollers are used to maintain the shape of the components. The electrodes and workpieces are heated to the desired temperature by electrical resistance. A specialized control system monitors the temperature of the workpieces. Welding temperature usually is between 982°C and 1090°C (1800°F and 2000°F) for titanium, and between 1090°C and 1200°C (2000°F and 2200°F) for nickel-base superalloys. The hot electrodes apply pressure in the range of 7 MPa to 138 MPa (1 ksi to 20 ksi) on the seam. The actual pressure depends on the type of metal being joined, the joint design, the temperature, and the welding speed.

Combined Forming and Welding

Two diffusion welding techniques, creep isostatic pressing and superplastic-forming diffusion welding, can be used to take advantage of the superplastic properties of certain metals or alloys to perform the dual functions of forming and welding in one operation. Some alloys deform or flow significantly at elevated temperatures under very small applied loads without yield or fracture. Titanium alloys exhibit this superplastic behavior. Complex shapes can be formed before

7. This diffusion welding technique, called *mokume gane*, is used in Japan to weld alloys together with simple tools. Mokume gane (translated as "creating a wood-grain pattern in metal") is accomplished with a series of steps to form a billet of diffusion-welded alloys. A design is created by exposing the contrast of different metals. The billet is then used to make jewelry with a wood-grain pattern.
8. Midgett, S., 2000, *Mokume Gane–A Comprehensive Study*, Franklin, N.C.: Earthshine Press.

welding using moderate gas pressures, or they can be welded before forming.

The creep isostatic press process is a two-step procedure combining creep or superplastic forming of titanium sheet structures with hot isostatic pressing to produce diffusion-welded structures, such as those used in various aerospace applications.

Inherent in this application is bringing two titanium sheets (skins) into intimate contact to form the external skin of the structure. First, one skin is creep-formed by gas pressure to the contour of a welding die. Then shaped inserts are placed at locations on the skin and a second skin is creep-formed by gas pressure over the first skin and the inserts. Diffusion welding of the formed sheets and the inserts is achieved by hot isostatic pressing in an autoclave. This method eliminates the need for precision-machined welding die sets and close dimensional tolerances of the workpieces.

Superplastic-Forming Diffusion Welding. Another technique, superplastic-forming diffusion welding (sometimes referred to as SPF/DFW) uses the same properties of titanium and titanium alloys as described previously; however, the welding is performed under low-pressure conditions. Because superplastic forming and diffusion welding of selected titanium alloys can be accomplished at the same temperature, the two operations can be combined in a single fabrication cycle.

For titanium alloys that exhibit superplastic properties, superplastic-forming diffusion welding considerably extends the range of low-cost, structurally efficient titanium aerospace components that can be manufactured. Titanium components fabricated by superplastic forming and diffusion welding may be substituted for conventionally fabricated aluminum alloy components.

Continuing developments in the superplastic diffusion welding of high-strength aluminums and metal matrix composites have increased the use of this method for the diffusion welding of aluminum.

The superplastic forming of the aluminum sheet may be done first, followed by welding, or the steps can be reversed. The order depends on the design of the component. If forming is required to bring the faying surfaces of the workpieces together for welding, forming is done first. If the faying surfaces are in contact, they are welded in the first process and then the workpiece is formed to final shape. A suitable nonmetallic agent such as Stopoff™ can be applied to prevent welding in selected areas.

Superplastic forming of Ti-6%Al-4%V alloy sheet can be accomplished by applying argon at low pressure and at a temperature of 925°C (1700°F) in a sealed die. Gas pressure of about 1035 kPa (150 psi) is used for both forming and welding. Preparation for the welding of titanium alloy sheets usually is limited to degreasing and acid etch.

EQUIPMENT

A wide variety of equipment and tooling is employed for diffusion welding. The basic requirement is to provide the means of applying and maintaining pressure and temperature in a controlled environment. The equipment is designed to meet these general process requirements and can be modified for specific applications. A general description of three types of diffusion welding equipment follows.

Autoclaves

The pressure required for hot isostatatic gas pressure welding can be applied uniformly to all joints in an assembly using gas pressure in a cold-wall autoclave. Autoclaves are pressure vessels and must be designed to meet applicable code requirements. It is important that all air be removed from the assembly prior to welding. The assembly itself may be evacuated and sealed by fusion welding, if this is practical. Otherwise, the assembly must be sealed in a thin, gas-tight envelope that is evacuated and sealed. Electron beam welding equipment, since it provides a vacuum chamber, is a convenient process for evacuating and sealing in one operation.

Gas pressure is applied externally against the evacuated assembly at welding temperature. Very high pressures can be applied using an autoclave, but the assembly must be capable of withstanding the applied pressure without macrodeformation. Some assembly designs may require internal support tooling with provisions for removing it after welding.

The primary component of hot isostatic equipment is a cold-wall autoclave, which can be designed for gas pressures up to 1035 MPa (150 ksi) and for workpiece temperatures in excess of 1650°C (3000°F). A schematic drawing of a typical autoclave is shown in Figure 12.4.

The workpieces are placed in the heated cavity of the autoclave, where openings on each end provide access to the vessel. Water-cooling usually is provided internally to maintain low temperature in the autoclave wall. Utilities and instrumentation are brought into the vessel through high-pressure fittings located in the end closures. The high temperatures are produced with an internal heating system; resistance heaters of various designs are used. Alumina or silica insulation is used to reduce heat losses to the cold wall, and the temperature is monitored and controlled by thermocouples located throughout the furnace and vessel. Pressurization is achieved by pumping inert gas into the autoclave with a compressor. Temperature and pressure are controlled independently, and any combination of heating and pressurizing rates can be programmed.

The most important consideration is the gas-tight envelope or container in which the workpieces are

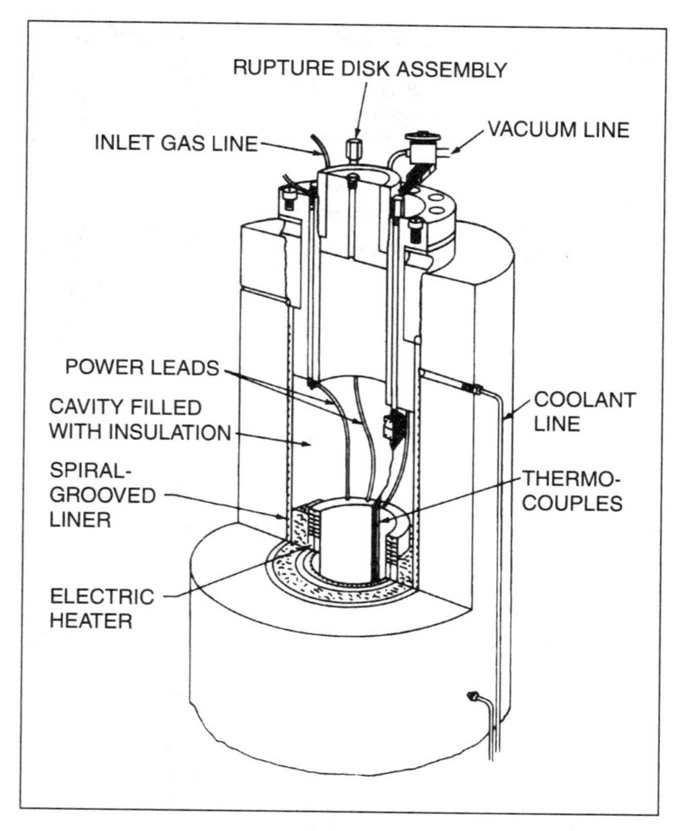

Figure 12.4—A Typical High-Temperature, Cold-Wall Autoclave

placed. If a leak develops in the container, pressure cannot be applied to or maintained on the joint. Sufficient gas pressure is applied so that local plastic flow will occur at the faying surface and all void space will be filled as a result of local deformation. With proper conditions, essentially no macrodeformation or changes in workpiece dimensions will occur during welding.

The gas pressure process variables are selected to suit the base metals. Usually, joints are made at the highest possible pressure to minimize the temperature needed. This method is well suited for welding brittle metals to ceramics or cermets because the isostatic pressure eliminates tensile stresses in the materials.

The principal advantage of the autoclave is the capability of handling complex shapes. It is well suited also to batch operations where large quantities of relatively small assemblies can be welded simultaneously. The major drawbacks are the capital equipment costs and the size limitations imposed by the internal dimensions of the autoclave. Autoclaves are available in sizes up to 92 cm (36 in.) inside diameter, and inside lengths of up to 275 cm (108 in.).

Presses

Mechanical or hydraulic presses are commonly used in diffusion welding. The basic requirements for a press are the following:

1. Sufficient load and size capacity,
2. A source of heat, and
3. The capability of maintaining uniform pressure for the required time.

Often it is necessary to weld inside a chamber that provides a protective atmosphere chamber around the weldment. An example of press equipment that can be adapted to diffusion welding applications is shown in Figure 12.5.

A standard press design does not exist for diffusion welding. The press can be as simple as a C-clamp or as complex as a hydraulic press that provides a vacuum or an inert atmosphere around the workpieces. Radiant, direct-fire, induction, and resistance types of heating can be used. Two advantages of presses are ease of operation and excellent process control. A disadvantage is the practical limitation of press size when large weldments are considered. Presses do not accommodate high production rates or batch operations.

Some of the limitations on size can be overcome by operating in a large forming or forging press without an inert atmosphere chamber. In this method, heated platens apply both heat and pressure to the components. The platens may be metallic or ceramic, depending on the temperature and pressure employed. Castable ceramics are particularly useful because contours can be accommodated easily without extensive machining. Heating elements can be cast into a ceramic die to provide uniform heat during welding. Close tolerances must be maintained between the die and the workpiece so that uniform pressure will be applied to the joint. This is a major problem with press-type equipment. It is difficult to maintain uniform pressure on the joint, and variations in weld quality can result.

Tooling requirements vary with the application. If lateral restraint is not provided, excessive upsetting may occur during welding. In such cases, lower pressure or temperature usually is required. Heated welding dies are required and die materials must be carefully selected. The die must withstand both the temperature and the pressure and must be compatible with the base metal. Interaction between the workpiece and the die can be controlled by stopoff agents and sometimes by oxidizing the die surface. Protection from the atmosphere is often achieved by sealing the workpieces in evacuated metal cans designed to conform to the die shape.

Retorts. Retorts can be used in conjunction with presses for the diffusion welding of titanium. Tooling blocks and spacers may be used to fill any voids between the titanium workpieces to maintain the shape

Photograph courtesy of Advanced Vacuum Systems, Inc.

Figure 12.5—1000-Ton Vacuum Hot Press for Diffusion Welding

of the workpieces. Presses with side- and end-restraining jacks can exert up to 13.8 MPa (2 ksi) pressure on the retort in all directions. In actual production, the completed assembly pack (retort, heating pads, and insulation) is heated before it is placed in the press. Large structures may require preheating as long as 40 hours. Several packs may be in assembly and preheat at one time. The actual time in the press will vary from 2 to 12 hours, depending on the shape of the structure and the mass of the titanium. The assembly pack is cooled to room temperature, dismantled, and the retort is then cut open. This approach is quite slow and is not readily adapted to high production rates.

Resistance Welding Machines

Resistance welding equipment can be used to produce diffusion spot welds between sheet metal workpieces. In general, conventional resistance welding equipment can be used without modification to achieve successful diffusion welds. The weld interface is resistance-heated under pressure with this equipment. The welding cycle is designed to avoid the melting of metal at the interface. Weld times generally are less than 1 second.

As in conventional resistance welding, the selection of the correct electrode material is important. The electrodes must be electrical conductors, possess high strength at welding temperatures, resist thermal shock, and resist sticking to the workpieces. A universal electrode material does not exist because of potential interaction with varied workpiece materials. Therefore, each combination must be carefully evaluated from the standpoint of metallurgical compatibility.

In some applications, a small chamber surrounding the electrodes is used to provide an inert atmosphere or vacuum during welding.

An advantage of this type of resistance welding equipment is the speed at which diffusion welds can be made. Each weld is made in a very short time; however, only a small area is welded in each cycle and many welds are needed to join a large area.

TOOLING

Selecting tooling equipment involves several important considerations. The main criteria are the following:

1. Ease of operation,
2. Reproducibility of the welding cycle,
3. Weld cycle time,
4. Operational maintenance, and
5. Capital cost.

In addition, the equipment must be capable of maintaining proper positions and shapes of the workpieces throughout the heating cycle.

The material from which the fixtures are made is another consideration. Suitable fixture materials are limited when welding temperatures are above 1320°C (2400°F). Only the refractory metals and certain nonmetallic materials have sufficient creep strength at such high temperatures. For example, tantalum and graphite may be the only suitable materials for fixtures for the diffusion welding of tungsten. Fixtures made of ceramic materials can be used, provided they are completely outgassed prior to welding.

Fixtures can be designed to take advantage of the difference in thermal expansion between the base metals and the fixture material. It is possible to generate some, if not all, of the pressure required for welding by the appropriate selection of the fixture material, base metal and the clearances between the fixture and the workpiece.

METALS WELDED

Titanium, nickel, and aluminum alloys are among the materials commonly welded by the diffusion welding process. Steels can be welded in certain applications. Diffusion welding is an excellent alternative process for joining many dissimilar materials that cannot be welded with other welding processes.

Titanium Alloys

Welding temperature is probably the most influential condition in determining weld quality in titanium welds; the temperature is set as high as possible without causing irreversible damage to the base metal. For the commonly used alpha-beta type of titanium alloys, this temperature is about 24°C to 38°C (75°F to 100°F) below the beta transus temperature. Thus, Ti-6%Al-4%V alloy with a beta transus of approximately 996°C (1825°F) normally is welded using temperatures between 925°C and 955°C (1700°F and 1750°F).

The time required to achieve high weld strength can vary considerably with other factors, such as faying surface roughness, welding temperature, and pressure. Welding times of 30 minutes to 60 minutes should be considered a practical minimum, with 2 hours to 4 hours being more desirable.

Nickel Alloys

Many nickel alloys, specifically the high-strength, heat-resistant alloys, are more difficult to weld than most other metals. These alloys must be welded at temperatures close to their melting temperatures, and because of their high-temperature strengths, relatively high pressures are required. In addition, extra care must be exercised when preparing the faying surfaces to ensure cleanliness and mutual conformity. Surface oxides that form on these alloys are stable at high temperatures and will not dissolve or diffuse into the base metal. During welding, the ambient atmosphere must be carefully controlled to prevent contamination of the faying surfaces.

Nickel Filler Metals. Pure nickel or a soft nickel alloy is commonly used as filler metal when joining nickel alloys by diffusion welding. These filler metals, generally from 2.5 micrometers (µm) to 25 µm (0.0001 in. to 0.001 in.) in thickness, serve several functions. Their relatively low yield strengths allow surface conformity to take place at relatively low welding pressures. More importantly, they are used during welding to prevent the formation of stable precipitates, such as oxides, carbides, or carbon nitrides at the weld interface. The welding time must be long enough to allow sufficient interdiffusion to occur at the weld interface.

Table 12.1
Typical Diffusion Welding Conditions for Some Nickel-Base Alloys

Base Metal (Trade Names)	Filler Metal	Welding Temperature		Pressure		Time, h
		°C	°F	kPa	Psi	
Inconel 600	Ni	1090	2000	690–3450	100–500	0.5
Hastelloy X	Ni	1120	2050	690–3450	100–500	4
Wrought Udimet 700	Ni-35%Co	1170	2140	6900	1000	4
Cast Udimet 700	Ni-35%Co	1190	2175	8275	1200	4
Rene 41	Ni-Be	1180	2150	10690	1550	2
Mar-M 200	Ni-25%Co	1205	2200	6900–13800	1000–2000	2

Welding conditions for diffusion welds in some nickel-base, heat-resistant alloys are shown in Table 12.1.

The pressure required for satisfactory welding is strongly influenced by the geometry of the joint members. Therefore, the required pressure for each application must be determined empirically.

The significance of filler metal and its composition was demonstrated by a series of diffusion welds in a wrought and cast proprietary nickel alloy. Welds were made without filler metal and then with filler metals 5 µm (0.0002 in.) of both pure nickel and nickel-35% cobalt alloy. The welding conditions were the same as those listed for this alloy in Table 12.1.

The microstructure of the welds in a proprietary wrought alloy is shown in Figure 12.6. When no filler metal was used, fine TiC (carbon)-and $NiTiO_3$ precipitates formed at the weld interface during welding and pinned the interfacial boundary, causing very poor mechanical properties in the weld. The nickel filler metal consisted of an electroplated layer on each surface. These layers probably welded together early in the cycle, and no precipitates were present to interfere with welding. Subsequent diffusion and grain boundary movement resulted in much improved mechanical properties. The pure nickel filler metals, however, resulted in preferential diffusion of aluminum and titanium into the nickel. This led to the formation of excessive amounts of the strengthening precipitate Ni_3 (Al, Ti) in the joint. The use of a filler metal of nickel-35% cobalt (Co) alloy prevented the diffusion of aluminum and titanium and resulted in a homogeneous joint.

Aluminum Alloys

Successful diffusion welds can be made in aluminum alloys as long as some means is employed to avoid, disrupt, or dissolve the tenacious surface oxide. A wide range of temperatures, pressures, and weld times may be utilized to weld aluminum alloys. For example, with alloy Type 6061 (AMS-A-22771) aluminum alloy, welding conditions as divergent as 385°C (725°F) and a pressure of 26 MPa (3800 psi) for several hours or 538°C (1000°F) and a pressure of 7 MPa (1000 psi) for one hour have been satisfactory.[9] However, the definitive condition is the melting point of the base metal. Welding normally is carried out in a vacuum or an inert gas atmosphere, although aluminum-boron fiber composites can be diffusion welded in ambient air. If no local deformation of the workpieces can be tolerated, the faying surfaces should be coated with a thin layer of silver or gold-copper alloy by electrolytic or vapor deposition. The coating will prevent surface oxidation during welding.

Steels

Steels normally are not welded by diffusion because in most applications they are more easily joined by conventional brazing or fusion welding processes. Diffusion welding can be utilized successfully for special applications in steel in which high-quality joints are required between large, flat surfaces. For example, low-carbon steels have been welded without filler metal over a wide range of conditions. Two sets of conditions that produced excellent welds in American Iron and Steel Institute (AISI) 1020 steel are 982°C to 1204°C (1800°F to 2200°F) with a pressure of 7 MPa (1 ksi) for 1 minute to 15 minutes and 1093°C to 1204°C (2000°F to 2200°F) with a pressure of 35 kPa (5 psi) for 2 hours. Welding can be accomplished either in a protective atmosphere or in ambient air, provided the joint is first seal-welded around the periphery to exclude air.

9. Aerospace Materials Specifications (AMS), a division of the Society of Automotive Engineers (SAE). AMS D Nonferrous Alloys Committee, 1999, *Aluminum Alloy Forgings, Heat Treated.* Warrendale, Pennsylvania: Society of Automotive Engineers.

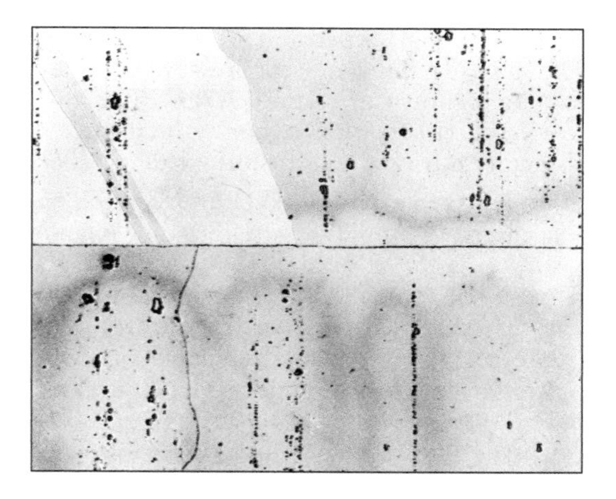

(A) No Interlayer

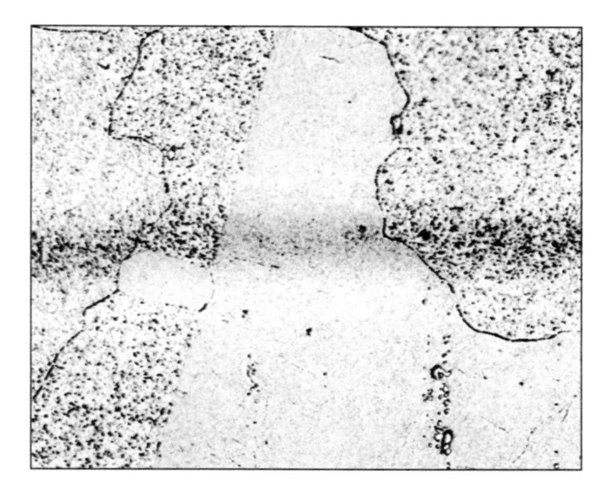

(B) Nickel Interlayer

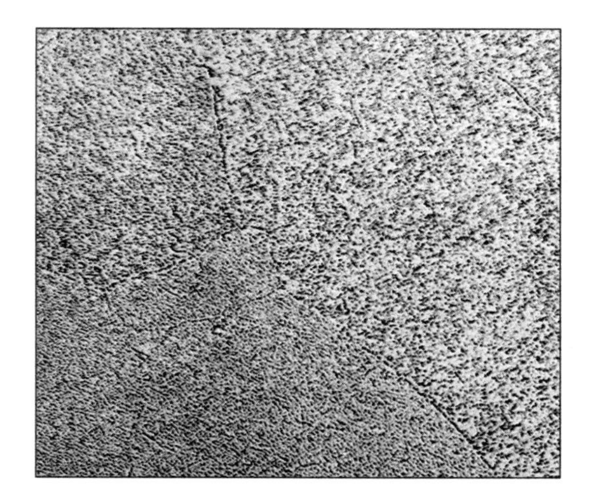

(C) Ni-35%Co Alloy Interlayer

Figure 12.6—Diffusion Welds in a Wrought Alloy (X250)

Stainless steels can be diffusion welded using conditions similar to those used for carbon steel; however, these steels are normally covered by a thin adherent oxide that must be removed prior to welding. Oxide removal can be accomplished either by welding at high temperatures in a dry hydrogen atmosphere or by copper-plating the faying surfaces after anodic cleaning. Copper oxide on the plating is relatively easy to reduce in hydrogen during heating to welding temperature. For illustrative purposes, sound welds were made in AMS 5630 martensitic stainless steel at 1093°C (2000°F) with a pressure of 690 kPa (100 psi) for 1.5 hour using a 2.5 µm (0.0001 in.) thick copper filler metal.

Dissimilar Metal Combinations

Diffusion welding is particularly well suited for joining many dissimilar metal combinations, even when the melting points of the two base metals differ widely or when they are not metallurgically compatible. In such cases, conventional fusion welding is not practical because it would result in either excessive melting of one of the metals or in the formation of brittle weld metal. Diffusion welding also is suitable when the high temperatures of fusion welding processes would cause an alloy to become brittle or drastically lower its strength, as is the case with some refractory metal alloys. Filler metals are sometimes used to prevent the formation of brittle intermetallic phases between certain metal combinations.

When determining conditions and filler metal requirements for the diffusion welding of a particular dissimilar metal combination, the effects of interdiffusion between the two base metals must be considered. Interdiffusion can cause certain problems, listed below with preventive actions, as a result of the following metallurgical phenomena:

1. An intermediate phase or a brittle intermetallic compound may form at the weld interface. The selection of an appropriate filler metal usually can prevent such problems;
2. Low melting phases may form (for example, eutectic melting). Sometimes this effect is beneficial;
3. A type of porosity known as *Kirkendall porosity* may form due to unequal rates of metal transfer by diffusion in the region adjacent to the weld interface.[10] Proper welding conditions or the use of an appropriate filler metal, or both, may prevent this problem.

A problem that often exists (not unique to diffusion welding) is the difference in the thermal expansion

10. *Kirkendall Porosity* is a phenomenon that occurs when atoms of one element diffuse in one direction faster than they can be replaced by other elements diffusing in the opposite direction.

characteristics of the two base metals. Any combination of dissimilar metals that is heated and cooled during welding or brazing will develop shear stresses in the joint if the coefficients of thermal expansion are not identical. The severity of the problem varies depending on the temperature span, the difference between the expansion coefficients, the size and shape of the joint members, and the nature of the weld formed between them. This becomes a design problem, in part, since distortion can result. Cracking can result when the joint strength or ductility, or both, are low and the stresses are high.

Many dissimilar metal combinations can be joined by the diffusion welding process but result in brittle intermetallic phases, and in some cases the reaction proceeds very rapidly due to the development of a liquid phase at the welding temperature. Although welds in these combinations produce brittle intermittent phases, acceptable joints can be obtained by compensating for the results of brittle phases in the design of the component. A combination of Zircaloy 2™ with Type 304 stainless steel is a good example of the situation in which a strong, useful joint can be made despite the presence of brittle phases. Figure 12.7 shows the joint designs employed for joining Type 304 (QQ-S-763F) stainless tubing to Zircaloy 2.

The differential expansion of the two materials is used to provide the required pressure for the conical, tapered joint shown in Figure 12.7(A). The joint shown in Figure 12.7(B) requires a longitudinal force to maintain pressure during the diffusion welding process.

Joints between stainless steel and Zircaloy 2 tubing with a diameter of 22.2 mm (7/8 in.) diameter and wall thickness of 3.2 mm (1/8 in.) can withstand from 83 MPa to 117 MPa (12 000 psi to 17 000 psi) internal pressure in hydraulic fracture tests. The fracture initiates by longitudinal splitting of the Zircaloy tube. Similar joints have withstood 100 pressure cycles between 690 kPa and 24 000 kPa (100 psi and 3500 psi) at 260°C (500°F) and 200 temperature cycles between 38 °C and 316°C (100°F and 600°F) without failure.

Representative conditions used for the diffusion welding of some dissimilar metal combinations are presented in Table 12.2. Often the temperature and the time used for a particular combination are selected as part of the necessary heat treatment for one of the alloys to develop design properties for a specific application. Several successful applications are shown in Table 12.3.

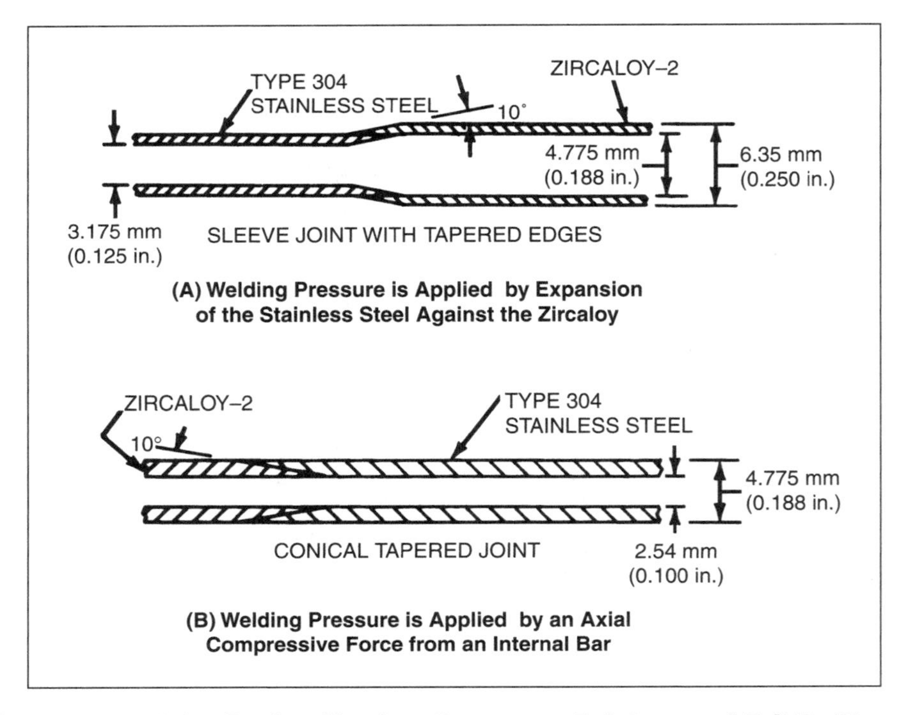

Figure 12.7—Joint Design Used to Overcome Existence of Brittle Phases

Table 12.2
Diffusion Welding Conditions for Some Dissimilar Metal Combinations

Base Metal Combinations	Filler Metal	Temperature		Time, h	Pressure[a]		Atmosphere
		°C	°F		MPa	ksi	
Cu to Al	—	510	950	0.25	7	1	Vacuum
Cu to 316 stainless steel	Cu	982	1800	2	Note a		Vacuum
Cu to Ti	—	849	1560	0.25	5	0.7	Vacuum
Cu to Cb-1%Zr	Cb-1%Zr	982	1800	4	Note a		Vacuum
Cu-10%Zn to Ti-6%Al-6%V-2%Sn	—	482	900	8	Note a		Vacuum
4340 steel[c] to Inconel 718[b]	—	943	1730	4	200	29	Vacuum
Nickel 200 to Inconel 600[b]	—	927	1700	3	7	1	Not reported
P Pyromet X-15[b] to T-111 Ta alloy	Au-Cu	593	1100	4	207	30	Not reported
Cb-1%Zr to 316 stainless steel	Cb-1%Zr	982	1800	4	Note a		Vacuum
Zircaloy-2[b] to 304 stainless steel	—	1021–1038	1870–1900	0.5	10	Note a	Vacuum

a. Pressure is applied with differential thermal expansion tooling.
b. Trade names.
c. Outgassing of the 4340 steel at 10(–5) torr (130 mPa) and 1850°F (1010°C) for 24 hours prior to welding was critical to the formation of satisfactory welds.

Table 12.3
Applications of Diffusion Welding in Various Industries

Product	Base Metal	Reason for Using DFW	Previous Method
F-15 Fighter fitting	Ti-6Al-4V	Cost reduction	Machining from forging or plate
Impeller for liquid-fuel rocket	Ti-5Al-2.5Sn	Higher quality	
Jewelry	18K gold alloy	Higher quality	Brazing
Electrode	Cu-316L stainless steel	Joining possible only with DFW	
Electrical connections	Cu to SiC	Joining possible only with DFW	None
Tube sheet	Cupro Ni-mild steel-316L stainless steel	Cost reduction	Rolling or explosion welding
Chock liner for steel rolling mill	Brass-mild steel	Cost reduction	Metal changed from solid Cu alloy
Cooling plate for cyclotron accelerator	Cu-316L stainless steel	Joining possible only with DFW	
Heat exchangers, large and small	Any combination of Ti, stainless steel, steel and Cu	Cost reductions and in some cases joining possible only with DFW	Solder/Braze or previously not capable
Continuous casting mold	Wear-resistant material, Cu alloy-304 stainless steel	Joining possible only with DFW	

DIFFUSION BRAZING

Diffusion brazing is similar to conventional brazing; various heating methods, atmospheres, joint designs, and equipment used in conventional brazing generally can be used interchangeably in diffusion brazing.[11] Diffusion-brazed joints can be produced with physical and mechanical properties almost identical to those of the base metal. To do this, it is necessary to completely diffuse the melting point depressant (solute) into the base metal. Diffusion brazing involves a number of stages, including the following:

1. Heating the braze joint, usually to the melting temperature of the filler metal;
2. Dissolution of the base metal;
3. Liquid phase solidification; and
4. Solute homogenization.

11. For more information on brazing, see Chapter 12 of American Welding Society (AWS) Committee on Brazing and Soldering, A. O'Brien, ed. 2004, Volume 2 of *Welding Handbook*, *Welding Processes, Part 1*, 9th ed., Miami: American Welding Society.

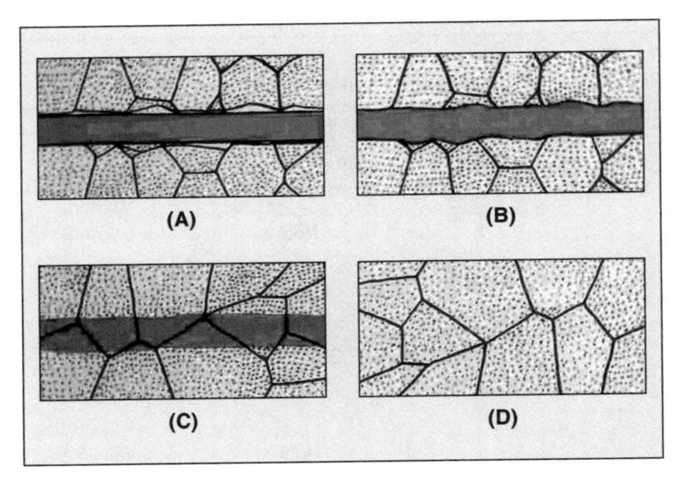

Figure 12.8—Diffusion Brazing Sequence

Figure 12.8 shows four stages in the brazing sequence. In the first stage, shown in Figure 12.8(A), the assembly is heated from room temperature to the melting point of the filler metal. Interdiffusion occurs between the filler metal and base material during heating. Figure 12.8(B) shows the second phase the dissolution stage, in which the base metal dissolves into the liquid and hence the width of the liquid zone increases. In the third stage, shown in Figure 12.8(C), the isothermal solidification stage, the liquid zone solidifies as a result of solute diffusion into the base metal at the brazing temperature. This stage generally is considered to be the most important since the completion time required for the entire process is largely determined by the time required for the completion of isothermal solidification. When solidification is complete, the assembly loses the identity of a brazed joint. In the last stage, shown in Figure 12.8(D), solid-state solute redistribution occurs. The homogenization temperature does not need to be that used during Stages (B) and (C), and this operation generally is terminated when the maximum solute concentration at the joint centerline reaches the pre-selected value.

METHODS

Two diffusion brazing methods are described in this section. The first method uses brazing filler metal with a chemical composition approximately the same as the base metal but with a lower melting temperature. The melting temperature is lowered by adding certain alloying elements, known as *melting point depressants* (MPDs), to the base metal composition or to a similar alloy composition. For example, the melting temperature of a nickel (Ni)-base high-temperature alloy can be lowered by a small addition of silicon or boron. In this case, the brazing filler metal melts and wets the faying surfaces of the base metal during the brazing cycle. This method is sometimes called *activated diffusion brazing* or *transient liquid-phase (TLP) brazing*.

The second method uses a metal interlayer that will alloy with the base metal to form one or more eutectic or peritectic compositions. When the brazing temperature is slightly higher than the eutectic or peritectic temperature, the metal interlayer and base metal form an alloy that produces a composition with a low melting point. The interlayer itself does not melt, but a low-melting alloy is formed *in situ*. An example of eutectic diffusion brazing is the joining of titanium alloys with copper at a brazing temperature below the melting point of copper. This method also is known as *eutectic brazing* when it is not followed by the isothermal solidification stage but cools and solidifies in the conventional manner.

With either method, the assembly is held at brazing temperature for a sufficient time for diffusion to produce a nearly uniform alloy composition across the joint. As this takes place, the melting temperature of the braze metal and the strength of the joint increase. The brazing time depends on the degree of homogeneity desired, the thickness of the initial filler metal layer, and the brazing temperature.

The relationship of the heating rate to the brazing temperature also may be important. A slow heating rate will allow more solid-state diffusion to take place, and more filler metal will be required to provide sufficient liquid to fill the joint. Conversely, if a large quantity of filler metal and a fast heating rate is used, the molten metal may run out of the joint and erode the base metal. The thick joint formed in this manner will require a longer diffusion time to achieve a suitable composition gradient across the joint.

The composition of the braze metal may be important with respect to its response to subsequent heat treatment. This is particularly true for metals that undergo phase transformation during heating and cooling. The composition of the alloy determines the transformation temperature and rate of transformation. Therefore, the phase morphology and mechanical properties of the joint can be controlled by the joint design and the brazing cycle.

Unlike conventional brazing, however, the assembly is held at the peak brazing temperature for an extended period of time. This allows for more extensive diffusion between the liquid and the base metal, with preferential diffusion of melting point suppressants (solutes) out of the liquid into the base metal. As a result, the liquid filler solidifies isothermally, or at constant temperature. This process offers some distinct advantages over conventional brazing, such as increased remelt temperature and improved joint properties, which in some cases can approach those of the base metal.

One of the most frequently cited examples is the copper-silver (Cu-Ag) combination. A thin layer of silver placed between pieces of copper will form a thin layer of eutectic liquid with a melting temperature of 780°C (1436°F) via interdiffusion at elevated temperature. Alternatively, a thin foil of the eutectic composition may simply be placed in the joint. In either case, if the assembly is held above 780°C (1436°F) for an extended time, the silver content of the eutectic liquid filler will diffuse away from the joint into the copper base metal and eventually result in isothermal solidification of the joint. In this situation, the remelt temperature and properties of the joint will approach that of the copper base metal. The material combinations are not limited to those that form solid solutions such as copper and silver. Combinations that form intermetallic compounds or new phases also may be used to produce joints with increased remelt temperatures.

PROCESS VARIABLES

Several variables, including temperature and heating rate, brazing time, pressure, metallurgical factors, and filler metals must be correctly selected to produce high-quality diffusion-brazed joints.

Temperature and Heating Rate

The temperature cycle used for diffusion brazing depends on the base metal and filler metal used. When a conventional low-melting filler metal is used, the assembly must be heated above the melting temperature of the filler metal, as in conventional brazing. As the brazing alloy melts, it wets the base metal and fills the voids in the joint; then the temperature is maintained to isothermally solidify the braze metal.

As previously described, eutectic diffusion brazing relies on the *in-situ* formation of the filler metal. Metal combinations generally are selected to form a molten eutectic that flows and fills the voids in the joint at brazing temperatures above the associated eutectic temperature. For example, a plating of copper on the faying surface of a silver-base metal will form a eutectic when heated to 815°C (1500°F). The eutectic melting temperature is 780°C (1435°F).

In systems where several eutectic and peritectic reactions take place at different temperatures, both the brazing temperature and the heating rate are important. Although a liquid phase can form at the lowest eutectic temperature, diffusion rates will be faster at higher temperatures. The heating rate will determine whether a molten eutectic is formed. If the heating rate is too low, solid-state diffusion will prevent the formation of a molten eutectic and consequently there will be inadequate liquid to properly fill voids at the faying surface.

The maximum brazing temperature may be established by the characteristics of the base metal; for example, incipient melting in most metals is not desirable. Brazing temperature also may be limited by the effect of temperature on the final metallurgical structure or by the heat treatment requirements of the brazement.

Time

The duration of the diffusion brazing cycle depends on the following variables:

1. Brazing temperature;
2. Diffusion rates of the filler metal and the base metal at brazing temperature; and
3. Maximum melting point suppressant, or solute, concentration of filler metal permissible at the joint.

The alloy composition at the joint may influence the response to heat treatment or the resulting mechanical properties of the joint. Therefore, the joint must be held at high temperature for an established minimum time to reduce the solute concentration of filler metal to an acceptable value.

Pressure

Similar to conventional brazing, diffusion brazing typically requires little or no pressure across the joint. In some cases, fixturing may be necessary to avoid excessive pressure, particularly when the molten filler metal must flow into the joint by capillary action. When the filler metal is placed in the joint before brazing, excessive pressure may force low-melting constituents to flow out of the joint before brazing temperature is achieved. In that case, the molten filler metal may not be sufficiently fluid to fill interface voids.

Metallurgical Factors

The metallurgical events that take place during diffusion brazing are similar to those that occur during diffusion welding. An additional factor is the variation in chemical composition across the joint. Compositional variations can significantly affect the response of a particular alloy to heat treatment. For metals that exhibit an allotropic transformation, the chemical composition affects both the transformation temperature and the rate of transformation. Thus, the response to heat treatment across a diffusion-brazed joint varies with the local chemical composition. For example, copper stabilizes the beta phase in titanium and decreases the beta-to-alpha transition temperature.

Filler Metals

Diffusion brazing filler metals may be conventional, low-melting alloys or metals that react with the base metal to form an *in-situ* low-melting liquid. The filler metal may be in powder, foil, or wire form, or it may be plated onto the surface of the base metal. Close control of the amount of filler metal in the joint is essential for consistent results. Detailed information on filler metals for brazing is published in the American Welding Society document *Specification for Filler Metals for Brazing and Braze* Welding, A5.8/A5.8M:2004.[12]

The application of pure metals and simple alloys by electroplating or vapor deposition can be accurately controlled. Films of desired thickness can be deposited on the faying surfaces. However, these processes are not always economical. Metal foil or wire formed into suitable shapes are better for many applications. An assortment of brazing filler metals in various shapes and forms is shown in Figure 12.9.

Nickel-base and cobalt-base alloys are commonly added to brazing filler metals to depress the melting temperature, but these elements also increase alloy hardness and brittleness. Consequently, filler metals with these elements are most commonly produced as powders. Powders can be a problem when precise amounts of filler metal are required. Boron in the range of 2% to 3.5% is used in nickel-base filler metals. Boron can be diffused into the surfaces of nickel alloy foil or wire shapes to produce filler metal preforms. Several variations of nickel-based alloys also are produced commercially as rapidly solidified, amorphous foils. These preforms provide good control of filler metal placement and volume for diffusion brazing applications.

12. American Welding Society (AWS) Committee on Filler Metals and Allied Materials, (1992, reaffirmed 2003) AWS A5.8, *Specification for Filler Metals for Brazing and Braze Welding*, A5.8/A5.8M:2004, Miami: American Welding Society.

DIFFUSION BRAZING TECHNIQUES

As in conventional brazing, an excessive joint clearance can be a limiting factor in diffusion brazing applications. Too wide a joint clearance can minimize or prevent capillary action of the liquid filler metal and consequently prevent proper wetting and filling of the joint. A wide-clearance diffusion brazing technique has been developed to overcome this problem.

Wide-Clearance Diffusion Brazing

Wide-clearance diffusion brazing uses a dense powder metal insert material that fills the volume of the joint clearance. The insert material is a pressed and sin-

Photographs courtesy of Lucas-Milhaupt, Inc.

Figure 12.9—Various Forms and Shapes of Filler Metal for Diffusion Brazing

tered composite of a high-melting additive metal and a brazing filler metal. The additive metal is selected to be similar in composition to the base metal. For example, a nickel-chrome alloy is used as an additive metal for brazing nickel-based alloys. At brazing temperature the liquid filler metal component of the insert reacts with the additive component in the same way it would with the base metal, resulting in isothermal solidification within the insert and consequently within the joint. Another approach to the wide clearance diffusion brazing technique is to use a porous powdered metal insert comprised of the additive metal only. Filler metal powder is applied to the outside of the joint and allowed to flow into the porous insert by capillary action during brazing.

An advantage of using powered metal composite inserts is that the process times typically are much shorter than conventional diffusion brazing. This is because the effective joint surface area of the composite insert is increased by several orders of magnitude compared to that of the faying surfaces in conventional applications. This reduces the necessary diffusion path of the filler metal solute and consequently the time for isothermal solidification and solid-state homogenization.

EQUIPMENT AND TOOLING

The equipment and tooling used for diffusion brazing are essentially the same as that used for conventional brazing. Vacuum- or controlled-atmosphere furnace brazing is by far the most common process. In some cases it may be more economical and convenient to braze with one piece of equipment and then follow with a diffusion heat treatment with other equipment. For example, the brazing could be done with resistance welding or induction heating equipment, and the diffusion heat treatment could be performed in a furnace. A typical vacuum furnace is shown in Figure 12.10.

Photograph courtesy of Advanced Vacuum Systems, Inc.

Figure 12.10—Typical Vacuum Furnace

MATERIALS AND APPLICATIONS

Diffusion brazed titanium alloys, nickel alloys, and aluminum alloys are used in the fabrication of aerospace and aircraft components, structural members, and numerous specialty items for which high quality is the most important requirement.

Following are several examples of diffusion brazing applications:

1. Cast nickel-base superalloys for high-temperature service;
2. Beryllium (Be) alloys;
3. Some dissimilar metal combinations;
4. Assemblies in which a combination joining and heat-treating cycle is desirable to minimize distortion;
5. Elevated temperature applications, such as high-strength titanium alloys in aircraft assemblies;
6. Large, complicated assemblies, such as honeycomb structures, when it is economical to produce many strong joints simultaneously;
7. Temperature-sensitive electronic components to be used in high-temperature service; and
8. Optoelectronic devices that must be rigidly fixed to maintain proper alignment even at elevated service temperatures.

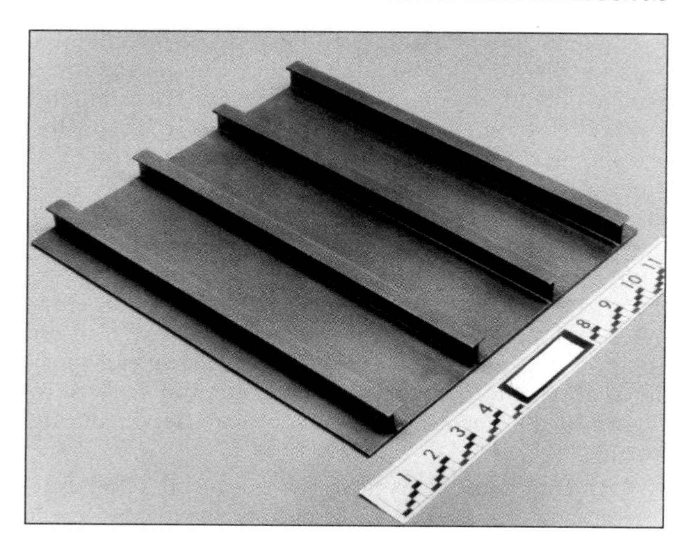

Figure 12.11—A Titanium-Alloy Stiffened Sheet Structure Fabricated by Continuous Seam Diffusion Brazing

Titanium Alloys

Continuous-seam diffusion brazing of titanium alloy sheet is used to fabricate stiffened skins for an integral one-piece structure. An example is shown in Figure 12.11. One of the first applications of this method was the fabrication of curved Ti-6%Al-4%V alloy I-beams used as structural members to support boron-aluminum composites on a fighter airplane. These beams were made from 0.64 mm (0.025 in.) sheet. Superplastic forming and diffusion brazing of titanium parts also were used. An augmenter flap fabricated by the process is shown in Figure 12.12.

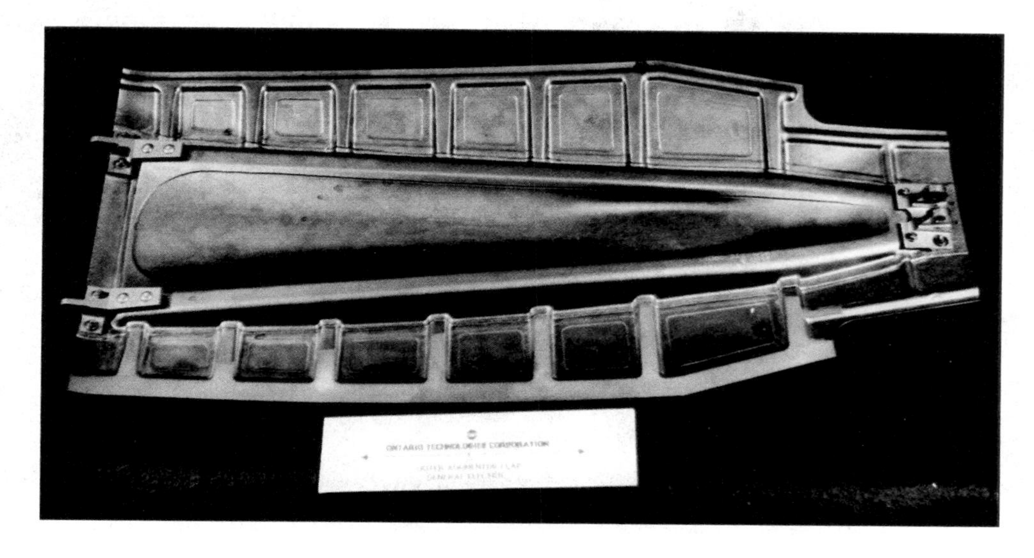

Figure 12.12—Augmenter Flap for a Jet Engine Fabricated by Superplastic Forming and Diffusion Brazing of Titanium

Conventional diffusion brazing techniques also are used for joining titanium alloys. Brazing times, temperatures, and prebrazing cleaning procedures are much the same as for diffusion welding. Pressure may be an amount just sufficient to hold the joint members in contact, and faying-surface preparation requirements are not as stringent.

The faying surfaces of the titanium alloy are electrolytically plated with a thin film of either pure copper or a series of elements, such as copper and nickel. When heated to the brazing temperature of 900°C to 925°C (1650°F to 1700°F), the copper layer reacts with the titanium alloy to form a molten eutectic at the braze interface. The brazement is then held at the brazing temperature for at least 1.5 hours. The assembly also may be given a subsequent heat treatment at the brazing temperature for several hours to reduce the composition gradient in the braze metal. Diffusion-brazed joints made with a copper filler metal and a cycle of 925°C (1700°F) for 4 hours exhibited tensile, shear, unnotched-fatigue, and stress-corrosion properties equal to those of the base metal. However, they had slightly lower notch-fatigue and corrosion-fatigue properties, and significantly lower fracture toughness.

A typical application of diffusion brazing is the fabrication of lightweight cylindrical cases of titanium alloys for jet engines. In this application, the titanium core is plated with a very thin layer of copper and nickel that reacts with the titanium to form a eutectic. During brazing in a vacuum of 10^5 torr, a eutectic liquid forms at 900°C (1650°F). This liquid performs the function of brazing filler metal between the core and face sheets. The eutectic quickly solidifies due to rapid diffusion at the braze interface.

In the past, the copper-nickel filler metal was electrodeposited on the edge of the core in a lamellar fashion, but currently, joints are produced using a thin homogeneous copper-nickel foil as the filler metal. The use of foil allows more precise control of the filler metal thickness and composition and eliminates several complicated steps that are required in the plating process. In addition, because the foil is a homogeneous layer, it produces its available liquid all at once as soon as the ternary eutectic point is reached. This technique is an improvement over the lamellar formation of liquid that is produced from the electroplating method.

Diffusion-brazed assemblies are held at the brazing temperatures for one to four hours to reduce the composition gradient at the braze interface by diffusion. A typical diffusion-brazed titanium alloy honeycomb structure 206 cm (81 in.) long and 2.7 kg (6 lb) total weight is shown in Figure 12.13.

Nickel Alloys

Nickel-base, heat-resistant alloys can be diffusion brazed to produce high-strength joints that resemble the base metal in both microstructure and mechanical properties.

Ni-base brazing filler metal contains melting-point depressants, such as silicon, boron, manganese, aluminum, titanium, and niobium. The filler metal contains sufficient amounts of depressants so that the resulting alloy is melted at a temperature that does not impair the

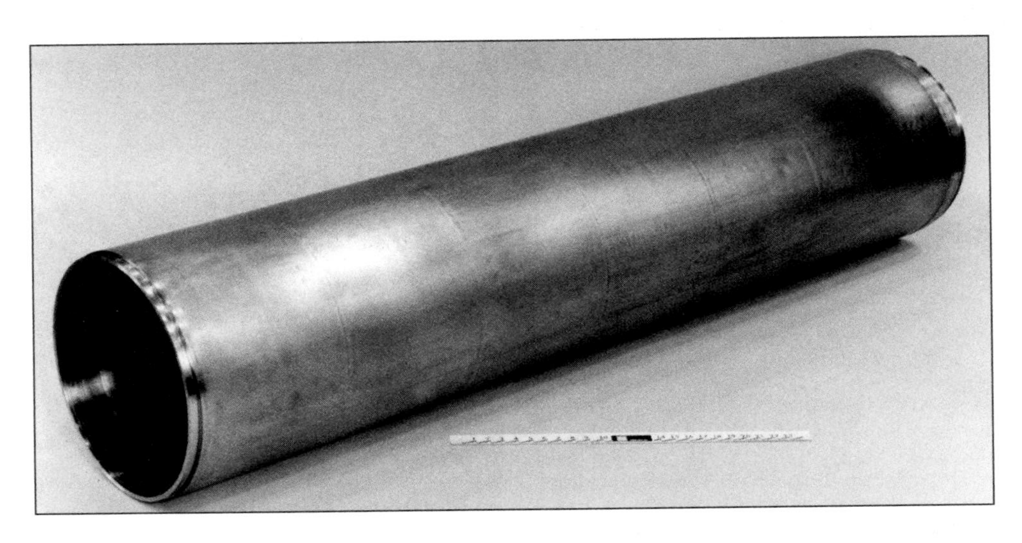

Figure 12.13—Diffusion-Brazed Titanium Alloy Honeycomb Structure

properties of the base metal. Ideally, brazing is accomplished at the normal solution-heat-treating temperature for a given base metal. Figure 12.14 shows a diffusion-brazed joint made in a proprietary wrought nickel alloy using the first procedure described. A filler metal of 0.08 mm (0.003 in.) thick Ni-15%Cr-15%Co-5%Mo-3%Be was used in the joint with a processing cycle of 1170°C (2140°F) for 24 hours in vacuum. A microprobe chemical analysis across a joint showed a uniform chemical composition, essentially that of the base metal. Stress rupture tests at 870°C and 980°C (1600°F and 1800°F) showed that the diffusion-brazed joints had essentially the same properties as the base metal.

Diffusion-brazed joints produced at lower temperatures and with a shorter time at brazing temperature may not be uniform in composition. As a result, some mechanical properties of the joints at elevated temperature may be lower than those of the base metal, particularly under stress-rupture conditions.

Single-Crystal Metals

Two metals of single-crystal microstructure can be joined by diffusion brazing to form a joint without grain boundaries and with the same crystal orientation as the base material. The two components must have the same crystal orientation. A braze alloy of essentially the same composition as the base metals but with melting point depressants is required. The assembly is heated to a temperature above the liquidus of the braze alloy and held at this temperature to allow the melting point depressant to diffuse from the filler metal into the base alloy. The base metal then solidifies isothermally. The solidification grows epitaxially from the base metal surfaces, and because the base metals are single-crystal with the same orientation, the solidifying braze grows as a single crystal with the same orientation as the base metal.

A diffusion-brazed joint between two single-crystal metals is shown in Figure 12.15. This joint was brazed with B-Ni2 foil (Ni-7%Cr-3%Fe-4.5%Si-3.2%B) for 16 hours at 1150°C (2100°F) followed by 22 hours at 1245°C (2275°F). Grain boundaries in the joint area are noticeably absent. The fractured surface of a similarly brazed joint is shown in Figure 12.16(A). The microstructure across the fracture of the tensile specimen is shown in Figure 12.16(B). The joint was brazed using a nickel-base foil containing 15% Cr and 4% B at 1175°C (2150°F) for 16 hours plus 1245°C (2275°F) for 22 hours. The tensile strength at 1095°C (2000°F) was 321 MPa (46.6 ksi). The reduction in area was 12.5%.

At this stage the joint has good properties, although not fully equivalent to those of the base metal. By permitting the brazement to remain at the brazing temperature for a longer time, the braze metal can be homogenized both in composition and structure until it is essentially equivalent to the base metal.

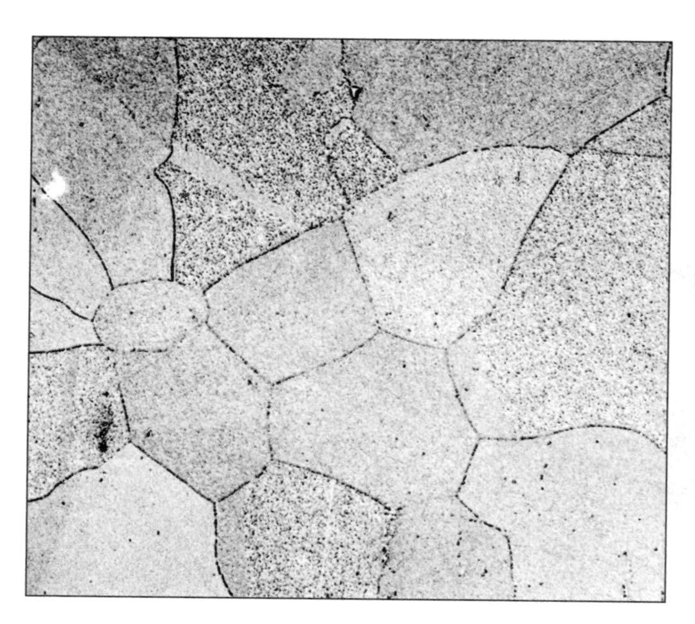

Figure 12.14—A Diffusion-Brazed Joint in Wrought Nickel Alloy

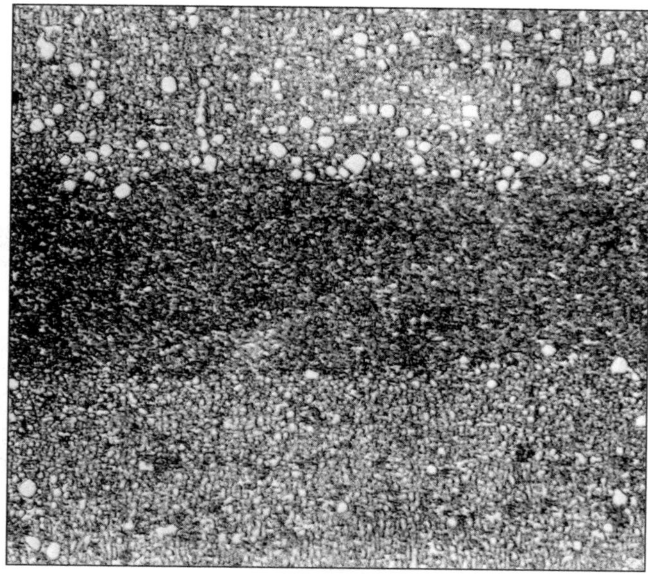

Figure 12.15—Diffusion-Brazed Single-Crystal Joint

(A) Fracture Surface (X8)

(B) Microstructure (X100)

Figure 12.16—Fracture Surface and Microstructure of Single-Crystal Brazement

Aluminum

Aluminum and aluminum alloys can be diffusion brazed using copper filler metal. Sound, strong joints can be produced in aluminum by limiting the copper thickness to 0.5 µm (2×10^{-5} in.) and restricting the brazing temperature between 554°C and 571°C (1030°F and 1060°F). The time at temperature should not exceed 15 minutes at the lower temperature limit or 7 minutes at the upper limit. Type A356.0 aluminum-7% silicon casting alloy can be diffusion brazed by electroplating one of the joint members with copper that will form a eutectic with the aluminum and silicon in the casting alloy when heated to 524°C (975°F).

To ensure optimum joint properties, the thickness of the copper filler metal, the brazing temperature, and the brazing time must be selected to promote isothermal solidification during brazing and thereby prevent the formation of the compound $CuAl_2$. Proper balancing of these conditions results in strong joints that can withstand quenching from the solution temperature required for heat-treating Type A356.0 alloy to the T61 condition. Electroplating the cover sheets with 0.38 µm to 0.5 µm (1.5×10^{-5} in. to 2.0×10^{-5} in.) of copper and holding between 527°C and 538°C (980°F and 1000°F) for one hour are satisfactory conditions. After quenching and aging, the joint strength will equal that of the casting itself, and the microstructure of the brazed joint will be indistinguishable from the casting.

DIFFUSION SOLDERING

Diffusion soldering is identical to diffusion brazing except that the filler metals used have a melting point below the established delineation temperature of 450°C. One example is silver (Ag) diffusion soldered with tin (Sn) at approximately 250°C. This system forms the intermetallic phase Ag_3Sn that provides a remelt temperature of 480°C or even higher with more extended process times and diffusion.[13, 14]

DIFFUSION SOLDERING APPLICATIONS

The primary goal in most diffusion soldering applications is to achieve a high remelt temperature while maintaining the relatively low process temperature of a solder joint. A low soldering temperature helps to minimize process-related thermal stresses or excessive heating of temperature-sensitive components, or both. These conditions are often found in electronics applications. Diffusion soldering often results in the formation of one or more intermetallic phase. These intermetallics usually have high strength and low ductility, which can limit the use of diffusion soldering for purely structural applications. Some of the more common applications include die or chip attachment to electronic substrates, packaging of high-temperature electronic devices, packaging of micro-electro-mechanical systems (MEMS) devices, and mounting optoelectronic devices. A variety of filler-metal alloy systems have been used or successfully demonstrated. Table 12.4 lists some of these systems and the associated temperatures.

Table 12.4
Diffusion Soldering Filler Metal Systems

Alloys	Typical Process Temperature °C (°F)	Potential Remelt Temperature °C (°F)
Au-Sn	225–250 (455–482)	280 (536)
Au-In	157–200 (306–392)	457 (850)
Cu-Sn	280–700 (536–1292)	415–800 (775–1472)
Cu-In	200 (392)	310–677 (590–1250)
Ag-In	220 (392)	750 (1382)

INSPECTION OF DIFFUSION-WELDED AND DIFFUSION-BRAZED JOINTS

Establishing the quality of a diffusion-welded or diffusion-brazed joint is difficult with existing nondestructive examination procedures because of the nature of the joints. Usually, little or no porosity exists if the joint is made with properly developed procedures. The most common discontinuity or flaw in a diffusion weldment or brazement is the lack of grain growth across the original interface. Efforts to distinguish intimate contact without grain growth across the interface from a perfect bond have not been successful.

Radiography, eddy current, and thermal examination methods are relatively unsatisfactory for inspection of most applications. Dye penetrant methods are useful for edge inspections.

Ultrasonic examination has proved the most useful for internal inspection, especially if a hairline separation exists. The sensitivity varies with the metal being tested, the ultrasonic frequency, the skill of the operator, and the degree of sophistication of the equipment. In general, discontinuities of less than 2.5 mm (0.1 in.) diameter are difficult to locate and a practical limit of about 1.0 mm (0.04 in.) exists. With special methods and very sophisticated equipment, discontinuities equivalent to 0.13 mm (0.005 in.) diameter can be detected in some metals. These testing methods cannot be considered routine and they work only under special conditions.

Various types of diffusion weld discontinuities representing possible production defects for superplastic forming and diffusion brazing processes have been evaluated in numerous titanium specimens. These specimens were used to evaluate ultrasonic examination techniques with regard to defect resolution. Table 12.5 summarizes the results. Higher frequency focusing probes may produce the best results.

Ultrasonic inspection equipment does not differentiate between complete intimate contact and an actual diffusion weld. Only metallographic examination can assure complete welding or brazing. Because this is a destructive test, it cannot always be performed on the workpiece in question. Fortunately, joints made by diffusion welding and diffusion brazing are reproducible when good process control is exercised. Random destructive sampling coupled with ultrasonic examination will provide a high confidence level. This approach has been successfully used in production.

A reliable nondestructive examination method for the inspection of diffusion-welded structures is desirable. However, conventional nondestructive examination procedures and equipment do not adequately differentiate between acceptable and unacceptable diffusion-welded or diffusion-brazed joints. Therefore it is necessary to supplement the nondestructive tests with destructive tests. Conventional radiography is not suitable for detecting the extremely small discontinuities

13. For detailed information on soldering refer to Vianco, P. T., American Welding Society (AWS) Committee on Soldering and Brazing, *Soldering Handbook,* 3rd ed. Miami: American Welding Society.
14. Refer to American Welding Society (AWS) Welding Handbook Committee, A. O'Brien, ed., 2003, *Welding Handbook,* 9th ed. Vol. 2, *Welding Processes, Part 1,* Miami: American Welding Society, Chapter 13.

Table 12.5
Detectability of Diffusion Brazing Defects with Various Ultrasonic Testing Methods

	Ultrasonic Testing Methods			
Diffusion Brazing Defects	Transmission-Focusing, Non-Focussing Probes 5 and 10 MHz	Pulse/Echo, Non-Focusing Probes 10 up to 20 MHz	Pulse/echo Non-Focusing Probes 30 up to 110 MHz	Pulse/Echo Point-Focusing Probes > 30 MHz
Course-dispersive macro defects (single defect size φ = 200μm; w = 3 mm)	D	WD	WD	WD
Course-dispersive macro defects (single defect size φ = 5μm; w = > 5 mm)	PD	D	WD	WD
Fine-dispersive macro defects (single defect size φ = 3μm; 0,1 <w< 5 mm)	PD	PD	WD	WD
Micro defect configurations (single defect size w = 1 μm; φ <15 μm)	ND	ND	PD	D

Notes:
ND = Not detected
PD = Partially detected
D = Detected
WD = Well detected

involved in diffusion welding, but the use of X-ray micro-focus techniques coupled with digital image enhancement provide an improvement in resolution.

ECONOMICS

Depending on the application, the cost of materials, labor and overhead can vary drastically. Exotic diffusion weldments may require large capital investments for items such as an autoclave, while simple diffusion weldments could be preformed with existing or less expensive equipment.

Diffusion welding is a difficult process to automate, especially since the small number of similar weldments typically needed may prohibit an investment in automation. However, depending on the size and complexity of the weldments, it is possible to process assemblies in batches, which would allow for automation and would result in cost savings. Since diffusion welding usually involves an unknown complexity of configuration and a variety of materials, the economics of diffusion welding is completely dependent on the application and its circumstances. Costs must be calculated on an individual basis for each application.

SAFE PRACTICES

Hazards encountered with diffusion welding and brazing are similar to those associated with other welding and cutting processes. Personnel must be protected against hot materials, gases, fumes, electrical shock, and chemicals.

The operation and maintenance of diffusion welding and brazing equipment should conform to the provisions of American National Standard Z49.1, *Safety in Welding, Cutting, and Allied Processes.*[15] This standard provides detailed procedures and instructions for safe practices that will protect personnel from injury or illness, and property and equipment from damage by fire or explosion arising from diffusion welding.

It is essential that adequate ventilation be provided so that personnel will not inhale gases and fumes generated while diffusion welding and brazing is in progress. Some filler metals and base metals contain toxic materials such as cadmium, beryllium, zinc, mercury, or lead, which are vaporized during brazing.

15. American National Standards Institute (ANSI) Accredited Standards Committee Z49, *Safety in Welding, Cutting, and Allied* Processes, ANSI Z49.1, Miami: American Welding Society.

Solvents such as chlorinated hydrocarbons, and cleaning compounds such as acids and alkalies, may be toxic or flammable or cause chemical burn when present in the brazing environment.

Requirements for the purging of furnaces or retorts that will contain a flammable atmosphere also are provided in the standard. In addition, to avoid suffocation, atmosphere furnaces must be carefully purged with air to ensure safe breathing conditions before personnel enter it.

Presses and pressure vessels that may be used in diffusion welding have their own associated hazards, and must be carefully assured that they are adequately inspected and properly operated.

Mandatory safe practices are contained in *Occupational Safety and Health Standards for General Industry*, in *Code of Federal Regulations (CFR)*, Title 29 CFR 1910.[16] Other safety codes, standards and publications are listed in Appendices A, B, and C of this volume.

CONCLUSION

The use of diffusion welding and brazing has allowed industrial fabricators to join materials previously viewed as impossible to join. Diffusion joining in its simplest form can be inexpensive and easy to accomplish, while at its most complex, it may be costly. However, in some instances, cost may be secondary to producing high-integrity welds in critical applications. The process still remains relatively easy as long as the appropriate procedures are followed.

BIBLIOGRAPHY

American National Standards Institute (ANSI). Accredited Standards Committee Z49. 2005. *Safety in welding, cutting, and allied processes*, ANSI Z49.1:2005, Miami: American Welding Society.

American Welding Society (AWS) Welding Handbook Committee. A. O'Brien, ed. 2004. *Welding Handbook*, Vol. 2. *Welding processes, Part 1*. Miami: American Welding Society, Chapters 12 and 13.

American Welding Society (AWS) Welding Handbook Committee. Jenney, C. L. and A. O'Brien, eds. 2001. *Welding Handbook,* Vol. 1, 9th ed. *Welding Science and Technology*, Miami: American Welding Society, Chapter 4.

American Welding Society (AWS) Committee on Brazing and Soldering, C. L. Jenney, ed., (forthcoming) *Brazing Handbook*, 5th ed., Miami: American Welding Society.

American Welding Society (AWS) Welding Handbook Committee. Oates, W. R. and A. M. Saitta, eds. 1998. *Welding Handbook*, Vol. 4, 8th ed. Miami: American Welding Society.

American Welding Society (AWS) Committee on Definitions and Symbols. 2001. *Standard welding terms and definitions*, AWS A3.0:2001. Miami: American Welding Society.

American Welding Society (AWS) Committee on Filler Metals and Allied Materials. 2004. *Specification for filler metals for brazing and braze welding*, A5.8/A5.8M:2004. Miami: American Welding Society.

Midgett, S. 2000. *Mokume gane—A comprehensive study*, Franklin, NC: Earthshine Press.

Vianco, P. T., American Welding Society (AWS) Committee on Brazing and Soldering. 1999. *Soldering Handbook*, 3rd ed. Miami: American Welding Society.

SUPPLEMENTARY READING LIST

Adam, P. and Steinhauser, L. 1986. Bonding of superalloys by diffusion welding and diffusion brazing. MTU, 61st Meeting of Structural and Materials Panel of AGARD, Oberammergau, Germany, 11–13 September 1985, AGARDCP-398, (7): 9.1–9.6.

Army Material Development and Readiness Command. 1981. Improved fabrication of fluidic laminates: fine blanking and semisolid-state diffusion bonding promise to improve fluidic components. *Journal Vol-U8202,* (10).

Arvin, G. H. et al. 1979. Evaluation of superplastic forming and co-diffusion bonding of Ti-6Al-4V titanium alloy expanded sandwich structures. NASA-CR-165827, NA-81-185-1, Cont. No. NASI-15788, Rockwell International (5).

Arvin, G. H. et al. 1980. Evaluation of superplastic forming and co-diffusion bonding of Ti-6Al-4V titanium alloy expanded sandwich structures. NASA-CR-165827, NA-81-185-1, Cont. No. NASI-15788. Rockwell International (12).

Blackburn, L. B. 1987. Effect of LID® processing on the microstructural and mechanical properties of Ti-6Al-4V and Ti-6Al-2Sn-4Zr-2Mo titanium foil-gauge materials. NASA Langley Research Center, NASA TP-2677.

16. Occupational Safety and Health Administration (OSHA), (latest edition), *Occupational Safety and Health Standards for General Industry*, in *Code of Federal Regulations (CFR)*, Title 29 CFR 1910, Subpart Q, Washington D.C.: Superintendent of Documents, U.S. Government Printing Office.

Blair, W. 1981. Fabrication of titanium multiwall thermal protection system (TPS) curved panel. Rohr Industries, Inc., NASA-CR-165754, Cont. No. NASI-15646.

Boire, M. and Jolys, P. 1985 and 1986. Application du soudage par diffusion associe au formage superplastique (SPF/DB) a la realization de structures en toles minces de TA6V. Aerospatiale, 61st Meeting of Structural and Materials Panel of AGARD, Oberammergau, Germany. AGARD-CP-398. (9): 10.4–10.12.

Calderon, P. D. et al. 1985. An investigation of diffusion welding of pure and alloyed aluminum to type 316 stainless steel. *Welding Journal* 64(4): 1045-s–1125-s.

Davé, V. R., Beyerlein, I. J., Hartman, D. A. and Barbieri, J. M. 2003. A probabilistic diffusion weld modeling framework. *Welding Journal* 82(7): 170-s–178-s.

Dini, J. W. et al. 1984. Use of electrodeposited silver as an aid in diffusion welding. *Welding Journal* 63(1): 285-s–345-s.

Doherty, P. E. and D. R. Harraden. 1977. New forms of filler metal for diffusion brazing. *Welding Journal* 56(10): 37-s–39-s.

Dunkerton, S. B. and C. J Dawes. 1986. The application of diffusion bonding and laser welding in the fabrication of aerospace structures. The Welding Institute, 61st Meeting of Structural and Materials Panel of AGARD, Oberammergau, Germany, 11–13 September 1985, AGARD-CP-398(7): 3.1–3.12.

Duvall, D. S., W. A. Owczarski, and D. F. Paulonis. 1974. TLP bonding: a new method for joining heat-resistant alloys. *Welding Journal* 53(4): 302-s–314-s.

Elmer, J. W. et al. 1988. The behavior of silver-aided diffusion-welded joints under tensile and torsional loads. *Welding Journal* 67(7): 1575.

Godziemba-Malisqewski, J. 1987. Thermal surge in diffusion welding-generation, inrush characteristic, and effects. *Welding Journal* 66(6): 1745.

Isserow, S. 1980. Diffusion welding of copper to titanium by hot isostatic pressing (HIP). AMMRC, Final Rept. AMRRC-TR-80-85, Journal Vol-U8111.

Luo, J-G, and V. L. Acoff. 2000. Interfacial reactions of titanium and aluminum during diffusion welding. *Welding Journal* 79(9): 239-s–243-s.

Kamat, G. R. 1988. Solid state diffusion welding of nickel to stainless steel. *Welding Journal* 67(6): 44.

Kapranos, P. and R. Priestner. 1987. NDE of diffusion bonds. University of Manchester/UMIST, *Metals and Materials* (4): 194-s–198-s.

Lee, C. C., and W. W. So. 2000. High-temperature silver-indium joints manufactured at lowtemperature. *Thin Solid Film*. 366: 196–201.

Leodolter, W. 1981. *Tool sealing arrangement and method*. Department of Air Force and McDonnell Douglas, Pat-App1-6-300767, Filed.

Lison, R. and J. F. Stelzer. 1979. Diffusion welding of reactive and refractory metals to stainless steel. *Welding Journal* 58(10): 3065-s–3145-s.

Lugscheider, E., T. Schittny, and E. Halmoy. 1989. Metallurgical aspects of additive-aided wide-clearance brazing with nickel-based filler metals. *Welding Journal* 68(1): 9-s–13-s.

McQuilkin, F. T. 1982. Feasibility of SPF/DB titanium sandwich for LFC wings. Rockwell International, NASA-CR-165929, Cont. No. NASI-16236, (6): 62.

Midgett, S., *Mokume Gane*. 2000. *A comprehensive study*, Franklin, NC. Earthshine Press.

Moore, T. J. and T. K. Glasgon. 1985. Diffusion welding of MA6000 and a conventional nickel-base superalloy. *Welding Journal* 64(8): 2195–2265.

Morley, R. A. and J. Caruso. 1980. Diffusion welding of 390 aluminum alloy hydraulic valve bodies. *Welding Journal* 59(8): 29-s–34-s.

Munir, Z. A. 1983. A theoretical analysis of the stability of surface oxides during diffusion welding of metals. *Welding Journal* 62(12): 3335–3365.

Naimon, E. R. et al. 1981. Diffusion welding of aluminum to stainless steel. *Welding Journal* 60(11): 17-s–20-s.

Niemann, J. T. and R. A. Garrett. 1974. Eutectic bonding of boron-aluminum structural components, Part 1. *Welding Journal* 53(4): 175-s–184-s.

Niemann, J. T. and R. A. Garrett. 1974. Eutectic bonding of boron-aluminum structural components, Part 2. *Welding Journal* 53(8): 351-s–359-s.

Niemann, J. T., and G. W. White. 1978. Fluxless diffusion brazing of aluminum castings. *Welding Journal* 57(10): 285-s–291-s.

Norris, B. and F. Gojny, 1987. Joining processes used in the fabrication of titanium and Inconel™ honeycomb sandwich structures. First SAMPE International Metals Symposium, August 18–20. Cherry Hill, N. J.: Rohr Industries, Inc.

O'Brien, M., C. R. Rice, and D. L. Olson. 1976. High-strength diffusion welding of silver coated base metals. *Welding Journal* 55(1): 25–27.

Osawa, T. 1995. Changes in the interface structure and strength of diffusion brazed joints of Al-Si systems alloy castings. *Welding Journal* 74(6): 206-s–212-s.

Orel, S. V., L. C. Parous, and W. F. Gale. 1995. Diffusion brazing of nickel aluminides. *Welding Journal* 74 (9): 319-s–324-s.

Owezarski, W. A. and D. F. Daulonis. 1981. Application of diffusion welding in the USA. *Welding Journal* 60(2): 22–33.

Partridge, P. G., J. Harvey, and D. V. Dunford. 1986. Diffusion bonding of Al-alloys in the solid state. Royal Aircraft Establishment, 61st Meeting of Structures and Materials Panel of AGARD, Oberammergau, Germany, 11–13 September 1985, AGARD-CP-398: 8.1–8.23.

Rosen, R. S. et al. 1986. The properties of silver-aided diffusion welds between uranium and stainless steel. *Welding Journal* 65(4): 835.

Sangha, S. P. S., D. M. Jacobson, and A. T. Peacock. 1998. Development of the copper-tin diffusion-brazing process. *Welding Journal* 77(10): 432-s–438-s.

Sheetz, H. A., P. L. Coppa, and J. Devine. 1979. Ultrasonically activated diffusion bonding forfluidic control assembly. Sonobond Corp., Cont. No. DAAA21-76-C0186, RLCD CR-79005, Final Report (2).

Schwartz, M. M. 1978. Diffusion brazing titanium sandwich structures. *Welding Journal* 57(9): 35–38.

Schwartz, M. M. 1979. *Metals Joining Manual.* McGraw-Hill Book Co.

Sharples, R. V. and I. A. Bucklow. 1986. Diffusion bonding of aluminum alloys to titanium. *The Welding Institute* 7836.01/85/448.3, 307/1986. (7).

Signes, E. G. 1968. Diffusion welding of steel in air. *Welding Journal* 47(12): 571-s–574-s.

Society of Automotive Engineers (SAE) Nadcap Management Council. *Nadcap requirements for diffusion welding.* AS 7110/8. 2003. Troy, Michigan: Society of Automotive Engineers.

Somani, M.C., N. C. Birla, and A. Tekin. 1998. Solid-state diffusion welding of wrought AISI 304 stainless steel to nimonic AP-1 superalloy powder by hot isostatic pressing. Welding Journal 77(2). 59-s–65-s.

Stephen, D. and S. J. Swadling. 1985. Diffusion bonding in the manufacture of aircraft structure. British Aerospace, 61st Meeting of Structures and Materials Panel of AGARD, Oberammergau, Germany, (9) AGARD-CP-398: 7.1–7.17.

Stephen, D. and S. J. Swadling. 1986. Diffusion bonding in the manufacture of aircraft structure. British Aerospace, 61st Meeting of Structures and Materials Panel of AGARD, Oberammergau, Germany, AGARD-CP-398: 7.1–7.17.

Sullivan, P. G. 1976, 1977, and 1978 Reports. Elevated temperature properties of boron/aluminum composites. NASA-CR-159445, Cont. No. NAS320079. Nevada Engineering and Technology Corp.

Tanzer, H. J. 1982. Fabrication and development of several heat pipe honeycomb sandwich panel concepts. Cont. No. NASI-16556, NASA CR-165962. Hughes Aircraft Co.

Tobor, G. and S. Elze. 1986. Ultrasonic testing techniques for diffusion-bonded titanium components. MBB 61st Meeting of Structures and Materials Panel of AGARD, Oberammergau, Germany, 11–13 September 1985, AGARDCP-398. (7).

Vianco, P. T. 1999. *Soldering handbook*, 3rd ed., Miami: American Welding Society.

Weisert, E. D. and G. W. Stacher. 1976. Fabricating titanium parts with SPF/DB process. *Metal Progress* 111(3): 11-1–11-107.

Wells, R. R. 1976. Microstructural control of thin-film diffusion brazed titanium. *Welding Journal* 55(1): 20-s–28-s.

Wigley, D. A. 1981. The structure and properties of diffusion-assisted bonded joints in 17-4 PH, type 347, 15-5 PH and Nitronic 40 stainless steels. Southampton University, NASA-CR-165745, Cont. No. NASI-16000.

Wilson, V. E. 1980. Superplastic formed and diffusion bonded titanium landing gear component feasibility study Cont. No. F33615-79-C3401, TR-80-3081, Final Report. Rockwell International.

Witherell, C. E. 1978. Diffusion welding multifilament superconducting components. *Welding Journal* 57(6): 153-s–60-s.

Yeh, Y.H., Tseng, Y.H. and Chuang, T.H. 1999. Effects of superplastic deformation on the diffusion welding SuperDux 65 stainless steel. *Welding Journal* 78(9): 301-s–304-s.

Zorc, B. and L. Kosec. 2000. A new approach to improvising the properties of brazed joints. *Welding Journal* 79(1): 24-s–31-s.

CHAPTER 13

ELECTRON BEAM WELDING

Prepared by the Welding Handbook Chapter Committee on Electron Beam Welding:

D. D. Kautz, Chair
Los Alamos National Laboratory

P. W. Hochanadel
Los Alamos National Laboratory

J. O. Milewski
Los Alamos National Laboratory

D. E. Powers
PTR—Precision Technologies, Inc.

K. J. Zacharias
Hamilton Sundstrand Special Systems International

Welding Handbook Volume 3 Committee Member:

P. F. Zammit
Brooklyn Iron Works, Inc.

Contents

Photograph courtesy of PTR—Precision Technologies, Inc.

CHAPTER 13

ELECTRON BEAM WELDING

INTRODUCTION

Electron beam welding (EBW) is a welding process that produces coalescence with a concentrated beam, composed of high-velocity electrons impinging on the joint. The process is used without shielding gas, except in some nonvacuum applications, and without the application of pressure. Variations of this process are high-vacuum electron beam welding (EBW-HV), medium-vacuum electron beam welding (EBW-MV), and nonvacuum electron beam welding (EBW-NV).[1,2] Since nonvacuum EBW is accomplished at near atmospheric pressure it is sometimes called *atmospheric electron beam welding*.

Since the commercial introduction of electron beam welding in the late 1950s, the benefits of the process have been used to great advantage by industrial manufacturers. The process was first employed by the nuclear industry, and shortly thereafter by the aircraft and aerospace industries. It was quickly recognized that the process had the capacity to enhance the quality and reliability of welds that would meet the highly critical requirements of components used by these industries. An added benefit of the process was reduced manufacturing costs.

During the initial period of commercial application, the electron beam welding process was limited to operating totally under high-vacuum conditions. However, new modes of operation soon were developed that did not require high vacuum to be used for the whole process, but only in the beam generation portion. This permitted the option of welding either in a medium-vacuum or a nonvacuum environment. This advancement led to the acceptance of electron beam welding by commercial, automotive, and consumer product manufacturers. As a consequence, EBW has been employed in a broad range of industries worldwide since the late 1960s and has successfully demonstrated the capability of producing both very shallow and extremely deep single-pass autogenous welds (that is, without the use of filler metal) with a minimum amount of thermal distortion of the workpiece.

This chapter provides an overview of the electron beam welding process and its variations, including fundamental information and the characteristics of welds produced by the process. Welding procedures, welding variables, and details of selecting welding conditions are discussed. A brief section on electron beam equipment is included in this chapter, and sections on the metals welded by the process, typical applications, and the economics of electron beam welding are presented. A discussion of safe practices specific to electron beam welding is provided. The chapter concludes with the bibliography and a list of sources of additional information on electron beam welding.

1. Welding terms and definitions used throughout this chapter are from *Standard Welding Terms and Definitions*, AWS A3.0:2001, Miami: American Welding Society.
2. At the time this chapter was prepared, the referenced codes and other standards were valid. If a code or other standard is cited without a publication date, it is understood that the latest edition of the document applies. If a code or other standard is cited with the publication date, the citations refer to that edition only, and it is understood that any future revisions or amendments to the code or other standard are not included. As codes and standards undergo frequent revision, the reader is advised to consult the most recent edition.

FUNDAMENTALS

Electron beam welding produces a fusion weld with heat obtained by impinging a beam composed of high-energy electrons onto the workpiece. Electrons are fundamental particles of matter, characterized by a negative charge and a very small mass. As used in electron beam welding, the electrons are raised to a high-energy state by acceleration to velocities in the range of 30% to 70% of the speed of light.

Basically, an electron beam welding system functions in much the same manner as a cathode ray tube (CRT), which uses a low-intensity electron beam to continuously scan the surface of a luminescent screen to produce an image. Electron beam welding systems use a high-intensity electron beam to continuously bombard a weld joint; the input energy that results when the electrons collide with the metal produces localized heating and melting of the weld joint. In both of these cases, the beam of electrons is created with an electron gun that typically contains some type of thermionic electron emitter, a bias voltage control electrode, and an anode. The electron emitter normally is referred to as the *cathode* or *filament* of the gun, and the bias voltage control electrode generally is referred to as the *grid* of the gun, or *grid cup*. Various supplementary devices, such as focus and deflection coils, also are provided downstream of the gun to focus and deflect this beam.

In EBW, the total beam-generating system (gun and electron optics) is called either the electron beam gun/column assembly, or simply the electron beam gun column.

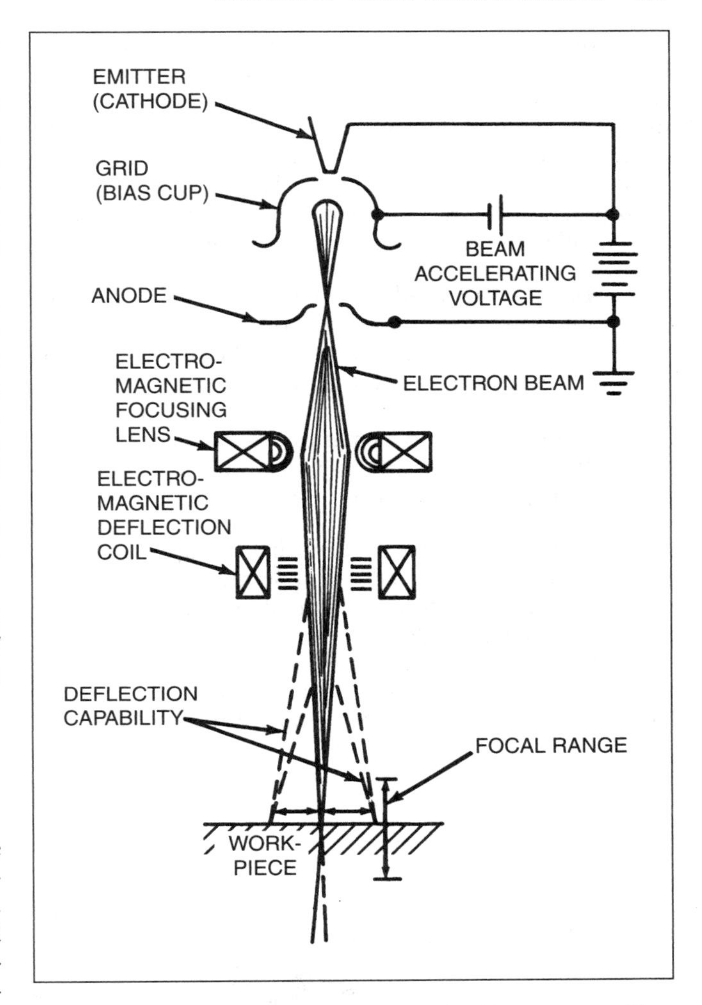

Figure 13.1—Simplified Schematic Representation of a Triode Electron Beam Gun Column

PRINCIPLES OF OPERATION

The heart of the electron beam welding process is the electron beam gun/column assembly. A simplified representation of the gun column is shown in Figure 13.1. Electrons are generated by heating a negatively charged emitting material to its thermionic emission temperature range, thus causing electrons to "boil off" the emitter (or cathode/filament) and to be electrostatically accelerated toward the positively charged anode. The precisely configured grid (bias cup) surrounding the emitter provides an electrostatic field geometry that accelerates the electrons and shapes them into the beam. The beam exits this region of the gun through an opening in the anode.

In a diode (cathode-anode) gun, the beam-shaping electrode and the emitter are both at the same electrical potential, and together are referred to as the *cathode*. In a triode (cathode-grid-anode) gun, the emitter and beam-shaping electrode are at different potentials; consequently the beam-shaping electrode can be biased to a slightly more negative value than the emitter in order to control the flow of the beam current. In this case, the emitter alone is called the *cathode* (or *filament*) and the beam-shaping electrode is called the *grid*. Because the anode is incorporated into the electron gun in both cases, beam generation (acceleration and shaping) is accomplished completely independent of the workpiece.

As it exits the gun, the beam of electrons has been accelerated to speeds in the range of 30% to 70% of the speed of light when gun voltages in the range of 25 kilovolts (kV) to 200 kV are employed. The beam then continues on toward the workpiece. Once the beam exits the gun, it gradually broadens as travel distance increases, as illustrated in Figure 13.1. This divergence is a result of the fact that all the electrons in the beam have some amount of radial velocity due to their thermal energy, and in addition, all experience some degree of mutual electrical repulsion. Some small effects also are created by the interaction of electrons with the remaining gas atoms and molecules in the beam path. While electrons at much higher energy levels will charge the particles, causing a self-focusing effect, the lower energy levels used in welding applications do not cause this phenomenon to occur. Therefore, in order to counteract this inherent divergence effect, an electromagnetic lens system is used to converge the beam and focus it into a small spot on the workpiece. The divergence and convergence angles of the beam are relatively small, which gives the concentrated beam a usable focal range (or depth of focus) extending over a distance of about 25 mm (1 in.), as shown in Figure 13.1.

In practice, the rate of energy input to the weld joint is controlled by four basic variables, as follows:

1. *Beam current*—the number of electrons per second impinging upon the workpiece;
2. *Beam accelerating voltage*—the magnitude of velocity of these electrons;
3. *Focal beam spot size*—the degree to which this beam is concentrated at the workpiece; and
4. *Welding speed*—the travel speed at which the workpiece or electron beam is being moved.

The maximum beam accelerating voltages and currents that can be achieved with commercially available electron beam gun/column assemblies vary over the ranges of 25 kV to 200 kV for the gun and 1 milliampere (mA) to 1000 mA for the current. The electron beams produced by these systems generally can be focused to minimum diameters in the range of 0.25 millimeter (mm) to 0.76 mm (0.01 inch [in.] to 0.03 in.). The resulting power level attainable from these units can reach values as high as 100 kilowatts (kW). Power density can reach values of 1.55×10^4 W/mm^2 (10^7 W/in.^{2}). These power densities are significantly higher than those possible with arc welding processes and are similar to those achievable by laser beam welding.

The potential welding capability of an electron beam system is indicated by the maximum power density that the system is capable of delivering to the workpiece. This comparison factor depends on the maximum beam power (current multiplied by voltage) and the minimum focal spot size attainable with the system. Electron beam welding systems with beam power levels up to 300 kW and power densities in excess of 1.55×10^5 W/mm^2 (10^8 W/in.^{2}) have been built but have never been used commercially.[3]

As illustrated in Figure 13.2, at power densities on the order of 1.55×10^2 W/mm^2 (10^5 W/in.^{2}) and greater, the electron beam is capable of instantly penetrating into a solid workpiece or a butt joint and

3. The Welding Research Institute of Osaka University in Japan has a 300-kW electron beam welding machine, which they have used to investigate the single-pass joining of extremely heavy sections.

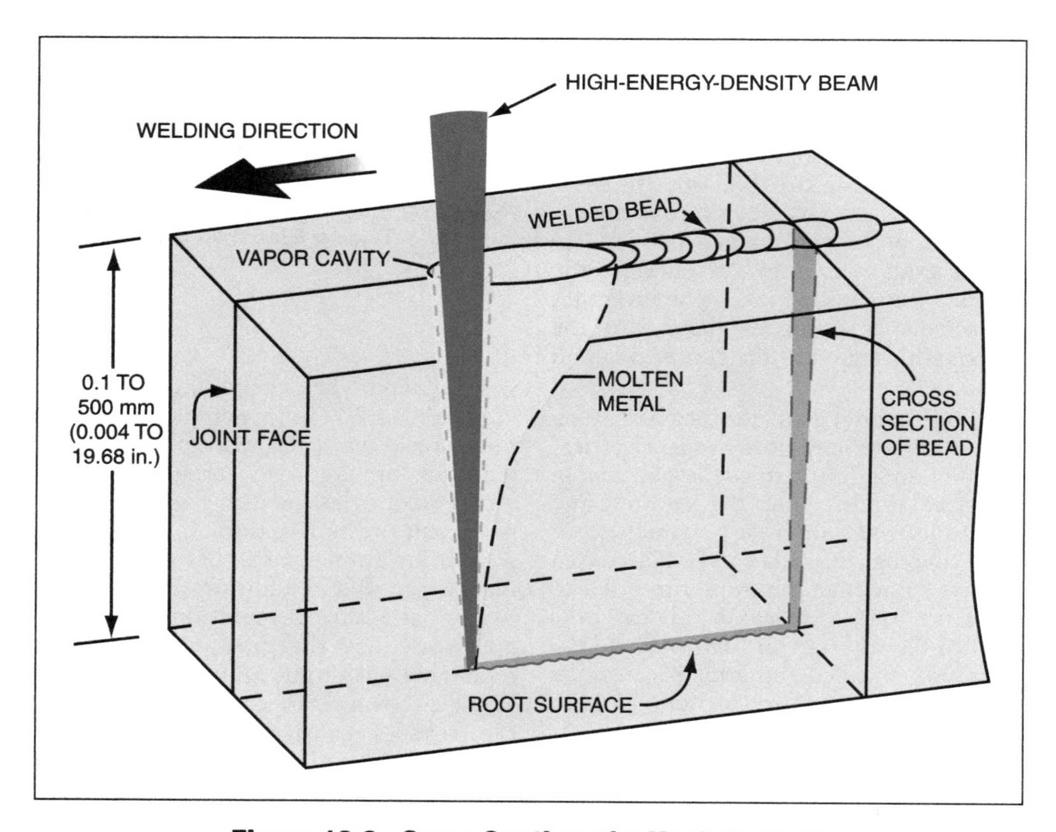

Figure 13.2—Cross Section of a Keyhole Weld

forming a vapor cavity, called a *keyhole,* which is surrounded by molten metal. Molten metal flows around the keyhole as the beam advances along the joint and solidifies at the rear to form weld metal. In keyhole applications, the joint penetration is much deeper than it is wide, and the heat-affected zone produced is very narrow. For example, the width of a weld in a butt joint in 13 mm (0.5 in.) thick steel may be as small as 0.8 mm (0.030 in.) when made in vacuum. This presents a remarkable contrast to the weld metal zone produced in arc-welded and gas-welded joints, in which penetration is achieved primarily through conduction melting.

Because the electron beam weld results from a keyhole formed by the beam, the angle of incidence at which the beam impinges on the surface of a workpiece can affect the final angle at which the keyhole is produced with respect to that surface. The angle of incidence also affects the resulting weld metal zone.

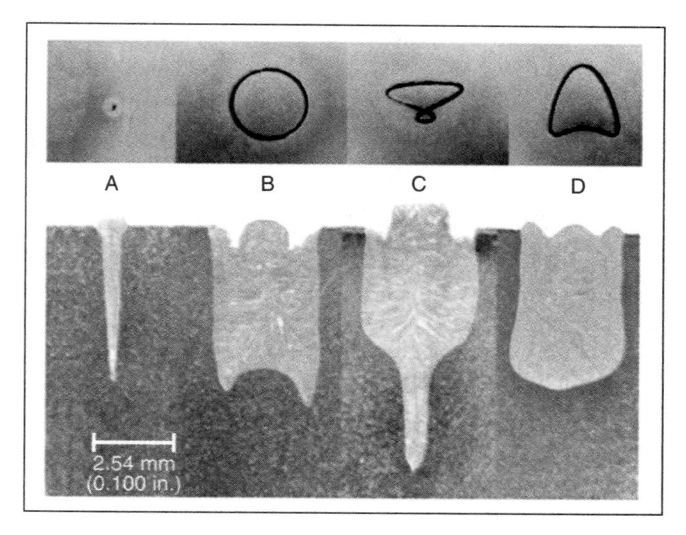

Key
A = No deflection
B = Circular
B = Lazy-8
D = Parabolic

Note: Top views of deflection patterns are enlarged for clarity and are not to scale with corresponding weld profiles.

Figure 13.3—Deflection Patterns of an Electron Beam Column and Corresponding Weld Profiles

Beam Motion Patterns

An electron beam can be readily moved about by electromagnetic deflection. This allows the creation of specific beam spot motion patterns, such as circles, ellipses, or parabolic shapes to be generated on the surface of a workpiece when an electronic pattern generator is used to drive the magnetic deflection coil system. Several of the various deflection patterns that can be generated are illustrated in Figure 13.3. The deflection capability sometimes can be used to provide beam travel motion; although in most instances, deflection is used to adjust the beam-to-joint alignment, or to apply a deflection pattern. The deflection modifies the average power density input to the joint and results in a change in weld characteristics. However, as previously noted, modifications always must be made with care when using any type of beam deflection to ensure that the angle of incidence of the beam does not adversely affect the final weld results, such as causing part of the weld joint to be missed by the electron beam.

PROCESS VARIATIONS

Three basic variations of electron beam welding commonly are used: high-vacuum (EBW-HV), medium-vacuum (EBW-MV), and nonvacuum (EBW-NV). The principal difference between these process modes is the ambient pressure at which welding is done. In the high-vacuum variation, often referred to as *hard vacuum welding,* the process operates below the pressure of 0.13 Pa (10^{-3} torr).[4] For medium-vacuum welding, the pressure range is 0.13 Pa to 3.3×10^3 Pa (10^{-3} torr to 25 torr). Within this range, the pressure span from about 0.13 Pa to 1.3×10^2 Pa (10^{-3} torr to 1 torr) is often called *partial vacuum* or *soft vacuum*, and the pressure span from about 1.3×10^2 Pa to 3.3×10^3 Pa (1 torr to 25 torr) is called *quick vacuum.* In all cases, the pressure in the electron beam gun column must be held at high vacuum, 1.3×10^{-2} Pa (10^{-4} torr) or lower, for stable and efficient operation.

High-vacuum and medium-vacuum electron beam welding is done inside a vacuum chamber. This imposes an evacuation time penalty while creating the high-purity environment. The medium-vacuum welding machine retains most of the advantages of high-vacuum welding, with the added advantage that the chamber requires shorter evacuation times. These features result in higher production rates. Although it incurs no pump-down time penalty, nonvacuum electron beam welding is not suitable for all applications because the welds produced are generally wider and shallower than equal-power electron beam welds produced in a vacuum. Also, the molten weld metal may interact with the atmosphere unless inert gases are used for shielding.

With medium-vacuum operation, the beam is generated in high vacuum and then projected into a welding chamber operating at a higher pressure. This is accomplished by providing an orifice below the beam

4. A *torr* is the accepted industry term for a pressure of 1 mm of mercury. Standard atmospheric pressure can be expressed as 760 torr or 760 mm Hg (2.992 in. Hg).

generation column that is large enough to emit the beam, yet small enough to impede any significant back-diffusion of gases into the gun chamber.

In nonvacuum electron beam welding equipment, the beam is generated in high vacuum and then projected through several specially designed orifices separating a series of differentially pumped chambers before finally emerging into a work environment near or at atmospheric pressure. When nonvacuum electron beam welding is performed directly in the atmosphere, beam-accelerating voltages greater than 150 kV normally are required. However, if the atmospheric pressure environment around the weldment is a gas such as helium, beam-accelerating voltages lower than 150 kV sometimes can be employed.

Figure 13.4 shows the three basic modes of electron beam welding. A fixed electron beam gun column is mounted on the exterior of the high- and medium-vacuum enclosures to illustrate these two modes. A mobile electron beam gun column, as shown in Figure 13.5, may be mounted on the interior of high- and medium-vacuum enclosures. This arrangement commonly is employed to provide the gun column with a higher degree of motion capability.

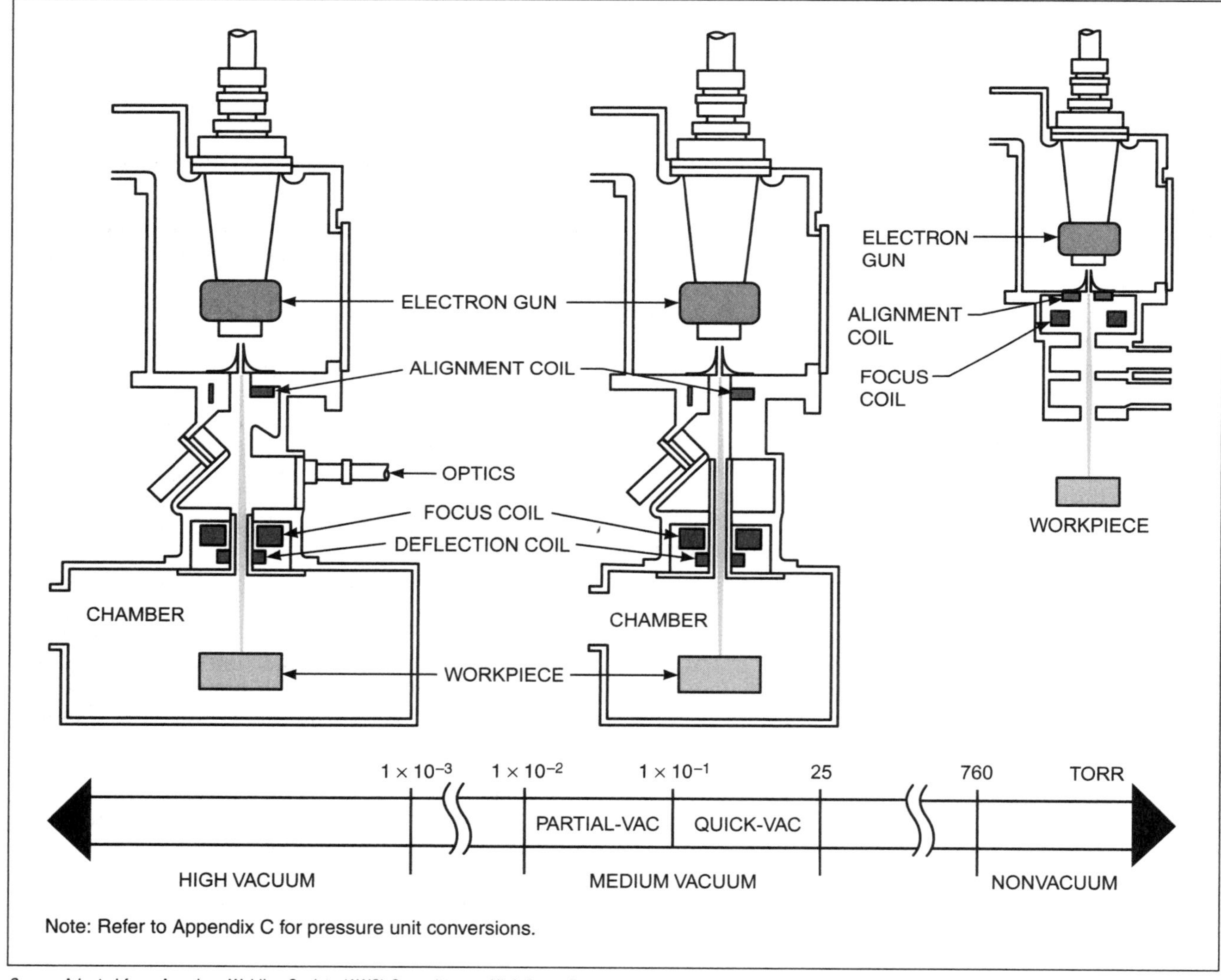

Source: Adapted from American Welding Society (AWS) Committee on High Beam Energy and Cutting, IC7.1M/C7.1:2004, Figure 5.

Figure 13.4—Basic Variations of Electron Beam Welding, with Corresponding Vacuum Scale

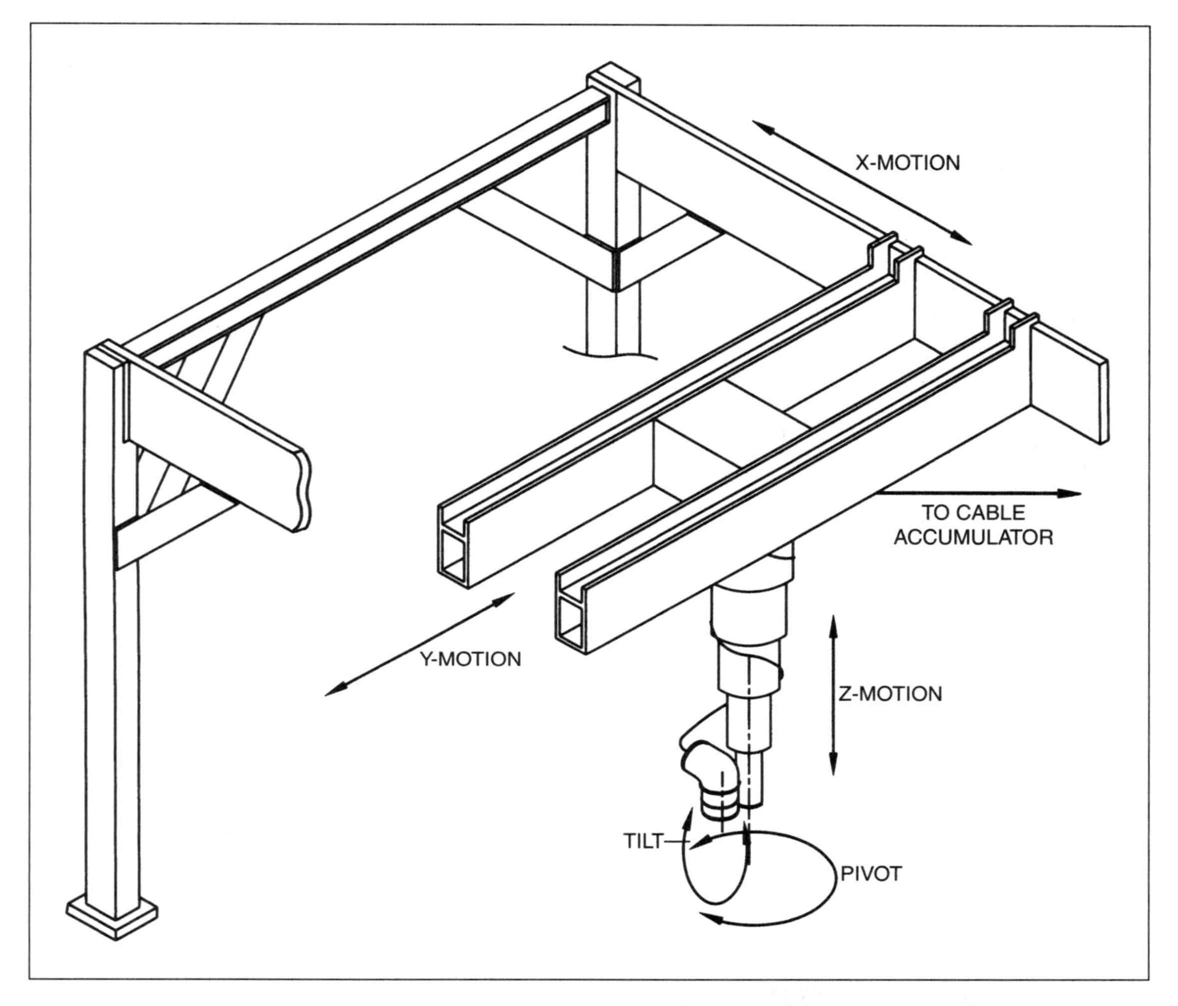

Figure 13.5—Mobile Electron Beam Welding Gun Column with Multi-Axis Motion System

High-Vacuum Welding

A high vacuum (1.3×10^{-2} Pa [10^{-4} torr] or lower) is the required environment for all electron beam guns. Although special methods allow the beam to enter environments of higher pressure, the gun itself will not operate effectively at pressures much greater than 1.3×10^{-2} Pa (10^{-4} torr).

The principal advantages of electron beam welding in a high-vacuum atmosphere are the following:

1. Maximum joint penetration and minimum weld width can be achieved, thereby incurring minimal weld shrinkage and distortion.
2. A high depth-to-width ratio is achieved in welds because of the high energy density of the beam and the resultant keyhole mode of melting;
3. Maximum purity of the weld metal is made possible by the clean environment provided by a high vacuum; and
4. The long gun-to-workpiece distances that are possible in a high vacuum improve the operator's ability to observe the welding process and also to weld joints having limited accessibility.

A high-vacuum environment minimizes the scatter of electrons produced by the collision of the electron beam with any residual gas molecules that might be present in the path of the beam; the frequency of these collisions varies directly with the concentration of gas molecules and the total distance traveled. The minimized scatter of electrons is particularly helpful when long travel distances must be employed.

The high vacuum minimizes exposure of the hot weld zone to oxygen and nitrogen contamination, and concurrently causes the gases evolved during welding to move rapidly away from the weld metal, thereby improving the purity of the weld metal. For this reason, high-vacuum welding is better suited for welding highly reactive metals than the medium-vacuum and nonvacuum process variations.

The production of a high vacuum in the chamber requires pumping times that significantly limit production rates. This pump-down limitation can be offset somewhat if a number of assemblies can be welded in a single load in a small-volume chamber. The number of components that can be welded per batch load is limited by the size of the chamber. As a result, high-vacuum welding generally is more suitable when relatively low production rates are involved. Various types of "air-to-air" workpiece-transfer schemes have been developed that allow component assemblies to be moved in and out of a high-vacuum region without venting the chamber. These procedures make it possible to use high-vacuum EBW on certain high-production joining applications, such as welding bimetallic saw blades.

Medium-Vacuum Welding

A principal feature of medium-vacuum electron beam welding is the capability of welding without pumping the welding chamber to a very low pressure (high vacuum). If the chamber is small, the required pumping time may be a matter of only a few seconds, which is of major importance in economical processing. When a minimum-volume welding chamber can be used, medium-vacuum welding is ideally suited for use in the mass production of assemblies that involve repetitive welding tasks. For example, gears in their final machined or stamped condition can be successfully welded to shafts, eliminating subsequent finishing while maintaining close dimensional tolerances. This application is shown in Figure 13.6.

Since medium-vacuum welding is done at pressures with a significant concentration of air (100 parts per million), this mode of EBW is less advantageous than high-vacuum EBW for welding reactive metals because they require an ultra-pure welding environment to retain properties. The higher concentration of air also scatters the beam electrons, enlarging the beam diameter and decreasing the power density. This results in welds that are slightly wider and more tapered, with less joint penetration than similar welds produced under high-vacuum conditions.

Nonvacuum Welding

The major advantage of nonvacuum (atmospheric) electron beam welding is that the workpieces do not

Photograph courtesy of PTR—Precision Technologies, Inc.

Figure 13.6—Gears Welded with Medium-Vacuum Electron Beam Welding

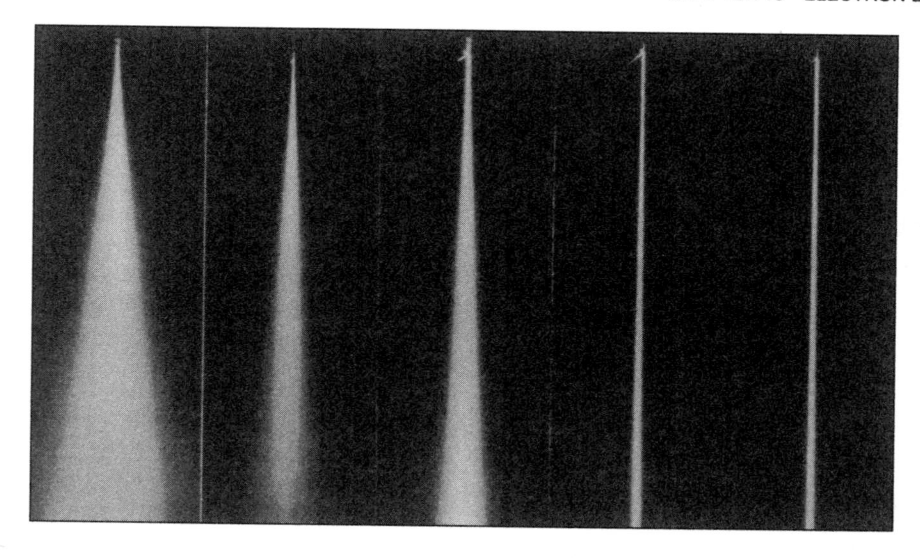

Figure 13.7—Dispersion Characteristics of the Electron Beam at Various Pressures

need to be enclosed in a vacuum chamber. Elimination of the time it takes to evacuate a chamber results in higher production rates and a lower cost per piece. Also, the size of the weldment is not limited by the size of the chamber. These advantages, however, are gained at the expense of not being able to achieve the depth-to-width ratio of the weld, the depth of fusion, or gun-to-workpiece distance attainable in a vacuum. The welding atmosphere is not as pure as it is with high- and medium-vacuum welding, even when inert gas shielding is employed. Although the use of a vacuum enclosure for the workpiece is not required, some type of radiation shielding must be provided to protect personnel from the X-rays generated when the electron beam strikes the workpiece.

Operating conditions for nonvacuum welding differ from the high- and medium-vacuum variations. Beam dispersion increases rapidly with ambient pressure, as shown in Figure 13.7. The maximum nonvacuum gun-to-workpiece distance employable, even when a helium gas environment is utilized, could be less than about 38 mm (1.5 inches). This restricts the shape of the workpieces to those that do not interfere with the gun column.

The joint penetration achieved in nonvacuum electron beam welding is affected by the power level of the beam, travel speed, gun-to-workpiece distance, and the ambient atmosphere through which the beam passes. Figure 13.8 shows weld joint penetration as a function of travel speed for three levels of beam power. It can be noted from the sample welds in Figure 13.8 that an increase in travel speed is gained for a given joint penetration as the power level is increased. These welds were made with 175 kV in air. Nonvacuum electron beam welding appears to demonstrate more efficient joint penetration at power levels above 50 kW. This result is attributed to the decreased density of the gas produced by local heating by the high-energy electron beam.

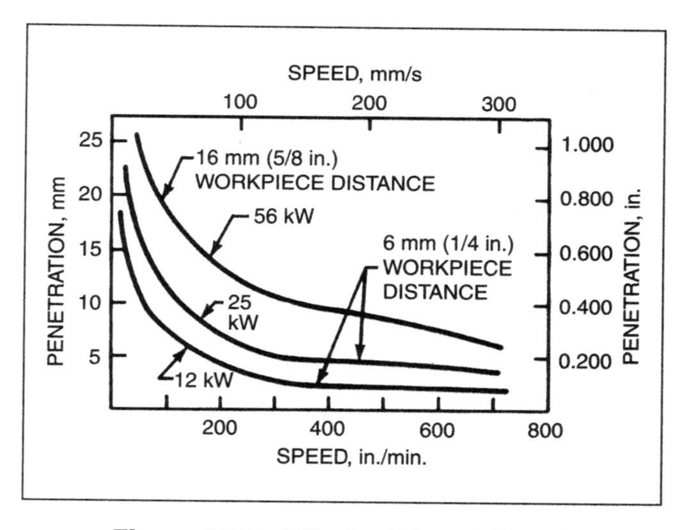

Figure 13.8—Effect of Travel Speed on Joint Penetration of Nonvacuum Electron Beam Welds in Steel

The graph in Figure 13.9 illustrates the effect on joint penetration of the ambient atmosphere, gun-to-workpiece distance, and travel speed in nonvacuum electron beam welds in AISI 4340 steel in atmospheres of helium, air, and argon. The depth of penetration is greater with helium, which is lighter than air, and lower with argon, which is heavier than air. Higher travel speeds can be achieved in a helium shielding gas for a given depth of penetration and gun-to-workpiece distance.

Many materials have been successfully welded using the nonvacuum technique, including carbon steels, low-alloy and stainless steels, high-temperature alloys, refractory alloys, copper alloys, and aluminum alloys. Some of these metals can be welded directly in air while

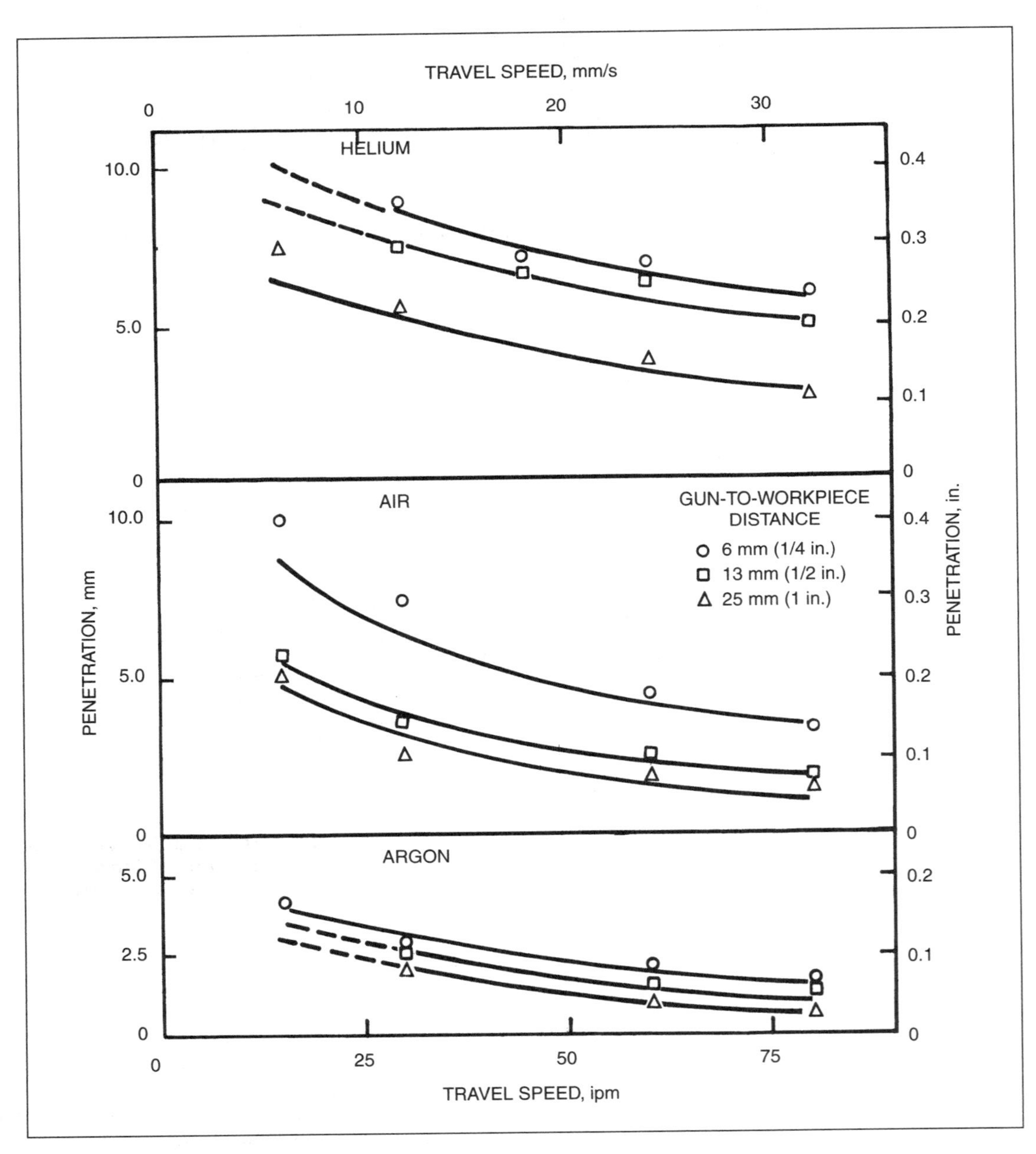

Figure 13.9—Three Gun-to-Workpiece Distances Illustrate Penetration versus Travel Speed for Nonvacuum Electron Beam Welds

others require a protective shield of inert gas to avoid excessive oxygen and nitrogen contamination.

With 60-kW nonvacuum equipment, it is possible to produce single-pass welds in many metals 25.4 mm (1 in.) thick at relatively high speeds. Figure 13.10 is a cross section of a nonvacuum weld in 19 mm (3/4 in.) Type 304 stainless steel plate made in air with 12 kW of power.

PROCESS ADAPTATIONS

Other adaptations of the electron beam process include electron beam braze welding (EBBW) and electron beam machining. Machining encompasses electron beam cutting (EBC) and drilling, the deposition of supplementary weld metal (surfacing, cladding, and hardfacing) and the free-form fabrication of components.

Electron Beam Braze Welding

Electron beam braze welding (EBBW) is a variation of braze welding (BW), a joining process that uses a filler metal with a liquidus above 450°C (840°F) and below the solidus of the base metal. The base metal is not melted. Unlike conventional braze welding, where the filler metal is distributed in the joint by capillary action, electron beam braze welding uses an electron beam as a heat source for simultaneously producing a brazing action and a welding action.

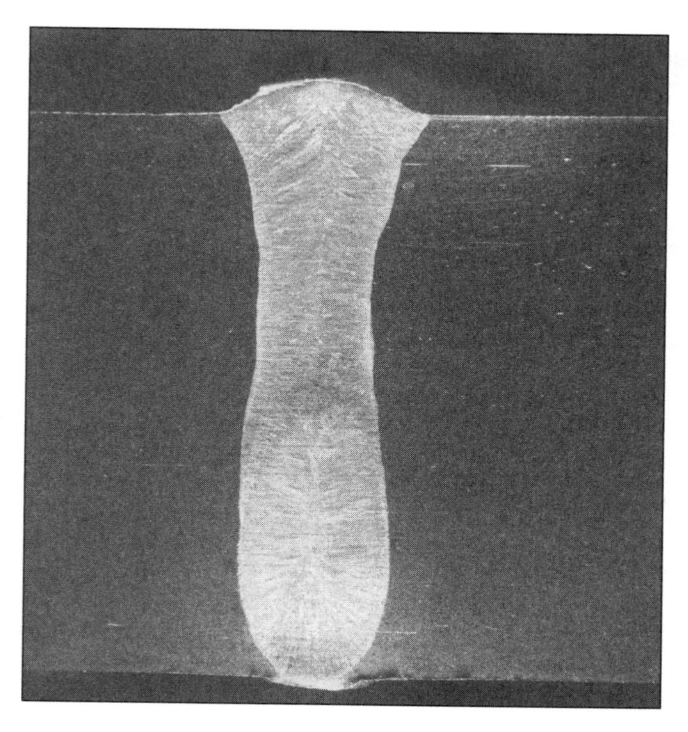

Figure 13.10—Transverse Section of a Nonvacuum Electron Beam Weld in 19 mm (3/4 in.) Stainless Steel Plate

A specific application of this hybrid technique uses an electron beam welding machine with computer numerical control capability to make it possible to produce high-quality welds in assemblies with complex designs. It can be used to eliminate the notch effect caused by incomplete fusion in some areas of T-joints where EBW alone is used. The stress riser normally resulting from the notch effect is appreciably reduced, and additional joint strength is achieved because of the bonding of the brazing alloy with the two workpieces.

Electron beam braze welding was initially developed to make T-joints in components of centrifugal compressor impellers for the oil, gas, and petrochemical industry. An example is shown in Figure 13.11. To join the impeller covers to the blades, a piece of brazing foil was inserted between the faying surfaces of the cover and the blade; then the assembly was welded with EBW. Figure 13.11(A) shows the assembly of the joint between the impeller blade and the cover. Figure 13.11(B) shows the placement of brazing foil, and (C) shows the completed joint. Complete fusion is achieved, as shown in the cross section of the joint in Figure 13.12.

Electron Beam Cutting

Electron beam cutting (EBC) is a thermal cutting process that severs metals by melting them with the heat from a concentrated beam, composed primarily of high-velocity electrons, impinging on the workpiece. Electron beam cutting is not discussed in detail in this chapter because of its limited use, but is mentioned as a variation of the electron beam process.

Electron Beam Drilling

The sequence of electron beam drilling is illustrated in Figure 13.13, from the impingement of the electron beam on the workpiece to the expulsion of molten material and finished hole. A highly concentrated electron beam spot on the order of 10^8 W/cm^2 (10^9 W/in.^{2}) or greater is pulsed in a highly reproducible fashion and is employed as means of rapidly creating a relatively parallel-sided vapor channel. Concurrently, some form of backing material is used to assist in ejecting the molten material surrounding the vapor channel. By combining high-speed beam deflection and CNC workpiece

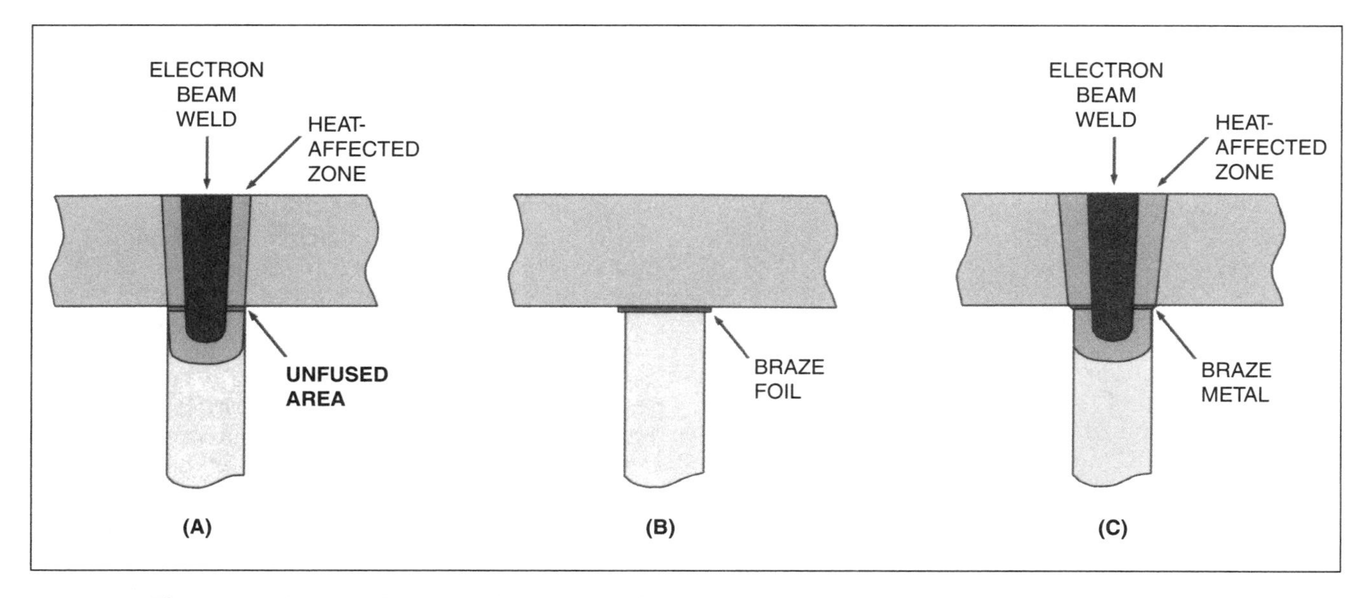

Figure 13.11—(A) Impeller Blade and Cover Assembly, (B) Placement of Braze Foil, and (C) Cross Section of Completed Braze-Welded Joint

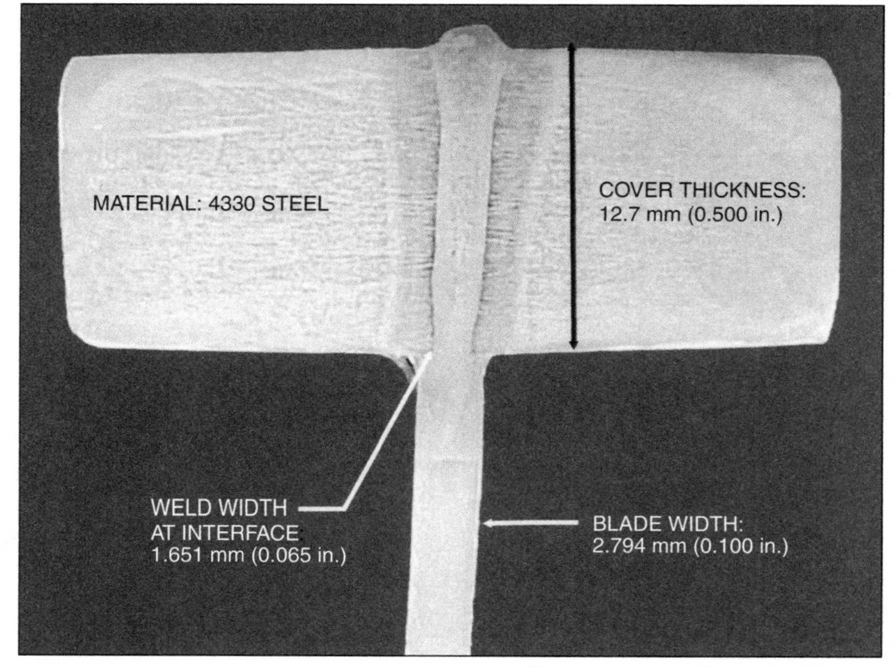

Photograph courtesy of Dresser-Rand Company

Figure 13.12—Finished T-Joint of Thin Impeller Blade and Thick Cover Joined with a Combination of Electron Beam Welding and Brazing

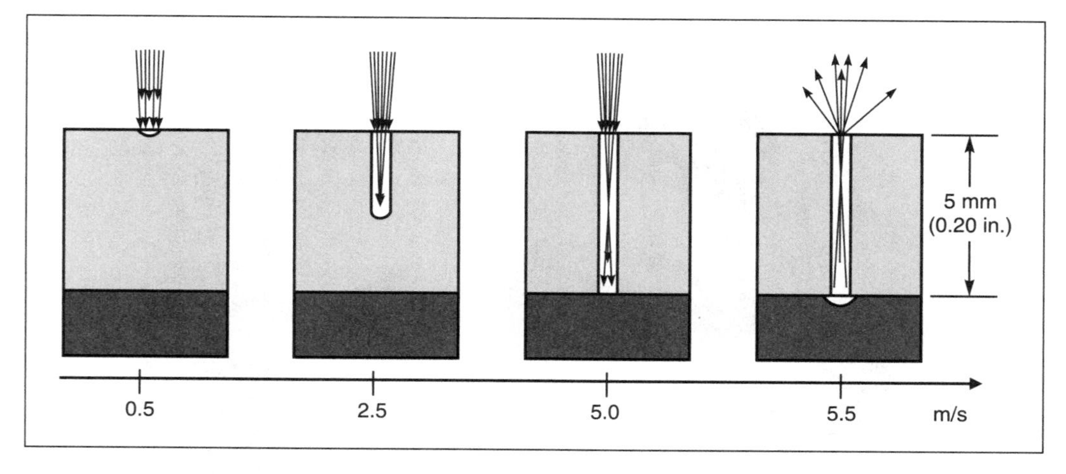

Figure 13.13—Sequence of Electron Beam Drilling with Speed of Beam Travel, Expulsion of Metal, and Completed Hole

motion control, serially drilled holes can be produced at extremely high rates (approximately 3,000 holes/s). Figure 13.14 shows an application of electron beam drilling, a spinner used in the production of fiberglass that has over 25,000 holes 0.55mm (0.02 in.) in diameter.

Electron Beam Surfacing, Cladding, and Hardfacing

In electron beam weld cladding, surfacing, and hard facing, metal (usually in wire form) is fed into the electron beam impingement spot to cause the metal to be rapidly deposited on the surface of the workpiece in a highly controlled fashion, thus quickly building up the desired quantity of metal in the precise region desired. This method traditionally has been employed to accomplish knife-edge seal repair tasks. Figure 13.15 shows a repair weld made with titanium wire on a titanium drum rotor section of an aircraft engine. The section on the left shows the initial metal buildup as deposited; the section on the right shows the welded product, finished by postweld machining.

Electron Beam Free-Form Fabrication

The principles of electron beam weld depositing are employed in special applications not only for repairs but also for manufacturing components of various products. This aspect of the process is generally referred to as *electron beam free-form fabrication*. In the early 2000s the National Aeronautics and Space Administration (NASA) began investigating this technique as a means of creating original components or replacement parts (large and small) while at zero gravity in space.[5]

To produce a component, the process uses a programmed three-axis CNC table to move a substrate under the wire-fed electron beam. Formation of the component begins as the beam deposits the wire on the substrate. As each level is completed, the z-axis moves to allow the next level to be deposited until the component is completed.

ADVANTAGES OF ELECTRON BEAM WELDING

Electron beam welding provides unique performance capabilities. The high-quality environment, high power density, and outstanding control capability solve a wide range of joining problems. The advantages of electron beam welding include the following:

1. The process is extremely efficient because EBW directly converts electrical energy into beam output energy;

5. Taminger, K. M. B. et al, 2002, Solid Freeform Fabrication: An Enabling Technology for Future Space Missions. Presented at the 2002 International Conference on Metal Powder Deposition for Rapid Manufacturing, San Antonio, Texas.

Photograph courtesy of Steigerwald Strahltechnik GmBH

Figure 13.14—Spinner with 25,600 Holes 0.55 mm (0.022 in.) Diameter Drilled by the Electron Beam Process

Figure 13.15—Repair Welds on Titanium Drum Rotor Part of an Aircraft Engine, As-Welded (Upper Left Section) and Finished Product After Machining (Upper Right Section)

2. Single-pass welds can be made in thick joints because electron beam welds exhibit a high depth-to-width ratio;
3. The heat input per unit length for a given depth of penetration can be much lower than that of arc welding, resulting in a narrow weld zone, less distortion, and fewer deleterious thermal effects in the workpiece;
4. A high-purity (vacuum) environment for welding minimizes contamination of the weld metal by oxygen and nitrogen;
5. The ability to project the beam over a distance of several feet in vacuum often allows welds to be made in otherwise inaccessible locations;
6. Rapid travel speeds are possible because of the high melting rates associated with the concentrated heat source, thus reducing welding time and increasing productivity and energy efficiency;
7. Butt joints that are reasonably square can be welded in one pass without the addition of filler metal in thick plate and relatively thin plate;
8. Hermetic closures can be welded with the high- or medium-vacuum modes of operation while retaining a vacuum inside the weldment;
9. The beam of electrons can be magnetically deflected to produce welds of various shapes, and also can be magnetically oscillated to improve weld quality or increase joint penetration;
10. The focused beam of electrons has a relatively long depth of focus, which will accommodate a broad range of beam-to-workpiece distances;
11. Compared to laser beam welding, materials with high thermal conductivity or high reflectivity can be welded easily without unwanted energy reflections and energy losses;

12. Single-pass welds with complete joint penetration, nearly parallel sides, and nearly symmetrical shrinkage can be produced; and
13. Dissimilar metals and metals with high thermal conductivity, such as copper, can be welded.

COMPARATIVE CHARACTERISTICS OF WELDS

The geometry of weld metal in electron beam welds differs significantly from those of conventional arc welding processes. In Figure 13.16, cross sections of typical transverse welds made with electron beam welding (A) and welds made with gas tungsten arc welding (B) are compared. The geometry of a typical electron beam weld exhibits a very large depth-to-width ratio compared to that of an arc weld. This feature results from the high power density of the electron beam. The beam is concentrated in a small area, and the beam power density can exceed the power densities available in arc welding by several orders of magnitude. However, when required, a defocused electron beam weld can be made that has depth-to-width ratio characteristics similar to a gas tungsten arc weld.

Depth-to-Width Ratios

The ability to produce welds with high depth-to-width ratios using electron beam welding accounts for two important advantages of the process. First, relatively thick joints can be welded in a single pass. Joints in base metals over approximately 6 mm (0.25 in.) thick that require multiple-pass arc welding procedures can be welded in a single pass by electron beam welding in considerably less time. An example of this is illustrated in Figure 13.17, which shows the cross section of a single-pass electron beam weld made in 100 mm (4 in.) thick carbon steel with a beam power of 33 kW and a travel speed of about 2 millimeters per second (m/s) (5 inches per minute [in./min]). Second, for a given thickness, the rate of travel speed at which welding can be accomplished is much greater than can be obtained with arc welding. An additional advantage is that the electron beam welding process introduces less distortion and fewer thermal effects than arc welding.

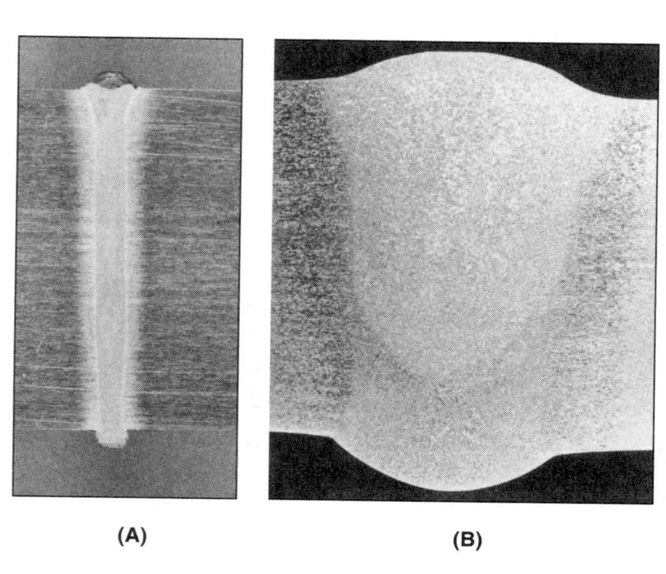

(A) (B)

Figure 13.16—Comparison of (A) Electron Beam Weld and (B) Gas Tungsten Arc Weld in 12.7 mm (1/2 in.) Type 2219 Aluminum Alloy Plate

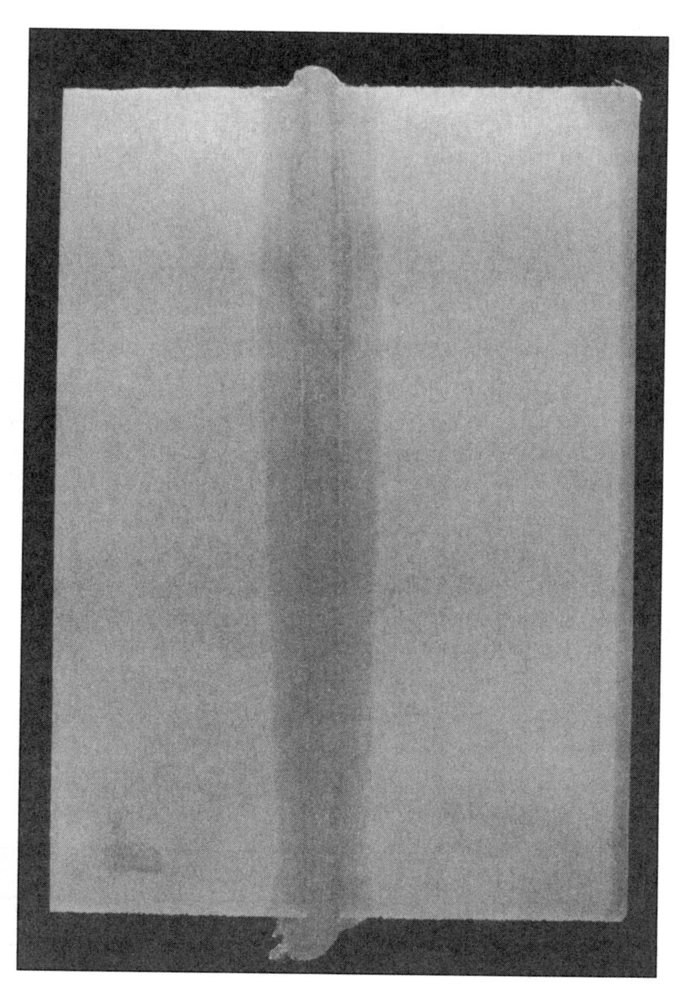

Figure 13.17—Single-Pass Electron Beam Weld in a 100 mm (4 in.) Thick Carbon Steel Section

High-vacuum welds with depth-to-width ratios of 50:1 are possible in a number of alloys. The welding of heavy sections in a single pass is practical using a square-groove butt joint. In aluminum, plates up to 46 cm (18 in.) thick can be welded in this manner.

Some problems may be encountered when joining thick sections with electron beam welding, but production applications in steel extend to workpieces more than 152 mm (6 in.) thick. The medium-vacuum mode sacrifices some of the joint penetration capability of the high-vacuum mode. In the partial-vacuum (soft-vacuum) region of the medium-vacuum mode, approximately 5% less depth of penetration is experienced. In the nonvacuum mode, the maximum depth of penetration attainable in steel joints is less than 51 mm (2 in.).

The ability to produce welds with these characteristics depends on the process mode, whether it is high-vacuum, medium-vacuum, or nonvacuum. In all cases, the welds are highly dependent on the focal spot size and the total beam power.

Joint Penetration

The plot in Figure 13.18 is representative of how penetration decreases with increasing ambient pressure, due to the beam diffusion brought about by increased pressure. It should be noted that the data in Figure 13.18 are normalized relative to data achievable under high-vacuum conditions. The extent of the beam spread shown in this data plot indicates that operating variables other than pressure, such as beam voltage and travel distance, also affect the joint penetration achieved at any given ambient pressure. The final depth-to-width ratio also is critically dependent on the physical properties of the base metal, especially the melting point, heat capacity, thermal diffusivity, and vapor pressure.

Dissimilar Metals

Electron beam welding, particularly in the high-vacuum mode, is an excellent tool for welding dissimilar metals, metals of different mass, and for the repair welding of components impossible to salvage when using other processes. Depending on the joint thickness and the type of base metal, the low total heat input to the workpiece can noticeably minimize distortion of the weld joint. In general, the high- and medium-vacuum modes are the most advantageous, although the non-

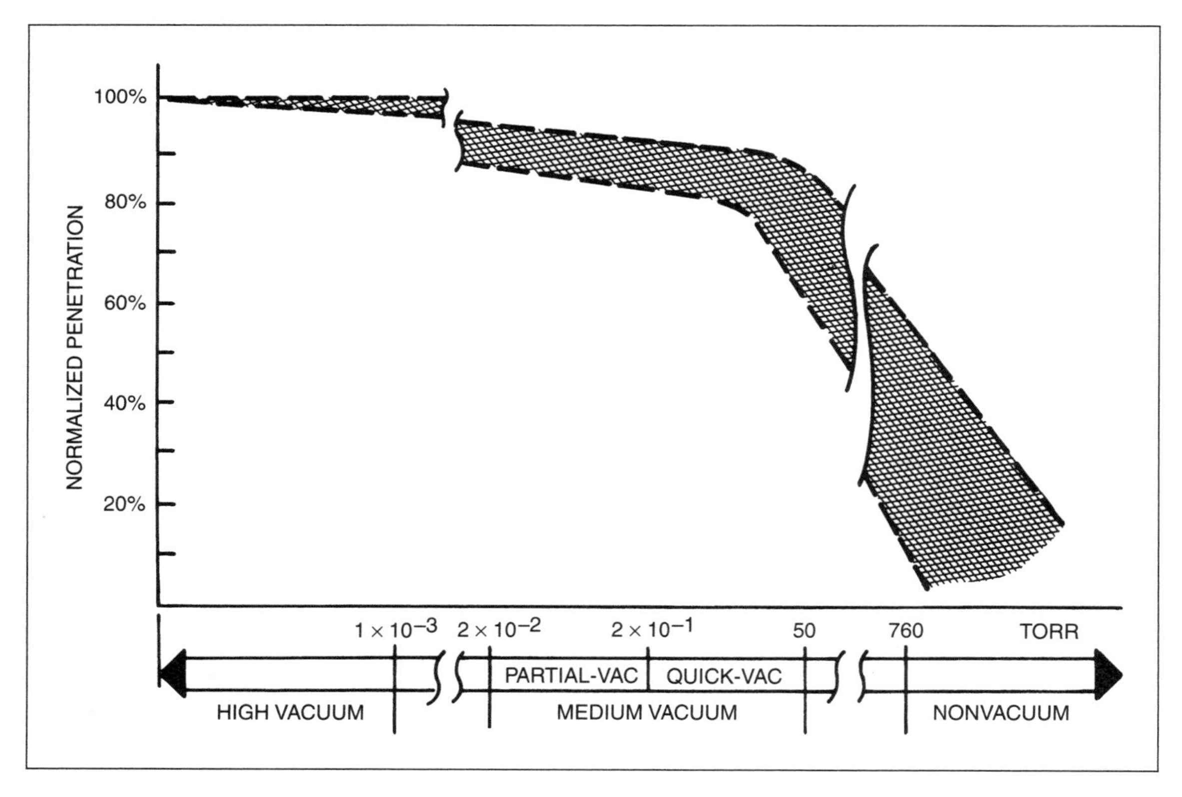

Figure 13.18—Penetration as a Function of Operating Pressure

vacuum mode offers some notable advantages over conventional arc welding processes.

Because a high power density beam produces welds that are not controlled by thermal conduction, metals of significantly different thermal conductivity can be welded. A joint between two pieces of different thickness has unequal heat losses into the thinner and thicker sections, but this is not a significant problem with electron beam welding. In thin metals or metals with low melting points, fusion can be accomplished without significant need for the backup heat sinks that would be required for arc welding. For unequal masses, the beam energy usually is concentrated on the thicker section, and the power is adjusted for penetration through the thin section.

Arc welding procedures often require preheating thick sections of metals that have high thermal conductivity, such as aluminum or copper. Little or no preheat is required for the electron beam welding of these metals because of the high power density available.

Reactive and Refractory Metals

Reactive and refractory metals, including tungsten, molybdenum, niobium, tantalum, zirconium and titanium, are difficult to weld with most welding processes. These metals are detrimentally affected by very small amounts of atmospheric contaminants, such as oxygen, nitrogen, and hydrogen. They can be welded with the electron beam process without introducing these contaminants, although it must be carefully verified that contaminants are not introduced on the faying surfaces of the joint. The high-vacuum mode is most suitable for joining these metals. The other two modes have decreasing weld performance capabilities, however, they still may be satisfactorily applied in specific cases.

Although electron beam welding is a high-power-density process, it is also a low-energy input process. The total energy input needed to weld a joint of a given thickness is considerably less than that required by conventional arc welding processes. Two advantages result from the low-energy input: first, it minimizes distortion and reduces the size of the heat-affected zone; second, the fast cooling rates of narrow electron beam welds can avoid metallurgical reactions, such as phase changes, although the fundamental rules of metallurgy regarding cooling rates and the resulting microstructure still apply. Nevertheless, the weld metal will have mechanical properties approaching those of the base metal.

Process Control

Another aspect of electron beam welding involves control of the process. As the process is pushed to the limit of its capabilities, the operating variables that influence the final results require much greater control. Accurate control of the electron beam process yields a high degree of reliability. The incorporation of CNC, minicomputers, and microprocessors provides additional control capability over welding conditions.

Control of the welding environment can control the composition of the weld metal. Electron beam welding in a high vacuum permits gases to escape and high-vapor-pressure metals to evaporate. This produces more refined weld metal, but also may cause the loss of certain alloying elements. At the other pressure extreme, nonvacuum electron beam welding in an air atmosphere may increase the nitrogen and oxygen content of the weld metal.

Less weld distortion is produced because of the narrow shape of the weld metal. In this type of weld, the weld metal is essentially parallel-sided except where the electron beam first impinges on the top surface of the abutted workpieces. (Refer to Figure 13.16[A].) Contraction of the metal during cooling is fairly uniform throughout the joint. When weld metal has the characteristic V-shape of arc welding, there usually is significant warping from the unequal thermal contraction at the top and bottom of the joint.

Since the electron beam can penetrate extremely thick sections, beveling or chamfering one or both edges of abutting members is not necessary. However, for such welds, tighter machining tolerances normally are required for high-vacuum electron beam welding than for arc welding. In medium-vacuum welding, cross sections of the weld metal are similar to those of high-vacuum welding, but the depth-to-width ratios are somewhat less. The weld bead produced with nonvacuum electron beam welding may be nearly as wide as a typical gas tungsten arc weld. The electron beam weld metal may possess all the characteristics of the weld metal produced by conventional welding processes because insufficient power or long gun-to-workpiece distances were employed. However, at high power and speed, depth-to-width ratios on the order of 5 to 1 are feasible with the nonvacuum welding mode.

LIMITATIONS

Some limitations of electron beam welding are the following:

1. Capital costs for EBW equipment are substantially higher than those of arc welding equipment, but in high-volume production, the final per-piece costs can be highly competitive;
2. Preparation for welds with high depth-to-width ratios requires precision machining of the joint edges, exact joint alignment, and good fit-up, with some exceptions;

3. The root opening must be minimized to take advantage of the small size of the electron beam;
4. The rapid solidification rates achieved can cause cracking in highly constrained, low-ferrite austenitic stainless steel;
5. For high- and medium-vacuum welding, the vacuum chamber must be large enough to accommodate the assembly, thus the size of the chamber and the time needed to evacuate it will influence production costs;
6. Partial-penetration welds with high depth-to-width ratios are susceptible to voids and porosity in the weld root;
7. Because the electron beam is deflected by magnetic fields, nonmagnetic or properly degaussed metals should be used for tooling and fixturing near the beam path, which will influence production costs;
8. With the nonvacuum mode of electron beam welding, the restriction on the distance from the bottom of the electron beam gun column to the workpiece may limit the product design in areas directly adjacent to the weld joint.
9. With all modes of EBW, radiation shielding must be maintained to ensure that there is no exposure of personnel to X-rays generated by EBW;
10. Adequate ventilation is required with nonvacuum EBW to ensure proper removal of ozone and other noxious gases formed during this mode of EBW or during other modes of welding when hazardous materials are generated.

It should be noted that the precision requirements for joint preparation mentioned in No. 2 and No. 3 are not mandatory if high depth-to-width ratio welds are not required.

EQUIPMENT

High-vacuum, medium-vacuum, and nonvacuum electron beam welding equipment employs an electron beam gun/column assembly, one or more vacuum pumping systems, and a power source. High-vacuum and medium-vacuum systems operate with the workpieces in an evacuated welding chamber. Although the nonvacuum mode does not require workpieces to be placed in a chamber, a vacuum environment is necessary for the electron beam gun column.

All three variations of electron beam welding can be performed using high-voltage equipment, that is, equipment using gun columns with beam-accelerating voltages greater than 60 kV. Nonvacuum electron beam welding performed directly in air requires beam-accelerating voltages greater than 150 kV. High-vacuum and medium-vacuum welding also can be performed with low-voltage equipment, for example, equipment with gun columns that employ beam-accelerating voltages of 60 kV and lower.

Because high-voltage gun columns generally are fairly large, they usually are mounted on the exterior of the welding chamber and are either fixed in position or provided with a limited amount of tilting or translational motion, or both. Low-voltage gun columns normally are small. Some units are fixed externally; others are internally mounted mobile units with up to five axes of combined translational motion. Figure 13.19 shows a high-vacuum system, an air-to-air bimetallic strip-welding machine in which individual strips can be continuously fed through the high-vacuum weld chamber by means of a series of graded pressure zones on both the inlet and outlet sides, thus allowing welding of the strips to be totally accomplished in a high-vacuum environment.

ELECTRON BEAM GUNS

In general, electron beam welding guns are operated in a space-charge-limited condition. When a gun is operated in this condition, the beam current produced at any accelerating voltage is proportional to the 3/2 power of the accelerating voltage ($I = KV^{3/2}$), where the constant of proportionality, K, is a function of gun geometry expressed as follows:

$$I = KV^{3/2} \tag{13.1}$$

where

I = Beam current, mA,
K = Dimensionless gun geometry constant, and
V = Beam voltage, kV.

Besides acceleration voltage, a broad range of conditions must be satisfied to enable an electron gun to deliver the required power magnitude and power density.

Optimum gun performance depends on the configuration of the gun, the characteristics of the emitter, total power capability, and provisions for focusing. For a given metal and joint thickness, characteristically narrow welds can be made if sufficient beam power is available to permit rapid travel speed and if the beam power density is great enough to develop and continuously maintain a keyhole to the depth of joint penetration required.

An electron beam gun generates, accelerates, and collimates electrons into a directed beam. The components and functions of the gun can be divided into the following two categories:

Photograph courtesy of PTR—Precision Technologies, Inc.

Figure 13.19—A High-Vacuum Welding Machine Designed for Continuous Processing of Bimetallic Strip Welds

1. The emitter, which generates electrons, uses directly heated wire or ribbon filament, or uses a rod or disk filament indirectly heated by electron bombardment or induction; and
2. Beam-shaping and forming elements, the grid and anode.

The specific emitter design chosen for an application will affect the characteristics of the final beam spot produced on the workpiece. Only self-accelerated guns similar to the telefocus-style guns invented by Pierce, Rogowski, and Steigerwald are used for electron beam welding. They have superior focusing and power capabilities, and also permit placing the gun anode and workpiece at earth ground potential.

The Pierce gun was originally designed as a diode capable of producing a rapidly converging beam with the primary focal spot close to the exit side of the anode, and having a uniform beam divergence thereafter. In comparison, the Rogowski gun was originally designed to produce a primary focal spot close to the emitter and to provide a uniformly divergent beam from that point outward. The Steigerwald telefocus gun was originally designed as a triode that produced a gradually converging beam with the primary focal spot some distance from the anode. Current models of the

Pierce, Rogowski, and Steigerwald guns are modifications of the original designs.

In a diode gun, when a change in the maximum beam current achievable at a given accelerating voltage is desired, the cathode-to-anode spacing in the gun must be changed. This changes the proportionality constant, K, of the gun. Several spacers usually are supplied by the manufacturer to provide a wide range of operating conditions. Each spacer has a range of beam power, beam spot size, and control sensitivity.

Diode guns vary the beam current output at a given voltage by controlling the power to the electron emitter, and thus the temperature of the emitter. Electron emission is related to both emitter temperature and accelerating voltage.

The triode gun is similar to the diode gun except that the beam-forming electrode (grid) is biased with a variable negative voltage relative to the emitter. In this type of gun, the emitter is referred to simply as the *filament*. The bias-voltage grid makes it easy to vary the beam current at any constant accelerating voltage. Thus, within limits, the accelerating voltage and beam current can be varied independently.

The ability to control beam current with a bias voltage allows rapid changes in beam current. Electronic switching circuits permit users to apply pulsed current to the beam. Pulsed beam current helps to minimize heat input to the workpiece while still achieving deep penetration. This capability is more extensively used for drilling than it is for welding. Accurate control of the slope rates of the beam current is extremely useful in many welding applications, particularly on circular welds at the start and finish overlap regions.

All gun/column assemblies employ an electromagnetic lens to focus the electron beam into a small spot on the workpiece after it exits the gun. Electromagnetic deflection coils usually oscillate the beam in either a repetitive or nonrepetitive fashion. These deflection coils generally are positioned immediately below the electromagnetic focusing lens, and are used to deflect the electron beam from its normal axial position. Two sets of deflection coils at 90° are capable of tracing classic Lissajous figures on the surface of the workpiece.[6] Sinusoidal beam deflection perpendicular to the direction of welding broadens the weld bead to simplify manual tracking of weld seams. Circular and elliptical deflection tends to reduce weld porosity. More complex deflection patterns also may be employed that will affect both depth of fusion and weld quality. (Refer to Figure 13.3.)

The modified Pierce, Rogowski, and Steigerwald guns weld with both high and low beam accelerating voltages. Similar power levels are available with high- and low-voltage guns. Beam power is the product of the accelerating voltage and the beam current. Therefore, operation at high voltage requires lower beam current than operation at low voltage for an equivalent beam output power. In high- or medium-vacuum applications, both low-voltage and high-voltage equipment produce welds of similar quality in most metals. However, differences will exist in the weld cross section produced at the same beam power, because one system operates with a low voltage and high current and the other with high voltage and low current. Low-voltage electron beam welds tend to be wider than those produced by high-voltage EBW equipment.

6. Jules A. Lissajous, French physicist, 1880.

POWER SOURCES

The high-voltage power used in electron beam welding machines is supplied by an assembly consisting of a high-voltage power source section and one or more auxiliary power source segments. The high-voltage power source segment provides the required gun beam accelerating voltage, and the auxiliary power source segments supply the emitter heating and beam current output control. Depending on whether the gun is a diode type or a triode type, the high-voltage power source consists of two or more of the following components:

1. A primary high-voltage direct-current (dc) power source that provides the constant beam-accelerating voltage and total beam current,
2. An emitter (filament) power source with either ac (alternate current) or dc output, and
3. A grid electrode power source that impresses a voltage between the emitter and the bias electrode to control beam current.

The primary high-voltage power source and auxiliary beam-generating power sources frequently are placed together in an oil-filled tank. High-purity, electrical-grade transformer oil serves as an electrical insulating medium and as a coolant that carries heat from the electrical components in the tank to the tank walls. These components typically are supported from the cover plate of the tank so that they can be removed from the tank with the cover plate; however, removal of the components is rarely required.

Another high-voltage insulating material occasionally used in EBW power sources is sulfur hexafluoride (SF_6) gas at pressures up to 3.1×10^2 Pa (45 pounds per square inch [45 psi]). A power source with this gas insulation is considerably more compact and lighter than an oil-insulated unit of the same rating. The components in both the primary high-voltage power source and the auxiliary power sources are primarily transformers, diodes (rectifiers), capacitors, and resistors. Electron tube diodes were initially used in the primary power source by some manufacturers, but tube diodes have been superseded by solid-state diodes, usually silicon. Cost, regulation, physical size, ability to absorb transients of voltage and current, and thermal capability are some of the considerations that affect the choice of components and the insulating medium.

Primary High-Voltage Power Source

The primary power source converts line input power to high-voltage dc power for the electron beam gun. Power ratings commercially available are in the range of 3 kW to 100 kW, but up to 300 kW is readily available. The power sources are designed for a particular type of electron beam gun (high or low voltage). Some typical ratings of these machines are listed in Table 13.1.

The maximum allowable voltage fluctuation (ripple) in the dc output varies, depending on the desired quality of beam focus. Excessive voltage fluctuation will produce an undesirable beam current fluctuation that may affect weld quality. Voltage fluctuation is usually controlled below 1%.

The inherent decrease in output voltage in proportion to increasing load typically is in the range of 15% to 20%. Various controls and regulators compensate for this voltage decrease and minimize the effects of line voltage variations, the effects of temperature, and other factors that influence the stability of the output voltage. Sophisticated controls eliminate all of these effects and maintain a stable accelerating voltage to within 1% of the selected value. These controls are especially useful for welds in highly critical applications. Other less costly controls eliminate only some of the effects, but are adequate for less critical applications. Some of the controls used, in approximate order of the level of sophistication and performance, include the following:

1. Line-voltage regulator (constant-voltage transformer);
2. Servo-operated variable transformer with feedback from the high-voltage output;
3. Modern SCR (silicon controlled rectifier) units, or the older motor-generator with electronic exciter and feedback from the high-voltage output; and
4. Electronic regulation of current and voltage feedback.

Electron beam welding equipment manufacturers can supply welding systems with integrated high-frequency, high-voltage power sources. This modern variety of switch-mode units replaced the low-frequency, three-phase rectified transformer power sources formerly used and also replaced the primary steering controls they required. Switch-mode power sources employ high-voltage generating sources that generally use frequencies greater than 20 kHz in value, and as an added advantage, provide a high-voltage source that involves very low stored energy values. The solid-state switching electronics employed in these units allow continuous monitoring of both beam voltage and current. The switching controls are fast-acting enough to sense a pending beam arc-over and extinguish it before it occurs. Also, switch-mode power sources use smaller enclosures than the transformer types, thus requiring less insulating medium. The lower-voltage, lower-power units (60 kV up to 7.5 kW) do not require any oil or gas insulating medium, and thus basically operate at atmospheric conditions.

Table 13.1
Typical Electron Beam Welding Machine Ratings

	Output	
Rating, kW	kV, max	mA, max
3	30	100
3	60	50
6	30	200
7.5	150	50
15	60	250
15	150	100
25	175	144
25	150	167
30	60	500
35	200	175
45	60	750
60	175	345
100	100	1000

Emitter Power Source

Power sources with a directly heated wire or ribbon (filament) emitter are the most commonly used. Filaments may be hairpin-shaped or may be a more complex shape. The current that heats the filament can be ac or dc, but dc is preferred because the magnetic field created by the filament-heating current may affect the direction of the beam. The cyclic nature of ac heating currents causes the beam spot to oscillate with a small, but significant, amplitude about a fixed point.

Since the magnitude of magnetic effects will increase with the increasing heating current, filtering must be used even with dc heating currents to reduce to 3% any ripple that might be present.

Current and voltage ratings of a filament power source depend on the type and size of the directly heated filament. For tungsten wire filament diameters of 0.5 mm (0.020 in.), the power source would be rated for 30A at 20V. Ribbon-type filaments have a much larger emitting area than wire filaments. Ribbons require power sources rated for higher currents and lower voltages (30 A to 70 A at 5 V to 10 V). High heating currents in dc filaments will produce a certain amount of initial beam deflection, but this is a static feature that usually can be corrected by beam alignment devices normally built into the gun/column assembly.

Indirectly heated emitters, such as rod and disk types also are used. In this type, an auxiliary bombardment or inductive-type heat source indirectly heats the gun emitter to electron emission temperatures. The power source for driving the auxiliary electron gun used to heat a disc emitter by bombardment would be rated for 100 mA to 200 mA at several kilovolts. The power source for indirectly heating a rod-type emitter, heated in a radial mode, would be rated for 2 A to 3 A at 400 V to 600 V.

Bias Voltage Power Source

The bias voltage power source for a triode gun usually is designed to give complete control of beam current from zero to maximum. To do this, the dc power source applies a variable voltage on the beam-shaping electrode, making it negative with respect to the emitter. A voltage in the range of 1500 V to 3000 V is needed to turn off the beam current. For maximum beam current, the voltage is 100 V to 300 V. Again in this instance, the bias voltage power source must have no greater than 1% beam-current ripple. Various electronic input power control devices are used to provide functions such as pulsing and ramping of the beam current.

Some electron beam equipment uses a self-biasing system. The bias voltage is derived partially from the primary accelerating voltage through a voltage divider and partially from the voltage across a series resistor in the main power circuit. This type of system does not have a separate bias voltage source.

Electromagnetic Lens and Deflection Coil Power Sources

The electromagnetic lens generally is powered by a solid-state constant-current power source (refer to the focusing lens shown in Figure 13.1). The strength of the magnetic field varies with the current flowing through the coil. The current provided to the coil must remain constant to produce a beam focal spot with consistent size, even when the voltage drop across the coil changes with temperature variations.

Beam deflection coils also are powered by solid-state devices (refer to Figure 13.1). Two sets of coils perpendicular to one another are usually placed at the base of the gun column for X and Y deflection of the beam. The programming of the power sources for the two sets of coils can provide beam movement along either axis singularly or both axes simultaneously. Complex geometric beam patterns, such as circle, ellipse, square, rectangle, and hyperbola can be produced by electronic control. The ripple on the dc input to the deflection coils and the electromagnetic lens must be low to minimize adverse effects of beam instability on weld quality.

VACUUM PUMPING SYSTEMS

Vacuum pumping systems are required to evacuate the electron beam gun chamber and the welding chamber (workpiece chamber) for high- and medium-vacuum welding and the orifice assembly used on the beam exit portion of gun/column assemblies for medium-vacuum and nonvacuum welding. Two basic types of vacuum pumps are used. One is a mechanical piston or vane type used to reduce pressure from 1.0×10^2 to 1.0×10^{-2} kPa (1 atmosphere to 0.1 torr). For medium-vacuum welding, these mechanical pumps generally are operated in conjunction with another kind of mechanical pump, a roots-type blower. The other pump is an oil-diffusion type used to reduce the pressure to 1.3×10^{-2} Pa (10^{-4} torr) or lower. Sequencing these pumps to produce the needed vacuum can be accomplished by manual or automatic operation of valves in the system. Automatic valve sequencing is standardized on commercial electron beam welding equipment. Appendix C of this volume provides a table showing conversions of Standard International and U.S. Customary pressure units that may be useful to users of electron beam welding systems.

The vacuum system for an electron beam gun chamber generally consists of a mechanical roughing pump backing either a diffusion pump or a turbo molecular pump. Systems used to evacuate a high-vacuum work chamber generally consist of a mechanical roughing pump backing either a diffusion pump or a cryogenic pump. Both systems may be ducted to the work chambers through a water-cooled or liquid nitrogen-cooled (optically dense) baffle, if necessary, to prevent any oil from backstreaming out of the diffusion pump into the gun or work chambers. However, most modern diffusion pumps have some form of integrated cold cap to capture backstreaming pump oil, so baffles rarely are needed. The use of a turbo molecular pump or a cryogenic pump will entirely eliminate any possibility that pump oil will enter the gun or chambers. A combination of diffusion and mechanical pumping is shown in Figure 13.20.

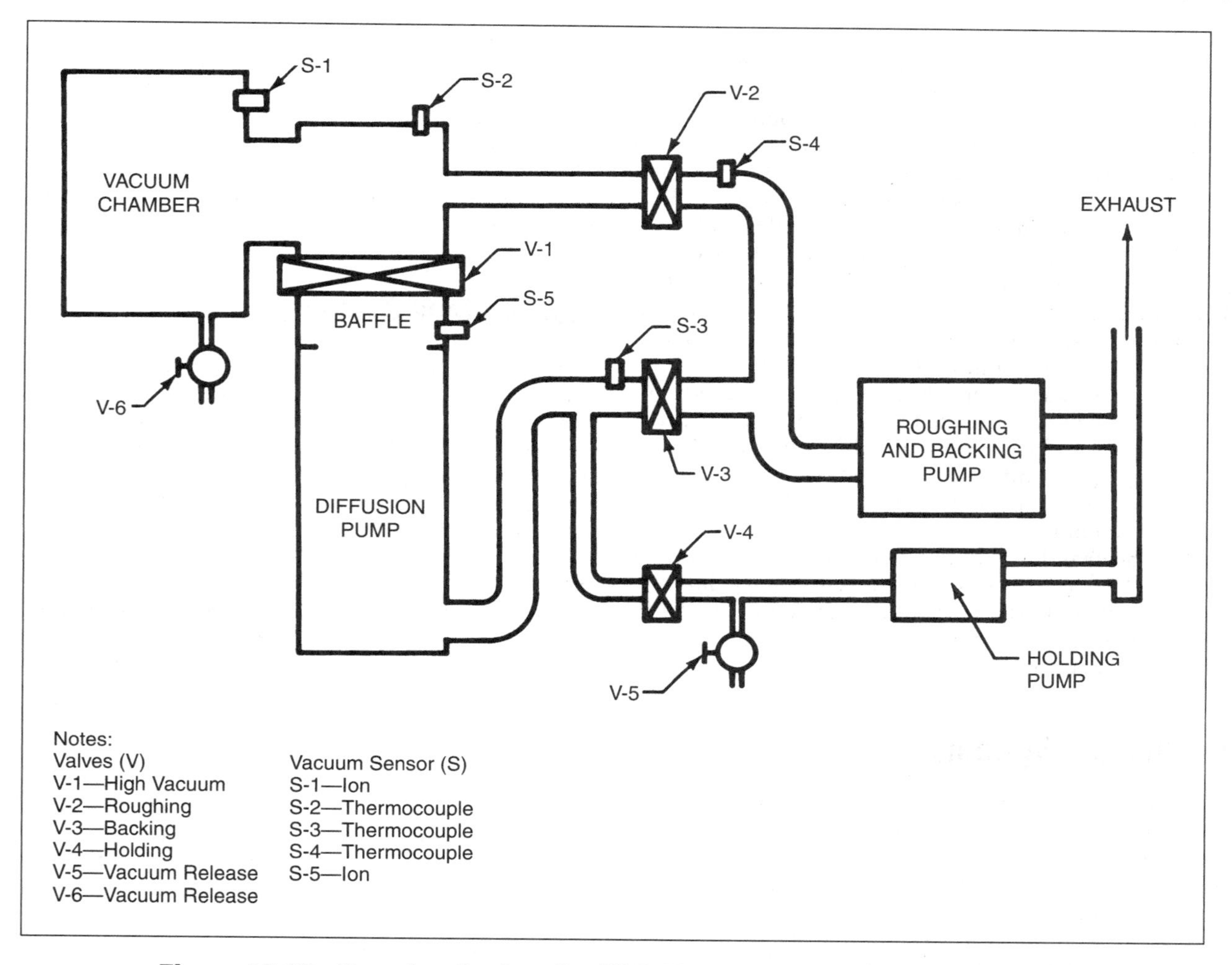

Figure 13.20—Pumping System for High-Vacuum Electron Beam Welding

The vacuum pumping system can be mounted on the same base as the vacuum chamber and connected to it with rigid ducting. The primary exception to this arrangement is the mechanical roughing pump, which must be connected to the system with a flexible tube to minimize any chance of transmitting vibration to the welding chamber. A large-diameter vacuum valve isolates the diffusion pump from the chamber during the roughing portion of the pumping cycle. A small mechanical pump keeps the isolated diffusion pump at low pressure.

The roughing and diffusion pumping periods normally are controlled by automatic sequencing of pneumatic or electric vacuum valves. Automatic evacuation cycles are accomplished with pressure-sensing relays that activate the appropriate valves in the preprogrammed sequence. The control units are designed to protect the vacuum system in case of an accidental pressure rise in the chamber.

For medium-vacuum welding, the work chamber is evacuated with a high-capacity mechanical vacuum pumping system. The types and sizes of mechanical pumps used in the system will depend on the size of the work chamber, the workload, and the desired production rate. Automatic evacuation cycles can be used to speed production.

When using nonvacuum welding equipment, the electron beam gun chamber is evacuated with a combination mechanical-diffusion pumping system. The various pressure stages in the orifice assembly through which the electron beam exits the gun/column assembly are pumped with a series of mechanical vacuum pumps.

In high- and medium-vacuum electron beam welding, the evacuation process and its rate depend on the capabilities of the pumps, the workpiece and fixturing load, the size of the chamber, and the total leakage rate of the system. The total leakage rate is the increase in chamber pressure per unit of time attributed to both real leaks and virtual leaks in the system. Real leaks are actual holes or voids in the chamber capable of passing air or gas. A virtual leak is the semblance of a real leak somewhere in the vacuum system; in actuality, this type of leak results from the outgassing of absorbed or occluded gases on the interior surfaces of the system when under vacuum. For satisfactory system operation, a helium mass spectrometer leak detector having a sensitivity of 1×10^{-4} standard cm^3/s of helium should reveal no in-leakage.[7] In addition, a pressure rate-of-rise test should be conducted to ensure that no detrimental virtual leaks are present. This test is conducted by isolating the chamber to be tested from the pumping system (without exposing it to atmosphere) immediately after completing a four-hour preparatory pump-down of the chamber into the high-vacuum range (1.3×10^{-2} Pa [10^{-4} torr] or lower). A customary limiting value for a rate-of-rise test is in the range of 0.13 Pa to 0.26 Pa (1 to 2×10^{-2} torr) per hour, averaged over a 10-hour test period.

Low-Voltage Systems

Low-voltage electron beam welding systems operate below 60 kV, and generally are designed with beam powers up to 100 kW. Work chambers in low-voltage systems usually are made of carbon steel plate. The thickness of the plate is designed to provide adequate X-ray protection and the structural strength necessary to withstand atmospheric pressure. Lead shielding may be required in certain areas of the chamber to ensure the total radiation tightness of the system.

The weldment inside the chamber may be observed by direct viewing through leaded glass windows. However, the effectiveness of this technique depends on the distance between the operator and the weld joint and on the shape of the weldment. When direct viewing is difficult, an optical viewing system may be provided to give the operator a magnified view of the weld joint. The optical viewing system can be used for setup operations, inspection of the weld, alignment of the weld joint relative to the electron beam, and positioning the gun to center the sharply focused electron beam on the weld joint.

Closed-circuit television provides another method of viewing. The light source and television camera may be mounted in a readily accessible location outside of the chamber, or both items may be located inside the chamber. An optical protection system of some type normally is employed to shield the viewing equipment from metal spatter and metal vapor deposition. The closed-circuit television system provides continuous monitoring of welds and permits minimum operator exposure to the intense light from the weld.

High-Voltage Systems

High-voltage electron beam welding systems operate above 60 kV, and generally are designed to operate at voltages between 100 kV and 200 kV (with beam powers up to 100 kW). The electron beam gun/column assembly of a high-vacuum welding machine is housed in an external vacuum chamber, mounted either on the top or side of the welding chamber, as illustrated in Figure 13.21. This mounting may involve either a stationary seal or sliding seal. When the sliding type is used, the motion of the gun/column assembly normally will be limited to a single axis of motion (X or Y). Any other required axis of motion must be provided by the weldment, for example, the type of X, Y, Z, or rotary motion of the workpiece that is commonly employed.

The external location of the high-voltage gun reduces its maneuverability, but provides ready access to gun components for service. This arrangement also provides the operator with a view of the beam spot and the weld through optics that are relatively coaxial with the electron beam. Direct viewing also may be provided through leaded glass windows in the chamber walls. Work chambers for this equipment usually are welded carbon steel boxes with a ribbed design, externally clad with lead for X-ray protection of personnel.

Nonvacuum electron beam welding machines do not require a vacuum chamber around the workpiece. The electron beam gun column may be fixed on top of a box (shielded from X-radiation) containing the workpiece and travel carriage. Another arrangement is to place both the gun column and the workpiece in a room (shielded from X-radiation) making both gun and workpiece capable of transverse motion; then the equipment is operated remotely from outside the room.

SEAM TRACKING METHODS

With electron beam welding, as with other automatic and semi-automatic welding processes, the relative positions of the beam spot and the weld joint must be established accurately before welding is started. Then this relationship must be accurately maintained throughout the entire welding cycle.

7. Refer to American Society for Testing and Materials (ASTM), 2004, *Test Methods for Leaks Using the Mass Spectrometer Leak Detector or Residual Gas Analyzer in the Tracer Probe Mode*, E0498 95R00, West Conshohocken, Pennsylvania: American Society for Testing and Materials.

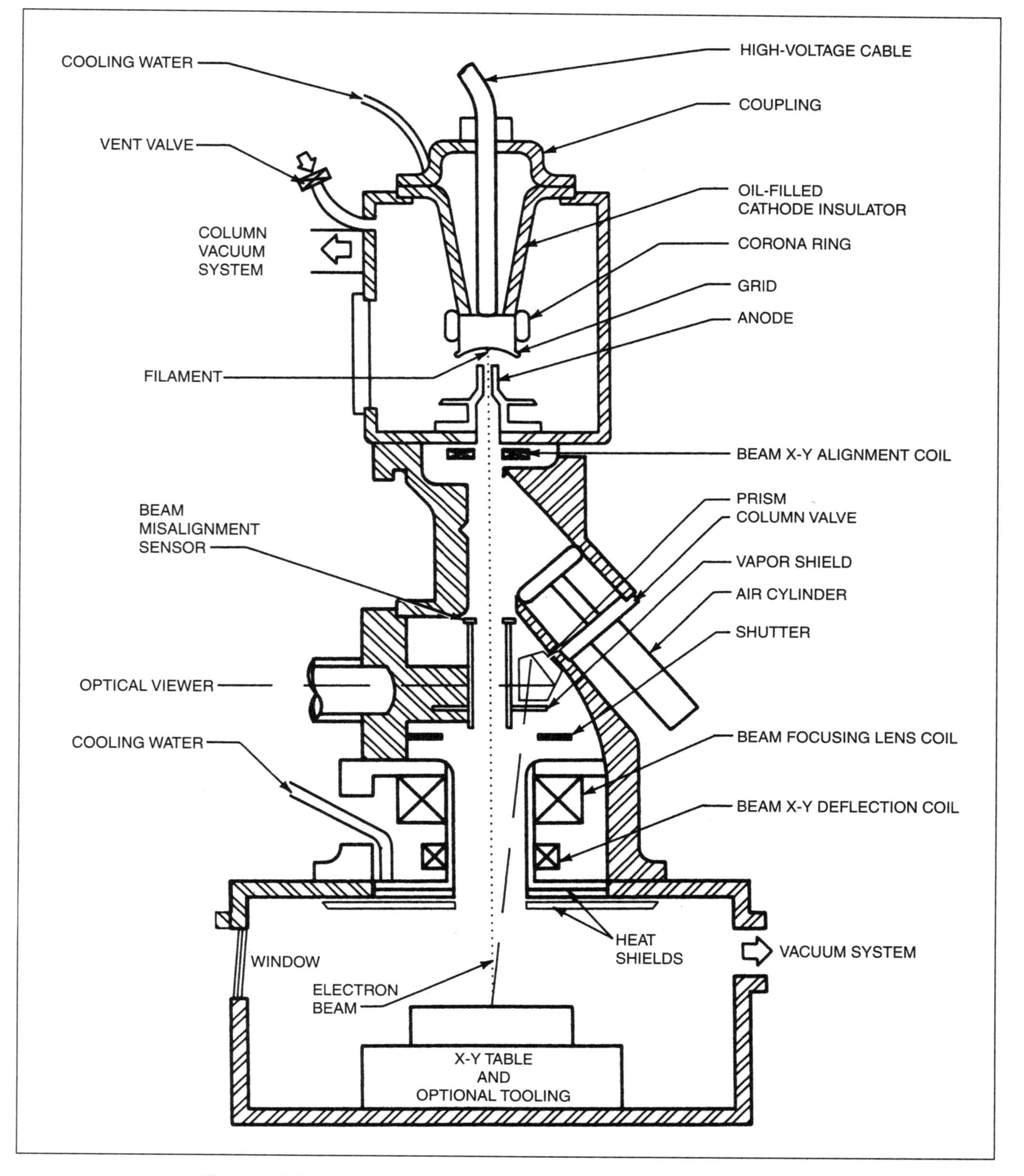

Figure 13.21—Cross Section of Gun Column and Work Chamber

This total requirement is somewhat complicated in electron beam welding for the following reasons:

1. The focal spot of the beam is very small and produces a relatively narrow weld bead;
2. Welding is performed at relatively high travel speeds; and
3. The workpiece is contained in a vacuum chamber or radiation enclosure, making continuous observation difficult.

As previously noted, most high-vacuum electron beam systems are equipped with a viewing system that permits the operator to observe the weld joint and the beam spot. The initially correct position of the electron beam in relation to the joint can be established easily through the use of a viewing system. Operator viewing normally is not provided in medium-vacuum and nonvacuum systems, so the initial beam-to-joint alignment is accomplished through precise handling of the workpiece by means of accurate adjustments to tooling and fixturing.

A means for automatically maintaining proper beam-to-joint alignment is desirable when welding long or slightly irregular joints. Optical viewing of welding and manual correction for deviations in the joint path is difficult at best, although some optical equipment is used in this fashion.

Two methods are used to maintain beam position along a nonlinear joint. The first involves programming by analog or continuous-path numerical controls. This method is applied when workpieces have been machined precisely to a required contour and are accurately positioned for welding. The second method uses an adaptive electromechanical control. This control has a tracking device that follows the weld joint and signals the control to adjust the position of the workpiece or gun to keep the beam on the joint. Both stylus and contour seam tracking devices employ the same electrical circuitry and may be quickly interchanged to accomplish various tracking requirements.

The stylus-type seam tracking system has a probe, or stylus, that rides in the joint. Lateral (cross-seam) movements of the probe resulting from a change in joint position are converted to electrical signals by a transducer. The electrical signals drive a positioning servomotor that moves the workpiece or gun to maintain the preset alignment. The electrical signals from this system define a right-error, a left-error, and the null or correct gun position. Alternatively, the electron beam itself can be deflected electronically to accommodate changes in the location of the joint.

A contour seam tracking system involves a simple modification of the stylus-type seam tracking system, which permits the edge welding of certain types of assemblies by using the weldment as a cam. A preloaded ball-type stylus rides against the edge of the weldment as the workpiece is rotated or driven linearly. As a proximity control to accommodate changes in the vertical position of the weld joint, the stylus is used to maintain a constant gun-to-workpiece distance by feeding the tracking signal to a servomotor drive on the Z-axis.

An electronic joint-finding system also is available, as illustrated in Figure 13.22, which can be used for seam locating and seam tracking functions. This unit taps into the same electron beam used for welding and uses it as a means of sensing the joint position. It finds the joint location by recognizing the absence of rays reflected by the joint in the midst of rays from the workpiece surface, thereby eliminating the need to calibrate an auxiliary joint-sensing device.

Once the vacuum chamber has been closed and evacuated for welding, a finely focused electron beam is aimed at the weld joint and scanned back and forth across it. This action produces a secondary electron emission or backflow that can be "collected" and continuously monitored. As the beam is traversed back and forth, the magnitude of secondary electron backflow being measured will decrease each time the beam passes across the joint. Thus, if the monitor signal is displayed as a visible oscilloscope trace, the resulting discontinuity in the oscilloscope line trace will indicate the location of the joint relative to the beam column centerline. Consequently, this method provides an easy means for initially aligning the beam column centerline with the joint. By initially scanning the entire joint in discrete steps, probable misalignments can be anticipated and corrected during welding.

The stylus, contour, and electronic seam tracking devices often are used in conjunction with a data storage system, in which the joint can be traced and its location recorded. Then the joint is welded, using the stored control program of the beam or workpiece position.

On systems equipped with CNC controls, programmed periodic scans across the joint during welding by the welding beam indicate the exact location of the joint, which the system repositions. The CNC control assumes on-line, real-time seam tracking.

POSITIONING EQUIPMENT

The response of work-handling mechanisms must be accurate and well defined to maintain the relative positions of the electron beam and the weld joint during the entire welding operation. The design and manufacture of positioning equipment should follow good machine tooling practices. Ruggedness, repeatability, smoothness, accuracy, and suitability for operation in a vacuum (if required) are prime requirements. Also, the magnetic susceptibility of the materials must be considered. Since travel speed affects weld geometry, this vari-

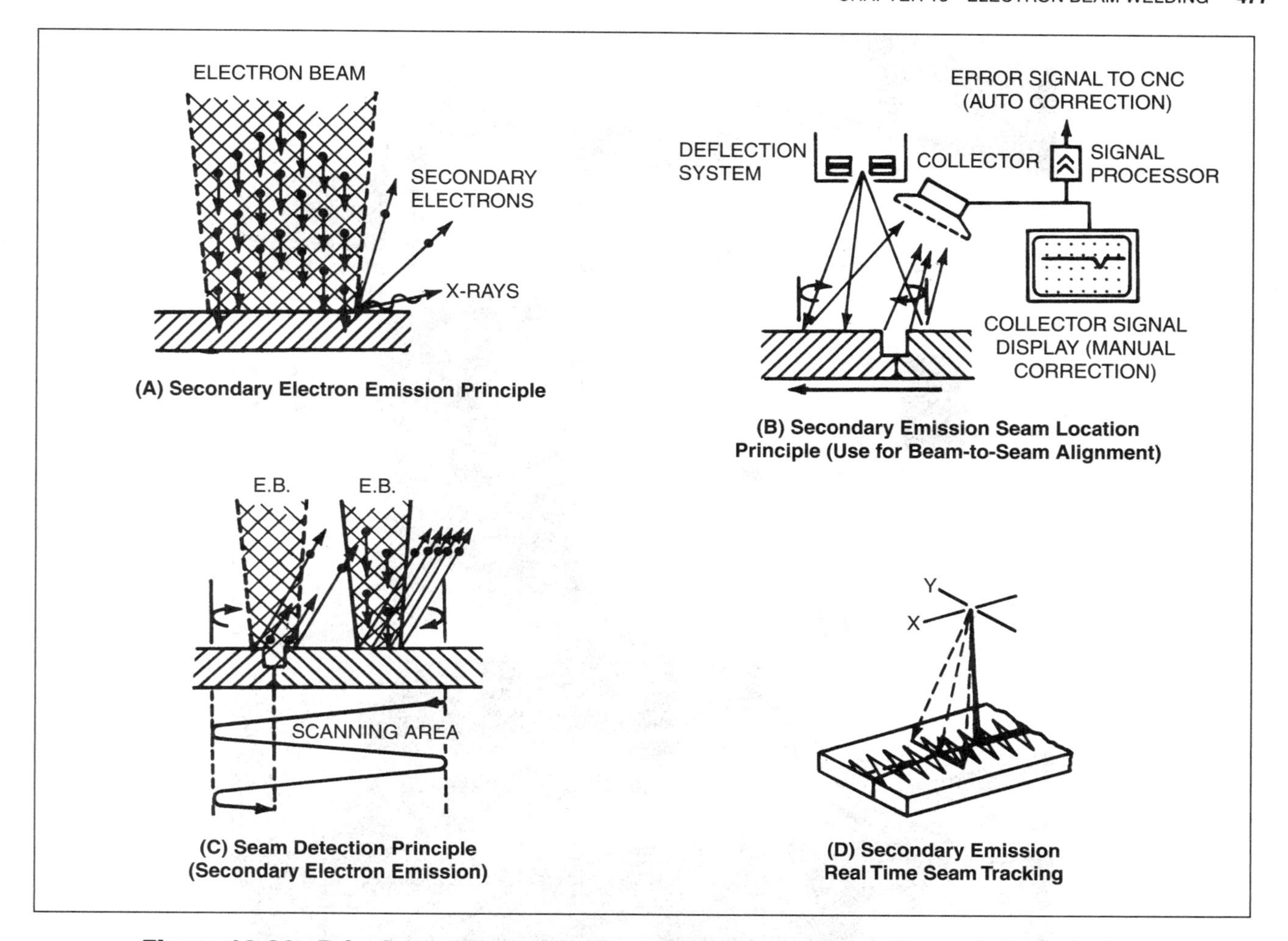

Figure 13.22—Principle of Seam Finding and Tracking Using Secondary Emission

able must be controlled accurately and repeatedly. In general, electric motor drives having an accuracy of about ±2% of prescribed speed are adequate.

Most electron beam welding machines provide standard mechanisms for linear and rotary motion of the workpiece relative to the electron beam. Horizontal linear motion usually is provided by movement of a worktable or by movement of the electron gun. Rotary motion about a vertical axis is achieved with a motor-driven horizontal rotary table. Chambers can be equipped with external platform devices that allow the worktable (and any work-handling mechanisms) to be withdrawn from the vacuum chamber for ease of loading and fixturing the workpieces. Modern computerized numerical control (CNC) systems provide control of precision motion systems and achieve full sequencing and control of the motion.

Figure 13.23 shows an X-Y worktable on its external platform. An adjustable (0° to 90°) power-driven rotary workpiece positioner with Z-motion capability has been mounted on the table. Rotary motion about a horizontal axis can be provided using this rotary tilt-positioner, which also allows the workpiece location to be adjusted. Linear, rotary, and helical joints can be aligned with the electron beam using a combination of the five motion axes (X/Y/Z/rotary/tilt) of workpiece motion.

Often it is desirable to weld circular joints in several segments during a single loading of the chamber. In this case, the components are arranged on an eccentric table attached to a motor-driven horizontal rotary table. The eccentric table holding the workpieces can position each piece in turn under the electron beam. The circular motion for the weld is made with the eccentric table.

Figure 13.23—Table with X-Y Drive and Rotary Positioner for Electron Beam Welding Chamber

Programmed indexing of the eccentric table from segment to segment can be added.

Multiple-spindle rotary fixtures also are used when making circumferential welds in a group of same components. The components are batch loaded, and then successively indexed into welding position by a motor drive. The joint on each piece can be positioned for welding by linear movements of the worktable on which the rotary fixture is mounted. It is possible to automate the entire operation.

CONTROLS

Since all the operating variables of an electron beam welding system are directly controllable, the process is readily adaptable to computerized numerical control (CNC). Movement of the workpieces or the gun and electron beam deflection can be preprogrammed in any combination. The beam current also is programmable. Thus, the beam can be easily changed from one discrete level to another, or changed at a specified rate. The easy and accurate CNC control of upslope/downslope and the capability of producing various beam deflection patterns enhance the capacity of EBW to produce welds of extremely high quality.

Other variables, including the accelerating voltage, beam focus, emitter power, chamber pressure and other auxiliary functions also can be part of the program for control or monitoring. Electron beam welding systems perform computerized contour welding of intricately shaped components, in which beam power and travel speed must be varied as a function of position along the weld path. The system also is used for components requiring multiple-pass weld programs. Modern CNC controllers continue to replace manual and programmable logic controller (PLC) systems.

MEDIUM-VACUUM EQUIPMENT

The equipment used for medium-vacuum electron beam welding basically is a modification of standard high-vacuum equipment. An orifice, or aperture tube, is added into the gun column assembly. This allows beam passage but impedes gas flow, thereby allowing the separately pumped gun region to remain under high vacuum when the chamber is operated at a medium-vacuum level. As on high-vacuum equipment, a column valve is used to isolate and maintain the gun region under high vacuum during chamber venting, and the added aperture helps to maintain a vacuum of 1.3×10^{-2} Pa (10^{-4} torr) or better in the gun region during beam operation, while still allowing the beam to impinge on a workpiece located in a medium-vacuum environment. Thus, on medium-vacuum equipment, the chamber is cyclically vented as new workpieces are loaded, and then rapidly pumped down to the specified medium-vacuum welding level without exposing the gun region to atmosphere. This permits high-volume weld production. Both low- and high-voltage EBW systems are available for medium-vacuum welding.

General-purpose medium-vacuum systems, such as the one shown in Figure 13.24, are used advantageously in short production runs. However, most medium-vacuum units are especially tooled for specific weld assemblies. In each case, the work chamber and tooling are an integrated assembly designed for a single application. Figure 13.25 illustrates typical medium-vacuum tooling concepts.

Various medium-vacuum welding systems are used for high-production applications. For example, a machine with a single welding station and multiple-loading stations can have a production capability of welding approximately 200 components per hour. A dual welding station machine could increase that production capability up to 500 components per hour. In the final analysis, production rates are dependent on the design of the weldment.

One method of achieving high production with medium-vacuum equipment is shown in Figure 13.26. This equipment uses a sliding seal to provide intermediate vacuum zones before and after the separately pumped zones in the medium-vacuum welding chamber. This method maintains a series of continuously pumped vacuum zones that eliminate the need to pause for evacuation, thus utilizing the high-production capability of an indexed feed table and allowing production rates in excess of 500 weldments per hour.

A modern and efficient sequence of operations is employed by automotive industry sectors in Europe and Asia for high-production, medium-vacuum welding of components. A schematic illustration of this system is shown in Figure 13.27. This method eliminates any relative motion between seals and sealing surfaces (a mandatory requirement on the sliding-seal method) and thus avoids the wear that inherently occurs as a result of the sliding motion between these surfaces, but still utilizes the advantages of the index-style of fast transfer of components. As illustrated in Figure 13.27, a segmented chamber is used. Multiple workpieces are placed in the load/unload segment of the chamber. The chamber can be unloaded and loaded and pumped down to a partial-vacuum level while components previously indexed into the processing segment of the chamber are being welded. Thus the chamber is loaded and pre-pumped while the workpieces in the processing chamber are being welded. Then, while under vacuum, the workpieces in the load/unload chamber and the workpiece assemblies in the processing chamber simultaneously lower, index, and lift, thereby interchanging workpieces in these two chambers without exposing the processing chamber to any environment other than a vacuum. While workpieces newly introduced to the processing chamber are being welded, the load/unload chamber can be vented, the weldments removed, and new assemblies loaded. The sequence continues as the chamber is pre-pumped and everything readied for the next full-vacuum interchange of the chamber segments. A finished-product version of the system schematically depicted in Figure 13.27 is shown in Figure 13.28, with a close-up view of the load/unload chamber portion pictured in Figure 13.29.

NONVACUUM EQUIPMENT

The use of nonvacuum EBW means that the beam must pass through a gas atmosphere to perform a welding

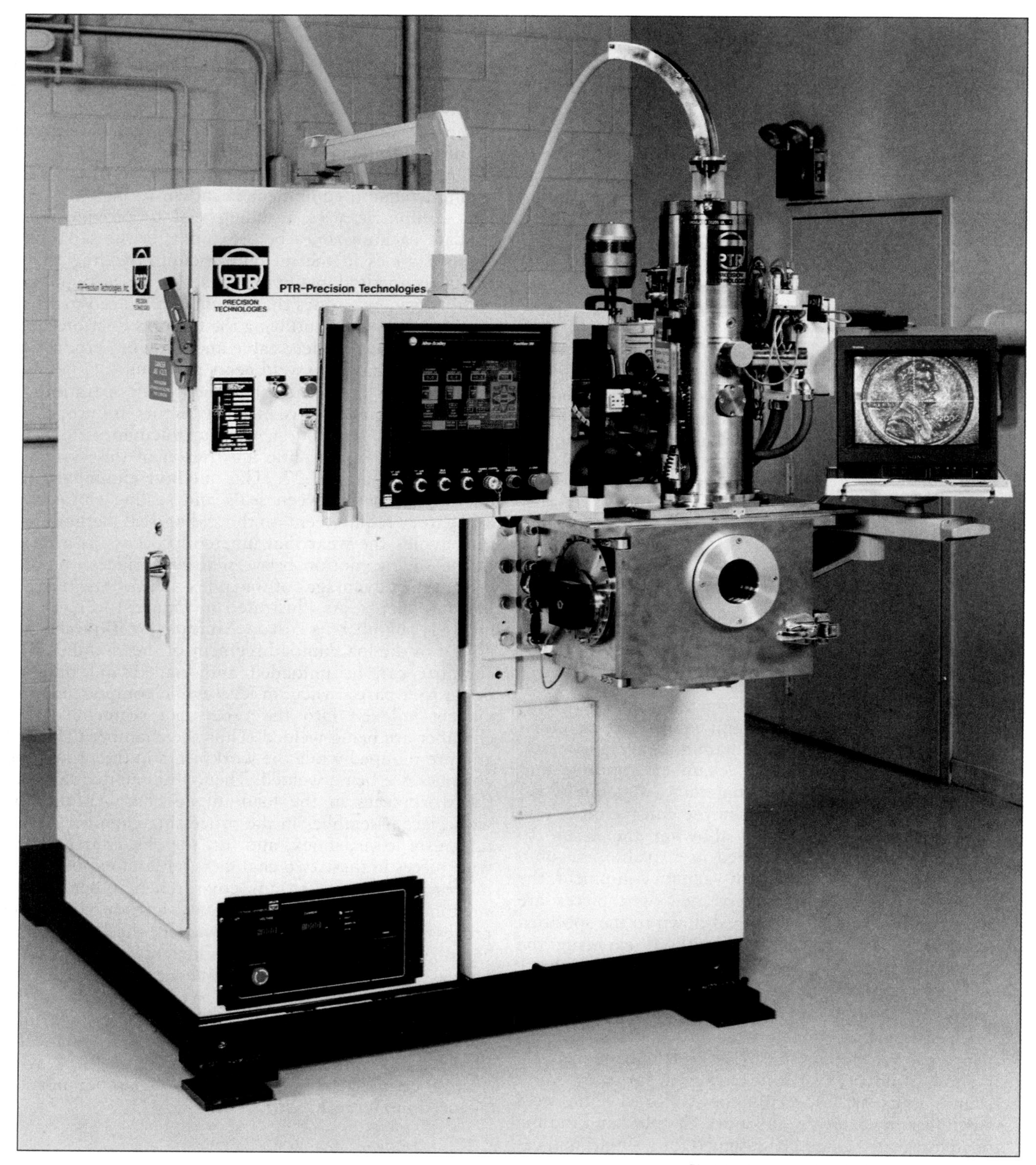

Photograph courtesy of PTR—Precision Technologies, Inc.

Figure 13.24—A General-Purpose Medium-Vacuum Electron Beam Welding Machine

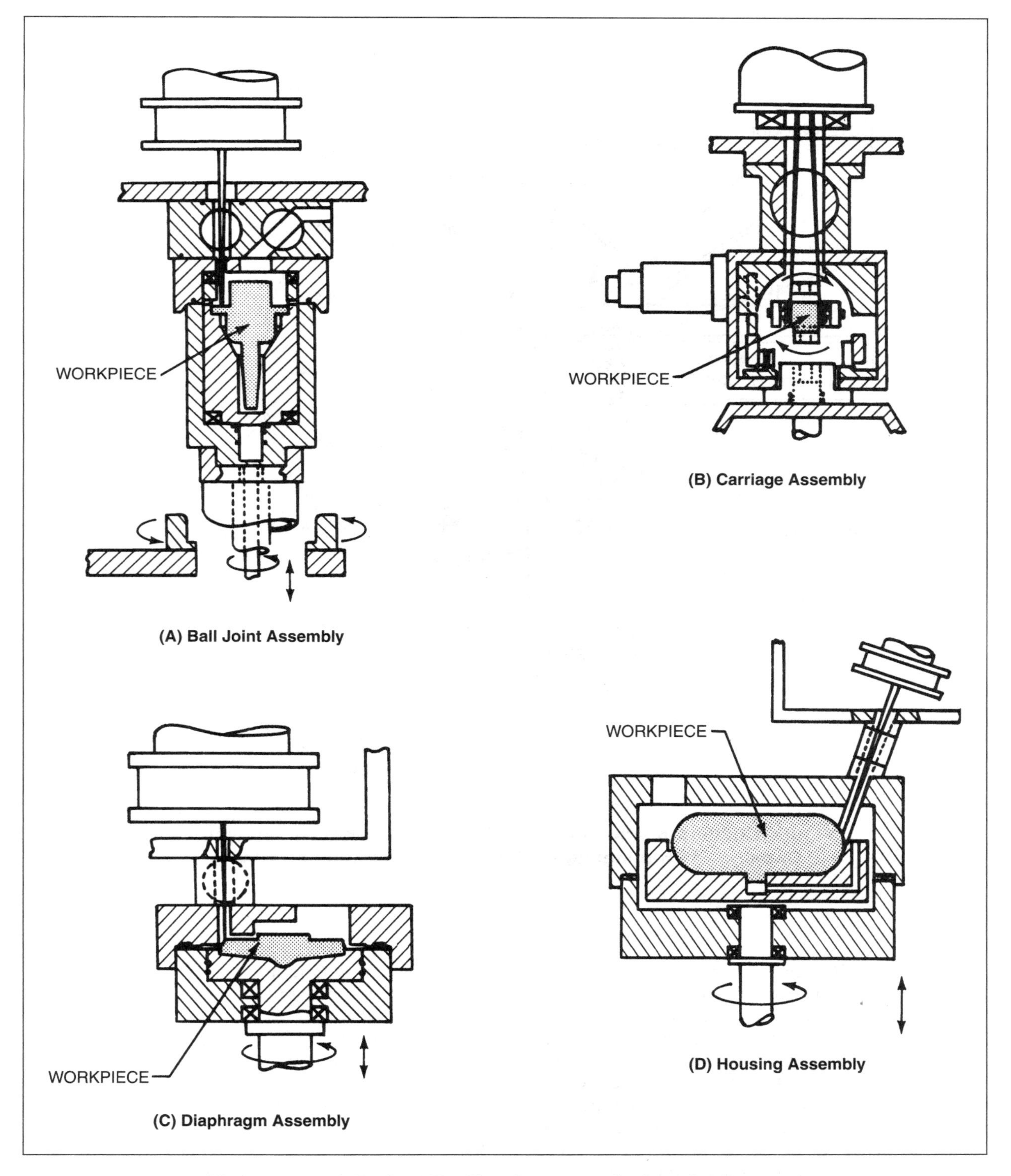

Figure 13.25—Typical Tooling Concepts in Special-Purpose Medium-Vacuum Electron Beam Welding Machines

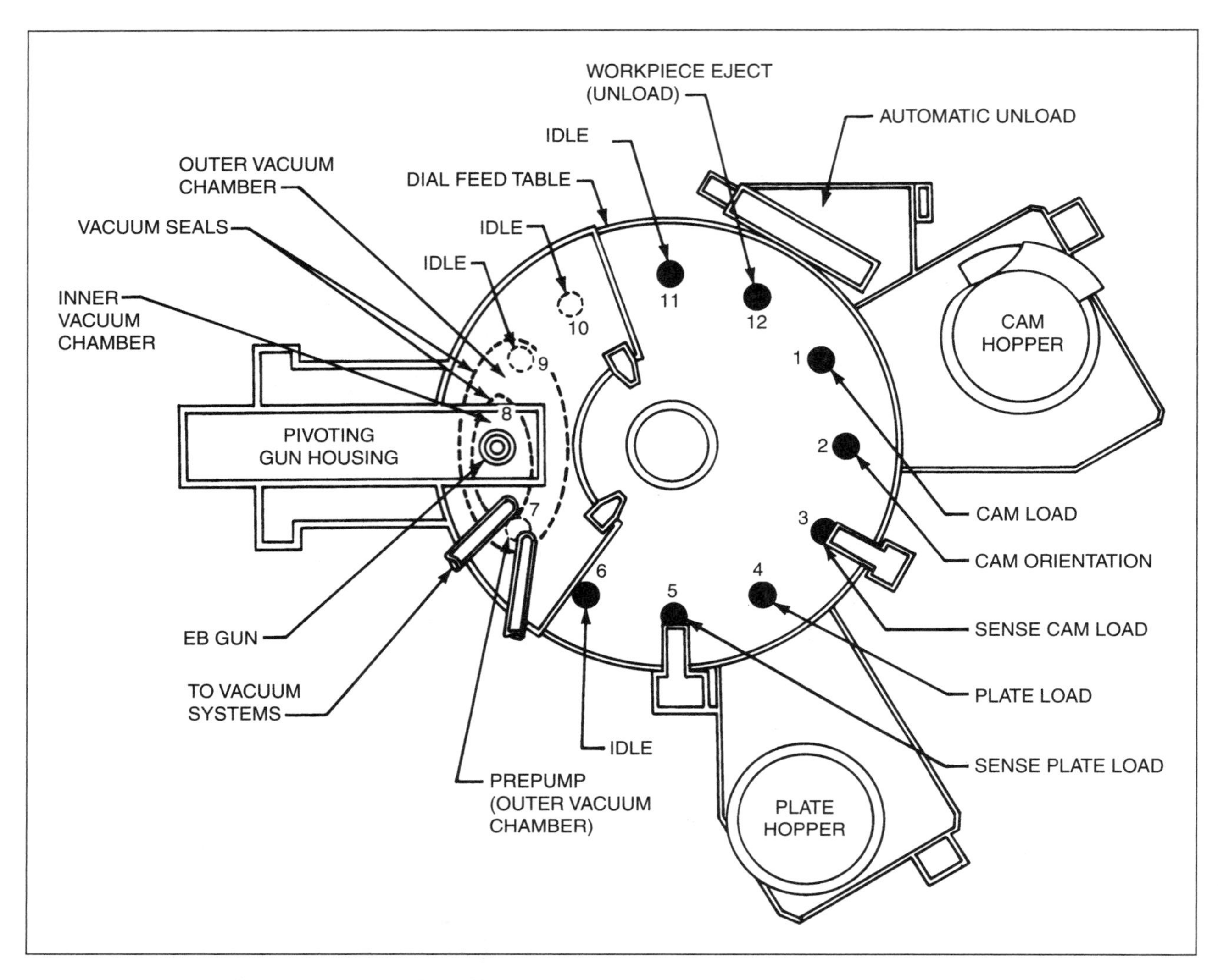

Figure 13.26—A Medium-Vacuum Electron Beam Welding System Capable of Prepumping for Continuous Workpiece Feeding

operation. In applications where the metal is nonreactive, a small shield gas arrangement may be used; in other cases, a larger area of inert gas may be needed to complete welds. A beam of electrons passing through a gas primarily is scattered by the shell electrons of the gas atoms or molecules. As the gas pressure increases, scattering becomes more severe (refer to Figure 13.7). This produces a noticeable broadening of the beam profile and a decrease in the power density of the beam, but not necessarily a loss in total beam power.

An electron beam must be generated in a high vacuum. In addition, the electron velocity (accelerating voltage) must be high enough to minimize the scattering effect of the atmosphere. As a result, to weld with the beam at atmospheric pressure, the beam is passed through a series of chambers or stages operating at progressively higher pressures. The series of chambers operating at successively higher pressures is obtained by *staging* (for example, by differentially pumping a number of chambers). A series of apertures is provided to permit the electron beam to pass through the wall of one chamber into the next, while restricting the gas flow in the opposite direction. This orifice-and-pumping system must be designed to maintain the atmospheric-to-high-vacuum gradient required. The electron beam must be accelerated through a high voltage. If the last stage is in air, this voltage must be a minimum of 150 kV in order to provide a practical working dis-

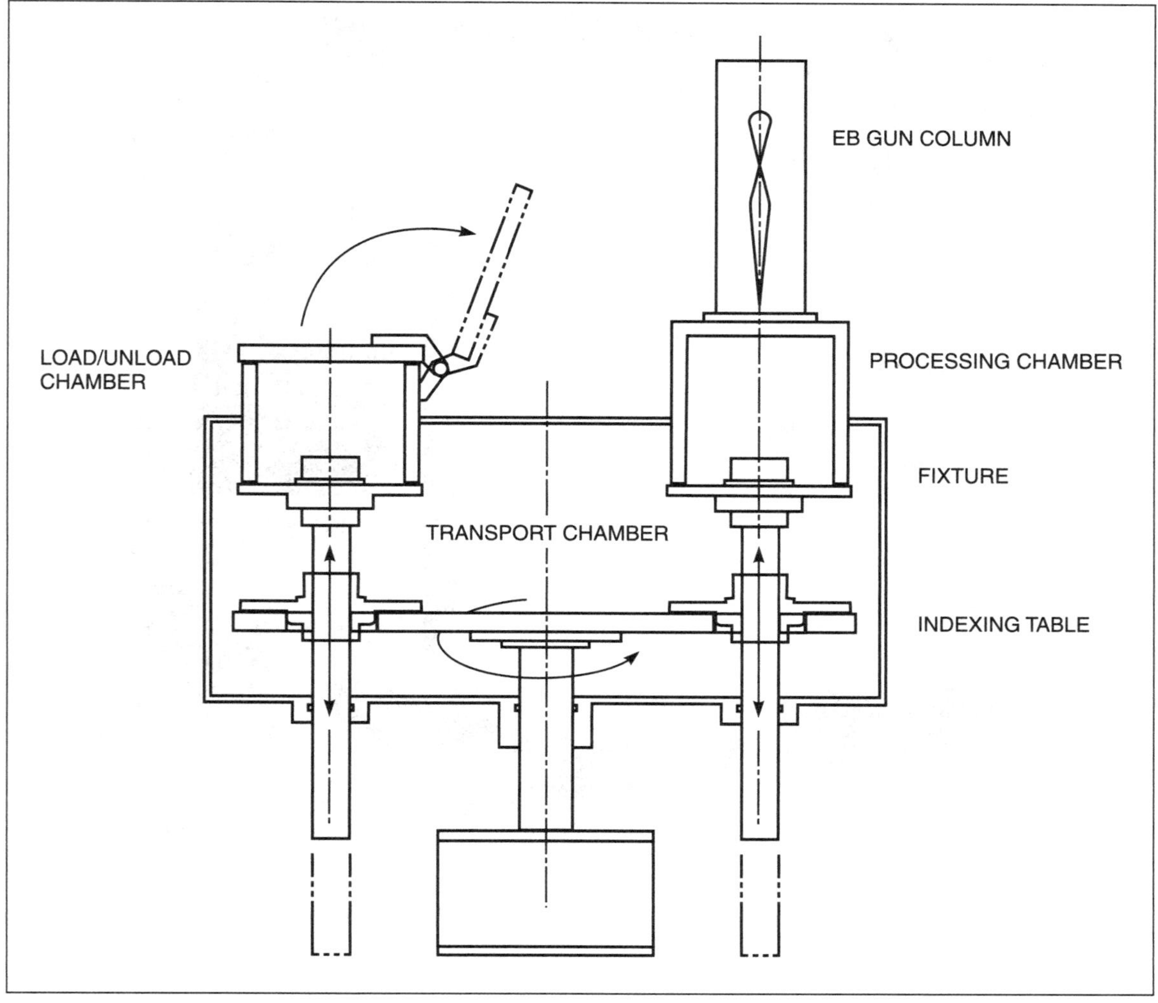

Source: PTR—Praezisionstechnik GmBH

Figure 13.27—Schematic of a Pre-Pumped, Drop-Bottom Indexing System Used in Medium-Vacuum Electron Beam Welding

tance between the final orifice and the workpiece. The beam power level used and the type of gas comprising the atmosphere through which the beam eventually passes can greatly influence the useful working distance.

Figure 13.30 is a diagram of a conventional nonvacuum electron beam gun/column assembly, including an orifice system. The electron gun in Figure 13.30 is typical of those used with the other modes of electron beam welding, and is capable of operating at accelerating voltages in the range of 150 kV to 200 kV. Beam current, and thus the power, is controlled by the voltage on the bias electrode of the gun. The beam is focused by an electromagnetic lens to the minimum diameter of the orifice system, shown at the bottom of Figure 13.30. It emerges from the vacuum environment into air at atmospheric pressure through the lower orifice. Inert gas shielding can be added, if desired. The workpiece is placed near the lower orifice.

During operation, a high vacuum is continuously maintained in the upper gun area by using an oil diffusion pump or turbo molecular pump on this region. Lesser vacuum levels are maintained in the interim pressure stages by mechanical pumps. In most cases, the workpiece is moved horizontally in front of the gun column, but the entire gun column can be moved if desired. As with the high-vacuum and medium-vacuum

Photograph courtesy of PTR—Praezisionstechnik GmBH

Figure 13.28—Finished-Product Version of the EBW-MV System Diagrammed in Figure 13.27

modes, the gun can be placed in either a vertical or a horizontal position.

Another type of nonvacuum electron beam welding gun unit features a gas-filled, high-voltage power source that can be mounted directly on the gun/column assembly. The unit then can be traversed along the weld joint during operation.

As with the sliding-seal and the modern concepts of medium-vacuum equipment, in which the time for evacuation of the work chamber is effectively eliminated, production rates in excess of 500 weldments per hour are readily attainable with the nonvacuum EBW mode. In addition, since the workpiece need not be enclosed in a chamber and a special atmosphere is not required, workpiece size and surface condition requirements are greatly alleviated.

The welding area for all electron beam processes must be shielded to protect personnel from the X-radiation produced during welding. Health hazards from electron beam radiation are discussed at the end of this chapter in the "Safe Practices" section.

WELDING PROCEDURES

Specifying welding procedures for an electron beam welding application must consider the many variables previously discussed, such as specification of equipment, movement of the gun or workpiece, magnitude of current, accelerating voltage, beam focus, beam deflection, travel speed and distance, emitter power and chamber pressure. Regarding chamber pressure, guidance for converting Standard International and U.S. Customary units of pressure is presented in Appendix C, "Cross-Reference Chart for Various Pressure Units."

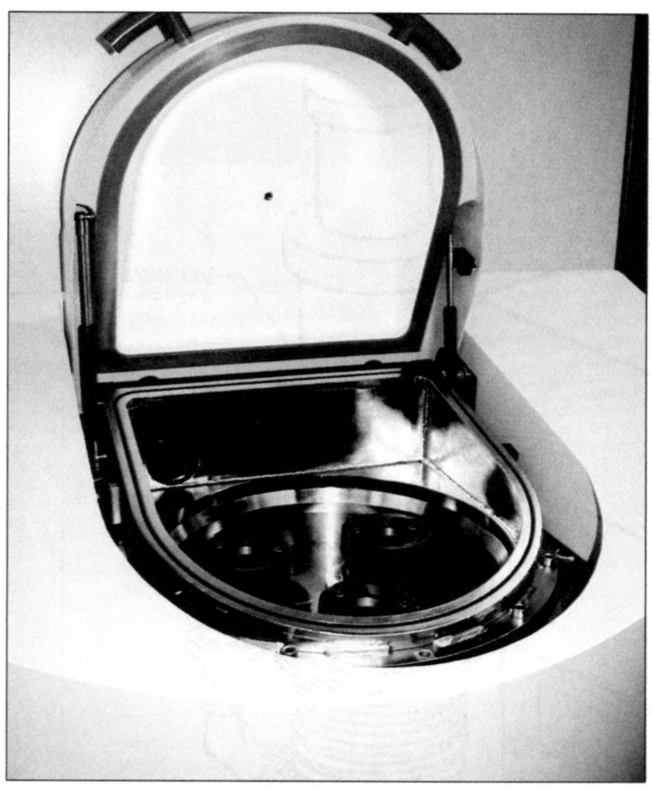

Photograph courtesy of PTR—Praezisionstechnik GmBH

Figure 13.29—View into the Load-Unload Chamber of the EBW-MV System Shown in Figure 13.28

Additional welding conditions that must be considered for electron beam welding include joint design, joint preparation, the fitup and fixturing of the workpieces, and the choice of base metals and filler metals. Recommendations for these variables are presented in this section.

JOINT DESIGN

Butt joints, corner joints, lap joints, edge joints, and T-joints can be made by electron beam welding using square butt joints or seam welds. Fillet welds are difficult to make and generally are not attempted. Typical joint designs for electron beam welding are shown in Figure 13.31. Modifications of these designs frequently are made for particular applications.

Square butt welds require fixturing to maintain fitup and alignment of the joint. They can be self-aligning if a rabbet joint design is used. The weld metal area can be increased using a bevel joint, but fitup and joint alignment for bevel joints are more difficult than for square butt joints. Edge, seam, and lap fillet welds are primarily used to join sheet metal of varying gauges.

Joint Preparation and Fitup

When no welding wire is added, fitup of the workpieces must be more precise than for arc welding processes, because poor fitup would result in incomplete filling of the weld joint. The beam must impinge on both members and melt them simultaneously, except for seam welds in which the beam penetrates through the top sheet. Underfill or incomplete fusion will result from poor fitup, and lap joints that are not clamped sufficiently will melt through.

A metal-to-metal fit of the workpieces is desirable but difficult to obtain. The acceptable root opening for a particular application depends on the mode of EBW employed, the type of base metal, the thickness and configuration of the joint, and the required weld quality. Thus, sheet sections being welded in the vacuum mode may require a fitup with tolerance of less than 0.1 mm (0.004 in.), plate sections being welded in the nonvacuum mode may tolerate a fitup more than five times greater. Aluminum alloys can tolerate somewhat larger root openings than steel. Beam deflection or oscillation is used in high- and medium-vacuum welding to widen the fusion zone, but when used in nonvacuum welding, it may permit larger root openings. Consequently, the maximum acceptable root opening and the tolerance for each particular application should be determined and qualified in order to avoid unnecessary joint preparation costs.

In general, roughness of the faying surfaces is not critical as long as the surfaces can be properly cleaned and all contamination removed. Burrs on the sheared edges of sheet are not detrimental unless they separate the faying surfaces of lap joints.

Cleaning

Cleanliness is a prime requisite for high-quality welding. The cleanliness level required depends on the end use of the welded product. Contamination of the weld metal may cause porosity or cracking, or both, and deterioration of mechanical properties. Improper cleaning of the workpieces may excessively lengthen chamber evacuation time, depending on the vacuum mode being employed.

In the past, acetone and methylethylketone were considered to be excellent solvents for cleaning electron gun components and workpieces. However, these

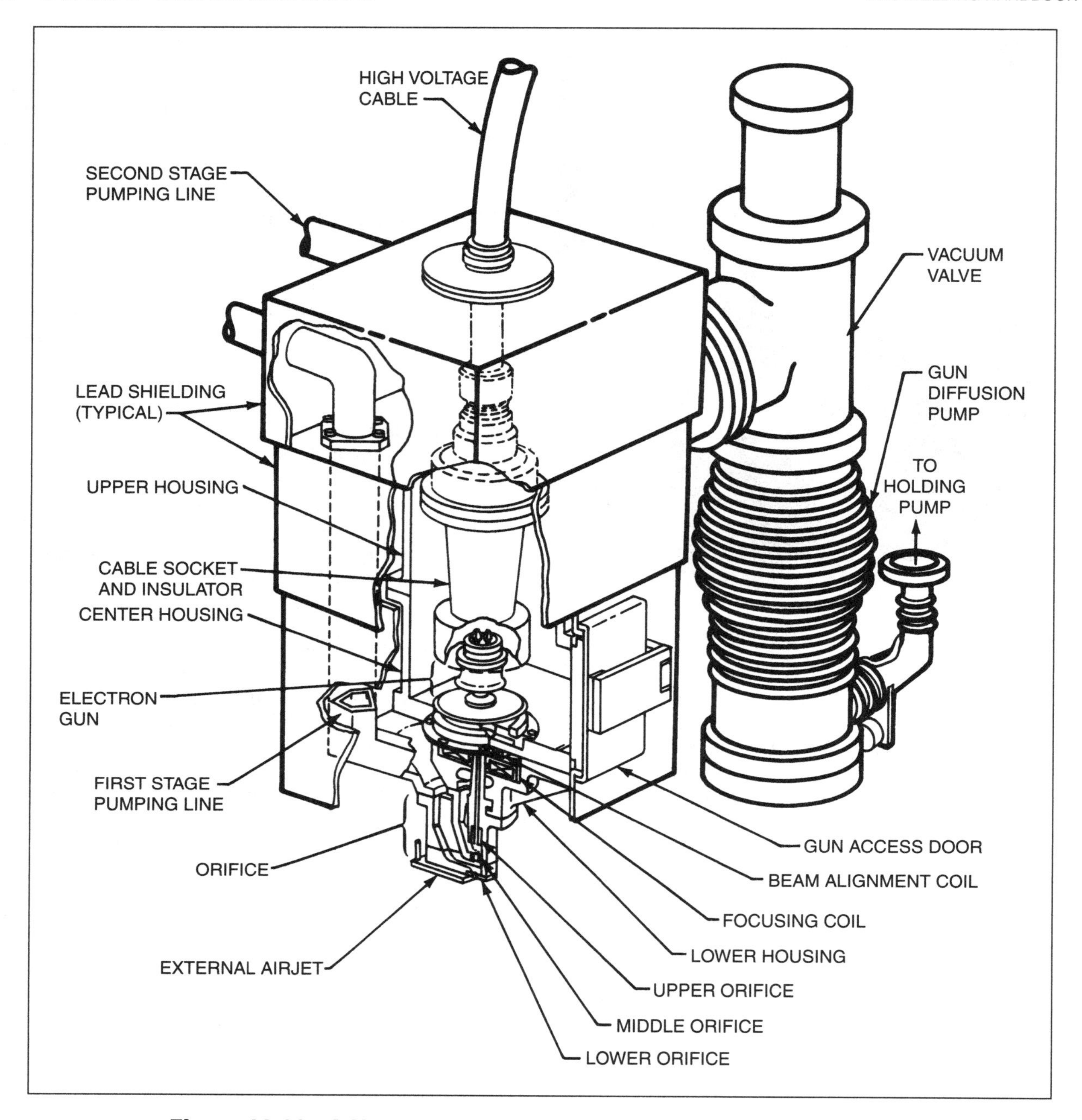

Figure 13.30—A Nonvacuum Electron Beam Gun Column Assembly

chemicals have since been considered to be possible hazardous substances, thus most facilities use pure alcohol instead.

Chlorinated hydrocarbon solvents should definitely not be used because of their detrimental effect on the operation of high-voltage equipment and because of their potential for forming phosgene gas when exposed to ultraviolet light. If a vapor degreaser containing a chlorinated hydrocarbon solvent must be used for heavy degreasing tasks, the components must be thoroughly washed in pure alcohol afterward. An alternative would be to use a fluorocarbon-type solvent for degreasing. After final cleaning, the joint area should not be touched by hands or tools.

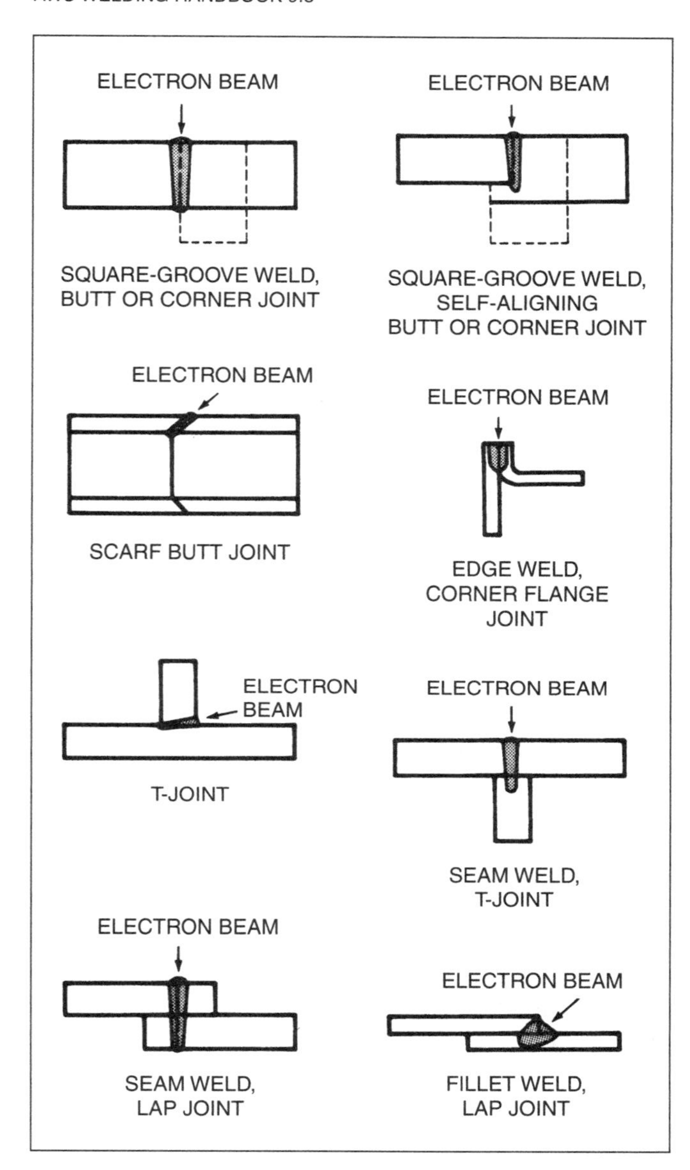

Figure 13.31—Typical Joint Designs for Electron Beam Welds

Surface oxides and other forms of contamination that cannot be dissolved by solvents should be removed by mechanical or chemical means, such as brushing, scraping, machining, or chemical etching. Flat surfaces of soft metals, for example, magnesium, aluminum, or copper can be scraped by hand. Machining without a coolant is preferred for all but very hard metals, where grinding must be used. Surfaces that are not prepared by machining should be chemically cleaned.

Grit blasting and grinding are not recommended for soft metals, including soft steels, because grit may become embedded in the surfaces. Wire brushing generally is not recommended because it also tends to embed contaminants in the metal surface. Nonvacuum welding generally will require less stringent precleaning than vacuum welding.

FIXTURING

Electron beam welding can be accomplished by manually or automatically controlling the functional operation of the system. The workpieces must be fixtured to align the joint, unless the design is self-fixturing; then either the assembly must move or the electron beam gun column must be moved to accomplish the weld.

Self-fixturing joints should be used when practical. A pressed or shrink fit can position circular components for welding. However, these methods require close-tolerance machining, which may not be economical for high-production welding.

Fixturing for electron beam welding need not be as strong and rigid as that required for automatic arc welding. The reason is that electron beam welds generally are made with much lower heat input per unit length of weld than arc welds. Therefore, stresses in the weldment caused by thermal gradients extend over a smaller volume of metal. However, fixturing used for EBW must not introduce magnetic effects that adversely affect the beam. The close fit-up and alignment required for joints in electron beam welds generally call for fixturing made to the same tolerances. Copper chill blocks plated with nickel can be used to remove heat from the joint.

Worktables and rotating positioners should have smooth and accurate motion at the required travel speeds. All fixturing and tooling should be made of nonmagnetic metals to prevent magnetic deflection of the beam. All magnetic metal workpieces should be demagnetized before welding.

The entry and exit of the electron beam tends to produce underfill at both ends of the welded joint. To minimize or eliminate underfill, weld tabs of the same metal as the workpieces should be fitted tightly against both ends of the joint so that the beam can be initiated on the starting tab, traversed along the weld joint, and terminated on the runoff tab. When the weld is completed the tabs can be removed flush with the ends of the workpiece.

FILLER METAL

Filler metal normally is not needed to obtain a weld with complete joint penetration when the faying surfaces of butt joints are fitted together with acceptable tolerances. As welding progresses along the joint, weld metal flows from the leading edge to the trailing edge of the keyhole. As the weld progressively freezes, thermal contraction usually produces a welded joint free of underfill when proper welding procedures are used. Certain joint designs use the thermal contraction of the weldment to produce an autogenous weld from multiple weld passes; these welding procedures use a narrow tapered root opening and a low-power-density beam to produce welds with complete joint penetration. These welds tend to exhibit few of the discontinuities sometimes encountered with single-pass autogenous welds.

For some applications it is desirable or necessary to add filler metal to obtain an acceptable welded joint. Filler metal may be needed to obtain certain physical or metallurgical characteristics in the weld metal. Characteristics of the weld metal that may be altered or improved by the addition of filler metal include ductility, tensile strength, hardness, and crack resistance. For example, preplacing a thin aluminum shim in the joint can produce a deoxidizing action in mild steel, which will reduce porosity in the weld.

When filler metal is added to the joint for metallurgical purposes, filler metal in the form of welding wire is often used, but not employed exclusively. The dilution obtained from a dissimilar filler metal added as wire at the joint surface does not occur uniformly from top to bottom of the weld. For a single-pass weld in heavy plate, filler metal may take the form of a thin shim. The presence of the filler shim requires that beam oscillation or a large-diameter spot be used to melt the shim and the base metal on both sides of the joint. This is not the case with weldments in thin metal, where welding wire can be added at the surface and dilution will occur throughout the entire joint. Typical examples of filler metal additions for metallurgical reasons are the welding of Type 6061 aluminum alloy using Type 4043 aluminum filler metal, and the welding of beryllium using aluminum or silver filler metal.[8]

Filler metal may be added at the surface to fill the joint during a second pass after the penetration pass has been made. This is done to obtain complete joint penetration in thick plate. Welding wire-feeding equipment usually is either a modified version of that used for gas tungsten arc welding or a unit specially designed for use in a vacuum chamber. Welding wire diameters generally are small, 0.8 mm (0.030 in.) and under, because the wire feeder must uniformly feed the wire into the leading edge of a small weld pool. The wire-feeding nozzle should be made of a heat-resistant metal.

When welding in a vacuum chamber, the welding wire drive motor must be sealed in a vacuum-tight enclosure or otherwise designed for use in a vacuum. Outgassing from an open motor will greatly increase the work chamber evacuation time. Provisions must be made for adjusting the wire-feed nozzle so that the welding wire is positioned relative to the electron beam and to the weld joint over the entire length of the joint.

SELECTION OF WELDING VARIABLES

The rate of energy input to the workpiece during EBW commonly is expressed in joules per inch, or joules per second.[9] The formula for this expression is the following:

$$\text{Energy input, J/mm (J/in.)} = EI/S = P/S \qquad (13.2)$$

where:

E = B-Beam accelerating voltage, V;
I = Beam current, A;
P = Beam power, W or J/s; and
S = Travel speed, mm/s (in./s).

Data for welding various thicknesses of a specific material can be plotted to permit interpolation of welding variables for that material over the range of values covered by the data. A curve relating energy input with thickness for a particular group of alloys can be determined from a few tests to establish the welding conditions for untested metal thicknesses. Figure 13.32 provides sample curves for several metals. These figures are particularly useful to determine starting-point conditions. The following factors make this possible:

1. Electron beam welding machine settings usually are regulated by closed-loop servo controls that ensure stability and reproducibility; and
2. The adjustment of each variable is independently controlled to permit flexibility in selection.

Assuming the vacuum level and electron beam gun-to-workpiece distance are held constant, only four basic variables need to be adjusted: accelerating voltage, beam current, travel speed, and beam focus. Beam deflection may constitute a fifth variable, if an oscillating beam motion is employed. These variables combine to make the process of establishing the welding schedule relatively simple.

8. Aluminum classification designators are defined by the Aluminum Association, 900 19th Street N.W., Washington, D.C. 20006.

9. Energy input to the weld from a heat source is discussed in more detail in American Welding Society (AWS) Welding Handbook, 2001, *Welding Science and Technology*, Miami: American Welding Society, Vol. 1, Chapter 2.

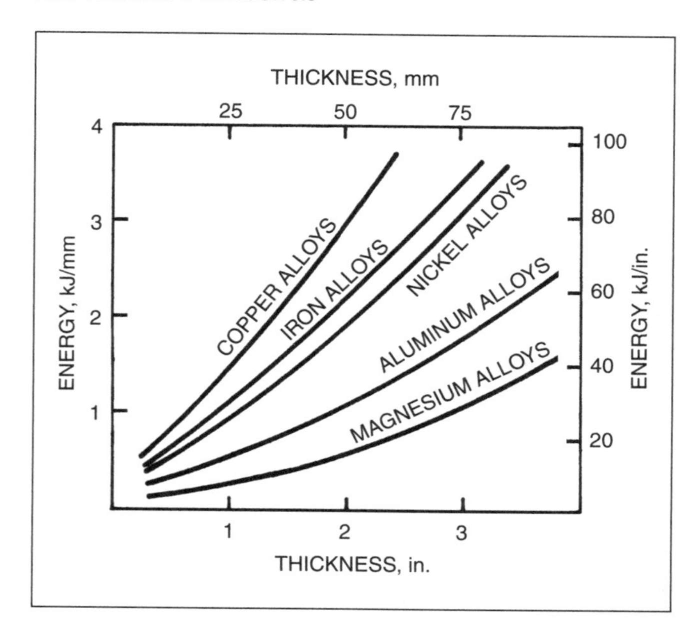

Figure 13.32—High-Vacuum Electron Beam Welding Energy Requirements for Complete Penetration Welds as a Function of Joint Thickness

Once the required energy input per unit length is determined for a given metal thickness, the travel speed can be selected and the required welding power defined, or the order can be reversed. The beam voltage and current can then be selected to produce the required power.

The beam size selected will depend on the desired geometry of the weld bead. To maintain a selected beam spot diameter at the surface of the workpiece, it is necessary to correspondingly increase the focus coil current as the accelerating voltage is increased, since the beam spot size is a dependent function of the accelerating voltage. Many electron beam welding machines automatically perform this compensation task. If the accelerating voltage is maintained constant but the gun-to-workpiece distance is increased, a corresponding decrease in focus coil current is necessary to maintain a selected beam spot diameter at the surface of the workpiece.

Changes in individual welding variables—accelerating voltage, beam current, travel speed, and beam spot size—will affect the joint penetration and bead geometry in the following manner:

1. *Accelerating voltage*: as the accelerating voltage is increased, the depth of penetration achievable also increases. For long gun-to-workpiece distances or the production of narrow, parallel-sided welds, the accelerating voltage should be increased and the beam current decreased to obtain maximum focal range. (Refer to Figure 13.1);
2. *Beam current*: for any given accelerating voltage, the joint penetration achievable will increase with increased beam current;
3. *Travel speed*: for any given beam power level, the weld bead will become narrower and joint penetration will decrease as the travel speed is increased; and
4. *Beam spot size*: sharp focus of the beam will produce narrow, parallel-sided weld geometries because the effective beam power density will be at its maximum. Defocusing the beam, by overfocusing or by underfocusing, will increase the effective beam diameter and reduce beam power density; consequently, this will tend to produce a shallow or V-shaped weld bead. These effects are shown in Figure 13.33.

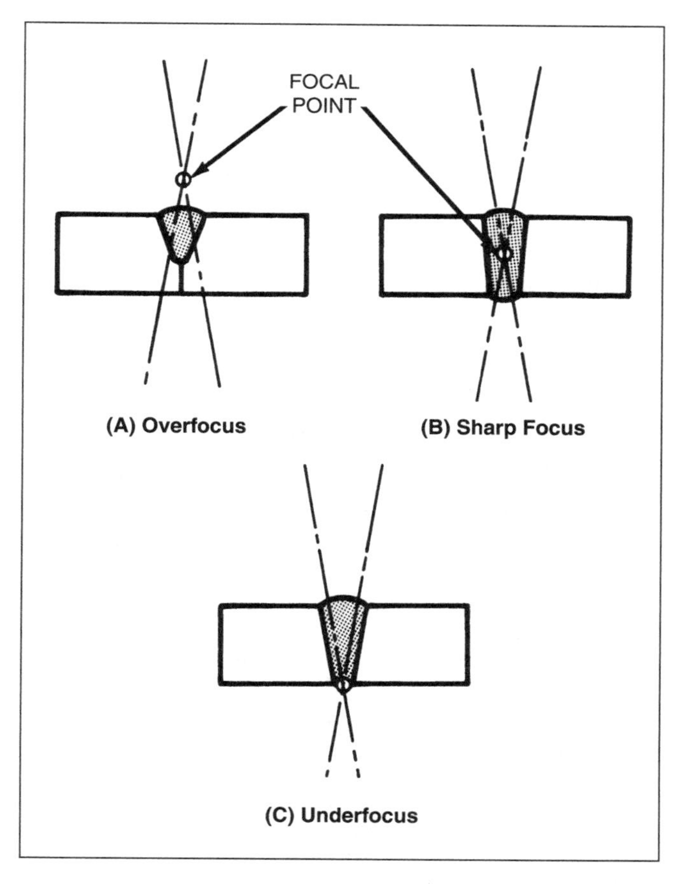

Figure 13.33—Effect of Electron Beam Focusing on Weld Bead Geometry and Joint Penetration

Underfocusing is often used for heavy-section welding in order to produce the highest possible effective aspect ratio. However, it should be carefully verified that the depressed beam focal spot does not produce a weldment with a large "nail head" or a bottle shape, since both conditions may lead to weld cracking.

METALS WELDED

In general, metals and alloys that can be welded by other fusion welding processes also can be joined by electron beam welding.[10] This includes similar and dissimilar metal combinations that are metallurgically compatible. The narrow weld metal geometry and thin heat-affected zones produce joints with better mechanical properties and fewer discontinuities than arc-welded joints. However, electron beam welds in alloys that are subject to hot cracking or porosity often will contain such discontinuities. The weldability of a particular alloy or combination of alloys depends on the metallurgical characteristics of the specific alloy or combination, and also depends on workpiece configuration, joint design, process variation, and the selected welding procedure.

Steels

Among the types of steels welded with electron beam welding are rimmed and killed steels, alloy steels, austenitic stainless steels, martensitic steels, and precipitation hardening steels.

Rimmed and Killed Steels. In ingot-cut rimmed steel, the chemical reaction that occurs between carbon and oxygen to form carbon monoxide gas (CO) will occur in the weld pool. As a result, violent weld pool action, spatter, and porosity in the solidified weld metal are expected with this type of steel.

Electron beam welds in rimmed steel can be improved if deoxidizers, such as manganese, silicon, or aluminum, are incorporated through filler metal additions. Deoxidizers also can be added locally to the joint area by painting, spraying, or inserting shims.

Running a low-power, shallow-penetration weld pass before doing a required high-power, deep-penetration pass often can reduce the violence of the weld pool action. The welding of rimmed steel can be improved by the careful selection of electron beam welding conditions, such as slow travel speed, to produce a wide and shallow weld cross section. The slow travel speed allows gases the time needed to escape from the molten weld metal. Then a weld of reasonable quality can be obtained. Various beam deflection patterns may help expedite gas escape, and thus may be effective in reducing weld porosity. Continuous-cast mild steels are silicon-aluminum killed, and therefore porosity is not a problem.

Alloy Steels. Without preheat, thick sections of alloy steels may crack when welded with the electron beam process. Very rapid cooling in the fusion and heat-affected zones will result in the formation of brittle martensite. The combination of a hard, brittle microstructure and residual stresses may create cracks. Cracking can be prevented by preheating. Preheat can be supplied with a defocused electron beam in many applications, relying on careful programming and monitoring to achieve the proper preheat temperature.

Austenitic Stainless Steels. The high cooling rates typical of electron beam welds help to inhibit carbide precipitation in stainless steels because of the short time that the weld zone is in the sensitizing temperature range. However, the high cooling rate may cause cracking in highly restrained, low-ferrite grades of material. Small quantities of sulfur, phosphorus, and boron in combination with a crack-susceptible microstructure usually are the cause of solidification cracking in these grades of stainless steel, because these elements have a tendency to form low-melting-point compounds with iron.

Martensitic Stainless Steels. Although martensitic stainless steels can be welded in almost any heat-treated condition, a hard martensitic heat-affected zone will result. Hardness and susceptibility to cracking increase in proportion to increasing carbon content and cooling rate. Rapid cooling can be prevented by preheating the base materials before welding.

Ferritic and Duplex Stainless Steels. While ferritic and duplex stainless steels are not commonly welded with the electron beam process, they may be welded with the careful use of postweld heat treatments to reduce the high susceptibility to cold-cracking. The heat treatment must be accomplished before the workpiece cools to ambient conditions, which makes processing difficult due to the need for complicated fixtures.

Precipitation-Hardening Stainless Steels. Precipitation-hardening stainless steels generally can be welded with the electron beam process and can produce good mechanical properties in the joint. The semi-austenitic types, such as 17-7PH and PH14-8Mo can be welded as readily as the 18-8 types of austenitic stainless steel. The weld metal becomes austenitic during

10. The weldability of various metals is discussed in American Welding Society (AWS) Welding Handbook, 1996, *Materials and Applications—Part 1*, Miami: American Welding Society, Vol. 3, 8th ed., and American Welding Society (AWS) Welding Handbook, 1998, *Materials and Applications—Part 2*, Miami: American Welding Society, Vol. 4, 8th ed.

welding and remains austenitic during cooling. In the martensitic types, such as 17-4PH and 15-5PH, the low carbon content precludes the formation of hard martensite in the weld metal and heat-affected zone. However, not all combinations of PH alloys can be welded without some cracking. Some precipitation-hardening stainless steels, such as 17-10P, have poor weldability because of the high phosphorus content of these steels.

Aluminum Alloys

In general, aluminum alloys that can be readily welded by gas tungsten arc and gas metal arc welding can be joined by electron beam welding. Two problems that may be encountered in some alloys are hot cracking and porosity.

The nonheat-treatable series of aluminum alloys (1xxx, 3xxx, and 5xxx) can be welded with the electron beam process without difficulty. When welded, these joints will possess mechanical properties similar to annealed base metal.

The heat-treatable alloys (2xxx, 6xxx, and 7xxx) are crack-sensitive to varying degrees when welded with EBW. Some may also be prone to weld porosity. Aluminum alloy AA6061-T6 and AA6066-T6, which are difficult alloys to join by other processes, can be successfully welded by the electron beam process. Best results with these alloys are obtained by incorporating small amounts of AA4043 aluminum filler metal or AA4047 aluminum brazing foil in the weld, although 5XXX series filler alloys also may be used.

As-welded joints in 38 mm (1.5 in.) thick AA7075-T651 aluminum alloy exhibit degraded mechanical properties compared to unwelded plate. The lowered weld properties are caused by overaging in the heat-affected zone. Postweld solution-treating and aging will produce heat-treated properties in the welded joint. At high travel speeds, weld porosity may result from the vaporization of certain elements in this alloy, the loss of which may change the properties of the weld metal. This effect should be taken into consideration before welding AA7075. The high zinc content of AA7075 aluminum alloy is responsible for vapor formation. At low travel speed, the vapor escapes to the surface before the weld metal solidifies.

Zinc-free aluminum alloys can be welded at higher speeds without developing severe porosity in the weld. It is advantageous to weld thermally hardened aluminum alloys at high travel speed to minimize the width of the softer weld and heat-affected zones.

Nickel-Base Alloys

Many nickel-base alloys commonly are welded with the electron beam welding process. Pure nickel may be welded, but the quality of welds usually will require post-weld material removal. Pure nickel also is susceptible to cracking, unless sulfur and other materials that react to form low-melting compounds are controlled. Nickel-copper alloys are similar to pure alloys, except that cracking susceptibility is reduced, while porosity susceptibility is increased due to the addition of copper. High-temperature aerospace alloys commonly are welded with the electron beam process, but precautions must be taken with the material condition when welding, as the large grain sizes associated with many of these alloys substantially reduce weldability.

Titanium and Zirconium

Titanium and zirconium absorb oxygen and nitrogen rapidly at welding temperatures, which reduces ductility in these metals. Acceptable levels of oxygen and nitrogen are quite low. Therefore, titanium, zirconium and alloys of these metals must be welded in an inert environment. High-vacuum electron beam welding is best for both metals, but medium-vacuum and nonvacuum welding with inert gas shielding may be acceptable for some titanium applications. Most zirconium applications require that welding be performed in a vacuum or an inert gas environment to preserve the corrosion resistance of the metal.

Refractory Metals

Rhenium, tantalum, niobium, vanadium, and alloys of these metals are readily weldable with electron beam welding. Tungsten, molybdenum and alloys of these, while weldable, are more difficult. Electron beam welding is an excellent process for joining the refractory metals, because the high power density allows the joint to be welded with minimum heat input. This is especially important with molybdenum and tungsten, because fusion and recrystallization raise the ductile-to-brittle transition temperatures above room temperature. The short time at temperature associated with electron beam welding minimizes grain growth and other reactions that raise transition temperatures.

Electron beam welding may be used successfully to join molybdenum and tungsten, provided the joints are not restrained during welding. Thin sections are easy to handle, and if design allows, it may be better to fabricate a composite structure by welding several thin sections rather than welding a single thick section. Alloys of metals containing rhenium are better suited for welding than pure tungsten or molybdenum because they remain more ductile at lower temperatures. Small additions of titanium or zirconium also improve the weldability of molybdenum. Freedom from impurities such as oxygen, nitrogen, and carbon is important because of the significant drop in ductility when impurities are present in any of the refractory materials.

Dissimilar Metals

Whether two dissimilar metals or alloys can be successfully welded depends on the physical properties of the metals, such as melting point, thermal conductivity, atomic size, and thermal expansion. Weldability usually is predicted by empirical experience in this area. A generalization about weldability can be made by examining the alloy phase diagram of the metals to be joined. If intermetallic compounds are formed by the metals to be joined, the weld will be brittle.

Information on the relative weldability of some dissimilar metals may be found in the welding literature. However, the available information about each particular application must be reviewed with regard to joint restraint and service environment. The problem of metallurgical incompatibility can sometimes be solved by the use of a filler metal shim or by welding each of the materials to a compatible transition piece. Examples are given in Table 13.2.

Table 13.3 presents a summary of the weldability of various metal combinations derived from phase diagram information and accumulated practical experience.

Often the electrical couple formed at the weld interface when two dissimilar materials are being welded

Table 13.2
Examples of Filler Metal Shims for Electron Beam Welding

Metal A	Metal B	Filler Shim
Tough pitch copper	Tough pitch copper	Nickel
Tough pitch copper	Mild steel	Nickel
Hastelloy X™	SAE 8620 steel	321 stainless steel
304 stainless steel	Monel™	Hastelloy B™
Inconel 713™	Inconel 713™	Udimet 500™
Rimmed steel	Rimmed steel	Aluminum

Table 13.3
Weldability of Dissimilar Metal Combinations

	Ag	Al	Au	Be	Co	Cu	Fe	Mg	Mo	Nb	Ni	Pt	Rh	Sn	Ta	Ti	W
Al	2																
Au	1	5															
Be	5	2	5														
Co	3	5	2	5													
Cu	2	2	1	5	2												
Fe	3	5	2	5	2	2											
Mg	5	2	5	5	5	5	3										
Mo	3	5	2	5	5	3	2	3									
Nb	4	5	4	5	5	2	5	4	1								
Ni	2	5	1	5	1	1	2	5	5	5							
Pt	1	5	1	5	1	1	1	5	2	5	1						
Rh	3	4	4	5	1	3	5	4	5	5	3	2					
Sn	2	2	5	3	5	2	5	5	3	5	5	5	3				
Ta	5	5	4	5	5	3	5	4	1	1	5	5	5	5			
Ti	2	5	5	5	5	5	5	3	1	1	5	5	5	5	1		
W	3	5	4	5	5	3	5	3	1	1	5	1	5	3	1	2	
Zr	5	5	5	5	5	5	5	3	5	1	5	5	5	5	2	1	5

Notes:
1 = Very desirable; solid solubility in all combinations.
2 = Probably acceptable; complex structures may exist.
3 = Use with caution; insufficient data for proper evaluation.
4 = Use with extreme caution; no data available.
5 = Undesirable combinations; intermediate compounds formed.

can induce an electromagnetic force (EMF) at elevated temperatures. If large circulating currents and magnetic fields are produced, they may cause the electron beam to be deflected from the joint centerline in medium-to-heavy-section weldments. This undesirable effect may be corrected by broadening the beam spot, by providing a slight bias to the angle of impingement of the beam, or by both techniques.

APPLICATIONS

Electron beam welding primarily is used for two distinctly different types of applications: high-precision and high-production.

High-precision applications require that the welding be accomplished in a high-purity (high-vacuum) environment to avoid contamination by oxygen or nitrogen, or both, and with minimum heat effects and maximum reproducibility. These applications mainly are in the nuclear, aircraft, aerospace, and electronic industries. Typical products include nuclear fuel elements, special-alloy jet engine components, pressure vessels for rocket propulsion systems, and hermetically sealed vacuum devices.

If a high-purity environment is not required, high-production applications take advantage of the low heat input, high speed, and the high reproducibility and reliability of electron beam welding. These relaxed conditions permit the welding of components in the semi-finished or finished condition using either medium-vacuum or nonvacuum equipment. Typical examples are gears, frames, steering columns, thin-walled tubing, and transmission and drive-train components for automobiles.

Another mode of high-production EBW is the "air-to-air" means used for welding bimetal strip used in band-saw and hacksaw blades, relays and other bimetal strip applications. In this mode, product is fed in a continuous manner from atmosphere through a vacuum region and back to atmosphere again, allowing welding to be accomplished under both high-vacuum and ambient conditions.

The major application of nonvacuum electron beam welding is in the high-volume production of industrial components, in which size, composition, or required hourly production rate precludes effective welding in a vacuum. This is exemplified in the automotive industry, where nonvacuum welding is employed in numerous applications. A drive ring welded to a torque converter turbine bowl assembly is shown in Figure 13.34(A). A profile of the nonvacuum electron beam weld produced in this application is shown in Figure 13.34(B).

(A)

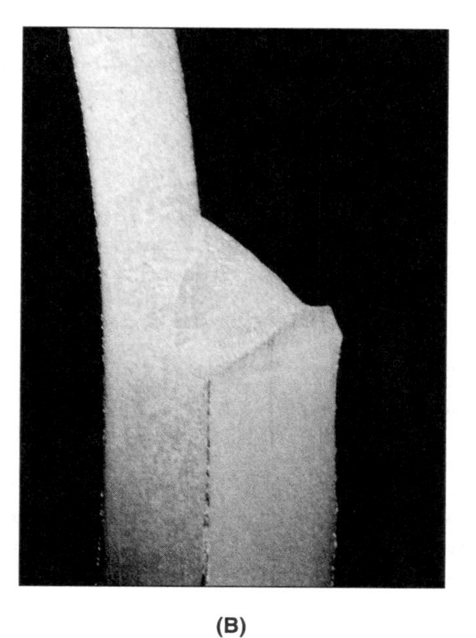

(B)

Photographs courtesy of DaimlerChrysler Transmission Division

Figure 13.34—(A) Drive Ring Welded to a Torque Converter Turbine Bowl Assembly and (B) Micrographic Profile of the Weld

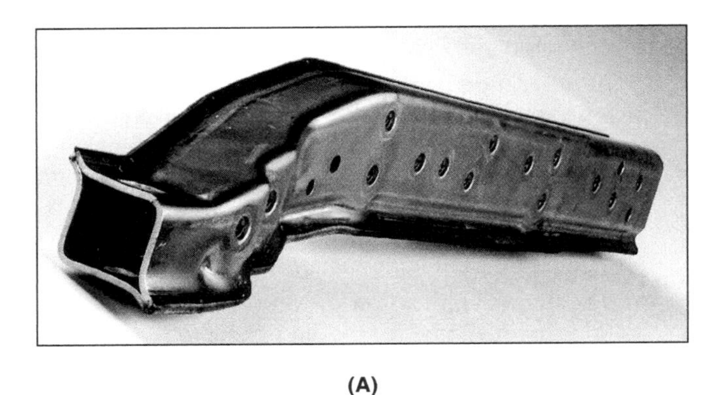

(A)

(B)

Photographs courtesy of Audi/VW Group

Figure 13.35—Aluminum Structural Member of an Automobile Dashboard (A) Welded with Nonvacuum Electron Beam Welding and (B) a Photomicrograph of the Weld

Another example of nonvacuum electron beam welding is shown in Figure 13.35. The application is a reduced-weight crossbar piece of an automobile instrument panel showing the edge weld in two formed half-shells manufactured from an aluminum alloy, shown in Figure 13.35(A), and a close-up view of the weld, shown (B). This weldment provides a component of equal structural strength to the steel item it replaces. The advantages are that it weighs 40% less than the steel item and was welded at an average speed of 12 m/min (450 in./min).

WELD QUALITY

To meet the quality specifications standardized by the welding industry, it is necessary to control the following three factors primarily responsible for the quality of electron beam welds:

1. Joint preparation, including cleaning;
2. Welding procedure, including provisions for keeping the beam on the joint; and
3. The characteristics of the metals being welded.

Joint preparation and welding procedure are discussed in previous sections of this chapter. The third factor relates to the physical properties, mechanical properties, and the metallurgical characteristics of the metals being welded. Weld discontinuities of metallurgical origin include cracking and porosity.

METALLURGICAL CHARACTERISTICS

The weld zone constitutes regions of different microstructure within the base metal structure. Unlike a cast ingot, weld metal grains usually grow from partially melted grains at the weld interface. The phenomenon is called epitaxial solidification. The nature of the weld metal structure is controlled by the size and orientation of the base metal grains, the thermal gradients, the weld pool, and the shape of the weld pool. The nature of the stress resulting from fusion welding is also important. Metal immediately adjacent to the moving weld pool is heated first; it expands against the restraining forces of the surrounding cold base metal; then it cools and contracts. During this process, the weld metal is plastically deformed (upset) during the heating cycle and restrained in tension during cooling. Residual tensile and compressive stresses surround the weld zone, often resulting in the warping of the welded assembly.

In consideration of these factors, the following unique characteristics of electron beam welding can be used to control weld joint properties:

1. Base metal recrystallization and grain growth can be minimized;
2. Beam oscillation and travel speed controls the shape of temperature gradients in the weld pool;
3. Low-heat input results in low thermal stress in the base metal and, hence, less distortion; and
4. Residual stresses are symmetrically distributed due to the characteristic two-dimensional symmetry (parallel sides) of the electron beam weld zone.

Unfortunately, it is not always possible to realize the full potential of the process, since the weldability of a metal is ultimately controlled by its own metallurgical factors. For this reason, electron beam welds may exhibit most of the common discontinuities associated with other fusion welding processes. A possible exception is hydrogen-induced cold cracking of carbon steel weldments, because normally there is no source of hydrogen in an autogenous joint welded with the high-vacuum electron beam process unless the materials have not been properly cleaned.

One type of discontinuity sometimes found in welds with incomplete joint penetration is the presence of large voids at the bottom of the weld metal. When examined by microscope, a large number of these voids typically will be aligned and will appear as linear porosity rather than scattered porosity. When the weld barely penetrates through the joint, root porosity will appear as incomplete filling, accompanied by spatter on the back side of the weld.

Another occurrence peculiar to the vacuum mode of welding is the release of trapped gas through the molten weld metal. This sometimes creates a discontinuity and results in rejection of the weld. It happens during an attempt to weld a gas-filled container that is not properly vented to the chamber vacuum.

Other discontinuities generally are the same as those found in other types of fusion welds, including the following:

1. Porosity,
2. Shrinkage voids,
3. Cracking,
4. Undercutting,
5. Underfill,
6. Missed joints, and
7. Incomplete fusion.

The probability of encountering these discontinuities is more pronounced when welding thick sections. Knowledge of the causes of the discontinuities and means for avoiding them is essential for the production of high-quality welds. As an example, when welding thick sections in the horizontal position, holes and porosity can be avoided by tilting the beam axis a few degrees out of the plane of welding. Equally important is a reliable nondestructive testing method, such as ultrasonic inspection, to determine the presence of certain types of discontinuities that are not detectable by radiography.

The narrow weld beads created by electron beam welding make radiographic inspection difficult. Certain joint designs incorporate a feature called a *radiographic window.* As shown in Figure 13.36, this provides a void within the joint that is easily resolved by the radiographic technique, when the void is not completely consumed by the weld. This window can be located at any position in the joint, and its absence in the radiograph after welding assures that the correct depth of penetration has been achieved.

Porosity and Spatter

Porosity in electron beam welds is caused by the evolution of gas as the metal is melted by the beam. The gas may form as a result of one or more of the following actions:

1. The volatilization of high vapor pressure elements in the alloy,
2. The escape of dissolved gases, or
3. The decomposition of compounds such as oxides and nitrides.

Copper-zinc alloys (brasses) and aluminum-magnesium alloys are difficult to weld with the electron beam

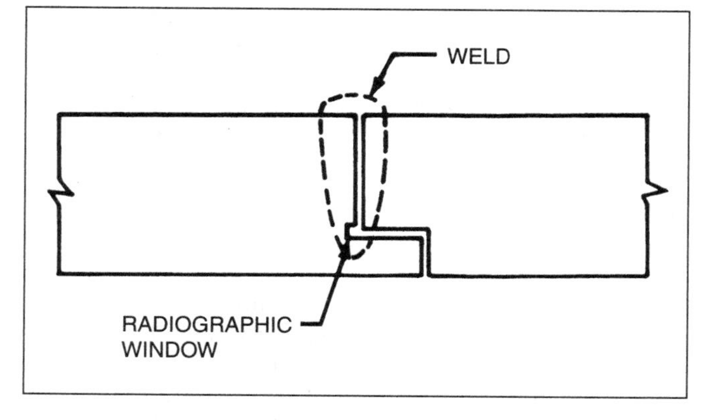

Figure 13.36—Radiographic Window Feature Often Used to Simplify the Reading of Weld Joint Radiographs

process because metal vapors evolve. Both zinc and magnesium have low boiling points. Dissolved gases and compounds are likely to be present in alloys originally melted in air or under protective gas atmospheres.

Spatter is caused by the same factors as porosity. The rapid evolution of gas or metal vapor causes the ejection of drops of molten weld metal that scatter over the workpiece surface and within the chamber. Spatter and porosity can even occur in vacuum-remelted alloys when a residual phase volatilizes under the intense heat of the electron beam.

An effective means of preventing porosity and spatter is to weld only vacuum-melted or fully deoxidized metals. When gas-emitting metals or alloys with high vapor pressure must be welded, special techniques are required to minimize porosity. Filler metal containing a deoxidizer may be added when welding metals are not completely deoxidized. Lowering the welding speed will provide time for gas bubbles to escape from the molten metal.

An oscillatory beam deflection may be effective in reducing porosity. In extreme cases, remelting the joint a second or third time will reduce it. However, these techniques reduce joint strength in age-hardening alloys that are heat treated prior to welding.

Shrinkage Voids

Shrinkage voids may occur between dendrites near the center of the weld metal. These voids are characterized by irregular outlines of porosity. Shrinkage voids usually occur in alloys having high volumetric shrinkage on solidification. In electron beam welds where the bond lines are essentially parallel, solidification proceeds uniformly from the base metal to the center of the weld. When solidification shrinkage of the metal is great, voids will form if the face and root surfaces freeze before the center of the weld freezes. An example of shrinkage voids in an electron beam weld in 15-7Mo PH stainless steel is shown in Figure 13.37. Low travel speed or beam oscillation may minimize or eliminate shrinkage voids by increasing the volume of molten metal and decreasing the solidification rate. However, these conditions generally will widen the fusion zone.

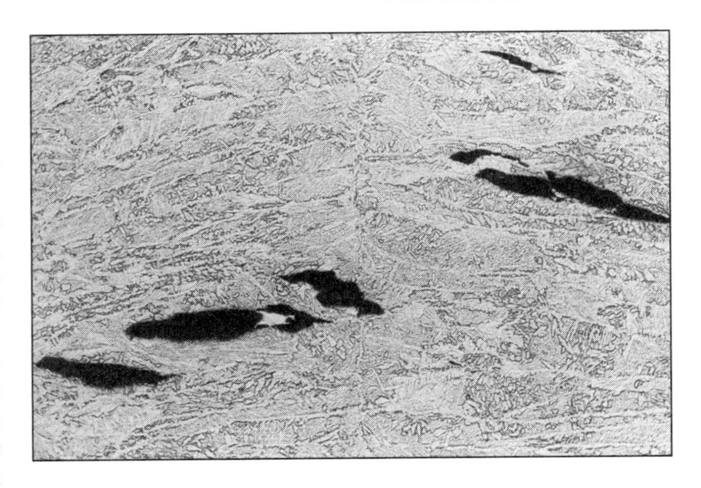

Figure 13.37—Shrinkage Voids in an Electron Beam Weld

Cracking

Solidification (hot) or cold cracks may form in electron beam welds in alloys that are subject to these types of cracking. Solidification cracking generally is intergranular and cold cracking is transgranular. Solidification cracks form in low-melting grain boundary phases during solidification of the weld metal. Cold cracks form after solidification as a result of high internal stresses produced by thermal contraction of the metal during cooling. A crack originates at some imperfection or point of stress concentration in the metal and propagates through the grains by cleavage.

Solidification cracking may be minimized by welding at high travel speeds with minimum beam energy or by controlling the composition of the materials to be welded. Cold cracking may be overcome by redesigning the joint to eliminate points of stress concentration. Quench-hardenable steels should be preheated to a suitable temperature to control the formation of martensite in the weld zone.

Undercutting

Electron beam welds with good bead geometry have essentially parallel bond lines with a uniform crown or a buildup of weld metal on the top surface, as shown in Figure 13.38(A). Undercutting refers to grooves produced in the base metal at the edges of the weld bead, as shown in Figure 13.38(B). Undercut occurs when the weld metal does not wet the base metal. Undercutting is promoted by very high travel speeds, characteristics of the workpiece material, improper cleaning procedures, or beam asymmetry (undercutting usually occurs on one side only). Alloy additions that reduce surface tension or increase fluidity, such as lili additions to carbon steel welds, have a beneficial effect.

Undercutting on the top surface of the weld can sometimes be filled by making a cosmetic pass. Usually this is performed at lower power levels relative to the penetration pass, and can be made more effective by beam deflection or a defocused beam used to widen the top of the bead. Certain joint designs provide extra metal above the desired finished surface of the weld-

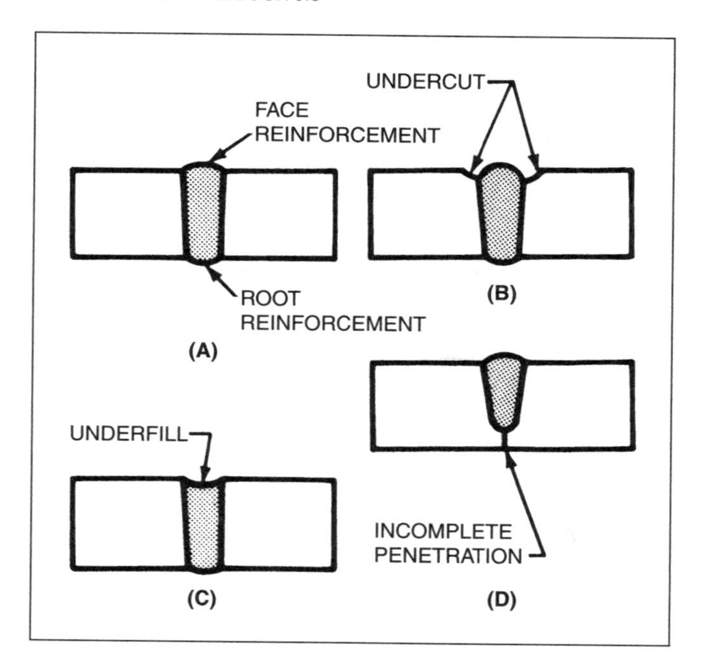

Figure 13.38—Electron Beam Weld Geometries: (A) Correct and (B), (C), and (D) Incorrect

ment that must be machined off after welding. Undercut is removed during the machining operation.

Underfill

Welds with complete joint penetration can develop either a uniform or irregular root surface, depending on the welding parameters and the base metal. The width of the root surface is dependent on welding conditions. In thick sections of metal, such as 76-mm (3-in.) stainless steel, the face and root surface shapes are dependent on the surface tension supporting the column of molten metal as it is being transported along the weld joint. At low welding speeds there will be a relatively large mass of molten weld metal, and the bead will tend to sag due to insufficient surface tension and the force of gravity. This will form extremely heavy root reinforcement and the weld face may show severe underfill (concavity), as shown in Figure 13.38(C). Various techniques can be used to eliminate this condition. These include the use of a backing strip, a step joint, or welding in the horizontal or the vertical position.

Excessive sagging of the root surface usually results when the beam energy is too high or the weld pool is too wide. This can be reduced by proper adjustment of the welding parameters. If underfilling persists at the best operating conditions of the beam, filler metal must be added to fill the groove. A number of techniques are effective in providing the required filler metal. One is to place a narrow strip of filler metal over the face of the joint and then weld through it. The thickness of the strip must be slightly greater than the depth of any undercut, so that the undercut will be entirely in the strip. Filler metal in the form of welding wire similarly may be added to the leading edge of the weld as it is being made, or during a subsequent smoothing pass made with a defocused beam. On circular welds, beam power ramping (upslope and downslope) may be used to minimize discontinuities in the overlap area.

Incomplete Joint Penetration

Numerous applications of electron beam welding do not require complete penetration of the joint. These applications generally involve seal welds or welds subjected to shearing forces only. In these cases, the sharp notch at the root of the weld is acceptable. However, when a welded joint must support a transverse tensile stress at the root of the weld, complete joint penetration is required. Incomplete penetration may be caused by low beam power, high travel speed, or improper focusing of the beam. This condition is shown in Figure 13.38(D).

Missed Segments of Joints

When a small-diameter electron beam is used to make a long joint in a thick section, the beam axis must be in the same plane as the joint faces and remain aligned with the joint along the entire length of travel. Otherwise, there is a great possibility of missing the joint at some location.

Beam Deflection. Electrostatic or magnetic forces can cause beam deflection even when the beam is properly aligned with the joint, which results in portions of the joint being missed. An electrostatic field can be generated by the accumulation of an electrical charge on an insulated surface, such as the glass in the vacuum chamber windows. The electron beam will be deflected away from or toward the charged surface if the beam passes close to it.

Residual magnetism in a ferromagnetic base metal or in the fixturing can cause unexpected beam deflection. For example, a steel component may be magnetized during grinding if it is held by a magnetic chuck, and the residual magnetism in the component will cause the beam to deflect and miss the joint. This can be avoided by demagnetizing all ferromagnetic components before welding, and by using nonmagnetic materials for fixturing.

Unexpected beam deflection can occur when welding dissimilar metals, especially when one is ferromagnetic. An example of this is shown in Figure 13.39, a weld

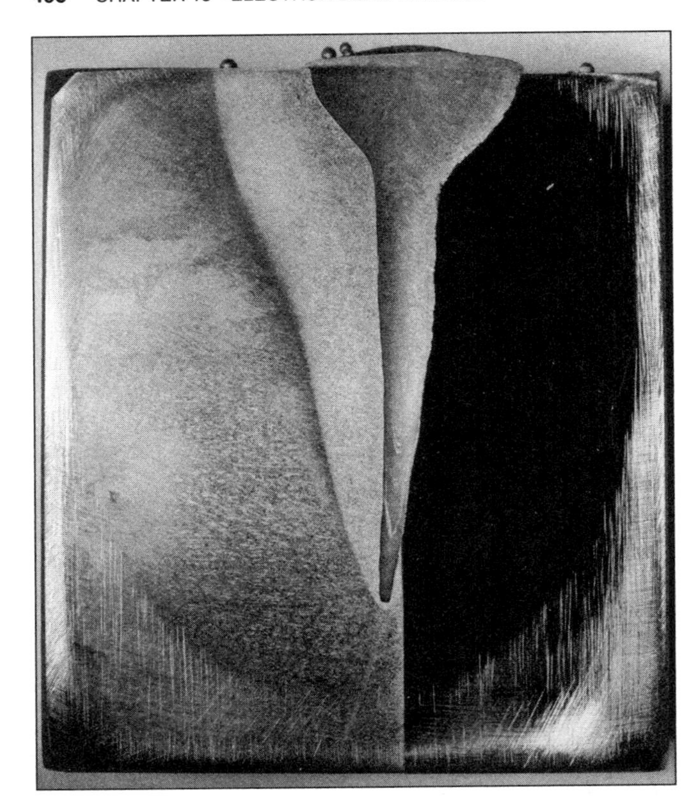

Figure 13.39—Beam Deflection When Welding Dissimilar Metals

between nonmagnetic nickel-base alloy and magnetic maraging steel. In this case, residual or induced magnetism in the steel deflected the electron beam and caused incomplete fusion at the root of the joint. If dissimilar materials are to be welded in production, it is important that test welds be made and examined to determine whether beam deflection will occur, and if so, to plan corrective action.

Witness Lines. The occurrence of missed joints can be verified by using joint designs that include witness lines and radiographic windows. Witness lines are scribed parallel to the joint on the face or root side of the joint, or both. The weld lies between these lines, and its position relative to the joint can be determined by postweld examination, as illustrated in Figure 13.40.

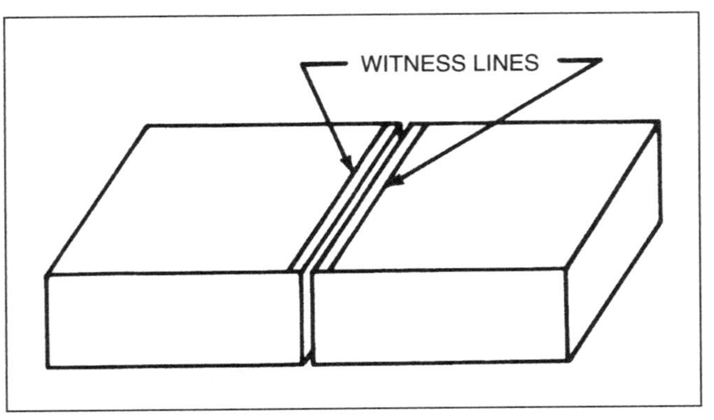

Figure 13.40—Witness Lines Scribed Parallel to Joint

Incomplete Fusion

Incomplete fusion, or spiking, occurs mostly in welds with incomplete penetration. However, it also can occur near the root of welds with complete penetration made with insufficient beam power. Figure 13.41 shows an example of this in an electron beam weld in a titanium alloy.

Incomplete fusion generally can be avoided by using properly adjusted welding parameters. There are circumstances, however, in which a partial joint penetration weld is required. One example is the welding of circular joints in which the beam power and penetration must be decreased as the end of the weld overlaps the starting point to avoid crater formation. A partial-penetration weld is formed in the overlap and incomplete fusion can occur. Another example is the welding of thick sections. Two partial-penetration weld passes, one from each side of the joint, may be required when the metal is too thick to be penetrated in a single pass. The second pass must reach the root of the first pass.

Welding with a slightly defocused beam and low travel speed (to compensate for the lower energy density) is effective in eliminating incomplete fusion. Beam oscillation, either circular or transverse, is sometimes effective. Preheating is helpful because it reduces the thermal gradients at the root of the weld.

Testing Methods for Incomplete Fusion. Incomplete fusion is difficult to locate with nondestructive examination because it is similar to fine cracks and usually cannot be detected by X-ray inspection. Penetrant tests are ineffective because the unfused area usually does not extend to the surface.

Ultrasonic testing is the only nondestructive examination method that can reliably detect incomplete fusion in electron beam welds. Nondestructive examinations require experienced personnel to perform the test and interpret the results. Even then, the test method is not suitable for many applications. Since certain joint designs can be easier to inspect ultrasonically, nondestructive examination personnel should be consulted prior to designing joints for critical assemblies.

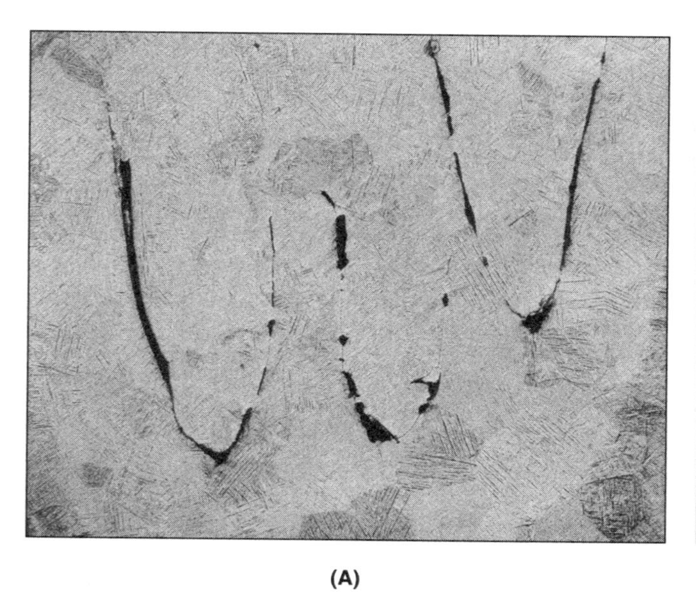

(A)

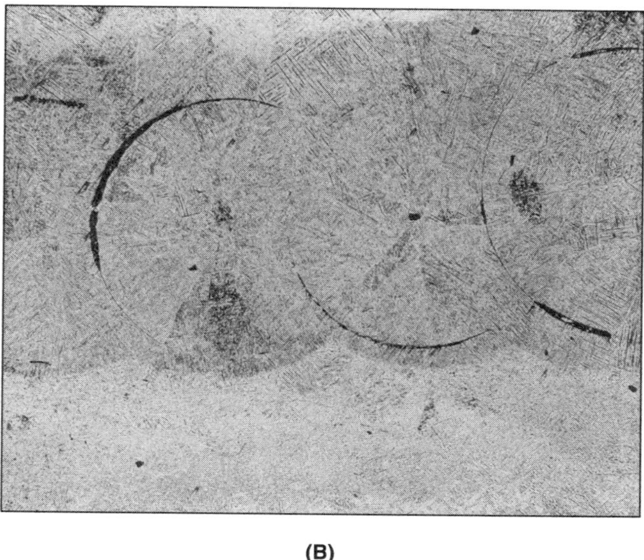

(B)

Figure 13.41—Incomplete Fusion in (A) Vertical and (B) Horizontal Sections through an Electron Beam Weld in a Titanium Alloy

ECONOMICS OF ELECTRON BEAM WELDING

The use of electron beam welding has opened entirely new categories of materials that can be welded. Aside from the benefits of single-pass, deep-penetration welding, electron beams also may be used to join extremely thin sections. While electron beam welding traditionally is associated with several niche markets, it should be considered that many of these markets could not exist without the use of this process.

While the original capital costs for electron beam welding may be high, in many cases the process allows throughput increases that, when amortized, may be extremely competitive with other welding processes. Laser beam welding is the process that traditionally competes with electron beam welding. As previously noted, each has niche markets in which the other has difficulty in competing, such as the welding of materials that are highly reflective, and applications requiring welds with very deep penetration.

From a capital cost standpoint, the laser beam welding and electron beam processes are similar in that costs of operating vacuum systems in EBW tend to be approximately the same as the costs of the large gas-handling systems in LBW.

The first cost consideration often is a decision on the viability of the capital investment required for electron beam welding equipment. Overhead expenses, usually calculated by a firm's accounting division, must be considered. In production, the primary costs associated with operating electron beam welding systems are for consumable items: electricity, vacuum pumps, and filaments. As with most industrial processes, welding operator or labor cost is higher than the cost of consumables. One way to ensure the viability of the capital invested in an electron beam welding machine is to set up a preventive maintenance schedule for the equipment. This would include such things as regular filament and grid cup cleaning and changes, vacuum leak checks and calibration of the gauges.

SAFE PRACTICES

Since electron beam welding machines employ a high-energy beam of electrons, the process requires users to observe several safety precautions normally not necessary with other types of fusion welding equipment. The primary hazards associated with electron beam equipment are electric shock, X-radiation, fumes and gases, and visible radiation.

In addition to the hazards associated with welding specific materials, such as beryllium, magnesium, and radioactive materials, health risks also may be associated with collateral materials, such as solvents and greases used in operating the equipment. Precautionary measures must be taken to ensure that all required safety procedures are strictly observed. Minimum standards of safe practice are published in AWS C7.1M/C7.1:2004,[11] *Recommended Practices for Electron Beam Welding*, and ANSI Z49.1:2005, *Safety in Welding, Cutting, and Allied Processes*.[12]

Other safety standards pertinent to the welding industry and facts of publication are listed in Appendices A and B of this volume.

ELECTRIC SHOCK

Every electron beam welding system operates with a voltage level high enough to cause fatal injury, regardless of whether the system is referred to as being a low-voltage or a high-voltage unit. Manufacturers of electron beam equipment must meet various national and independent testing laboratory requirements to ensure that their machines are well insulated against the hazards of contact with the high voltage. However, all precautions required when working around high voltages should be observed when working with EBW machines.

RADIATION

The X-rays generated by an electron beam welding machine are produced when electrons, traveling at a high velocity, collide with matter. The majority of these X-rays are produced when the electron beam impinges on the workpiece. Substantial levels of radiation also are produced when the beam strikes gas molecules or metal vapor in both the gun column and work chamber. National and independent testing laboratories and federal regulations have established and published firm rules for permissible X-radiation exposure levels, and producers and users of equipment must observe these rules. Examples of these publications are the NSB Handbook 114, *General Safety Standards for Installations using Non-Medical X-Ray and Sealed Gamma Ray Sources, Energies up to 10 Megaelectron Volts (MeV)*,[13] which contains rules governing the safe operation of X-ray machines, and *Occupational Health Standards—Ionizing Radiation*, Title 29, Chapter 17, Part 1910, Section 96.[14]

Assuming proper design, the vacuum enclosure should be manufactured from steel approximately 25-mm (1-in.) thick. This will satisfy the X-ray shielding requirements for beam systems using accelerating voltages of up to 60 kV. For units with higher beam-accelerating voltages, the use of a much thicker steel or a lead covering on top of the steel is needed to satisfy the X-ray shielding requirements for these units. Leaded glass windows are employed in both high- and low-voltage electron beam systems. In general, commercially available shielded vacuum chamber walls and leaded glass windows provide sufficient radiation protection for operators.

In nonvacuum systems, some type of radiation enclosure must be provided to ensure the safety of the operator. Thick walls of high-density concrete (or a similar material) may be selected instead of steel and lead, especially if a large radiation enclosure is required. Special safety precautions must be imposed to prevent personnel from accidentally entering or being trapped inside these enclosures when equipment is in operation.

A complete X-radiation survey of the electron beam equipment must be made at the time of installation and at regular intervals thereafter. These surveys should be conducted by personnel who are knowledgeable about the use of radiation survey equipment and are trained to perform radiation surveys to assure initial and continued compliance with all radiation regulations and standards applicable to the site where the equipment is installed.

FUMES AND GASES

It is unlikely that the very small amount of air left in a high-vacuum electron beam chamber would be sufficient to produce ozone and oxides of nitrogen in harmful concentrations. However, nonvacuum and medium-vacuum electron beam systems are capable of producing these byproducts and other types of airborne contaminants in concentrations well above acceptable levels.

11. American Welding Society (AWS) Committee on High Energy Beam Welding and Cutting, 2004, *Recommended Practices for Electron Beam Welding*, AWS C7.1M/C7.1:2004. Miami: American Welding Society.
12. American National Standards Institute (ANSI) Accredited Standards Committee Z49, 2005, *Safety in Welding, Cutting, and Allied Processes*, Miami: American Welding Society.
13. National Institute of Standards and Technology, NSB Handbook 114, *General Safety Standards for Installations using Non-Medical X-Ray and Sealed Gamma Ray Sources, Energies up to 10 MeV*, Gaithersburg, Maryland: National Institute of Standards and Technology.
14. Occupational Safety and Health Administration (OSHA), *Occupational Health Standards—Ionizing Radiation*, Title 29, Chapter 17, Part 1910, Section 96. Washington D.C.: U.S. Government Printing Office.

Adequate area ventilation around the equipment must be employed to reduce concentrations of any airborne contaminants below the maximum allowable exposure levels. Proper exhausting equipment and techniques should be employed to maintain residual concentrations in the chamber or enclosure below these limits. When hazardous materials are welded, precaution should be exercised to ensure that operators are protected from excessive exposure to fine particulate generated during welding.

Solvents

Chlorinated hydrocarbon solvents have a potential for forming phosgene gas when exposed to ultraviolet light. If a vapor degreaser containing a chlorinated hydrocarbon solvent must be used for heavy degreasing tasks, the components must be thoroughly washed in pure alcohol afterward. An alternative would be to use a fluorocarbon-type solvent for degreasing. After final cleaning, the joint area should not be touched by hands or tools.

The Occupational Safety and Health Administration (OSHA) has established *Permissible Exposure Limits (PEL)*, which can be found in the Occupational Safety and Health Standards, 29 CFR 1910.1000 (Subpart Z) Toxic and Hazardous Substances.[15]

The American Conference of Governmental Industrial Hygienists (ACGIH) has established recommended *Threshold Limit Values (TLV)*, which can be found in *Threshold Limit Values for Chemical Substances and Physical Agents in the Workroom Environment.*[16]

VISIBLE RADIATION

Direct viewing of the visible radiation emitted by the molten weld metal can be harmful to eyesight. In the presence of intense light sources, proper eye protection is necessary. Optical viewing should be done through filters selected in accordance with the latest edition of *Occupational and Educational Eye and Face Protection,* ANSI Z87.1.[17]

15. Occupational Safety and Health Administration (OSHA), *Title 29—Labor, In Code of Federal Regulations (CFR)*, 1910.1000, Subpart Z, *Toxic and Hazardous Substances*, Washington D.C.: Superintendent of Documents, U.S. Government Printing Office.
16. The American Conference of Governmental Industrial Hygienists (ACGIH), *Threshold Limit Values (TLV) for Chemical Substances and Physical Agents in the Workroom Environment,* Cincinnati, Ohio: American Conference of Industrial Hygienists.
17. American National Standards Institute (ANSI), *Occupational and Educational Eye and Face Protection,* ANSI Z87.1, Des Plaines, Illinois: American Society of Safety Engineers.

CONCLUSION

Initially developed as a sophisticated laboratory tool in the 1940s, electron beam welding first became commercially available in the 1950s. Early applications of the process focused on single-pass, deep-penetration welding, but present applications include the welding of extremely thin sections and weldments in which very tight tolerances of the finished product must be maintained. While the linear footage of materials welded with electron beam welding will never approach that of various arc processes, the flexibility of the process and general properties of electron beam weldments will provide a steady use for the process for the foreseeable future. Many materials that are readily weldable with electron beams are difficult or impossible to weld with other fusion processes.

BIBLIOGRAPHY

American Conference of Governmental Industrial Hygienists (ACGIH). *Threshold limit values for chemical substances and physical agents in the workroom environment.* Cincinnati, Ohio: American Conference of Industrial Hygienists.

American National Standards Institute (ANSI) Accredited Standards Committee Z49.1. 2005. ANSI Z49.1: 2005. *Safety in welding, cutting, and allied processes.* Miami: American Welding Society.

American National Standards Institute (ANSI), Accredited Standards Committee Z87. *Occupational and educational eye and face protection,* ANSI Z87.1, Des Plaines, Illinois: American Society of Safety Engineers.

American Society for Testing and Materials (ASTM). 2004. *Test methods for leaks using the mass spectrometer leak detector or residual gas analyzer in the tracer probe mode,* E0498 95R00. West Conshohocken, Pennsylvania: American Society for Testing and Materials.

American Welding Society (AWS) Committee on Definitions and Symbols. 2001. *Standard welding terms and definitions,* AWS A3.0:2001. Miami: American Welding Society.

American Welding Society (AWS) Committee on High Energy Beam Welding and Cutting. 2004. AWS C7.1M/C7.1:2004. *Recommended practices for electron beam welding.* Miami: American Welding Society.

American Welding Society (AWS) *Welding Handbook.* 2001. *Welding science and technology.* Miami: American Welding Society. Vol. 1, Chapter 2.

American Welding Society (AWS) *Welding Handbook*. 1998. *Materials and applications—part 2*. Miami: American Welding Society. Vol. 4.

National Institute of Standards and Technology. NSB Handbook 114. *General safety standards for installations using non-medical X-ray and sealed gamma ray sources, energies up to 10 MeV*. Gaithersburg, Maryland: National Institute of Standards and Technology.

Occupational Safety and Health Administration (OSHA). *Occupational health standards—ionizing radiation*, Title 29, Chapter 17, Part 1910, Section 96. Washington D.C.: U.S. Government Printing Office.

SUPPLEMENTARY READING LIST

Baujat, V. and C. Charles. 1990. Submarine hull construction using narrow groove GMAW. *Welding Journal* 69(8): 31–35.

Caroll, M. J., and D. E. Powers. 1985. Automatic joint tracking for CNC-programmed electron beam. *Welding Journal* 64(8): 34–38.

Dixon, R. D., J. O. Milewski, and S. Fetzko. 1987. Electron beam welding data acquisition system using a personal computer. *Welding Journal* 66(4): 41–46.

Elmer, J. W., and A. T. Teruya. 2001. An enhanced faraday cup for rapid determination of power density distribution in electron beams. *Welding Journal* 80(12): 288-s–295-s.

Farrell, W. J. and J. D. Ferrario. 1987. A computer-controlled, wide-bandwidth deflection system for EB welding and heat treating. *Welding Journal* 66(10): 41–49.

King, J. F., S. A. David, J. E. Sims, and A. M. Nasreldin. 1986. Electron beam welding of heavy-section 3Cr-1.5 Mo alloy. *Welding Journal* 65(7): 39–47.

LaFlamme, G. R., and D. E. Powers. 1991. Diagnostic device quantifies, defines geometric characteristics of electron beams. *Welding Journal* 70(10): 37–40.

LaFlamme, G. R., and D. E. Powers. 1994. Electron beam welding of copper containers to encapsulate nuclear waste. *Welding Journal* 73(12): 37–40.

Metzbower, E. A. 1990. Laser beam welding: thermal profiles and HAZ hardness. *Welding Journal* 69(7): 272s.

Murphy, J. L., T. M. Mustaleski, and L. C. Watson. 1988. Multipass autogenous electron beam welding. *Welding Journal* 67(9): 187-s to 195-s.

Mustaleski, T. M., R. L. McCaw, and O. E. Sims. 1988. Electron beam welding of nickel—aluminum bronze. *Welding Journal* 67(7): 53–59.

Patterson, R. A., P. L. Martin, B. K. Damkroger, and L. Christodoulou. 1990. Titanium aluminide: electron beam weldability. *Welding Journal* 69(1): 39-s–44-s.

Powers, D. E. 1997. Nonvacuum electron beam welding enhances automotive manufacturing. *Welding Journal* 76(11): 59–62.

Powers, D. E. and R. K. Colegrove. 1986. A new mobile EB gun/column assembly. *Welding Journal* 65(9): 47–51.

Powers, D. E. and G. R. LaFlamme. 1988. EBW vs. LBW—A complete look at the cost and performance traits of both processes. *Welding Journal* 67(3): 25–31.

Powers, D. E. and G. Schubert. 2000. Electron Beam Welding: A useful tool for the automotive industry. *Welding Journal* 79(2): 35–38.

Schulze, K. R., and D. E. Powers, 2004. EBW of aluminum breaks out of the vacuum, *Welding Journal* 83(2): 32–38.

CHAPTER 14

LASER BEAM WELDING, CUTTING, AND ASSOCIATED PROCESSES

Prepared by the Welding Handbook Chapter Committee on Laser Beam Welding and Cutting:

D. Kautz, Chair
Los Alamos National Laboratory

V. E. Merchant
Consultant

J. O. Milewski
Los Alamos National Laboratory

D. E. Powers
PTR Precision Technologies Inc.

Welding Handbook Volume 3 Committee Member:

P. F. Zammit
Brooklyn Iron Works Inc.

Contents

Photograph courtesy of ESAB

CHAPTER 14

LASER BEAM WELDING, CUTTING, AND ASSOCIATED PROCESSES

INTRODUCTION

Laser beam welding (LBW) produces coalescence with the heat from a laser beam impinging on the joint. Filler metal occasionally may be used, but the process is primarily used autogenously.

Laser beam cutting (LBC) is a thermal cutting process that utilizes highly localized melting or vaporizing to sever metal with the heat from a laser beam. The process is used with or without a gas to aid in the removal of molten and vaporized material.[1,2]

Laser is an acronym for *light amplification by stimulated emission of radiation.* A laser is a device that uses an optical resonating system incorporating a crystal or gas medium and reflective mirrors or focusing lenses to amplify and synchronize light waves into a coherent beam. Coherence of the beam is produced by stimulated electronic or molecular transitions to lower levels of energy. The laser emits this concentrated beam as energy that can be focused on the weld joint or cutting site and applied as heat to make the weld or cut. The engineering disciplines involved in laser beam material processing include laser mechanics, optics, fluid dynamics, and materials science.

The first laser was introduced in 1960. It used a ruby crystal electrically excited (pumped) by a flashlamp to produce a laser beam. By the late 1960s, the beam was successfully performing the first laser material processing application: the drilling of diamond dies used in wire drawing. Solid-state lasers of this type produce only short pulses of light energy at repetition frequencies limited by the heat capacity of the crystal. Consequently, even though individual pulses exhibit instantaneous peak power levels in the megawatt range, pulsed ruby lasers are limited to low average power output levels. They have largely been replaced by continuous wave (CW) solid-state lasers, many of which use neodymium-doped, yttrium aluminum garnet (Nd:YAG) crystal rods to produce a continuous, monochromatic beam output in the average power range of 0.5 kilowatt (kW) to 4 kW. Other lasers use carbon dioxide (CO_2) and other gases as the lasing medium, with output power ranges of 0.5 kW to 45 kW.

Among laser material-processing applications, cutting is the most common, with many types of machines used in industrial production systems worldwide. It is estimated that laser cutting and related equipment for drilling, trimming, machining, scribing, and laser transformation hardening represent well over 50% of industrial laser installations. Laser beam welding is widely

1. Welding terms and definitions used throughout this chapter are from *Standard Welding Terms and Definitions*, AWS A3.0:2001, Miami: American Welding Society.
2. At the time this chapter was prepared, the referenced codes and other standards were valid. If a code or other standard is cited without a publication date, it is understood that the latest edition of the document referred to applies. If a code or other standard is cited with the publication date, the citation refers to that edition only, and it is understood that any future revisions or amendments to the code or other standard are not included. As codes and standards undergo frequent revision, the reader is advised to consult the most recent edition.

used for the high-quality welds required by the automotive, aerospace, shipbuilding, pipeline, and air conditioning industries. Examples of other applications are the fabrication and hermetic sealing of relay containers, cases for electronic, medical, and other devices, and the production of aluminum tubing.

Many of the laser techniques associated with welding, cutting, marking, and surfacing processes also are associated with other industries. Different types and power levels of lasers are used for industrial, medical, construction, and office applications. In medicine, lasers are used for welding and cutting applications such as eye repair welding and self-cauterizing surgical cutting and repairing. In offices around the world laser printers use laser surfacing techniques to apply dry ink to paper; in plastics-painting applications lasers are used to prepare the surface for better adhesion. It is not unusual to hear typical welding terms applied to many uses of lasers in non-traditional applications.

This chapter is devoted to the fundamentals of the laser beam welding and cutting processes, modes of operation, and equipment. Other topics include laser beam welding operating systems, applications, joint design and preparation, weld quality, and the economics of using these processes. Similar information is presented for laser beam cutting, drilling, and related processes. The chapter concludes with a discussion on safety issues specific to laser beam operations and the safety codes that must be followed to provide the best working environment where lasers are in use.

FUNDAMENTALS OF LASER BEAM WELDING AND CUTTING

From an engineering standpoint, the laser is an energy conversion device that transforms energy from a primary source such as electrical, chemical, thermal, optical, or nuclear, into a beam of electromagnetic radiation at a specific frequency of light such as ultraviolet, visible, or infrared. This transformation is facilitated by certain solid, liquid, or gaseous mediums which, when excited by various techniques on either a molecular or atomic scale, will produce a coherent and relatively monochromatic form of light (that is, light consisting of a very narrow band of frequencies): a beam of laser light.

In practice, when the medium between the end mirrors of an optical resonator cavity is excited or pumped to the point of population inversion, a condition occurs in which the majority of active atoms (or molecules) in the medium are put into a higher-than-normal energy state. This results in a source of coherent light that can then reflect back and forth between the end mirrors of the cavity. When light waves are coherent they are in phase; that is, the time variation of the electric field in all the light waves is synchronized. A cascade effect causes the level of this coherent light to reach a threshold point (the point at which the gain in light amplification begins to exceed any losses in light that simultaneously occur), thereby allowing the device to start emitting a beam of laser light.

Because laser beams are coherent and monochromatic, both low-power and high-power laser light beams have low divergence angles. Consequently, the beams of light can be transported over relatively long distances before being highly concentrated through the use of either transmissive optics (lenses) or reflective optics (mirrors) to provide the level of beam power density, i.e., the power per unit area, measured in watts per square meter (W/m^2) (watts per square inch [$W/in.^2$]) needed to do a variety of material-processing tasks. Among these tasks are welding, various forms of cutting, and laser transformation hardening.

Basic laser technology is used in both welding and cutting. In laser beam welding, the heat obtained from the concentrated beam of light is impinged on the joint and produces a fusion weld. Laser beam cutting is accomplished by impinging the beam on the surface to be cut, drilled, scored or machined.

The two types of lasers predominantly used for industrial welding, cutting, and other material processing applications are the 1.06 µm (41.7 µin.) wavelength Nd:YAG laser, which uses the neodymium (Nd) ion as the active element, and the 10.6 µm (417 µin.) wavelength CO_2 laser, which uses the CO_2 molecule as the active element. Several types of equipment, such as electrically pumped, pulsed, and continuous-wave gas lasers use alternating current (ac), direct current (dc), or radio frequency (rf) as the means of excitation. Carbon dioxide (CO_2) lasers with beam power outputs of up to 12 kW also are in general use. Although no longer commercially produced, lasers with power levels up to 45 kW are still in use for a wide variety of industrial material processing tasks. These high-power lasers are capable of making single-pass welds with complete joint penetration in steel up to 15 mm (0.6 in.) thick and cuts in materials up to 25 mm (1 in.) in thickness.

The laser beam is directed to the workpiece by flat optical elements, such as mirrors, and then focused onto a small spot, which creates high power density at the weld joint or cut location using either reflective focusing elements or lenses. Although the bulk laser power is important, it is predominantly the beam power density, or the amount of power per unit area, that governs the interactive mechanisms between the beam and workpiece.

Laser beam welding and cutting are non-contact processes; no part of the apparatus is in contact with the workpiece surface. An inert gas generally is employed to shield the weld pool to prevent oxidation and to provide control of the plasma plume that is generated. In some laser beam cutting techniques, an inert gas is used as part of the lasing medium and also to remove cutting debris, although the gases used are not from the same supply. In others systems, oxygen or air is used to enhance the cutting process.

LASER BEAM WELDING

Laser beam welding has numerous advantages over other processes; however, it also has several limitations that should be considered when selecting the welding process for a particular application.

ADVANTAGES

The major advantages include the following:

1. Heat input is close to the minimum required to fuse the weld metal. Heat-induced distortion of the workpiece and metallurgical effects in the heat-affected zone (HAZ) are minimized;
2. Single-pass laser beam welding procedures have been qualified for metals up to 3.2 mm (1.25 in.) thick, although more typically joints up to 19 mm (0.75 in.) may be welded. This reduces the time needed to weld thick sections and reduces or eliminates the need for welding wire and elaborate joint preparation;
3. Electrodes are not required to conduct current to the workpiece, thereby eliminating electrode contamination, indentation, or damage from the high currents used in other welding processes;
4. Tool wear is essentially eliminated because LBW is a non-contact process;
5. Laser beams are readily focused, aligned, and directed by optical elements, permitting welding in areas not easily accessible by other processes and allowing the laser to be conveniently located relative to the workpiece or redirected around tooling and obstacles in the workpiece;
6. The process allows workpieces with internal volumes to be hermetically welded to leave a vacuum or a controlled atmosphere in the finished product;
7. The laser beam can be focused on a small area, permitting the joining of small, closely spaced components with extremely small welds;
8. A wide variety of materials and many combinations of different types of materials can be welded, including those with dissimilar physical properties, such as electrical resistance, and several that are electrically insulating;
9. The laser can be readily mechanized for automated, high-speed welding, including the use of computer numerical controls (CNC) or computer-controlled welding;
10. Welds in thin metal and small-diameter wire are less susceptible to incomplete fusion than arc welds;
11. Laser welds are not influenced by the presence of magnetic fields, as are arc welds and electron beam welds;
12. No vacuum is required and no X-rays are generated;
13. Aspect ratios (depth-to-width ratios) on the order of 10:1 are attainable when a keyhole weld is made by forming a cavity in the metal; and
14. The laser beam can be transmitted to more than one workstation using beam-switching optics, which allows beam timesharing.

LIMITATIONS

Compared to other welding methods, laser beam welding has certain limitations, including the following:

1. Joints must be accurately positioned laterally under the laser beam and at a controlled position with respect to the laser beam focal spot;
2. When weld surfaces must be mechanically forced together, the clamping mechanism must ensure that the final joint position is accurately aligned with the laser beam impingement point;
3. The maximum joint thickness is somewhat limited, as weld penetrations greater than 19 mm (0.75 in.) generally are considered impractical for LBW production applications;
4. The high reflectivity and high thermal conductivity of some metals, such as aluminum and copper alloys, may adversely affect weldability with the laser;
5. When performing moderate-to-high-power laser beam welding, appropriate plasma and plume control devices must be employed to ensure that welds are reproducible;
6. Lasers tend to have low energy conversion efficiency;
7. As a consequence of the rapid solidification characteristic of laser beam welds, some weld porosity and brittleness can be expected in many common engineering alloys; and
8. Laser equipment and fixturing costs may be high.

LASER BEAM SYSTEMS

Laser beam welding systems include solid-state lasers, direct-diode lasers, fiber lasers, and gas lasers. Solid-state Nd:YAG lasers and gas CO_2 lasers are the most widely used in industry.

Solid-State Lasers

Nd:YAG lasers utilize an impurity in the lasing material, or host, as the active medium. Thus the neodymium ion (Nd+++) is used as a dopant, an impurity purposely added to either a glass or a YAG crystal. The 1.06 µm (41.7 µin.) output wavelength is characteristic of the neodymium ion. The lasing material is in the form of a cylinder, called a rod, of about 150 mm (6 in.) long by 9 mm (0.35 in.) in diameter. Both ends of the cylinder are ground to close-tolerance flat parallel surfaces, then polished to a high optical finish and silver-coated to create reflective surfaces. The crystal is excited by using an intense krypton or xenon lamp, or in recent designs, by using diode lasers. A simplified arrangement of the rod, lamp, and mirrors in a solid-state Nd:YAG laser is shown in Figure 14.1.

The selection of the host material for the neodymium ion depends on several factors. These include the availability of large quantities of good optical-quality rods (i.e., having an acceptable hardness and the capability of being polished), with acceptable levels of thermal conductivity, fluorescent lifetime, efficiency, and optical absorption bands. All of these factors enable the system to emit reasonable amounts of energy in a single pulse. Successful materials are those from which large amounts of energy can be extracted. The ideal characteristics of the Nd:YAG crystal make it an excellent host material.

The output characteristics of Nd:YAG lasers depend on the excitation method used, which may be either continuous wave or pulsed. A pulsed laser is defined as a laser with an output controlled to produce a pulse with duration of 25 milliseconds or less. A continuous-wave laser has an output that operates in a continuous rather than a pulsed mode. A laser operating with a continuous output for a period greater than 25 milliseconds is regarded as a continuous-wave laser.[3]

Continuous-wave Nd:YAG lasers using multiple lasing rods can achieve power levels of greater than one kilowatt. For pulsed lasers, the output characteristics depend on lamp or diode configuration. Table 14.1 lists the characteristics of Nd:YAG lasers and suggests possible trade-offs that might be made between the average power, pulse energy, pulse duration, and pulse repetition rates for these lasers.

As referenced in Table 14.1, TEM_{00} is defined as laser radiation of the fundamental transverse electric

3. American Welding Society (AWS) Committee on High-Energy Beam Welding and Cutting, 1998, *Recommended Practices for Laser Beam Welding, Cutting, and Drilling*, C7.2:1998, Miami: American Welding Society.

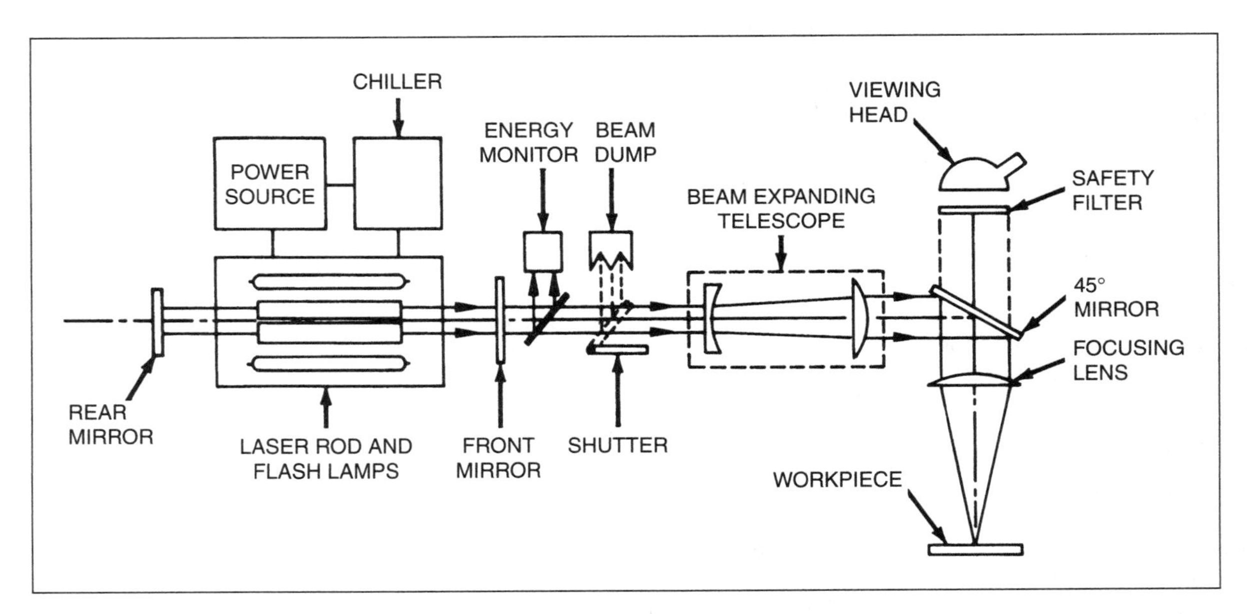

Figure 14.1—The Elements of an Nd:YAG Laser

Table 14.1
Output of an Nd:YAG Laser

Average Power	
Continuous wave	>4500 W
Multiple mode	<20 W (TEM_{00})*,
Pulsed	1000 W
Divergence	
Multiple mode	1 mrad to 20 mrad
TEM_{mn}**	3 mrad (diffraction-limited beam divergence for this wavelength and typical beam)
Beam Diameter	
Continuous wave	0.1 mm to 10.2 mm (0.004 in. to 0.4 in.)
Pulsed	5.1 mm to 10.2 mm (0.2 in. to 0.4 in.)
Pulse Length	
Welding	0.1 millisecond to 20 milliseconds at 0.1 Hz to 100 Hz
Drilling and marking	0.01 millisecond to 1 millisecond at 0.1 Hz to 30 000 Hz
Output Energy	
Multiple mode	<500 J/pulse
Pulsed (TEM_{00})	5 mJ to 70 J/pulse
Repetition Rate	0.1 to 30 000 Hz in pulsed mode to systems that are continuous wave
Peak Power	10 kW–50 kW

*TEM_{00} = Mode of laser operation where the beam has no nodes in the energy distribution in the beam. An approximately Gaussian or normal energy distribution within the beam.

**TEM_{mn} = Transverse electromagnetic mode, where the subscripts *m* and *n* specify the transverse modal lines across the emerging laser beam. Generally, beams used for welding, cutting and drilling have a TEM_{00} mode.

mode. The spatial intensity has a Gaussian distribution. Gaussian distribution can be expressed as a symmetric two-dimensional equation that approximately describes the spatial power distribution of many laser beams, as follows:

$$P(r) = P_o e^{\frac{-2r^2}{w^2}} \tag{14.1}$$

where

$P(r)$ = Power density

P_o = Maximum power density,

e = 2.718 (Euler's number),

r = Radial distance out from the center of distribution, and

w = Radius at which power density is 0.135 times maximum.

The relatively narrow frequency band exhibited by an Nd:YAG laser facilitates continuous-wave operation at room temperature, making the CW Nd:YAG laser second only to the CW gas lasers in terms of power generation. Overall, however, lamp-pumped Nd:YAG lasers are considerably less efficient (typically 2%, versus 10% for gas lasers), resulting in lower power output. Diode-pumped Nd:YAG lasers are considerably more efficient than lamp-pumped Nd:YAG lasers, with quoted efficiencies of more than 30%. The higher efficiency is achieved as the laser rod is pumped at frequencies most absorptive by the lasing rod, since other frequencies emitted by flashlamps are not present. This reduces the heat load in the rods, allowing them to generate more light without excess thermal distortion.

Laser diode bars are more efficient and operate at cooler temperatures than lamp-pumped systems. Improvements in laser beam quality and a reduction in the size of output focusing optics are additional benefits of this type of laser. Figure 14.2 shows a multiple-kW diode-pumped laser that may be used for welding, cutting, or thermal treatments.

In the pulsed mode, the active medium of an Nd:YAG laser is pumped intermittently instead of continuously by using a pulsed power source to drive the flashlamp. Figure 14.3 shows the time relationship of the flashlamp and laser output pulses of a typical pulsed solid-state laser. The beginning of the flashlamp pulse establishes a population inversion in the active medium. When the net gain reaches 1.0 (when each photon impact produces another photon), lasing begins and continues as a series of closely spaced spikes for the duration of the flashlamp pulse. These spikes are produced by gain switching in the active medium. The gain rises quickly to a high value because of the intense pumping level. This results in a high-intensity standing wave in the optical cavity, which quickly depletes the population inversion for that particular wavelength, and lasing stops. Thus, the laser momentarily switches off by using up all of its gain. The population inversion and the gain are regenerated as long as the flashlamp pump is ongoing, leading to a sequence of short spikes of laser output. Sophisticated diagnostic equipment is required to resolve these short spikes, although normally only the envelop of the short spikes is observed.

Because of the spiking in the output, the peak power of a pulsed solid-state laser is difficult to determine and tends to vary from pulse to pulse, even though the overall energy and duration of each pulse may remain constant. For these reasons, specifications of pulsed solid-state laser beam welding systems usually do not include the maximum output power. Instead, pulsed energy and pulse duration are specified. Peak output can be approximated by dividing the energy of the output pulse (the duration of the sequence of the short pulses) by the duration of the pulse. The duration of the output pulse of solid-state lasers used for metalworking varies from as short as 50 microseconds per second (μ/s) to as long as 50 milliseconds per second (ms/s) with a usual pulse duration of about 1 ms. Only Nd:YAG systems

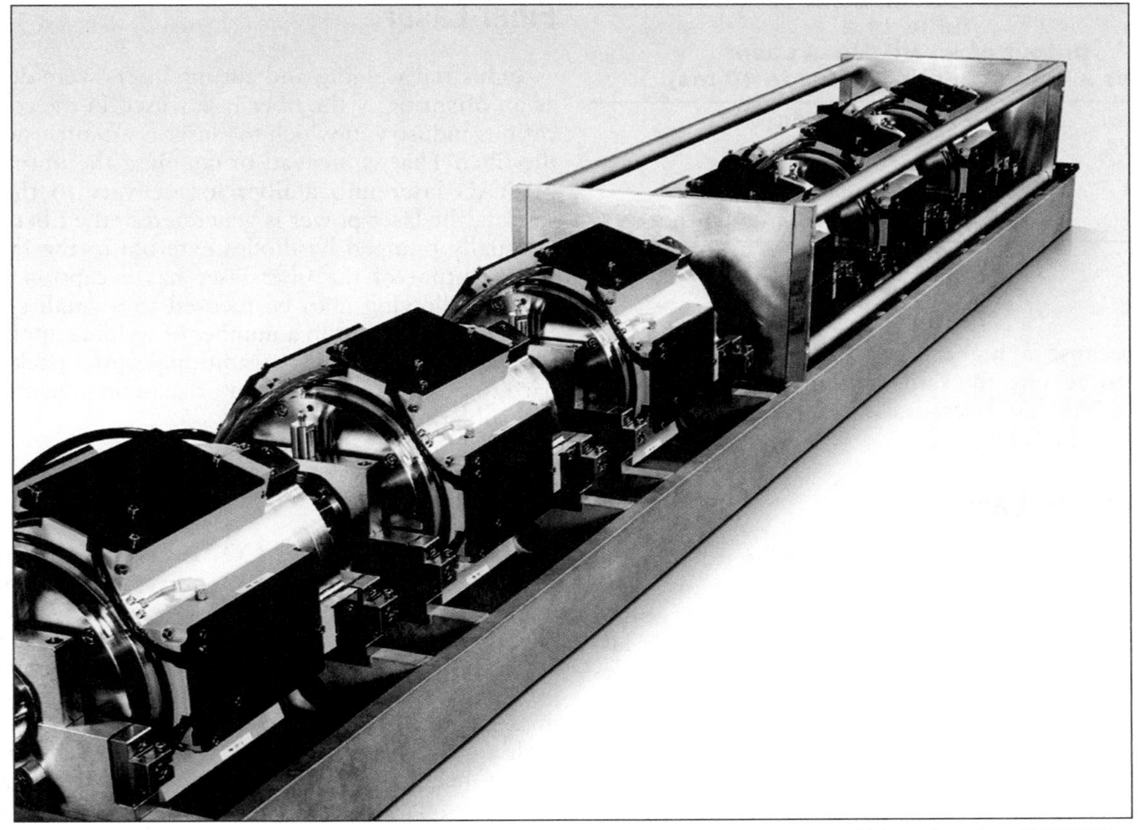

Photograph courtesy of TRUMPF, Inc.

Figure 14.2—Solid-State Diode-Pumped CW Multiple-kW Laser

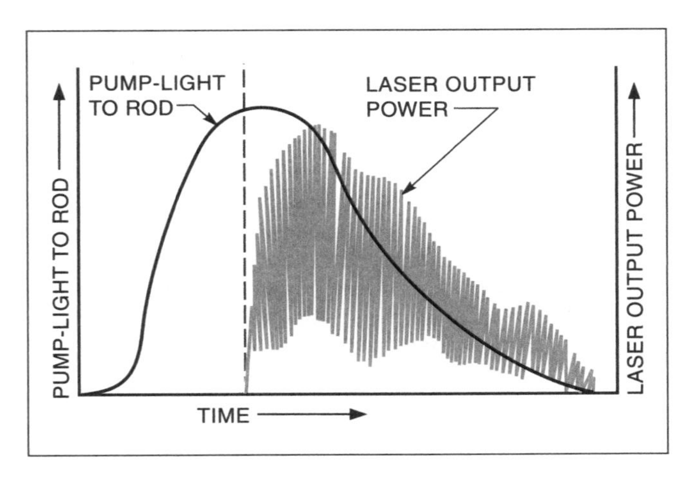

Figure 14.3—Output of a Typical Pulsed Solid-State Laser Compared to Pump-Light Input to the Rod as a Function of Time

are capable of pulse durations much greater than 2 ms, and some Nd:YAG laser drilling machines use pulse durations of 5 ms to 8 ms.

Improvements in laser beam quality can be realized by using diode pumping, which allows the lasers to be pumped more efficiently. This permits the use of smaller-diameter lasing rods, which results in higher beam quality.

Glass also has certain desirable characteristics as a laser host material. One of these features is that large pieces of high-quality optical glass fibers can be fabricated into a variety of sizes and shapes with diameters that range from a few microns to rods with diameters of 100 mm (4 in.) and lengths of up to 2 m (6-1/2 ft). Since the thermal conductivity of glass is lower than that of most crystalline hosts, the slow cooling of the glass limits the maximum repetition rate at any given level of pulse energy. Also, the emission lines of neodymium ions in glass are broader than those in crystalline materials. This raises the threshold of the glass for laser

Table 14.2
Output of an Nd-Glass Laser
(for a Pulse Length of 1 ms to 10 ms)

Output energy	20 J/pulse (multiple mode)
Repetition rate	10 Hz
Divergence	5 mrad to 10 mrad
Beam diameter	0.4 mm to 10.3 mm (0.02 in. to 0.4 in.)

action, because a higher ion population inversion is required to achieve the same gain. The output characteristics of Nd:glass lasers suitable for laser beam welding are listed in Table 14.2.

Direct-Diode Laser

Developments in high-power-output laser diodes and diode-array cooling allow the arrangement of arrays that can provide over 3 kW of direct-laser output power, instead of using this power to pump another lasing medium. Diode lasers work on the same principles that standard light-emitting diodes (LED) work, except that they generate much higher power levels and the light generated is of a single wavelength. High electrical conversion efficiency can be achieved with direct-diode lasers and lower maintenance results from significantly fewer system components. These laser systems are extremely compact, but have limitations associated with large asymmetrical focal spots and power densities. Because of the lower energy densities produced by direct-diode laser systems, they cannot be used for welding in the keyhole mode, but they can provide high-speed, conduction-mode weld processing. Figure 14.4 shows a typical direct-diode laser.

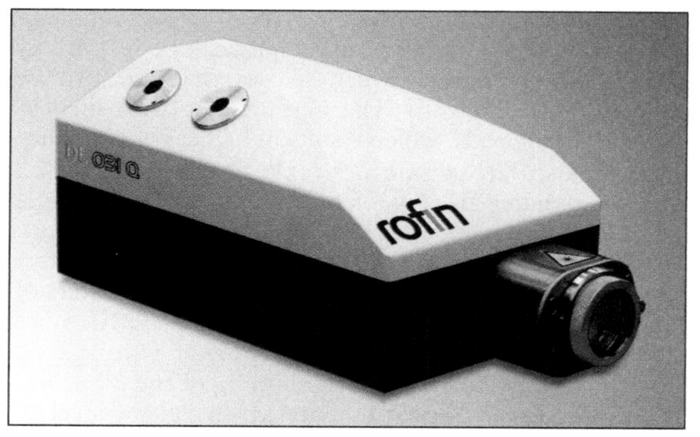

Photograph courtesy of ROFIN-SINAR Laser GmbH

Figure 14.4—Direct-Diode Laser

Fiber Laser

Industrial welding and cutting lasers were developed as an offspring of the fiber lasers used in the communications industry in which the laser is an intrinsic part of the fiber. That is, instead of coupling the output of an Nd:YAG laser into a fiber for delivery to the workstation, the laser power is generated in the fiber itself. It is usually pumped by diodes external to the fiber. The beam output of the fiber laser has exceptionally high quality, allowing it to be focused to a small spot size. Fiber lasers are used in a number of welding applications, especially in areas where traditional optics packages are hard to position and in miniaturized-component welding and soldering.

Gas Laser

The electric-discharge CO_2 lasers are the most efficient types available for high-power laser beam welding. Electric-discharge gas lasers generally have three basic subassemblies: the optical cavity, the gas flow loop, and the electric discharge system that includes the power source and electrodes. These lasers use a gas mixture of primarily nitrogen (N_2) and helium (He) with a small percentage of CO_2. A very common mixture is helium with 15% N_2 and 5% CO_2 added. A radio frequency (rf), or "electric glow" discharge is used to pump this laser medium, that is, to excite the CO_2 molecule. Gas heating produced in this fashion is controlled by maintaining a continuous flow of the gas mixture through the optical cavity. CO_2 lasers usually are categorized according to the type of gas flow system they use: slow axial, fast axial, transverse, or diffusion-cooled designs. Figure 14.5 shows the CO_2 diffusion-cooled laser beam welding of a transmission gear.

Slow Axial-Flow Laser. The slow axial-flow laser is the simplest of the CO_2 lasers. The gas flows in the same direction as that of the optical axis of the laser resonator and the electric excitation field, or gas discharge path, as shown in Figure 14.6. The axial gas flow is maintained through the tube to replenish molecules depleted from the effects of the gas discharge being used for excitation. This causes the carbon dioxide to be reduced to carbon monoxide and oxygen by electron bombardment. Catalytic devices are used in the recycling gas flow path to recombine the dissociation products, thus replenishing the laser gases.

The cooling of the laser gas in a slow axial-flow system is achieved by conduction through the walls of the discharge tubes to a liquid coolant in the cooling mantle. Some form of external heat exchange system is then used to dissipate the heat that is continuously extracted by this process.

Photograph courtesy of TRUMPF, Inc.

Figure 14.5— CO_2 Diffusion-Cooled Laser Beam Welding a of Transmission Gear

A mirror is located at each end of the discharge tubes to complete the resonator cavity. Typically, one mirror, the rear mirror, is totally reflective and the other mirror, the output coupler, is partially transmissive. Slow axial-flow resonators are capable of generating laser beams with a continuous power rating of about 80 watts (W) for every meter (3 ft) of discharge length. A folded tube configuration is used to achieve output power levels of 50 W to 3 kW, rather than extending the length of the resonator cavity to attain these power levels.

Fast Axial-Flow Laser. Fast axial-flow lasers have a component arrangement similar to the slow axial-flow laser. The difference is that a roots blower or turbo pump is used to circulate the laser gas at high speed through the discharge region and corresponding heat exchangers, as shown in Figure 14.7. Cooling is enhanced within the confines of the laser by forcing the hot laser gas through gas-to-liquid heat exchangers. This results in a higher rate of heat extraction than is available with slow axial-flow lasers. It also provides the capacity for output power levels greater than 2 kW per 1 meter (3 ft) of discharge length. As with slow axial-flow lasers, most fast axial-flow lasers can be pulsed. Fast axial-flow lasers with continuous-wave output power levels from 500 W to 6 kW are available.

Transverse-Flow Laser. Transverse-flow lasers operate by continuously circulating gas across the resonator cavity axis by means of a blower with a high-speed fan. An electrical discharge is maintained perpendicular to the direction of the gas flow and the optical axis of the laser beam. Because the resonator volume is large relative to its length, mirrors can be placed at each end to reflect the light beam several times through the discharge region before it is transmitted through an output coupler. The ability to achieve a long optical path within a short resonator structure allows transverse-flow lasers to be compact, but capable of generating high output power. Transverse-flow lasers with continuous-wave output power levels between 1 kW and 12 kW are

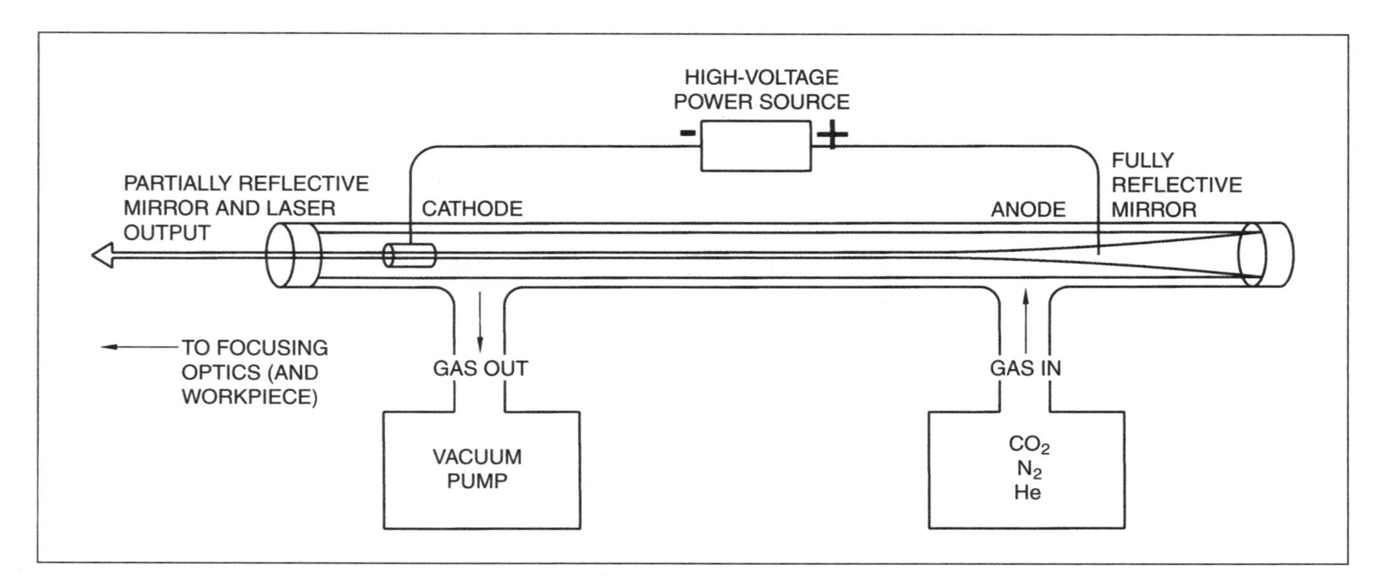

Figure 14.6—Schematic View of a Slow Axial-Flow Laser

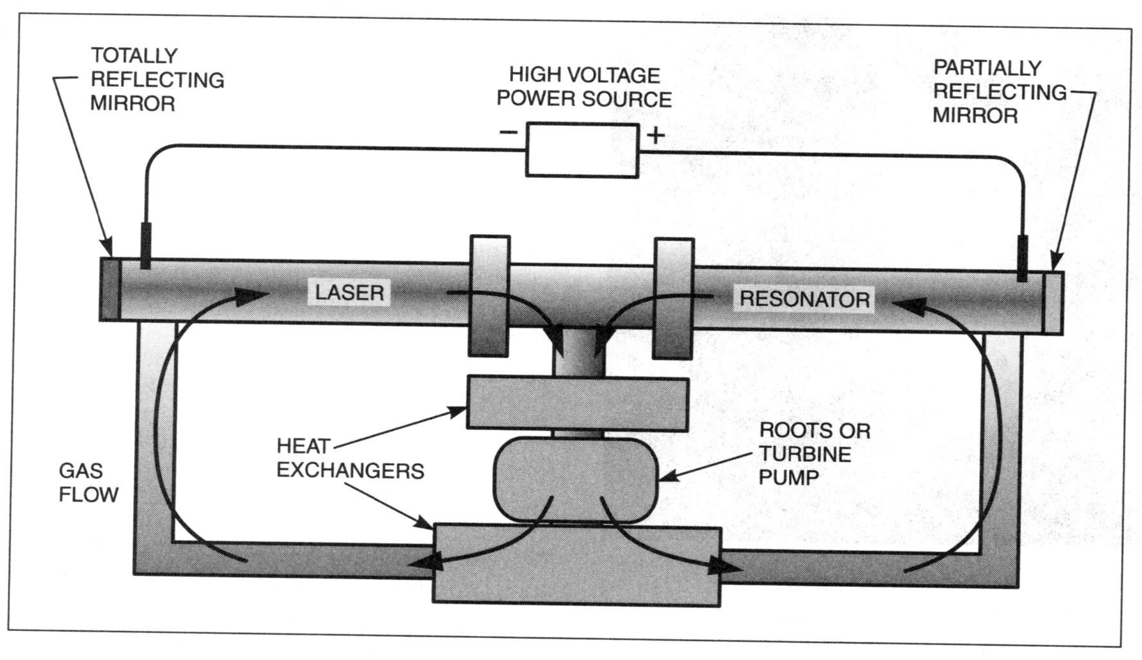

Figure 14.7—Schematic View of a Fast Axial Flow Laser

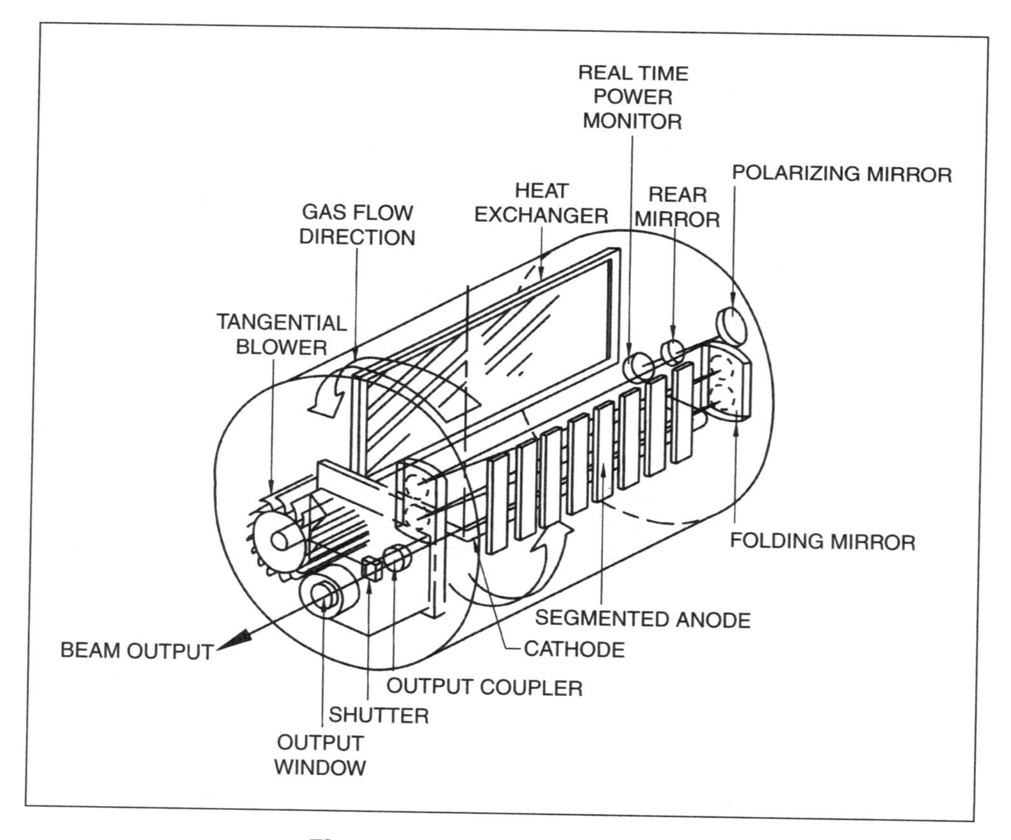

Figure 14.8—Schematic View of a Transverse-Flow Laser

commercially available, but have been built with power levels up to 45 kW. A schematic of a transverse-flow laser is shown in Figure 14.8.

A typical transverse-flow laser system operates in the range of 5% to 10% of atmospheric pressure. The gas-flow loop contains the laser gas, comprised of about 95% He and N_2 and 5% CO_2. The gas is driven around an enclosed loop by a large axial vane or similar pump. The gas is electrically excited in the laser cavity where the optical power is produced; unused power in the form of heat is removed by a gas-to-water heat exchanger located just downstream of the laser cavity, and the gas is re-circulated to be electrically excited again. The electrical discharge system contains a high-voltage dc, ac, or rf power source, which is connected to an electrode array to provide excitation to the volume of gas throughout the laser cavity. For the continuous-wave laser beam, ballast resistors sometimes are placed in series with the electrodes to provide a smoother electrical discharge.

The optical cavity usually is formed by the precise location of water-cooled metal mirrors of specific radius of curvature and spacing. This determines the laser beam mode. The mirrors are mounted on a truss or similar structure designed to minimize distortion due to temperature variations.

The beam quality of high-power industrial lasers, although not as good or as focusable as near-Gaussian low-power lasers, provides the focusing capability needed for the keyhole welding technique. High-power industrial lasers, however, normally do not have the desired pulsing capability needed for some applications. Cathode maintenance may be required periodically in some high-powered lasers. When optical-grade windows are used to transmit the laser beam from the laser cavity to the ambient environment, the limited service life and cost of this consumable item also must be considered.

Diffusion-Cooled Gas Laser. Diffusion-cooled gas lasers have a compact cavity design compared to the other lasers, as shown in Figure 14.9. The rf electrodes are water-cooled slabs placed parallel to one another, as shown in Figure 14.10. Excitation occurs in the laser gas between the electrodes when an rf signal is passed through the gas. This type of cavity generates a laser beam of higher quality than other cavity designs, with laser beam divergence typically under one milliradian. Cooling occurs by heat conduction through the gas to the water-cooled electrodes. This eliminates the requirement for a gas circulation system. These lasers have output power levels up to 5 kW.

Photograph courtesy of Cincinnati Incorporated

Figure 14.9—Resonator Cavity of a Diffusion-Cooled Gas Laser

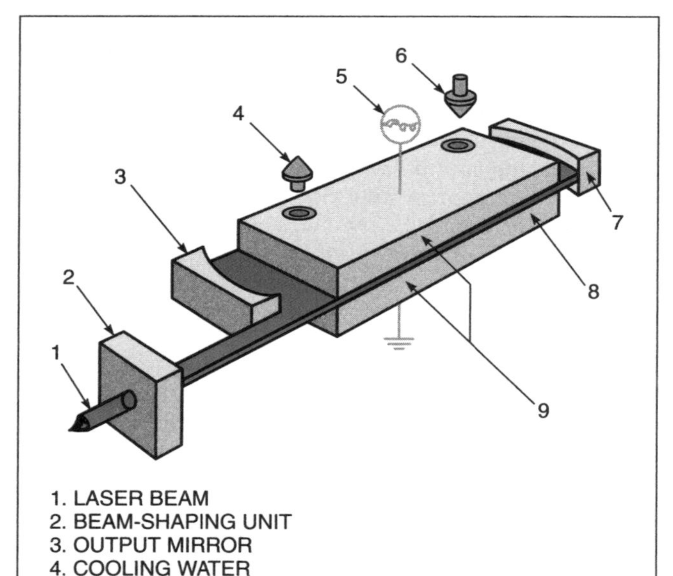

Source: Rofin-Sinar Technologies.

Figure 14.10—Design of a Diffusion-Cooled Resonator

Hybrid Laser-GMAW Systems

Hybrid systems that combine laser beam and arc welding processes have been developed to improve efficiency and productivity. The gas metal arc welding (GMAW) hybrid system is the most common adaptation. The laser beam improves the wetting of the molten material in the weld pool, which promotes much higher travel speeds than are attainable with the gas metal arc process alone. An added benefit is that weld pool stability is greatly improved. Typical engineering materials, including steel and aluminum alloys, may be welded with the hybrid laser-arc process. These systems are employed in such industries as heavy equipment and vehicle fabrication, shipbuilding, and construction. Micrographs in Figure 14.11 showing a GMAW weld, a laser weld, and a GMA-Hybrid weld illustrate the improvements that may be gained when a hybrid system is used. The welds are V-groove butt joints in steel plate, 12.7 mm (0.050 in.) thick, ASTM A-36 Grade A, with 8.8 mm (0.35-in.) land and beveled with a 90-degree included angle. The chamfer at the root is 3.2 mm (0.125 in.).

LASER BEAM DELIVERY AND FOCUSING OPTICS

Laser beams must be focused to a small diameter in order to produce the high power density required for welding. Focusing is accomplished with transmitting lenses or reflective mirrors, as shown in Figures 14.12, 14.13, and 14.14. The spot size can be varied by the design of the optics and the choice of focal length. For a given laser beam, the final focused spot size is directly proportional to the focal length. The resultant power density is inversely proportional to the square of the focal length, while the depth of focus attained varies directly with focal length. Therefore, laser beams focused with short focal-length lenses require greater precision in maintaining the lens-to-workpiece distance than when longer focal length lenses are used.

The shortest practical focal length used for CO_2 laser beam welding is about 125 mm (5 in.) because of the adverse effects on focusing that result from spatter and vapor produced during welding. Since the spot size at the focal plane varies inversely with the diameter of the laser beam incident on the focusing optic element, a

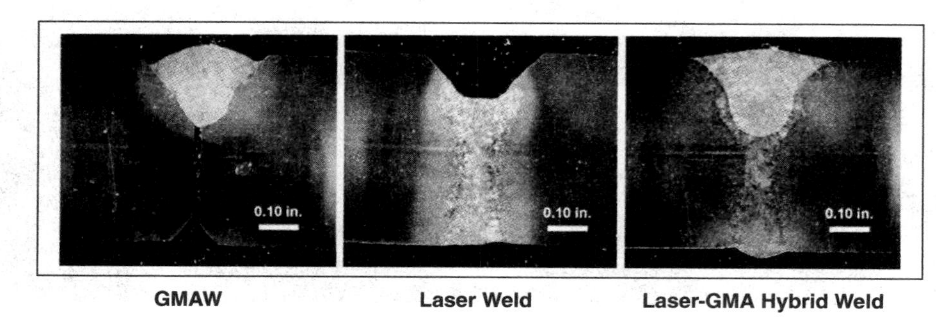

GMAW **Laser Weld** **Laser-GMA Hybrid Weld**

Source: Ruetzel, E. W., D. A. Mikesic, A. Nujesucm, Applied Research Laboratory, Pennsylvania State University, State College, PA.

Figure 14.11—Micrographs of GMAW, Laser Beam, and GMA-Hybrid Welds in 12.7 mm (0.50 in.) Steel Plate

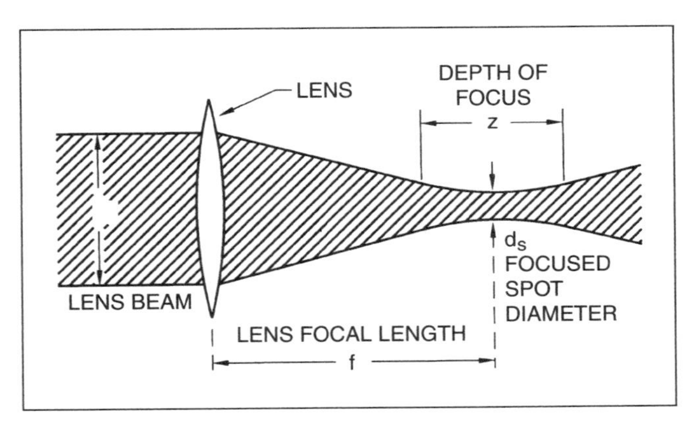

Figure 14.12—A Gaussian Laser Beam Focused with a Simple Lens

laser beam expander, i.e., an optical system used to increase the diameter of the beam may be used prior to focusing. This allows longer focal lengths to be used without loss of power density. Figure 14.12 illustrates the ratio of the focal length of a lens to the diameter of a laser beam passing through the lens, in which diameter (f/d_o) is referred to as the F-number. The spot size of the focused laser beam, i.e., the focused beam spot diameter (d_s), varies directly in proportion to the wavelength (λ) of the laser beam and the F-number of the focusing system. This is expressed in Equation 14.2.

$$d_s = K(\text{F-number})\lambda = K(f/d_o)\lambda \quad (14.2)$$

where

d_s = Focused beam spot diameter,
K = Quality measure specifying the ability to focus the laser beam,
F-number = Ratio of focal length of the lens to the diameter of the beam passing through the lens,
f/d_o = Diameter of optic (F-number), and
λ = Wavelength of the laser beam.

The power density of the laser beam at the focus spot is inversely proportional to the laser beam diameter, squared. As the F-number used for a particular system decreases, so does the focused spot diameter, resulting in greater power density.

Fiber optic delivery is possible for Nd:YAG wavelengths and may provide additional flexibility in laser beam delivery systems. Up to 6 kW of Nd:YAG wavelength energy can be delivered through a single quartz fiber as small as 300 µm (0.012 in.). A small percentage of energy is lost with the fiber, but transmission losses are negligible in distances of 100 m (328 ft) or less. During transmission within a fiber, the beam path is fully enclosed and impervious to dust and contamination. Optical switching mechanisms can divert all or part of the laser beam energy to multiple fibers serving multiple processing locations.

Focusing Systems

The transmissive and reflective optics used in laser beam systems are lenses and mirrors. Solid-state laser systems, lamp-pumped, diode-pumped, or direct-diode systems generally use a lens to focus the laser beam on the workpiece. The type of lens ranges from a simple plano-convex lens with an anti-reflective coating to a compound lens, cylindrical lens, or a lens with a graded index of refraction.

Gas lasers generally employ mirrors as reflective optics. The mirrors usually are made of highly polished metal and are water-cooled to withstand high-incident power. The mirrors may be either bare or coated to enhance reflectivity. Compared to transmissive lenses, these mirrors are less sensitive to damage from spatter and fumes, and are easier to maintain in production welding. Highly polished copper or molybdenum mirrors are common, but gold-coated mirrors provide the highest reflectivity, and thus the least amount of laser beam attenuation. However, they are expensive and are susceptible to surface damage because gold is relatively soft.

Molybdenum-coated mirrors, although more expensive than uncoated mirrors, have good reflectivity and are less susceptible to fume and spatter damage. Thus, a laser system may use gold-coated mirrors to transmit the laser beam to the workstation, but then may employ molybdenum-coated mirrors within the workstation. Several types of reflective-style focus heads used for high-power (multiple-kilowatt) laser beam welding are shown in Figure 14.13.

The primary advantage of optical mirrors made of metal is that they may be water-cooled by flowing a liquid through the passages beneath the reflecting surface or around the periphery. Compared to transmissive-type optics, mirrors allow more efficient cooling and ultimately provide more repeatable results. Mirrors and mirror focusing systems can be purchased in many of the different configurations needed for most welding manufacturing requirements. Many of these can be purchased directly off the shelf; special orders may require a modest lead-time. Additionally, some optical fabrication facilities provide services for re-polishing or refurbishing mirrors at reasonable prices.

Programmable focusing systems can precisely route the laser beam to various locations on the workpiece without requiring the movement or repositioning of the

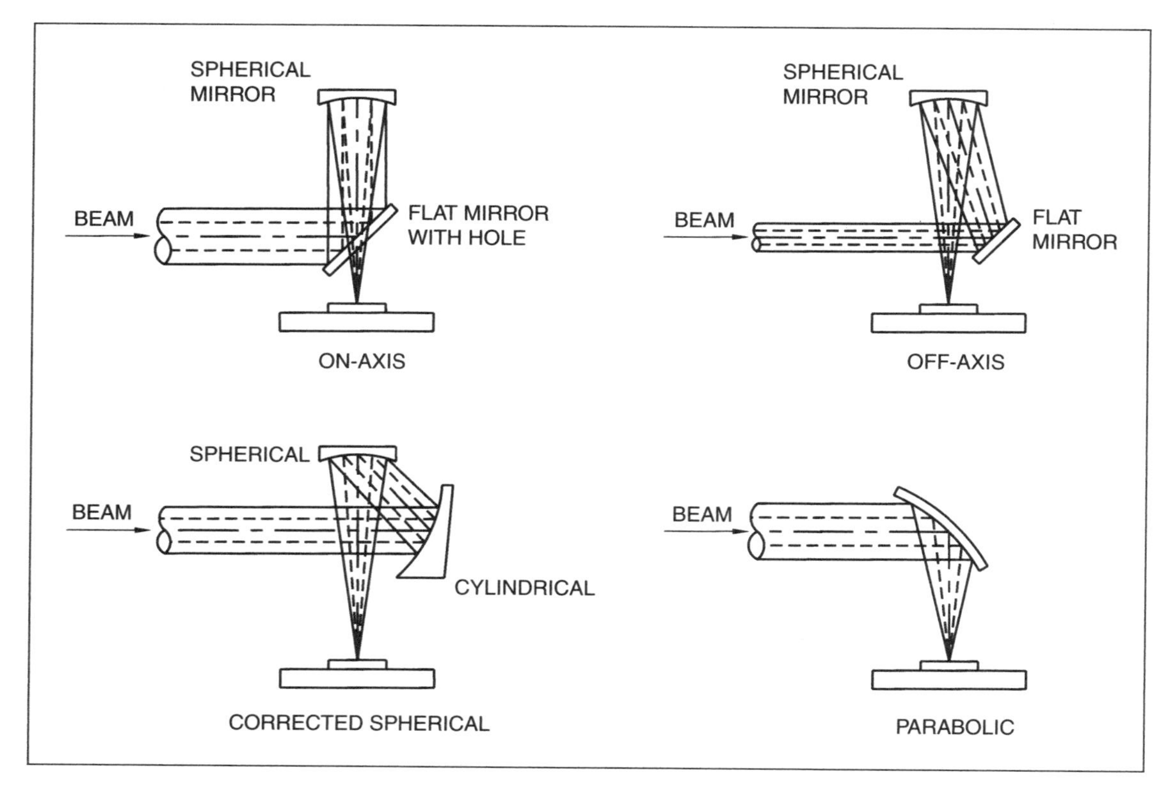

Figure 14.13—Laser Beam Focusing Heads

workpieces. This is especially convenient when spot and seam welds are to be repeated. Figure 14.14 shows a programmable focusing optic that uses a galvanometer scanner with two mirrors to rapidly direct the laser beam to designated points on the workpiece.

Beam Quality. The quality of a laser beam is measured by its focusing capability, which is a function of the transverse mode of the laser beam and the extent of aberrations and divergence introduced by the optics. The radiant energy oscillating from one end of the laser cavity to the other forms an intense electromagnetic field. This field can assume many different cross-sectional shapes, called *transverse electromagnetic modes* (TEM), which establish the radial energy distribution of the laser beam. The TEM mode is expressed as TEM_{mn}, where the subscripts m and n specify the transverse modal lines across the emerging laser beam cross section. As a result, the laser beam in cross section is segmented into two distinct planes at right angles to one another. This is illustrated in Figure 14.15. The number of energy density modes, or "valleys," in each

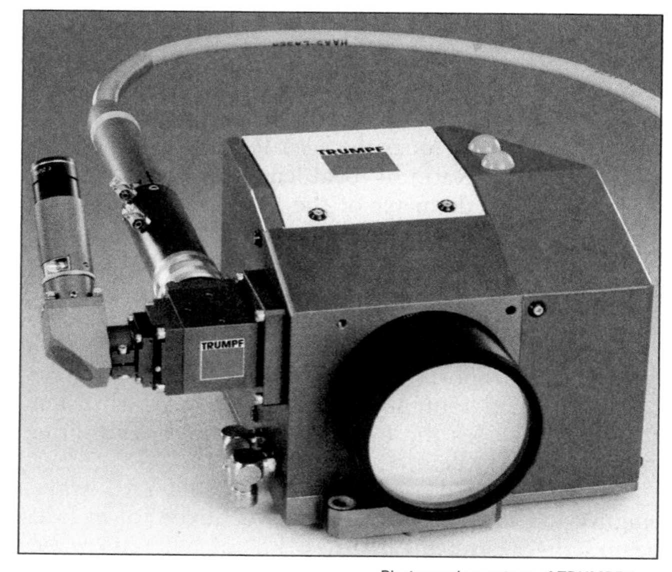

Photograph courtesy of TRUMPF, Inc.

Figure 14.14—Programmable Focusing Optic

of these directions is expressed as a subscript. Therefore, the notation TEM_{00} is representative of the lowest-order mode (or purest beam), and the power distribution across this laser beam is Gaussian. A pure TEM_{00} laser beam is the highest "absolute quality" laser beam attainable, and thus has the most focusing capability; however, it may not be the ideal mode to use for welding. The mode used depends on the specific welding task to be performed. Figure 14.15 illustrates a variety of TEM modes that can be generated by the CO_2 lasers used in industry.

At laser output power greater than 2 kW, the ability to generate a high-quality output beam could be limited by several factors. Discontinuities in the lasing media, more prevalent in transverse-flow units than in axial flow systems, affect beam quality. Thermally induced changes of diffraction in materials used as output couplers and windows also can significantly affect the output quality of the laser beam and result in degradation of weld performance. This phenomenon commonly is referred to as thermal focusing or thermal lensing. The phenomenon also can be produced by the presence of

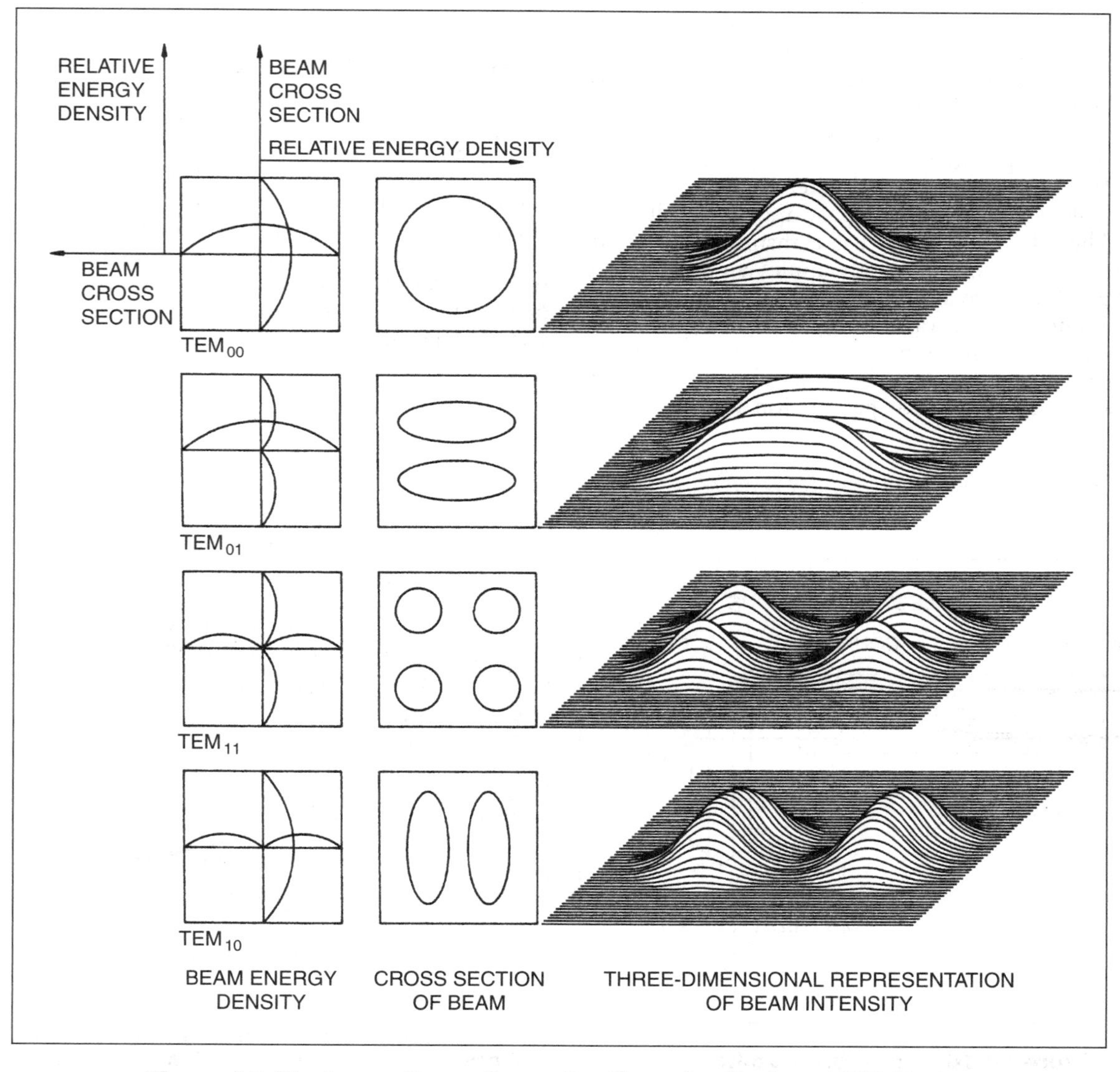

Figure 14.15—Laser Beam Cross Sections for Various TEM Modes

humidity, Freon residues, or other heavy molecular gases somewhere along the transmission path of the laser beam. In an effort to preserve the consistency of the process and to maintain a reasonably long service life of laser optics, which are consumable items, many laser manufacturers use a laser beam of higher order than TEM_{00}, thus producing less beam distortion. Beam distortion may be due to uneven heating of the transmissive optical elements, which changes laser beam shape, or it may be due to the temperature dependence of the refractive index.

The TEM_{00} and some of the higher-order modes usually are produced using a stable oscillator configuration similar to that of a slow axial flow laser. (Refer to Figure 14.7.) In contrast, the unstable oscillator configuration depicted in Figure 14.16 illustrates another means of generating beams greater that 1 kW. This method focuses the beam to power densities high enough to produce deep-penetration fusion welds. The intensity profile of the laser beam produced with the unstable oscillator configuration shown in Figure 14.16 will be annular. Multiple-kilowatt lasers using this resonator style are used in production welding at power levels up to 12 kW. Some lasers with higher power capability than 12 kW were used in the past and may still be in service. These lasers may not use transmissive elements but instead may use an aerodynamic window, which uses a "gas knife" to limit entry of contaminants into the laser cavity, and to transmit the laser beam from the reduced-pressure environment in the laser cavity to atmospheric pressure. Aerodynamic windows are almost always mandatory for production welding above 6 kW because optical-grade windows have a very short operating life.

As illustrated in Figure 14.16, the most significant unstable oscillator parameter affecting laser beam quality (for example, focusing capability) is magnification M, which is defined as the ratio of the outer diameter of the near-field annular output beam to the inner diameter. As magnification increases, both focusing capability and welding performance improve. The improved welding performance of M = 4 versus M = 2 laser cavity optics is shown in Figure 14.17. This effect also is illustrated in Figure 14.18, in which fusion-zone cross sections of welds in carbon steel made with M = 2, M = 3, and M = 4 laser cavity optics are compared. Welding conditions were 10 kW, 2.03 m/s (80 in./min), bead-on-plate welds performed with argon as the shielding gas.

Laser Beam Polarization

In some cases, laser beam welding speed is dependent on the alignment of the plane of polarization of the laser beam incident on the workpiece relative to the travel direction of the laser beam. Highest welding speeds with the narrowest weld bead geometries result when the polarization plane matches the direction of welding. Conversely, welding in a direction perpendicular to the polarization plane results in the lowest welding speed. Due to this effect, circular polarization

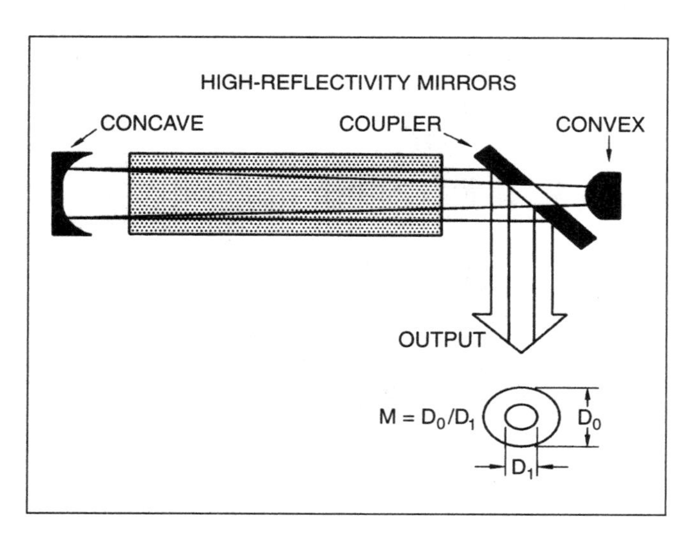

Figure 14.16—Unstable Oscillator Laser Cavity Optics

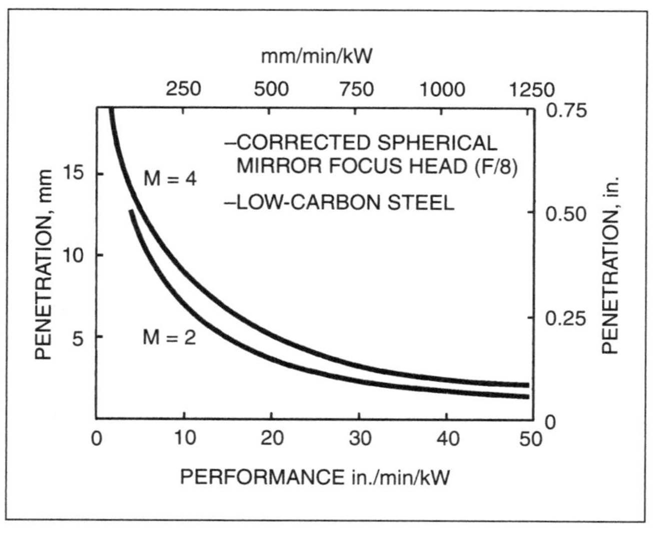

Figure 14.17—Welding Performance Comparison for M = 2 and M = 4 Laser Cavity Optics for Welds in Low-Carbon Steel

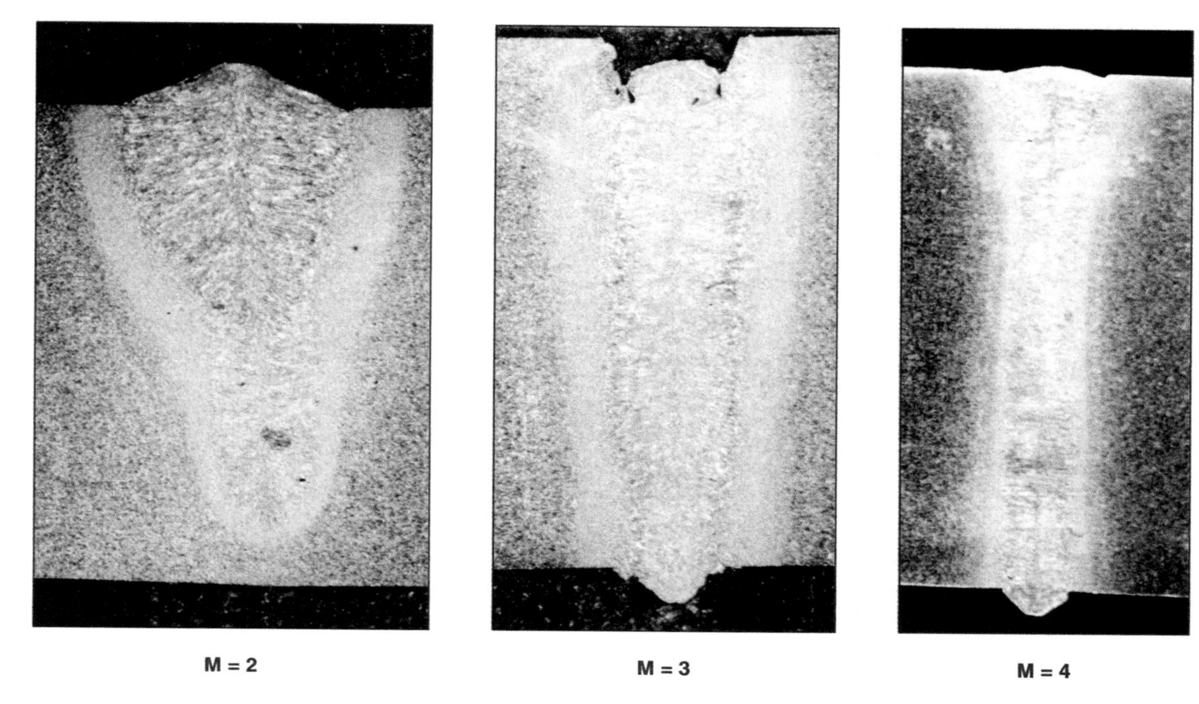

Figure 14.18—Comparison of Bead-on-Plate Weld Profiles in 9.5 mm (3/8 in.) Carbon Steel Using M = 2, M = 3, and M = 4

(i.e., equal in all directions) of the laser beam often is used to obtain consistent results regardless of the orientation of the polarization plane of the laser output with respect to the direction of welding.

The output of a laser frequently is characterized as being either randomly polarized or linearly polarized. In a randomly polarized output, the design of the resonator in the laser allows the polarization plane to drift in a random fashion. This often happens when welding at relatively high speeds. Welds made with randomly polarized lasers generally do not exhibit any effects of polarization. Linearly polarized output beams are produced by laser resonator designs in which the plane of polarization is locked and cannot rotate. In this system, the orientation of the plane of polarization of the incident laser beam relative to the direction of welding must be maintained constant in order to guarantee repeatable results. This characteristic in the design of the welding system must be considered, and a choice must be made whether to use circular polarization optics in the laser beam delivery system. In typical production welding systems, the optical system for laser beam delivery is fixed and the workpiece is manipulated under the focused beam. Polarization generally is not an issue when welding with multi-mode laser beams such as those produced by the Nd:YAG laser.

AUXILIARY SYSTEMS

High-power industrial lasers use many auxiliary systems that usually are controlled by a programmable logic controller, which results in quick and easy operation of the laser system. Control systems that automate laser startup and operation require minimum operator skill.

Following are some of the auxiliary systems used in industrial laser beam welding:

1. Gas supply system from which either premixed gas or separate bulk gases are supplied to the laser;
2. Vacuum system to maintain the laser pressure at its operating level or to evacuate the laser cavity when a fresh mixture of gas is required;
3. Bulk water supply for the heat exchanger and mirror cooling;
4. Solid or aerodynamic window system to transport the high-power beam from the reduced-pressure laser cavity to the work environment at atmospheric pressure;
5. Power meters to monitor beam power; and
6. A beam-shuttering system that delivers the beam to the workstation on demand via operator or workstation computerized numerical controls.

Laser beam welding normally is done inside an enclosure that protects the operator from stray laser beam radiation, such as that reflected from the workpiece. High-power lasers normally use all reflective, water-cooled optical elements that require a minimum of periodic maintenance and are ruggedly built to withstand the rigors of the manufacturing environment.

Laser Beam Switching

The ability to transmit the laser beam directly through the atmosphere over distances of several meters (yards) makes it feasible to use a single laser source to serve several workstations. In production, one laser may serve multiple workstations by employing laser beam-switching mirrors. Beam switching of Nd:YAG lasers can deliver the beam through one or more optical fibers to multiple workstations.

The primary advantage of operating in this manner is that it helps increase the usage factor of the laser beam. While the laser beam is welding in one station, workpieces can be prepared for welding in another. When the task in the first station is completed, the beam can be switched immediately to the second station.

Robotic Laser Beam Delivery

Although laser beam welding was initially performed in much the same manner as electron beam welding by manipulating the joint to be welded under a stationary laser beam, robotic devices can be used to control the laser beams over the joints in a stationary workpiece, or in conjunction with a multi-axis work table for additional positioning flexibility. Consequently, modern robotics with highly flexible laser beam motion systems with the capability of performing three-dimensional welding and cutting tasks are in common use. Figure 14.19 shows 5-axis robot system with a 2-axis positioner using an Nd:YAG laser to weld stainless steel tanks.

Two different methods are used to provide robotic capability to laser beam welding (LBW) and laser beam cutting (LBC). The first involves mounting beam-focusing optics on the end of an articulated mechanical arm and using a series of beam-directing mirror combinations situated in the various joints of the arm or in a delivery fiber to direct the laser beam to these optics. The second method is to mount the laser beam-focusing lens on a Z-travel axis suspended from an X-Y gantry

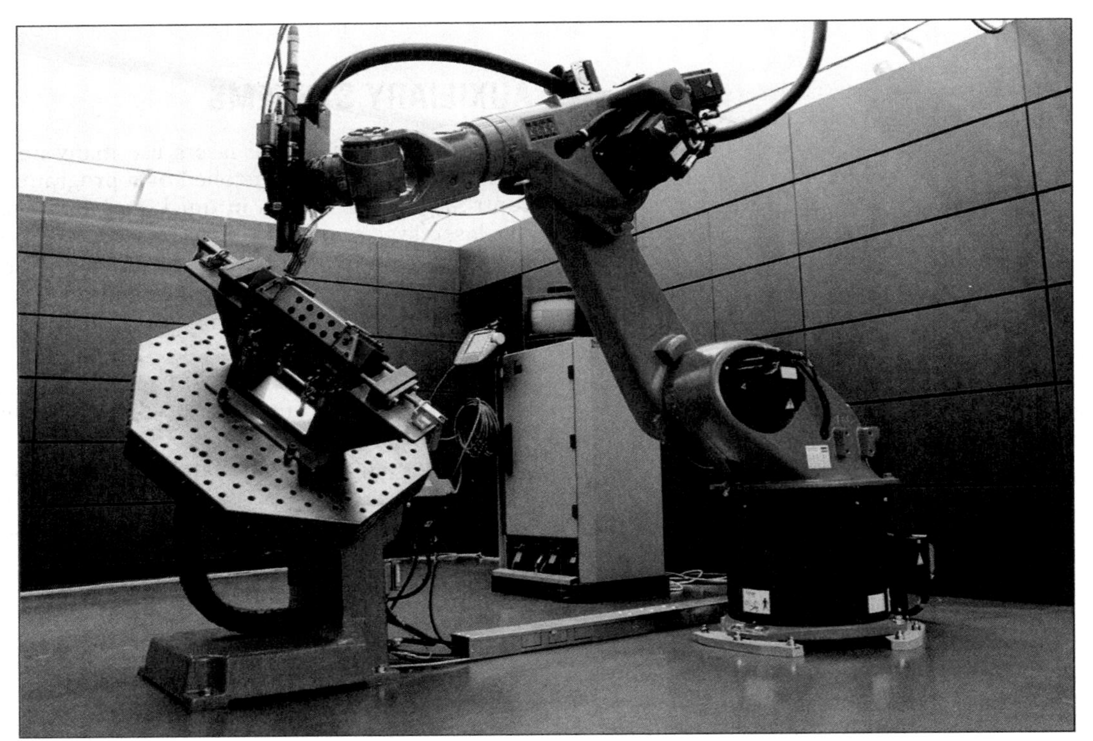

Photograph courtesy of TRUMPF, Inc.

Figure 14.19— Nd:YAG Laser with a 5-Axis Robot and a 2-Axis Positioner

motion system, and then use a series of singular laser beam-directing mirrors (or fiber) to direct the beam to the workpiece. Both of these robotic laser concepts are commonly used in industrial applications to accomplish a variety of welding and cutting tasks.

AUXILIARY GAS SHIELDING

When the weld metal must be shielded until it cools below the oxidation temperature, auxiliary inert shielding gas can be provided using gas-flow hardware compatible with the laser beam welding process. As with conventional welding processes, the main objective is to provide an inert environment over the length of the joint, which is at or above the oxidation temperature, without disturbing the weld pool. In some cases, it also is necessary to provide underbead shielding. Inert gas shielding also can be obtained by placing the workpiece in a glovebox, which can be purged and backfilled with inert gas.

ENERGY ABSORPTION

Effective laser beam welding depends on the absorption of laser energy by the workpiece. However, shiny metal surfaces at room temperature are highly reflective of laser light, particularly at a wavelength of 10.6 µm (417 µin.). For example, the absorption of low-intensity 10.6 µm (417 µin.) CO_2 laser light beam is 40% for stainless steel and as low as 1% for polished aluminum or copper. Absorption levels are higher for Nd:YAG and ruby laser beams. The absorption in most metals increases with rising temperature, and surface temperature increases rapidly at the laser beam impingement point when the metal is exposed to high-power-density laser radiation. At power densities on the order of 1.55 W/mm^2 × 10^3 W/mm^2 (106 W/in.^{2}), the absorption level for most steels and superalloys is approximately 90%. For aluminum and copper, the absorption level occurs at intensities of about 1.55 W/mm^2 × 104 W/mm^2 (1 W/in.^{2} × 10^7 W/in.^{2}) and for tungsten at 1.55 W/mm^2 × 10^5 W/mm^2 (1 W/in.^{2} × 10^8 W/in.^{2}).

Coupling problems occur when metals such as copper, aluminum, and silver are highly reflective of the laser wavelength. The problem usually is solved in one of three ways: changing to a different wave length laser, etching or painting the surface to reduce reflectivity, or using a keyhole weld in which the energy density is great enough to overcome reflectivity.

WELDING TECHNIQUES

The four commonly used techniques of laser beam welding described in this section are conduction-mode, deep-penetration (keyhole), shallow-penetration, and pulsed laser beams used in thin-section welding.

Conduction Welding

Conduction welding occurs when the power density of the beam lacks sufficient intensity to produce a keyhole. In this mode, welds made with laser beams have aspect ratios similar to those made by arc welding processes.

Keyhole Welding

Keyhole welding, or deep penetration welding, is a technique in which a concentrated heat source (the laser beam) penetrates partially or completely through a workpiece, forming a hole, or keyhole, at the leading edge of the weld pool. As the laser beam progresses, molten metal fills in behind the hole to form the weld bead and results in a homogeneous weld.

When laser beam power densities on the order of 1 W/cm^2 × 10^6 W/cm^2 (6.5 W/in.^{2} × 10^6 W/in.^{2}) or greater are achieved, deep-penetration fusion welding is accomplished by a keyhole energy transfer mechanism. At this level of power density, the energy input of the impinging beam is so intense that energy absorbed by the base metal cannot be removed by normal conduction, convection or radiation mechanisms, causing the spot where the laser beam is impinged to instantaneously vaporize. This particular keyhole welding phenomenon is common only to laser beam and electron beam welding, indicating that it is primarily a function of power density and is not dependent on wavelength. Figure 14.20 illustrates the power densities associated with the transition from conduction welding to keyhole welding, which are the same for laser beam welding and electron beam welding. While the welds shown in this figure were produced by electron beam welding, a similar transition and weld profiles are produced by high-power laser beams.

When the material at the interaction point melts and vaporizes, the vapor recoil pressure creates a deep cavity, or keyhole, as shown in Figure 14.21. The keyhole is a vapor column surrounded by a thin cylinder of molten metal. When the workpiece moves relative to the beam, the vapor pressure of the metal sustains the keyhole, and the molten metal surrounding the keyhole flows in the opposite direction of welding, where it rapidly solidifies and forms a weld with a narrow fusion zone. Depth-to-width ratios in the range of 10 to 1 can be obtained with laser beam welding in the keyhole mode. The narrow fusion zones with high depth-to-

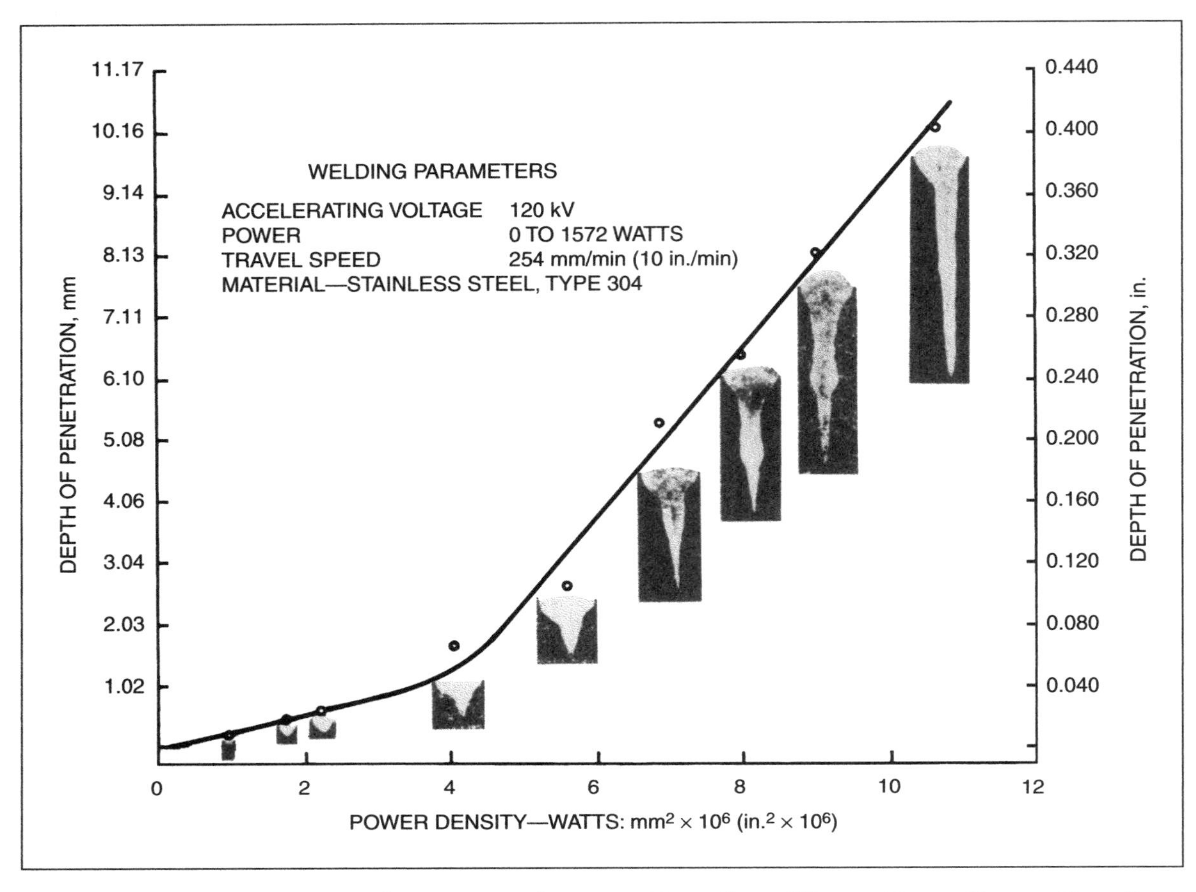

Figure 14.20—Depth of Joint Penetration as a Function of Laser Beam Power Density

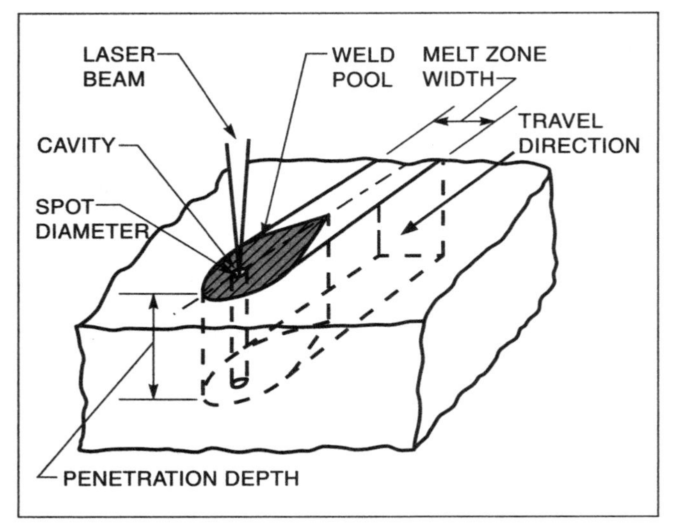

Figure 14.21—Schematic View of a Keyhole Weld

width ratios formed by LBW at atmospheric pressure are similar to electron beam welds made in a vacuum.

Plasma Suppression. When intense heat generated by the laser beam melts the workpiece, some of the liquid metal vaporizes into a gaseous state. A fraction of this gas is ionized by the high-energy beam and becomes plasma. The presence of plasma is detrimental because it absorbs and attenuates the laser beam.

Since plasma can cause the laser beam to be severely attenuated, failure to control or significantly suppress the plasma will result in reduced welding performance and diminished depth of fusion, as well as the occasional collapse of the keyhole. Fortunately, plasma can be suppressed by blowing it away with a stream of gas transverse to the laser beam axis.

For laser power less than 5 kW, helium, argon, or various mixtures of the two may be effectively used to suppress plasma. Nitrogen and CO_2 also may be effectively used if they are compatible with the molten metal and vapor. Helium is preferred for high power levels because it is less likely to break down under the influence of the high-energy-density laser beam.

Flow rates and flow velocities must be adjusted to suppress the plasma, but should not be so great that they disturb the weld pool. Many plasma-suppression jet designs have been effectively used. One of the simplest techniques is to flow the gas across the beam impingement point, where the beam strikes the workpiece, as shown in Figure 14.22. For production welding, the size of the plasma suppression jet can be minimized and precisely aligned to reduce the volume of the flow. For experimental work, a larger plasma suppression jet may be useful because small changes in process parameters such as workpiece distance can be evaluated independently of the precise location of the plasma suppression jet.

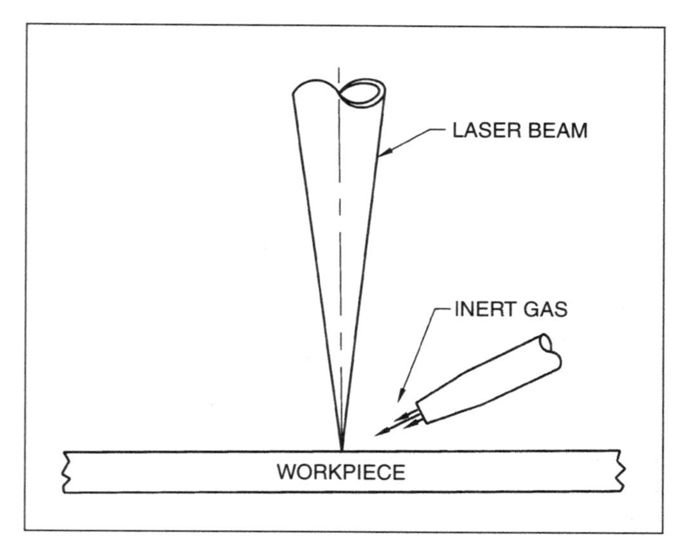

Figure 14.22—Plasma Suppression Using a Transverse Jet of Inert Gas

Shallow-Penetration Welding

When levels of laser output power lower than 1 kW are being utilized and welding speeds approach the depth-of-fusion limit dictated by the lower laser output power being used, the effects of thermal conduction become more prevalent than the deep welding effects described for keyhole welding. The result of this transition generally can be recognized as a broadening of the weld bead from the deep keyhole (high-aspect-ratio weld shape) to the characteristic wineglass shape illustrated in Figure 14.23.

Shallow-penetration welds, such as closure welds for hermetic sealing applications performed with a pulsed laser, typically are conduction-mode welds, with little or no keyhole formation. These welds are used to seal electronic enclosures and batteries. Surface plasma generation aids in transferring the laser beam energy into the workpiece, giving the desired melt zone required to form the closure weld.

In some applications, it may be necessary to reduce the welding speed in order to obtain a wider weld through conduction effects. This method is commonly used in sheet metal applications for butt welding, in which fitup tolerance forces the laser into accepting lower welding speeds. This ensures the reliability and repeatability of the weld.

Pulsed Laser Beam Welding

Pulsed laser beam welding is an excellent process for welding thin sections. Stainless steel as thin as 0.002 mm (0.0001 in.) is successfully welded by pulsed Nd:YAG and pulsed CO_2 laser beam welding machines. These systems are especially well suited for most thin-section welding. As with other processes, welds with complete joint penetration are preferable to welds with partial penetration.

The distinct advantages and limitations of pulsed laser beam welding should be evaluated. The process has the following advantages:

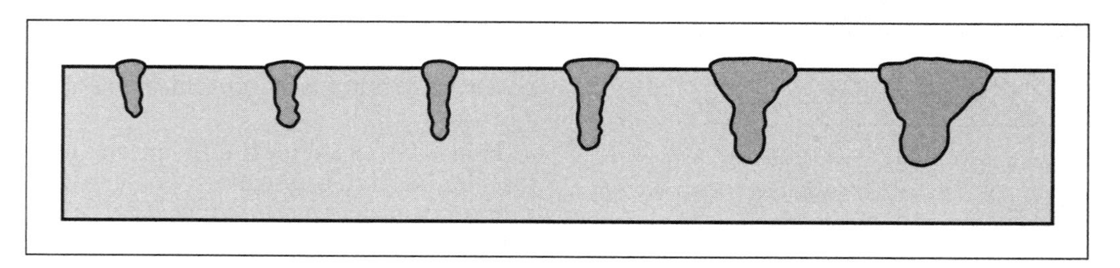

Figure 14.23—Variation of Weld Penetration as Travel Speed Changes Using Continuous-Wave Output Power

1. Produces small fusion and small heat-affected zones,
2. Uses low heat input,
3. Allows execution of precision tack welds, and
4. Uses unique properties of laser heat transfer.

The first two advantages of laser beam welding listed are self-explanatory. The ability to make precision tack welds in a joint is aided by the pulsed mode of welding. A single pulse may be used to tack the joint before the final weld is made. The heat-transfer properties during laser beam welding differ dramatically from welding processes that depend on electrical conductivity to make welds. These properties allow the laser beam welding of many materials that are not electrically conductive. Materials transparent to light of a specific wavelength, however, can be used for hold-down fixturing during laser beam welding, as shown in Figure 14.24.

The following limitations of pulsed laser beam welding should be considered:

1. Extremely high cooling rates may lead to weld defects,
2. Sensitivity to material chemistry may lead to weld defects, and
3. Problems with highly reflective materials may occur.

Disadvantages of the pulsed laser beam process when welding thin sections generally are related to metal cracking or laser coupling problems. Cracking typically is caused by fast cooling rates, which may lead to undesirable brittle phases in some metals. Hot cracking also can occur because of material chemistry problems. These problems may be solved, in most cases, by preheating the workpiece or using a different laser wavelength to reduce the cooling rate, or by changing to a more suitable material to prevent hot cracking.

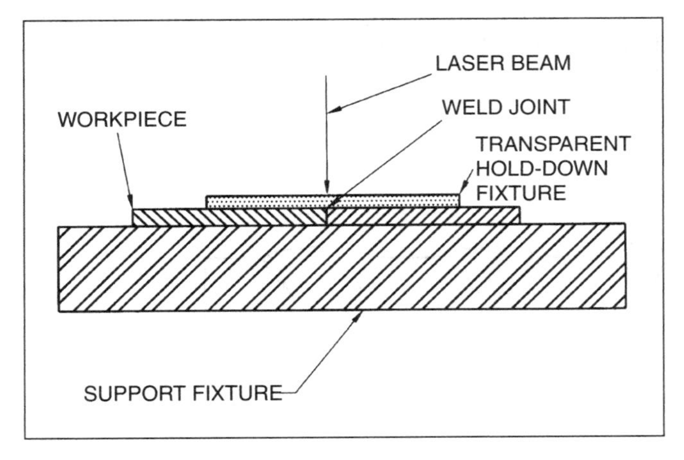

Figure 14.24—Diagram of Transparent Hold-Down Fixturing for Flat Thin-Section Welding

Fixturing is extremely important for thin-section laser beam welding. Close tolerances must be held to maintain joint fitups without allowing mismatch, the misalignment of the workpieces, or excessive root openings. Standing-edge joints are preferable for thin-section welds since the actual weld-joint cross section is enlarged. Butt joints are difficult to design for laser beam welding, and distortion during welding may cause joint fitup problems.

APPLICATIONS

Laser beam welding is used in an extensive variety of applications in the automotive industry, such as in the production of automotive transmissions, air conditioners, and roof assemblies. Annular and circumferential rotary welds are used to join transmission components such as clutch assemblies, clutch housings, synchronization gears and drive gears. These welds need from 3 kW to 6 kW of laser beam power, depending on the weld speed. They also require joint penetration that typically does not exceed 3.2 mm (0.125 in.). Figure 14.25 shows an Nd:YAG laser integrated into a robotic laser system to weld automobile roofs.

Materials welded in automotive applications are usually carbon steels, alloy steels or aluminum alloys. In some cases, components such as gear teeth are selectively hardened before welding. There are many advantages to the laser beam welding of such assemblies. The low-heat input provided by the laser has no effect on the pre-hardened zones adjacent to the weld. Also, the low heat input results in minimal distortion, so that precision stampings often can be welded to finished dimensions. Tailored blanks consisting of different thicknesses or grades of steel can be joined by laser beam welding prior to stamping or forming operations. The high-speed capability of laser beam welding is easily adapted to automation, making it ideal for the high production rates required by the automotive industry. Modern automobiles use large quantities of aluminum alloys to reduce weight and increase fuel efficiency. Laser weldability is important in the selection of these alloys.

Figure 14.26 shows a fully automated, 3-kW system for welding clutch assemblies. This system incorporates a laser beam switch to use one laser for sequential welding at two separate but in-line workstations. While welding in one station, components are loaded into the other station, thus helping to maximize the production capability of this dual-station system. As illustrated in

Photograph courtesy of TRUMPF, Inc.

Figure 14.25—Automobile Roof Welded with a Robotic Laser Beam System

Figure 14.26—Automated 3-kW System Employed for Welding Automotive Clutch Assemblies

Figure 14.27, individual hub and housing components are brought to each station, assembled, pressed to proper dimensions, and then welded. This is accomplished under the control of a central unit. Welding speed is 228 cm/min. (90 in./min) for these assemblies.

Figure 14.28 shows a weld of another transmission component, involving welding a threaded annular boss to a circular ring. In this application a 2.5 kW (CO_2) laser is used to make a 4.75 mm (0.187 inch)-deep weld at 152 cm/min. (60 in./min), employing helium as the shielding gas.

Figure 14.29(A) shows a recuperator plate pair for a heat exchanger, which is joined by welds made around each of the air holes [Figure 14.29(B)] in the cutout pattern. The metal used for these plates is 0.2 mm (0.008 in.) Inconel 625™. They are welded with a CO_2 laser rated for 750 W (continuous output power), operating in an enhanced pulse mode (for example, 1.5 milliseconds per pulse duration and a 200 pulse per second repetition rate). Weld type is lap weld, made at 3 m/min (120 in./min) with a 127 mm (5 in.) focal-length lens.

Figure 14.27—Laser Welding Station Used to Press and Weld Individual Hub and Housing Components into Finished Parts

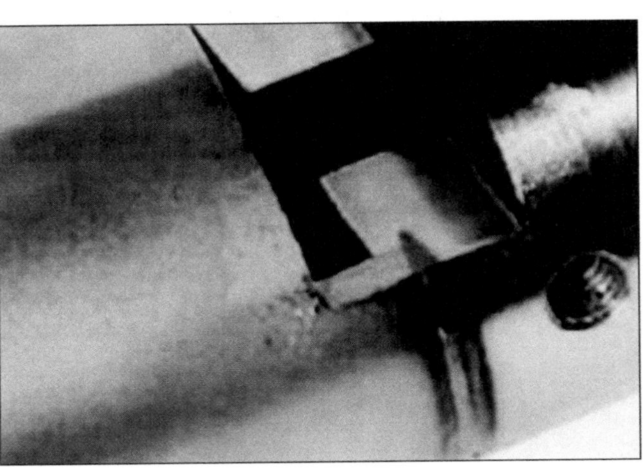

Figure 14.28—Cross Section of a Laser Beam Weld Joining a Boss to a Ring

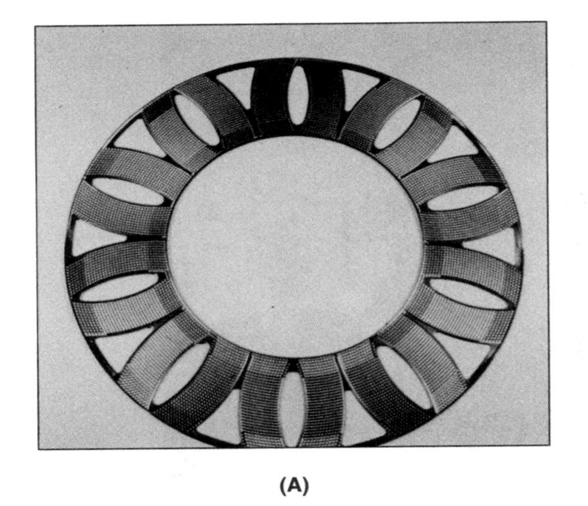

(A)

(B)

Figure 14.29—(A) Recuperator Plate Pair for a Heat Exchanger and (B) Closeup of Laser Beam Welds Around Each of the Air Holes

Figure 14.30 shows the cross section of a Type 416 stainless steel cap welded onto a Type 310 stainless steel body, using a 750-W laser at a welding speed of 114 cm/min. (45 in./min). Joint penetration of the weld into the body component is 1.27 mm (0.050 in.).

Laser beam welding also is used in the production of electrical relays and relay containers, and for sealing electronic devices and heart pacemaker cases. Other applications include the continuous welding of aluminum tubing for thermal windows and for refrigerator doors. The laser is used for welding offshore pipelines and in shipbuilding.

High-Speed Assembly Line Production

Laser beam welding is used in areas where resistance and projection seam welding, and electron beam welding have been the preferred processes. In tube and pipe seam welding mills, laser beam welding machines are used to produce high-quality welds with very few heat-affected zone problems. This type of welding is especially useful for products like stainless steel tubing and pipe welding, where higher-cost grades of stainless steel or expensive heat treating operations are required to ensure that corrosion of the material will not cause problems during service.

Metals Welded

In addition to carbon or alloy steels, laser beam welding can be used to join most similar metals and many dissimilar metals that are metallurgically compatible.

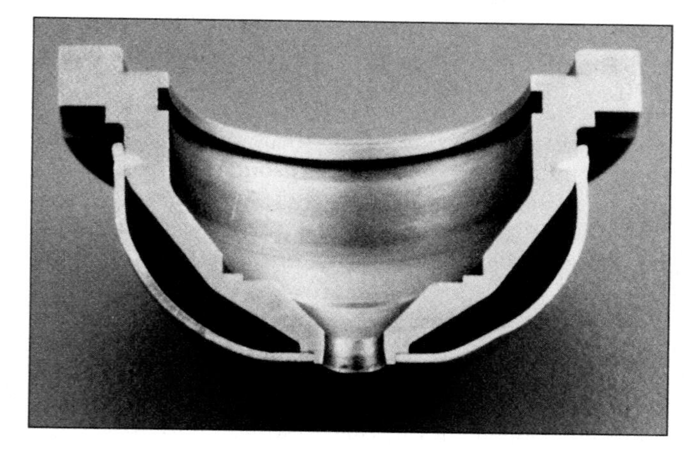

Figure 14.30—Cross Section of Laser Beam Weld in a Type 416 Stainless Steel Cap

Carbon Steel. Low-carbon steels are readily weldable, but if the carbon content exceeds 0.25%, martensitic transformation may cause brittle welds and cracking. Pulsed welding helps minimize the tendency for cracking. Fully killed or semi-killed steels are preferable, especially for structural applications, because welds in rimmed steel may have voids. Steels containing high amounts of sulfur and phosphorus may be subject to hot cracking during welding. Also, porosity may occur in free-machining steels containing sulfur, selenium, cadmium, or lead.

Difficulty has been encountered in welding carburized or nitrided steels. Welds in these alloys are generally porous and exhibit cracks. Occasionally, small amounts of nickel in the form of shims are added to these metals and some alloy steels to increase toughness. Aluminum in small quantities is sometimes added to joints of rimmed steel to reduce porosity caused by entrapped gases.

Stainless Steel and Other Iron and Nickel-Base Alloys. Many stainless steels are considered good candidates for laser beam welding. The low thermal conductivity of these metals permits forming narrower welds with deeper joint penetration than is possible with carbon steels. Stainless steel of the AISI 3XX series, with the exception of free-machining AISI Types 303 and 303Se and stabilized AISI Types 321 and 347, are readily welded. Welds made in some of the AISI 4XX series stainless steels can be brittle and may require post-weld annealing. Many heat-resistant nickel-base and iron-base alloys can be welded successfully with laser beams.

Copper and Brass. Copper-to-copper and brass-to-brass often are welded and they can be welded to other metals with the specialized joint designs used in conduction-mode welding.

Aluminum. Aluminum and its weldable alloys can be joined to make partial-penetration assembly welds. These welds are commonly made by pulsed conduction-mode laser beams to produce items such as hermetically sealed electronic packages. In this type of application, the joint design must retain the aluminum in tension.

Refractory Metals. Refractory metals such as tungsten often are joined by laser beam welding in the conduction mode to produce electronic assemblies, but welding these metals requires higher power than other materials. Special care is required for welds made to seal nickel-plated Kovar™ electronic components to ensure that the plating does not contain phosphorous. Residual phosphorous is common in the electroless nickel-plating used on Kovar components that are to be resistance welded. Titanium alloys and other refractory alloys can be welded, but an inert atmosphere is required to prevent oxidation.

Dissimilar Metals. Dissimilar metal joints commonly are encountered in conduction-mode welds where the twisting of conductors form a mechanical support that minimizes the bending of potentially brittle joints. Dissimilar metals that have different physical properties (reflectivity, conductivity and melting temperature) often are joined in the welding of electrical conductors. Special techniques such as adding extra turns of one material to the joint as opposed to the other may be required to balance the melting characteristics of the materials. Some of these concepts also can be applied to structural and assembly welds, but the possibilities are much more limited.

JOINT DESIGN

Joints designed for laser beam welding must meet the criteria of the manufacturing engineer. Strength and safety specifications must be incorporated. Joints must be accessible to a focused laser beam and must prove economical in consideration of machining operations before and after welding. A good joint design can enhance a laser beam welding production system because tooling design, manufacture, and maintenance are affected by weld joint design. An optimum weld joint design may facilitate the assembly of a weldment before welding. The weld joint design should allow easy access and inspection.

Types of Joints

Welded joints used in laser beam fabrications normally are designed for structural, assembly, sealing, or similar purposes. Butt joints, corner joints, spot welds, lap joints, and variations and combinations of these designs can be welded with the laser beam process. Some of the joint configurations are shown in Figure 14.31. Joints that trap and concentrate optical energy through multiple reflections may be useful when welding highly reflective materials.

Butt Joint. The geometry of a butt joint may be annular, circumferential or linear. Joint cleanliness must be maintained and, as with any welding operation, rust and mill scale must be removed so that fusion-zone integrity is not degraded. A butt joint is illustrated in Figure 14.31(A).

Fitup is an important consideration in joint preparation for laser beam butt welds. In some cases, root openings up to 3% of the metal thickness can be tolerated. However, underfill occurs if the root opening is too wide.

When employing a butt joint design for laser beam welding, hold-down tooling should be considered, especially if the weld is to be repeated in an automated high-volume production application. Butt joints adapt well to production welding operations, but intimate contact must be assured through appropriate fixturing design and dimensional control of the workpiece. When annular butt joints are to be welded, subassemblies can be manufactured to include an interference fit, allowing assembly to take place independently of the welding process. Preassembly can simplify the overall design of a laser beam welding system. A separate press or assembly station also adds another measure of quality control to the production system, because workpiece size tolerances can be gauged before welding.

Butt joints are used in structural, assembly, and sealing applications. Single-pass, deep-penetration butt welds in 31.8 mm (1.25 in.)-thick base metal can be welded with high-power 25-kW lasers using the keyhole penetration technique. A majority of successful laser beam welding applications in the automotive transmission industry use single-pass welds with complete joint penetration of 2.3 mm to 5.1 mm (0.090 in. to 0.200 in.), which require laser power of 5 kW to 9 kW.

Keyhole welding is readily performed with sharply focused laser beams of 1 kW or greater. Assembly or structural joints requiring limited or partial joint penetration also may be welded with lower-power lasers. Keyhole formation does not take place at low power levels, but a weld pool is created via heat conduction from the workpiece surface in a manner similar to arc welding processes.

Corner Joint. Corner joints, illustrated in Figure 14.31(B), often are used in the laser beam welding of assemblies and for sealing applications. The use of a corner joint design is limited by the thickness of the plate or sheet. The amount of power needed is proportionate to the thickness. Thinner materials need less power, but the location of the focus spot is critical and must be reliable throughout the production cycle.

The corner joint has the advantage of good accessibility when tooling and fixturing are important to maintaining the integrity of the fitup.

T-Joint. A laser beam can be aimed at the weld root of an accessible T-joint, shown in Figure 14.31(C). At an optimum directed angle, a focused beam may follow the root opening between the intersecting workpieces. Depending on plate thickness and the amount of laser power used, fusion takes place at the interface between the workpieces. The stress load is transferred from one member to another primarily through the joint root.

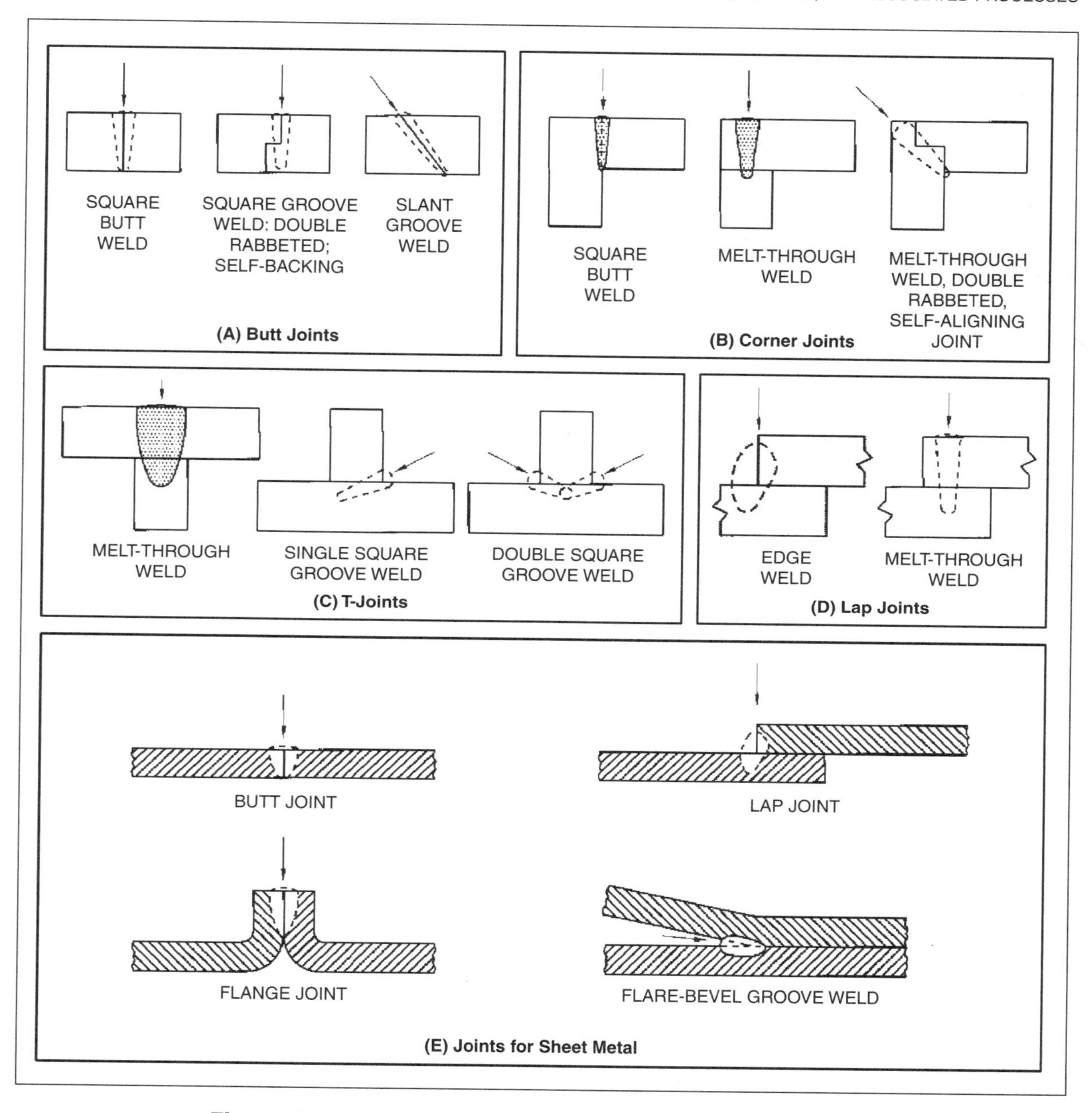

Figure 14.31—Joint Configurations for Laser Beam Welds

If a fillet is formed, it will act to reduce stress. However, laser-formed fillets usually are not as pronounced as typical arc-welded fillets.

Lap Joint. Lap joints, as shown in Figure 14.31(D), typically are used in sheet metal assembly applications. The focused laser beam can be impinged on the top surface, causing weld penetration into or through one or more metal sheets that are in contact.

Intimate contact is not necessarily required in the melt-through weld of a lap joint because molten metal will bridge a limited amount of space, or gapping, between layers. A gap may be advantageous in some applications that use coated steels because it may avoid the entrapment of products outgassed from the coating. The gap tolerance has a relatively narrow band, which is related to workpiece thickness and laser beam spot size. For example, for 0.8 mm to 1.0 mm

(0.030 in. to 0.040 in.) metal welded with a laser beam diameter of about 0.5 mm (0.020 in.), a gap of about 0.08 mm (0.003 in.) is required to prevent porosity resulting from outgassing. However, if the gap exceeds approximately 0.15 mm (0.006 in.), complete fusion of the joint may not occur.

As with any welded lap joint, the fusion zone interface is the stress carrier. To increase the interface area, the laser beam may be directed in a circular or linear pattern by moving the beam delivery optics. Although laser-welded lap joints are not as sensitive to fitup variances as other joints, workpiece fitup generally must be maintained by using suitable fixtures or tooling.

Laser-welded lap joints usually are characterized by a slight reinforcement at the fusion zone faces. For melt-through lap welds, a slight reinforcement of the weld root also may be achieved. When the lower member of the assembly is not penetrated, a deformation resulting from heat distortion usually can be observed on the bottom surface. This may indicate an incomplete joint penetration defect; however, distortion would not be as pronounced as that found in a complete joint penetration. It should be noted that the lap joint is less dependent on the focused laser beam location tolerance than butt joints, corner joints, or T-joints.

Joint Preparation

Successful laser beam welding depends on the careful cleaning of the weld zone and the precise fitup of the workpieces.

Cleaning. All laser-welded joints must be cleaned to ensure that the weld zone is free of rust, mill scale, lubricants or other contaminants. Workpiece cleaning systems are easily integrated with laser beam welding production processes via materials-handling conveyors. As in all welding processes, the type of cleaning system used must comply with local and state environmental laws.

Fitup Requirements. The sensitivity of lap welds to gaps in fitup is discussed in the section on "Joint Design." Butt joints, T-joints, and similar joints also are sensitive to fitup problems due to the small size of the laser beam used for welding. The fitup tolerance for these joints is dependent on material thickness, weld speed, beam diameter, and beam quality. Fitup tolerance normally increases with material thickness. However, as the joint mismatch or root opening increases, the reinforcement normally associated with line-on-line fitup of laser welds decreases. When the root opening is too large for weld bead reinforcement, underfill will occur. If the root opening continues to increase, underfill becomes increasingly severe and reaches the point that fusion does not occur. This condition is characterized by the laser beam actually channeling through the joint without being absorbed by the workpiece at the faying surfaces.

For laser beams used to weld materials up to 12.7 mm (0.5 in.) thick, typical root opening tolerance would be up to 3% of material thickness. The minimum allowable tolerance specified should be based on all fitup tolerance factors and the required fusion-zone geometry.

Weld Joint Mismatch

Laser beams are considerably more tolerant of root misalignment of the joint members than they are of root openings that are too large. Mismatch as great as half the material thickness can be tolerated; however, assembly finish requirements may necessitate much tighter control. Joint specifications and integrity should be used to designate the mismatch tolerance.

LASER BEAM CUTTING

Laser beam cutting is a vital thermal cutting process used in industrial production internationally not only for cutting, but also for other material-removal tasks such as drilling, machining, trimming, and scoring.

The basic technology presented in previous sections of this chapter on laser beam welding also is used in laser beam cutting and related applications. The latest edition of the AWS publication *Recommended Practices for Laser Beam Welding, Cutting, and Drilling*, AWS C7.2:1998 contains comprehensive information.[4]

Laser beam cutting and allied processes usually are performed with electrically pumped, pulsed or continuous-wave Nd:YAG lasers or CO_2 lasers with alternating current, direct current or radio frequency as the means of excitation.

Laser beam cuts are made by locally melting or vaporizing metal or other material with the heat generated by impinging a laser beam on the surface of the workpiece. The process is most often used with an assist gas. The heat from the reaction of the assist gas with the workpiece aids the cutting process. The assist gas also serves to carry molten material away from the cut zone and carries fumes and spatter away from the focusing lens of the laser. To illustrate the function of the assist gas, laser beam cutting with oxygen as the assist gas can be compared to oxyfuel gas cutting, except that the heat that causes the oxygen to react with the workpiece comes from the laser beam rather than the oxygen-fuel reaction.

4. See Reference 3.

The output power of a laser beam cutting machine (described in the Equipment section) is measured in watts or kilowatts. For pulsed lasers, it is necessary to consider not only the average laser power measured in watts but also the energy available in a pulse. The pulsed energy is measured in joules, where one joule is one watt-second. The capability of the laser to cut often depends not on the power or energy of the laser, but the power or energy density at the focus spot created by the lens or mirror. The power density on the workpiece is measured in W/mm^2 (W/in.^{2}) and energy density on the workpiece is measured in J/mm^2 (J/in.^{2}).

Relatively modest amounts of laser energy can be focused to a very small spot size, resulting in high power density. Cutting and drilling require power density in the range of 10^4 W/mm^2 to 10^6 W/mm^2 (6.5 W/in.^{2} × 10^6 W/in.^{2} to 6.5 W/in.^{2} × 10^8 W/in.^{2}). This high concentration of energy causes the melting and vaporization of the workpiece material. The molten material is removed by a jet of gas. Depending on the material, a jet of reactive gas such as oxygen (for an exothermic reaction) can be applied coaxially with the beam, improving the speed of the process speed and of the cut edge quality. The physical mechanisms involved in the cutting and material-removing processes are complex and are influenced by the properties of the material and several process variables.

A high-power CO_2 laser can make cuts in carbon steel exceeding 25 mm (1 in.) in thickness. CO_2 lasers in the range of 400 W to 4000 W are the predominant types used. Nd:YAG lasers are also used for laser cutting. Figure 14.32 shows the laser beam shape-cutting of carbon steel components. Another laser type, the excimer laser, produces short pulses of ultraviolet radiation that are useful for fine machining.

PRINCIPLES OF LASER BEAM CUTTING AND DRILLING

Laser beam cutting and laser beam drilling are two different processes used for cutting, drilling, and other operations related to removing metal or other material. Both processes can use either pulsed or continuous-wave lasers as the primary energy source. Although laser cutting can be achieved with either a pulsed or continuous beam, laser drilling always uses a pulsed beam. Laser beam cutting requires the cutting gas to flow completely through the material to make a cut. Therefore, to start the cut, a hole is drilled through the workpiece in a technique known as piercing. Piercing permits the gas to pass completely through the material and allows the cutting process to take place. Consequently, even continuous-wave lasers used for cutting must have a pulsed mode to achieve piercing. Many factors, such as those listed in Table 14.3, are variables that must be coordinated in laser cutting and drilling.

The laser beam cutting process produces a kerf in the workpiece by the simultaneous action of a focused laser beam with power density greater than 10^4 W/mm^2 (6.5 W/in.^{2} × 10^6 W/in.^{2}) and an assist gas jet. The laser beam impingement functions as a heat source from which the energy is absorbed throughout the thickness of the material. The assist gas jet ejects the molten

Photograph courtesy of ESAB

Figure 14.32—Laser Beam Cutting of Carbon Steel Components

Table 14.3
Factors Influencing Laser Beam Cutting and Drilling Processes

I. Laser Type		
Solid State	**Gas**	
Nd:YAG (pulsed & CW)	Carbon dioxide (pulsed & CW)	
Direct diode	Excimer (pulsed)	
II. Optical Considerations		
Laser Output	**Focusing Optics**	**Focused Beam**
Power	Lens or reflecting optics	Focal length
Divergence	Quality, F-number	Beam diameter
Wavelength	Focal length	Optical mode
Mode structure	Optical material	Wavelength
Polarization	Aperture	Power density
Size	Coating	Depth of focus
Energy per pulse	Cooling	
III. Material Properties		
Surface	**Bulk**	
Condition	Thickness	
Reflective properties	Density	
Absorptive properties	Heat of fusion	
	Heat of vaporization	
	Diffusivity	
IV. Assist Gas Jets		
Inert Gas	**Exothermic Gas**	
Pressure	Pressure	
Orifice size	Orifice size	
Orifice contour	Orifice contour	
Momentum	Momentum	
Depth of cut	Depth of cut	
Standoff distance	Mass flow	
Type	Type	
Additive (water)	Additive (water)	

material created within the keyhole through the root of the kerf. In most cases, the exothermic chemical reaction of an active gas is used to improve cutting efficiency. For example, in the laser beam cutting of steel, about half of the heat input comes from the laser and the other half from the energy generated by the reaction between the steel and the oxygen in the assist gas. Commonly used assist gases are listed in Table 14.4.

Table 14.4
Assist Gases Used for the Laser Beam Cutting of Various Materials

Assist Gas	Material	Comments
Air	Aluminum	Good result up to 0.060 in. (1.5 mm)
	Plastic	Depends on material, some work well, others exhibit burning and chemical breakdown
	Wood	Some burning on surface occurs
	Composites	Depends on formulation, epoxy-containing composites exhibit some edge breakdown
	Alumina	All gases react similarly; air is the least expensive
	Glass and Quartz	Works well, but need to radius any corners
Oxygen	Carbon Steel	Good finish, high speed; oxide layer on surface
	Stainless Steel	Heavy oxide on surface
	Copper	Good surface up to 1/8 in. (3 mm)
Nitrogen	Stainless Steel	
	Aluminum	Clean, oxide-free edges to 1/8 in. (3 mm)
	Nickel Alloys	
Argon	Titanium	Inert assist gas required to produce good cutting of various alloys

PROCESS VARIATIONS

Variations of the laser beam cutting process include laser beam air cutting (LBC-A), laser beam evaporative cutting (LBC-EV), laser beam inert gas cutting (LBC-IG), and laser beam oxygen cutting (LBC-O).

Laser Beam Air Cutting

Laser beam air cutting is a laser beam cutting process variation that melts the workpiece and uses an air jet to remove molten and vaporized material. Using air as an active assist gas is the most common variation of LBC. This is due to increased material removal rates and the inexpensive nature of the gas supply.

Laser Beam Evaporative Cutting

Laser beam evaporative cutting is a laser beam cutting process that vaporizes the workpiece with or without an assist gas, typically inert gas, to aid the removal of vaporized material.[8] This variation utilizes keyhole formation in thick sections to cut materials with smaller kerfs using less power and fewer consumables than other cutting processes. This variation may be used in conjunction with the other three variations.

Laser Beam Inert Gas Cutting

Laser beam inert gas cutting is a laser beam cutting process variation that melts the workpiece and uses an inert assist gas to remove molten and vaporized material. Due to the expense of inert assist gases, this variation is only used on very reactive materials such as titanium and tantalum alloys, in which oxygen and nitrogen contamination could harm the properties of the metal on the kerf edges of the cut.

Laser Beam Oxygen Cutting

Laser beam oxygen cutting is a laser beam cutting process variation that uses the heat from the chemical reaction between oxygen and the base material at elevated temperatures. The necessary temperature for reaction is maintained with a laser beam. The use of this assist gas increases material removal rates, but the cost of consumables is increased. Comparison studies should be completed before switching to oxygen to ensure the economy of the application. Care must be exercised when using oxygen so that excess levels, which could promote fires, do not build up in enclosures or work areas.

ADVANTAGES OF LASER BEAM CUTTING

The laser beam cutting process can be used to advantage in many industrial applications. One important advantage is that the shape of the workpiece can be changed without the extensive reworking that would be required if mechanical tools were used. No tool wear or change-out is involved, and finishing operations usually are not required. Within its optimum base metal thickness range, laser beam cutting is an alternative to punching or blanking, and to the oxyfuel gas and plasma arc cutting processes. Laser beam cutting is especially advantageous for prototyping studies and for short production runs, although it can be used for mass-produced blanking of components. Compared to most conventional cutting processes, noise, vibration, and fume levels involved in laser beam cutting are quite low.

Laser beam cutting results are highly reproducible, and laser systems have achieved operating uptimes greater than 95%. Relative movement between the beam and the workpiece can be easily programmed using CNC workstations. High precision and good edge quality are commonly achieved even in three-dimensional laser cutting.

Other advantages of the laser beam cutting process include the following:

1. Produces a narrow kerf width,
2. Leaves a narrow heat-affected zone,
3. Operates at high cutting speeds,
4. Produces good cut quality,
5. Adapts well to automation, and
6. Requires no mechanical contact between the cutting device and the workpiece.

Modern laser systems with fully integrated robotic systems that readily interface with personal computers provide good control of laser systems and operating variables. The level of operator skill required for laser beam cutting is lessened by the availability of comprehensive software packages and easy-to-learn programming.

Manufacturers continue to reduce the size and weight of CO_2 lasers and to improve beam quality and the pulsing capability of multiple-kilowatt CO_2 lasers. Single-mode Nd:YAG lasers and Nd:YAG lasers with outputs at the multiple-kilowatt level are available. These improvements in system designs provide a high level of accuracy and repeatability of the process. Efficient automatic plate-loading and scrap-removal systems are available that result in increased operating uptime.

LASER BEAM DRILLING

Laser beam drilling is a pulsed operation involving higher power densities and shorter dwell times than laser cutting. Holes are produced by single or multiple pulses. Laser beam drilling is a cost-effective alternative to mechanical drilling, electro-chemical machining, and electrical-discharge machining for making holes of relatively shallow depth. Although lasers designed for cutting are used for drilling, it is most efficiently done with lasers specifically designed for drilling. With these high-energy repetitively pulsed lasers, cutting can be performed by using the laser to drill a series of very closely adjoining holes. Hole diameters produced by laser beam drilling typically range from about 0.075 mm to 1.5 mm (0.003 in. to 0.060 in.).

Examples of laser drilling on jet engine blades and a rotor component are shown in Figure 14.33.

Laser beam drilling produces clean holes with small layers of recast, that is, metal melted during drilling that may cling to the inside surfaces of the hole. When large holes are required, a trepanning technique is used

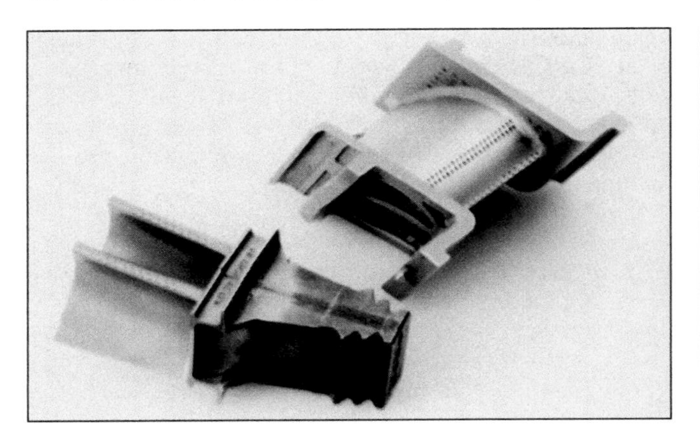

Figure 14.33—Jet Engine Blades and a Rotor Component with Laser-Drilled Holes

Photograph courtesy of TRUMPF, Inc.

Figure 14.34—Nd:YAG Laser Drilling a Lubrication Hole in a Piston Rod of an Engine

in which the laser beam is used in a cutting mode to produce a circle with the required diameter. However, the beam is first used in a drilling mode to drill a hole from which the cutting process starts. This hole-drilling or piercing operation uses a higher gas pressure and a higher peak-power repetitively pulsed beam than is required after the piercing operation, when the beam may revert to a lower peak power or even to continuous operation to perform the cutting operation.

The high-intensity pulsed outputs from solid-state lasers with short wavelengths such as those produced by Nd:YAG, Nd:glass, and ruby are more suitable for drilling. In industrial operations, the Nd:YAG laser predominates in the laser drilling of metals. Figure 14.34 shows an Nd:YAG laser drilling a lubrication hole in a piston rod of an engine. CO_2 lasers are commonly used for drilling nonmetallic materials such as ceramics, composites, plastics, and rubber.

The laser drilling of metal requires a pulsed laser with the beam focused to power densities greater than 10^5 W/mm^2 (6.5 W/in.^{2} × 10^7 W/in.^{2}). When the focused beam strikes a surface, material is melted and volatilized, and the molten and vaporized material is ejected, forming a hole through the workpiece. The depth of the hole normally achieved is approximately six times the diameter of the hole. Multiple pulses may be required to completely penetrate the thickness of the material. Materials up to 25 mm (1 in.) thick can be drilled.

Advantages of Laser Beam Drilling

Laser beam drilling shares most of the advantages of laser cutting. Laser drilling is especially advantageous when the required diameters of the holes are smaller than 0.5 mm (0.020 in.) and when holes are to be made in areas inaccessible to conventional tools. Beam entry can be at glancing angles to the surface, a situation in which mechanical tools would be susceptible to breakage.

Additional advantages of laser beam drilling include the following:

1. Short drilling times;
2. Adaptability to automation;
3. Ability to penetrate difficult-to-drill materials; and
4. No tool wear resulting from contact with the workpiece, as is the case in mechanical drilling.

LASER BEAM FOCUSING

In both the cutting and drilling processes it is necessary to achieve power densities from 10^4 W/mm^2 to 10^6 W/mm^2 (6.5 W/in.^{2} × 10^6 W/in.^{2} to 6.5 W/in.^{2} × 10^8 W/in.^{2}). This is accomplished by focusing the beam with lenses or reflective optics, depending on the laser type and wavelength. In either case, the beam spot size is defined the same way and is expressed by the following relationship:

$$d_s = M^2 \lambda \, F/D \tag{14.3}$$

where

d_s = Focused spot diameter μm (μin.),

M^2 = Magnification,

λ = Laser optical wavelength, μm (μin.).

F = Focal length of lens or mirror mm (in.),

D = Aperture diameter of optical focusing device directing the beam on focusing mirror mm (in.), and

Table 14.5 shows some typical characteristics for three of the most common continuous laser beams, which are dependent on the optical beam mode structure and its divergence.

Cutting

In most cutting applications the laser is operated continuously at power levels between 400 W and 4000 W, somewhat lower than the peak powers of the pulsed lasers described for drilling applications. The lower power densities required for cutting with continuous-wave lasers generally are in the range of 10^4 W/mm² to 105 W/mm² (6.5 W/in.² × 10^6 W/in.² to 6.5 W/in.² × 10^7 W/in.²). This requires spot sizes on the order of 0.5 mm (0.020 in.) at the CO_2 laser wavelength in order to achieve the required power densities.

Drilling

For drilling, short focal-length lenses are used to focus the high-peak-power optical beams from the pulsed lasers to spot sizes on the order of 0.6 mm (0.024 in.) in diameter to achieve power density levels exceeding 10^5 W/mm² (6.5 W/in.² × 10^7 W/in.²). Under these conditions, the material is volatilized and ejected from the workpiece, leaving a partially drilled hole. Complete penetration is achieved with multiple pulses. Low beam divergence can be achieved by specifically configured laser resonators. Low beam divergence enhances drilling characteristics and depth by improving the reflective propagation of the laser beam into the hole being drilled. Beam diameter is controlled by changing the aperture of the focusing device. The aperture may be used to increase the energy density and improve the beam intensity distribution of the focused beam, which may be beneficial for drilling applications.

Table 14.5
Effect of Beam Mode on Focusability

Type of Laser Beam	Magnification (M2)
Gaussian*	1
Unstable resonator*	
M = 2**	4.0
M = 4	3.5

*Magnifications 'M' most used.
**Magnification ratio of an annular beam: M = Beam OD ÷ Beam ID.

LASER AND MATERIAL INTERACTIONS

An important fundamental feature of laser beam cutting and laser beam drilling is the interaction of the laser beam with the material surface. Figure 14.35 shows the relationship between the optical beam, the focusing system, the assist gas jet and the workpiece. In Figure 14.36, this relationship is actualized in the laser-beam cutting of a carbon-steel sheet. The optical lens or mirror focuses the input laser beam to a spot size, as described in Equation 14.3. The location of the spot size (generally aimed within the thickness of the workpiece) of the focal plane relative to the workpiece surface depends on several factors, all governed by the relationships in Equation 14.3. In practice, the optimum location is determined experimentally to suit the application.

Most metals are reflective at the laser wavelengths under consideration. The coupling of the beam and workpiece is inefficient and absorption of the laser energy is initially low. However, the absorption coefficient is a function of the temperature of the material, which changes during the transient phase of the process. This relationship is shown in Figure 14.37, for a laser beam perpendicularly incident on a workpiece surface.

In spite of the initial weak absorption, the light absorption at the workpiece begins to increase with the workpiece temperature at the location of beam impingement and the reflectivity decreases quite rapidly. Temperature and absorption increase until melting occurs, and vaporization even occurs if power levels appropriate for drilling are used. That permits a keyhole to form, which acts as a radiation trap. The laser beam then acts as energy in a distributed heat source within the material and forms a pool of molten metal. When the pool is exposed to vaporization pressures or to a high-pressure gas jet, or both, molten metal is ejected through the root of the workpiece to start the cut. Once this pierced hole has been established, the leading edge of the cut is heated by the beam and chemical reaction (if reactive gas is used), and cutting debris is removed by the high-pressure stream of gas.

To initiate the keyhole cutting process, it is essential that the power density be high enough to overcome the reflection barrier. The depth of the cut is then

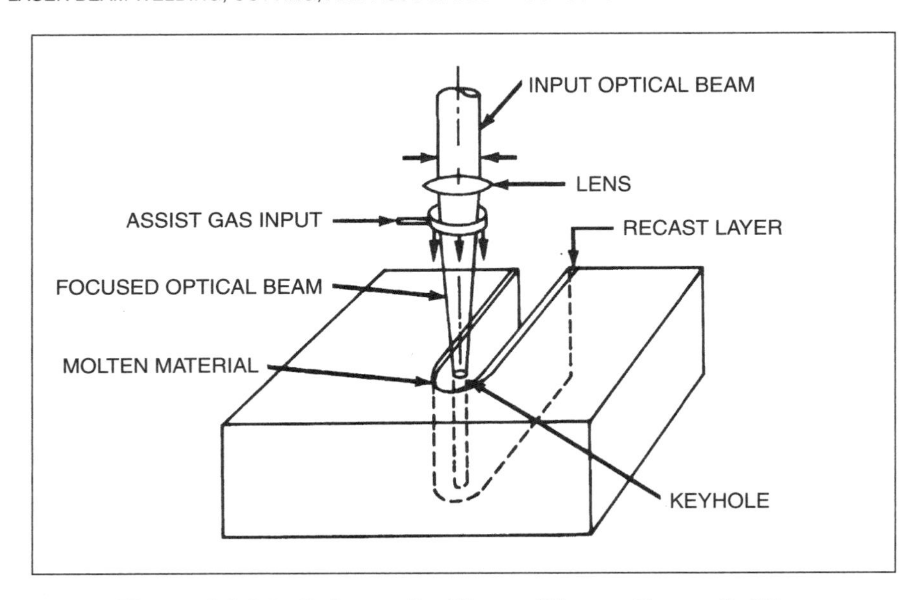

Figure 14.35—Schematic View of Laser Beam Cutting

Photograph courtesy of ESAB

Figure 14.36—Laser Beam Cutting of a Carbon-Steel Sheet

controlled by the melting and vaporization relationships illustrated in Figure 14.38. At power density levels below 5 W/mm^2 × 103 W/mm^2 (3.25 W/in.^{2} × 106 W/in.^{2}), only surface melting is achieved. To develop a keyhole, power densities in the range of 104 W/mm^2 to 105 W/mm^2 (6.5 W/in.^{2} × 106 W/in.^{2} to 6.5 W/in.^{2} × 107 W/in.^{2}) are required. Both melting and vaporization occur within the keyhole range. The complete vaporization required for drilling, including the piercing mode of the laser cutting process, is achieved above this range.

In the laser beam cutting process, the beam is absorbed by a molten layer on the front surface of the kerf, as shown in Figure 14.35. This layer is at a sharp angle of incidence of the beam and is at an elevated temperature. Both of these factors enhance the absorption of the beam in the material, so a high fraction of the beam is absorbed on the cut front. Consequently,

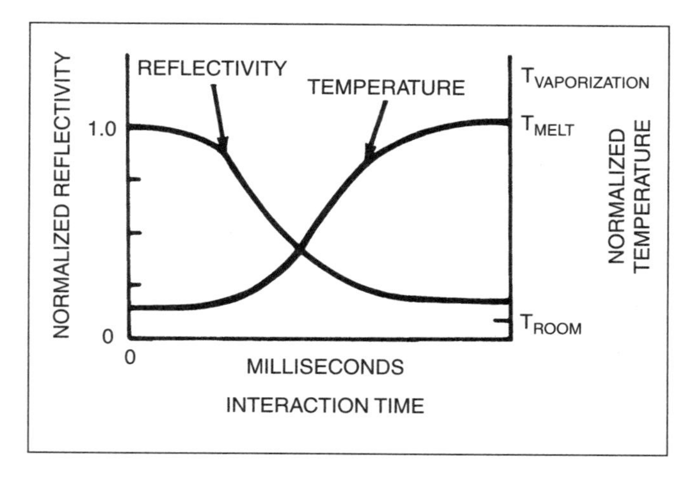

Figure 14.37—Reflectivity and Temperature Transient Time for Typical Metals

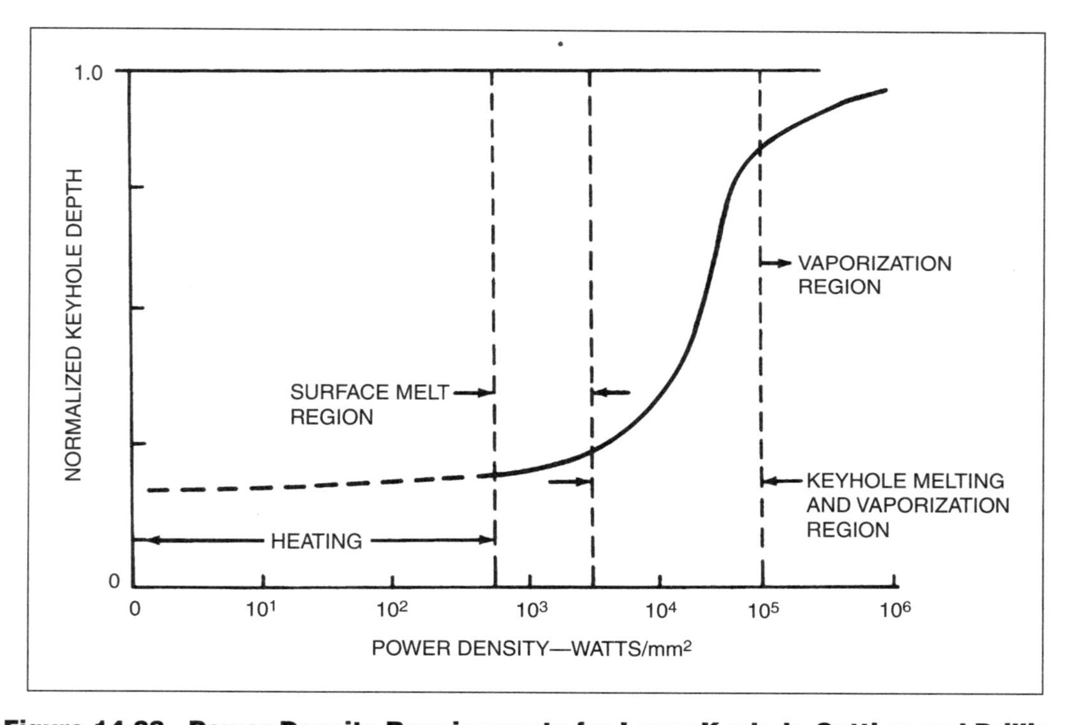

Figure 14.38—Power Density Requirements for Laser Keyhole Cutting and Drilling

once the pierced hole has been produced, cutting can occur at lower laser power densities than those required for drilling.

Assist Gas Jet

The liquid column formed by the laser on the front surface of the kerf can be removed by the force of an assist gas jet. The momentum of the gas from the jet such as that shown in Figure 14.36 ejects a large percentage of the molten material through the root of the kerf. A very thin recast layer is left along the sidewall of the kerf.

As previously mentioned, the gas jet often can be a reactive gas; in this case the cutting capability of the laser system is enhanced because of the additional heat from the reaction of the gas with the material being cut. The gas jet also removes smoke and spatter from the nozzle and focusing lens, enhancing the service life of these components. A beam delivery system for laser cutting with gas assist is shown in Figure 14.39.

PHYSICAL MECHANISMS OF LASER CUTTING

The primary factors that influence the laser cutting process (refer to Table 14.3) are the power level, mode, and polarization. Additional considerations are optical variables such as the focal length, aperture diameter, depth of focus, F-number, and location of the focal plane relative to the workpiece.

The energy balance of the laser beam cutting process is shown in Figure 14.40. The energy sources are the laser and the reactive gas. The primary reasons for energy loss from the beam-workpiece material interaction region are heat conduction, reflection from the cut front, heat of vaporization, convection, radiation, and the energy contained in material ejected at the root of the cut.

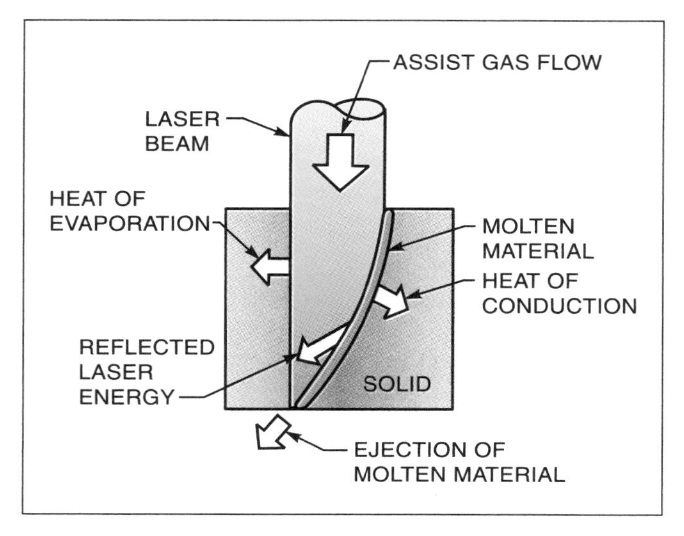

Figure 14.40—Schematic View of Energy Balance of Laser Cutting

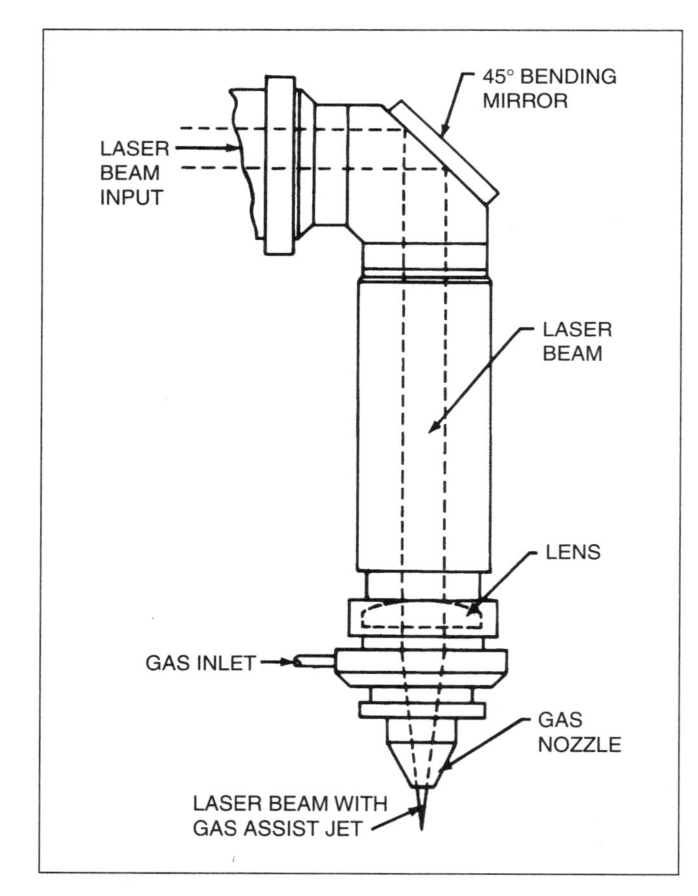

Figure 14.39—Beam Delivery System for Laser Cutting With Assist Gas

The most critical mechanism that occurs is the absorption of the incident radiation on the cut front. Cutting would not be possible without efficient absorption of the impinging beam. Absorption efficiency depends on several factors, including the cut width, the instantaneous slope of the cut front, polarization of the input optical beam and distribution of the optical beam intensity in the longitudinal and radial directions.

The cut front, sketched in Figure 14.40, is the interface between the impinging beam with its assist gas and the molten layer of metal or material. The cut front is the surface along which the molten material is ejected by the momentum of the assist gas from the root of the base metal kerf. Because of the shape of the cut front, the molten material is ejected from the lower surface of the sheet at an angle sloped toward the previously cut material. The angle of the incident laser beam relative to the molten layer along the cut front contributes to the efficiency of energy absorption.

Basic cutting mechanisms and the type of laser used in each case are described in Table 14.6. A subsequent section describes the characteristics of the lasers. For the pulsed Nd:YAG laser, the pressure of vaporized materials can be so great during cutting that some material is also ejected in the solid or liquid state. The continuous-wave and pulsed CO_2 laser removes most material by first melting it and then blowing it from the kerf with an assist gas. If a reactive gas such as oxygen is used, the process becomes exothermic and additional energy is supplied by the oxidation of the material. This

Table 14.6
Basic Cutting Mechanisms

Mechanism	Description
Solid-State Laser-Nd:YAG	
Evaporation	Material removal by volatilization at 10^5 W/mm^2 (6.5 W/in.^{2} × 10^7 W/in.^{2})
	Pulsed only
Gas Laser—CO_2	
Molten material flow	Most material removed in liquid state by means of inert assist gas with beam intensities of 10^4 W/mm^2 (6.5 W/in.^{2} × 10^6 W/in.^{2})
Exothermic	Most material removed in liquid state at beam intensities of 10^4 W/mm^2 (6.5 W/in.^{2} × 10^6 W/in.^{2})
	Additional energy supplied by oxygen gas assist
Excimer Laser	
Photo ablation	Material removed by photo ablation when used on polymers having bond energies below the excimer photon energy level

sequence of events leaves an oxide layer on the cut surface of the material. If the reactive gas is replaced by an inert gas such as nitrogen, no oxide layer is produced when the gas stream removes the molten material. However, this process does not benefit from the extra heat from the endothermic reaction. Cutting speeds with a reactive gas typically are twice those of maximum cutting speeds attained when cutting with an inert gas.

Contour cutting is usually performed with a circularly polarized laser beam. If the cutting were done with a linearly polarized beam, cutting speed and the nature of the cut surface would depend on whether the beam was moving parallel or perpendicular to the direction of polarization. This would result in intermittent quality, which is avoided by using a circularly polarized beam. Most laser cutting systems have a device that converts the output of the laser, which is usually linearly polarized, into a circularly polarized beam.

Photograph courtesy of ESAB

Figure 14.41—CO_2 Laser Bevel Cutting System

LASER BEAM CUTTING EQUIPMENT

Laser beam cutting and drilling require a precisely focused, coherent laser beam. Two primary types of laser sources, the pulsed Nd:YAG laser, operating at a wavelength of 1.06 microns (µm) and the CO_2 laser, operating either a pulsed or continuous mode at a wavelength of 10.6 µm are predominately used for cutting and drilling. Figure 14.41 shows a CO_2 laser integrated into a bevel cutting system.

Another type, the excimer laser, emits ultraviolet energy from molecules called excimers that are created in pulsed electrical discharges and exist only for a very short period of time, on the order of nanoseconds. Excimer lasers use combinations of a gas such as argon, xenon, or krypton with halogens such as fluorine. They operate at wavelengths of 248 µm and are used as a beam source for drilling some materials. When this laser is used on polymers that have bond energies below the excimer photon energies, drilling is accomplished by the mechanism of photo ablation.

CO_2 Lasers

The CO_2 laser, the most common beam source for contour-cutting applications, is a gas-discharge device: it operates by sending an electric current through a mixture of helium, nitrogen, and carbon dioxide. The electrically excited gas produces laser emission at 10.6 µm in the far infrared spectrum. The gas typically is passed through a heat exchanger where it is cooled before being recycled. In some lower-power lasers used for cutting, the heat is extracted through the walls of the laser container. Figure 14.38 shows a laser beam cutting application using a CO_2 laser.

Temporal Characteristics. CO_2 lasers can operate in a continuous-wave mode or in a variety of pulsed modes. The pulse frequency may be as high as 10 kilohertz (kHz). The most common types of pulsing are termed *gated* and *enhanced*. In the gated mode, the laser operates at a peak power level that is within its normal continuous wave (CW) range. The laser input, and hence the output, is turned on and off to produce an output at a controllable duty cycle. Gated pulses can be any length of time less than the period corresponding to the repetition rate of the pulses. In some of the enhanced laser designs, turning on the power results in a more powerful pulse that can have a peak power several times higher than the CW rating. Enhanced pulses, usually lasting about 100 microseconds regardless of repetition rate, are especially useful in cutting reflective material.

Spatial Characteristics. Most high-powered CO_2 lasers have good optical quality controlled by the mirror system of the laser. Lasers with beam outputs up to 1500 W are close approximations of the fundamental Gaussian mode. These beams may be focused to the limit set by the diffraction of light. Higher-power lasers have a somewhat lower-quality mode; nevertheless focused spot diameters of 0.5 mm (0.020 in.) typically can be achieved.

Slow-Flow Lasers

The earliest industrial CO_2 lasers consisted of glass tubes with mirrors at both ends. The laser gas flowed through the tube while electricity was applied near each mirror. These simple and reliable devices produced about 50 W/m (15 W/ft) of discharge length. Slow-flow laser systems become difficult to design if more than 1500 watts are required. Although they may be obsolete, some slow-flow lasers are still in use because they can produce stable, high-quality outputs, and because the large volume of active medium allows for substantial pulse enhancement.

Fast Axial-Flow Lasers

Fast axial-flow lasers use specially designed compressors to flow the gas rapidly down the discharge tube. The gas is cooled in a heat exchanger instead of cooling by conducting it to the wall of the tube, as occurs in a slow-flow laser. Fast axial-flow lasers, excited by either direct current or radio frequency, are available in the multi-kilowatt power range and are commonly used in cutting systems.

Transverse-Flow Lasers

The transverse-flow laser was developed to produce high power in a small package. It does this by circulating the laser gas through the discharge region at high speed and then cooling it with a heat exchanger. Transverse lasers may have asymmetric modes because the electrically excited region is relatively short and wide, and there are variations in gas temperature and current density across this region. Despite these limitations, transverse-flow machines have been highly successful as cutting lasers.

Diffusion-Cooled Lasers

Diffusion-cooled lasers are compact laser systems in which the gas is not circulated or exchanged. The lasing medium is excited by radio frequency energy because it minimizes decomposition of the gas better than electrically excited lasers. Excess heat produced by the excitation energy diffuses to the walls of the resonator, cooling the gas. These lasers are not capable of enhanced pulsing because they operate at a higher gas pressure. They are less complex because they do not require vacuum pumps or replenishment of the gas. Diffusion-cooled lasers are often selected because of this simpler design.

Nd:YAG Lasers

The Nd:YAG laser is the standard laser used for industrial drilling. Continuous-wave Nd:YAG lasers are also used for contour cutting. An Nd:YAG laser contains a crystalline Nd:YAG rod surrounded by xenon or krypton lamps, light-emitting diodes, or laser diodes. Light from the diodes or lamps pump the neodymium atoms to an excited state, in which they emit light at a wavelength of 1.06 µm (0.41 µin.) in the near infrared spectrum. Active cooling of the rod is required to control heat buildup. Multiple laser rods can be placed within the laser resonator to allow multiple-kilowatt power output.

Temporal Characteristics. Industrial Nd:YAG lasers can be operated either in a continuous wave or pulsed mode. If pulsed, the repetition rate is generally below 200 Hz. Control of the power going into the lamps allows precise adjustment of the shape and duration of the laser pulse. The solid lasing medium has a high concentration of light-emitting atoms, so the peak power can be very high. High-energy pulses of short duration evaporate the molten debris from the material being cut or drilled.

Spatial Characteristics of Nd:YAG Beams. Laser rods generate heat in the center and are cooled on the outside. When substantial power is produced, a temperature gradient develops across the diameter of the rod. This gradient induces changes in the refractive index of the rod, which degrades the optical performance of the laser. High-power Nd:YAG lasers have multiple-mode outputs with high divergence, which limits the diameter of the focus spot that can be produced by the laser and the distance that the beam can be propagated. Either the laser generator must be close to the workstation, or the beam must be focused into a fiber optic by which it is delivered to the workpiece. However, fiber optic delivery is used more often for welding than for cutting applications because of the focusing characteristics of the beam from the fiber. Design modifications in the pumping mechanisms of Nd:YAG lasers have improved beam quality by reducing excess heat. Using diode lasers to pump the ND:YAG material improves the beam quality and the overall efficiency of the process.

Other Laser Types

Several other types of lasers use different mediums, including glass, ruby, noble gas, and an electrically modified Nd:YAG.

Nd:glass Laser. Nd:glass lasers are similar to Nd:YAG lasers, except that the laser rod is made of neodymium-doped glass rather than garnet. When glass rather than YAG is used as a matrix, a higher concentration of neodymium atoms can be incorporated in the laser rod. This allows glass lasers to produce stronger pulses than Nd:YAG lasers, which makes them more appropriate for deep drilling. The poor thermal conductivity of glass limits the pulse repetition rate of glass lasers to about one pulse per second, which is too slow for the laser to make cost-effective cuts.

Ruby Laser. Ruby was the first material from which laser emission was observed. The ruby laser is a flashlamp-pumped, solid-state device like the Nd:YAG and Nd:glass laser. Although the ruby laser has characteristics similar to Nd:glass lasers, the ruby laser emits visible light rather than infrared light. It has been largely replaced by other cutting systems, but is still used for drilling applications.

Excimer Laser. The term *excimer* is a contraction of the words "excited dimer." Excimer lasers are pulsed high-pressure gas lasers that emit light at ultraviolet wavelengths. A dimer is basically a molecule that exists only in the excited state, such as in krypton fluoride (KrF). Krypton fluoride molecules are formed when the appropriate gas mixture (typically a noble gas and a halogen) is excited in a pulsed electrical discharge. Lasing occurs when the excited molecule returns to the lower state. The short pulses of excimer lasers combined with their typically short wavelengths make them useful for the ablation of non-metallic materials.

Frequency-Tripled Nd:YAG Laser. The frequency-tripled Nd:YAG lasers are Nd:YAG lasers modified with a special frequency-shifting crystal inserted into the beam path. Like excimer lasers, these lasers also produce short-pulsed ultraviolet output beams. They are used in many of the same applications as excimer lasers.

AUXILIARY SYSTEMS

For production cutting, drilling, machining and other laser beam material-removing tasks, the laser must be integrated with mechanical devices to deliver the beam to the workpiece and provide a means of orienting the workpiece relative to the beam. Computer-controlled laser contour cutters, laser drillers, and machine tools are among the auxiliary machines used. The most commonly used control system is computerized numerical control, which reads numerical data and transforms it into axis motion and laser system commands. The cutting head, consisting of a focusing lens and provision for an assist gas, must be kept at a certain distance from the workpiece. The component subsystems are enclosed in a laser system or laser workstation to provide efficient and safe operation for personnel.

There is considerable variety in the design of these systems. Standard machines are available for a variety of tasks, such as the contour cutting of sheet metal, drilling turbine blades, and cutting shaped or formed components. Special units can be obtained for other tasks, such as slitting sheet materials in production lines.

Beam Delivery Systems

For optimum cut quality, the laser optics must be held motionless, because any vibration or misalignment in the beam delivery system will result in inaccurate or poor-quality cuts. Fixed optics, however, require that the workpiece move and this becomes more complicated when the application involves large sheets of material. The minimum floor area for a fixed-beam system is four times the maximum sheet size, which becomes a problem with large workpieces. Automatic sheet feeding and removal are difficult, as is accurate contour cutting with widely varying loads.

Under these conditions, moving the optics simplifies the laser system. With a moving beam system, sheets move only when they are being loaded onto the cutting table or removed from it. The drive system always handles the same load, allowing optimum servo response. However, with moving optics, steps must be taken to preserve beam divergence, alignment, rigidity, and beam path cleanliness. Figure 14.42 shows a large cutting system configured with a moving CO_2 laser.

Robot systems with several axes provide a sophisticated extension of a moving beam system. These systems allow the beam to impinge on the workpiece from any angle. Figure 14.43(A) shows a 6-axis robot laser cutting system preparing an edge for welding. The cutting head in this system also can be used for welding, as shown in 14.43(B).

Beam Divergence. Laser beams change as they are propagated through space. Diameters and other properties of the beam vary as a function of distance from the source. Since the laser beam in a moving beam system intercepts the focusing lens at different locations, variations occur in the location of the focal spot, causing the focus spot size to vary. The result is that cutting conditions may vary at different locations on the table, requiring either optical or computer-programming compensation.

Alignment. With fixed optics, it is necessary to deliver the laser beam through the optical delivery system without clipping, alteration, or interference by any optical components, contaminants, or apertures. For a moving beam to function properly, the focused beam must translate across the entire workpiece without change in alignment.

Rigidity. A fixed cutting head can be made rigid by using a massive support structure, and this is an option. However, when the cutting head is moving, suppression of vibration and deflection requires careful engineering to eliminate surface roughness or possible deviations from the programmed path, especially when the cut path requires sharp corners.

Beam Path Cleanliness. Dust particles that settle on lenses and mirrors are heated by the beam and can cause damage to these components. For this reason, the beam path of all industrial laser systems must be sealed

Photograph courtesy of ESAB

Figure 14.42—Large Cutting System Featuring a Moving CO_2 Laser

(A)

(B)

Figure 14.43—6-Axis Robot Laser System

against the contaminants that exist in shop environments. This may be simple for a fixed beam but is more complex when the elements are moving. In many moving-beam systems, the laser optics share enclosures with gears, motors, and other sources of contaminants, shortening the service life of the optics.

Focusing Heads

The focusing head concentrates the laser output beam on the workpiece, and also incorporates gas nozzles to provide the assist gases. The head also may contain devices for precise positioning of the beam on the workpiece.

The relation between the surface of the workpiece and the focal spot of the lens is one of the most important variables in laser cutting. Control of this distance is essential to maintaining consistency in the process. The depth of focus for CO_2 lasers is a function of the optical characteristics of the system. In many cases, variation in workpiece thickness can be greater than the depth of focus in the system. It is important, then, for the system to provide a way to hold focus when cutting uneven materials.

Machines that cut flat sheets often have focusing heads with ball bearings that ride on the surface of the workpiece. This approach works well but may mar the finish on some workpieces and it is unsuitable for contoured workpieces. A more sophisticated approach is to attach a drive motor to the focusing head and control it with a sensor. Capacitive probes work well in all orientations and do not protrude from the cutting head. The use of capacitive probes is restricted to electrically conductive workpieces. Contact probes, consisting of a fork or cup around the cutting nozzle, work on any material but function only in the vertical position.

Polarization Control

Polarization of the laser beam is important, particularly when cutting ferrous and other reactive metals with oxygen. Laser light may be polarized in several different ways: linearly, elliptically, circularly, or randomly, depending on the design of the laser. The best results in oxygen-assisted cutting of metals are obtained by using circular polarization. Linear and elliptical polarizations will not produce the same cut in all directions of travel, and they tend to produce a slanted cut edge in some directions. Random polarization will produce an acceptable cut only if it remains consistently random. A laser that produces linearly polarized light can be used to make good cuts by inserting optical devices (known as phase shifters or circular polarizers) into the beam path, that convert the linear polarization to circular polarization.

MATERIALS

Laser cutting is a thermal process: the laser beam cuts materials by heating them until they melt, decompose, or vaporize. Thus the thermal properties of materials determine the response to laser radiation. The optical properties of the materials are equally important, since the energy is transferred in the form of light. In many cases, reactive or inert gases are used to assist cutting, so the chemical response of the material is also important.

Metals

Metals have high thermal diffusivity and optical reflectivity compared to other types of materials. They also melt without decomposing and have very high boiling points. Therefore, laser beam cutting requires high power densities to deliver energy into the metal faster than it can be conducted away and requires an assist gas to remove the liquid metal from the kerf. Within this broad characterization, significant variations exist among metals that affect the suitability of a metal for laser cutting. Typical laser cutting conditions for various materials are shown in Table 14.7.

Carbon Steel. Carbon steel is one of the easiest metals to cut with a laser. When oxygen is used as the assist gas, an examination of the energy balance during cutting shows that most of the heat comes from the exothermic reaction of iron and oxygen, with the laser beam serving as a pilot, or preheat, energy source. The metal heated by the laser burns in the oxygen stream, leaving the area surrounding the kerf unaffected. The cut edge can be extremely smooth, with finishes better than 0.8 µm (32 µin.) achievable in 1.5-mm (0.06-in.) thick sheet.

When cutting poor-quality steel with inclusions, an explosive interaction can occur between the laser radiation and the inclusions, resulting in a rough and unsightly cut edge. Killed steels, which are produced with very low inclusion contents and tightly controlled chemical compositions, consistently produce better quality kerf edges than lower-quality rimmed steels.

Alloy Steel. The term *alloy steel* covers a wide range of metals. Low-alloy steels, such as AISI 4140 and AISI

Table 14.7
Typical Conditions for Laser Beam Cutting

	Thickness		Travel Speed			
Material	mm	in.	mm/s	in./min	Power Watts	Assist Gas
Carbon steel	1.5	0.060	63.5	150	400	O
	3.2	0.125	50.8	120	800	O
	6.0	0.250	33.9	801	200	O
	9.5	0.375	21.2	501	500	O
Stainless steel	1.5	0.060	63.5	150	1500	N
	3.2	0.125	16.9	40	1500	N
	6.3	0.250	16.9	40	650	O
	9.5	0.375	12.7	30	800	O
Titanium Ti6Al4V	1.5	0.060	63.5	150	1500	Ar
Kevlar-epoxy®	3.2	0.125	105.8	250	400	Air
	6.3	0.250	105.8	250	1500	Air
G10 glass-polyester	1.5	0.060	254.0	600	1000	Air
Boron-aluminum	1.0	0.040	127.0	300	1500	Air
Silicon carbide	1.8	0.030	10.6	25	150	Ar

O = Oxygen
N = Nitrogen
Ar = Argon

8620, cut much like carbon steel. The lower impurity levels generally found in low-alloy steels result in improved cut quality compared to commercial cold-rolled carbon steel. Increasing quantities of alloying elements change the behavior of the steel. Tool steels with high tungsten additions cut slowly and with some slag adherence. The addition of chromium reduces the reactivity of the steel with oxygen, which produces adherent scale on the cut edge.

Stainless Steel. Stainless steels are corrosion-resistant alloy steels with two primary classifications: austenitic (300 series) and ferritic/martensitic (400 series). The alloying elements that give corrosion-resistant properties to stainless steels make them resistant to oxidation. As a result, stainless steels react quite differently to laser energy than carbon steels.

The 400 series stainless steels that have chromium as the primary alloying element can be cut cleanly with oxygen as the assist gas, but a tenacious chromium oxide layer remains on the cut edge.

The austenitic stainless steels, which have nickel and chromium additions, tend to produce tenacious slag on the bottom of the kerf edges in addition to the chromium oxide layer. The slag and oxide are serious problems in production, since they require additional operations such as grinding or machining to produce a finished workpiece. If a laser-cut component is to be welded, oxide must be removed prior to welding. Stainless steel can be cut without oxide or slag adherence by using a non-reactive assist gas (nitrogen or argon) instead of a reactive gas (air or oxygen). The edges produced in this manner can be welded without grinding, machining, or other preparation. This clean-cutting operation requires a sacrifice in cutting speed, which is generally recouped by a reduced need for processing after cutting.

Aluminum. Because aluminum at room temperature is very reflective at the CO_2 laser wavelength, the cutting of aluminum requires higher power than the cutting of steel and closer attention to accurate focusing. Reflected energy when the beam is improperly focused on the workpiece can cause damage to the focusing optics or other laser components. Higher power and precise focusing provide enough intensity on the surface to overcome the initially high reflectivity. Laser cuts in aluminum generally leave slag on the lower edge of the kerf that has to be removed by a secondary operation. The use of an enhanced mode of operation may reduce the amount of slag. The laser beam cutting of aluminum results in a rougher surface than that of steel, and microcracking may be present in the laser-cut edges of some aluminum alloys. These two factors may impede the process in some applications.

The output of the Nd:YAG laser is more readily absorbed in aluminum, hence Nd:YAG lasers are most often used in aluminum applications.

Copper and Copper Alloys. Copper, which has higher diffusivity and reflectance than aluminum, is very difficult to cut with low-power lasers. However, copper is easily cut with kilowatt-class CO_2 lasers, as long as they have good TEM_{00} modes, i.e., laser radiation of the fundamental transverse electric mode (the spatial intensity has a Gaussian mode), and the system keeps the beam focused on the workpiece. Nd:YAG lasers, with their high-pulse power and shorter wavelength, are able to cut copper with little difficulty.

Results obtained with the laser beam cutting of copper alloys, such as brass and bronze, are similar to those of cuts in aluminum.

Nickel-Base Alloys. Most nickel-base alloys are intended for use in severe service conditions, for example, high temperature or corrosive environments. While these metals are easily cut by the laser, it is usually necessary to examine the cut component for metallurgical defects, such as microcracking and grain growth, to ensure that it will perform properly. Tests have shown that cuts made with an inert assist gas have higher kerf quality than those produced by oxygen-assisted cutting.

Titanium. Titanium and titanium alloys react with oxygen and nitrogen to form brittle compounds at the cut edge, which is generally an unacceptable condition. Therefore, it is necessary to use argon as the assist gas for titanium cutting. However, argon ionizes easily under CO_2 laser cutting conditions and this can lead to plasma formation above the workpiece. When this happens, the laser output must be revised to obtain consistent results. One effective method to reduce plasma formation is to direct the laser beam at a slight offset from vertical.

Other Materials

An impressive variety of organic, inorganic, and composite materials may be cut by the laser beam cutting processes, ranging from alumina, quartz, and glass to polymers, cloth, paper, and high-performance metal-matrix materials.

Organic Materials. Organic materials commonly cut by the laser beam process include cloth, paper, and many types of polymers. Organic materials generally are decomposed by laser light. The energy required to do this usually is much lower than that required to melt inorganic substances, so cutting can often be done at high speeds or with lower-power lasers. The large volume of decomposition products causes some problems: the process speed and edge quality is limited by the difficulty of removing gases from the kerf. In addition, many organic materials emit toxic compounds during laser cutting. These byproducts must be handled in a manner that eliminates hazards to operators and to the environment.

Cloth presents few problems for laser cutting because it is so thin. Commercially available cloth cutting systems operate with the roll of cloth as feedstock and are capable of extremely high cutting speeds. This requires lightweight optical components to withstand the acceleration and deceleration of the focusing head.

A wide variety of polymers are cut with lasers. The beam causes melting, vaporization, and decomposition of the material. Thermoplastics, such as polypropylene and polystyrene, are cut by shearing the molten material. Thermosets, such as phenolics or epoxies, are cut by decomposition. Materials that decompose in the laser beam during cutting leave a carbon residue on the cut edge. This often must be removed in another operation, such as bead blasting, before the workpieces can be used. Decomposition products of laser-cut polymers are potential health hazards, thus operating personnel must take protection precautions.

Inorganic Materials. Many common inorganic materials, such as alumina, quartz, and glass, have high melting points and poor thermal shock resistance. As a result, special techniques have to be used to process these materials. Localized heat input must be kept at a minimum.

Alumina (Al_2O_3) is often cut or scribed by lasers. Cutting is performed using high-powered pulses to vaporize the material, since recast melted material is a problem. The high melting point of alumina coupled with the low average power of lasers operating in the enhanced pulse mode result in low cutting speeds.

The process of scribing is the standard method of preparing alumina substrates for hybrid microcircuits. Scribing is performed by drilling rows of holes partially though the material. These perforations make it possible to snap the ceramic apart along the drilled lines. For typical 0.64 mm (0.025 in.) thick alumina substrates, holes are drilled to a depth of 0.2 mm (0.008 in.) and 0.18 mm (0.007 in.) apart. For these conditions, laser pulsing at 1000 Hz can scribe at 175 mm/s (7 in./s).

Quartz can be cut in much the same way as metal because it has a high resistance to thermal shock. Continuous-wave CO_2 radiation is used, since quartz is transparent to the 1.06 μm 0.41 μin. light emitted by Nd:YAG lasers. Strains caused by thermal stresses often must be relieved by annealing the workpieces after cutting.

The laser cutting of glass is limited by the poor thermal shock resistance of most glass compositions. This causes complex glass workpieces to crack apart after cutting. Glass also tends to form recast material on the cut edge because it does not have a well-defined melting point.

Composites. Composites are materials consisting of two or more distinct constituents. Usually, one component is fibrous, and the other forms a surrounding matrix. By selecting appropriate matrices and reinforcing elements, composite materials can be engineered to optimize the properties required for a specific use. From the standpoint of laser cutting, the main differences between composites are whether the matrix or the fiber, or both, are organic.

The laser has little difficulty cutting organic fibers set in an organic matrix. An example is a commonly used high-performance composite, Kevlar® (aramid fibers in an epoxy matrix), that is readily cut in thicknesses up to 6.4 mm (0.25 in.). Thicker sections exhibit considerable charring of the cut edge.

The presence of inorganic materials changes the response of composites to laser heating. To cut fiberglass-epoxy, the laser must melt the fiberglass. This takes much more energy than decomposing the epoxy, so the fiberglass component controls the processing rate. Graphite-epoxy is extremely difficult to cut because graphite vaporizes at 3593°C (6500°F). Since graphite has fairly good thermal conductivity, the epoxy near the cutting zone is exposed to high temperatures that decompose it for a significant distance from the cut edge. Laser cutting of graphite-epoxy is thus limited to relatively thin sections of 1.6 mm (0.06 in.) or thinner.

Some of the highest performing materials available are the metal-matrix composites. The addition of refractory fibers to this type of superalloy matrix produces tremendous strength and high toughness at high temperatures. Unfortunately, these same characteristics make it very difficult to machine these materials. Lasers have successfully cut several types of metal-matrix composites. One effect that must be controlled is the melting back of the matrix from the cut edge, which leaves exposed fibers. The use of high-energy pulses, such as those produced by Nd:YAG lasers, minimizes this problem.

PROCESS VARIABLES

Variables that affect the laser cutting process can be categorized as material-related, laser-related, and process-related. These categories include the assist-gas aspects of the process.

Material-Related Variables

The thermal and optical properties of a material determine how it interacts with a laser beam. Any specific material, however, can react differently in various applications, depending on the properties of the material.

Thickness. Thickness is the most important variable affecting how a given material can be cut with a laser. In general, cutting speed is inversely proportional to thickness.

Surface Finish. The surface finish of highly reflective materials, such as pure aluminum or copper, can affect the initial coupling of laser energy. Sheet metal with extremely reflective surfaces may not cut consistently.

Carbon steel often has surface rust and scale. These oxides interfere with the oxygen-assisted cutting process and cause poor edge quality. Because of the contaminated surface and inclusions in the steel, hot-rolled plate is less amenable to laser cutting than cold-rolled plate. For decorative applications, highly finished sheets of stainless steel and aluminum often are provided with coatings of paper or plastic to protect the surface from scratches during manufacturing. These layers do not have much direct effect on the cutting process. However, the assist gas can get under the protective coating, lift the coating off the surface, and cause damage to mechanical parts of the laser, such as focusing heads.

Laser-Related Variables

Most lasers are characterized by their maximum continuous power output. Lasers that operate only in the continuous mode are defined by their maximum power specification. Other lasers can be pulsed to high peak powers but produce low average powers. Nd:YAG lasers, which typically operate in the pulsed mode, will deliver their rated average powers only under specific pulse conditions. Modern Nd:YAG lasers have the capability to shape the power output during individual pulses, and to ramp on and off a sequence of pulses.

Continuous or Pulsed Operation. Some materials, such as steel in thicknesses greater than 6.4 mm (0.25 in.), require high continuous power, while others, such as alumina, must be cut with short high-energy pulses. As power is increased, cutting speed for a specific material also increases. Some lasers produce a sequence of pulses on top of a low-level continuous output. This mode is suitable for cutting reflective materials such as aluminum.

The ability to vary laser power using computerized numerical control is important when cutting intricate shapes, because the motion system often cannot maintain constant speed for all configurations of a workpiece.

Mode. In laser beam cutting, the term *mode* describes the shape of the beam in a direction perpendicular to the direction of propagation. The ideal laser output for cutting is the fundamental Gaussian mode, TEM_{00}. This can be focused on the smallest spot and has the greatest depth of focus (the least change in power density with distance of all possible modes). CO_2 lasers with power lower than 2000 W can be made to produce beams that closely approximate the Gaussian profile.

The mode can be evaluated by electronic mode monitors, involving spinning wires or tubes, or charged coupled device cameras. The shape of the beam can be evaluated by observing burns produced by CO_2 lasers in acrylic material or by Nd:YAG lasers in burn-test paper. The method of evaluating the beam is important. In fact, lasers that appear to have a good beam can be unstable on a millisecond time scale. These short time variations in the mode are common and have significant effects on cut quality.

Beam stability and focusing capability are especially important in the contour cutting of thin carbon steel sheet (3.2 mm [0.125 in.] or less). A stable, low-order beam is required to produce surface finishes of 0.81 µm (32 µin.) or better. Since this is a significant application for lasers, considerable attention should be devoted toward achieving the required level of stability. When the material thickness exceeds the depth of focus of the laser system, the focusing quality of the beam has less effect on edge quality.

Duty Cycle. The highest speed (and often the highest quality) is achieved using a continuous-wave beam. In many cutting applications, however, it is preferable or necessary to use a pulsed beam. When cutting intricate steel components with a CO_2 laser, the motion system may be unable to maintain the linear speeds that are appropriate for good cut quality. Reducing the continuous-wave power by slowing the travel speed is useful, but is ineffective at speeds below about 8.5 mm/s (20 in./min) because of the bulk heating of the workpiece. The solution to this difficulty is to maintain continuous-wave power and pulse the beam to reduce the percentage of time that the beam is on. A typical schedule is to have the beam on 25% of the time with a repetition rate of 500 Hz. The actual repetition rate is dependent on the ability of the laser to generate clean, well-defined pulses. There should be no laser emission during the off part of the cycle, because that heats the material and reduces the benefit of pulsing. The pulses should be uniform in duration and power.

As described above, continuous-wave lasers used for the laser cutting of metals must have a pulsed mode to pierce the material to start the cut. Otherwise, the cutting system program always must start the cut at the edge of the workpiece to produce cosmetically pleasing cut edges. The piercing process delivers a sequence of pulses to the focus spot on the workpiece, vaporizes the material as a hole is formed, and ejects the molten metal produced by the laser.

Certain lasers that can be pulsed electronically have high "simmer" levels, which keep the population inversion in the laser cavity ready to produce a beam very quickly. Simmer current is applied to ensure uniform response to the pulse current, and can result in significant CW output during periods when the beam should be off. This type of beam does not produce satisfactory cuts in thin materials. Mechanical choppers, which mechanically block the beam path, actually have an advantage in this instance, since they produce highly uniform pulses and reduce output to absolutely nothing between the pulses. The main disadvantages of mechanical pulsing are the limited repetition rate and slow response to commands to change the cycle.

Another type of pulse used in cutting is termed enhanced pulsing or superpulsing. This type involves circuitry designed to trigger a pulse with preset duration and power. The pulse is usually repeated at a frequency of 10 Hz to 200 Hz for Nd:YAG lasers and 100 Hz to 5000 Hz for CO_2. Nd:YAG lasers generally operate with enhanced pulsing, and many CO_2 lasers can be modified to operate in this manner. Slow-flow CO_2 lasers can produce several times their CW outputs when operated in enhanced pulse mode. Fast-flow lasers, because of the small volume of active lasing medium used, cannot deliver the same degree of enhancement and lose effectiveness at high repetition rates.

Applications such as ceramic scribing and the cutting of refractory materials are usually performed with enhanced pulsing. Short, high-intensity pulses vaporize substances before they have time to conduct heat away, thus reducing the volume of molten material and minimizing recast. The same technique produces good results in metal-matrix composites.

Beam Propagation. The focus distance and spot size that result when a lens focuses a laser beam are well-defined functions of the distance between the lens and the laser. Because of high divergence (the beam spreads out rapidly as it leaves the laser head), Nd:YAG lasers usually are set within 1.2 m (4 ft) of the focusing lens and maintained at that distance. The small size of an Nd:YAG laser head allows it to be set on a moving axis, so that there are no significant in-process variations in the focus.

CO_2 laser heads are large and operate best when kept stationary. The low divergence of the beam allows it to propagate 10 m (30 ft) or more. A potential problem arises when the process variables are set correctly for a specific laser-to-lens distance and then the distance changes, as with a moving-beam system. Large changes in distance will change the focal spot location, possibly causing loss of process quality.

Process-Related Variables

Once a laser system is built, many of the variables are fixed. However, a large number of variables remain that must be controlled to get reliable cutting. These include control of the focusing lens, and many variables associated with the assist gas.

Focusing Lens. The focusing lens controls the spot size and depth of focus. When cutting metals in thicknesses of 0.25 mm to 10 mm (0.010 in. to 0.38 in.), the use of a lens with a 125 mm (5 in.) focal length produces a combination of optimum spot size and depth of focus.

For thin material, a focal length of 64 mm (2.5 in.) produces better results. The smaller focal length results in a smaller spot size, which allows higher travel speeds. This lowers the heat input into the workpiece, and hence distortion or buckling that otherwise might occur near the cut edge. The depth of focus, however, is only about 25% of that achieved with a 125-mm (5-in.) lens, which limits the utility of the 64 mm (2.5 in.) lens to materials 3.2 mm (0.125 in.) or thinner.

For thick metal or organic materials, a 190 mm or 250-mm (7.5-in. or 10-in.) lens is sometimes used. The increased depth of focus provided by these lenses results in straighter kerfs than those made by shorter lenses and also reduces contamination of the lens by vaporized materials.

Assist Gas Variables

Almost all laser cutting is gas-assisted. Gas-related variables have a significant effect on cutting results. Oxygen reacts with most metals and many nonmetals. Carbon steel usually is cut with oxygen to get the best surface finishes and process rates. Acrylic plastic may be cut with oxygen to achieve very high cutting speeds.

Air is used for cutting aluminum and alumina. Since it is the least expensive assist gas available, it is commonly used for nonmetals, where the gas composition has little effect on finish.

Nitrogen is used with good results when cutting aluminum, stainless steel, and nickel-base alloys. Nitrogen is used rather than oxygen in cutting mild steel when it is desired to apply certain finishes. The slower cutting rates can be tolerated because they avoid a secondary oxide-cleaning operation on the kerf. Nitrogen is reactive to titanium, and should not be used on that metal. Argon, which is inert, must be used to produce clean edges on titanium.

Molten material is removed from the cut by gas pressure. The pressure varies from 35 kilopascals (kPa) (5 pounds per square inch [psi]) to 830 kPa (120 psi) for cuts made in acrylic material with an inert gas.

In general, the effectiveness of the gas-sweeping action improves as the pressure is increased.

For certain applications, however, the assist gas pressure cannot exceed specific limits. For example, in the oxygen-assisted cutting of carbon steel, excess pressure causes uncontrolled burning of the material. Thick plate is usually cut at pressures of 70 kPa to 140 kPa (10 psi to 20 psi) as measured in the cutting head.

In thick organic materials, the use of an assist gas at high pressure results in incandescent decomposition products in the kerf. These radiate energy and widen the kerf in the middle of the cut face.

Nozzles. The gas pressure in the laser head is transmitted to the workpiece through a nozzle that is coaxial with the laser beam. Cutting nozzle diameters vary from 0.75 mm to 3.2 mm (0.030 in. to 0.125 in.). The smaller sizes are used with thin materials. Cutting 6.4 mm (1/4 in.) steel with a nozzle smaller than 1.5 mm (0.06 in.) in diameter gives unsatisfactory results because the pressure profile of a small nozzle does not extend far enough from the beam centerline to clean up the bottom of the kerf. A nozzle too large for a given material uses excessive amounts of assist gas.

Nozzle damage has serious effects on cut quality. Asymmetry in the opening causes changes in performance as the direction of cutting varies. It is not possible to get good results in metal cutting with a dented, spatter-coated, or burned nozzle.

The distance between the nozzle and the workpiece controls the pressure in the kerf. The relationship is not linear because most laser cutting is done at supersonic flow velocities, and the resulting shock waves produce complex pressure patterns. The pressure at the workpiece can, in fact, decrease as the nozzle is brought closer to it. The typical standoff distance is approximately the same as the diameter of the nozzle. It is often more critical to control the nozzle standoff distance than it is to maintain beam focus.

Cutting Speed. The cost effectiveness of laser beam cutting is related to high processing rates. In contour cutting, the processing rate is the same as travel speed. For a given material, thickness, and laser power, there is a range of cutting speeds that produce satisfactory results. If operated above the maximum speed, the cut only intermittently penetrates the sheet thickness, or has excessive slag blocking the kerf. If operated below the minimum, the heat from the cutting process destroys the edge of the workpiece.

For most materials, cutting speed at constant laser power is more or less inversely proportional to thickness. There is a characteristic maximum thickness, above which no cutting will occur at any travel speed, and there are dynamic effects that reduce process efficiency at very high speeds.

It is often impossible to maintain the linear speed that gives the best results. For example, 16-gauge or 1.5 mm (0.060 in.) cold-rolled steel should be cut at about 64 mm/sec (150 in./min) with 500 watts of continuous-wave (CW) power from a CO_2 laser. Typical laser-cut components, however, are too intricate for most motion systems to trace out at this speed. Corners, for example, require that one axis decelerate to zero and the other accelerate to the cutting speed. If motion accelerates at 0.1 g (3.2 ft/sec^2), the table must travel 2.0 mm (0.080 in.) before it reaches 64 mm/sec (150 in./min). The reduced speed in the corner can cause burning of the workpiece.

Laser systems incorporate several ways of dealing with this. One way is to vary CW power as a function of speed. This is very effective when the right relationship is used and the laser responds quickly to power commands. Another method is to change to pulsed operation and cut at low speed. While simpler to implement than power control, pulsing has the obvious disadvantage of increasing processing time.

Controlling the duty cycle as a function of speed has the potential of maximizing speed and quality: at full speed, the laser runs in CW mode. As speed drops in corners or small radii, the laser is pulsed at a high repetition rate. The percent of time that the beam is on is varied to suit the instantaneous speed. The range of travel speeds accommodated by a variable duty cycle is much greater than the range that varying CW power can handle. With a suitable schedule of duty cycle vs. speed, optimum quality can be achieved on any geometry.

CHARACTERISTICS OF LASER BEAM CUTS

Laser systems are preferred for cutting because of the high quality of the cuts produced. The attributes of laser cutting are narrow kerf width, smooth surface finish, clean edges, and good dimensional accuracy.

Kerf widths produced by CO_2 lasers range from 0.1 mm to 1.0 mm (0.004 to 0.040 in.). The usual goal is to generate the narrowest kerf possible, since that minimizes the amount of material that is removed. This has two advantages: the heat input is reduced and accuracy is increased. Lenses with short focal length that have small focused spot sizes are used to produce narrow kerfs. As material thickness increases, the kerf tends to widen. Narrow kerfs in thick material impede the flow of cutting assist gas and the ejection of cut material. Another benefit of narrow kerfs in laser cutting is the small heat-affected zone (HAZ) in the material on the cut edge.

Surface Finish of Laser Cuts

One measure of cut quality is the degree of surface roughness. The ability to produce finished components can depend on maintaining acceptable smoothness. It is possible to cut 20-gauge or 0.92 mm (0.036 in.) carbon steel sheet with an average roughness (Ra) less than 0.81 µm (32 µin.) This type of finish is adequate for most purposes. The stability of the laser, the smoothness of the motion system, and rigidity of the beam delivery all must be optimized to achieve such results. The roughness of the cut edge increases in proportion to the thickness of the steel. The best finish achievable on 9.5 mm (3/8 in.) plate is on the order of 6.4 µm (250 µin.) Ra. Inert gas cutting, used on many metals to obtain weld-ready edges, uses high pressure. The turbulence created by the high pressure increases surface roughness to about 1.6 µm (63 µin.) on material 1.6 mm (0.063 in.) in thickness.

Other materials have different characteristics. Acrylic plastic, which vaporizes during cutting, can have a 0.20 µm (8 µin.) finish on a 25 mm (1 in.) section if the assist-gas flow is low enough to avoid turbulence. Surfaces of plastics, such as polycarbonate, which decompose in the beam, are much rougher. It is hard to produce finishes better than 6.4 µm (250 µin.) on polycarbonate.

The assist gas jet used in laser cutting pushes molten material out of the narrow channel created by a focused laser beam. Under some circumstances a portion of this material adheres to the bottom of the cut edge. This slag or dross is always undesirable and often unacceptable. With carbon steel, dross appears when the focus is incorrect, the gas pressure is too low, or the travel speed too high. Cuts in stainless steel and aluminum are prone to slag adherence; extremely high-assist gas pressures are often needed to eliminate it, even in thin sections. Anti-spatter coatings such as graphite can be used to reduce the adhesion of recast material to the bottom of a laser-cut sheet.

Dimensional Accuracy

The accuracy attained by laser-cut components is a function of the following conditions:

1. Accuracy of the cutting table,
2. Ability of the CNC to contour to the programmed path,
3. Stability of the laser beam, and
4. Control of distortion induced in the workpiece by the cutting process.

Efficient motion systems are available that are capable of maintaining accuracy at the high speeds achieved by laser cutting. The construction of the beam delivery system must be sturdy enough to inhibit vibration and deflection during cutting, because any changes in focal position or focused spot size will change the kerf dimensions, which in turn will alter the dimensions of the workpiece. Movement of the workpiece itself is a significant source of dimensional error. If the workpiece moves because of thermal expansion during cutting, the cut parts will not match the tool path. As laser cutters approach accuracies of 0.0025 mm (0.0001 in.), thermal effects will be more apparent. The best-known way to deal with this thermal distortion is to alter the cutting program in the opposite direction of the distortion.

SETTING UP FOR LASER CUTTING

In addition to the variables mentioned in the previous section, other variables must be considered when setting up for consistent quality cutting. Correct alignment, beam focusing, standoff distance, gas combinations and pressure are among cutting conditions that must be integrated into the cutting schedule.

Correct Alignment

The beam coming from the laser goes through several optical elements before it impinges the workpiece. Correct alignment of the beam-delivery system is essential for proper operation.

It is relatively easy to align a fixed-beam system. As long as the beam does not clip (hit something opaque like the side of a mirror housing) and goes through the center of the focusing lens and gas nozzle, the system is aligned. The stationary elements of a fixed-beam delivery system tend to stay aligned because they are not subject to shaking or vibration.

A moving-beam system is aligned if there is no change in the beam location when the axes are run through their range of motion. This is usually checked for each axis at both extremes of its travel, and mirrors are adjusted until the beam stays in place. Alignment requirements are more stringent on moving-beam systems because they have many mirrors, long beam paths, and moving parts.

Gantry systems are aligned much like moving beam systems. Rotational axes may add some difficulty because they require that the beam be parallel to the axis within 0.2 milliradian (mrad) to maintain nozzle alignment when the axis rotates.

Beam Focusing

Laser cutting quality depends on the focusing of the beam. The relation of the focal spot to the surface of the work is one of the most important variables in the process.

Since there is significant variation between different lenses of the same nominal focal length, it is necessary to test each lens under power. There are several tests for focal spot. One method is to make a flat-position weld along a sloping plate and measure the distance from the nozzle to the narrowest part of the weld bead. Another is to make a series of cuts in thin metal while changing focus and designate the thinnest kerf. Whatever method is used, it is important to be consistent so that continuity is maintained in the process. In performing these tests, it may be necessary to move the cutting nozzle upward from the workpiece, possibly using a larger-diameter orifice than is optimum for the laser cutting process.

The focus spot position for most metal cutting should be at or slightly below the surface of the work. With inert gas cutting, slag is minimized by locating the focus deeper into the material. The focus can be set with calipers, feeler gauges, or through CNC commands.

It is important to keep the focus in the same place throughout the cutting process. This is easily accomplished when cutting flat sheets, but most materials are not flat in comparison to the laser optics depth of focus. The cutting system must have some form of focus control to accommodate out-of-flat sheets.

Standoff Distance

The distance between the tip of the nozzle and the workpiece is an important processing parameter. This usually can be adjusted during initial set-up by screwing the nozzle a certain distance into the focusing head and securing it with a locknut. Individual equipment suppliers may have other means of making the adjustment. Normally, a nozzle standoff distance equivalent to the nozzle orifice size is used. That is, a nozzle with a 1 mm (0.040 in.) orifice usually is adjusted so that the tip of the nozzle is 1 mm (0.040 in.) from the workpiece. This adjustment should take place after setting the position of the focus relative to the surface. This standoff distance provides a satisfactory pressure of assist gas at the surface and allows enough space for ejected material to escape without entering the focusing head.

Assist Gas Pressure

Table 14.4 shows commonly used combinations of assist gases and the materials on which they are used. The focusing lens is part of the pressurized head and is limited in the pressure that it can withstand. A standard 2.5 mm × 28 mm (0.10 in. × 1.1 in.) diameter zinc selenide lens will withstand 550 kPa (80 psi). Higher pressures, such as those used in the inert gas cutting of metals, require thicker lenses. At high pressures, the cost of operating the system increases because of increased gas consumption. In addition, there is more chance of leakage and seal damage.

The focused laser beam must go through the center of the assist gas nozzle to get uniform cutting performance in all directions. All laser systems provide some means of adjusting for concentricity, and there are several ways to check it. One of the most accurate ways is to pierce a hole in thin [0.75 mm (0.030 in.)] steel while observing the material to see the direction that metal is ejected. The lens or nozzle is then adjusted to make the ejected metal form a uniform starburst around the nozzle. This will occur when the beam and nozzle are concentric to within 50 μm (0.002 in.), which is the order of accuracy needed.

TROUBLESHOOTING

Because CO_2 lasers are used to process a wide variety of metallic and nonmetallic materials with varying properties, it can often be difficult to identify the variable that might be the cause of poor-quality cuts. Deterioration in cutting performance usually can be attributed to one of the following conditions:

1. Incorrect cutting speed;
2. Distorted laser output coupler and absorption of radiant energy by vapors in the beam path;
3. Incorrect cutting gas or cutting gas pressure;
4. Incorrect nozzle height;
5. Incorrect lens, focal length, or beam-focus setting;
6. Defective or dirty lens;
7. Incorrect alignment of the beam in the cutting head; or
8. Damaged nozzle tip.

Incorrect Cutting Speed

The effect of cutting speed on cut quality for specific materials is discussed in the Materials section. Often the speed that gives the best quality is somewhat slower than the maximum speed, but slowing down beyond a certain point also will reduce the quality. Consistent results can be obtained when the optimum speed is determined empirically.

When cutting with oxygen as the assist gas, relatively small changes in the chemical composition of ferrous metals can produce significant changes in optimum cutting speed. This implies a need for test cuts when a new batch of material is received.

Generally, cutting speed is directly related to laser power and the power density at the workpiece. If it becomes necessary to reduce the cutting speed from a previously determined optimum, then a fault involving loss of power or power density should be suspected. Loss of power from the laser usually is indicated by a lower reading on an internal power meter of the laser. Loss of power also could occur along the path of the beam between the laser and the focusing lens if any of the reflecting mirrors become dirty. If laser power has not changed, and the material being cut has not changed, then the need to reduce cutting speed may result from reduced power density caused by a larger focus spot on the workpiece surface. The larger spot usually produces a wider kerf than was previously obtained.

Distorted Laser Output Coupler and Absorbing Vapor in the Beam Path

A distorted laser output coupler and organic or other absorbing vapors in the beam path may cause reduced power density. Trichloroethylene, paint solvents, and polymer plasticizing agents are examples of absorptive vapors that frequently are encountered during processing. A small, positive flow of clean dry air or nitrogen into one end of the beam path between the laser and the focusing lens usually is sufficient to remove vapors such as these from the beam path.

Incorrect Cutting Gas or Cutting Gas Pressure

When changing from cutting one type of material to another, it may be necessary to change the type of assist gas. Attempting to cut flammable materials with pure oxygen is a fire hazard. Attempting to cut most metals with air or inert gas when the original parameters were developed with oxygen as the assist gas would produce the same symptoms as cutting with insufficient power.

A deterioration of cut quality also can be noted when the pressure or flow of the assist gas varies from its optimum level. One example of this occurs as a gas cylinder empties. The noticeable effect would be a greater accumulation of oxide slag when cutting metal.

Incorrect Nozzle Height

When cutting metal, the nozzle should be relatively close to the surface, approximately 0.5 mm to 2 mm (0.02 in. to 0.08 in.), to ensure maximum removal of molten slag. When cutting materials that do not produce molten cutting residue, the spacing is less critical. When cutting plastics, such as acrylics that are softened by heat, a frosted effect on the cut edge might be produced by the gas flow from the nozzle. This effect can be minimized by increasing the nozzle-to-workpiece distance and by using minimum gas flow.

A height-control probe can be used to maintain a constant nozzle-to-workpiece distance. Both contact and noncontact sensors are available to detect workpiece undulations. Noncontact devices, such as capacitive sensors, are best suited to metals.

Incorrect Lens Focal Length or Beam-Focus Setting

An incorrect lens focal length or beam-focus setting is most likely to occur after changing a lens. If the point of focus is considerably above or below the nozzle tip, the nozzle will intercept part of the beam and hence the nozzle will become very hot. This will allow less power to reach the workpiece, resulting in a reduction of cutting performance. Reflections off the bore of the nozzle can cause burn marks at the side of the cut. This may be particularly noticeable in thermally sensitive materials, such as paper and plastics.

If the focus is inside or just above the nozzle tip, the beam may pass safely through the nozzle orifice, but the beam will diverge when it reaches the surface of the workpiece. This will result in a wider and more poorly shaped kerf than normal, and because of the loss of power density, lower cutting speed may result.

Defective or Dirty Lens

If the lens becomes defective or dirty it will cause the position of the focal spot to change during cutting operations due to thermal lensing. If this happens during cutting, the effects would be the same as those described for incorrect beam focus setting.

It should be noted that a reduction in focal length also might occur due to a thermally focusing laser output coupler. The output coupler should be routinely checked for cleanliness.

Incorrect Alignment of the Beam in the Cutting Head

If the laser beam is not concentric with the gas jet as it exits the nozzle, an asymmetric cutting action may take place. If the misalignment causes the beam to clip the nozzle, overheating of the nozzle will occur.

The effect of asymmetric metal cutting is that a burnout action may be induced preferentially on one side of the cut, leaving asymmetric dross adherence on the bottom surface or producing a cut in which the kerf is not perpendicular to the surface. When preferential burnout occurs, it is due to the beam being offset in the nozzle toward the side where burnout occurs.

Damaged Nozzle Tip

Damage to the nozzle tip can occur as a result of molten oxide blown onto the nozzle during metal piercing or when attempting to cut metal too fast. The effect is the same as a misalignment of the beam in the nozzle because the gas jet profile will be permanently asymmetric as a result of the damage.

INSPECTION

Inspection criteria for laser cuts are largely dependent on the type of material that has been cut. Visual inspection, nondestructive examination, and sometimes mechanical (destructive) testing are the techniques used. Three areas of concern when inspecting laser cut materials are physical appearance, dimensional accuracy, and thermal alterations.

Visual Inspection

Visual inspection is the first and often the only inspection method used in laser cutting. A laser-cut surface is visually inspected for dross (solidified, oxidized metallic material adhering to the workpiece at the bottom of the cut), which usually is unacceptable. Surface roughness is viewed qualitatively to determine if the cut is similar to previous acceptable cuts produced in the same metal. The color of the cut metal edge also is a consideration.

Physical Appearance. Some metals, such as titanium, stainless steels, and nickel-base alloys usually are cut with inert gas to produce oxide-free cuts with a bright silver appearance. Oxide-free cuts are advantageous when the cut component subsequently will be welded, or when the cut surface is exposed in the end product. The angle of the striations in the laser cut is viewed because of its relationship to the cutting speed. If rates are near the maximum speed, the vertical striations deflect at the root of the cut. Slower cutting speeds yield striations that are completely vertical.

The appearance of the cut surface of plastics, ceramics, wood, and composites varies greatly. Cuts in thermoplastics made with proper conditions have a fire-polished edge. Thermoset plastics are cut with the objective of minimizing charring or discoloration. Ceramics are visually inspected for cracks that result from low ductility and toughness.

Dimensional Accuracy. Dimensional accuracy is another factor in cut quality. Components can be inspected with traditional measuring devices and accuracies of ±25 μm (±0.001 in.) are commonly achieved. A controlling factor in dimensional accuracy is the surface finish of the cut.

The surface roughness on laser-cut metals varies through the thickness of the cut. Typically, the top surface is smoother than the bottom surface. Therefore, the surface roughness measurements should always be taken in the same location.

Taper, or parallelism, is another dimensional value on which laser cuts are evaluated. The minimum value for parallelism is dependent on the type of material cut. Parallelism in metal cutting can be held within 1.5 milliradians to 7.3 milliradians for sheet metal.

Nondestructive Examination

Nondestructive examination (NDE) of metals cut with lasers are inspected for the size of the HAZ, the amount of solidified metal on the cut surface, and the length and number of microcracks penetrating into the recast, the HAZ, and base metal. Common NDE for laser cuts include penetrant, ultrasonic, and in ferromagnetic materials, magnetic particle methods.

The HAZ in laser-cut metals varies with the composition and thickness of the metal. The width of the HAZ is usually between 0.025 mm to 0.25 mm (0.001 in. and 0.010 in.). The HAZ is uniform along the face of the cut. Dross on the bottom of the cut can increase the HAZ at the root.

The laser cutting of metals produces a liquid phase in the metal, creating molten metal that is removed with a coaxial gas jet. Some of the molten metal from this phase clings to the base metal and resolidifies on the walls of the cut surface. This resolidified metal is known as recast or remelt. The depth of the recast usually is only a few micrometers or tens of microinches in laser cutting.

Destructive Examination

During development, cuts made in crack-sensitive materials should be analyzed destructively to determine whether the parameters developed cause cracking in the materials, which may propagate during service. Microcracks can result from the thermal input of laser cutting. The laser cutting process can produce high thermal stresses at the cut edge, which may result in the nucleation of microcracks. These small cracks can be detrimental to the service life of the laser-cut component if the material has poor toughness. Some materials, such as heat-treatable aluminum alloys, hardenable steel alloys, refractory metals, or ceramics, will form microcracks easier than others. For instance, heat-treatable aluminum alloys lose ductility at elevated temperatures, a phenomenon known as hot shortness. These metals are particularly sensitive to the formation of microcracks.

Microcracks are quantified by metallographic cross-section to determine maximum crack length, average crack length, or the total number of cracks. The location of the microcracks is also pertinent. Microcracks in the recast layer may be acceptable, but microcracks extending into the HAZ or base metal may not be acceptable. Acceptability of the size, number, and location of microcracks is dependent on the toughness of the metal, the intended service for the laser-cut component, and industry specifications.

Thermal alterations to nonmetals may be advantageous or detrimental. A laser cut in a fibrous thermoplastic material is an advantage because it seals the edge, while mechanical cuts are detrimental because they leave a frayed edge. Another example is the delamination of composite materials caused by laser cutting, which can lead to premature failure in service.

Mechanical Testing

Thermal alterations to the base metal may have dramatic effects on the service life of a laser-cut component. Inspection for thermal alterations is usually accomplished by mechanical (destructive) testing. Depending on the service conditions for the finished product, typical mechanical testing methods include tensile, torsional, or impact tests.

QUALITY

High-quality laser beam cuts can be produced when proper procedures are followed. The high energy density achievable with this process allows materials to be severed with minimum heat input and minimum alteration of the cut surface. Good cut quality can be achieved in a given application by controlling variables, particularly in selecting the laser mode, the focal length of the lens, and positioning the focus of the laser beam.

Selection of a Laser Mode

A key factor in obtaining good quality with minimum heat input to the material is the selection of the correct laser mode. The laser mode usually is determined at the time the laser is purchased. The mode is subject to change, however, due to accidental misalignment or perhaps damage to the internal laser optics, which leads to degradation of cut quality. The output mode of the laser should be checked periodically, as directed in the maintenance manual from the laser equipment manufacturer, and maintenance should be performed as recommended.

Focal Length of the Lens

The focal length of the lens also affects quality. Usually, as the material thickness is increased, the focal length also should be increased for a given beam diameter. A lens with a longer focal length projects a greater depth of field, which maintains the proper power density to cut the material and minimize kerf taper.

Focal Positioning

The correct focal position in the material is important to maintaining consistent results. Often this is the only variable that can be controlled in real time by using autofocusing techniques. The two most common autofocusing methods are mechanical and capacitive sensors. The focus position and the operation of the autofocus control, if used, should be checked periodically.

ECONOMICS OF LASER BEAM WELDING AND CUTTING

The use of laser beam welding has opened whole new categories of materials that can be welded and cut. An area in which lasers have proven extremely valuable is in the precision cutting of parts that are then joined to form a fabricated product. In addition to the benefits of single-pass, deep-penetration welding, lasers also may be used to join extremely thin sections. Lasers may be used to weld materials that lack electrical conductivity, such as many types of plastics, and weld assemblies that contain materials that are too fragile to be exposed to the electrical currents associated with most other welding processes.

The first cost consideration often is a decision on the viability of the capital investment required for laser equipment. Overhead expenses, usually calculated by a firm's accounting division, must be considered. In production, the primary costs associated with operating laser systems are for consumable items: electricity, optics, flashlamps (solid-state only), and gases (used to generate CO_2 light and to assist in cutting). As with most industrial processes, welding and cutting operator or labor costs are higher than the cost of the consumables.

WELDING

Laser beam welding is traditionally associated with several niche markets, but it should be recognized that many of these markets could not exist without the use of this process.

While the original capital costs may be high for lasers, in many cases lasers may be used to replace costly machining equipment needed as a precursor to welding the finished product. The traditional competitor of laser beam welding is electron beam welding. As noted, each has niche markets in which the other has difficulty in competing, such as the welding of materials with high magnetic susceptibility and non-metallics.

From a capital cost standpoint they are similar because the costs of operating vacuum systems versus large gas handling systems tend to be similar.

CUTTING

Plasma arc cutting and oxygen cutting processes are traditional competitors of laser beam cutting. While capital costs are substantially reduced for competing processes, neither allows the precision cut surfaces or small kerfs and heat-affected zones (HAZs) associated with laser beam cutting. However, LBC usually is cost effective only when very small, highly reactive or thermally sensitive materials must be fabricated. An exception is the use of laser beam cutting in the development and fabrication of dies for prototype blanks, where LBC results in significant cost savings.

When LBC is used for drilling, the only competitive processes are mechanical milling and drilling, and electron beam drilling. Laser beam cutting provides greater speed and smaller dimensional features than available with the mechanical processes, but once again the order of magnitude of the capital cost differential drives LBD to niche applications. Electron beam drilling, while extremely fast, may be applied only to metallic materials, making it competitive in only a portion of the microdrilling market.

Some sample costs for linear cuts in mild steel are listed in Table 14.8. The operator cost accounted for 60% of these costs, machine amortization (which depends on whether the machine is used in one, two, or three shifts per day) accounts for 25% of the total cost in this example.

OPERATING COSTS OF ND:YAG LASER DRILLING

As in laser beam cutting, the primary costs in most Nd:YAG laser beam drilling are for electricity, optics, gas, and flashlamp replacement. The service life of a lamp is in the range of 1 to 10 million pulses, depending on the power used. Typical hourly costs (calculated in the mid-2000s) for the process were between $7 and $10. Operator costs, as in most welding, cutting, and drilling operations, are higher than the cost of consumables.

Table 14.8
Sample Cost Calculations for Linear Cuts in Mild Steel

Metal	Thickness	Cost per Linear Unit
Mild Steel	2.3 mm (0.090 in.)	$0.005 to $0.007 per 25.4 mm (1 in.)
	6.3 mm (0.25 in.)	$0.009 per 25.4 mm (1 in.)
	12.7 mm (0.5 in.)	$0.25 per 25.4 mm (1 in.)
	22.2 mm (0.875 in.)	$0.042 per 25.4 mm (1 in.)

*Based on 2005 prices.

SAFE PRACTICES

Safe practices common to all welding, cutting, and allied processes must be followed in addition to the safety considerations specific to laser beam operations. General aspects of safety in welding and cutting are covered in the AWS standard *Safety in Welding, Cutting, and Allied Processes*, ANSI Z49.1.[5] Safety instructions in the operation manual from the equipment manufacturers must be strictly followed.

The misuse of laser beam welding and cutting equipment can result in permanent damage to the eyes and skin of operators and nearby personnel. Specific precautionary measures must be exercised to avoid other hazards sometimes associated with using lasers. Hazards include servicing high-voltage power sources and exposure to harmful fumes that can be released when certain metals or materials are welded or cut. In some instances, these hazards can be far more significant than the laser-beam-related hazards. Laser safety information is detailed in the American National Standards Institute (ANSI) publication *American National Standard for the Safe Use of Lasers*, ANSI Z136.1;[6] the United States Food and Drug Administration (FDA) publication, *Performance Standard for Light-Emitting Products*,[7] and the Laser Institute of America publication, *Safe Use of Lasers*.[8] Reference to these documents is strongly recommended. Detailed safety training is recommended for personnel working with lasers, including technicians and members of the technical support staff.[9]

5. American National Standards Institute (ANSI) Accredited Standards Committee Z49, (latest edition), *Safety in Welding, Cutting, and Allied Processes*, ANSI Z49.1, Miami: American Welding Society.
6. American National Standards Institute, (ANSI), Accredited Standards Committee Z136, 2006, *American National Standard for the Safe User of Lasers*, ANSI Z136.1, New York: American National Standards Institute.
7. Food and Drug Administration, 1999, Title 21, Volume 8, Parts 800 to 1299, Section 1040 in *Code of Federal Regulations: Federal Performance Standard for Light-Emitting Products*, Washington D.C.: U.S. Government Printing Office.
8. The Laser Institute of America (LIA) *Safe Use of Lasers*, LIA 136.1, Orlando, Florida: Laser Institute of America.
9. Sliney and Wolbarsht, 1980, *Safety with Lasers and Other Optical Sources*, New York: Plenum.

LASER BEAM-RELATED HAZARDS

The American National Standards Institute standard divides lasers into four major classes with some subclasses that define the beam-related hazards associated with each type of laser. These categories are summarized in *Recommended Practices for Laser Beam Welding, Cutting, and Drilling*, AWS C7.2:1998.[10]

Class 1: Denotes exempt lasers or laser systems that cannot, under normal operating conditions, produce a hazard. An example is the type of laser used at grocery store checkout counters to read bar codes. Many material-processing laser systems that are completely enclosed and that are placed in protective interlocked housings may also qualify as Class 1 lasers.

Class 2: Denotes low-power visible lasers or laser systems, which, because of the natural human response of squinting in bright light, normally do not present a hazard, but may produce a hazard if viewed directly for extended periods.

Class 3: Class 3R denotes lasers or laser systems that pose an eye hazard under direct viewing conditions, but have reduced safety requirement because of low risk. Class 3B denotes lasers or laser systems that can produce a hazard if viewed directly, including intrabeam (direct) viewing of specular (concentrated) reflections. Except for high-power Class 3B lasers and special viewing conditions, Class 3 lasers usually do not produce a hazardous diffuse reflection.

Class 4: Denotes lasers or laser systems that can produce a hazard, not only if the laser beam or specular reflections are viewed directly, but also from direct viewing of diffused reflections. In addition to eye damage, the laser beam and its reflections may produce skin burns and fire hazards with flammable materials. Almost all material-processing laser systems are Class 4 unless the beam is totally enclosed.

Laser Safety Officer

All facilities with open-beam, materials-processing laser installations must designate a Laser Safety Officer (LSO). The duties of the LSO include administrative and procedural controls, ensuring that the correct engineering controls are in place and functional. The engineering controls include but are not limited to safety interlocks, signs, signals and light-tight beam enclosures. The LSO is required to determine and inform operators and other pertinent personnel of the maximum permissible exposure for the laser system installed in the facility. The maximum permissible exposure is calculated using the formulas and tables in *American National Standard for the Safe Use of Lasers*, ANSI Z136.1.[11] Limits are given for intra-beam viewing, extended source viewing, and skin exposure. It should be noted that a high-power collimated or unfocused laser beam is more hazardous over long distances than a focused beam, which diverges rapidly. Other hazards of laser beam welding include electrical, ultraviolet and infrared plasma radiation, visible radiation, harmful fumes and gasses, and noise associated with the process.

In welding, cutting, and drilling applications, the visible light emitted from the metal surface can approach that of arc welding processes. In these cases, visible shade- and wavelength-specific filters must be employed.

10. See Reference 3.

ELECTRICAL HAZARDS

The voltages used in lasers are sufficient to cause fatal injuries to personnel; electrical accidents account for most laser-related fatalities. All electrical equipment associated with laser beam materials processing should be installed in conformance to National Fire Protection Association (NFPA) 70, *National Electric Code*[12] and NFPA 79, *Industrial Machinery.*[13] Detailed information on the safe use of welding power sources is presented in Volume 2 of *Welding Handbook, Welding Processes—Part 1.*[14]

All doors and access panels must be secured, either electrically or mechanically, to prevent access to electrical components by unauthorized personnel. All personnel working on or around high-voltage components should be trained in the proper safety techniques for electrical systems, as well as in the technique of removing a victim from an electrical circuit and administering cardiopulmonary resuscitation (CPR). Personnel should be aware of any additional electrical safety requirements of the laser system installed in their facility and adhere to them carefully. Wherever access to high voltage is possible, the area should be properly posted in conformance to the latest editions of *Safety Color Code—Environmental Facility Safety Signs—Criteria for Safety Symbols, Product Safety Signs and Labels,*

11. See Reference 8.
12. National Fire Protection Association (NFPA), 2005, *National Electric Code*, NFPA 70, Quincy, Massachusetts: National Fire Protection Association.
13. National Fire Protection Association (NFPA), 2002, *Industrial Machinery*, NFPA 79, Quincy, Massachusetts: National Fire Protection Association.
14. American Welding Society (AWS) Welding Handbook Committee, C. Jenney and A. O'Brien, Eds., 2001, Welding Handbook, 9th Edition, *Welding Science and Technology, Part 1*, Chapter 1, Miami: American Welding Society.

and Accident Prevention Tags, ANSI Z535, and *Color Chart*, ANSI 535.1.[15]

LASER RADIATION HAZARDS

The two hazards associated with laser radiation are eye and skin damage. Depending on the laser beam wavelength, damage to either the cornea or the retina, or both, are possible. Exposure to radiation from a CO_2 laser 10.6 µm (417 µin.) typically results in corneal damage. The radiation from an ND:YAG laser, at 1.06 µm (41.7 µin.), is much closer to the visible spectrum. Visible and near-infrared radiation can be transmitted by the cornea and lens; the lens will focus the laser light on the retina and can cause retinal damage.

Requirements for eye protection are described in *Practice for Occupational and Educational Eye and Face Protection*, ANSI Z87.1[16] and the Laser Institute of America (LIA) publication *Guide for the Selection of Laser Eye Protection*.[17] It is important that the selected eye protection be clearly marked to ensure that it is only used for the laser wavelength and power levels for which it is intended. It should be noted that eye protection is only intended to protect operators from reflected light and should never be used for protection from either a direct collimated or focused beam.

Skin damage mainly concerns burns. Burns resulting from exposure to high-powered lasers can be deep and can cause severe and permanent damage. ANSI Z136.1, *American National Standard for the Safe Use of Lasers*[18] should be consulted for definitions for all laser wavelengths, including the typical industrial lasers and alignment lasers.

Visible Radiation Hazards

Visible and ultraviolet radiation emitted during laser beam welding, cutting, and materials processing also can be harmful to eyesight. During welding, a bright plume (similar to a welding arc) is generated from the interaction between the laser beam and the workpiece. The size and intensity of this plume is a function of the material being processed, the power level, and the shielding gas being used; therefore, safety guidelines should include these variables.

Injuries resulting from optical radiation during laser beam welding may include skin erythema (sunburn) from the ultraviolet emitted, and photokeratitus (arc eye) through the ultraviolet light to the infrared. These may contribute to the formation of cataracts. Particular care must be exercised when using short-wave-length lasers, such as an excimer laser, in an open-beam application because UV radiation presents a greater risk of delayed effects such as skin cancer and cataracts if the skin and eyes are not fully protected.

The optical viewing system should provide filtering in conformance to ANSI Z87.1 and should include provision for filtering the visible, infrared, and ultraviolet radiation from the plume, as well as the laser radiation. All persons involved with laser beam materials processing should be instructed in the use of proper optical filtering and must be required to use such protection, unless the window enclosure of the laser system filters the energy.

FUMES AND GASES

Using lasers to weld, cut, drill or perform related materials-removing processes may generate fumes, dust and gases that can be hazardous to personnel. Airborne contaminants may include metal particles and oxides, ozone, breakdown from polymeric materials, and other toxic gases. As described in *Safety Requirements for Confined Spaces,* ANSI Z117.1, the excessive buildup of laser discharge gases, shielding gases, and assist gases should be carefully avoided, especially in enclosed spaces where oxygen can be displaced.[19]

Adequate protection to personnel should be provided in conformance to *Safety in Welding, Cutting, and Allied Processes* ANSI Z49.1.[20] The Material Safety Data Sheets (MSDS) available from the material supplier should be consulted to determine what hazards exist when using the product.

CONTROL MEASURES

Control measures center on enclosing as much of the laser beam path as possible and baffling the target area to reduce the chance of hazardous reflections. Care should be taken to install dark welding filters to reduce the level of visible light to a comfortable level. Robotic systems should be designed and installed to limit laser beam traverse to ensure that the laser beam is not directed at personnel. A qualified laser safety officer

15. American National Standards Association, (ANSI) Accredited Standards Committee 535, 2002, *Safety Color Code—Environmental Facility Safety Signs—Criteria for Safety Symbols, Product Safety Signs and Labels, and Accident Prevention Tags*, ANSI 535 and 535.1, *Color Chart*, New York: American National Standards Association.
16. American National Standards Association, (ANSI) Accredited Standards Committee Z87, 2003, *Practice for Occupational and Educational Eye and Face Protection*, ANSI Z87.1, New York: American National Standards Association.
17. Laser Institute of America (LIA), *Guide for the Selection of Laser Eye Protection*. Orlando, Florida: Laser Institute of America.
18. See Reference 6.

19. American National Standards Institute (ANSI) Accredited Standards Committee Z117, *Safety Requirements for Confined Spaces*, ANSI Z117.1, New York: American National Standards Institute.
20. See Reference 5.

should review each laser processing operation before it is placed into service.

OTHER HAZARDS

The hazards of welding not associated with the laser beam include items such as electrical shock, toxic gases and other occupational hazards. The proper safety precautions are clearly defined in ANSI Z49.1. The general safety requirements expressed in this publication, as well as those provided by ANSI Z136.1 should be strictly followed at all times.

The United States Food and Drug Administration (FDA) Center for Devices and Radiological Health (part of the U.S. Department of Health and Human Services) provides a manufacturers' performance characteristics standard, *Performance Standard for Laser Products*.[21]

LASER BEAM CUTTING HAZARDS

The standard used in the United States to design a laser facility is ANSI Z136.1, *American National Standard for the Safe Use of Lasers*. Safety personnel in locations where laser equipment is installed should be familiar with the engineering and operational controls specified in these regulations. This section discusses only those aspects of laser safety that particularly relate to the laser cutting operation. Laser safety guidelines should be stressed and understood by all persons who operate or work in the vicinity of lasers.

Exposure to Direct or Reflected Light

Beam exposure is the most common safety hazard associated with laser cutting. Lasers that can cut engineering materials also can cause great damage to the human body. Laser beam exposure can cause eye damage, including burning of the cornea or the retina, or both. Lasers also can cause severe skin and tissue damage on unprotected areas of the body. Protective measures, as described in the referenced ANSI standards, should be undertaken to ensure that employees are protected from light reflected from the workpiece.

Fumes from Workpiece Materials

The source of fumes in laser beam cutting involves the materials being cut. Some metals and materials emit toxic vapors, dusts, or fumes during the cutting process. The laser beam cutting of polymethyl methacrylate, polyvinyl chloride, Kevlar® and other organic polymeric materials produces byproducts containing toxicants and carcinogenic compounds. Precautions must be exercised to ensure that proper ventilation is supplied in the area of laser operation. Before cutting any material, Material Safety Data Sheets should be consulted to determine associated health hazards and precautionary techniques. Fire extinguishers must be available in case a fire is started by the laser cutting process.

21. U.S. Department of Health and Human Services, Code of Federal Regulations, Title 21, Subchapter J, Part 1040, Latest Edition, *Performance Standards for Laser Products*, Silver Spring, Maryland: U.S. Department of Health and Human Services.

CONCLUSION

Laser beam welding has made significant progress since first becoming available in the 1960s. Although early applications for the process focused on single-pass, deep-penetration welding, present applications include the welding of extremely thin sections and weldments in which very tight tolerances of the finished product must be maintained. While the linear footage of materials welded with laser beam welding will never approach that of various arc processes, the flexibility and general properties of using photons instead of electrons for some types of welding makes laser beam welding a process that always has many uses in commercial fabrication.

Laser beam cutting and drilling also have progressed significantly. The wide range of materials that may be cut or drilled by lasers makes the laser beam a very flexible tool for many precise applications. A fairly new, but promising, use for lasers is for surface treatments. Due to the very high power densities achieved and the physical properties of photons when impinged on many materials, previously unobtainable properties may be imparted on a wide array of materials, including metals, ceramics, and polymers.

BIBLIOGRAPHY

American National Standards Institute (ANSI) Accredited Standards Committee Z136. 2000. *Safe use of lasers*. ANSI Z136.1. New York: American National Standards Institute.

American National Standards Institute (ANSI) Accredited Standards Committee Z117. *Safety requirements for confined spaces*. ANSI Z117.1. 2003. New York: American National Standards Institute.

American National Standards Institute (ANSI) Accredited Standards Committee Z49. 2005. *Safety in welding, cutting, and allied processes.* ANSI Z49.1. Miami: American Welding Society.

American National Standards Institute (ANSI), Accredited Standards Committee Z535. 2002. *Safety color code.* Z535.1. New York: American National Standards Institute.

American National Standards Institute (ANSI) Accredited Standards Committee Z535. 2002. *Accident prevention tags.* Z535.4. New York: American National Standards Institute.

American National Standards Institute (ANSI) Accredited Standards Committee Z87. 2003. *Practice for occupational and educational eye and face protection.* ANSI Z87.1. New York: American National Standards Institute.

American National Standards Institute (ANSI), Accredited Standards Committee Z136. 1999. *Standard for the safe user of lasers.* ANSI Z136.1. Miami: American Welding Society.

American Welding Society (AWS) Committee on Definitions and Symbols. 2001. *Standard welding terms and definitions.* AWS A3.0:2001. Miami: American Welding Society.

American Welding Society (AWS) Committee on High-Energy Beam Welding and Cutting. 1998. *Recommended practices for laser beam welding, cutting, and drilling.* AWS C7.2. American Welding Society: Miami.

American Society for Metals (ASM International) Wegner, C. 1985. *Nontraditional machining: conference proceedings, 2–3 December.* Materials Park, Ohio: ASM International.

Association for Iron and Steel Technology (AIST). 2001. *Standard wrought steels.* Warrendale, PA: Association for Iron and Steel Technology.

American Welding Society (AWS) Welding Handbook Committee. Jenney, C. L., and A. O'Brien, eds. 2001. Welding Handbook, Volume 1, 9th ed. *Welding science and technology.* Miami: American Welding Society. Chapter 9.

Laser Institute of America (LIA). Laser Safety Committee. *Guide for the selection of laser eye protection.* 5th Edition. Orlando, Florida: Laser Institute of America.

Laser Institute of America (LIA). Ready, J. F., and D. F. Farson, eds. *Handbook of laser materials processing.* Orlando, Florida: Laser Institute of America.

Laser Institute of American (LIA). Sliney, D. M., and W. J. Marshall *Laser safety guide.* 10th Edition. Orlando, Florida: Laser Institute of America.

Laser Institute of America (LIA). 2004. *Laser safety information bulletin.* Orlando, Florida: Laser Institute of America.

National Fire Protection Association (NFPA). 2005. *National electric code.* NFPA 70. Quincy, Massachusetts: National Fire Protection Association.

National Fire Protection Association (NFPA). 2002. *Industrial machinery.* NFPA 79. Quincy, Massachusetts: National Fire Protection Association.

Occupational Safety and Health Administration (OSHA). (Latest annual edition). Code of federal regulations (CFR) Title 29, Part 1910 (in its entirety). *Occupational safety and health standards.* Washington, D.C.: Superintendent of Documents. U.S. Government Printing Office.

Sliney and Wolbarsht. 1980. *Safety with lasers and other optical sources.* New York: Plenum.

United States Department of Health and Human Services, Code of Federal Regulations, Title 21, Subchapter J, Part 1040. *Performance standards for laser products.* Silver Spring, Md: U.S. Department of Health and Human Services.

United States Food and Drug Administration. 1999. Title 21, Volume 8, Parts 800 to 1299, Section 1040 in Code of Federal Regulations. *Federal performance standard for light-emitting products.* Washington D.C.: U.S. Government Printing Office.

SUPPLEMENTARY READING LIST

AISI, Advanced High-Strength Welding Information http://www.asp.org/publications.htm

Anthony, P. 1989. Choosing the right CO_2 laser. *Industrial Laser Review.* 4(2).

Banas, C. M. 1978. High-power laser welding. *Optical Engineering.* 17(3): 2410–2416.

Brown, C., and C. M. Banas. 1986. High-power laser beam welding in reduced-pressure atmospheres. *Welding Journal.* 65(7): 48–53.

Çam, G., S. Erim, C. Yeni, and M. Koçak. 1999. Determination of mechanical and fracture properties of laser beam welded steel joints. *Welding Journal.* 78(6): 193-s–201-s.

Crafer, R. C. 1978. Improved welding performance from a 2-kW axial flow CO_2 laser welding machine. *Advances in welding process, 4th International Conference.* Harrogate, England, 9–11 May 1978, Cambridge, England: The Welding Institute.

Dodd, A., and J. Bialach. 2003. Lasers repair turbine blades. *Welding Journal.* 82(8): 43–45.

Firestone, R. F. 1987. Lasers and other nonabrasive machining methods for ceramics. *Advanced Ceramics Conference.* February 1987, Cincinnati, Ohio. Hubbard Woods, IL: Metals Science Co.

Frewin, M. R., and D. A. Scott. 1999. Finite element model of pulsed laser welding. *Welding Journal.* 78(1): 15-s–22-s.

Graf, T., and H. Staufer. 2003. Laser-hybrid drives VW improvements. *Welding Journal.* 82(1): 43–48.

Harry, J. E. 1974. *Industrial lasers and their applications.* New York: McGraw-Hill.

Holbert, R. K., T. M. Mustaleski, and L. D. Frye. 1987. Laser beam welding of stainless steel sheet. *Welding Journal.* 66(8): 21–25.

Jon, M. C. 1985. Non-contact acoustic emission monitoring of laser beam welding. *Welding Journal.* 64(9): 43–48.

Kern, M., P. Berger, and H. Hugel. 2000. Magneto-fluid dynamic control of seam quality in CO_2 laser beam welding. *Welding Journal.* 79(3): 72-s–78-s.

Li, M. Y., and E. Kannaey-Asibu, Jr. 2002. Monte Carlo simulation of heat-affected zone microstructure in laser-beam-welded nickel steels. *Welding Journal.* 81(3): 37-s–44-s.

Mazumder, J., and W. M. Steen. 1981. Laser welding of steels used in can-making. *Welding Journal.* 60(6): 19–25.

Messler, R. W. Jr., J. Bell, and O. Craigue. 2003. Laser beam weld-bonding of AA5754 for automobile structures. *Welding Journal.* 82(6): 151-s–159-s.

Missori, S., F. Murdolo, and A. Sili. 2004. Single-pass laser beam welding of clad steel plate. *Welding Journal.* 83(2): 65-s–71-s.

Missori, S., and A. Sili. 2000. Structural characterization of C–Mn steel laser beam welded joints with powder filler metal. *Welding Journal.* 79(11): 317-s–323-s.

Morgan-Warren, E. J. 1979. The application of laser welding to overcome joint asymmetry. *Welding Journal.* 58(3): 76s–82s.

Pastor, M., H. Zhao, R. P. Martukanitz, and T. DebRoy. 1999. Porosity, underfill and magnesium loss during continuous-wave Nd: YAG laser welding of thin plates of aluminum alloys 182 and 5754. *Welding Journal.* 78(6): 207-s–216-2.

Powers, D. E., and G. R. LaFlamme. 1988. EBW vs. LBW: A comparative look at the cost and performance traits of both processes. *Welding Journal.* 67(3): 25–31.

Ram, V., G. Kohn, and A. Stern. 1986. CO_2 laser beam weldability of zircaloy 2.™ *Welding Journal.* 65(7): 33–37.

Rupp, E. W. 1985. Water-cooling of laser: design considerations and techniques. *Laser and Applications.* (3): 91.

Russo, A. J. 1990. Thermocapillary flow in pulsed laser beam weld pools. *Welding Journal.* 69(1): 23-s.

Schwartz, M. M. 1971. Laser welding and cutting. *Welding Research Council Bulletin.* New York: (11): No. 167.

Seretsky, J., and E. R. Ryba. 1976. Laser welding dissimilar metals: titanium to nickel. *Welding Journal.* 55(7): 208-s–1-s.

Sharp, C. M., and C. J. Nilsen. 1988. High-speed laser beam welding in the can-making industry. *Welding Journal.* 67(1): 25–28.

Sherwell, J. R. 1977. Design for laser beam welding. *Welding Design and Fabrication.* 50(6): 106–110.

Smith II, R. D., G. P. Landis, I. Maroef, D. L. Olson, and T. R. Wildeman. 2001. The determination of hydrogen distribution in high-steel weldmelts—part 1: Laser ablation methods. *Welding Journal.* 80(12): 115-s–121-s.

Xie, J. 2002. Dual beam laser welding. *Welding Journal.* 81(10): 223-s–230-s.

Xie, J. 2002. Weld morphology and thermal modeling in dual-beam laser welding. *Welding Journal.* (81)12: 283-s–290-s.

Zhang, S. 2002. Stresses in laser-beam-welded lap joints determined by outer surface strains. *Welding Journal.* (81)1: 14-s–18-s.

Zhao, H., and T. DebRoy. 2001. Pore formation during laser beam welding of die-cast magnesium allow Am60B-mechanism and remedy. *Welding Journal.* (80)8: 204-s–210-s.

CHAPTER 15

OTHER WELDING AND CUTTING PROCESSES

Photograph courtesy of Orgo-Thermit, Inc.

Prepared by the Welding Handbook Chapter Committee on Other Welding and Cutting Processes:

J. E. Gould, Chair
Edison Welding Institute

H. R. Castner
Edison Welding Institute

M. Kimchi
Edison Welding Institute

J. V. Kristan
Trane Corporation

S. B. Maitland
Goodrich Aerospace

J. Reynolds
Process Welding Systems

P. Zhang
Edison Welding Institute

Welding Handbook Committee Member:

D. W. Dickinson
The Ohio State University

Contents

CHAPTER 15

OTHER WELDING AND CUTTING PROCESSES

INTRODUCTION

This chapter covers several welding and cutting processes that differ from the more commonly used processes described in the other chapters of this volume. Some processes discussed in this chapter may not be widely used, but are important in their fields of special application. Among these processes are cold welding, hot pressure welding, thermite welding, and percussion welding. In addition, modern research and development applied to technologies of the past have resulted in two emerging processes: magnetic pulse welding and electrospark deposition. These are also included in this chapter.

Two processes from the distant past are covered in the chapter. Forge welding has been used for centuries and still has limited modern use. Water jet cutting uses the ancient concept of a stream of water as a tool; however, coupled with modern technology, this process is highly effective as a cutting tool for a growing range of modern industrial applications.

Early arc welding processes, although obsolete, are included in this chapter because of their historical significance. They were precursors of the shielded arc welding processes that have become highly important to modern metal fabricating.

FORGE WELDING

Forge welding (FOW) is a solid-state welding process that produces a weld by heating the workpieces to welding temperature and applying blows sufficient to cause permanent deformation at the faying surfaces.[1,2]

Forge welding is the earliest known welding process, and it remained in common use until well into the nineteenth century. Blacksmiths used this process. Pressure vessels and steel pipe were among the industrial items once fabricated by forge welding.

FUNDAMENTALS

The process is used in some applications with modern methods of applying the heat and pressure necessary to achieve a weld. The main present-day applications are in the production of tubing and clad metals.

The metals to be joined by forge welding may be heated in a forge, a furnace, or by other appropriate means until the metal is highly malleable. A weld is accomplished by removing the workpieces from the heat source, superimposing them, and then applying pressure or hammer blows to the joint.

Heating time is the major variable that affects joint quality. Insufficient heat fails to bring the surfaces to the proper degree of plasticity, and welding does not take place. If the metal is overheated, a brittle, low-strength joint may result. The overheated joint is likely to have a rough, spongy appearance where the metal is severely oxidized. The temperature must be uniform throughout the joint interfaces to yield a satisfactory weld.

1. Welding terms and definitions used throughout this chapter are from *Standard Welding Terms and Definitions*, AWS A3.0:2001, Miami: American Welding Society.

2. At the time this chapter was prepared, the referenced codes and other standards were valid. If a code or other standard is cited without a publication date, it is understood that the latest edition of the document referred to applies. If a code or other standard is cited with the publication date, the citation refers to that edition only, and it is understood that any future revisions or amendments to the code or other standard are not included. As codes and standards undergo frequent revision, the reader is advised to consult the most recent edition.

Forge Welding Techniques

One method of forge welding, known as *hammer welding,* is accomplished by applying hammer blows to achieve deformation. Coalescence is produced by heating the workpieces in a forge or other furnace and then applying pressure by hammering. Manual hammer welding is the oldest technique. Pressure is applied to the heated workpieces by repeated high-velocity blows with a comparatively light sledge hammer. Hammering is accomplished in modern automatic and semiautomatic operations by a heavy power-driven hammer operating at low velocity. The hammer may be powered by steam, hydraulic, or pneumatic equipment.

The size and quantity of parts to be fabricated determines the choice of either manual or power-driven hammering. This method of forge welding has been replaced in industry by other welding processes, but it is still used for some maintenance or repair applications, and also is used in the fabrication of blades, sculptures and other artistic objects.

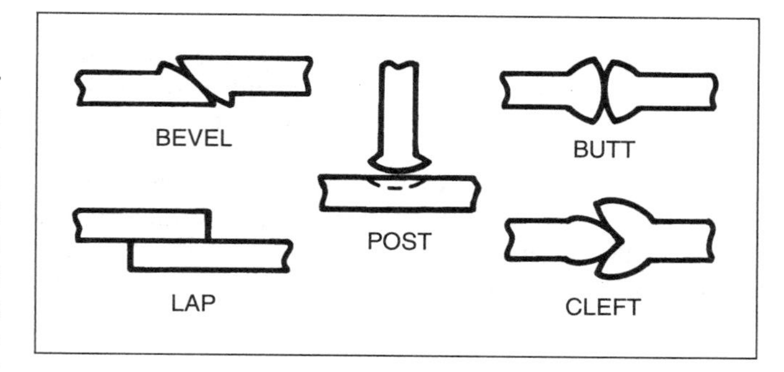

Figure 15.1—Typical Joint Designs Employed for Manual Forge Welding

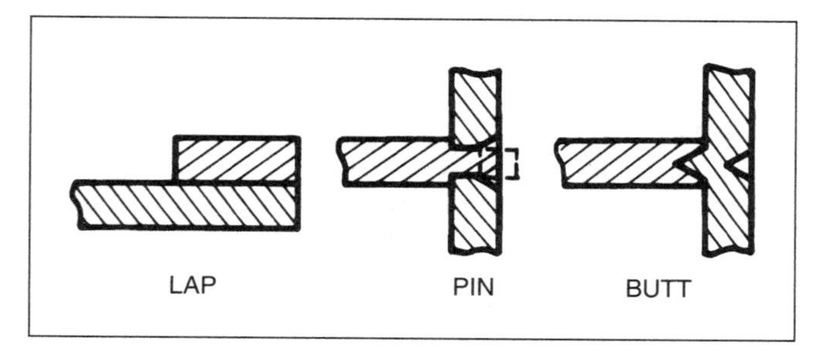

Figure 15.2—Typical Joint Designs Used for Automatic Forge Welding

METALS WELDED

Low-carbon steels are the metals most commonly joined by forge welding. Sheets, bars, tubes, pipes and plates of these materials are readily welded.

The major influences on the grain structure of the weld and heat-affected zone are the amount of forging applied and the temperature at which the forge welding takes place. A high temperature generally is necessary for the production of a sound forge weld. Annealing can refine the grain size in a forge-welded steel joint and can improve joint ductility.

Thin, extruded sections of aluminum alloy are joined edge-to-edge by forge welding using automatic equipment to form integrally stiffened panels. The panels are used for lightweight truck and trailer bodies. The success of the operation depends on the use of the correct temperature and pressure, effective positioning and clamping devices, edge preparation, and other factors. The forge welding of aluminum for this application is very similar to hot pressure welding because the edges to be joined are heated to welding temperature and then upset (causing bulk deformation) by the application of pressure.

JOINT DESIGN

The five joint designs applicable to manual forge welding are the bevel, lap, butt, post, and cleft joints shown in Figure 15.1. The joint surfaces for these welds are slightly rounded or crowned. This shape ensures that the center of the pieces will weld first so that any slag, dirt, or oxide on the surfaces will be forced out of the joint as pressure is applied. The faying surfaces are prepared by beveling (scarfing) or otherwise removing metal to shape one surface to fit into the other. The lap, pin, and butt joints used for automatic forge welding are shown in Figure 15.2.

Scarfing (creating a corresponding notch on the edges of both workpieces) is the term applied to the preparation of the workpieces for the pin joint and butt joints illustrated in Figure 15.1. Each workpiece must be upset for an adequate distance from the scarfed surface to provide metal for mechanical working during welding.

FLUXES

In the forge welding of certain metals, a flux must be used to prevent the formation of oxide scale. Preexisting flux and the oxides on the surfaces of the metal combine to form a protective coating on the heated surfaces. This coating lowers the melting point of the existing oxide and prevents the formation of additional oxide.

Two commonly used fluxes for steels are silica sand and borax (sodium tetraborate). Flux is not required for very low-carbon steels (ingot iron) and wrought iron, because the oxides of these metals have low melting points. The flux most commonly used in the forge welding of high-carbon steels is borax. Because it has a relatively low fusion point, borax may be sprinkled on the metal while it is in the process of heating. Silica sand is suitable as a flux in the forge welding of low-carbon steel.

EARLY ARC WELDING PROCESSES

This section discusses three arc welding processes that are seldom used or considered obsolete: carbon arc welding (CAW), bare metal arc welding (BMAW), and atomic hydrogen welding (AHW). They are included in this chapter for reference and for their historical interest because they are the predecessors of the shielded arc welding processes in modern use. Additional information on these processes can be found in the previous editions of the American Welding Society *Welding Handbook*.

CARBON ARC WELDING

Carbon arc welding (CAW) is an obsolete process that used an arc between a carbon electrode and the weld pool. The process was used with or without shielding and without the application of pressure.[3] The earliest welding process that used an electric arc to heat and melt metal is attributed to Benardos and Olszewski, who patented carbon arc welding in both Europe and the United States in the mid-1880s.[4] Figure 15.3 illustrates the carbon arc welding apparatus developed by Benardos and Olszewski. The carbon arc was used only as a heat source, so filler metal had to be added separately when needed. Carbon arc welding was used to join steels, cast iron, and copper alloys and also for weld surfacing, soldering, and the braze welding of galvanized steel.[5]

Variations of this process included twin carbon arc welding (CAW-T), which used an arc between two carbon electrodes; shielded carbon arc welding (CAW-S), which used shielding from the combustion of solid metal fed into the arc or from a blanket of flux on the workpieces, or both; and gas carbon arc welding (CAW-G), which used a shielding gas. Carbon arc welding and its variations no longer are commercially significant in modern welding applications.

Figure 15.4 shows a twin carbon arc welding torch with two adjustable arms that hold the carbon electrodes. Heat was produced by an arc between these two carbon electrodes and the workpiece was not part of the electrical circuit. The twin carbon arc torch was used for welding, brazing, surfacing, or soldering, and also for heating.

3. American Welding Society (AWS), 1942, *Welding Handbook*, second edition, Miami: American Welding Society.
4. American Welding Society (AWS), *Welding Handbook*, 1971, *Welding, Cutting, and Related Processes*, Vol. 3b, sixth edition, Miami: American Welding Society.
5. Simonson, R. D., 1969, *The History of Welding*, Morton Grove, Illinois: Monticello Books, Inc.

BARE METAL ARC WELDING

Bare metal arc welding (BMAW) is an obsolete arc welding process that used an arc between a bare or lightly coated electrode and the weld pool. This process was the next advance in arc welding technology, succeeding the carbon arc process. Development of the metal electrode for welding is attributed to C. L. Coffin, an American who, in 1890, was awarded a United States patent for the process. The weld filler metal came from the electrode that was melted by the arc and transferred to the workpiece. The arc was not shielded. Bare metal arc welding was suitable for only a few metals; the strength and ductility of the welds rarely matched those of the base metal. The chief disadvantage of welding with a bare electrode was that the molten filler and weld metal were exposed to nitrogen, oxygen, and moisture in the atmosphere. The resulting welds often had low ductility, poor fusion, and sometimes contained porosity or were subject to hydrogen cracking. The process also required considerable skill because the arc was unstable and hard to control. Experiments as early as 1904 showed that a light coating of oxide (rust) or lime drawing compound on the bare metal electrode improved arc stability and overall operability of the process. Continued development of electrode coatings resulted in the introduction of heavily covered welding electrodes in the late 1920s. These "shielded metal arc" electrodes improved both arc stability and weld properties compared to using bare electrodes; this breakthrough led to the development of shielded metal arc welding (SMAW). Shielded metal arc welding became the process of choice and replaced bare metal arc welding for a wide range of applications.

ATOMIC HYDROGEN WELDING

Atomic hydrogen welding (AHW) is a process that uses an arc between two metal electrodes in a shielding atmosphere of hydrogen and without the application of pressure. It is an obsolete or seldom-used process.

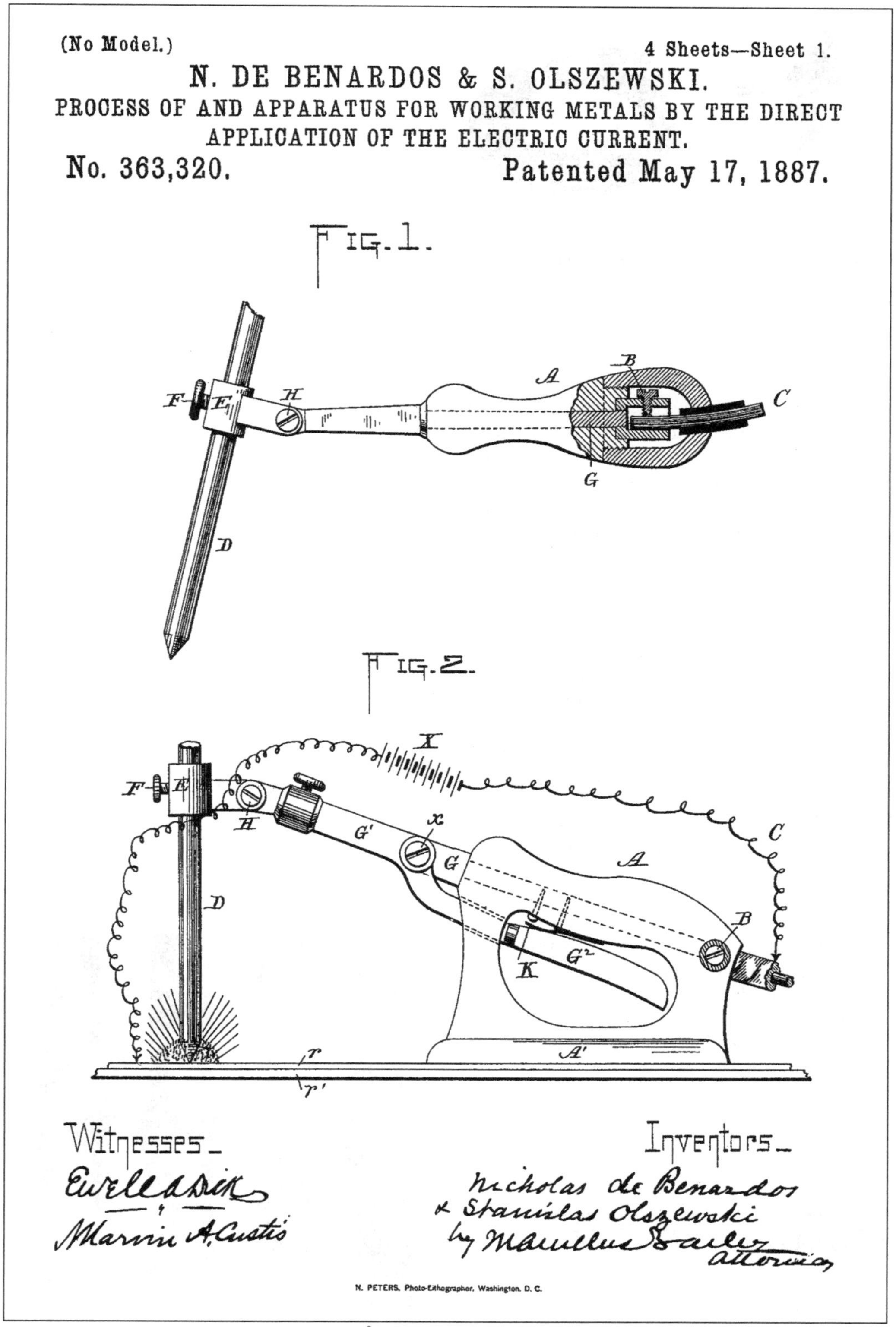
(No Model.)
4 Sheets—Sheet 1.
N. DE BENARDOS & S. OLSZEWSKI.
PROCESS OF AND APPARATUS FOR WORKING METALS BY THE DIRECT APPLICATION OF THE ELECTRIC CURRENT.
No. 363,320.
Patented May 17, 1887.
FIG. 1.
A
B
C
F
E
H
G
D
FIG. 2.
X
F
E
H
x
G'
G
A
C
D
B
K
G²
A'
r
r'
Witnesses_
Inventors_
Nicholas de Benardos
& Stanislas Olszewski
N. PETERS, Photo-Lithographer, Washington, D. C.

Figure 15.3—Patent Application for the Original Arc Welding Process

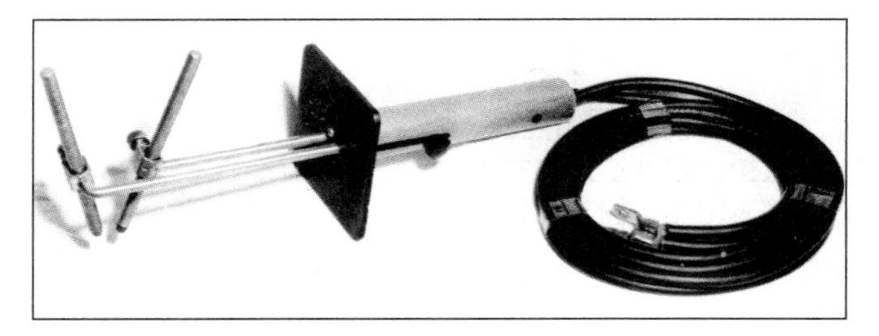

Figure 15.4—A Twin Carbon Arc Welding Torch

The limitations of welding with the carbon arc and bare metal arc led not only to the development of coated electrodes as previously described, but also to experiments with the gas shielding of welding arcs. During the 1920s, the American chemists and physicists, P. Alexander and I. Langmuir, made carbon arc welds in chambers filled with hydrogen. They replaced the carbon electrodes with tungsten and made welds outside of the chamber by blowing hydrogen through the arc, which formed a hot flame of atomic hydrogen that liberated additional heat. The hydrogen also cooled the electrodes and protected them and the molten metal from the atmosphere. The resulting atomic hydrogen welding process, using an arc between two tungsten electrodes in a hydrogen shielding atmosphere, was patented by Langmuir in 1926. The process could be used with or without filler metal, which was added separately when needed.

The arc in AHW is maintained between the two tungsten electrodes, independent of the workpieces. The workpiece is a part of the electrical circuit only to the extent that a portion of the arc comes in contact with it. The advantage of atomic hydrogen welding was the ability to manipulate the arc separately from the work being welded, and therefore control heat input over a fairly wide range. While AHW was not widely used, it has found special applications, including welding nonferrous metals, tool steels, and for repair welding. Atomic hydrogen welding has been replaced by other gas shielded arc welding processes such as gas tungsten arc welding (GTAW) and gas metal arc welding (GMAW).

HOT PRESSURE WELDING

Hot pressure welding (HPW) is a process that produces a joint simultaneously over the entire faying surfaces. The process generally is used with oxyfuel gas heating, the application of pressure, and without filler metal.

Hot pressure welding currently is used for applications in which the high strength and integrity of the joints are critical, for example, for the manufacturing of airplane landing gear. The basic elements of the process assisted in the development of similar processes, such as flash welding and friction welding, which use other sources of energy.

FUNDAMENTALS

The two techniques used in hot pressure welding are the closed-joint and open-joint methods. In the closed-joint method, the clean weld interfaces of the workpieces to be joined are abutted under pressure and heated, typically by oxyfuel flames, until a predetermined deformation of the joint occurs. In the open-joint method, the faying surfaces are individually heated to the melting temperature by gas flames and then brought into contact and placed under load for upsetting (bulk deformation resulting from pressure). Both methods are easily adapted to mechanized operation. Hot pressure welding can be used for welding low- and high-carbon steels, low- and high-alloy steels, and several nonferrous metals and alloys.

In the closed-joint method, the metal along the weld interface does not reach the melting point; thus, the mode of welding is different from that of fusion welding. In general, welding takes place by the action of grain growth, diffusion, and grain coalescence across the interface under the impetus of high temperature (about 1200°C [2200°F] for low-carbon steel) and upsetting pressure. The welds are characterized by a smooth-surfaced bulge as a result of upsetting, as shown in Figure 15.5, and by the general absence of cast metal at the weld interface.

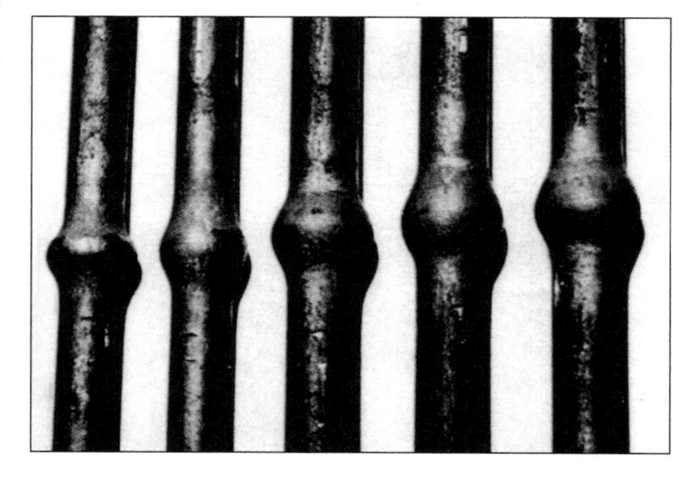

Figure 15.5—Typical Hot Pressure Welds in 25-mm (1-in.) and 32-mm (1-1/4-in.) Diameter Steel Rods

The open face method uses a gap between the workpieces at initiation of the process. In this case, the faces are heated, alternately to the proper forging temperature or melting point. At peak temperature, the workpieces are advanced together, expelling any liquid that may be present and forging the components together. These welds resemble flash welds in general appearance.

Details of the Closed-Joint Method

Most hot pressure welding is done in the closed-joint method. In this method, the workpieces are butted together under initial pressure to ensure intimate contact. Heating generally is done with one or more water-cooled, multiple-flame oxyfuel torches, which are designed to generate sufficient heat and distribute it uniformly throughout or around the periphery of the entire section to be welded. For sections over 25 mm (1 in.) thick, it is advisable to heat the joint uniformly from all sides. A torch arrangement for a circumferential weld is shown in Figure 15.6. Figure 15.7(A) shows the weld in progress and (B) the completed weld.

Figure 15.6—Torch Arrangement for a Circumferential Hot Pressure Weld in Tubing

(A) In Progress

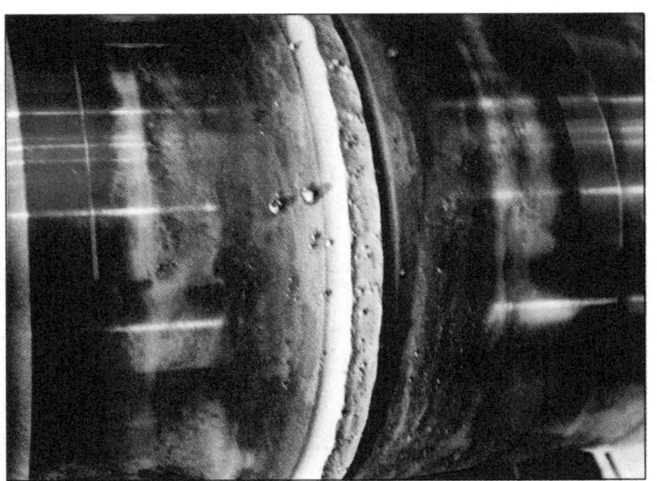

(B) Finished Weld

Figure 15.7—Hot Pressure Circumferential Weld

Solid or hollow round sections, such as shafts or pipes, usually are welded with circular ring torches. The torch head may be a split type for easy loading and removal of the workpieces from the welding machine. More elaborate heating heads are required for complicated shapes because they must conform to the shape of the workpiece to provide uniform heating.

Hot pressure welding machines must be designed to apply the desired pressure and maintain alignment during welding; provision for maintaining uniform pressure and alignment is essential.

Joint Preparation

The quality and type of edge preparation of steel workpieces depends on the type of steel. In general, the abutting ends should be machined or ground to a smooth, clean surface. Eliminating oil, rust, grinding dust, and other foreign material from the weld interface is of great importance.

The geometry of the abutting faces depends on the application and the alloy. Beveling one or both workpieces allows some control over the shape of the upset metal. Figure 15.8 shows typical designs for joint preparation for hot pressure welding and the effect of beveling on the shapes of the completed welds.

For illustration, if it is assumed that two 125-mm (5-in.) diameter steel pipes with wall thickness of 6.4 mm (1/4 in.) are to be hot-pressure welded end to end using a butt joint, the general procedures would be as follows:

A split torch head should be selected that would provide small oxyfuel flames around the full circumference of the joint. The head should be mounted in the same plane as the weld interface, with provision for axial oscillation. The abutting ends of the pipe should be beveled to a groove angle of 6° to 10° and should have a

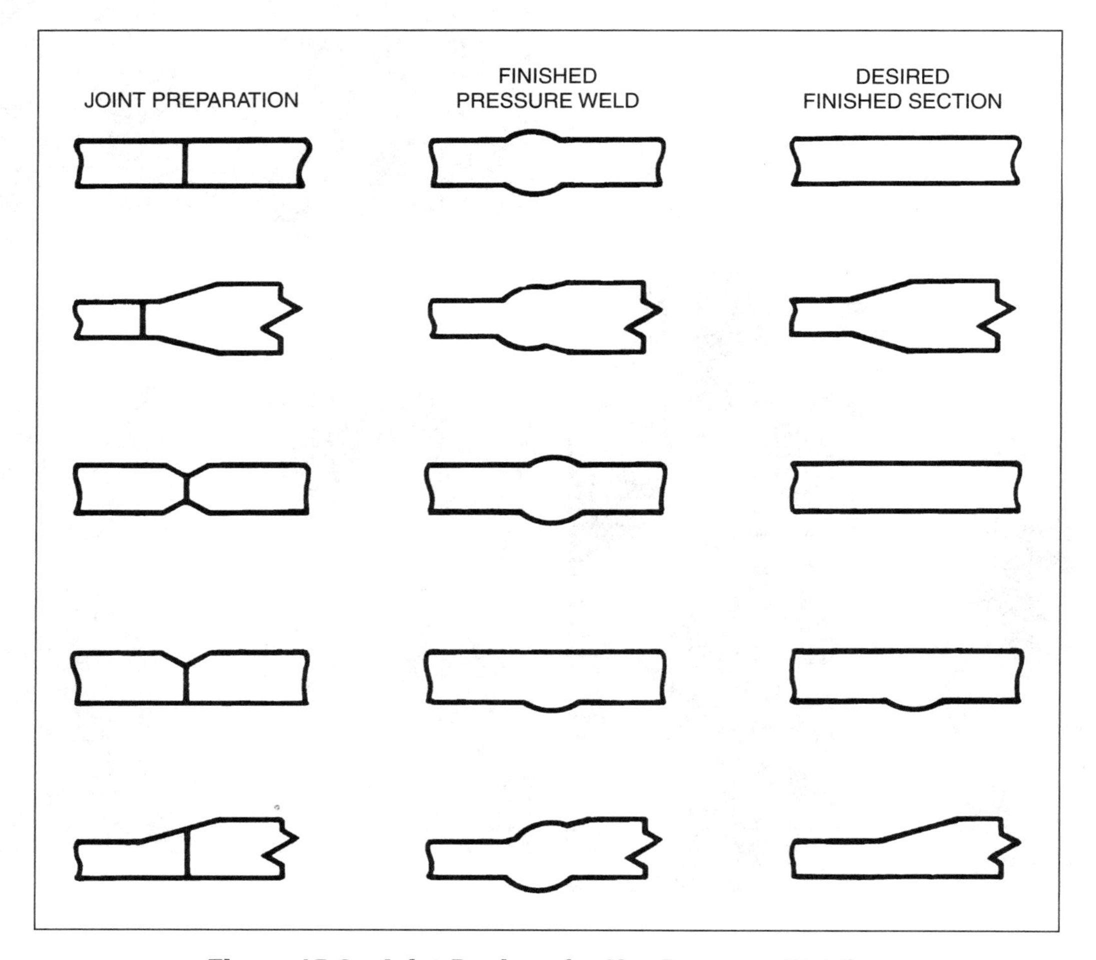

Figure 15.8—Joint Designs for Hot Pressure Welding

smooth, clean finish. The pipes should be placed in the machine and aligned; then a force of 2650 kg (5850 lbs) should be applied to produce a low compressive pressure of 10 MPa (1500 psi).

While this force is maintained, the torch flames should be oscillated axially a short distance across the weld joint. As the joint heats up, the metal upsets. The joint interfaces close together, preventing oxidation at higher temperatures. As the metal temperature increases, the compressive strength of the steel decreases and the joint begins to upset uniformly. When the metal is at welding temperature throughout the full thickness, an upsetting pressure of 28 MPa (4000 psi) should be applied until the weld zone is upset for a distance of 4.8 mm (0.188 in.). The torches should then be extinguished and the assembly removed from the machine.

Welding Parameters

Several parameters (welding variables) can be adjusted to control the basic hot pressure welding procedure. However, it should be noted that the principal variable is the sequence of pressure application. These variables can be introduced to meet special requirements of certain metals, such as high-carbon steels, high-chromium steels, and nonferrous metals. For example, the constant-pressure method is recommended for welding joints in high-carbon steels.

Another technique can be applied to high-chromium steels and some nonferrous metals. A high initial pressure in the range of 40 MPa to 70 MPa (6000 psi to 10 000 psi) is applied before heating is started. The high pressure is maintained until the metal in the weld zone starts to upset (bulk deformation resulting from pressure). Pressure is then decreased until welding temperature is reached; then the high pressure is again applied to upset the joint. The high pressure forces the joint interfaces together and prevents oxidation.

Examples of pressure cycles used for the hot pressure welding of several metals are listed in Table 15.1. Table 15.2 provides the average dimensions of closed-joint pressure welds in metals of various thicknesses.

The quality of a weld depends to an important degree on proper upsetting during the welding operation. The upset distance or shortening of the weld zone increases with metal thickness. The recommended amounts of upset provided in Table 15.2 usually are measured from fixed points on the workpieces or the welding machine.

Table 15.1
Typical Upset Pressure Cycles for Pressure Gas Welds

Type of Metal	Method	End Pressure, MPa (psi)		
		Initial	Intermediate	Final
Low carbon steel	Closed joint	3 to 10 (500 to 1500)	...	28 (4000)
High carbon steel	Closed joint	19 (2700)	...	19 (2700)
Stainless steel	Closed joint	69 (10 000)	34 (5000)	69 (10 000)
Monel alloy	Closed joint	45 (6500)	...	45 (6500)
Steel (carbon and alloy)	Open joint	...	...	28 to 34 (4000 to 5000)

Table 15.2
Joint Dimensions of Pressure Gas Welds, Squared End Preparation, and Closed-Joint Method

Metal Thickness (T), mm (in.)	Length of Upset (L), mm (in.)	Approx. Upset Height (H), mm (in.)	Total Upset, mm (in.)
3 (1/8)	5 to 6 (3/16 to 1/4)	2 (1/16)	3 (1/8)
6 (1/4)	8 to 13 (5/16 to 1/2)	2 (3/32)	6 (1/4)
10 (3/8)	14 to 16 (9/16 to 5/8)	3 (1/8)	8 (5/16)
13 (1/2)	19 to 22 (3/4 to 7/8)	5 (3/16)	10 (3/8)
19 (3/4)	27 to 30 (1-1/16 to 1-3/16)	6 (1/4)	13 (1/2)
25 (1)	32 to 38 (1-1/4 to 1-1/2)	10 (3/8)	16 (5/8)

Equipment

A hot pressure welding machine includes the following components:

1. Equipment for applying upsetting force;
2. Suitable heating sources such as torches and tips, or alternatively, induction coils designed to provide uniform and controlled heating of the weld zone; and
3. Indicators and measuring devices necessary for regulating the process during welding.

The complexity of the machine depends on the configuration and size of the workpieces and the degree to which the process is mechanized. In most cases, it is advisable to use special heating arrangements and a special apparatus for gripping and applying force to the workpieces.

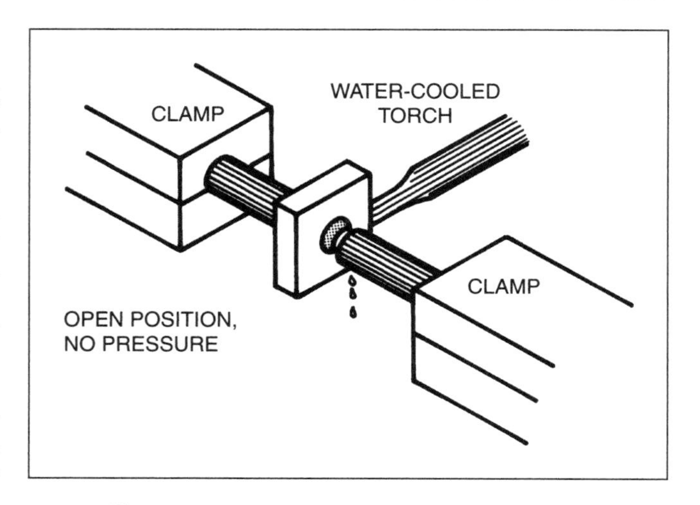

Figure 15.9—Torch and Typical Setup for Open-Joint Hot Pressure Welding

Auxiliary Equipment. Hot pressure welding machines require auxiliary equipment for operation. For use with burner rings, the gas supply equipment must adequately provide the maximum flow requirement, and the gas regulators must be capable of maintaining a uniform flame adjustment. Quick-acting gas shut-off valves are highly recommended. In many instances, needle valves are advantageous for fine adjustment of the flame. The best control of gas flow and heat input is obtained when the pressure gauges are located close to the torch. This permits the operator to readily check gas pressures. Flow meters can be used to assure uniform gas flow. For induction heating machines, the power source can be used to provide specific process waveforms (for example, pulsing waveforms) to facilitate uniform heating.

An ample supply of water is needed for cooling the heating elements and, in some cases, the clamps and parts of the press. Fixtures for aligning and supporting the workpieces are needed. Automatic control units for regulating the upset force and heating cycles and then terminating the operation can be incorporated in the welding machine.

Open-Joint Machines

Open-joint hot pressure welding machines must provide more accurate alignment and be ruggedly constructed to withstand rapidly applied upset forces. Welding machines similar to those used for flash welding are suitable.

The most common heating head is a flat, multiple-flame burner, such as the one shown in Figure 15.9, which produces a uniform flame pattern conforming to the cross section of the workpieces. Of note, induction heating is becoming more popular as a replacement for the burner. Good alignment of the heating head with the faces of the joint is important to minimize oxidation and to obtain uniform heating and subsequent upsetting. A removable spacer block can be used to aid alignment.

Saw-cut surfaces are satisfactory for welding since the abutting ends are commonly melted before the weld is consummated. A thin layer of oxide on the joint faces has little effect on weld quality, but significant amounts of foreign substances, such as rust or oil, should be removed before welding.

The general procedure for open-joint hot pressure welding is to align the workpieces so that a suitable torch tip can be properly spaced between the joint faces, as shown in Figure 15.10. When thin sections are welded, the torch tip is placed just outside the joint with the flames directed at the joint faces. The tip is designed to heat the full cross section of the weld faces. The flames are maintained in this position until a molten film entirely covers both faces. The torch is then withdrawn, and the workpieces are rapidly brought together with a force that will produce a constant pressure of 28 MPa to 35 MPa (4000 psi to 5000 psi) at the weld interface. This step is shown in Figure 15.10. Pressure is maintained until upsetting of the metal ceases. The total upset is controlled by adjusting both the applied pressure and the temperature of the hot metal; it is not preset on the equipment.

Applications

Hot pressure welding has been successfully applied to plain-carbon steel, low-alloy and high-alloy steels, and to several nonferrous metals, including nickel-copper, nickel-chromium, and copper-silicon alloys. The process also has been very useful for joining dissimilar metals.

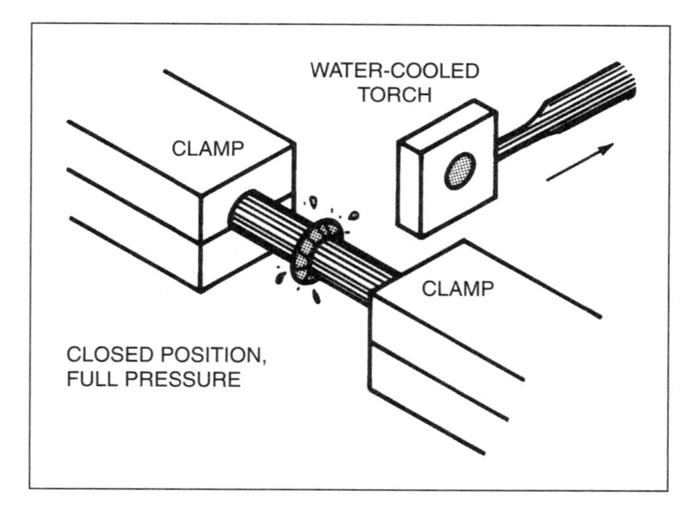

Figure 15.10—An Open-Joint Hot Pressure Weld as Upsetting Starts

Rails. The first commercial application of this process was the welding of railroad rails. However, this process has largely been replaced by the flash welding process. When hot pressure welding is used, the closed-joint method usually is employed with equipment designed specifically for the application.

The rail ends are carefully prepared by power sawing and then are cleaned. The rails are gripped by special clamps and a force is applied to produce about 20 MPa (2800 psi) on the joint. The joint is then heated with specially shaped heating tips or heads that are oscillated automatically across the joint until the metal reaches welding temperature. Adequate pressure is applied to produce the required upset. Most of the upset metal on the ball and edges of the base is removed by oxygen cutting. The welded joint is then indexed or moved to the next position, where the weld zone is normalized to refine the grain size and restore normal hardness. Finally, the weld is ground to the contour of the rail, examined by magnetic particle inspection, and oiled for protection against rusting. Rails that have been welded, normalized, and inspected in this manner have given satisfactory service under heavy and fast traffic for extended periods of time.

Mechanical and Physical Properties

In general, hot pressure welding has a minimum effect on the mechanical and physical properties of the base metals. Because of the relatively large mass of hot metal in the weld zone, the cooling rate usually is quite low.

In the closed-joint method, the maximum temperature of the metal is below the temperature at which overheating and rapid grain growth occur. In the open-joint method, the melted metal film is squeezed out of the joint during upset. These characteristics are advantageous for welding high-carbon steels and some nonferrous alloys that are hot-short or affected by overheating.

Another important factor is the absence of deposited metal. The entire weld zone is base metal, therefore it responds to heat treatment in the same manner. This includes the effect of the heat of welding on the corrosion resistance of welded stainless steels. If unimpaired corrosion resistance is necessary, stabilized stainless steel must be used or the welded assembly must be given a heat treatment for stabilization after welding.

Heat Treatment. Hot pressure welds in low-carbon steels seldom require heat treatment or stress relief, because the heat-affected zone in these steels usually is normalized and relatively stress free. Hot pressure welding is used with low-alloy and high-carbon steels for fabricating assemblies subject to high service stresses, and postweld heat treatment is necessary. Heat treatment frequently can be done with the same heating heads used for welding.

In rails, the annealed zone on each side of the weld may be too soft. To overcome this problem, the weld zone can be heated to normalizing temperature using heating heads and then air-cooled to restore the desired hardness. Similarly, heat treatment with the welding flame may be suitable for developing desired mechanical properties in welded joints in some low-alloy steels such as those used for oil well drilling tools. This type of heat treatment, which is essentially a normalizing operation, refines the grain size in the weld zone and improves ductility and toughness. For highly hardenable steels, annealing or slow cooling after welding may be necessary to prevent hardening or surface cracking in the weld zone. To develop optimum properties in welds in heat-treatable steels, furnace heat treatment commonly is used.

Mechanical Properties. Since there is no deposited metal in pressure welds, the mechanical properties of welds depend on the composition of base metals, the cooling rate, and the quality of the weld. When dissimilar steels are joined, the properties of the welded joint will be more nearly like those of the weaker member.

Microstructure. The location of the original interface in hot pressure welds in plain carbon steels and many alloy steels is very difficult to detect in a metallographic cross section using normal etchants. However, it is possible to locate the weld line with special polishing and etching techniques. A typical photomicrograph of a hot pressure weld in steel is shown in Figure 15.11.

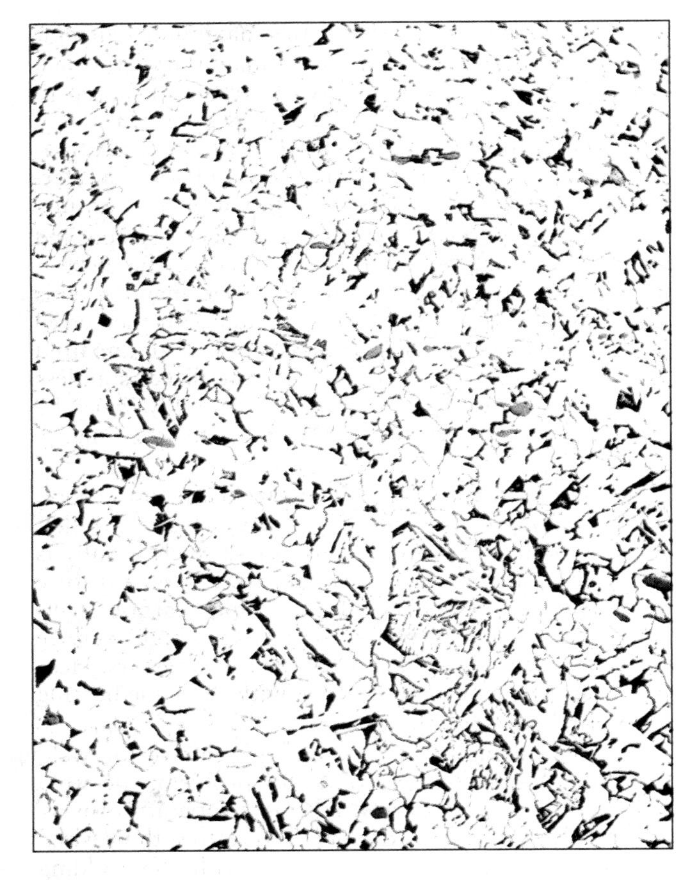

Figure 15.11—Photomicrograph of a Hot Pressure Weld in 1020 Steel, in the As-Welded Condition

Although not readily apparent, it can be observed that the weld interface extends vertically through the center of the photomicrograph.

QUALITY CONTROL

Successful hot pressure welding by the closed-joint method requires positive and continuous control of the variables that influence the quality of a weld. The variables include the following:

1. Degree of roughness and cleanliness of edge preparation,
2. Pressure cycle,
3. Alignment of the workpieces,
4. Welding cycle time,
5. Performance of the heating unit,
6. Desired upset or shortening, and
7. Cooling time in the machine after upset.

Process Control

Examination of the pressure-upsetting cycle can indicate whether the welding conditions are conforming to the prescribed procedure. With constant heat input into the weld zone, constant width of heated zone, and uniform pressure sequence, the entire cycle of heating and upsetting should be completed with a variation in welding time of not more than 25%. On this basis, if a weld requires an unduly long or short time, the conditions that prevailed during welding should be evaluated. A large time variation from nominal would indicate that some factor other than time was controlled improperly and that the weld might be of questionable quality. The malfunctioning of the pressure system or the heating heads and the slippage of the workpieces in the clamps are examples of conditions that might cause poor-quality welds.

Data acquisition and other means of record keeping have proven valuable for maintaining good control of the following variables:

1. Pressure cycle,
2. Total time or times for certain stages of the procedure,
3. Gas flow rates (for burner systems) or welding currents (for induction systems), and
4. Total upset distance.

Some conditions that are important to the closed-joint method are not as important in the open-joint method. For the open-joint method, the cleanliness of the weld interface is not critical unless excessive amounts of contaminants are present. Melting at the weld interface offsets the need for thorough surface preparation. The amount of upset or shortening is not necessarily constant; therefore, it is not an index of weld quality. However, due attention must be given to the pressure cycle and the performance of the heating system.

Inspection

The following general characteristics usually are evaluated by visual inspection:

1. Presence or absence of excessive melting,
2. Contour and uniformity of upset, and
3. Position of the weld line with respect to the midpoint of the upset zone.

In general, if the controls are adequate and there is no appreciable variation from an accepted standard, it may be concluded that the pressure weld is of normal quality.

In highly stressed assemblies, added assurance of weld consistency and quality may be needed. Sample welds selected either at random or at fixed intervals should be destructively tested. This procedure will serve as a positive and continuous check on the welding cycle, process controls, and the properties of the welded assemblies.

Magnetic-particle inspection can be used for the nondestructive inspection of hot pressure welds.[6] The nick-break test can be used as a convenient quality check for soundness.[7] A fracture of a sample weld along the weld interface will show the extent of metallic bonding, grain size, and evidence of the overheating of the weld interface. Changes in the welding cycle can be checked quickly with this test. Experience has proven that when the nick-break tests show satisfactory crystalline fracture throughout the weld cross section, all other tests usually will prove that the weld is satisfactory.

Proof testing may be used as an alternative test or as a compliment to destructive testing. Proof testing is designed to disclose defective welds and pass acceptable welds. A welded joint is subjected to either a tensile or a bending load, or both, to produce a maximum tensile stress just below the yield strength of the metal. A poor quality weld will fail in this test.

SAFE PRACTICES

The potential hazards of hot pressure welding include mechanical and possibly electrical risks, as well as those from open flames and fumes. Personnel must be aware of these hazards and wear protective gear.

Mechanical

The basic systems are mechanical presses, and must be treated with appropriate safety practices. This includes safety devices to prevent injury to the hands of the operator or other parts of the body. Initiating devices, such as push buttons or foot switches, should be arranged and guarded to prevent inadvertent actuation.

Welding machine guards, fixtures, or operating controls must prevent the operator from entering the area of joining during operation. Dual hand controls, latches, presence-sensing devices, or similar devices should be employed to prevent operation in an unsafe manner.

Electrical

Some variants of the process use induction heating. Induction heating systems incorporate high voltages, and must be handled accordingly. All doors and access panels on such systems and controls must be kept locked or interlocked to prevent access by unauthorized personnel. A lock-out procedure should be followed prior to working with the electrical or hydraulic systems.

Personal Protection

The operator should wear flash guards of suitable fire-resistant material as protection from sparks and flames. In addition, the operator should wear personal eye shields with suitable shaded lenses. Operating personnel should wear ear protection when the welding operations produce high noise levels. Metal fumes produced during welding operations should be removed by local ventilating systems.

Additional information on safe practices for welding is provided in the latest edition of ANSI *Safety in Welding, Cutting, and Allied Processes*, Z49.1 published by the American Welding Society. This document is available on line at www.aws.org.[8]

THERMITE WELDING

Thermite welding (TW) is a process that produces coalescence of metals by heating them with superheated molten metal from an aluminothermic reaction between a metal oxide and aluminum. Filler metal is obtained from the liquid metal.

Thermite welding was developed near the end of the nineteenth century when Hans Goldschmidt of Germany discovered that the exothermic reaction between aluminum powder and a metal oxide can be initiated by an external heat source. The reaction is highly exothermic, and consequently self-sustaining once started.

Utility, portability, and minimal capital cost are the substantial advantages that motivate the selection of thermite welding over other processes. This welding process generally does not require an external power source, with the exception of a gas torch which may be necessary for preheating under certain conditions. In addition, preparatory cutting of the workpieces and completion procedures (such as shearing and grinding of the thermite weld), may require a generator or hydraulic pump. In the basic process, however, the aluminothermic reaction negates the need for an external

6. For a description of this test, refer to American Welding Society (AWS) Committee on Methods of Inspection, 2000, *Welding Inspection Handbook*, Miami: American Welding Society.

7. For a description of this test, refer to American Welding Society (AWS) B4 Committee on Mechanical Testing of Welds, 2000, *Standard Methods for Mechanical Testing of Welds*, AWS B4.0M:2000, Miami: American Welding Society.

8. American National Standards Institute (ANSI), Accredited Standards Committee Z49, (latest edition), *Safety in Welding, Cutting, and Allied Processes*, ANSI Z49.1, Miami: American Welding Society.

power source to perform the weld. This makes thermite welding appropriate for use in remote areas where access to the welding site is limited and electric power may not be available.

FUNDAMENTALS

The thermochemical reaction that produces a thermite weld takes place according to the following general formula:

Metal oxide + aluminum (powder) →
aluminum oxide + metal + heat (15.1)

The reaction can be started and completed only if the oxygen affinity of the reducing agent (aluminum) is higher than that of the metal oxide to be reduced. The heat generated by this exothermic reaction results in a liquid product (slag) consisting of metal and aluminum oxide. If the density of the slag is lower than that of the metal, as is the case with steel and aluminum oxide, they separate immediately. The slag floats to the surface and the molten steel drops into the cavity to be welded.[9]

Typical thermochemical reactions and the thermal energies produced are the following:

$3Fe_3O_4 + 8\ Al\ 9Fe + 4Al_2O3$ H = 3350 kJ
$3FeO + 2\ Al\ 3Fe + Al_2O_3$ H = 880 kJ
$Fe_2O_3 + 2\ Al\ 2Fe + Al_2O_3$ H = 850 kJ
$3CuO + 2\ Al\ 3Cu + Al_2O_3$ H = 1210 kJ
$3Cu_2O + 2\ Al\ V\ 6Cu + Al_2O_3$ H = 1060 kJ

Aluminum is the reducing agent in these reactions. Theoretically, magnesium, silicon, and calcium also could be used; however, magnesium and calcium have limited use for general applications. Silicon often is used in thermite mixtures for heat treatment, but it is rarely used in welding. In some cases, an aluminum-silicon alloy is used as the reducing agent.

The first of the reactions listed in the equation above is the one most commonly used as a basic mixture for thermite welding. The proportions of these mixtures usually are about three parts by weight of iron oxide to one part of aluminum. The theoretical temperature created by this reaction is about 3100°C (5600°F). The addition of non-reacting constituents, heat loss to the reaction vessel, and radiation reduce this temperature to about 2480°C (4500°F). This temperature is about the maximum that can be tolerated, because aluminum vaporizes at 2500°C (4530°F). However, the minimum temperature should not be much lower because the aluminum slag (Al_2O_3) solidifies at 2040°C (3700°F).

Heat loss is highly dependent on the quantity of thermite being reacted. When large quantities are reacted, the heat loss per pound of thermite is considerably lower and the reaction more complete than when small quantities of thermite are reacted.

Alloying elements in the form of ferroalloys to match the chemistry of the workpieces can be added to the thermite compound. Other additions are used to increase the fluidity and lower the solidification temperature of the slag.

The thermite reaction is non-explosive and requires less than one minute for completion, regardless of quantity. To start the reaction, a special ignition powder or ignition rod is required; both can be ignited with an ordinary match. The ignition powder or rod will produce enough heat to raise the thermite powder in contact with the rod to the ignition temperature, roughly 1200°C (2200°F). The workpieces should be aligned properly and the weld interface should be free of rust, loose dirt, moisture, and grease. A proper size root opening must be provided at the weld interface, the size depending on the width of the joint. Wider joints normally require a larger root opening. A mold, which may be built up on or premanufactured to conform to the workpieces, is placed around the joint to be welded.

To weld a butt joint, the joint faces should be adequately preheated to promote complete fusion between the thermite deposit and the base metal. Even though it is a welding process, thermite welding resembles metal casting, in which proper gates and risers are needed to perform the following functions:

1. Compensate for shrinkage during solidification,
2. Eliminate typical defects that appear in cast metal,
3. Provide proper flow of the molten steel, and
4. Avoid turbulence as the metal flows into the joint.

APPLICATIONS

The primary application of thermite welding is for rail joining, particularly in the field installation or repair of rail tracks for railroad, mining, and crane operations. Another application of the thermite process is the welding of copper conductors to rail for proper operation of the signal system, in which copper oxide is used instead of iron oxide.

Rail Welding

The welding of rail sections into continuous lengths is an effective method of minimizing the number of

9. Refer to American Welding Society (AWS) 2001, *Welding Handbook, Welding Science and Technology,* Chapter 2, Physics of Welding, Vol. 2, 9th Ed., Miami: American Welding Society, 61–62.

bolted joints in the track structure. In coal mines, the main hauling track is often welded to minimize maintenance and to reduce excessive coal spillage caused by uneven track. Crane rails in buildings usually are welded to minimize joint maintenance and vibration of the building as heavily loaded rail car wheels pass over the joint.

Thermite mixtures are available for all types of rail steels. The majority of rails are carbon-manganese (C-Mn) steels. However, the high–hardness and high-strength rail steel used to reduce wear in curved track often contain additional alloying elements, such as chromium (Cr), molybdenum (Mo), vanadium (V), and silicon (Si). The addition of rare earth metals or alloys may decrease the amount of sulfur and phosphorous in the weld deposit, resulting in improved mechanical properties. Specially designed thermite molds that allow the insertion of an alloy plug at the head of the rail are used to increase local hardness while maintaining desirable properties in the web and base of the rail. Figure 15.12 shows a weld in progress using a disposable crucible.

Welding with Preheat. Premanufactured two-piece or three-piece molds designed for single use generally are preferred for the thermite welding of standard rail sizes. The installation of a thermite weld in a rail is shown in Figure 15.13 (A through F). The mold is aligned so that the center coincides with that of the root opening, which typically is 25 mm (1 in.) between the rail ends. The molds are fitted together (A), and packed (B). The rail ends are preheated (C) in the range of 600°C to 1000°C (1100°F to 1800°F) with a gas torch flame directed into the mold. A refractory-lined crucible containing the thermite charge is positioned above the mold halves after preheating is completed. The crucible also is designed for a single use, which minimizes size and weight for better ergonomics. The charge is then ignited to initiate the thermite reaction, (D). Tapping takes place (E) and the thermite reaction continues. Additional time is allowed for the steel and aluminum oxide to separate. The molten steel flows into the joint while the residual oxide spills into slag pans, and the pour is complete, (F). In most procedures, the metal is center-poured (fed into the middle of the root opening), and then diverted to the outer legs to enter and fill the root opening, beginning at the rail base by way of a diverter plug.

A self-tapping seal (thimble) is used in the bottom of the crucible. Approximately 20 seconds to 30 seconds after

Photograph courtesy of Transportation Technology Center, Inc.

Figure 15.12—Thermite Welding of Rail with a Single-Use Crucible

(A) Mold Fitup

(B) Packing the Mold

(C) Preheat

(D) Thermite Reaction

Photographs courtesy of Transportation Technology Center, Inc. and Johnson County Community College

Figure 15.13—Installation of a Thermite Weld in a Railroad Track

the thermite reaction is complete, the molten metal melts the seal and pours out of the bottom of the crucible into the joint root between the two rail sections. The lower-density liquid slag floats to the top of the thermite metal in the crucible. The liquid slag does not reach the mold cavity until all of the molten steel has entered and filled both the cavity between the rail sections and the mold. The slag remains on top of the weld and solidifies in slag pans. When the metal has solidified, the mold halves are removed and discarded. The excess metal is removed by hand grinding or by hydraulic or manual shearing devices. Preheating times and temperatures may be reduced by using a larger thermite charge. The heat dissipated into the rails during welding requires a larger mass of molten steel.

Welding without Preheat. Although not used extensively, the self-preheating method is designed to

(E) Tapping

(F) Pour Complete

Photographs courtesy of Transportation Technology Center, Inc. and Johnson County Community College

Figure 15.13 (Continued)—Installation of a Thermite Weld in a Railroad Track

eliminate the variables associated with torch preheating and the equipment needed for the welding operation. The rail ends are preheated by a portion of the molten metal produced by the thermite reaction. The crucible and mold are a one-piece design. The thermite molds, commonly known as *shell molds*, are premanufactured of sand bonded with phenolic resins. They are very light, non-hygroscopic, and moisture-free. They have a long shelf life, typically one to two years. After the thermite reaction is completed, the molten steel automatically flows from the crucible into the joint rather than passing through the atmosphere, as is the case with a separate crucible.

A cross section of this thermite mold is displayed in Figure 15.14, showing the shape of the cavity in which the molten filler metal flows. A hollow chamber in the mold under the weld area receives the first molten metal, called the *preheat metal*, and allows it to preheat the rail ends. By the time the chamber is filled, sufficient molten metal should have passed over the rail ends to preheat them to the required temperature to assure

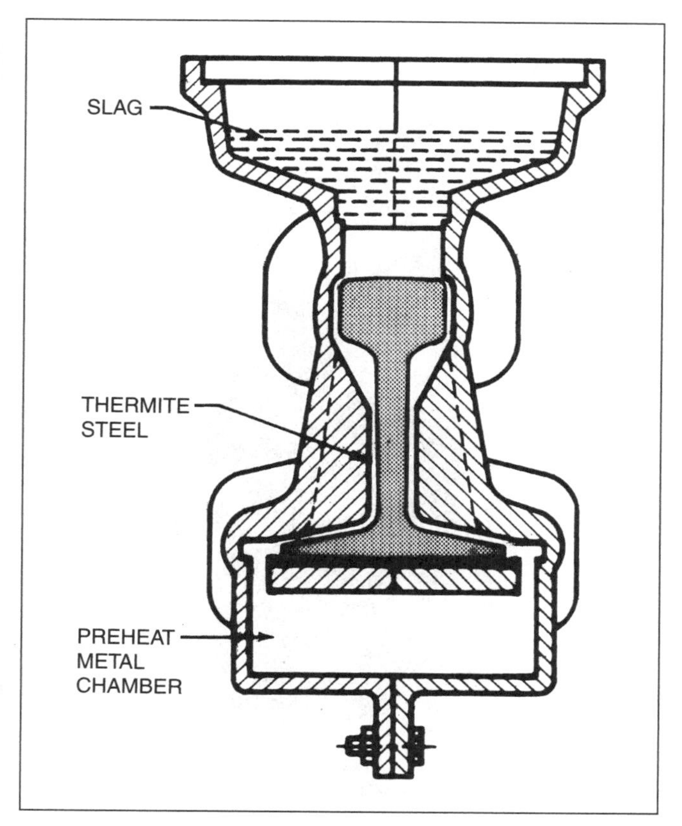

Figure 15.14—Cross Section of a Mold-Crucible with a Preheat Metal Chamber

complete fusion with the base metal. The portions of the thermite mixture for this process are about twice the size of those used for the external preheat method.

The heat-affected zones (HAZ) in the adjacent rail sections are considerably smaller than when external preheating is used. Figure 15.15 shows a typical section through a thermite rail weld made with the self-preheating process.

Rail Repair Welding

Rail repair welds are made in much the same way as new rail is welded. Premanufactured molds designed specifically for each project are used, facilitating the installation and improving weld performance. Molds in half-sections are used to conform to the shape of the rail ends to compensate for vertical mismatch between the opposing rails and the reduction in height of the railhead that might have occurred due to wear.

Joint Preparation. The rails to be joined should be properly positioned and aligned for welding. The component ends should be clean and free of grease and cutting debris. The oxide that may be present on all other surfaces to be covered by the thermite mold or in contact with it should be cleaned to provide an optimal surface for welding. When repairing a fractured rail or defective weld, the rail metal typically is cut with an abrasive wheel but also may be prepared with a cutting torch along the line of fracture or defect to provide a parallel-sided root opening. The width of the root opening typically is fixed, independent of the size of the rail section. Optimal weld practices, however, would require a root opening appropriate to the size of the section to be welded, as shown in Table 15.3. All loose oxide and slag from torch cutting as well as dirt and grease should be removed from the workpieces where the mold will be located.

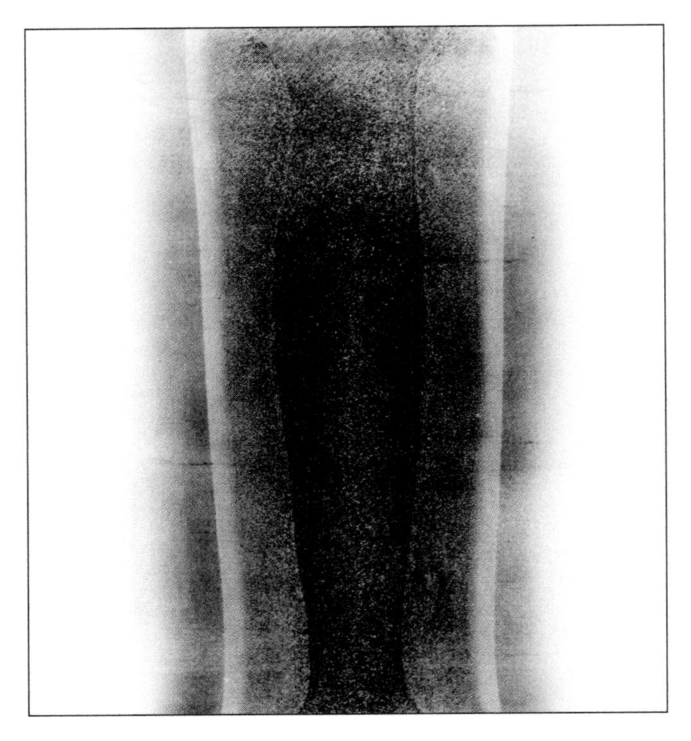

Figure 15.15—Photomacrograph of a Vertical Section of a Typical Thermite Rail Weld

Mold designs for repair welds also include risers for feeding the molten metal back into the thermite weld allowing for contraction of the weld during cooling. The exact increase depends on the size of the weld and the length of the root opening.

Applying the Mold. The premanufactured molds are applied in two or three pieces. The common practice, independent of mold design, is to fit the pieces onto the rail, ensuring that each piece conforms to the profile. Excess mold material is removed by filing. Various materials are used to fill gaps between the mold and the workpiece if the rail is in a severely worn condition.

The top of the mold is hollowed out to provide a basin for the slag produced, while runways (runners) in the mold allow excess slag to be caught in slag pans installed at both sides of the rail. The mold must be adequately vented to facilitate the escape of moisture and gases during preheating and welding. Figure 15.16 shows a typical thermite repair mold pattern. The geometry of the mold between the weld and the rail must minimize stress raisers to prevent weld fatigue and allow optimal performance.

The quality of the thermite mold requires special attention. It must be made of material with a high melting temperature, high permeability, and adequate shear strength. To ensure proper functionality of the mold, the sand should be free of clay components that have low melting points.

Preheating. Preheating is accomplished by directing a gas flame into the chamber through the heating gate. A torch designed specifically for the purpose may burn propane, natural gas, kerosene, or gasoline.

The purposes of preheating are to remove moisture and contaminants from the surfaces of the workpieces and to ensure that proper fusion takes place. The heat is typically applied for a period of 3 minutes to 5 minutes, depending on the preheat method used. The preheat is complete when the rail interfaces have reached optimal temperature and the mold has been thoroughly dried. The mold must be completely dried to avoid weld

Table 15.3
Examples of Weld Dimensions, Mold, and Thermite Requirements (in Inches)

Section Size or Diameter mm (in.)	Root Opening mm (in.)	Collar Opening mm (in.)	Risers Number Required	Risers Diameter Opening mm (in.)	Pouring Gates No.	Pouring Gates Diameter of Opening mm (in.)	Heating Gates No.	Heating Gates Diameter mm (in.)	Connecting Gates No.	Connecting Gates Diameter mm (in.)	Thermite Required[a] kg (lb)
Rectangular Sections											
50.8 × 50.8 (2 × 2)	11.1 (7/16)	38.1 × 11.1 (1-1/2 × 7/16)	1	19 (3/4)	1	19 (3/4)	1	31.7 (1-1/4)	—	—	13.2 (6)
50.8 × 101.6 (2 × 4)	14.3 (9/16)	33.3 × 14.3 (1-5/16 × 9/16)	1	19 (3/4)	1	25.4 (1)	1	31.7 (1-1/4)	—	—	26.4 (12)
101.6 × 101.6 (4 × 4)	17.5 (11/16)	66.7 × 17.5 (2-5/8 × 11/16)	1	25.4 (1)	1	25.4 (1)	1	31.7 (1-1/4)	—	—	55.1 (25)
101.6 × 203.2 (4 × 8)	22.2 (7/8)	77.3 × 22.2 (3-7/16 × 7/8)	1	25.4 (1)	1	25.4 (1)	2[b]	31.7 (1-1/4)	—	—	110.2 (50)
203.2 × 203.2 (8 × 8)	28.6 (1-1/8)	117.5 × 28.6 (4-5/8 × 1-1/8)	1	43.4 (1-3/4)	1	31.7 (1-1/4)	2[b]	31.7 (1-1/4)	—	—	275.5 (125)
203.2 × 304.8 (8 × 12)	31.7 (1-1/4)	139.7 × 31.7 (5-1/2 × 1-1/4)	1	43.4 (1-3/4)	1	31.7 (1-1/4)	1	31.7 (1-1/4)	1	31.7 (1-1/4)	385.8 (175)
304.8 × 304.8 (12 × 12)	36.5 (1-7/16)	165.1 × 36.5 (6-1/2 × 1-7/16)	1	63.5 (2-1/2)	1	38.1 (1-1/2)	2[b]	38.1 (1-1/2)	1	38.1 (1-1/2)	661.3 (300)
304.8 × 457.2 (12 × 18)	42.9 (1-11/16)	196.8 × 42.9 (7-3/4 × 1-11/16)	1	63.5 (2-1/2)	1	38.1 (1-1/2)	2[b]	38.1 (1-1/2)	1	38.1 (1-1/2)	1102.3 (500)
406.4 × 406.4 (16 × 16)	43.4 (1-3/4)	203.9 × 43.4 (8-15/16 × 1-3/4)	1	69.9 (2-3/4)	2	50.8 (2)	2	38.1 (1-1/2)	2	38.1 (1-1/2)	1543.2 (700)
406.4 × 610.0 (16 × 24)	50.8 (2)	252.4 × 50.8 (9-15/16 × 2)	1	69.9 (2-3/4)	2	50.8 (2)	2	38.1 (1-1/2)	2	38.1 (1-1/2)	2535.3 (1150)
610.0 × 610.0 (24 × 24)	58.6 (2-5/16)	290.0 × 58.6 (11-13/16 × 2-5/16)	2	63.5 (2-1/2)	2	50.8 (2)	2	44.5 (1-3/4)	2	44.5 (1-3/4)	4133.7 (1875)
610.0 × 914.4 (24 × 36)	66.7 (2-5/8)	358.8 × 66.7 (14-1/8 × 2-5/8)	2	63.5 (2-1/2)	2	50.8 (2)	2	50.8 (2)	4	50.8 (2)	6889.4 (3125)
Round Sections											
50.8 (2)	11.1 (7/16)	34.9 × 11.1 (1-3/8 × 7/16)	1	19 (3/4)	1	19 (3/4)	1	31.7 (1-1/4)	—	—	11.0 (5)
101.6 (4)	15.9 (5/8)	60.3 × 15.9 (2-3/8 × 5/8)	1	25.4 (1)	1	25.4 (1)	1	31.7 (1-1/4)	—	—	55.1 (25)
203.2 (8)	25.4 (1)	106.4 × 25.4 (4-3/16 × 1)	1	38.1 (1-1/2)	1	31.7 (1-1/4)	1	31.7 (1-1/4)	—	—	165.3 (75)
304.8 (12)	33.3 (1-5/16)	149.2 × 33.3 (5-7/8 × 1-5/16)	1	44.5 (1-3/4)	1	38.1 (1-1/2)	1	38.1 (1-1/2)	1	38.1 (1-1/2)	440.9 (200)
406.4 (16)	41.3 (1-5/8)	190.5 × 41.3 (7-1/2 × 1-5/8)	1	50.8 (2)	1	38.1 (1-1/2)	1	38.1 (1-1/2)	1	38.1 (1-1/2)	936.9 (425)

Notes:

a. Quantities of thermite mixture in this table include provision for a 10% excess of steel in the slag basin for a single pour and a 20% excess for a double pour.

b. Includes one separate back-heating gate.

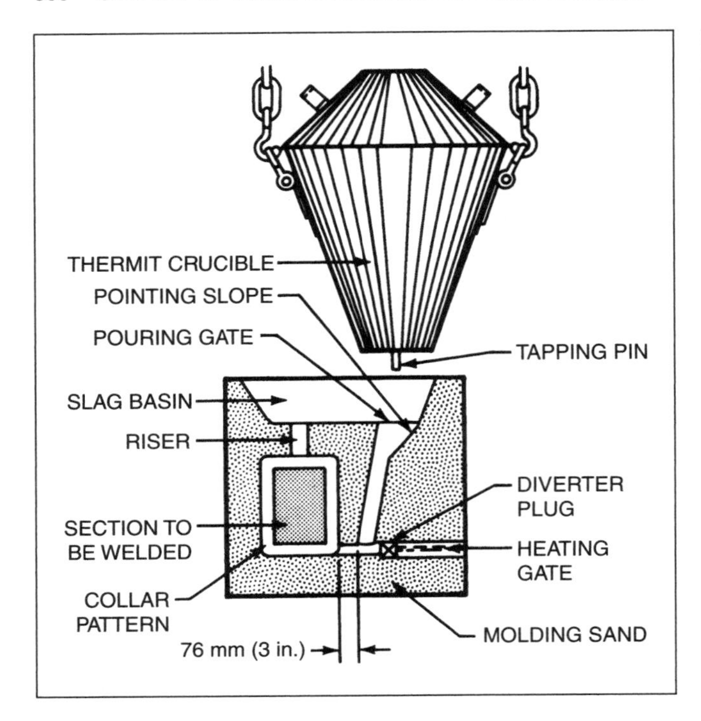

Figure 15.16—Cross Section of a Typical Thermite Mold for Repair Welding With External Preheat

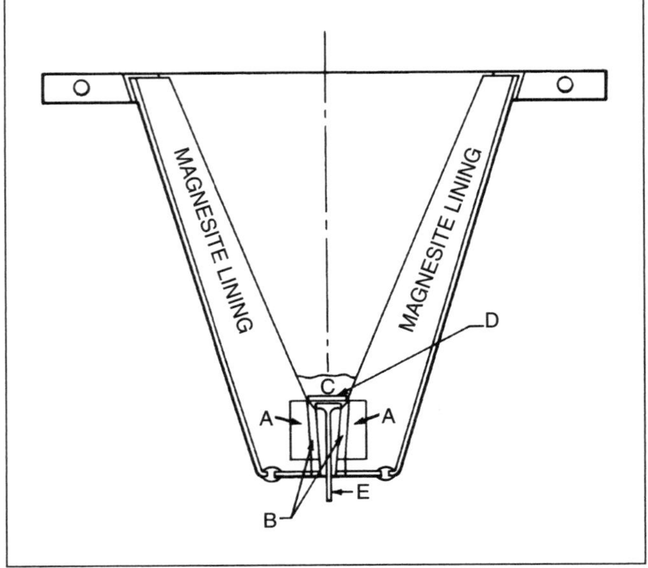

Figure 15.17—Cross Section of a Thermite Crucible

porosity generated by residual moisture in the molding sand. Preheating is continued until the ends of the workpieces are approximately cherry red in color, an indication that the temperature is between 800°C and 1000°C (1500°F and 1800°F).

On completion of preheating, the heating gate or opening is plugged by a diverter plug (a short piece made of the same material as the mold). This ensures that the pour begins at the bottom and flows to the top, and also assists in reducing turbulent flow.

Charging the Crucible. When welding preheated rail, the thermite reaction takes place in a refractory-lined crucible, as shown in Figure 15.17 (A through E). A hard refractory stone, often magnesite, at the bottom of a multiple-use crucible shown in (A) holds a replaceable refractory orifice or thimble, (B). The thimble is plugged by inserting a tapping pin (E) through it and then placing a metal disk (D) on top of the pin. The disk is covered with a layer of refractory sand (C). The thermite mixture should be placed in the crucible in a manner that will not dislodge the sand layer. Although a multiple-use crucible can be used if numerous welds are to be made, very few crucibles of this type are used in rail welding. Most railroads use the modern single-use version. In a single-use crucible, there is no need for manipulation or modification of the thimble, as it has been prepared at the factory for the intended single use. The single-use crucible has a safety bypass thimble in case it does not tap within the specific time frame.

Making the Weld. The reaction typically is initiated by an ignition rod using the gas preheat torch or the latent heat from the preheated rail ends to ignite the rod. After the reaction is complete and the motion of the molten metal subsides, the crucible is tapped by the melting of the crucible thimble. The molten steel flows into the mold and fills the joint. The top portion of the mold protruding above the rail is removed subsequent to solidification of the metal, and a hydraulic shear is used to trim the weld close to the head of the rail. This minimizes the time required to finish-grind the railhead flat for traffic. After the weld metal has solidified and cooled to an acceptable temperature, the mold is stripped away.

Other Thermite Welding Applications

Thermite welding is employed in the marine field for the repair of heavy sections of ferrous metal such as broken stern frames, rudder parts, shafts, and struts.

Broken pinions, pinion teeth, and necks of sheet and plate rolls are replaced with entirely new pieces, cast or forged slightly oversize to permit machining. They are thermite-welded to the main section. Similarly, badly worn wobblers on the ends of steel mill rolls may be replaced with a sufficiently tough thermite metal deposit that can be machined. The method is particularly well suited for repairs involving large volumes of metal, where the heat of fusion cannot be raised satisfactorily or efficiently by other means or when fractures or voids in large sections require a large quantity of weld metal.

Thermite welding can be used for the repair of ingot molds at significant savings over replacement. In one method, an eroded cavity in the bottom of the mold can be filled with thermite metal, but this repair has to be repeated after every second or third pouring. In another method, the eroded bottom of the mold can be cut off and completely rebuilt with thermite metal. This method of repair is more sophisticated and requires larger quantities of the thermite mixture, but the life of the ingot mold can be more than doubled.

When working with large dredge cutters, the blades may be thermite-welded to a center ring. Quantities up to several thousand pounds can be poured at one time. In this case, thermite welding can be considered a production tool rather than a repair method.

Figure 15.18—Thermite-Welded Reinforcing Steel Bars After Tensile and Bend Tests

Reinforcing Bar Welding. Thermite welding without preheat provides a method of splicing concrete-reinforcing steel bars. Continuous reinforcing bars permit the design of concrete columns or beams smaller in section than when the bars are not welded together. In this method, two premanufactured mold halves are positioned at the joint in the aligned bars and sealed to them with an adhesive compound and sand to avoid the loss of molten metal.

A closure disk is located in a well at the base of the thermite crucible section of the mold. The thermite powder is placed in the crucible and the reaction initiated. After completion of the reaction, the molten steel melts through the closure disk and fills the root opening between the bars. The initial molten steel entering the mold chamber preheats the bar ends as it flows over them into a preheated metal chamber. The molten steel fills the root opening and completes the weld.

Reinforcing bars can be welded with this process in any position with properly designed molds. Thermite-welded reinforcing bar specimens tested in tension and bending are shown in Figure 15.18

As an alternative to welding, reinforcing bars can be joined end-to-end by depositing thermite metal between a steel sleeve and the enclosed bars. This is primarily a mechanical joint. Arrangements for horizontal and vertical connections are shown in Figure 15.19. The sleeve is placed around the abutted bars. The inside diameter of the sleeve is somewhat larger than the diameter of the bars; this is necessary to provide space for the cast steel. Both the inner surface of the sleeve and the surfaces of the bars are serrated. A graphite or carbon dioxide-cured sand mold is mounted over an orifice in the sleeve through which the molten steel flows into the annular space between the bars and the sleeve. The thermite mixture is placed above a metal closure disk within the mold.

On ignition of the thermite powder, molten steel is produced. It melts the metal disk, flows into the annular space between the bars and the sleeve, and then solidifies. About five minutes are required to make a joint with this technique.

Welding of Electrical Components. A thermite mixture of copper oxide and aluminum is used to weld joints in copper conductors. The reaction between the two materials produces superheated molten copper and slag within in 1 second to 5 seconds. Other metals in the form of slugs or powder can be added to produce alloys for particular applications of thermite welding.

The process is primarily used for welding copper bars, cables, and wires; also for welding dissimilar metals, for example, copper conductors to steel rails for

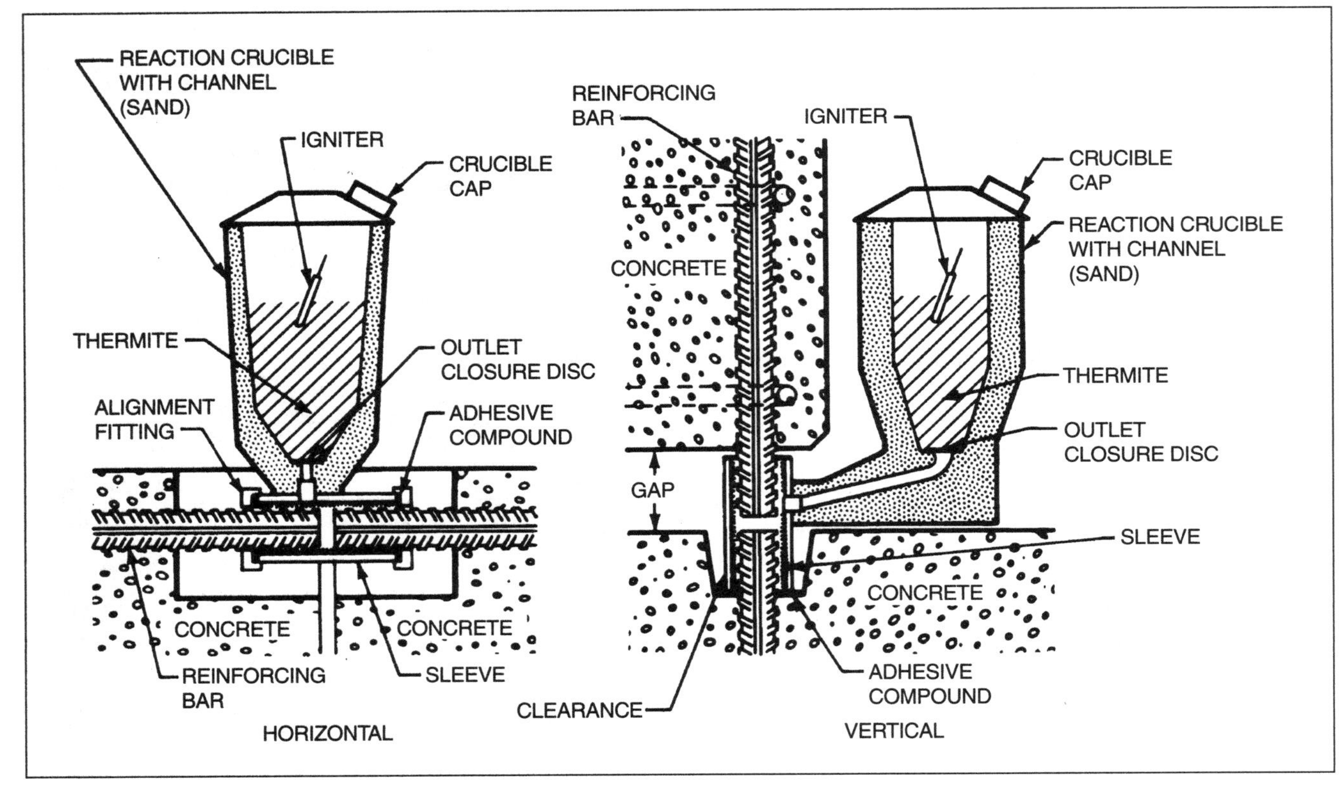

Figure 15.19—Thermite Sleeve Joint for Reinforcing Bars

grounding. For the latter application, a graphite mold is clamped to the rail section at the joint. As soon as the thermite reaction is complete, the molten copper melts the disk and flows into the joint cavity. The molten copper solidifies in a few seconds and creates a weld between the base metal and the copper cable. Then the mold is removed. The mold can be used again after removing the slag from the reaction chamber.

Heat Treatment of Welds. Thermite welding of high-alloy steels requires special thermite mixtures. These mixtures produce only heat; molten metal is not produced. This thermite mixture is designed to create thermal energy for heat-treating purposes. However, since using thermite for heat treating is the cost equivalent of producing a second weld, this approach has not been extensively used.

Heat treatment using thermite is done in the following way. Using special binders, the thermite mixture to be heat-treated is formed to the configuration of the workpieces. The binders keep their exact shape during and after the thermite reaction. The maximum temperature produced in the workpiece can be adjusted by the composition of the thermite compound.

SAFE PRACTICES

Moisture is a major concern in thermite welding. The presence of moisture in the thermite mixture, in the crucible, or on the workpieces can lead to the rapid formation of steam when the thermite reaction takes place. Steam pressure may cause the violent ejection of molten metal from the crucible. Therefore, the thermite mixture should be stored in a dry place, the crucible should be dry, and moisture should not be allowed to enter the system before or during welding.

The work area should be free of combustible materials that may be ignited by sparks or small particles of molten metal. The area should be well ventilated to avoid the buildup of fumes and gases from the reaction. Starting rods should be protected from accidental ignition.

Personnel should wear appropriate protection against hot particles or sparks. This includes gloves, full-face shields with filter lenses for eye protection, and headgear. Safety boots are recommended to protect the feet from hot sparks. Clothing should not have pockets or cuffs that might catch hot particles.

Preheating should be done using safety precautions applicable to all oxyfuel gas equipment and operations.

Additional information on safe practices for welding is provided in the latest edition of ANSI *Safety in Welding, Cutting, and Allied Processes*, Z49.1 published by the American Welding Society. This document is available on line at www.aws.org.[10]

COLD WELDING

Cold welding (CW) is a solid-state process in which pressure is used to cause substantial deformation, and through this form a weld. An important characteristic of the process is the absence of heat, either applied externally or generated by the welding process.

FUNDAMENTALS

The force applied during cold welding causes the metal between the electrodes to upset laterally as illustrated in Figure 15.20. The lateral flow of metal brings about the following effects:

1. Breaks up the oxide film present on the abutting surfaces and carries most of it away from the joint surface (Figure 15.20[A]);
2. Forces oxide-free metal on one side of the weld interface into intimate contact with oxide-free metal on the other side, (15.20 [B]); and
3. Maintains interfacial contact forces between these clean materials, allowing the formation of metallic bonds (15.20 [C], thus creating the weld (15.20 [D]).

Cold welds usually are insensitive to the rate of upsetting of the metal, within limits. Regardless of upset speed, welding will take place if there is sufficient upset.

A fundamental requisite for satisfactory cold welding is that at least one of the metals to be joined is highly ductile and does not exhibit extreme work-hardening. Both butt joints and lap joints can be cold welded.

10. See reference 8.

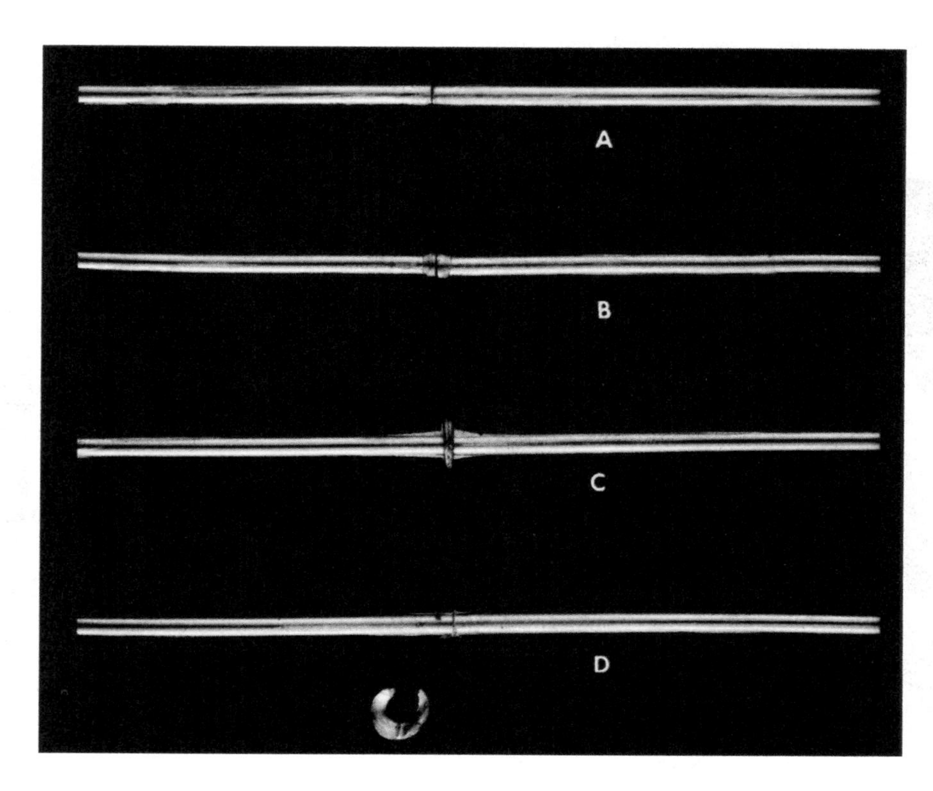

Figure 15.20—States of Upset During the Cold Welding of a Butt Joint

EQUIPMENT

Cold welding can be accomplished through any means of applying high pressures at the weld interface. The types of equipment used include hydraulic or mechanical presses, rolls, or special manually or pneumatically operated tools. These tools can be used for both butt and lap welds. A hand tool such as a toggle cutter is suitable for very light work. A typical manually operated press, as shown in Figure 15.21, is used for light work such as wire welding. Heavy work requires power-operated machines. The rate at which pressure is applied usually does not affect the strength or quality of the weld.

MATERIALS WELDED

Metals with a face-centered cubic (fcc) lattice structure are best suited for cold welding, provided they are not susceptible to rapid work-hardening. Soft tempers of metals such as aluminum and copper are most easily welded with this process. It is more difficult to weld cold-worked or heat-treated alloys of these metals. Other fcc-structured metals that may be readily welded by the cold welding process are gold, silver, palladium and platinum.

The joining of copper to aluminum by cold welding is a good example of an application of the process. This type of cold weld is characterized by substantially greater deformation of the aluminum than the copper because of the difference in the yield strength and work-hardening behavior between the two metals.

Dissimilar Metals

Numerous dissimilar metals can be joined by cold welding, whether or not they are intrinsically compatible. Since cold welding is carried out at room temperature, there is no significant diffusion between dissimilar metals during welding. The alloying characteristics of the metals being joined do not affect the manner in which the cold welding operation is carried out. However, interdiffusion at elevated temperatures can affect the choice of postweld thermal treatments and the performance of the weld in service.

Welds made between metals that are essentially insoluble in one another usually are stable. However, diffusion can form intermetallic compounds at elevated service temperatures. In some cases, this intermetallic layer may be brittle and will cause a marked reduction in the performance of the weld. These welds are particularly sensitive to bending or impact loading after this intermetallic layer has formed.

The rates at which intermetallic compounds form are partly determined by the interdiffusion characteristics of the two species. This includes exposure times and temperatures, as well as the specific diffusion constants for the respective material combination. Thus, bimetal cold weldments require careful consideration of the specific metal combination as well as the service environment. For example, Figure 15.22 shows a layered structure at the interface of an aluminum-copper weldment after

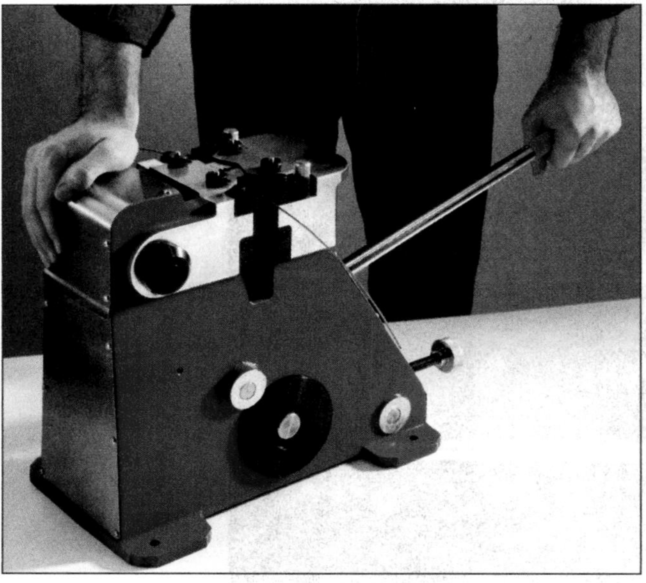

Photograph courtesy of Huestis Industrial

Figure 15.21—A Manually Operated Cold Welding Machine

Figure 15.22—Layered Structure in an Aluminum-Copper Cold Weld after Exposure at 260°C (500°F) for 60 Days

exposure at elevated temperatures. The layered structure contains a brittle Al-Cu intermetallic compound that effectively weakens the weldment. Figure 15.23 shows how rapidly the thickness of the diffused zone increases at high service temperatures. Mechanical tests have shown that the strength and ductility of the joint decrease when the thickness of the interfacial layer exceeds about 0.05 mm (0.002 in.). Consequently, aluminum-copper cold welds should be used only in applications where service temperatures are low and peak temperatures seldom, if ever, exceed 65°C (150°F).

Microstrucure

In butt joints, the lateral flow of metal between the welding dies during upset produces a cross-grained structure adjacent to the interface of the weld. This structure is detailed in Figure 15.24. The lateral flow brings about a change in grain orientation, with material in the upset region nominally aligned parallel to the resulting joint. This change in orientation generally does not affect the joint performance of isotropic materials (for example, aluminum), although networks of planar grain boundaries or realigned interior contamination may be a concern.

Cold welding requires that clean metal faces come into intimate contact for a satisfactory joint. Proper surface preparation is necessary to ensure joints of maximum strength. Dirt, absorbed gas, oils, or oxide films on the surface interfere with metal-to-metal contact and must be removed to obtain strong welds.

The best method of surface preparation for lap welds is wire brushing at a surface speed of about 15 m/s (3000 ft/min). A motor-driven rotary brush of 0.1-mm (0.004-in.) diameter stainless steel wire is commonly used. It should be noted that softer wire brushes may burnish the surface, while coarser types may remove too much metal and roughen the surface. The surfaces should be degreased prior to brushing to avoid contamination of the brush. It is important that the clean surface not be touched with the hands before welding. Hand contact may transfer grease or oil on the faying surface, impairing the formation of a strong joint. Welding should take place as soon as practical after cleaning to avoid any oxidation at the weld interface. When welding aluminum, for example, welding should take place within about 30 minutes of cleaning.

Chemical and other abrasive cleaning methods have not proven satisfactory for cold welding. The residue from chemical cleaning or abrasive particles on the sur-

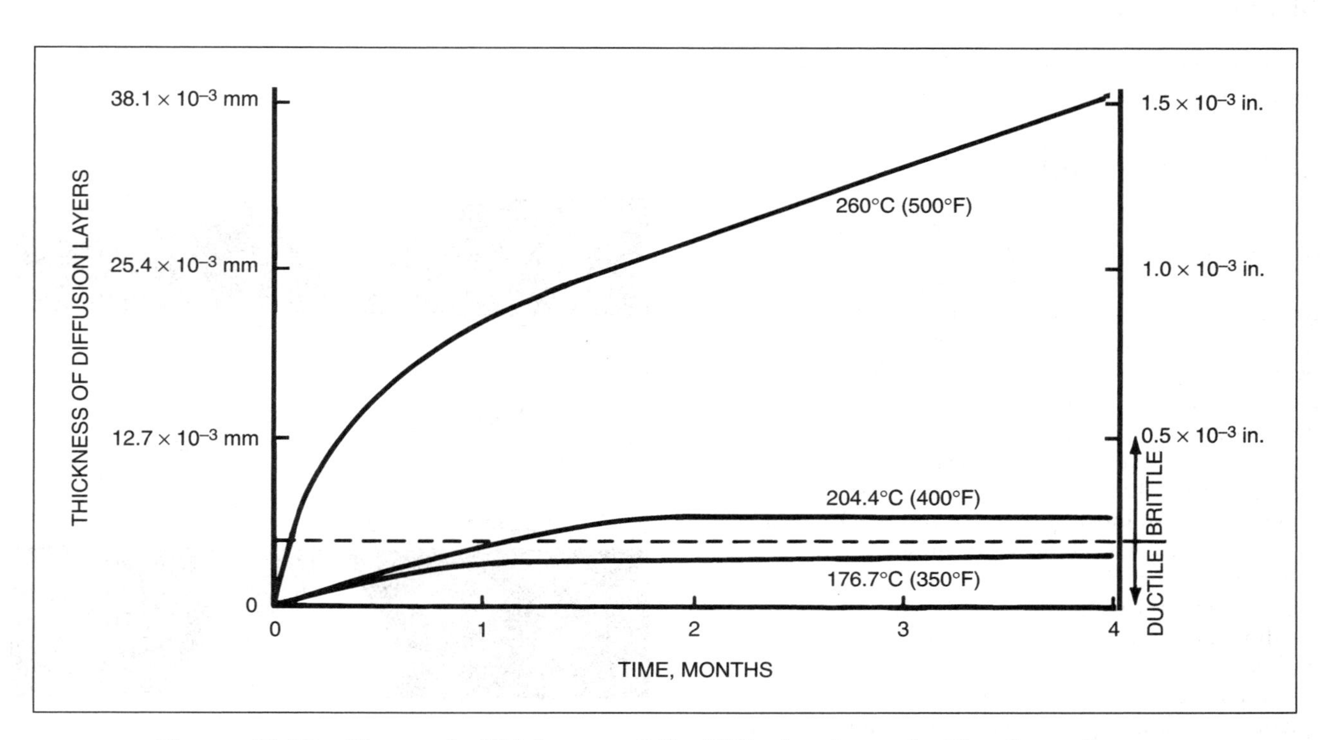

Figure 15.23—Change in Thickness of the Diffusion Layer in Aluminum-Copper Cold Welds with Exposure Time at Three Elevated Temperatures

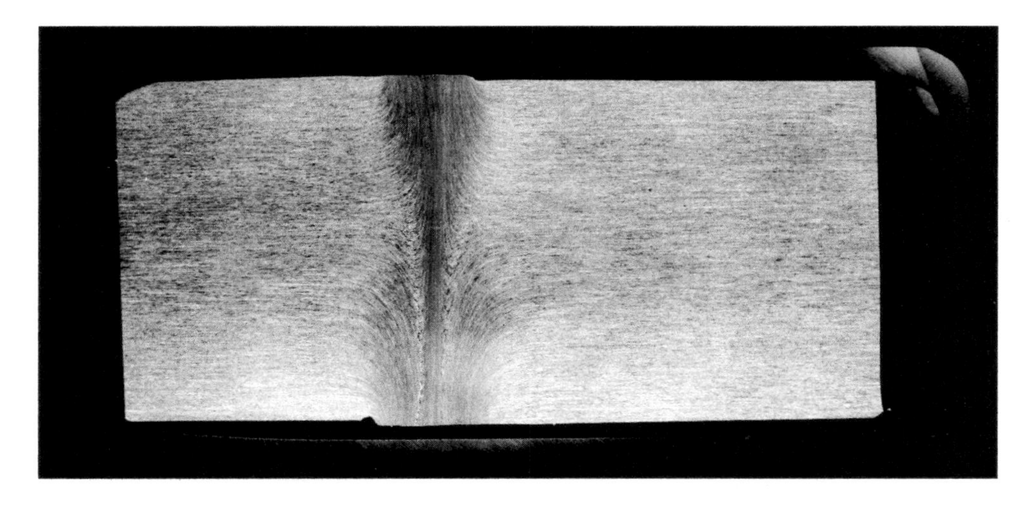

Figure 15.24—Transverse Flow Lines in the Surface of a Cold-Welded Butt Joint

face or embedded in it may prevent the formation of a sound weld.

APPLICATIONS

Cold welding is mainly used in the production of wire and the welding of bars, tubes, and plate metals.

Butt Joints in Wire

Cold-welded butt joints are used in the manufacture of aluminum, copper, gold, silver, and platinum wire. The most common use is to join successive reels of wire for continuous drawing to smaller diameters. Butt joints also are used to repair breaks in the wire that occur during the drawing operation. Diameters ranging from 0.06 mm to 12.7 mm (0.0025 in. to 0.50 in.) have been successfully welded. The aluminum alloys that have been welded with good results include Types EC, 1100, 2319, 3003, 4043, all of the 5000 series, 6061, and 6201. With most of these alloys, the as-welded wire can be drawn successfully to smaller diameters after removal of the flash. Flash is made up of material expelled from a weld prior to the upset portion of the welding cycle. Cold-welded joints in wire containing high-strength alloys, such as Types 2014 and 7178, usually must be annealed to prevent breaks during subsequent drawing operations. Aluminum alloys that contain lead and bismuth (Types 2011 and 6262) are difficult to join with cold welding.

Single-upset welding normally is not used for butt joints in wires smaller than 4.8 mm (3/16 in.) in diameter. A cross section of a single-upset butt joint in Type 1100 aluminum wire is shown in Figure 15.25.

This weld compares favorably with the multiple-upset weld illustrated in Figure 15.26. The multiple upset, offset-flash technique commonly is used in wire

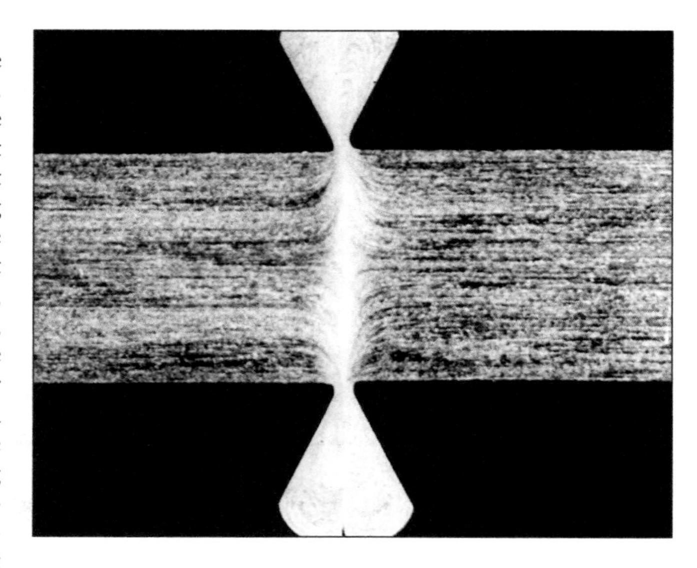

Figure 15.25—Single-Upset Cold Weld in Type 1100 Aluminum Wire

Figure 15.26—Multiple-Upset Cold Weld in Type 1100 Aluminum Wire Using an Offset Flash Technique

drawing to splice 0.64-mm to 3.25-mm (0.025-in. to 0.128-in.) diameter wires and also for aluminum alloy wires that cannot be welded effectively with single upset.

Electrical Wire Welding

If welding is permitted in the applicable ASTM International specification, cold welding may be used to join successive lengths of wire to facilitate the stranding of long lengths of multiple-strand electrical conductors. The weld flash is removed, and the weld is finished with a file or suitable abrasive to obtain a smooth, uniform appearance.

Joints in annealed wire of any of the weldable aluminum alloys exhibit tensile strengths exceeding 95% of the base metal strength. In cold-worked Type EC or 5005 aluminum wire and heat-treated Type 6201 wire, weld efficiencies of 92% to 100% are attained. In bend testing, the welded joint can be bent or twisted about half as many times as unwelded wire of the same alloy before failure.

Tests using Types EC, 5005, and 6201 aluminum alloys showed that seven-strand No. 4 American Wire Gauge (AWG) aluminum alloy conductors with a cold weld in one strand had the same breaking strength as similar conductors without welded strands.

For copper wire, work hardening at the weld interface increases the metal strength to that of the drawn wire.

Butt Joints in Rods and Tubing

In addition to butt joints in wire, cold welding commonly is used in the production of rod, tubing, and simple extruded shapes of similar and dissimilar metals. Figures 15.27(A) and 15.27(B) illustrate butt joints in tubing and solid forms. The weld can be as strong as the base metal when correct procedures are used.

When welding butt joints in tubing and bars or other solid forms, a short section usually is sheared from the ends of the workpieces to expose fresh, clean surfaces. This eliminates the need for the brush cleaning required for lap joints. The shear should be designed to provide square ends so that the workpieces do not deflect from axial alignment as welding force is applied. During

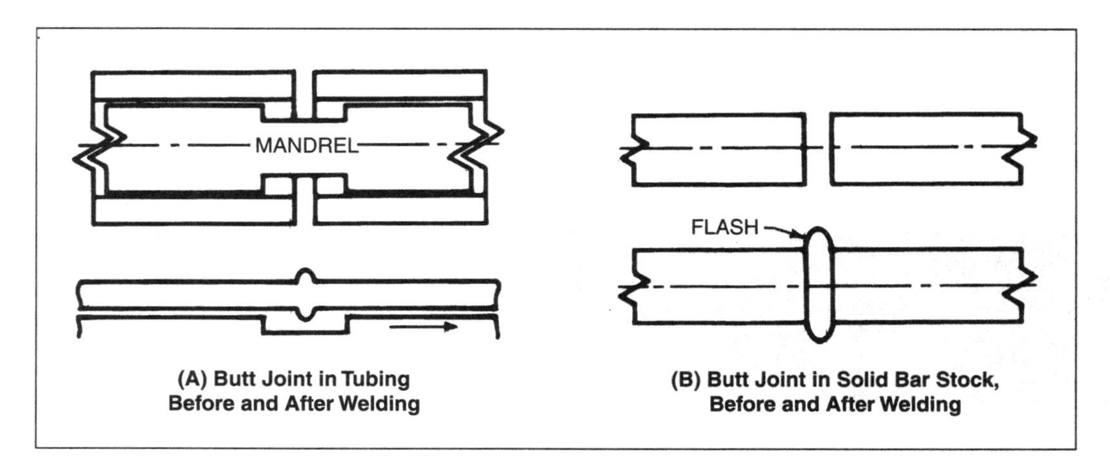

(A) Butt Joint in Tubing Before and After Welding

(B) Butt Joint in Solid Bar Stock, Before and After Welding

Figure 15.27—Butt Joints in Tubing and Solid Forms

shearing, a thin film of the particular metal being cut may accumulate on the shear blades. If the shear subsequently is used to cut a different metal, accumulated metal on the blade may transfer to the cut surface and inhibit welding. Therefore, the shear blades should be cleaned before use on another metal for cold welding.

Cold welding usually does not require degreasing of the workpieces before shearing if the residual film of lubricant usually present on metal surfaces is very thin. However, if there is a heavy oil film on the metal, degreasing may be necessary to avoid contamination of the cut surfaces and to prevent the workpieces from slipping in the clamping dies. The best practice, therefore, is always to clean before shearing.

To form cold welds in the butt configuration, the workpieces should be positioned in the clamping dies with sufficient initial extension to ensure adequate upset. However, extension of the workpieces should not be excessive or they may bend instead of upsetting. Similarly, trying to upset a workpiece with an excessively large die opening will cause the workpieces to bend or buckle, as shown in Figure 15.28. If the combined projecting distance exceeds approximately four times the thickness or diameter of the workpieces, the ends may deflect and slide past one another when force is applied. This distance should be equivalent to the maximum total upset that can be used to accomplish the weld. The minimum upset distance varies with the alloy being joined.

The welding dies should firmly grip the workpieces to prevent slippage when the upset force is applied. Any slippage will reduce the amount of upset and consequently diminish the final quality of the joint. For a firm grip, the dimensions of the workpieces are critical so that the dies can achieve maximum contacting loads for welding. The allowable tolerance depends on the design of the die and holder, as well as the finish on the gripping surface.

Deep knurling on the gripping surfaces aids in preventing slippage. However, knurling also may result in marking or scarring the workpiece surfaces. The allowable indentation tolerance for round sections is about 3% of the workpiece diameter. Somewhat wider tolerances are permissible for rectangular-shaped parts. This is because the dies usually bear on only two sides; however, the parallelism of the closed grips must be maintained in order to obtain the uniform upsetting of metal. Alignment of the dies should be within about 10% of the workpiece thickness.

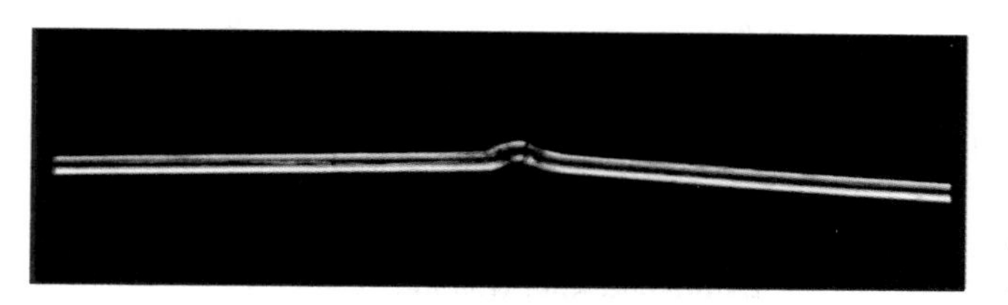

Figure 15.28—Bending Produced During Upset Caused by Excessive Projecting Lengths

The amount of upset required to produce a full-strength weld in some alloys sometimes exceeds that which can be provided in one step. However, if an initial upset can produce a bond of sufficient strength to hold the workpieces together, additional upset can be applied through multiple stages. This is done by repositioning the weldment in the dies and repeating the upsetting operation as many times as necessary to achieve a joint with the required quality characteristics. Surface preparation prior to welding is relatively unimportant when a multiple upsetting technique is used. This technique produces greater surface strains, resulting in more complete displacement of contaminants from the weld interface.

Figure 15.29 illustrates the various stages involved in making a multiple-upset weld in a butt joint between strips.

Applications occasionally arise in which flash cannot be aligned on two sides of the joint. Figure 15.30 illustrates a cold weld being made in a die designed to produce an offset flash. This technique produces a discontinuous flash that is easy to remove and results in a weld joint that is at an angle to the wire axis. The weld joint positioned at an angle to the axis is less influenced by discontinuities in the weld.

Lap Joints

Commercial uses of cold-welded lap joints include packaging applications and the fabrication of electrical devices, probably the field in which cold-welded lap joints are most commonly used. This process is especially well suited to the fabrication of electrical components in which transitions from aluminum windings to copper terminations are required. These applications range from small electronic devices to large power distribution transformers.

Another application for cold-welded lap joints is the sealing of commercially pure aluminum, copper, or nickel tubing by pinching it between two dies. This is accomplished with the flange welding technique. In this application, the radius and width of the die face must be designed to accomplish cold welding and then cut the tubing in two pieces, thus sealing each end. The opposing dies must be carefully aligned for welding. The face radius is the key to successful cold welding and must be determined experimentally for the type metal and the tube wall thickness.

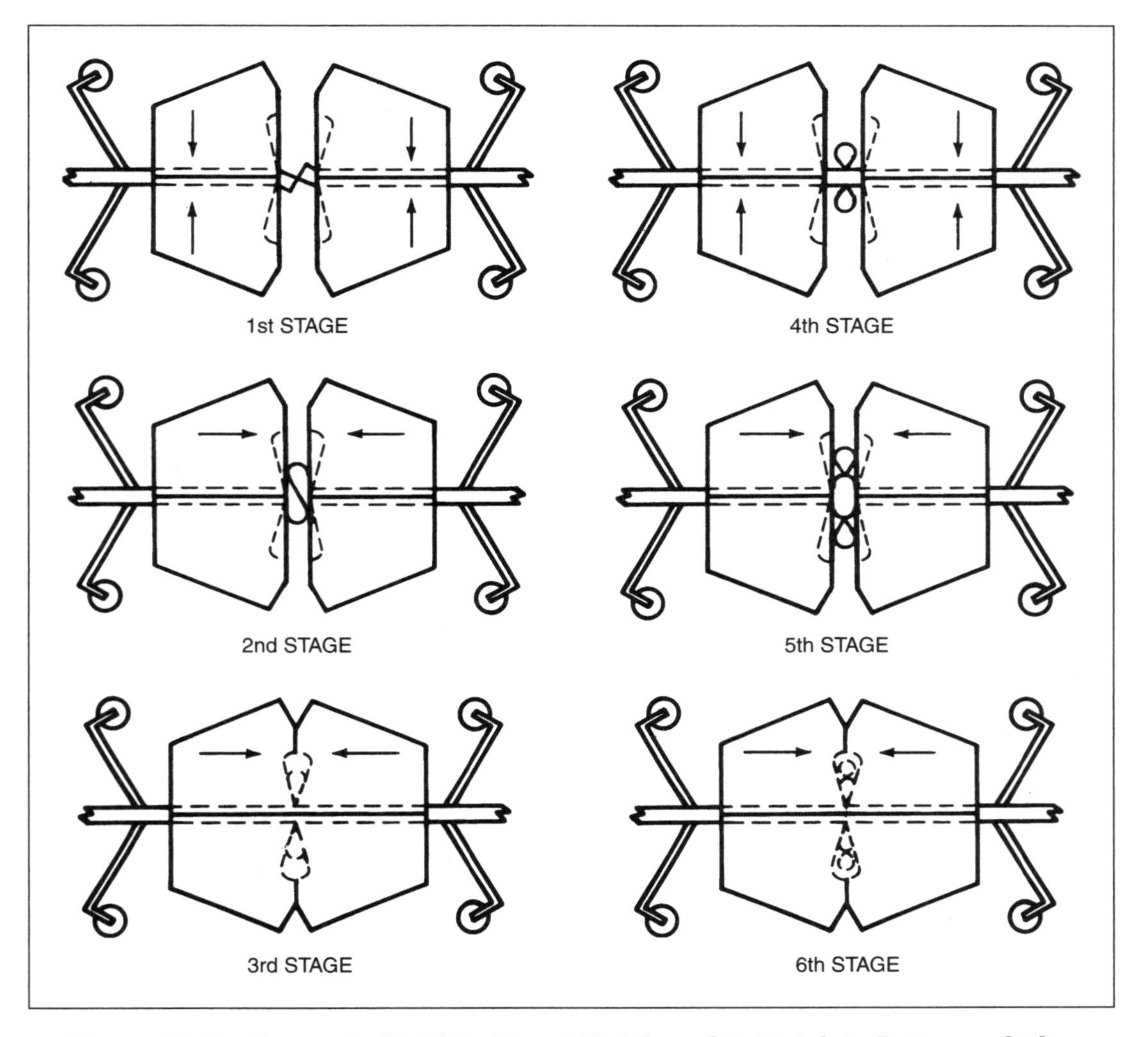

Figure 15.29—Stages in Multiple-Upset Welding of Butt Joints Between Strips

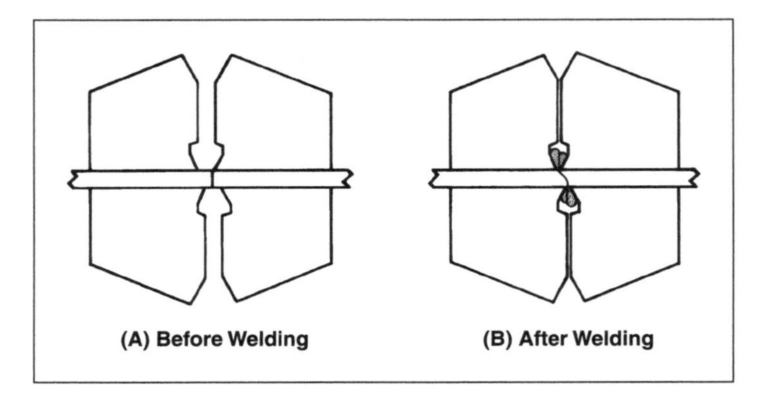

Figure 15.30—Cold Welding of Wire with an Offset-Flash Technique

Lap welds can be used for joining sections of aluminum sheet or foil to one another or to copper. Pressure is applied to the lapped workpieces by dies that indent the metal and cause it to flow at the weld interface. For aluminum, this pressure ranges between 1034 MPa and 3477 MPa (150 ksi and 500 ksi), depending on the compressive yield strength of the aluminum alloy being welded. Excellent lap welds can be produced in non-heat-treatable aluminum alloys, such as Types EC, 1100, and 3003. However, lap joints in aluminum alloys containing more than 3% magnesium (the 2000 and 7000 series of wrought alloys) and castings are not readily welded.

Table 15.4 provides recommended deformations and typical joint efficiencies for lap welds in several common aluminum alloys. For most alloys, the maximum joint strength is reached when using deformations between 60% and 70%. It is apparent that the intrinsic strength of the weld may increase at deformations exceeding 70%, but the overall strength of the assembly is decreased. Lap welds exhibit good shear and tensile strengths but have poor resistance to bending or peel loading.

Equal deformations can be achieved when lap welding aluminum to copper or welding dissimilar aluminum alloys by using dies with the bearing areas approximately in inverse proportion to the compressive yield strength of each metal.

As noted, proper indentation is essential for achieving high-quality lap welds. The indentation may take the form of a narrow strip, a ring, or a continuous seam. Typical weld-indenter configurations used for cold welding are shown in Figure 15.31. The selection of indentation configuration is determined largely by the appearance and performance characteristics required of the product.

The bar-type indentation, shown in Figures 15.31(A) and (B), causes lateral deformation in both directions, nominally perpendicular to its major axis. Indentation in the form of a ring, shown in Figure 15.31(C), may cause undesirable curvature of the sheet surfaces; for example, a ring weld may have a smooth convex dome of metal within the ring. This dome is formed when the metal is forced from between the dies as pressure is applied. Continuous and intermittent seam welding, shown in Figure 15.31(E) and (F), can be employed in the manufacture of thin-wall tubing or for making lap welds in sheet.

Symmetrical dies that indent both sides of the joint, shown in Figure 15.32(A) and (B), generally are used. If one surface must be free of indentation, a flat plate or anvil may be used on one side to produce the weld shown in Figure 15.32(C). Thinner gauges of sheet metal or wire can be cold-welded using simple dies mounted on hand-operated tools.

Draw welding is a form of lap welding used to seal cans or containers. In this application, both the lid and the can are flared before welding. The components are placed in a close-fitting die. A punch forces the components into the die, and a cold weld forms in the flared metal as it is drawn down over the punch. Figure 15.32(D) illustrates this type of joint.

Dies usually are subjected to high pressures and should be made of tool steel hardened to about 60 on the Rockwell C scale. Pressures from 1000 MPa to 3400 MPa (150 ksi to 500 ksi) are required to weld aluminum, depending on the composition and temper of the alloy. Copper requires pressures that may be two times to four times greater than those required for aluminum. Aluminum can be cold welded to copper using specially designed dies to compensate for the difference in yield strengths of these metals.

Table 15.4
Lap Joints Made by Cold Welding in Selected Aluminum Alloys

Alloy	Temper	Recommended Deformation, %	Joint Efficiency, %
3003	0	50	85
3003	H14	70	70
3003	H16	70	60
3003	H18	60	55
3004	0	60	60
3004	H34	55	40
5052	0	60	65
5052	H34	60	45
6061	T6	60	50
7075	T6	40	10

SAFE PRACTICES

The potential hazards of cold pressure welding are primarily mechanical. Personnel must be aware of these hazards and take appropriate safety precautions.

Mechanical

The basic systems are mechanical presses, and must be treated with appropriate safety practices. This includes safety devices to prevent injury to the operator's hand or other parts of the body. Initiating devices, such as push buttons or foot switches, should be arranged and guarded to prevent inadvertent actuation.

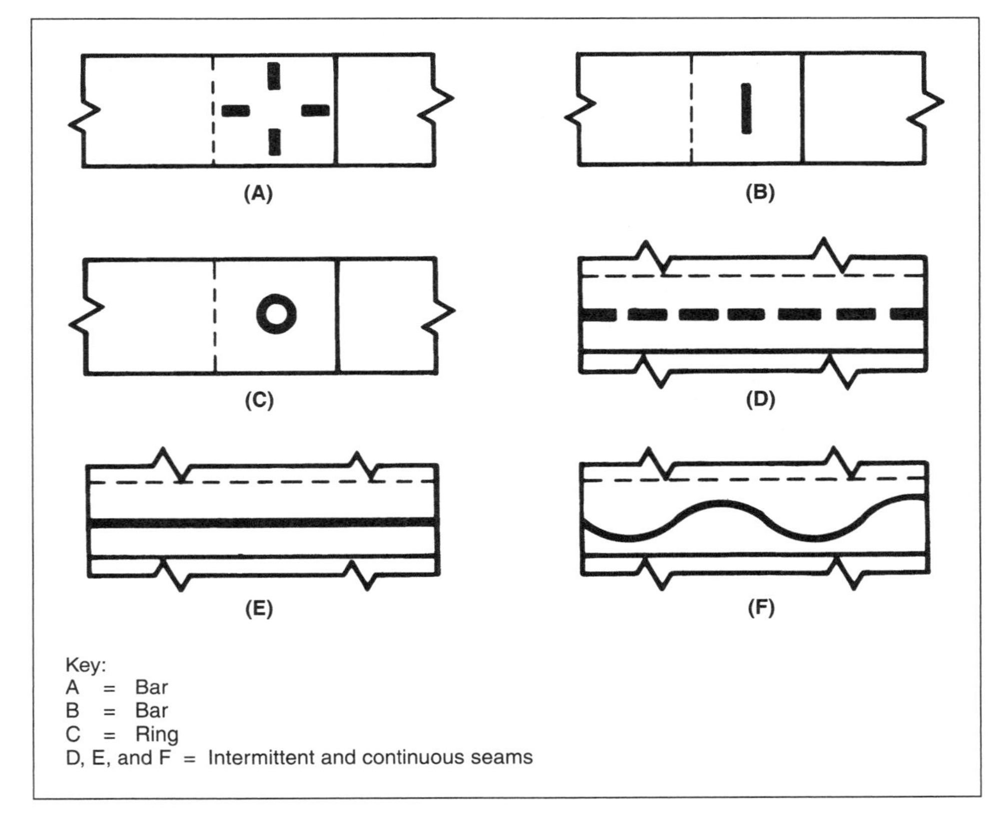

Figure 15.31—Typical Lap Weld Indentor Configurations Used in Cold Welding

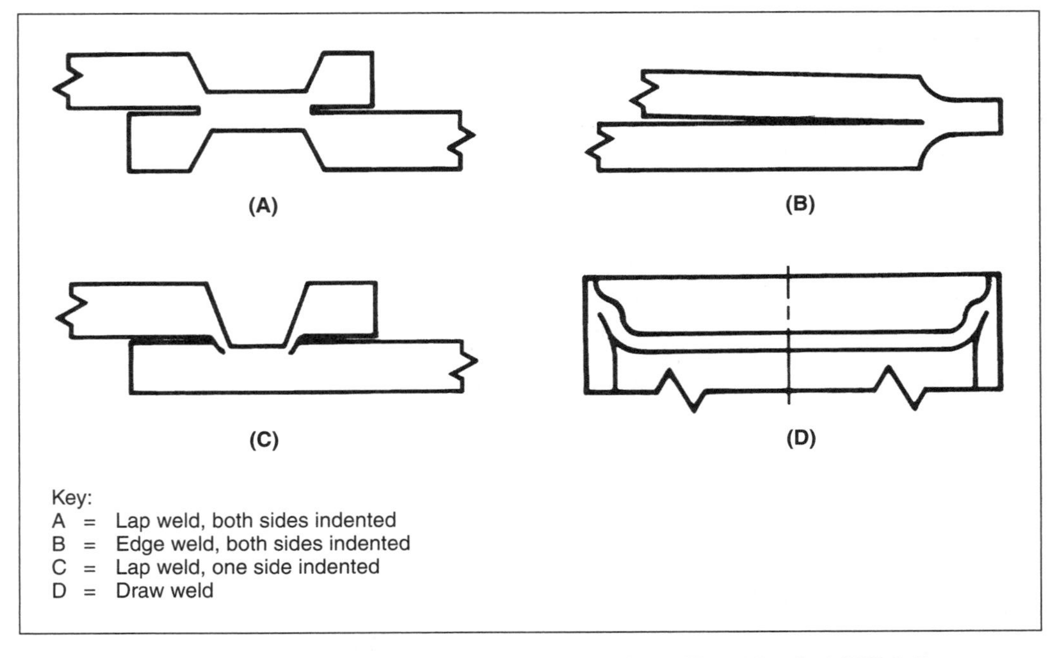

Figure 15.32—Contour of Several Lap Joints Used in Cold Welding

Machine guards, fixtures, or operating controls must prevent the operator from entering the area of joining during operation. Dual hand controls, latches, presence-sensing devices, or similar devices should be employed to prevent operation in an unsafe manner.

Personal Protection

The operator should wear personal eye shields when working with this process. Operating personnel may require ear protection when the welding operations produce high noise levels.

Additional information on safe practices for welding is provided in the latest edition of ANSI *Safety in Welding, Cutting, and Allied Processes*, Z49.1 published by the American Welding Society. This document is available on line at www.aws.org.[11]

MAGNETIC PULSE WELDING

Magnetic pulse welding is a solid state process that uses electromagnetic pressure to accelerate one workpiece to produce an impact against another workpiece. The metallic bond created by this process is similar to the bond created by explosion welding. Magnetic pulse welding, also known as electromagnetic pulse or magnetic impact welding, is highly regarded for the capability of joining dissimilar materials.

PHYSICS OF THE PROCESS

Electromagnetic metal processing was developed in the late 1800s, and in succeeding years most applications for this technology were in metal forming. It was not recognized as a viable welding process, but a substantial renewal of interest has occurred recently in the further development of this technology for welding.

Fundamentally, both metal forming and welding use the same underlying physics. The process is driven by the primary circuit. A significant amount of energy, usually between 5 kilojoules (kJ) and 200 kJ, (1124 lb force and 44 962 lb force) is stored in capacitors charged to a high voltage that may range between 3000 V and 30 000 V. The capacitors are then discharged through low-inductance and highly conductive bus bars into a coil, or actuator. The resulting current takes the form of a damped sine wave, characterized as a ringing inductance-resistance-capacitance (LRC) circuit. Peak currents during this process range between tens of thousands and millions of amperes (A), with pulse widths on the order of tens of microseconds. This creates an extremely intense transient magnetic field in the vicinity of the coil. The magnetic field induces eddy currents in any conductive materials nearby, in the opposite direction to the primary current. The opposing fields in the coil and workpiece result in a high repulsion force. This force drives the flyer, or driver, workpiece (the workpiece closest to the driving coil) at high velocity toward the target, the stationary workpiece, resulting in a high impact between the two metals. The impact pressure drives away the surface contaminants and provides for the intimate contact of clean surfaces across the weld interface. Metallic bonding results from this contact. A schematic of the process is shown in Figure 15.33.

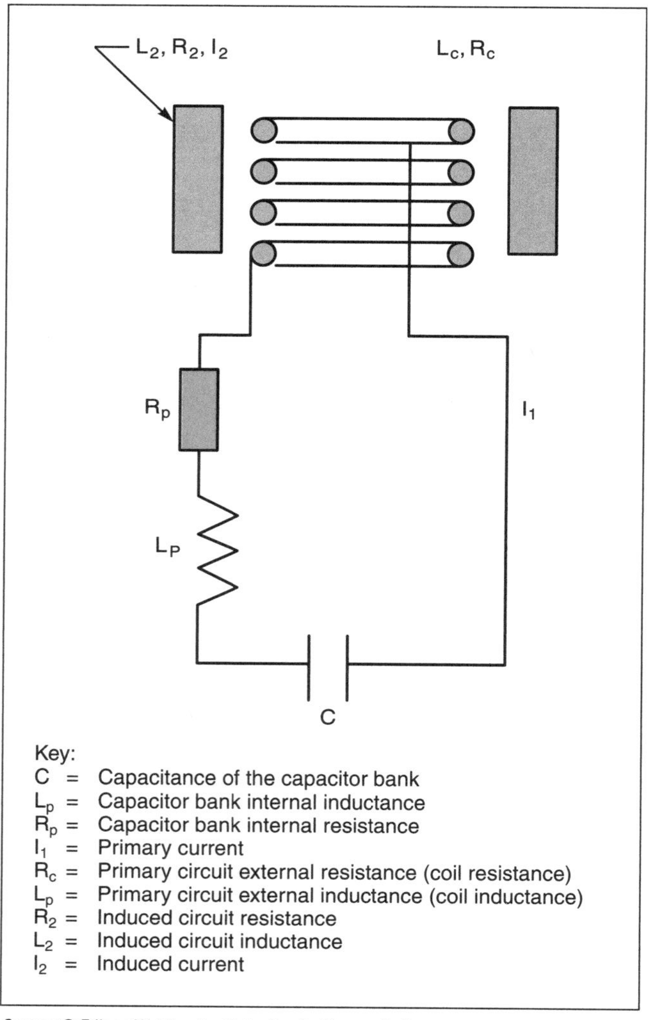

Source: © Edison Welding Institute. Used with permission.

Figure 15.33—Basic Diagram of the Magnetic Pulse Welding Process

11. See reference 8.

The following three elements are fundamental to achieving good magnetic pulse-welded joints:

1. Correct welding machine parameters,
2. Consideration of metal or material properties, and
3. Relative positioning of the flyer and the target workpieces.

Welding machine parameters determine the frequency and magnitude of the current waveform. High frequencies typically are favored for magnetic pulse welding. If the frequency is too low, the buildup of eddy currents in the flier workpiece will not be sufficient to achieve the velocities necessary for impact joining. The frequency is directly related to the electrical characteristics (LRC) of the circuit, including the capacitors and coil. Low system capacitances and inductances favor high-frequency characteristics.

The properties of the workpiece metal, particularly of the flier, also contribute to determining the weldability of a given metal. Properties to be considered include electrical conductivity and strength. Metals with high electrical conductivity and low strength are most easily welded with the magnetic pulse process. Higher electrical conductivity facilitates greater induced currents in the flier workpiece, with correspondingly greater magnetic pressures. Lower yield strengths facilitate displacements of the flier at lower magnetic pressures, and are easier to accelerate to the required speed for welding. Aluminum and copper are very well suited to electromagnetic welding, both aluminum-to-aluminum, copper-to-copper and to dissimilar metals. Carbon steel also can be welded when adjustments are made to the system power and frequency. Metals with relatively low electrical conductivity, such as titanium and austenitic stainless steels, are almost impossible to directly weld with the magnetic pulse process. They are readily welded, however, with the use of a driver plate. The driver plate is essentially a band of conductive material (typically copper) wrapped around the low-conductivity flier. During welding, the driver reacts with the coil, pushing the actual flier to the necessary velocities for metallic bonding.

POWER SOURCE

The essential component of a magnetic pulse welding system is a capacitor bank. The energy stored in the system can be determined from the size (capacitance) of the bank and the charge voltage using the following equation:

$$E = \frac{CV^2}{2} \qquad (15.1)$$

where

E = Energy,
C = Capacitance, and
V = Voltage

The energy is provided to the capacitors by a dedicated charging system. The capacity of the charging system largely controls the time required to charge the bank between subsequent welds. The charging circuit generally is actively cooled, allowing repeated use during production applications.

As previously mentioned, energy is transferred from the capacitors to the coil with an assembly of bus bars. Two considerations are key in the design of the bus bar assembly: it must have low inductance (in general the majority of the system inductance should be at the coil), and low resistance contacts. When in use, the capacitors are charged relatively slowly to a predefined voltage. Once this voltage is reached, a fast-action switch is used to allow current flow to the coil. Switching typically is done using solid-state silicon controlled rectifiers (SCRs).

Coil Designs

Various methods exist for fabricating electromagnetic welding coils. The various approaches provide different inductance characteristics and can potentially require different capacitor bank system characteristics for optimum functionality. Some examples of coil designs are single-turn and machined coils, multiple-turn coils, and magnetic field coils.

Single-Turn and Machined Coils. The simplest electromagnetic compression coil consists of a slot and tapered hole in a conductive plate, as shown in Figure 15.34. Metals with high conductivity and high strength

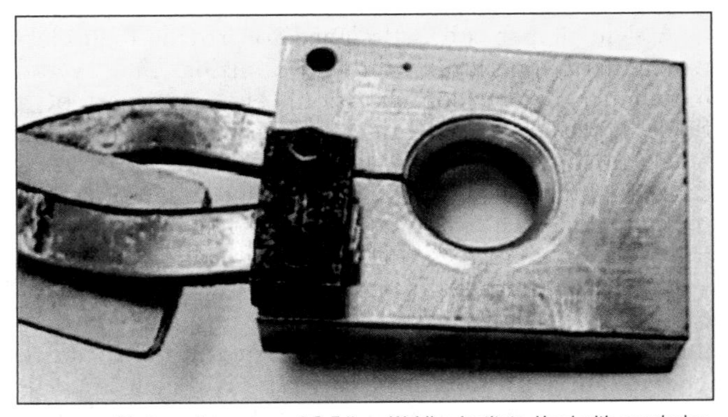

Photograph courtesy of © Edison Welding Institute. Used with permission.

Figure 15.34—A Single-Turn Compression Coil used in Magnetic Pulse Welding

generally are used. High conductivity improves system efficiency; metals with high yield strength provide a strong, robust coil with a long service life. High-strength aluminum alloys, copper-beryllium alloys, brasses, and materials strengthened with oxide dispersion are candidate coil materials. Polyamide films such as Kapton® are used to provide insulation both at the coil slot and between the coil and workpiece.

Multiple-Turn Coils. Simple multiple-turn coils for electromagnetic welding can be fabricated by wrapping conductive wire, for example, copper magnetic wire, over a strong nonconductive mandrel, such as one made from a Grade G-10 laminated phenolic composite. Reinforced epoxy or urethane often is used as a structural and insulating overlay. Compression, expansion or flat coils can be made in this way. There are two important limitations of this type of coil. First, since the forces on the workpiece and coil are equal and opposite, the coil construction materials determine the pressures that can be tolerated. For this reason, this type of coil construction usually is limited to pressures on the order of 48 MPa (7000 psi). Second, the winding pitch limits the local field intensity that can be achieved. This, in turn, limits the local pressures that can be generated. As a result, multiple-turn coils are often used in short-run productions, where they are considered disposable items, called *one-shot coils.*

Magnetic Field-Shaper Coils

In general, the highest electromagnetic pressures can be generated when the working surface of the coil is made from a monolithic block of a high-strength, high-conductivity metal. This is accomplished with single-turn coils. However, these coils often are quite inefficient. To improve efficiency, field-shaper coils can be used.

A field-shaper coil is designed to provide high electromagnetic pressures while minimizing the overall inductance of the configuration. The operation of a field-shaper coil is illustrated in Figure 15.35(A), with a machine shown as Figure 15.35(B). Field-shaper coils can be used in conjunction with either single or multiple-turn coils. The main coil of a field shaper acts as a transformer, increasing the current in proportion to the apparent windings ratio (main coil windings per 1 turn of the field shaper), and focusing this current over the area defined by the face area of the field shaper. The advantage of a coil of this type is that it concentrates large magnetic pressures in a desired area. This is done at the cost of reduced efficiencies (losses at the main coil and shaper gap) compared to a well-designed single-stage coil.

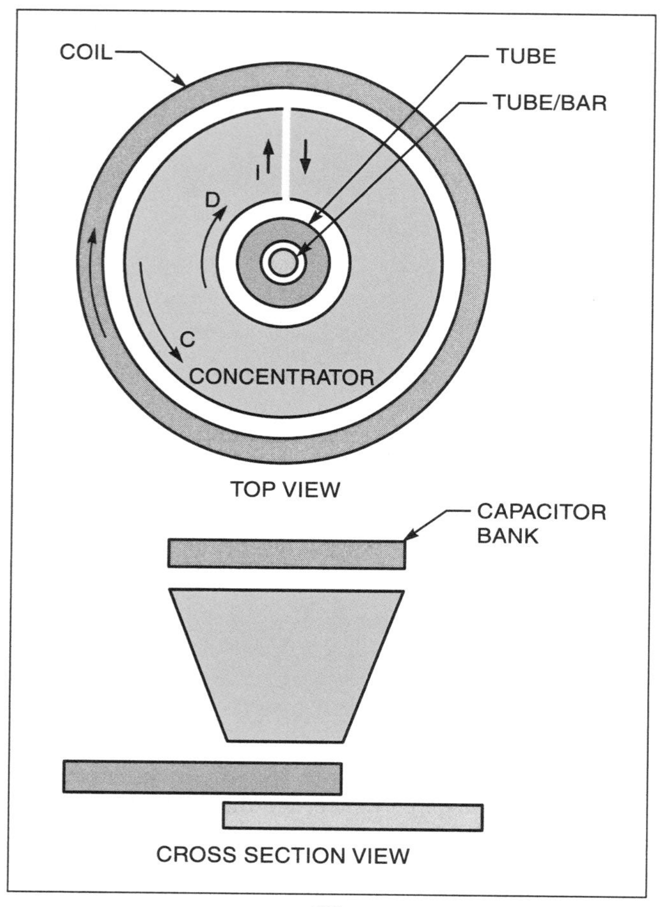

(A)

(B)

Source: © Edison Welding Institute. Used with permission.

Figure 15.35—(A) Cross Section of Coil and Field Shaper and (B) Magnetic Pulse Welding Machine

JOINT CONFIGURATIONS

The magnetic pulse welding process is used mainly for joints in tubular products. The flyer tube thickness generally varies between 0.1 mm to 3 mm (0.004 in. to 0.118 in.); the target tube must be thick enough to withstand the impact forces. If a target tube is not strong enough to tolerate the electromagnetic impact, an internal mandrel support is used. In these applications, the flyer tube can be expanded or compressed against the target tube or rod. The compression variation is shown in Figure 15.36. In this case, the driver coil is located around the outer tube, (the flier tube). The outer tube is then driven inward (compressed) at high speed, and then impacts the inner tube to create a weld. A single-turn coil with tapered opening works for most metals. In cases where higher efficiency is needed, multiple-turn coil with a field shaper is preferred. An expansion-type coil is placed inside of the inner tube as shown in Figure 15.37. The coil then drives the inner tube outward (with an expansion coil), contacting the outer tube again with sufficient velocity to create a joint on impact. Expansion coils typically are multi-turn windings. The coils are increasingly difficult to make as the tube size decreases, although tubes that can be expansion-welded usually are less than 50 mm (2 in.) in diameter.

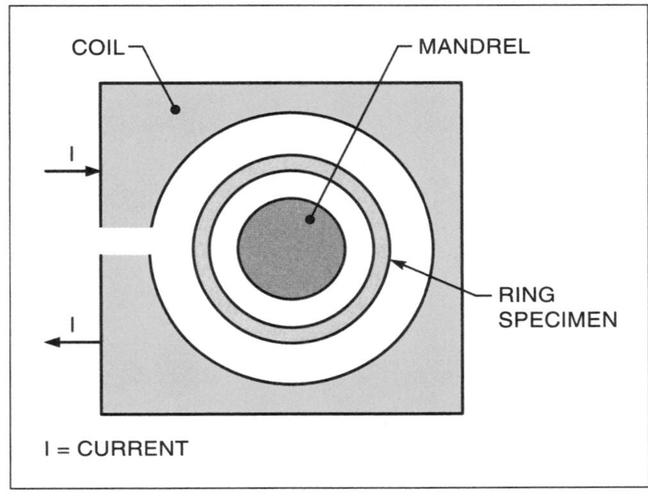

Source: OSU, Material Science and Engineering

Figure 15.36—Compression Welding Arrangement for Tubing

Magnetic pulse welding also can be used to join sheet metal to plate. However, research on this type of work is preliminary, and at the time of publication of this book, only a few material combinations had been joined in this way. Generally, the sheet metal used in these applications is thinner than 1 mm (0.039 in.). Magnetic impulse welding of sheet metal also is hampered by the difficulty in making appropriate coils. The types of coils often used are not energy-efficient, and often cannot tolerate the forces necessary for welding.

Sheet Metal

Two types of joint configurations are used in sheet metal welding. In one configuration, shown in Figure 15.38(A), one sheet is fixed against a mandrel. The other sheet is placed adjacent to the coil. The interaction between this sheet and the coil accelerates the sheet and impacts the stationary sheet at sufficient velocity to create a weld. Coils for this type of configuration can be

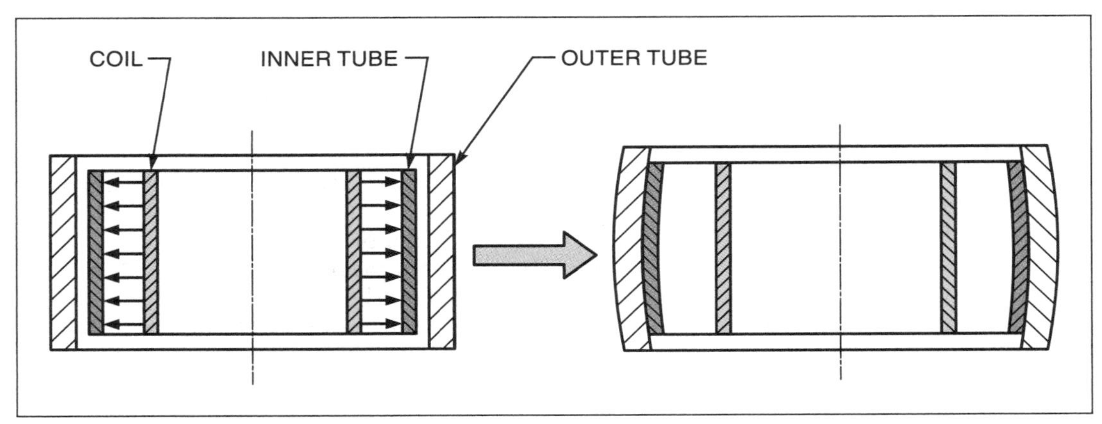

Source: OSU, Material Science and Engineering

Figure 15.37—Expansion Welding Arrangement for Tubing

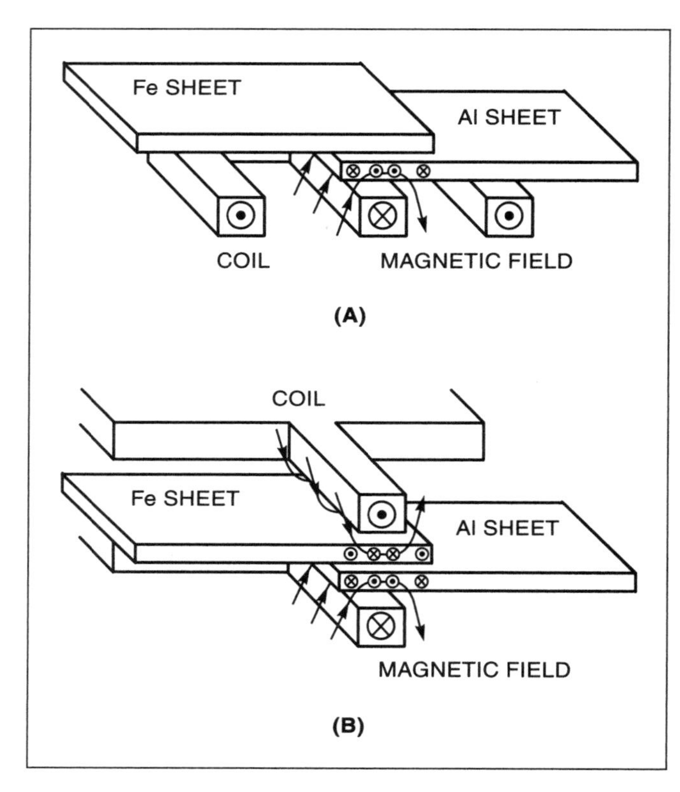

Figure 15.38—Magnetic Pulse Welding of Sheet Metal with (A) Coil on One Side and (B) Coil on Both Sides of Sheet

varied, depending on the joint geometry. For example, a bar-type coil is well suited for linear seam welds. The second type of configuration is shown in 15.39(B). During welding, both sheets are driven by adjacent coils, achieving the necessary relative velocity to accomplish bonding. Coils of this type are designated as "H" coils.

APPLICATIONS

Magnetic pulse welding has been successfully applied to various similar and dissimilar metal combinations. Materials with high conductivity, such as aluminum and copper, are the easiest to weld with the magnetic pulse process. Aluminum has been successfully welded to aluminum, steel, stainless steel, brass, copper, tungsten, and magnesium, and can conceptually be welded to many other metals. Copper has been successfully welded to copper, aluminum, steel, stainless steel, nickel, brass, and can presumably be welded to other metals. Other materials that have lower electrical conductivity, such as steel, stainless steel, nickel, and titanium require higher power to achieve a weld. Sometimes conductive drivers are used for magnetic pulse welding to reduce power requirements. Material combinations that have been successfully welded include steel to steel, steel to stainless steel, stainless steel to stainless steel, and nickel to titanium. Figure 15.39 shows some of the weld interface structures for similar and dissimilar metal combinations.

Tubular Structures

Magnetic pulse welding has great potential for joining tubular structures for automotive and aerospace applications and for fluid-carrying tubes. The process has several advantages that can significantly reduce manufacturing costs, summarized as follows:

1. High-strength joints can be produced that are stronger than the base metal;
2. Leak-tight welds can be made;
3. High welding speeds, in the millisecond range, make the process readily adapted to automation;
4. Dissimilar metals and difficult-to-weld materials, such as 303 stainless steel, can be joined;
5. Cold processing enables the immediate handling of components;
6. Welds are made with no heat-affected zone and minimum distortion;
7. Post-cleaning operations and post-weld heat treatments are unnecessary; and
8. The process is cost efficient because no filler metals or shielding gases are needed, and environmental costs are reduced.

One of the earliest applications of magnetic pulse welding was the joining of end closures to nuclear fuel pins. Various combinations of materials were welded, including 316 stainless steel, Inconel 706®, PE 16, and RA 330. The diameter of the outer tube in this application was 5.84 mm (0.230 in.). A mild-steel driver was used. Figure 15.40 shows a welded sample of the tube.

Another application of the magnetic pulse process is the welding of seals on pressure vessels and accumulators. The materials used in this type application include mild steel and aluminum. Figure 15.41 shows a weld in an aluminum pressure vessel 51 mm (2 in.) in diameter. In this application, Type 6061 T0 was used for the body, and 6061 T6 and 1010 steel were used as closures.

One of the most significant applications for magnetic pulse welding in the automotive industry has been the production of drive shafts. Drive shafts usually are made of steel or aluminum, depending on the application. Improvements in performance can be obtained through the use of composite aluminum-steel shafts utilizing aluminum tubes and steel yokes. Production of

Figure 15.39—(Top) Magnetic Pulse Welds and (Bottom) Corresponding Interface Structures for Various Metal Combinations

Photograph courtesy of Magneform

Figure 15.40—End Closure of a Nuclear Fuel Pin Welded with the Magnetic Pulse Process

(A)

(B)

Photographs courtesy of Magneform

Figure 15.41—Magnetic Pulse Welds in an Aluminum Pressure Vessel (A) Before Welding and (B) the Completed Weld

composite shafts of this type is limited by the difficulty of welding the aluminum tube to the steel yoke. The main concern is the formation of detrimental intermetallic phases at the weld interface associated with heat from conventional welding processes. However, magnetic pulse welding techniques have been developed to avoid these problems.

Magnetic pulse welding has been used to join fuel pipes, fuel filters, exhaust system components, power cables, and for the construction of automotive body parts. The development of new applications of the magnetic pulse welding process continues, with the goal of advancing these applications to mass production. The process is achieving increased recognition for applications across the industrial spectrum.

SAFE PRACTICES

The potential hazards of magnetic pulse welding include mechanical and electrical risks, noise, flash and fumes.

Mechanical

The welding machine should be equipped with appropriate safety devices to prevent injury to the operator's hands or other parts of the body. Initiating devices, such as push buttons or foot switches, should be arranged and guarded to prevent inadvertent actuation.

Machine guards, fixtures, and operating controls must prevent the operator from coming in contact with the coil and workpiece, and must block or deflect the weld jet associated with the process.

Electrical

All doors and access panels on machines and controls must be kept locked or interlocked to prevent access by unauthorized personnel. The interlocks should interrupt the power and discharge all the capacitors through a suitable resistive load when the panel door is open.

Personal Protection

Appropriate guards should be in place to isolate the operator from the process. Operating personnel should wear ear protection when the welding operations produce high noise levels.

Additional information on safe practices for welding is provided in the latest edition of ANSI *Safety in Welding, Cutting, and Allied Processes*, Z49.1 published by the American Welding Society. This document is available on line at www.aws.org.[12]

ELECTRO-SPARK DEPOSITION

Electro-spark deposition (ESD) is a microwelding process capable of depositing any metallic material to any other metallic material, provided proper parameters and conditions are observed. The process primarily is used for surfacing, cladding, and other metal modifications, but not for joining. Applications of electro-spark deposition fostered other names for the process, such as *spark hardening, electric spark toughening*, and *electro-spark alloying*.[13]

12. See reference 8.

13. Adapted with permission. Reynolds, J. L., R. L. Holdren and L. E. Brown, 2003, Electro-spark Deposition, *Advanced Materials and Processes* 161(3), Materials Park, Ohio: ASM International.

Although ESD can deliver a number of benefits, the process often is limited by the ability to produce the desired deposition rates. The thickness of electro-spark deposits often is limited to less than 250 µm, with deposition rates usually less than 1 gram per hour.[14]

Research on electro-spark deposition began in the early 1900s, with the origin of the process linked to studies of phenomena associated with arced materials.[15, 16] Electro-spark depositing was introduced in 1924 by H. S. Rawdon of the United States Bureau of Standards.[17] Rawdon discovered that iron, when sparked with another similar material, became very hard. His further research proved that the increase in hardness was the result of martensite formation due to the rapid solidification of the steel during sparking.

Similar research determined that the hardness of a deposit could be influenced by the spark atmosphere. A study of deposits on titanium under oil using a tungsten electrode revealed that the resulting deposit was tungsten carbide, and that electrodes of tungsten carbide could be applied to metals to increase hardness above that achieved by other methods. During the discoveries by Rawdon and others, the process also was under investigation in the Union of Soviet Socialist Republics, where several versions of ESD were being studied and implemented.

By the mid 1970s, the process showed viability for nuclear applications and further research continued to improve deposit quality.

EQUIPMENT

Electro-spark deposition equipment has two main components: the power source, and the electrode holder, or torch, as shown in Figure 15.42. The power source primarily consists of a direct current (dc) rectifier and discharge circuit. The electrode holder has an integrated mechanism that produces electrode motion and also may be designed to deliver a shielding gas.

Electro-spark deposition produces deposits that have little effect on substrate properties.[18] This is achieved by

14. Johnson, R. N., 1988, Principles and Applications of Electro-spark Deposition, *Proceedings, 117th Metallurgical Society Meeting*, Phoenix, Arizona: The Metallurgical Society.

15. Li, Zhengwei, et al, 2000, Electro-Spark Deposition Coatings for High Temperature Oxidation Resistance, *High-Temperature Materials and Processes* 19(6): 443–458.

16. Sheldon, G. L., and R. N. Johnson, 1985, Electro-Spark Deposition—A Technique for Producing Wear-Resistant Coatings, *The International Conference on Wear Materials,* Vancouver, B.C. Canada. New York: The American Society of Mechanical Engineers.

17. Johnson, R. N., and G. I. Sheldon, 1986, Advances in the Electro-spark Deposition Coating Process, *Journal of Vacuum Science A*, 4(6): 2740–2746.

18. Wang, P. Z. et al, 1997, Accelerated Electro-Spark Deposition and Wear Behavior of Coatings, *Journal of Materials Engineering and Performance,* 6(6): 780–784.

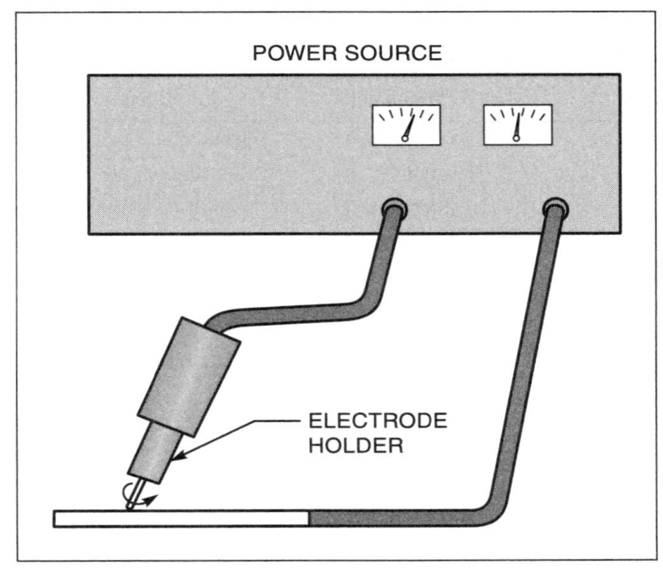

Source: Advanced Surfaces and Processes Incorporated

Figure 15.42—Arrangement of Electro-Spark Deposition Equipment

using a moving electrode energized by a series of capacitors, creating a minute deposit as the electrode is momentarily shorted with the substrate. The momentary short consists of an instantaneous high-current pulse, which creates a solid-state bond between the electrode and the substrate that is then broken away by the velocity of the moving electrode. By controlling the discharge rate and pulse width through the use of a firing circuit, the overall heat input can be minimized.[19]

Power Source

The pulsing circuit of an electro-spark power source is composed of a series of capacitors that are discharged by a firing circuit. The firing circuit is most commonly controlled by either a resistor-capacitor circuit, referred to as an *R-C circuit*, or by a microprocessor.

Torch or Electrode Holder

The primary purpose of the torch is to allow for the safe energizing of the electrode during operation. The torch body typically consists of an outer non-conductive material that minimizes the risk of electrical shock, and has an inner metallic lining that provides mechanical stability.

19. I. V. Galinov and R. B. Luban, 1996, Mass Transfer Trends during Electro-spark Alloy, *Surface and Coatings Technology* 76.

Electrode movement provides the velocity needed to break contact between the electrode and the workpiece. The simplest form of electrode manipulation is a vibrating motion that moves the electrode in and out with respect to the torch. Most commonly, however, the electrode is rotated by a motor-driven system. Electrical current is transferred from the power source to a moving shaft that holds the electrode. Electrical current is supplied from the power cable to the electrode by several methods, including the use of techniques such as a braided cable wrapped around the rotating shaft and the use of rotary contactors.

The torch also may be designed to deliver shielding gas to provide the proper atmospheric conditions at the weld area. A gas nozzle similar to the type used for gas tungsten arc welding (GTAW) may be affixed to the electrode end of the torch.

PROCESS VARIABLES

Electro-spark deposition produces minute deposits at extremely low rates. However, the disadvantage of low productivity is offset by the fact that the deposits occur with extremely low heat input, resulting in virtually no dilution of the substrate. To produce deposits and maintain very low heat input, several process variables must be controlled. The variables that can be controlled are the electrode, substrate, environment, and electrical characteristics. They can be grouped into several categories, as listed in Table 15.5.

Electrodes

The geometry of the electrode generally is round to facilitate electrode rotation. Several electrode geometries can be used, including a straight round electrode or a disc.

The selection of electrode material is based on the substrate properties and the desired properties of the deposit. Deposits made in repair applications often require the electrode to be the same material as the substrate. However, if the deposit is applied to provide wear resistance, wear resistive materials such as tungsten carbide or chromium carbide should be used.[20,21,22]

20. Johnson, Rodger N., 1984, Tribilogical Coatings for Liquid Metal and Irradiation Environments, *Proceedings Conference on Coatings and Bimetallics for Energy Systems and Chemical Process Environments*, Hilton Head, South Carolina, Materials Park, Ohio: ASM International, 113–123.

21. Sheldon, G. L., 1995, Galling Resistant Surfaces on Stainless Steel Trough Electro-spark Alloying, *Journal Of Tribology* 117: 343–349.

22. Sheldon, G. L., and R. N. Johnson, 1985, Electro-spark Deposition—A Technique for Producing Wear Resistant Coatings, *The International Conference on Wear Materials,* Vancouver, B.C. Canada: The American Society of Mechanical Engineers, 388–396.

Table 15.5
Process Variations of Electro-Spark Welding

Electrode	Substrate	Electrical	Environment	Other
Material	Material	Spark energy	Gas composition	Spark time per unit area
Geometry	Surface finish	Voltage	Flow rate	Number of passes
Motion	Cleanliness	Current	Flow geometry	Overlap of passes
Speed	Temperature	Capacitance		
Contact force	Geometry	Pulse duration		
		Inductance		
		Spark frequency		

The electrode deposition is affected by motion, velocity, and contact pressure. Direction of motion affects the contact velocity and process stability. Velocity determines the relative time that electrode and substrate are in contact during a discharge. Electrode pressure affects the stability of the process by controlling the arc length between the electrode and the substrate. The type of substrate often has the most influence on the properties of the deposited metal. Deposits often are easier to make on hard materials. The surface finish affects the process through material preparation of the surface and substrate cleanliness. For example, machined surfaces normally are preferred to those not machined.

The environment in which a deposit is made can affect the quality and the rate of deposition. Typically, a shielding gas is used to prevent atmospheric contamination of the deposit. Argon is the most widely used shielding gas used for electro-spark deposition.

The electrical characteristics of ESD exert the most influence on deposit quality and deposition rate. Spark energy has the most effect on the process and is interdependent on all other electrical factors, with higher spark energies usually providing higher deposition rates.

Spark energy W_p is defined in the following equation:

$$W = \int_{0}^{t_p} U(t)I(t)dt \qquad (15.2)$$

where

$U(t)$ = Voltage across the arc length,
$I(t)$ = Current, and
t_p = Duration of pulse.

The variables of spark energy are controlled by the capacitance, charging voltage, inductance, and resistivity of the circuit.

The Deposit

An electro-spark deposit is the result of multiple micro deposits referred to as *splats*. A microscopic view of the splats is shown in Figure 15.43. The splats form layers that typically range from 3 μm to 250 μm in thickness, although 1-mm (0.002-in.) thick deposits have been achieved. A cross-section of an IN625 deposit is shown in Figure 15.44. Deposition rates are extremely slow and time consuming due to the micro size of the splats.

There is virtually no dilution of the base metal and a very small heat-affected zone (HAZ) because there is very little heat input into the substrate. Discontinuities associated with the process include deposit cracking and the development of voids between splats. These discontinuities often can be reduced by the proper selection of parameters. Deposit inspection is typically limited to destructive evaluation by metallography because of the micro-size of defects that may exist.

APPLICATIONS

The first use of electro-spark deposition was the application of a steel overlay that formed martensite caused by a spark between a steel electrode and a steel substrate. Many other overlays have been developed to coat wear areas with hard-facing material.[23]

The electrodes used to make the hard-facing deposits often consist of a carbide composition. Figure 15.45 displays the hardness profile of an ESD hard-facing deposit. The service life of cutting edges can be improved by applying a coating to one side of a cutting edge. The softer material will erode from the uncoated edge, leav-

23. Brown, E. A., G. L. Sheldon, and A. E. Bayoumi, 1990, A Parametric Study of Improving Tool Life by Electro-Spark Deposition, *Wear* 138: 137–151.

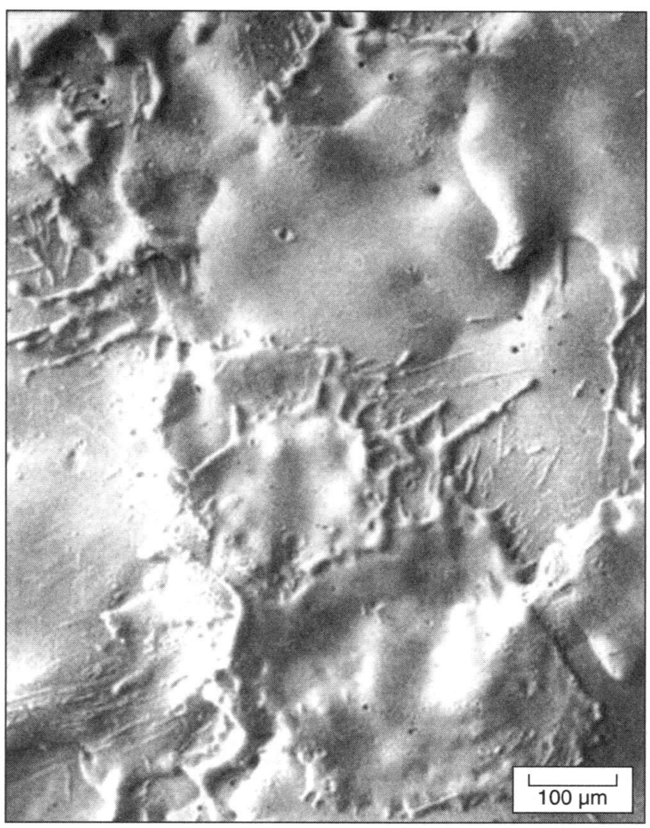

Micrograph courtesy of Advanced Surfaces and Processes Incorporated

Figure 15.43—Coating Formed by Overlapped Splats

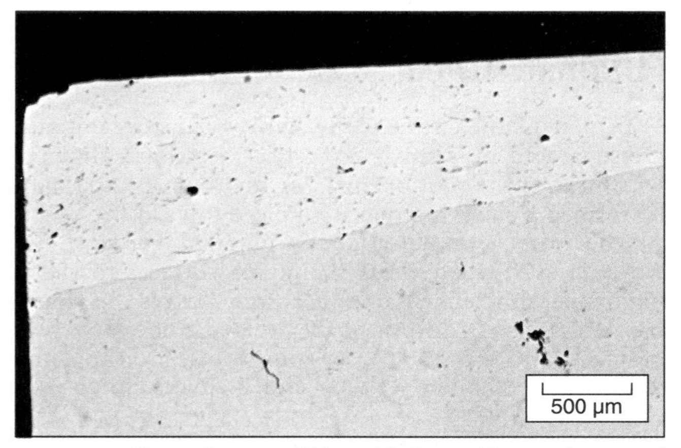

Micrograph courtesy of Advanced Surfaces and Processes Incorporated

Figure 15.44—Electro-Spark Deposited IN625

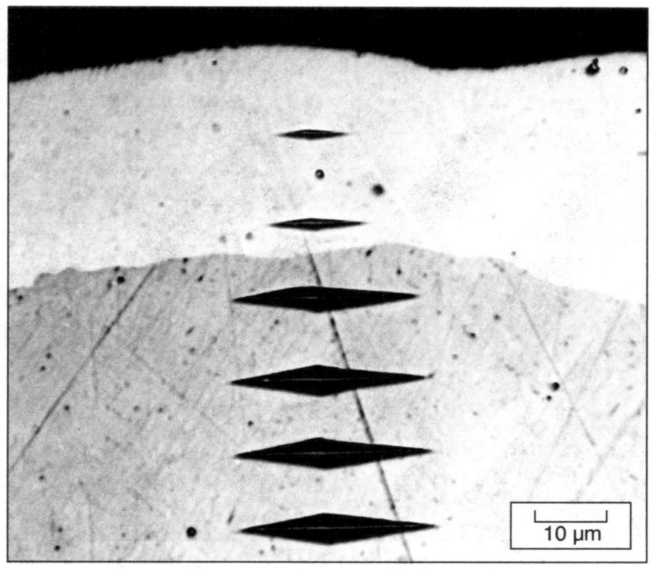

Photograph courtesy of Advanced Surfaces and Processes Incorporated

Figure 15.45—Hardness Profile of a Carbide Composition Deposit

ing a sharpened edge of the coated material. Another application of ESD is the hard facing of surface contact areas to increase the service life of the deposited item. Electro-spark-deposited coatings also can be applied to electrical contact areas to increase conductivity.

Another application of electro-spark deposition of hard facing material is to repair a pre-existing coating. For example, to repair chrome plating, the existing coating would have to be stripped and the plating restored by conventional methods. Electro-spark-deposited material can be used to coat the defective area locally, providing monetary savings in addition to reducing the negative environmental impact of the plating process.

Electro-spark deposition also has the capability of restoring machine parts considered difficult to weld. Applications include the repair of tooling materials to prolong tool life and the repair of items such as turbine engine components. Deposits often can be made on these difficult-to-weld materials because the process inputs very little heat. As a result, there is virtually no dilution of the base metal and a very small heat-affected zone. Figure 15.46 shows platinum-coated turbine blades; Figure 15.47 shows a cross-section of the repair deposit.

SAFE PRACTICES

The potential hazards of electro-spark deposition include mechanical and electrical risks, as well as those

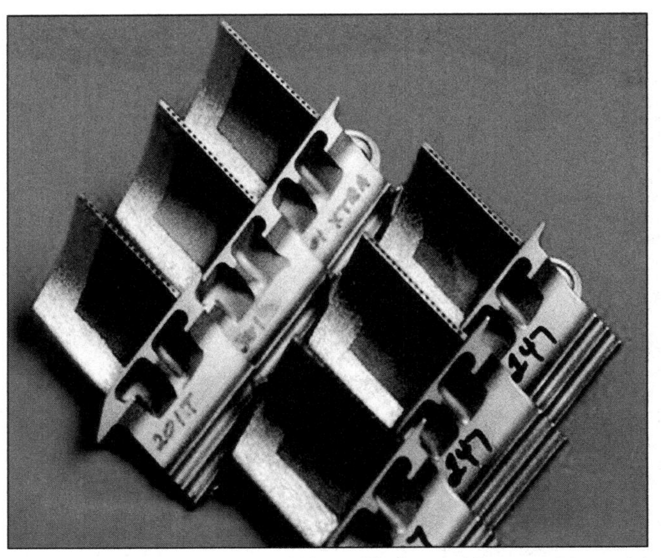

Photograph courtesy of Advanced Surfaces and Processes Incorporated

Figure 15.46— Electro-Spark Deposition of Platinum on Turbine Blades

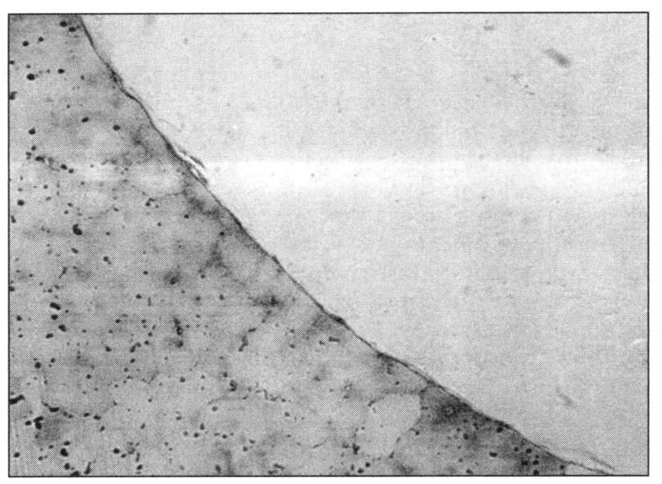

Micrograph courtesy of Advanced Surfaces and Processes Incorporated

Figure 15.47—Cross Section of a Platinum-Coated Repair using Electro-Spark Deposition

from flash and fumes. Personnel must be aware of these hazards and wear protective gear.

Electrical

This process utilizes capacitors for energy storage. A manually operated switch or other positive device should also be provided in addition to the mechanical interlock or contacts. The use of this device will assure complete discharge of the capacitors.

Personal Protection

The welding machine should be equipped with appropriate safety devices to prevent injury to the operator's hand or other parts of the body. Care must be taken to protect the operator when manipulating the torch. Precautions include gloves for thermal protection, as well as to electrically isolate the operator from the torch. The torch includes an electrode rotating at high speed. Care must be taken to avoid injury from this electrode. The process also produces a high degree of sparks. Proper eye and skin protection is necessary when using this process.

Additional information on safe practices for welding is provided in the latest edition of ANSI *Safety in Welding, Cutting, and Allied Processes*, Z49.1 published by the American Welding Society. This document is available on line at www.aws.org.[24]

PERCUSSION WELDING

Percussion welding (PEW) is a welding process that produces coalescence with an arc resulting from a rapid discharge of electrical energy. Pressure is applied percussively during or immediately following the electrical discharge.

Percussion welding is used in the electronics industry to join wires, contacts, leads, and similar items to a flat surface.

FUNDAMENTALS

In percussion welding, the two workpieces initially are separated by a small projection on one workpiece, which forms the root opening, or the root opening may be formed by moving one workpiece toward the other. At the proper time, an arc is initiated between the workpieces. The arc heats the faying surfaces to welding temperature. Then an impact force drives the workpieces together to produce the welded joint. The two basic variations of the percussion process are capacitor discharge percussion welding and magnetic force percussion welding.

Although the steps may differ in the variations of the process and in certain applications, a welding operator

24. See Reference 8.

usually would perform the following sequence in making a percussion weld:

1. Load and clamp the workpieces into the machine,
2. Apply a low force on the workpieces or release the driving system,
3. Establish an arc between the faying surfaces with high voltage to ionize the gas between the workpieces or with high current to melt and vaporize a projection on one workpiece,
4. Move the workpieces together percussively with an applied force to extinguish the arc and complete the weld,
5. Turn off the current,
6. Release the force,
7. Unclamp the welded assembly, and
8. Unload the machine.

Percussion welding is similar to capacitor discharge stud welding. The differences between the two processes are in the type of application and the type of power source each requires. Percussion welding can be used to join wires, rods, and tubes of the same cross section. Welding current is supplied by a capacitor storage bank in these applications. The process also can be used to weld wires or contacts to large flat areas with power from a capacitor bank or a transformer.

Welding heat is generated by a high-current arc between the two workpieces. The high current density melts a thin layer of metal on the faying surfaces within a few milliseconds. Then the molten surfaces are brought together in a percussive manner to complete the weld.

CAPACITOR DISCHARGE PERCUSSION WELDING

Power for the capacitor discharge method is furnished by a capacitor storage bank. The arc is initiated by the voltage across the terminals of the capacitor bank (charging voltage) or a superimposed high-voltage pulse. Motion may be imparted to the movable workpiece by mechanical or pneumatic means.

Equipment

Two types of machines are used for joining sections of wire. One machine uses a high-voltage, low-capacitance system, with a charging voltage range from 1 kV to 3 kV. With this system, special preparation of the wire ends is not critical since the applied potential is sufficient to ionize the air in the root opening and start the arc. The other system uses a low-voltage, high-capacitance source. This system has the advantages of a safe working voltage (about 50 V), and a simple power source that produces minimum weld spatter. In some designs, the high-voltage power is discharged through a transformer with low-voltage output.

A low-voltage system requires a 600-V arc-starting circuit and special preparation of wire ends. Once the air gap is ionized with the 600-V (low-amperage) circuit, the arc is sustained by the 50-V circuit. The arc initiation circuit causes no appreciable melting.

One type of low-voltage capacitor discharge welding machine used for welding electrical wires to terminals consists of a hand-held gun and a portable power source. The gun is designed to weld the workpieces (the wire and the terminal) by holding a small flat or square terminal in one set of stationary jaws and the wire to be welded in a set of movable jaws. When the gun is triggered, springs move the wire toward the terminal at a high velocity. A feathered edge on the end of the wire greatly improves arc starting. The arc is initiated at the point of contact of the wire and terminal. The welding current melts the feathered edge on the wire faster than the wire is moving toward the terminal. The arc spreads over the wire area and melts a layer of about 0.050 mm to 0.076 mm (0.002 in. to 0.003 in.) in thickness in each wire. The arc is extinguished after about 150 microseconds to 600 microseconds as the two wires come in contact.

Another version of a portable, low-voltage welding machine employs a high-frequency pulse to initiate the arc. This feature eliminates the need for a special shape on the wire end. The machine uses an electromechanical actuator to accelerate the wire and to provide the necessary forging force.

Low-voltage semiautomatic and automatic machines are used to weld assemblies similar to the one shown in Figure 15.48. Component leads usually are precoated annealed copper. Terminals may be brass, precoated brass, or nickel-silver alloys. Wires and leads with diameters of 0.2 mm to 2.6 mm (0.006 in. to 0.102 in.)

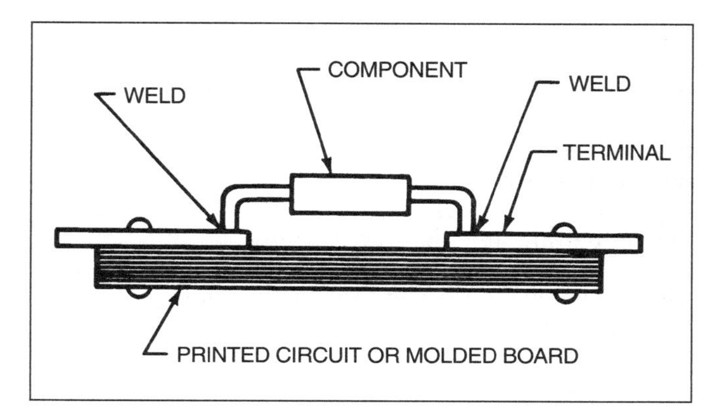

Figure 15.48—A Typical Percussion-Welded Electronic Assembly

can be welded to terminals and plates of various thicknesses above 0.2 mm (0.006 in.).

Capacitor discharge equipment usually includes controls for welding voltage, capacitance, and high-frequency voltage (when high frequency is used). Control of the motion mechanism also is provided.

MAGNETIC FORCE PERCUSSION WELDING

Power is supplied by a welding transformer for magnetic force percussion welding. The arc is initiated by vaporizing a small projection on one workpiece with high current from the transformer. The vaporized metal provides an arc path. The percussive force is applied to the joint by an electromagnet that is synchronized with the welding current. Magnetic force percussion welds are made in less than one half-cycle of 60 Hz. Consequently, the timing between the initiation of the arc and the application of magnetic force is critical.

Welding Machines

Magnetic force welding machines use a low-voltage power source (20 V to 35 V) from a transformer, a projection-type arc starter, and an electromagnetic system to produce the weld force. A unit generally consists of a modified press-type resistance welding machine with specially designed transformer, controls, and tooling.

Magnetic force percussion welding machines usually have an independent power source for the electromagnet so that the force magnitude and time of application can be varied to coincide with the initiation of welding current. This is accomplished by using two separate transformers, one for welding power and one for the electromagnet power. The acceleration of the force member can be controlled by adjusting the magnitude of the electromagnet current, thereby providing a control for the duration of arc time.

Since welding is done during one-half cycle of 60 Hz, current is unidirectional. In some cases, the polarity of the two workpieces may have some effect on weld quality. In general, the same conditions with respect to polarity that prevail in direct current (dc) arc welding also are in effect in percussion welding. The current is always passing through the transformer in the same direction and the core may become partially saturated. Consequently, the electrical controls should provide a low-amplitude one half-cycle pulse in the opposite direction to remove remaining flux from the transformer and electromagnet. This can be done during the loading time.

Joint Design

For welding two flat surfaces together, a projection similar to that used in resistance welding must be formed on one piece. An example is shown in Figure 15.49, in which a contact assembly is welded. The diameter and height of the projection must be developed for each application. The diameter must be large enough to support the initial force applied to the workpieces, but too small to carry the welding current. The height determines the root opening between the faying surfaces and, thus, the initial arc voltage. When large-area contacts are welded, two projections may be required.

The surfaces to be joined must be flat and parallel during welding so that arcing will occur over the entire area. Areas that are not melted probably will not weld when impacted together.

Voltage and Current

It is necessary to establish and maintain the desired magnitude of voltage and current required for the weld area. These are determined by the projection desired, the capacity of the welding transformer, and the impedance of the secondary circuit. The transformer should have low impedance with secondary voltages higher than those commonly used in resistance welding.

Arc Time

Arc time can be considered as the time beginning with the explosion of the projection and ending when the two workpieces come together and the arc is extinguished. The timing between initiating the arc and applying magnetic force is very critical.

The arc time is a function of the following factors:

1. Magnitude of magnetic force,
2. Timing of the magnetic force with relation to welding current,
3. Inertia or mass of the moving parts in the force system,
4. Height of the projection, and
5. Magnitude of the welding current and the diameter of the projection.

The acceleration of the movable head is directly proportional to the magnetic force applied and inversely proportional to the mass. The acceleration of the movable head with the two-transformer system can be controlled by adjusting the magnitude of the force current, thereby providing a control for the duration of arc time, within limits.

ADVANTAGES AND LIMITATIONS OF PERCUSSION WELDING

The extreme brevity of the arc in both versions of percussion welding limits melting to a very thin layer on the faying surfaces. Consequently, there is very little

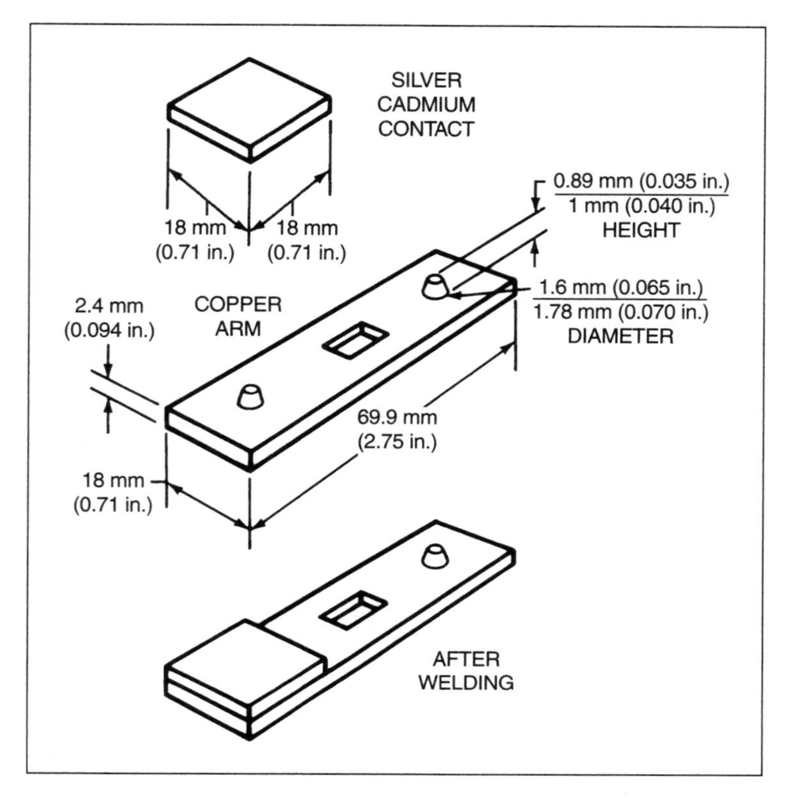

Figure 15.49—Typical Design of a Magnetic-Force, Percussion-Welded Contact Assembly

upset or flash on the periphery of the welded joint, but enough to remove impurities from the joint. Heat-treated or cold-worked metals can be welded without annealing. Dissimilar metals are readily welded with this process.

Filler metal is not used and there is no cast metal at the weld interface. A percussion-welded joint usually has higher strength and conductivity than a brazed joint. Unlike brazing, no special flux or atmosphere is required.

A particular advantage of the capacitor discharge method is that the capacitor charging rate is low and easily controlled compared to the discharge rate. The line-power factor is better than with a single-phase alternating-current machine. Both these factors promote good operating efficiency and low line-power demand.

Percussion welding can tolerate a slight amount of contamination on the faying surfaces because expulsion of the thin molten layer tends to carry contaminants out of the joint.

However, countering these advantages are several limitations to percussion welding. The percussion welding process is limited to butt joints between two similar sections and to flat pads or contacts joined to flat surfaces. Also, the total area that can be joined is limited because control of an arc path between two large surfaces is difficult.

Joints between two similar sections usually can be accomplished more economically by other processes. Percussion welding is confined mainly to the joining of dissimilar metals normally considered unweldable by other processes, and to the production of joints in which it is imperative to avoid upset.

Another limitation is that two separate pieces must be joined. For example, the process cannot be used to weld a ring made from one piece.

APPLICATIONS OF PERCUSSION WELDING

The magnetic force method is primarily used for joining electrical contacts to contactor arms. Weldable metal combinations include copper-to-copper, silver-tungsten to copper, silver oxide to copper, and silver-cadmium oxide to brass. Areas from 26 mm^2 to 820 mm^2 (0.040 in^2 to 1.27 in^2) can be welded in production.

Some metal loss occurs at the weld interface, and in most instances some flash must be removed from the periphery of the weld.

The capacitor discharge method usually is employed to produce the following types of joints:

1. Butt joints between wires or rods,
2. Lead wire ends to flat conductors or terminals, and
3. Contacts to relay arms.

The wire usually is made of copper and may be solid or stranded, bare or precoated. The rods usually are copper, brass, or nickel-silver. Other alloys such as steel, Alumel,™ Chromel,™ aluminum, and tantalum may be welded to one another or to other materials. The method also is applicable to welds in reactive, refractory, and dissimilar metals, because the short weld time limits contamination of the reactive metals and the formation of low-strength intermetallic zones in the joints.

Industrial Applications

Applications of percussion welding are mainly in the electrical contact or component industry. Large contact assemblies for relays and contactors usually are made with magnetic force percussion welding machines. These machines can be automated for high-speed production.

Hand-held capacitor discharge equipment can be used to weld wires to pins. This application is particularly useful in manufacturing aerospace or other equipment that is subject to shock and vibration. The process also is used to weld electronic components to terminals.

EFFECTS ON METALS WELDED

A percussion weld is made in an extremely short time. This can range from microseconds to milliseconds. Because of this short time, the heat-affected zones of percussion welds usually are shallow, less than 0.25 mm (0.010 in.). There is little oxidation of the faying surfaces and a minimum of alloying between dissimilar metals. Since the heat-affected zone is so small, heat-treated metals may be welded without softening the metal. The heat input is so concentrated and of such short duration that heat-sensitive components near the weld area are not affected by the welding cycle.

Heat balance between workpieces usually is not a factor of concern. Because percussion welding is essentially a direct-current process, the polarity of the two workpieces involved may be important in some cases, as it is in arc welding.

Metal Loss

The metal loss that occurs during a percussion weld is not as great as the loss incurred in arc stud welding. The loss varies with the area of the weld and the type of welding machine. Metal loss generally can be ignored for components to be joined by capacitor discharge percussion welding. However, metal loss should be considered in magnetic force percussion welding.

Flash

Flash is the metal that is expelled at high velocity from the weld interface during a percussion weld. It can damage adjacent tooling and may affect the accuracy of the tooling assembly. Any flash attached to the weld joint should be removed so that it will not cause a problem in service.

WELD QUALITY

The quality of percussion welds can be determined by metallographic examination and mechanical testing. Metallographic examination will show the weld interface and the widths of the heat-affected zones. In dissimilar metals, metallography may reveal the degree of alloying at the weld interface. Microhardness tests on a metallographic section may indicate the effect of welding on the base metal.

Welded joints may be tested for tension, bending, or shear, depending on the joint design. The effect of vibration may be important in some applications. The test method should be designed to qualify the welding procedures and weld joint properties for the intended application.

SAFE PRACTICES

The potential hazards of percussion welding include mechanical and electrical risks, noise, flash and fumes. Personnel must be aware of these hazards and wear protective gear.

Mechanical

The welding machine should be equipped with appropriate safety devices to prevent injury to the hand of the operator or other parts of the body. Initiating devices, such as push buttons or foot switches, should be arranged and guarded to prevent inadvertent actuation.

Machine guards, fixtures, or operating controls must be designed to prevent the hands of the operator from entering between the work-holding clamps or the workpieces. Dual hand controls, latches, presence-sensing devices, or similar devices should be employed to prevent operation in an unsafe manner.

Electrical

All doors and access panels on machines and controls must be kept locked or interlocked to prevent access by unauthorized personnel. When the equipment utilizes capacitors for energy storage, the interlocks should interrupt the power and discharge all the capacitors through a suitable resistive load when the panel door is open. A manually operated switch or other positive device should also be provided in addition to the mechanical interlock or contacts. The use of this device will assure complete discharge of the capacitors.

A lock-out procedure should be followed prior to working with the electrical or hydraulic systems.

Personal Protection

The operator should wear flash guards of suitable fire-resistant material as protection from sparks and flames. In addition, the operator should wear personal eye shields with suitable shaded lenses. Operating personnel should wear ear protection when the welding operations produce high noise levels. Metal fumes produced during welding operations should be removed by local ventilating systems.

Additional information on safe practices for welding is provided in the latest edition of ANSI *Safety in Welding, Cutting, and Allied Processes*, Z49.1 published by the American Welding Society. This document is available on line at www.aws.org.[25]

WATER JET CUTTING

Water jet cutting uses a high-velocity water jet to produce cuts in a wide variety of metals and other material. The jet is formed by water forced by high pressure through an orifice in a synthetic sapphire. The orifice is approximately 0.1 mm to 0.6 mm (0.004 in. to 0.024 in.) in diameter; the high pressure ranges from 207 MPa to 414 MPa (30 000 psi to 60 000 psi). The velocity of the jet may vary from 520 meters per second (m/s) to 914 m/s (1700 feet per second [ft/s] to 3000 ft/s). When the jet of water is applied to the cut site at these speeds and pressures it rapidly erodes the material and performs much like a saw blade. The water stream, with a flow rate of 0.4 liters per minute (L/min) to 19 L/min (0.1 gallon per minute [gal/min] to 5 gal/min) usually is manipulated by a robot or gantry system, but small workpieces may be guided manually past a stationary water jet. The typical range of nozzle-to-workpiece distance is 0.25 mm to 25 mm (0.010 in. to 1.0 in.). Distances less than 6.4 mm (1/4 in.) are preferred.

Metals and other hard materials are cut by adding an abrasive in powder form to the water stream. In this method, called *hydroabrasive machining* or *abrasive-jet machining*, the abrasive particles (often garnet) are accelerated by the water and accomplish most of the cutting. Higher flow rates of water are required to accelerate the particles.

Materials are cut cleanly, without ragged edges (unless the travel speed is too high), without heat, and generally faster than bandsaw cutting. A smooth, narrow (0.8-mm to 2.5-mm [0.030-in. to 0.100-in.]) kerf is produced. If the process is properly applied, there is no thermal damage, thus eliminating deformation or delamination of the workpieces. No dust is produced.

The ancient Egyptians used sand combined with water for mining and cleaning. Sandblasters in the present century use a pressurized stream (3400 kPa [500 psi]) for cleaning and paint removal. In 1968, Franz patented a concept for a very high-pressure water jet cutting system.[26] His patent for producing a coherent cutting stream involved the addition of a long-chain liquid polymer to the water stream to prevent it from breaking up as it exited the orifice of the pressurized chamber.

Prior to its application as a cutting tool in metalworking and fabricating industries, high-pressure water was used in forestry and mining. In the 1970s, high-pressure water jet cutting technology with pressures in the range of 207 MPa to 379 MPa (30 000 psi to 55 000 psi) was developed to cut nonmetallic materials.

The first commercial water jet cutting system was sold in 1971 to cut furniture shapes from laminated paper stock that band saws, reciprocating saws, and routers could not handle well. In 1983, the process was modified by the addition of abrasives (such as silica and garnet particles) to the water stream to cut metals, composites, and other hard materials.

FUNDAMENTALS

The water used in water jet cutting first passes through a booster pump to pressurize it to about 1300 kPa (190 psi) and to filter it. Then an intensifier pump (a hydraulically driven double-acting reciprocating pump) creates a water pressure of 207 MPa to 414 MPa (30 000 psi to 60 000 psi) with a flow rate of up to 13.3 L/min (3.5 gal/min]). The stream is forced

25. See Reference 8.

26. Dr. Norman Franz, forestry engineer, invented the process initially to cut lumber.

through a sapphire orifice and forms a water jet. The velocity of the jet depends on the water pressure.

For abrasive cutting, dry abrasives can be fed from a hopper into a mixing chamber. There the water accelerates the particles to supersonic velocities. The high-speed slurry is focused and then exits the nozzle in a stream 0.5 mm to 2.3 mm (0.020 in. to 0.090 in.) in diameter. Water jets can be made with jet diameters as small as 80 micrometers (µm) (0.003 in.), suitable for cutting paper. Abrasive jets generally are not manufactured in smaller sizes than 0.23 mm (0.009 in.) in diameter.

Depending on the properties of the workpiece material, the actual cutting is a result of erosion, shearing, or failure under rapidly changing localized stress fields. Water jet cutting does not produce thermal or mechanical distortions. There is a slight work-hardening of metals at the cut surface. Downstream of the kerf, the water or water-abrasive stream is collected in a tank or *catcher.* Figure 15.50 shows a water jet head cutting through 9.5-mm (3/8-in.) thick carbon steel.

APPLICATIONS

Water jet and abrasive water jet cutting systems compete with processes such as bandsaw and reciprocating-knife cutting, flame cutting, plasma arc cutting and laser beam cutting. Water jet cutting systems can handle materials that would be damaged by heat from thermal processes or that would adhere to mechanical cutting tools and cause malfunction. In some cases, water jet cutting systems can cost-effectively replace three operations: the rough cutting, milling, and deburring of contoured shapes.

The extremely wide range of materials that may be cut are listed in Table 15.6. Although water jet cutting and abrasive jet machining often are perceived as sheet-material processing systems, thick metals and other materials also can be cut. Examples of cuts made to test the limits of the process involved 190 mm (7.5 in.) carbon steel, 75 mm (3 in.) 7075 T-6 aluminum, 64 mm (2.5 in.) 470-ply graphite/epoxy, and 250 mm (10 in.) titanium.

Hundreds of factory water jet systems, including water-jet-equipped robots, are in place in dozens of countries. Among the diverse industries that use the technology include automotive, aerospace and defense; manufacturers of building supplies, circuit boards, packaging, paper, glass, rubber products, and oil and gas equipment; also fabrication shops, foundries, food processors, job shops, mines, shipyards, and steel service centers.

Aerospace applications include the abrasive jet cutting of advanced composite structures; titanium, nickel,

Photograph courtesy of ESAB

Figure 15.50—Water Jet Head Cutting 9.5-mm (3/8-in.) Thick Carbon Steel

Table 15.6
Water Jet Cutting Speeds on Various Materials

	Thickness		Travel Speed	
Material	mm	in.	mm/s	(in./min)
ABS plastic	2.0	0.080	34	80
Cardboard	1.4	0.055	102	240
Corrugated cardboard	6.4	0.250	51	120
Circuit board	2.6	0.103	423	100
Leather	1.6	0.063	1600	3800
Plexiglass®	3.0	0.118	15	35
Rubber	1.3	0.050	1500	3600
Rubber-backed carpet	9.5	0.375	2500	6000
Wood	3.2	0.125	17	40

and cobalt superalloys; and the stack cutting of metals and fiberglass. Abrasive water jet systems are particularly useful for cutting composites because cuts can be made without delamination and heat damage in cut components.

Automotive companies and their suppliers use water jets and abrasive jets for trimming composite panels, bumpers, glass, door panel linings and carpeting.

Foundries use abrasive jets to remove exterior burned-in sand from iron castings and monoshell ceramic coatings from investment castings. Degating and definning also are common applications.

A steel, circular saw-blade cut using hydroabrasive machining is shown in Figure 15.51.

Figure 15.51—A Steel Circular Saw Blade Cut with Hydroabrasive Machining

ADVANTAGES AND LIMITATIONS

Two major advantages of water jet cutting are its wide range of applications and the capability of making cuts without heat. The versatility of the process is demonstrated by the simultaneous cuts through carbon steel, brass, copper, aluminum and stainless steel shown in Figure 15.52. An abrasive jet is particularly good for cutting laminates of different materials, including sandwiches of metals and nonmetals. Since the abrasive jet can penetrate most materials, no predrilling is required to start, and cutting may be omnidirectional.

Multiple shapes can be nested and cut, depending on the limits of the control system and the workpiece size. Tapering the kerf generally is not a problem unless the cutting speed is too high, the workpieces are too thick, or worn nozzles are involved. Little or no deburring is required. The process is easily adapted to robotic control.

Figure 15.52—Abrasive Water Jet Stack Cutting of Various Metals

Low maintenance is an advantage. Other than the orifice and the nozzle, there are no tools to wear out, although the robot mechanism and the pumps may show signs of wear. Minimal lateral forces are generated, which allows simplified fixture designs.

Tolerances depend on the equipment and the workpiece material and thickness, but can be as close as ±0.1 mm (0.004 in.) on dimensions and ±50 µm (0.002 in.) on positioning. Laser cutting achieves closer tolerances.

Finishes vary widely. Abrasive water jet finishes on aerospace components have been reported in the range (R_a) of 1.6 µ/mm to 6 µ/mm (63 µ/in. to 250 µ/in.).

In basic water jet cutting, the kerf width usually is 0.13 mm (0.005 in.) or wider; in abrasive water jet cutting it usually is 0.8 mm (0.032 in.) or wider. The water jet tends to spread as it leaves the nozzle, so the kerf is wider at the bottom than at the top. Kerf tapering can be reduced by adding long-chain polymers, such as polyethylene oxide, to the water or by reducing cutting speed.

With the exception of sophisticated systems such as those required for aerospace applications, most abrasive and water jet cutting equipment can be operated with computer numerical control (CNC) systems, which are relatively easy to program.

Table 15.7
Cutting Speeds on Various Materials with Abrasive Water Jet

Material	Thickness		Travel Speed	
	mm	in.	mm/s	(in./min)
Aluminum	3.2	0.125	17	40
Aluminum	12.7	0.50	8	18
Aluminum	19.0	0.75	2	5
Brass	3.2	0.125	8.5	20
Brass	10.8	0.425	2	5
Bronze	25.4	1.0	0.5	1
Copper	1.6	0.063	15	35
Copper	15.9	0.625	3	8
Lead	50.8	2.0	3	8
Carbon steel	19.1	0.75	3	8
Cast iron	38.1	1.5	0.5	1
Stainless steel	2.5	0.1	25	25
Stainless steel (304)	25.4	1.0	2	4
Stainless steel (304)	101.6	4.0	0.5	1
Armor plate	19.1	0.75	4	10
Inconel®	15.9	0.625	3	8
Inconel 718	31.8	1.25	0.5	1
Titanium	0.6	0.025	25	60
Titanium	12.7	0.500	5	12
Tool steel	6.4	0.250	4	10
Ceramic (99.6% aluminum)	0.6	0.025	2.5	6
Fiberglass®	2.5	0.100	85	200
Fiberglass	6.4	0.250	42	100
Glass	6.4	0.250	42	100
Glass	19.1	0.75	17	40
Graphite/epoxy	6.4	0.250	34	80
Graphite/epoxy	25.4	1.0	6	15
Kevlar®	9.5	0.375	17	40
Kevlar	25.4	1.0	1.3	3
Lexan®	12.7	0.5	5	12
Metal-matrix composite	3.2	0.125	13	30
Pheonolic	12.7	0.5	4	10
Plexiglass®	4.4	0.175	21	50
Rubber belting	7.6	0.300	85	200

Limitations

The main limitation of the water jet cutting process is its relatively low cutting speed. Typical cutting speeds are shown in Table 15.7. Another limitation is that a device must be provided to collect the exhaust liquid from the cutting stream. Also, initial capital costs are high because of the pumps and pressure chamber required to propel and direct the water jet.

The material to be cut must be softer than the abrasive used. Very thin ductile metals are prone to bending stress from an abrasive jet and often show exit burrs. Ceramics cut with a water jet show a decrease in as-fired strength.

Nozzles must be replaced every two to four hours (sometimes even more frequently) in abrasive water jet systems. The abrasive grit wears the carbide nozzles to an out-of-round condition, causing the jet to lose symmetry and thus the deterioration of cut quality.

Optimally, the water supplied for the system should be deionized water filtered to a particle size of 0.5 µm, to reduce maintenance, but it is possible that other water treatment methods may be used. Many systems operate successfully with simple line filters on the municipal water supply, if the water is relatively soft. Proper disposal of waste water and slurry from the cutting operation must be a part of the system.

The fatigue life of abrasive water-jet-cut edges in critical applications, such as aerospace structures, can be lower than that of a raw sheared edge if a coarse 60-grit abrasive particle is used. Decreasing the particle size to 150 grit increases fatigue life 50% or more, but there is a corresponding decrease in cutting speed.

EQUIPMENT

The key pieces of equipment for a water jet or an abrasive water jet system are the following:

1. A special high-pressure pump or intensifier to provide the stream of water;
2. Plumbing and a tank, or catcher unit, to handle the water;
3. A gantry, robotic, or other delivery system to traverse and guide the water jet;
4. The nozzle assembly unit that forms the jet; and
5. An abrasive particle delivery system.

The particle delivery system includes a hopper, a metering valve, and a mixing unit that mixes the abrasive particles into the water stream. A typical arrangement of a water jet cutting system is shown in Figure 15.53.

Equipment is available in a range from individual components to turnkey systems. Complex systems, such as 5-axis robotic motion systems, usually are custom built. In some instances, flame-cutting machines have been converted to water jet cutting systems. Figure 15.54 shows a four-head gantry system.

Auxiliary Equipment

Auxiliary equipment for loading and unloading workpieces, such as cranes, gantry robots, or pedestal-mounted robots may be necessary. This work-handling equipment generally is separate from the robotic or other system that drives the water jet cutting head.

When contour cutting with five-axis systems of motion, it may be necessary to have a special catcher device to stop the water jet and dissipate its energy. Hard water may require a water-treatment system.

Periodic cleaning of the water table to remove abrasive grit and metal particles generated during cutting is necessary.

CONSUMABLE MATERIALS

The sapphire orifice is the component of a cutting system that is subject to most wear. In abrasive systems, it is the carbide nozzle. On pure water jet systems, a

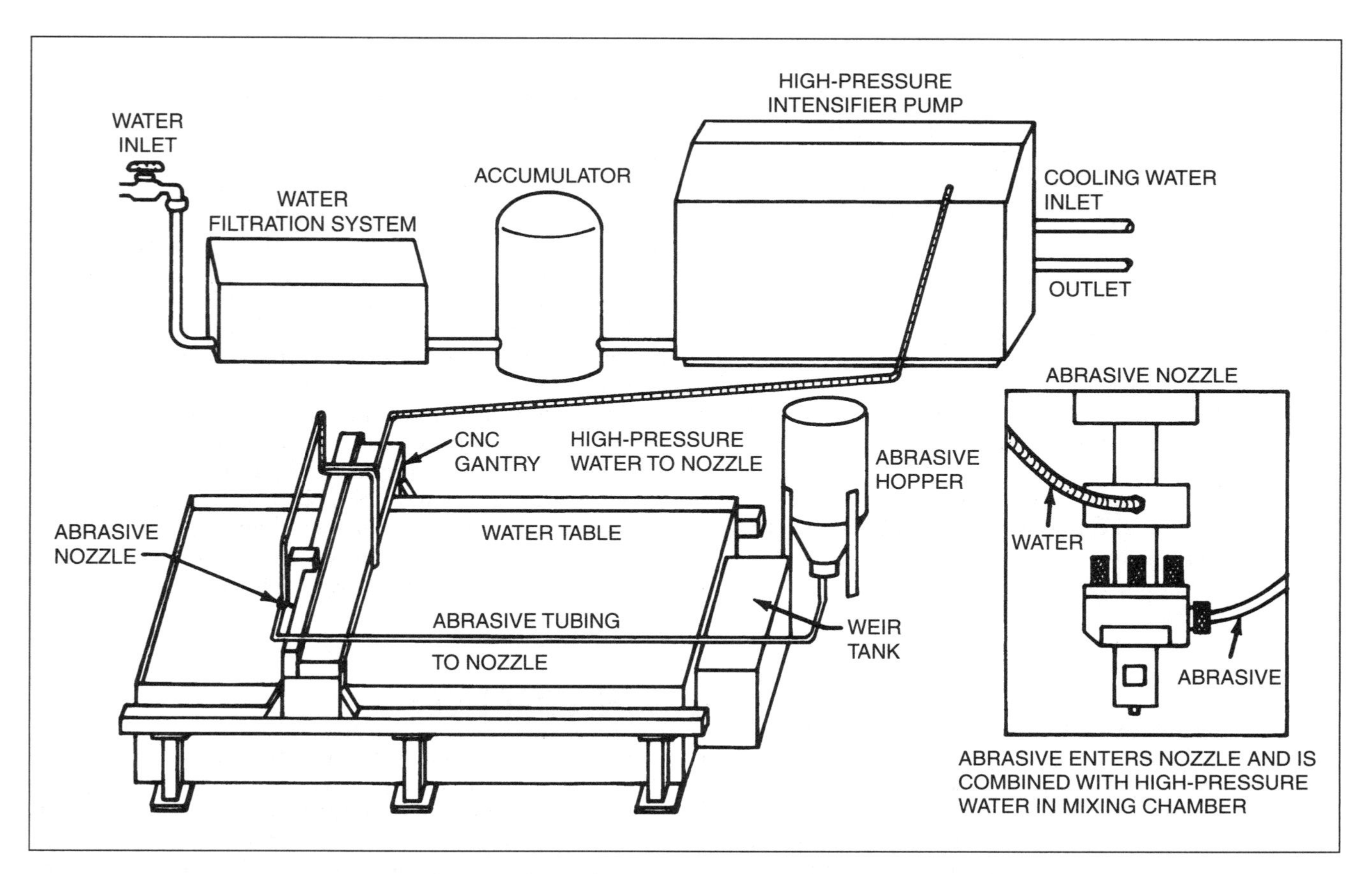

Figure 15.53—Typical Abrasive Water Jet Cutting System

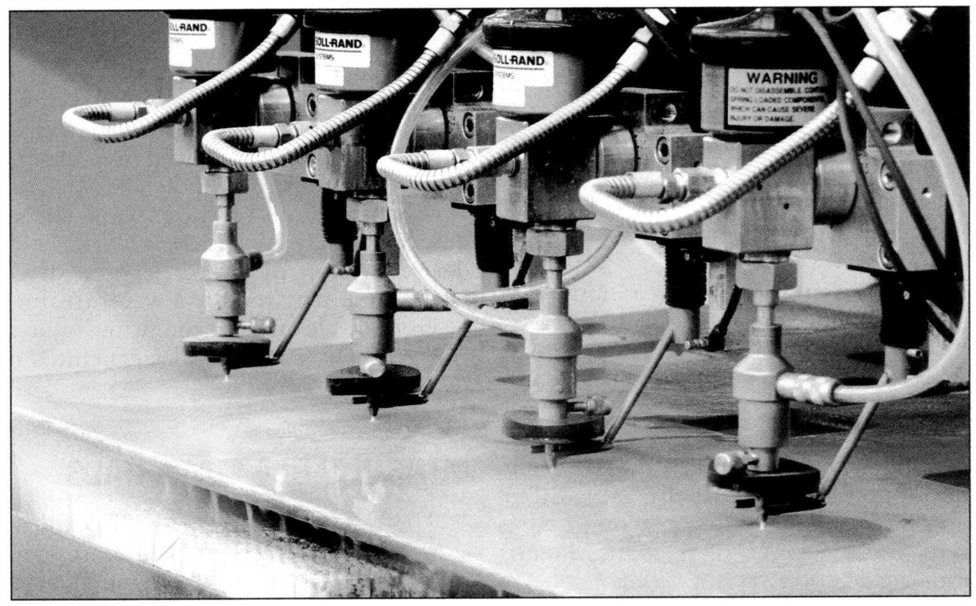

Photograph courtesy of ESAB

Figure 15.54—Water Jet Cutting Machine Using a Four-Head Gantry System

synthetic sapphire may last up to 200 hours. In abrasive systems, the carbide abrasive nozzles last from two to four hours. Other consumables are water, abrasive particles, and electricity. Abrasive particles are used at the approximate rate of 0.1 kg to 1.4 kg (0.25 lb to 3.0 lbs) per minute.

PROCESS VARIABLES

Several variables must be considered to achieve the desired cutting depth and surface characteristics of the cut, including the following:

1. Water jet pressure and diameter;
2. Size, type, and flow rate of the abrasive material;
3. Travel speed;
4. Angle of cutting; and
5. Number of passes.

Typical thicknesses cut by water jet cutting range from 2.5 µm to 30 cm (0.0001 in. to 12 in). The thickness and density of materials that can be cut with a water jet can be increased by increasing the pressure, increasing the jet diameter, and lowering the travel speed.

The cutting speed of an abrasive jet can be increased by increasing the flow rate of the water or the size of the abrasive particles, or both. The use of smaller sizes of abrasive particles and slower cutting speeds will improve the edge quality of the cuts. Increasing the water pressure in abrasive jet cutting increases the capability of the jet to cut thick plate up to 305 mm (12 in.) because of higher velocities of the particle. The optimum pressure tends to remain in the range of 207 MPa to 310 MPa (30 000 psi to 45 000 psi), because higher pressures result in increased equipment maintenance costs with only slight process advantages.

Fine abrasive particles below 150 mesh are relatively ineffective; the most effective general purpose size for cutting metals is 60 mesh or 80 mesh. For very hard ceramics, boron carbide abrasives sometimes are used.

High abrasive flow rates result in high cutting costs: a 0.91 kilogram/min (kg/min) (2 pounds/min [lb/min]) flow rate at a nominal $0.12/lb would result in an hourly cost of $14.40, not including cleaning and disposal costs. This represents a large portion of the total hourly cost. These high flow rates also result in rapid wear of the mixing nozzles.

SAFE PRACTICES

Because the water jet or abrasive jet easily could cut flesh or bone, operator protection is mandatory. Maintenance personnel must be trained in the safety aspects of handling high-pressure equipment and water lines. Each cutting installation should be designed to provide shielding from an accidental discharge of high-pressure water if the high pressure should rupture any of the tubing. Pressure sensors should be in place to shut down the system in case of tubing failure.

Noise generated during cutting typically is in the range of 80 decibels (dB) to 95 dB, but may reach 120 dB. Safety enclosures provided to protect the operator from the cutting operation are designed to deaden sound, but operators should use ear protection.

Additional information on safe practices for welding is provided in the latest edition of ANSI *Safety in Welding, Cutting, and Allied Processes*, Z49.1 published by the American Welding Society. This document is available on line at www.aws.org.

CONCLUSION

The technologies included in this chapter range from the oldest known methods of welding to those only recently made available for production applications. In each case, a general description of the process has been provided, along with details of the necessary equipment and applications demonstrating how the technology is used in manufacturing. The Supplementary Reading List at the end of this chapter is provided for readers requiring a more detailed knowledge of these processes.

BIBLIOGRAPHY

American National Standards Institute (ANSI) Accredited Standards Committee Z49. 2005. *Safety in welding, cutting, and allied processes*. ANSI Z49.1:2005. Miami: American Welding Society. www.aws.org

American Welding Society (AWS). 1942. *Welding Handbook*, 2nd Edition. Miami: American Welding Society.

American Welding Society (AWS) Welding Handbook Committee. 1991. *Welding handbook, Welding processes*. Vol. 2, 8th ed. Miami: American Welding Society.

American Welding Society (AWS) Committee on Definitions and Symbols. 2001. *Standard welding terms and definitions,* AWS A3.0:2001. Miami: American Welding Society.

American Welding Society (AWS) Committee on Methods of Inspection. 2000. *Welding inspection handbook*. Miami: American Welding Society.

American Welding Society (AWS) Committee on Mechanical Testing of Welds. 2000. *Standard methods for mechanical testing of welds,* AWS B4.0M:2000. Miami: American Welding Society.

American Welding Society (AWS) Welding Handbook Committee. 2001. *Welding handbook, welding science and technology*. Chapter 2, Physics of welding. Vol. 2, 9th Ed. Miami: American Welding Society. 61–62.

Brown, A. E., G. L. Sheldon, and A. E. Bayoumi. 1990. A parametric study of improving tool life by electro-spark deposition. *Wear* (*138*): 137–151.

Cueman, M. K., and R. Williamson. 1989. Process model for percussion welding. *Welding Journal* 68(9): 372s–376s.

Galinov I. V., and R. B. Luban. 1996. Mass transfer trends during electro-spark alloying. *Surface and Coatings Technology* 76: 9–18.

Johnson, R. N. 1988. Principles and applications of electro-spark deposition. *Proceedings of the 117th metallurgical society meeting*. Phoenix, Arizona: The Metallurgical Society. 189–213.

R. N. Johnson and G. L. Sheldon. 1986. Advances in the electro-spark deposition coating process. *Journal of vacuum science & technology A*. Nov./Dec. 4(6): 2740–2746.

Johnson, R. N. 1984. Tribological coatings for liquid metal and irradiation environments. *Proceedings of the conference on coatings and bimetallics for energy systems and chemical process environments*. Hilton Head, South Carolina: American Society of Metals. 113–123.

Li, Z. et al. 2000. Electro-spark deposition coatings for high temperature oxidation resistance. *High-temperature materials and processes* 19(6):43–45.

Sheldon, G. L., and R. N. Johnson. 1985. Electro-spark deposition-a technique for producing wear resistant coatings. *The international conference on wear materials*. Vancouver, B. C., Canada: The American Society of Mechanical Engineers. 388–396.

Sheldon, G. L. 1995. Galling resistant surfaces on stainless steel through electro-spark alloying. *Journal of tribology* 117: 343–349.

Wang, P. Z., et al. 1997. Accelerated electro-spark deposition and wear behavior of coatings. *Journal of materials engineering and performance* (6): 780–784.

SUPPLEMENTARY READING LIST

EARLY ARC WELDING PROCESSES

American Welding Society (AWS) Welding Handbook Committee. 1991. *Welding handbook*. Vol. 2. *Welding processes*. Miami: American Welding Society. 892–922.

Simonson, R. D. 1969. *History of welding*. Morton Grove, Illinois: Monticello Books, Inc.

COLD WELDING

Gould, J. E. 2002. Mechanisms of bonding for solid-state welding processes. *Joining of advanced and specialty materials IV*. Materials Park, Ohio: ASM International: 89–97.

Jellison, J. L., and F. J. Zanner. 1985. Solid-state welding. *Metals handbook*, Vol. 6, 9th ed. Materials Park. Ohio: ASM International.

Houldcraft, P. T. 1977. *Welding process technology*. London: Cambridge University Press. 217–21.

Milner, D. R., and G. W. Rowe. 1962. Fundamentals of solid phase welding. *Metallurgical review*. 28(7): 433–480.

Mohamed, H. A. and J. Washburn. 1975. Mechanism of solid-state pressure welding. *Welding Journal* 54(9): 302-s–10-s.

Tylecote, R. F. 1968. *The solid-state welding of metals*. New York: St. Martin's Press.

HOT PRESSURE WELDING

Bryant, W. A. 1975. A method for specifying hot isostatic pressure welding parameters. *Welding Journal* 54(12): 433s–35s.

Guy, A. G. and A. L. Eiss. 1957. Diffusion phenomena in pressure welding. *Welding Journal* 36(11): 473-s–480-s.

Hastings, D. C. 1955. An application of pressure welding to fabricate continuous welded rails. *Welding Journal* 34(11): 1065–69.

Jellison, J. L., and F. J. Zanner. 1985. Solid-state welding. *Metals Handbook*, Vol. 6, 9th Ed. Materials Park, Ohio: ASM International.

Lage, A. P. 1956. Application of pressure welding to the aircraft industry. *Welding Journal* 35(11): 1103-s–1109-s.

Lessmann, G. G. and W. A. Bryant. 1972. Complex rotor fabrication by hot isostatic pressure welding. *Welding Journal* 51(12): 606-s–614-s.

McKittrick, E. S., and W. E. Donalds. 1959. Oxyacetylene pressure welding of high-speed rocket test track. *Welding Journal* 38(5): 469–74.

Metzger, G. E. 1978. Hot pressure welding of aluminum alloys. *Welding Journal*. 57(1): 37–43.

THERMITE WELDING

Ailes. A. S. 1964. Modern applications of thermite welding. *Welding and metal fabricating* 32(9): 335–43. 414–19.

Cikara, M. 1961. Repair of rails by thermite welding and some observations on the testing of welded joints. *Welding and allied processes in maintenance and repair work*. New York: Elsevier Publishing Co. 318–34.

Fricke, H. D. 1985. Thermite welding. *Metals handbook*, Vol. 6, 9th Ed. Materials Park, Ohio: ASM International.

Fricke, H. D., H. Guntermann, and N. Jacoby. 1976. Thermite welding process for rails of special quality. *ETR*. 25(4). (In German).

Guntermann, H. 1975. The applications of the thermite process in areas besides rail welding. *ZEV-Glaser Annalen*. (In German).

Jacoby, N. 1977. Special processes of the thermite welding technique. *Der eisenbahningenieur*. No. 3. (In German).

Kubaschewski, E., L. L Evans, and C. B. Alcock. 1967. *Metallurgical thermochemistry*, 4th ed. London-New York: Pergamon Press.

Rossi, B. E., 1954. *Welding engineering*. New York: McGraw-Hill.

PERCUSSION WELDING

Cueman, M. K., and R. Williamson. Process model for percussion welding. *Welding Journal* 68(9): 372-s–376-s.

Holko, Kenneth H. 1970. Magnetic force upset welding dissimilar thickness stainless steel T-joints. *Welding Journal* 49(9): 427–439-s.

Kotecki, D. J., D. L. Cheever, and D. G. Howden. 1974. Capacitor discharge percussion welding: microtubes to tube sheets. *Welding Journal* 53(9): 557-s–560-s.

Wilson, R. D. 1994. Explore the potential of capacitor discharge welding. *Advanced materials processing*. 145(6): 93–94.

ELECTRO-SPARK DEPOSITION

Galinov, I. V., and R. B. Luban. 1996. Mass transfer trends during electrospark alloying. *Surface coatings and technology* 79: 9–18.

Johnson, R. N. 1988. Principals and applications of electro-spark deposition. *Surface modification technologies*. Warrendale, Pennsylvania: TMS.

Johnson, R. N. and Sheldon, G. L. 1986. Advances in the electrospark deposition coating process. *Journal of vacuum science & technology A*. 4(6): 2740–2746.

Rawdon, H. S. 1924. *Transactions of the American Institute of Mechanical Engineering* 79: 37.

Sheldon, G. L. and Johnson, R. N. 1985. Electro-spark deposition—a technique for producing wear resistant coatings. *Wear of materials 1985*. New York: ASME.

Wand, P.-Z., G-S. Pan, Y. Zhou, J-X. Qu, and H-S. Shao. 1997. Accelerated electrospark deposition and the wear behavior of coatings. *Journal of materials engineering and performance* 6(6): 780–784.

WATER JET CUTTING

ASM Nontraditional machining. 1986. *Conference proceedings, December 1985*. Materials Park, Ohio: ASM International.

American Society of Mechanical Engineers. 1987. *Proceedings of the fourth U.S. water jet conference*. Berkeley, California. New York: American Society of Mechanical Engineers.

Behringer-Ploskonka, C. A., 1987. Waterjet cutting—a technology afloat on a sea of potential. *Manufacturing engineering* 11.

Firestone, R. F. 1987. Lasers and other nonabrasive machining methods for ceramics. *Advanced Ceramics Conference, February 1987*. Cincinnati, Ohio. Hubbard Woods, Illinois: Metals Science Co.

Hashih, M. 1984. Abrasive water jet cutting studies. Kent, Washington: Flow Industries Inc.

Holland, C. L. 1985. Implementing abrasive water jet cutting. *Fabtech conference, Chicago*. SME technical paper #MF85-875. Chula Vista, California. Dearborn, Michigan: Society of Manufacturing Engineers.

Jones, E. P. 1986. Water jet and abrasive water jet and their application in the automotive industry. Presented at the Tracking Robotic Applications in Automotive Manufacturing Conference, Detroit, Michigan, September 1986. Kent Washington: Flow Systems.

Martin, J. M., ed. 1980. Using water as a cutting tool. *American machinist* (4).

Ohlsson, L., J. Powell, A. Ivarson, and C. Magnusson. 1991. Comparison between abrasive water jet cutting and laser cutting. *Laser applications* (3): 46–50.

Slattery, T. J. 1987. Abrasive water jet carves out metalworking niche. *Machine & tool blue book* (4).

Sprow, E. E., ed. 1987. Cutting composites: three choices for any budget. *Tooling and production* (12).

Steinhauser, J. 1985. Abrasive water jets: on the cutting edge of technology. Presented at Fabtech Conference, Chicago. September 1985. Kent Washington: Flow Systems.

Wightman, D. F. 1986. Water jets on the cutting edge of machining. FMS Conference, Chicago. March 1986. Tech Paper MS86-171. American Society of Mechanical Engineers.

Wightman, D. F. 1988. Hydroabrasive near-net shaping of titanium parts and forgings. *Westec '88 conference*. Los Angeles, Technical paper MR88-141. Dearborn, Michigan: Society of Manufacturing Engineers.

Appendix A

AMERICAN WELDING SOCIETY SAFETY AND HEALTH STANDARDS LIST

Appendix A is adapted from Annex A, *Safety in Welding, Cutting, and Allied Processes,* ANSI Z49.1:2005.[1,2] It is not part of that standard, but it is included for informational purposes only.

The following standards contain information that may be useful in meeting the safety and health requirements of welding, cutting, and allied processes.

A3.0 *Standard Welding Terms and Definitions*

F1.1 *Method for Sampling Airborne Particulates Generated by Welding and Allied Processes*

F1.2 *Laboratory Method for Measuring Fume Generation Rates and Total Fume Emission of Welding and Allied Processes*

F1.3 *Evaluating Contaminants in the Welding Environment: A Sampling Strategy Guide*

F1.5 *Methods for Sampling and Analyzing Gases from Welding and Allied Processes*

F1.6 *Guide for Estimating Welding Emissions for EPA and Ventilation Permit Reporting*

F2.2 *Lens Shade Selector*

F2.3 *Specification for Use and Performance of Transparent Welding Curtains and Screens*

F3.2 *Ventilation Guide for Weld Fume*

F4.1 *Recommended Safe Practices for the Preparation for Welding and Cutting Containers and Piping*

1. American National Standards Institute (ANSI) Committee Z49, *Safety in Welding, Cutting, and Allied Processes*, ANSI Z49.1:2005, American Welding Society.
2. At the time of preparation of this appendix, the codes and other standards were valid; however, as codes and standards undergo frequent revision, the reader is advised to consult the most recent edition.

Appendix B

SAFETY CODES AND OTHER STANDARDS

Appendix B is adapted from Annex B of *Safety in Welding, Cutting, and Allied Processes*, ANSI Z49.1:2005.[1,2] It is not part of that standard, but it is included for informational purposes only.

The following codes, standards, specifications, pamphlets, and books contain information that may be useful in meeting the safety and health requirements of welding, cutting, and allied processes.

ACGIH		*Threshold Limit Values (TLVs®) for Chemical Substances, TLVs® for Physical Agents, and Industrial Ventilation Manual*
AGA		*Purging Principles and Practices*
ANSI	A13.1	*Scheme for the Identification of Piping Systems*
	B11.1	*Safety Requirements for Construction, Care and Use of Mechanical Power Presses*
	B15.1	*Safety Standard for Mechanical Power Transmission Apparatus (with ASME)*
	B31.1	*Power Piping (with ASME)*
	Z535.4	*Standard for Product Safety Signs and Labels*
	Z87.1	*Occupational and Educational Eye and Face Protection Devices*
	Z88.2	*Practice for Respiratory Protection*
	Z89.1	*Industrial Head Protection*
API	STD 1104	*Welding of Pipelines and Related Facilities*
	PUBL 2009	*Safe Welding, Cutting, and Other Hot Work Practices in the Petroleum and Petrochemical Industries*
	PUBL 2013	*Cleaning Mobile Tanks in Flammable or Combustible Liquid Service*
	STD 2015	*Safe Entry and Cleaning of Petroleum Storage Tanks*
	PUBL 2201	*Safe Hot Tapping Practices in the Petroleum and Petrochemical Industries*
AVS		*Vacuum Hazards Manual*
CGA	C-7	*Guide to Preparation of Precautionary Labeling and Marking of Compressed Gas Containers*
	E-1	*Standard Connections for Regulator Outlets, Torches, and Fitted Hoses for Welding and Cutting*
	E-2	*Hose Line Check Valve Standards for Welding and Cutting*

1. American National Standards Institute (ANSI) Committee Z49, 2005, *Safety in Welding, Cutting, and Allied Processes*, ANSI Z49.1:2005, American Welding Society.
2. At the time of preparation of this appendix, the codes and other standards were valid; however, as codes and standards undergo frequent revision, the reader is advised to consult the most recent edition.

	G-7.1	*Commodity Specification for Air*
	P-1	*Safe Handling of Compressed Gases in Containers*
	V-1	*Standard for Compressed Gas Cylinder Valve Outlet and Inlet Connections*
NEMA	EW1	*Electric Arc Welding Power Sources*
	EW4	*Graphic Symbols for Arc Welding and Cutting Apparatus*
	EW6	*Guidelines for Precautionary Labeling for Arc Welding and Cutting Products*
NFPA	50	*Standard for Bulk Oxygen Systems at Consumer Sites*
	51	*Standard for the Design of Oxygen-Fuel Gas Systems for Welding, Cutting, and Allied Processes*
	51B	*Standard for Fire Prevention During Welding, Cutting, and Other Hot Work*
	70	*National Electrical Code®*
	79	*Electrical Standard for Industrial Machinery*
	306	*Control of Gas Hazards on Vessels*
	327	*Standard Procedures for Cleaning or Safeguarding Small Tanks and Containers Without Entry*
	701	*Standard Methods of Fire Tests for Flame Propagation of Textiles and Films*
NIOSH	78-138	*Safety and Health in Arc Welding and Gas Welding and Cutting*
	80-144	*Certified Equipment List*
NSC		*Accident Prevention Manual*
		Fundamentals of Industrial Hygiene
OSHA		*Occupational Safety and Health Standards for General Industry (29 CFR Part 1910, Subpart Q)*
		Occupational Safety and Health Standards for Construction (29 CFR Part 1926, Subpart J)
RMA	IP-7	*Specification for Rubber Welding Hose*
RWMA		*Resistance Welding Manual*
UL	252	*Standard for Compressed Gas Regulators*
	551	*Standard for Transformer-Type Arc Welding Machines*

PUBLISHERS OF SAFETY CODES AND OTHER STANDARDS

ACGIH **American Conference of Governmental Industrial Hygienists**
1330 Kemper Meadow Drive
Cincinnati, OH 45240
www.acgih.org

AGA **American Gas Association**
400 N. Capitol Street, N.W., Suite 450
Washington, DC 20001
www.aga.org

ANSI **American National Standards Institute**
25 West 42nd Street, 4th Floor
New York, NY 10036
www.ansi.org

API **American Petroleum Institute**
1220 L Street N.W.
Washington, DC 20005
www.api.org

ASME **American Society of Mechanical Engineers**
Three Park Avenue
New York, NY 10016-5990
www.asme.org

ASTM **ASTM International**
100 Bar Harbor Drive
P.O. Box C700
West Conshohocken, PA 19428-2559
www.astm.org

AVS **American Vacuum Society**
120 Wall Street, 32nd Floor
New York, NY 10005
www.avs.org

AWS **American Welding Society**
550 N.W. LeJeune Road
Miami, FL 33126
www.aws.org

CGA **Compressed Gas Association**
4221 Walney Road, 5th Floor
Chantilly, VA 20151-2923
www.cganet.com

MSHA **Mine Safety and Health Administration**
1100 Wilson Boulevard
Arlington, VA 22209
www.msha.gov

NEMA **National Electrical Manufacturers Association**
1300 North 17th Street, Suite 1847
Rosslyn, VA 22209
www.nema.org

NFPA **National Fire Protection Association**
One Batterymarch Park
Quincy, MA 02269-9101
www.nfpa.org

NIOSH **National Institute for Occupational Safety and Health**
4676 Columbia Parkway
Cincinnati, OH 45226
www.cdc.gov/niosh

NSC **National Safety Council**
1121 Spring Lake Drive
Itasca, IL 60143-3201
www.nsc.org

OSHA **Occupational Safety and Health Administration**
200 Constitution Avenue N.W.
Washington, DC 20210
www.osha.gov

RWMA **Resistance Welding Manufacturing Alliance**
550 N.W. LeJeune Road
Miami, FL 33126
www.rwma.org

RMA **Rubber Manufacturers Association**
1400 K Street N.W.
Washington, DC 20005
www.rma.org

UL **Underwriters Laboratories, Incorporated**
333 Pfingsten Road
Northbrook, IL 60062
www.ul.com

US **U.S. Government Printing Office**
732 North Capitol Street N.W.
Washington, DC 20401
www.gpo.gov

Appendix C

CONVERSIONS FOR VARIOUS PRESSURE UNITS

The following table, adapted from Annex A of *Recommended Practices for Electron Beam Welding*, AWS C7.1M/C7.1:2004, provides equivalents of pressure units in the International System of Units and U.S. Customary Units. The table is not a part of that document but is included for informational purposes only.

Pressure Conversions—International System of Units/U.S. Customary

ATM	µm Hg	mm of Water	Torr or mm Hg	psi	Pa	Bar	in. Hg (absolute)	in. Hg (gauge)	% Vacuum	ppm Dry Air*	ppm Oxygen*	ppm Nitrogen*	ppm Carbon Dioxide*
6.80	5,171,496	70,359.57	5,171.50	100.00	689,476	6.895	203.602	173.68		6,804,600	1,422,161	5,314,393	2,382
5.44	4,137,197	56,287.65	4,137.20	80.00	544,410	5.444	162.882	132.96		5,443,680	1,137,729	4,251,514	1,905
4.08	3,102,898	42,215.74	3,102.90	60.00	400,448	4.004	122.161	92.24		4,082,760	853,297	3,188,636	1,429
2.72	2,068,599	28,143.83	2,068.60	40.00	254,003	2.540	81.441	51.52		2,721,840	568,865	2,125,757	953
1.36	1,034,299	14,071.91	1,034.30	20.00	108,799	1.088	40.720	10.80		1,360,920	284,432	1,062,879	476
1.00	760,000	10,340.00	760.00	14.70	101,325	1.013	29.921	0.00	0.000	1,000,000	209,000	781,000	350
0.987	750,120	10,205.58	750.12	14.50	100,008	1.000	29.532	–0.39	1.300	987,000	206,283	770,847	345
0.968	735,680	9,879.00	735.68	14.23	98,083	0.981	28.964	–0.96	1.900	968,000	202,312	756,008	339
0.921	699,960	9,098.56	699.96	13.53	93,320	0.933	27.558	–2.36	7.900	921,000	192,489	719,301	322
0.789	599,640	7,178.76	599.64	11.60	79,945	0.799	23.608	–6.31	21.000	789,000	164,901	616,209	276
0.658	500,080	4,723.63	500.08	9.670	66,672	0.667	19.688	–10.23	34.000	658,000	137,522	513,898	230
0.526	399,760	2,484.63	399.76	7.730	53,297	0.533	15.739	–14.18	47.000	526,000	109,934	410,806	184
0.500	380,000	1,242.31	380.00	7.348	50,663	0.507	14.961	–14.96	50.000	500,000	104,500	390,500	175
0.395	300,200	490.71	300.20	5.805	40,023	0.400	11.819	–18.10	61.000	395,000	82,555	308,495	138
0.264	200,640	129.55	200.64	3.880	26,750	0.267	7.8992	–22.02	74.000	264,000	55,176	206,184	92.40
0.132	100,320	17.10	100.32	1.940	13,375	0.134	3.9496	–25.97	87.000	132,000	27,588	103,092	46.20
0.118	89,680	2.02	89.68	1.734	11,956	0.120	3.5307	–26.39	88.000	118,000	24,662	92,158	41.30
0.105	79,800	0.21	79.80	1.543	10,639	0.106	3.1417	–26.78	89.500	105,000	21,945	82,005	36.75
0.0921	69,996	0.02	70.00	1.353	9,332	0.0933	2.7558	–27.17	90.800	92,100	19,249	71,930	32.24
0.0789	59,964	0.00	59.96	1.160	7,995	0.0799	2.3608	–27.56	92.100	78,900	16,490	61,621	27.62
0.0680	51,680	0.00	51.68	0.9993	6,890	0.0689	2.0346	–27.89	93.000	68,000	14,212	53,108	23.80
0.0658	50,008	0.00	50.01	0.9670	6,667	0.0667	1.9688	–27.95	93.500	65,800	13,752	51,390	23.03
0.0526	39,976	0.00	39.98	0.7730	5,330	0.0533	1.5739	–28.35	94.800	52,600	10,993	41,081	18.41
0.0395	30,020	0.00	30.02	0.5805	4,002	0.0400	1.1819	–28.74	96.100	39,500	8,256	30,850	13.83
0.0340	25,840	0.00	25.84	0.4997	3,445	0.0345	1.0173	–28.90	96.600	34,000	7,106	26,554	11.90
0.0264	20,064	0.00	20.06	0.3880	2,675	0.0267	0.7899	–29.13	97.400	26,400	5,518	20,618	9.24
0.0132	10,032	0.00	10.03	0.1940	1,337	0.0134	0.3950	–29.53	98.700	13,200	2,759	10,309	4.62
0.0100	7,600	0.00	7.600	0.1470	1,013	0.0101	0.2992	–29.62	99.000	10,000	2,090	7,810	3.50
0.00132	1,003.2	0.000	1.003	0.01940	133.7	0.00134	0.0395	–29.882	99.900	1,320	276	1,031	0.462
0.000987	750.12	0.000	0.7501	0.01450	100.0	0.00100	0.0295	–29.892	99.900	987	206	771	0.345
0.000132	100.32	0.0000	0.1003	0.001940	13.37	0.000134	0.00395	–29.9174	99.990	132	28	103	0.0462
0.0000132	10.032	0.0000	0.01003	0.0001940	1.337	0.0000134	0.000395	–29.9209	99.999	13.20	2.76	10.31	0.00462
0.00000132	1.0032	0.00000	0.001003	0.00001940	0.1337	0.00000134	0.0000395	–29.92126	99.99990	1.320	0.276	1.031	0.000462
0.000000132	0.1003	0.000000	0.0001003	0.00000194	0.01337	0.000000134	0.00000395	–29.921296	99.999990	0.1320	0.0276	0.1031	0.0000462

*Assumes absolutely dry air and compares to standard atmospheric pressure at sea level and air composition of 78.1% nitrogen, 20.9% oxygen and 0.035% carbon dioxide.

Notes:

(1) 1 torr = 1 mm Hg; 1 micron = 10^{-3} torr (10^{-3} mm Hg).

(2) 1 atm (= standard atmosphere) = 101 325 Pa (standard pressure reference).

(3) % vacuum: a pressure increase (or decrease) of approximately 1Mbar corresponds to a change of vacuum of 0.1%.

(4) 1 mm water (column) = 10^{-4} at = 0.1 Mbar.

MAJOR SUBJECT INDEX

Volumes 1, 2, and 3—Ninth Edition, and Volumes 3 and 4—Eighth Edition

C

D

E

P

Q

R

S

T

U

Z

INDEX

A

B

C

D

E

F

G

H

I

J

K

L

M

N

O

P

Q

R

S

T

U

V

W

Z